BIOSENSORS
AND MOLECULAR TECHNOLOGIES
FOR CANCER DIAGNOSTICS

Series in Sensors

Series Editors: Barry Jones and Haiying Huang

SERIES IN SENSORS

BIOSENSORS
AND MOLECULAR TECHNOLOGIES
FOR CANCER DIAGNOSTICS

Edited by **Keith E. Herold** • **Avraham Rasooly**

CRC Press
Taylor & Francis Group
Boca Raton London New York

CRC Press is an imprint of the
Taylor & Francis Group, an **informa** business

CRC Press
Taylor & Francis Group
6000 Broken Sound Parkway NW, Suite 300
Boca Raton, FL 33487-2742

First issued in paperback 2016

Version Date: 20120418

ISBN 13: 978-1-138-19853-1 (pbk)
ISBN 13: 978-1-4398-4165-5 (hbk)

Library of Congress Cataloging-in-Publication Data

Biosensors and molecular technologies for cancer diagnostics / editors, Keith E. Herold, Avraham Rasooly.
 p. ; cm. -- (Series in sensors)
 Includes bibliographical references and index.
 ISBN 978-1-4398-4165-5 (hardback : alk. paper)
 I. Herold, K. E. II. Rasooly, Avraham. III. Series: Sensors series.
 [DNLM: 1. Neoplasms--diagnosis. 2. Biosensing Techniques. 3. Early Diagnosis. 4. Immunoassay--methods. 5. Molecular Diagnostic Techniques. 6. Tumor Markers, Biological. QZ 241]

 616.99'4075--dc23

 2012013520

Visit the Taylor & Francis Web site at
http://www.taylorandfrancis.com

and the CRC Press Web site at
http://www.crcpress.com

Contents

Part IV Optical Technologies for Cancer Detection and Diagnostics: Spectrometry for Cancer Analysis

Part V Optical Technologies for Cancer Detection and Diagnostics: Optical Imaging for Cancer Analysis

Part IX Electronic and Magnetic Technologies for Cancer Analysis

Part X Thermometric Sensing

Part XI Cantilever-Based Technology

Preface

The field of biosensor-based cancer molecular diagnostics is a new, interdisciplinary field with considerable promise. Biosensors have the potential to play a major role in early detection and treatment of cancer through molecular diagnostics as these technologies and our knowledge of cancer molecular markers advance. While many types of biosensors have been developed and used in a wide variety of biomedical and other analytical settings, biosensors are largely absent today in cancer detection and diagnostics. The aim of this book is to help bridge the gap between research and clinical application. This is accomplished by describing a range of biosensor technologies for cancer detection and diagnostics with the hope of facilitating the integration of biosensor technologies into cancer clinical practice.

Cancer Detection and Diagnostics

Cancer consists of more than 200 distinct diseases affecting more than 60 human organs. It is the second largest cause of death in developed countries. The highest number of deaths results from lung, colon, breast, pancreas, and prostate cancers as well as leukemia.

Cancer is characterized by uncontrolled and unlimited cell division, which leads to invasion of nearby tissues and metastasis to distant organs. To develop technologies for cancer diagnostics, it is important to understand the complexity and the basic elements of cancer biology, which can help define molecular detection targets. Cancer is a genetic disease caused by changes or modification of the DNA sequences of key genes. These changes result in altered gene or protein expression, resulting in altered protein composition of tumor cells. As DNA alterations accumulate, cancer cells become progressively disregulated, eventually leading to uncontrolled proliferation of cancer. Thus, the development of effective cancer detection and diagnostics relies on the utilization of unique cancer molecular markers (e.g., DNA mutations, protein modification, or alteration of gene expression). Moreover, the emergence of companion and personal medicine, which tailors treatment to the individual (e.g., prescribing cancer drugs based on the presence or absence of specific mutations), highlights both the need for the identification of specific molecular targets and the technologies to detect these targets.

Biosensors for Cancer Detection and Diagnostics

Cancer molecular target detection is critical for effective detection, diagnosis, and treatment. While in recent years many types of biosensors have been developed and used in a wide array of biomedical and other analytical settings, biosensors are largely absent in cancer detection and diagnostic practice despite the fact that cancers are probably the most complicated of diseases and are very prevalent. Biosensors are analytical devices that use biological ligands, such as antibodies, peptides, and aptamers, for detection. They have two main components: (1) a biological recognition element (ligand) that facilitates specific binding, via a biochemical reaction of a target to a ligand, and (2) a signal conversion unit (transducer) that measures the ligand binding to the target molecule. Transducers, based on various physical principles (such as optical, electrochemical, and optomechanical), convert the biochemical interactions between the target and the ligand into a measurable signal. Other elements of biosensors are the input/output systems to operate the device and fluidics systems to manage samples and reagents needed for the analysis.

Biosensors have several potential advantages over other biodetection methods for cancer analysis, especially for performing complex analysis, due to their increased assay speed and flexibility. Rapid, real-time analysis can provide immediate interactive information to health-care providers, which can be incorporated into the planning of cancer patient care. Many biosensors allow multitarget analyses, which are critical for the analysis of complex and heterogeneous cancer samples. Other beneficial features include automation and reduced cost of testing. Thus, biosensor-based diagnostics might facilitate improved cancer screening and increase the rates of early detection with attendant improved prognosis.

There are numerous types of biosensors utilizing various transduction modes, and the choice of a suitable detector is very complex and is based on many factors, such as the nature of the application, the labeled molecule (if used), the sensitivity required, the number of channels (or area) measured, cost, technical expertise, and the speed of detection needed.

However, despite promising biosensor capabilities, biosensors are essentially absent today for cancer clinical analysis. This is due to a combination of factors, including a lack of knowledge of biosensor capabilities on the part of decision makers, regulatory barriers, high cost, and entrenched industry standard methods.

The Book's Aims and Approaches

The aim of the book is to describe the major elements of cancer detection and diagnostics, the basic elements of biosensors and their applications to cancer detection and diagnostics, hoping, ultimately, to facilitate the integration of biosensor technologies into cancer detection and diagnostic practice. It addresses cancer molecular diagnostics (e.g., genomic or proteomic) from the perspective of biosensors and biodetection. It also explains how to measure and understand such molecular markers using biosensors and provides a discussion of the medical advantages of rapid and accurate cancer diagnostics. It then goes on describe major biosensor technologies (including optical, electrochemical, and optomechanical), focusing on cancer analysis and the clinical utility of these technologies for cancer detection and diagnostics, prognostics, and treatment. The book is designed to make biosensor technology more accessible and understandable to molecular biologists, oncologists, pathologists, and engineers so that it can play a major role in the early molecular diagnosis of cancer.

The Book's Organization

The book includes an introductory chapter describing some of the basic elements of cancer relevant to detection and diagnostics. The remaining chapters treat biosensor technology, including specific cancer applications.

The book chapters are organized according to the biosensor mode of detection, starting with a part on various modes of optical-based biosensors that are probably the most common approach for biosensing. The optical biosensors part describes a broad array of optical technologies, including surface plasmon resonance (SPR), evanescent wave and waveguides, spectrometry, optical imaging, fluorescence, luminescence, refractive index detection technologies, and photoacoustic analysis.

The part on electrochemical detection describes several modes of electrochemical biosensors, including potentiometric and amperometric biosensors and their applications for cancer detection and diagnostics.

The part on electronic and magnetic technologies for cancer analysis includes chapters on field effect transistors, magnetic resonance, and electric field–based biosensors.

In addition to these better known biosensors, the book includes parts on thermometric sensing and optomechanical cantilever-based technologies. To make biosensors accessible and understandable to a nonengineering audience, the chapters include a background on the technology and schematics of the devices.

We hope that this book will introduce biosensors to oncologists, molecular biologists, and clinicians, and that biosensors will play an instrumental role in the early detection, diagnostics, and treatment of cancer through molecular cancer analysis, as our knowledge of cancer molecular markers advances.

MATLAB® is a registered trademark of The MathWorks, Inc. For MATLAB® and Simulink® product information, please contact:

The MathWorks, Inc.
3 Apple Hill Drive
Natick, MA, 01760-2098 USA
Tel: 508-647-7000
Fax: 508-647-7001
E-mail: info@mathworks.com
Web: www.mathworks.com

Keith E. Herold
Fischell Department of Bioengineering
University of Maryland
College Park, Maryland

Avraham Rasooly
Division of Cancer Treatment and Diagnosis
National Cancer Institute
Rockville, Maryland

and

Division of Biology
Center for Devices and Radiological Health
Food and Drug Adminstration
Silver Spring, Maryland

Contributors

Sharmila Anandasabapathy
Division of Gastroenterology
Mount Sinai Medical Center
New York, New York

Aaron S. Anderson
Chemistry Division
Los Alamos National Laboratory
Los Alamos, New Mexico

Brian Athos
BioElectroMed Corporation
Burlingame, California

Vassilios I. Avramis
Division of Hematology/Oncology
Children's Hospital Los Angeles
Keck School of Medicine
University of South California
Los Angeles, California

Shabbir B. Bambot
Guided Therapeutics Inc.
Norcross, Georgia

Andrew C. Barton
Cranfield Health
Cranfield University
Bedfordshire, United Kingdom

Adela Ben-Yakar
Department of Biomedical Engineering
and
Department of Mechanical Engineering
The University of Texas at Austin
Austin, Texas

Shekhar Bhansali
Alcatel-Lucent Professor and chair
Electrical and Computer Engineering
Florida International University
Miami, Florida

Ramaswamy Bhuvaneswari
National Cancer Centre Singapore
Singapore, Singapore

Celia Bonaventura
Alderon Biosciences, Inc.
and
Nicholas School of the Environment
Duke University Marine Laboratory
Beaufort, North Carolina

Norman D. Brault
Department of Chemical Engineering
University of Washington
Seattle, Washington

François Breton
CyToCap
Ecole Normale Supérieure de Cachan
Institut d'Alembert
Cachan, France

Quincy Brown
Department of Biomedical Engineering
Duke University
Durham, North Carolina

Richard H. Bruce
Palo Alto Research Center
Palo Alto, California

Riccardo Castagna
Department of Applied Science and Technology
Politecnico di Torino
Torino, Italy

Cesar M. Castro
Center for Systems Biology
and
Cancer Center
Massachusetts General Hospital
and
Harvard Medical School
Boston, Massachusetts

Junseok Chae
School of Electrical, Computer, and Energy
    Engineering
Arizona State University
Tempe, Arizona

Madhumita Chatterjee
Department of Oncology
Karmanos Cancer Institute
Wayne State University School of Medicine
Detroit, Michigan

Soo khee Chee
National Cancer Centre Singapore
Singapore, Singapore

Wei Chen
Department of Oncology
Karmanos Cancer Institute
Wayne State University School of Medicine
Detroit, Michigan

Yu Chen
Department of Physics
Boston University
Boston, Massachusetts

Yuanzhong Chen
Fujian Institute of Hematology
Union Hospital
Fujian Medical University
Fuzhou, China

Chang Kyoung Choi
Department of Mechanical Engineering-
    Engineering Mechanics
Michigan Technological University
Houghton, Michigan

Seokheun Choi
School of Electrical, Computer, and Energy
    Engineering
Arizona State University
Tempe, Arizona

R.F. Chuaqui
National Institute of Health
National Cancer Institute
Rockville, Maryland

Ahmet F. Coskun
Department of Electrical Engineering
University of California, Los Angeles
Los Angeles, California

Frank Davis
Cranfield Health
Cranfield University
Bedfordshire, United Kingdom

Venkataraman Dharuman
Department of Bioelectronics and Biosensors
Alagappa University
Karaikudi, India

Yiwu Ding
Resonant Sensors Incorporated
Arlington, Texas

Yuhong Du
Department of Pharmacology
and
Emory Chemical Biology Discovery Center
School of Medicine
Emory University
Atlanta, Georgia

Steven P. Dudas
Department of Oncology
Karmanos Cancer Institute
Wayne State University School of Medicine
Detroit, Michigan

Nicholas J. Durr
Department of Biomedical Engineering
The University of Texas at Austin
Austin, Texas

V. Egorov
Artann Laboratories
Trenton, New Jersey

Marica B. Ericson
Department of Physics
University of Gothenburg
Gothenburg, Sweden

and

Department of Mechanical Engineering
The University of Texas at Austin
Austin, Texas

Shyamsunder Erramilli
Department of Physics
Boston University
Boston, Massachusetts

Xudong Fan
Department of Biomedical Engineering
University of Michigan
Ann Arbor, Michigan

Ye Fang
Science and Technology Division
Department of Biochemical Technologies
Corning Incorporated
Corning, New York

Ann M. Ferrie
Science and Technology Division
Department of Biochemical Technologies
Corning Incorporated
Corning, New York

Alex Fragoso
Departament d'Enginyeria Quimica
Universitat Rovira i Virgili
Tarragona, Spain

Haian Fu
Department of Pharmacology
and
Department of Hematology and Medical Oncology
and
Emory Chemical Biology Discovery Center
School of Medicine
Emory University
Atlanta, Georgia

Ann Gillenwater
Department of Head & Neck Surgery
MD Anderson Cancer Center
The University of Texas
Houston, Texas

Benjamin S. Goldschmidt
Department of Biological Engineering
University of Missouri
Columbia, Missouri

Shaoqin Gong
Department of Biomedical Engineering
and
Wisconsin Institutes for Discovery
University of Wisconsin-Madison
Madison, Wisconsin

W. Kevin Grace
Chemistry Division
Los Alamos National Laboratory
Los Alamos, New Mexico

Alex E. Grill
Department of Pharmaceutics
University of Minnesota
Minneapolis, Minnesota

Jong Hoon Hahn
Department of Chemistry
BK School of Molecular Science
Pohang University of Science and Technology
Pohang, South Korea

Nile Hartman
nGimat™
Atlanta, Georgia

Robert Henkens
Alderon Biosciences, Inc.
and
Nicholas School of the Environment
Duke University Marine Laboratory
Beaufort, North Carolina

Keith E. Herold
Fischell Department of Bioengineering
University of Maryland
College Park, Maryland

Ewa Heyduk
Edward A. Doisy Department of Biochemistry
    and Molecular Biology
School of Medicine
St. Louis University
St. Louis, Missouri

Tomasz Heyduk
Edward A. Doisy Department of Biochemistry
    and Molecular Biology
School of Medicine
St. Louis University
St. Louis, Missouri

Séamus P.J. Higson
Cranfield Health
Cranfield University
Bedfordshire, United Kingdom

Mi K. Hong
Department of Physics
Boston University
Boston, Massachusetts

H. Ben Hsieh
Palo Alto Research Center
Palo Alto, California

Yi-Heui Hsieh
Department of Chemistry
National Chung Hsing University
Taichung, Taiwan

Xu Hun
Key Laboratory of Eco-chemical Engineering
Ministry of Education
and
State Key Laboratory Base of Eco-chemical
    Engineering
and
Shandong Provincial Key Laboratory of
    Biochemical Analysis
and
College of Chemistry and Molecular Engineering
Qingdao University of Science and Technology
Qingdao, China
and
School of Materials and Chemical Engineering
Xi'an Technological University
Xi'an, China

Alireza Javadi
Department of Biomedical Engineering
and
Wisconsin Institutes for Discovery
University of Wisconsin-Madison
Madison, Wisconsin

Shaoyi Jiang
Department of Chemical Engineering
University of Washington
Seattle, Washington

Charles Kearney
Resonant Sensors Incorporated
Arlington, Texas

Ivan Keogh
School of Physics
National University of Ireland, Galway
and
University College Hospital, Galway
Galway, Ireland

Peter Koulen
Vision Research Center
Departments of Ophthalmology and Basic
    Medical Science
School of Medicine
University of Missouri-Kansas City
Kansas City, Missouri

Gregory J. Kowalski
Department of Mechanical and Industrial
    Engineering
Northeastern University
Boston, Massachusetts

Mark Kreis
BioElectroMed Corporation
Burlingame, California

Lee-Jene Lai
Scientific Research Division
National Synchrotron Radiation Research Center
Hsinchu, Taiwan

Dale Larson
Charles Stark Draper Laboratory
Cambridge, Massachusetts

Veronica Leautaud
Department of Bioengineering
Rice University
Houston, Texas

Hakho Lee
Center for Systems Biology
Massachusetts General Hospital
and
Harvard Medical School
Boston, Massachusetts

Genxi Li
Department of Biochemistry
and
National Key Laboratory of Pharmaceutical
    Biotechnology
Nanjing University
Nanjing, China
and
Laboratory of Biosensing Technology
School of Life Science
Shanghai University
Shanghai, China

Wei Liao
School of Dentistry
Dental Research Institute
University of California, Los Angeles
Los Angeles, California

Chwee Teck Lim
Graduate School for Integrative Sciences and
    Engineering
and
Division of Bioengineering
and
Department of Mechanical Engineering
and
Mechanobiology Institute
National University of Singapore
Singapore, Singapore

Wan Teck Lim
Department of Medical Oncology
National Cancer Center Singapore
and
Duke-NUS Graduate Medical School
Singapore, Singapore

Xinhua Lin
Faculty of Pharmacy
Department of Pharmaceutical Analysis
Fujian Medical University
Fuzhou, China

Ailin Liu
Faculty of Pharmacy
Department of Pharmaceutical Analysis
Fujian Medical University
Fuzhou, China

Shih-Jen Liu
Vaccine Research and Development Center
National Health Research Institutes
Miaoli, Taiwan

Xiaohe Liu
Palo Alto Research Center
Palo Alto, California

KaYing Lui
BioElectroMed Corporation
Burlingame, California

Robert Magnusson
Resonant Sensors Incorporated
and
Department of Electrical Engineering
University of Texas at Arlington
Arlington, Texas

Gary K. Maki
Integrated Molecular Sensors
Coeur d'Alene, Idaho

Wusi C. Maki
Integrated Molecular Sensors
Coeur d'Alene, Idaho

Ángel Maquieira
Departamento de Química
Instituto de Reconocimiento Molecular y
    Desarrollo Tecnológico
Universidad Politécnica de Valencia
Valencia, Spain

Niranka Mishra
University of Idaho
Moscow, Idaho

Jianwei Mo
Kumetrix, Inc.
Union City, California

and

UC Biodevices Corp.
Fremont, California

Pritiraj Mohanty
Department of Physics
Boston University
Boston, Massachusetts

Harshini Mukundan
Chemistry Division
Los Alamos National Laboratory
Los Alamos, New Mexico

Pamela Nuccitelli
BioElectroMed Corporation
Burlingame, California

Richard Nuccitelli
BioElectroMed Corporation
Burlingame, California

Malini Olivo
School of Physics
National University of Ireland, Galway
Galway, Ireland

and

Department of Pharmacy
National University of Singapore
Singapore, Singapore

Ciara K. O'Sullivan
Departament d'Enginyeria Quimica
Universitat Rovira i Virgili
Tarragona, Spain

and

Institució Catalana de Recerca i Estudis Avançats
Passeig Lluis Campanys
Barcelona, Spain

Aydogan Ozcan
Department of Electrical Engineering
and
California NanoSystems Institute
University of California, Los Angeles
Los Angeles, California

Gregory Palmer
Department of Radiation Oncology
Duke University
Durham, North Carolina

Jayanth Panyam
Department of Pharmaceutics
University of Minnesota
Minneapolis, Minnesota

Giljun Park
Department of Microbiology
The University of Tennessee, Knoxville
Knoxville, Tennessee

Mark C. Pierce
Department of Biomedical Engineering
Rutgers, the State University of New Jersey
Piscataway, New Jersey

Dorielle Price
Bio-MEMS and Microsystems Laboratory
Department of Electrical Engineering
University of South Florida
Tampa, Florida

Min Qui
Emory Chemical Biology Discovery Center
School of Medicine
Emory University
Atlanta, Georgia

Abdur Rub Abdur Rahman
Bio-MEMS and Microsystems Laboratory
Department of Electrical Engineering
University of South Florida
Tampa, Florida

Nimmi Ramanujam
Department of Biomedical Engineering
Duke University
Durham, North Carolina

Avraham Rasooly
Division of Cancer Treatment and Diagnosis
National Cancer Institute
Rockville, Maryland

and

Division of Biology
Center for Devices and Radiological Health
Food and Drug Adminstration
Silver Spring, Maryland

Björn M. Reinhard
Department of Chemistry
and
Photonics Center
Boston University
Boston, Massachusetts

Alexander Revzin
Department of Biomedical Engineering
University of California, Davis
Davis, California

Carlo Ricciardi
Department of Applied Science and Technology
Politecnico di Torino
Torino, Italy

Rebecca Richards-Kortum
Department of Bioengineering
Rice University
Houston, Texas

Emilie Roger
Department of Pharmaceutics
University of Minnesota
Minneapolis, Minnesota

Guoxin Rong
Department of Chemistry
and
Photonics Center
Boston University
Boston, Massachusetts

Kyle D. Rood
Department of Biological Engineering
University of Missouri
Columbia, Missouri

Carol L. Rosenberg
School of Medicine
Boston University
Boston, Massachusetts

A. Sarvazyan
Artann Laboratories
Trenton, New Jersey

N. Sarvazyan
Artann Laboratories
Trenton, New Jersey

Mehmet Sen
Department of Mechanical and Industrial
    Engineering
Northeastern University
Boston, Massachusetts

Ikbal Sencan
Department of Electrical Engineering
University of California, Los Angeles
Los Angeles, California

John E. Shively
Beckman Research Institute
City of Hope
Duarte, California

Lynell R. Skewis
Department of Chemistry
and
Photonics Center
Boston University
Boston, Massachusetts

George Somlo
Department of Medical Oncology and
    Therapeutics Research
City of Hope
Duarte, California

Tim E. Sparer
Department of Microbiology
The University of Tennessee, Knoxville
Knoxville, Tennessee

Ting-Wei Su
Department of Electrical Engineering
University of California, Los Angeles
Los Angeles, California

Brian J. Sullivan
Kumetrix, Inc.
Union City, California

Basil I. Swanson
Chemistry Division
Los Alamos National Laboratory
Los Alamos, New Mexico

Michael A. Tainsky
Department of Oncology
Karmanos Cancer Institute
Wayne State University School of Medicine
Detroit, Michigan

Min-Han Tan
Department of Medical Oncology
National Cancer Center Singapore
Singapore, Singapore

and

Genomic Medicine Institute
Cleveland Clinic
Cleveland, Ohio

Swee Jin Tan
Graduate School for Integrative Sciences and
    Engineering
National University of Singapore
and
Institute of Materials Research and Engineering
A*STAR (Agency for Science Technology and
    Research)
Singapore, Singapore

Liang Tiao
School of Materials and Chemical Engineering
Xi'an Technological University
Xi'an, China

Kevin Tran
BioElectroMed Corporation
Burlingame, California

Phuong-Lan Tran
CyToCap
Ecole Normale Supérieure de Cachan
Institut d'Alembert
Cachan, France

JrHung Tsai
Kumetrix, Inc.
Union City, California

John A. Viator
Department of Biological Engineering
and
Department of Dermatology
University of Missouri
Columbia, Missouri

Karthik Vishwanath
Department of Biomedical Engineering
Duke University
Durham, North Carolina

Hongyun Wang
Department of Chemistry
and
Photonics Center
Boston University
Boston, Massachusetts

Jing Wang
Department of Biochemistry
and
National Key Laboratory of Pharmaceutical
    Biotechnology
Nanjing University
Nanjing, China

and

Department of Chemistry
and
Photonics Center
Boston University
Boston, Massachusetts

Kun Wang
Faculty of Pharmacy
Department of Pharmaceutical Analysis
Fujian Medical University
Fuzhou, China

Xihua Wang
Department of Physics
Boston University
Boston, Massachusetts

Debra Wawro
Resonant Sensors Incorporated
Arlington, Texas

David T.W. Weaver
Ninth Sense, Inc.
Boston, Massachusetts

Fang Wei
School of Dentistry
Dental Research Institute
University of California, Los Angeles
Los Angeles, California

Kho Kiang Wei
National Cancer Centre Singapore
Singapore, Singapore

Ralph Weissleder
Center for Systems Biology
and
Cancer Center
Massachusetts General Hospital
and
Harvard Medical School
Boston, Massachusetts

Michael S. Wilson
EIC Laboratories, Inc.
Norwood, Massachusetts

David T.W. Wong
School of Dentistry
Dental Research Institute
University of California, Los Angeles
Los Angeles, California

Bo Yan
Department of Chemistry
and
Photonics Center
Boston University
Boston, Massachusetts

Jun Yan
Department of Biomedical Engineering
University of California, Davis
Davis, California

Minghui Yang
Department of Mechanical Engineering
University of Wisconsin-Milwaukee
Milwaukee, Wisconsin

Qiuming Yu
Department of Chemical Engineering
University of Washington
Seattle, Washington

Zhujun Zhang
School of Materials and Chemical Engineering
Xi'an Technological University
Xi'an, China

Hongying Zhu
Department of Biological Engineering
University of Missouri
Columbia, Missouri

Shelby Zimmerman
Resonant Sensors Incorporated
Arlington, Texas

Part I

Introduction

1

Cancer and the Use of Biosensors for Cancer Clinical Testing

R.F. Chuaqui, Keith E. Herold, and Avraham Rasooly

CONTENTS

1.1 Introduction

One purpose of this chapter is to describe the basic concepts and terminology of cancer to provide an introduction to bioengineers that are developing technologies for cancer detection and diagnostics. This chapter also includes a very general introduction of biosensors. Cancers are complex diseases, so our aim is to give an overview of the major issues that might impact biosensor technology development. The main elements required to adapt biosensors for cancer clinical testing are summarized in Figure 1.1. To develop cancer biosensor technology, cancer biomarkers identified from basic and clinical research and from genomic and proteomic analyses must be first validated. Identification of high-sensitivity, high-specificity biomarkers is the greatest challenge for cancer diagnostics. A key goal of this chapter is to

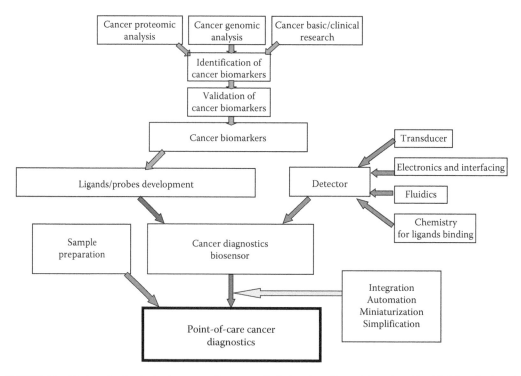

FIGURE 1.1 Strategy for developing biosensors for cancer clinical testing. Cancer biomarkers identified from basic and clinical research and from genomic and proteomic analyses must be validated. Ligands and probes for these markers can then be combined with detectors to produce biosensors for cancer-related clinical testing. Point-of-care cancer testing requires integration and automation of the technology as well as development of appropriate sample preparation methods.

address the issues of cancer biology which can lead to discovery of new cancer biomarkers. Ligands and probes are needed for the identification of validated biomarkers which can be combined with transducers to produce biosensors for cancer-related clinical testing. Other important elements are sample preparation and the "technical" format for testing. As shown in Figure 1.1, point-of-care cancer testing requires integration and automation of the technology as well as development of appropriate sample preparation methods.

1.1.1 General Tumor Features

Neoplasias are tissue growths that escape regular proliferation control mechanisms. Malignant neoplasia, that is, cancer, results from the gradual accumulation of multiple molecular alterations at the genomic (genetic and epigenetic), transcriptional, translational, and posttranslational level. Cancer consists of more than 200 distinct diseases affecting over 60 human organs. Malignant tumors are capable of invading adjacent tissues, as well as detaching groups of cells from the primary growth followed by implanting and proliferating at distant sites (the process of metastasis). Cancer cells undergo uncontrolled abnormal cell division, increasing the total number of cancer cells and leading to growth of the primary tumor. Cancer cells often acquire the ability to infiltrate adjacent normal tissues. In addition, some cancer cells acquire the ability to penetrate the walls of blood or lymphatic vessels and to move through the blood/lymph circulation (circulating tumor cells [CTCs]). Such cells can repenetrate through the vessel walls entering remote tissues where the cells continue to multiply forming a new, metastatic (or secondary) tumor (lymphatic or hematogenous spread). The adaptation to new sites is a long process (may take several years) and requires several genetic changes.

During the multistep carcinogenesis process, genes and protein expression in key pathways are altered, leading to dysregulation of cellular proliferation control mechanisms. The clinical aggressiveness of a malignant tumor depends on these changes, which provide growth advantages to the tumor over adjacent normal tissues as well as invasive and metastatic potential. These features include (1) unlimited cell division; (2) invasion of nearby tissues via degradation of extracellular matrix (ECM); (3) increased cellular motility, favoring metastasis to distant organs; (4) resistance to antigrowth signaling; (5) evasion of apoptosis (programmed cell death) leading to self-sufficient growth [1]; and (6) sustained vessel formation (angiogenesis) providing nutrients and oxygen to tumor cells [2]. A general feature of cancer is its complexity: tumors are histologically composed of a mixed cell population, including tumor cells, premalignant cells, normal epithelium, stroma, and inflammatory infiltrate (see Figure 1.2). Another

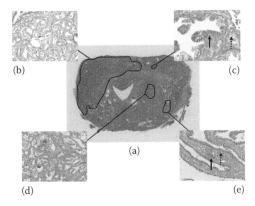

FIGURE 1.2 Example of cancer histopathology. Prostate cancer showing heterogeneous cell populations in the same tissue section. (a) Whole mount section of the prostate (H&E, 4×). Independent foci of invasive carcinoma can be seen, such as well-differentiated (b, 40×) and moderately differentiated tumors (d). In addition, nearby microscopic areas of prostate intraepithelial neoplasia (c, 40×), the preinvasive precursor of the tumor can be seen, with epithelial cells (arrow) and stromal cells (dashed arrow). Fibromuscular stroma, normal epithelium, and inflammatory infiltrate can also be found (e, 40×) with vessel (arrow) and lymphocytes (dashed arrow). When preparing clinical tissue samples for molecular analysis, it is key to consider this cellular heterogeneity. Pure cell populations from each component can be procured by performing tissue microdissection.

dimension of cancer complexity is the fact that it contains multiple molecular alterations, with a large number of genes and mutations leading to cancer.

1.1.2 Impact of the Disease

Cancer is the second largest cause of death in developed countries, with a total of 562,875 cancer deaths recorded in the United States in 2007, the most recent year for which actual data are available, accounting for approximately 23% of all deaths in the United States (http://caonline.amcancersoc.org/cgi/content/full/caac.20073v1?ijkey=c6e8e191143f64c78e36b7d4ce9efb872f538147#BIB13). According to these statistics, the expected number of deaths from the various cancers projected for 2010 among men is as follows: cancers of the prostate (32,060, 11% of cases), lung and bronchus (86,200, 29%), colorectum (26,580, 9%), and pancreas (18,770, 6%). The projected deaths from the four most common cancers among women in 2010 are as follows: breast (39,840, 29%), lung and bronchus (71,080, 26%), colorectum (24,760, 9%), and pancreas (18,030, 7%). Although cancer treatments have progressed impressively since the war on cancer began in 1971, cancer remains a potent killer, and the most effective aspect of the current medical arsenal is early detection.

1.1.3 Biosensors for Cancer Detection

Biosensors consist of a recognition element (ligand or receptor), a signal conversion unit (transducer), and an output interface. Biosensors have the potential to play a key role in reducing the impact of cancer through early detection, which may lead to improved cancer detection and diagnostics and thus reduce the impact of the disease and improve public health through early and more personalized treatment.

1.2 Main Elements of Cancer Biology

1.2.1 Cancer Typing, Classification, and Nomenclature

Cancers can originate in various tissue types, and the tissue of origin is used to define the malignant cell type. The most common type is carcinoma, malignant tumors originating from epithelial cells (an example is shown in Figure 1.2). Sarcomas are cancers originating from supporting tissue cells (including bone and soft tissue such as cartilage, fat, and muscle). Lymphomas are cancers that originate in the lymph nodes and cells of the lymphatic system. Leukemias are cancers of the immature blood stem cells that grow in the bone marrow and normally accumulate in large numbers in the bloodstream. Cancer can originate in every organ in the body. The highest number of deaths results from cancers in the lung, colon, breast (females), pancreas, prostate (males), and blood (leukemia).

Histologically, malignant tumors are characterized by general architectural distortion, cellular atypia, and lack of differentiation. Cancers are described by their pathological grade and clinical stage. These histological classification systems describe how well the cells that make up the tumor are differentiated and how far cancer has spread anatomically. Grading (how abnormal the cells appear under the microscope) and staging systems (the clinical stage of tumors from small localized cancers to metastatic cancer) vary significantly for different tumor types depending on the organ from which the tumor originated. In general, tumor classification correlates with the patient's prognosis although there is significant heterogeneity in patient outcomes within a given tumor stage and grade.

Figure 1.2 shows a micrograph of a prostate cancer section stained with hematoxylin and eosin (H&E). Prostate cancer presents a particularly complex histology since several histological foci and cell types are usually admixed, including higher-grade tumor (Figure 1.2b), lower-grade tumor (Figure 1.2c) with epithelial cells (arrow) and stromal cells (dashed arrow), intraepithelial neoplasia, that is, preinvasive prostate lesion (Figure 1.2d), as well as normal tissue, including stroma and epithelium (Figure 1.2e) with blood vessels (arrow), and lymphocytes (dashed arrow). This cell heterogeneity represents a challenge when molecular analysis is performed on prostate tissue, especially if pure populations of cells are required for the correct interpretation of the data.

1.2.2 Staging of the Disease

Malignant diseases progress through clinical stages, which describe how much a cancer has spread in a patient. Tumor stage is the leading prognostic indicator of the disease. The probability of a cure is much higher with early stage disease. One of the most commonly used staging systems is the TNM classification [3]. TNM refers to the local invasion of the tumor (T), involvement of regional lymph nodes (N), and presence of distant metastasis (M). As an example, a T1, N1, M0 stage corresponds to a small primary tumor (T1), presence of tumor cells in only a few regional lymph nodes (N1), and no distant metastases (M0). A T3, N2, M1 would correspond to a tumor with deeper invasion, more distant lymph node involvement, and presence of distant metastases. Specific TNM classifications vary depending on the organ site of the tumor.

1.2.2.1 Preinvasive or Intraepithelial Neoplasias

In general, cancer arises in individual cells which show some histologically malignant features before invasion of adjacent tissues takes place. Detection and treatment of the disease at this early stage provides a high probability of curing the disease.

1.2.2.2 Primary Tumor

It may take years for a preinvasive lesion to progress to the stage of invasion of adjacent tissues. Frequently preinvasive and invasive microscopic regions coexist in the same tissue (Figure 1.2). Once that initial invasion takes place, the disease evolves through progressive spreading stages: local invasion, regional metastasis (mostly lymph nodes), distant metastasis.

1.2.2.3 Local Invasion

Local invasion is a complex process that depends on a series of steps. Some of the cellular functions (and their corresponding genes) involved in tumor invasion are attachment (integrins, CD44, VLA-4, ICAM-1, OPN), dissociation (cadherins), basement membrane disruption (collagens, fibronectin, elastin, proteoglycans), proteolysis of ECM (matrix metalloproteinases, interstitial collagenases, stromelysins, gelatinases, TIMPs), and motility (HGF, IGF-II, ATX, vitronectin, fibronectin, laminin, type I and IV collagen, thrombospondin, IGF-I, IL-8, histamine) [4]. Although the specifics of the invasion depend strongly on which of these many disease pathways are taken, the end result is that the tumor breaks out of the tissue where it originated and begins to disrupt its host on a larger scale.

1.2.2.4 Metastatic Disease

Metastasis is a multistep process which results in the growth of tumor cells at sites distant from the primary neoplasm. Metastatic disease is critical to clinical management of cancer since the majority of cancer mortality is associated with disseminated disease rather than a localized, primary tumor. To successfully colonize a secondary site, a cancer cell must complete a sequential series of steps. First, the malignant cells have to break away from the primary tumor, followed by attachment, degradation, and breaching of the surrounding ECM. This is followed by penetration of the walls of circulatory vessels and movement through the blood/lymph circulation (lymphatic or hematogenous spread), and these cells are called CTCs. Repenetration through the vessel walls allows access to remote tissues where the cancer cells adapt to the new site and multiply forming a new metastatic (or secondary) tumor.

Metastasis may occur years or even decades after from the initial diagnosis of the primary tumor which significantly complicates clinical monitoring of the disease. In general, metastasis is an extremely inefficient process because the body has defenses against such rouge cells. To successfully colonize a distant site, a cancer cell must complete all of the steps of the metastatic cascade before it is destroyed by the immune system. Tumors can shed many millions of cells into the bloodstream daily [5], yet very few clinically relevant metastases are formed [6]. Cancer cells in the bloodstream have been shown to arrest in capillary beds and extravasate with high efficiency [7]. Most such cells exist in dormancy at the secondary site for long periods of time [7] even for years [8,9]. A small fraction of CTCs that extravasate form micrometastases, but the majority of these lesions appear to regress [7], probably due to apoptosis [10]. Although only a very small number of CTCs form metastatic tumors, such tumors often elude early diagnosis and tend to be hard to treat. Microenvironmental factors are thought to play important roles at the secondary site, as well as at the primary tumor, in determining the state (i.e., dormant or proliferative) of the cancer cells.

Several models describing mechanisms of metastasis were recently reviewed [11]. The most commonly accepted one is the progression model which suggests that a series of mutations are needed either in subpopulations of the primary tumor or in disseminated cells, resulting in a small fraction of cells that acquire full metastatic potential. The inefficiency of metastasis is explained by the low probability that any given cell within the primary tumor will acquire all of the multiple alterations required for the successful implementation of the metastatic cascade. An alternative early oncogenesis model suggests that the majority of cells within a primary tumor possess an inherent metastatic capacity. In this model, a prometastatic gene signature expressed within only a small subpopulation of cells within the primary tumor initiates metastasis. The transient metastatic compartment model suggests that not all cells within a tumor retain the capacity to colonize secondary sites due to random or microenvironmentally induced epigenetic events or inadequate access to vasculature. The cellular fusion model suggests that cellular fusions are a source of genomic instability contributing to metastatic progression. A second, related hypothesis for the acquisition of metastatic capacity is based on the role of horizontal gene transfer.

Many molecular alterations implicated in metastasis have been described, including epidermal growth factor receptor (EGFR) ligand epiregulin (EREG), cyclooxygenase-2 (COX2), and matrix metalloproteinases 1 and 2 (MMP-1 and MMP-2) that have been shown to be associated with metastasis of breast cancer to the lung [12]. A recent review [13] describes more than 45 genes associated with metastasis, including more than 16 promoter genes implicated in invasion and metastasis, several genes involved in transcriptional regulation (*CTNNB1, EZH2, MTA1, MYC, SMAD2*, and *SNAI1*), and various receptors (*HER2* and *HGFR*). Nearly 30 other genes were identified as suppressor genes implicated in invasion and metastasis that serve at various steps of cell signaling (autocrine and paracrine ligands), receptors, signal transducers, transcription factors, or modulators and mediators of cellular response [13].

1.2.3 Biological Processes in Carcinogenesis and Tumor Progression

Carcinogenesis is a complex multiphase biological process involving many genes and affecting numerous molecular and cellular functions. The biological development and progression of a tumor depends on these processes, including initial neoplastic conversion, local invasion, metastasis, and angiogenesis. A tumor at the original site (primary tumor) contains a heterogeneous population of cancer cells and normal cells at a ratio which depends on the stage of the tumor. Early detection and treatment of the primary tumor before metastatic spread to secondary sites is key to the control and treatment of cancer. CTCs play an important role in metastasis. The biggest opportunity in controlling cancer is early identification of the primary tumor. Unfortunately, the primary tumor can often grow without outward symptoms leading to initial diagnosis at a late stage. Cancer of unknown primary site (CUP), which corresponds to metastasis in the absence of an identifiable primary tumor, is the fourth most common cause of cancer death [14]. Biological features which play a role in the development and progression of tumors are discussed in this section.

1.2.3.1 Cell Proliferation

The central biologic element in cancer is the uncontrolled growth of abnormal cells. Cancerous mutations at the genomic (genetic and epigenetic), transcriptional, translational, and posttranslational modification level usually impact one of two broad categories of genes: oncogenes, promoting cell proliferation, and tumor suppressor genes (TSGs), which control cellular growth.

Oncogene: Oncogenes are genes that, when mutated or when their expression is altered, can turn a normal cell into a tumor cell. In their normal state, oncogenes are called proto-oncogenes. Mutations in proto-oncogenes can lead to an abnormally active protein, either via protein modification (e.g., phosphorylation) or increased protein concentration. Either case can lead to abnormal cell survival and division. An example is RAS [15], a small GTPase which resides on the inner surface of the plasma membrane and serves to link receptor binding to kinases like RAF. RAS and RAF are key components of the mitogen-activated protein kinase (MAPK/ERK) pathway, a signal transduction pathway that couples intracellular responses to the binding of extracellular growth factors such as epidermal growth factor (EGF) to the membrane EGFRs. A chain of tyrosine kinases are activated, which transfer the signal into the nucleus activating DNA transcription factors (promoters), which turn on their associated genes. MAPKs activate many transcription factors such as c-myc, c-fos, and CREB; MAPK is dysregulated in at least one-third of all cancers [16]. The high frequency of activating mutations centered around the RAS–RAF axis suggests that this is the regulatory hotspot of the pathway [16]. Identification of such RAS mutations is important especially for targeted therapeutics. For example, the *KRAS* oncogene is constitutively activated by a small set of specific mutations which almost all occur in codons 12 and 13 of exon 2. The presence of somatic mutations in the *KRAS* gene is associated with poor prognosis and lack of response to anti-EGFR therapy in colorectal (CRC) and non-small-cell lung cancer (NSCLC) patients [17]. Six of the most common mutations of *KRAS*, all in codon 12, have been associated with lack of response to EGFR-targeted therapies (e.g., Erbitux or Vectibix) in both CRC and NSCLC patients [18]. Effective therapeutics require the analysis of these mutations.

Tumor suppressor genes: TSGs are genes which suppress tumor formation. TSG function can be lost in cancer resulting in dysregulated cell growth. TSGs include several well-known genes such as *p53*, which is involved in cell cycle control and regulation of apoptosis [19,20]. Loss of p53 is found in 70% of colon cancers, 30%–50% of breast cancers, and 50% of lung cancers [21]. In small cell lung cancers, 75%–100% exhibit p53 mutations [22]. The so-called two-hit theory [23] is the most common mechanism proposed to explain inactivation of TSG, whereby both alleles must be inactivated in order to disrupt the gene's function. Most frequently, one allele is mutated, and the second allele is inactivated via deletion.

1.2.3.2 Circulating Tumor Cells

CTCs are cancer cells that shed from tumors and are circulating in the peripheral blood. CTCs, which are typically present at concentrations less than a few cells per milliliter of blood, constitute seeds for metastasis and are instrumental for the spread of the disease. The number of CTCs has been shown to correlate with clinical outcome. Various studies have shown that the existence of CTCs in blood and/or bone marrow samples is of prognostic significance in different types of solid tumors and that CTC numbers also help monitor response to treatment [24–28]. It was shown that in patients with metastatic breast cancer, the median survival of patients with values of 5 or more CTCs per 7.5 mL of blood was 2.7 months compared with survival of 7.0 months in patients with fewer than 5 CTCs per 7.5 mL of blood [28]. Analysis of CTCs may allow early detection of metastasis processes prior to the invasive stage, and the ability to remove CTCs from circulation could potentially limit metastases after surgery. However, accurate detection of CTCs is difficult due to their small numbers. Furthermore, the presence of CTCs usually means that the disease has progressed to a late stage. CTCs are generally not found in early stage cancer.

1.2.3.3 Angiogenesis

Angiogenesis is a process involving the growth of new blood vessels which is essential to tumor maintenance and growth. A negative correlation is observed between patient survival and the degree of vascularization of several tumor types, including gastric [29], breast [30–32], prostate [33], esophageal [29], vulvar [34], and melanoma [35]. Several steps are important in the process of angiogenesis, including (1) proliferation of endothelial cells, (2) breakdown of the ECM, and (3) migration of endothelial cells [4]. These steps can be promoted by growth factors secreted by tumor cells. These factors are termed angiogenic factors and include the heparin-binding growth factor or fibroblast growth factor family, transforming growth factor-A, angiogenin, vascular permeability growth factor (VPF), vascular endothelial growth factor (VEGF), and others [36].

1.2.3.4 Apoptosis (Programmed Cell Death) Genes

Apoptosis plays a major role in homeostasis by removing cells to make way for new cells so as to maintain cell populations in tissues. The mechanism of apoptosis is highly complex, involving a cascade of molecular events, including (a) an extracellular inducer pathway, (b) an intrinsic mitochondrial pathway, and (c) a T-cell-mediated cytotoxicity [37]. All converge on the same terminal, execution pathway: caspase activation leading to DNA fragmentation. Several genes and pathways are involved in each of these three apoptosis pathways. To date, the best-characterized ligands and corresponding death receptors include FasL/FasR, TNF-α/TNFR1, Apo3L/DR3, Apo2L/DR4, and Apo2L/DR5 [37]. Several stimuli are factors that participate in the intrinsic stress-induced apoptosis pathway, such as radiation, toxins, hypoxia, hyperthermia, viral infections, and free radicals. Changes in the inner mitochondrial membrane result in the loss of the mitochondrial transmembrane potential and release of proapoptotic proteins (cytochrome c, Smac/DIABLO, and the serine protease HtrA2/Omi, as well as AIF, endonuclease G, and CAD) [37]. Cytotoxic T lymphocytes (CTLs), as well as specific caspases and granzymes A and B, play a central role in T-cell-mediated cytotoxicity.

1.2.3.5 Tumor and Cancer Cell Dormancy

Tumor dormancy has been defined clinically as "the disease-free period" between clinical "cure" of the primary cancer and its subsequent local or distant recurrence/metastasis [38]. Cancer cell dormancy is a significant clinical problem because primary treatment of a cancer may be apparently successful, but the tumor may recur years or decades later, introducing uncertainty regarding treatment. Clinical trials in breast cancer have shown that some patients develop metastases years after removal of the primary tumor implying dormancy and that these dormant cancer cells may in some cases be effectively treated with long-term therapy [8]. The sites harboring the dormant cells are not well known although bone marrow is often cited and the signals that initiate and maintain dormancy are poorly understood. Proposed mechanisms include angiogenic metastatic dormancy due to reduced expression of angiogenic promoters such as VEGF and/or high expression of inhibitors such as thrombospondin [39,40]. Disseminated tumor cells (DTCs) in the bone marrow of cancer patients were shown to be negative for proliferation markers such as PCNA or Ki67 [41–43]. Several other mechanisms which may be involved in dormancy were reviewed recently [38], which include a G_0–G_1 arrest with high expression of p21 and p27, reduced signaling through adhesion molecules such as α5β1, and adhesion signaling pathways such as focal adhesion kinase (FAK) observed in dormant tumor cells.

1.2.3.6 Drug Resistance and Cancer Stem Cells

Cells with a capacity for long-term renewal are known in normal compartments such as epithelial linings and hemopoietic systems [44]. These multipotent progenitors can undergo rapid cell division and repopulate their compartments. The discovery of a stem cell compartment in tumors has strongly influenced the understanding of carcinogenesis as well as chemotherapy. Tumor stem cells may not only represent the origin of tumors but also are capable of expressing high levels of drug transporters that confer drug resistance. The most common drug transporters are ATP-binding cassette (ABC) transporters, such as ABCB1, which is an efflux pump with broad substrate activity. These transporters can create drug-resistant pluripotent cells that can survive chemotherapy and potentially reinitiate the tumor.

1.2.3.7 DNA Repair

DNA damage can occur due to both normal metabolic processes and external factors, such as UV-light exposure or chemical mutagens. Therefore, normal cells have a series of mechanisms that not only detect DNA damage but also are able to repair the damage, creating an important protection mechanism. Several genes are involved in these detection and repair processes. These key genes can be altered in cancer. Not only does inactivation of these genes impair the ability of the cell to repair DNA damage, but also the resulting additional mutations can be part of tumor initiation and accelerated tumor progression [45,46].

1.2.4 Inherited and Somatic Changes in Cancer

Both inherited genetic traits and somatic cell mutations can contribute to cancer. Somatic cell mutations arising in individual cells account for most cancers. Environmental factors (such as diet, lifestyle, and exposure to ultraviolet radiation or carcinogenic pollutants) are thought to play a major role in the development of most cancers. In general, exposure to specific carcinogens is at low levels, and each person's defense mechanisms are unique. Combined with the multistep nature of cancer initiation, this makes cancer incidence appear sporadic. It is estimated that only 5%–10% of cancers are hereditary [47]. Hereditary cancers arise in individuals with specific germline mutations that convey a much higher risk of developing cancer. Examples of hereditary cancers include breast-ovarian cancer associated with mutations in the *BRCA1/2* genes [48,49], and CRC associated with mutations in the *HNPCC* or *FAP* genes [50].

There are many types of genetic alterations that can lead to cancer initiation. Cancer-related mutations can result from changes in single bases within the DNA sequence, from DNA replication errors (RER) that add or delete bases within mononucleotide or dinucleotide repeat sequences [51], from malfunction of the DNA mismatch repair genes leading to alterations in the length of simple repetitive sequences (microsatellite instability), or from deletion or insertion of DNA sequences ranging from short sequences to large portions of chromosomal arms. In addition, many chromosomal changes, including increased or decreased number of chromosomes, and chromosomal rearrangements have been identified in cancer cells.

1.2.4.1 Epigenetic Changes in Cancerous Cells

Epigenetic changes are modifications of DNA that affect gene expression without altering DNA nucleotide sequences. The predominant epigenetic changes are cytosine methylation, histone deacetylation, histone methylation, and loss of imprinting (LOI) [52].

Aberrant cytosine methylation: The most common epigenetic modification in cancer is methylation of the C5 position of cytosine within the nucleoside deoxycytosine [53–56]. Methylation is observed most often in promoter-associated CpG islands. Methylation silences genes by interfering with sequence-specific binding of transcription factors or by producing more general effects on chromatin. For example, more than half of sporadic cases of colon cancer with microsatellite instability exhibit promoter hypermethylation of a DNA mismatch repair gene, the *hMLH1* gene [57]. Transcriptional silencing through methylation may occur in combination with other inactivating mechanisms [58], such as a combination of hypermethylation and loss of heterozygosity (LOH) of the *BRCA1* in certain breast and ovarian tumors [59–63]. LOH is the loss of the functional allele of a gene.

Alterations in histone acetylation: Variant histone acetylation patterns are present in many tumors [64,65]. One striking example is the class of acute leukemia cases in which a fusion of the MLL (mixed lineage leukemia) and histone acetyltransferase (CBP) genes leads to disrupted histone acetylation and gene silencing [66].

Aberrant histone methylation: Histone methyltransferases methylate lysines within the tails of histones H3 and H4 [67] which contain a conserved SET domain [68]. SET domain proteins are multifunctional chromatin regulators. Some cancers (e.g., CRC and hepatocellular carcinomas affecting the liver cells) are associated with overexpression of SET-methylating enzymes [69,70].

Loss of imprinting: Imprinting is an epigenetic change that occurs in the gamete and causes reduced expression of one of the two alleles of a gene in somatic cells. LOI results in expression of both alleles. The result may contribute to tumorigenesis either by activating a transcriptionally repressed allele or by inactivating an expressed allele of an imprinted TSG [71]. LOI-mediated silencing of expression of the single functional allele has been demonstrated in ovarian cancer [72] and breast cancer. There is some evidence that LOI works through altered methylation [73].

1.2.4.2 RNA Interference

In addition to the well-known genetic and epigenetic roles in cancer, in recent years, the role of RNAi has become a new focus of research. RNAi involves small noncoding RNAs that regulate target gene expression posttranscriptionally. This newly discovered mechanism for gene-expression inhibition [74,75] acts by degradation of specific RNA transcripts, thus interfering with mRNA translation [76], or epigenetic silencing [77,78] through RNA-directed DNA methylation. Short interfering RNA (siRNA) recognizes specific mRNA targets by homologous base pairing [74,75] and results in their degradation. Micro-RNA (miRNA) silences target genes either by degrading mRNA molecules or by inhibiting their translation [79]. This genetic interference can influence invasion and metastasis, and miRNA has been shown to modulate the translation of several suppressor genes, including PTEN, CD44, SERPINB5, or of promoter genes, including RAS [80].

miRNAs operate along multiple pathways, making them a powerful cancer diagnostic tool for early detection, risk assessment, prognosis, and therapeutics. For example, a microarray profile of 94 chronic lymphocytic leukemia (CLL) patients identified a panel of nine miRNAs (*miR-181a, miR-155, miR-146, miR-24–2, miR-23b, miR-23a, miR-222, miR-221,* and *miR-29c*) whose expression was associated with prognosis and progression in CLL [81,82]. *miR-141* was shown to be present at a much higher level in patients with prostate cancer [83] suggesting utility as a stable blood-based marker for cancer detection. Similarly, several circulating miRNAs in plasma of patients with gastric cancers (GCs) have shown potential utility as complementary tumor markers for GC [84] and specifically to identify the origin of unknown primary cancers (UPC) [85].

1.2.5 Proteomics

Proteomics is the large-scale study of proteins using new technologies based on protein separation (e.g., 2D gels), protein identification (e.g., mass spectrometry [MS]), and bioinformatics (e.g., the capability to analyze and reconstruct protein sequences). Proteomics enables the study of the proteome (the entire set of proteins expressed by a genome), enabling the identification of proteins involved in cancer. The number of such cancer-relevant proteins is much larger than the number of promoter and suppressor genes [13], many of which encode transcriptional and translational regulators. For example, the transcription of CDH1, encoding the invasion-suppressor E-cadherin, is regulated by Snail, Slug, and bHLH, receiving signals from more than 20 autocrine and paracrine ligands via multiple signal transduction pathways [86]. Moreover, each ligand is regulated posttranscriptionally (through phosphorylation and sumoylation). Such complex pathways, involving cascades of interrelated proteins (with many redundancies), are the hallmark of cancer which can be deciphered only by proteomics because

- Proteins are the endpoint of biological processes, and thus, proteins reflect pathological changes much more directly than changes in DNA or RNA.
- Gene-expression analysis does not always correlate well with protein expression due to myriad complexities in the translation process.
- Posttranscriptional modifications (e.g., phosphorylation, glycosylation, and sumoylation) play critical roles in protein activity and can be detected only with proteomic studies.
- The vast number of protein–protein interactions essential for biological processes can be detected only with proteomic studies.
- Proteomics allows direct monitoring of dynamic changes in the state of a cell or tissue.

Proteomic analysis is important clinically for diagnostic biomarker discovery and validation, discovery of molecular targets for treatment, therapy decisions, and for patient monitoring. Unlike genomics, which is a relatively mature field, proteomics is only slowly producing clinically verified biomarkers. However, proteomics has great potential to develop such markers for cancer diagnostics.

1.2.6 Cancer Genes, Pathways, and Clusters

It is critical to understand the complexity of pathway alterations in tumorigenesis when developing biosensors for cancer testing. The genes involved in carcinogenesis usually belong to one of two broad categories: oncogenes, promoting cell proliferation, and TSGs, which control cellular growth. However, changes can involve other groups of genes controlling functions not directly related with cell proliferation, but nevertheless key in cancer cell survival and progression. These genes can affect relevant processes such as invasion, angiogenesis, DNA repair, apoptosis (programmed cell death), or drug resistance (see Section 1.2.3). To understand the complexity of the molecular changes involved in cancer, it is important to keep in mind that many genes may play a role in more than one of these functions. As an example, p53 is a TSG that is inactivated in a significant proportion of cancers and plays a role in several p53-dependant apoptosis pathways [87] or is involved in p53-mediated inhibition of tumor angiogenesis [88]. In addition, it has been shown that p53 plays central roles in DNA repair [89] regulation of cell motility [90]. Multiple examples could be cited, so it is important to keep in mind that most of the oncogenes and TSGs can, besides proliferation, influence several cellular processes that provide the tumor advantages over neighboring tissues.

Large-scale sequencing methods have enabled analyses of mutated genes that cause common human cancers. Such mutation analyses of common human tumors have revealed cancer genomic maps composed of a few frequently mutated genes clustered in a limited number of central molecular pathways. Systems biology approaches are being used to identify and better define these core pathways. There are many classes of metabolic pathways such as signaling, cell cycle, cell death, and DNA repair. A few examples of cancer pathways are provided in the following text, including topics in DNA repair and EGF signaling. However, in almost all cancers, multiple pathways are involved, and such examples are provided as follows for pancreatic cancer and glioblastomas.

1.2.6.1 Examples of Specific Cancer Pathways

1.2.6.1.1 DNA Repair Pathways and Mechanisms

The publication of the human genome revealed 130 genes whose products participate in DNA repair. DNA damage can occur due to external agents, such as ionizing radiation, ultraviolet rays, highly reactive oxidants, or chemicals, including many hydrocarbons and many chemotherapeutic agents. DNA damage can also result from normal metabolic activities. Therefore, a constant DNA repair process (Section 1.2.3.7) is in place in mammalian cells. Failure of this system can lead to permanent DNA damage and accumulation of potentially carcinogenetic mutations [91]. On the other hand, DNA repair pathways can provide malignant cell tools to overcome DNA damage induced by chemotherapeutic agents and therefore survive the treatment. Therefore, DNA repair proteins have also become targets for cancer therapy [92].

The three major DNA repair mechanisms are base excision, nucleotide excision, and mismatch repair. These mechanisms are best understood in *E. coli*, although they are thought to be similar in eukaryotic cells. In base excision, the bases may be modified by deamination or alkylation. The modified base is recognized and removed by a DNA glycosylase and endonuclease, and the gap is filled by DNA polymerase I using the other strand as the template. Finally, the two ends of the repaired strand are joined via DNA ligase. In nucleotide excision in *E. coli*, proteins UvrA, UvrB, and UvrC are involved in removing the damaged nucleotides, and the gap is then filled by DNA polymerase I and DNA ligase. Mismatch can occur due to internal errors in DNA synthesis (replication), and the repair mechanism recognizes the mismatch. Key genes are MSH2, MLH1, and DNA polymerases. Recognition of a mismatch requires several proteins, including the MSH2 product. MLH1 is one of the critical proteins involved in cutting the mismatch. The repair is performed by the DNA polymerase delta. All of these processes are critical for normal cell metabolism, and all of these processes are implicated in various cancers when they stop functioning properly.

1.2.6.1.2 Epidermal Growth Factor Receptor Signaling Pathway

The EGFR signaling pathway is one of the most important pathways that regulate mammalian cell growth, proliferation, survival, and differentiation [6,7]. The EGFR pathway represents a typical signaling system with multiple input and output mechanisms/molecules interconnected to targets with high redundancy, feedback control loops, and extensive cross talk with other pathways. As an example, aberrant regulation of the EGFR pathway due to amplification of HER2/neu is implicated in about 20% of breast cancers [1]. Cross talk with other growth factor receptor pathways is found in breast cancer, including the estrogen (EP) and progesterone (PR) receptors, which are involved in 75% of breast cancers [2,3]. Another example of cross talk is with the DNA repair pathways mediated by poly (ADP-ribose) polymerase PARP4 [93].

1.2.6.2 Examples of Multipathway Cancers

1.2.6.2.1 Pancreatic Cancer Signaling Pathways

A comprehensive DNA analysis of 23,219 transcripts, representing 20,661 protein-coding genes from 24 pancreatic cancers, was performed [94]. The DNA analysis included sequencing and a search for homozygous deletions and amplifications in the tumor DNA by microarray. The results suggest that pancreatic cancers contain an average of 63 genetic alterations, the majority of which are point mutations. The analysis defined a core set of 12 cellular signaling pathways and processes that were genetically altered in 67%–100% of the tumors. The core signaling pathways and processes genetically altered in most pancreatic cancers are apoptosis DNA damage control regulation of G1/S phase transition, Hedgehog signaling, homophilic cell adhesion, integrin signaling, c-Jun N-terminal kinase signaling, KRAS signaling, regulation of invasion, small GTPase–dependent signaling (other than KRAS), TGF-b signaling, and Wnt/Notch signaling. Dysregulation of these core pathways can explain the major features of pancreatic tumorigenesis.

1.2.6.2.2 Glioblastoma Pathways

An analysis of data (DNA copy number, gene expression, and DNA methylation) from the Cancer Genome Atlas (TCGA) from 206 glioblastomas (the most common type of primary adult brain cancer) shows DNA sequence mutations in 91 of the 206 glioblastomas. This analysis provides new insight into the roles of known cancer-related genes (such as ERBB2, NF1, and TP53). The analysis found frequent mutations of the phosphatidylinositol-3-OH kinase regulatory subunit gene PIK3R1 and provided a more integrated understanding of the multiple pathways altered in the development of glioblastomas. Inclusion of mutation, DNA methylation, and clinical treatment data reveals a link, with potential clinical implications, between MGMT (methylated-DNA–protein-cysteine methyltransferase gene) promoter methylation and a hypermutator phenotype related to mismatch repair deficiency in treated glioblastomas [95,96]. Somatic nucleotide substitution, homozygous deletion, and focal amplification analysis identified a highly interconnected network of aberrations, including three major pathways: RTK signaling, p53, and retinoblastoma (RB) protein suppressor. The same pathways emerged from copy number data analysis alone and from sequencing data.

1.3 Cancer Analysis, Detection, and Diagnostics

Earlier cancer detection and accurate diagnosis are critical for improving the outcomes for cancer patients. Currently, morphological and histological characteristics of tumors are the most important diagnostic and prognostic indicators of cancer along with individual biomarkers, such as prostate-specific antigen (PSA). However, given the complexity of cancer, the utility of such single markers, is often limited. As we have seen, there is increasing use of molecular tools, both genomic and proteomic, to profile multiple molecular alterations in tumors and produce "molecular signatures" based on many variables. Since tumor development involves many biological changes, the molecular signatures can be highly complex. Correlation of these signatures with clinical observations can yield information useful to physicians in support of patient management decisions such as selection of effective cancer treatment. In addition, the ability to rapidly and efficiently identify a reliable set of disease-specific biomarkers is key to enable a future of personalized medicine. A significant challenge for the medical community is to devise practical means to discover these complex molecular signatures in the clinical setting.

1.3.1 Cancer Biomarkers

The FDA defines a biomarker as follows: "A characteristic that is objectively measured and evaluated as an indicator of normal biological processes, pathogenic processes, or pharmacologic responses to a therapeutic intervention" [97]. Cancer biomarkers are molecular characteristics elements that can be detected in samples from cancer patients. These biomarkers (molecular signatures) are produced when cancer is present, either by the tumor itself or as a response to the tumor. They may include DNA, DNA modifications, RNA, proteins or protein modifications, or other biological molecules. Cancer biomarkers can be used in several ways [98], including risk screening, detection, diagnosis, prognosis, and monitoring. Biosensor technologies have great potential for identification, characterization, and clinical application of cancer biomarkers.

Risk screening and detection markers: These markers can be used to screen asymptomatic individuals who are at risk of cancer such as testing for mutations in BRCA1. The U.S. Preventive Services Task Force issued recommendations for screening for BRCA1 and BRCA2 gene mutations in [99,100]. Even though routine screening for the two BRCA genes is not recommended, the Task Force does recommend screening women with specific family medical histories.

Diagnostic and disease monitoring markers: Markers are used for identification, characterization, and monitoring progression of cancer, including markers used to identify diseased individuals among those presenting with vague symptoms or to distinguish among subclasses of disease. This includes PSA for detection and monitoring for recurrence of prostate cancer or CA-125 monitoring for recurrence of ovarian cancer.

Prognostic and predictive markers: Prognostic markers provide information that correlates to outcome (e.g., tumor recurrence or patient survival) independent of the treatment the patient receives. An example is the spread of cancer cells to lymph nodes which correlates with increased likelihood of tumor recurrence. Predictive markers correlate with patient response to a specific therapy. For example, breast cancers that overexpress the estrogen receptor (ER) tend to respond to receptor antagonist therapies such as tamoxifen.

Biomarker assays: The key performance measures of a biomarker assay are sensitivity of detection of the disease (measures the proportion of actual positives which are correctly identified), specificity of the assay (measures the proportion of negatives which are correctly identified), and the predictive value of the assay (proportion of patients positive for the biomarker who are correctly diagnosed). Although many studies have searched for definitive correlations between clinical endpoints and cancer biomarkers, there are still very few clinically proven markers that are used by oncologists to make patient management decisions. This has been attributed to variable patient populations, tumor characteristics, treatments, and assay methods in the studies to discover, characterize, and validate cancer biomarkers [101]. The National Cancer Institute (NCI; http://cadp.cancer.gov/) is placing significant emphasis on research to confirm and validate biomarkers for clinical applications.

1.3.2 Single Target Biomarkers

Single biomarkers used today for cancer clinical testing are primarily specific proteins detected in higher-than-normal amounts in patients with certain types of cancer. Although, in general, single biomarkers lack the sensitivity and the specificity needed, they have some clinical utility, and several biomarkers for cancer diagnostics were reviewed recently [102]. Examples of such tumor markers include PSA and prostatic acid phosphatase (PAP) for prostate cancer; CA 15–3 or CA 27–29 for breast cancer; mucin-like carcinoma-associated antigen (MCA) for breast (metastatic) cancer; CA 125 for ovarian cancer; CA 19–9 for gastric, pancreatic, or biliary tract cancers; carcinoembryonic antigen (CEA) for colorectal, pancreatic, or gastric cancers; human chorionic gonadotropin-b (HCG) for pancreatic, lung, or bladder, and rarely, gastrointestinal, lymphoma, or breast cancers; HER2/neu for breast (primary or metastatic) cancer; neuron-specific enolase (NSE) for small-cell cancer of the lung and neuroblastoma; thyroglobulin (hTG) for follicular or papillary carcinoma of the thyroid; and tissue polypeptide antigen (TPA) for lung, bladder, prostate, ovarian, liver, pancreatic, thyroid, head and neck, or colon cancers.

Biomarkers of metastatic cancer: In addition to biomarkers used for cancer detection and diagnostics, several biomarkers have utility for analysis of metastatic cancers:

Estrogen and progesterone receptors: Normal breast cells express estrogen and progesterone receptors on the cell surface. In some breast cancers, these receptors are overexpressed. Hormone receptor overexpression is prognostic of rapid tumor growth. Hormone receptors are predictive markers for the response of breast cancer to hormonal therapy (therapy that blocks body hormones from stimulating growth of breast cancer).

HER2 expression: HER2 is a member of a family of transmembrane tyrosine kinase receptors involved in several major cellular processes, including control of cell growth, cell survival, differentiation, and migration [103,104]. *HER2* overexpression occurs in approximately 20%–25% of breast cancers [105] and gastric tumors (approximately 9%–38%) [106]. *HER2* overexpression is prognostic of very aggressive breast cancer tumor growth which can be blocked by anti-HER2 agents (trastuzumab and lapatinib), and increased *HER2* gene copy number is associated with response to gefitinib therapy in EGFR-positive NSCLC patients [107].

EGFR: EGFR is a member of the HER family of tyrosine kinases involved in proliferation, angiogenesis, invasion, and metastasis through initiation of a mitogenic signaling cascade. Overexpression of EGFR has been found to occur in approximately 40%–80% of NSCLC, conferring a poor prognostic outcome [108,109]. EGFR mutations function as both prognostic and predictive biomarkers among patients with metastatic NSCLC.

KRAS: The *KRAS* gene is present in almost all malignant tumors, and it was shown to be critical for cell growth and tumor development in 35%–45% of colorectal adenocarcinomas and in up to 30% of lung adenocarcinomas [110,111]. As discussed earlier, wild-type *KRAS* is predictive of response to treatment with anti-EGFR antibodies in colorectal tumors, and the presence of a *KRAS* mutation is indicative of bad prognostic outcome among colorectal and non-small-cell lung tumors.

Due to the complexity of cancer, single biomarkers usually have limited specificity and selectivity which limits their clinical utility. Exceptions include biomarkers developed as diagnostic markers to predict the effectiveness of targeted treatments (see Section 1.3.3.2). The complexity of cancer and the number of proteins/genes/pathways/mutations involved in cancer typically require multitarget panels for cancer diagnostics.

1.3.3 Gene or Protein Signatures and Multigene Profiling

Unlike single biomarkers based on a single molecular characteristic, genetic or protein signatures rely on many genes or proteins, typically analyzed by high-throughput technologies such as microarrays. Such signatures can be developed from data which identify combinations of molecular characteristics with statistical correlation to disease.

1.3.3.1 Gene or Protein Signatures for Breast Cancer Diagnosis

One example is breast cancer in which expression profiling studies have identified a number of molecular signatures that correlate with cancer in different cell types [112]. These include (a) luminal types of tumors expressing epithelial-like genes such as E-cadherin and cytokeratins 8 and 18 [112–115]; (b) basal-like breast tumors, which are positive for cytokeratins 5 and 6 [114,115], typically present in normal myoepithelial cells; (c) HER2/neu-positive tumors; and (d) the so-called "normal" variants of breast tumors.

- Luminal cancers of the mammary gland tissues are predominantly hormone receptor–positive and express cytokeratins 8 and 18. Luminal cancers can be divided into luminal A (mostly histologically low grade) and luminal B (high grade with a worse prognosis).
- HER2-positive cancers which, as discussed earlier, are characterized by amplification and overexpression of the HER2/neu gene (hormone receptors) are of poor prognosis.

Basal-like are typically "triple-negative," that is, do not express ER, PR, or HER2, and they correlate with poor prognosis. Such subtyping enabled prediction of the clinical outcome of 49 women with locally advanced breast cancer [114], and it was shown that relapse-free survival was poorest among women whose tumors were of HER2 or basal-like subtypes. However, a study using a different set of samples [116] suggests that the luminal A and basal are the most robust phenotypic subtypes of breast cancer, that is, they most consistently stratify different sets of samples among multiple laboratories.

Gene-expression profiles were used to develop a 128-gene signature which distinguishes between primary and metastatic adenocarcinomas from the same tumor types [117]. When the 128-gene signature was applied to 279 primary solid tumors of diverse origin, it was able to predict metastases and prognostic outcome ($p < 0.03$). Of the 128 original genes, a subset of 17 genes within the primary tumors (e.g., breast and prostate adenocarcinomas) enable prediction of the future development of metastases.

Similar to the issues of single biomarkers, the complexity of cancer compounded with large genetic variation and numerous mutations limits specificity and selectivity of multitarget panels for cancer diagnostics.

1.3.3.2 Companion Diagnostic Tests to Predict the Effectiveness of Treatment

One trend in cancer diagnostics is the development of laboratory tests designed to screen patients as candidates for treatment with particular drugs. Targeting treatments to patients who are more likely to see benefits is an effective strategy to increase overall treatment effectiveness. If a particular drug is target specific, a single biomarker can be used to verify the presence or the state of the target (although multiple biomarkers can also be used). Companion diagnostics is a critical area in which biosensors can play an important clinical role as these tests are increasingly important because of the current trend to develop anticancer drugs that target gene mutations, including:

1. Herceptin, a monoclonal antibody approved for treatment of HER2-positive metastatic breast cancer
2. Gleevec, a tyrosine kinase inhibitor (TKI) (protein) for patients with Philadelphia chromosome-positive chronic myeloid leukemia (CML) and gastrointestinal stromal tumors
3. Erbitux, an EGFR inhibitor for CRC patients
4. Iressa, an EGFR inhibitor for patients with NSCLC
5. Tarcevas, another EGFR inhibitor for the treatment of advanced or metastatic NSCLC

Examples of biomarkers used for companion diagnostic tests are described next.

Cell surface receptor CD20: CD20 is a protein expressed in the surface of normal B-lymphocytes. However, it can also be present in some lymphomas and leukemias. Its presence in these tumors predicts efficacy of the monoclonal antibody rituximab treatment, which targets the CD20 antigen [118]. It is also used for transplant rejection and some autoimmune disorders [119].

Test to identify patients who may respond to Herceptin: In breast cancer, overexpression of HER2 is an indication for targeted therapy with Herceptin (trastuzumab), an antibody directed against the HER2 receptor. Overexpression of HER2 on the cell surface results primarily from abnormal DNA replication leading to extracopies of genes (gene amplification). Because only patients with tumors which overexpress the HER2 protein respond to Herceptin, a DNA hybridization assay for the HER2 gene locus on chromosome 17q11.2–21 can be used to determine HER2 gene amplification. Such an assay (SPoT-Light HER2 CISH) was approved by the FDA in 2008 to identify patients who may respond to Herceptin. This is a direct test to determine the existence of the Herceptin target.

Predictor of benefit from tamoxifen: Other companion diagnostic tests utilize biomarkers which are not directly related to the target but are instead correlated to the responsiveness to the treatment. An example is the Oncotype DX breast cancer test for women with early stage breast cancer who are ER-positive and whose lymph nodes are negative. It utilizes a panel of 21 genetic markers to determine a recurrence score (RS) (likelihood of breast cancer recurrence within 10 years) and is a predictor of benefit from tamoxifen treatment (see following text) which is an ER antagonist. The Oncotype DX test is a polymerase chain reaction (PCR)-based assay designed to analyze both fresh and fixed tissue samples.

MammaPrint is a microarray-based prognostic breast cancer gene-expression profiling test of 70 genes developed using 78 patients [120]. This test provides information about the likelihood of tumor recurrence (metastases within 5 years) and enables to classify patients as high- or low-genomic risk (good or poor prognosis categories). A second validation also confirmed the prognostic capacity of the gene signature [121]. The U.S. Food and Drug Administration (FDA) approved MammaPrint for clinical use (2007) as a prognostic test (for lymph node–negative breast cancer patients under 61 years of age with tumors of less than 5 cm) with the aim to determine the need for adjuvant chemotherapy or hormone therapy (e.g., tamoxifen). MammaPrint is designed to analyze only fresh or frozen tissues from stage I or stage II and lymph node–negative (in the United States) or three or fewer lymph nodes (in Europe).

Prediction of chemotherapy benefit in endocrine-responsive, early breast cancer using Oncotype DX and MammaPrint multigene assays was studied recently [122]. In this study, both assays define groups of patients with low scores who do not appear to benefit from chemotherapy and a second group with very high scores who derive major benefit from CMF or CAF chemotherapy. The use of these assays led to a change in treatment decision in 30% of the cases, most of the times avoiding chemotherapy that would not have been beneficial for these patients.

Tests to monitor Gleevec resistance: The molecular basis of CML (chronic myelogenous leukemia) is a mutation (*BCR-ABL*) which leads to a fusion protein of *ABL* with *BCR*. The fused BCR-ABL protein is a constitutively active tyrosine kinase. This enzyme is the specific target for Gleevec (imatinib mesylate) which inhibits *BCR-ABL* activity as well as other tyrosine kinases. One hypothesis for TKI resistance is TKI-induced selection of subclones differentiating into immature B-cell progenitors that pass on TKI resistance.

Many relapses are associated with a secondary mutation in the ABL portion of the gene which correlates with treatment failure. However, in about 50% of patients, clinical resistance does not appear to be linked to known mutations and is attributed to a multifactorial array of mechanisms. Resistance testing can help guide alternative CML treatment. CML patients on Gleevec can be monitored with quantitative reverse-transcriptase–polymerase-chain-reaction (qRT-PCR), and patients with a log increase in BCR-ABL levels should be assayed for the presence of Gleevec-resistant mutations [123–126]. Depending on the results, an increased dosage of Gleevec (or other kinase inhibitor) may control the levels of *BCR-ABL*. Alternative therapies for CML patients with Gleevec resistance depend on mutation type (e.g., the T315I mutation confers resistance to multiple targeted TKI) [127]. Mutation-specific PCR assays that use hydrolysis probes and an array of allele-specific primers enable analysis of nine Gleevec resistance mutations associated with the substitution of six amino acid residues [128].

Direct sequencing, alone and in combination with denaturing high-performance liquid chromatography (LC), and two high-sensitivity allele-specific oligonucleotide PCR approaches were compared for analysis of BCR-ABL mutations [129], and it was found that the combination is a reliable screening technique for the detection of BCR-ABL kinase domain mutations.

Predictor of benefit from Erbitux or Vectibix: Another companion diagnostic test utilizes genetic markers which are not directly related to the target but are instead correlated to the responsiveness to the treatment. EGFR is overexpressed in a large majority of colon cancers. Two anti-EGFR monoclonal antibody drugs (Erbitux or Vectibix) are indicated for treatment of metastatic CRC [17,18]. Activation of EGFR triggers multiple downstream signaling pathways resulting in promotion of tumor growth, inhibition of apoptosis, vascular proliferation, invasion, and metastasis. One of the most important pathways appears to be the RAS signaling pathway [130].

Developing companion diagnostics: Developing companion diagnostics requires understanding of both the mechanism of the anticancer drug and the biology of its specific target. For example, KRAS is known to be involved in the development and progression of colon cancers where KRAS oncogene is constitutively activated by a small set of specific mutations which almost all occur in codons 12 and 13 of exon 2 in about 40% of colon cancers independent of EGFR-mediated signaling. Thus, a constitutively expressed KRAS protein would not likely be affected by inhibition of the upstream EGFR by anti-EGFR drugs [17]. Companion diagnostics for EGFR inhibitors thus must include KRAS mutation analysis. PCR and gene sequencing are widely used for KRAS mutation analysis. However, the challenge in such mutation analysis is the heterogeneity of the samples; often (especially in early stages) samples are a mixture which contains also a large amount of the wild-type gene; effective detection method must enable the detection of the mutant gene in the presence of access (e.g., 1/100) of the wild-type gene.

1.3.4 Genomic and Proteomic Approaches for Cancer Analysis, Detection, and Diagnostics

Proteomic and genomic analyses generate vast amounts of data, but these approaches are primarily research tools that may not be practical for routine cancer clinical testing because of their complexity. In addition, as discussed earlier, cancer samples are most often heterogeneous (e.g., Figure 1.2) containing both cancer and noncancer cells and both mutant and wild-type genes. To overcome this problem, laser capture microdissection (LCM) technology was developed which enables the selection of only cancer cells for analysis. In addition to sample and genetic variability, the cumbersome and lengthy protocols introduce issues of reproducibility and must be carried out by highly skilled technicians.

However, it is often possible to get clinically useful information from a smaller, more manageable number of biomarkers. For example, an assay that quantifies the likelihood of breast cancer recurrence in women with newly diagnosed, early stage breast cancer was developed using only 16 target genes analyzed using an RT-PCR assay [131]. This small group of target genes was chosen after analysis of the expression of 250 candidate genes selected from a variety of sources. The assay was able to predict risk of disease recurrence and to predict whether or not patients will benefit from chemotherapy for cases with node-negative, ER–positive breast cancer.

When a set of useful biomarkers can be reduced to a small number of assays, biosensor-based detection becomes practical and advantageous for cancer clinical testing because it is faster, more user friendly, less expensive, and less technically demanding than microarray or proteomic analyses. However, significant technical development is still needed, particularly for protein-based biosensors.

As we have seen, carcinogenesis is the result of an accumulation of events involving multiple pathways and groups of genes (proteins), including TSGs, oncogenes, and/or DNA repair genes. Traditionally, one of the main goals of molecular studies using clinical tumor samples has been the identification of individual markers with potential clinical utility. Even though several such clinical markers have been identified (see Section 1.3.2) and are in use today, the genomic era has spawned efforts toward the study of the overall pattern of the genome, transcriptome, and proteome in each sample [132–134]. Using techniques that can rapidly study a large number of analytes per sample (i.e., high-throughput screening [HTS]), the goal is to identify patterns of transcripts or proteins that have prognostic, diagnostic, or therapeutic utility [135]. Many HTS techniques can analyze multiple samples in each experiment. Such HTS techniques are at the center of current cancer biomarker discovery and validation efforts. Tumor signatures using HTS techniques have been identified at all molecular levels, including DNA, RNA, and protein.

1.3.4.1 High-Throughput Screening for the Identification of DNA Signatures

Cancer-related mutations can result from changes in single bases within the DNA sequence or from deletion or amplification (i.e., duplication) of DNA sequences ranging from individual bases to short sequences to large portions of chromosomal arms. In addition, many chromosomal changes including an increased or decreased number of chromosomes and chromosomal rearrangements have been identified in cancer cells.

DNA sequencing has significantly evolved from a slow, manual base-by-base procedure to an impressively high-throughput, automated technique in the past 30 years. Originally described by Sanger et al.

in the 1970s, the early approach allowed the determination of single DNA fragment sequences on a gel [136,137]. Sequencing has moved through several generations of technology development and is progressing rapidly, with the so-called Next Gen Sequencing approaches being developed with the capability to sequence the entire human genome 100 times faster than the technology used to sequence the first human genome [138]. Most of these methods are based on amplification of genetic material by PCR, ligation of amplified material to a solid surface, and sequencing of hundreds of thousands of targets using sequence by synthesis or sequence by ligation. The sequencing is done in a massively parallel fashion, and a large number of data are captured by a computer. Based on these technologies, sequencing is becoming a low-cost commodity service industry enabling many current biomarker discovery studies.

Sequencing can be performed by hybridization with a DNA chip (http://www.illumina.com/documents/products/technotes/technote_denovo_assembly.pdf) [139]. A variation on the theme is the application of comparative genomic hybridization (CGH), a technique that allows the comparison between two sources of DNA, usually normal vs. tumor DNA. Unbalanced chromosomal changes (gains/losses) in all chromosomes can be identified simultaneously, generating a pattern of changes in a single experiment. Microarray-based CGH has been able to generate cancer signatures in a variety of tumors, including various solid malignancies [140], prostate cancer chromosomal deletions and additions [57,58], and leukemia [141]. Although CGH techniques have proven very useful, Next Gen Sequencing may soon allow direct sequencing of cancer cell genomes and comparison to healthy cell genomes using the computer.

The identification of systematic deletions in tumor genomes at one locus suggests the presence of a TSG that is eliminated or inactivated by the deletion [23]. The deletions can be detected by comparing tumor vs. matched normal DNA. When the normal DNA is heterozygous at the TSG locus, a single deletion (or mutation) can disable the TSG. Such deletions are termed LOH and were originally analyzed using single microsatellite amplification [142]. Microsatellites represent well-documented, easily recognized signposts in the genome that provide a convenient deletion assay. However, the application of multiplex PCR has allowed the analysis of multiple microsatellites in a single experiment, even from minute LCM samples [143,144]. More recently, single nucleotide polymorphisms (SNPs) have overtaken microsatellites as the preferred genomic signposts for deletion analysis using DNA microarrays which allow for high-resolution genotyping of as many as a half million SNPs in a single assay [145–147]. Imbalances such as gene amplification [148] as well as genome-wide LOH can be detected using this method [149]. These methods have been used to identify deletion signatures in several tumor types [150], some of them with direct clinical implications such as clinical response to therapy in breast, ovary, and hematological malignancies [151–153]. It remains to be seen if these methods will be supplanted by direct sequencing.

The MitoChip is another example of an HTS DNA platform. This array-based sequencing chip was developed for the rapid and high-throughput analysis of mitochondrial DNA (mtDNA) [154–157]. Single nucleotide changes, insertions, and deletions in the mitochondrial genome can play an important role in carcinogenesis and are detected using this method [157]. Interestingly, there is evidence that mtDNA mutations could be a reliable tool for molecular assessment of cancer risk. Dasgupta identified massive clonal mtDNA mutation patches that develop in lifetime smokers and ultimately give rise to clinically significant cancers [158].

Epigenetic signatures can also be identified using high-throughput array-based analysis. As discussed in Section 1.2.4.1, epigenetic features are modifications of DNA that affect gene expression without altering DNA nucleotide sequences. The predominant epigenetic features are cytosine methylation, histone deacetylation, histone methylation, and LOI [52]. Currently, platforms using "BeadChip" technology can generate a comprehensive genome-wide profile of human DNA methylation [159].

1.3.4.2 High-Throughput Screening for the Identification of RNA Signatures

Quantitative RT-PCR (qRT-PCR) is routinely used in initial validation of RNA biomarkers [135]. qRT-PCR allows for the precise measurement of the level of expression of the transcript of interest. Multiple samples can be analyzed in a single experiment. Comparison to so-called housekeeping genes allows for increased precision through normalization of analyzed samples. Oncotype DX is a successful example of a qRT-PCR assay with clinical implications and whose development depended on the availability of clinical trial specimens. The Oncotype Dx™ was tested on FFPE breast cancer tissue [131] from the Trial Assigning Individualized Options for Treatment (TAILORx) breast cancer trial. The development went through several phases, including the study of 250 genes on archival material by high-throughput qRT-PCR. These were then reduced to a 21-gene signature. RS identified patients with low, intermediate, or high recurrence risk (10-year 6.8%, 14.3%, and 30.5%, respectively, $p < 0.001$). Significant benefit from chemotherapy was demonstrated in high-risk RS patients (decrease of 26.7% in absolute risk), while no benefit from chemotherapy was shown in low-risk RS patients. The multicenter TAILORx (ECOG) is integrating the 21-gene assay into the clinical decision-making process.

MammaPrint is a microarray-based prognostic breast cancer gene-expression profiling test of 70 genes as described in Section 1.3.3.2. The inherent high-throughput features of a microarray assay are harnessed to assay for a gene-expression signature. This microarray analysis provides information about the likelihood of tumor recurrence (distant metastases within 5 years) and enables accurate prognostic classification of tumors as well as information about sensitivity to available chemotherapy (tamoxifen).

Analysis of whole transcriptome has been a goal of many of the HTS platforms that have been developed [135,160,161]. Microarrays are currently among the most robust analysis tools for measuring the transcriptome and are excellent at generating data sets for transcriptome comparison (e.g., tumor vs. normal tissue, between cancer types, histologic and phenotypic subtypes) [135,162]. Usually signatures or profiles of a few hundred differentially expressed genes define each cell or tumor type. These "molecular signatures" are potential biomarkers for several types of cancer, including breast, lung, ovarian, oral, brain, CRC, and prostate cancers, as well as melanomas, lymphomas, and leukemias [163], and may be useful in classifying tumors and predicting response to therapies. In addition, microarray data on differential expression can yield clues about the role of both individual genes and pathways in cancer. Some striking examples of signatures with clinical implications have been identified in subtypes of lymphomas. For example, two classes of diffuse large B-cell lymphoma (DLBCL), with distinct prognostic and therapeutic implications, have been identified. One type is responsive to current therapies, and new therapies are being sought for the other [164,165].

This demonstrates that a molecular classification of tumors based on gene-expression profile can help identify clinically significant subtypes of cancer. Molecular classification could be a useful supplement to current histological methods. Molecular profiles with clinical implications have also been identified in Burkitt and mantle-cell lymphomas [166,167].

Coupling fine microdissection techniques (LCM) with array technology has enabled the identification of expression signatures of individual histological compartments. As an example, Richardson et al. were able to compare epithelial and stromal cells between tumor and adjacent normal areas in prostate cancer sections [168]. Signatures composed of 35–44 genes were identified that characterize the different compartments, including recognizable changes occurring specifically in the stroma surrounding tumor cells. In a study of breast tumor stroma, a new stroma-derived prognostic predictor (SDPP) was identified, allowing stratification of disease outcome independent of standard clinical prognostic factors [169]. The 200 most variable genes between normal and tumor areas were studied and used to generate various predictors, but the one with the best performance was composed of only 26 genes.

miRNAs are a large class of small noncoding RNAs that regulate protein expression in eukaryotic cells and therefore play a key role in carcinogenesis. Recently, technologies have been developed to facilitate high-throughput analysis of these molecules in array-based approaches [170,171]. RNA signatures can be identified that can help to molecularly define classes and subclasses of tumors. The data can increase the understanding of pathogenesis of diseases such as the relationship between various lung cancers [172]. In addition, these profiles may ultimately be useful in the management of cancer. For example, a recent study identified nine miRNA signatures able to discriminate between clinically relevant DLBCL categories [173]. Each signature was composed of a few (3–15) key miRNAs.

1.3.4.3 High-Throughput Screening for the Identification of Protein Signatures

High-throughput proteomic approaches have been used to generate protein expression maps (PEMs) of tumors. In a study of breast cancer [174], it was found that among a total of 43,302 proteins detected across 20 samples, 170 were elevated twofold or more in the cancer cells compared to normal cells. When normal squamous epithelium and corresponding tumor cells from two patients with esophageal cancer were compared, 17 out of approximately 675 distinct proteins (or isoforms) exhibited tumor-specific alterations [175]. Although similar proteomic studies have been carried out on many tumor types [176], as with all new technologies, there is no widely accepted standardization, and there are continuing issues of reproducibility.

The development of HTS techniques for the analysis of proteins is a continuing challenge for researchers, since no method currently exists to amplify proteins, in contrast to PCR which is a primary tool in nucleic acid analysis. Protein microarrays (PMAs) were developed to provide a high-throughput method of analyzing the tissue proteome by generating large expression data sets. Comparisons to gene-expression transcript data can elucidate the protein/transcript expression ratio. PMAs can be used both for initial biomarker discovery as well as for patient screening [177,178]. New proteomic techniques are being applied to identify the temporal changes in protein levels associated with tumor development. The main technologies used in proteomic studies are various chromatography separation technologies, two-dimensional gel electrophoresis, MS, and protein arrays [93].

Separation and identification of proteins: Two-dimensional denaturing polyacrylamide gel electrophoresis (2D-PAGE) is one of the basic proteomic technologies for biomarker discovery. In the first dimension, separation is carried out by isoelectric focusing (IEF) in a pH gradient, followed by separation according to molecular mass (SDS-PAGE) using electrophoresis. The proteins are then detected by staining the proteins in the gel with silver or Coomassie blue. MS techniques can be used for protein identification by precise determination of the molecular weight of proteins or protein fragments from individual spots on the gel combined with bioinformatic analysis.

MS is a technique used for measuring mass to charge ratio (m/z) of gas-phase ions and is the primary method of protein identification. An MS instrument consists of an ion source which charges the molecules, a mass analyzer that uses an electromagnetic field to accelerate and separate the ions according to their m/z ratio, and a detector. MS analysis enables identification and characterization of proteins, analysis of protein interactions, determination of posttranslational modifications, and for quantification of proteins. Two main techniques are used for ionization of proteins: matrix-assisted laser desorption/ionization (MALDI), utilizing a laser to vaporize the sample molecules dispersed in a matrix, and electrospray ionization (ESI), in which the liquid sample is ionized by applying an electrical charge as the liquid is pumped into a vacuum chamber as a jet.

To reduce sample complexity prior to MS, several separation technologies have been used. The most widely used is 2D-PAGE as described in this section. Other approaches include coupling the MS to LC or capillary electrophoresis (CE). Tandem MS (MS/MS) is used for protein/peptide identification and sequencing applying multiple steps of MS separated by a fragmentation step.

Cancer proteomics is based on the comparison of proteins in diseased and normal tissues and utilizes MS to identify the proteins. One approach involves combining two protein samples, with differential labeling using stable isotopes, followed by MS analysis. An alternative label-free approach involves separate analysis of individual samples using MS.

Several examples of proteomic analysis of cancer samples can be found in the literature. One involves profiling of serum protein in pancreatic cancer patients [179] which identified 16 differentially expressed proteins. Protein cluster discrimination analysis, obtained from MS, for the classification of ovarian cancer serum samples showed sensitivity, specificity, and positive predictive values all above 97% [180]. Proteomic patterns were developed for classification of ovarian cancer and cutaneous T-cell lymphoma (CTCL) serum samples using very simple algorithms for classification of cancer and normal samples [181]. A seven-peak proteomic pattern enabled effective (sensitivity and specificity greater than 93%) discrimination of 33 advanced ovarian cancer patients from 31 noncancer controls [182]. Quantitative proteomics based on size separation and MS was used to compare serum of women with ovarian cancer to healthy women and women with benign ovarian tumors. The analysis resulted in the identification of 19 validated markers [183].

Protein microarrays: PMA technology involves the simultaneous parallel analysis of a large number of different proteins in one single experiment. Protein arrays are emerging as an important tool of proteomic analysis including studies of protein expression, posttranslational modifications, and various protein–protein interactions. There are several PMA types: (a) label-based antibody array, where the protein sample is pretreated with labels to allow detection [135,184–186]; (b) sandwich antibody array, where two antibodies (one bound to the array and one labeled) are used to detect the analyte; (c) reverse phase array, where the protein sample is immobilized on the surface [135,187]; and (d) single-protein array, where only one protein of interest is used and immobilized on the surface. These approaches have been successfully used to elucidate potential early phase biomarkers from microdissected prostate cancer tissues [135].

Another widely used HTS method for protein analysis is the construction of tissue microarrays (TMAs) [188]. TMAs are generally used for in situ protein analysis using conventional immunohistochemical techniques. TMAs are excellent for validating gene-expression array data in large sets of tissue samples [135]. They are typically constructed from cores of formalin-fixed paraffin-embedded (FFPE) tissue samples inserted in a high-density array where hundreds of cases can be studied in one staining process. They can be constructed to compare cancer-relevant biomarkers; for example, multiple biomarkers were found by TMA to correlate with prostate cancer progression [135,189–192]. Prostate cancer prognosis markers identified using TMAs include Ki-67, cyclin D1, Tag, E-cadherin, alpha- and beta-catenin, Bcl-2, hMSH2, and Stat5 [193–201]. Several valuable clinical features can be assessed using TMAs. For example, at the NCI, various TMAs (breast, colon, and melanoma) were constructed to specifically enable comparison of clinically relevant subgroups of tumors. These TMAs were statistically designed to permit comparisons of biomarker expression across stages of disease and to predict survival and recurrence outcomes.

Layered peptide arrays correspond to a novel high-throughput proteomics technique [132,202]. In this approach, protein samples can be analyzed by passing the samples through various ultrathin membranes coated with capture peptides. The proteome contained in the fluid samples (e.g., multiwell plates), gels, or even tissues can be moved through the layers, keeping their two-dimensional resolution. Up to 5000 measurements per experiment can be performed [202–204].

1.3.5 Analysis of Clinical Samples

Various models have traditionally been used to perform molecular analysis of cancer, including human cells in vitro (cell lines), laboratory animals, and human tissue (ex vivo) specimens. Cell lines are generally straightforward to grow in the laboratory yielding easy access to scalable amounts of sample. However, cell lines may not necessarily maintain all the molecular features of the tissues from which they were derived ([205–207] http://cgapmf.nih.gov/ProstateExample/ProstateTissueProcessing/CellLines.html). The direct analysis of clinical samples can provide the best opportunity to closely reflect the biological alterations as they occur in humans in vivo. However, molecular analysis of clinical tissue samples involves several challenges, including regulatory issues, tissue sample acquisition and handling which can affect the quality of biomolecules, limited amounts of sample, and limited number of cases for a particular study [205]. Finally, tissues are complex, composed of a mixed cell population, including tumor cells, premalignant cells, normal epithelium, stroma, and inflammatory infiltrate. Tissue microdissection has become a key tool to procure the desired cell population for studies (e.g., normal vs. tumor cells). However, microdissected material usually yields small amounts of DNA, mRNA, or proteins, generally in the nanogram (ng) range [205].

For any of these models, subsequent analysis requires that a sample be taken and processed in some manner to prepare it for whatever assay is needed. The sample processing depends on what biomolecules are being detected as discussed next.

1.3.5.1 Sample Preparation and Cancer Cell Enrichment

Sample preparation represents a major stumbling block to the widespread use of biosensors in clinical applications. Traditionally, cancer diagnosis is carried out using paraffin-embedded tumor tissue and/

or analysis of individual biomarkers in blood. However, paraffin is not the ideal preservation method for the DNA, RNA, and protein used in many biosensor-based analyses. Nucleic acids recovered from paraffin-embedded tissues are often significantly degraded compared to those recovered from fresh tissue although PCR has been performed successfully in many cases [61,208–212]. Similarly, mRNA has been successfully isolated from archival tissue and was found, in some cases, to be suitable for accurate, real-time quantitative PCR [131,209,211,213–217]. Proteins are more difficult to recover from paraffin-embedded samples than nucleic acids, although several methods for extraction and analysis of diagnostically useful proteins have been described [218]. Protein analysis is one of the most important applications for biosensors, and the fact that proteins are not amenable to analysis following conventional fixation methods means that fresh or frozen samples must be analyzed.

1.3.5.2 Laser Capture Microdissection

As mentioned earlier in the discussion of Figure 1.2, cancer tissues are often heterogeneous containing both normal and cancer cells. If the goal is to understand the biology of a single cell type, it is important to obtain pure cell populations for analysis. Tissues are composed of a variety of normal epithelial, stromal, inflammatory, and/or pathological cell populations (Figure 1.2). LCM is one of the most commonly used techniques to procure pure or near pure cell populations for analysis. This ability has been critical in the identification of several tumor deletion sites that could otherwise be masked with contamination by nonneoplastic cells [205]. At the RNA level, several characteristic changes in tumor cells have been identified using LCM samples. Harrell et al. identified more than 1000 genetic changes between the primary lesions and the metastasis using LCM samples [219]. Less than 1% of these changes were identifiable in a nonmicrodissected analysis. LCM was developed as an infrared (IR) laser-based system for isolating pure cell populations from tissue sections [220]. Subsequently, several ultraviolet-based dissection systems have also been developed [205]. These laser cutting systems provide semiautomated features to the microdissection process. More recently, expression laser microdissection (xMD) was developed as a high-throughput, operator-independent microdissection system. As opposed to traditional laser microdissection, where desired cells are recognized morphologically under microscopic visualization, xMD targets cells with antibodies, similar to conventional immunohistochemistry [221]. A large number of cells can be procured by the laser in seconds. In addition, very fine targets can be obtained, as was shown in the molecular analysis of pure endothelial cells obtained from tumor and normal adjacent areas in prostate cancer tissues [222].

1.3.5.3 Other Cancer Enrichment Methods in Nondissected Material

It may be necessary to isolate a few cancer cells from among a large volume of normal cells. Various enrichment methods to select cancer cells have been developed, including density gradient centrifugation [223,224], which enriches ~632-fold and provides mean tumor cell recovery rates of ~86% [224]. Counterflow centrifugal elutriation (CCE) has also been used [225]. Immunomagnetic enrichment of cancer cells is more specific [226–234] and allows 2300-fold [235] or 8139-fold [231] enrichment. Flow cytometry has also been used to sort cancer cells using several biomarkers simultaneously [235–238]. Cell enrichment technologies for isolating rare cells from biological fluids are also being developed, for example, separation of tumor cells based on their mechanical characteristics using a massively parallel microfabricated sieving device [239]. Another approach for cell enrichment is sorting and separating breast cancer cells in peripheral blood in a microfluidic system based on their electrophysiological characteristics or the use of flow cytometry [240].

1.3.5.4 Methods for Obtaining Tissue Samples

Two important sources for clinical samples are fine needle aspiration (FNA) and saliva.

Fine needle aspiration: Clinical tissue samples from cancer patients are usually obtained via open surgical procedures or through FNAs. Surgical procedures can be part of the patient's treatment (excisional), such as in CRC, breast, lung, and other cancers. Surgical biopsies can also aim to take only part of the tumor for the pathologist to examine (incisional). Surgical resections yield larger amounts of tissue, which can include neoplastic and surrounding normal tissue. In the last few decades, finer methods have been used to obtain tissues for diagnostic purposes for cancer patients [241]. FNA uses very thin needles to obtain fluid, cells, or small tissue amounts for diagnostic purposes. Core needle biopsies, on the other hand, have a special cutting edge and yield larger amounts of material than aspirates. Currently, needle biopsy procedures are frequently guided through radiological imaging, increasing the effectiveness and reducing the risks of the procedure.

Salivary-based diagnostics: Saliva, a biofluid that can be obtained noninvasively and which is readily available, can be used as an inexpensive source of diagnostic sample. Saliva sampling methods generate minimal subject discomfort for cancer testing.

1.4 Biosensor Technologies for Cancer Analysis

Biosensor technologies have a great potential for cancer detection, diagnosis, monitoring, and guiding treatment. Biosensors consist of a recognition element (ligand or receptor), a signal conversion unit (transducer), and an output interface (the electronic component for interacting with the instrument). This general design (Figure 1.3) is common to all biosensors. There are two types of biosensors, direct detection and indirect detection biosensors.

1.4.1 Current Biosensor Technologies

Direct detection biosensors such as surface plasmon resonance (SPR) are biosensors in which the recognition reaction is directly measured using various physical properties. Direct detection methods are label free, a factor that simplifies and speeds up the detection, keeping down the assay cost. Indirect detection biosensors, such as electrochemical sensors, require a secondary labeled ligand for detection of the target. The secondary labeled ligands are often antibodies conjugated to an enzyme such as alkaline phosphatase. The sensor detects the catalytic reaction carried out by the enzyme. The need for secondary labeled ligand and the two-step "sandwich" assay (binding the target to the primary ligand and the binding of the secondary ligand to the captured target) complicate the detection.

As discussed earlier, the trend over the past few years is toward high-throughput systems with increased integration and miniaturization approaching the molecular level. This trend has several inherent advantages because it is leading to the production of higher-capacity devices (more channels), simpler support systems, less expensive device manufacture, and more rapid response time. An example of this emerging technology is antibody arrays, in which antibodies are placed in an orderly arrangement in a

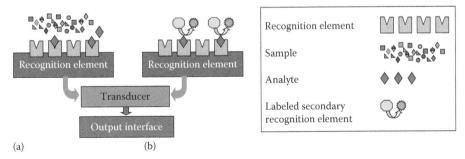

FIGURE 1.3 Schematic general design of biosensors. Biosensors consist of a recognition element (receptor), a signal conversion unit (transducer), and an output interface. Biosensors can be divided into two general types: (a) noncatalytic (label-free) biosensors used for direct detection and (b) indirect detection biosensors which often utilize catalytic assay for detection. In both cases, the analyte first binds specifically to the recognition element. In indirect detection biosensors, a secondary labeled molecule then binds to the target analyte, and the sensor detects the effects of the labeled molecule.

two-dimensional format on a solid support. Using antibody arrays [242,243], various antigens can be identified by their binding to specific antibodies at known positions. Such array biosensors, combined with a microfluidic system and a detection element, enable automated multianalyte analysis. The use of such integrated biosensor chips may allow a comprehensive analysis of a complex biospecimen, such as blood or other body fluids. Biosensors have several potential advantages over other methods of clinical cancer testing, including increased speed and flexibility. Rapid, real-time methods can provide more timely information to health care providers that can be used for diagnosis. In addition, biosensors enable automated multiple measurements on a single sample leading to reduced costs. Biosensor-based diagnostics have the potential to expand cancer screening at reasonable cost and to improve the rates of early detection with attendant improvements in prognosis. Biosensor technologies are expected to enhance health care delivery in the community setting and to benefit underserved populations.

Ligands for cancer testing: The recognition element, or ligand, is the biological component of the biosensor that produces the signal. There are various types of recognition elements, ranging from whole cells to relatively small molecules. Recognition elements can be divided into two general categories: noncatalytic elements (Figure 1.3a) and catalytic (Figure 1.3b). Noncatalytic recognition elements, such as cell receptors or antibodies, are often used for direct detection (i.e., label-free) biosensors in which the interaction is directly measured and the output signal is produced immediately. Many biosensors have been developed for protein analysis, indicating their potential for detecting protein biomarkers of cancer. However, one of the bottlenecks for multitarget cancer testing is the availability of recognition ligands. While antibodies are excellent ligands in terms of binding strength and selectivity, they are expensive to produce, incompatible with some high-throughput approaches, and often have limited shelf life. The critical role of antibodies as the recognition ligand in most biosensor systems makes this bottleneck critical. There has been considerable interest in research on alternatives to antibodies, and those efforts are ongoing today. Several very promising avenues of investigation have been pursued, including aptamers, peptides, scaffolded peptides, combined binding agents derived from low-affinity ligands, and combinatorial chemistry ligands. Aptamers are protein-binding nucleic acids that bind other molecules with high affinity [244–246] ($K_D \sim 10^{-9}$ M). Peptides can be selected from phage display libraries [247,248] with high affinity for a target. The ligand equilibrium dissociation constant (K_D) for antibodies and some high-affinity aptamers is on the order of 10^{-9} M. In contrast, other ligands such as peptides, scaffolded peptides, and combined binding agents typically have K_Ds that are two to three orders of magnitude lower ($\sim 10^{-6}$ to 10^{-7} M). A new generation of high-affinity and versatile ligands compatible with high-throughput systems is a key element for biosensor development and utilization.

Transducers: Transducers are the physical components of the sensor that respond to the signal of the biosensing process and convert the response to measurable form. Transducers are based on a variety of physical principles, including electrochemistry, electrical conductivity, fluorescence, optics, mass detection, and spectrometry. Biosensors can be classified as either direct or indirect transducers. Direct detection transducers, such as SPR or quartz resonator transducers, detect the physical presence of the target bound to a recognition element. Direct detection methods are label free, a factor that simplifies and speeds up the detection, keeping down the assay cost. Indirect detection biosensors, such as electrochemical sensors, require a secondary labeled ligand for detection of the target. Given the complexity of cancer, transducers for cancer analysis must be capable of multiplexing which will enable analysis of multiple cancer associated analytes (cell types, proteins, mutations, epigenetic alterations, etc.) simultaneously.

Direct detection transducers: The main types of direct detection transducers are direct optical biosensors, mechanical biosensors, and field-effect transistors (FETs).

Direct detection optical transducers: Direct optical biosensors are usually based on evanescent wave physics and measure changes in light interactions with the surface of the biosensor. Molecular interactions in the immediate vicinity of the transducer surface, such as binding of the antibody and antigen, result in refractive index changes which can be detected in several ways. The main direct measurement optical transducers are SPR and resonant mirror biosensors. In SPR biosensors, the recognition ligand (often an antibody) is immobilized on the prism surface. Changes within the range of the plasmon field (such as antibody–antigen binding) cause changes in plasmon resonance. Resonant mirror biosensors also measure changes in the refractive index but are based on layer of high refractive index instead of a gold film sensing layer used by SPR. Light entering the resonant layer from the prism is coupled to binding events on the surface through the evanescent field. As in SPR biosensors, binding events on the sensor surface can be measured.

Direct detection mechanical transducers: In addition to optical direct transducers, mechanical transducers such as resonant crystal and cantilever both based on sensor surface dynamics used as direct sensing transducers. Resonant crystal biosensors (also known as quartz crystal microbalance (QCM), piezoelectric or acoustic wave) measure changes in acoustic resonant frequency of a quartz crystal due to bound mass on the crystal surface. Cantilever-based biosensors are newly developed transducers measuring mass-dependent vibrational frequency or bending stress changes on the cantilever surface. Ligands can be covalently attached to the surface of the cantilever, and the mass change induced by target-ligand binding events can be detected by changes in the resonant frequency or by measuring the cantilever position using optical methods (e.g., by focusing a laser beam on the tip of the cantilever to measure the position).

Direct detection field-effect transistors: FETs are transistors that are gated by changes in the electric field. A special class of FETs, called ion-sensitive FETs (ISFETs), are useful as biosensors. FETs have a source, a drain, and a gate. Recently, FETs have been fabricated using nanomaterials (e.g., carbon nanotubes) to enhance performance. In an FET, the ligands are immobilized on the gate surface. Single-walled carbon nanotube (SWNT)-based FETs have shown a large conductance change in response to binding events on the surface.

1.4.1.1 Indirect Detection Transducers

As discussed earlier, indirect detection requires that the target analyte is first bound by a labeled secondary molecule, for example, an antibody conjugated to an enzyme such as alkaline phosphatase. The main types of indirect transducers are optical and electrochemical. In both types, the sensor detects the catalytic reaction carried out by the enzyme as change in light or electric changes. Indirect detection biosensors are very sensitive, relatively simple, and generally less expensive than direct detection sensors, although they use a labeled secondary molecule and typically exhibit a slower response than direct detection sensors. There are several types of indirect detection biosensors, including fluorescence-label, chemiluminometric, electrochemical, and light-addressable potentiometric sensors.

Fluorescent-label biosensors: In fluorescent sandwich immunoassays, fluorochrome molecules are used to label the secondary antibodies. When the fluorochrome is excited, the emission can be detected by the biosensor transducer. Generally, fluorescent-label biosensors can detect analytes at concentrations as low as 1–10 ng/mL. Several multichannel fluorescent-label biosensors have been developed to test for the presence of multiple pathogens and toxins simultaneously. This is a versatile type of instrument suitable for antibody-based detection of microbes and toxins. Fluorescence can be measured in several ways, including photodiodes, photomultipliers (PMT), phototransistors, or charge-coupled devices (CCD) which measure emission intensity over an area.

Electrochemical biosensors: Electrochemical detectors measure electron transfer due to oxidation/reduction reactions catalyzed by an enzyme such as alkaline phosphatase or chemical reaction measured with a suitable electrode. The common electrochemical transducers are amperometric devices that detect changes in current at constant potential, conductimetric devices that detect changes in conductivity between two electrodes (at constant voltage), impedimetric devices that measure the ratio of voltage to alternating current, and potentiometric devices that detect changes in potential at constant current.

1.4.2 Emerging Technologies

"Lab-on-a-chip" biosensors containing integrated microfabricated fluidic systems enable manipulation of fluids moving in small channels [249–252]. Such devices can be designed to perform multistep high-resolution biological or chemical assays. Many are fabricated using molding or photolithographic processes using composite materials, quartz, silica, or glass substrates. Lab-on-a-chip research is part of a trend in biosensor development toward high-throughput systems with increased integration and miniaturization. Examples include real-time DNA amplification for detection of human papillomavirus (HPV) [253], microfluidic platform for the isolation of specific cells from blood samples [254], cell isolation and purification of low-abundance cells (about 1:10,000 cells) [255], and electrochemical array detection of single-base mismatch mutations [256]. Enthalpy arrays based on analysis of thermodynamics of molecular interactions which enable direct detection [257] were applied for detection of a large range of biomolecular interactions, including DNA and protein binding and enzymatic reactions. The biological compact disk (BioCD), an interferometric biosensor that uses self-referencing surface-normal interferometry to detect molecules bound to the surface of a spinning disk [258], may enable high-throughput (more than a million spots per disk) protein or DNA analysis. Fiber-optic array scanning technology (FAST) applies laser-printing optics to enable rare-cell detection at rates as high as 300,000/s [259], enabling a 500-fold speedup over conventional detection with comparable sensitivity and superior specificity.

Nanotechnology offers promise in several areas, including transducers, recognition ligands, labeling technology, and manufacturing. Nanotransducers offer new potential capabilities such as miniaturization to nanoscale size, higher sensitivities to one or few molecule detection, reduced cost of production, high throughput, and a large number of channels. More importantly, the use of quantum effects, high surface-to-volume effects, or both provided by nanotechnology may allow the creation of new approaches for cancer detection, diagnostics, targeting, imaging, and drug delivery for cancer. Several applications of nanotechnology are currently being developed for biosensing. This includes microcantilevers which measure bending due to changes in surface mass [260–263]. It also includes quantum dots (QDs) which are nanometer-sized semiconductor crystals that absorb white light and reemit light at a specific color. Changes in the optical properties of QDs result from changes on their surface (e.g., antibody–antigen binding) [264–266]. QD systems can be used for biological labeling and biosensing. Metal nanoparticles which exhibit colloidal SPR (nano-SPR) [267] are another example of utilizing the optical properties of nanoparticles. Carbon nanotubes and noncarbon nanowires change their electric superconducting, insulating, semiconducting, or conducting properties because of changes on surface such as binding antigens by attached antibodies. The NCI is developing a research program for utilization of nanotechnology for cancer diagnosis and treatment.

1.4.3 Challenges in Moving Biosensors to Point-of-Care Cancer Testing

One eventual goal for biosensor application is point-of-care testing (POCT). As suggested by the wide use of biosensors for measuring blood sugar levels, POCT applications have great clinical potential for cancer clinical testing. There are several challenges in moving biosensors to POCT for cancer, including:

- The nature of cancer samples—cancer tissues are often heterogeneous containing both normal and cancer cells.
- The complexity of cancer—many genes and pathways are known to be involved in cancer development.
- The large genetic variability of cancer—numerous mutations for each gene are associated with cancer.
- Development of reproducible biomarker assays.
- Need for new recognition ligands, including nonantibody recognition ligands.
- Development of multichannel biosensors which will enable analysis of multiple cancer associated analytes.
- Need for advances in sample preparation and cancer cell enrichment to overcome sample heterogeneity and the need to analyze small samples.
- Miniaturization and integration.
- Development of new, more sensitive transducers.

- Microfluidic integration.
- Need for the development of advanced manufacturing techniques.
- Need for cost reduction.

1.5 Summary

Although many biosensors have been developed for cancer-related testing, few biosensors have penetrated the market to become widely used. Cancer complexity requires the development of molecular tools, both genomic and proteomic, to profile tumors and produce "molecular signatures" (based on genetic and epigenetic signatures, changes in gene expression and protein profiles, and protein post-translational modifications) has opened up new opportunities for utilizing biosensors in cancer testing. Harnessing the potential of biosensors is challenging because of cancer's complexity and diversity which requires assaying multiple analytes in parallel. The strategy for the development of biosensors, shown in Figure 1.1, includes utilization of cancer biomarkers identified from genomic and proteomic analyses and clinical research. Successful development of biosensor-based cancer testing will require continued development and validation of biomarkers and development of ligands for those biomarkers, as well as continued development of sample preparation methods and multichannel biosensors able to analyze many cancer markers simultaneously. The use of biosensors for cancer clinical testing is expected to reduce assay time to increase flexibility by enabling multitarget analyses, automation, and cost reduction. Biosensors have the potential to deliver molecular testing to the community health care setting and to underserved populations.

REFERENCES

1. Ukraintseva, S.V. and A.I. Yashina, Cancer as "rejuvenescence". *Ann NY Acad Sci*, 2004, **1019**, 200–205.
2. Hanahan, D. and R.A. Weinberg, The hallmarks of cancer. *Cell*, 2000, **100**(1), 57–70.
3. Sobin, L.H., M.K. Gospodarowicz, and C.H. Wittekind, *TNM Classification of Malignant Tumors*, 7th edn., 2009. Wiley-Blackwell, Oxford, U.K.
4. Woodhouse, E.C., R.F. Chuaqui, and L.A. Liotta, General mechanisms of metastasis. *Cancer*, 1997, **80**(8 Suppl), 1529–1537.
5. Butler, T.P. and P.M. Gullino, Quantitation of cell shedding into efferent blood of mammary adenocarcinoma. *Cancer Res*, 1975, **35**(3), 512–516.
6. Tarin, D. et al., Mechanisms of human tumor metastasis studied in patients with peritoneovenous shunts. *Cancer Res*, 1984, **44**(8), 3584–3592.
7. Luzzi, K.J. et al., Multistep nature of metastatic inefficiency: Dormancy of solitary cells after successful extravasation and limited survival of early micrometastases. *Am J Pathol*, 1998, **153**(3), 865–873.
8. Riethmuller, G. and C.A. Klein, Early cancer cell dissemination and late metastatic relapse: Clinical reflections and biological approaches to the dormancy problem in patients. *Semin Cancer Biol*, 2001, **11**(4), 307–311.
9. Pantel, K. et al., Frequency and prognostic significance of isolated tumour cells in bone marrow of patients with non-small-cell lung cancer without overt metastases. *Lancet*, 1996, **347**(9002), 649–653.
10. Wong, C.W. et al., Apoptosis: An early event in metastatic inefficiency. *Cancer Res*, 2001, **61**(1), 333–338.
11. Hunter, K.W., N.P. Crawford, and J. Alsarraj, Mechanisms of metastasis. *Breast Cancer Res*, 2008, **10**(Suppl 1), S2.
12. Eltarhouny, S.A. et al., Genes controlling spread of breast cancer to lung "gang of 4". *Exp Oncol*, 2008, **30**(2), 91–95.
13. Mareel, M., M.J. Oliveira, and I. Madani, Cancer invasion and metastasis: Interacting ecosystems. *Virchows Arch*, 2009, **454**(6), 599–622.
14. Pentheroudakis, G., E. Briasoulis, and N. Pavlidis, Cancer of unknown primary site: Missing primary or missing biology? *Oncologist*, 2007, **12**(4), 418–425.

15. Nandan, M.O. and V.W. Yang, An update on the biology of RAS/RAF mutations in colorectal cancer. *Curr Colorectal Cancer Rep*, 2011, **7**(2), 113–120.
16. Dhillon, A.S. et al., MAP kinase signalling pathways in cancer. *Oncogene*, 2007, **26**(22), 3279–3290.
17. Ruzzo, A. et al., Molecular predictors of efficacy to anti-EGFR agents in colorectal cancer patients. *Curr Cancer Drug Targets*, 2010, **10**(1), 68–79.
18. Dahabreh, I.J. et al., Systematic review: Anti-epidermal growth factor receptor treatment effect modification by KRAS mutations in advanced colorectal cancer. *Ann Intern Med*, 2011, **154**(1), 37–49.
19. Perry, M.E. and A.J. Levine, Tumor-suppressor p53 and the cell cycle. *Curr Opin Genet Dev*, 1993, **3**(1), 50–54.
20. Mercer, W.E., Cell cycle regulation and the p53 tumor suppressor protein. *Crit Rev Eukaryot Gene Expr*, 1992, **2**(3), 251–263.
21. Lane, D.P., On the expression of the p53 protein in human cancer. *Mol Biol Rep*, 1994, **19**(1), 23–29.
22. Leslie, N.R. and C.P. Downes, PTEN function: How normal cells control it and tumour cells lose it. *Biochem J*, 2004, **382**(Pt 1), 1–11.
23. Knudson, A.G., Jr., Mutation and cancer: Statistical study of retinoblastoma. *Proc Natl Acad Sci USA*, 1971, **68**(4), 820–823.
24. Allard, W.J. et al., Tumor cells circulate in the peripheral blood of all major carcinomas but not in healthy subjects or patients with nonmalignant diseases. *Clin Cancer Res*, 2004, **10**(20), 6897–6904.
25. Nagrath, S. et al., Isolation of rare circulating tumour cells in cancer patients by microchip technology. *Nature*, 2007, **450**(7173), 1235–1239.
26. Hayes, D.F. et al., Circulating tumor cells at each follow-up time point during therapy of metastatic breast cancer patients predict progression-free and overall survival. *Clin Cancer Res*, 2006, **12**(14 Pt 1), 4218–4224.
27. Cristofanilli, M., Circulating tumor cells, disease progression, and survival in metastatic breast cancer. *Semin Oncol*, 2006, **33**(3 Suppl 9), S9–S14.
28. Cristofanilli, M. et al., Circulating tumor cells, disease progression, and survival in metastatic breast cancer. *N Engl J Med*, 2004, **351**(8), 781–791.
29. Tanigawa, N. et al., Tumor angiogenesis and mode of metastasis in patients with colorectal cancer. *Cancer Res*, 1997, **57**(6), 1043–1046.
30. Gasparini, G., Clinical significance of the determination of angiogenesis in human breast cancer: Update of the biological background and overview of the Vicenza studies. *Eur J Cancer*, 1996, **32A**(14), 2485–2493.
31. Karaiossifidi, H. et al., Tumor angiogenesis in node-negative breast cancer: Relationship with relapse free survival. *Anticancer Res*, 1996, **16**(6C), 4001–4002.
32. Heimann, R. et al., Angiogenesis as a predictor of long-term survival for patients with node-negative breast cancer. *J Natl Cancer Inst*, 1996, **88**(23), 1764–1769.
33. Silberman, M.A. et al., Tumor angiogenesis correlates with progression after radical prostatectomy but not with pathologic stage in Gleason sum 5 to 7 adenocarcinoma of the prostate. *Cancer*, 1997, **79**(4), 772–779.
34. Obermair, A. et al., Influence of microvessel density and vascular permeability factor/vascular endothelial growth factor expression on prognosis in vulvar cancer. *Gynecol Oncol*, 1996, **63**(2), 204–209.
35. Srivastava, A. et al., The prognostic significance of tumor vascularity in intermediate-thickness (0.76–4.0 mm thick) skin melanoma. A quantitative histologic study. *Am J Pathol*, 1988, **133**(2), 419–423.
36. Denijn, M. and D.J. Ruiter, The possible role of angiogenesis in the metastatic potential of human melanoma. Clinicopathological aspects. *Melanoma Res*, 1993, **3**(1), 5–14.
37. Elmore, S., Apoptosis: A review of programmed cell death. *Toxicol Pathol*, 2007, **35**(4), 495–516.
38. Ossowski, L. and J.A. Aguirre-Ghiso, Dormancy of metastatic melanoma. *Pigment Cell Melanoma Res*, 2010, **23**(1), 41–56.
39. Naumov, G.N., J. Folkman, and O. Straume, Tumor dormancy due to failure of angiogenesis: Role of the microenvironment. *Clin Exp Metastasis*, 2009, **26**(1), 51–60.
40. Naumov, G.N., L.A. Akslen, and J. Folkman, Role of angiogenesis in human tumor dormancy: Animal models of the angiogenic switch. *Cell Cycle*, 2006, **5**(16), 1779–1787.
41. Pantel, K. and R.H. Brakenhoff, Dissecting the metastatic cascade. *Nat Rev Cancer*, 2004, **4**(6), 448–456.
42. Pantel, K., C. Alix-Panabieres, and S. Riethdorf, Cancer micrometastases. *Nat Rev Clin Oncol*, 2009, **6**(6), 339–351.

43. Slade, M.J. et al., Comparison of bone marrow, disseminated tumour cells and blood-circulating tumour cells in breast cancer patients after primary treatment. *Br J Cancer*, 2009, **100**(1), 160–166.

44. Dean, M., Cancer stem cells: Implications for cancer causation and therapy resistance. *Discov Med*, 2005, **5**(27), 278–282.

45. Robert, V. et al., High frequency in esophageal cancers of p53 alterations inactivating the regulation of genes involved in cell cycle and apoptosis. *Carcinogenesis*, 2000, **21**(4), 563–565.

46. Gallie, B.L. and R.A. Phillips, Retinoblastoma: A model of oncogenesis. *Ophthalmology*, 1984, **91**(6), 666–672.

47. Lindblom, A. and M. Nordenskjold, Hereditary cancer. *Acta Oncol*, 1999, **38**(4), 439–447.

48. Eeles, R.A. et al., The genetics of familial breast cancer and their practical implications. *Eur J Cancer*, 1994, **30A**(9), 1383–1390.

49. Easton, D., D. Ford, and J. Peto, Inherited susceptibility to breast cancer. *Cancer Surv*, 1993, **18**, 95–113.

50. Strate, L.L. and S. Syngal, Hereditary colorectal cancer syndromes. *Cancer Causes Control*, 2005, **16**(3), 201–213.

51. Thibodeau, S.N., G. Bren, and D. Schaid, Microsatellite instability in cancer of the proximal colon. *Science*, 1993, **260**(5109), 816–819.

52. Balch, C. et al., New anti-cancer strategies: Epigenetic therapies and biomarkers. *Front Biosci*, 2005, **10**, 1897–1931.

53. Jones, L.K. and V. Saha, Chromatin modification, leukaemia and implications for therapy. *Br J Haematol*, 2002, **118**(3), 714–727.

54. Jones, P.A., DNA methylation and cancer. *Oncogene*, 2002, **21**(35), 5358–5360.

55. Liang, G. et al., Identification of DNA methylation differences during tumorigenesis by methylation-sensitive arbitrarily primed polymerase chain reaction. *Methods*, 2002, **27**(2), 150–155.

56. Nguyen, C.T., F.A. Gonzales, and P.A. Jones, Altered chromatin structure associated with methylation-induced gene silencing in cancer cells: Correlation of accessibility, methylation, MeCP2 binding and acetylation. *Nucleic Acids Res*, 2001, **29**(22), 4598–4606.

57. Cunningham, J.M. et al., Hypermethylation of the hMLH1 promoter in colon cancer with microsatellite instability. *Cancer Res*, 1998, **58**(15), 3455–3460.

58. Jones, P.A. and P.W. Laird, Cancer epigenetics comes of age. *Nat Genet*, 1999, **21**(2), 163–167.

59. Esteller, M., Epigenetic lesions causing genetic lesions in human cancer: Promoter hypermethylation of DNA repair genes. *Eur J Cancer*, 2000, **36**(18), 2294–2300.

60. Esteller, M. et al., Analysis of adenomatous polyposis coli promoter hypermethylation in human cancer. *Cancer Res*, 2000, **60**(16), 4366–4371.

61. Trojan, J. et al., 5′-CpG island methylation of the LKB1/STK11 promoter and allelic loss at chromosome 19p13.3 in sporadic colorectal cancer. *Gut*, 2000, **47**(2), 272–276.

62. Esteller, M. et al., Inactivation of the DNA repair gene O6-methylguanine-DNA methyltransferase by promoter hypermethylation is associated with G to A mutations in K-ras in colorectal tumorigenesis. *Cancer Res*, 2000, **60**(9), 2368–2371.

63. Esteller, M. et al., Promoter hypermethylation and BRCA1 inactivation in sporadic breast and ovarian tumors. *J Natl Cancer Inst*, 2000, **92**(7), 564–569.

64. Claus, R. and M. Lubbert, Epigenetic targets in hematopoietic malignancies. *Oncogene*, 2003, **22**(42), 6489–6496.

65. Kalebic, T., Epigenetic changes: Potential therapeutic targets. *Ann NY Acad Sci*, 2003, **983**, 278–285.

66. Lavau, C. et al., Chromatin-related properties of CBP fused to MLL generate a myelodysplastic-like syndrome that evolves into myeloid leukemia. *Embo J*, 2000, **19**(17), 4655–4664.

67. Bannister, A.J. and T. Kouzarides, Histone methylation: Recognizing the methyl mark. *Methods Enzymol*, 2004, **376**, 269–288.

68. Yeates, T.O., Structures of SET domain proteins: Protein lysine methyltransferases make their mark. *Cell*, 2002, **111**(1), 5–7.

69. Bracken, A.P. et al., EZH2 is downstream of the pRB-E2F pathway, essential for proliferation and amplified in cancer. *Embo J*, 2003, **22**(20), 5323–5335.

70. Hamamoto, R. et al., SMYD3 encodes a histone methyltransferase involved in the proliferation of cancer cells. *Nat Cell Biol*, 2004, **6**(8), 731–740.

71. Reid, L.H. et al., Genomic organization of the human p57KIP2 gene and its analysis in the G401 Wilms' tumor assay. *Cancer Res*, 1996, **56**(6), 1214–1218.

72. Yu, Y. et al., Epigenetic regulation of ARHI in breast and ovarian cancer cells. *Ann NY Acad Sci*, 2003, **983**, 268–277.

73. Ransohoff, D.F., Developing molecular biomarkers for cancer. *Science*, 2003, **299**(5613), 1679–1680.

74. Fire, A. et al., Potent and specific genetic interference by double-stranded RNA in Caenorhabditis elegans. *Nature*, 1998, **391**(6669), 806–811.

75. Kennerdell, J.R. and R.W. Carthew, Use of dsRNA-mediated genetic interference to demonstrate that frizzled and frizzled 2 act in the wingless pathway. *Cell*, 1998, **95**(7), 1017–1026.

76. Lopez, P.J. et al., Translation inhibitors stabilize Escherichia coli mRNAs independently of ribosome protection. *Proc Natl Acad Sci USA*, 1998, **95**(11), 6067–6072.

77. Ekwall, K., The roles of histone modifications and small RNA in centromere function. *Chromosome Res*, 2004, **12**(6), 535–542.

78. Lippman, Z. et al., Distinct mechanisms determine transposon inheritance and methylation via small interfering RNA and histone modification. *PLoS Biol*, 2003, **1**(3), E67.

79. Bartel, D.P. and C.Z. Chen, Micromanagers of gene expression: The potentially widespread influence of metazoan microRNAs. *Nat Rev Genet*, 2004, **5**(5), 396–400.

80. Ma, L. and R.A. Weinberg, Micromanagers of malignancy: Role of microRNAs in regulating metastasis. *Trends Genet*, 2008, **24**(9), 448–456.

81. Nicoloso, M.S. et al., MicroRNAs in the pathogeny of chronic lymphocytic leukaemia. *Br J Haematol*, 2007, **139**(5), 709–716.

82. Calin, G.A. et al., A microRNA signature associated with prognosis and progression in chronic lymphocytic leukemia. *N Engl J Med*, 2005, **353**(17), 1793–1801.

83. Mitchell, P.S. et al., Circulating microRNAs as stable blood-based markers for cancer detection. *Proc Natl Acad Sci USA*, 2008, **105**(30), 10513–10518.

84. Tsujiura, M. et al., Circulating microRNAs in plasma of patients with gastric cancers. *Br J Cancer*, 2010, **102**(7), 1174–1179.

85. Rosenfeld, N. et al., MicroRNAs accurately identify cancer tissue origin. *Nat Biotechnol*, 2008, **26**(4), 462–469.

86. van Roy, F. and G. Berx, The cell-cell adhesion molecule E-cadherin. *Cell Mol Life Sci*, 2008, **65**(23), 3756–3788.

87. Shen, Y. and E. White, p53-dependent apoptosis pathways. *Adv Cancer Res*, 2001, **82**, 55–84.

88. Teodoro, J.G. et al., p53-mediated inhibition of angiogenesis through up-regulation of a collagen prolyl hydroxylase. *Science*, 2006, **313**(5789), 968–971.

89. Liu, Y. and M. Kulesz-Martin, p53 protein at the hub of cellular DNA damage response pathways through sequence-specific and non-sequence-specific DNA binding. *Carcinogenesis*, 2001, **22**(6), 851–860.

90. Guo, F. et al., p19Arf-p53 tumor suppressor pathway regulates cell motility by suppression of phosphoinositide 3-kinase and Rac1 GTPase activities. *J Biol Chem*, 2003, **278**(16), 14414–14419.

91. Wood, R.D. et al., Human DNA repair genes. *Science*, 2001, **291**(5507), 1284–1289.

92. Helleday, T. et al., DNA repair pathways as targets for cancer therapy. *Nat Rev Cancer*, 2008, **8**(3), 193–204.

93. Munoz-Gamez, J.A. et al., Inhibition of poly (ADP-ribose) polymerase-1 enhances doxorubicin activity against liver cancer cells. *Cancer Lett*, 2011, **301**(1), 47–56.

94. Jones, S. et al., Core signaling pathways in human pancreatic cancers revealed by global genomic analyses. *Science*, 2008, **321**(5897), 1801–1806.

95. The Cancer Genome Atlas Research Network, Comprehensive genomic characterization defines human glioblastoma genes and core pathways. *Nature*, 2008, **455**(7216), 1061–1068.

96. Ding, L. et al., Somatic mutations affect key pathways in lung adenocarcinoma. *Nature*, 2008, **455**(7216), 1069–1075.

97. Biomarkers Definitions Working Group, Biomarkers and surrogate endpoints: Preferred definitions and conceptual framework. *Clin. Pharmacol. Ther.*, 2001, **69**(3), 89–95.

98. Rasooly, A. and J. Jackson, Development of biosensors for cancer clinical testing. *Biosensors and Bioelectronics*, 2006, **21**, 1851–1858.

99. Nelson, H.D. et al., Genetic risk assessment and BRCA mutation testing for breast and ovarian cancer susceptibility: Recommendation statement. *Ann Intern Med*, 2005, **143**, 355–361.

100. Nelson, H.D. et al., Genetic risk assessment and BRCA mutation testing for breast and ovarian cancer susceptibility: Systematic evidence review for the U.S. Preventive Services Task Force. *Ann Intern Med*, 2005, **143**(5), 362–379.

101. McShane, L.M. et al., Reporting recommendations for tumor marker prognostic studies. *J Clin Oncol*, 2005, **23**(36), 9067–9072.

102. Hudler, P., M. Gorsic, and R. Komel, Proteomic strategies and challenges in tumor metastasis research. *Clin Exp Metastasis*, 2010, **27**(6), 441–451.

103. van de Vijver, M.J. et al., A gene-expression signature as a predictor of survival in breast cancer. *N Engl J Med*, 2002, **347**(25), 1999–2009.

104. Yarden, Y. and M.X. Sliwkowski, Untangling the ErbB signalling network. *Nat Rev Mol Cell Biol*, 2001, **2**(2), 127–137.

105. Slamon, D.J. et al., Use of chemotherapy plus a monoclonal antibody against HER2 for metastatic breast cancer that overexpresses HER2. *N Engl J Med*, 2001, **344**(11), 783–792.

106. Gravalos, C. and A. Jimeno, HER2 in gastric cancer: A new prognostic factor and a novel therapeutic target. *Ann Oncol*, 2008, **19**(9), 1523–1529.

107. Cappuzzo, F. et al., Increased HER2 gene copy number is associated with response to gefitinib therapy in epidermal growth factor receptor-positive non-small-cell lung cancer patients. *J Clin Oncol*, 2005, **23**(22), 5007–5018.

108. Rusch, V. et al., Overexpression of the epidermal growth factor receptor and its ligand transforming growth factor alpha is frequent in resectable non-small cell lung cancer but does not predict tumor progression. *Clin Cancer Res*, 1997, **3**(4), 515–522.

109. Hirsch, F.R. et al., Epidermal growth factor receptor in non-small-cell lung carcinomas: Correlation between gene copy number and protein expression and impact on prognosis. *J Clin Oncol*, 2003, **21**(20), 3798–3807.

110. Lynch, T.J. et al., Activating mutations in the epidermal growth factor receptor underlying responsiveness of non-small-cell lung cancer to gefitinib. *N Engl J Med*, 2004, **350**(21), 2129–2139.

111. Barault, L. et al., Mutations in the RAS-MAPK, PI(3)K (phosphatidylinositol-3-OH kinase) signaling network correlate with poor survival in a population-based series of colon cancers. *Int J Cancer*, 2008, **122**(10), 2255–2259.

112. Perou, C.M. et al., Molecular portraits of human breast tumours. *Nature*, 2000, **406**(6797), 747–752.

113. Turner, N.C. and J.S. Reis-Filho, Basal-like breast cancer and the BRCA1 phenotype. *Oncogene*, 2006, **25**(43), 5846–5853.

114. Sorlie, T. et al., Gene expression patterns of breast carcinomas distinguish tumor subclasses with clinical implications. *Proc Natl Acad Sci USA*, 2001, **98**(19), 10869–10874.

115. Sorlie, T. et al., Repeated observation of breast tumor subtypes in independent gene expression data sets. *Proc Natl Acad Sci USA*, 2003, **100**(14), 8418–8423.

116. Alexe, G. et al., Data perturbation independent diagnosis and validation of breast cancer subtypes using clustering and patterns. *Cancer Inform*, 2007, **2**, 243–274.

117. Ramaswamy, S. et al., A molecular signature of metastasis in primary solid tumors. *Nat Genet*, 2003, **33**(1), 49–54.

118. Soria, J.C. et al., Added value of molecular targeted agents in oncology. *Ann Oncol*, 2011, **22**(8), 1703–1716.

119. Fuchinoue, S. et al., The 5-year outcome of ABO-incompatible kidney transplantation with rituximab induction. *Transplantation*, 2011, **91**(8), 853–857.

120. van't Veer, L.J. et al., Gene expression profiling predicts clinical outcome of breast cancer. *Nature*, 2002, **415**(6871), 530–536.

121. Buyse, M. et al., Validation and clinical utility of a 70-gene prognostic signature for women with node-negative breast cancer. *J Natl Cancer Inst*, 2006, **98**(17), 1183–1192.

122. Albain, K.S., S. Paik, and L. van't Veer, Prediction of adjuvant chemotherapy benefit in endocrine responsive, early breast cancer using multigene assays. *Breast*, 2009, **18**(Suppl 3), S141–S145.

123. Ou, J., J.A. Vergilio, and A. Bagg, Molecular diagnosis and monitoring in the clinical management of patients with chronic myelogenous leukemia treated with tyrosine kinase inhibitors. *Am J Hematol*, 2008, **83**(4), 296–302.

124. Hughes, T. and S. Branford, Molecular monitoring of BCR-ABL as a guide to clinical management in chronic myeloid leukaemia. *Blood Rev*, 2006, **20**(1), 29–41.

125. Martinelli, G. et al., Monitoring minimal residual disease and controlling drug resistance in chronic myeloid leukaemia patients in treatment with imatinib as a guide to clinical management. *Hematol Oncol*, 2006, **24**(4), 196–204.

126. Cea, M. et al., Molecular diagnosis and monitoring of chronic myelogenous leukemia: BCR-Abl and more. *J BUON*, 2009, **14**(4), 565–573.

127. Manrique Arechavaleta, G. et al., Rapid and sensitive allele-specific (AS)-RT-PCR assay for detection of T315I mutation in chronic myeloid leukemia patients treated with tyrosine-kinase inhibitors. *Clin Exp Med*, 2010, **11**(1), 55–59.

128. Gruber, F.X. et al., Detection of drug-resistant clones in chronic myelogenous leukemia patients during dasatinib and nilotinib treatment. *Clin Chem*, 2010, **56**(3), 469–473.

129. Ernst, T. et al., A co-operative evaluation of different methods of detecting BCR-ABL kinase domain mutations in patients with chronic myeloid leukemia on second-line dasatinib or nilotinib therapy after failure of imatinib. *Haematologica*, 2009, **94**(9), 1227–1235.

130. Guo, A. et al., Signaling networks assembled by oncogenic EGFR and c-Met. *Proc Natl Acad Sci USA*, 2008, **105**(2), 692–697.

131. Paik, S. et al., A multigene assay to predict recurrence of tamoxifen-treated, node-negative breast cancer. *N Engl J Med*, 2004, **351**(27), 2817–2826.

132. Gannot, G. et al., Layered peptide arrays: High-throughput antibody screening of clinical samples. *J Mol Diagn*, 2005, **7**(4), 427–436.

133. Visintin, M., M. Quondam, and A. Cattaneo, The intracellular antibody capture technology: Towards the high-throughput selection of functional intracellular antibodies for target validation. *Methods*, 2004, **34**(2), 200–214.

134. Braunschweig, T., J.Y. Chung, and S.M. Hewitt, Perspectives in tissue microarrays. *Comb Chem High Throughput Screen*, 2004, **7**(6), 575–585.

135. Erickson, R.P., Somatic gene mutation and human disease other than cancer: An update. *Mutat Res*, 2010, **705**(2), 96–106.

136. Sanger, F. et al., Nucleotide sequence of bacteriophage phi X174 DNA. *Nature*, 1977, **265**(5596), 687–695.

137. Sanger, F., S. Nicklen, and A.R. Coulson, DNA sequencing with chain-terminating inhibitors. *Proc Natl Acad Sci USA*, 1977, **74**(12), 5463–5467.

138. Schuster, S.C., Next-generation sequencing transforms today's biology. *Nat Methods*, 2008, **5**(1), 16–18.

139. Fox, S., S. Filichkin, and T.C. Mockler, Applications of ultra-high-throughput sequencing. *Methods Mol Biol*, 2009, **553**, 79–108.

140. Costa, J.L. et al., Array comparative genomic hybridization copy number profiling: A new tool for translational research in solid malignancies. *Semin Radiat Oncol*, 2008, **18**(2), 98–104.

141. Rucker, F.G. et al., Molecular characterization of AML with ins(21;8)(q22;q22q22) reveals similarity to t(8;21) AML. *Genes Chromosome Cancer*, 2011, **50**(1), 51–58.

142. Zhuang, Z. et al., Premeiotic origin of teratomas: Is meiosis required for differentiation into mature tissues? *Cell Cycle*, 2005, **4**(11), 1683–1687.

143. Acevedo, C.M. et al., Loss of heterozygosity on chromosome arms 3p and 6q in microdissected adenocarcinomas of the uterine cervix and adenocarcinoma in situ. *Cancer*, 2002, **94**(3), 793–802.

144. Chuaqui, R., M. Silva, and M. Emmert-Buck, Allelic deletion mapping on chromosome 6q and X chromosome inactivation clonality patterns in cervical intraepithelial neoplasia and invasive carcinoma. *Gynecol Oncol*, 2001, **80**(3), 364–371.

145. Bignell, G.R. et al., High-resolution analysis of DNA copy number using oligonucleotide microarrays. *Genome Res*, 2004, **14**(2), 287–295.

146. Zhao, X. et al., An integrated view of copy number and allelic alterations in the cancer genome using single nucleotide polymorphism arrays. *Cancer Res*, 2004, **64**(9), 3060–3071.

147. Gunderson, K.L. et al., A genome-wide scalable SNP genotyping assay using microarray technology. *Nat Genet*, 2005, **37**(5), 549–554.

148. Huang, J. et al., Whole genome DNA copy number changes identified by high density oligonucleotide arrays. *Hum Genomics*, 2004, **1**(4), 287–299.

149. Dumur, C.I. et al., Genome-wide detection of LOH in prostate cancer using human SNP microarray technology. *Genomics*, 2003, **81**(3), 260–269.

150. Yavas, G. et al., An optimization framework for unsupervised identification of rare copy number variation from SNP array data. *Genome Biol*, 2009, **10**(10), R119.

151. Wiechec, E. and L.L. Hansen, The effect of genetic variability on drug response in conventional breast cancer treatment. *Eur J Pharmacol*, 2009, **625**(1–3), 122–130.

152. Despierre, E. et al., The molecular genetic basis of ovarian cancer and its roadmap towards a better treatment. *Gynecol Oncol*, 2010, **117**(2), 358–365.

153. Heinrichs, S., C. Li, and A.T. Look, SNP array analysis in hematologic malignancies: Avoiding false discoveries. *Blood*, 2010, **115**(21), 4157–4161.

154. Palanichamy, M.G. and Y.P. Zhang, Potential pitfalls in MitoChip detected tumor-specific somatic mutations: A call for caution when interpreting patient data. *BMC Cancer*, 2010, **10**, 597.

155. Subramaniam, V. et al., MITOCHIP assessment of differential gene expression in the skeletal muscle of Ant1 knockout mice: Coordinate regulation of OXPHOS, antioxidant, and apoptotic genes. *Biochim Biophys Acta*, 2008, **1777**(7–8), 666–675.

156. Desai, V.G. and J.C. Fuscoe, Transcriptional profiling for understanding the basis of mitochondrial involvement in disease and toxicity using the mitochondria-specific MitoChip. *Mutat Res*, 2007, **616**(1–2), 210–212.

157. Maitra, A. et al., The Human MitoChip: A high-throughput sequencing microarray for mitochondrial mutation detection. *Genome Res*, 2004, **14**(5), 812–819.

158. Dasgupta, S. et al., Following mitochondrial footprints through a long mucosal path to lung cancer. *PLoS One*, 2009, **4**(8), e6533.

159. Weisenberger, D.J. et al., DNA methylation analysis by digital bisulfite genomic sequencing and digital MethyLight. *Nucleic Acids Res*, 2008, **36**(14), 4689–4698.

160. Russo, G. et al., pRB2/p130 target genes in non-small lung cancer cells identified by microarray analysis. *Oncogene*, 2003, **22**(44), 6959–6969.

161. Russo, G., C. Zegar, and A. Giordano, Advantages and limitations of microarray technology in human cancer. *Oncogene*, 2003, **22**(42), 6497–6507.

162. Singh, D. et al., Gene expression correlates of clinical prostate cancer behavior. *Cancer Cell*, 2002, **1**(2), 203–209.

163. Macgregor, P.F. and J.A. Squire, Application of microarrays to the analysis of gene expression in cancer. *Clin Chem*, 2002, **48**(8), 1170–1177.

164. Alizadeh, A.A. et al., Distinct types of diffuse large B-cell lymphoma identified by gene expression profiling. *Nature*, 2000, **403**(6769), 503–511.

165. Staudt, L.M. and S. Dave, The biology of human lymphoid malignancies revealed by gene expression profiling. *Adv Immunol*, 2005, **87**, 163–208.

166. Dave, S., Gene expression profiling and outcome prediction in non-Hodgkin lymphoma. *Biol Blood Marrow Transplant*, 2006, **12**(1 Suppl 1), 50–52.

167. Fu, K., J. Iqbal, and W.C. Chan, Recent advances in the molecular diagnosis of diffuse large B-cell lymphoma. *Expert Rev Mol Diagn*, 2005, **5**(3), 397–408.

168. Richardson, A.M. et al., Global expression analysis of prostate cancer-associated stroma and epithelia. *Diagn Mol Pathol*, 2007, **16**(4), 189–197.

169. Finak, G. et al., Stromal gene expression predicts clinical outcome in breast cancer. *Nat Med*, 2008, **14**(5), 518–527.

170. Cummins, J.M. and V.E. Velculescu, Implications of micro-RNA profiling for cancer diagnosis. *Oncogene*, 2006, **25**(46), 6220–6227.

171. Cummins, J.M. et al., The colorectal microRNAome. *Proc Natl Acad Sci USA*, 2006, **103**(10), 3687–3692.

172. Du, L. and A. Pertsemlidis, MicroRNAs and lung cancer: Tumors and 22-mers. *Cancer Metastasis Rev*, 2010, **29**(1), 109–122.

173. Culpin, R.E. et al., A 9 series microRNA signature differentiates between germinal centre and activated B-cell-like diffuse large B-cell lymphoma cell lines. *Int J Oncol*, 2010, **37**(2), 367–376.

174. Page, M.J. et al., Proteomic definition of normal human luminal and myoepithelial breast cells purified from reduction mammoplasties. *Proc Natl Acad Sci USA*, 1999, **96**(22), 12589–12594.

175. Emmert-Buck, M.R. et al., Molecular profiling of clinical tissue specimens: Feasibility and applications. *Am J Pathol*, 2000, **156**(4), 1109–1115.

176. Cai, Z., J.F. Chiu, and Q.Y. He, Application of proteomics in the study of tumor metastasis. *Genomics Proteomics Bioinformatics*, 2004, **2**(3), 152–166.

177. Zhang, L. et al., Discovery and preclinical validation of salivary transcriptomic and proteomic biomarkers for the non-invasive detection of breast cancer. *PLoS One*, 2010, **5**(12), e15573.

178. Guan, P. et al., Expression and significance of FOXM1 in human cervical cancer: A tissue micro-array study. *Clin Invest Med*, 2011, **34**(1), E1.

179. Rong, Y. et al., Proteomics analysis of serum protein profiling in pancreatic cancer patients by DIGE: Up-regulation of mannose-binding lectin 2 and myosin light chain kinase 2. *BMC Gastroenterol*, 2010, **10**(1), 68.

180. Hong, Y.J. et al., Discrimination analysis of mass spectrometry proteomics for ovarian cancer detection. *Acta Pharmacol Sin*, 2008, **29**(10), 1240–1246.

181. Liu, C. et al., Proteomic patterns for classification of ovarian cancer and CTCL serum samples utilizing peak pairs indicative of post-translational modifications. *Proteomics*, 2007, **7**(22), 4045–4052.

182. Wang, J. et al., Proteomic studies of early-stage and advanced ovarian cancer patients. *Gynecol Oncol*, 2008, **111**(1), 111–119.

183. Amon, L.M. et al., Integrative proteomic analysis of serum and peritoneal fluids helps identify proteins that are up-regulated in serum of women with ovarian cancer. *PLoS One*, 2010, **5**(6), e11137.

184. MacBeath, G. and S.L. Schreiber, Printing proteins as microarrays for high-throughput function determination. *Science*, 2000, **289**(5485), 1760–1763.

185. Zhu, Q. et al., Enzymatic profiling system in a small-molecule microarray. *Org Lett*, 2003, **5**(8), 1257–1260.

186. Lal, S.P., R.I. Christopherson, and C.G. dos Remedios, Antibody arrays: An embryonic but rapidly growing technology. *Drug Discov Today*, 2002, **7**(18 Suppl), S143–S149.

187. Espina, V. et al., Protein Microarrays: Molecular profiling technologies for clinical specimens. *Proteomics*, 2003, **3**(11), 2091–2100.

188. Battifora, H., The multitumor (sausage) tissue block: Novel method for immunohistochemical antibody testing. *Lab Invest*, 1986, **55**(2), 244–248.

189. Lu, Y. et al., CCR2 expression correlates with prostate cancer progression. *J Cell Biochem*, 2007, **101**(3), 676–685.

190. Fromont, G. et al., BCAR1 expression in prostate cancer: Association with 16q23 LOH status, tumor progression and EGFR/KAI1 staining. *Prostate*, 2007, **67**(3), 268–273.

191. Davis, M.T. et al., Cancer biomarker discovery via low molecular weight serum proteome profiling—Where is the tumor? *Proteomics Clin Appl*, 2007, **1**(12), 1545–1558.

192. Ma, S. et al., The significance of LMO2 expression in the progression of prostate cancer. *J Pathol*, 2007, **211**(3), 278–285.

193. Aaltomaa, S. et al., Expression of Ki-67, cyclin D1 and apoptosis markers correlated with survival in prostate cancer patients treated by radical prostatectomy. *Anticancer Res*, 2006, **26**(6C), 4873–4878.

194. Aaltomaa, S. et al., Reduced alpha- and beta-catenin expression predicts shortened survival in local prostate cancer. *Anticancer Res*, 2005, **25**(6C), 4707–4712.

195. Ray, G. et al., Modulation of cell-cycle regulatory signaling network by 2-methoxyestradiol in prostate cancer cells is mediated through multiple signal transduction pathways. *Biochemistry*, 2006, **45**(11), 3703–3713.

196. Ray, M.E. et al., Nadir prostate-specific antigen within 12 months after radiotherapy predicts biochemical and distant failure. *Urology*, 2006, **68**(6), 1257–1262.

197. Ray, M.E. et al., E-cadherin protein expression predicts prostate cancer salvage radiotherapy outcomes. *J Urol*, 2006, **176**(4 Pt 1), 1409–1414; discussion 1414.

198. Ray, M.R. et al., Cyclin G-associated kinase: A novel androgen receptor-interacting transcriptional coactivator that is overexpressed in hormone refractory prostate cancer. *Int J Cancer*, 2006, **118**(5), 1108–1119.

199. Abdulkader, I. et al., Cell-cycle-associated markers and clinical outcome in human epithelial cancers: A tissue microarray study. *Oncol Rep*, 2005, **14**(6), 1527–1531.

200. Prtilo, A. et al., Tissue microarray analysis of hMSH2 expression predicts outcome in men with prostate cancer. *J Urol*, 2005, **174**(5), 1814–1818; discussion 1818.

201. Li, H. et al., Activation of signal transducer and activator of transcription-5 in prostate cancer predicts early recurrence. *Clin Cancer Res*, 2005, **11**(16), 5863–5868.

202. Gannot, G. et al., Layered expression scanning: Multiplex molecular analysis of diverse life science platforms. *Clin Chim Acta*, 2007, **376**(1–2), 9–16.

203. Gannot, G. et al., Layered peptide array for multiplex immunohistochemistry. *J Mol Diagn*, 2007, **9**(3), 297–304.

204. Gannot, G. et al., Layered peptide arrays: A diverse technique for antibody screening of clinical samples. *Ann NY Acad Sci*, 2007, **1098**, 451–453.

205. Rodriguez-Canales, J. et al., *High-Throughput Image Reconstruction and Analysis: Intelligent Microscopy Applications*, Eds. B. Ravi and C. Guillermo, 2009. IBM Research, Yorktown Heights, New York.

206. Bright, R.K. et al., Generation and genetic characterization of immortal human prostate epithelial cell lines derived from primary cancer specimens. *Cancer Res*, 1997, **57**(5), 995–1002.

207. Ornstein, D.K. et al., Proteomic analysis of laser capture microdissected human prostate cancer and in vitro prostate cell lines. *Electrophoresis*, 2000, **21**(11), 2235–2242.

208. Bielawski, K. et al., The suitability of DNA extracted from formalin-fixed, paraffin-embedded tissues for double differential polymerase chain reaction analysis. *Int J Mol Med*, 2001, **8**(5), 573–578.

209. Gloghini, A. et al., RT-PCR analysis of RNA extracted from Bouin-fixed and paraffin-embedded lymphoid tissues. *J Mol Diagn*, 2004, **6**(4), 290–296.

210. Diaz-Cano, S.J. and S.P. Brady, DNA extraction from formalin-fixed, paraffin-embedded tissues: Protein digestion as a limiting step for retrieval of high-quality DNA. *Diagn Mol Pathol*, 1997, **6**(6), 342–346.

211. Weirich, G. et al., Fixed archival tissue. Purify DNA and primers for good PCR yield! *Mol Biotechnol*, 1997, **8**(3), 299–301.

212. Wang, H.Y. et al., A genotyping system capable of simultaneously analyzing >1000 single nucleotide polymorphisms in a haploid genome. *Genome Res*, 2005, **15**(2), 276–283.

213. Abrahamsen, H.N. et al., Towards quantitative mRNA analysis in paraffin-embedded tissues using real-time reverse transcriptase-polymerase chain reaction: A methodological study on lymph nodes from melanoma patients. *J Mol Diagn*, 2003, **5**(1), 34–41.

214. Specht, K. et al., Quantitative gene expression analysis in microdissected archival tissue by real-time RT-PCR. *J Mol Med*, 2000, **78**(7), B27.

215. Shibutani, M. et al., Methacarn fixation: A novel tool for analysis of gene expressions in paraffin-embedded tissue specimens. *Lab Invest*, 2000, **80**(2), 199–208.

216. Goldsworthy, S.M. et al., Effects of fixation on RNA extraction and amplification from laser capture microdissected tissue. *Mol Carcinog*, 1999, **25**(2), 86–91.

217. Stanta, G., S. Bonin, and R. Perin, RNA extraction from formalin-fixed and paraffin-embedded tissues. *Methods Mol Biol*, 1998, **86**, 23–26.

218. Ikeda, K. et al., Extraction and analysis of diagnostically useful proteins from formalin-fixed, paraffin-embedded tissue sections. *J Histochem Cytochem*, 1998, **46**(3), 397–403.

219. Harrell, J.C. et al., Contaminating cells alter gene signatures in whole organ versus laser capture microdissected tumors: A comparison of experimental breast cancers and their lymph node metastases. *Clin Exp Metastasis*, 2008, **25**(1), 81–88.

220. Emmert-Buck, M.R. et al., Laser capture microdissection. *Science*, 1996, **274**(5289), 998–1001.

221. Tangrea, M.A. et al., Novel proteomic approaches for tissue analysis. *Expert Rev Proteomics*, 2004, **1**(2), 185–192.

222. Grover, A.C. et al., Tumor-associated endothelial cells display GSTP1 and RARbeta2 promoter methylation in human prostate cancer. *J Transl Med*, 2006, **4**, 13.

223. Choesmel, V. et al., Enrichment methods to detect bone marrow micrometastases in breast carcinoma patients: Clinical relevance. *Breast Cancer Res*, 2004, **6**(5), R556–R570.

224. Rosenberg, R. et al., Comparison of two density gradient centrifugation systems for the enrichment of disseminated tumor cells in blood. *Cytometry*, 2002, **49**(4), 150–158.

225. Dlubek, D. et al., Enrichment of normal progenitors in counter-flow centrifugal elutriation (CCE) fractions of fresh chronic myeloid leukemia leukapheresis products. *Eur J Haematol*, 2002, **68**(5), 281–288.

226. Zieglschmid, V. et al., Combination of immunomagnetic enrichment with multiplex RT-pCR analysis for the detection of disseminated tumor cells. *Anticancer Res*, 2005, **25**(3A), 1803–1810.

227. Woelfle, U. et al., Bi-specific immunomagnetic enrichment of micrometastatic tumour cell clusters from bone marrow of cancer patients. *J Immunol Methods*, 2005, **300**(1–2), 136–145.

228. Ulmer, A. et al., Immunomagnetic enrichment, genomic characterization, and prognostic impact of circulating melanoma cells. *Clin Cancer Res*, 2004, **10**(2), 531–537.

229. Lara, O. et al., Enrichment of rare cancer cells through depletion of normal cells using density and flow-through, immunomagnetic cell separation. *Exp Hematol*, 2004, **32**(10), 891–904.

230. Zigeuner, R.E. et al., Isolation of circulating cancer cells from whole blood by immunomagnetic cell enrichment and unenriched immunocytochemistry in vitro. *J Urol*, 2003, **169**(2), 701–705.

231. Weihrauch, M.R. et al., Immunomagnetic enrichment and detection of isolated tumor cells in bone marrow of patients with epithelial malignancies. *Clin Exp Metastasis*, 2002, **19**(7), 617–621.

232. Weihrauch, M.R. et al., Immunomagnetic enrichment and detection of micrometastases in colorectal cancer: Correlation with established clinical parameters. *J Clin Oncol*, 2002, **20**(21), 4338–4343.

233. Kielhorn, E., K. Schofield, and D.L. Rimm, Use of magnetic enrichment for detection of carcinoma cells in fluid specimens. *Cancer*, 2002, **94**(1), 205–211.

234. Zigeuner, R.E. et al., Immunomagnetic cell enrichment detects more disseminated cancer cells than immunocytochemistry in vitro. *J Urol*, 2000, **164**(5), 1834–1837.

235. Hu, X.C. et al., Immunomagnetic tumor cell enrichment is promising in detecting circulating breast cancer cells. *Oncology*, 2003, **64**(2), 160–165.

236. Allan, A.L. et al., Detection and quantification of circulating tumor cells in mouse models of human breast cancer using immunomagnetic enrichment and multiparameter flow cytometry. *Cytometry A*, 2005, **65**(1), 4–14.

237. Kraemer, P.S. et al., Flow cytometric enrichment for respiratory epithelial cells in sputum. *Cytometry A*, 2004, **60**(1), 1–7.

238. Feuerer, M. et al., Enrichment of memory T cells and other profound immunological changes in the bone marrow from untreated breast cancer patients. *Int J Cancer*, 2001, **92**(1), 96–105.

239. Mohamed, H. et al., Development of a rare cell fractionation device: Application for cancer detection. *IEEE Trans Nanobiosci*, 2004, **3**(4), 251–256.

240. de Tute, R.M., Flow cytometry and its use in the diagnosis and management of mature lymphoid malignancies. *Histopathology*, 2011, **58**(1), 90–105.

241. Nassar, A., Core needle biopsy versus fine needle aspiration biopsy in breast-A historical perspective and opportunities in the modern era. *Diagn Cytopathol*, 2010, **39**(5), 380–388.

242. Taitt, C.R., G.P. Anderson, and F.S. Ligler, Evanescent wave fluorescence biosensors. *Biosens Bioelectron*, 2005, **20**(12), 2470–2487.

243. Ligler, F.S. et al., Array biosensor for detection of toxins. *Anal Bioanal Chem*, 2003, **377**(3), 469–477.

244. Famulok, M. and A. Jenne, Oligonucleotide libraries—Variatio delectat. *Curr Opin Chem Biol*, 1998, **2**(3), 320–327.

245. Shaikh, K.A. et al., A modular microfluidic architecture for integrated biochemical analysis. *Proc Natl Acad Sci USA*, 2005, **102**(28), 9745–9750.

246. Collett, J.R. et al., Functional RNA microarrays for high-throughput screening of antiprotein aptamers. *Anal Biochem*, 2005, **338**(1), 113–123.

247. Chang, C.Y. et al., Development of peptide antagonists for the androgen receptor using combinatorial peptide phage display. *Mol Endocrinol*, 2005, **19**(10), 2478.

248. Shukla, G.S. and D.N. Krag, Selection of tumor-targeting agents on freshly excised human breast tumors using a phage display library. *Oncol Rep*, 2005, **13**(4), 757–764.

249. Mitchell, P., Microfluidics—Downsizing large-scale biology. *Nat Biotechnol*, 2001, **19**(8), 717–721.

250. Abrantes, M. et al., Adaptation of a surface plasmon resonance biosensor with microfluidics for use with small sample volumes and long contact times. *Anal Chem*, 2001, **73**(13), 2828–2835.

251. Stokes, D.L., G.D. Griffin, and T. Vo-Dinh, Detection of *E. coli* using a microfluidics-based antibody biochip detection system. *J Anal Chem*, 2001, **369**(3–4), 295–301.

252. Wang, P.C., J. Gao, and C.S. Lee, High-resolution chiral separation using microfluidics-based membrane chromatography. *J Chromatogr A*, 2002, **942**(1–2), 115–122.

253. Gulliksen, A. et al., Parallel nanoliter detection of cancer markers using polymer microchips. *Lab Chip*, 2005, **5**(4), 416–420.

254. Kurita, R. et al., Differential measurement with a microfluidic device for the highly selective continuous measurement of histamine released from rat basophilic leukemia cells (RBL-2H3). *Lab Chip*, 2002, **2**(1), 34–38.

255. Furdui, V.I. and D.J. Harrison, Immunomagnetic T cell capture from blood for PCR analysis using microfluidic systems. *Lab Chip*, 2004, **4**(6), 614–618.

256. Hashimoto, K. and Y. Ishimori, Preliminary evaluation of electrochemical PNA array for detection of single base mismatch mutations. *Lab Chip*, 2001, **1**(1), 61–63.

257. Torres, F.E. et al., Enthalpy arrays. *Proc Natl Acad Sci USA*, 2004, **101**(26), 9517–9522.

258. Varma, M.M. et al., Spinning-disk self-referencing interferometry of antigen-antibody recognition. *Opt Lett*, 2004, **29**(9), 950–952.

259. Krivacic, R.T. et al., A rare-cell detector for cancer. *Proc Natl Acad Sci USA*, 2004, **101**(29), 10501–10504.

260. Fritz, J. et al., Translating biomolecular recognition into nanomechanics. *Science*, 2000, **288**(5464), 316–318.

261. Baller, M.K. et al., A cantilever array-based artificial nose. *Ultramicroscopy*, 2000, **82**(1–4), 1–9.

262. Hansen, K.M. et al., Cantilever-based optical deflection assay for discrimination of DNA single-nucleotide mismatches. *Anal Chem*, 2001, **73**(7), 1567–1571.

263. Wu, G. et al., Bioassay of prostate-specific antigen (PSA) using microcantilevers. *Nat Biotechnol*, 2001, **19**(9), 856–860.

264. Chan, W.C. and S. Nie, Quantum dot bioconjugates for ultrasensitive nonisotopic detection. *Science*, 1998, **281**(5385), 2016–2018.

265. Bruchez, M., Jr. et al., Semiconductor nanocrystals as fluorescent biological labels. *Science*, 1998, **281**(5385), 2013–2016.

266. Medintz, I.L. et al., Self-assembled nanoscale biosensors based on quantum dot FRET donors. *Nat Mater*, 2003, **2**(9), 630–638.

267. Nath, N. and A. Chilkoti, Interfacial phase transition of an environmentally responsive elastin biopolymer adsorbed on functionalized gold nanoparticles studied by colloidal surface plasmon resonance. *J Am Chem Soc*, 2001, **123**(34), 8197–8202.

Part II

Optical Technologies for Cancer Detection and Diagnostics: Surface Plasmon Resonance

2

Surface Plasmon Resonance Biosensor Based on Competitive Protein Adsorption for the Prognosis of Thyroid Cancer

Seokheun Choi and Junseok Chae

CONTENTS

2.1   Introduction

Thyroid cancer is a cancerous growth of the thyroid gland and the most common endocrine malignancy, accounting for approximately 1% of malignancies worldwide [1]. Thyroglobulin (Tg) concentration in blood is a key parameter in the follow-up of patients treated for differentiated thyroid cancer (DTC) [2]. The power of Tg detection lies in the fact that Tg can only be made by the thyroid gland. Tg is a very specific protein biomarker and can be used for prognosis after the eradication of normal thyroid tissue in thyroid cancer patients. Up to 20% of patients with initial treatment of thyroidectomy and radioablation therapy show subsequent persistence or recurrence of the disease, and 8% of them eventually die [3]. Thus, highly sensitive and selective tools to detect Tg are useful for the persistent and recurrent disease.

Competitive and sandwich immunoassays are common types of assays used to detect Tg [4]. Competitive immunoassays use a limited amount of antibody and Tg, labeled with visible markers. By introducing known amounts of unlabeled Tg, a calibration curve is created in which the amount of labeled Tg bound to the antibody is inversely related to the amount of unlabeled Tg in the sample tested. Sandwich immunoassays begin by having primary Tg antibodies adsorbed onto a surface, which then capture Tg, and the additional labeled secondary Tg binds the captured Tg to amplify signal. Both immunoassay methods are well established for Tg measurement, and several kits are commercially available to perform these immunoassays [4].

However, there are two main challenges associated with the immunoassay methods. First of all, both immunoassay methods depend on fluorescence or radioisotope labeling or enzymatic amplification to visualize target-antibody binding. The chemical labeling may modify the target proteins' characteristics, altering their behavior [5]. Moreover, the labeling procedure is time-consuming and labor-intensive, and it is often difficult to achieve accurate quantification due to variable labeling efficiency for different proteins. On the other hand, the label-free technique can eliminate the labeling process and is becoming an attractive alternative for biological analysis [6]. Among various label-free mechanisms, SPR (surface plasmon resonance) has been one of the leading techniques due to its extremely high sensitivity which offers detection limits up to few ppt (pg mL^{-1}) [7].

The second main challenge involves creating highly active and robust sensing surfaces, which is crucial to the performance of the immunoassays. However, developing high-quality antibodies is extremely laborious and expensive. Therefore, the key limiting factor in immunoassays is to develop specific antibodies with sufficient affinity to analytes. Besides these limitations, integrating antibodies on to the surface of a transducer is also a time-consuming and labor-intensive process, often becoming the bottle neck of high-throughput sensing systems [8].

Here, we report a fundamentally different protein detection method that relies on the competitive nature of protein adsorption onto a surface. The method can be used as a complement to the conventional immunoassay for Tg measurement. It is known that, in general, a smaller-size protein initially covering the surface will be displaced by a larger-size protein, a phenomenon discovered by Vroman and Adams in 1969 [9]. The Vroman effect has been studied extensively in the biochemistry community for competitive protein adsorption to materials having different surface chemistries and surface energies [10]. The adsorption/exchange reaction depends on the size and concentration of proteins, solution pH, surface properties, and environmental temperature. Recently, Dr. Rotello and his group demonstrated the use of arrays of green fluorescent protein (GFP) and nanoparticles (NPs) in the detection of proteins at biorelevant concentrations in human serum. They showed that the binding equilibrium between GFP and NPs was altered because of competitive protein adsorption which thus modulated the fluorescence response [11]. Our group has also shown that the adsorption/exchange of proteins occurs and their adsorption strength varies according to the surface properties using SPR [8]. We used four different nonhuman proteins (aprotinin, lysozyme, streptavidin, and isolectin) to rank their strength of affinity by monitoring adsorption/exchange of each protein. Based upon the measurements, we present that Tg in a protein mixture can be selectively detected based on the exchange reaction. The basic concept may be realized by engineering a pair of surfaces, each covered by relatively larger proteins based on their molecular weight to the target protein. The pair of surfaces forms a highly selective protein sensor since the target protein only displaces proteins on one surface (small molecular weight) and not the other (large molecular weight) (Figure 2.1). By using this technique, specific target proteins may be identified without using conventional immunoassay methods. In this sense, detection techniques based on the Vroman effect obviates the need to rely on the antibodies and their attachment to the sensing surface because it simply uses physical adsorption of proteins to the surface.

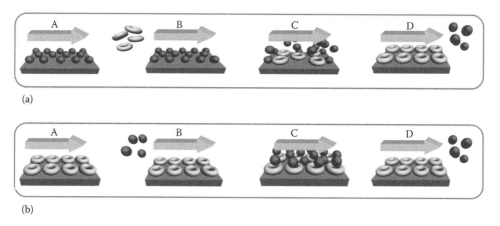

(a)

(b)

FIGURE 2.1 Illustration of the competitive adsorption/exchange of proteins (Vroman effect) on a surface: (a) when small molecular weight proteins adsorb first and then large molecular weight proteins arrive to the surface later, protein displacement occurs. (b) The reverse sequence (large molecular weight proteins adsorb first and small molecular weight proteins are introduced later) does not occur.

In order to demonstrate the practicality of the sensor, we first show that five different human serum proteins, albumin, haptoglobin, immunoglobulin G (IgG), fibrinogen, and Tg, have different adsorption strengths to the surface and that competitive adsorption/exchange controls the displacement of individual types of proteins. Then, we demonstrate that Tg in a protein mixture can be selectively detected based on the exchange reaction. The protein sensing surfaces are integrated with microfluidic systems, and all adsorption/exchange behavior is monitored in the microfluidic channels using SPR in real time.

2.2 Materials and Methods

2.2.1 Surface Plasmon Resonance

SPR is an optical tool that uses the excitation of surface plasmons, and it is a surface-sensitive analytical tool responding to slight changes in refractive index occurring adjacent to the metal film (Figure 2.2). In this way, binding of proteins on the surface and their affinity interactions can be monitored in real time without labeling. SPR can be configured in various ways to monitor the molecular interactions. So far, there are three different optical systems to excite surface plasmons: systems with (1) prisms, (2) gratings, and (3) optical waveguides. Among them, SPR using a prism coupler is most widely adopted and also called the "Kretschmann configuration" because of enhanced stability and sensitivity. The SPR using a prism consists of several components: a light source, a prism, a sensing surface, a flow regulator, and a photodetector. A glass slide covered by a thin gold film is optically coupled to a prism via a

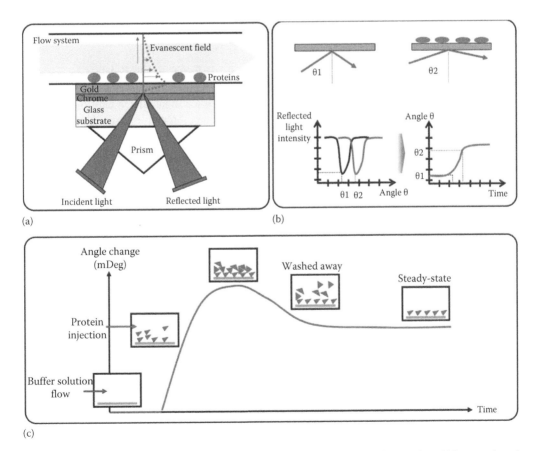

FIGURE 2.2 (a) Schematic view of the SPR-based biosensor. (b) Molecular interactions on the gold layer produce the SPR angle changes. (c) Typical SPR sensorgram.

refractive index matching liquid. Plane polarized light is directed through a prism to the gold film over a wide range of incident angles, and the intensity of the reflected light is detected by a photodetector. At a certain incident light angle, surface plasmon reaches a maximum, and the reflected light becomes minimum, which is denoted as the SPR angle. The propagation constant of the surface plasmon wave propagating at the interface between a dielectric and a metal is given by

$$k_{sp} = k_0 \sqrt{\frac{\varepsilon_{metal}\, \varepsilon_{sample}}{\varepsilon_{metal} + \varepsilon_{sample}}} \qquad (2.1)$$

where

k_0 is the free space wave vector of the optical wave

ε_{metal} and ε_{sample} are the complex dielectric constants of the metal and the sample medium respectively

The enhanced wave vector of incident light (k_x) is given by the following:

$$k_x = k_0 n_{glass} \sin \theta_i \qquad (2.2)$$

where

n_{glass} is the refractive index of the prism
θ_i is the angle of incidence

In order to excite the plasmons at resonance, the two vectors should be matched, i.e.,

$$k_{sp} = k_x \qquad (2.3)$$

Therefore, a small variation of ε_{sample}, which is the dielectric properties of the medium adjacent to the sensing surface, will lead to a large change of SPR angle (θ_i). Figure 2.2c shows a typical SPR sensorgram. Initially, PBS (phosphate buffered saline) is circulated for about 20 min until the angle shift stabilizes. Once the angle shift stabilizes, the protein sample flows through the microfluidic channels, reaches the surface, and generates an angle shift that is proportional to the molecular interactions on the surface. When protein adsorption completes, we let PBS wash the surface to remove excess weakly bound proteins to obtain a steady-state angle value.

2.2.2 Agents

We used five human serum proteins to characterize the sensor: albumin (66 kDa), haptoglobin (86 kDa), IgG (150 kDa), fibrinogen (340 kDa), and thyroglobulin (660 kDa) (Sigma-Aldrich and Calbiochem). All the chemicals were received as lyophilized powders and used without further purification. The proteins were made up to 0.05% (w/v) concentrations in PBS 1× (1.15 g/L-Na_2HPO_4, 0.20 g/L-KCl, 0.20 g/L-KH_2PO_4, 8.0 g/L-NaCl, pH 7.4) immediately prior to analysis. The concentration of mixed proteins is 0.05% (w/v).

2.2.3 Sensing Surface Formation

Glass slides (BK7, $n = 1.517$, 150 μm) were first cleaned in a piranha solution (a 3:1 ration of H_2SO_4 and H_2O_2) for 10 min. The slides were then rinsed with water and ethanol sequentially and were dried under a N_2 stream. A "lift-off" method was used to pattern the sensing surfaces. AZ4330 positive photoresist layer was spin coated and patterned by photolithography on the cleaned slide. After the patterned photoresist was developed in AZ 400 K developer, a thin layer of Cr (2 nm) was deposited as an adhesion layer, followed by Au deposition to a thickness of 48 nm using a sputter. Finally, the slide was immersed with agitation in AZ 400T stripper to "lift-off" the resist with unwanted metal film. Then, the slides were cleaned with oxygen plasma (Harrick Plasma Inc.) at 18 W for 1 min.

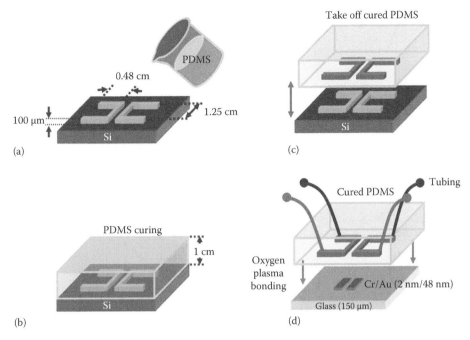

FIGURE 2.3 Fabrication process of the microfluidic device. (a) Photoresist is spin coated on a silicon wafer and patterned for channels. Then the pattern is transferred to silicon by deep RIE. (b) PDMS is poured and cured on the patterned silicon wafer. (c) The molded PDMS is peeled off manually. (d) Inlet and outlet tubes are inserted through the molded PDMS. Then the molded PDMS and the glass slide having a patterned Cr/Au layer are bonded by using oxygen plasma treatment.

2.2.4 Fabrication of Microfluidic Device

We used a soft-lithography technique to fabricate the microfluidic device (Figure 2.3). For the top layer, photoresist AZ4330 was spin coated on a silicon wafer and patterned for channels. The pattern was transferred to the silicon wafer by DRIE (deep reactive ion etching) to etch about 100 μm vertically. The width of the channel is 2.1 mm, and the two channels are separated by 1.3 mm. PDMS solution was poured and cured onto the silicon wafer and then the molded PDMS was peeled off. The thickness of the layer is approximately 1 cm. Inlet/outlet tubes (Upchurch Scientific) (inner diameter: 25 μm; outer diameter: 360 μm) were inserted through the molded PDMS using a syringe needle and fixed by an adhesive. A bottom glass slide has patterned Cr/Au (2 nm/48 nm) pads, which are two sensing surfaces for detection of the protein exchange. The size of the sensing surfaces is 1.8 mm wide and 8.0 mm long. The width of the channel is slightly wider than that of the gold surface to ensure bonding of the two substrates using oxygen plasma at 18 W for 1 min.

2.2.5 SPR Setup

As shown in Figure 2.4, the fabricated microfluidic device was mounted to the semicylindrical prism of SPR (BI-SPR 1000, Biosensing Instrument Inc.) by using a refractive index matching liquid. The experimental setup, shown in Figure 2.5, is equipped with a computer-controlled data acquisition system (BI-2000 control program). The SPR produces two sensorgrams in real time; one on a reference channel and the other on a sensing channel. All experiments were performed at 25°C.

2.2.6 Thermodynamic Process of Protein Adsorption/Exchange

The exchange process is led by thermodynamics: proteins with different dimensions and morphologies adsorb differently to a surface based upon their thermodynamic energy preferences and behave in ways that

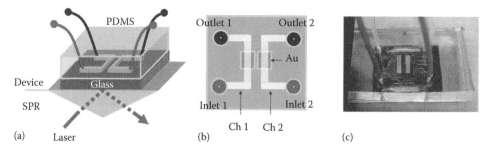

FIGURE 2.4 The microfluidic device consisting of two channels. A schematic of (a) the microfluidic device on SPR, (b) its top view, and (c) a photo image of the microfluidic device mounted on SPR.

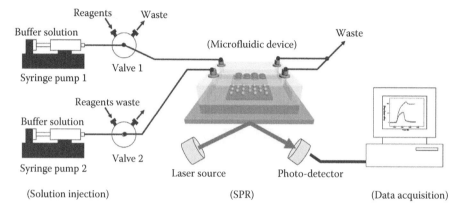

FIGURE 2.5 Characterization setup of the Vroman effect–based biosensor; the fabricated microfluidic device is mounted to SPR. The setup is equipped with a computer-controlled data acquisition system. The angle shift is monitored in real time as sample solution flows through the channels driven by external syringe pumps. Two sensorgrams are generated in real time: one on reference and the other on sensing channels, respectively.

minimize the overall system energy. This can be interpreted as a natural outcome of surface reorganization to achieve the equilibrium interphase composition. Therefore, the change in the free energy of the protein adsorption process on a certain surface reflects the protein exchange reaction, which can be expressed as

$$\Delta G = \Delta G^\circ + RT \ln\left(\frac{[AB]}{[A][B]}\right) = -RT \ln K_A = RT \ln K_D \tag{2.4}$$

where

ΔG and ΔG° represent the changes in adsorption free energy and in standard-state adsorption free energy

R is the ideal gas constant [1.985 cal/K·mol]

T is the absolute temperature [298 K]

$[A]$ is the molar concentration of the protein in solution

$[B]$ and $[AB]$ are the mole fraction of surface sites occupied by bare surface and adsorbed protein, respectively

The equilibrium association constant K_A and dissociation constant K_D can be experimentally acquired from SPR kinetic analysis. For example, K_D of albumin and fibrinogen were 13.2 and 2.0 μM, respectively, and their respective ΔG values were −6.6 and −7.8 kcal/mol. Albumin has relatively small Gibbs free energy, which means that the albumin preadsorbed surface has a high chance of being displaced by fibrinogen.

2.3 Results and Discussion

In its first implementation, we examine five human serum proteins, albumin (66 kDa), haptoglobin (86 kDa), IgG (150 kDa), fibrinogen (340 kDa), and thyroglobulin (660 kDa), for their different adsorption strengths onto the hydrophobic gold surface (contact angle measurement: 83.2° ± 0.75°). The different adsorption strengths induce surface exchange reactions among the proteins. Most proteins have amphiphilic surfaces; they have hydrophobic residues buried within their core and their hydrophilic residues facing outside or vice versa [12]. The adsorption of any protein onto a solid surface is highly dependent on hydrophobic attraction. Hydrophilic residues determine the orientation of the adsorbed protein but do not take part in the adsorption process itself. Therefore, proteins rapidly adsorb onto a hydrophobic surface by unfolding and spreading their hydrophobic core over the surface [12]. When more strongly interacting proteins arrive at the surface, they displace preadsorbed proteins. However, the reverse sequence does not occur (Figure 2.1b) [8]. This process is led by thermodynamics; proteins that have different dimensions and morphologies adsorb differently to a surface based upon their thermodynamic energy preferences and behave to minimize the overall system energy.

First, a protein is preadsorbed onto the surface. Then another protein subsequently reaches the surface to interact with the preadsorbed protein (Table 2.1). If protein exchange occurs, a permanent angle change of SPR profiles is generated. During the preadsorption procedure, the surface needs to be saturated by the preadsorbed proteins before the protein exchange; otherwise, a subsequently introduced protein will adsorb onto the space between the preadsorbed proteins if they do not form a fully packed protein monolayer. The surface was saturated using 0.05% (w/v) proteins to form a fully packed monolayer, verified by reinjecting the same sets of protein (Table 2.1). Since we used neutral PBS (pH 7.4) and all the proteins used for experiments have the *pI* (isoelectric point) less than pH 7.0, multilayer formation due to protein-protein electrostatic interaction does not occur. For example, when albumin is preadsorbed on the surface, the other proteins, including haptoglobin, IgG, fibrinogen, and thyroglobulin, displace the albumin, producing 40, 90, 320, and 200 mDeg angle changes, respectively. However, the SPR angle change is almost negligible when these proteins are exposed to albumin, confirming that the measured angle changes are not due to the multilayer formation but indeed due to the protein displacement. Table 2.1 shows all the cases of the displacement and the monolayer formation among five different proteins. The displacement strength is ranked in the following order: fibrinogen (340 kDa) > thyroglobulin (660 kDa) > IgG (150 kDa) > haptoglobin (86 kDa) > albumin (66 kDa). Fibrinogen can displace all other proteins, while haptoglobin can displace only albumin.

As shown from the affinity rank, the Vroman effect is not always associated with the size of proteins; for instance, thyroglobulin is larger than fibrinogen and yet is displaced by fibrinogen. We may

TABLE 2.1

Summary of Protein Displacements of Five Serum Proteins

Protein Injection	SPR Angle (mDeg)	Protein Injection	SPR Angle (mDeg)	Protein Injection	SPR Angle (mDeg)	Protein Injection	SPR Angle (mDeg)	Protein Injection	SPR Angle (mDeg)
Alb→Alb	X	Hp→Alb	X	IgG→Alb	X	Fib→Alb	X	Tg→Alb	X
	1		3		−2		2		−4
Alb→Hp	O	Hp→Hp	X	IgG→Hp	X	Fib→Hp	X	Tg→Hp	X
	40		−1		2		−1		0
Alb→IgG	O	Hp→IgG	O	IgG→IgG	X	Fib→IgG	X	Tg→IgG	X
	90		75		3		0		−2
Alb→Fib	O	Hp→Fib	O	IgG→Fib	O	Fib→Fib	X	Tg→Fib	O
	320		249		142		4		90
Alb→Tg	O	Hp→Tg	O	IgG→Tg	O	Fib→Tg	X	Tg→Tg	X
	200		87		62		3		3

Preadsorbed proteins react with ones arriving subsequently. O: displacement occurs; X: no displacement.

possibly explain this phenomenon because of their structures and potential conformation at the surface. Fibrinogen has a triad structure tethering two D domains to a central E domain by triple-stranded α-helical-coiled coils (αC) [13]. On the neutralized Au surfaces, D and E domains are involved in the strong adsorption, and in the neutralized PBS, positively charged αC domains have no electrostatic repulsion against the surface. On the other hand, thyroglobulin has a rather high degree of spherical symmetry and compactness. We expect that thyroglobulin has a smaller hydrophobic attraction than fibrinogen because thyroglobulin (Tg) acts as a compact particle [14].

Figure 2.6 depicts how to detect Tg using a pair of surfaces preadsorbed by two known proteins (fibrinogen and IgG) in a microfluidic system. When Tg flows on the surface preadsorbed by IgG, Tg displaces IgG in channel 1 but flows through the fibrinogen-covered surface (channel 2) without any displacement reaction. The differential measurement (Δθ) of the SPR angle changes from channels 1 and 2 enables the detection of Tg (Figure 2.6c). When a mixture of the four proteins solution (albumin, haptoglobin, IgG, and Tg) was injected, the sensor only responded to the specific target protein, Tg, which is a response similar to conventional affinity-based sensing techniques (Figure 2.7). The surfaces were preadsorbed with IgG and fibrinogen. Upon the mixture reaching the surfaces, channel 1 showed 64 mDeg angle change, and channel 2 had little angle change (2 mDeg). This set of results demonstrates that fibrinogen remained on the surface, whereas IgG was displaced by the mixture. Since albumin, haptoglobin, and IgG cannot displace IgG and a multilayer surface is not formed in the process, the only protein that can

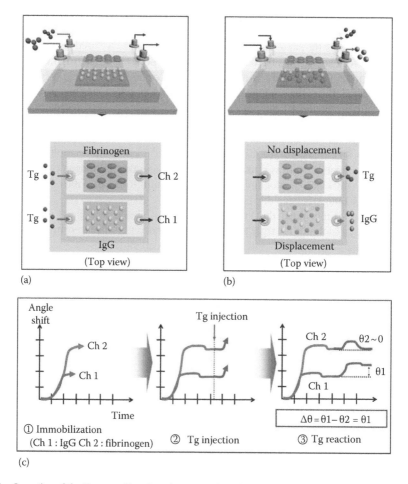

FIGURE 2.6 Operating of the Vroman effect–based sensor and a schematic of the SPR profiles. (a) The two sensing surfaces are adsorbed by IgG and fibrinogen, respectively. (b) Tg is injected into both channels. Tg displaces IgG in channel 1, while no displacement occurs in channel 2. (c) The sensor produces a permanent differential angle change (Δθ).

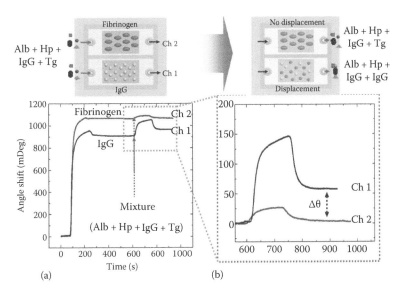

FIGURE 2.7 (a) SPR sensorgrams of Tg detection in a mixture of proteins. First, IgG and fibrinogen are adsorbed on the sensing surfaces in channels 1 and 2, respectively. The SPR angle shifts of IgG and fibrinogen are 909 mDeg and 1090 mDeg, respectively. Once the angle shift stabilizes, the mixture is injected into both channels. (b) Normalized SPR sensorgram after the mixture injection. The differential measurement of the SPR angle change from channels 1 (64 mDeg) and 2 (2 mDeg) allows the detection of Tg ($\Delta\theta = 62$ mDeg).

displace IgG is Tg. Although target molecules do not displace preadsorbed proteins on the surface in channel 2, SPR shows a small increase with moderate slope. However, their interactions to the surface are very weak, and they are easily detached by subsequent PBS flow, resulting in a very small SPR angle change. Figure 2.7b shows normalized SPR angle shift upon the injection of the mixture of proteins. After stabilization, the permanent angle change is almost negligible.

Figure 2.8 shows high selectivity of Tg on the preadsorbed surfaces with IgG and fibrinogen. The selectivity was measured against albumin, haptoglobin, and IgG. When Tg was injected, a 59 mDeg

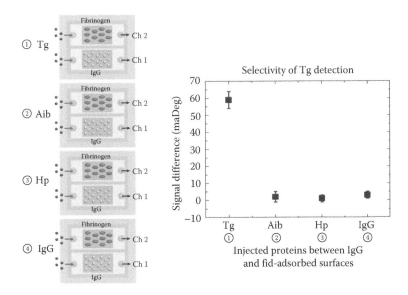

FIGURE 2.8 Selectivity of the Vroman effect–based biosensor of Tg detection. The two surfaces are preadsorbed by IgG and fibrinogen, respectively. Monitoring the differential SPR angle shifts of such configuration allows selective detection of Tg in a mixture of proteins.

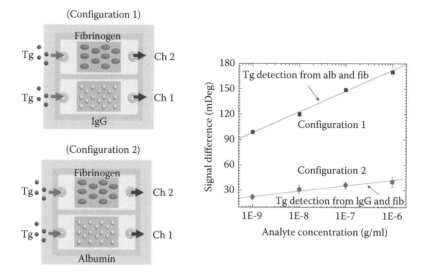

FIGURE 2.9 Sensitivity of the Vroman effect–based biosensor of Tg detection. Sensitivity can be controlled by preadsorbed proteins pair. The two sets of configurations demonstrate that the biosensor can detect Tg of at least 1 ng/ml.

angle was generated, while others had insignificant angle changes, all less than 5 mDeg. This suggests that the Vroman effect–based biosensor can be configured to detect Tg with high selectivity. Figure 2.9 shows a calibration curve of Tg from 1 ng/mL to 1 μg/mL. The calibration curves show that the sensitivity (the slope) is a function of the preadsorbed protein configuration. The sensitivity of albumin-fibrinogen pair, the slope of the curves, is higher than that of IgG-fibrinogen configuration. We verify the ability of competitive protein adsorption in allowing detection of Tg in four different protein mixtures (albumin, haptoglobin, IgG, and Tg) without the need for antibody immobilization and labeling. The work is still a proof-of-concept protein sensor, yet it delivers several advantages over existing immunoassay methods to detect Tg. Rather expensive, time-consuming, and labor-intensive immobilization processes may be substituted for the proposed sensor. Also, this method can be applied to detect other target molecules by choosing an appropriate combination of preadsorbed protein pairs.

2.4 Future Direction

Conventional approaches to biosensors make use of a "lock-and-key" design, wherein a specific bioreceptor is immobilized on the sensing surface. In theory, this process strongly and highly selectively binds target analytes. In practice, however, most biosensors suffer from interference caused by molecules that are structurally or chemically similar to the desired analyte. This is an inevitable consequence of the "lock" being able to fit a number of imperfect "keys" [15]. Additionally, biomarkers themselves pose an imperfect measure, as most biomarkers are nonspecific to particular diseases, and most diseases have more than one biomarker associated with their incidence. Furthermore, the concentration of other proteins in serum may change the concentration of biomarkers. These challenges lead to adopt rapid and efficient identification of total protein imbalances in serum for disease diagnosis. The current method of observing competitive protein adsorption/exchange has limitations of having high selectivity in a complex mixture, such as undiluted serum. Differentially responsive arrays that take advantage of competitive protein adsorption may have potential to overcome the limitation. Our future work is to develop a sensor array based on a distinct pattern of responses which provide a better fingerprint of the analyte—one that allows for classification and identification.

ACKNOWLEDGMENT

The authors would like to thank the U.S. National Science Foundation for supporting this work under grant #0846961.

REFERENCES

1. Park, J. S., K. K. Oh, E. Kim, H. Chang, and S. W. Hong. 2006. Sonographic screening for thyroid cancer in females undergoing breast sonography. *American Journal of Roentgenology* 186:1025–1028.
2. Krahn, J. and T. Dembinski. 2009. Thyroglobulin and anti-thyroglobulin assays in thyroid cancer monitoring. *Clinical Biochemistry* 42:416–419.
3. Frasoldati, A., M. Pesenti, M. Gallo, A. Caroggio, D. Salvo, and R. Valcavi. 2002. Diagnosis of neck recurrences in patients with differentiated thyroid carcinoma. *Cancer* 97:90–96.
4. Wartofsky, L. and D. V. Nostrand. 2005. *Thyroid Cancer: A Comprehensive Guide to Clinical Management*, 2nd edn. Humana Press, Totowa, NJ.
5. Haab, B. B. 2003. Methods and applications of antibody microarrays in cancer research. *Proteomics* 3:2116–2122.
6. Yu, X., D. Xu, and Q. Cheng. 2006. Label-free detection methods for protein microarrays. *Proteomics* 6:5493–5503.
7. Ince, R. and R. Narayanaswamy. 2006. Analysis of the performance of interferometry, surface plasmon resonance and luminescence as biosensors and chemosensors. *Analytica Chimica Acta* 569:1–20.
8. Choi, S., Y. Yang, and J. Chae. 2008. Surface plasmon resonance protein sensor using Vroman effect. *Biosensors and Bioelectronics* 24:893–899.
9. Vroman, L. and A. L. Adams. 1969. Findings with the recording ellipsometer suggesting rapid exchange of specific plasma proteins at liquid/solid interfaces. *Surface Science* 16:438–446.
10. Horbett, T. A. and J. L. Brash. 1995. *Proteins at Interfaces II: Fundamentals and Applications*. American Chemical Society, Washington, DC, pp. 112–128.
11. De, M., S. Rana, H. Akpinar, O. R. Miranda, R. R. Arvizo, U. H. F. Bunz, and V. M. Rotello. 2009. Sensing of proteins in human serum using conjugates of nanoparticles and green fluorescent protein. *Nature Chemistry* 1:461–465.
12. Latour, Jr. R. A. 2005. Biomaterials: Protein-surface interactions. *Encyclopedia of Biomaterials and Biomedical Engineering*, 1–15.
13. Holden, M. A. and P. S. Cremer. 2005. Microfluidic tools for studying the specific binding, adsorption, and displacement of proteins at interfaces. *Annual Review of Physical Chemistry* 56:369–387.
14. Bloth, B. and R. Bergquist. 1968. The ultrastructure of human thyroglobulin. *Journal of Experimental Medicine* 128:1129–1136.
15. Albert, K. J., N. S. Lewis, C. L. Schauer, G. A. Sotzing, S. E. Stitzel, T. P. Vaid, and D. R. Walt. 2000. Cross-reactive chemical sensor arrays. *Chemical Reviews* 100:2595–2626.

3

Surface Plasmon Resonance Analysis of Nanoparticles for Targeted Drug Delivery

Emilie Roger, Alex E. Grill, and Jayanth Panyam

CONTENTS

3.1 Introduction

3.1.1 Nanotechnology

Nanotechnology is defined as "the understanding and control of matter at dimensions of 1–100 nanometers, where unique phenomena enable novel applications" [9,10]. The nanometer size range is suitable for manipulation at the molecular level, with potential applications in drug delivery, imaging, and diagnosis of diseases such as cancer [2,21]. Drugs and imaging agents can be encapsulated in, adsorbed on, or conjugated to nanometer-size delivery systems. Nanodelivery systems include lipid-based carriers such as liposomes (vesicles in which an aqueous volume is surrounded by bilayer membrane), micelles (self-assembling amphiphiles that form supramolecular core–shell structures), dendrimers (polymers that comprise a series of well-defined branches around an inner core), and polymeric nanoparticles, including nanospheres and nanocapsules [25].

Although nanoparticles have been investigated as drug carriers for different disease applications, there has been a particularly intense interest in their use for anticancer drug delivery. The specific anatomical and pathophysiological features of solid tumors provide a unique opportunity for the delivery of drugs and imaging agents to cancer cells using nanocarriers [2]. First, rapidly growing tumor cells have a higher requirement for nutrients, so solid tumors initiate and promote the growth of new vasculature (angiogenesis) to meet that need. The newly growing vessels are structurally deficient, have poorly formed junctions, and, thus, are unusually porous. This allows particles smaller than ~200 nm to extravasate (i.e., pass through vessel walls). Moreover, tumors are also characterized by poor lymphatic drainage. Consequently, particles that extravasate into the tumor tissue are cleared rather slowly [15]. This phenomenon is referred to as the enhanced permeation and retention (EPR) effect. To achieve selective tumor accumulation via the EPR effect, nanocarriers must have long circulating times. First-generation nanocarriers were rapidly eliminated from the systemic circulation by the mononuclear phagocyte system (MPS), which prevented these nanocarriers from reaching the tumor site by the EPR effect. The second-generation "stealth" nanocarriers were developed to evade the MPS; these typically have a coating of polyethylene glycol (PEG) or other hydrophilic polymer on the surface [34].

Accumulation of nanocarriers in the tumor tissue can be further improved by targeting specific antigens or receptors overexpressed on the cancer cells. For example, folate receptors are overexpressed in many ovary, mammary gland, lung, kidney, brain, colon, prostate, nose, and throat tumors. Indeed, elevated folate receptor expression has been found in nearly 89% of ovarian carcinomas [18]. However, folate receptors are only minimally distributed in normal tissues [13]. Therefore, folic acid–conjugated nanocarriers have attracted wide attention for targeting tumor cells. Targeted nanocarriers have several advantages. First, targeted nanoparticles can enhance the concentration of drug within the target malignant cell cytoplasm and reduce the intracellular clearance [10]. Second, nanoparticles have the potential to reduce the toxicity of anticancer drugs in normal tissue and minimize undesired side effects. Targeted nanoparticles have been investigated extensively for delivery of anticancer drugs like paclitaxel (a major antitumor agent widely used in the treatment of breast cancer, non–small cell lung cancer and ovarian carcinoma) [16,23].

The efficacy of targeted nanoparticles depends on (1) the specific and selective recognition ability of the targeting molecule, and (2) the properties of nanoparticles and their ability to interact with specific receptors. Several previous reports have discussed the choice of targeting agent and the methods for conjugating them to nanoparticles [4,7,20,24,32]. Targeting ligands can be incorporated during the manufacturing of nanoparticles or can be conjugated after the fabrication of nanoparticles (post–particle synthesis). An example of the former is the Interfacial Activity Assisted Surface Functionalization (IAASF) technique [19]. Most polymeric nanoparticles are prepared using some modification of an emulsion-solvent evaporation process. In this technique, polymer solution in organic solvent is emulsified in an aqueous surfactant solution, and the organic solvent is evaporated to form polymeric nanoparticles. Hydrophobic drugs such as paclitaxel are dissolved in the organic solvent along with the polymer. In the IAASF technique, a diblock copolymer such as poly(lactide)–polyethylene glycol (PLA–PEG), with or without a targeting ligand such as folic acid attached to the PEG terminus (PLA–PEG–folic acid), is added to the oil-in-water emulsion and stirred at ambient temperature. The hydrophobic region in PLA–PEG copolymer partitions into polymer containing oil phase, thus coating the nanoparticle surface with PEG–ligand (Figure 3.1).

The number of targeting molecules present on the surface of nanoparticles is a critical determinant of targeting effectiveness. Many methods are available to determine the presence and/or quantity of ligand molecules on nanoparticles (Table 3.1). However, these methods do not allow for the analysis of the specific activity (i.e., binding ability) of the ligands. Consequently, over the last few years, surface plasmon resonance (SPR) has become an important tool to analyze the specific interactions between ligand-conjugated nanoparticles and the corresponding cellular target.

3.1.2 Surface Plasmon Resonance

SPR is a versatile technique for measuring molecular interactions in real time. Surface plasmons are electromagnetic waves that propagate in a direction parallel to the metal/dielectric interface. SPR phenomenon occurs when polarized light, under conditions of total internal reflection, strikes an electrically conducting metal layer at the interface between media of different refractive indexes: glass (high refractive index) and a buffer (low refractive index). Since the electromagnetic wave is on the boundary of the metal and the external medium (e.g., air or water), the oscillations are very sensitive to any change of this

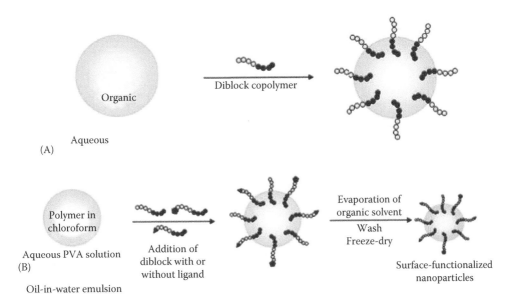

FIGURE 3.1   IAASF technique. (A) Behavior of amphiphilic block copolymers in an emulsion media. Hydrophobic regions (filled blocks) orient themselves with the organic phase resulting in outward facing hydrophilic regions (open blocks). (B) An example of the application of the IAASF technique. Introduction of PLA–PEG and PLA–PEG–ligand conjugates during the emulsification step results in nanoparticles with PEG and PEG–ligand on nanoparticle surface. (From Patil, Y.B. et al., *Biomaterials*, 30, 859, 2009.)

TABLE 3.1

Methods Used to Characterize Functionalized Nanoparticles

Surface characterization methods	Nuclear magnetic resonance
	Zeta potential
	FT-IR
	Dynamic light scattering
	Microscopy
	Size exclusion chromatography
Activity/efficacy methods	In vitro cytotoxicity
	Cellular uptake in vitro
	Antitumor efficacy in vivo

boundary, such as the adsorption of molecules on the surface. The SPR technique uses this phenomenon to provide kinetic and equilibrium data for interaction of biomolecules to other molecules bound to the sensor surface [33].

SPR can be used to study many different types of molecular interactions—antibody–antigen, receptor–ligand, etc. Typically, the target of interest (receptor, antibody) is immobilized on a sensor surface (metal film, typically gold or silver, acting as a mirror), and a flowing solution containing antigen or ligand molecules is injected across the sensor. When polarized light is shone through the sensor metal, the light gets reflected with a typical angle of incidence. At this angle of incidence, the light excites surface plasmon, inducing plasmon resonance and causing a dip in the intensity of the reflected light [5,29]. The angle at which the maximum loss of the reflected light intensity occurs is called resonance or SPR angle (Figure 3.2A).

Adsorption (or desorption) of molecules to (or from) the sensor surface is associated with a change in refractive index of the solvent near the surface. The change in refractive index, in turn, alters the SPR

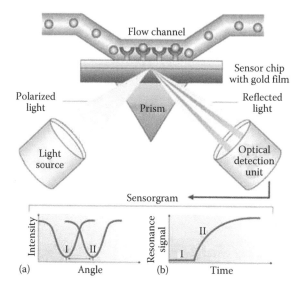

FIGURE 3.2   SPR technology. Schematic illustration of the components of biosensor-SPR technology: the optical unit generates and measures the SPR angle. (a) A change in refractive index at the surface will cause an angle shift from I to II when molecules bind to the surface and change the mass of the surface layer. (b) This change in resonance angle can be monitored in real-time through a plot of resonance signal as a function of time. (From Cooper, M.A., *Nat. Rev. Drug Discov.*, 1, 515, 2002.)

angle. The change in SPR angle is proportional to the mass of bound material and is recorded in a sensorgram. The change of angle is monitored in real time, and the time course is represented in the sensorgram (Figure 3.2B). If binding occurs as the solution passes over the sensor surface, the resonance signal in the sensorgram increases. On the other hand, the dissociation of bound molecules is characterized by a decrease of the resonance signal. Consequently, the amount of adsorbed species can be determined after injection of the original baseline buffer. The analysis of the kinetic parameters permits the evaluation of association constant (K_a) and dissociation constant (K_d). The affinity constant (K_D) can be also calculated from the ratio of these rate constants ($K_D = K_d/K_a$) (Figure 3.3).

The flowing solution can contain virtually any type of analyte ranging from proteins to small molecules, viruses, cells, or nanoparticles. To permit selective detection of binding of the analyte to a specific target, the surface needs to be modified with a target of interest suited for selective capturing of the analyte but is not prone to adsorbing any other component present in the sample or buffer. It is important to note that the SPR angle is dependent on the characteristics of the system, and the refractive index changes with modification of the medium [33].

3.2 Materials

3.2.1 SPR Materials

SPR biosensor analysis is usually performed according to the biosensor hardware used. A list of biosensor manufacturers is given in Table 3.2. The most widely used equipment is the BIAcore 1000 or 3000 (www.biacore.com). Most SPR instruments use an optical method to measure the refractive index near the sensor surface (within ~300 nm) [1]. Generally, SPR instruments are comprised of three units: the optical unit (the detector), the liquid handling unit (sample delivery system), and the sensor surface [3,30]. Choosing the sensor surface is critical to the experiment. Table 3.3 shows common sensor surfaces currently available. Running buffer is typically HEPES-buffered saline (10 mM HEPES pH 7.4, 150 mM NaCl, 3 mM EDTA, 0.005% surfactant P20).

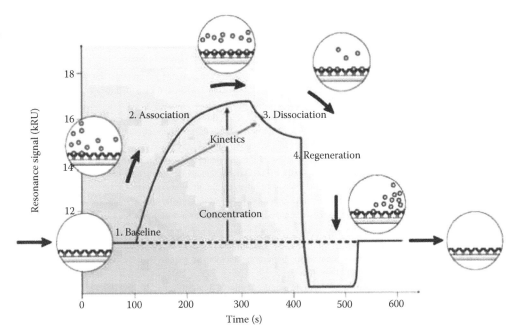

FIGURE 3.3 A typical sensorgram illustrating the different steps in an SPR experiment: (1) the baseline: only buffer. (2) Association phase: the solution of analyte in the running buffer is passed over the receptor, and binding of analyte on sensor surface chip starts, resulting in an increase of the resonance signal. The observed association rate (K_{obs}) can be determined, and when the concentration of analyte is known, the association rate constant (K_a) can be calculated. By definition, the amount of analyte that is associating and dissociating is equal at equilibrium. (3) Buffer containing no analyte is injected for dissociation of the receptor–analyte complex. Analysis of these data gives the dissociation rate constant (K_d). (4) The regeneration solution is injected to bring the surface back to starting conditions. The injections are typically repeated at a range of concentrations to generate a set of sensorgrams. (From Cooper, M.A., *Nat. Rev. Drug Discov.*, 1, 515, 2002.)

TABLE 3.2

Companies That Currently Manufacture Biosensor Hardware

Company	Website
Affinity Sensors	http://www.sierrasensors.com/
Aviv Instruments	http://www.avivbiomedical.com/
Biacore	http://www.biacore.com/lifesciences/index.html
IBIS	http://www.ibis-spr.nl
Nippon Lasers	http://www.rikei.com
Texas Instruments	http://www.ti.com
Windsor Scientific Limited	http://www.windsor-ltd.co.uk

3.3 Methods

3.3.1 SPR Methods

The different steps that comprise an SPR experiment are as follows:

- Preparation of target and analyte
- Selection of a suitable sensor chip

TABLE 3.3

Various Biacore Sensor Chips Currently Available (Biacore Interaction Analysis Product Guide)

Name	Biacore Surface	Molecule to Be Immobilized
CM5	Carboxymethyl dextran	The first choice for immobilization of proteins, tagged proteins, low-molecular-weight molecules (typically <1000 Da), nucleic acids, and carbohydrates via –NH2, –SH, –CHO, –OH, or –COOH groups
CM4		Used for immobilization of proteins, low-molecular-weight molecules (typically <1000 Da), nucleic acids, and carbohydrates when sample contaminants have a high positive charge
CM3		Used for immobilization of viruses, intact cells and proteins, low-molecular-weight molecules (typically <1000 Da), nucleic acids, and carbohydrates when the interaction partner in solution is very large, e.g., molecular complex
C1	Flat carboxymethylated	Used for immobilization of viruses and intact cells and proteins, nucleic acids, and carbohydrates when the interaction partner in solution is multivalent or very large
SA	Streptavidin	Used for immobilization of biotinylated peptides, proteins, nucleic acids, or carbohydrates
HPA	Hydrophobic monolayer	Used when working with model membrane systems (used to incorporate molecules into a lipid monolayer)
L1	Lipophilic dextran	Used to incorporate a molecule into a lipid bilayer
NTA	Nickel chelation	Used for immobilization of histidine-tagged molecules
Au	Gold surface	Used for a wide variety of coating techniques, including those using harsh conditions that the chip carrier would not withstand

- Immobilization of target on the sensor surface
- Injection of analyte over sensor surface
- Recording of response and analysis of data

A critical step in the SPR experiment is the attachment of the target molecules (such as a receptor or antigen) onto the surface of a sensor chip. It is essential that after immobilization onto the sensor surface, target molecules retain their native conformation and binding activity. Further, this attachment must be stable over the entire course of the binding assay, and sufficient binding sites should be presented to the solution phase to interact with the analyte (such as antibody-conjugated nanoparticles). The sensor surface chosen should also be resistant to nonspecific binding. As discussed previously, different sensor surfaces are available (Table 3.3), and the most suitable surface is chosen according to the nature of the target molecule to be coupled and the requirements of the analysis.

Different methods are available to immobilize target molecules on the sensor surface [12,35]. In general, there are two approaches to attaching biomolecules on sensor chips: covalent and noncovalent capture. The main covalent coupling methods utilize amine (e.g., amino terminus or surface lysine residues), thiol, or aldehyde functional groups on proteins and free carboxymethyl groups on the sensor chip surface. In the case of noncovalent capture, it is an indirect immobilization of the target molecule. This technique is most commonly used for biotinylated ligands. For example, covalently coupled streptavidin on the sensor chip surface can be used to capture biotinylated ligands (e.g., nucleic acids) using the multiple biotin-binding sites of streptavidin. Similarly, antibodies covalently conjugated to the sensor surface can be used to capture other target molecules on the surface. For instance, murine antifolate antibody amine coupled to the CM-5 surface can be used to capture folic acid on the sensor surface through antibody–antigen interaction [8,19,26]. Many examples of immobilization protocols are available on the major manufacturers' websites [30]. Additionally, annual surveys on SPR methods, applications, and appropriate experimental approaches by Myszka and colleagues provide many additional helpful suggestions on experimental protocols to obtain high-quality biosensor data [17,22].

The next step is the injection of analyte over the sensor surface. Typically, the analyte is dissolved or dispersed in HEPES-buffered saline. The flow rate and time depend on the nature of the experiment and the equipment [1]. Finally, the last step is the analysis of data using the integrated software. An SPR system is able to determine the specificity of an interaction, the kinetics of an interaction (i.e., K_a and K_d), and the affinity of an interaction.

3.3.2 Example Characterization of Targeted Nanoparticles by SPR

In this section, the use of SPR for surface analysis and targeting potential of folic acid– and/or biotin-conjugated poly(D,L-lactide-*co*-glycolide) (PLGA) polymer nanoparticles is described. The details of nanoparticle fabrication are first provided, followed by SPR analysis.

3.3.2.1 Fabrication of Polymers

3.3.2.1.1 Preparation of Biotin–PEG–OH and Folic Acid–PEG–OH

α-Amine-ω-hydroxy PEG was dissolved in acetonitrile. Methylene chloride and TEA were added to the solution and stirred. *N*-Hydroxysuccinimide (NHS)–biotin or NHS–folic acid was introduced to the aforementioned reaction mixture and stirred overnight under nitrogen. The reaction was stopped by the addition of diethyl ether. Precipitated polymer was filtered and washed with diethyl ether. The polymer was then dissolved in hot 2-propanol. This polymer was reprecipitated by cooling, and the unconjugated folic acid and biotin molecules were removed by dialysis. The purified polymer was then lyophilized [19].

3.3.2.1.2 Preparation of Biotin–PEG–PLA and Folic Acid–PEG–PLA

Graft polymerization of lactide onto folic acid–PEG–OH or biotin–PEG–OH was done by solvent polymerization technique. Glasswares were silanized and dried overnight at 130°C. Folic acid–PEG–OH or biotin–PEG–OH and L-lactide were added into a round bottom flask and diluted with toluene. Stannous-2-ethyl-hexanoate in toluene was added to the reaction mixture mentioned earlier and refluxed for 4 h under nitrogen. The remaining viscous material was heated to 140°C for 1 h under nitrogen. The reaction mixture was cooled and dissolved in dichloromethane. This polymer solution was then added slowly to chilled diethyl ether. The final polymer conjugate was isolated by vacuum filtration and then lyophilized [19].

3.3.2.1.3 Fabrication of Targeted Paclitaxel-Loaded Polymeric Nanoparticles

To prepare nanoparticles, PLGA and paclitaxel were first dissolved in chloroform. The organic solution was emulsified on ice for 5 min in an aqueous 2.5% PVA solution using a probe sonicator. PLA–PEG–biotin and/or PLA–PEG–folic acid was added dropwise to the resulting emulsion. Chloroform was evaporated overnight by continuous stirring at ambient conditions. Nanoparticles were washed three times with distilled water and collected through ultracentrifugation (150,000×g for 30 min). After washing, particles were lyophilized.

3.3.2.2 Characterizing the Functional Efficacy of Targeted Nanoparticles

3.3.2.2.1 SPR Analysis

For these studies, folate receptor (or folate-binding protein, FBP) was immobilized on the surface of carboxylated, activated, dextran-coated gold film (sensor chip CM-5) by amine coupling. First, a solution containing 0.1 M NHS and 0.4 M N'-(3-dimethylaminopropyl)-N-ethylcarbodiimide (NHS/EDC) (1/1, v/v) was injected to activate the carboxylated dextran. Following this, 100 mM potassium phosphate buffer (pH 7.0) containing FBP, 4 mM mercaptoethanol, and 10% v/v glycerol was injected. Finally, ethanolamine in water (pH 8.5) was injected to deactivate residual NHS-esters on the sensor chip. Biacore instrument was equilibrated with HBSS buffer, and the samples were then successively injected into the system. The binding of nanoparticles to the sensor chip was performed, and the nanoparticles were allowed to interact with the FBP for 10 min. The specific interaction with the protein shifted the wavelength at which the SPR occurred. This shift was registered as response units (RUs). Nonconjugated nanoparticles were used as control. It is important to test all the control groups on newly immobilized FBP [26–28].

3.3.2.2.2 *Accumulation of Nanoparticles in Tumor Cells* In Vitro

MCF-7 (drug-sensitive human breast carcinoma), NCI-ADR (drug-resistant ovarian adenocarcinoma), and 4T1 (drug-sensitive mouse mammary carcinoma) cells were seeded in 24-well plates. Following attachment, cells were treated for 30 min with 6-coumarin (fluorescent dye)-labeled nanoparticle formulations with or without excess free folic acid and biotin. The treated cells were washed twice, and then lysed. Cell lysates were lyophilized and then extracted with methanol. The extracts were centrifuged at 5000 rpm for 10 min, and the supernatants were subjected to HPLC analysis. Nanoparticle concentrations were determined based on 6-coumarin loading in nanoparticles, and then normalized to total cell protein content determined using Pierce protein assay kit [19].

3.3.2.2.3 *Efficacy of Paclitaxel-Loaded Nanoparticles in MCF-7 Tumor Xenograft Model*

MCF-7 cells were used to develop tumor xenografts in NCR–NU mice. Mice were implanted subcutaneously with a 90 day 17-β-estradiol pellet on the dorsal side before tumor cell injection. Following implantation, each animal received one to two million MCF-7 cells suspended in PBS. Tumors were allowed to grow to a size of 100 mm³ in diameter before treatments were administered. Animals were injected intravenously with paclitaxel dissolved in Cremophor® or encapsulated in different nanoparticle formulations. Animals that received Cremophor® or targeted blank nanoparticles were used as vehicle-treated controls. Tumor volumes were assessed regularly by measuring two perpendicular diameters with calipers. Tumor volume on the day of treatment was normalized to 100% for all groups [19].

3.3.2.2.4 *Determination of Paclitaxel in Tumor Samples*

NCR–NU mice bearing MCF-7 tumor xenografts (~100 mm³) were injected intravenously with different paclitaxel treatments as described before. Animals were euthanized 2 h postinjection, and tumors were excised, homogenized, and lyophilized. Extraction with tert-butyl methyl ether was used for sample preparation. Paclitaxel accumulation was quantified using LC-MS/MS with docetaxel as the internal standard. Paclitaxel concentration was normalized to the tissue weight [19].

3.4 Results and Discussion

3.4.1 Specificity of Binding

The first utilization of SPR in targeted drug delivery was to study the affinity of targeted nanoparticles for a specific receptor. In the example case study described in Section 3.2, SPR analysis was used to characterize the targeting potential of PLGA nanoparticles with surface-bound biotin and/or folic acid. These studies indicated a significant difference in binding affinity to respective targets for nanoparticles with and without ligands (Figure 3.4A and B). A three to five-fold increase in RUs was observed for biotin-functionalized nanoparticles for binding to streptavidin compared to that of nonfunctionalized nanoparticles (Figure 3.4B). A difference of 200 RU was observed for binding of folic acid–conjugated nanoparticles to anti–folic acid antibody compared to that for nanoparticles without folic acid (Figure 3.4A). The magnitude of differences in the RUs for biotin nanoparticles and folic acid nanoparticles was attributed to the differences in the binding affinities of biotin–streptavidin and folic acid–antibody combinations. Binding of folic acid to its target is weaker ($K_D \sim 10^{-10}$ M) [14] than the binding of biotin to streptavidin ($K_D \sim 10^{-15}$ M) [11]. Presence of both biotin and folic acid on the surface resulted in slightly weaker binding for the individual ligands, probably due to fewer number of ligand molecules present on the surface. For example, nanoparticles with both biotin and folic acid had about 13.0 ± 1.4 biotin molecules per particle while biotin-only nanoparticles had about 17.0 ± 1.6 biotin molecules per particle.

Other studies have used FBP (instead of antifolate antibody) as the target. For example, SPR analysis was used to determine the binding of poly[aminopoly(ethylene glycol)cyanoacrylate-*co*-hexadecyl cyanoacrylate] (poly(H_2NPEGCA-*co*-HDCA)) nanoparticles to FBP. As can be seen from Figure 3.5, folate conjugate nanoparticles demonstrated a significantly higher binding (translating into higher RUs on the sensorgram) to FBP than nonfunctionalized nanoparticles. These results suggested that nanoparticles

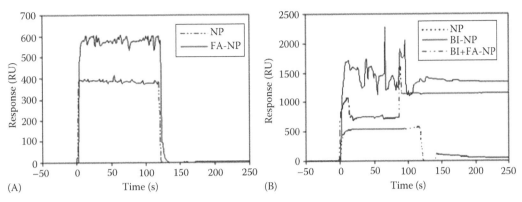

FIGURE 3.4 (A) SPR analysis of binding of folic acid–functionalized PLGA nanoparticles (FA-NP) and nonfunctionalized PLGA nanoparticles to antifolate antibody–coated sensor chip (10 mg/mL). (B) SPR-based binding analysis of biotin-functionalized nanoparticles (BI-NP), dual biotin and folic acid–functionalized nanoparticles (BI+FA-NP), and nonfunctionalized nanoparticles (NP) to immobilized streptavidin-coated sensor chip. (From Patil, Y.B. et al., *Biomaterials*, 30, 859, 2009.)

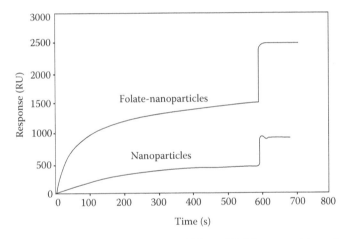

FIGURE 3.5 SPR analysis of binding of folate-poly(H$_2$NPEGCA-*co*-HDCA) nanoparticles and control nanoparticles (nonconjugated poly(H$_2$NPEGCA-*co*-HDCA) nanoparticles) (5 mg/mL) to immobilized FBP. (From Stella, B. et al., *J. Pharm. Sci.*, 89, 1452, 2000.)

were successfully functionalized with folic acid and were able to effectively recognize the target compared to control nanoparticles [19].

3.4.2 Correlation of SPR Data with Biological Performance

While SPR is an excellent in vitro characterization tool, the targeting effectiveness of ligand-conjugated nanoparticles needs to be confirmed in biological systems. Bioactivity can be determined through in vitro cell culture assays and through in vivo animal model studies. Cell culture studies typically include determination of the ability of targeted nanocarriers to be taken up selectively by target cells (uptake studies) and the ability of drug-loaded nanocarriers to selectively kill target cells (cytotoxicity studies). In vivo studies often involve determining the differences in the biodistribution of targeted and nontargeted nanocarriers as well as the enhancement in drug delivery to the target tissue. Finally, demonstration of greater therapeutic efficacy (such as tumor growth inhibition) is the most important indicator of the bioactivity of targeted nanocarriers.

In the example case study with targeted PLGA nanoparticles, the SPR data correlated with in vitro and in vivo performance of nanoparticles. Folic acid and biotin–functionalized nanoparticles showed a significant increase in intracellular accumulation when incubated with various cancer cell lines compared to controls (Figure 3.6). Further, when incubated with excess folic acid and biotin, the accumulation returned to basal levels, confirming the effects of surface functionalization. Paclitaxel-loaded biotin and folic acid–functionalized nanoparticles also showed increased efficacy in vivo against an MCF-7 xenograft model (Figure 3.7). A slower rate of tumor growth and a smaller final tumor volume were observed in mice treated with nanoparticles with dual surface functionalization compared to controls. Interestingly, these particles also performed better than nanoparticles functionalized with biotin or folic acid only. This suggests a possible additive or synergistic effect of multiple targeting moieties on the nanoparticle surface. Biodistribution studies were carried out to confirm that dual targeting leads to greater nanoparticle accumulation. Intratumoral paclitaxel concentrations were highest in the dual functionalized nanoparticles, compared to all other groups (Figure 3.8). These data further point to possible additive or synergistic effect between the targeting agents. Thus, the SPR analysis and the data from

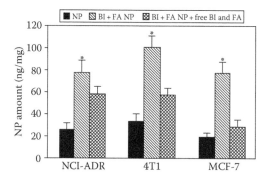

FIGURE 3.6 Intracellular accumulation of biotin and folic acid–functionalized nanoparticles in various breast cancer cell lines. Blank nanoparticles (NP), biotin and folic acid–functionalized nanoparticles (BI + FA NP), and BI + FA NP with excess free biotin and folic acid (BI+FA NP+free BI and FA) were incubated with NCI-ADR, 4T1, and MCF-7 breast cancer cell lines for 30 min at 37°C. Cells were lysed and nanoparticle accumulation was determined using HPLC, and the data were normalized to the total cell protein. Data as mean ± D ($n=6$). *$P<0.05$ vs. other groups. (From Patil, Y.B. et al., *Biomaterials*, 30, 859, 2009.)

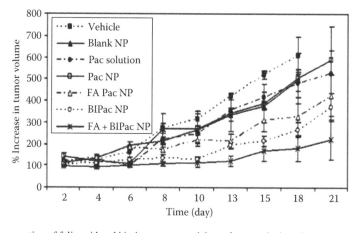

FIGURE 3.7 Incorporation of folic acid and biotin on nanoparticle surface results in enhanced anticancer effectiveness of nanoparticle-encapsulated paclitaxel. NCR–NU mice bearing MCF-7 xenografts were injected with paclitaxel in solution or encapsulated in nanoparticles (~3 mg/animal) with or without surface functionalization. Growth in tumor volume was determined over a period of 21 days. Data as mean ± SD ($n=6$). (From Patil, Y.B. et al., *Biomaterials*, 30, 859, 2009.)

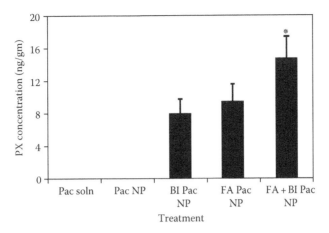

FIGURE 3.8 Effect of ligand functionalization on tumor accumulation of paclitaxel. NCR–NU mice bearing MCF-7 xenografts were injected with paclitaxel in solution or encapsulated in nanoparticles (~3 mg/animal) with or without surface functionalization. Animals were sacrificed 2 h postinjection, and paclitaxel concentration in the tumor tissue was determined by LC-MS/MS. Data as mean ± SD ($n = 4$). $^*P < 0.05$ vs. other treatment groups. (From Patil, Y.B. et al., *Biomaterials*, 30, 859, 2009.)

biological studies together confirm that the targeting moieties are on the surface of the nanoparticle, and that they are functional.

3.4.3 Quantitative Determination of Affinity Constants

The SPR technique can also be used to perform a variety of binding studies and to determine binding constants. By comparing the calculated affinity constants, it is possible to compare the affinity of different targeting molecules for a particular receptor. Stella et al. [28] have shown a greater affinity of folic acid–conjugated nanoparticles for FBP in comparison with free folic acid. An apparent dissociation constant of 800 ± 170 nM for folate-conjugated nanoparticles was calculated on the basis of 79 μM folate/2.5 mg folate-conjugated nanoparticles. Free folic acid had an apparent dissociation constant of only 11 ± 1 μM, necessitating a 108-fold higher molar concentration to completely inhibit the interaction of folic acid–conjugated nanoparticles with FBP. To explain the greater binding affinity of folic acid–conjugated nanoparticles, authors suggested that folic acid–conjugated nanoparticles probably display a multivalent interaction with the FBP on the sensor chip surface while free folic acid displays a monovalent interaction.

In a similar study, Tassa et al. [31] observed the affinity and binding kinetics of small molecule–conjugated nanoparticles with the FKBP12-GST protein (glutathione S-transferase fusion protein) immobilized to the sensor chip. The conjugation of small molecule ligands to nanoparticles was shown to increase the avidity of ligand–target interactions, resulting in differences in both association and dissociation kinetics between nanoparticle-conjugated and free ligand. They also concluded that the presumed multivalent conjugation of targeting ligands on the surface of nanoparticles can explain the enhancement of binding and, consequently, the greater affinity of nanoparticle-conjugated ligands.

3.4.4 Analysis of Properties of Targeting Molecules

By comparing the binding affinity of a native targeting molecule with that of a nanoparticle-conjugated targeting molecule, it is possible to draw conclusions about the activity of the targeting molecule. This is especially important since processes used to conjugate molecules on the surface of particles can affect the intrinsic properties of the targeting molecules. Thus, Debotton et al. [6] have used SPR to show that conjugation to nanoparticles did not alter the intrinsic specificity and affinity of the AMB8LK

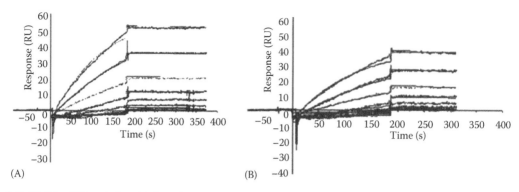

FIGURE 3.9   SPR analysis of binding of (A) AMB8LK antibody and (B) AMB8LK-functionalized nanoparticles at different concentrations to H-ferritin. (From Debotton, N. et al., *Eur. J. Pharm. Biopharm.*, 74, 148, 2010.)

TABLE 3.4

Affinity Constant Obtained from the Interaction between AMB8LK Antibody or AMB8LK Immunonanoparticles and H-Ferritin

	AMB8LK	**AMB8LK-Immunonanoparticles**
K_a (M^{-1} s^{-1})	3.45×10^4	3.3×10^4
K_d (s^{-1})	1.53×10^{-4}	1.27×10^{-4}
K_D (M)	6.67×10^{-9}	8.16×10^{-9}

Source: Debotton, N. et al., *Eur. J. Pharm. Biopharm.*, 74, 148, 2010.

monoclonal antibody which specifically recognizes H-ferritin. In the said study, the antibody was coupled by a thioether bond to maleimide groups on the surface of nanoparticles. Sensorgrams obtained with both native AMB8LK and AMB8LK-conjugated nanoparticles showed similar profiles (Figure 3.9). Calculated K_a, K_d, and K_D were close, which demonstrated the conservation of intrinsic properties of the antibody following conjugation to nanoparticles (Table 3.4).

3.4.5 Screening of Molecule Binding to a Specific Protein

SPR is also an interesting method to screen for molecules with optimal targeting properties. For example, Tassa et al. [31] have used SPR to study the affinity and binding kinetics of nanoparticles conjugated to different small molecules. FKBP12-GST was used as the target protein. Constant experimental conditions were used for all the molecules: a common protein target, shared nanoparticle scaffold, similar valency, and identical conjugation chemistry. It was concluded that the differences in binding affinities obtained with the different small molecules could translate into functional differences and could provide information regarding the structure activity relationship of targeted nanoparticles in both in vitro and in vivo systems.

3.5 Conclusions and Future Trends

SPR is a well-established method for characterizing binding interactions. Currently, the targeting ability of functionalized nanoparticles is generally investigated using in vitro cell culture and in vivo animal model studies. SPR provides a rapid and easy alternative to cell culture studies for screening new targeting ligands and as a quality control tool. Important considerations for the routine use of SPR are the cost of SPR equipment and the availability of a suitable biosensor surface. Overall, SPR studies could provide critical information regarding binding kinetics and target affinities to facilitate the design and synthesis of functionalized nanoparticles for imaging, diagnosis, and treatment.

REFERENCES

1. Anton van der Merwe, P. 2001. Surface plasmon resonance. In Stephen E. Harding and B. Z. Chowdhry (eds.), *Protein–Ligand Interactions: Hydrodynamics and Calorimetry*. Oxford University Press, New York, pp. 137–184.
2. Arruebo, M., M. Valladares, and A. González-Fernández. 2009. Antibody-conjugated nanoparticles for biomedical applications. *J Nanomater*.
3. Baird, C. L., E. S. Courtenay, and D. G. Myszka. 2002. Surface plasmon resonance characterization of drug/liposome interactions. *Anal Biochem* **310**:93–99.
4. Beduneau, A., P. Saulnier, and J. P. Benoit. 2007. Active targeting of brain tumors using nanocarriers. *Biomaterials* **28**:4947–4967.
5. Cooper, M. A. 2002. Optical biosensors in drug discovery. *Nat Rev Drug Discov* **1**:515–528.
6. Debotton, N., H. Zer, M. Parnes, O. Harush-Frenkel, J. Kadouche, and S. Benita. 2010. A quantitative evaluation of the molecular binding affinity between a monoclonal antibody conjugated to a nanoparticle and an antigen by surface plasmon resonance. *Eur J Pharm Biopharm* **74**:148–156.
7. Gref, R., P. Couvreur, G. Barratt, and E. Mysiakine. 2003. Surface-engineered nanoparticles for multiple ligand coupling. *Biomaterials* **24**:4529–4537.
8. Hong, S., P. R. Leroueil, I. J. Majoros, B. G. Orr, J. R. Baker, Jr., and M. M. Banaszak Holl. 2007. The binding avidity of a nanoparticle-based multivalent targeted drug delivery platform. *Chem Biol* **14**:107–115.
9. Jain, K. K. 2008. Recent advances in nanooncology. *Technol Cancer Res Treat* **7**:1–13.
10. Koo, O. M., I. Rubinstein, and H. Onyuksel. 2005. Role of nanotechnology in targeted drug delivery and imaging: A concise review. *Nanomedicine* **1**:193–212.
11. Laitinen, O. H., H. R. Nordlund, V. P. Hytonen, S. T. Uotila, A. T. Marttila, J. Savolainen, K. J. Airenne, O. Livnah, E. A. Bayer, M. Wilchek, and M. S. Kulomaa. 2003. Rational design of an active avidin monomer. *J Biol Chem* **278**:4010–4014.
12. Lofas, S., B. Johnsson, A. Edstrom, A. Hansson, G. Lindquist, R. M. Muller Hillgren, and L. Stigh. 1995. Methods for site controlled coupling to carboxymethyldextran surfaces in surface plasmon resonance sensors. *Biosens Bioelectr* **10**:813–822.
13. Lu, Y. and P. S. Low. 2002. Folate-mediated delivery of macromolecular anticancer therapeutic agents. *Adv Drug Deliv Rev* **54**:675–693.
14. Lu, Y., E. Sega, C. P. Leamon, and P. S. Low. 2004. Folate receptor-targeted immunotherapy of cancer: Mechanism and therapeutic potential. *Adv Drug Deliv Rev* **56**:1161–1176.
15. Maeda, H. and Y. Matsumura. 1989. Tumoritropic and lymphotropic principles of macromolecular drugs. *Crit Rev Ther Drug Carrier Syst* **6**:193–210.
16. Malika, S., R. M. Cusidób, M. H. Mirjalilic, E. Moyanod, J. Palazónb, and M. Bonfillb. 2011. Production of the anticancer drug taxol in *Taxus baccata* suspension cultures: A review. *Process Biochem* **46**:23–34.
17. Myszka, D. G. and R. L. Rich. 2000. Implementing surface plasmon resonance biosensors in drug discovery. *Pharm Sci Technol Today* **3**:310–317.
18. Parker, N., M. J. Turk, E. Westrick, J. D. Lewis, P. S. Low, and C. P. Leamon. 2005. Folate receptor expression in carcinomas and normal tissues determined by a quantitative radioligand binding assay. *Anal Biochem* **338**:284–293.
19. Patil, Y. B., U. S. Toti, A. Khdair, L. Ma, and J. Panyam. 2009. Single-step surface functionalization of polymeric nanoparticles for targeted drug delivery. *Biomaterials* **30**:859–866.
20. Popielarski, S. R., S. H. Pun, and M. E. Davis. 2005. A nanoparticle-based model delivery system to guide the rational design of gene delivery to the liver. 1. Synthesis and characterization. *Bioconjug Chem* **16**:1063–1070.
21. Retel, V. P., M. J. Hummel, and W. H. van Harten. 2009. Review on early technology assessments of nanotechnologies in oncology. *Mol Oncol* **3**:394–401.

22. Rich, R. L. and D. G. Myszka. 2000. Advances in surface plasmon resonance biosensor analysis. *Curr Opin Biotechnol* **11**:54–61.

23. Rowinsky, E. K., M. Wright, B. Monsarrat, G. J. Lesser, and R. C. Donehower. 1993. Taxol: Pharmacology, metabolism and clinical implications. *Cancer Surv* **17**:283–304.

24. Ruoslahti, E., S. N. Bhatia, and M. J. Sailor. 2010. Targeting of drugs and nanoparticles to tumors. *J Cell Biol* **188**:759–768.

25. Sahoo, S. K. and V. Labhasetwar. 2003. Nanotech approaches to drug delivery and imaging. *Drug Discov Today* **8**:1112–1120.

26. Salmaso, S., A. Semenzato, P. Caliceti, J. Hoebeke, F. Sonvico, C. Dubernet, and P. Couvreur. 2004. Specific antitumor targetable beta-cyclodextrin-poly(ethylene glycol)-folic acid drug delivery bioconjugate. *Bioconjug Chem* **15**:997–1004.

27. Sonvico, F., S. Mornet, S. Vasseur, C. Dubernet, D. Jaillard, J. Degrouard, J. Hoebeke, E. Duguet, P. Colombo, and P. Couvreur. 2005. Folate-conjugated iron oxide nanoparticles for solid tumor targeting as potential specific magnetic hyperthermia mediators: Synthesis, physicochemical characterization, and in vitro experiments. *Bioconjug Chem* **16**:1181–1188.

28. Stella, B., S. Arpicco, M. T. Peracchia, D. Desmaele, J. Hoebeke, M. Renoir, J. D'Angelo, L. Cattel, and P. Couvreur. 2000. Design of folic acid-conjugated nanoparticles for drug targeting. *J Pharm Sci* **89**:1452–1464.

29. Szabo, A., L. Stolz, and R. Granzow. 1995. Surface plasmon resonance and its use in biomolecular interaction analysis (BIA). *Curr Opin Struct Biol* **5**:699–705.

30. Tanious, F. A., B. Nguyen, and W. D. Wilson. 2008. Biosensor-surface plasmon resonance methods for quantitative analysis of biomolecular interactions. *Methods Cell Biol* **84**:53–77.

31. Tassa, C., J. L. Duffner, T. A. Lewis, R. Weissleder, S. L. Schreiber, A. N. Koehler, and S. Y. Shaw. 2010. Binding affinity and kinetic analysis of targeted small molecule-modified nanoparticles. *Bioconjug Chem* **21**:14–19.

32. Torchilin, V. P. 2010. Passive and active drug targeting: Drug delivery to tumors as an example. *Handb Exp Pharmacol* **197**:3–53.

33. Tudos, A. J. and R. B. M. Schasfoort. 2008. Introduction to surface plasmon resonance. In A. J. Tudos and R. B. M. Schasfoort (eds.), *Handbook of Surface Plasmon Resonance*. RCS Publishing, Cambridge, U.K., pp. 1–14.

34. Wang, X., L. Yang, Z. G. Chen, and D. M. Shin. 2008. Application of nanotechnology in cancer therapy and imaging. *CA Cancer J Clin* **58**:97–110.

35. Xu, F., B. Persson, S. Lofas, and W. Knoll. 2006. Surface plasmon optical studies of carboxymethyl dextran brushes versus networks. *Langmuir* **22**:3352–3357.

4

Dual-Functional Zwitterionic Carboxybetaine for Highly Sensitive and Specific Cancer Biomarker Detection in Complex Media Using SPR Biosensors

Norman D. Brault, Shaoyi Jiang, and Qiuming Yu

CONTENTS

4.1 Introduction

The past several decades have introduced many advances to the field of cancer biology which have significantly enhanced our fundamental knowledge of the disease [1,15,16,47]. In sharp contrast, even far-reaching comparable improvements in the clinic have not yet been achieved as current cancer diagnostics are frequently incomplete, leading to the use of treatments which offer either transient benefits or none at all [1,15]. This discrepancy has led to a large push for the discovery and validation of reliable biomarkers which can improve diagnosis and disease stratification as well as guide molecularly targeted therapy and monitor the therapeutic response [1,2,15,16,18,27,28,36,40,41,47,52]. While biomarkers (e.g., DNA, mRNA, proteins) [27] can be found in various types of human bodily fluids (e.g., urine, saliva) [45], the dynamic nature of the human circulatory system to reflect numerous physiological and abnormal processes, in addition to its ease of accessibility, makes human plasma and serum the best and most logical sources for biomarker analysis [18,27]. However, sensitive detection of these target analytes (at or below nanogram/milliliter quantities) buried in the comprehensive proteome of human blood (thousands of core proteins which span more than 10 orders of magnitude in concentration) has severely hindered the biomarker development process due to the inability to fully verify the marker as being clinically relevant [12,15,23,45].

In order to better enable the verification of candidate biomarkers, biosensors which are capable of highly specific and sensitive detection directly from complex media are highly desirable [50]. To this end, biosensing techniques have to resolve three major issues. First, due to the vast sea of proteins in human plasma (or serum), the platform must reduce or eliminate the presence of nonspecific protein adsorption (i.e., biofouling). Secondly, the surface must allow for efficient and convenient immobilization of molecular recognition elements (e.g., antibodies). Lastly, the biosensing method/device should be highly sensitive while allowing for the quantification of small levels of the desired specific interaction from complex media, preferably in real time, and with relatively simple detection protocols.

The nonspecific adsorption of proteins onto sensing surfaces has severally hindered the full exploitation of biosensors concerned with detection from complex media [4,50,53]. The presence of biofouling primarily results in an overwhelming background noise. This not only reduces the detection limits of biomarkers, already at low concentrations, but also leads to many false positives [45]. In the context of biosensors, there have been several solutions to cope with biofouling. The most common method involves the use of blocking agents (e.g., BSA, nonfat dry milk, fish gelatin) [32]. While this is the simplest method to reduce nonspecific adsorption from complex media, there are several drawbacks including lot-to-lot inconsistencies, interference with desired specific binding, as well as cross-reactivity with assay components.

Alternative methods to blocking agents have resulted in the development of protein-resistant surface coatings [8,43]. While many hydrophilic surfaces can reduce protein adsorption, few materials can meet the additional challenges of biosensing applications, all the while showing effective resistance to complex media [25]. The most commonly used sensor coatings are based on ethylene glycol (EG) and its derivatives, such as poly(ethylene glycol) (PEG) or oligoethylene glycol (OEG) [3]. It is believed that the surface hydration achieved via hydrogen bonding for EG and steric effects of the longer PEG chains play important roles in resisting protein adsorption [3,19,35]. Several studies have shown that the presence of a tightly bound water layer on an OEG surface results in a large repulsive force, leading to the resistance of nonspecific protein adsorption [20,63,64]. A typical example of this material used for nonfouling applications is shown in Figure 4.1a. Relatively complex chemical steps are used to generate reactive functional groups on OEG-based surface coatings for the immobilization of molecular recognition elements.

Oligo-ethylene glycol (OEG)
(hydrogen bonding)
(a)

Sulfobetaine (SB)
(electrostatic interactions)
(b)

Phosphorylbetaine (PC)
(c)   **(electrostatic interactions)**

Carboxybetaine (CB)
(d)   **(electrostatic interactions)**

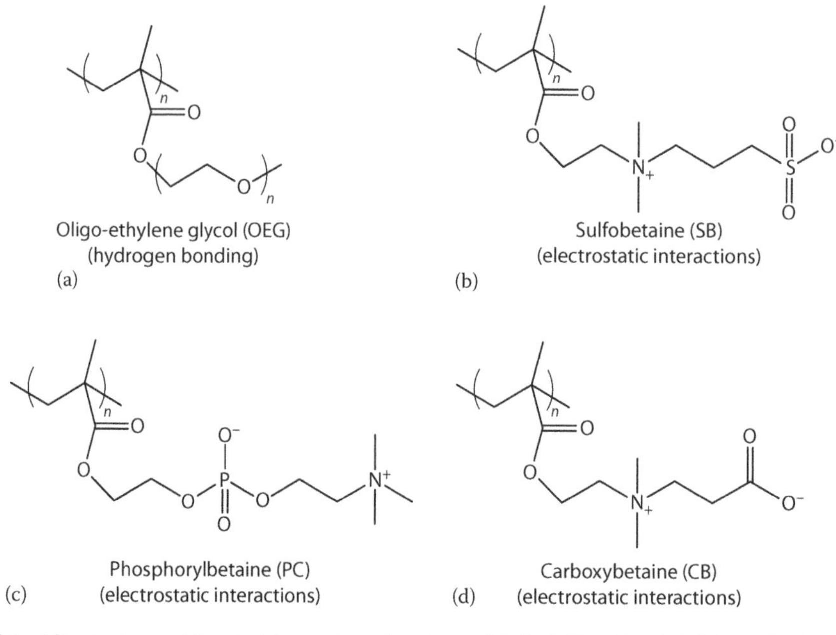

FIGURE 4.1 The structures of four highly protein-resistant materials including (a) poly(ethylene glycol methacrylate) (PEGMA), (b) SB, (c) PC, and (d) CB. Nonzwitterionic EG-based materials (a) achieve the surface hydration necessary for protein-resistant properties via hydrogen-bonding interactions. However, the three zwitterionic materials (b–d) create a stronger interfacial water structure via electrostatically induced hydration which results in superior protein resistance from complex media compared to EG. (Adapted from Jiang, S.Y. and Z.Q. Cao: Ultralow-fouling, functionalizable, and hydrolyzable zwitterionic materials and their derivatives for biological applications. *Adv. Mater.* 2010. 22. 920–932. Copyright Wiley-VCH Verlag GmbH & Co. KGaA. Reproduced with permission.)

As an alternative to the formation of a hydration layer via hydrogen bonding, zwitterionic materials (i.e., compounds containing both cationic and anionic groups in the same monomer), such as phosphorylbetaine (PC), sulfobetaine (SB), and carboxybetaine (CB), achieve much stronger hydration via ionic solvation [11]. The structures of these zwitterionic compounds are shown in Figure 4.1b through d. Simulation studies have indicated the electrostatically induced hydration of these materials to be stronger than the hydrogen-bonding-induced hydration for EG and hence provide the primary physical mechanism for improved protein resistance [21,48]. Numerous studies of zwitterionic materials have yielded surface coatings with nonfouling properties to both single protein solutions and complex media (i.e., undiluted human serum and plasma) measured using surface plasmon resonance (SPR) biosensors [55,58,59,61,62]. The results indicated the importance of densely packed polymer surfaces with controlled lengths for effectively reducing nonspecific adsorption. Furthermore, a unique property of CB over the other low-fouling materials is the presence of a carboxylate functional group in each monomer. This can be used in common amino-couple procedures to covalently attach molecular recognition elements, all the while maintaining postfunctionalized nonfouling properties, a vital characteristic for highly sensitive analyte detection [54,55,59]. This dual functionality of CB makes it a highly attractive material for numerous applications including cancer biomarker detection.

While reducing nonspecific adsorption helps to increase the signal-to-noise ratio (S/N) necessary for highly sensitive detection, the ability to immobilize active biorecognition elements onto the surface for the specific detection of target analytes is also important for obtaining low detection limits. Since the great majority of sensing applications detect protein biomarkers primarily using antibodies, this method will be discussed herein. However, other high affinity binding agents including aptamers, peptides, and several others have also been developed [44] but will not be covered. Typical immobilization strategies for antibodies fall into three categories: physical adsorption (e.g., hydrophobic surfaces), covalent linkage (e.g., EDC/NHS amino-coupling chemistry), or affinity based (e.g., avidin-biotin linkages) [23,46]. While proper orientation of the attached antibodies has been shown to improve the assay sensitivity [9,57] and can be achieved in certain circumstances or with various techniques, the added assay complexity is undesirable, thereby resulting in the majority of immobilization protocols involving random orientations of the ligands attached to the surface [46]. It is thus the combination of both efficient immobilization of proteins and the reduction of nonspecific adsorption that yield a highly sensitive assay.

The ultimate usefulness and applicability of a given assay, even with an optimized S/N, depend upon the specific biosensing and read-out methods used. Currently, label-free and real-time sensing technology has been on the rise due to its simplicity, facilitation in obtaining both kinetic data and binding affinities, and the capability of high-throughput screening in numerous applications [13,23,50]. One important example includes SPR biosensors. SPR biosensors have become one of the most widely adopted label-free and real-time instruments for quantifying biomolecular interactions. These instruments optically excite surface plasmons typically using gold substrates resulting in a highly sensitive detection platform with the capability of measuring changes in refractive index as small as 10^{-7} RIU which can be converted into protein surface coverage (e.g., ng/cm^2) [42]. The use of gold substrates also enables many different types of surface chemistry to be studied with SPR biosensors.

For demonstrating highly sensitive label-free detection in this review, ALCAM was chosen as the model cancer biomarker. Activated leukocyte cell adhesion molecule (ALCAM, CD 166) is a 105 kDa glycoprotein with a normal presence in human blood ($\sim$80 ng/mL) [14]. It is a member of the immunoglobulin superfamily and has shown to be critical in tumor development and progression. The potential use of this target for cancer detection has been studied in numerous tumor types including melanoma, prostate cancer, breast cancer, colorectal cancer, bladder cancer, and esophageal squamous cell carcinoma. While further work is necessary to confirm ALCAM as a reliable diagnostic biomarker, several tumors have indicated upregulation of the protein and several have indicated downregulation [38], illustrating the need for sensitive detection capabilities.

The methods and results presented in this summary will illustrate the use of functionalizable zwitterionic poly(carboxybetaine) (pCB) for the highly sensitive detection of a model cancer biomarker (ALCAM) directly from human plasma using an SPR biosensor. Important considerations and variables for the three important steps (i.e., protein resistance, immobilization, and biosensor selection) for this system will be presented along with a comparison of other competing technologies. This chapter's conclusion will briefly discuss the future trends in this field.

4.2 Materials

4.2.1 SPR Instrumentation and Substrates

SPR instruments are optically based sensor platforms capable of quantifying biomolecular interactions on a surface of a thin metallic film which supports surface plasmons [23]. A custom-built SPR sensor developed at the Institute of Photonics and Electronics, Prague, Czech Republic, was used for this work and is shown in Figure 4.2. The key components of the SPR biosensor include a polychromatic halogen light source (Ocean Optics, Model LS-1), collimator, the SPR unit, receiving optics, photospectrometer (Ocean Optics, Model S2000), temperature controller (ILX Lightwave, Model LDT-5525), a peristaltic pump (ISMATEC Model ISM 936), samples, and waste collection. The SPR unit consists of a gold-coated SPR substrate interfaced with a prism and a four-channel flow-cell (allowing for four independent measurements on the same sample). This sensor adapts the Kretschmann attenuated total reflection configuration with wavelength modulation to monitor biomolecular interactions in real-time [54].

Briefly, as illustrated in Figure 4.3a [51], SPR sensor chips are composed of a glass slide coated with an adhesion-promoting titanium film (~2 nm) followed by an SPR active gold layer (~48 nm) using an electron beam evaporator and 99.999% pure gold slugs from D.F. Goldsmith. These gold surfaces are typically modified with a thin layer of surface chemistry to provide for antibody immobilization and the protein-resistant background necessary for detection. After mounting the chip, a collimated polychromatic halogen light source is used to excite surface plasmons at the interface between the gold layer and the dielectric medium (e.g., water). This generates an electromagnetic field with an intensity that decays exponentially into both the gold film and the dielectric. The specific wavelength of light at which this excitation occurs (i.e., the surface plasmon resonant wavelength) is shown as a dip in the wavelength spectrum of the reflected light (Figure 4.3b). The resonant wavelength is dependent upon

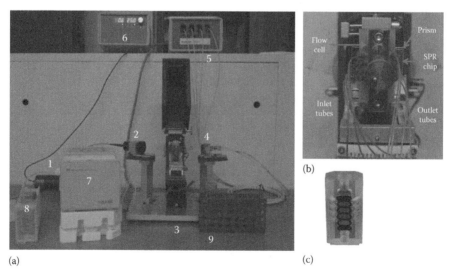

(a) (b) (c)

FIGURE 4.2 (a) The custom-built SPR biosensor used in this work. The key components include (1) a polychromatic halogen light source, (2) collimator, (3) SPR unit, (4) receiving optics, (5) photospectrometer, (6) temperature controller, (7) peristaltic pump, (8) samples, and (9) waste collection. (b) The close-up view of the SPR unit showing a gold-coated SPR chip interfaced with a prism and a four-channel flow-cell. (c) The front-side view of a four-channel flow-cell which enables four independent experiments to be conducted simultaneously.

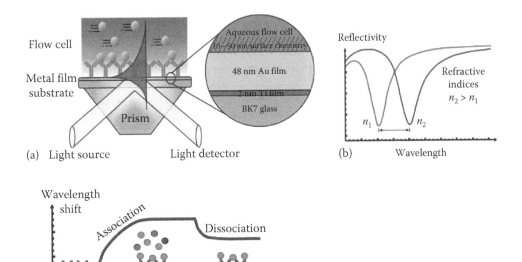

FIGURE 4.3 SPR biosensing based on wavelength modulation. (a) A schematic representing the fundamentals of SPR biosensors. SPR chips consist of BK7 glass coated with an adhesion-promoting layer of titanium (2 nm) and an SPR-active layer of gold (48 nm). The gold surface is typically modified with a thin layer of surface chemistry (e.g., polymers or self-assembled monolayers [SAMs]) for immobilizing probing molecules and providing a nonfouling background. First, the SPR chip is mounted onto the prism and a flow-cell is attached to the gold-coated side of the substrate. A collimated poly-chromatic halogen light source is then directed through the prism and used to generate surface plasmons which propagate along the interface between the gold and the dielectric medium (e.g., water). These surface plasmons produce an electro-magnetic field which decays exponentially into both the gold film as well as the dielectric. The reflected light from the chip gets directed through the prism and detected by a photospectrometer. (b) An illustration of the SPR wavelength shift as a function of the refractive index (RI). The specific wavelength of the incident light used to generate surface plasmons (the SPR wavelength) is detected as a dip in the spectrum of reflected light. Since the majority of the electromagnetic field (due to the propagating surface plasmons) is located within the dielectric, the position of the SPR wavelength is very sensitive to RI changes occurring within this medium, such as proteins adsorbing to the surface. (c) An example of an SPR sensorgram for a sensing chip immobilized with antibodies. After establishing a baseline using running buffer, a solution consisting of antigens and other proteins is flowed over the surface. Antigens are specifically captured by the antibody-immobilized surface during the "association" phase resulting in an increase in the wavelength shift. Upon injecting running buffer, any remaining or loosely bound antigen is removed from the flow-cell during the "dissociation" phase, and the wavelength decreases. The net amount of protein coverage can then be taken as the difference between buffer baselines and converted to a surface coverage. (Adapted from *Sensors Actuat. B Chem.*, 107, Taylor, A.D., Yu, Q.M., Chen, S.F., Homola, J., and Jiang, S.Y., Comparison of *E. coli* O157: H7 preparation methods used for detection with surface plasmon resonance sensor, 202–208, Copyright 2005, with permission from Elsevier B.V.)

the refractive index of the dielectric (i.e., solution) in contact with the chip within the probe range of the electromagnetic field (approximately 200–400 nm from the surface). By monitoring changes in the reso-nant wavelength of the reflected light, highly sensitive label-free detection in real-time is enabled [23]. Figure 4.3c depicts a typical SPR sensorgram showing the wavelength shift as a function of time for an antibody-functionalized surface. The figure demonstrates the different physical phenomena occurring and the corresponding sensor response.

4.2.2 Polymerization, Immobilization, and Detection

N-[3-(Dimethylamino)propyl] acrylamide (DMAPA, 98%, Cat. No. D1785) was purchased from TCI America (Portland, OR). β-Propiolactone (97%, Cat. No. B23197) was purchased from Alfa Aesar

(Ward Hill, MA). 11-Mercapto-1-undecanol (97%, Cat. No. 447528), copper (I) bromide (99.999%, Cat. No. 254185), copper (II) bromide (99.999%, Cat. No. 437867), 2,2′-bipyridine (BPY, 99%, Cat. No. D216305), phosphate-buffered saline (PBS, 0.01 M phosphate, 0.138 M NaCl, 0.0027 M KCl, pH 7.4, Cat. No. P3813), Tween 20 (Cat. No. P9416), and bovine serum albumin (BSA, Cat. No. A7030) were purchased from Sigma-Aldrich Corporation (St. Louis, MO). Bromoisobutyryl bromide (98%, Cat. No. 403091000), *N*-hydroxysuccinimide (NHS, 98%, Cat. No. 157270250), *N*-ethyl-*N*′-(3-diethylaminopropyl) carbodiimide hydrochloride (EDC, 98%, Cat. No. 171440500), and *N,N*-dimethylformamide (DMF, Cat. No. 27960) were purchased from Acros Organics (Morris Plains, NJ). Sodium borate (SB) (Cat. No. 3570), sodium phosphate (Cat. No. 3818), sodium chloride (Cat. No. 3624), sodium hydroxide (98%, Cat. No. 3722), and ether (Cat. No. 9259) were purchased from J.T. Baker (Phillipsburg, NJ). Anhydrous sodium acetate (SA) (Cat. No. 71179) was purchased from Fluka BioChemika (Switzerland). Anhydrous potassium carbonate (Cat. No. PX1390), acetone (Cat. No. AX0115), and tetrahydrofuran (THF, Cat. No. TX0282) were purchased from EMD Chemicals (Gibbstown, NJ). Pooled human plasma (male only, Cat. No. 752PR-CPD-PM) containing the anticoagulant citric phosphate dextran (CPD) and pooled human serum (male only, Cat. No. 751-NS-MP) were purchased from BioChemed Services (Winchester, VA). Human monoclonal antibodies against ALCAM (anti-ALCAM, Cat. No. MAB6561) and human recombinant ALCAM/Fc chimera (Cat. No. 656-AL) were purchased from R&D Systems (Minneapolis, MN). Rabbit polyclonal biotinylated antibody to *Salmonella sp.* (Cat. No. B65707R) was purchased from Meridian Life Science Inc. (Saco, MA). Ethanol (200 proof, Cat. No. 2701) was purchased from Decon Laboratories (King of Prussia, PA). The water used was purified using a Millipore water purification system with a minimum resistivity of 18.2 MΩ cm. Acetone was dried with anhydrous potassium carbonate before use. The thiol initiator for surface-initiated atom transfer radical polymerization (SI-ATRP) was ω-mercaptoundecyl bromoisobutyrate synthesized via the reaction of bromoisobutyryl bromide and 11-mercapto-1-undecanol using a method published previously [26].

4.3 Methods

4.3.1 Synthesis of CBAA Monomer

The monomer, (3-Acryloylamino-propyl)-(2-carboxy-ethyl)-dimethyl-ammonium (carboxybetaine acrylamide, CBAA), was synthesized by reacting 1.54 g DMAPA with 0.99 g of β-propiolactone in 50 mL of dried acetone at 0°C for 2 h under nitrogen protection. The crude product was then filtered, washed with ether, and then dried under vacuum to yield a white powder which was subsequently stored at 4°C. Yield: 81%.^{1}H NMR (Bruker 500 MHz, DMSO-d_6): 8.61 (t, 1H, N–H), 6.28 (t, 1H, CHH=C*H*), 6.13 (t, 1H, CH*H*=CH), 5.61 (t, 1H, C*H*H=CH), 3.44 (t, 2H, N–CH$_2$–CH$_2$–COO), 3.21 (m, 4H, NH–CH$_2$–CH$_2$–CH$_2$), 2.97 (s, 6H, N–(CH$_3$)$_2$), 2.25 (t, 2H, CH$_2$–COO), 1.87 (t, 2H, NH–CH$_2$–CH$_2$–CH$_2$). A vital step in the synthesis of the monomer is to ensure the complete removal of the unreacted DMAPA starting material. This reactant contains the acrylamide backbone, allowing it to get incorporated into the polymerization process. If this occurs, the resulting polyCBAA (pCBAA) film will be fouling due to the net positive charge created by the free tertiary amine at neutral pH. Proteins which carry a net negative charge will be attracted to and nonspecifically adsorb onto the surface.

4.3.2 Preparation of pCBAA Films on SPR Chips

The first step for the formation of nonfouling pCBAA films is to clean the initial gold-coated SPR substrates by rinsing with ethanol, UV ozone cleaning for 20 min, rinsing with water and ethanol, and then drying with a stream of compressed filtered air. The initiator self-assembled monolayer (SAM) necessary for SI-ATRP (Figure 4.4) can be formed by submersing the cleaned substrates into a 0.1 mM solution of ω-mercaptoundecyl bromoisobutyrate in pure ethanol for 24 h at room temperature. Subsequent rinsing with THF and ethanol followed by drying readies the substrate for the next step, ATRP. THF is necessary to remove any undissolved impurities from the initiator solution which could adversely affect the polymerization reaction.

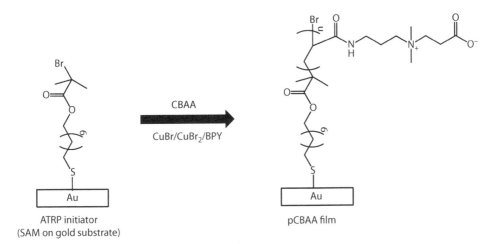

FIGURE 4.4 SI-ATRP was used to form dense polymer brushes with uniform and finely controlled thicknesses. First, a self-assembled monolayer (SAM) of ATRP initiators was formed onto a clean gold-coated SPR chip. The chip was then placed into a small tube reactor. Upon the addition of reaction materials (CBAA monomer, CuBr, CuBr$_2$, BPY, and solvent) under nitrogen protection, the polymerization proceeds for several hours and results in a thin film of pCBAA. (Adapted with permission from Yang, W., Xue, H., Li, W., Zhang, J.L., and Jiang, S.Y., Pursuing "zero" protein adsorption of poly(carboxybetaine) from undiluted blood serum and plasma, *Langmuir*, 25, 11911–11916, 2009. Copyright 2009 American Chemical Society.)

ATRP is a living free-radical polymerization and was chosen for generating nonfouling pCBAA surfaces due to its ability to obtain highly dense polymer brushes with uniform and finely controlled thicknesses. A key component of this procedure involves maintaining an oxygen-free environment by using nitrogen protection throughout the entire reaction. This is important because of the ability of oxygen to react with and consume the catalyst. Other important variables affect the quality of the film including the ratio of organic solvent to water, monomer concentration, ratio of deactivator to activator (e.g., CuBr$_2$ to CuBr as shown in the Results section), and the reaction time. The protocol presented here is the optimized condition for this system [59].

First, 28.53 mg of CuBr, 8.88 mg of CuBr$_2$, and 61.71 mg of BPY along with two SPR chips containing the ATRP initiator SAM were placed into a glass tube reactor which was then sealed and placed under a nitrogen atmosphere using the conventional Schlenk line technique. Nitrogen-purged DMF (2 mL) was then added to the reactor followed by slight mixing under streaming nitrogen until the solution darkened. Next, the CBAA monomer (320 mg) was measured and placed into a separate sealed glass tube under a continuous stream of nitrogen protection. After completely dissolving the monomer using nitrogen-purged DMF (1.2 mL) and water (0.8 mL) by stirring, a syringe was used to transfer the CBAA solution directly to the reactor. The glass tube reactor was then shaken very slightly for 15 min under nitrogen protection and placed in a shaker at 100 rpm and room temperature for 2.75 h (total reaction time was 3 h). After the polymerization was completed, the chips were removed, rinsed with water, and submerged in PBS overnight to remove any unreacted monomer. Upon the removal of the chip from the buffer solution, rinsing with water, and drying, the pCBAA substrate was then ready to be analyzed using the SPR biosensor.

4.3.3 Calibration of SPR Sensitivity

Due to the sensitivity of the SPR sensor being dependent upon the distance of the binding event from the original SPR-active gold layer and the resonant wavelength, a correction factor due to the pCBAA polymer thickness must be determined [23]. This allows for a direct comparison to other surfaces studied by SPR in terms of fouling, functionalization, and detection using common units (e.g., ng/cm^2). The calibration can be done in two ways. The first method is strictly experimental and involves measuring the sensor response to increasing NaCl (aq) concentrations (e.g., increments from 0.5% to 1.5% corresponding to a net change in refractive index of 1.79×10^{-3} RIU) on the pCBAA surface and taking the average net change. After performing the same experiment using a bare gold chip at the same resonant wavelength (i.e., 750 nm), the ratio between the two net averages provides the calibration factor [4,23]. The second method uses the measured value of the refractive index of the pCBAA film and the SPR wavelength shift of the pCBAA substrate relative to that of a bare chip, combined with theoretical simulations, to estimate the change in surface sensitivity [54]. The latter method was used for this work, yielding an average correction coefficient of ~1.3 for film thicknesses of 10–30 nm (i.e., a 1 nm shift on pCBAA-coated substrates equated to a 1.3 nm shift on uncoated chips). Hence, for the pCBAA films used in this study, a 1 nm shift corresponded to a protein surface coverage of ~22 ng/cm² [22], and the detection limit, corresponding to three standard deviations of baseline noise, was ~0.2 ng/cm² [54]. For thicker films, the correction factor is larger.

4.3.4 Functionalization of Surfaces

The pCBAA substrate is now ready to be functionalized *in situ* via amino-coupling surface chemistry and subsequently used for the direct detection of a target analyte from undiluted complex media [54,55]. The key steps of the functionalization process include surface activation, antibody immobilization, followed by surface deactivation and washing, as shown in Figure 4.5. Successful implementation allows the formation of an effective monolayer of immobilized and biologically active antibody (i.e., 220–280 ng/cm²)

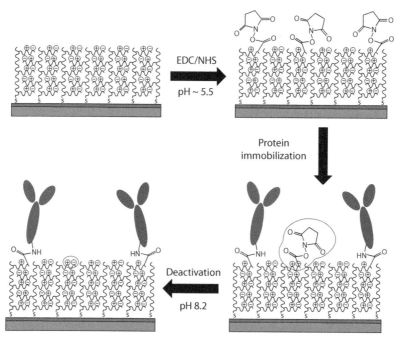

FIGURE 4.5 The covalent immobilization of antibodies onto the ultralow-fouling pCBAA surface. First, a low-pH solution of EDC/NHS in water is used to "activate" the surface by reacting with carboxylic groups to form NHS esters, which also generates a slightly positive net charge on the surface. Flowing a solution of antibodies at a pH greater than the iso-electric point enables the favorable electrostatic interactions (i.e., the positively charged surface and negatively charged protein) to be used for immobilizing an approximate monolayer of antibody. Subsequently injecting an aqueous pH 8.2 solution hydrolyzes any unreacted NHS esters back into the ultralow-fouling zwitterionic background, thereby enabling highly sensitive and specific detection from complex media. (Adapted with permission from Vaisocherova, H., Yang, W., Zhang, Z., Cao, Z.Q., Cheng, G., Piliarik, M., Homola, J., and Jiang, S.Y., Ultralow fouling and functionalizable surface chemistry based on a zwitterionic polymer enabling sensitive and specific protein detection in undiluted blood plasma, *Anal. Chem.*, 80, 7894–7901, 2008. Copyright 2008 American Chemical Society.)

in a nonfouling background. The running buffer for this procedure is 10 mM SA at pH 5.0, and the flow rate is 30 µL/min. The temperature of the flow-cell is controlled at 25°C throughout the entire procedure. After establishing an initial baseline using SA, the surface is then activated using conventional EDC/NHS chemistry. A fresh solution containing 0.05 M NHS and 0.2 M EDC in water (at ~pH 5.5) is flowed over the sensor surface for 7 min followed by SA buffer. Protonation of some of the carboxylate groups at this pH enables the formation of reactive NHS-esters that can react with primary amines on proteins. The immobilization of antibodies to the surface is performed by flowing the antibody solution, consisting of anti-ALCAM or anti-Salmonella at 50 µg/mL in 10 mM SB pH 8.2–9.0, for 14 min across the surface followed by washing with a solution of 10 mM sodium phosphate containing 300 mM sodium chloride (PBNa) at pH 8.2 for 7 min. The net amount of immobilization (in ng/cm^2) is then obtained by flowing SA buffer and taking the difference between the starting (i.e., SA buffer after activation) and ending baselines. The result is multiplied by the correction factor due to the polymer thickness and converted to a surface coverage, which resulted in ~250 ng/cm^2 of immobilized antibody.

There are several notable points to be made here. For the activation procedure, the presence of a carboxylate group in each monomer combined with the high density of polymer brushes by ATRP enables a relatively high degree of surface activation. However, due to the rapid hydrolysis of the NHS-esters with water as the pH increases, it is necessary to flow the low-pH 5.0 solution (SA) following activation in order to preserve the reactive groups. Furthermore, the ability to immobilize antibody using this method takes advantage of several important properties of the pCBAA surface. The partial consumption of protonated carboxylate groups on the film during the activation procedure results in a loss of a net neutral charge on some of the monomer branches. Due to the permanent presence of the quaternary amine, the surface takes on a net positive charge. By using antibody solutions with a pH 8.2–9.0 (typically above the isoelectric points of most proteins), the antibodies take on a net negative charge. Minimizing the total salt concentration maximizes the electrostatic interactions (i.e., the driving force) and provides for a relatively large local protein concentration near the surface. These proteins can outcompete the hydrolysis reaction and allow monolayer formation of covalently linked antibodies. Washing with PBNa solution then serves two purposes. Exposure of the activated pCBAA surface to a pH 8–9 solution for ~21 min results in complete surface deactivation. Thus, the 14 min immobilization step at pH ~8.5 is followed by a rinsing step at pH 8.2 for 7 min to hydrolyze any residual activated groups back into the original nonfouling background. This is emphasized in the last step of Figure 4.5. PBNa also helps the removal of any noncovalently and loosely attached antibodies. No further deactivation or blocking steps are then necessary before moving on to detection.

4.3.5 Measurements of Nonspecific Adsorption

The nonspecific protein adsorption of both the original pCBAA surface as well as the functionalized surface can be tested using undiluted human plasma and serum. For any surface to be used for cancer biomarker detection, it is absolutely vital to maintain protein-resistant properties following ligand immobilization. The general fouling experiments are performed as follows. After establishing a stable baseline using PBS buffer, complex media is flowed over the surface for 10 min at a flow rate of 50 µL/min followed by washing with buffer for an additional 10 min. The difference between the starting and ending baselines is then converted to a net protein surface coverage. In order to accurately test the fouling properties of a postfunctionalized surface, it is recommended to immobilize an antibody that lacks any target analyte naturally present in the complex media (e.g., anti-Salmonella as used in this work).

4.3.6 Detection of Cancer Biomarkers

The highly sensitive detection of a target analyte directly from undiluted human plasma using an antibody-functionalized pCBAA surface was demonstrated for the cancer biomarker ALCAM [54]. For pCBAA to be applicable in cancer diagnostics, it must be able to measure small changes in protein concentration. Here, anti-Salmonella was used as a reference surface because it has no specific binding partner present in human plasma. The binding buffer for analyte detection consisted of 10 mM sodium phosphate, 0.3 M NaCl, 0.1% Tween 20, and 0.1% BSA at pH 7.4. Together, BSA and Tween 20 helped to minimize the loss of analytes to nonspecific adsorption onto both the tubing and flow-cell surfaces in the SPR microfluidics. They also aided in the removal of loosely bound target analytes attached to the immobilized antibodies [32].

Following the immobilization of both anti-ALCAM and the reference antibody (anti-Salmonella) on the pCBAA surface, a stable baseline was obtained using the detection buffer. In order to determine the amount of native antigen present in the plasma (which can be subtracted from the spiked samples), unspiked media was flowed over both antibody surfaces for 15 min at a flow rate of 50 μL/min followed by washing with buffer. Using *de novo*–functionalized substrates, the standard detection curve was obtained by flowing antigen spiked into undiluted human plasma at a concentration range of 7.8–1000 ng/mL over both antibody surfaces for 15 min followed by washing for 15 min. The net response for both the measuring and reference channels was calculated as the difference between the starting and ending buffer baselines. After subtracting the reference response, the specific amount of spiked antigen was then determined by subtracting the native ALCAM response and converting to a protein surface coverage.

4.4 Results

One way for controlling the film thickness of pCBAA polymers made via ATRP is by varying the ratio of Cu(II)/Cu(I) (deactivator to activator). Films in the range of 10–55 nm were obtained when the ratio of Cu(II)/Cu(I) was varied from 10% to 30% while holding the reaction time at 3 h. These films were then exposed to undiluted human serum, undiluted human plasma, and undiluted aged human serum (serum which has been frozen for greater than 1 year at −80°C and tends to result in more severe protein adsorption than fresh serum) at both 25°C and 37°C. Protein adsorption versus film thickness at both temperatures (Figure 4.6) revealed an optimal thickness of ~21 nm. Figure 4.7 shows the SPR sensorgrams for undiluted human plasma and serum with exposure times of 10 and 60 min for a ~21 nm thick pCBAA film at 25°C. The data showed that all four-sensor responses returned to the original PBS baseline and illustrate the ultralow-fouling properties for this zwitterionic material. Additionally, less than 5 ng/cm² of adsorption was obtained for film thicknesses of 15–26 nm and 15–40 nm for experiments performed at 25°C and 37°C, respectively. This indicates that the electrostatically induced hydration of pCBAA

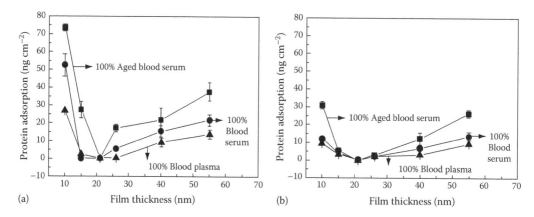

FIGURE 4.6 The biofouling of pCBAA surfaces with different film thickness for three different types of complex media measured by an SPR biosensor at 25°C (a) and 37°C (b). For both temperatures, an optimal film thickness of 21 nm yielded undetectable protein adsorption from the three samples. Furthermore, the low-fouling abilities of 15–40 nm thick films at 37°C indicate a wide usable range of pCB for a variety of applications. (Adapted with permission from Yang, W., Xue, H., Li, W., Zhang, J.L., and Jiang, S.Y., Pursuing "zero" protein adsorption of poly(carboxybetaine) from undiluted blood serum and plasma, *Langmuir*, 25, 11911–11916, 2009. Copyright 2009 American Chemical Society.)

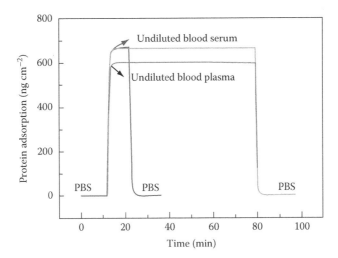

FIGURE 4.7 The SPR response to undiluted human serum and plasma for 21 nm thick films of pCBAA-coated gold substrates. After obtaining a stable baseline with PBS at 50 μL/min and 25°C, human serum and plasma were then injected into the sensor for 10 and 60 min followed by PBS. The data revealed that all four-sensor responses returned to the original buffer baseline, thereby illustrating the ultralow-fouling properties of this zwitterionic material for both short and long exposure times. (Adapted with permission from Yang, W., Xue, H., Li, W., Zhang, J.L., and Jiang, S.Y., Pursuing "zero" protein adsorption of poly(carboxybetaine) from undiluted blood serum and plasma, *Langmuir*, 25, 11911–11916, 2009. Copyright 2009 American Chemical Society.)

polymers is maintained even at elevated temperatures, enabling a wide range of film thicknesses which are protein resistant [59]. In contrast, commonly used EG-based materials lose their nonfouling properties as the temperature increases due to a reduction in hydration caused by the breaking of hydrogen bonds [34]. Therefore, CB polymers offer an additional advantage for body temperature and *in vivo* applications.

The method for the immobilization of antibodies onto the pCBAA surface utilized EDC/NHS amino-coupling chemistry. In order to exploit the nonfouling properties of a pCBAA surface functionalized with an effective monolayer of protein, the appropriate reaction conditions had to be determined [54,55]. Table 4.1 shows a summary of the activation and immobilization conditions tested using anti-Salmonella and the resulting protein-fouling properties to 50% and 100% plasma as determined by the SPR biosensor. The data reveal that the pH of both the activation and immobilization solutions affected the levels of functionalization and/or subsequent protein resistance. Excellent nonfouling properties for all postfunctionalized surfaces activated at pH ~5.5 were observed.

A pCBAA surface functionalized with anti-ALCAM and a reference antibody (anti-Salmonella) was then selected to demonstrate the specific and sensitive detection abilities of this novel biosensing platform. The SPR sensorgrams for detecting ALCAM from both undiluted human plasma and serum are shown in Figure 4.8a and b, respectively. Here, a net response for anti-ALCAM and a zero response for anti-Salmonella were observed, demonstrating the specificity of detection and lack of any nonspecific protein adsorption. This subsequently enabled the standard detection curve to be carried out. The reference-compensated sensor response curves for the direct detection of ALCAM from undiluted human plasma are shown in Figure 4.9a. The SPR response shown for these spiked samples includes the subtracted signals from both the native ALCAM present in human plasma as well as the response from the control anti-Salmonella surface. Thus, the *y*-axis shows the increase in protein surface coverage due to only the spiked ALCAM from the desired specific interaction of the antigen/ antibody pair. The vertical lines indicate three stages of the detection: the initial buffer baseline, detection from human plasma, and buffer wash, identical to the steps shown in Figure 4.8a and b. The absence of background noise allowed the binding kinetics to be clearly monitored directly from the complex media. The limit of detection was found to be ~10 ng/mL, and the complete detection curve is shown in Figure 4.9b.

TABLE 4.1

Summary of the Activation and Immobilization Conditions Tested Using pCBAA Substrates and the Resulting Protein-Fouling Properties to 50% and 100% Human Plasma as Determined by an SPR Sensor

No.	Activation pH	Immobilization/ Deactivation pH	Protein Conc. (μg/mL)	Immob. Level (ng/cm²)	Nonspecific Protein Adsorption (50% Plasma) (ng/cm²)	Nonspecific Protein Adsorption (100% Plasma) (ng/cm²)
1	**5.5**	9.0	50	**230.3**	1.2	**2.6**
2	**4.9**	9.0	50	**418.5**	7.1	**15.0**
3	**4.0**	9.0	50	**398.1**	22.1	**29.4**
4	**2.2**	9.0	50	**4.2**	0.7	**0.9**
5	5.5	**5.2**	50	**76.5**	0.8	**1.2**
6	5.5	**7.2**	50	**164.9**	2.2	**3.2**
7	5.5	**8.2**	50	**257.9**	1.6	**3.6**
8	5.5	**8.9**	50	**254.8**	0.5	**2.4**
9	5.5	**10.8**	50	**70.9**	0.1	**1.0**
10	5.5	9.0	**10**	**88.8**	0.9	**3.4**
11	5.5	9.0	**50**	**239.2**	1.1	**3.6**
12	5.5	9.0	**200**	**241.8**	2.0	**0.1**

Source: Adapted from Biosens. Bioelectron., 24, Vaisocherova, H., Zhang, Z., Yang, W., Cao, Z.Q., Cheng, G., Taylor, A.D., Piliarik, M., Homola, J., and Jiang, S.Y., Functionalizable surface platform with reduced nonspecific protein adsorption from full blood plasma-material selection and protein immobilization optimization, 1924–1930, Copyright 2009, with permission from Elsevier B.V.

Note: The human plasma was diluted with PBS buffer. The variation of conditions and the final biofouling levels are highlighted in bold.

In summary, a new surface chemistry platform based on ultralow-fouling zwitterionic CB polymers was demonstrated to be capable of highly sensitive and specific detection of cancer biomarkers directly from undiluted human plasma using an SPR biosensor, which provides fast, robust, label-free, and real-time analysis of biomolecular interactions. Effective protein immobilization and deactivation steps using protein-resistant pCBAA required no additional blocking methods. This offered a significant improvement over other surface chemistry platforms and enabled the direct detection of the cancer biomarker, ALCAM, from undiluted human plasma down to 10 ng/mL [54].

4.5 Discussion

While the benefits of zwitterionic CB polymers made via SI-ATRP have already been discussed, several other nonfouling materials are also frequently used on sensor platforms. As mentioned previously, EG-based films are the most common material utilized for obtaining protein-resistant surface coatings [3]. Short OEG SAMs allow simple sample preparation, but the nonfouling properties are not sufficient to meet the challenges of undiluted complex media [31,54]. These surfaces generate a hydration layer via hydrogen bonding, and as a result, their protein resistance decreases at elevated temperatures (e.g., 37°C) [34]. Although OEG SAMs with chemically modified reactive groups (e.g., COOH) can be conjugated to proteins and used for detection, the surface chemistry is more severe, requiring multiple reaction steps, and usually results in the worsening of the subsequent nonfouling properties [24]. Each OEG chain also only contains a single reactive group for protein immobilization which ultimately results in a lower degree of functionality compared to that of zwitterionic pCB films.

Despite the relative protein resistance of short OEG-based and mixed-charged SAMs, thicker zwitterionic polymer films are necessary for more complex solutions such as blood. As shown in Figure 4.10,

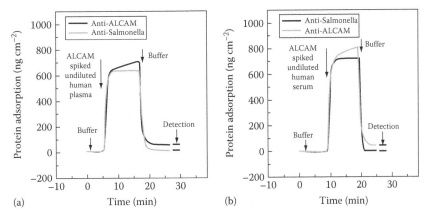

FIGURE 4.8 SPR sensorgrams for the detection of ALCAM directly from undiluted human plasma (a) and serum (b). Following the immobilization of anti-ALCAM and control anti-Salmonella onto the pCBAA surface, a running buffer consisting of 10 mM phosphate, 0.3 M NaCl, 0.1% Tween 20, and 0.1% BSA at pH 7.4 was used to obtain a stable baseline at 50 μL/min. Samples consisting of antigen-spiked human plasma and serum were then flowed over the detection and control antibody–immobilized surfaces followed by running buffer. Both media resulted in a zero response for the control antibody, indicating the absence of any nonspecific binding. Therefore, the difference between the starting and ending baselines for the detection antibody accounted for only the desired specific interaction between the ALCAM/anti-ALCAM pair. The ultralow-fouling background of the functionalized pCBAA surface thus allows for highly sensitive and specific detection of analytes directly from complex media. (Adapted with permission from Vaisocherova, H., Yang, W., Zhang, Z., Cao, Z.Q., Cheng, G., Piliarik, M., Homola, J., and Jiang, S.Y., Ultralow fouling and functionalizable surface chemistry based on a zwitterionic polymer enabling sensitive and specific protein detection in undiluted blood plasma, *Anal. Chem.*, 80, 7894–7901, 2008; Yang, W., Xue, H., Li, W., Zhang, J.L., and Jiang, S.Y., Pursuing "zero" protein adsorption of poly(carboxybetaine) from undiluted blood serum and plasma, *Langmuir*, 25, 11911–11916, 2009. Copyright 2008 and 2009 American Chemical Society.)

SAMs of OEG, mixed trimethylamine ($-N(CH_3)^+$) and sulfonic acid ($-SO_3^-$), and mixed trimethylamine ($-N(CH_3)^+$) and carboxylic acid ($-CO_2^-$) are not protein resistant to undiluted human serum or plasma, although these surfaces are very effective at resisting adsorption from single protein solutions (e.g., fibrinogen in PBS) or 10% blood serum [10,31]. However, the thicker zwitterionic polymer films created via ATRP have improved nonfouling properties to complex media, with CB being the most effective.

Carboxymethylated dextran is another type of nonfouling material and has been widely used in Biacore's SPR sensor technology. The three-dimensional matrix is capable of extremely high levels of immobilization due to more accessible functionalization sites [23]. However, in addition to limited nonfouling properties from complex media, such as undiluted blood plasma or serum [7,49], these functionalized surfaces also require chemical blocking to passivate unreacted activated sites before detection [37].

While polymers composed of PC, SB, and CB have been shown to be protein resistant, it should be pointed out that not all zwitterionic materials can be assumed to have nonfouling properties. Previous work has shown the importance of the size and hydrophobicity of the spacer group between the two charged moieties of the monomer [55]. This is due to the requirement of densely packed polymer surfaces for reducing nonspecific binding, as mentioned previously, in addition to minimizing the hydrophobicity, a characteristic well known to cause fouling [39].

Besides covalent protein attachment, as demonstrated in this work, other approaches for the immobilization of molecules include physical adsorption (via hydrophobic and electrostatic interactions) and bioaffinity reagents (e.g., avidin/biotin linkage or DNA hybridization). While the specific method chosen will vary from system to system and is dependent upon the chemical properties of the surface and protein, applied immobilization strategies should take two important factors into account. The immobilization should not affect the conformation or function of the ligand (perhaps via steric hindrance or

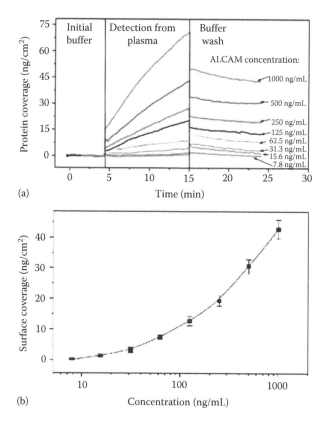

(a)

(b)

FIGURE 4.9 (a) The reference-compensated SPR sensorgrams for the direct detection of ALCAM from undiluted human plasma using an antibody-functionalized pCBAA substrate. The SPR response shown for these spiked samples includes the subtracted signals from both the native ALCAM present in human plasma (obtained by running pure plasma samples) as well as the response from the control anti-Salmonella surface. Thus, the *y*-axis indicates the increase in protein surface coverage due to only the spiked ALCAM from the desired specific interaction of the antigen/antibody pair. The vertical lines indicate three stages of the detection: the initial buffer baseline, detection from human plasma, and buffer wash, identical to the steps shown in Figure 4.8a and b. The ultralow-fouling pCBAA background enabled the binding kinetics to be clearly monitored directly from the complex media. (b) A plot of the net SPR response versus spiked antigen concentration in human plasma resulted in a standard detection curve with a limit of detection of ~10 ng/mL. (Adapted with permission from Vaisocherova, H., Yang, W., Zhang, Z., Cao, Z.Q., Cheng, G., Piliarik, M., Homola, J., and Jiang, S.Y., Ultralow fouling and functionalizable surface chemistry based on a zwitterionic polymer enabling sensitive and specific protein detection in undiluted blood plasma, *Anal. Chem.*, 80, 7894–7901, 2008. Copyright 2008 American Chemical Society.)

structural deformation) nor worsen the postfunctionalized nonfouling properties of the resulting surface platform [46,55].

Lastly, while SPR biosensing has become an important tool for the study of biomolecular interactions, other biodetection assays are still used frequently. One of the most popular, which is also used for ALCAM detection [14], is the label-based enzyme-linked immunosorbent assay (ELISA). Despite ELISA representing a low-cost and sensitive method for detecting target analytes from solution, it, along with other label-based sandwich assays (e.g., fluorescence, radio, or enzymatic based), requires the use of two different antibodies with different binding epitopes in order to first capture and then detect the target of interest. The costs associated with obtaining multiple high-quality antibodies for a single analyte have resulted in limited developments for these techniques [32]. A recent study showed that SPR biosensors serve as a good alternative to conventional ELISA for cancer biomarker detection. Specifically, the sensitivity and reproducibility of SPR were shown to match that of ELISA, all the while enabling shorter assay times and without the need for signal amplification. Eight human serum samples (four positive for pancreatic cancer and four negative controls) were

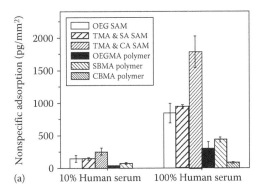 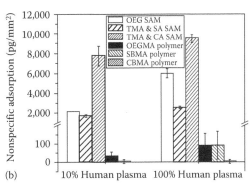

FIGURE 4.10 Nonspecific protein adsorption measured for six different surfaces to (a) 10% human serum in PBS and undiluted human serum and (b) 10% human plasma in PBS and undiluted human plasma using an SPR biosensor. The six surfaces include SAMs of OEG, mixed trimethylamine ($-N(CH_3)^+$, TMA) and sulfonic acid ($-SO_3^-$, SA), and mixed TMA and carboxylic acid ($-CO_2^-$, CA), as well as SI-ATRP-created polymers using methacrylate (MA) monomers of OEG, SB, and CB. Short SAMs composed of OEG or mixed-charged alkanethiols were unable to prevent nonspecific adsorption from complex media. The thicker zwitterionic polymer films were needed to obtain protein resistance to undiluted human blood plasma and serum, with CB being the most effective. (TMA + SA and TMA + CA are the mixed-charged SAM versions of SB and CB, respectively.) (Adapted with permission from Ladd, J., Zhang, Z., Chen, S., Hower, J.C., and Jiang, S., Zwitterionic polymers exhibiting high resistance to nonspecific protein adsorption from human serum and plasma, *Biomacromolecules*, 9, 1357–1361, 2008. Copyright 2008 American Chemical Society.)

analyzed for a single cancer biomarker. Statistically different results were observed between the two groups using both direct and amplified (i.e., sandwich assay) SPR detection which correlated well with ELISA results [53]. Similar trends were also observed between SPR and ELISA for the detection of autoantibodies from human serum in colon and ovarian cancer samples [29]. The use of a nonfouling surface chemistry helped to improve the SPR assay sensitivity and aid in making direct SPR measurements more reliable, illustrating the importance of developing protein-resistant materials for cancer diagnostics. It should be mentioned that mass spectroscopy is also a rapidly evolving technique to identify and characterize isolated proteins (including ALCAM). However, cost aside, recent improvements in the sensitivity and mass accuracy have not enabled this technique to fully resolve individual protein constituents of the complex blood proteome. To overcome this, extensive fractionation is required prior to analysis [14].

Techniques for label-free biomarker detection based on other sensing mechanisms have also been under development. One example is suspended microchannel resonators (SMRs) which take advantage of micro- and nanoscience and technology [5]. These sensors (shown in Figure 4.11a) monitor absorbed mass on the surface of a microcantilever via changes in vibrational frequency of the cantilever tip. It was demonstrated that the SMR is highly sensitive, thus representing great possibilities for increasing the detection limits of cancer biomarkers from complex media when integrated with zwitterionic pCB surface chemistry [56].

4.6 Future Trends

The zwitterionic pCB technology combined with SPR biosensing represents a new type of detection platform with highly desirable properties for cancer diagnostics. The extraordinarily high diversity and heterogeneity within cancer further requires the detection of multiple biomarkers for improved stratification and treatment [1,30,32]. As a result, antibody microarrays on pCB are being developed by implementing the methods presented here with high-throughput analysis using SPR imaging sensors. The ability to detect multiple disease markers simultaneously from a single sample in a protein-resistant background will offer significant improvements for cancer diagnostics. In addition to serving as a technology for cancer detection in the clinic, pCB films can also be used as a sensing platform for other applications

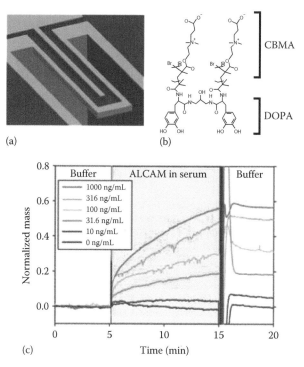

(a)   (b)

(c)

FIGURE 4.11   (a) An SMR. This detection platform consists of a U-shaped sensor channel within a resonating microcantilever contained in a vacuum. As fluid continuously flows through the device, absorbed mass on the walls of the sensing channel can be monitored by changes in vibrational frequency, allowing for label-free detection with subfemtogram resolution. (Adapted from *Nature*, 446, Burg, T.P., Godin, M., Knudsen, S.M., Shen, W., Carlson, G., Foster, J.S., Babcock, K., and Manalis, S.R., Weighing of biomolecules, single cells and single nanoparticles in fluid, 1066–1069, Copyright 2010, with permission from Elsevier B.V.) (b) The molecular structure of DOPA-pCBMA conjugates. (c) The SMR sensorgrams for direct detection of ALCAM from undiluted serum. Grafting of DOPA-pCBMA polymer conjugates to the sensor surface followed by antibody immobilization enabled the direct detection of ALCAM from undiluted serum using the SMR. (Adapted with permission from von Muhlen, M.G., Brault, N.D., Knudsen, S.M., Jiang, S.Y., and Manalis, S.R., Label-free biomarker sensing in undiluted serum with suspended microchannel resonators, *Anal. Chem.*, 82, 1905–1910, 2010. Copyright 2010 American Chemical Society.)

involving complex media such as food safety, environmental monitoring, and national security (e.g., biowarfare agent detection).

The decreasing size of biosensor platforms in order to increase sensitivity and minimize sample volume requirements [12] has raised additional challenges. One of them is to create protein-resistant coatings *in situ* because conventional surface-initiated polymerization techniques are difficult to be implemented in such small and confined structures. Zwitterionic polymers containing an adhesive moiety which can attach "to" the surface from a solution have been developed [4,17,33]. As shown in Figure 4.11, DOPA-pCBMA polymer conjugates (Figure 4.11b) have been successfully used with SMRs to detect target analytes directly from serum (Figure 4.11c) [56]. This illustrates the potential impact pCB materials hold in the field of nanoscale biosensors. Methods for improving the surface density of these attached polymers for such devices are still highly desirable. In addition, the development of zwitterionic CB nanoparticles for *in vivo* cancer diagnostics has also been progressing [6,60]. For example, pCB-functionalized gold nanoparticles were successfully prepared and exhibited excellent stability in undiluted human serum, indicating its potential to increase *in vivo* circulation time. Following EDC/NHS activation, antibody immobilization resulted in the ability to specifically detect the corresponding antigen [60]. In summary, the nonfouling and functionalizable properties of pCB make it highly desirable for many different biomedical applications and represent the future of the next-generation biomaterials.

ACKNOWLEDGMENTS

This work was supported by the University of Washington and the National Oceanic and Atmospheric Administration (Oceans and Human Health Initiative/NOS) to Q.Y. This work was also supported by the Office of Naval Research (N000140711036 and N000140910137), the National Science Foundation (DMR-0705907 and CBET 0827274/0625829), and the National Cancer Institute via SAIC-Frederick (28XS119) to S.J. support for N.D.B was provided by the Center for Nanotechnology at the University of Washington via funding from an IGERT Fellowship Award NSF #DGE-0504573.

REFERENCES

1. Altieri, D. C. 2008. Opinion: Survivin, cancer networks and pathway-directed drug discovery. *Nature Reviews Cancer* **8**:61–70.
2. Baker, M. 2005. In biomarkers we trust? *Nature Biotechnology* **23**:297–304.
3. Blattler, T. M., S. Pasche, M. Textor, and H. J. Griesser. 2006. High salt stability and protein resistance of poly(L-lysine)-g-poly(ethylene glycol) copolymers covalently immobilized via aldehyde plasma polymer interlayers on inorganic and polymeric substrates. *Langmuir* **22**:5760–5769.
4. Brault, N. D., C. L. Gao, H. Xue, M. Piliarik, J. Homola, S. Y. Jiang, and Q. M. Yu. 2010. Ultra-low fouling and functionalizable zwitterionic coatings grafted onto SiO2 via a biomimetic adhesive group for sensing and detection in complex media. *Biosensors and Bioelectronics* **25**:2276–2282.
5. Burg, T. P., M. Godin, S. M. Knudsen, W. Shen, G. Carlson, J. S. Foster, K. Babcock, and S. R. Manalis. 2007. Weighing of biomolecules, single cells and single nanoparticles in fluid. *Nature* **446**:1066–1069.
6. Cao, Z. Q., Q. M. Yu, H. Xue, G. Cheng, and S. Y. Jiang. 2010. Nanoparticles for drug delivery prepared from amphiphilic PLGA zwitterionic block copolymers with sharp contrast in polarity between two blocks. *Angewandte Chemie—International Edition* **49**:3771–3776.
7. Carlsson, J., C. Gullstrand, G. T. Westermark, J. Ludvigsson, K. Enander, and B. Liedberg. 2008. An indirect competitive immunoassay for insulin autoantibodies based on surface plasmon resonance. *Biosensors and Bioelectronics* **24**:876–881.
8. Chapman, R. G., E. Ostuni, L. Yan, and G. M. Whitesides. 2000. Preparation of mixed self-assembled monolayers (SAMs) that resist adsorption of proteins using the reaction of amines with a SAM that presents interchain carboxylic anhydride groups. *Langmuir* **16**:6927–6936.
9. Chen, S. F., L. Y. Liu, J. Zhou, and S. Y. Jiang. 2003. Controlling antibody orientation on charged self-assembled monolayers. *Langmuir* **19**:2859–2864.
10. Chen, S. F., F. C. Yu, Q. M. Yu, Y. He, and S. Y. Jiang. 2006. Strong resistance of a thin crystalline layer of balanced charged groups to protein adsorption. *Langmuir* **22**:8186–8191.
11. Chen, S. F., J. Zheng, L. Y. Li, and S. Y. Jiang. 2005. Strong resistance of phosphorylcholine self-assembled monolayers to protein adsorption: Insights into nonfouling properties of zwitterionic materials. *Journal of the American Chemical Society* **127**:14473–14478.
12. Cheng, M. M. C., G. Cuda, Y. L. Bunimovich, M. Gaspari, J. R. Heath, H. D. Hill, C. A. Mirkin, A. J. Nijdam, R. Terracciano, T. Thundat, and M. Ferrari. 2006. Nanotechnologies for biomolecular detection and medical diagnostics. *Current Opinion in Chemical Biology* **10**:11–19.
13. Cooper, M. A. 2003. Label-free screening of bio-molecular interactions. *Analytical and Bioanalytical Chemistry* **377**:834–842.
14. Faca, V. M., K. S. Song, H. Wang, Q. Zhang, A. L. Krasnoselsky, L. F. Newcomb, R. R. Plentz, et al. 2008. A mouse to human search for plasma proteome changes associated with pancreatic tumor development. *Plos Medicine* **5**:953–967.
15. Ferrari, M. 2005. Cancer nanotechnology: Opportunities and challenges. *Nature Reviews Cancer* **5**:161–171.
16. Fine, B. M. and L. Amler. 2009. Predictive biomarkers in the development of oncology drugs: A therapeutic industry perspective. *Clinical Pharmacology and Therapeutics* **85**:535–538.
17. Gao, C. L., G. Z. Li, H. Xue, W. Yang, F. B. Zhang, and S. Y. Jiang. 2010. Functionalizable and ultra-low fouling zwitterionic surfaces via adhesive mussel mimetic linkages. *Biomaterials* **31**:1486–1492.
18. Hanash, S. M., S. J. Pitteri, and V. M. Faca. 2008. Mining the plasma proteome for cancer biomarkers. *Nature* **452**:571–579.

19. Harder, P., M. Grunze, R. Dahint, G. M. Whitesides, and P. E. Laibinis. 1998. Molecular conformation in oligo(ethylene glycol)-terminated self-assembled monolayers on gold and silver surfaces determines their ability to resist protein adsorption. *Journal of Physical Chemistry B* **102**:426–436.

20. He, Y., Y. Chang, J. C. Hower, J. Zheng, S. F. Chen, and S. Jiang. 2008. Origin of repulsive force and structure/dynamics of interfacial water in OEG-protein interactions: A molecular simulation study. *Physical Chemistry Chemical Physics* **10**:5539–5544.

21. He, Y., J. Hower, S. Chen, M. T. Bernards, Y. Chang, and S. Jiang. 2008. Molecular simulation studies of protein interactions with zwitterionic phosphorylcholine self-assembled monolayers in the presence of water. *Langmuir* **24**:10358–10364.

22. Homola, J. 2006. *Surface Plasmon Resonance Based Sensors*, Vol. 4. Springer, Berlin, Germany.

23. Homola, J. 2008. Surface plasmon resonance sensors for detection of chemical and biological species. *Chemical Reviews* **108**:462–493.

24. Hucknall, A., S. Rangarajan, and A. Chilkoti. 2009. In pursuit of zero: Polymer brushes that resist the adsorption of proteins. *Advanced Materials* **21**:2441–2446.

25. Jiang, S. Y. and Z. Q. Cao. 2010. Ultralow-fouling, functionalizable, and hydrolyzable zwitterionic materials and their derivatives for biological applications. *Advanced Materials* **22**:920–932.

26. Jones, D. M., A. A. Brown, and W. T. S. Huck. 2002. Surface-initiated polymerizations in aqueous media: Effect of initiator density. *Langmuir* **18**:1265–1269.

27. Kulasingam, V. and E. P. Diamandis. 2008. Strategies for discovering novel cancer biomarkers through utilization of emerging technologies. *Nature Clinical Practice Oncology* **5**:588–599.

28. Kulasingam, V., M. P. Pavlou, and E. P. Diamandis. 2010. Integrating high-throughput technologies in the quest for effective biomarkers for ovarian cancer. *Nature Reviews Cancer* **10**:371–378.

29. Ladd, J., H. L. Lu, A. D. Taylor, V. Goodell, M. L. Disis, and S. Y. Jiang. 2009. Direct detection of carcinoembryonic antigen autoantibodies in clinical human serum samples using a surface plasmon resonance sensor. *Colloids and Surfaces B—Biointerfaces* **70**:1–6.

30. Ladd, J., A. D. Taylor, M. Piliarik, J. Homola, and S. Y. Jiang. 2009. Label-free detection of cancer biomarker candidates using surface plasmon resonance imaging. *Analytical and Bioanalytical Chemistry* **393**:1157–1163.

31. Ladd, J., Z. Zhang, S. Chen, J. C. Hower, and S. Jiang. 2008. Zwitterionic polymers exhibiting high resistance to nonspecific protein adsorption from human serum and plasma. *Biomacromolecules* **9**:1357–1361.

32. Lausted, C., Z. Y. Hu, and L. Hood. 2008. Quantitative serum proteomics from surface plasmon resonance imaging. *Molecular & Cellular Proteomics* **7**:2464–2474.

33. Li, G. Z., G. Cheng, H. Xue, S. F. Chen, F. B. Zhang, and S. Y. Jiang. 2008. Ultra low fouling zwitterionic polymers with a biomimetic adhesive group. *Biomaterials* **29**:4592–4597.

34. Li, L. Y., S. F. Chen, and S. Y. Jiang. 2007. Protein interactions with oligo(ethylene glycol) (OEG) self-assembled monolayers: OEG stability, surface packing density and protein adsorption. *Journal of Biomaterials Science—Polymer Edition* **18**:1415–1427.

35. Li, L. Y., S. F. Chen, J. Zheng, B. D. Ratner, and S. Y. Jiang. 2005. Protein adsorption on oligo(ethylene glycol)-terminated alkanethiolate self-assembled monolayers: The molecular basis for nonfouling behavior. *Journal of Physical Chemistry B* **109**:2934–2941.

36. Ludwig, J. A. and J. N. Weinstein. 2005. Biomarkers in cancer staging, prognosis and treatment selection. *Nature Reviews Cancer* **5**:845–856.

37. Masson, J. F., T. M. Battaglia, P. Khairallah, S. Beaudoin, and K. S. Booksh. 2007. Quantitative measurement of cardiac markers in undiluted serum. *Analytical Chemistry* **79**:612–619.

38. Ofori-Acquah, S. F. and J. A. King. 2008. Activated leukocyte cell adhesion molecule: A new paradox in cancer. *Translational Research* **151**:122–128.

39. Ostuni, E., R. G. Chapman, R. E. Holmlin, S. Takayama, and G. M. Whitesides. 2001. A survey of structure-property relationships of surfaces that resist the adsorption of protein. *Langmuir* **17**:5605–5620.

40. Paulovich, A. G., J. R. Whiteaker, A. N. Hoofnagle, and P. Wang. 2008. The interface between biomarker discovery and clinical validation: The tar pit of the protein biomarker pipeline. *Proteomics Clinical Applications* **2**:1386–1402.

41. Petricoin, E. F., C. Belluco, R. P. Araujo, and L. A. Liotta. 2006. The blood peptidome: A higher dimension of information content for cancer biomarker discovery. *Nature Reviews Cancer* **6**:961–967.

42. Piliarik, M. and J. Homola. 2009. Surface plasmon resonance (SPR) sensors: Approaching their limits? *Optics Express* **17**:16505–16517.

43. Prime, K. L. and G. M. Whitesides. 1991. Self-assembled organic monolayers: Model systems for studying adsorption of proteins at surfaces. *Science* **252**:1164–1167.

44. Rasooly, A. 2006. Moving biosensors to point-of-care cancer diagnostics. *Biosensors and Bioelectronics* **21**:1847–1850.

45. Rifai, N., M. A. Gillette, and S. A. Carr. 2006. Protein biomarker discovery and validation: The long and uncertain path to clinical utility. *Nature Biotechnology* **24**:971–983.

46. Rusmini, F., Z. Y. Zhong, and J. Feijen. 2007. Protein immobilization strategies for protein biochips. *Biomacromolecules* **8**:1775–1789.

47. Sawyers, C. L. 2008. The cancer biomarker problem. *Nature* **452**:548–552.

48. Shao, Q., Y. He, A. D. White, and S. Jiang. Difference in hydration between carboxybetaine and sulfobetaine. *The Journal of Physical Chemistry B* **114**:16625–16631.

49. Situ, C., A. R. G. Wylie, A. Douglas, and C. T. Elliott. 2008. Reduction of severe bovine serum associated matrix effects on carboxymethylated dextran coated biosensor surfaces. *Talanta* **76**:832–836.

50. Stern, E., A. Vacic, N. K. Rajan, J. M. Criscione, J. Park, B. R. Ilic, D. J. Mooney, M. A. Reed, and T. M. Fahmy. 2010. Label-free biomarker detection from whole blood. *Nature Nanotechnology* **5**:138–142.

51. Taylor, A. D., Q. M. Yu, S. F. Chen, J. Homola, and S. Y. Jiang. 2005. Comparison of *E. coli* O157: H7 preparation methods used for detection with surface plasmon resonance sensor. *Sensors and Actuators B—Chemical* **107**:202–208.

52. Ullah, M. F. and M. Aatif. 2009. The footprints of cancer development: Cancer biomarkers. *Cancer Treatment Reviews* **35**:193–200.

53. Vaisocherova, H., V. M. Faca, A. D. Taylor, S. Hanash, and S. Y. Jiang. 2009. Comparative study of SPR and ELISA methods based on analysis of CD166/ALCAM levels in cancer and control human sera. *Biosensors and Bioelectronics* **24**:2143–2148.

54. Vaisocherova, H., W. Yang, Z. Zhang, Z. Q. Cao, G. Cheng, M. Piliarik, J. Homola, and S. Y. Jiang. 2008. Ultralow fouling and functionalizable surface chemistry based on a zwitterionic polymer enabling sensitive and specific protein detection in undiluted blood plasma. *Analytical Chemistry* **80**:7894–7901.

55. Vaisocherova, H., Z. Zhang, W. Yang, Z. Q. Cao, G. Cheng, A. D. Taylor, M. Piliarik, J. Homola, and S. Y. Jiang. 2009. Functionalizable surface platform with reduced nonspecific protein adsorption from full blood plasma–Material selection and protein immobilization optimization. *Biosensors and Bioelectronics* **24**:1924–1930.

56. von Muhlen, M. G., N. D. Brault, S. M. Knudsen, S. Y. Jiang, and S. R. Manalis. 2010. Label-free biomarker sensing in undiluted serum with suspended microchannel resonators. *Analytical Chemistry* **82**:1905–1910.

57. Wang, H., D. G. Castner, B. D. Ratner, and S. Y. Jiang. 2004. Probing the orientation of surface-immobilized immunoglobulin G by time-of-flight secondary ion mass spectrometry. *Langmuir* **20**:1877–1887.

58. Yang, W., S. F. Chen, G. Cheng, H. Vaisocherova, H. Xue, W. Li, J. L. Zhang, and S. Y. Jiang. 2008. Film thickness dependence of protein adsorption from blood serum and plasma onto poly(sulfobetaine)-grafted surfaces. *Langmuir* **24**:9211–9214.

59. Yang, W., H. Xue, W. Li, J. L. Zhang, and S. Y. Jiang. 2009. Pursuing "zero" protein adsorption of poly(carboxybetaine) from undiluted blood serum and plasma. *Langmuir* **25**:11911–11916.

60. Yang, W., L. Zhang, S. L. Wang, A. D. White, and S. Y. Jiang. 2009. Functionalizable and ultra stable nanoparticles coated with zwitterionic poly(carboxybetaine) in undiluted blood serum. *Biomaterials* **30**:5617–5621.

61. Zhang, Z., S. F. Chen, and S. Y. Jiang. 2006. Dual-functional biomimetic materials: Nonfouling poly(carboxybetaine) with active functional groups for protein immobilization. *Biomacromolecules* **7**:3311–3315.

62. Zhang, Z., H. Vaisocherova, G. Cheng, W. Yang, H. Xue, and S. Y. Jiang. 2008. Nonfouling behavior of polycarboxybetaine-grafted surfaces: Structural and environmental effects. *Biomacromolecules* **9**:2686–2692.

63. Zheng, J., L. Y. Li, S. F. Chen, and S. Y. Jiang. 2004. Molecular simulation study of water interactions with oligo (ethylene glycol)-terminated alkanethiol self-assembled monolayers. *Langmuir* **20**:8931–8938.

64. Zheng, J., L. Y. Li, H. K. Tsao, Y. J. Sheng, S. F. Chen, and S. Y. Jiang. 2005. Strong repulsive forces between protein and oligo (ethylene glycol) self-assembled monolayers: A molecular simulation study. *Biophysical Journal* **89**:158–166.

5

Surface Plasmon Resonance (SPR) and ELISA Methods for Antibody Determinations as Tools for Therapeutic Monitoring of Patients with Acute Lymphoblastic Leukemia (ALL) after Native or Pegylated Escherichia coli and Erwinia chrysanthemi Asparaginases

Vassilios I. Avramis

CONTENTS

5.1 Introduction

5.1.1 Nutrients, Gene Expression, and Progression of Malignancies

Malignancies have their etiology in malformations of DNA such as point mutations, chromosomal deletions, inversions, copy number variants (CNV), loss of heterozygocity (LOH), and other chromosomal
abnormalities. However, protein biosynthesis is fundamental in all living cells because it is the mechanism that drives the cellular growth and development of all normal (healthy) and cancerous cells. For
the most part, malignancies are expressing variants of protein biosynthesis due to mutations of abnormal
upregulation of protein biosynthesis. Furthermore, many determinant causative proteins act in concert
with hormones (growth hormone [GH], androgens, estrogens, etc.), about 20 families of growth factors
and their ligands (VEGF, EGF, PDGF, IGF, etc.), which along with ample nutrients (glucose, amino
acids [AAs], lipids, minerals, vitamins, etc.) constitute an important role in the control of gene expression leading to cellular growth and proliferation and/or differentiation. AAs, along with other nutrients,
serve as the building blocks of protein biosynthesis as well as signal transduction messengers to transmit
the nutritional status of the entire organism to individual cells. Given the critical roles AA exerts in cell
survival, growth, metabolism, and signaling, it is not surprising that mammalian tumors have evolved
mechanisms to adapt to protein malnutrition and related AA deprivation for short time intervals, i.e., during chemotherapy. The steady state of the cellular pool for each of the nonessential AA is the result of a
balance between de novo biosynthesis plus nutrient input and utilization and removal (degradation). Diet
and proteolysis provide all the essential and the majority of the nonessential AA pools. Certain key nonessential AAs like L-glutamine (Gln) and L-asparagine (Asn), which are necessary for neuronal function
and survival of malignant cells, are in great demand for protein biosynthesis and as sources of carbon
and nitrogen in highly proliferative conditions. Therefore, the molecular basis of gene regulation by AA
is an important field of research for study of the regulation of global protein inhibition under physiological and pharmacological conditions. However, we must leave the broad spectrum of cell survival aside,
and now, we will focus on the aspects of the lymphoproliferative malignancies, which is an exciting field
in oncology.

The era of cancer chemotherapy was initiated in earnest in the past 50-plus years. Cytotoxic chemotherapy drugs, along with improved surgical techniques and radiation therapy, are the principal means of
effectively controlling malignancies. Many classes of antineoplastic drugs were investigated and developed in the 1960s and 1970s, including the bacterial L-asparaginases (ASNases), which are still in use
today. Naturally, major improvements have been made in the treatment outcomes of many of these malignancies, especially in leukemias and lymphomas, with the use of classical and novel anticancer drugs.
Recently, the development of specific target-oriented molecules, inhibiting specific proteins and/or their
antibodies (Abs), has provided further successes in specific malignancies. As a result, significant steps
have been made in prolonging patients' lives and/or curing many cancer patients. One of these classes
of drugs was the bacterial-origin global protein inhibitor L-ASNase used in its native or pegylated form.
ASNases inhibit global protein biosynthesis of many cell cycle and antiapoptotic proteins in cells undergoing chemoradiation; hence, they have become the drug of choice in many induction and reinduction
combination regimens against leukemia and lymphomas. To this end, assays for the determination of
the pharmacodynamic effectiveness of these bacterial proteins and their Abs have been developed and
applied in leukemia evaluations.

5.1.2 Useful Bacteria in Oncology

E. coli is gram negative, facultative anaerobic, and nonsporulating. Cells are typically rod shaped and
are about 2 µm long and 0.5 µm in diameter, with a cell volume of 0.6–0.7 µm³. *E. coli* is commonly

found in the lower intestine of warm-blooded organisms (endotherms). Most *E. coli* strains are harmless (saprophytes), but some, such as serotype O157:H7, can cause serious food poisoning in humans and are occasionally responsible for food product recalls.

Erwinia chrysanthemi (*Dickeya chrysanthemi*) is a gram-negative bacillus that belongs to the family Enterobacteriaceae. *Erwinia* (ErW) is a genus of Enterobacteriaceae bacteria containing mostly plant pathogenic species. It is a close relative of *E. coli* and other animal pathogens that include *Salmonella*, *Shigella*, *Klebsiella*, *Proteus*, and *Yersinia* (*Y. pestis*). Members of this family are facultative anaerobes, able to ferment sugars to lactic acid, contain amidase and nitrate reductase but lack oxidases. Even though many clinical pathogens are part of the Enterobacteriaceae family, most members of this family are plant pathogens. Both the *E. coli* and *Erwinia chrysanthemi* bacteria produce an amidase (ASNase with glutaminase activity), which for the past 50 years has been used successfully against lymphomatous malignancies.

5.1.3 Immune Responses to (Bacterial) Biology Oncology Products

Beneficial new biology oncology drugs have been proven useful against cancers in the past many decades. They provide promising results by prolonging the outcomes and, hence, the lives of cancer patients. These promising anticancer biology products provide disease-controlling medical interventions. Although carcinogenesis has its origins in DNA malformations, which are based on chromosomal abnormalities, translocations, inversions, including point mutations, missing or extra chromosomal structures, or other abnormalities, all of which eventually lead to cancer initiation and development. However, the growth of the initial cancer cell depends primarily on altered or mutated protein biosynthesis. Therefore, the majority of the oncology drugs are small molecules targeting DNA with a few targeting protein biosynthesis inhibitions. Hence, once the first few malignant cells have been generated, in order to grow, they need to synthesize proteins in a continuous and more rapid manner than healthy cells. Therefore, the uses of the few protein inhibitor drugs, which are bacterial proteins themselves, like ASNases, have been shown to be very effective in combination therapies.

Classical immunology teaches us that protein therapeutics may elicit an immune response upon a single or multiple administrations [1–3]. When cross-reactive polyclonal Abs are developed, they may cause unexpected adverse effects, including the discontinuation of biologic-drug treatments. For example, Ab(+) to ASNases in ALL patients will cause discontinuation of ASNase therapy, which most likely will have deleterious effects on long-term event-free survival (EFS) [3,4]. Therefore, it is essential that immunological assays of therapeutic proteins be developed and applied during clinical trial for better safety and efficacy treatment assessments [1–3].

5.1.4 Asparaginase, Asparagine Synthetase, and Protein Synthesis

Protein synthesis is fundamental in all living cells dividing and resting, both healthy and malignant. Translational control in eukaryotic cells is critical for gene regulation during nutrient deprivation stress, development and differentiation, neuronal function, disease progression, and aging. Therefore, translation of the genetic code to proteins and its regulation matter. AAs serve as signal transduction messengers to transmit the nutritional status of the entire organism to individual cells. One of the mechanisms by which AAs mediate this signaling is altered transcription for specific genes *via* a *signal transduction process* referred to as the *amino acid response* (*AAR*) pathway. Under these conditions, increased expression occurs of genes containing AA response elements (i.e., Asn synthetase [ASNS], glutamine synthetase [GS], c-Myc, thiopurine methyltransferase, Bcl-2, dihydrofolate reductase, and many others), as several enzymes favor glutathione (GSH) biosynthesis [5].

ASNases are large-molecular-weight (134–150 kDa) bacterial proteins that have certain advantages and disadvantages [2]. ASNase is an antileukemia enzyme, which is used in treatment of hematopoietic malignancies, in particular in acute lymphoblastic leukemia (ALL) and other lymphoid malignancies such as non-Hodgkin's lymphoma (NHL) in induction or in reinduction in relapsed patients. Their mechanism of action is by depriving these tumor cells from the AAs L-asparagine (Asn) and L-glutamine (Gln) [2,3].

5.1.5 Asparaginases in Lymphoid Malignancies

Intermittent or continuous use of ASNase has been shown to be active against ALL through the depletion of serum Asn and to a lesser extent, Gln. However, after multiple ASNase administrations, patients develop Ab against this bacterial protein [2,3]. Patients who relapse, mostly after the development of neutralizing Ab(+) against this antigen, develop drug resistance to these drugs. This form of ASNase resistance may be exacerbated and possibly correlated with the upregulation of the Gln-dependent ASNS [1–3,6–8].

To ameliorate this insidious condition, a noncross Ab(+) reacting with *Erwinia chrysanthemi* ASNase (ErW ASNase) is a preferred alternate formulation for depletion of serum Asn and Gln during reinduction therapies. ErW ASNase has ~10× better glutaminase activity; hence, it is a better product to deaminate Gln, the cosubstrate of ASNS. This becomes more important since Gln deprivation (<2 h) triggers intracellular events affecting the mitochondria, leading to an irreversible commitment for apoptosis [9]. Thus, many investigators have developed the hypothesis that patients in long-term remission had optimal treatment with ASNases, at least in part, due to limited or absence of Ab(+) and ASNS expressions.

Many Phase II and Phase III clinical studies have reported a wide range of the presence of human antibacterial ASNase Ab (HABA) [Ab(+)] formation. However, these were assayed by an Ab-capture enzyme-linked immunosorbent (ELISA) Ab(+) assay, which were not always specifying the immunoglobulin subtype [2,3,10–15].

5.1.6 Use of Asparaginases in Lymphoid Malignancies

ASNases are administered either intramuscularly (IM) or, preferably, intravenously (IV) to patients [2,3,14,15]. Although ASNases do not diffuse into the extracellular space, the helper T-cells in the circulation along with the dendritic cells are exposed to this bacterial antigen protein, thus, initiating a host response. Once the immune response is activated, most often the ASNase formulation has to be either discontinued or a non-cross-reacting bacterial ASNase (ErW ASNase, *Erwinase*) has to be administered when available [2,3]. As many as 70% of high-risk ALL patients treated with the native *E. coli* ASNase may develop anti-ASNase Ab(+) after IM administrations, many of whom becoming Ab positive without demonstrating any clinical evidence of Ab(+) formation (silent hypersensitivity), and other with higher Ab(+) levels after the presentation of clinical allergy reaction [3,11–15]. In patients who develop hypersensitivity post-ASNase treatments, there appears to be an inverse relationship between high Ab(+) levels and low or undetected enzymatic activity due to neutralization of the antigen-ASNase [2,3,14].

5.1.7 Unmet Needs of Asparaginase Ab(+) Determination

ALL is a very complex malignancy of multiple etiologies. Thus, the mechanisms of treatment failure in ALL are poorly understood [4]. However, recently, attention has been focused on the Ab(+) development to ASNase; hence, an intrinsic form of drug-antigen-induced resistance is utilized by the leukemic blasts to ASNases, drugs which then become ineffective to improve efficacy and long-term outcomes in ALL. However, there are many unmet needs in deciphering who and why do patients develop anti-ASNase Ab(+). Therefore, a search for a global bioassay for the determination of pharmacodynamics post-ASNase therapy was undertaken. Indeed, despite the ELISA Ab(+) assay availability, no coordinated efforts have been made for a centralized application of the Ab(+) determination in sera from the many thousands of ALL patients diagnosed ever year in the United States. Moreover, replacement of the multiple injections of native *E. coli* ASNase with the long half-life PEG-ASNase improved the complete remission rates [2,3]. However, replacement of native *E. coli* ASNase with an equal dose and schedule of the shorter half-life ErW ASNase eroded outcomes in the context of multiagent therapy [14,16]. Recently, in relapsed or Ab(+) ALL patients, much higher doses of ErW ASNase were administered IV with equal or perhaps improved efficacy without an apparent evidence of increased toxicity of this drug [17].

Clearly, the immune response to ASNase in ALL patients is very heterogeneous; thus, methods must be able to detect different Ab(+) isoforms, affinities, and/or neutralizing specificities. Thus far, serum samples have been used for screening of anti-ASNase Ab(+) using an inverse sandwich Ab(+)-antigen bioassay (ELISA) in both pediatric [2,3] and adult ALL patients [15]. IgG immunoglobulin is the most abundant isotype, whereas IgE is the least abundant class of immunoglobulins circulating in human serum, respectively [18,19]. Only a limited number of publications reporting ELISA-quantifiable results have been published, and even fewer have been reported comparing ELISA with other assays for Ab(+) in patients who receive biological oncology drugs for their immunogenicity determination [20].

5.1.8 ELISA and Surface Plasmon Resonance (Biacore) Assays

When anti-ASNase Ab(+) is detected by an ELISA Ab(+) assay, then a biochemical enzymatic assay determining the percent (%) neutralization of ex vivo–added antigen (1 IU ASNase) over a short time frame is performed at high expense of time and effort to validate the neutralization of enzymatic activity by the Ab(+) [2,3,14,15]. Therefore, a simpler, accurate, and rapid immunological method is needed to determine the Ab(+) type and its antigen neutralization. Surface plasmon resonance (SPR-Biacore) has emerged as a biosensor assay that allows for the sensitive detection of both low- and high-affinity Ab(+), their isoforms, and the binding characteristic to the antigen drug [21,22].

5.1.9 Elements in Biosensor Technologies

Smart biosensor technology (SBT) is a comprehensive, interdisciplinary guide to extending the capabilities of biosensors for a broad range of fields. It is demonstrating how to apply the latest knowledge to solve problems of protein–protein and protein–macromolecule (protein–DNA or protein–mRNA) interactions. This can be a photoelectric biosensing, chemical sensing, and/or color imaging. These biosensor applications can detect food pathogens, monitoring toxicities of natural products, antigen–Ab reactions and related toxicities such as allergens, and viruses.

The development of the biosensor-based SPR technology became available about 10 years ago. During this time, the use of these techniques has increased steadily. The most widely used one is the Biacore produced by Biacore AB, which has developed into a range of instruments. Biacore offers advantages for analyzing both weak and strong macromolecular interactions, thus allowing measurements not possible by old techniques. The underlying physicochemical principles of SPR are complex. It suffices to know that SPR-based instruments use an optical method to measure refractive index (RI) near a sensor surface (within 300 nm). In these instruments, the surface forms on the floor of a flow cell (20–60 nL in volume) through which an aqueous solution (running buffer) passes under continuous flow of a few nanoliters per minute.

The ligand (any antigen) is immobilized onto the carboxymethylated dextran matrix sensor surface. The reactant—the analyte—is injected in aqueous solution (sample buffer) though the flow cell under pressure and continuous flow. As the analyte binds to the ligand over time, the accumulation of protein on the surface results in an increase in the RI. The change of the RI is measured in real time, and the result is projected (plotted) as response resonance units (RU) vs. time (sensorgram). At the same time, the background response is also generated if there is a difference in the RIs of running sample buffers. This background response is subtracted from the sensorgram to obtain the actual protein–protein binding response. One RU represents approximately 1 pg/protein/mm^2 buffer. In reality, the lowest cutoff limit of SPR of about $\geq$50 pg protein/mm^2 is needed. As a result of the binding, the association rate constant (kon) is measured. Then the flow cell is washed with clean buffer; the analyte is being washed from the binding sites in the flow cell. As a result, the dissociation rate constant (apparent koff) can be determined after it reaches equilibrium in binding.

There is an important limitation of SPR largely because of the mass transport limitations when the analyte is in high or very high concentrations. Then it is difficult to measure accurately kon values faster than a nanomolar per second (nM/s). This is the upper limit of the analyte binding determination. If there are high concentrations of analytes, then a dilution of the sample may be needed to be tested. Therefore, obtaining accurate kinetic data is a very demanding and time-consuming task, and it requires a thorough understanding of the type of binding (e.g., antigen–Ab), their kinetics, and potential sources of artifact (nonspecific immunoglobulins or allo- or mutant proteins). Other limitation on this biosensor method is the size of the analyte, i.e., when the analyte has low MW, it gives very small responses. Other potential problem can be the purity of the protein that is bound to the surface of the flow cell. Any impurities of the target protein which are incorporated into the flow cell surface can and will cause misleading readings of RU.

5.1.9.1 Regeneration of the Cell Surface

Once a covalently immobilized protein has been shown to be active with respect to binding to its ligand or monoclonal Ab, regeneration of the flow cell can be attempted. The goal is to elute with clean buffer in order to remove all the bound material(s) from the immobilized protein. Thus, regeneration of the flow cell allows surfaces to be reused and regenerated many times, saving both time and money.

Based on the principles presented earlier, a new biosensor anti-ASNase Ab(+) assay was developed using SPR on a Biacore T-100 unit to detect Ab(+) binding to immobilized ASNase proteins in a series of protein chips capturing three different ASNase antigens. Appropriate negative controls, as well as patients' sera with Ab(−) and Ab(+), were tested for the presence of neutralizing IgG-4 or IgE Ab(+) [23]. Described in the following are the comparative SPR and ELISA Ab(+) detection methods in ASNase Ab(+) and Ab(−) patients and in volunteers' sera specimens against both native *E. coli*, PEG-ASNase, and ErW ASNase formulations [22].

5.2 Materials, Equipment, and Assay Methods

5.2.1 Equipment and Materials

A Biacore T-100 processing unit (GE Life Sciences Uppsala, Sweden) was used.

Biacore CM5 sensor chip(s) (part # BR-1006-68 GE Life Sciences Uppsala, Sweden) and 1× Biacore HBS-EP + part # BR-1008-26 (0.1 M HEPES, 1.5 M NaCl, 30 mM EDTA, and 0.5% v/v surfactant P20, pH = 7) were used. In addition, the GE Life Sciences Uppsala, Sweden, amine coupling kit 1-ethyl-3-(3-dimethylaminopropyl) carbodiimide hydrochloride (EDC) (part # BR-1000-50) was used, as well as N-hydroxysuccinimide (NHS), 1.0 M ethanolamine-HCl, pH = 8.5 [22].

5.2.2 Biacore-SPR Ab(+) Assay: Creating a Sensor Surface

Using Biacore's CM5 sensor chips and following standard amine chemistry protocols, ErW ASNase was diluted to 150 μg/mL in 10 mM sodium acetate, pH = 4.5, and immobilized to a level of 4000 RU into flow cell 2. Native *E. coli* ASNase was also diluted to 150 μg/mL in 10 mM sodium acetate, pH = 4, and immobilized to a level of 2000 RU into flow cell 4. Flow cells 1 and 3 were activated and blocked and assigned as BLANK controls. In flow cell 5, the same process as in flow cell 4 was applied for PEG-ASNase (with not very good binding, and eventually, it was removed from the series of protein chips). A reference unit (RU) is a measure of 0.0001° shift in RI a light beam form a solid surface (gold foil) embedded with the antigen. An RU unit is approximately equivalent to 1 pg/mm^2 in the dextran. RUs for each serum specimens, both control and experimental samples, were obtained over time (20 h) and after multiple cycles of analysis (>10) in an automated process. There was a superimposable reproducibility for these Ab(−) and Ab(+) values over time and upon reassaying at room temperature, which provided the validation for these assays [22].

5.2.3 Antigen–Ab(+) Binding Analysis

The assay was performed using Biacore's HBS-EP+ running buffer to the protein-embedded chips, in series. Serum was diluted 1:10 with running buffer and was injected over all four sensor surfaces (SPR) for 60 s at 30 μL/min followed by a 60 s injection of anti-isotype-specific Ab. The complex Ab(+) was removed, and the chip was regenerated back to immobilized protein-drug (ASNase) using 10 mM glycine pH = 2.0 [22]. Biacore software was used to analyze the quantitative results. Biacore software can be constrained to comply with the FDA's regulation 21 CFR part 11, so that initially unskilled users may operate the instrument without having the need to program or analyze the data manually. These functions can be set up by the PI or project administrator on the unit's computer.

5.3 Patients and Treatment

5.3.1 Patients

Between May 2005 and June 2007, patients with newly diagnosed ALL and HR features not enrolled on CCG studies received chemotherapy treatments with similar protocols which included *E. coli* native ASNase or PEG-ASNase after the appropriate institutional IRB review and the individual informed consents were obtained [22,23]. Patients, who developed clinical allergy to native ASNase or PEG-ASNase treatments, if symptoms persisted, were treated with alternative chemotherapy after discontinuation of ASNase treatments, including the use or ErW ASNase, when it was available in the United States for a "compassionate use." Of interest is that the ErW ASNase has become available on a special agreement by the FDA to be purchased for personal use in patients with obvious clinical allergy to the *E. coli* ASNase formulations. The majority of these patients who reacted with obvious clinical allergy post-ASNase treatments and sent us serum specimens to be analyzed for anti-ASNase Ab(+) on a compassionate basis and under separate IRB approval were mostly male >10 years of age (HR ALL) patients.

5.3.2 ASNase Formulations

All preparations of ASNase were given to these patients by IM injection. The doses were $6000\,\text{IU}/\text{m}^2 \times 3$ per week for 2 weeks postinduction, for native *E. coli*-ASNase (ELSPAR, Merck & Co., Inc. West Point, PA), and $2.5\,\text{IU}/\text{m}^2$ 2 or 4 weeks for PEG-ASNase, as in previous clinical trials [2,3,14,15]. The laboratory investigators were "blinded" on many clinical issues including the timing or the exact dosing schedule these patients had when they reacted, and the serum specimen(s) were sent to the ASNase reference laboratory for Ab(+) and ASNase activity determinations. The results were relayed to the referring institution free of cost.

5.3.3 Clinical Sample Testing

Enzymatic activity assay and Ab(+) ELISA assay quantification assays for the determination of ASNase therapeutic activity had been developed with quality control for *E. coli* or ErW ASNase activity, anti–*E. coli,* anti-PEG-ASNase, and/or anti-ErW ASNase Ab(+) in human sera [3,14–16]. Briefly, serum *E. coli* or ErW enzymatic activity is measured using an enzymatic reaction that converts L-Asn to L-aspartate and ammonia in the presence of ASNase (Nesslerization of ammonia) [2,3,14–16]. Pre-ASNase (control) and post-treatment-ASNase serum specimens from ALL patients were collected during various phases of therapy and placed immediately in ice-water bath (0°C) to prevent ex vivo AA deamination. The linearity of the calibration lines, the lower limit of quantification (LLOQ), and the inter-and intrabatch accuracy and precision of these assays were excellent [2,3,14]. Anti–*E. coli* PEG-ASNase or anti-*Erwinase* Ab(+) titers were measured using an ELISA assay in serum, which is a multistep assay in a similar manner to that of native *E. coli* ASNase with minor modifications for the accommodation of the new protein size and computerized equipment [2,3,11–16]. This Ab(+) assay uses the VECTASTAIN ABC kit, which was obtained from VECTOR Laboratories, Burlingame, CA. The ELISA Ab(+) assay used the same negative control serum as Ab(−); thus, all of these Ab(+) specimens were expressed as ratios of Ab(+) over the same denominator of Ab(−). Therefore, the within-day, within-month, and within-year variance of ratios of known Ab(+) over Ab(−) was extremely small. Details of this ELISA Ab(+) assay on accuracy and precision, upper and lower limits of quantification, can be seen elsewhere [22].

5.4 Results

5.4.1 Response and Function of Linearity between ELISA and Biacore Ab(+) Assays

In order to compare SPR and ELISA Ab detection methods of life-threatening human-antibacterial Abs [Ab(+)], sera from six individuals were analyzed for anti–*E. coli* ASNase Ab(+). A typical depiction of Ab(−) sera from six individuals who have very low RU—Ab negative as well as its wide Ab(+) range between low and high titer anti–*E. coli* ASNase Ab(+)—is shown in Figure 5.1. Among them are two patients with ALL who received pegaspargase and then developed pancreatitis, a toxicity associated with higher levels of ASNase enzymatic activity, resulting in Ab(−) status; two healthy volunteers never having been exposed to ASNase; and two patients with other forms of cancer, never having been exposed to ASNase. After 20 assay cycles, all of these samples had extremely low RU (Figure 5.1). Ab(+) patients had much higher and wider RU range (Figure 5.1). Patients represented with the solid triangles and solid square points who had the lowest and highest Ab(+) at 2750 RU, respectively, are demonstrating the dynamic range of the SPR Ab(+) assay. Moreover, data from these patients show the reproducibility of the wide dynamic range of Ab(+) detection by this method over the same number of cycles. Thus, the SPR Ab(+) assay can decipher at very low RU values an Ab(+) from the Ab(−) sera (Figure 5.1), whereas the ELISA assay has limitations.

Figure 5.2a shows the excellent linearity between $\log_{10}$ Ab ratio (ELISA assay) vs. the $\log_{10}$ Ab(+) at dilution of 1:1000 of original serum specimen. The ELISA Ab(+) assay line is determined in triplicate in $n = 3$ separate experiments of the same stock but with separate Ab(+) dilutions or total $n = 9$ OD observations. The 95% confidence intervals (CI) of the "mean of means" are superimposable with the best-fit line indicating the high accuracy and reproducibility of the ELISA Ab(+) assay. Figure 5.2b shows the relationship between the $\log_{10}$ RU by Biacore Ab assay vs. the $\log_{10}$ Ab ratio obtained from the abscissa of Figure 5.2a. With the exception of one set of data which fall outside the 95% CI, the linearity between ELISA and SPR (Biacore) methodologies is validated.

5.4.2 Immunoassay "Cutoff Point" Determination

It is clearly understood that there are no perfect assays of any kind. As mentioned previously, all assays and especially bioassays have inherited errors. These come from imperfections of the electrical instruments, the variability of the power voltage, and the accepted measurement error in making these

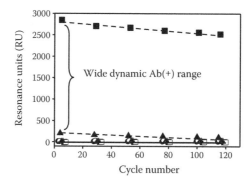

FIGURE 5.1 Biacore Ab assay raw RU data showing the repeated evaluation of sera specimens from very-high-Ab(+) data (solid squares) assayed 20 times. At the same time period in many sera from a low Ab value determined by ELISA assay, from six control sera—two patients with pancreatitis (verified to be Ab[−] by ELISA), two patients with solid tumors not having received ASNase, and two healthy volunteers—were determined in sequence to be Ab(−) (open circles and squares–filled triangle symbols near the bottom of the figure). Of note is the reproducibility of RU by the SPR assay on these six Ab(−) sera, which were repeated 20 separate times. Of interest is that although these Ab(−) sera specimens were assayed and intermingled with many Ab(+) sera, the results indicated that there was no residual no Ab(+) "memory" by the SPR protein chips after the "wash out" phase post-Ab(+) sera determinations.

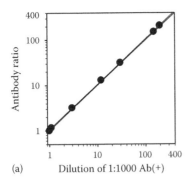
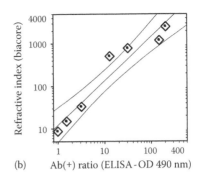

(a)   Dilution of 1:1000 Ab(+)   (b)   Ab(+) ratio (ELISA - OD 490 nm)

FIGURE 5.2 (a) The excellent linearity between $\log_{10}$ Ab(+) ratio over negative control by ELISA assay vs. the $\log_{10}$ Ab(+) at dilution of 1:1000 of original serum specimen. The best-fit line is determined in triplicate of $n = 3$ or total $n = 9$ OD observations. The 95% confidence intervals (CI) of the mean of means are superimposable with the best-fit line of mean of means, indicating the high accuracy and reproducibility of the ELISA Ab(+) assay. Of interest is that the Ab(+) assay at 1:500 or 1:100 dilutions had no effect on the linearity or reproducibility of the assay. (b) The relationship between the $\log_{10}$ RU by Biacore SPR Ab(+) assay from the same Ab(+) stock solution vs. the $\log_{10}$ Ab(+) ratio obtained from the abscissa of (a). The linearity between ELISA and Biacore SPR Ab(+) methodologies is well established.

biological samples, measuring their volumes, etc. This is the reason for using statistics in particular probability determinations to minimize these errors to a manageable level so that this can provide accurate and reproducible data.

The cutoff point is a statistical point where the linearity of the calibration line is lost or changes in direction. During this process, the slope of the line declines in its steepness and becomes a curve or the slope: Δoptical density/Δconcentration changes. In other words, it is the point where the experimentalist makes a decision point, below which a lower biological value has changed the slope significantly; hence, the point is considered "negative." These assays, once developed, are used by technologists who need to know where to declare a "cutoff" value positive or negative. When all values are high or very high, this is of no concern; however, when the values of an analyte by an assay are relatively low, then the statistical probabilities guide us as to the 95% probability of being correct, either positive or negative value. This is the significance of the cutoff point, which must be determined for each analyte and for each assay.

Thus, the important value of "cutoff point" potentially differentiates positive from negative samples in any bioassay. The ELISA Ab(+) assay response from these 84 Ab(+) patients ranged from <1 (0.85) to >500 ratio over negative control Ab(−) serum. For the SPR method, the RU ranged from ~8.1 to 2800 RU. Three of 84 patients (3.6%) who were Ab(+) by ELISA were found not to be IgG anti-ASNase Ab(+) by the SPR method. Of interest is that these samples were not neutralizing ASNase enzymatic activity ex vivo.

Moreover, sera donors with Ab(−) values were resolved precisely for both SPR and ELISA assays against the respective proteins (Figures 5.1 and 5.3). As in every comparative set of assays, very few specimens were shown to contain low affinity Ab(+) against both *E. coli* and ErW ASNase antigens. These sera from one patient were excluded. Results from the remaining samples were transformed ($\log_{10}$–$\log_{10}$) as necessary in order to make the data easier to be understood (Figure 5.4). In this figure, three serum samples which had been determined to be Ab(+) by the ELISA assay (black squares) appear to be Ab(−) by the SPR anti–*E. coli* ASNase assay. There appears to be a very narrow window of ambiguity between the Ab(−) (open circles) and the Ab(+) values by the ELISA assay (black squares below the 8.0 RU line), three of which are Ab(+) by the SPR method (gray squares) with values well exceeding 8 RU units.

The "cutoff point" of Ab(+), or the lowest limit of quantification (LLOQ), below which all values were considered as Ab(−), was then calculated as 8.1 RU and 1.003 ratio over (−) control for the SPR and ELISA AB(+) assays, respectively (Table 5.1, Figure 5.4). These values were statistically significant from the Ab(+) at $p > 0.05$ (Table 5.1). Of interest is that in the past evaluations, a ratio of <1.05 by ELISA assay method was considered a "cutoff point" for Ab(−) determinations [2,3,9,10].

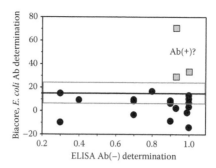

FIGURE 5.3 Comparison of the ELISA and Biacore SPR antigen–Ab interaction methods in Ab(−) sera (solid black circles) for the native anti–*E. coli* ASNase Ab. In this experiment, all sera specimens determined to be Ab(−) by ELISA assay were evaluated by SPR assay. The majority of them were found to be Ab(−) by this assay method as well. However, there are three serum samples (gray squares) which had been determined to be Ab(−) by the ELISA assay, but they appear to be Ab(+) by the Biacore anti–*E. coli* ASNase assay. These sera specimens (gray squares) were neutralizing ex vivo 1 IU of ASNase antigen over 15 min. Therefore, the SPR Ab(+) method is clearly segregating the Ab(+) from the many Ab(−) values.

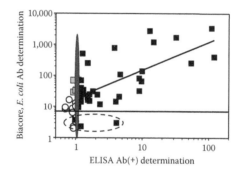

FIGURE 5.4 Comparison of the ELISA and Biacore SPR antigen–Ab interaction methods in Ab(+) sera for the native anti–*E. coli* ASNase Ab. The cutoff point is the statistical value, above of which there are Ab(+) and below of which are Ab(−) points. The "cutoff" point for the SPR method was at 8 RU units indicated as a straight line. For the ELISA was 1.003 ratio over negative control. The majority of the Ab(+) sera by ELISA assay (solid black square symbols) were also Ab(+) by SPR assay. Three serum samples which had been determined to be Ab(+) by the ELISA assay (black squares) appeared to be below the "cutoff" line; thus, they are Ab(−) by the Biacore anti–*E. coli* ASNase assay. There appears to be a very narrow window of ambiguity between the Ab(−) (open circles) and the Ab(+) values by the ELISA assay (black squares below the 8 RU line), three of which are Ab(+) by the SPR method (gray squares) with values well exceeding 8 RU units.

This Ab(−) value is very close to the mean + SDEV of the current Ab(+) determination by ELISA assay (ratio of 1.003 + 0.069).

5.4.3 Comparison of the ELISA and Surface Plasmon Resonance for Assessing Clinical Immunogenicity of IgG Antibodies

Every bioassay has inherent limitations and so do ELISA determinations of macromolecules (antigen-Abs). In order to be as accurate as possible in the "true" determination of a patient's serum being anti-ASNase Ab(+), we had to resort to determining the Ab assay by ELISA and the percent neutralization of the ex vivo–added antigen (1 IU of ASNase), a process which requires kinetic biochemical determinations; hence, it is very time consuming. Therefore, when the SPR method became available, it was felt that it had to be used to either validate or reject the existing database obtained by ELISA Ab assay. To this end, the comparison of the two protein–Ab interaction methods is shown in Figure 5.3 for the native anti–*E. coli* ASNase Ab(+) (Figure 5.3 and Table 5.1). The mean (±SDEV) Ab ratio in the Ab(−) patient samples (black circles—Figure 5.3) of 1.003 ± 0.69 by ELISA compares with 8.1 ± 16.17 RU

TABLE 5.1

Comparisons of ASNase-Antibody Ab(−) vs. Ab(+) and RU and Off-Rates from the Embedded
Immobilized Samples in Ab(+) and Ab(−) to *E. coli* (a) and ErW (b) ASNase

	ELISA (OD)	Biacore (RU)	Magnitude *E. coli* ASNase (RU)	Off-Rate *E. coli* ASNase (1/s)	Magnitude ErW ASNase (RU)	Off-Rate ErW ASNase (1/s)
(a) *E. coli*						
Ab(+) $n = 53$		*E. coli*	*E. coli*	*E. coli*	ErW	ErW
Mean	4.84	255.74	241.86	0.007	13.88	0.012
SEDEV	10.16	767.43	771.73	0.046	16.85	0.018
%CV	210.05	300.08	319.08	662.34	121.41	146.78
Variance	103.22	588941.8	595557.7	0.0021	283.77	0.00033
Ab(−) $n = 24$						
Mean	1.003	8.1	20.89	0.0025	3.13	0.0075
SEDEV	0.69	16.17	57.25	0.013	136.75	0.012
%CV	68.77	199.75	274.08	523.01	4374.158	163.06
Variance	0.476	261.51	3277.27	0.00017	18701.75	0.00015
T-test, $p =$	>0.05	0.043	0.043	>0.05	>0.05	>0.05
(b) *ErW*						
Ab(+)+ERW $n = 17$						
Mean		41.37	25.41	0.0019	29.29	0.005
SEDEV		50.92	54.93	0.008	8.65	0.0018
%CV		123.08	216.16	418.26	29.52	39.12
Variance		2592.71	3017.36	6.39E − 05	74.77	3.13E − 06
Ab(−)+ERW $n = 2$						
Mean		6.85	3.5	0.006	6.85	0.0003
SEDEV		2.47	2.97	0.003	2.47	0.0035
%CV		36.13	84.85	47.14	36.13	1288.51
Variance		6.13	8.82	6.77E − 06	6.13	1.21E − 05
T-test, $p =$		0.013	>0.05	0.048	0.0003	>0.05
T-test, $p =$		0.047	0.046	>0.05	>0.05	0.086

Note: T-test$_1$ = test between Ab(+) and Ab(−); T-test$_2$ = T-test between ASNase Ab(−) in 24 patients (*E. coli*) and in 2
patients in ErW.

from the SPR method. In contrast, the Ab(+) sera specimens in this group of patients' sera had a mean
of 4.84 ± 10.16 ELISA assay and 255.74 ± 767.43 RU for the SPR methods, respectively, which, despite
the wide variance, were significant values ($p = 0.043$). For ErW ASNase, the mean Ab(+) values were
41.37 + 50.92 with the SPR method, and for the Ab(−), the RU values were 6.85 ± 2.47 (Table 5.1). In
two other patients who had been treated with both *E. coli* and ErW ASNase formulations and had an
obvious clinical allergic reaction, the RU of 6.85 ± 2.47 by SPR was not indicative of an anti-ErW Ab(+).
However, the RU for *E. coli* ASNase is strongly suggesting that this Ab(+) may be responsible for the
clinical allergic reactions (Table 5.1). Eighteen out of 21 (18/21 = 85.71%) samples which had been determined to be Ab(−) by ELISA method (<1.05 ratio) are well below an RU value of 8.1. Moreover, there
are three serum samples (gray squares) which had been determined to be Ab(−) by the ELISA assay,
but they appear to be Ab(+) by the Biacore anti–*E. coli* ASNase assay. These sera specimens were neutralizing ex vivo 1 IU of ASNase antigen over 15 min. However, 3 out of 21 specimens have had higher
RU value from 28 to 70 units [Ab(+) by SPR method], strongly suggesting that there was Ab(+) in these
specimens (Figure 5.3).

Figure 5.4 represents the comparison between Ab(+) ratio as determined by ELISA assay and RU units
as determined by the SPR method in the same Ab(−) and Ab(+) samples [$n_1 = 24$ Ab(−) and $n_2 = 54$ Ab(+)]

(Table 5.1). In addition to the three Ab(−) specimens by SPR as depicted in Figure 5.3, there are three additional specimens which had been determined to be Ab(+) by ELISA, and they are shown to be Ab(−) by SPR method as noted in the graph. Despite the narrow range of ambiguity around the 1.05–1.1 Ab(+) ratio values, the vast majority of the specimens (93.2%) were determined to be Ab(+) by either ELISA or SPR determination as illustrated by being above the LLOQ in the best-fit line for the square symbols, thus demonstrating a positive Ab(+) by SPR method, and beyond the gray ellipsis indicating Ab(+) by ELISA method (Figure 5.4).

5.4.4 ALL Patient Ab(+) Determination

Many Ab(−) patients by ELISA and volunteer samples were Ab(−) via the SPR assay over 120 cycles run overnight. The Ab(−) values are shown in Figure 5.1 and Table 5.1.

Preexisting Ab(+) sera (ELISA assay method) were selected and detected by the SPR method in 84 patients with evaluable baseline sample [ELISA Ab(+) assay]. All subjects exhibited IgG anti–*E. coli* ASNase Ab(+), which were reactive and tightly bound with the *E. coli* antigen as determined by the values of the association rate constant (K_a). All these samples cross-reacted with PEG-ASNase. Of interest is that only 1 out of 84 patients' sera cross-reacted with the ErW ASNase antigen as well (1.2%, Figure 5.5). These ALL patients had received more than three cycles of chemotherapy that contained either native or PEG-ASNase and were presented with obvious moderate or severe clinical allergic reactions.

5.4.5 Cross-Reaction between IgG and Neutralization of ASNase Antigen

The interaction of anti-ErW Ab(+) and anti–*E. coli* ASNase Ab(+) with the respective antigens is shown in Figure 5.5. The dissociation or the "off-rate" of the ASNase-Ab complexes from the immobilized proteins in the SPR sensor protein chips, expressed in inverse seconds, clearly demonstrates that the anti–*E. coli* ASNase Ab(+) is very tightly bound to its antigen as illustrated by the low degree slope of RU over time and up to 300 s. The near-horizontal line after the peak Ab(+)-antigen binding suggests a very tight interaction with low rates of random dissociation. At that time, the injection of the IgG subtype human Ab was introduced into the system which identified the Ab(+) as IgG type only. The time-dependent RU

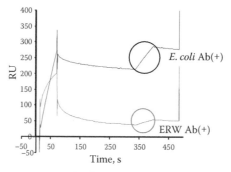

High affinity to *Erwinia*
IgG neutralizing *Erwinia*

FIGURE 5.5 The depiction of the time-dependent RU units for a serum sample by the Biacore Ab(+) assay after the injection of a specific human IgG-4 Ab at 300 s, showing the immediate increase in the RU values, strongly suggesting that the IgG Ab(+) to *E. coli* ASNase is reacting (neutralizing) and that the same IgG serum also cross-reacts and neutralizes the ErW ASNase immobilized in the protein chip (weaker signal, lower line). Based on the relative slope of the increased post-IgG injection into the Biacore unit, the Ab(+) in the patient's serum specimen is raised primarily against *E. coli* ASNase, which apparently evolves and cross-reacts with the ErW ASNase antigen. However, after the end of the IgG injection, the RU values during the wash-phase for either ASNase formulation is parallel, indicating that the off-rates are similar to either *E. coli* or ErW ASNase Ab bound complex.

units for a serum sample by the SPR Ab(+) assay method after the injection of a specific human IgG Ab (IgG-4) at 300 s is shown by the increase in the RU values, strongly suggesting that the IgG Ab(+) to *E. coli* ASNase is neutralizing (Figure 5.5). The neutralization had been validated by the loss of ex vivo *E. coli* ASNase enzymatic activity. In addition, the same IgG serum also cross-reacted and neutralized the ErW ASNase immobilized in the follow-up SPR protein chip (Figure 5.5, lower line). In other words, based on the relative slope of the increased post-IgG injection into the Biacore T-100 unit, the Ab(+) in the patient's serum specimen is raised primarily against *E. coli* ASNase, which also cross-reacted with the ErW ASNase antigen. However, after the end of the IgG injection, the RU values during the wash-phase for either formulation were parallel, indicating that the dissociation or the "off-rates" are similar to either *E. coli* or ErW ASNase Ab(+) bound complexes.

5.4.6 Ab(+) Determination in Patients Treated with Erwinase

In addition to the *E. coli* ASNase–treated patients, there were a handful who received ErW ASNase ($n = 17$). In this population, the anti-*Erwinase* IgG Ab(+) is statistically higher than in the Ab(−) to *E. coli* who received ErW ASNase or in the Ab(−) patients' sera who did not. This means that ErW ASNase administration is contributing to the production of almost immediate anti-ErW ASNase IgG Ab(+) formation by the already activated immune response system of the patients toward this bacterial protein with relatively high degree of AA homology to *E. coli* ASNase. However, in these few patients, the Ab(+) was non-IgG and, hence nonneutralizing. In addition, in the Ab(−) patients who receive ErW ASNase, the dissociation (expressed as Koff) range of a nonspecific IgG immunoglobulin was dissociated 3.16-fold faster than in the Ab(+) to *E. coli* formulations. In spite of the wide number of observations, the lower Koff rate for ErW ASNase was statistically significant ($p = 0.04$).

Similarly, the RU magnitude and the Koff of the off-rate of ErW ASNase from its antigen were 4.4 and 2.8 times faster in the Ab(−) than Ab(+), respectively. Of interest is that none of these patients received ErW ASNase antigen ($p > 0.05$). This may indicate that once an IgG Ab(+) against one bacterial ASNase formulation is developed by the host, the same or an evolved isotype IgG Ab(+) is most likely to recognize and bind to variable degrees (higher or lower) to a similar family of bacterial antigens, to which these patients have not been exposed. Most importantly, one patient among the 53 Ab(+) to *E. coli* ASNase (<2%) had a strong binding (neutralizing) to IgG against the ErW ASNase antigen. This patient had not received ErW ASNase to our knowledge in the clinic. Overall, the anti–*E. coli* ASNase IgG Ab(+) and Ab(−) sera compared very well with either the ELISA and the SPR Biacore Ab methods. However, the later method was rapid, accurate, 10-fold more sensitive, and reproducible, hence, a better bioassay in identifying the neutralization of the IgG Ab(+).

5.4.7 Dissociation of Ab(+) from Its Antigen

The dissociation or the "off-rate" of Ab(+) from its Ab(+)-antigen complex binding site is very slow as depicted by the slow rate of release (Table 5.1a and b). Moreover, after the IgG subtype of human Ab(+) was injected into the system, the anti-ErW ASNase Ab(+) is bound relatively tightly to its antigen with a very slow dissociation of this Ab(+) from its Ab(+)-Antigen complex site (Figure 5.5, Table 5.1). This is expressed as 1.7-fold tighter binding between the ErW ASNase and *E. coli* ASNase IgG Ab(+) (not significant). This experiment demonstrated the high affinity of anti–*E. coli* ASNase Ab(+) toward ErW ASNase, indicating a rare yet possible cross-reaction with this antigen in patients with severe clinical allergy postexposure to *E. coli* formulations. This evidence suggests that there is a maturation or random evolution of Ab(+) from one ASNase protein tetramer to another from a different bacterial species. This occurred in one patient who had a severe to life-threatening clinical allergic reaction to PEG-ASNase. Lastly, this IgG Ab(+) is neutralizing both *E. coli* and ErW ASNase. Therefore, it is not surprising that the dissociation or "off-rate" values are not substantially different (Table 5.1), in other words, it is just as a tight bound with low dissociation. Perhaps the dissociation rate is not an ideal biomarker tool for the SPR Ab(+) assay, unless one wants to identify low-affinity binding Abs in order to show early allergic clinical response to ASNase therapy.

5.5 Discussion

The resolution and advances of DNA arrays have enriched scientists in the many intricacies of pathway-specific complexities of tumor growth and proliferation. To counter this, many advances on biology oncology products, either in the form of purified native or pegylated proteins like growth factor Abs (PEG_ASNase, Avastin™, etc.), have made measurable improvements in the lives of oncology patients. ASNases, like the recombinant DNA-produced proteins or monoclonal antibodies (mAbs) against many growth factor ligands and/or their receptors, require a new approach in developing accurate pharmaco-analytical methods in determining the key biomarkers for the evaluation of these oncology products in clinical trials. These antigen or Ab assays must be developed and validated with different assay methods in support of the biomarker assessment in clinical trials and eventual proof of the biology product's usefulness.

The ELISA anti-ASNase Ab(+) qualified assay has been used to evaluate and monitor HABA Ab(+) in patients post *E. coli* or ErW ASNase administration in ALL patients [2–4,11–16,22]. The presence of the Ab(+) in patients with hypersensitivity reactions to either of these ASNase formulations leads to a more rapid clearance of the enzymatic activity (neutralization); hence, the half-life of both native *E. coli* and PEG-ASNase preparations are reduced substantially [3,22–26]. The vast majority of these assays determine the IgG isoform of the polyclonal HABA Ab(+) and all show a clear advantage of Ab(+) determination vs. obvious clinical allergic observation. Today, only one study has correlated the Ab(+) subpopulation of ALL patients with inferior outcome [3].

The clinical presentation in these patients' allergic symptoms ranged from localized skin rash and swelling to respiratory difficulties and, very rarely, anaphylaxis. Neutralizing anti-PEG-ASNase Ab(+)s were seen in 85% of these patients [23]. The results were examined with respect to age, gender, ASNase formulation received, time of post-ASNase administration, and the physician's clinical observations. Therefore, the observations made by the clinical staff were 2σ (2 *sigma*) accurate.

A wide range of IgG Ab(+) sensitivity from 1 to 2800 RU was achieved by the SPR Ab(+) assay, although the majority of the Ab(+) in this group of serum specimens was in the 20–500 RU range. There was a clear distinction between the Ab(−) serum values over 20 h of repeated assay time and the low Ab(+) RU values (Figure 5.1). The linearity of either bioassay method was excellent, as was their direct comparison even at 1:1000 standard Ab(+) dilutions (Figure 5.2), whereas Figures 5.3 and 5.4 show the results from the cross-validation of these assays.

Protein depletion experiments with human serum (washing step) revealed that the observed clinical responses in the majority of these patients were predominantly caused by specific IgG isoforms binding to its ligand. This isoform of Ab(+) is theorized to be the cause for the allergic clinical reactions that had been documented post-*E. coli* ASNase formulation administrations [2,3,11–16]. The variable immune type of clinical allergic responses was defined by a peak response from a few to many hours' post-ASNase administrations (Type II immune reactions). These were managed by standard symptomatic care in the various clinics. In a few occasions, these responses did not reappear even after PEG-ASNase repeat administrations a few days or weeks later when the antigen was still present in some fashion in their system. In other patients, the repeat administration of the pegylated ASNase antigen caused more severe clinical allergic reactions, requiring discontinuation of therapy with this drug. Therefore, these severe types of HABA Ab(+) responses to *E. coli* ASNase were regarded as critical for the patient's safety and well-being, thus requiring appropriate intervention, and for the continuation of the antileukemia combination regimen treatment with the alternative ErW ASNase [2,3]. The vast majority of the IgG Ab(+) (85%) by the ELISA method was neutralizing, suggesting that the circulating Ab(+) negated the pharmacodynamic (PD) efficacy against ALL cells [2,3].

Among the Ab(+), 14 patients had cutaneous manifestations (rash, hives, urticaria). All of these had neutralizing moderate to low Ab(+) ratios by ELISA assay followed by lack of enzymatic activity or Asn and Gln depletion in the same specimens. These patients' sera had low RU values by SPR method. All other patients who had localized or generalized swelling had higher Ab(+) ratios by ELISA and higher RU values by the SPR methods. Neutralizing Ab(+) was correlated with much lower or absent ASNase enzymatic activity. Lower Ab(+) ratios may be associated with earlier collection time points in the anamnestic response. In the remaining 15% of patients with an apparent clinical reaction, we found no Ab(+) nor substantial loss of ASNase activity. Therefore, monitoring ASNase activity and Ab(+) may be a useful tool for guiding oncologists with optimal ASNase therapy. Patients with no Ab(+) and substantial enzymatic activity might be rechallenged with the same product despite an apparent allergic reaction [3,23].

In the comparison, the advantage of SPR vs. ELISA Ab(+) assays is that SPR is a novel protein-chip technology which provides fast, reproducible, accurate, and ~10 times more sensitive results than the ELISA Ab(+) assay. Hence, SPR is a better bioassay in identifying the neutralizing IgG Ab(+). Rather than using a secondary Ab to detect the patient response to ErW ASNase as it is needed in the ELISA assay, the SPR technology allows for detection directly in the patients' serum without the need for secondary signals. This is an important advantage that relates to SPR protein chip's ability to detect weak-affinity Ab(+) by approximately 10-fold lower sensitivity that would be washed away in an ELISA Ab(+) assay. Moreover, the Biacore T-100 unit is an automated computer-controlled instrument allowing for analysis of many samples, thus reducing the cost per individual specimen assay, in a programmed manner for up to 48 h.

In this study, we evaluated the fact that the SPR method detected as Ab(−) nonneutralizing Ab, which had been determined Ab(+) by the ELISA method (false-negative Ab determinations) (Figure 5.4). However, a combination of ELISA and % neutralization of ASNase antigen assay was equally as successful in determining the neutralizing Ab(+) even at miniscule circulating relative concentrations as was the SPR Ab(+) method [22]. However, the cost of the second assay was a serious matter for consideration. Thus, these analyses were approximately equal. This evidence provided the external validation for either method, with the SPR being more sensitive and able to identify clearly the isotype of the neutralizing Ab(+) formed. During cross-validation, using the same Ab(+) and Ab(−) serum controls (blinded samples to the experimentalists), the SPR assay was shown to be again more sensitive in RU values than the Ab(+) ratio over negative control per ELISA Ab(+) assay. However, this cross-validation of these assays has yet to be evaluated in a prospective clinical trial. This unmet need, hopefully, will be fulfilled in the near future.

Of interest is that the difference in performance between these two Ab(+) assays is related to the real time analysis occurring in the SPR method compared with the fact that ELISA is an end-point analysis. Therefore, due to their faster dissociation rate constants (Koff), low-affinity Ab(+) are more likely to passively bind and release the protein-drug during the wash cycle of an ELISA Ab(+) assay. However, this is not possible to occur in the SPR assay, which detects the Ab/antigen complexes throughout the binding and release. Moreover, the ELISA assay favors high-affinity Ab(+), which is more likely to maintain a successful bridge formation between protein–antigen and its human antibacterial Ab(+). Conversely, low-affinity Ab(+) are more likely to either not bind or, if they bind to antigen, to dissociate from the surface-bound protein-drug and return to the soluble phase, hence resulting in a decreased sensitivity and/or increased Ab(+)-antigen interference, thus resulting in a false-negative Ab assay result.

Furthermore, and despite being a relatively 6 month old serum samples (stored properly) for the detection of polyclonal Ab(+) against *E. coli* ASNase, the SPR assay identified more neutralizing anti-ASNase Ab(+) than the ELISA method, especially at the low end of the Ab(+) quantification (Figure 5.4). Due to the very low "noise" of the Biacore T-100 unit, the range between LLOQ and below the LLOQ (BLLOQ) RU signal is very wide so that the probability of a false-negative result is diminished. This evidence suggested that screening immunoassays for biology oncology products may have been detecting different isotypes of Abs (IgM or IgA but not IgG). Therefore, appropriate human immunoglobulin standards of these isoforms can resolve this issue. Thus, the hypothesis that can be put forward is that the SPR assay detects more Ab(+) samples because of its ability to detect low-affinity Ab(+) against ASNase, which was demonstrated by the data (Figure 5.4).

One of these ALL patients who had been reported to have had severe to life-threatening clinical allergy reactions post-PEG-ASNase, but with moderate IgG anti–*E. coli* ASNase Ab(+) levels, showed an IgG AB(+) binding response against ErW ASNase antigen as well, even thought the patient had not been exposed clinically to this ASNase formulation (Figure 5.5). All of the Ab(+) sera by SPR method clearly showed that this was an IgG Ab(+) to *E. coli* ASNase formulations, but it remains to be tested for the IgM and/or IgE cross-Ab(+) reactions with these specific antigens. Naturally, the later Ab(+) isotype is of greater interest in that its presence is of serious concern for Type I immunological reaction upon reexposure to the ASNase antigen, which the patient did not exhibit.

In recent years, monoclonal Ab and recombinant DNA pure protein oncology antigens are often used in high concentrations in oncology patients, i.e., for ErW ASNase $25,000 IU/m^2 \rightarrow$ for a large-BMI patient that approximately equals or exceeds $50,000 IU/dose$ ($\times 6$ per treatment) and/or typically the new linker pegylated ASNases that have long half-lives of elimination. Therefore, protein–antigen–Ab interactions are of great interest. Most interestingly, at the same time, they present with a challenge when one evaluates the immunogenicity of these oncology products, thus making the precise clinical evaluation of the Ab(+) a necessity rather than a cost-cutting luxury in the clinics. In our immunosurveillance evaluations, the ability of the ELISA method has failed to explain fully the observed clinical results [22,23]. In seeking more advanced methods of Ab(+) detection, we delved into the protein-chip SPR method. To our pleasant surprise, the advantage of the SPR Ab(+) assay is that it was able to detect and characterize Ab(+) with different subtypes of immunoglobulins rapidly, reliably, and with greater sensitivity and accuracy. Moreover, it is possible to distinguish different antigen–Ab(+) affinities, expressed as RU values, which may be more likely to be correlated and possibly explain the severe clinical allergic presentations.

Lastly, as it has already been recommended [20,21], our results [22] suggest that careful consideration must be given both by the clinical investigators and the regulatory agencies to the assay formats and procedures applied to immunogenicity testing in support of biological drug development. The use of SPR Ab(+) method platform assures that low-affinity polyclonal Ab(+) will be detected as well as their isotype, although cross-validation with ELISA methods should occur to enable accurate and reliable Ab(+) detection. Since time is critical in determining Ab(+) or Ab(−) after a clinical allergic reaction in a patient is presented, the SPR assay holds an advantage in that "real-time" accurate RU values of a potentially threatening IgG Ab(+) can be produced.

Finally, the knowledge of Ab–antigen dissociation is of great importance, especially now that we have access to additional protein formulations of ASNases. The future of determining immunogenicity to biology oncology and similar antigenic oncology products by the very sensitive and specific SPR Ab(+) detection method is now available for use in future clinical trials. Time only will tell if this expensive research effort really translates into benefits of ALL patients in the clinic.

5.6 Where the Technology Is Headed and Future Applications

Cancer cell proliferation depends on protein synthesis via specific protein biosynthesis pathways. This is especially important in homeostatic, cell cycle, and oncogene gene expression into proteins, which then drive the cellular machinery for growth and metastasis. Considering the significance that protein expression exerts to cancer cell growth and metastasis, unfortunately, there are few drugs licensed against protein biosynthesis in the chemotherapy-stressed malignant cells. One such class of important antileukemia and antilymphoma drugs is the bacterial ASNases from *E. coli* and *Erwinia chrysanthemi*. The important contributions of ASNases have been to increase the outcomes and improve the quality of life of leukemia patients, a subject which has been documented in many publications. However, the unmet need in this field of oncology is the difficulty in developing useful biomarkers in order to determine their pharmacological usefulness and success during treatments. One of the invidious position a patient can be is the one with silent hypersensitivity due to anti-ASNase Ab(+) without obvious clinical allergic symptoms. The presence of the Ab(+) nullifies the pharmacodynamics of the ASNase drugs; however, there are alternative intense chemotherapy regimens one can offer to these patients. Thus, knowing the Ab(+) status to ASNases by the clinical attending can be instrumental to the patient's well-being and survival.

Moreover, new recombinant DNA ASNase proteins are being tested in patients with excellent results [27–29]. Hopefully, these purer recombinant-DNA proteins will diminish the Ab development and, hence, the need for determining anti-ASNase Ab(+) in the clinic. However, until these new products become widely available in leukemia and lymphoma clinics, one has to deal with the best possible manner in determining Ab(+) for the potential improvement of patient treatment and outcomes. To address this unmet need, the ELISA anti-ASNase Ab(+) was developed long time ago and used in a number of Phase III clinical trials. However, because of the high cost in performing these assays "on demand" and "in real time," they became very expensive for routine use, therefore ineffective to be offered to the oncologists at large. In order to address the imperfection of the ELISA Ab(+) assay, the protein-chip SPR Ab(+) method was developed and the two methods were compared. Unfortunately, the cost of possessing the T-100 or an equivalent SPR unit is very high for an academic institution. Taking all the evidence together, one can conclude that the SPR protein-chip method with its improved sensitivity and reproducibility is the new way for the future Ab(+) determinations. This can only take place when one assumes that priorities get reciprocal to the cost of saving the lives in a considerable number of leukemia-lymphoma patients, especially those with clinical allergies and those with silent hypersensitivity. Thus, this manner of thinking is especially more important with the advent of multiple biology oncology products, many of which seem to induce Ab(+) in patients. All of these Ab(+) can be detected by SPR protein-chip technology. Therefore, it will be a wonderful day when this will become a reality for all patients.

ACKNOWLEDGMENTS

I would like to thank the many patients and hospital personnel who obtained the many patient samples; and all those who assayed the ELISA Ab(+); and the hospital volunteer who tabulated the data for this work to be finalized in its present form. Also, I would like to thank Children's Hospital Los Angeles, Division of Hematology/Oncology for the support and trust they bestowed on me, without which none of this would have been possible.

REFERENCES

1. Wadhwa W, Bird C, Dilger P, Gaines-Das R, Thorpe R. Strategies for detection, measurement, and characterization of unwanted antibodies induced by therapeutic biologicals. *J Immunol Methods*, 2003, 278:1–17.
2. Avramis VI, Panosyan EH. Pharmacokinetic/pharmacodynamic relationships of asparaginase formulations: The past, the present and recommendations for the future. *Clin Pharmacokinet*, 2005, 44(4):367–393.
3. Panosyan EH, Seibel NL, Martin-Aragon S, Gaynon PS, Avramis IA, Sather H, Franklin J et al. Children's Cancer Group Study CCG-1961. Asparaginase antibody and asparaginase activity in children with higher-risk acute lymphoblastic leukemia: Children's Cancer Group Study CCG-1961. *J Pediatr Hematol Oncol*, 2004, 26(4):217–226.
4. Gaynon PS. Childhood acute lymphoblastic leukaemia and relapse. *Br J Haematol*, 2005, 131(5):579–587.
5. Lee JI, Dominy JE Jr., Sikalidis AK, Hirschberger LL, Wang W, Stipanuk MH. HepG2/C3A cells respond to cysteine deprivation by induction of the amino acid deprivation/integrated stress response pathway. *Physiol Genomics*, 2008, 33(2):218–229.
6. Meiser A (Ed). *Biochemistry of Amino Acids*, 2nd edn., Vol. 1, pp. 457–460, Academic Press, New York, 1965.
7. Horowitz B, Madras BK, Meister A, Old LJ, Boyes EA, Stockert E. Asparagine synthetase activity of mouse leukemias. *Science*, 1968, 160(827):533–535.
8. Horowitz B, Meister A. Glutamine-dependent asparagine synthetase from leukemia cells. Chloride dependence, mechanism of action, and inhibition. *J Biol Chem*, 1972, 247(20):6708–6719.
9. Paquette JC, Guérin PJ, Gauthier ER. Rapid induction of the intrinsic apoptotic pathway by L-glutamine starvation. *J Cell Physiol*, 2005, 202:912–921.
10. Kishi S, Evans WE, Relling MV. Pharmacokinetic, pharmacodynamic, and pharmacogenetic considerations. In: *Childhood Leukemias*, C.-H. Pui, Ed., 2nd edn., pp. 391–413, Cambridge University Press, Cambridge, MA, 2006.
11. Woo MH, Hak LJ, Storm MC, Evans WE, Sandlund JT, Rivera GK, Wang B, Pui CH, Relling MV. Anti-asparaginase antibodies following *E. coli* asparaginase therapy in pediatric acute lymphoblastic leukemia. *Leukemia*, 1998, 12(10):1527–1533.

12. Woo MH, Hak LJ, Storm MC, Sandlund JT, Ribeiro RC, Rivera GK, Rubnitz JE, Harrison PL, Wang B, Evans WE, Pui CH, Relling MV. Hypersensitivity or development of antibodies to asparaginase does not impact treatment outcome of childhood acute lymphoblastic leukemia. *J Clin Oncol*, 2000, 18(7):1525–1532.

13. Wang B, Relling MV, Storm MC, Woo MH, Ribeiro R, Pui CH, Hak LJ. Evaluation of immunologic cross-reaction of anti-asparaginase antibodies in acute lymphoblastic leukemia (ALL) and lymphoma patients. *Leukemia*, 2003, 17(8):1583–1588.

14. Avramis VI, Sencer S, Periclou AP, Sather H, Bostrom BC, Cohen LJ, Ettinger AG et al. A randomized comparison of native *Escherichia coli* asparaginase and polyethylene glycol conjugated asparaginase for treatment of children with newly diagnosed standard-risk acute lymphoblastic leukemia: A Children's Cancer Group study. *Blood*, 2002, 99(6):1986–1994.

15. Douer D, Yampolsky H, Cohen LJ, Watkins K, Levine AM, Periclou AP, Avramis, VI. Pharmacodynamics and safety of intravenous pegylated L-asparaginase during remission induction in adult newly diagnosed acute lymphoblastic leukemia. *Blood*, 2007, 109:2744–2750.

16. Avramis VI, Martin-Aragon S, Avramis EV, Asselin BL. Pharmacoanalytical assays of Erwinia asparaginase (Erwinase) and pharmacokinetic results in high-risk acute lymphoblastic leukemia (HR ALL) patients: Simulations of Erwinase population PK-PD models. *Anticancer Res*, 2007, 27(4C):2561–2572.

17. Salzer W, Asselin B, Supko J, Devidas M, Kaiser N, Plourde PV, Winick N, Reaman G, Raetz E, Carroll W, Hunger S. Administration of Erwinia asparaginase (Erwinase®) following allergy to PEG-asparaginase in children and young adults with acute lymphoblastic leukemia treated on AALL07P2 achieves therapeutic Nadir serum asparaginase activity: A report from the children's oncology group (COG). In: *Proceedings of 52nd ASH Annual Meeting and Exposition*, Abstr# 2134, presented in December at ASH-2010, Orlando, FL, *Blood*, November 19, 2010, 116, p. 882.

18. Gurbaxani B. Mathematical modeling as accounting: Predicting the fate of serum proteins and therapeutic monoclonal antibodies. *Clin Immunol*, 2007, 122(2):121–124.

19. Jakobsen CG, Bodtger U, Kristensen P, Poulsen LK, Roggen EL. Isolation of high-affinity human IgE and IgG antibodies recognising Bet v 1 and Humicola lanuginosa lipase from combinatorial phage libraries. *Mol Immunol*, 2004, 41(10):941–953.

20. Lofgren JA, Dhandapani S, Pennucci JJ, Abbott CM, Mytych DT, Kaliyaperumal A, Swanson SJ, Mullenix MC. Comparing ELISA and surface plasmon resonance for assessing clinical immunogenicity of panitumumab. *J Immunol*, 2007, 178(11):7467–7472.

21. Swanson SJ, Mytych D, Ferbas J. Use of biosensors to monitor the immune response. *Dev Biol*, 2002, 109:71–78.

22. Avramis VI, Avramis EV, Hunter W, Long MC. Immunogenicity of native or pegylated *E. coli* and Erwinia asparaginases assessed by ELISA and surface plasmon resonance (SPR-biacore) assays of IgG antibodies (Ab) in sera from patients with acute lymphoblastic leukemia (ALL). *Anticancer Res*, 2009, 29(1):299–302.

23. Tiwari PN, Zielinski M, Quinn JJ, Siegel SE, Gaynon PS, Wakamatsu P, Seibel NL, Avramis VI. Assessment of anti-asparaginase (ASNase) antibodies (Ab) and ASNase activity after suspected clinical allergy. *Blood*, 2006, 108(11):532a, (Abstr # 1878).

24. Klug Albertsen B, Schmiegelow K, Schrøder H, Carlsen NT, Rosthøj S, Avramis VI, Jakobsen P. Anti-Erwinia asparaginase antibodies during treatment of childhood acute lymphoblastic leukemia and their relationship to outcome: A case-control study. *Cancer Chemother Pharmacol*, 2002, 50(2):117–120.

25. Albertsen BK, Schrøder H, Jakobsen P, Avramis VI, Müller HJ, Schmiegelow K, Carlsen NT. Antibody formation during intravenous and intramuscular therapy with Erwinia asparaginase. *Med Pediatr Oncol*, 2002, 38(5):310–316.

26. Asselin BL, Whitin JC, Coppola DJ et al. Comparative pharmacokinetic studies of three asparaginase preparations. *J Clin Oncol*, 1993, 11:1780–1786.

27. Wikman LE, Krasotkina J, Kuchumova A, Sokolov NN, Papageorgiou AC. Crystallization and preliminary crystallographic analysis of L-asparaginase from Erwinia carotovora. *Acta Crystallogr Sect F Struct Biol Cryst Commun*, 2005, 61(Pt 4):407–409.

28. Kotzia GA, Labrou NE. L-asparaginase from Erwinia chrysanthemi 3937: Cloning, expression and characterization. *J Biotechnol*, 2007, 127(4):657–669.
29. Pieters R, Appel I, Kuehnel HJ, Tetzlaff-Fohr I, Pichlmeier U, van der Vaart I, Visser E, Stigter R. Pharmacokinetics, pharmacodynamics, efficacy, and safety of a new recombinant asparaginase preparation in children with previously untreated acute lymphoblastic leukemia: A randomized phase 2 clinical trial. *Blood*, 2008, 112(13):4832–4838.

Part III

Optical Technologies for Cancer Detection and Diagnostics: Evanescent Wave and Waveguide Biosensors

6

Photonic Biochip Sensor System for Early Detection of Ovarian Cancer

Debra Wawro, Peter Koulen, Shelby Zimmerman, Yiwu Ding, Charles Kearney, and Robert Magnusson

CONTENTS

6.1 Introduction

Among gynecologic tumors, ovarian carcinoma has the highest lethality rate with about 14,600 deaths from ovarian cancer in the United States in 2009 [1]. Ovarian cancer is the fifth leading cause of cancer deaths among women and often goes undetected until it reaches an advanced stage. Ovarian cancer causes more deaths in the United States than all other female reproductive cancers combined [1]. Due to the availability of early detection tests and improved treatments, survival rates are much higher for other cancers such as breast or cervical cancer. However, less than 20% [1,2] of ovarian cancers are found at an early stage due to lack of reproducible and definitive diagnostic methods. Early symptomatic diagnosis is difficult as the typical symptoms (such as bloating, pelvic pain, and/or nausea) are common to a host of causes.

Determining the disease stage is very important since there are different prognoses and treatments for ovarian cancer as it progresses. Typically, the cancer stage is determined by the surgical removal of tissue samples from the pelvis and abdomen. Survival rates of 5 years (or more) are likely for women diagnosed with early-stage ovarian cancer. If not properly staged, cancer that has spread outside of the ovaries may be overlooked and subsequently not treated. Ideally, an accurate, nonsurgical diagnostic tool is needed to diagnose the presence of ovarian cancer and accurately discriminate between early and late stages such as metastatic versus primary forms.

In this work, we describe high-accuracy sensor chips and systems that provide effective detection of an array of biomarker proteins in blood and serum to accurately diagnose ovarian cancer as well as provide advanced information regarding the disease stage. While there are currently no established clinical diagnostic tools using urinalysis or seranalysis, experimentally and clinically identified targets [3,4] can be categorized into two groups: group 1 consists of biomarker proteins that are upregulated twofold or higher in metastatic over primary ovarian serous papillary carcinoma (such as collagen type I, calreticulin, fibronectin, and epidermal growth factor receptor), and group 2 consists of biomarker proteins that are upregulated twofold or higher in primary over metastatic ovarian serous papillary carcinoma (such as apolipoprotein A-1 and ryanodine receptor). This differentiation yields accurate diagnosis of the disease and staging information that can be used to monitor presymptomatic aspects of the disease, disease progression, and the efficacy of therapies.

6.1.1 Guided-Mode Resonance Sensor Technology

The technology is based on an optical resonance effect that occurs in periodic waveguides (also referred to as waveguide gratings) called the guided-mode resonance (GMR) effect. This effect occurs in subwavelength waveguide gratings (whereby the period of the thin-film waveguide is smaller than the wavelength of incident light). Fundamentally, this device operates as a high-resolution optical filter, reflecting or transmitting a narrow band of wavelengths at a specific angle [5]. As shown in Figure 6.1, illuminating a GMR sensor with a laser diode emitting a diverging beam results in a reflective peak at a particular angle. A slight variation in resonance conditions alters the angular position. Thus, interaction of a target analyte with a biochemical layer on the sensor surface yields measurable angular shifts that directly identify the binding event. The sensor element is prepared with standard surface chemistries to chemically attach a selective layer, such as antibodies. It is multifunctional as only the sensitizing antibody layer needs to be chemically altered to detect different species. Arrays of antibody spots (100–200 microns in diameter) can be utilized to detect hundreds of analytes on a single biochip element. Sensors responsive to picomolar (pM) concentrations have been demonstrated in our labs for antibody–antigen interactions. Repeatable fabrication processes are in place to produce the resonant grating sensor element in low-cost polymer materials. The entire system (including the resonant sensor elements) is designed to operate in a low-power, integrated format.

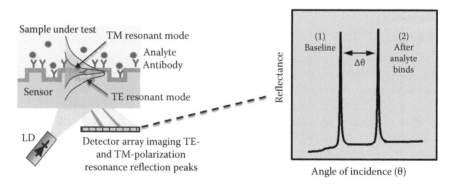

FIGURE 6.1 Schematic of a label-free GMR sensor chip element operating in reflection mode. The diverging (expanding) beam from the laser diode (LD) is incident on the sensor with a continuous range of angles. As binding events occur at the sensor surface, resonance peak changes can be tracked as a function of incident angle ($\Delta\theta$). The resonance occurs at different angles for TE and TM polarization states of the input light enabling high-accuracy, cross-referenced detection. The incident light excites photonic surface states shown as TE and TM modes. As schematically indicated, these modes interact differently with the surrounding media enabling the polarization-based differentiation.

Compelling attributes of this diagnostic technology include:

1. Label-free operation minimizes chemical processing.
2. Highly sensitive operation allows for a wide range of biomarker chemistries.
3. Sharp, well-defined resonance reflection peaks provide accurate, high-resolution data.
4. High signal-to-noise ratio enables the use of low-power light sources and detectors.
5. Capability for ultracompact, highly integrated sensor elements and systems.
6. Dense biochip format available for parallel multispecies detection with sensor arrays.
7. Capability for multiparametric sensing via polarization and modal diversity.

This sensor technology is broadly applicable to medical diagnostics, drug development, industrial process control, genomics, and environmental monitoring. In this chapter, the focus is on compact, rapid monitoring of disease markers in biological fluids.

6.1.2 Label-Free GMR Sensing Approach

Our resonance elements have compelling applications for biosensors as they are tunable on changes in device parameters such as sample density changes and/or thickness variations associated with a biochemical interaction. As the antibody binds with its matched antigen, the buildup of the attaching biolayer can be monitored directly without the use of chemical labels (such as fluorescent tags); instead, the corresponding resonance angular shift can be followed with a detector array [6–9] as shown in Figure 6.1. Before the sample containing the analyte is introduced on the sensor element, a baseline resonance response is measured (shown as peak [1] in Figure 6.1). This baseline can be measured in either neat buffer or air. After the sample to be measured is introduced and the analyte binds to the antibody layer on the sensor surface, the resonance shifts to a new angular location (shown as peak [2] in Figure 6.1). The resonance shift is monitored until the biochemical interaction stabilizes (typically between 15 min and 2 h). Using this approach, diagnostic tests can be conducted in a rapid manner; the test time is only limited by the binding rate between the selective layer (such as antibodies) and the targeted biomarker protein being detected. Standard immunoassay tests (such as enzyme-linked immunosorbent assays, or ELISA) involve extensive and complicated incubation and washing steps with results not obtained until 4–24 h after test initiation. By using GMR sensor technology, results can be obtained rapidly with no required washing steps. This greatly simplifies medical diagnostics, enabling doctor offices and hospitals to perform routine screening on a much larger scale with dramatically less labor, time, and cost.

The resonance response is sensitive to the incident light polarization (polarization is a fundamental property of light that describes the orientation of the wave oscillations). Thus, separate resonance peaks occur for incident TE (transverse-electric light: laser-light electric field that oscillates normal to the plane of incidence) and TM (transverse-magnetic light: magnetic field that oscillates normal to the plane of incidence) polarization states. This property can provide cross-referenced data that can be used to calibrate for variations such as temperature or sample background density simultaneously in the same sensor spot. Moreover, very conveniently, the layer can be designed to support additional resonant leaky modes, thereby providing additional resonance peaks for further increased detection accuracy and reduction of the probability of false readings [9,10]. Concurrent, colocalized data acquisition via this polarization and modal diversity eliminates errors generated by the use of separate reference sites. This property is a major advantage of GMR sensor technology. Our systems have target sensitivity in the low ng/mL (or pM) range to detect disease-induced alterations in protein concentration indicative of disease progression.

6.1.3 GMR Sensors versus Other Photonic Biosensors: Brief Review

The coupling of a freely propagating beam of light to a state of confinement at a surface is presently the subject of considerable research activity. Periodic structures with subwavelength features provide effective means of achieving such coupling. The resulting strong localization of energy at a dielectric (or metallic) layer is the basis for effective biosensors. Thus, subwavelength periodically modulated

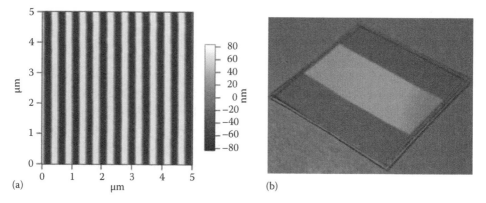

FIGURE 6.2 Submicron resonant grating. (a) Atomic force microscope (AFM) picture of a ~520 nm period grating contact printed in an optical polymer. (b) A submicron molded grating. The grating is coated with a thin high-index layer (TiO$_2$ or HfO$_2$) to realize a GMR sensor element.

dielectric layers can be applied to realize high-performance GMR sensors [5–12]. We fabricate GMR filters and sensors in dielectric media such as moldable polymers, fused silica, silicon dioxide, hafnium dioxide (Figure 6.2), silicon nitride, etc. The GMR biosensors used in this work are based upon a single-layer filter design fabricated using low-cost submicron molding methods. We utilize polymers that are imprinted with submicron grating patterns (~520 nm grating periods) and coated with a high-index dielectric material (such as TiO$_2$ or HfO$_2$) to realize resonant sensors. Figure 6.2 shows an example of a GMR sensor [8,9].

Numerous optical sensors for bio- and chemical detection have been developed commercially and in the research literature. Key label-free technologies include the surface-plasmon resonance (SPR) sensor, MEMS-based sensors, nanosensors (rods and particles), the resonant mirror, Bragg grating sensors, waveguide sensors, waveguide interferometric sensors, ellipsometry, and grating coupled sensors [13–16]. Other methods include immunomagnetic separation, polymerase chain reaction, and standard immunoassay approaches that incorporate fluorescent, absorptive, radioactive, and luminescence labels [15,16].

Although dramatically different in concept and function, the SPR sensor [13,14] comes closest in features and operation to the GMR sensor proposed here. The term surface plasmon (SP) refers to an electromagnetic field charge-density oscillation that can occur at the interface between a conductor and a dielectric (e.g., gold/glass interface). An SP mode can be resonantly excited by TM-polarized incident light but not TE-polarized light. Phase matching occurs by employing a metallized diffraction grating or by using total internal reflection from a high-index material such as in prism coupling or from a guided wave in an optical fiber. When an SPR surface wave is excited, an absorption minimum occurs in a specific wavelength band. While angular and spectral sensitivity is very high for SPR sensors, the resonance linewidth is rather large. Since only a single polarization (TM) can physically be used for detection, the refractive index and thickness attachments cannot simultaneously be resolved in one measurement. This is a particularly significant problem for portable diagnostic applications where thermal variations are probable.

With GMR sensor technology, multiple system architectures are possible. Systems can be designed to operate at any convenient wavelength and can be constructed using low-cost materials. By monitoring the angular resonance, very compact, low-power systems can be realized. Due to inherent polarization diversity, for each resonant mode, two narrow resonances shift their positions on the sensor surface when a bioreaction occurs, thereby providing the possibility of cross-referenced data. This information can be combined with rigorous theoretical models to backfit data to distinguish between environmental variations and the targeted biochemical binding events [9,10]. Additionally, due to a large dynamic range available during detection, this sensor can operate in gas or liquid environments. The basic sensor can be fashioned into dense arrays for effective identification of hundreds (or even thousands) of species in a single sample.

6.2 Materials

Resonant Sensors Incorporated supplies the sensor elements and the prototype sensor system. We purchase the optical components in the prototype system (850 and 650 nm laser diodes, polarizer, mounts, and lenses) from Thorlabs (Newton, NJ). We purchase some of the microwell upper structures from Nunc (Rochester, NY). In the experiments, we initially coat the sensor elements with the commercially available silane, 3-aminopropyltriethoxysilane (Pierce, Rockford, IL). We use disuccinimidyl suberate (Pierce, Rockford, IL) as a homobifunctional cross-linker to chemically attach the antibodies to the amino-silanated surface. We purchase the antibodies and proteins used in this work from several commercial sources. We prepare all antibodies and standards 30 min before use in neutral phosphate buffered saline (PBS). We use bovine serum albumin (BSA) as the blocking buffer. We purchase the cell lines TOV-112D (#CRL-11731) and TOV-21G (#CRL-11730) from ATCC (Manassas, VA) and culture as recommended by the manufacturer. We purchase the cell medium components (MCDB 105, Medium 199) from Sigma-Aldrich (St. Louis, MO), and PAA Laboratories (Dartmouth, MA) supplies the heat-inactivated fetal bovine serum. We purchase all other buffers and chemicals used in the experiments from Lonza (Basel, Switzerland) or Sigma-Aldrich (St. Louis, MO).

6.3 Methods

6.3.1 Biomarker Chemistries

We choose the ovarian cancer antibodies based on the following criteria:

1. Antibodies can be produced in large quantities without changes in functional properties.
2. Antibodies have proven epitope specificity, i.e., they recognize the targeted ovarian cancer biomarker and do not cross-react with other proteins or compounds.
3. Antibodies have high relevance for ovarian cancer, i.e., their targets are reliable biomarkers for ovarian cancer with a proven upregulation of at least twofold or higher in patients with ovarian serous papillary carcinoma [3–4].
4. In addition to the properties described in criterion 3, antibodies have high relevance for unequivocally distinguishing ovarian cancer from other types of cancer and diseases and, equally important, for distinguishing between metastatic and primary ovarian serous papillary carcinoma [3–4].

By using GMR sensor technology to detect multiple target concentrations simultaneously, a differential diagnosis can potentially be achieved. The specificity in target recognition (criteria 2–4) is particularly relevant because, ultimately, the GMR sensor technology allows expedient decisions on additional diagnostics and therapy choices for clinicians in the field.

6.3.2 In Vitro Cell Model

Human cell lines are used for the detection of ovarian cancer biomarker proteins and for the testing of feasibility of sensor operation in complex samples. In order to reach the highest possible clinical relevance (for the most financially viable research plan), the in vitro models for ovarian cancer are chosen based on human cell lines that derived directly from patients with ovarian cancer and are not from other types of cancer with ovarian side effects/metastases. Additionally, the in vitro models are established (used by ovarian cancer researchers in peer-reviewed publications) and reproducible (available through ATCC).

The two cell lines used in this assay are TOV-112D (ATCC, CRL-11731) and TOV-21G (ATCC, CRL-11730). Cell line TOV-112D was initiated in 1992 from a patient with early-onset ovarian cancer (grade 3, stage IIIC primary malignant adenocarcinoma; endometrioid carcinoma). Cell line TOV-21G was initiated in 1991 from a patient with primary malignant adenocarcinoma and clear cell carcinoma (grade 3, stage III). We culture both cell lines detailed as follows.

6.3.2.1 Cell Culture Start

We thaw and transfer cells to a 15 mL conical tube. We spin cells at 200 ×g for 1 min. We remove the supernatant and replace it with 1 mL of complete medium (MCDB 105 and medium 199 with fetal bovine serum). A cell count is done on the Nexcelom T4 Cellometer (Nexcelom Bioscience LLC, Lawrence, MA). We then seed cells in two 75 cm² flasks per vial of cells.

6.3.2.2 Subculturing or Passage

We remove the media and collect them for supernatant. We replace the media with 0.25% trypsin/EDTA, and we place the flask in an incubator for approximately 3–5 min. Once cells are detached, we remove the suspension and place it in a 15 mL falcon tube. We spin the cell suspension at 200×g for 1 min. We remove the trypsin and resuspend the cell pellet in complete medium (amount varies depending on confluence). We then seed the suspension into a fresh flask.

6.3.2.3 Supernatant Collection

To collect supernatant, we remove the media from the culture flask, place it in a 50 mL falcon tube, and spin it at 300×g for 1 min. We remove the supernatant and place it in a fresh 50 mL falcon tube. The tubes are then frozen at −80°C. For detection experiments, the thawed supernatant is diluted 1:100 in PBS/Tween 20.

6.3.3 Biosensor System

In this work, we develop a sensor system to analyze multiple biochemical reactions in a microarray format. In this embodiment, the expanding beam of a low-power laser diode is incident on the sensor chip element from the substrate side as shown in Figure 6.1. We choose the incident wavelength to optimize detection sensitivity and minimize optical absorption in the sample. For biochemical measurements, this is most convenient in the red or near-IR range. Multianalyte testing is accomplished by scanning the sensor element across the system reader using a small translation stage (schematic in Figure 6.3). A bottomless microwell plate (modified for 6 or 12 channels) is bonded to the sensor element to create an array (as shown in Figure 6.4). This system can be miniaturized by integrating a lens array whereby >100 spots can be interrogated simultaneously without moving parts. The sensor element can be configured with a microarray of antibody spots (100–200 μm in diameter) such that multiple analytes can be tested in a single sample.

A prototype system built in our labs is depicted in Figure 6.4. As shown in the schematic in Figure 6.3, light from a laser diode is passed through a beam-splitting element to a focusing lens. This beam-shaping lens provides an angular spread of the incident beam that is typically less than ±5°. The focused beam is incident on the GMR sensor element, and light is reflected (or transmitted) at specific angle(s) corresponding to exited resonant leaky modes. The resonantly reflected light is redirected through the beam-splitting element to a high-density detector array for readout. A lens is used before the detector array to optimize the angular range of interest (or dynamic range) of the system. As biochemical binding events occur on the sensor surface, the resonance reflection shifts in angle, and a measured peak is tracked as a function of position on the detector array. An alternate configuration (not shown) can be used by placing the detector above the GMR sensor element to measure the transmitted resonance null(s). Data are acquired via USB and downloaded into an ASCII text file for processing. Angular resolutions for this system are ∼10^{-4} degrees with measureable refractive index sensitivity to ∼10^{-5} RIU (refractive index units). Currently, system noise contributors include laser speckle (from the coherent source) and spurious reflections from optical surfaces. The hardware is controlled via a software program written in C++. This program was developed to control the data acquisition and provide the user with means to analyze the resonance shifts

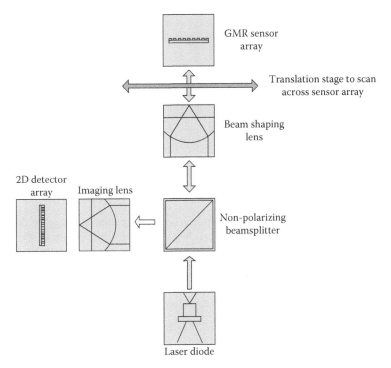

FIGURE 6.3 Schematic of a GMR reader system based on the fundamental sensor concept shown in Figure 6.1. The sensor reader is scanned across the bottom of the sensor element for multianalyte detection.

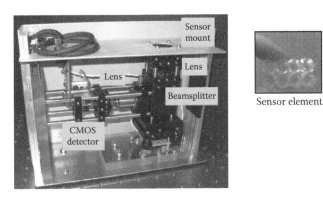

FIGURE 6.4 Illustration of a single-channel GMR sensor system prototype. This format can be extended to high-density arrays for multispecies analysis. Power is provided by USB port interface. Translation stage not shown.

in order to quantify a sample. The user can monitor the binding reactions in real time to obtain a measured concentration and/or binding kinetic results. Low-power laser diodes and photodetector elements can be used due to the high signal response reflected by, or transmitted through, the sensor element.

Figures 6.5 and 6.6 depict GMR sensor resonance minima (or nulls) measured in transmission with a 2D detector array (setup similar to that shown in Figure 6.4). In this geometry, the imaging detector is placed above the GMR sensor plate to measure the transmitted beam from an incident diverging laser diode (850 nm). The sensor elements are characterized by varying the cover refractive index (n_C is the refractive index adjacent to the sensor surface) in dilutions of the solvent dimethyl sulfoxide (DMSO) [18]. As the concentration of DMSO increases in the sample (from 10% to 50%), the refractive index also increases. The subsequent resonance shifts for both TE- and TM-resonance minima are measured and compared to the theoretically predicted values.

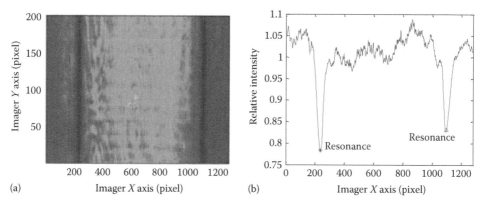

FIGURE 6.5   (a) Image snapshot taken with a 2D detector array depicting the resonant minima for TE polarization with $n = 1.33$ cover refractive index. (b) Corresponding line-scan cross section clearly illustrating the TE resonance minima.

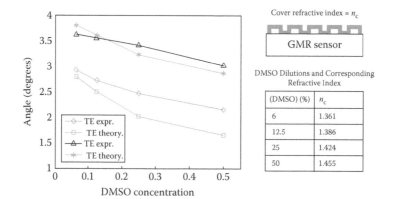

FIGURE 6.6   Measured (expr.) and calculated (theory) transmitted resonance minima angular location shifts under variations of refractive index. The TE and TM resonance null locations shift as the cover refractive index (n_C) varies in response to changing concentrations of the solvent DMSO in water. (From Ning, H. and Wiegand, S., *J. Chem. Phys.*, 125, 221102, 2006.) The measurement setup includes an incident diverging laser diode (850 nm) and a CCD detector array to measure the transmitted resonance minima. (From Wawro, D. et al., Guided-mode resonance sensors for rapid medical diagnostic testing applications, in *Optical Fibers and Sensors for Medical Diagnostics and Treatment Applications IX*, ed. I. Gannot, *Proceedings of the SPIE*, San Jose, CA, Vol. 7173, p. 717303, 2009.) The sensor structure is a ~520 nm period grating covered with a layer of sputtered HfO_2.

One advantage of measuring the GMR sensor response in transmission is that a compact system architecture requiring minimal optics is realized. However, in applications where the sample of interest is nonideal (such as raw blood), it can be advantageous to avoid transmitting the signal beam through the sample fluid. By measuring the sensor response in reflection (as depicted in Figures 6.1 and 6.3), the detection signal interacts only with the sample at the sensor surface, thereby minimizing any scattering due to the fluid. Figures 6.7 through 6.11 provide biosensor test results obtained using the reflection geometry; these will be discussed as follows.

6.3.4 Detection Experiments

Experiments detecting biomarkers for ovarian cancer are performed using an optical setup similar to that depicted in Figure 6.1. A laser with wavelength 650 nm is used as the light source in a reflection configuration. The incident wavelength is chosen to optimize detection sensitivity and dynamic range for biomarker detection in these experiments using the current sensor element design (described in Figure 6.2). The input light is shaped with a cylindrical lens to provide a focused beam (rather than a diverging beam

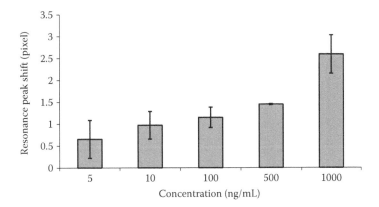

FIGURE 6.7 Measured GMR peak shift versus concentration for the detection of apolipoprotein A-1 in buffer (2 h incubation). Buffer is used as a reference and subtracted from data. All results are repeated in triplicate (outliers are removed) and averaged.

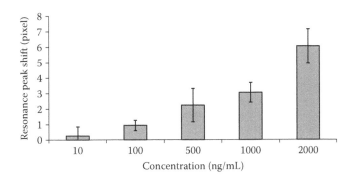

FIGURE 6.8 Measured GMR peak shift versus concentration for the detection of collagen type I in buffer after a 45 min incubation. Neat buffer is used as a reference and subtracted from measured data. All results are repeated in triplicate (outliers are removed) and averaged.

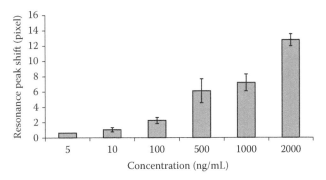

FIGURE 6.9 Experimental sensing results for detection of known concentrations of fibronectin in buffer. All results are repeated in triplicate (outliers are removed) and averaged.

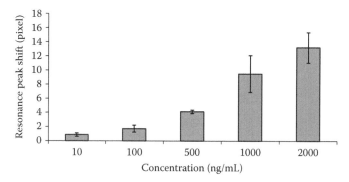

FIGURE 6.10   Resonance peak shift response for detection of known concentrations of fibronectin in cell culture media (2 h incubation). Diluted cell medium (1:100) is used as a reference, and subtracted from measured data. All results are repeated in triplicate (outliers are removed) and averaged.

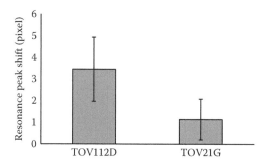

FIGURE 6.11   Detection of fibronectin in samples collected from cell culture supernatant in relevant ovarian cancer cell lines TOV112D and TOV21G. Supernatant was diluted 1:100 in PBS before sample detection. All results are repeated in triplicate (outliers are removed) and averaged.

as depicted in Figure 6.1) approximately 200 microns wide, while the reflected signal is directed to a 2D CCD device (Sony, 782 × 582 pixels). A reflected resonance peak shift as small as 0.2 pixels can be resolved with this system. Sensor elements are bound to a bottomless microwell plate to create an array, and each element is interrogated separately with the aid of a linear translation stage. The sensor plates are initially coated with a commercially available silane, 3-aminopropyltriethoxysilane, to provide a means to covalently bond the antibody/aptamer to the sensor surface. All antibody-based detection chemistries (Figures 6.7 through 6.11) are chemically attached to the silane-coated sensor surface using the homo-bifunctional cross-linking agent, disuccinimidyl suberate. The antibody-coated sensor is blocked and washed with PBS/Tween 20 in preparation for sample measurement. An initial baseline is taken in buffer (as a reference), and then approximately 75 μL of sample is dispensed by a pipette onto the activated sensor surface. As the immobilized antibody binds to the matched biomarker protein, the reflected peak response changes in angle and shifts across the CCD camera. After incubating the sample for 45 min (up to 2 h depending on binding kinetics), the sensor is washed 3× in PBS, and a postbinding baseline is measured. All results are repeated in triplicate (outliers are removed) and averaged.

6.4 Results

Figure 6.6 depicts the measured and calculated transmittance null locations as the refractive index is modified in the cover region of the sensor element (system layout shown in Figure 6.4). The experimental values are compared to the theoretical calculations using in-house rigorous coupled-wave analysis (RCWA) programs [17]. There is an agreement between computed and measured values, particularly for TM polarization.

Discrepancies between the theory and experiment are likely due to fabrication variations in the sensor element parameters. The measured sensitivity of the TE and TM resonances to the refractive index changes is approximately equal in this experiment. In general, sensitivity of TE and TM resonances can be different and is fundamentally dependent on device design. The DMSO dilutions correspond to the refractive indices [18] as shown in the inset of Figure 6.6. The transmitted resonant nulls (there are two that are visible due to angular symmetry) tracked in this experiment are clearly seen in the images illustrated in Figure 6.5. The angular spread of the incident and reflected beams is carefully measured (tracking change in beam size versus distance) in order to quantify the angular locations shown in Figure 6.6. The resonance peaks/nulls can shift to lower or higher values during a detection event depending on sensor design and the biomolecular reaction properties. The absolute angular location (in degrees) is important when backfitting the data to theory. However, in most cases, it is the relative resonance shift that is important.

Figures 6.7 through 6.9 depict results detecting various concentrations of the ovarian cancer biomarkers apolipoprotein A-1, collagen type I, and fibronectin in backgrounds of buffer using a reflection mode setup (described in Section 6.3). All tests are performed at room temperature and without thermal control. In these experiments, the TM polarization resonance is tracked during detection. The incubation time varies for each analyte depending on reaction times, which slow down when the room is cooler. Neat buffer is used as a reference, and resonance shifts (due to thermal variations) are subtracted from the data. In all cases, a blocking agent (BSA) is used to minimize nonspecific binding. Limits of detection for the biomarkers apolipoprotein A-1, collagen type I, and fibronectin are measured to be 5–10 ng/mL, respectively. These sensitivities are expected to improve as we optimize the attachment chemistries and system reader noise limits. To explore performance using realistic sample backgrounds, known concentrations of fibronectin are also detected in a background of cell media with minimal sensitivity degradation (Figure 6.10). A known standard concentration of 2000 ng/mL fibronectin is used as the high standard. Dilutions of fibronectin are prepared using the cell medium in a 1:100 dilution in PBS/Tween 20 (pH 7.4) to obtain the desired concentrations. A 1:100 dilution of cell medium is used as a reference blank. Figure 6.11 depicts two measured cell culture supernatant samples taken from ovarian cancer cell lines (TOV112D and TOV21G) known to produce the biomarkers under study, including fibronectin. After comparing the measured cell supernatant sample response with the corresponding standards curve (generated from a linear fit of data taken in Figure 6.10), it is determined that supernatant samples from cell lines TOV112D and TOV21G contain ~440 and ~55 ng/mL fibronectin, respectively. Work is ongoing to extend the testing in unprocessed serum samples. Early experiments indicate minimal sensitivity degradation (not measureable) in complex backgrounds such as serum by use of commercially available BSAs. Since the measured reflected resonance signal only penetrates the fluid by a few hundred nanometers at the sensor surface, there is minimal impact on sensor performance due to scattering or signal loss. Additionally, we have analyzed sample volumes as small as 20 μL (with a sensor area of ~100 μm diameter) without impact on sensitivity. By using the GMR sensor approach, samples from patients can be measured rapidly for a range of biomarker analytes.

6.5 Discussion

There is a great need for fast, easy-to-use, cost-effective analytical test methods that have high sensitivity, specificity, and reliability for in vivo and in vitro applications. Conventional medical diagnostic tools are typically based on immunoassay approaches that require sophisticated processing steps, and they are usually performed by highly trained personnel in a laboratory environment. These techniques typically involve the use of a chemical label, such as fluorescent or radioactive markers that add complexity, labor, and materials cost. Using this conventional approach, results are obtained in ~4–24 h. In contrast, we use a label-free detection system where arrays of predictive biomarkers are screened with minimal processing steps required for operation. Results can typically be obtained in less than 30 min (limited only by the binding dynamics of the antibody–antigen interactions). This greatly simplifies medical diagnostic testing and enables medical practitioners to effectively perform routine screening on a much larger scale with dramatically less labor and cost.

6.6 Conclusions

A new point-of-care diagnostic technology has been developed to rapidly screen an array of clinically relevant ovarian cancer biomarkers in blood or serum. This photonic resonance sensor technology has the capability to rapidly measure multiple agents simultaneously with minimal sample processing without requiring the use of chemical labels. A prototype sensor system is developed using common low-cost electronic components such as laser diodes and CCD arrays. Sensor performance is quantified by the detection of apolipoprotein A-1, collagen type I, and fibronectin in buffer and cell culture supernatant. Future work will extend the biosensor array for simultaneous detection of hundreds (or even thousands) of analyte biomarkers and will utilize multiple resonance peaks to increase detection accuracy. This will include analysis of ovarian cancer patient serum samples to measure a wide range of relevant biomarker concentrations such that regulatory requirements can be established for clinical use. This advanced system will provide highly accurate means to monitor presymptomatic aspects, disease progression, and effectiveness of treatments in ovarian cancer therapies.

ACKNOWLEDGMENTS

The project described was supported in part by the National Cancer Institute award #R43CA135960 and the State of Texas Emerging Technology Fund. The content is solely the responsibility of the authors and does not necessarily represent the official views of the National Cancer Institute, the National Institutes of Health, or the State of Texas Emerging Technology Fund.

REFERENCES

1. 2010. Ovarian Cancer National Alliance, Reference literature. http://www.ovariancancer.org/about-ovarian-cancer/statistics/
2. DiSaia, P. J. and W. T. Creasman, eds. 2002. *Epithelial Ovarian Cancer: Clinical Gynecologic Oncology.* Mosby–Year Book, Inc., St. Louis, MO, pp. 185–206.
3. Liotta, L. A., M. Lowenthal, A. Mehta, T. P. Conrads, T. D. Veenstra, D. A. Fishman, and E. F. Petricoin, III. 2005. Importance of communication between producers and consumers of publicly available experimental data. *J. Natl. Cancer Inst.* **97**:310–314.
4. Bignotti, E., R. A. Tassi, S. Calza, A. Ravaggi, E. Bandiera, E. Rossi, C. Donzelli, B. Pasinetti, S. Pecorelli, and A. D. Santin. 2007. Gene expression profile of ovarian serous papillary carcinomas: Identification of metastasis-associated genes. *Am. J. Obstet. Gynecol.* **196**:245.
5. Magnusson, R. and S. S. Wang. 1992. New principle for optical filters. *Appl. Phys. Lett.* **61**:1022–1024.
6. Wawro, D., S. Tibuleac, R. Magnusson, and H. Liu. 2000. Optical fiber endface biosensor based on resonances in dielectric waveguide gratings. *Proc. SPIE* **3911**:86–94.
7. Wawro, D., S. Tibuleac, and R. Magnusson. 2006. Optical waveguide-mode resonant biosensors. *Opt. Imag. Sens. Syst. Homeland Security Appl.* **2**:367–384.
8. Wawro, D., Y. Ding, S. Gimlin, S. Zimmerman, C. Kearney, K. Pawlowski, and R. Magnusson. 2009. Guided-mode resonance sensors for rapid medical diagnostic testing applications. In *Optical Fibers and Sensors for Medical Diagnostics and Treatment Applications IX*, ed. I. Gannot, *Proceedings of the SPIE*, San Jose, CA, Vol. 7173, p. 717303.
9. Wawro, D., P. Koulen, Y. Ding, S. Zimmerman, and R. Magnusson. 2010. Guided-mode resonance sensor system for early detection of ovarian cancer. In *Optical Diagnostics and Sensing X: Toward Point-of-Care Diagnostics, Proceedings of the SPIE*, San Francisco, CA, Vol. 7572, p. 75720D.
10. Magnusson, R., D. Wawro, S. Zimmerman, and Y. Ding. 2011. Resonant photonic biosensors with polarization based multiparametric discrimination in each channel. *Sensors.* **11**:1476–1488.
11. Kikuta, H., N. Maegawa, A. Mizutani, K. Iwata, and H. Toyota. 2001. Refractive index sensor with a guided-mode resonant grating filter. *Proc. SPIE* **4416**:219–222.
12. Cunningham, B., P. Li, B. Lin, and J. Pepper. 2002. Colorimetric resonant reflection as a direct biochemical assay technique. *Sens. Actuat. B* **81**:316–328.

13. Homola, J. 2003. Present and future of surface plasmon resonance biosensors. *Anal. Bioanal. Chem.* **377**:528–539.

14. Raether, H. 1988. *Surface Plasmons on Smooth and Rough Surfaces and on Gratings*. Springer-Verlag, Berlin, Germany.

15. Cooper, M. Summer 2006. Current biosensor technologies in drug discovery. *Drug Discov. World* **7**:68–82.

16. Cunningham, A. 1998. *Introduction to Bioanalytical Sensors*. John Wiley & Sons, New York.

17. Gaylord, T. K. and M. G. Moharam. 1985. Analysis and applications of optical diffraction by gratings. *Proc. IEEE* **73**:894–937.

18. Ning, H. and S. Wiegand. 2006. Experimental investigation of the Soret effect in acetone/water and dimethylsulfoxide/water mixtures. *J. Chem. Phys.* **125**:221102.

7

Label-Free Optofluidic Ring Resonator Biosensors for Sensitive Detection of Cancer Biomarkers

Hongying Zhu and Xudong Fan

CONTENTS

7.1 Introduction

Cancer is a leading cause of death worldwide, and the incidence of cancer continues to increase. The clinical treatment outcome for a cancer patient highly relies on the stage at which the cancer is diagnosed, and therefore, the early diagnosis and effective treatment against it are critical to increasing the survival rate. Traditional cancer diagnostic methods, such as various medical imaging techniques (e.g., computed tomography [CT], magnetic resonance imaging [MRI]) [1,2], are not sensitive enough for the early cancer detection and also lack the ability to predict the patient's response to the treatment. In addition, those tests are costly and time consuming, which limits their applications in mass screening for large population.

With the increased understanding of the cancer at molecular level in the last few decades, there is growing interest in using cancer biomarkers for cancer diagnosis [3,4]. Cancer biomarkers are biomolecules presented in specific tissues or body fluids (e.g., blood, urine, or saliva) of a cancer patient. Cancer biomarkers can be nucleic acids, proteins, sugars, lipids, small metabolites, or whole cells [5]. They may be molecules secreted by the cancer itself, or a specific body response to the presence of the cancer [5]. The presence and concentration of those biomarkers provide important molecular information for early diagnosis and prognosis. Most cancer biomarkers used today in clinical diagnosis are proteins presented in sera, such as prostate-specific antigen (PSA) for prostate cancer [6], CA15-3 and Her-2/neu for breast cancer [7,8], and CA19-9 for pancreatic cancer [9]. The concentrations of those protein biomarkers are typically measured by various antibody-based immunoassays, such as enzyme-linked immunosorbent assay (ELISA) [10], with high sensitivity and specificity. Those tests are usually performed by the well-trained personnel in centralized laboratories with automated immunoassay analyzers, which inevitably increases the patient turn-around time and test cost. Consequently, an easy-to-use, cheap, sensitive, and portable analytical device that can directly perform the test in doctor's office (or even patient bedside) is highly desirable.

The optofluidic ring resonator (OFRR), within the category of evanescent-wave-based label-free refractive index (RI) sensors, is an emerging sensing platform and has experienced considerable growth in recent years [11–20]. As shown in Figure 7.1, the OFRR employs a glass capillary with around 100 μm in diameter. The capillary serves as a microfluidic channel for liquid sample delivery, whereas the capillary circular cross section forms an optical ring resonator. The light can be evanescently coupled into the ring resonator through a fiber taper [15] or a waveguide [21]. The light circulates along the ring resonator curved surface through total internal reflection (TIR). This circulating mode is called whispering gallery mode (WGM). The capillary wall is sufficiently thin (<4 μm) so that the evanescent field of the WGM extends into the capillary interior surface to interact with the analytes. Essentially, the OFRR sensor performs label-free sensing by detecting RI changes near the inner surface induced by the binding of biomolecules. The relation between the WGM spectral position and the RI is shown as follows [22]:

$$2\pi n_{eff} r = m\lambda, \tag{7.1}$$

where
   n_{eff} is the effective RI experienced by the WGM
   r is the capillary radius
   m is an integer representing the WGM angular momentum
   λ is the WGM resonant wavelength (or spectral position)

Antibodies (or other biorecognition elements) are immobilized on the capillary interior surface to provide the detection specificity. The WGM spectral position shifts when target molecules are captured by the antibodies. The real-time WGM spectral shift provides the quantitative and kinetic information about molecular interaction on the OFRR sensing surface. The resonant nature of the WGMs creates extremely long effective light-matter interaction length, thus greatly improving the OFRR sensitivity [11]. The OFRR sensor has demonstrated a mass detection limit on the order of 1 pg/mm², competitive with commercialized surface plasmon resonance (SPR) sensors [11]. Due to its advantageous features

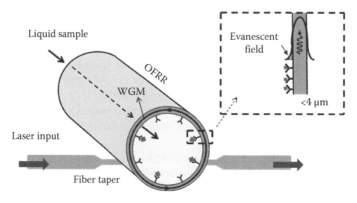

FIGURE 7.1 Conceptual illustration of the OFRR sensing principle.

such as label-free, excellent sensitivity, specificity, low fabrication cost, and easy sample handling, the OFRR sensor is a very promising analytical tool for cancer biomarker detection.

In this chapter, the OFRR sensor fabrication method and sensing experimental setup will be described in details. The protocol of using the OFRR sensor for the detection of cancer biomarkers will be presented. Breast cancer serum biomarker CA15-3 will be employed as a model system in the study. The future prospects of the OFRR sensor development will be discussed.

7.2 Materials

7.2.1 Materials and Equipment for the OFRR Sensor and Fiber Taper Fabrication

1. Two CO_2 lasers such as #F48-2-285W/2538 with maximum power 40 W or #J48-1S-7644 with maximum power 30 W (Synrad, Inc., Mukilteo, WI)
2. Silica capillary tubing: #700850 (857 µm outer diameter [OD] and 699 µm inner diameter [ID]) (Polymicro Technologies, Phoenix, AZ)
3. Fiber optic cable: single mode fiber, #SMF28 (Corning, Inc., Corning, NY)
4. UV-curable glue: Norland optical adhesive, #8101 (Norland, Cranbury, NJ)
5. Plain microscope slides: Fisherbrand, #12-550A (Fisher Scientific, Pittsburgh, PA)

7.2.2 Experimental Setup Equipment

1. Syringe pump: #55-1140 (Harvard Apparatus, Holliston, MA).
2. Tunable laser: various tunable lasers can be used, such as 1550 nm distributed feedback (DFB) laser (JDSU #CQF935, JDS Uniphase Corp., Milpitas, CA) and 980 nm external cavity laser (Velocity #6309, New Focus, San Jose, CA).
3. Photodetector: large area IR photoreceiver, #2033 (New Focus, Santa Clara, CA).
4. Tygon microbore tubing: 0.01″ ID, 0.03″ OD, #EW-06418-01 (Cole-Parmer, Vernon Hills, IL).

7.2.3 Buffers and Chemical Reagents

1. 18 MΩ water (referred to as water throughout the chapter): generated by the EASYpure UV, #D7401 (Barnstead/Thermolyne Corp., Dubuque, IA)
2. Ethanol: absolute, 200 proof, #E7023 (Sigma-Aldrich, St. Louis, MO)
3. Methanol: absolute, #M1775 (Sigma-Aldrich, St. Louis, MO)
4. Phosphate-buffered saline (PBS) tablets: #P-4417 (Sigma-Aldrich, St. Louis, MO)
5. Hydrochloric acid (HCl): #A144-212 (Fisher Scientific, Pittsburgh, PA)
6. Hydrofluoric acid (HF): 48%, #244279 (Sigma-Aldrich, St. Louis, MO)
7. 3-Aminopropyl-trimethoxylilane (3-APTMS): 97%, #281778 (Sigma-Aldrich, St. Louis, MO)
8. Cross-linker: glutaraldehyde, 50%, #G7651 (Sigma-Aldrich, St. Louis, MO)
9. Tween-20 surfactant: 10%, #28320 (Thermo Scientific, Rockford, IL)
10. Amine-PEG-amine: molecular weight = 1000 Da (Laysan Bio Inc., Arab, AL)

7.2.4 Biomolecule Samples

1. Bovine serum albumin (BSA): minimum 96%, #A2153 (Sigma-Aldrich, St. Louis, MO)
2. Mouse monoclonal anti-CA15-3 antibody: clone number 2F16 and product number #C0050-22A (USBiological, Marblehead, MA)
3. Mouse monoclonal anti-ER antibody: clone number 4E73 and product number #E3565-09 (USBiological, Marblehead, MA)

4. Purified CA15-3 antigen: #150-12A (Lee Biosolutions, St. Louis, MO)
5. Purified CA12-5 antigen: #150-11(Lee Biosolutions, St. Louis, MO)
6. Fetal calf serum (FCS): #10437-077 (Invitrogen, Carlsbad, CA)

7.3 Methods

This section provides detailed information regarding the OFRR sensor fabrication, fiber taper fabrication, sensing experimental setup assembly, and experimental procedures of using the OFRR sensor for cancer biomarker detection.

7.3.1 OFRR Sensor Fabrication Method

In order to expose the evanescent field into the capillary core, the capillary wall must be sufficiently thin (typically less than 4 µm). The glass capillaries with such a thin wall are not commercially available yet. To make such a thin-walled capillary, an automated pulling station is assembled in the lab, as shown in Figure 7.2A. The thin-walled capillary is formed by quickly stretching the center section of a glass capillary (OD: 857 µm and ID: 699 µm) under the heat provided by CO_2 lasers. The lasers and mechanical motion stations are controlled by a computer. The whole setup is enclosed in an acrylic box to reduce air flow, which may cause temperature fluctuations at the heating zone. The procedures for the OFRR sensor fabrication are described as follows:

1. The glass capillary is mounted between two mechanical motion stages, and the capillary positions are tightly held by two pieces of 3M double-sided tapes.
2. Two CO_2 lasers are placed on the opposite side of the capillary to heat the capillary evenly. The laser is turned on and the laser beams are focused onto the capillary. The pulling and feed-in stages start to move after several seconds when the capillary at the heating zone becomes soft. Laser power level, pulling speed, and feed-in speed are controlled by the computer through a

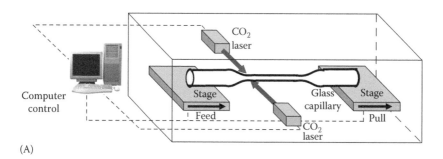

(A)

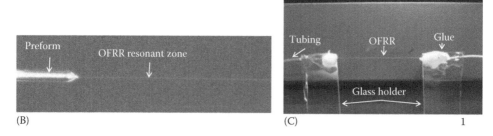

(B)   (C)   1

FIGURE 7.2 (A) OFRR sensor fabrication setup. (B) Picture of capillary after pulling. (C) Mounting the OFRR sensor on a glass slide holder and connecting it with the tubing.

Labview program. The program operates through the data acquisition card and the PCI-based motion controller to generate the voltage-based outputs to control the laser output power and the pull/feed-in stages.

3. The pull and feed-in stages stop moving when the desired capillary length is achieved. By controlling the pull and feed-in speed ratio, we are able to control the final OFRR OD and wall thickness [11]. Typically, the OFRR OD is around 50–100 μm with a wall thickness from 5 to 10 μm.

4. The capillary is removed from the stage. The capillary tapering section is the resonant region as indicated in Figure 7.2B.

5. Cut the capillary tapered region into small pieces for experiments. The last step is to mount the capillary onto a glass slide holder and connect the Tygon tubing to both ends of the OFRR capillary, as shown in Figure 7.2C. The connection parts are sealed with a small amount of UV-curable glue.

6. Note that the OFRR wall thickness achieved by the method mentioned earlier may still not be sufficiently thin. A subsequent HF etching step is needed to further thin the wall, as discussed in Section 7.3.4. To eliminate this HF etching step, we can use an initial capillary with a proper design (i.e., larger ratio between the initial OD and the wall thickness). Furthermore, to avoid potential collapsing effect (i.e., the wall thickness does not reduce proportionately while the OD decreases during the pulling), the capillary can be pressurized.

7.3.2 Fiber Taper Fabrication Method

Fiber taper is employed to evanescently couple the light into the OFRR sensor. The fiber taper is fabricated by stretching an SMF-28 fiber slowly under heat provided by the flame from a hydrogen torch. Similar to the capillary pulling setup described earlier, the mechanical stages for the fiber taper pulling are also controlled by a computer through a Labview program. The fiber taper fabrication setup is illustrated in Figure 7.3A and B. The fabrication procedure is described as follows:

1. Prepare 1 m long optical fiber and strip away a couple of centimeter-long fiber polymer jacket near the fiber center. Clean the stripped region using Kim wipes with ethanol solution.

2. The optical fiber is mounted onto the mechanical motion stages and is held tightly with magnetic clamps on both sides.

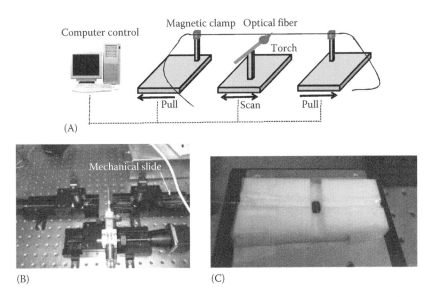

FIGURE 7.3 (A) Fiber taper fabrication illustration, (B) fiber taper fabrication setup picture, and (C) fiber taper holder.

3. Position the torch close to the center of the stripped region. Turn on the oxygen/hydrogen gas and light the torch.

4. Scan the flame along the stripped region back and forth for several seconds to preheat the fiber.

5. Start the taper pulling program and the stages start to pull the fiber slowly in opposite directions.

6. When the desired taper diameter is achieved, the stages are stopped and the torch is turned off. The fiber taper diameter is dependent on the laser wavelength used in experiments. Short wavelength will have a shorter evanescent field penetration depth, and therefore the fiber taper has to be sufficiently thin to expose enough evanescent field for coupling. Typically the tapered region on the fiber taper is around 2–3 μm in diameter.

7. The fiber taper is very fragile. For easy handling, it is mounted onto a plastic module (see Figure 7.3C).

7.3.3 Sensing Experimental Setup Assembly

The experimental setup is used to characterize the OFRR sensor bulk refractive index sensitivity (BRIS) and to detect the biomolecules binding to the OFRR sensing surface. The setup is shown in Figure 7.4, and it is assembled following the steps described in the following:

1. Mount the OFRR sensor on a three-dimensional mechanical stage.

2. Mount the fiber taper on another three-dimensional mechanical stage.

3. Connect the fiber taper input to a tunable diode laser (980 or 1550 nm) and place a photodetector at the fiber taper output to measure the transmission light intensity.

4. Start to scan the laser (~100 pm).

5. Bring the OFRR sensor in contact with the fiber taper. Monitor the photodetector signal. The laser periodically scans across ~100 pm wavelength range. When the scanning laser passes through a resonant wavelength, the light is coupled into the ring resonator and causes the transmission light intensity to drop. The spectral dip is used to indicate the WGM spectral position.

6. Move the OFRR sensor along the fiber taper while monitoring the WGM spectral position. Find out the best coupling position on the fiber taper for the experiment.

7. Pump the liquid sample through the OFRR sensor using a syringe pump at a constant speed.

8. The measurement system is controlled by a computer through a data acquisition card, and the WGM spectral position is recorded for postanalysis.

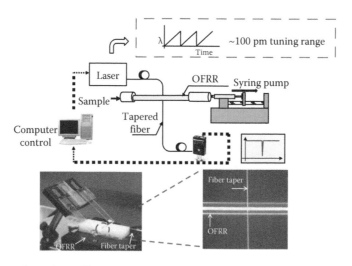

FIGURE 7.4 The experimental setup diagram.

7.3.4 OFRR Sensor Preparation and Sensitivity Characterization

The theoretical analysis has shown that the OFRR sensor sensitivity is directly dependent on the capillary wall thickness [11]. In order to expose the WGMs evanescent field into the capillary core to interact with the sample flowing inside it, the capillary wall needs to be sufficiently thin (less than 4 µm). However, as described earlier, the capillary wall is typically around 5–10 µm after mechanical pulling. Therefore, it is necessary to further reduce the wall thickness. To do this, diluted HF is flowed through the capillary to etch the capillary inner wall, which can give us the control over the wall thickness reduction with the nanometer accuracy. After etching, the OFRR sensor sensitivity to the bulk solution RI change is characterized with ethanol–water mixtures. The OFRR sensor preparation and sensitivity characterization are guided by the following steps:

1. Prepare 10%, 5%, and 0.5% HF solution in water.
2. Prepare ethanol solution, 1%, 0.75%, 0.5%, and 0.25% (v/v), in water. The RI of the ethanol–water mixture is known [23].
3. Bring the OFRR sensor in contact with the fiber taper. Find the best coupling position on the taper and start to record the WGM spectrum. Flow 10% HF solution through the capillary to etch the capillary wall and monitor the WGM spectrum during the whole etching process.
4. When the OFRR sensor becomes sensitive, the WGM spectral position will shift to the shorter wavelength (blueshift).
5. Stop the HF etching process and flow the water through the capillary to rinse off the HF residues.
6. Test the OFRR's BRIS with various ethanol solutions. Initially fill the OFRR sensor with the water to establish a stable detection baseline. Then the solution is changed from water to the ethanol solution (e.g., 1% ethanol). Since the bulk RI of the ethanol solution is higher than the water, based on Equation 7.1, the WGM spectral position will shift to the longer wavelength (redshift). After the WGM spectral position is stabilized in the ethanol solution, the flow is switched back to the water. Record the spectral position during the whole process. If the spectral redshift is invisible, continue the HF etching process. If the spectral shift is visible, start to perform step 7.
7. Repeat step 6 with various ethanol concentrations (1%, 0.75%, 0.5%, and 0.25%). Record the real-time WGM spectral shift for each ethanol concentration, which is called sensorgram as shown in Figure 7.5A. Plot the amount of WGM spectral shift versus the RI change. Perform a linear fit for those data points. The slope of the curve is the BRIS, which is in units of nanometers per RI unit (i.e., nm/RIU). An example of the calibration curve is shown in Figure 7.5B.

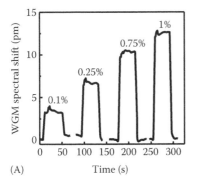

(A)

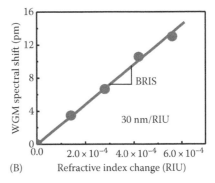

(B)

FIGURE 7.5   (A) Sensorgrams for various bulk RI changes induced by different concentrations of ethanol solutions. (B) Bulk RI sensitivity obtained from the slope of the spectral shift versus the RI change.

8. When the desired the sensitivity is achieved for a particular experiment, the HF etching and sensitivity calibration process are finished.

9. Clean the OFRR sensor with HCl-methanol mixture at volume ratio 50:50 for 10 min. Rinse the OFRR sensor with water thoroughly and then refill it with DI water. After this, the OFRR is ready for biosensing application.

7.3.5 Protocol for Breast Cancer Biomarker CA15-3 Detection Using the OFRR Sensor

Specificity is critical for the OFRR sensor to detect the cancer biomarker in serum. Here, antibodies are employed as the biorecognition element immobilized on the OFRR interior surface to provide the specificity. In order to reduce the nonspecific binding resulting from serum proteins, it is necessary to block the unoccupied activated functional groups on the sensor surface after antibody immobilization. The protocol for the detection of CA15-3 in human serum is listed as follows:

1. Prepare the OFRR sensor as described in Section 7.3.4.

2. Prepare 1% silane solution (3-APTMS) in 10% ethanol and flow 3-APTMS through the OFRR sensor for 30 min. Rinse the OFRR sensor with water.

3. Prepare 5% homobifunctional cross-linker glutaraldehyde in PBS buffer. Pass the glutaraldehyde solution through the OFRR sensor for 30 min and rinse it with PBS buffer thoroughly.

4. Prepare 50 µg/mL anti-CA15-3 antibody solution in PBS buffer. Introduce the antibody solution into the OFRR sensor at a flow rate of 5 µL/min for 20–30 min. Then rinse the OFRR sensor with PBS buffer. Antibodies are covalently bound to the activated OFRR sensor surface. Monitor the WGM spectral position change during the whole process.

5. Prepare 1 mg/mL amine-PEG in PBS buffer and 10% FCS in PBS buffer containing 0.5% Tween-20. Flow amine-PEG solution through the OFRR sensor for 30 min and rinse it with PBS buffer. Following this, switch the flow to 10% FCS in PBS buffer containing 0.5% Tween-20 for 30 min to further block the sensing surface. Rinse the OFRR sensor with the running buffer (3% FCS-PBS/0.5% Tween-20).

6. Fill the OFRR sensor with the running buffer (3% FCS-PBS/0.5% Tween-20), and the OFRR sensor is ready to use for the detection of CA15-3 in human serum.

7. Centrifuge the blood sample and collect the serum. Note that for long-term use, serum samples can be stored under −80°C.

8. Dilute the serum with running buffer. Flow the diluted serum sample through the OFRR sensor and record the WGM spectral shift for postanalysis.

7.3.6 Biomolecule Surface Density Estimation Method

The surface density of captured biomolecules can be estimated using a linear relationship between the wavelength shift ($\delta\lambda$), BRIS (S), and the biomolecule surface density (σ_p), which is established from a simple evanescent field model [18]:

$$\frac{\delta\lambda}{\lambda} = \sigma_p \alpha_{ex} \frac{2\pi\sqrt{n_2^2 - n_3^2}}{\varepsilon_0 \lambda^2} \cdot \frac{n_2}{n_3^2} S, \qquad (7.2)$$

where
  α_{ex} is the excess polarizability of the analyte, and it is proportional to the molecular weight [24]
  n_2 (1.45) and n_3 (1.33) are the RI of the OFRR sensor wall and the aqueous medium in the capillary core
  ε_0 is the vacuum permittivity
  λ is the wavelength of the WGM (~980 or ~1550 nm)

Equation 7.2 allows us to quantify the biomolecule surface density on the OFRR inner wall using the experimentally measurable sensing signal, i.e., the WGM spectral shift.

7.4 Results

Breast cancer biomarker CA15-3 is chosen as a model system in our study to demonstrate the feasibility of using the OFRR sensor for cancer biomarker detection. CA15-3 is a mucin-like glycoprotein with molecular weight of 300–450 kDa. In a healthy human, the CA15-3 concentrations are below 30 units/mL [25,26]. Patients with primary breast cancer or metastatic breast cancer show elevated CA15-3 levels. CA15-3 is an important biomarker used for breast cancer prognosis and monitoring of breast cancer recurrence. In this section, the results of using the OFRR sensor for CA15-3 detection are presented and discussed, including the antibody immobilization, minimizing nonspecific serum protein adsorption, and the detection of the CA15-3 in human serum with high specificity.

7.4.1 Monoclonal Antibody Immobilization

First, a series of experiments are performed to optimize the antibody immobilization (e.g., incubation time and antibody concentration). Figure 7.6A is a representative sensorgram for anti-CA15-3 antibody immobilization. Upon the injection of 50 μg/mL antibody solution (~323 nM), the WGM spectral position quickly shifts to a longer wavelength within the first few seconds. This huge and rapid shift is mainly due to the bulk RI change inside the OFRR sensor because the antibody solution contained 40% glycerol. The WGM spectral continuously shifts due to the actual antibody binding to the surface. After certain amount of time, the flow is changed back to PBS buffer for surface rinsing. The difference between the final stabilized WGM spectral position and the initial baseline is the actual amount of spectral shift due to antibody immobilized on the OFRR interior surface. Note that in Figure 7.6A, the WGM spectral shift is converted to the antibody surface density using Equation 7.2, in which $(\alpha_{ex})_{antibody} = 4\pi\varepsilon_0 \times (9.04 \times 10^{-21})$ cm^3 [24]. Figure 7.6B shows that the antibody surface density is proportional to the immobilization time for the first 40 min. The antibody surface density is about 4–5×10^{11}/cm^2 after 30 min of immobilization. This result is similar to the typical antibody density in ELISA ($\sim 5.8 \times 10^{11}$/cm^2), and 30–40 min of antibody immobilization time is used for all the experiments.

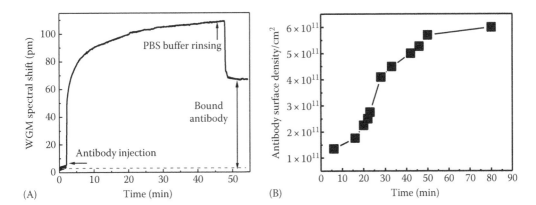

FIGURE 7.6   (A) Real-time antibody immobilization sensorgram. (B) Antibody surface density versus the immobilization time obtained with 11 different OFRR sensors. Antibody surface density is calculated using Equation 7.2. (Reprinted with permission from Zhu, H., Dale, P.S., Caldwell, C.W., and Fan, X., Rapid and label-free detection of breast cancer biomarker CA15-3 in clinical human serum samples with optofluidic ring resonator sensors, *Anal. Chem.*, 81, 9858–9865, 2009. Copyright 2009 American Chemical Society.)

7.4.2 Minimizing Nonspecific Serum Protein Adsorption

Essentially, the OFRR sensor detects the RI change near the sensing surface. Therefore, nonspecific serum protein adsorption is a major challenge for using the OFRR sensor to detect the protein biomarkers in complex biological media such as serum or plasma. The nonspecific adsorption can be greatly reduced by using dual surface blocking method described in Section 7.3.5. As shown in Figure 7.7, the dual surface blocking method yields lowest nonspecific protein adsorption density compare to the other surface blocking methods. The nonspecific protein surface density is estimated with Equation 7.2 from the measured WGM spectral shift [27]. The nonspecific protein densities for threefold and sixfold diluted FCS are only about $4.5 \times 10^9/cm^2$ (or $0.5\,ng/cm^2$) and $2.7 \times 10^9/cm^2$ (or $0.3\,ng/cm^2$), respectively, which meet the ultralow fouling surface criteria [28]. The nonspecific protein densities from other higher dilutions 8- to 21-fold are not detectable. Consequently, with the dual surface blocking method, the OFRR sensor is capable of detecting CA15-3 in diluted serum directly without any additional reference channels or amplification steps.

7.4.3 Detection of CA15-3 in PBS Buffer

Figure 7.8A displays a representative sensorgram for the direct detection of CA15-3 in PBS buffer at concentrations of 1 unit/mL and 5 units/mL. The CA15-3 surface density gradually increases after the sample injection. After 20 min, the density reaches over 90% of the saturation signal, meaning that the whole detection can be completed in less than 30 min even taking into account subsequent rinsing step. The OFRR sensor specificity and cross-reactivity are also studied. Figure 7.8B shows the response of the anti-ER antibodies coated OFRR sensor to a high concentration of CA15-3 (100 units/mL) in PBS buffer. No noticeable shift is observed in the WGM spectral position, indicating that only anti-CA15-3 antibody can bind CA15-3 specifically. Figure 7.8C shows the cross-reactivity study results, in which anti-CA15-3 antibodies are immobilized and the surface is blocked. Interfering antigen CA12-5, a glycoprotein used as an ovarian cancer biomarker, is flowed through the OFRR sensor at a concentration of 100 units/mL for 20 min. No noticeable WGM spectral shift is observed. Immediately after rinsing the OFRR surface with PBS buffer, 100 units/mL CA15-3 solution is introduced into the OFRR sensor and a clear WGM spectral shift is observed, suggesting that anti-CA15-3 antibody has high specificity toward CA15-3 and low cross-reactivity to other interfering antigens.

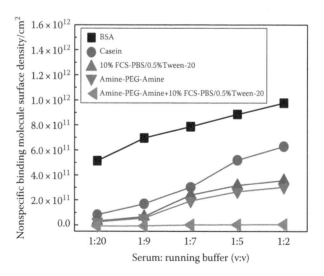

FIGURE 7.7 Surface densities of nonspecifically adsorbed serum protein molecules versus FCS concentration using different surface blocking methods. FCS concentrations are 33%, 16.7%, 12.5%, 10%, and 4.8% for 1:2, 1:5, 1:7, 1:9, and 1:20 dilutions, respectively. (Reprinted with permission from Zhu, H., Dale, P.S., Caldwell, C.W., and Fan, X., Rapid and label-free detection of breast cancer biomarker CA15-3 in clinical human serum samples with optofluidic ring resonator sensors, *Anal. Chem.*, 81, 9858–9865, 2009. Copyright 2009 American Chemical Society.)

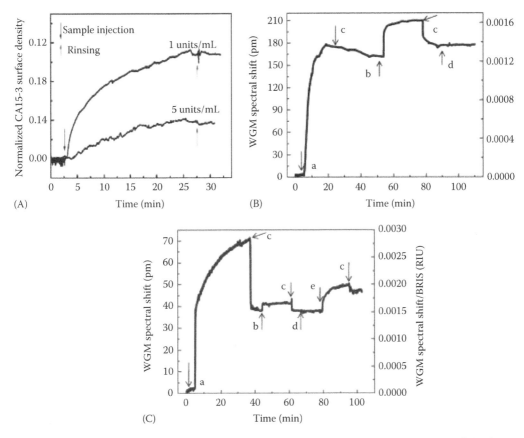

FIGURE 7.8 (A) Real-time sensorgram for CA15-3 in PBS buffer at concentrations of 1 unit/mL and 5 units/mL. (B) Demonstration specificity of the CA15-3 antigen. Sensor responses to (a) immobilization of anti-ER antibody (20 μg/mL), (b) surface blocking with amine-PEG-amine, (c) PBS buffer rinsing, and (d) CA15-3 antigen (100 units/mL). (C) Demonstration of no cross-reactivity of anti-CA15-3 antibody. Sensor responses to (a) immobilization of anti-CA15-3 antibody (50 μg/mL), (b) surface blocking with amine-PEG-amine, (c) PBS buffer rinsing, (d) CA12-5 (100 units/mL), and (e) CA15-3 (100 units/mL). (Reprinted with permission from Zhu, H., Dale, P.S., Caldwell, C.W., and Fan, X., Rapid and label-free detection of breast cancer biomarker CA15-3 in clinical human serum samples with optofluidic ring resonator sensors, *Anal. Chem.*, 81, 9858–9865, 2009. Copyright 2009 American Chemical Society.)

7.4.4 Detection and Quantification of CA15-3 in Human Serum

To quantify the concentration of CA15-3 in an unknown blood sample, the net WGM spectral shift should be compared to a calibration curve that plots the net WGM shift versus CA15-3 concentration. Purified CA15-3 is spiked into the FCS to prepare different concentrations of CA15-3 solution as "calibrators." The net WGM spectral shift for each corresponding concentration is recorded and converted to the normalized CA15-3 surface density (CA15-3 density/antibody density). Figure 7.9 shows the calibration curves for 6- and 21-fold dilutions. The calibration curves are fitted by Equation 7.3:

$$\sigma = \frac{\sigma_{max} \cdot [CA15\text{-}3]}{K_d + [CA15\text{-}3]}, \tag{7.3}$$

where
    σ is the normalized CA15-3 surface density
    σ_{max} is the maximum of the normalized CA15-3 surface density
    K_d is the dissociation constant for antibody–antigen binding
    [$CA15$-3] is the CA15-3 concentration

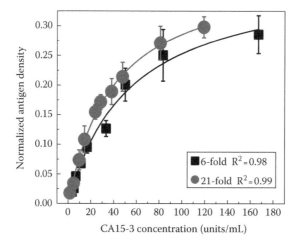

FIGURE 7.9 OFRR sensor response for the detection of CA15-3 spiked into FCS. Samples are diluted 6 fold or 21 fold with 3% FCS-PBS/0.5% Tween-20 buffer. Each point is measured five times. The solid curves are fitted with Equation 7.3, where $\sigma_{max} = 0.389$ (0.399) and $K_d = 54$ units/mL (40 units/mL) for 6 (21) fold dilution. (Reprinted with permission from Zhu, H., Dale, P.S., Caldwell, C.W., and Fan, X., Rapid and label-free detection of breast cancer biomarker CA15-3 in clinical human serum samples with optofluidic ring resonator sensors, *Anal. Chem.*, 81, 9858–9865, 2009. Copyright 2009 American Chemical Society.)

The dissociation constant is 54 units/mL for six fold dilution and 40 units/mL for 21-fold dilution. Figure 7.10 shows some of the real-time sensorgrams for the patient serum samples test. The healthy male serum does not generate an observable signal (Figure 7.10A) after buffer rinsing. The signals from three stage IV breast cancer patients' sera are significantly greater than that of the healthy male serum (Figure 7.10B through D). These results suggest that the OFRR sensor can easily differentiate breast cancer patient from healthy people. Additionally, the detection can be completed in 20 min, much quicker than the detection using a commercialized CA15-3 ELISA kit, which usually takes approximately 3 h from serum sample incubation to getting the final test results. On the basis of the calibration curves provided in Figure 7.9, it is able to estimate the CA15-3 concentration for those patients. Figure 7.11A compares the results obtained with the OFRR tests using 21-fold dilutions with those with clinical standard tests performed at ARUP Lab (Salt Lake City, UT). Figure 7.11B shows that good correlation is achieved between the OFRR test and the standard test.

7.5 Discussion and Future Trends

This chapter introduces the development and applications of the OFRR sensor for cancer biomarker detection in human serum using breast cancer biomarker, CA15-3, as a model system. The OFRR sensor demonstrates the label-free, direct, sensitive, and rapid (<30 min) detection of cancer biomarkers. Compared to the traditional fluorescent-based "sandwich" immunoassay, the OFRR sensor avoids tedious experimental procedures, reduces the reagent cost (no secondary antibody and no labeling), and greatly shortens the detection time. Compared to other label-free optical biosensor–based biomarker detection methods (e.g., SPR [29–32] and optical diffraction gratings [33]), the OFRR sensor has simpler optical alignment without any bulky optical components. The capillary nature of the OFRR sensor allows simple and safe liquid (e.g., blood) handling. The microsized fluidics channel requires only submicroliter sample volume which can be obtained by a finger prick. In addition, the OFRR sensor can also be mass produced at low cost and can be used as a disposable component.

Despite those advantageous features, the OFRR sensor platform still has to be further developed to meet the requirements for actual applications in cancer diagnosis. Clinically, no single biomarker can effectively predict the breast cancer stage and outcome. As a consequence, the detection of a biomarker

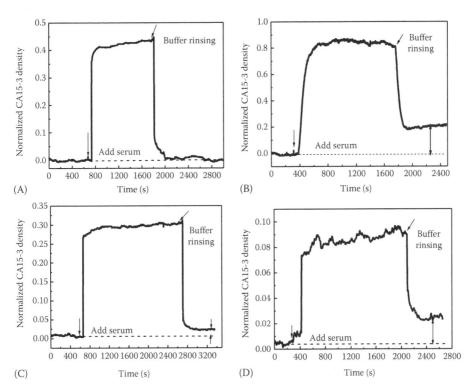

FIGURE 7.10 OFRR sensor response to diluted human sera. (A) No observable signal from sixfold dilution of healthy male serum. (B) Sensor response from sixfold dilution of stage IV breast cancer Patient #4 serum. Original concentration: 211 units/mL (lab test result). (C) Sensor response from 21-fold dilution of stage IV breast cancer Patient #1. Original concentration: 19 units/mL (lab test result). (D) Sensor response from 21-fold dilution of stage IV breast cancer Patient #2. Original concentration: 25 units/mL (lab test result). (Reprinted with permission from Zhu, H., Dale, P.S., Caldwell, C.W., and Fan, X., Rapid and label-free detection of breast cancer biomarker CA15-3 in clinical human serum samples with optofluidic ring resonator sensors, *Anal. Chem.*, 81, 9858–9865, 2009. Copyright 2009 American Chemical Society.)

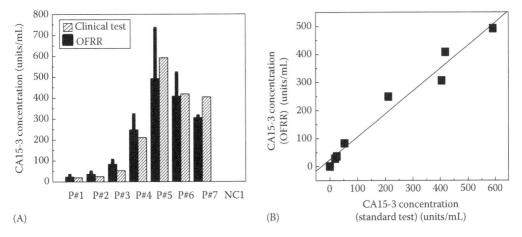

FIGURE 7.11 (A) Comparison of CA15-3 levels of breast cancer patients (Patients #1–#7) and a negative control (NC1) measured by the OFRR with 21-fold dilution and standard clinical test. Each point is measured three times. (B) Correlation between the OFRR sensor test results and the standard clinical test results. (Reprinted with permission from Zhu, H., Dale, P.S., Caldwell, C.W., and Fan, X., Rapid and label-free detection of breast cancer biomarker CA15-3 in clinical human serum samples with optofluidic ring resonator sensors, *Anal. Chem.*, 81, 9858–9865, 2009. Copyright 2009 American Chemical Society.)

panel is a requisite of any diagnostic strategy that utilizes them. Future work will concentrate on the improvement of the OFRR multiplexing detection capability. For example, multiple OFRR capillaries can be coated with different antibodies to detect different biomarkers. They can be further embedded in a low-RI polymer–based package to improve the mechanical robustness [13]. With those improvements, we envision that the OFRR sensor will eventually become a portable device in the clinical use for better understanding of how serum biomarkers function and how they are related to different diseases.

ACKNOWLEDGMENT

We acknowledge the financial support from the Wallace H. Coulter Foundation.

REFERENCES

1. Frangioni, J. V. 2008. New technologies for human cancer imaging. *J. Clin. Oncol.* **26**:4012–4021.
2. Wagner, H. N. and P. S. Conti. 1991. Advances in medical imaging for cancer diagnosis and treatment. *Cancer* **67**:1121–1128.
3. Hanash, S. M., S. J. Pitteri, and V. M. Faca. 2008. Mining the plasma proteome for cancer biomarkers. *Nature* **452**:571–579.
4. Ludwig, J. A. and J. N. Weinstein. 2005. Biomarkers in cancer staging, prognosis and treatment selection. *Nat. Rev. Cancer* **5**:845–856.
5. Diamandis, E. P., H. A. Fritche, H. Lilja, D. W. Chan, and M. K. Schwartz. 2002. *Tumor Markers: Physiology, Pathobiology, Technology and Clinical Applications*. AACC Press, Washington, DC.
6. Caplan, A. and A. Kratz. 2002. Prostate-specific antigen and the early diagnosis of prostate cancer. *Am. J. Clin. Pathol.* **117**:S104–S108.
7. Pegram, M. D., G. Pauletti, and D. J. Slamon. 1998. HER-2/neu as a predictive marker of response to breast cancer therapy. *Breast Cancer Res. Treat.* **52**:65–77.
8. Duffy, M. J. 2005. CA15-3 and related mucins as circulating markers in breast cancer. *Ann. Clin. Biochem.* **51**:494–503.
9. Ferrone, C. R., D. M. Finkelstein, S. P. Thayer, A. Muzikansky, C. F. Castillom, and A. L. Warshaw. 2006. Perioperative CA19-9 levels can predict stage and survival in patients with resectable pancreatic adenocarcinoma. *J. Clin. Oncol.* **24**:2897–2902.
10. Zangar, R. C., D. S. Daly, and A. M. White. 2006. ELISA microarray technology as a high-throughput system for cancer biomarker validation. *Expert Rev. Proteomics* **3**:37–44.
11. Fan, X., I. M. White, H. Zhu, J. D. Suter, and H. Oveys. 2007. Overview of novel integrated optical ring resonator bio/chemical sensors. *Proc. SPIE* **6542**:64520M 1–20.
12. Fan, X., I. M. White, S. I. Shopova, H. Zhu, J. D. Suter, and Y. Sun. 2008. Sensitive optical biosensors for unlabeled targets: A review. *Anal. Chim. Acta.* **620**:8–26.
13. Suter, J. D., Y. Sun, D. J. Howard, J. A. Viator, and X. Fan. 2008. PDMS embedded opto-fluidic microring resonator lasers. *Opt. Express* **16**:10248–10253.
14. Suter, J. D., I. M. White, H. Zhu, H. Shi, C. W. Caldwell, and X. Fan. 2008. Label-free quantitative DNA detection using the liquid core optical ring resonator. *Biosens. Bioelectron.* **23**:1003–1009.
15. White, I. M., H. Oveys, and X. Fan. 2006. Liquid-core optical ring-resonator sensors. *Opt. Lett.* **31**:1319–1321.
16. Yang, G., I. M. White, and X. Fan. 2008. An opto-fluidic ring resonator biosensor for the detection of organophosphorus pesticides. *Sens. Actuators B* **133**:105–112.
17. Zhu, H., J. D. Suter, I. M. White, and X. Fan. 2006. Aptamer based microsphere biosensor for thrombin detection. *Sensors* **6**:785–795.
18. Zhu, H., I. M. White, J. D. Suter, P. S. Dale, and X. Fan. 2007. Analysis of biomolecule detection with optofluidic ring resonator sensors. *Opt. Express* **15**:9139–9146.
19. Zhu, H., I. M. White, J. D. Suter, and X. Fan. 2008. Phage-based label-free biomolecule detection in an opto-fluidic ring resonator. *Biosens. Bioelectron.* **24**: 461–466.
20. Zhu, H., I. M. White, J. D. Suter, M. Zourob, and X. Fan. 2008. Opto-fluidic micro-ring resonator for sensitive label-free viral detection. *Analyst* **133**:356–360.

21. White, I. M., J. Gohring, Y. Sun, G. Yang, S. Lacey, and X. Fan. 2007. Versatile waveguide-coupled opto-fluidic devices based on liquid core optical ring resonators. *Appl. Phys. Lett.* **91**:241104.
22. Chang, R. K. and A. J. Campillo. 1996. *Optical Processes in Microcavities*. World Scientific Publishing Co. Pte Ltd., Singapore.
23. Ghoreyshi, A. A., F. A. Farhadpour, M. Soltanieh, and A. Bansal. 2003. Transport of small polar molecules across nonporous polymeric membranes. *J. Membr. Sci.* **211**:193–214.
24. Arnold, S., M. Khoshsima, and I. Teraoka. 2003. Shift of whispering-gallery modes in microspheres by protein adsorption. *Opt. Lett.* **28**:272–274.
25. Kumpulainen, E. J., R. J. Keskikuru, and R. T. Johansson. 2002. Serum tumor marker CA15-3 and stage are the two most powerful predictors of survival in primary breast cancer. *Breast Cancer Res. Treat.* **76**:95–102.
26. Shering, S. G., F. Sherry, and E. W. McDermott. 1998. Preoperative CA15-3 concentrations predict outcome of patients with breast carcinoma. *Cancer* **83**:122521–122527.
27. Zhu, H., P. S. Dale, C. W. Caldwell, and X. Fan. 2009. Rapid and label-free detection of breast cancer biomarker CA15-3 in clinical human serum samples with optofluidic ring resonator sensors. *Anal. Chem.* **81**:9858–9865.
28. Ladd, J., Z. Zhang, S. Chen, J. C. Hower, and S. Jiang. 2008. Zwitterionic polymers exhibiting high resistance to nonspecific protein adsorption from human serum and plasma. *Biomacromolecules* **9**:1357–1361.
29. Teramura, Y. and H. Iwata. 2007. Label-free immunosensing for a-fetoprotein in human plasma using surface plasmon resonance. *Anal. Biochem.* **365**:201–207.
30. Cui, X., F. Yang, Y. Sha, and X. Yang. 2003. Real-time immunoassay for ferritin using surface plasmon resonance biosensor. *Talanta* **60**:53–61.
31. Vaisocherova, H., V. M. Faca, A. D. Taylor, R. Hanash, and S. Jiang. 2009. Comparative study of SPR and ELISA methods based on analysis of CD166/ALCAM levels in cancer and control human sera. *Biosens. Bioelectron.* **15**:2143–2148.
32. Vaisocherova, H., W. Yang, Z. Zhang, Z. Cao, G. Cheng, M. Piliarik, J. Homola, and S. Jiang. 2008. Ultralow fouling and functionalizable surface chemistry based on a zwitterionic polymer enabling sensitive and specific protein detection in undiluted blood plasma. *Anal. Chem.* **80**:7894–7901.
33. Acharya, G., C. Chang, D. D. Doorneweerd, E. Vlashi, W. A. Henne, L. C. Hartmann, P. S. Low, and C. A. Savran. 2007. Immunomagnetic diffractometry for detection of diagnostic serum markers. *J. Am. Chem. Soc.* **129**:15824–15829.

8

Resonant Waveguide Grating Biosensor for Cancer Signaling

Ye Fang and Ann M. Ferrie

CONTENTS

8.1 Introduction

Cancer is a collection of distinct genetic diseases. Central to cancer development is a single cell to progressively acquire genetic mutations that lead to the activation of oncogenes and/or inactivation of tumor suppressor genes [1–3]. The quest to discover the full complement of cancer development has led to identification of a wide range of oncogenes as well as genetic and epigenetic abnormalities, many of which encode signaling proteins preferentially involved in a small number of pathways [4–6]. Since signaling proteins often operate through a large and complex network, crucial to cancer biology is to elucidate the mechanisms how mutated proteins alter and govern signaling of cancer cells in the context of intracellular and/or intercellular signaling networks [7–10]. Integrative cellular assays now in development, such as label-free resonant waveguide grating (RWG) biosensor cellular assays, are making it possible to study various signaling pathways and their network interactions in a variety of disease states including cancers [11–15].

Central to the RWG biosensor is a nanograting waveguide, which consists of a thin film of Nb_2O_5 on a glass substrate having a nanograting structure (Figure 8.1). The Nb_2O_5 thin film acts as a waveguide, since it has higher refractive index (~2.36) than either the glass substrate (~1.5) or live cell does (~1.37). The diffraction grating allows the resonant coupling of light into the waveguide thin film, thus creating an electromagnetic field (i.e., evanescent wave) at the sensor surface [16].

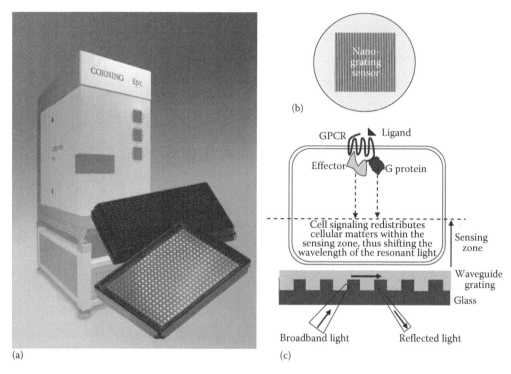

FIGURE 8.1 RWG biosensor and system for cancer signaling. (a) Corning Epic® system and Epic 384-well biosensor tissue-culture-treated microplate. (b) A scanning electron microscopic image of the nanograting waveguide embedded in the bottom of each well. (c) A schematic drawing showing the principle of the biosensor for whole-cell sensing. The activation of a GPCR leads to cell signaling mediated via both G protein– and effector protein–dependent mechanisms, which causes dynamic redistribution of cellular matters within the sensing zone, thus shifting the wavelength of the reflected light. Movement of biomass into the sensing zone (~150nm) leads to increase in local refractive of index, which, in turn, results in a positive shift in resonant wavelength (P-DMR); vice versa, biomass leaving the sensing volume results in a decrease in resonant wavelength (N-DMR).

The evanescent wave is sensitive to perturbations in local refractive index at or near the sensor surface [17], which can be induced by the presence of live cells or a stimulus-triggered cellular response [12]. Such a perturbation leads to a shift in resonant wavelength [12]. The biosensor detection system can noninvasively and continuously monitor the changes in resonant wavelength of the biosensor having live cells, leading to a real-time kinetic response of the cells upon stimulation. The resultant kinetic response is often termed as dynamic mass redistribution (DMR). This is because the local index of refraction within a cell is known to be directly proportional to the density and distribution of biomass (e.g., proteins, molecular complexes)—the biophysical characteristic behind the contrast of cells under light microscopy [18]. More importantly, there is significant redistribution of cellular matters in many, if not all, cell signaling and cellular processes [19,20]. Such a redistribution is often not random; instead, it is tightly regulated and is often dynamic both spatially and temporally [20,21]. Cell signaling, particularly mediated through receptors, often involves protein trafficking, microfilament remodeling, cell adhesion alterations, and morphological changes of cells, all of which can lead to DMR. Depending on the receptor and cellular background, different cellular event(s) may dominate a receptor DMR signal in a specific cell.

Since DMR is common to many cell signaling and cellular processes, it is not surprising to see in recent years that RWG biosensor cellular assays have found broad applications in a diverse array of cellular processes including adhesion [22,23], viral infection [24], proliferation [25], and apoptosis [26] of cells, as well as in cell signaling mediated through diverse types of receptors including G protein–coupled receptors (GPCRs) [27–29], ion channels [30], kinases [21,31], enzymes [32], and structural proteins [33]. Here, using human epidermoid carcinoma cell line A431 as a model system, we describe how to use RWG biosensor to mine functional receptors using a focused library of GPCR agonists and to determine the critical pathways downstream epidermal growth factor (EGF) receptor.

8.2 Materials

8.2.1 Reagents

1. Epinephrine (Sigma Chemical Co., St. Louis, MO)
2. Clonidine (Sigma)
3. Dopamine (Sigma)
4. Acetylcholine (Sigma)
5. Angiotensin (Bachem, King of Prussia, PA)
6. Bradykinin (Bachem)
7. Epidermal growth factor (Bachem)
8. SFLLR-amide (Bachem)
9. SLIGKV-amide (Bachem)
10. Substance P (Bachem)
11. Urotensin II (Bachem)
12. Vasoactive intestinal peptide (VIP) (Bachem)
13. Adenosine (Tocris Chemical Co., St. Louis, MO)
14. IB-MECA (Tocris)
15. R-(-)-phenylephrine (Tocris)
16. Histamine (Tocris)
17. ATP (Tocris)
18. CCPA (Tocris)
19. UTP (Tocris)
20. Prostaglandin D2 (Tocris)
21. Prostaglandin E2 (Tocris)
22. Lysophosphatidic acid (LPA) (Tocris)
23. Sphingosine-1-phosphate (S1P) (Tocris)
24. U0126 (Tocris)
25. AG1478 (Enzo Life Science, Plymouth Meeting, PA)
26. BML-265 (Enzo Life Science)
27. Rottlerin (Enzo Life Science)
28. Protease-free and lipid-free bovine serum (Sigma)

8.2.2 Tissue Culture Medium and Cell Line

1. Human epidermoid carcinoma A431 cells (American Type Cell Culture (ATCC), Manassas, VA)
2. Dulbecco's modified Eagle's medium (DMEM) (Invitrogen, Carlsbad, CA)
3. Fetal bovine serum (FBS) (Invitrogen)
4. Penicillin-streptomycin liquid (100×) (Invitrogen)
5. Complete cell culture medium (DMEM supplemented to a final concentration of 10% with FBS, 4.5 g/L of glucose, 2 mM of glutamine, 50 U/mL of penicillin, and 50 µg/mL of streptomycin)
6. Serum-free starvation medium (DMEM supplemented to a final concentration of 4.5 g/L of glucose, 2 mM of glutamine, 50 U/mL of penicillin, and 50 µg/mL of streptomycin)
7. Trypsin-EDTA (0.25% trypsin with EDTA 4Na$^+$) (Invitrogen)

8.2.3 Cell Culture Vessels, Microplates, and Instruments

1. Corning® 75 cm^2 tissue-culture-treated polystyrene flask with vented cap (T-75) (Corning Inc., Corning, NY)
2. Corning 96-well and 384-well polypropylene compound storage plates (Corning Inc.)
3. Corning Epic 384-well tissue-culture-treated microplate (Corning Inc.)
4. BioTek cell plate washer ELx405 (BioTek Instruments Inc., Winooski, VT)
5. Corning Epic® system (Corning Inc.)
6. Matrix 16-channel electronic pipettor (Thermo Fisher Scientific, Hudson, NH)
7. Beckman Multimek 96 liquid dispenser (Beckman Coulter, Brea, CA)

8.2.4 RWG Biosensor Cellular Assay Solutions

1. Hank's balanced salt solution (HBSS), 1× with calcium and magnesium but no phenol red (Invitrogen)
2. 1 M HEPES buffer solution, pH 7.1 (Invitrogen)
3. Assay buffered solution (1× HBSS, 10 mM HEPES, pH 7.1)

8.2.5 Data Analysis Softwares

1. Epic® Offline Viewer (Corning Inc.)
2. Microsoft Excel (Microsoft Inc., Seattle, WA)
3. Prism Software (Graph Pad Software Inc., La Jolla, CA)
4. TreeView (http://rana.lbl.gov/eisen/)

8.3 Methods

Figure 8.2 outlines the flowchart of RWG biosensor cellular assays for characterizing functional receptors in cancer cells. This one-step assay is often termed an agonism assay, which includes five critical steps: (1) preparation of a compound source plate, (2) preparation of a biosensor cell assay plate, (3) recording of the baselines, (4) addition of compounds, and (5) recording of cellular responses. Incorporating a second stimulation step, wherein a specific receptor agonist (e.g., the EGFR agonist EGF) is used to further stimulate the compound-treated cells, can be used to determine the impacts of different compounds on the agonist-induced DMR signal, thus mapping the core pathway(s) of the agonist-interacting receptor. Such a two-step assay is commonly termed an antagonism assay.

8.3.1 Preparation of GPCR Agonist Focused Library

GPCRs represent the largest family of cell surface receptors. Many GPCRs are up-regulated in cancer, leading to increased activation of G proteins and their associated signaling pathways. To mine the functional receptors in A431 cell line, a GPCR agonist focused library is prepared as follows:

1. All peptides and lipids are directly dissolved into 100% dimethyl sulfoxide (DMSO) to make 1 mM stock solutions.
2. All small organic molecules are directly dissolved into 100% DMSO to make 10 mM stock solutions.
3. The stock solutions are aliquoted into a 96-well microplate to make a mother plate, each well containing 50 μL of a stock compound solution.

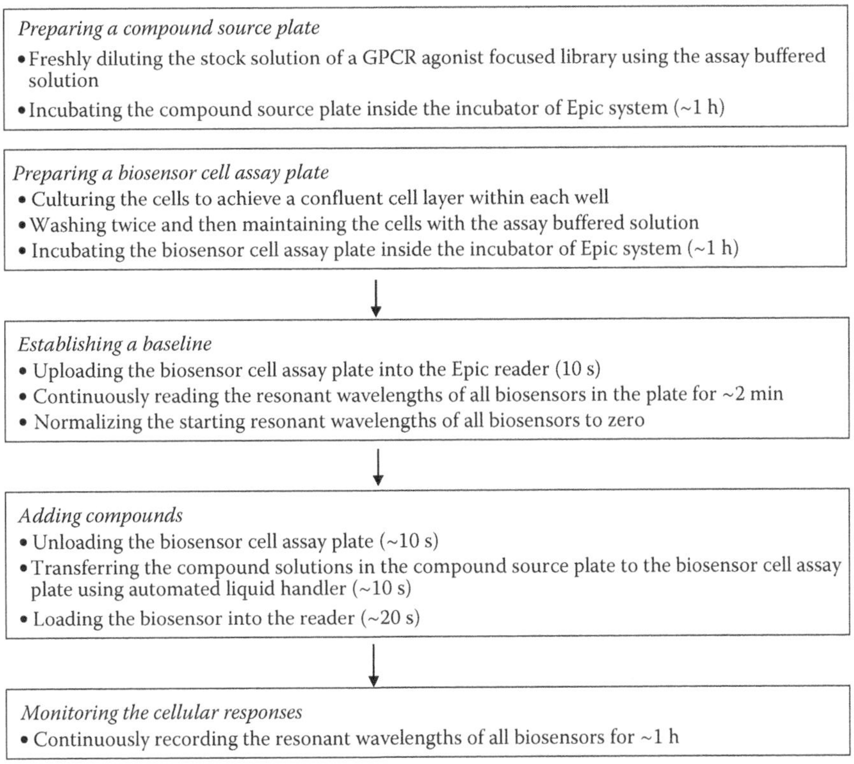

Preparing a compound source plate
- Freshly diluting the stock solution of a GPCR agonist focused library using the assay buffered solution
- Incubating the compound source plate inside the incubator of Epic system (~1 h)

Preparing a biosensor cell assay plate
- Culturing the cells to achieve a confluent cell layer within each well
- Washing twice and then maintaining the cells with the assay buffered solution
- Incubating the biosensor cell assay plate inside the incubator of Epic system (~1 h)

Establishing a baseline
- Uploading the biosensor cell assay plate into the Epic reader (10 s)
- Continuously reading the resonant wavelengths of all biosensors in the plate for ~2 min
- Normalizing the starting resonant wavelengths of all biosensors to zero

---

Adding compounds
- Unloading the biosensor cell assay plate (~10 s)
- Transferring the compound solutions in the compound source plate to the biosensor cell assay plate using automated liquid handler (~10 s)
- Loading the biosensor into the reader (~20 s)

Monitoring the cellular responses
- Continuously recording the resonant wavelengths of all biosensors for ~1 h

---

FIGURE 8.2 A flowchart of typical agonism assay using Epic system.

4. A mother compound plate is further aliquoted into a series of daughter plates, each well containing 1 μL of a stock compound solution.

5. Both mother and daughter plates after sealed are stored at −80°C.

8.3.2 Preparation of Cells for RWG Biosensor Cellular Assays

For RWG biosensor cellular assays, anchorage-dependent cells can be directly cultured onto the biosensor surface to form an adherent layer of cells, while suspension cells (e.g., lymphoblastic leukemia cells) can be brought to be closely in contact with the biosensor surface via physical sedimentation or specific binding between immobilized ligands and a cell surface receptor [12,15]. Since A431 is an anchorage-dependent cancerous cell line and cell signaling is often sensitive to cellular status (e.g., cell cycle, and proliferating and quiescent state), the following protocol employs standard cell culture and starvation techniques to prepare cells for biosensor cellular assays. Simple modifications of this protocol can be used for studying cell signaling in different types of cells including other cancerous cells. These modifications include different cell culture medium and duration, initial seeding numbers of cells, cell synchronization, and specific coatings of biosensor surfaces.

1. A431 cells received from ATCC are passed in T-75 flasks using the recommended protocol. The cells up to at least 15 passages after being received from ATCC are appropriated for the biosensor cellular assays in our laboratory. The native A431 cells when approaching confluence are passed with trypsin/EDTA to provide new maintenance cultures on T-75 flasks and experimental cultures on Epic 384-well tissue-culture-treated microplates. A 1:10 split of the cells will provide maintenance cultures on T-75 flasks such that they are approaching confluence after 5 days. Cells on one T-75 flask are sufficient for experimental cultures on at least one biosensor microplate.

2. Once the cells reach confluence on the T-75 flask, the cells are harvested using 1× trypsin-EDTA solution. The cells are then centrifuged down to form cell pellet. After the supernatant is removed, the cell pellet is washed once to remove any remaining trypsin, and then resuspended in freshly prepared complete cell culture medium.

3. ~2×10^4 cells at passages 3–15 suspended in 50 µL of the complete medium are added into each well of an Epic 384-well biosensor microplate. All biosensor microplates are tissue culture treated using standard manufacturing process and packaged under sterilization condition, and thus can be used directly for cell culture. The cell plating is carried out using a 16-channel electronic pipettor. Air bubbles trapped in the bottoms of the wells, if any, are manually removed by the pipettor.

4. After incubation inside the laminar flow hood for ~20 min, the cells are cultured for ~24 h at 37°C under air/5% CO_2. At this point, the cells usually reach ~80%–90% confluency. The cells are rinsed twice with the serum-free starvation medium, and then incubated for another 24 h in the serum-free medium at 37°C under air/5% CO_2. During this starvation step, the cells continue growing until approaching ~100% confluency and finally reach a quiescent state via the serum withdrawal and contact inhibition [34].

8.3.3 Mining Functional Receptors with RWG Biosensor Cellular Assays

To mine functional GPCRs in A431 cells, the one-step agonism assay is used. This agonism assay utilizes a GPCR agonist focused library to stimulate the cells. The DMR signals of all GPCR agonists are recorded in real time using Corning Epic system. This system consists of a RWG detector, an external liquid handling accessory, and a scheduler enabled with internal and external robotics, such that it can process large numbers of microplates using end-point measurements for high-throughput screening (HTS) or using kinetic measurements for high information content assays. Central to the system is a scanning wavelength interrogation detector (Figure 8.3), enabling microtiter plate–based biomolecular interaction analysis and cell-based assays [12,35–37]. The agonism assay starts with a baseline recording and proceeds with a compound addition step (~40 s) and a subsequent real-time kinetic measure of cellular responses (Figure 8.4):

1. A daughter plate of the focused GPCR agonist library is freshly thawed and used once.

2. The assay buffered solution without and with 0.1% high-grade bovine serum albumin (BSA) is freshly made.

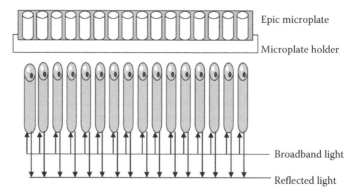

Epic microplate

Microplate holder

Broadband light

Reflected light

FIGURE 8.3 A schematic drawing of scanning wavelength interrogation Epic reader system. The system uses a linear array of 16 illumination/detection heads to scan a 384-well microplate, one column at a time. In each bundled head, there is an optical fiber to send a laser light to illuminate a biosensor located in each well and a second optical fiber to receive and record the reflected light. The illumination wavelengths are centered on ~830 nm. Time to scan the whole plate is ~6 s. The microplate sits on a microplate holder. The movement of microplate in and out the system is handled using both internal and external built robotics.

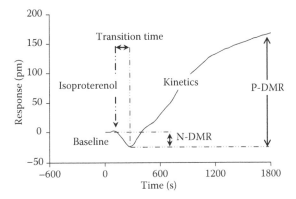

FIGURE 8.4 Characteristics of the DMR response of a quiescent A431 cell adlayer upon stimulation with isoproterenol of 0.1 nM. Multiparameters can be extracted from the DMR signal. Isoproterenol is a full agonist of β_2AR, which endogenously expresses in A431.

3. The compounds in the daughter plate are diluted 250-fold to make 4× compound solutions and then transferred into a 384-well polypropylene compound storage plate using Beckman Multimek 96 liquid dispenser to prepare a compound source plate. Except for peptides and lipids that are diluted with 0.1% BSA, all other compounds are directly diluted using the assay buffered solution. The DMSO concentration is ~0.4%. The wells in the columns 1 and 2 only contain the assay buffered solution having 0.4% DMSO, while the wells in columns 23 and 24 are the assay buffered solution having 0.4% DMSO and 0.1% BSA. The assay vehicle–induced DMR can be subtracted from compound-induced DMR signals to remove background interferences, if any. The presence of high-grade BSA (0.1%) generally does not cause any obvious DMR signal.

4. The starved cells are washed twice with the assay buffered solution using the BioTek washer and maintained in 30 µL of the assay buffered solution to prepare a cell assay plate.

5. Both the cell assay plate and the compound source plate are incubated in the incubator of the Epic system for ~1 h to minimize artifacts induced by temperature difference between solutions.

6. Afterward, the baseline wavelengths of all biosensors in the cell assay microplate are recorded and normalized to zero, followed by a 2 min continuous recording to establish a baseline. At this point, the baseline should reach a steady state, leading to a net-zero response (Figure 8.4).

7. Cellular responses are triggered by transferring 10 µL of the 4× compound solutions from the compound source plate into the cell assay plate using the external liquid handling accessory. Afterward, the shifts in resonant wavelength are continuously monitored for ~1 h to generate DMR signals.

8. The real wavelength shifts at specific time points poststimulation, as indicated in Figure 8.5, are extracted from each agonist DMR signal and used as a basis for calculating the similarity of any given pairs of DMR signals using hierarchical Euclidean cluster methods. The heat map is further generated using TreeView software developed by Prof. M.B. Eisen and his colleagues at University of California at Berkeley (http://rana.lbl.gov/eisen/).

8.3.4 Determining Pathways Downstream EGFR Signaling Using Biosensor Cellular Assays

Pathway deconvolution using RWG biosensor can be achieved via examining the impacts of chemical or biochemical (e.g., RNAi) interventions of various signaling cascade proteins on a receptor agonist–induced DMR. Although the agonism assay can also be used in costimulation, sequential two-step assay is the most appropriate assay format for pathway deconvolution. This is partly because the biosensor is

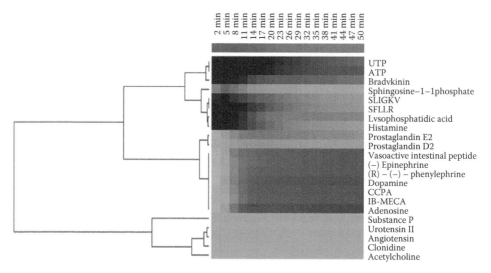

FIGURE 8.5 Heat map of diverse GPCR agonist DMR signals of quiescent A431 cells. The heat map was generated using the hierarchical Euclidean cluster analysis and TreeView. (From Eisen, M.B. et al., *Proc. Natl. Acad. Sci. USA*, 95, 14863, 1998; Fang, Y., *Drug Discov. Today Technol.*, 7, e5, 2010.) The real responses of all DMR signals at discrete time points, as indicated, were used as the basis for similarity analysis. The amplitude and direction is indicated by color— the black indicates a positive value, the white a negative value, and the light gray a value close to zero.

multiplexing in nature and partly because many compounds display polypharmacology. The sequential assays can effectively separate a compound-induced cellular response from the impact of the compound on the receptor agonist–induced response. The following is an example of pathway deconvolution of EGFR signaling in A431 cells using a chemical biology approach:

1. A compound source plate containing various kinase inhibitors, each at four replicates, is freshly made. Each inhibitor in 10 mM 100% DMSO is diluted using the assay buffered solution into 4× 10 μM solution containing 0.4% DMSO.

2. The starved A431 cells in the biosensor microplate are washed twice with the assay buffered solution and maintained in 30 μL of the buffered solution.

3. Both the compound source plate and the cell assay plate are incubated in the incubator of the reader system for ~1 h. The inhibitor solutions are then transferred into the cell assay plate.

4. Fifty minutes after incubation with the inhibitors, the cell assay plate is assayed using the biosensor system to establish a baseline. Afterward, a 10 μL solution containing 5× 32 nM EGF in 0.1% BSA is transferred into the cell assay plate and mixed once. The EGF responses are then recorded in real time for about 1.5 h.

5. The impacts of each inhibitor is examined by comparing the EGF DMR signal in the absence and presence of the inhibitor and used as an indicator whether the cellular target(s) or pathway(s) with which the inhibitor intervene are involved in the EGFR signaling.

8.4 Results

RWG biosensor cellular assay utilizes a label-free optical biosensor to measure dynamic redistribution of cellular matters, equivalent to DMR, upon receptor activation. The DMR is manifested by the shift in resonant wavelength (in picometer) of the biosensor [16]. The wavelength shift may vary in magnitude and direction (positive or negative) over time depending on how different activated signaling pathways cause various intracellular molecules to relocate within the sensing zone of the biosensor. Thus, receptor activity is translated into a complex optical trace ("fingerprint") that represents the generic response of living cells, reminiscent of the holistic responses obtained in tissue or organ bath experiments. A DMR signal often contains certain characteristics, including the phases, amplitudes, kinetics, and durations of distinct DMR events (e.g., positive DMR [P-DMR] and negative DMR [N-DMR]). Each characteristic can be analyzed and used to study receptor signaling and ligand pharmacology [21,29]. As shown in Figure 8.4, the β2-adrenergic receptor full agonist isoproterenol triggered a DMR signal in quiescent A431 cells that consists of two phases—an initial N-DMR and a sequential P-DMR. The amplitudes of both N-DMR and P-DMR events define the efficacy of isoproterenol to activate the β2-adrenergic receptor. Fitting the P-DMR event with nonlinear regression leads to a characteristic kinetics of receptor signaling. Profiling a large panel of β2-adrenergic receptor–specific agonists in A431 cells using DMR assays suggests that different agonists lead to distinct kinetics, which is related to the ability of agonists to cause receptor internalization [29].

Cell surface receptors comprise more than a third of all cellular proteins and are important mediators in signal transduction. Cancer cells are often characterized by the up-regulation and/or mutations of these receptors [38]. Since RWG biosensor is multiplexing in nature, we were interested in mining the functional GPCRs in A431 cells using a GPCR agonist focused library. Stimulation of quiescent A431 cells led to robust DMR signals for a subset of agonists in the library. Representative examples were shown in Figure 8.6. Results showed that the DMR measurements are highly reproducible, and the DMR signals obtained exhibited complicated patterns, many of which differ in the overall shape,

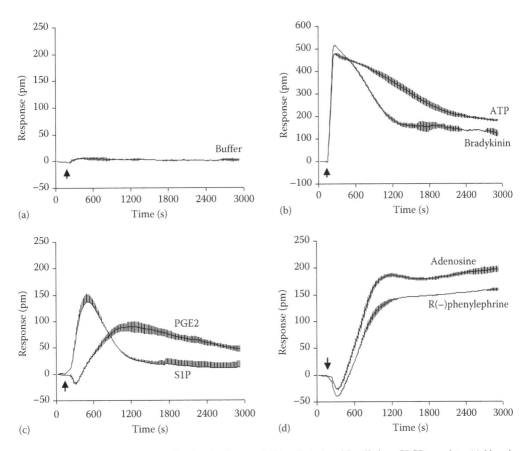

FIGURE 8.6 Representative DMR signals of quiescent A431 cells induced by distinct GPCR agonists. (a) Negative control (i.e., buffer). (b) The Gq-coupled P2Y2 receptor agonist ATP, and the the Gq-dominant bradykinin B2 receptor agonist bradykinin; (c) The Gq-dominant S1P receptors agonist sphinosine-1-phosphate, and the G_s-coupled prostaglandin EP4 receptor agonist PGE2; (d) The Gs-coupled adenosine A2B receptor agonist adenosine, and the Gs-coupled beta2-adrenergic receptor agonist R(-)phenylephrine. Each graph represents an averaged response of 8 replicates. Each agonist was assayed at 10 micromolar. The arrows indicate the time when agonists are introduced.

amplitudes, kinetics, and durations. It is obviously difficult to classify different agonists based on the kinetic parameters of all agonist-induced responses in a large library, particularly when these responses are complicated and diverse. Thus, to study the relationships among different DMR signals, we adopted similarity analysis using hierarchical Euclidean clustering—a technology widely used in gene expression analysis [39,40]. The heat map obtained showed that these DMR signals can be classified into two classes at the lowest resolution (Figure 8.5). The first class includes substance P, urotensin II, angiotensin, clonidine, and acetylcholine, all of which did not result in any obvious DMR. All other DMR signals can be further classified into two categories, each consisting of several subclusters. The G_s-type DMR signal [29] was observed for the three known G_s-coupled β_2-adrenergic receptor agonist epinephrine, dopamine, and phenylephrine [29], and the three G_s-coupled adenosine A_{2B} receptor agonist adenosine, IB-MECA, and CCPA [27], the G_s-coupled VIP_1 agonist vasoactive intestinal peptide, and the G_s-coupled EP_4 agonist prostaglandin E_2 and prostaglandin D_2. The distinct G_q-like DMR signal [27,41] was observed for the G_q-coupled $P2Y_2$ receptor agonist ATP and UTP [27], the G_q-coupled histamine receptor H_1R agonist histamine [32], the G_q-dominant bradykinin B_2 receptor agonist bradykinin [41], the G_q-dominant S1P receptor agonist sphingosine-1-phosphate, the G_q-dominant LPA receptor agonist LPA [27], and the G_q-dominant protease-activated receptor PAR_1 agonist SFLLR-amide and PAR_2 agonist SLIGKV-amide [42]. These DMR signals are not only correlated well with the expression of their corresponding receptor(s) in A431 (data unpublished) but also reflect the major pathways of distinct receptors. These results suggest that similarity analysis and heat map visualization are effective means to classify different agonists based on DMR assays.

EGF receptors belong to a family of receptor tyrosine kinases and are one of the most frequently mutated and/or dysregulated proto-oncogenes in many cancers [38]. A431 is well known for its overexpression of EGFR. Thus, we were interested in studying the core pathways downstream EGFR signaling in A431. Previously, we used RWG biosensor cellular assays to map the signaling and its network interactions of EGFR in A431 cells [21]. The EGFR signaling was found to require its intrinsic tyrosine kinase activity and to be mostly originated from the internalized receptors. The EGFR signaling also led to actin remodeling, dynamin- and clathrin-dependent receptor internalization, and MEK pathway–mediated cell detachment (possibly via FAK). To further determine the core pathways of EGFR signaling, a judicious selection of kinase inhibitors was made to examine their impacts on the EGF DMR signal in quiescent A431 cells. Figure 8.7 shows the different sensitivity of the EGF DMR signal in quiescent A431 cells to distinct kinase modulators. As expected, the EGFR tyrosine kinase inhibitors, A1478 and BML-265, almost completely blocked the EGF signal. However, the MEK1/2 inhibitor U0126 selectively attenuated the late DMR event, and the protein kinase C (PKC) inhibitor rottlerin selectively blocked the early DMR event. These results suggest that distinct pathways preferentially occur at different time domains during signaling—the PKC pathway plays important role in the early cellular response upon the activation of EGFR, but the MAPK pathways dominate in the late response. These results also suggest that cell detachment dominates in the late response mainly via MEK pathway, since MEK activation is required for the EGF-induced cell detachment [21]. Cell detachment causes a decrease in mass density and distribution within the sensing zone of the biosensor, leading to an N-DMR event.

8.5 Discussions

Conventional genomic approaches have been very fruitful for the discovery of genetic and epigenetic mutations [1–6]. Since it is proteins, but not genes, that fulfill most biological functions of cells, the functional consequences of these genetic abnormalities are still largely uncharted by these approaches. The wide adoption of recombinant DNA technologies has made molecular characterization assays possible to delineate many oncogenic pathways. However, such linear view of cancer signaling is considered to be insufficient nowadays, since the signaling proteins rarely operate in isolation through linear pathways but rather through a large and complex network. Thus, cellular assays that are integrative in measurements, as promised by RWG biosensors, are advantageous and desired.

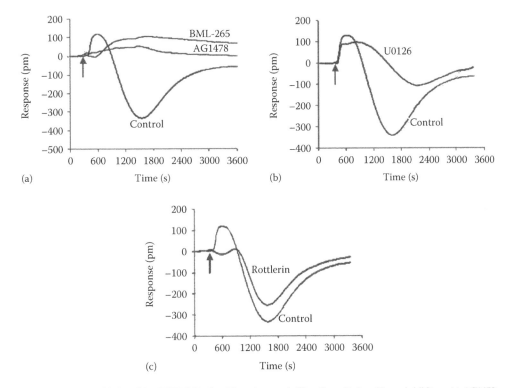

FIGURE 8.7 The sensitivity of the EGF DMR signal in quiescent A431 cells to distinct kinase inhibitors: (a) AG1478 and BML-265, (b) U0126, and (c) rottlerin. The control is the EGF response of cells pretreated with the assay vehicle only. In all experiments, EGF was at 32 nM, while all other compounds were at 10 μM. The arrows indicate the time when EGF is introduced.

RWG biosensor cellular assays have numerous potential applications in cancer signaling, ranging from mining the cell surface proteome, discovering the pathways downstream a receptor and the core pathways in a cancerous cell, and examining the drug responsiveness. This is made possible due to several critical characteristics of the biosensor cellular assays. First, the biosensor is label free and non-invasive, thus permitting multiple assay formats without any manipulations and alterations to live cells [40]. Second, the biosensor offers high sensitivity due to the amplification potential of redistribution of cellular matters downstream receptor signaling, thus permitting the direct measurement of endogenous receptors [16]. Third, the biosensor is pathway unbiased but pathway sensitive, thus permitting large scale mapping of functional receptors and pathways in a cancer cell [27]. Lastly, the DMR signals obtained offer high information content and contain contributions from any signaling pathways and events having significant redistribution of cellular matters within the sensing volume of the biosensor, thus allowing the systems cell biology study of the core pathways and network interactions downstream a receptor [21,41].

However, RWG biosensor, label-free biosensor in general, is largely nonspecific, due to the collective outcome of several factors. First, the biosensor is multiplexing in nature, meaning that any cellular events related to DMR can be detected. Second, many molecules often display polypharmacology, suggesting that the exact cellular mechanisms that lead to detectable DMR signal need to be deconvoluted. Third, the types of possible DMR signals are mathematically much less than possible biological spaces in a cell, suggesting that the modulations of distinct classes of cellular targets may lead to similar DMR. Thus, novel assay design and methodologies are further required to deconvolute the cellular events underlying the DMR profiles, thereby linking distinct receptor-mediated signaling pathways and process to each DMR event. Similar to any other molecular characterization assays, conventional approaches (e.g., chemical biology, RNAi knockdown, gene transfection, and cell biology) can be used to further validate the results obtained via DMR assays.

8.6 Perspective

Cancer remains to be one of the most devastating diseases today. Understanding the origin and development of cancer is crucial for discovering the next-generation anticancer therapies and for developing novel means to achieve effective diagnosis. Advances in functional genomics have identified a great variety of genetic abnormalities associated with cancerogenesis, while molecular imaging and quantification assays have further elucidated the complex interactions within pathways. Understanding the systems biology and pharmacology are the next frontiers in cancer research. Integrative cellular assays, as demonstrated by RWG biosensors, are finding their advantages in cancer research. The ability to assay endogenous receptors without any manipulations makes it possible to characterize cancer cells including primary cells and to identify responsive therapeutics for specific cancers. Next-generation RWG biosensors will have higher spatial resolution [43,44], enabling noninvasively measure cancer signaling in single cells including rare cancer cells and in mixed populations of cancer cells such as tissue cells, reprogrammed cells, and unpurified primary cells. Together with the development of novel methodologies, RWG biosensor, label-free biosensors in general, will provide new perspectives in cancer biology.

REFERENCES

1. Vogelstein, B. and K.W. Kinzler. 2004. Cancer genes and the pathways they control. *Nat. Med.* 10:789–799.
2. Weinstein, I.B. 2002. Cancer. Addiction to oncogenes—The Achilles heal of cancer. *Science* 297:63–64.
3. Weinstein, I.B. and A. Joe. 2008. Oncogene addiction. *Cancer Res.* 68:3077–3080.
4. Wood, L.D., D.W. Parsons, S. Jones, J. Lin, T. Sjoblom, R.J. Leary, D. Shen, S.M. Boca, T. Barber, J. Ptak et al. 2007. The genomic landscapes of human breast and colorectal cancers. *Science* 318:1108–1113.
5. Jones, S., X. Zhang, D.W. Parsons, J.C. Lin, R.J. Leary, P. Angenendt, P. Mankoo, H. Carter, H. Kamiyama, A. Jimeno, S.M. Hong, B. Fu, M.T. Lin, E.S. Calhoun, M. Kamiyama, K. Walter, T. Nikolskaya, Y. Nikolsky, J. Hartigan, D.R. Smith, M. Hidalgo, S.D. Leach, A.P. Klein, E.M. Jaffee, M. Goggins, A. Maitra, C. Iacobuzio-Donahue, J.R. Eshleman, S.E. Kern, R.H. Hruban, R. Karchin, N. Papadopoulos, G. Parmigiani, B. Vogelstein, V.E. Velculescu, and K.W. Kinzler. 2008. Core signaling pathways in human pancreatic cancers revealed by global genomic analyses. *Science* 321:1801–1806.
6. Velculescu, V.E. 2008. Defining the blueprint of the cancer genome. *Carcinogenesis* 29:1087–1091.
7. Hanahan, D. and R.A. Weinberg. 2000. The hallmarks of cancer. *Cell* 100:57–70.
8. Kroemer, G. and J. Pouyssegur. 2008. Tumor cell metabolism: Cancer's Achilles' heel. *Cancer Cell* 13:472–482.
9. Luo, J., N.L. Solimini, and S.J. Elledge. 2009. Principles of cancer therapy: Oncogene and non-oncogene addiction. *Cell* 136:823–837.
10. Hahn, W.C. and R.A. Weinberg. 2002. Modelling the molecular circuitry of cancer. *Nat. Rev. Cancer* 2:331–341.
11. Fang, Y. 2006. Label-free cell-based assays with optical biosensors in drug discovery. *Assays Drug Dev. Technol.* 4:583–595.
12. Fang, Y. 2007. Non-invasive optical biosensor for probing cell signaling. *Sensors* 7:2316–2329.
13. Rocheville, M. and J.C. Jerman. 2009. 7TM pharmacology measured by label-free: A holistic approach to cell signaling. *Curr. Opin. Pharmacol.* 9:643–649.
14. Kenakin, T. 2009. Cellular assays as portals to seven-transmembrane receptor-based drug discovery. *Nat. Rev. Drug Discov.* 8:617–626.
15. Fang, Y., A.G. Frutos, and R. Verklereen. 2008. Label-free cell assays for GPCR screening. *Comb. Chem. HTS* 11:357–369.
16. Fang, Y., A.M. Ferrie, N.H. Fontaine, J. Mauro, and J. Balakrishnan. 2006. Resonant waveguide grating biosensor for living cell sensing. *Biophys. J.* 91:1925–1940.

17. Tiefenthaler, K. and W. Lukosz. 1989. Sensitivity of grating couplers as integrated-optical chemical sensors. *J. Opt. Soc. Am. B* 6:209–220.
18. Barer, R. and S. Joseph. 1954. Refractometry of living cells. Part I. Basic principles. *Quart. J. Microsc. Science* 95:399–423.
19. Shumay, E., S. Gavi, H.Y. Wang, and C.C. Malbon. 2004. Trafficking of β2-adrenergic receptors: Insulin and β2-agonists regulate internalization by distinct cytoskeletal pathways. *J. Cell Sci.* 117:593–600.
20. Kholodenko, B.N. 2003. Four-dimensional organization of protein kinase signaling cascades: The roles of diffusion, endocytosis and molecular motors. *J. Exper. Biol.* 206:2073–2082.
21. Fang, Y., A.M. Ferrie, N.H. Fontaine, and P.K. Yuen. 2005. Characteristics of dynamic mass redistribution of EGF receptor signaling in living cells measured with label free optical biosensors. *Anal. Chem.* 77:5720–5725.
22. Ramsden, J.J. and R. Horvath. 2009. Optical biosensors for cell adhesion. *J. Recept. Signal Transduct.* 29:211–223.
23. Fang, Y. 2010. Label-free biosensor cellular assays for cell adhesion. *J. Adhesion Sci. Technol.* 24:1011–1021.
24. Owens, R.M., Q. Wang, J.A. You, J. Jiambutr, A.S.L. Xu, R.B. Marala, and M.M. Jin. 2009. Real-time quantitation of viral replication and inhibitor potency using a label-free optical biosensor. *J. Recept. Signal Transduct.* 29:195–201.
25. Ramsden, J.J., S.Y. Li, J.E. Prenosil, and E. Heinzle. 1995. Optical method for measurement of number and shape of attached cells in real time. *Cytometry* 19:97–102.
26. Voros, J., R. Graf, G.L. Kenausis, A. Bruinink, J. Mayer, M. Textor, E. Wintermante, and N.D. Spencer. 2000. Feasibility study of an online toxicological sensor based on the optical waveguide technique. *Biosens. Bioelectr.* 15:423–429.
27. Fang, Y., G. Li, and A.M. Ferrie. 2007. Non-invasive optical biosensor for assaying endogenous G protein-coupled receptors in adherent cells. *J. Pharmacol. Toxicol. Methods* 55:314–322.
28. Henstridge, C.M., N.A.B. Balenga, R. Schroder, J.K. Kargl, W. Platzer, L. Martini, S. Arthur, J. Penman, J.L. Whistler, E. Kostenis, W. Waldhoer, and A.J. Irving. GPR55 ligands promote receptor coupling to multiple signalling pathways. *Br. J. Pharmacol.* 160:604–614.
29. Fang, Y. and A.M. Ferrie. 2008. Label-free optical biosensor for ligand-directed functional selectivity of acting on β₂ adrenoceptor in living cells. *FEBS Lett.* 582:558–564.
30. Flemin, M.R. and L.K. Kaczmarek. 2009. Use of optical biosensors to detect modulation of Slack potassium channels by G protein-coupled receptors. *J. Recept. Signal Transduct.* 29:173–181.
31. Du, Y., Z. Li, L. Li, Z. Chen, S.Y. Sun, P. Chen, D.M. Shin, F.R. Khuri, and H. Fu. 2009. Distinct growth factor-induced dynamic mass redistribution (DMR) profiles for monitoring oncogenic signaling pathways in various cancer cells. *J. Recept. Signal Transduct.* 29:182–194.
32. Tran, E. and Y. Fang. 2009. Label-free optical biosensor for probing integrative role of adenylyl cyclase in G protein-coupled receptor signaling. *J. Recept. Signal Transduct.* 29:154–162.
33. Fang, Y., A.M. Ferrie, and G. Li. 2005. Probing cytoskeleton modulation by optical biosensors. *FEBS Lett.* 579:4175–4180.
34. Coller, H.A., L. Sang, and J.M. Roberts. 2006. A new description of cellular quiescence. *PLoS Biol.* 4:e83.
35. Tran, E. and Y. Fang. 2008. Duplexed label-free G protein-coupled receptor assays for high throughput screening. *J. Biomol. Screen.* 13:975–985.
36. Peters, M.F., F. Vaillancourt, M. Heroux, M. Valiquette, and C.W. Scott. 2010. Comparing label-free biosensors for pharmacological screening with cell-based functional assays. *Assay Drug Dev. Technol.* 8:219–227.
37. Lee, P.H., A. Gao, C. van Staden, J. Ly, J. Salon, A. Xu, Y. Fang, and R. Verkleeren. 2008. Evaluation of dynamic mass redistribution technology for pharmacological studies of recombinant and endogenously expressed G protein-coupled receptors. *Assay Drug Dev. Technol.* 6:83–93.
38. Sharma, S.V., D.W. Bell, J. Settleman, and D.A. Haber. 2007. Epidermal growth factor receptor mutations in lung cancer. *Nat. Rev. Cancer* 7:169–181.
39. Eisen, M.B., P.T. Spellman, P.O. Brown, and D. Botstein. 1998. Cluster analysis and display of genome-wide expression patterns. *Proc. Natl. Acad. Sci. USA* 95:14863–14868.
40. Fang, Y. 2010. Label-free receptor assays. *Drug Discov. Today Technol.* 7: e5–e11.

41. Fang, Y., G. Li, and J. Peng. 2005. Optical biosensor provides insights for bradykinin B_2 receptor signaling in A431 cells. *FEBS Lett.* 579:6365–6374.
42. Fang, Y. and A.M. Ferrie. 2007. Optical biosensor differentiates signaling of endogenous PAR_1 and PAR_2 in A431 cells. *BMC Cell Biol.* 8:e24.
43. Horvath, R., K. Cottier, H.C. Pedersen, and J.J. Ramsden. 2008. Mutlidepth screening of living cells using optical waveguide. *Biosens. Bioelectr.* 24:799–804.
44. Ziblat, R., V. Lirtsman, D. Davidov, and B. Aroeti. 2006. Infrared surface plasmon resonance: A novel tool for real time sensing of variations in living cells. *Biophys. J.* 90:2592–2599.

9

Optical Waveguide-Based Biosensors for the Detection of Breast Cancer Biomarkers

**Harshini Mukundan, John E. Shively, Aaron S. Anderson,
Nile Hartman, W. Kevin Grace, and Basil I. Swanson**

CONTENTS

9.1 Introduction

Breast cancer is the second most common form of cancer in the United States, with 200,000 new cases each year. The 5-year survival rate associated with the disease is >98% if diagnosed early (i.e., when localized). Survival rate markedly decreases with disease progression and is <26% for end-stage (stage IV, metastatic) disease [1–3]. It is therefore apparent that early detection of breast cancer is essential for survival. However, current methods of diagnosis cannot guarantee early detection of breast cancer. In women under 40 years of age, clinical self-exams are the only method used for early diagnosis. Mammography is used in women over 40 years of age but is associated with high false-positive rates. Indeed, the possibility of having breast cancer after "indicative" mammogram is only 9% [4,5]. Magnetic resonance imaging (MRI) is more sensitive than mammography for early detection. However, MRI is expensive and intensive and requires skilled technical assistance. Most importantly, it is not efficient at the detection of ductal carcinoma in situ. In general, current strategies for breast cancer diagnosis must be complimented by invasive biopsy for confirmation [6]. Accordingly, there is a clear need for reliable methods for the early detection of breast cancer.

9.1.1 Biomarkers for Breast Cancer Detection

Quantitative detection of biomarkers has promise for aiding the early diagnosis of many cancers. A biomarker has been defined as any characteristic that can objectively be measured and evaluated as an indicator of normal biological processes, pathogenic processes, or pharmacological responses to a therapeutic intervention [7]. Molecular biomarkers are said to present a tremendous opportunity to improve the outcome of people with cancer by enhancing detection and treatment approaches. Nass and Moses have said that "Biomarkers will be instrumental in making that transition" toward more reliable diagnosis of cancer. Indeed, carbohydrate antigen-125 has been valuable in tracking progression and relapse in ovarian cancers [8]. Perhaps the most recognized of all biomarkers used in cancer detection is the prostate-specific antigen. More recently, the use of other biomarkers such as human kallikrein 2, urokinase-type plasminogen activator receptor, and others along with the prostate-specific antigen has been suggested to improve predicative value of the diagnosis [9]. Several biomarkers, such as the carcinoembryonic antigen (CEA) [10,11], carbohydrate antigen 15-3 [10,11], human epidermal receptor 2 (HER2/Neu) [12], and the estrogen receptor [13], may also have value in detection, prognosis, and prediction of relapse in breast cancer. However, the American Cancer Society does not recommend the use of biomarkers for diagnosis of breast cancer [14]. Their reasoning is listed here, as well as counter arguments that defend biomarkers as excellent targets for diagnostics:

1. The aforementioned biomarkers are overexpressed in many types of cancers and also in some noncancerous conditions. For instance, CEA is overexpressed in blood, liver, lung, and ovarian cancers [15], as well as in smokers [16]. This problem can be resolved by sampling for the biomarker in the organ of interest. For example, expression of CEA in serum is not an indicator of a specific type of cancer. However, its presence in the aspirate fluid collected from the breast is a positive indicator of breast cancer [17].

2. Many biomarkers (e.g., estrogen receptors) are also expressed in the normal human host. Knowledge of basal expression levels and a quantitative measurement of the change in concentration with disease are therefore imperative for accurate diagnosis.

3. No single biomarker is expressed during the entire course of disease. Quantitative determination of biomarker profiles during the course of disease, and multiplex detection of a limited suite of such molecules, is required before their use in early detection.

4. Secreted concentrations of biomarkers in the host are very low, thereby requiring a very sensitive diagnostic strategy. In the case of samples such as the NAF, detection should also be achieved in very low sample volumes ($\sim$10–50 μL).

Conventional immunoassay platforms (e.g., ELISA) cannot address all of these requirements. Novel signal transduction technologies, such as optical sensing using waveguide-based platforms, are capable of addressing each of these requirements and can potentially be applied to biomarker-based early detection of cancer.

9.1.2 Optical Principles of Waveguide-Based Sensors

An optical waveguide guides electromagnetic waves in the optical spectrum. The intense optical field generated by waveguides has been exploited for sensitive biodetection by several investigators and commercial entities. The sensor team at Los Alamos National Laboratory (LANL) has developed an optical waveguide biosensor for the detection of biomarkers utilizing the evanescent field of single-mode planar optical waveguides (Figure 9.1). Single mode planar optical waveguides produce a high intensity field at the bioactive surface as well as offer discrimination from background fluorescence due to the exponential decay in the intensity of the evanescent field (Figure 9.1C). The team at LANL has developed a benchtop

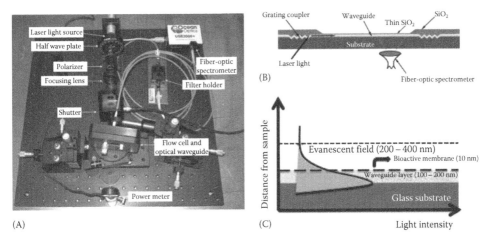

FIGURE 9.1 (A) Waveguide-based, benchtop optical biosensor instrument developed at LANL and its components. (B) Diagram of a single-mode Si_3N_4 planar optical waveguide structure. Grating couplers are used to couple in the incident excitation light into the thin Si_3N_4 waveguide film. The light is confined to the guided mode of the waveguide due to total internal reflection (TIR) with ~120 reflections per millimeter. A small amount of light leaks at each one of the reflections creating an evanescent field at the surface of the waveguide. The intensity of the evanescent field decays exponentially with distance to near complete loss of intensity at one half the wavelength of the excitation light (~250–350 nm). A fiber-optic spectrometer is oriented normal to the surface of the waveguide and is used to collect the emitted fluorescence from the sample. (C) Diagram showing the exponential decay of the evanescent field from the surface of a single-mode planar optical waveguide.

system, interfaced with a spectrometer, for accurate quantitative detection of biological molecules [18]. We have adapted this system for use with both single-channel and multichannel waveguides to increase sample throughput. In addition, the team has also developed a hand-held fieldable instrument for rapid detection of such biomolecules. The use of optical waveguides, with proper functionalization and transduction, offers significantly higher sensitivity for detection in complex biological samples, as will be described in this chapter.

Physical properties of single-mode planar optical waveguides as applicable to pathogen detection have been reviewed [18]. Single-mode planar waveguides are typically comprised of a very thin film dielectric material with a high refractive index (e.g., silicon oxynitride, tantalum pentoxide) that is deposited on a low index substrate [19]. They are more suitable for biosensor platforms, compared to multimode planar and fiber waveguides, because of several reasons: Single-mode planar optical waveguides support several thousand reflections per centimeter of beam propagation for visible wavelengths, two orders of magnitude higher than multimode planar and fiber waveguides. Thus, these waveguides have a greater sensitivity due to high evanescent field intensity at the surface. They also possess a high contrast that allows for the rapid decay of the evanescent field with increasing distance from the surface of the waveguide, with no appreciable intensity beyond one-half the wavelength of the excitation light (~250–300 nm). This allows for spatial resolution between surface-bound assay components and sample solution contaminants.

9.1.3 Surface Chemistry for Waveguide Functionalization

A sensing film provides the interface between the waveguide and the biological sample being sensed. Several different sensing films such as phospholipid bilayers, alkyl silanes with a distal reactive group (amines, thiols, epoxides, carboxylic acids, and aldehydes), polyethylene glycol chains, polyamino acids, and proteins adsorbed to a surface have been used in consort with waveguide-based biosensor platforms [18]. The sensor team at LANL has developed silane-based self-assembled monolayers (SAMs) that are very effective in resisting nonspecific binding (NSB) on the surface associated with complex biological samples such as serum and urine [20,21]. Our approach involves the functionalization of silica surfaces by self-assembly of an amine-terminated silane film using vapor deposition of

3′-aminopropylmethyldiethoxysilane (APDMES). The amine-terminated films are chemically modified with a mixture of carboxylic acid–terminated poly(ethylene glycol) (PEG) chains. A fraction of the PEG chains (0.1–10 mol %) terminate in biotin, which produces a surface with an affinity toward streptavidin. SAMs so generated are routinely characterized by a combination of methods such as ellipsometry and atomic force microscopy [20,21] and have been found to be stable for over a year in air, resistant to washing with detergents and harsh chemicals and are potentially reusable. SAMs are critical for efficient transition of our sensor platform to a point-of-care diagnostic application.

9.1.4 Commercially Available Planar Optical Waveguide-Based Technology

A comprehensive review of commercial technology is available [18]. Corning has developed EPIC®, a label-free platform, which uses resonant waveguide sensors. Here, binding of the antigen to the surface results in a change in the index of refraction, which is then measured as a shift in the wavelength. This technology has been applied to the detection of biological molecules including, but not limited to, specific proteins, biochemical assays, cell-based assays, and fibronectin measurements [22,23]. The ZeptoREADER™ is a fluorescence-based microarray readout system (532 and 635 nm lasers), which utilizes ZeptoCHIPs™ planar waveguides optimized for DNA and protein arrays and can be adapted for the detection of genes associated with breast cancer such as BRCA1. Recently, MicroVacuum® launched the OWLS 120™, a label-free biosensor that utilizes optical waveguide light mode spectroscopy (OWLS). The system measures refractive index changes with antigen binding and can be used for the investigation of binding processes in real time. The system has been evaluated for the multiplexed detection of biomarkers leading to early cancer diagnosis [24].

9.1.5 Novel Waveguide-Based Technology in Development

Several investigators have been evaluating the application of waveguides to biomarker detection. However, a detailed discussion of all these technologies is beyond the scope of this chapter. Two examples, from several, are discussed. Xu et al. have demonstrated a microfluidic chip with an integrated planar waveguide fabricated in poly(methylmethacrylate) and applied for the evaluation of DNA hybridization. The technology was evaluated in the detection of low abundance point mutations in BRCA1 genes associated with breast cancer. The polymorphisms present at less than 1% of DNA content were successfully detected using this approach, demonstrating excellent sensitivity within a short time (5 min) [25]. Another notable technology is the use of high-quality-factor microcavity resonators where light is coupled into the cavity via an adjacent waveguide positioned within the evanescent field. One such category is the silicon on insulator optical microring resonators applicable to label-free multiplexed biomolecular application. Washburn et al. have applied this technology to the detection of CEA in spiked bovine fetal serum. Compared to ELISA, a significantly better limit of detection (2 ng/mL) is achieved using this method within a much shorter time (15 min) with a dynamic range spanning three orders of magnitude [26]. This is comparable to the limit of detection achieved using our sensor platform and much superior than that achieved by plate-based immunoassays. Yet another example is the stacked planar affinity regulated resonant optical waveguide biosensor (SPARROW) that is currently being applied to the molecular screening of cancer drug design [27].

The rest of the chapter will focus on the development of a sensitive assay for CEA using the waveguide-based technology developed at LANL.

9.2 Materials

Antigens and antibodies: Humanized mouse monoclonal antibodies T84.1 and T84.66 were from the City of Hope Medical Center, as was recombinant CEA. *Optical*: The waveguide-based optical biosensor (Figure 9.1) was developed at LANL. SiON$_x$ planar optical waveguides were fabricated at nGimat Inc. (Atlanta). *Antibody processing*: EZ-link Sulfo-NHS-LC-LC-Biotin and streptavidin were procured from Pierce. Amino-functionalized quantum dots (QD; 655 nm, QD655), Qdot® 655 ITK™ carboxyl QDs, and AF647-*N*-hydroxysuccinimide ester labeling kits were from Invitrogen. Miniature G-25 sephadex columns were from Harvard Apparatus, and bovine serum was from Hyclone Laboratories. *Functional surfaces*: 1,2-Dioleoyl-sn-glycero-3-phosphocholine (DOPC) and 1,2-dioleoyl-sn-glycero-3-phosphoethanolamine-*N*-(cap biotinyl) were from Avanti Polar Lipids, Inc. 3-(APDMES) was from Gelest. All discrete, monodisperse polyethyleneglycol reagents were obtained from Quanta Biodesign, Ltd. Benzotriazole-1-yl-oxy-tris-pyrrolidino-phosphonium hexafluorophosphate (PyBOP) and *N*-(biotinyloxy)succinimide (Biotin-OSu) were purchased from NovaBiochem. *Other*: Buffers, chemicals, and all other components were from Sigma-Aldrich or Fisher Scientific unless otherwise specified. Patient samples (serum and nipple aspirate fluid [NAF]) were collected at the City of Hope under an IRB-approved protocol. Patient sample numbers indicated in the text (Figures 9.5 and 9.6) are as received from the City of Hope and have not been modified.

9.3 Methods

9.3.1 Choice of Antibodies

Selection of antibodies that bind orthogonal epitopes of an antigen is one of the most critical steps for development of a sensitive sandwich immunoassay. Dr. Shively's group at the City of Hope has identified several antibodies for CEA and developed an epitope map of antigen binding, which was used to guide antibody selection. Indeed, many of the issues associated with the detection of biomarkers in complex samples have been attributed to antibodies with poor sensitivity and high nonspecific interactions. Therefore, choice of antibodies is critical in determining the efficiency of the assay. In addition, a relatively long shelf life (>1 year) and stability after labeling (e.g., biotinylation) are desired. Preliminary immunoassay experiments were performed to determine that T84.1 (binds the *N* domain of CEA) is the most effective choice for capture antibody, whereas T84.66 (binds the A3 domain of CEA) was chosen as the reporter antibody.

9.3.2 Labeling, Purification, and Characterization of Antibodies

For enhancing sensitivity, minimizing NSB, and improving specificity, it is essential that the assay reagents be well characterized. The capture antibody (T84.1) is biotinylated, and the reporter antibody (T84.66) is labeled with a fluorescent dye for use in our platform. Labeling procedures are well characterized [28,29] and are only briefly described in this chapter. For biotinylation, the capture antibody was incubated with Sulfo-NHS biotin (20 mol excess) for 1 h at room temperature (RT) on a rotator. MacroSpin gel filtration columns were used for the separation of the labeled antibody from free biotin. Characterization of the biotinylated antibody involved determination of (1) degree of biotinylation using an EZ Biotin Quantitation Kit that uses 4'-hydroxyazobenzene-2-carboxylic acid, (2) functionality of the biotinylated antibody (i.e., ability to bind antigen) by immunoblot analysis on nitrocellulose filter paper using a goat anti-mouse secondary antibody labeled with alkaline phosphatase as the reporter, and (3) evaluation of biotinylation of the antibody binding of streptavidin labeled with alkaline phosphatase reporter to it on a nitrocellulose surface (immunoblot). Optimal labeling and uncompromised functionality of the antibody are essential for a sensitive assay. The protein concentration of the labeled material was measured using absorbance at 280 nm.

We have evaluated the use of both organic dyes (Alexa Fluor 647, AF647) and QDs as the fluorescent label. An AF647-NHS ester labeling kit was used for labeling with the dye, and the reaction was carried out according to the manufacturer's recommendation. Briefly, the antibody was reacted with AF647-NHS ester in the dark for 1 h and purified by gel filtration. Degree of labeling was determined by UV/Vis spectroscopy, based on the corrected ratio of absorbance at 650 nm (dye) and 280 nm (protein). For effective functionality, long-term stability, and preventing aggregation, it is essential that the extent of labeling is between ~4 and 8 mol of dye for every mole of protein. Higher degree of labeling facilitates formation of protein aggregates, increased NSB, and background fluorescence.

The use of QDs as a fluorescence reporter is driven by the need for a multiplex platform, capable of the simultaneous detection of several biomarkers using the same excitation source. QDs have been extensively used in biological applications and simply put, consist of nanoscale atom clusters with a semiconductor core and shell and an amphiphilic polymer coating that provides a protective, water-soluble surface that can be differentially modified [30,31]. A direct, predictable relationship between the physical size of QDs and the energy of the exciton (and therefore, the wavelength of emitted fluorescence) makes them tunable and, hence, highly attractive for use in multiplex detection. In addition, QDs exhibit excellent brightness and much greater photostability than traditional organic fluorophores [32], making long-term excitation possible under conditions that would lead to the photoinduced deterioration (photobleaching) of other types of fluorophores. These features have facilitated the use of QDs in several bioapplications such as immunofluorescence assays, live cell imaging, biotechnology detection, single-molecule biophysics, and in vivo animal imaging [30]. We have also examined both NH and hinge labeling of antibodies using QDs [28]. For labeling of antibodies with carboxyl-coated QDs with emission at 655 nm (QD655), they were mixed with the antibody and N-ethyl-N'-dimethylaminopropyl-carbodiimide hydrochloride (EDC, 100 mM) was added. Following a 2 h conjugation (RT), the reaction was quenched and the material was purified by electrophoresis followed by UV illumination (366 nm). The conjugate was separated by spin filtration followed by at least five exchanges with borate buffer. Absorbance of QD655 was measured to calculate its concentration ($\varepsilon_{638} = 8 \times 10^5$ M^{-1} cm^{-1}). For labeling of T84.66 using this strategy, a molar ratio of antibody to QD of 3:1 was typically observed in the conjugate. We have also evaluated hinge labeling of antibodies using QDs, and a detailed description of this procedure is previously published [28].

9.3.3 Choice of Waveguide and Functional Surfaces

The LANL waveguide biosensor is based on fluorescence measurements from single-mode planar optical waveguides. The waveguide biosensor instrument consists of a laser light source required for coupling in light into the planar optical waveguides (Figure 9.1A). Then a combination of a half wave plate and polarizer are used as an attenuator to adjust the laser power and orient the input polarization to the waveguide. A 150 mm focal length plano-convex lens focuses the laser light into the waveguide and increases coupling efficiencies. The shutter is used to control the laser exposure time to the sample and mitigate photobleaching of the sample. We use an Ocean Optics fiber-optic spectrometer to collect the fluorescence signal from the sample. The power meter is used for setting the laser intensity and measuring the coupling efficiency of the waveguide.

We have used conventional single-mode planar optical waveguides, which include both single channel and multichannel. The single-channel waveguide is fabricated from a 50.8 mm diameter 1 mm thick fused silica (SiO$_2$) optically polished substrate, and the multichannel is fabricated from a 76.2 mm diameter 2 mm thick fused silica (SiO$_2$) optically polished substrate. These substrates typically have a 10^{-5} scratch-dig specification and an rms surface roughness of 0.5 nm or less. The optically polished fused silica substrates have a nominal refractive index of $n = 1.46$ (low index). The grating is etched into the substrate and then a 110 nm ± 5 nm thick waveguide film of SiO$_2$ with a refractive index of $n = 1.81$ (high index) is deposited by ion-beam sputtering (IBS). Grating couplers are utilized to couple laser light into the thin waveguide film of Si$_3$N$_4$. This method of coupling was chosen because of its relaxed positional alignment tolerances and high coupling efficiencies and because it allows the ability to input couple light from either surface of the waveguide. To physically protect and isolate the gratings from ambient environmental effects, which could cause detuning of the coupling angle, the grating coupler region is overcoated with a thick layer of dense SiO$_2$ that is 1100 nm or thicker. This dense SiO$_2$ overcoat also defines the sensing regions of the waveguide. The defined sensing region of the waveguide is coated with a thin 10 nm thick SiO$_2$ film, which provides the functional groups required for covalent attachment of an appropriate sensing chemistry required for use in bioassays. The physical properties of single-mode planar waveguides have been reviewed [18]. After the waveguide coatings have been deposited on to the substrate, the substrates are cut to 50.8 mm × 25.5 mm for the single channel and 76.2 mm × 25.4 mm for the multichannel. Cutting the

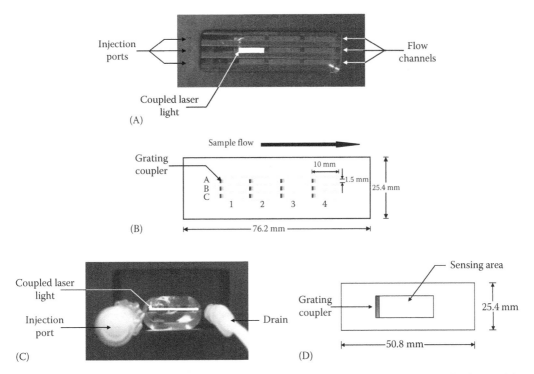

FIGURE 9.2 (A) Multichannel waveguide mounted in a flow cell. (B) Schematic representation of the fabrication of the multichannel waveguide. As indicated, there are three flow channels and four waveguide gratings within each channel in this prototype. (C) Single-channel waveguide mounted in flow cell. (D) Schematic representation of single-channel waveguide. The white line in either case indicates the streak observed when incident light is coupled into a grating.

substrate to these dimensions permits the substrate to be incorporated into a flow cell and mounted into a holder for the addition of reagents.

The single-channel waveguide is fabricated with a single grating coupler which is 2 mm × 10 mm and has a sensing area of 10 mm × 30 mm (Figure 9.3A and B). The flow cell for the single-channel waveguide consists of the single-channel waveguide, a 0.5 mm thick silicone rubber gasket with a laser-cut single channel, and a glass cover (76.2 mm × 25.4 mm × 1 mm) with two 1 mm diameter through holes bored through it which are then registered with the channel of the silicone gasket and sensing area of the waveguide. The flow cell is then captured into a holder to allow the addition of reagent via a septum and allow the integration into the experimental apparatus for measurement and detection. The total volume of the flow cell and holder is 70 µL.

An ideal biomarker-based detection platform should be capable of the simultaneous detection of a suite of such molecules in a sample. To achieve this goal, our team has also developed multichannel waveguides, consisting of three channels with four sensing elements per channel (Figure 9.2A and B). Each sensing element is 10 mm long and 1.5 mm wide and has a dedicated 1.5 mm × 1.5 mm grating coupler element, thus enabling each sensing element the ability to be excited independently of each other. All the channels and sensing elements have a chrome absorbing material surrounding them, thus allowing each element to be excited discretely without cross talk or inadvertent coupling between the sensing elements. The flow cell consists of the multichannel waveguide, a laser-cut three-channel 0.5 mm thick silicone rubber gasket and a glass cover (76.2 mm × 25.4 mm × 1 mm) with six 1 mm diameter holes bored into it at precise locations to align with the silicone rubber gasket, and the three individual sensing channels. This flow cell is then captured in a holder, which incorporates septum material to form a seal with the glass cover and allow the addition of reagents via a syringe and to be integrated into our experimental apparatus for measurement and detection. This configuration minimizes reagent volumes (100 µL/channel), is self-sealing, and allows the addition of reagents.

The LANL waveguide biosensor is based on fluorescence measurements of biomarkers in the evanescent field of single-mode planar optical waveguides. The waveguide biosensor instrument consists of a laser light source required for coupling in light into the planar optical waveguides (Figure 9.1). Then, a combination of a half wave plate and polarizer are used as an attenuator to adjust the laser power and orient the input polarization to the waveguide. A 150 mm focal length plano-convex lens focuses the laser light into the waveguide and increases coupling efficiencies. The shutter is used to control the laser exposure time to the sample and mitigate photobleaching. We use an Ocean Optics fiber-optic spectrometer to collect the fluorescence signal from the sample. The power meter is used for setting the laser intensity and measuring the coupling efficiency of the waveguide.

We have validated the feasibility of application of multichannel waveguides to the detection of protective antigen and lethal factor toxins of *Bacillus anthracis* with excellent sensitivity and specificity [29] and anticipate translating this technology to the detection of cancer biomarkers in the near future.

We have used two different surfaces with our waveguide-based platform: supported phospholipid bilayers and SAMs. Before either functionalization, waveguides are cleaned by sonication in chloroform and ethanol, followed by exposure to ozone in a UV-ozone cleaner [20]. For functionalization using a supported lipid bilayers, thin films of DOPC with 1% cap-biotin-PE are deposited on test tubes by evaporation of chloroform. Following hydration in PBS and multiple freeze-thaw cycles in liquid nitrogen, micelles are sonicated with a probe tip sonicator. A supported phospholipid bilayer is formed on the surface of cleaned waveguides by injecting the vesicle solution and allowing it to stabilize for 12 h. The bilayer surface is subsequently blocked with PBS containing 2% bovine serum albumin for 1 h before use. Phospholipid bilayers mimic cell membranes and are intrinsically resistant to NSB, and multiple sandwich assays can be consecutively performed on waveguides functionalized with lipid bilayers [33]. However, the lipid surfaces are not regenerable for repeat use.

For coating of waveguides with SAMs [20,21], the substrates are submerged in water, blown dry, and placed in a vacuum desiccator with a small volume of APDMES. After coating, substrates are baked, allowed to cool in a desiccator, rinsed with ethanol, dried, and analyzed by contact angle measurements that provide a reliable and simple way to test the relative hydrophilicity of the surface film, which in turn provides a good measure of film integrity and composition. PEG reagents containing a carboxylic acid at one end and either a methoxy group (99%–99.99% of the total PEG volume) or a protected amine (0.01%–1% of the PEG volume) at the other are coupled to the amine surfaces overnight using standard peptide coupling conditions, again followed by contact angle analysis. The protecting groups (Fmoc) are removed by immersion in a piperidine/*N*-methylpyrrolidinone (NMP) solution, and biotin is attached to the resulting free-amine via its commercially available *N*-hydroxysuccinimide ester. All steps after amino silanization are carried out in NMP as the solvent, and the slides are rinsed thoroughly between steps. After biotinylation, slides are rinsed with NMP, acetone, and ethanol and dried under argon. SAM-coated waveguides and coverslips are mounted into flow cells, but the SAMs are not blocked and are therefore available for immediate use. Waveguides coated with SAMs can be stored in air for >1 year, washed with harsh chemicals and detergents, and are potentially reusable. SAMs also efficiently reject NSB associated with complex biological samples, especially urine [19,21].

9.3.4 Detection of CEA

The waveguide-based optical biosensor was used for all measurements discussed here, and the fluorescence signal was measured using an Ocean Optics spectrometer interface. The assays described in this section have been previously published [28,33]. The schema for a typical sandwich immunoassay on the waveguide-based platform is shown in Figure 9.3A. All experiments were carried out using the general scheme outlined here, although experimental parameters (choice of functional surface, fluorescence reporter, concentration of the antigen, and others) were varied depending on the experiment. For all experiments, the waveguides were functionalized for use in biological assays and assembled into a flow cell to accommodate the addition of reagents. Both SAMs and supported lipid bilayers contained biotin (1%), which allowed the use of biotin-avidin chemistry to attach the capture antibody to the waveguide surface. Time course experiments were performed to determine

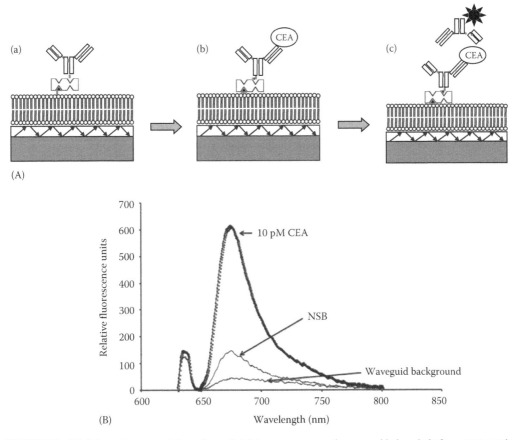

FIGURE 9.3 (A) Schematic representation of a sandwich immunoassay on the waveguide-based platform, not to scale. (a) Waveguides are functionalized, in this case with a supported lipid bilayer. (b) Biotin-avidin chemistry is used to trap the capture antibody on the surface. Addition of the sample containing the antigen allows for binding of antigen to the entrapped capture antibody. (c) Subsequent addition of the reporter antibody allows for the specific measurement of reporter fluorescence on the Ocean Optics spectrometer. (B) Specific detection of CEA (10 pM) using the waveguide-based biosensor in a sandwich immunoassay format. The waveguide-associated background, as indicated, is very low and is simply a measure of intrinsic impurities. The NSB associated with the antibody is low (200 RFU). The specific signal associated with the detection of 10 pM CEA is 600 RFU.

that an incubation time of 5–10 min for the antibodies and 10 min for CEA allow for saturation binding. The concentration of the sample in the low-volume flow cell, specific detection of the surface components, and high binding affinities of the antibodies to CEA all contribute to rapid binding. A typical experimental result for detection of CEA is indicated in Figure 9.3B. Once assembled in the flow cells, the waveguide-associated background was measured in all experiments. This is an intrinsic measure of impurities associated with the waveguide itself and is typically <100 relative fluorescence units (RFU). Then, streptavidin is injected (200 μL) into the flow cell and binds to the biotin on the functional surface. Subsequently, a biotinylated capture antibody (T84.1) was added to the flow cell and incubated for 5 min. Unbound antibody is removed by washing (PBS, 1.5 mL, 60× flow cell rinse). The waveguide is now prepped for sample measurement. Before sample measurement, NSB associated with control serum and the fluorescently labeled reporter antibody (T84.66) is measured in each experiment (Figure 9.3B). Then, the antigen or an unknown sample is added to the flow cell. Following incubation, the reporter antibody is added again and the specific signal associated with binding to the antigen is then measured on the spectrometer interface. The signal/NSB is then calculated for the experiment [33]. Once experimental parameters were established, assay performance was measured with change in one of many functional parameters such as the surface (lipid bilayers vs. SAMs) or fluorescence labels (QDs vs. organic dyes).

For the quantitative determination of CEA in patient samples, a standard measurement using recombinant CEA was first made. Subsequently, patient sample was added and the associated signal measured. A second standard CEA measurement was also done at the end of each experiment. The results from the two standards were always consistent (not shown). The concentration of CEA in the patient sample is extrapolated from the standard measurements made on the same waveguide. For NAF, a special flow cell was developed for assaying low-volume samples. NAF was diluted 10-fold in PBS before testing. For each patient, CEA concentrations from both breasts were measured and calibrated to the total protein concentration to correct for differences between breasts and among samples.

9.4 Results

9.4.1 Comparison of the Use of Phospholipid Bilayers vs. Self-Assembled Monolayers on the Optical Biosensor Platform

For an effective comparison, NSB and specific measurement of CEA was performed on waveguides functionalized with either SAMs of lipid bilayers in the absence of blocking. A typical experimental result for measurement of CEA on a waveguide functionalized with supported lipid bilayers after blocking is indicated in Figure 9.3B. As indicated in Figure 9.4B, NSB is greatly increased in the absence of blocking with the use of supported lipid bilayers, thereby compromising signal/background (S/B). In contrast, a low NSB and high S/B is measured with the use of SAMs (Figure 9.4A). This, together with increased stability and robustness of SAMs, makes them better surfaces for conducting sandwich immunoassays. In other cases, e.g., with urine samples, the use of SAMs offers an even greater advantage in stability over lipid bilayers [21].

9.4.2 Comparison of the Use of Organic Dyes with QDs in the Detection of CEA

Detection of CEA on phospholipid bilayers functionalized waveguides with T84.66 labeled with either AF647 (Figure 9.3B) or QD-655 (Figure 9.5) is shown. There is a small decrease in the S/B for 100 pM CEA with QDs, but the assay is still extremely sensitive. Very little photobleaching was observed with repeat excitation (>2 min, 3 s integrations, 440 μW input power) with the use of the QD-labeled reporter antibody [28]. This photostability, combined with tunability (ability to excite multiple QDs that emit at different wavelengths using a single excitation source) of QDs, is ideal for multiplex strategies.

We have previously evaluated NH-domain labeling of antibodies vs. labeling in the hinge region. These results, while interesting, are beyond the scope of this chapter. A detailed description can be found in the literature [28].

9.4.3 Concentration-Dependent Detection of CEA on the Waveguide-Based Optical Biosensor

Detection of CEA on lipid bilayers using T84.66-AF647 as the reporter is shown in Figure 9.3B. In all experiments, waveguide-background (an intrinsic measure of impurities associated with the waveguide itself) and NSB associated with complex samples were measured. Results are plotted as RFU, measured on the spectrometer interface, as a function of wavelength. Our limit of detection (<0.5 pM, S/BG 2.2) is below the reported physiological concentration of CEA in serum (~5 ng/mL or 5–10 pM) from normal individuals. The sensitivity achieved for detection of CEA on the planar optical waveguide platform is an order of magnitude better than traditional immunoassays using the same antibody, and a concentration curve has been reported with a limit of detection of 500 fM [33], and our assay is linear over two orders of magnitude. We have found this to be true for other assays on our system as well (influenza [34], anthrax [29]).

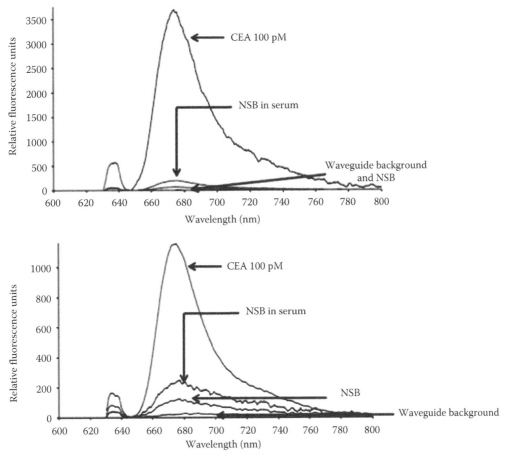

FIGURE 9.4 Detection of CEA [100 pM] on a waveguide functionalized with (A) SAMs and (B) supported lipid bilayers without blocking. Waveguide-associated background is low in both cases but is indistinguishable from the nonspecific increase associated with the bovine serum in (A). Some increase in signal is seen in waveguides functionalized with bilayers (B). This is followed by measurement of NSB associated with the fluorescently labeled reporter antibody. The specific signal associated with 100 pM CEA is indicated, clearly showing a greater signal and lower nonspecific background with the use of SAMs (A) over lipid bilayers (B).

9.4.4 Detection of CEA in Serum and NAF

Detection of CEA in NAF is a more definitive indicator of breast cancer than serum [35]. However, the use of NAF is difficult because of the low volume of samples collected, thereby requiring an assay with exquisite sensitivity. All samples were from patient with abnormal mammograms. Further information on patient disease was not obtained during the course of this study. We were able to quantitatively measure CEA in a small cohort of patient serum (Figure 9.6A) and NAF (Figure 9.6B). In NAF, we obtained a 100% corroboration with disease progression in the patient in that only one individual who had very high concentration of CEA in NAF actually progressed to develop breast cancer. All other individuals had false-positive mammograms and did not develop breast cancer. It is interesting to note that we measured CEA in NAF from both breasts, irrespective of the one that showed abnormal mammograms. This has implicit applications to the detection of ductal carcinoma in situ. The threshold concentration of CEA was assumed to be 7 ng/mL (39 pM) in serum based on literature [36]. Based on this assumption, 7/15 individuals probed had a positive measurement of CEA in serum using our method. Only one of the five patients had very high concentrations of CEA in aspirate fluid and this patient did develop breast cancer, suggesting a positive correlation.

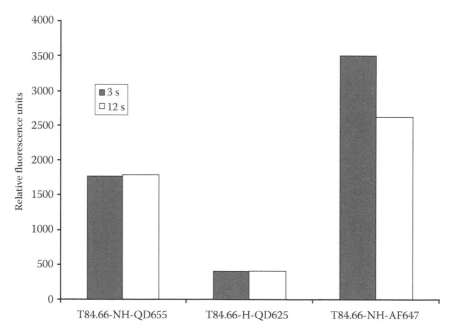

FIGURE 9.5   Detection of CEA with QD-conjugated antibodies. The signal intensity measured for the detection of CEA after a 3 s integration (black bars) and 12 s integration (clear bars) is shown when the reporter antibody T84.66 is labeled with either a QD in the NH position, QD in the hinge position, or with AF647 (organic dye). Labeling with the QDs results in a lower signal intensity than labeling with organic dyes, but there is no photobleaching with extended integration as indicated. There is 15%–20% decrease in signal intensity with repeat integration when the reporter antibody is labeled with aF647.

9.5 Discussion

Application of planar optical waveguides to the sensitive, rapid, quantitative detection of CEA is demonstrated. The work also describes the development of essential components for such an assay, especially labeling strategies and functional surfaces. We have successfully measured CEA in a small cohort of patient samples using this approach, demonstrating feasibility. However, no single biomarker is an accurate indicator of cancer, prognosis, or relapse. Hence, our team has developed multiplex detection strategies for the simultaneous detection of a limited suite of biomarkers using photostable QDs as the fluorescence reporter. Indeed, the lessons learned from labeling anti-CEA T84.66 described in this chapter were essential to this development. We have recently described the simultaneous detection of protective antigen and lethal factor from anthrax using QDs as the reporters. The assay is quantitative (with an internal standard for comparison of inter- and intra-assay differences), sensitive (LOD 1 pM for each antigen), and rapid (15 min). We have further expanded our capability with the development of multichannel waveguides (Figure 9.2) capable of the simultaneous detection of a suite of biomarkers in three samples in quadruplicate. Efforts are underway to adapt this technology to the detection of breast cancer biomarkers.

9.6 Future Directions

We are currently working on the development of sensitive assays for three other breast cancer biomarkers on our optical biosensor platform. Once complete, the assays will be integrated into a multiplex assay and used for the evaluation of multiple biomarkers in serum and aspirate fluid collected from patients at different time points during the course of the disease. This will allow for the development of longitudinal

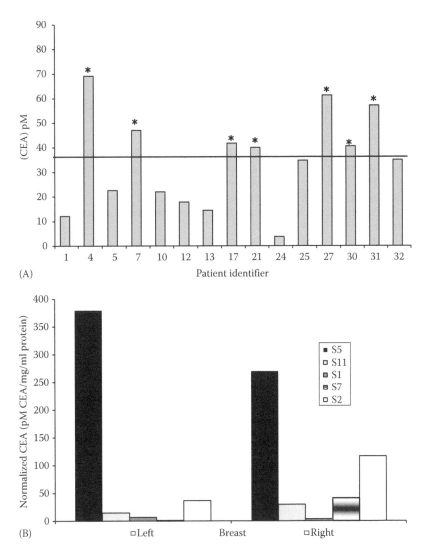

FIGURE 9.6 (A) Detection of CEA in serum or (B) NAF from patients with abnormal mammograms. Patient numbers are indicated as received from the collection source, City of Hope, and have no bearing with the results reported. CEA (pM) measured in patient serum is plotted as a function of patient number (random) in (A). * in (A) indicates samples with serum CEA concentrations >39 pM (assumed threshold for abnormal CEA, based on the literature). (B) Shows measurement of CEA in NAF from left and right breasts of five different patients with abnormal mammograms. Only patient S5 progressed to develop breast cancer; all others had false-positive mammogram readings. CEA was measured from both breasts irrespective of which breast showed the abnormal mammogram.

expression profiles for these biomarkers and evaluate their applicability in predicting breast cancer, its relapse, and prognosis.

REFERENCES

1. Cheng, H.D., J. Shan, W. Ju, Y. Guo, and L. Zhang. 2009. Automated breast cancer detection and classification using U.S. images: A survey. *Pattern Recogn.* 43: 299–317.
2. Lippman, M.E. 2008. In *Harrisons Principles of Internal Medicine*, 17th edn., Chapter 86, Eds.: Fauci, A.S., E. Braunwald, D.D. Kasper, S.L. Hauser, D.L. Longo, J.L. Jameson, and J. Loscalzo. McGraw-Hill Book Company, New York.

3. Kearney, A.J. and M. Murray. 2008. Breast cancer screening recommendations: Is mammography the only answer? *J. Midwife Women's Health* 54: 393–400.

4. Miller, A.B. 2003. Is mammography screening for breast cancer really not justifiable? *Recent Results Cancer Res.* 163: 115–128, discussion 246.

5. Gotzche, P.C. and M. Nielson. 2006. Screening for BC with mammography, Cochrane database systems review 4: CS001877.

6. Kriege, M. et al. 2004. Efficacy of MRI and mammography for breast-cancer screening in women with familial or genetic predisposition. *NEJM* 351(5): 427–437.

7. Nass, S.J. and H. Moses, Eds. 2007. In *Cancer Biomarkers: The Promises and Challenges for Improving Detection and Treatment, Committee for Developing Biomarker-Based Tools for Cancer Screening and Diagnosis and Treatment.* The National Academies Press, Washington, DC, pp. 1–18.

8. Bast Jr., R.C. 2004. Early detection of ovarian cancer: New technologies in pursuit of a disease that is neither common nor rare. *Trans. Am. Clin. Climatol. Assoc.* 115: 233–248.

9. Nogueira, L., R. Corradi, and J.A. Eastham. 2009. Other biomarkers for detecting prostate cancer. *BJU Int.* 105: 166–169.

10. Ebeling, F.G., P. Stieber, M. Untch, D. Nagel, G.E. Konecny, U.M. Schmitt, A. Fateh-Moghadam, and D. Seidel. 2002. Serum CEA and CA15-3 as prognostic factors in primary breast cancer, *Brit. J. Cancer.* 86(8): 1217–1222.

11. Dnistrian, A., M.K. Schwartz, E.J. Greeberber, C.A. Smith, and D.C. Schwartz. 1991. CEA and CA15-3 in the clinical evaluation of breast cancer. *Clin. Chim. Acta* 200: 81–93.

12. Ross, J.S. and J.A. Fletcher. 1998. The Her2/neu oncogene in breast cancer: Prognostic factor, predictive factor are target for therapy. *Oncologist* 3(4): 237–252.

13. Ring, B.Z., R.S. Seitz, R. Beck, W.J. Shasteen, S.M. Tarr, and M.C.U. Cheang. 2006. Novel prognostic, immunohistochemical biomarker panel for estrogen receptor positive breast cancer. *J. Clin. Oncol.* 24(19): 3039–3047.

14. Smith, R.A., V. Cokkinides, and H.J. Eyre. 2004. American chemical society guidelines for early detection of cancer. *CA Cancer J. Clin.* 54: 41–52.

15. Grunert, F., G.A. Luckenback, B. Haderlie, K. Schwartz, and S. VonKleist. 1983. Comparison of colon, lung and breast derived CEA and cross-reacting antigens by monoclonal antibodies and fingerprint analysis. *Ann. NY Acad. Sci.* 417: 75–85.

16. Jothy, S., S.A Brazinsky, M. Chin-A-Loy, A. Haggarty, M.J. Krantz, M. Cheung, and A. Fuks. 1986. Characterization of monoclonal antibodies to CEA with increased tumor specificity. *Lab. Invest.* 54: 108–117.

17. Foretova, L., J.E. Garber, N.L. Sadowsky, S.J. Verselis, D.M. Joseph, A.F. Andrade, P.G. Gudrais, D. Fairdough, and F.P. Li. 1998. Carcinoembryonic antigen in breast nipple aspirate fluid. *Cancer Epidemiol. Biomarker Prev.* 7(3): 195–198.

18. Mukundan, H., A.S. Anderson, K. Grace, W.K. Grace, N. Hartman, J. Martinez, and B.I. Swanson. 2009. Waveguide-based sensors for pathogen detection. *Sensors* 9(7): 5783–5809.

19. Nishihara, H., M. Haruna, and T. Suhara. 1985. In *Optical Integrated Circuits*. McGraw-Hill Book Company, New York, pp. 41–49.

20. Anderson, A.S., A.M. Dattelbaum, G.A. Montaño, D.N. Price, J.G. Schmidt, J.S. Martinez, W.K. Grace, K.M. Grace, and B.I. Swanson. 2008. Functional PEG modified thin films for biological detection. *Langmuir* 24(5): 2240–2247.

21. Anderson, A.S., A. Dattelbaum, H. Mukundan, D. Price, W.K. Grace, and B.I. Swanson. 2009. Robust sensing films for pathogen detection and medical diagnostics. *Proceedings of the SPIE* 7167: 71670Q. doi:10.1117/12.809383.

22. Fang, Y., A.M. Ferrie, N.H. Fontaine, J. Mauro, and J. Balakrishnan. 2006. Resonant waveguide grating biosensor for living cell sensing. *Biophys. J.* 91: 1925–1940.

23. Cunningham, B., P. Li, B. Lin, and J. Pepper. 2002. Colorimetric resonant reflection as a direct biochemical assay technique. *Sens. Actuat. B Chem.* 81: 316–328.

24. Grieshaber, D., E. Reimhult, and J. Voros. 2007. Enzymatic biosensors towards a multiplex electronic detection system for early cancer diagnostics. Nano/micro engineered and molecular systems. *IEEE NEMS* 1: 402–405.

25. Xu, F., P. Datta, H. Wang, S. Gurung, M. Hashimoto, S. Wei, J. Goettert, R.L. McCarley, and S.A. Soper. 2007. Polymer microfluidic chips with integrated waveguides for reading microarray. *Anal. Chem.* 79(23): 9007–9013.

26. Washburn, A.L., L.C. Gunn, and R.C. Bailey. 2009. Label free quantitation of a cancer biomarker in complex media using silicon photonics microring resonators. *Anal. Chem.* 81: 9499–9506.

27. Feng, K. 2007. Biolayer modeling and optimization for the SPARROW biosensor. PhD. dissertation, Department of Physics, West Virginia University, Morgantown, WV, pp. 17–18.

28. Mukundan, H., H. Xie, A.S. Anderson, W.K. Grace, J.E. Shively, and B.I. Swanson. 2009. Optimizing a waveguide-based sandwich immunoassay for tumor biomarkers: Evaluating fluorescent labels and functional surfaces. *Bioconj. Chem.* 20(2): 222–230.

29. Mukundan, H., H. Xie, D. Price, J.Z. Kubicek-Sutherland, W.K. Grace, A.S. Anderson, J.S. Martinez, N. Hartman, and B.I. Swanson. 2009. Quantitative multiplex detection of pathogen biomarkers on multichannel waveguides. *Anal. Chem.* 82(1): 136–144.

30. Jamieson, T., R. Bakhshi, D. Petrova, R. Pocock, M. Imani, and A.M. Seifalian. 2007. Biological applications of quantum dots. *Biomaterials* 28(31): 4717–4732.

31. Michalet, X., F.F. Pinaud, L.A. Bentolila, J.M. Tsay, S. Doose, J.J. Li, G. Sundaresan, A.M. Wu, S.S. Gambhir, and S. Weiss. 2005. Quantum dots for live cells, in *vivo* imaging and diagnostics. *Science* 307(5709): 538–544.

32. Medintz, I.L., H.T. Uyeda, E.R. Goldman, and H. Matoussi. 2005. Quantum dots bioconjugates for imaging labeling and sensing. *Nat. Mater.* 4(6): 435–446.

33. Mukundan, H., J.Z. Kubicek, A. Holt, J.E. Shively, J.S. Martinez, K. Grace, W.K. Grace, and B.I. Swanson. 2009. Planar optical waveguide-based biosensor for the quantitative detection of tumor markers. *Sens. Actuat. B Chem.* 138(2): 453–460.

34. Kale, R., H. Mukundan, D. Price, J. Foster-Harris, D.M. Lewallen, B.I. Swanson, J.G. Schmidt, and S.S. Iyer. 2008. Detection of intact influenza viruses using biotinylated biantennary s-sialosides. *J. Am. Chem. Soc.* 130(26): 8169–8171.

35. Zhao, Y., S.J. Verselis, N. Klar, N.L. Sadowsky, C.M. Kaelin, B. Smith, L. Foretova, and F.P. Li. 2001. Nipple fluid carcinoembryonic antigen and prostate specific antigen in cancer bearing and tumor-free breasts. *J. Clin. Oncol.* 19: 1462–1467.

36. Hansen, H.J., G. LaFontaine, E.S. Newman, M.K. Schwartz, A. Malkin, K. Mojzisik, E.W. Martin, D.M. Goldenberg. 1989. Solving the problem of antibody interference in commercial sandwich type immunoassays of carcinoembryonic antigen. *Clin. Chem.* 35: 146–151.

10

Label-Free Resonant Waveguide Grating (RWG) Biosensor Technology for Noninvasive Detection of Oncogenic Signaling Pathways in Cancer Cells

Yuhong Du, Min Qui, and Haian Fu

CONTENTS

10.1 Introduction 171
10.2 Materials 173
    10.2.1 Reagents 173
        10.2.1.1 Reagents for Cell Culture 173
        10.2.1.2 Pathway Modulators 174
    10.2.2 Epic Biosensor System 174
        10.2.2.1 Biosensor Cell Plates 174
        10.2.2.2 Biosensor Detection Reader 174
10.3 Methods 174
    10.3.1 Assay Design 174
    10.3.2 Cell Preparation and Seeding 175
    10.3.3 Baseline DMR Signal Measurement 175
    10.3.4 Ligand-Induced DMR Signal Measurement in Response to Test Compounds 176
        10.3.4.1 Compound Treatment 176
        10.3.4.2 Ligand Addition 176
    10.3.5 Data Collection and Analysis 176
        10.3.5.1 DMR Response Curve 176
        10.3.5.2 Defining Optical Signature Parameters 176
        10.3.5.3 Determining the Effect of Pathway Modulators on Ligand-Induced
                DMR Response 177
10.4 Results 177
    10.4.1 EGF Induces a Unique DMR Response in a Cell Type–Dependent Manner 177
    10.4.2 DMR Response Induced by EGF Is Mediated by EGFR 178
    10.4.3 EGF-Induced DMR Signals in SCCHN Cells Are Mediated by PI3K 178
    10.4.4 DMR Signatures as a Tool for Evaluating the Efficacy of EGFR-Targeting Drugs 179
10.5 Discussion 180
10.6 Future Trends 182
Acknowledgments 183
References 183

10.1 Introduction

Dysregulated signaling pathways that control cell proliferation, immortalization, apoptosis, invasion, and angiogenesis have been correlated with cancer development and progression [1]. Many growth control pathways are initiated by signals at the cell surface. Extracellular growth signals are transmitted into cells through several types of transmembrane receptors. Upon ligand binding, a transmembrane receptor is activated, which initiates a cascade of intracellular signaling events, thereby regulating cell function. Thus, technologies that can monitor signal transduction are essential for the understanding of various cellular processes and for drug discovery.

Traditional methods for monitoring ligand-induced cell signaling pathways typically involve the use of intact cells, cell lines, or primary cells. In order to enhance detection sensitivity and selectivity, many cell-based assays require artificially manipulated cell lines. For example, cells are often engineered to express fluorophore-tagged and/or overexpressed targets or a reporter system. End-point assays are usually performed to simplify the detection of final products. Conventional assays generally measure a single cellular event one at a time, such as the activation of a signaling molecule or cell growth status in an end-point assay. These assays provide mature technology platforms and are widely used for studying cell signaling pathways. However, the cellular response to a specific stimulus is a complex process and involves a network of interactions among many molecules. It is becoming increasingly apparent that innovative technologies that are capable of monitoring an integrated cellular response in living cells are required for effectively dissecting complex cellular signaling pathways. Label-free technologies for cell-based assays offer such powerful tools and allow integrated cellular signaling to be detected [2,3].

A label-free detection system generally consists of a biosensor and a reader. A biosensor is an analytical device that typically utilizes a transducer to convert a molecular recognition event or a ligand-induced alternation of a cell monolayer into a measurable signal. A reader is then used to collect, amplify, and process signals from the transducer. Depending on the transducing mechanism used, many types of biosensors are available, including resonant, optical detection, thermodetection, ion-sensitive field-effect transistor, and electrochemical biosensors [4]. Among various label-free technologies, an electrical impedance–based biosensor and optical biosensors have been successfully used to study ligand-induced cell signaling through endogenous receptors [5–11]. Informative results have been obtained from cell-based functional assays using the label-free technology conducted on various instruments, such as CellKey (impedance-based; MDS Analytical Technologies), Bind (optical-based; SRU Biosystems), and Epic (optical-based; Corning Inc.) [12]. Utilizing the inherent morphological and adhesive characteristics of cells as a physiologically relevant and quantitative readout in various cellular assays, these technologies allow the detection of temporal signaling events under physiological conditions. Label-free biosensors are available in 96-well, 384-well, or 1536-well format for both basic research and for drug discovery.

This chapter focuses on the Epic optical biosensor system, which utilizes resonant waveguide grating (RWG) mechanism for monitoring biomolecular interactions [2,5,13]. The RWG biosensor is an evanescent wave–based sensor and is capable of detecting refractive index changes in the vicinity of the sensor surface [14]. The change in refractive index can be detected by a "reader" and is directly proportional to mass redistribution on the surface of the biosensor. The basic principle of the Epic optical biosensor for cell-based functional assays is illustrated in Figure 10.1 [2,13]. The RWG biosensor system consists of three layers, a glass substrate with a diffractive grating, a high index of refraction waveguide thin film coating, and a cell layer. Living cells can be directly cultured and coupled onto the waveguide surface to form a cell layer. When unstimulated cells are illuminated with broadband light, the RWG biosensor only reflects a specific wavelength that is sensitive to the waveguide local index of refraction and is directly proportional to the density and distribution of biomass (e.g., proteins, molecular complexes) in living cells. In response to a ligand stimulation, for example, growth factor, any biomass alternations from cell micromotion, such as protein translocation, microfilament remodeling, cell adhesion alternations, and morphological changes of cells, can lead to a change in the local index of refraction and therefore a change in reflected wavelength. The measured reflected wavelength difference in picometer (pm) unit before and after stimulation indicates the intracellular dynamic mass redistribution (DMR) triggered by a given stimulus and can be used as a functional readout for ligand-induced cellular signaling. As the energy of the broadband evanescent electromagnetic field on the RWG sensor surface decays over distance from the surface, only the changes in DMR within a limited depth from the biosensor surface (~150 nm) are detected, which is referred to as detection zone. Therefore, only partial changes in cellular events at the bottom portion of the cells in close contact with the biosensor surface are measured by the

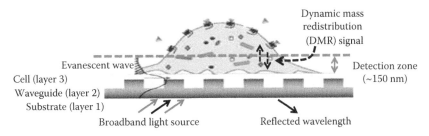

FIGURE 10.1 The cell-based RWG biosensor system for monitoring ligand-induced DMR signal in living cells. The system consists of three layers: a glass substrate (layer 1), a high index of refraction waveguide thin film with a grating structure (layer 2), and a cell layer (layer 3) attached above the layer 2 through cell culture. The electromagnetic field, termed evanescent wave, is generated by the diffraction grating–coupled waveguide resonance and extends above the waveguide surface. Any refractive index change in this region will perturb the evanescent field, resulting in a change of the waveguide's effective refractive index. When illuminated with broadband light at a fixed angle of incidence, the sensor only reflects a narrow band of wavelength that is a sensitive function of biosensor local refraction index. The biomass (e.g., proteins and molecular complex) redistribution of the cells in response to a stimulus leads to the change of the local effective refraction index and the shift of reflected wavelength measured as pm unit. As the amplitude of the evanescent wave decays exponentially with increasing distance, only the DMR at the bottom portion (around 150 nm) of the cells can be detected, which is referred to as "detection zone." The RWG biosensor can therefore noninvasively detect stimulus-induced DMR signal in living cells as a functional readout for cell signaling.

RWG biosensor. The reflected wavelength shift as expressed in DMR signal after a ligand stimulation of the cells may vary in magnitude and direction (positive or negative) over time. Such a change of DMR often reflects various intracellular molecular movements corresponding to either activated or inactivated signaling pathways. For example, positive DMR (P-DMR) may indicate that the movement of biomolecules is toward the detection zone, while negative DMR (N-DMR) may be an indication of biomolecules moving away from the biosensor detection zone. The noninvasive feature of the Epic RWG biosensor or other cell-based label-free technologies allows direct and real-time detection of complex and integrated cellular signaling events in native cells. The measured DMR is an integrated cellular response to a particular environmental stimulus. Many cell signaling events, including protein trafficking, microfilament remodeling, cell adhesion alternations, and morphological changes of cells, may contribute to the detected overall DMR signal [15]. Because DMR can be used as a universal readout to reflect cellular response to a variety of signals, label-free biosensors are expected to have broad applications in cell biology.

Cell growth is tightly regulated by a network of signal transduction pathways. Epidermal growth factor (EGF), or other growth factors, and their receptors play a fundamental role in cell growth control [6,16]. A better understanding of growth factor–mediated signaling pathways and their role in tumorigenesis is vital for the discovery of new therapeutic strategies and for the effective use of existing anticancer therapies. The EGF receptor (EGFR) is dysregulated in a variety of human cancers, including head and neck cancer, and lung cancer. In this chapter, using EGFR signaling as a model system, we describe the application of the Epic label-free optical biosensor for the development of a biosensor assay to reveal distinct responses of different cancer cells to EGF.

10.2 Materials

10.2.1 Reagents

10.2.1.1 Reagents for Cell Culture

RPMI1640 medium (Cat. 10-092-CM), DMEM/F12 medium (Cat. 10-092-CM), 0.25% trypsin-EDTA solution (Cat. 25-052-CI), and penicillin-streptomycin (100×) (Cat. 15140-155) were purchased from Mediatech Inc. (Manassas, VA) and stored at 4°C before use. Heat-inactivated fetal bovine serum (FBS) was obtained from Atlanta Biologicals Inc. (Lawrenceville, GA) and stored at −20°C. Hank's balanced salt solution (HBSS) (Cat. 14025-126) and HEPES buffer (1 M) (Cat. 15630-080) were purchased from Invitrogen (Carlsbad, CA).

10.2.1.2 Pathway Modulators

Recombinant human EGF expressed in *E. coli* was purchased from Sigma Chemical Co. (St. Louis, MO; Cat. E9644) and reconstituted in 10 mM acetic acid as a 1 mg/mL stock and stored at −80°C before use. Various pathway inhibitors were purchased from LC Laboratories (Woburn, MA), including gefitinib (Iressa) (Cat. G-4408), erlotinib (Tarceva) (Cat. E-4007), PD98059 (Cat. P-4313), wortmannin (Cat. W-2990), U0126 (Cat. U-6770), and LY 294002 (Cat. L-7962). All compounds were dissolved in DMSO as 10 mM stocks and stored at −80°C before use.

10.2.2 Epic Biosensor System

10.2.2.1 Biosensor Cell Plates

Corning Epic 384-well fibronectin-coated optical biosensor cell assay microplates (Cat. 5042) were obtained from Corning Inc. (Corning, NY) and stored at 4°C. The plates have black walls and a clear bottom. Cells seeded at the bottom of biosensor plates can be observed using an optical microscope.

10.2.2.2 Biosensor Detection Reader

The Corning Epic® system at the Emory Chemical Biology Discovery Center from Corning Inc. consists of a temperature-controlled chamber with build-in stackers for up to 20 cell plates. Its optical detection system utilizes a linear array of fiber optic heads for scanning the biosensor plate and recording the DMR signal. The system is operated by the Epic Quest software. Depending on the type of biosensor plates used (cell plate or biochemical plate), the system can be applied for monitoring bimolecular interactions in both cell-based assays and biochemical assays in a 384-well plate format for high-throughput screening (HTS). As the Epic biosensors are sensitive to changes in temperature, all experiments should be carried out at room temperature (25°C) to avoid temperature fluctuation.

10.3 Methods

10.3.1 Assay Design

The general process of using the Epic optical biosensor for monitoring a ligand-induced DMR signal is illustrated in Figure 10.2. Cells are cultured and attached onto the biosensor surface to form a cell layer. The Epic reader is used to illuminate the biosensor surface with the cell layer and to capture the reflected wavelength change. The reflected wavelength shift in pm is referred to as the DMR signal. Under unstimulated conditions, the local mass density in cells within the detection zone can reach an equilibrium state. The recorded DMR of these cells is defined as "baseline DMR signal." To monitor ligand-induced pathways in living cells, the first step is to establish the baseline DMR signal. Then, a ligand, such as a growth factor, is added to the cells. Ligand-induced DMR events of the cell layer leads to a change in the index of the refraction at the sensor surface. DMR could be due to a number of cellular events, such as relocation of intracellular proteins, reorganization of structural proteins, and changes in cell adhesion or detachment. The measured DMR signal difference between baseline DMR and ligand-stimulated signal is defined as the ligand-induced DMR signal [2,13]. The 384-well Epic biosensor plate is scanned for DMR signal over a period of time. The recorded DMR signals are plotted against the measurement time, which is referred to as real-time DMR response. The shape of DMR response curves, such as magnitude and directions (increase or decrease) over time, is used to define the optical signatures of a particular cell type in response to a specific stimulus.

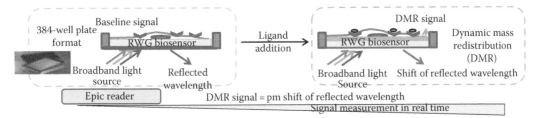

FIGURE 10.2 Assay design with the Epic RWG optical biosensor system. Cells are placed and grown on the 384-well Epic biosensor plate for overnight. Cells are subject to serum starvation for 24 h for studying growth factor signaling or as needed. The plate is thermoequilibrated inside the Epic reader for 1 h, and the DMR signal is measured every 1 min for 10 min with Epic readers to obtain the baseline DMR signal. A ligand, such as EGF, is then added to cells to initiate a signaling cascade. The DMR signal is immediately monitored in live cells every 1 min for a desired period of time with the Epic reader. The buffer-only wells without ligand are included in each plate as blank DMR signal controls. The DMR signal from each well is calculated by subtracting the basal DMR signal before ligand addition from DMR signal recorded at each time point after ligand addition in the same well. The real-time DMR response of the cells stimulated with ligand or buffer as background control is obtained by plotting the DMR signal against the measurement time from each well.

10.3.2 Cell Preparation and Seeding

The lung adenocarcinoma cancer cell line A549 was maintained in RPMI1640 medium supplemented with 10% FBS and 1% penicillin-streptomycin. UPCI-37B squamous cell carcinoma of the head and neck (SCCHN) cells were grown in DMEM/F12 medium (1:1) supplemented with 10% FBS and 1% penicillin-streptomycin. All cells were cultured in T-75 or T-175 tissue culture flasks at 37°C in a humidified atmosphere with 5% CO_2.

Cells growing in tissue culture flasks at 90% confluency were harvested with trypsin-EDTA solution (0.25%) and resuspended in culture medium with FBS (10%) and penicillin-streptomycin (1%). About ~10,000 cells (40 μL) were dispensed onto each well of a 384-well Epic cell plate using a Multidrop Combi dispenser (Thermo Fisher Scientific). The plates were incubated at 37°C in a humidified incubator with 5% CO_2 for 24 h to allow the cells to attach to the plate. Because the status of cells, such as cell health and confluency, can significantly influence the measurement of the DMR signal, the quality of cells was observed at each step using a microscope. A cell layer of high confluency is usually used to achieve optimal results.

10.3.3 Baseline DMR Signal Measurement

To place cells under quiescent conditions for basal DMR signal measurement prior to stimulation, the culture medium was first replaced with serum-free medium in 348-well Epic plate using a buffer-exchanging procedure. Briefly, the medium on the plate was gently aspirated using a plate washer (BioTek) with predefined settings to allow 10 μL of medium to remain in each well without disturbing the cell monolayer. Serum-free medium (50 μL) was then dispensed to the cell plate using a Multidrop Combi dispenser. This process was repeated for four times. After the buffer exchange process, integrity of a uniform monolayer of cells was confirmed by microscopy. The plates were then incubated at 37°C in a humidified atmosphere with 5% CO_2 for another 24 h.

Before measurement, the medium in the biosensor plate was replaced again with assay buffer (HBSS with 20 mM HEPES) using the buffer-exchanging procedure described earlier. After two cycles of assay buffer exchange, the assay buffer was adjusted to bring the total volume to 30 μL in each well. The plate was then loaded onto the Epic reader and incubated for at least 1 h for thermoequilibration. The baseline DMR signal was recorded in real time at every 1 min for 10 min to obtain the baseline DMR response signal from each well.

10.3.4 Ligand-Induced DMR Signal Measurement in Response to Test Compounds

10.3.4.1 Compound Treatment

All compounds in DMSO were diluted in assay buffer to achieve the desired concentrations. DMSO is a high index of refraction solvent. To reduce the bulk reflective index effect upon compound addition, buffer matching was performed for all experiments involving compound treatment. For buffer matching, assay buffer containing 0.5% DMSO was used for buffer exchange and for dilution of compounds as well as ligand (EGF).

To test the effect of specific inhibitors or pathway modulators on ligand-induced DMR, cells were treated with each compound for a defined period of time before ligand addition. For compound addition, 10 μL of each diluted compound was transferred from the compound plate to the Epic cell plate using a Sciclone liquid handler with a 384-channel low-volume cannula array (Caliper Life Sciences) to ensure uniform dispensing to each well. The final DMSO concentration was set at 0.5% (v/v). Vehicle (DMSO) controls without test compounds were included in each plate. After compound addition, the DMR signal was monitored continuously every minute for 30 min. The effect of compounds on the DMR signal was obtained.

10.3.4.2 Ligand Addition

The DMR signal of cells in response to EGF was tested using serum-starved cells in the presence or absence of a test compound. First, a ligand plate was prepared. A 20 μL of buffer alone or EGF diluted in assay buffer as dispensed to a 384-well round bottom plate to serve as the ligand plate. An Epic cell plate with or without compound was used for the experiments. A 10 μL volume was transferred from the ligand plate to the Epic cell plate using a Sciclone liquid handler with a 384-channel low-volume cannula array to ensure simultaneous ligand addition to each well. Buffer-only control wells without EGF were included in each plate. To obtain the DMR signal of cells in response to EGF, the Epic cell plate was scanned using an Epic reader. The DMR signal was recorded every 1 min for 45 min or as otherwise indicated.

10.3.5 Data Collection and Analysis

The DMR signal was measured at each time point as pm unit (recorded pm). All values of recorded pm indicate the shift of the reflected wavelength at different time points after subtracting the measured value at time zero (the first reading of the baseline measurement). Data analysis was performed to obtain DMR optical signatures of cells in response to a ligand. Data collected by the Epic reader were first converted from text format file to a Microsoft Excel (Microsoft Corporation, Redmond, WA) format using the Corning Microplate Analyzer v2.0. Data in Excel format were used to obtain the optical signatures of cells upon ligand stimulation.

10.3.5.1 DMR Response Curve

DMR response curves were used to define the DMR signatures of the cells. To obtain a real-time DMR response curve after ligand addition, ΔResponse was calculated as *ΔResponse = Recorded pm (at the indicated time point) − Recorded pm (at the time point before ligand addition)*. The DMR response curve is plotted as ΔResponse against measurement time after ligand addition.

10.3.5.2 Defining Optical Signature Parameters

The characteristic cellular DMR in response to a specific stimulus, such as a growth factor, is defined by the shape of the DMR response curve. An increase in the response curve over time is termed a P-DMR event, and a decrease in the response curve is referred to as an N-DMR event [17]. An N-DMR event indicates a decrease of mass at the biosensor detection zone, while a P-DMR event indicates an increase of mass at the biosensor detection zone. The DMR signal is defined as the maximal ΔResponse at the peak of a P-DMR or N-DMR event.

10.3.5.3 Determining the Effect of Pathway Modulators on Ligand-Induced DMR Response

The effect of compound treatment on ligand-induced DMR is expressed as % of control and calculated based on the following equation:

$$\% \text{ of control} = (DMR_{compound} - DMR_{blank})/(DMR_{ligand} - DMR_{blank}) \times 100$$

where
  DMR_{blank} is the DMR signal from wells with vehicle (DMSO) alone without EGF, which defines the minimal DMR signal
  $DMR_{compound}$ is the DMR signal from wells with ligand (e.g., EGF) in the presence of a compound
  DMR_{ligand} is the DMR signal from wells with ligand (e.g., EGF) in the presence of vehicle but without compound, which defines the maximal DMR signal

The IC_{50} value of a compound on EGF-induced DMR was obtained using Prism 4.0 (Graphpad Software, San Diego, CA). All experiments were performed at least in triplicate and repeated at least three times. Representative response curves are shown.

10.4 Results

10.4.1 EGF Induces a Unique DMR Response in a Cell Type–Dependent Manner

Lung cancer A549 cells were chosen as a model system to evaluate the EGF-induced DMR response in living cells using Epic optical biosensor. A549 cells were seeded in 384-well Epic plate for overnight. After serum starvation for 24 h, basal DMR signal was measured. Cells were then exposed to EGF (100 ng/mL), and DMR response was continuously monitored by the Epic reader. ΔResponse was calculated by subtracting the basal DMR signal from the detected DMR signal after EGF addition at each measurement time point. The real-time response curve was then obtained by plotting the ΔResponse against measurement time. As shown in Figure 10.3A, when serum-starved A549 cells were exposed to buffer only, no DMR response was detected. Upon EGF addition, however, a unique DMR response was observed which contained two distinct events: a rapid initial decrease in the DMR (N-DMR) signal followed by an increase of the DMR (P-DMR) signal. The N-DMR signal reached maximum about 15 min after EGF addition, which corresponds to a DMR signal change of about 70 pm. The P-DMR event immediately followed the N-DMR event in A549 cells. The maximum of N-DMR and P-DMR signal, in response to addition of EGF, increased in a dose-dependent manner [17] (data not shown). It is possible that the N-DMR response may be in part due to the relocation of biomolecules away from the bottom portion of cells attached to the biosensor surface. It could be due to receptor endocytosis or cell detachment [18]. The detailed mechanism behind the EGF-induced N-DMR signal in A549 cells requires further investigation.

Interestingly, when the effect of EGF on DMR signal was evaluated in UPCI-37B SCCHN cancer cells, a DMR signature which was completely different from that of A549 lung cancer cells was observed (Figure 10.3B). In response to EGF, only a rapid increase in the DMR response (P-DMR event) was detected. The DMR signal peaked about 18 min after the addition of EGF, which corresponds to a detected DMR signal of about 100 pm. These results suggest that the molecular events of A549 cells in response to EGF are different from those of UPCI-37B SCCHN cells.

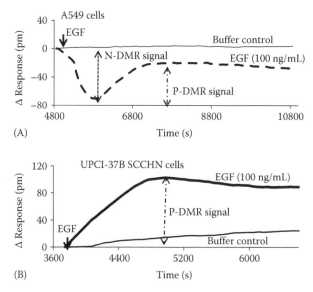

FIGURE 10.3 Distinct DMR response signatures obtained from different cancer cells stimulated with the same ligand, an example with EGF. (A) The unique real-time DMR response of A549 lung cancer cells to EGF. Upon EGF (100 ng/mL) addition to serum-starved A549 cells, two DMR events were observed—an N-DMR event followed by a P-DMR event—while the cells stimulated with the buffer only without EGF gave rise to almost no DMR response. P-DMR signal: the maximum increase (positive) of DMR response. N-DMR signal: the maximum decrease (negative) of DMR response. (B) The completely different real-time DMR response of UPCI-37B SCCHN cancer cells compared to A549 cells to EGF. Upon exposed to EGF (100 ng/mL), UPCI-3B cells exhibited only P-DMR event. Black arrow indicates the time point of EGF addition after baseline measurement. Baseline measurement which started at time zero was not shown here.

10.4.2 DMR Response Induced by EGF Is Mediated by EGFR

It has been well established that EGF binds to EGFR to trigger a cellular growth regulatory response. To determine the biological relevance of the DMR signal, we examined the effect of EGFR inhibition on EGF-induced DMR. Pretreatment of SCCHN cells with AG 1478, an EGFR tyrosine kinase inhibitor, resulted in a dose-dependent inhibition of EGF-induced DMR signal (Figure 10.4A). The estimated IC_{50} of AG 1478 on the EGF-induced P-DMR signal was 0.16 μM (Figure 10.4B). This result is consistent with a previously reported value obtained in A431 cells using the same biosensor technology [18], even though the DMR signal signature of UPCI-37B cells was different from that of A431 cells. These results suggest that EGF-induced DMR signals in SCCHN cells are dependent on EGFR tyrosine kinase activity, which supports the use of the established DMR signature as a biological readout for EGF signaling.

10.4.3 EGF-Induced DMR Signals in SCCHN Cells Are Mediated by PI3K

Two major downstream pathways are involved in EGFR signaling: the PI3K/Akt pathway and the Ras/Raf/MAPK pathway. To further validate the biological relevance of the DMR response, we examined the effect of blocking these two pathways on the generation of EGF-induced DMR signals. While LY 294002 and wortmannin are small molecule PI3K inhibitors, U0126 and PD98059 inhibit MEK1/MEK2. Addition of LY 294002 (25 μM) or wortmannin (125 nM) before EGF treatment significantly reduced the magnitude of DMR signals (Figure 10.5A). However, inhibition of MEK1/MEK2, with U0126 (1 μM) or PD98059 (10 μM), did not suppress the DMR signal (Figure 10.5B). These data suggest that the EGF-induced DMR response in UPCI-37B SCCHN cancer cells is primarily mediated by the PI3K pathway.

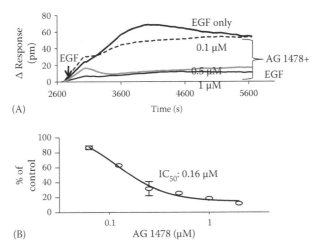

(A)

(B)

FIGURE 10.4 The DMR signal induced by EGF in UPCI-37B SCCHN cancer cells is specific for EGFR. The effect of an EGFR tyrosine kinase inhibitor, AG 1478, on the generation of the EGF-induced DMR signal in UPCI-37B cells was evaluated. (A) Pretreatment of UPCI-37B cells with AG 1478 for 30 min before EGF addition dose-dependently blocked the DMR signal in response to EGF (100 ng/mL). Black arrow indicates the time point of EGF addition after baseline measurement followed by the compound treatment for 30 min. The basal DMR signal from baseline measurement and compound treatment without EGF are not shown here. (B) Dose–response curve of AG 1478 derived from data in (A). The inhibitory effect of AG 1478 on the P-DMR signal induced by EGF was normalized to the vehicle control wells in the presence of EGF but without compound. The IC_{50} was calculated using Prism 4.0 software.

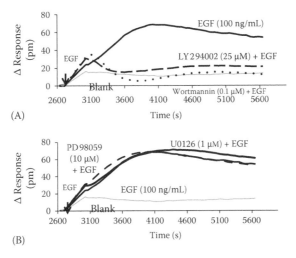

(A)

(B)

FIGURE 10.5 The PI3K pathway plays a critical role in mediating EGF-induced DMR signal in SCCHN cancer cells. (A) Pretreatment of UPCI-37B cells with two PI3K inhibitors, LY 294002 (25 μM) and wortmannin (125 nM), for 30 min, blocked the DMR signal in response to EGF (100 ng/mL). (B) Preincubation of two MAPK inhibitors, U0126 (1 μM) and PD98059 (10 μM), with UPCI-37B cells, had no effect on the EGF-induced DMR signal. Black arrow indicates the time point of EGF addition after baseline measurement followed by the compound treatment. The basal DMR signal from baseline and compound treatment without EGF are not shown here.

10.4.4 DMR Signatures as a Tool for Evaluating the Efficacy of EGFR-Targeting Drugs

EGFR is frequently activated in SCCHN, lung, and other cancer types and is a validated cancer target. To explore potential applications of the Epic RWG biosensor as a tool for efficacy evaluation of targeted therapies, the effect of two small molecule FDA-approved EGFR kinase inhibitors, gefitinib (Iressa) and erlotinib (Tarceva), on EGF-induced DMR signals was tested. Pretreatment of UPCI-37B

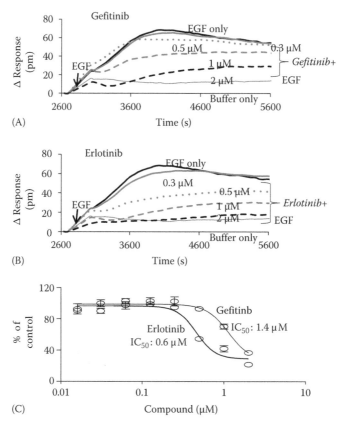

FIGURE 10.6 Application of the RWG biosensor assay for efficacy evaluation of anticancer drugs. The effect of gefitinib and erlotinib, two FDA-approved drugs targeting EGFR, on EGF-induced DMR signal in UPCI-37B SCCHN cancer cells was evaluated. (A) Gefitinib pretreatment of the cells dose-dependently inhibited the EGF-induced DMR response. (B) Erlotinib pretreatment of the cells inhibited the DMR response of UPCI-37B cells to EGF in a dose-dependent manner. (C) The dose–response curves were obtained, and the IC_{50} values were calculated using Prism 4.0 software using data derived from (A) and (B). The inhibitory effect of gefitinib and erlotinib on EGF-induced DMR signal was normalized to the EGF vehicle (DMSO) control wells without compound and expressed as % of control.

SCCHN cells with increasing concentrations of gefitinib (Figure 10.6A) or erlotinib (Figure 10.6B) dose-dependently blocked DMR response compared with that of vehicle-treated control wells. The estimated IC_{50} values of gefitinib and erlotinib on P-DMR signal were 1.4 and 0.6 μM, respectively (Figure 10.6C). The efficacy of gefitinib and erlotinib measured using the RWG technology is consistent with previous results from cell viability assays [19]. These results indicate that the Epic RWG biosensor technology may be used as an effective tool to evaluate the efficacy of EGFR-targeted therapies. Furthermore, the 384-well format of the biosensor enables the application of this technology as a cell-based HTS tool for the discovery of new therapeutic agents to target abnormal pathways involved in human diseases.

10.5 Discussion

Among various label-free biosensor technologies, surface plasmon resonance (SPR) has been the most extensively explored and developed tool to study bimolecular interactions [7,20]. The surface plasmon is a special type of the electromagnetic field propagating along a metal and dielectric interface, and its electromagnetic field decays evanescently into both the metal and dielectric. Metals such as gold and silver support the surface plasmons at visible frequencies. The label-free SPR technologies coupled with a microfluidic system have the advantage of monitoring biomolecular interactions in solution for kinetic studies. The binding kinetics (association and dissociation constants), affinity, and concentration of targets can be quantified. RWG-based technologies are also capable of detecting biomolecular interactions. The high-density plate format as used in the Epic system allows this technology to be used for HTS. However, such a plate format does not support the experiments required to obtain kinetic parameters for the biomolecular interactions, a strength of the current SPR technology. On the other hand, no cell-based technologies have been developed based on SPR. Because an RWG-based surface has a distinct physical composition relative to the metal/glass sensor surface used in SPR, this offers the unique advantage of the RWG technology in allowing both biochemical and cell-based assays to be carried out in a single platform [2].

Current label-free technologies for cell-based assays make use of various aspects of integrated cellular and morphological changes in response to stimulants to obtain a multiparametric and biologically relevant readout [2,21–24]. They do not require fluorescence-tagged molecules or an overexpression system. The signal change is typically a composite readout of multiple cellular changes; thus, it is a reflection of integrated cellular biology and pharmacology, in contrast to conventional pathway- or reaction-specific and label-based detection technologies. Cell-based label-free technologies have now emerged as highly valuable tools to allow the detection of signaling events under physiological conditions without the need for cell line manipulations.

The Corning Epic optical biosensor system for cell-based assays is based on the use of RWG optical biosensors to measure the DMR signal of cells in response to a specific stimulus. This technology does not require labeling or manipulation of cellular targets and offers a highly sensitive and versatile detection platform for profiling ligand-receptor cell signaling pathways. The resulting DMR signal could be due to many cellular events, such as receptor internalization, recruitment of cellular components to activated receptors, cell movement, spreading, and detachment [18,25]. Therefore, the measured DMR signal is an integrated cellular response to a particular environmental stimulus. The detected DMR signatures may be defined by various parameters, such as magnitude or direction (P-DMR or N-DMR events). These parameters depend on how activated/suppressed signaling pathways cause various intracellular molecules to redistribute at the bottom portion of cells which is within the range of the biosensor detection zone. Although the sensing depth of current RWG biosensors may not penetrate through the entire volume of the cells attached to the sensor surface, a collective response from integrated cellular signaling pathways enables the signal propagation within the entire cytoplasm and permits the detection of a cellular response within a partial volume of the cell. This is demonstrated by the consistent behavior of well-characterized drugs captured by either RWG biosensors or other label- or non-label-based technologies [26]. Because of its unique feature, RWG biosensor technology has been widely used in a range of cell-based assays. They include functional assays for the activation of receptors, such as G protein–coupled receptors [15,27–30] and RTKs [17,18], cell proliferation and growth inhibition assays [10,22,31], signal pathway identification and deconvolution [4,25,26], hit identification and pharmacological profiling, compound screening assays for the identification of specific ligand-induced pathway modulators [17], probing cytoskeleton modulations [18], monitoring viral infection [32], and ion channel assays [33].

Using the Epic RWG optical biosensor, we have identified and characterized distinct optical signatures of EGF-induced signaling events in lung and SCCHN cancer cells [12]. Two DMR events, N-DMR followed by P-DMR, have been obtained from the DMR profiles of A549 lung cancer cells in response to EGF. However, EGF only induced a P-DMR event in UPCI-37B SCCHN cells. Unlike our observed DMR response of A549 lung cancer and UPCI-37B SCCHN cancer cells to EGF addition, the well-established DMR response of epidermoid carcinoma A431 cells to EGF exhibited an initial rapid increase of DMR signal followed by a slowly decreased DMR signal as previously reported [18]. These data suggest that the unique optical signatures measured by the RWG biosensor are cell type dependent. The distinct DMR signatures are consistent with the fact that pathways and mechanisms involved in the regulation of cell signaling are cell type specific. The characteristic DMR response signatures, as defined by the Epic biosensor, may serve as a promising approach for cell line characterization and for monitoring cell-specific oncogenic signaling pathways. High-content information may also be obtained from the real-time DMR response in living cells and enable the detection of integrated signaling pathways under physiological conditions with endogenous receptors.

The use of the tyrosine kinase inhibitor, AG 1478, has validated the specificity of the EGFR-mediated DMR signal. The data support the importance of EGFR tyrosine kinase activity in transmitting EGF-induced cellular signals to the cell/sensor interface to trigger a DMR response. Evaluation of a panel of specific inhibitors on the DMR signal of UPCI-37B SCCHN cells revealed that the PI3K/Akt pathway, but not the Ras/Raf/MAPK pathway, plays a critical role in EGF-induced signaling in these cells. The possible application of the DMR profiles for efficacy evaluation of existing drugs has been examined using two FDA-approved drugs, gefitinib and erlotinib. These data suggest that RWG biosensor technology may have potential for examining the action of EGFR-targeting agents in clinical settings. With the established optical signatures and the availability of a high-throughput 384-well plate format, RWG biosensors offer valuable tools for the discovery of novel therapeutic agents for the treatment of various cancers.

Continued probing of EGFR and other growth factor receptor–induced signaling pathways with the use of the Epic RWG biosensor is expected to reveal distinct DMR signal signatures for various oncogenic pathways in a variety of cancer types. Understanding of the molecular basis of these distinct DMR profiles and their correlation with therapeutic response will aid in their application as in vitro predictive tools for in vivo efficacy. Characterization of these optical signatures for a variety of cell types will help establish a cell optical signature "library." Such information will be invaluable for cell line characterization, pathway studies, and future therapeutic development.

10.6 Future Trends

Application of label-free technologies in cell-based assays is still in the early stage. Because cell-based assays use complex cell systems, many changes will occur when the test system is exposed to an exogenous signal. Certain resistance to applying label-free technology in cell-based assays still exists, in part because data interpretation can be complicated given our current limited knowledge. However, with increased usage and accumulated knowledge, we will gain further insight into the biological meanings of observed DMR signals, which will aid in the interpretation of assay results. It is envisioned that such label-free pathway analysis technologies will have substantial impact on both the dissection of complex biological signaling networks in basic research and the understanding of mechanisms of drug action in drug discovery.

The ability to detect a cellular response within native cells will broaden the use of label-free technology in diverse cell types including primary cells and challenging cells. Furthermore, the pathway-unbiased feature may enable the use of these technologies for the discovery of therapeutics with novel mechanisms and uncover unexpected signaling events. The exquisite sensitivity of label-free technologies may even allow for the resolution of cellular processes that were previously undetectable. Such a property begs for expanded clinical applications of the technology in the future. Indeed, it has been used in identifying the presence of specific surface markers of cancer [34]. It is possible to use the label-free RWG biosensor to study the signaling events of stem cells or cancer stem cells and possibly their response to therapies. Current studies have indicated the importance of circulating tumor cells in early diagnosis and monitoring treatment response. It may not be totally unrealistic to speculate that the RWG biosensor would be employed to characterize these circulating tumor cells for expanded clinical applications. Another area of potential values is to examine the mechanism of drug resistance in a large number of diseases, from infectious disease to cancer. The unbiased cell-based label-free detection technologies may not only reveal pathways critical in mediating the resistance but may also offer a practical HTS approach to identify strategies and agents to overcome the drug resistance problem. As we gain further insights into the biological significance of the DMR signatures, it is anticipated that this label-free biosensor technology will find a wide-ranging powerful applications from basic research, drug discovery, to translational medicine.

ACKNOWLEDGMENTS

We thank Cheryl Meyerkord for her critical reading and editing of the manuscript. Yuhong Du is a recipient of Emory University's SPORE in Head and Neck Cancer Career Development award (P50 CA128613), the Emory URC award, and Emory Winship Cancer Institute Kennedy Seed Grant award. Haian Fu is a Georgia Research Alliance Distinguished Investigator and Georgia Cancer Coalition Distinguished Cancer Scholar.

REFERENCES

1. Hanahan, D. and Weinberg, R.A. (2000). The hallmarks of cancer. *Cell 100*, 57–70.
2. Fang, Y., Ferrie, A.M., Fontaine, N.H., Mauro, J., and Balakrishnan, J. (2006). Resonant waveguide grating biosensor for living cell sensing. *Biophys J 91*, 1925–1940.
3. Minor, L.K. (2008). Label-free cell-based functional assays. *Comb Chem High Throughput Screen 11*, 573–580.
4. Bosch, M.E., Sanchez, A.J., Rojas, F.S., and Ojeda, C.B. (2007). Optical chemical biosensors for high-throughput screening of drugs. *Comb Chem High Throughput Screen 10*, 413–432.
5. Fang, Y. (2006). Label-free cell-based assays with optical biosensors in drug discovery. *Assay Drug Dev Technol 4*, 583–595.
6. Scaltriti, M. and Baselga, J. (2006). The epidermal growth factor receptor pathway: A model for targeted therapy. *Clin Cancer Res 12*, 5268–5272.
7. Homola, J. (2003). Present and future of surface plasmon resonance biosensors. *Anal Bioanal Chem 377*, 528–539.
8. Cunningham, B.T. and Laing, L. (2006). Microplate-based, label-free detection of biomolecular interactions: Applications in proteomics. *Expert Rev Proteomics 3*, 271–281.
9. Cunningham, B.T., Li, P., Schulz, S., Lin, B., Baird, C., Gerstenmaier, J., Genick, C., Wang, F., Fine, E., and Laing, L. (2004). Label-free assays on the BIND system. *J Biomol Screen 9*, 481–490.
10. Lin, B., Qiu, J., Gerstenmeier, J., Li, P., Pien, H., Pepper, J., and Cunningham, B. (2002). A label-free optical technique for detecting small molecule interactions. *Biosens Bioelectron 17*, 827–834.
11. Wu, M., Coblitz, B., Shikano, S., Long, S., Spieker, M., Frutos, A.G., Mukhopadhyay, S., and Li, M. (2006). Phospho-specific recognition by 14-3-3 proteins and antibodies monitored by a high throughput label-free optical biosensor. *FEBS Lett 580*, 5681–5689.
12. Peters, M.F., Vaillancourt, F., Heroux, M., Valiquette, M., and Scott, C.W. (2010). Comparing label-free biosensors for pharmacological screening with cell-based functional assays. *Assay Drug Dev Technol 8*, 219–227.
13. Fang, Y., Li, G.G., and Peng, J. (2005). Optical biosensor provides insights for bradykinin B(2) receptor signaling in A431 cells. *FEBS Lett 579*, 6365–6374.
14. Tiefenthaler, K. and Lukosz, W. (1989). Sensitivity of grating couplers as integrated-optical chemical sensors. *J Opt Soc Am B 6*, 209–220.
15. Fang, Y., Li, G., and Ferrie, A.M. (2007). Non-invasive optical biosensor for assaying endogenous G protein-coupled receptors in adherent cells. *J Pharmacol Toxicol Methods 55*, 314–322.
16. Schlessinger, J. and Lemmon, M.A. (2006). Nuclear signaling by receptor tyrosine kinases: The first robin of spring. *Cell 127*, 45–48.
17. Du, Y., Li, Z., Li, L., Chen, Z.G., Sun, S.Y., Chen, P., Shin, D.M., Khuri, F.R., and Fu, H. (2009). Distinct growth factor-induced dynamic mass redistribution (DMR) profiles for monitoring oncogenic signaling pathways in various cancer cells. *J Recept Signal Transduct Res 29*, 182–194.
18. Fang, Y., Ferrie, A.M., Fontaine, N.H., and Yuen, P.K. (2005). Characteristics of dynamic mass redistribution of epidermal growth factor receptor signaling in living cells measured with label-free optical biosensors. *Anal Chem 77*, 5720–5725.
19. Muller, S., Su, L., Tighiouart, M., Saba, N., Zhang, H., Shin, D.M., and Chen, Z.G. (2008). Distinctive E-cadherin and epidermal growth factor receptor expression in metastatic and nonmetastatic head and neck squamous cell carcinoma: Predictive and prognostic correlation. *Cancer 113*, 97–107.
20. Hoa, X.D., Kirk, A.G., and Tabrizian, M. (2007). Towards integrated and sensitive surface plasmon resonance biosensors: A review of recent progress. *Biosens Bioelectron 23*, 151–160.

21. McGuinness, R. (2007). Impedance-based cellular assay technologies: Recent advances, future promise. *Curr Opin Pharmacol 7*, 535–540.
22. Atienza, J.M., Yu, N., Wang, X., Xu, X., and Abassi, Y. (2006). Label-free and real-time cell-based kinase assay for screening selective and potent receptor tyrosine kinase inhibitors using microelectronic sensor array. *J Biomol Screen 11*, 634–643.
23. Giaever, I. and Keese, C.R. (1993). A morphological biosensor for mammalian cells. *Nature 366*, 591–592.
24. Xi, B., Yu, N., Wang, X., Xu, X., and Abassi, Y.A. (2008). The application of cell-based label-free technology in drug discovery. *Biotechnol J 3*, 484–495.
25. Fang, Y., Ferrie, A.M., and Li, G. (2005). Probing cytoskeleton modulation by optical biosensors. *FEBS Lett 579*, 4175–4180.
26. Lee, P.H., Gao, A., van Staden, C., Ly, J., Salon, J., Xu, A., Fang, Y., and Verkleeren, R. (2008). Evaluation of dynamic mass redistribution technology for pharmacological studies of recombinant and endogenously expressed G protein-coupled receptors. *Assay Drug Dev Technol 6*, 83–94.
27. Schroder, R., Janssen, N., Schmidt, J., Kebig, A., Merten, N., Hennen, S., Muller, A., Blattermann, S., Mohr-Andra, M., Zahn, S., Wenzel, J., Smith, N.J., Gomeza, J., Drewke, C., Milligan, G., Mohr, K., and Kostenis, E. (2010). Deconvolution of complex G protein-coupled receptor signaling in live cells using dynamic mass redistribution measurements. *Nat Biotechnol 28*, 943–949.
28. Fang, Y. and Ferrie, A.M. (2008). Label-free optical biosensor for ligand-directed functional selectivity acting on beta(2) adrenoceptor in living cells. *FEBS Lett 582*, 558–564.
29. Fang, Y., Ferrie, A.M., and Tran, E. (2009). Resonant waveguide grating biosensor for whole-cell GPCR assays. *Methods Mol Biol 552*, 239–252.
30. Fang, Y., Frutos, A.G., and Verklereen, R. (2008). Label-free cell-based assays for GPCR screening. *Comb Chem High Throughput Screen 11*, 357–369.
31. Chan, L.L., Gosangari, S.L., Watkin, K.L., and Cunningham, B.T. (2007). A label-free photonic crystal biosensor imaging method for detection of cancer cell cytotoxicity and proliferation. *Apoptosis 12*, 1061–1068.
32. Owens, R.M., Wang, C., You, J.A., Jiambutr, J., Xu, A.S., Marala, R.B., and Jin, M.M. (2009). Real-time quantitation of viral replication and inhibitor potency using a label-free optical biosensor. *J Recept Signal Transduct Res 29*, 195–201.
33. Fleming, M.R. and Kaczmarek, L.K. (2009). Use of optical biosensors to detect modulation of Slack potassium channels by G protein-coupled receptors. *J Recept Signal Transduct Res 29*, 173–181.
34. Fang, Y. (2010). Label-free and non-invasive biosensor cellular assays for cell adhesion. *J Adhesion Sci Technol 24*, 1011–1021.

Part IV

Optical Technologies for Cancer Detection and Diagnostics: Spectrometry for Cancer Analysis

11

Noninvasive and Quantitative Sensing of Tumor Physiology and Function via Steady-State Diffuse Optical Spectroscopy

Karthik Vishwanath, Gregory Palmer, Quincy Brown, and Nimmi Ramanujam

CONTENTS

11.1 Introduction

It is well known that there is an abnormal growth of neovasculature in solid tumors. These vessels typically form an inefficient and leaky network of blood vessels, thereby constraining delivery of oxygen and other nutrients to the cancer cells [1]. This inefficient system of oxygen delivery combined with the high metabolic demand within the proliferating cancer cells drives a feedback cycle that further propagates aberrant angiogenesis. Both hypoxia and angiogenesis have been flagged as indicators of cancer, and hypoxic microenvironments have been identified in solid tumors of many different tissues [2].

Tumor hypoxia has shown to influence the growth rate and metabolism of tumors, and promote metastatic behavior [3,4]. These phenotypical changes are manifested via various signaling pathways stimulated by hypoxia. One of the most widely understood and studied signaling pathways is through the hypoxia inducible factor (HIF-1) which has been shown to upregulate dozens of genes which influence cellular processes such as proliferation, apoptosis metabolism, and angiogenesis [5,6]. The decreased oxygen concentration in solid hypoxic tumors also decreases the efficacy of radiation and chemotherapy [7,8]. Previously, several studies have investigated the link between clinical outcomes and hypoxia using a variety of different methods and have indicated that tumor hypoxia is an adverse prognostic factor for overall survival and local control, and that tumor hypoxia suggests worse disease-free survival for patients with soft tissue sarcomas, cervical cancer, and head and neck cancers [4,9,10]. These studies motivate the importance of measuring hypoxia as well as the need for being able to measure it, repeatedly and longitudinally, in vivo.

Although there has been an increasing amount of interest in knowing the actual oxygenation levels within solid tumors, there are only a limited number of methods available for the measurement of tumor hypoxia and even fewer of these techniques are applicable for use in vivo. The most well-established method to directly quantify tumor oxygenation is by sensing the partial pressure of tissue oxygenation using a platinum-based microelectrode. Electrode polarography yields point measurements of pO_2 at spatial locations along the path of (micro)electrode insertion and withdrawal within the tissue and provides the best direct estimation of oxygen content in tissue [11]. Tissue oxygen concentrations have also been assessed using perfluorinated compounds in combination with electron paramagnetic resonance spectroscopy, and using 2-nitroimidazoles with positron emission tomography (PET) or magnetic resonance (MR) spectroscopy [10]. Alternatively, tumor hypoxia is indirectly estimated using immunohistochemical (IHC) assays that detect intrinsic markers of hypoxia (such as carbonic anhydrase IX, osteopontin, or HIF-1) or through extrinsic agents (such as pimonidazole or EF5) that selectively stain hypoxic cells [10].

A principal shortcoming of all these aforementioned methods is that they cannot be used to dynamically, frequently, and/or repeatedly monitor changes in tumor oxygenation. Such longitudinal measurements of tumor oxygenation could be extremely useful, for instance, to monitor temporal changes of oxygenation patterns within solid tumors that are undergoing targeted molecular therapies, or in preclinical trials of new therapeutic drug regimens where drug efficacy is significantly influenced by the tumor's oxygenation status.

Optical diffuse reflectance spectroscopy (DRS) has emerged as a technology that can quantitatively and nondestructively quantify changes in several key biomarkers of carcinogenesis including vascular tumor oxygenation and total blood volume, and/or measure the amount of exogenous contrast agent such as an administered tracer agent or drug. Figure 11.1 shows a graphical schematic of how optical spectroscopy works. A quick, simple, and inexpensive optical spectroscopic implementation involves measurement of the tissue diffuse reflectance spectrum—this technique quantifies the wavelength dependence of the reflected light after it has undergone absorption and scattering interactions in tissue and can be analyzed to provide a quantitative measure of the absorption and scattering coefficients of the interrogated tissue [12–14]. These measures of absorption and scattering in turn provide

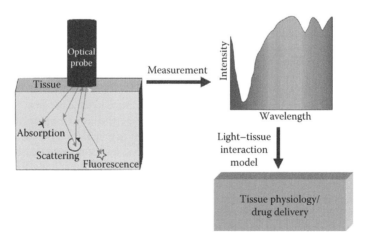

FIGURE 11.1 Optical measurements are acquired using a fiber-optic probe placed in contact with the tissue sample surface. Light is coupled into the tissue via illumination fibers, where it can undergo absorption, scattering, and fluorescence interactions with the tissue. Some of the light is diverted back to the tissue surface to be collected by the collection optical fibers. The collected signal can be quantified to obtain a diffuse reflectance spectrum as a function of wavelength. Modeling algorithms can then be used to extract the underlying optical properties for detection of physiologic indicators, probe update, and/or drug delivery. (Reprinted from *J. Controll. Release*, 142(3), Palmer, G.M., Boruta, R.J., Viglianti, B.L., Lan, L., Spasojevic, I., and Dewhirst, M.W., Noninvasive monitoring of intra-tumor drug concentration and therapeutic response using optical spectroscopy, 457–464, Copyright 2010, with permission from Elsevier.)

information about the composition of the interrogated tissues which are related directly to tissue function and morphology.

In this chapter, we introduce the methods of DRS, briefly describe the main physical principles behind this technology, provide an overview of the quantitative model used to extract physiological endpoints from the measurements, and finally present a selection of preclinical and clinical studies where we have applied this technology successfully to monitor vascular mediated physiology of solid tumors. We would like to emphasize that all of the work presented in this chapter will specifically discuss the applications of quantitative DRS in the ultraviolet-visible (UV-Vis) (300–700 nm) wavelength spectrum. Although several groups have reported on the use of NIR wavelengths for diffuse optical spectroscopy and tomography to image the presence of cancerous tissues in human patients [15–21], as well as to image the response of solid tumors to chemo- and radiotherapy [22–25], we will not cover that large volume of work here. The optical methods that use NIR wavelengths allow deep sensing of tissues, but only provide indirect measurements from lesions. The UV-Vis wavelengths on the other hand provide direct measurements of optical properties of the lesions since these techniques require the placement of the optical probes directly on the sites of measurements. Thus, these UV-Vis spectroscopic methods provide the essence of the underlying contrast that may be utilized by the deeper sensing NIR methods.

11.2 Materials and Methods

11.2.1 Overview of Optical Spectroscopy Techniques

Methods of optical spectroscopy have historically been employed for chemical analysis of biological samples. Over the last two decades, there has been increased interest in bringing the methods of optical spectroscopy to noninvasively and rapidly examine human tissues and in particular tissues in vivo. These efforts have been motivated by the exquisite chemical sensitivity and selectivity offered by optical techniques. Interactions of light with complex media, such as biological tissues, are characterized by a variety of processes that depend on the physical nature of the light (such as its intensity, frequency spectrum, pulse width, polarization state, overall power, etc.) and the specific tissue morphology and composition [26,27]. When light is incident on a biological sample, it is multiply scattered due to microscopic variations in the tissue refractive index (such as between the cytosol, the cell membrane, cellular organelles, and the surrounding extracellular matrix). The incident radiation can also be absorbed by chromophores present in the tissue, absorbed by tissue fluorophores to be remitted as fluorescence, and/or be inelastically scattered. These optical responses can be measured by a variety of spectroscopic techniques, including UV-Vis reflectance spectroscopy, near-infrared (NIR) spectroscopy (NIRS), elastic scattering spectroscopy (ESS), Raman spectroscopy, and/or fluorescence (or phosphorescence) spectroscopy.

In diffuse optical (reflectance or transmittance) spectroscopy, the wavelengths of the illumination source typically span the spectral range from the UV through the NIR wavelengths. These illumination light sources can be temporally invariant (such as broadband lamps and/or CW lasers), pulsed light sources (such as pulsed LED or lasers), or intensity-modulated sources (such as CW laser or LED sources whose driving currents are sinusoidally modulated) [28,29]. These different types of light sources, by definition, give rise to steady-state, time-resolved, and frequency-domain spectroscopies, respectively. For the remainder of this chapter, we will focus solely on the applications of steady-state DRS performed using wavelengths spanning the UV-Vis spectrum where absorption of tissue is comparable to or greater than its scattering.

DRS is sensitive to the optical absorption and scattering of soft tissues. The shape and magnitude of the absorption coefficient depends on the extinction coefficient and concentrations, respectively, of dominant tissue chromophores. Common and ubiquitous tissue chromophores include oxygenated hemoglobin (HbO$_2$) and deoxygenated hemoglobin (dHb), β-carotene, melanin, water, lipids, and other proteins in the UV-NIR spectrum [30–32]. Since DRS can measure both HbO$_2$ and dHb, one can estimate both the total blood concentration (THb = HbO$_2$ + dHb) and the percent oxygenation saturation in tumors

($SO_2 = 100 \times HbO_2/THb$). The optical scattering coefficient is known to be sensitive to the spatial architecture and organization of the tissue and therefore can be used as a means to track changes in cellular morphology and density [33,34].

11.2.2 Relationship of Diffuse Reflectance Spectroscopy to Other Surface Optical Scanning Methods

DRS techniques sense perturbations of the incident broadband spectrum induced by the optical absorption and scattering properties of the biological tissues. In general, DRS scanning techniques rely only on the spectral shape and intensities of the incident and reflected light and are insensitive to the coherent properties (such as polarization or phase) of light. This method detects multiply scattered light, and its sensing depth depends on the geometrical arrangement of the sources and detectors as well as the optical properties of the interrogated medium itself. Other surface scanning optical techniques include optical coherence tomography (OCT) and confocal and two-photon fluorescence methods—all of which provide much higher spatial resolution than DRS methods, but include trade-offs with respect to sensing depth, sources of contrast, complexity of instrumentation, scan time, and data analysis.

OCT is an interferometric imaging technique which provides high-resolution (~1–10 μm) images at depths of ~1–2 mm below the surface. In OCT, depth scans (A-scans) are acquired over a two-dimensional (2D) area (B-scans) to reconstruct 3D tissue volumes. Depth in the sample is gated by resolving the difference in path lengths in a Michelson interferometer between reflecting and scattering sites in the sample and that of a known reference delay. Reflection and scattering sites in the sample are resolved corresponding to the coherence length of the illumination source in the axial dimension, and the diameter of the focused spot size in the sample in the lateral direction. The primary source of contrast for OCT is the scattering coefficient of the tissue.

Confocal imaging uses a focused laser source to excite fluorescence at a point in the image plane of the sample and rejects light above and below the image plane by a pinhole that is positioned at the conjugate image plane. By controlling the size of the pinhole, the lateral resolution of a confocal microscope can be better than the Rayleigh limit. A confocal imaging system thus allows for depth sectioning capability. A 3D image is constructed by raster-scanning in the lateral *x-y* plane using motorized mirrors, and stepping through the *z* plane, commonly using a motorized stage focus.

Another surface scanning technique that is employed for in vivo imaging of biological tissues is multiphoton fluorescence. In this technique, an ultrashort, pulsed excitation light of half the absorption frequency is tightly focused using a high NA objective to achieve a high photon density at the focal plane such that two (or three) photons are simultaneously absorbed by tissue fluorophores. Since the probability of multiphoton excitation has a quadratic dependence on the photon intensity for two-photon excitation, fluorescence is generated only within the focal point and thus retains spatial resolutions that are comparable to that of laser scanning confocal imaging despite using longer excitation wavelengths. Additionally, the use of longer wavelengths allows for deeper penetration (on the order of several hundred microns) in tissue.

Although these surface scanning techniques can provide images with much higher spatial resolutions than DRS, they require specialized light sources, highly optimized optical components, complex detector elements, and electronics. OCT is only sensitive to tissue scattering and thus mainly provides structural contrast of the tissue. Confocal imaging is typically used for tissue fluorescence, has limited penetration depths in thick tissues, loses depth resolution in thick tissues, and involves a trade-off between high-resolution and sampling volume. High-resolution confocal images have low sampling volume or area, whereas DRS has lower spatial resolution but can sample larger volumes. Finally, these higher-resolution techniques are difficult to translate into clinical settings and are much more expensive compared to DRS methods.

11.2.3 Instrumentation for Measurement of Diffuse Reflectance Spectra

Measurement of diffuse reflectance requires illuminating the tissue with a broadband light source and recording the multiply scattered reflected light as a function of wavelength. This requires a few

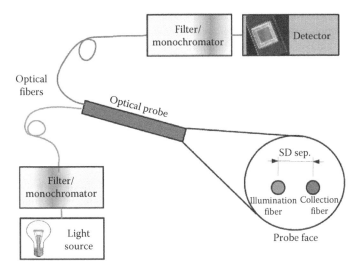

FIGURE 11.2 Schematic of a measurement device, consisting of a light source and monochromator/filter element. This delivers filtered light to the sample via a fiber-optic probe. Similarly on the detection end, a detector is shown with a separate monochromator/filter element, enabling independent selection of illumination and collection wavelengths (e.g., this enables flexibility to measure fluorescence). In general, at least one filtering or dispersive element is needed to measure diffuse reflectance spectra. At the optical probe interface, which couples light to the tissue, the fibers are terminated in a solid probe. This holds the illumination and collection fibers in a fixed geometry, characterized by the source-detector separation (SD Sep.). Larger SD separations cause the light to travel farther and deeper through the tissue on average, thus influencing the sensing volume.

components that are common to all spectroscopic instruments: a light source to illuminate the sample, a means of filtering or selecting wavelengths, and a detector to quantify the collected light intensity spectrum (see Figure 11.2). The illumination light source is typically coupled to a fiber-optic probe which acts as a conduit to deliver this light to the tissue site of interest. Common broadband light sources for steady-state UV-Vis DRS include standard high-power arc lamps such as the Xe-arc lamp or deuterium-tungsten lamps. Since DRS requires a broad spectral range of light to be viable, several lasers and LEDs would need to be combined or scanned serially to cover required spectral wavelengths. However, recent advances in solid-state electronics have brought about broadband white-LED sources (e.g., XR-E, Cree, Durham, NC) and photonic-crystal-fiber-based supercontinuum coherent laser sources (e.g., the SC-400 from Fianium, Inc., Southampton, UK) which are ideally suited as light sources for DRS. The filter/monochromator components depicted in Figure 11.2 allow passage of a given spectral range of wavelengths to be used for illumination and collection from the tissue. A monochromator isolates a narrow spectral band of the input source by using a diffraction grating mounted on a rotating turret. The ability to customize the grating and the entrance and exit slit widths gives wide flexibility to the spectral range and bandwidth that can be filtered by the monochromator, but suffers from the disadvantage of being relatively slow while scanning across a wide spectral range due to the inherent mechanical motion of the turret. Alternately, it is possible to insert a series of spectral bandpass filters into the light path (either from the source, or to the detector) in order to choose the spectral ranges of detection and illumination. Developments in liquid crystal (or acousto-optically) tunable filters (LCTF/AOTF) offer the possibility of using these devices to control the spectral ranges electronically, instead of mechanically, albeit at a much higher system cost. The final element needed to make a spectroscopic measurement device is the detector element. This element can either be a detector array such as a charge couple device (CCD) or a photodiode array, or a single channel detector such as a photomultiplier tube (PMT) or photodiode. Typically, a CCD-based detection scheme is used as a spectrograph which eliminates the need for the filter/monochromator element in the detection arm of the system. PMT-based detection on the other hand allows for the detection of signals with a much higher dynamic range than CCD-based systems, but are typically bulkier and slower than the CCD-based detectors.

There are several key considerations in designing such spectroscopic devices including spectral bandwidth, spectral resolution, wavelength range, measurement speed, and throughput, as well as practical concerns such as system cost, complexity, and size with trade-offs between them. Important aspects of these types of optical instrumentation are the geometry of illumination and collection of light (which determine the sensing volume of the device and signal throughput), methods for accurate and routine calibration, and algorithms for converting measured optical spectra to quantitative, physical endpoints that reflect the underlying tissue composition.

11.2.4 Fiber Probe Geometry to Illuminate and Collect Diffuse Reflectance from Tissues

The sensing depth of DRS varies from few millimeters in the UV-Vis spectrum to several centimeters in the NIR region [35]. In the UV-Vis region, tissues are strongly absorbing, which limits the overall tissue sensing depth. With increasing wavelength, the overall absorption coefficient decreases, and the ratio of scattering to absorption coefficients increases. Thus, in the red and NIR wavelengths, tissues are generally more transparent, and photons can migrate through several centimeters of tissue which allows NIR spectroscopy to interrogate subsurface solid tumors such as those in the breast and neck nodes. UV-Vis spectroscopy complementarily has a superficial sensing depth and can directly interrogate tumor growth in readily accessible sites such as the head and neck, anus, cervix, and recurrent chest wall disease in breast cancer. Optical spectroscopic probes can also be guided through endoscopes and biopsy needles to access tumors within body cavities such as colon cancer or within surrounding tissue such as breast cancers. This technology is also well suited for drug discovery in preclinical tumors in rodent models.

The probing volume and depth are affected by the optical properties (absorption and scattering coefficients) of the medium, as well as the probe geometry—particularly the source-detector separation. As the source-detector separation increases, the probing depth and volume increase which increases the sensitivity of the probe to deeper tissue [36]. However, the extent to which the source and detector can be separated is limited by throughput of light through the tissue or signal to noise ratio, which is influenced by the tissue absorption and scattering coefficients. Source-detector separations can be made significantly larger in the NIR (~centimeters) compared to the UV-Vis range (~millimeters) where the ratio of scattering to absorption is large compared to that in the UV-Vis range.

11.2.5 Inverse Models to Extract Underlying Tissue Composition from Diffuse Reflectance

Once measured, the diffuse reflectance must be processed through rigorous computational or theoretical models to obtain quantitative information about the absorption and scattering properties of the tissue [13,16,37]. Quantitative determination of the underlying absorption and scattering properties provides direct insight into tissue physiology, function, and/or morphology. However, extraction of these parameters from the measured diffuse reflectance spectra is a nontrivial problem. Given the complexity of real biological tissues, inverse operation by making assumptions about structure and/or the distribution of absorbers or scatterers in the tissue without which finding unique solutions to the inverse problem is not possible. It is therefore imperative that any developed algorithm be rigorously tested so that it can elucidate appropriate functional biochemical relationships when applied to living tissues. Most inverse methods used in optical spectroscopic and tomographic reconstructions assume that the optical properties of the medium under investigation can be characterized by uniform absorption and scattering properties (over predefined length scales). There are a wide variety of different approaches that have been developed by several researchers over the previous two decades [32]. These methods can roughly be subdivided into analytical and numerical approaches. Analytical approaches (generally simplified approximations of the Boltzmann radiative transport equation) have the advantage of low computational requirements, but are not widely applicable to a large variety of wavelength ranges or experimental conditions, particularly the UV-Vis spectral range. On the other hand, numerical approaches (such as stochastic Monte Carlo modeling) have high computational complexity but are more generally applicable over a wide range of tissue types or wavelengths and can accurately model real complex experimental

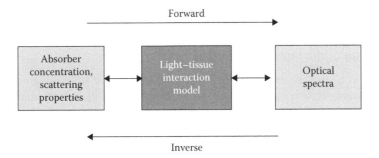

FIGURE 11.3 The forward model is a deterministic algorithm that generates the diffuse reflectance spectra, based on the tissue optical properties (absorption and scattering as a function of wavelength). The inverse model takes the measured spectra and extracts the unknown optical properties of the sample from which it was measured. This is an ill-posed problem since there is not a unique set of optical properties for a given diffuse reflectance intensity. Thus, other assumptions must be made about the nature of the absorbers and scatterers present in the medium.

conditions. In either case, there is distinction between the forward problem (determining tissue diffuse reflectance spectra based on the optical properties) and the inverse problem (extracting tissue optical properties from a measured spectrum), as shown in Figure 11.3. The forward problem is solved deterministically by making certain assumptions about the tissue model, while the inverse problem involves inferring the optical properties based on the combination of the measured data, the model assumptions, and knowledge of tissue optical properties (e.g., the absorption spectrum of hemoglobin). It is to be noted that the extracted optical properties, and thus the functional and physiological properties, of the medium are only as robust as the assumptions made in developing the forward techniques. We have developed a Monte Carlo approach to extract quantitative absorption and scattering properties of the medium being probed using the measured diffuse reflectance spectra [38,39] by employing scaling and similarity relationships that accelerate the inverse modeling. Application of this model involves a least-squares fitting approach, whereby the modeled (estimated) optical properties of the medium are varied such that they produce minimal error between the measured and predicted spectra generated using scaling of a baseline Monte Carlo simulation for different tissue optical properties.

For all of the data shown in this chapter, the data analyses that translated a measured diffuse reflectance spectrum into extracted tissue absorption and scattering coefficients (thereby allowing quantitative estimates of biologically relevant endpoints such as the total hemoglobin concentration, vascular oxygen saturation, etc.) used the inverse Monte Carlo developed previously and followed a fixed procedure. Each collected reflectance spectrum is divided by the reflectance obtained from the surface of a standard reflectance Spectralon standard (Spectralon Reflectance Standard, Labsphere, North Sutton, NH) to calibrate for light intensity throughput and wavelength calibration. These calibration spectra are obtained every time the light source is turned on (i.e., on each day of measurement). The calibrated reflectance data are then input into the inverse Monte Carlo model, and a selection is made regarding the type and number of absorbers expected to be present in the tissues—for instance, resected breast tissues in the tumor margin study (as described later) are analyzed by using oxy-, deoxyhemoglobin, lymphazurin (a dye used in the lumpectomy/mastectomy surgical procedures), and β-carotene (an absorber known to be present in adipose breast tissue) [40–42], while data from animal tumors only used hemoglobin spectra and an associated mouse-skin absorber [43,44]. Lastly, the diffuse reflectance measured (using the exact same instrument configuration as that used to collect the tissue reflectance spectra) from a calibrated reference phantom with known optical absorption and scattering properties is input into the inverse model. As described in detail elsewhere [39,45], this reference phantom allows the inverse model to scale the computed diffuse reflectance by the Monte Carlo method to that detected by the instrument, thereby allowing the least-squares approach to accurately fit the measured data. Once the inverse model reaches convergence, the outputs from the model give the concentrations of the absorbers, as well as the shape and magnitude of the absorption and scattering coefficients that best fit the measured data.

11.3 Results

11.3.1 Validation of the Optical Technology

We have conducted a series of extensive studies to test the ability of the inverse model in quantitatively extracting hemoglobin oxygen saturations (SO_2), total hemoglobin concentrations, as well as the wavelength-averaged optical scattering coefficients from experimentally measured diffuse reflectance when obtained using a variety of different instruments and fiber geometries in well-controlled, homogeneous tissue-simulating phantoms [45]. The performance of the instrument was tested by measuring the diffuse reflectance data from liquid tissue phantoms that were prepared by diluting stock solutions of hemoglobin (as absorber) and 1 μm diameter polystyrene microspheres (as scatterer) to a final fixed volume. Knowledge of the stock concentrations of hemoglobin and scatterer solutions allowed calculation of the expected optical properties of absorption and scattering in the final phantom. The measured diffuse reflectance was analyzed using the inverse Monte Carlo model described previously [39] to extract optical absorption and scattering from the phantoms. Figure 11.4a shows the comparison between the extracted values of the total hemoglobin concentration across a set of phantoms whose absorption coefficient (and hence the hemoglobin concentration) was increased with sequential additions of a stock hemoglobin solution to span total hemoglobin concentrations between 1 and 50 μM. Figure 11.4b shows the extracted vs. expected mean scattering coefficient for a second series of diffuse reflectance measurements obtained from phantoms that had varying scattering coefficients (scattering was varied by sequential additions of polystyrene microspheres suspension and calculated using Mie theory [45]). The values of both the extracted hemoglobin concentration and the mean scattering coefficient were less than 10% of their expected values across the range of values studied here.

Next, we tested the accuracy and sensitivity of the instrumentation and inverse model to extract the hemoglobin saturation in a phantom model [45]. A liquid phantom with polystyrene microspheres suspension as scatterer and HbO_2 as absorber was prepared and was desaturated using yeast. Concurrent optical and pO_2 measurements (using an oxygen needle electrode) were taken while the oxyhemoglobin in the phantom was gradually deoxygenated over a 1 h time period. Optical measurements of diffuse reflectance were obtained approximately every minute postaddition of yeast. The percent oxygen over the course of the optical measurements was independently and continuously monitored by an oxygen-sensitive electrode (MI-730, Microelectrodes, Inc., Bedford, NH) at the beginning and end of each optical measurement and averaged, for that time point. Oxygen partial pressure (pO_2) was determined from the percent oxygen measured by the electrode assuming a linear relationship, and the expected SO_2–pO_2 relationships were estimated using the Hill-curve relationship [46]. SO_2 values were derived from the optical measurements by extracting the concentration of oxyhemoglobin (HbO_2) and deoxyhemoglobin (dHb) and computing the % saturation, $SO_2 = 100 \times HbO_2/(HbO_2 + dHb)$. The changes in the pO_2–SO_2

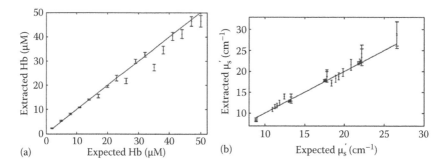

(a)   Expected Hb (μM)   (b)   Expected μ_s' (cm^{-1})

FIGURE 11.4 Comparisons between the expected and extracted values of total hemoglobin concentration (a) and mean scattering coefficients (b) in liquid tissue phantoms. The solid line shows perfect agreement. (Reproduced from Bender, J.E., K. Vishwanath, L.K. Moore, J.Q. Brown, V. Chang, G.M. Palmer, and N. Ramanujam, A robust-Monte Carlo model for the extraction of biological absorption and scattering in vivo. IEEE Trans Biomed Eng, 2009. 56(4): 960–968., copyright 2009 IEEE. With permission.)

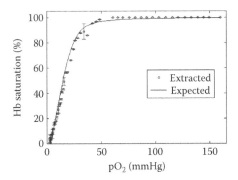

FIGURE 11.5 The congruence between the hemoglobin saturation (symbols) extracted from the optical measurements and pO_2–SO_2 Hill curve (line) calculated from measurements of percent oxygen from a needle electrode in a tissue phantom containing HbO_2. Desaturation was done using Baker's yeast. (Reproduced from Bender, J.E., K. Vishwanath, L.K. Moore, J.Q. Brown, V. Chang, G.M. Palmer, and N. Ramanujam, A robust-Monte Carlo model for the extraction of biological absorption and scattering in vivo. IEEE Trans Biomed Eng, 2009. 56(4): 960–968., copyright 2009 IEEE. With permission.)

values (Figure 11.5 symbols) were quantified using the Hill curve (Figure 11.5 line). These results show that the optical technology can provide robust, accurate, and real-time estimates of changes in hemoglobin saturation.

11.3.2 Applications to Preclinical Studies

We have carried out a series of preclinical studies with nude mice with our technology. Here, we present a set of four different nude-mouse xenograft studies that used visible DRS to quantitatively extract optical endpoints specifically as pertained to tumor hypoxia. The first two of these studies report on preliminary validation study conducted to ascertain the ability of optical methods to measure tumor oxygenation status in vivo by comparisons to well-established techniques using pO_2 microelectrodes and immunohistochemistry [43,44]. The objective of the last two sets of studies was to determine the association of tumor oxygenation status measured optically to tumor growth delay curves when the animals were treated with radiation and chemotherapy [47,48]. In the following text, we discuss the main findings of these studies to highlight the advantages of using these noninvasive measurements to extract functional information from solid tumors.

In the first set of these animal studies, we used ($N = 19$) nude mice to investigate the relationship between the vascular hemoglobin saturation and the established gold standard of needle sensors (OxyLite, Oxford Optronics) [43]. 4T1 mammary carcinoma xenografts were grown in the flank of each animal to a size of nearly 1 cm. The fiber-optic probe was placed adjacent to two-electrode needle sensors, and the tumor oxygenation was monitored over time. The optical probe used here had a mean sensing depth of ~1 mm under the skin of these mouse tumors. First these responses were measured in room air, then subsequently under carbogen (95% oxygen, 5% CO_2) inhalation. Significant increases in both the pO_2 values obtained using the OxyLite electrode and hemoglobin saturation were seen following carbogen breathing. The panel of figures in Figure 11.6 show the temporal oxygenation changes detected in a representative animal with the two pO_2 electrodes (Figure 11.6a and b) and the optical probe (Figure 11.6c). A direct comparison between the two measurements (optical and electrode) proved somewhat challenging due the issue of small sensing volumes of the OxyLite sensors and makes the reported pO_2 highly sensitive to positioning within the tumor of each animal (Figure 11.6a and b). Nevertheless, it is apparent on comparing Figure 11.6b and c that the optical sensor exhibited a faster response time relative to the pO_2 sensor and showed increases in SO_2 changes upon carbogen breathing. Across the 19 animals measured, significant increases in vascular hemoglobin saturation using the optical measurements were observed before carbogen inhalation relative to after carbogen inhalation: oxygen saturation increased from $30\% \pm 19\%$ at baseline to $49\% \pm 20\%$ with carbogen breathing (paired t-test: $p = 0.002$, unpaired: $p = 0.01$). It is to be noted that the

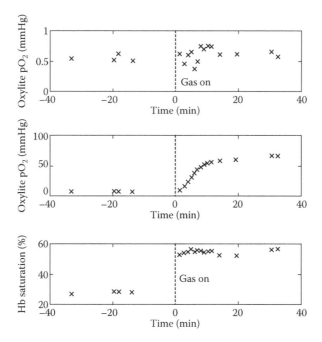

FIGURE 11.6 Representative time course data for showing pO_2 for each of the two OxyLite probes (top two panels) and the hemoglobin saturation. The animal was provided carbogen continuously after $t = 0$ min, as indicated by the dashed line. Hemoglobin saturation and pO_2 for one of the OxyLite probes (middle panel) can be seen to increase with carbogen breathing. The other OxyLite probe shows no change upon carbogen breathing (top panel). (Reproduced from Palmer, G.M., R.J. Viola, T. Schroeder, P.S. Yarmolenko, M.W. Dewhirst, and N. Ramanujam, Quantitative diffuse reflectance and fluorescence spectroscopy: Tool to monitor tumor physiology in vivo. J Biomed Opt, 2009. 14(2): 024010., copyright 2009 SPIE. With permission.)

optical measurements extracted oxygen saturation based on the relative amounts of oxy- and deoxyhemoglobin concentrations in the tissues, and thus measures vascular oxygen saturation.

In the second animal study, we studied the relationship between optical biomarkers and IHC-assessed hypoxic fractions (using pimonidazole), and necrotic fractions in 50 nude mice were subcutaneously inoculated with a suspension of 10^6 4T1 mouse breast carcinoma cells in their right flanks [44]. Once the tumor diameters reached 4–6 mm, the animals were evenly distributed (by tumor volume) into control and treatment groups (with each group having 25 animals). The treated group received a maximum tolerated dose (MTD; 10 mg/kg i.v.) of doxorubicin (DOX), while the control group received an equivalent volume of saline. Treatment day was labeled day 0. Diffuse reflectance measurements were obtained longitudinally from each animals using optical spectroscopy for 2 weeks on days 0, 2, 5, 7, 10, and 13. For IHC analyses, a total of 10 mice (5 from each group) were randomly selected and sacrificed at four time points (on days 0, 5, 10, and 13). The tumors from these animals were resected, snap-frozen, and subsequently sliced into 10 μm thick sections for IHC analysis. Each section was stained with pimonidazole and hematoxylin and eosin (H&E). Using fluorescence microscopy, images of the tumor sections were analyzed to extract the hypoxic fractions from pimonidazole staining and necrotic fractions from H&E staining, as described previously [44].

Quantitative assessments of the temporal trends displayed by the optical and IHC endpoints were estimated using multivariate statistical models. These models were built in an additive fashion to determine the effect of time (day) and treatment for both groups of data and indicated that the hypoxic and necrotic fractions (from IHC) and dHb concentrations and $<\mu_s'>$ (from the optical measurements) had correlated temporal trends ($p < 0.001$). Figure 11.7a through d shows these temporal relationships between the two IHC endpoints of hypoxic and necrotic fractions (plain bars, left axis) with the optical endpoints (shaded bars, right axis). Figure 11.7a and b shows these temporal variations in the IHC and optical endpoints for the treated animals, while Figure 11.7c and d shows these data for the control group. It is to be noted that

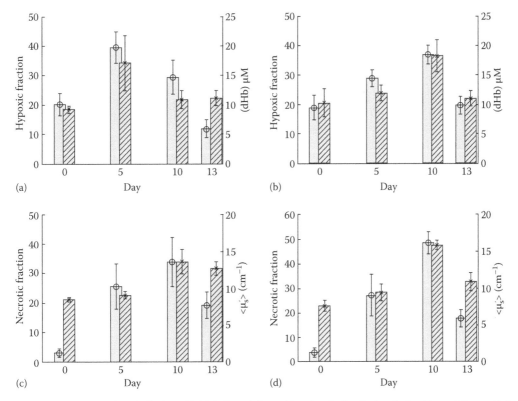

FIGURE 11.7 Temporal trends in the IHC (plain bars, left *y*-axis) and optical endpoints (striped bars, right *y*-axis) for animals treated with DOX (left column) and control (right column) animals. (a) and (b) show the relationships between the average hypoxic fractions assessed using IHC and optical extraction of the dHb concentration for the DOX treated (left) and control animals (right), respectively. (c) and (d) show the concordance between necrotic fraction from IHC (plain bars, left *y*-axis) and $<\mu_s'>$ (hatched bars, right *y*-axis) for the treated (left) and control (right) animals, respectively. (Reproduced from Vishwanath, K., H. Yuan, W.T. Barry, M.W. Dewhirst, and N. Ramanujam, Using optical spectroscopy to longitudinally monitor physiological changes within solid tumors. Neoplasia, 2009. 11(9): 889–900. With permission.)

the optical endpoints shown in Figure 11.7a through d correspond to the five animals that were selected to undergo IHC at each time point. These results demonstrate that the optical biomarkers could track important physiological endpoints as estimated using the "gold standard" of IHC and show the feasibility of these methods to dynamically measure tumor oxygenation in vivo.

In the third and fourth set of animal studies, we investigated how these functional endpoints within solid tumors obtained using optical methods changed as the animals were exposed to conventional radiation or chemotherapeutic treatments. The third animal study tested whether early measurements of diffuse reflectance obtained noninvasively from solid tumors that were exposed to radiation treatment could predict (or be associated with) final treatment outcomes [47]. Here, 34 nude mice were inoculated with 10^6 FaDu human hypopharyngeal squamous cell carcinoma cells, subcutaneously, on their right flanks. Once the average tumor diameters reached 5–8 mm, the animals were evenly distributed by tumor volume into control and treatment groups in a 1:2 ratio, respectively. $N = 23$ animals in the treatment group were exposed to 39 Gy of radiation, while $N = 11$ animals in the control group received sham irradiation. This dose of 39 Gy was chosen as it was previously reported as the TCD_{50} (dose which provides local control to 50% of the treated population) for this cell line in nude mice [49]. Prior to radiation exposure, all animals (whether treated or shammed) were anesthetized using Nembutal (i.p. diluted 1:5 by volume) and placed in a custom-built stereotactic holder in the irradiator. Treatment day was labeled day 0. All tumors were monitored optically before treatment to get baseline measurements on day 0 and then again on days 1, 3, 5, 7, 10, 12, 14, and 17. The animals were anesthetized via isoflurane breathing (1.5% isoflurane gas mixed with oxygen) throughout the course of the optical measurements. Tumor volumes were

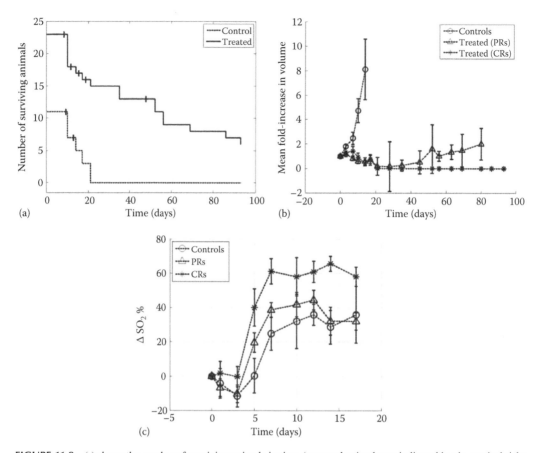

FIGURE 11.8 (a) shows the number of surviving animals in time (censored animals are indicated by the vertical ticks at the points of censoring); solid line: treated group; dotted line: control group. (b) shows the mean fold increase in tumor volume for the treated (triangles and asterisks) and untreated animals (controls, circles). The treated animals were classified as CRs (asterisks) or PRS (triangles) based on the presence or absence of lesions at the primary site 90 days posttreatment. (c) shows the average change in tumor blood oxygen saturation (relative to day 0, ΔSO_2) for each group (circles: CRs; triangles: PRs; asterisks: controls). (Reproduced from Vishwanath, K., D. Klein, K. Chang, T. Schroeder, M.W. Dewhirst, and N. Ramanujam, Quantitative optical spectroscopy can identify long-term local tumor control in irradiated murine head and neck xenografts. J Biomed Opt, 2009. 14(5): 054051. With permission.)

measured once using calipers each day over the course of the optical measurements and then one or two times per week until 90 days posttreatment.

Figure 11.8a shows the number of surviving animals for the treated (dotted line) and control (solid line) groups over the 90 day period posttreatment. Censored animals in the study are indicated by the customary vertical ticks (five control and seven treated animals were censored due to tumor ulcerations). Figure 11.8b shows the tumor growth delay curves for three groups: triangles and asterisks identify the treated group, while circles show the control animals. Symbols represent the mean value of the fraction increase in tumor volume calculated for all animals that were alive for each group, at each time point, while the error bars represent the standard error. Six of the treated animals showed complete cures with no local or distant tumor recurrences up to day 90 and were classified as complete responders (asterisks; CRs). The other 17 treated animals were classified as partial responders (PRs, triangles) since the treatment in these animals failed to achieve local control with regrowth occurring before 90 days (the last recurrence was scored on day 63 posttreatment). The measured diffuse reflectance spectra were inverted to obtain the tumor oxygen saturation (SO_2) in all animals, at each measurement time point. Figure 11.8c shows the group-averaged time trends in ΔSO_2 for the controls, PRs, and CRs. Here, ΔSO_2 on a given day represented the change in SO_2 on that day relative to the baseline measurement for each particular animal

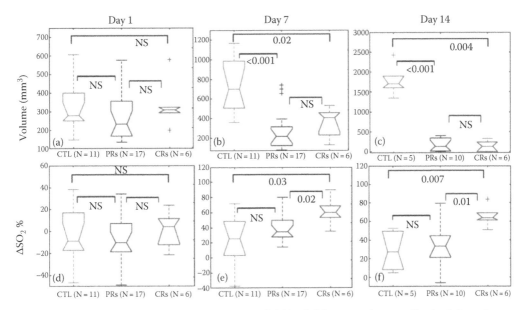

FIGURE 11.9 Changes in tumor volume (a–c, top row) and ΔSO_2 (d–f, bottom row) across all animals in each group, on days 1, 7, and 14 (columns) visualized through box plots (CTL: Controls; PRs: Partial responders; CRs: Complete responders). Each figure shows the *p* value for all pairwise Wilcoxon comparisons (NS: not significant). (Reproduced from Vishwanath, K., D. Klein, K. Chang, T. Schroeder, M.W. Dewhirst, and N. Ramanujam, Quantitative optical spectroscopy can identify long-term local tumor control in irradiated murine head and neck xenografts. J Biomed Opt, 2009. 14(5): 054051. With permission.)

and allowed comparison of the changes in SO_2 values relative to each subject's baseline. As in Figure 11.8b, the symbols are average values computed using all surviving animals, at each time point, while the vertical lines are standard errors. Figure 11.8c shows that the tumor oxygen saturation for the CRs had the steepest increase and the highest value relative to the controls and PRs 7–14 days posttreatment.

Figure 11.9 compares the changes in tumor volume (Figure 11.9a through c) and ΔSO_2 (Figure 11.9d through f) across animals in each group, on day 1 (Figure 11.9a and d), day 7 (Figure 11.9b and e), and day 14 (Figure 11.9c and f) using the notched form of box plots [50]. Figure 11.9a and d showed no differences in tumor volumes or ΔSO_2 values between all groups on day 1. By day 7, it was possible to isolate the treated animals (in both the PR and CR group) from the untreated controls on the basis of an ANOVA test on the tumor volumes, and these differences were enhanced on day 14. However, there was no statistically significant difference between the tumor volumes of the PRs and CRs either on day 7 or day 14. The changes in ΔSO_2 on days 7 and 14, for each group on the other hand, showed that the CRs were well separated from both the control group and the PR group, while there was no statistically significant difference (NS) between the PRs and control animals, on any of the days. This proof-of-concept study suggests that DRS could have the potential to differentiate CRs from partial and nonresponders in preclinical models early in the course of therapy.

The fourth preclinical study we present here demonstrates the use of optical spectroscopy to quantify tumor hypoxia and drug uptake and see how these parameters related to treatment outcome. In this study, dual SKOV3 ovarian tumors were grown in the flanks of nude mice. Upon reaching treatment volume (approximately 150–200 mm³), the animals were randomized into six groups and were treated with free DOX, low-temperature-sensitive liposomal-DOX (LTSL-DOX), or saline, with each treatment group with and without hyperthermia applied to the larger tumor [48]. This study was designed to test the ability of thermally targeted drug delivery to control distant metastases. Optical measurements were used to characterize the influence of tumor physiology on treatment outcome. First we established the ability of optical spectroscopy to quantify drug delivery, taking advantage of the intrinsic fluorescence properties of DOX. We found that there was a linear relationship between the intrinsic optical fluorescence in solid tumors and the DOX concentration measured by HPLC ($r = 0.88$, $p < 0.001$; data not shown). A dual

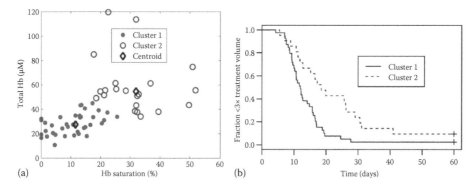

FIGURE 11.10 Prediction of treatment outcome following LTSL-DOX + HT treatment, based on optical spectroscopy prior to treatment. (a) Cluster analysis result. Time to three times treatment volume for (closed symbol) cluster was significantly shorter than for (open symbol) cluster $p = 0.004$. Lower [Hb] and SO_2 indicate that cluster 1 was more hypoxic. (b) Time to three times treatment volume for cluster 1 was significantly shorter than for cluster 2 ($p = 0.003$). (Reproduced from Palmer, G.M., R.J. Boruta, B.L. Viglianti, L. Lan, I. Spasojevic, and M.W. Dewhirst, Non-invasive monitoring of intra-tumor drug concentration and therapeutic response using optical spectroscopy. J Control Release, 2010. 142(3): 457–464. With permission.)

flank tumor model was used with the primary (larger) tumor receiving hyperthermia treatment for 1 h in a 42°C water bath. Free or LTSL-DOX was injected into the tail vein immediately before hyperthermia treatment. The primary (heated) side was observed to have the highest drug accumulation (as estimated through measurement of fluorescence measured via the optical probe [48]), with most of those with the highest drug concentration belonging to the LTSL-DOX + HT treatment group.

Next, we observed that the baseline (pretreatment) tumor oxygen saturation measured optically was statistically significantly associated with faster time to treatment failure (time to three times treatment volume). This was assessed by clustering the pretreatment data based on the pretreatment total hemoglobin and hemoglobin saturation. Two distinct clusters were seen (Figure 11.10a), one having low blood volume and oxygen saturation, and the other having higher oxygenation and higher hemoglobin content. Figure 11.10b shows the Kaplan-Meier plots for time to three times initial volumes for the two clusters shown in Figure 11.10a. It is clear that the more hypoxic cluster (cluster 1) had a significantly shorter time to three times treatment volume ($p = 0.003$ by log rank test).

11.3.3 Applications to Clinical Studies

As an example of the translational applicability of these optical methods in clinical applications, we have demonstrated the ability of DRS for in vivo measurement of vascular oxygenation in 37 patients with breast cancer [51]. Patients with a confirmed diagnosis of breast cancer undergoing surgical resection of their cancers were recruited for this study. Prior to resection, a needle biopsy cannula was introduced into normal and tumor tissues by the surgeon under ultrasound guidance. Importantly, this procedure was done before commencement of the surgical procedure, so the vascular supply to the tissue was not compromised by surgery. A specially designed optical probe was then interfaced with the tissue by inserting it into the open lumen of the biopsy cannula. Following collection of spectroscopic data, the tissue was then removed by an automated biopsy machine (Bard Vacora) through the same cannula used for spectroscopy. Pathologic analysis of the tissue biopsies was conducted to determine the diagnosis and molecular subtype of the tumor.

Figure 11.11 shows pertinent results from this study. As observed in Figure 11.11a, the median vascular hemoglobin saturation of malignant tissues is overall lower than normal tissues, as expected. However, the range of oxygenation values in the malignant vasculature spanned the normal range. This indicates that not all breast tumors in this patient cohort exhibited marked hypoxia, as seen by the large fraction of HbO_2 present in the vasculature of some tumors (data not shown). Examination of the molecular subtypes of the tumors, particularly regarding HER2/neu status, revealed that tumors which were positive

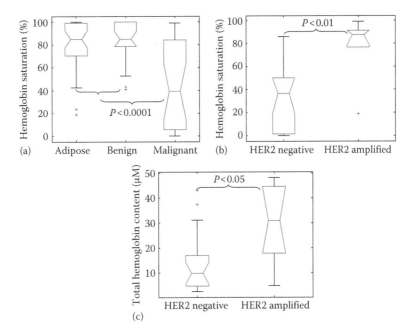

FIGURE 11.11 (a) Vascular hemoglobin saturation measured in the breast in vivo, stratified by pathologic diagnosis of the corresponding tissue biopsy. (b) Vascular hemoglobin saturation measured in vivo in malignant tissues only, stratified by HER2/neu receptor status. (c) Volume-averaged total hemoglobin concentration measured in vivo in malignant tissues only, stratified by HER2/neu receptor status. *P* values indicate significance of Wilcoxon rank-sum test. (Reproduced from Brown, J.Q., L.G. Wilke, J. Geradts, S.A. Kennedy, G.M. Palmer, and N. Ramanujam, Quantitative optical spectroscopy: A robust tool for direct measurement of breast cancer vascular oxygenation and total hemoglobin content in vivo. Cancer Res, 2009. 69(7): 2919–2926. With permission.)

for upregulation of the HER2/neu receptor were better oxygenated than their HER2/neu negative counterparts (Figure 11.11b). In addition, the amplification of the HER2/neu receptor was also associated with increased volume-averaged total hemoglobin content (Figure 11.11c). These observations are consistent with previous studies which used IHC markers of hypoxia and angiogenesis, which also demonstrated better oxygenation and vascular density in HER2/neu positive tumors. This is consistent with the understood model of HER2/neu upregulation of angiogenesis, which results in better tumor perfusion and increased vascular oxygenation [52–54]. However, the benefit of the optical approach is that these characteristics of the tumor microenvironment were measured in vivo with immediate availability of results.

11.3.3.1 Quantitative Diffuse Reflectance Imaging for Ex Vivo Breast Tumor Margin Analysis

Surgery remains the primary first-line intervention for women with a confirmed diagnosis of breast cancer. This surgical procedure can be a full mastectomy, in which the entire breast is removed along with the offending malignancy, or it may be a less radical procedure, in which much of the breast is retained to improve cosmetic outcomes. This procedure, called breast conserving surgery (or commonly, partial mastectomy), involves removal of the malignancy along with a surrounding rim of normal tissue and is appropriate for tumors which are small in size relative to the entire breast volume, or which have been reduced in size with presurgical (neoadjuvant) chemotherapy. Unfortunately, the ability of the surgeon to accurately evaluate whether the entire tumor has been removed successfully during the operation is limited, and a reported 20%–70% of partial mastectomy patients must undergo additional operations to remove cancer identified by histopathology as left behind during the first surgery [55]. Although methods for intraoperative histopathology exist (such as frozen section analysis or touch prep cytology) for determination of the presence of cancer cells at the surface of the specimen, or within 2 mm of the surface of the specimen, these techniques have not been widely adopted due to challenges in implementation (e.g., frozen section performs poorly on fatty samples) and pathology personnel cost. Thus, there is a compelling need for an intraoperative technology which provides an objective determination of the presence of residual disease at the specimen surface, or "margin."

We have recently reported on the development [40], characterization [42], and pilot clinical results [40,56] of an imaging implementation of our spectroscopy technology, in which multichannel fiber-optic probes are used to collect spectra systematically over the surface of a partial mastectomy specimen. In this way, the technology is able to reconstruct 2D maps of the tissue surface, with a third dimension being the spectral information from each location on the 2D map. This "hyperspectral cube" contains information about the underlying tissue structure and function down to a depth of 1–2 mm below the surface [40,42]. Using our quantitative spectral analysis technology (i.e., the inverse Monte Carlo inversion algorithm), we can convert these hyperspectral maps to quantitative images of the tissue which are reflective of parameters such as total hemoglobin concentration, β-carotene concentration, or the reduced light scattering coefficient. These images contain information that is indicative of the presence of cancer, and may potentially be used to alert the surgeon during the first operation whether additional tissue must be resected to avoid an additional operation.

Figure 11.12a contains a schematic of the quantitative diffuse reflectance imaging system. The system consists of a Xenon-arc lamp and monochromator for illumination, a custom fiber-optic probe with eight independent tissue measurement channels, and an imaging spectrograph and CCD for collection of reflectance signals from the tissue. The illumination monochromator is set to zero-order diffraction for white-light illumination for diffuse reflectance. The eight individual tissue channels are bundled at the distal (i.e., tissue) end into a 2×4 array using a custom aluminum chuck. The channels are separated by 10 mm center to center and are designed to interface to the tissue through a custom-designed tissue specimen holder, which consists of a Plexiglas container with an array of through holes drilled over its entire surface (Figure 11.9b). The spacing between the holes is 5 mm center to center. The probe is inserted into the box to interface with tissue, diffuse reflectance signals are collected from 400 to 600 nm, and then the probe is manually translated into an adjacent set of holes. In this way, two successive placements of the probe array can cover an area of approximately 3.4×1.4 cm, with 5 mm spatial resolution. The probe is systematically interfaced with the box in order to cover the entire tissue surface. The result of this exercise is the collection of spectra from each "pixel" on the margin surface. These spectral data are then converted into quantitative tissue optical properties using the inverse Monte Carlo model, from which relevant parameters such as absorber concentrations or reduced scattering coefficients may be computed.

Figure 11.13 shows sample tissue parameter maps collected with the spectral imaging device. Figure 11.13a and c are maps of the ratio of β-carotene concentration to the wavelength-averaged reduced scattering coefficient ($<\mu_s'>$). β-carotene is a dietary carotenoid primarily stored in fatty tissues; therefore, the β-carotene concentration can be used as a surrogate for the density of fat tissue present [57,58]. Likewise, the $<\mu_s'>$ is positively correlated with the amount of fibrous and cellular components in the tissue [57,58]. Therefore, these maps are reflective of the relative amount of fat to fibrous/cellular tissue. Since breast tumors involve a combined cellular and stromal component that displaces fat, we expect the presence of residual disease at the margin surface to be marked by a lower β-carotene/$<\mu_s'>$ ratio. Observation of Figure 11.13 demonstrates this, as Figure 11.13a is a map of a margin with path-confirmed residual disease at the margin, and Figure 11.13b is a map of a margin with no residual disease found by pathology. Figure 11.13a is marked by areas of low β-carotene/$<\mu_s'>$, whereas Figure 11.13b is characterized by higher β-carotene/$<\mu_s'>$ values overall. Interestingly, the central area in the map of Fig. 11.13a was found to be the location of residual disease upon closer postoperative pathologic examination.

The β-carotene/$<\mu_s'>$ map is just one of several informative parameter maps that can be reconstructed to represent the tissue surface. Other parameters include the total hemoglobin concentration (related to tissue vascularity), or the reduced albedo (which is a measure of the amount of light back-scattered compared to absorbed by the tissue). One attractive application of this technology would be the automated intraoperative determination of margin status, as positive or negative for residual disease. One strategy which we have employed toward this goal is to compute a full set of parameter maps for each margin

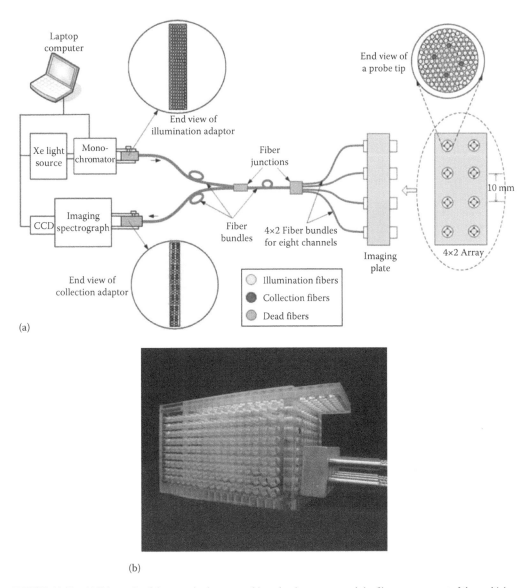

FIGURE 11.12 (a) Schematic of the quantitative spectral imaging instrument and the fiber arrangement of the multichannel probe and (b) photograph of mock tissue in the custom specimen container, which is used to interface the spectral imaging probe to the tissue. (Reproduced from Brown, J.Q., T.M. Bydlon, S.A. Kennedy, J. Geradts, L.G. Wilke, W. Barry, L.M. Richards, M.K. Junker, J. Gallagher, and N. Ramanujam, Assessment of breast tumor margins via quantitative diffuse reflectance imaging, In Biomedical Applications of Light Scattering IV, 2010, Proceedings of the SPIE, Vol. 7573, p. 75730O. With permission.)

and to reduce the dimensionality of each parameter map using an image-descriptive statistical measure (e.g., mean, median, or the percentage of image pixels below a predetermined threshold value) [56]. This results in a number of possible predictor variables for each margin, which can be used to train an appropriate margin status prediction algorithm. We have used a conditional inference tree predictive modeling approach, which has a number of attractive advantages over methods such as logistic regression [41]. Namely, conditional inference trees allow for automated selection of predictor variables, do not require a cross validation step for unbiased estimation of performance, make no assumptions about the distribution of the predictor variables, and are not sensitive to the use of multiple candidate predictor variables which have different absolute units or are of different form (integer, ordinal, binary, or nominal) [59].

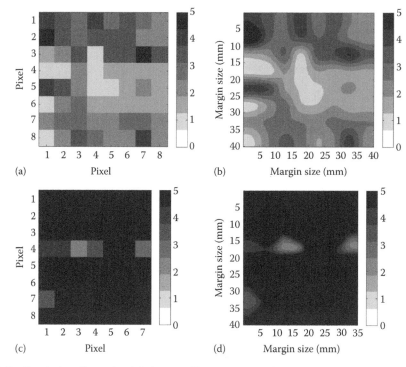

FIGURE 11.13   Raw (a, c) and interpolated (b, d) maps of β-carotene: <μ_s'> for a path-confirmed margin positive for residual invasive ductal carcinoma (a and b), and for a path-confirmed margin negative for residual disease (c and d). The parameter β-carotene: <μ_s'> is calculated from the quantitative absorption and reduced scattering coefficients which are extracted
from the spectral data at each pixel. Interpolated maps are for aid in visual analysis only; quantitative image-descriptive variables are computed from the raw parameter maps. (Reproduced from Brown, J.Q., T.M. Bydlon, S.A. Kennedy,
J. Geradts, L.G. Wilke, W. Barry, L.M. Richards, M.K. Junker, J. Gallagher, and N. Ramanujam, Assessment of breast
tumor margins via quantitative diffuse reflectance imaging, In Biomedical Applications of Light Scattering IV, 2010,
Proceedings of the SPIE, Vol. 7573, p. 75730O. With permission.)

Using this approach on an initial preliminary cohort of 55 margins from 48 patients, we found that the
device was able to detect residual disease (at or within 2 mm of the margin surface) with 79% positive
predictive value and 67% negative predictive value [40,41]. Data analysis on a larger 100 patient cohort is
underway at the time of this writing, and it is expected that inclusion of patient demographic and clinical
factors (such as mammographic breast density) as predictor variables in the conditional inference trees
will increase the accuracy of the device.

Additional technology improvements involve reducing the complexity, size, and cost of the device by
using a simplified detection scheme consisting of an array of silicon photodiodes at the tissue surface
[60,61]. The hyperspectral reflectance cubes are generated by scanning a reduced set of illumination
wavelengths and collecting the resulting reflectance maps at each wavelength. The technology has
been validated in tissue simulating phantoms, and additional development is aimed at increasing the
light efficiency of the device and refining the manufacturing methods, in preparation for future clinical
validation studies.

11.4 Discussion and Future Trends

The preclinical and clinical studies discussed earlier demonstrate the promise held by these novel,
noninvasive, and quantitative technologies to measure vascular mediated physiological changes
across a variety of organ sites. The first set of the preclinical studies revealed a clear concordance

between the optical biomarkers and physiological endpoints of tumor pO_2 (measured using electrode polarography) and tumor hypoxia as ascertained by IHC methods. The next set of preclinical studies indicate how these techniques may provide insight into changes in tumor physiology as the animals were exposed to radiation and chemotherapeutic treatments. In these studies, we found that the optical endpoint of vascular oxygen saturation and total hemoglobin predicted for treatment outcomes several days before volumetric measurements indicated these results. These preliminary findings are particularly interesting as they indicate the potential of obtaining longitudinal measurements from the individual subjects, thereby tracking how the underlying tumor physiology changes during treatment which in turn could provide insights to identify individual response. Lastly, we have also used the optical methods in a preliminary clinical study where we observed changes in the tumor oxygenation patterns between different molecular subtypes of breast cancer. Overall all we observed that malignant breast tissues were lower oxygen saturation relative to normal breast tissue, but there was a clear variation in the oxygenation pattern of the tumors dependent on the molecular subtype of the tissue itself. Vascular oxygenation patterns observed in the preclinical therapy studies lent further credence to the fact that better oxygen saturation in the tumors across treatment implied an overall better outcome in the individuals.

Optical reflectance spectroscopy holds an important place within the methods of biomedical optics and the larger field of medical imaging. Perhaps the single most attractive aspect of this technology lies in the simplicity of the instrumentation that can be used. These devices can operate in near real time, easily fit into the confines of clinics, and report on a variety of structural and functional endpoints. Given advances in field of semiconductor devices and fiber optics, it will be possible to even further reduce the footprint and power requirements of these devices. Further research is needed to harness the full potential of this technology in reporting about tissue structure and function along with larger clinical trials and commercialization efforts to make these methods a standard clinical tool.

REFERENCES

1. Folkman, J., What is the evidence that tumors are angiogenesis dependent? *J Natl Cancer Inst*, 1990. **82**(1): 4–6.
2. Hanahan, D. and R.A. Weinberg, The hallmarks of cancer. *Cell*, 2000. **100**(1): 57–70.
3. Vaupel, P., Hypoxia and aggressive tumor phenotype: Implications for therapy and prognosis. *Oncologist*, 2008. **13 Suppl 3**: 21–26.
4. Vaupel, P. and A. Mayer, Hypoxia in cancer: Significance and impact on clinical outcome. *Cancer Metastasis Rev*, 2007. **26**(2): 225–239.
5. Dewhirst, M.W., Y. Cao, C.Y. Li, and B. Moeller, Exploring the role of hif-1 in early angiogenesis and response to radiotherapy. *Radiother Oncol*, 2007. **83**(3): 249–255.
6. Semenza, G.L., Targeting hif-1 for cancer therapy. *Nat Rev Cancer*, 2003. **3**(10): 721–732.
7. Moeller, B.J., R.A. Richardson, and M.W. Dewhirst, Hypoxia and radiotherapy: Opportunities for improved outcomes in cancer treatment. *Cancer Metastasis Rev*, 2007. **26**(2): 241–248.
8. Vaupel, P. and M. Hockel, Tumor oxygenation and its relevance to tumor physiology and treatment. *Adv Exp Med Biol*, 2003. **510**: 45–49.
9. Evans, S.M. and C.J. Koch, Prognostic significance of tumor oxygenation in humans. *Cancer Lett*, 2003. **195**(1): 1–16.
10. Tatum, J.L., G.J. Kelloff, R.J. Gillies, J.M. Arbeit, J.M. Brown, K.S. Chao, J.D. Chapman, W.C. Eckelman, A.W. Fyles, A.J. Giaccia, R.P. Hill, C.J. Koch, M.C. Krishna, K.A. Krohn, J.S. Lewis, R.P. Mason, G. Melillo, A.R. Padhani, G. Powis, J.G. Rajendran, R. Reba, S.P. Robinson, G.L. Semenza, H.M. Swartz, P. Vaupel, D. Yang, B. Croft, J. Hoffman, G. Liu, H. Stone, and D. Sullivan, Hypoxia: Importance in tumor biology, noninvasive measurement by imaging, and value of its measurement in the management of cancer therapy. *Int J Radiat Biol*, 2006. **82**(10): 699–757.
11. Vaupel, P., M. Hockel, and A. Mayer, Detection and characterization of tumor hypoxia using po2 histography. *Antioxid Redox Signal*, 2007. **9**(8): 1221–1235.
12. Xu, R.X. and S.P. Povoski, Diffuse optical imaging and spectroscopy for cancer. *Expert Rev Med Devices*, 2007. **4**(1): 83–95.

13. Richards-Kortum, R. and E. Sevick-Muraca, Quantitative optical spectroscopy for tissue diagnosis. *Ann Rev Phys Chem*, 1996. **47**: 555–606.

14. Bigio, I.J. and S.G. Bown, Spectroscopic sensing of cancer and cancer therapy: Current status of translational research. *Cancer Biol Ther*, 2004. **3**(3): 259–267.

15. Gibson, A.P., J.C. Hebden, and S.R. Arridge, Recent advances in diffuse optical imaging. *Phys Med Biol*, 2005. **50**(4): R1–R43.

16. Hielscher, A.H., A.Y. Bluestone, G.S. Abdoulaev, A.D. Klose, J. Lasker, M. Stewart, U. Netz, and J. Beuthan, Near-infrared diffuse optical tomography. *Dis Markers*, 2002. **18**: 313–337.

17. Brooksby, B., B. Pogue, S. Jiang, H. Dehghani, S. Srinivasan, C. Kogel, T. Tosteson, J. Weaver, S. Poplack, and K. Paulsen, Imaging breast adipose and fibroglandular tissue molecular signatures by using hybrid MRI-guided near-infrared spectral tomography. *PNAS*, 2006. **103**(23): 8828–8833.

18. Pogue, B.W., S.P. Poplack, T.O. McBride, W.A. Wells, K.S. Osterman, U.L. Osterberg, and K.D. Paulsen, Quantitative hemoglobin tomography with diffuse near-infrared spectroscopy: Pilot results in the breast. *Radiology*, 2001. **218**(1): 261–266.

19. Perini, R., R. Choe, A.G. Yodh, C. Sehgal, C.R. Divgi, and M.A. Rosen, Non-invasive assessment of tumor neovasculature: Techniques and clinical applications. *Cancer Metastasis Rev*, 2008. **27**(4): 615–630.

20. Shah, N., A.E. Cerussi, D. Jakubowski, D. Hsiang, J. Butler, and B.J. Tromberg, The role of diffuse optical spectroscopy in the clinical management of breast cancer. *Dis Markers*, 2003. **19**(2–3): 95–105.

21. Tromberg, B.J., A.E. Cerussi, N. Shah, M. Compton, A. Durkin, D. Hsiang, J. Butler, and R. Mehta, Imaging in breast cancer: Diffuse optics in breast cancer: Detecting tumors in pre-menopausal women and monitoring neoadjuvant chemotherapy. *Breast Cancer Res*, 2005. **7**(6): 276–278.

22. Cerussi, A., D. Hsiang, N. Shah, R. Mehta, A. Durkin, J. Butler, and B.J. Tromberg, Predicting response to breast cancer neoadjuvant chemotherapy using diffuse optical spectroscopy. *Proc Natl Acad Sci USA*, 2007. **104**(10): 4014–4019.

23. Jakubowski, D.B., A.E. Cerussi, F. Bevilacqua, N. Shah, D. Hsiang, J. Butler, and B.J. Tromberg, Monitoring neoadjuvant chemotherapy in breast cancer using quantitative diffuse optical spectroscopy: A case study. *J Biomed Opt*, 2004. **9**(1): 230–238.

24. Choe, R., A. Corlu, K. Lee, T. Durduran, S.D. Konecky, M. Grosicka-Koptyra, S.R. Arridge, B.J. Czerniecki, D.L. Fraker, A. DeMichele, B. Chance, M.A. Rosen, and A.G. Yodh, Diffuse optical tomography of breast cancer during neoadjuvant chemotherapy: A case study with comparison to MRI. *Med Phys*, 2005. **32**(4): 1128–1139.

25. Sunar, U., H. Quon, T. Durduran, J. Zhang, J. Du, C. Zhou, G. Yu, R. Choe, A. Kilger, R. Lustig, L. Loevner, S. Nioka, B. Chance, and A.G. Yodh, Noninvasive diffuse optical measurement of blood flow and blood oxygenation for monitoring radiation therapy in patients with head and neck tumors: A pilot study. *J Biomed Opt*, 2006. **11**(6): 064021.

26. Welch, A.J. and M.J.C.V. Gemert, *Optical-Thermal Response of Laser-Irradiated Tissue*. 1995, New York: Plenum Press.

27. Vo-Dinh, T., ed. *Biomedical Photonics Handbook*, Vol. 1, 2003, New York: CRC Press.

28. Hebden, J.C., S.R. Arridge, and D.T. Delpy, Optical imaging in medicine: I. Experimental techniques. *Phys Med Biol*, 1997. **42**(5): 825–840.

29. Tuchin, V.V., *Handbook of Optical Biomedical Diagnostics*, 2002, Bellingham. WA: SPIE Press.

30. Bigio, I. and J. Mourant, Ultraviolet and visible spectroscopies for tissue diagnostics: Fluorescence spectroscopy and elastic-scattering spectroscopy. *Phys Med Biol*, 1997. **42**: 803–814.

31. Zuzak, K.J., M.D. Schaeberle, M.T. Gladwin, R.O. Cannon 3rd, and I.W. Levin, Noninvasive determination of spatially resolved and time-resolved tissue perfusion in humans during nitric oxide inhibition and inhalation by use of a visible-reflectance hyperspectral imaging technique. *Circulation*, 2001. **104**(24): 2905–2910.

32. Brown, J.Q., K. Vishwanath, G.M. Palmer, and N. Ramanujam, Advances in quantitative UV-visible spectroscopy for clinical and pre-clinical application in cancer. *Curr Opin Biotechnol*, 2009. **20**(1): 119–131.

33. Srinivasan, S., B.W. Pogue, S. Jiang, H. Dehghani, C. Kogel, S. Soho, J.J. Gibson, T.D. Tosteson, S.P. Poplack, and K.D. Paulsen, Interpreting hemoglobin and water concentration, oxygen saturation, and scattering measured in vivo by near-infrared breast tomography. *PNAS*, 2003. **100**(21): 12349–12354.

34. Bigio, I.J., S.G. Brown, G. Briggs, C. Kelley, S. Lakhani, D. Picard, P.M. Ripley, I.G. Rose, and C. Saunders, Diagnosis of breast cancer using elastic-scattering spectroscopy: Preliminary clinical results. *J Biomed Opt*, 2000. **5**(2): 221–228.

35. Taroni, P., A. Pifferi, A. Torricelli, D. Comelli, and R. Cubeddu, In vivo absorption and scattering spectroscopy of biological tissues. *Photochem Photobiol Sci*, 2003. **2**(2): 124–129.

36. Zhu, C., Q. Liu, and N. Ramanujam, Effect of fiber optic probe geometry on depth-resolved fluorescence measurements from epithelial tissues: A Monte Carlo simulation. *J Biomed Opt*, 2003. **8**(2): 237–247.

37. Arridge, S., Optical tomography in medical imaging. *Inv Prob*, 1999. **15**: R41–R93.

38. Kienle, A. and M.S. Patterson, Determination of the optical properties of turbid media from a single Monte Carlo simulation. *Phys Med Biol*, 1996. **41**(10): 2221–2227.

39. Palmer, G.M. and N. Ramanujam, Monte Carlo-based inverse model for calculating tissue optical properties. Part I: Theory and validation on synthetic phantoms. *Appl Opt*, 2006. **45**(5): 1062–1071.

40. Brown, J.Q., T.M. Bydlon, S.A. Kennedy, J. Geradts, L.G. Wilke, W. Barry, L.M. Richards, M.K. Junker, J. Gallagher, and N. Ramanujam, Assessment of breast tumor margins via quantitative diffuse reflectance imaging. In *Biomedical Applications of Light Scattering IV*, 2010. *Proceedings of the SPIE*, Vol. 7573, p. 75730O.

41. Brown, J.Q., T.M. Bydlon, L.M. Richards, B. Yu, S.A. Kennedy, J. Geradts, L.G. Wilke, M.K. Junker, J. Gallagher, W.T. Barry, and N. Ramanujam, Optical assessment of tumor resection margins in the breast. *IEEE J Sel Top Quantum Electron*, 2010. **16**(3): 530–544.

42. Bydlon, T.M., S.A. Kennedy, L.M. Richards, J.Q. Brown, B. Yu, M.K. Junker, J. Gallagher, J. Geradts, L.G. Wilke, and N. Ramanujam, Performance metrics of an optical spectral imaging system for intraoperative assessment of breast tumor margins. *Opt Express*, 2010. **18**(8): 8058–8076.

43. Palmer, G.M., R.J. Viola, T. Schroeder, P.S. Yarmolenko, M.W. Dewhirst, and N. Ramanujam, Quantitative diffuse reflectance and fluorescence spectroscopy: Tool to monitor tumor physiology in vivo. *J Biomed Opt*, 2009. **14**(2): 024010.

44. Vishwanath, K., H. Yuan, W.T. Barry, M.W. Dewhirst, and N. Ramanujam, Using optical spectroscopy to longitudinally monitor physiological changes within solid tumors. *Neoplasia*, 2009. **11**(9): 889–900.

45. Bender, J.E., K. Vishwanath, L.K. Moore, J.Q. Brown, V. Chang, G.M. Palmer, and N. Ramanujam, A robust-Monte Carlo model for the extraction of biological absorption and scattering in vivo. *IEEE Trans Biomed Eng*, 2009. **56**(4): 960–968.

46. Kelman, G.R., Digital computer subroutine for the conversion of oxygen tension into saturation. *J Appl Physiol*, 1966. **21**(4): 1375–1376.

47. Vishwanath, K., D. Klein, K. Chang, T. Schroeder, M.W. Dewhirst, and N. Ramanujam, Quantitative optical spectroscopy can identify long-term local tumor control in irradiated murine head and neck xenografts. *J Biomed Opt*, 2009. **14**(5): 054051.

48. Palmer, G.M., R.J. Boruta, B.L. Viglianti, L. Lan, I. Spasojevic, and M.W. Dewhirst, Non-invasive monitoring of intra-tumor drug concentration and therapeutic response using optical spectroscopy. *J Control Release*, 2010. **142**(3): 457–464.

49. Petersen, C., D. Zips, M. Krause, K. Schone, W. Eicheler, C. Hoinkis, H.D. Thames, and M. Baumann, Repopulation of FaDu human squamous cell carcinoma during fractionated radiotherapy correlates with reoxygenation. *Int J Radiat Oncol Biol Phys*, 2001. **51**(2): 483–493.

50. McGill, R., J.W. Tukey, and W.A. Larsen, Variations of box plots. *Am Statist*, 1978. **32**(1): 12–16.

51. Brown, J.Q., L.G. Wilke, J. Geradts, S.A. Kennedy, G.M. Palmer, and N. Ramanujam, Quantitative optical spectroscopy: A robust tool for direct measurement of breast cancer vascular oxygenation and total hemoglobin content in vivo. *Cancer Res*, 2009. **69**(7): 2919–2926.

52. Blackwell, K.L., M.W. Dewhirst, V. Liotcheva, S. Snyder, G. Broadwater, R. Bentley, A. Lal, G. Riggins, S. Anderson, J. Vredenburgh, A. Proia, and L.N. Harris, Her-2 gene amplification correlates with higher levels of angiogenesis and lower levels of hypoxia in primary breast tumors. *Clin Cancer Res*, 2004. **10**(12 Pt 1): 4083–4088.

53. Laughner, E., P. Taghavi, K. Chiles, P.C. Mahon, and G.L. Semenza, Her2 (neu) signaling increases the rate of hypoxia-inducible factor 1α (hif-1α) synthesis: Novel mechanism for hif-1-mediated vascular endothelial growth factor expression. *Mol Cell Biol*, 2001. **21**(12): 3995.

54. Hohenberger, P., C. Felgner, W. Haensch, and P.M. Schlag, Tumor oxygenation correlates with molecular growth determinants in breast cancer. *Breast Cancer Res Treat*, 1998. **48**(2): 97–106.

55. Jacobs, L., Positive margins: The challenge continues for breast surgeons. *Ann Surg Oncol*, 2008. **15**(5): 1271–1272.
56. Wilke, L.G., J.Q. Brown, T.M. Bydlon, S.A. Kennedy, L.M. Richards, M.K. Junker, J. Gallagher, W.T. Barry, J. Geradts, and N. Ramanujam, Rapid noninvasive optical imaging of tissue composition in breast tumor margins. *Am J Surg*, 2009. **198**(4): 566–574.
57. Kennedy, S., J. Geradts, T. Bydlon, J.Q. Brown, J. Gallagher, M. Junker, W. Barry, N. Ramanujam, and L. Wilke, Optical breast cancer margin assessment: An observational study of the effects of tissue heterogeneity on optical contrast. *Breast Cancer Res*, 2010. **12**(6): R91.
58. Zhu, C., G.M. Palmer, T.M. Breslin, J. Harter, and N. Ramanujam, Diagnosis of breast cancer using fluorescence and diffuse reflectance spectroscopy: A Monte-Carlo-model-based approach. *J Biomed Opt*, 2008. **13**(3): 034015.
59. Hothorn, L.A., Multiple comparisons and multiple contrasts in randomized dose-response trials-confidence interval oriented approaches. *J Biopharm Stat*, 2006. **16**(5): 711–731.
60. Fu, H.L., B. Yu, J.Y. Lo, G.M. Palmer, T.F. Kuech, and N. Ramanujam, A low-cost, portable, and quantitative spectral imaging system for application to biological tissues. *Opt Express*, 2010. **18**(12): 12630–12645.
61. Lo, J.Y., B. Yu, H.L. Fu, J.E. Bender, G.M. Palmer, T.F. Kuech, and N. Ramanujam, A strategy for quantitative spectral imaging of tissue absorption and scattering using light emitting diodes and photodiodes. *Opt Express*, 2009. **17**(3): 1372–1384.

12

Noble Metal Nanoparticles as Probes for Cancer Biomarker Detection and Dynamic Distance Measurements in Plasmon Coupling Microscopy

Hongyun Wang, Guoxin Rong, Jing Wang, Bo Yan,
Lynell R. Skewis, and Björn M. Reinhard

CONTENTS

12.1 Introduction

A mechanistic understanding of fundamental cancer processes requires a detailed knowledge not only of the underlying biochemistry but also of the accompanying structural changes in the biopolymers involved and of the interplay between "chemical" and "mechanical" changes. Some cancer-relevant molecular processes, for instance, in cell signaling, gain complexity from the fact that multiple components interact with each other in a time-dependent fashion.[1] The investigation of these intrinsically dynamic processes is difficult since it requires the ability to detect and quantify distance and position changes on nanometer length scales in real time. Ideally, these measurements are performed with single molecule sensitivity to bypass the need for ensemble synchronization and to access short-lived intermediates.[2,3] Since the wave nature of light limits the resolution of conventional light microscopy to approximately 400 nm in the visible, the investigation of nanometer distance changes in individual biopolymers in real time using an optical microscope requires special techniques, such as fluorescence resonance energy transfer (FRET),[4,5] two color colocalization,[6–8] or nanometal surface energy transfer[9] between noble metal nanoparticles and dyes. While some of these methods have proven indispensable to investigate the mechanistic aspects of complex and intrinsically dynamic biological processes, all of the methods mentioned earlier have specific advantages and disadvantages, and no single method addresses all the specific requirements of a diverse application range. For instance, FRET is limited to relatively short separations of approximately 8 nm and below, and, like all fluorescence dye–based assays, FRET suffers from the photophysical limitations of the dyes which blink and bleach.[4] Semiconductor quantum dots[10] have higher photostabilities and are brighter than organic dyes and are thus alternatives as fluorescent labels. Even FRET-like studies which utilize energy transfer between individual pairs of quantum dot donor and organic dye acceptor have been realized in aqueous solution.[11] Blinking remains, however, a significant concern with quantum dots, and although the blinking can be reduced in the presence of thiols and other small ligands that occupy surface trap sites,[12–14] the need for these surface ligands reduces the applicability of quantum dots.

Measurements of nanoscale distances with high temporal resolution over extended time periods in biological systems remain challenging. The latter motivates the development of new tools to access this distance range through optical microscopy.

In this chapter, we outline alternative, noble metal nanoparticle–based strategies for monitoring nanometer scale distances and for resolving subdiffraction limit contacts in particle tracking. The optical properties of noble metal nanoparticles are determined by particle plasmons, which are coherent collective electron oscillations of a nanoparticle's conduction band electrons.[15] Gold and silver nanoparticles have plasmon resonance frequencies in the visible range of the electromagnetic spectrum, and a resonant excitation of these plasmons results in large optical cross sections of the particles.[16,17] The precise resonance frequency of a particle plasmon depends on the dielectric function of the metal, the particle's size and shape, and the dielectric of the environment.[18] Excellent introductions to the optical properties of nanoparticles are given in Refs. [15,18,19].

Gold and silver nanoparticles with diameters ≥ 20 nm have sufficiently large scattering cross sections to facilitate their detection in a conventional dark field microscope. Since their signal is based on light scattering, noble metal nanoparticles do not blink or bleach and are thus labels with extraordinary photostability for imaging and biosensing. While the optical properties of individual noble metal nanoparticle are already remarkable, the assembly of gold and silver nanoparticles into higher order structures provides additional functionalities due to the particles' ability to interact in the near and far field.[20–33] For the purpose of resolving distances on nanometer length scales, the near-field interactions (interactions that occur on short distances below $S = \lambda/(2\pi n)$, where λ is the plasmon resonance wavelength and n is the refractive index of the medium) between individual nanoparticles are of particular interest. We restrict our discussion of the underlying electromagnetic coupling mechanism here to the dipole coupling between ideal spheres in the quasistatic approximation. In this approximate but for our purpose sufficient model unpolarized light excites two coupled plasmon modes: a longitudinal and a vertical mode.[34–38] Figure 12.1 summarizes the charge distribution, dipole orientation, and distance-dependent spectral response of coupled nanoparticle dimers. In the longitudinal mode (a.1–a.3), the coupled particle dipoles are aligned in a head-to-tail fashion along the long dimer axis, whereas in the vertical mode (b.1–b.3), the dipoles have a parallel alignment perpendicular to the long dimer axis. At large interparticle separations (c.1), the particle plasmons do not couple, and the longitudinal and vertical modes are degenerate. With decreasing interparticle separations, the plasmons in the individual nanoparticles start to interact. As a consequence of the relative orientations of the net dipole moments and the resulting charge distributions in the particles, the hybridization of the plasmon modes results in a splitting of the longitudinal and vertical plasmon modes in the dimer (c.2). With decreasing interparticle separation, the longitudinal plasmon mode continues to redshift, and the vertical mode further blueshifts (c.3).

Under unpolarized white light illumination, the longitudinal mode dominates the spectrum of the coupled dimer, but the vertical mode can be detected if it is selectively excited using linearly polarized light (see for instance Ref. [38]). An elegant theory that illustrates the interactions between the different modes in close-by noble metal nanoparticles in an intuitive fashion is the plasmon hybridization method developed by Nordlander and coworkers.[39,40] This method describes plasmon hybridization in an analogy to orbital hybridization in molecular orbital theory and provides useful predictions of the hybridized mode energies in coupled noble metal nanoparticle structures. The interparticle separation–dependent optical response in systems of coupled nanoparticles forms the foundation of plasmon coupling–based approaches to measure distances, such as plasmon rulers[34,35,41–45] and plasmon coupling microscopy (PCM).[46–48]

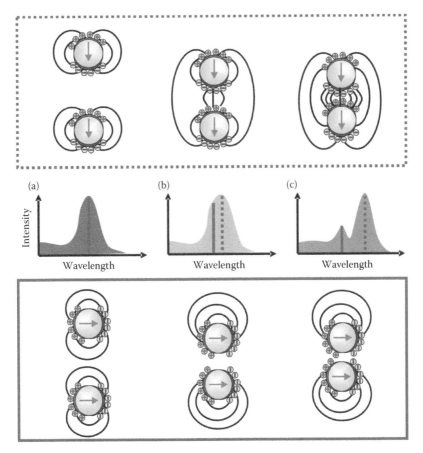

FIGURE 12.1 Charge distribution and dipole orientations in (a) longitudinal and (b) vertical plasmon modes. (c) Spectral shift of longitudinal and vertical mode. With decreasing interparticle separation, the longitudinal mode redshifts, whereas the vertical mode blueshifts. For clarity, the effect is exaggerated in (c).

This chapter is organized as follows. We review techniques to generate immunolabels for PCM and functional nanoparticle assemblies (plasmon rulers) and introduce the experimental setup used for PCM in Section 12.3. In Section 21.4, we review the current status of PCM for monitoring distances on nanometer length scales. We demonstrate the method's ability to resolve subdiffraction limit contacts between laterally diffusing nanoparticles. This ability can be used, for instance, for characterizing cancer-relevant receptor association processes. We also demonstrate that calibrated silver plasmon rulers can provide a two-dimensional structural profiling of cell membranes. Given the potential role of the cell surface morphology as cancer biomarker,[49–52] methods that can reveal structural details of the cell surface have high relevance for cancer diagnostics and research.

12.2 Materials

Material	Vendor	Catalog Number
A431 cells	ATCC	CRL-1555
Agarose	EMD	2125
Ag colloid, 20 nm	Ted Pella (BBI)	15705-5SC
Ag colloid, 40 nm	Ted Pella (BBI)	15707-5SC

(continued)

Material	Vendor	Catalog Number
Albumin, biotin-labeled bovine (BSA-biotin)	Sigma	A8549
Alkyl-PEG-acid ($HSC_{11}H_{22}(OC_2H_4)_6OCH_2COOH$)	ProChimia	TH 011-01
Antibiotin affinity isolated antigen specific antibody	Sigma	B3640
Anti-IgG biotin conjugate	Sigma	B7264
Au colloid, 40 nm	Ted Pella	15707
BSPP	Strem Chemicals	15-0463
Biotin-PEG-disulfide (biotin-$C_4H_8CONHC_2H_4(OC_2H_4)_8OC_2H_4NH$ $COC_2H_4SSC_2H_4CONHC_2H_4O(C_2H_4O)_8C_2H_4NHCOC_4H_8$-biotin)	Polypure	41151-0895
$CuSO_4 \cdot 5H_2O$	Sigma	203165
Cover glass no. 2	VWR	48382-128
DNA oligonucleotides	Integrated DNA Technologies	
D-tube dialyzer midi, MWCO 6–8 kDa	EMD	71507
EDC	Thermo Scientific	22980
Hank's balanced salt solution	Sigma	14025
HeLa cells	ATCC	CCL-2
HEPES	Sigma	H4034
Ascorbic acid	Aldrich	25556-4
Melon gel monoclonal IgG purification kit	Pierce	45214
MES monohydrate	Acros Organics	32776
Monoclonal antiepidermal growth factor receptor antibody (29.1)	Sigma	E2760
NeutrAvidin	Pierce	31000
PEG-acid disulfide ($HOOCC_2H_4O(C_2H_4O)_7C_2H_4SSC_2H_4(OC_2H_4)_7$ OC_2H_4COOH)	Polypure	37157-0795
Propargyl-dPEG-NHS ester	Quanta Biodesign	10511
Sodium phosphate dibasic	Sigma	S3264
Sodium phosphate monobasic	Sigma	S3139
Sulfo-NHS	Thermo Scientific	24510
Superblock blocking buffer in PBS	Pierce	37515
TEM binary finder grids	SPI Supplies	2002C-XA
Thiol-PEG-azide N_3-$(CH_2CH_2O)_{77}C_2H_4SH$	Nanocs	PG-AZTH-3K
1M Tris, pH 7	Ambion	AM9851
1M Tris, pH 8	Ambion	AM9856
Tris/borate/EDTA buffer	Sigma	T3913
Zeba spin desalting column (7 kDa weight cutoff)	Thermo Scientific	89882

12.3 Methods

12.3.1 Gel Matrix–Assisted Preparation of Stable Antibody-Functionalized Gold and Silver Immunolabels

Figure 12.2 outlines a preparative strategy for antibody-functionalized gold and silver nanoparticles that are stable in live cell imaging buffer conditions.[53] The outlined functionalization approach immobilizes nanoparticles in a low-density gel matrix which facilitates a convenient transfer of the particles between different reaction and washing buffers. In a typical preparation, 50 μL of a ~1 × 10^{12} particles/mL noble metal nanoparticle solution (nominal diameters: 40 nm [gold]; 20 nm [silver]) is incubated with 5 μL of a 10 mM alkyl-PEG-acid ($HSC_{11}H_{22}(OC_2H_4)_6OCH_2COOH$) solution for 12 h. The resulting pegylated particles are then run into a 1% agarose gel in a 100 mM, pH 7.4 phosphate running buffer. The pegylated silver nanoparticle band is cut from the gel and immersed in 100 mM 2-(N-morpholino)ethanesulfonic acid (MES) buffer, pH 5, for 15 min. The isolated gel slabs are subsequently transferred into a cross-linker solution: 2 mM 1-ethyl-3-[3-dimethylaminopropyl]carbodiimide hydrochloride (EDC) and 5 mM

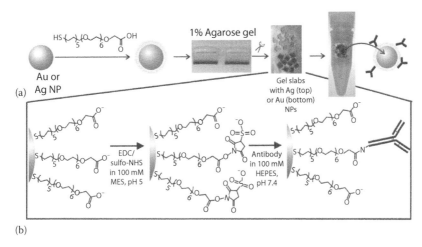

FIGURE 12.2 Gel matrix–assisted nanoparticle functionalization approach. (a) PEG-stabilized particles are run into a 1% agarose gel in 100 mM phosphate buffer, pH 7.4. The nanoparticle containing bands are cut and then exchanged between activation buffer, cross-linking reagent, and antibody containing coupling buffer. The antibody-functionalized particles are electroeluted and stored in 0.5× TBE buffer. (b) Details of cross-linking chemistry. (Reproduced with permission from Skewis, L.R. and Reinhard, B.M., *ACS Appl. Mater. Interfaces*, 2, 35–40, 2010. Copyright 2010 American Chemical Society.)

N-hydroxysulfosuccinimide (sulfo-NHS) in 100 mM MES, pH 5. After allowing the activation reaction to proceed for 1 h, the slabs are removed from excess cross-linking solution and rinsed once with coupling buffer, 100 mM 4-(2-hydroxyethyl)-1-piperazineethanesulfonic acid (HEPES) buffer, pH 7.4. The slabs are then immersed in a solution of desired antibody in the same coupling buffer. The coupling reaction is carried out for 12 h at 4°C. We found that antibody concentration of approximately 1 µg/mL are sufficient to provide nanoparticle labels with good binding affinities. It is important to purify antibodies received in ascites solution using a Pierce Melon Gel Monoclonal IgG Purification Kit prior to cross-coupling. After the coupling is complete, the antibody-functionalized nanoparticles are recovered from the gel slabs through electroelution in 0.5× Tris/borate/EDTA (TBE) buffer. The amines of the TBE buffer also serve to quench the cross-linking reaction. The recovered nanoparticles are washed and eventually concentrated through centrifugation. Excess antibodies that might have electroeluted from the gel together with the nanoparticles are removed with the supernatant.

12.3.2 Plasmon Ruler Assembly

Different assembly strategies have been developed for gold and silver plasmon rulers using single- or double-stranded DNA or RNA as tether molecules.[35,41,43,44] In Figure 12.3, we outline a general self-assembly strategy for single-stranded RNA tethered gold nanoparticles.[43] The general idea of this approach is to connect nanoparticles functionalized with different DNA handles through complementary DNA or RNA oligonucleotides. In the following text, we summarize the details of this process using commercial 40 nm citrate-stabilized gold colloid (9×10^{10} particles/mL); if the concentration or size of the used gold colloid is changed, the reaction conditions need to be adjusted. We describe here the assembly process for 29-nucleotide-long DNA handles and 100-nucleotide-long RNA tether. One of the handles is 3′; the other 5′ thiolated. Both handles have a 9-nucleotide-long poly-A sequence adjacent to the thiol modification and contain a 20-nucleotide-long sequence which is complementary to one of the ends of the RNA tether. Changes in the length of the handles or RNA might require slight adjustments of the procedure.

In the first step of the preparation, 6 mg bis(p-sulfonatophenyl)phenylphosphine dihydrate dipotassium salt (BSPP) is added to 11 mL of 40 nm gold colloid and stirred overnight at 45°C. The particles are then pipetted into 1 mL aliquots and spun down at 5500 rpm for 10 min. The supernatant is removed and exchanged with T40 (40 mM NaCl, 10 mM Tris:HCl, pH 8.0). The particles are washed once more

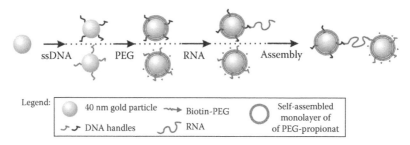

FIGURE 12.3 RNA-programmed self-assembly of plasmon rulers. The noble metal nanoparticles are functionalized with ssDNA and pegylated with short thiol-propionate-PEGs. One particle is functionalized with biotinylated PEGs as well. Then RNA is annealed to one of the particles. Finally, the particles are combined to self-assemble. (Reproduced with permission from Skewis, L.R. and Reinhard, B.M., *Nano Lett.*, 8(1), 214–220, 2008. Copyright 2008 American Chemical Society.)

through centrifugation and resuspension in T40. Then the particles are spun down again, and 950 μL supernatant is removed. The pellet is resuspended in the remaining 50 μL solution. Then 3 μL of a 1 μM solution of the respective DNA handles are added to the particles, corresponding to a nominal ratio of DNA handles per particle of ~20:1. Two batches (A and B) with DNA sequences complementary to the 3′ and 5′ ends of the RNA are prepared and incubated overnight. In the next step, the surface of the DNA-functionalized nanoparticles is passivated with a monolayer of short acid-PEGs ($HSC_2H_4(OC_2H_4)_7OC_2H_4COOH$). The PEG ligands are necessary to stabilize the particles in buffer conditions (≥80 mM NaCl, 10 mM Tris:HCl, pH 8) required for an efficient dimer formation. A 4.8 mM solution of acid-PEG disulfide is prepared in 5 mM aqueous solution of BSPP. A second solution with 4.8 mM acid-PEG disulfide and 0.2 mM biotin-PEG ($HSC_2H_4CONHC_2H_4O(C_2H_4O)_8C_2H_4NHCOC_4H_8$-biotin) in 5 mM BSPP is generated in parallel. Five microliters of these solutions are added to the nanoparticle batches A (acid-PEG) or B (PEG mix). The small amounts of biotinylated PEGs in batch B facilitate the immobilization of the assembled dimers to a NeutrAvidin-functionalized glass surface. Both batches A and B are incubated overnight. Then 950 μL T20 (20 mM NaCl, 10 mM Tris:HCl pH 8.0) is added to the samples, and the particles are subsequently spun down at 5500 rpm for 10 min. The supernatant is discarded. Batch A is washed once more with T60 before the RNA tether (RNA: nanoparticle ratio ~30:1) is added to a total volume of 50 μL nanoparticles. The RNA tether is annealed to the DNA handles on the nanoparticles by placing the mix into a water bath of ~70°C and allowing the mix to cool down to room temperature. Batches A and B are then washed through resuspension and centrifugation in T20 (batch A 2×, batch A 1×) and finally resuspended in T100. Subsequently, batches A and B are combined to anneal at room temperature overnight. After the annealing step, the assembled dimers need to be purified and separated from monomers or larger nanoparticle assemblies. This can be conveniently achieved using gel electrophoresis. The annealing mix and controls (batches A and B before annealing) are loaded into a 1% agarose gel (0.6 g agarose in 60 mL 0.5× TBE) and run at 170 V for 15 min in 0.5× TBE buffer. The dimer bands are then cut out of the gels and electroeluted at 170 V for 10 min using D-tube dialyzer midi.

A similar strategy as outlined earlier can be applied for single-stranded DNA (ssDNA) tethered dimers if the RNA tether is replaced through ssDNA with 3′ and 5′ ends complementary to the handles. Double-stranded DNA tethered plasmon rulers can be obtained through direct annealing of complementary DNA oligonucleotide handles.[41,44] In this case, the number of DNA molecules per nanoparticle can be reduced. We obtained a successful plasmon ruler assembly with stoichiometric ratios of one DNA molecule per particle.

Plasmon ruler assembly is also not restricted to gold nanoparticles. Silver plasmon rulers can be obtained through DNA-programmed self-assembly if several details of the assembly process are adjusted. For silver plasmon ruler assembly, the BSPP step is omitted. Instead, silver nanoparticles with nominal diameters of 40 nm (9×10^9 particles/mL) or 20 nm (7×10^{10} particles/mL) are pipetted into 1 mL aliquots and centrifuged at 5500 rpm (8500 for 20 nm particles) for 10 min. Then 980 μL of supernatant is removed, and 20 μL of 40 mM phosphate buffer (pH = 8) is added. After resuspension of the pellet, the

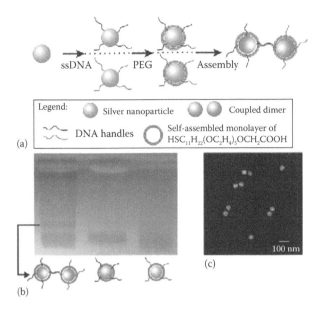

FIGURE 12.4 Assembly and characterization of silver plasmon rulers. (a) Silver nanoparticles are functionalized with complementary ssDNA and pegylated. Finally, the particles are combined to anneal. (b) Agarose gel (1%) of hybridization mix (left lane) and silver monomers with DNA handles (two right lanes). The hybrid mix contains new bands indicative of successful hybridization. The new band (arrow) running closest to the monomers is cut out of the gel, and the particles are recovered through gel electroelution. (c) A TEM micrograph of the recovered particles from this band confirms that the band is enriched in dimers. (Reprinted with permission from Yang, L., Wang, H., and Reinhard, B.M., *J. Phys. Chem. C,* 114, 4901–4908, 2010. Copyright 2010 American Chemical Society.)

samples are split into two batches, A and B, and aqueous solutions of thiolated DNA oligonucleotides are added. The sequence of the handles in A and B are complementary. For our applications, we typically use DNA:nanoparticle ratios in the range 1:1–10:1. After incubation of the particles with the oligonucleotides overnight, alkyl-PEG-acids are added to batch A, and a mix of alkyl-PEG-acids and biotin-PEGs (molar ratio: ~25:1) are added to batch B. For details of the pegylation step, see earlier text. After another night of incubation, the PEG-stabilized silver nanoparticle-oligonucleotides conjugates are cleaned by repeated centrifugation and resuspension (3×) and finally dispersed in 80 mM phosphate buffer, pH 8. In the last reaction step, the two batches are combined for <1 h to assemble into dimers. Formed dimers are then purified by gel electrophoresis in a 1% agarose gel, using 0.5× TBE as running buffer at 170 V for 15 min. Finally, the dimer band is isolated from the gel and recovered by electroelution and stored in TBE buffer at 4°C.

The silver plasmon ruler assembly approach is summarized in Figure 12.4a and a gel of silver plasmon rulers is shown in Figure 12.4b.[35] The reaction mix contains a ladder structure indicative of the formation of DNA-linked nanoparticle assemblies. The dimer band—which is absent in the precursor particles—is cut out of the gel, and the dimers in this band are recovered through gel electrophoresis. Figure 12.4c shows TEM images of some of the recovered dimers. We emphasize that the individual particles in the dimers are well separated. This observation confirms that the particles are not just nonspecifically attached but indeed linked by a polymer tether.

12.3.3 Plasmon Coupling Microscopy

PCM is based on conventional dark field microscopy where light is injected at oblique angles into the sample plane so that only the light that is scattered into the direction of the objective by objects lying in the focal plane is collected. Dark field illumination thus efficiently suppresses the excitation light and enables to image strong light scatterers with good signal-to-noise. Noble metal nanoparticles are efficient

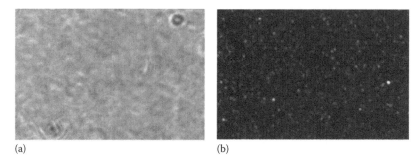

(a)   (b)

FIGURE 12.5   Comparison of a 40 nm diameter gold nanoparticle–decorated glass surface under bright field (a) and dark field (b) illumination. Whereas under dark field illumination, individual gold nanoparticles are clearly visible as (vivid green) dots, in bright field illumination, the nanoparticles are not recognizable.

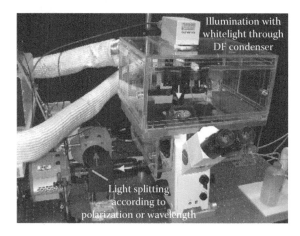

FIGURE 12.6   Picture of a plasmon coupling microscope. The setup is built around an inverted dark field (DF) microscope. The collected light is split either according to wavelength or polarization using a dichroic filter or a polarizing beam splitter, respectively. In the setup shown here, the split beams are recorded on two separate EMCCD detectors. The microscope is equipped with a cage incubator for temperature control during live cell studies.

light scatterers[16,17] and therefore bright labels in dark field microscopy.[53–58] Figure 12.5 demonstrates the striking contrast difference of gold nanoparticles under bright field and dark field illumination. The dark field image was recorded using an Olympus IX71 inverted microscope with an Olympus oil condenser (NA = 1.2 − 1.4) and a 60× objective (NA = 0.65). The sample was illuminated with unpolarized white light using a 100 W Tungsten halogen lamp. In PCM, the light scattered from the nanoparticle labels is split according to color or polarization through ratiometric imaging. This splitting is accomplished by placing an appropriate dichroic filter or polarizing beam splitter into the beam path. Figure 12.6 shows a plasmon coupling microscope in which the two beams are collected on two separate electron-multiplied charge-coupled devices (EMCCD). A simpler implementation of PCM can be realized using commercial dual-view systems, which refocus two light beams split according to wavelength or polarization on two separate areas of the same EMCCD detector.[47]

12.4 Results

12.4.1 Cancer Biomarker Labeling

Cancer-specific molecular features (so called biomarkers) are of great value for elucidating the molecular mechanisms of the disease as well as for improving the accuracy of cancer detection and

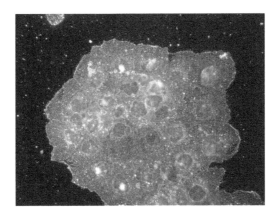

FIGURE 12.7   A patch of A431 cells that were stained for EGFR with 40 gold nanoparticle immunolabels (magnification: 60×). The nanoparticle labels on the surface create a vivid green color. See for instance reference 88.

staging. A reliable and selective detection of biomarkers at low concentrations requires new detection modalities. Gold and silver nanoparticles are very bright optical probes that enable to localize—in principle—individual biomarkers in living and fixed cells using dark field microscopy. Specificity in these labeling experiments is commonly conveyed through functionalization of the nanoparticles with an antibody that recognizes and binds to a particular biomarker.[59–61]

One approach for functionalizing nanoparticles is based on bifunctional PEG molecules that are end functionalized with azide moieties which can be cross-linked with alkyne-functionalized antibodies using click chemistry[62] (copper-catalyzed azide-alkyne cycloaddition). The PEGs used in this process are functionalized with a thiol residue for anchoring to the metal surface on one side and an azide residue for chemical cross-linking on the other end. The antibody that is to be tethered to the nanoparticle via the PEG linker needs to be functionalized with a propargyl residue. This residue can be covalently attached to the antibody by incubating the antibody with propargyl-dPEG-NHS ester on ice. The antibody is then purified through a size exclusion column (7 kDa molecular weight cutoff) and subsequently incubated with the azide-PEG-functionalized nanoparticles in the presence of a click chemistry catalyst (0.5 mM ascorbic acid and 0.1 mM CuSO4) at 4°C overnight. Figure 12.7 shows fixed A431 cells that were stained for the epidermal growth factor receptor (EGFR) using 40 nm gold nanoparticles. The labeling in Figure 12.7 was achieved by incubating the cells with anti-EGFR (antibody 29.1) and subsequently with biotin-labeled secondary antibody. Finally, 40 nm gold nanoparticles functionalized with antibiotin antibodies were bound to the biotin-labeled secondary antibodies. EGFR is a receptor tyrosine kinase whose over-expression is correlated with poor prognosis in cancer.[63–65] Figure 12.7 shows that the gold immunolabels achieve a distinct and easily detectable staining of this important cancer biomarker.

In PCM, immunolabels cannot only be used to detect a particular cancer biomarker but also to investigate its biological function through imaging its distribution and interaction with other cellular components as function of time. One complication for using noble metal nanoparticles as probes in these studies arises from the high salt concentrations in live cell studies. Typical live cell imaging buffers, such as Hank's buffer, have NaCl concentrations of ~150 mM and contain divalent ions. Under these conditions, gold and silver nanoparticles with sufficient scattering cross sections to enable single particle detection (diameters >20 nm) require a stabilizing surface modification that protects them from aggregation. The challenge for using nanoparticle in live cell imaging is therefore to generate small labels with a defined number of functional groups that are stable under live cell imaging conditions and that selectively bind to a specific cancer biomarker.

One approach to synthesize stable, yet functionalized, noble metal nanoparticles is the gel matrix assisted nanoparticle functionalization approach, described in Section 12.3.1.[53] This method creates a protecting self-assembled brush of short, negatively charged PEGs around the particles, which is then covalently decorated with the antibodies of interest. The stability of the particles obtained through this approach in Hank's buffer was tested by UV-Vis spectroscopy; significant particle agglomeration would

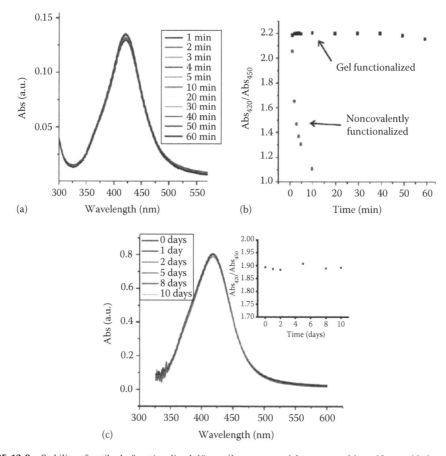

FIGURE 12.8 Stability of antibody-functionalized 40 nm silver nanoparticles protected by self-assembled monolayers of short PEG molecules. (a) UV-Vis absorbance spectra of anti-EGFR-functionalized silver nanoparticles in Hank's buffer. (b) Ratio of absorbances measured at 420 and 450 nm for antibody-functionalized nanoparticles obtained through the gel matrix–assisted functionalization and noncovalent attachment. (c) Absorbance of anti-EGFR-functionalized 40 nm silver nanoparticles in storage buffer (100 mM HEPES) over 10 days. (Reprinted with permission from Skewis, L.R. and Reinhard, B.M., *ACS Appl. Mater. Interfaces*, 2, 35–40, 2010. Copyright 2010 American Chemical Society.)

be indicated by a redshift and broadening of the plasmon resonance. The UV-Vis spectra of the labels in Figure 12.8a show, however, only negligible changes in the plasmon resonance as function of time during a total observation time of 60 min, indicating that the particles are stable. The observed stability of the PEG-functionalized particles is contrasted by a rapid precipitation of immunolabels that were obtained through noncovalent attachment of antibodies to nanoparticles without any additional stabilization. This is illustrated in Figure 12.8b where the ratio of the absorbances measured at 450 and 420 nm for the two different immunolabels are compared. Whereas the spectra of the PEG-stabilized particles remain nearly unchanged, the spectra of the immunolabels obtained through noncovalent antibody attachment show a strong, continuous redshift. In Figure 12.8c, we test the stability of the PEG-stabilized particles in 100 mM HEPES (pH 7.4) storage buffer. These data confirm that the particles can be stored at 4°C without aggregation for several days.

12.4.2 Plasmon Coupling Microscopy to Resolve Subdiffraction Contacts between Diffusing Cell Surface Species

Important cell surface receptors, such as receptor tyrosine kinases, require interactions between two or more receptors to trigger a signaling response.[66] Overexpression or aberrant activation of tyrosine

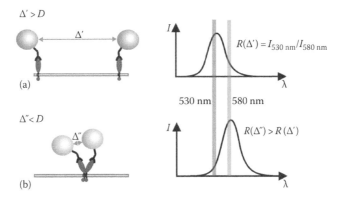

FIGURE 12.9 Plasmon coupling between cell surface–confined gold nanoparticle labels. (a) For interparticle separations Δ' larger than the particle diameter D, the resonance wavelength λ_{res} is that of an individual particle. (b) For interparticle separations $\Delta'' < D$, the plasmons in the particles couple, and the resonance wavelength λ_{res} redshifts with decreasing separation. This spectral shift is observable as an increase in the intensity ratio $R = I_{580\,nm}/I_{530\,nm}$. (Reprinted with permission from Rong, G., Wang, H., Skewis, L.R., and Reinhard, B.M., *Nano Lett.*, 8, 3386–3393, 2008. Copyright 2008 American Chemical Society.)

kinases is often associated with cancer. To reveal the molecular mechanisms underlying cell surface receptor association and to understand the role of these associates in cancerogenesis, new imaging modalities are required that can resolve nanoscale distances between laterally diffusing receptors with high temporal resolution. Given the distance-dependent optical properties of noble metal nanoparticles, the ability to label specific cell surface moieties provides an experimental avenue for these challenging investigations. At constant refractive index, the scattering resonance wavelengths of particle pairs and larger oligomers encode information about potential near-field interactions between the nanoparticles in situations when they are no longer discernible in the optical microscope. Figure 12.9 illustrates how these distance-dependent near-field interactions, which can be monitored in the far field through an analysis of the immunolabels' spectral responses, provide information about separations on nanometer length scales. For separations larger than approximately one particle diameter, the spectral response of the nanoparticle labels corresponds to that of the monomers. If the particles approach each other closer, the plasmons of the individual particles couple, and the "color" of the nanoparticles changes. The lateral mobility of the targeted receptors represents, however, a challenge for the spectral analysis of the light scattered from the nanoparticle labels. To overcome this problem, we have recently implemented PCM, which combines conventional particle tracking with a ratiometric analysis of the scattered light. A basic scheme of a PCM setup is shown in Figure 12.10a; an image of this microscope is shown in Figure 12.6. The dual color detection scheme in PCM facilitates a simultaneous recording of two monochromatic images of the same field of view. Spectral shifts induced through near-field interactions between interacting nanoparticles are detected in this scheme by ratiometric analysis of the collected light. Changes in the ratio R of the intensities of individual scatterers (or multiple colocalized scatterers)[48] monitored on two wavelength channels can indicate plasmon coupling–induced spectral shifts (Figure 12.9).

In coupled dimers of spherical noble metal nanoparticles, the scattering cross section of the longitudinal mode is larger than that of the vertical mode. Different than for monomers, the light scattered off dimers is therefore polarized. As we will discuss later in more detail (see Section 12.4.4), in some applications it is useful to probe both the scattering wavelength and the polarization of the scattered light. To that end, the excitation light is toggled between two different excitation wavelengths in polarization-resolved PCM, and the light is detected on two orthogonal polarization channels (Figure 12.10b). In each frame (n), two intensities I_1 and I_2 are recorded on polarization channels 1 and 2. We can thus calculate the reduced linear dichroism (P)[44,67,68] in each frame:

$$P(n) = \frac{I_1(n) - I_2(n)}{I_1(n) + I_2(n)}$$

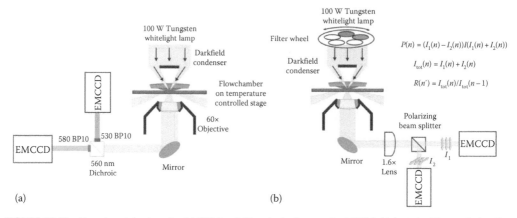

FIGURE 12.10 Experimental setups for (a) PCM and (b) polarization-resolved PCM. (Adapted with permission from Rong, G., Wang, H., Skewis, L.R., and Reinhard, B.M., *Nano Lett.*, 8, 3386–3393, 2008; Rong, G., Wang, H., and Reinhard, B.M., *Nano Lett.*, 10, 230–238, 2010. Copyright 2008 and 2010 American Chemical Society.)

P depends on the interparticle separation and the orientation of the long dimer axis; rotational motions of nanoparticle dimers on the membrane surface can be detected through anticorrelated changes in I_1 and I_2. Since subsequent images are recorded at different excitation wavelengths, polarization-resolved PCM also provides ratiometric spectral information. Using the total intensities $I_{tot}(n) = I_1(n) + I_2(n)$ for each frame *n*, an intensity ratio (*R*) is calculated from two subsequent images:

$$R(n) = \frac{I_{tot}(n)}{I_{tot}(n-1)}$$

In the setup in Figure 12.10b, the toggling of the excitation wavelength is achieved through a filter wheel. The temporal resolution in this case is fairly slow since it is limited by the mechanics of the filter wheel. Faster temporal resolutions are possible if the filter wheel is replaced with two shuttered monochromatic excitation lasers or a supercontinuous white light source in combination with appropriate filters and shutters. The latter has the practical advantage that the monitored wavelength channels can be easily varied and adjusted to a specific spectral range by choice of the filters. The freedom to adjust the monitored wavelength channels is instrumental in optimizing the sensitivity to a particular separation range, material (gold versus silver), or cluster size of interest.

Figure 12.11 illustrates how short-range interactions between individual gold nanoparticle labels are detected in PCM. The figure contains curve-fitted images, or point spread functions (PSF), for two gold nanoparticles diffusing on a cell membrane at three time points during aggregation. The nanoparticles are bound to integrins.[47] In Figure 12.11a, the two particles P1 and P2 are still discernible. In Figure 12.11b, P1 and P2 optically colocalize which results in a change of both the total intensity and the intensity distribution on the two channels. The peak intensity is now significantly higher on the 580 nm than on the 530 nm channel, and the intensity ratio has increased to $R \approx 1.3$. In Figure 12.11c, both the total intensity and the *R* value reach their maximum. The high *R* value of $R \approx 1.6$ reveals strong interparticle coupling between P1 and P2. The strong spectral shift and the prolonged colocalization of P1 and P2 are clear indications for nonreversible short-range interactions between the nanoparticles.

12.4.3 Quantification of Distance-Dependent Plasmon Coupling: Silver Plasmon Ruler Calibration

The ability to detect subdiffraction limit contacts in a conventional wide field microscope through PCM is already very useful for monitoring dynamic interactions between multiple nanoparticle-labeled cell surface components. In some applications, it might be desirable to convert the optical response of

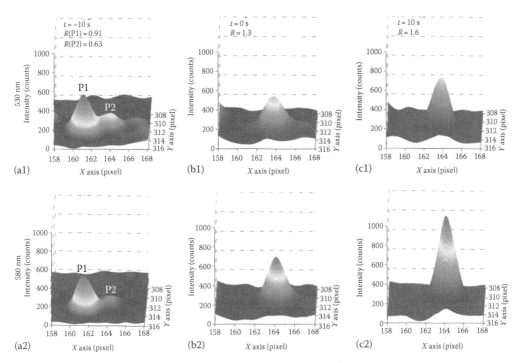

FIGURE 12.11 Collision between two particles (P1 and P2) monitored in PCM. The top row (a1-c1) contains the images recorded on the 530 nm channel; the bottom row (a2-c2) shows the 580 nm channel at t = −10s, 0s, +10s. Concurrent with the optical colocalization of the two particles, the relative intensities on the 580 nm channel increase. The strong gain in *R* indicates a strong redshift of λ_{res}. (Reprinted with permission from Rong, G., Wang, H., Skewis, L.R., and Reinhard, B.M., *Nano Lett.*, 8, 3386–3393, 2008. Copyright 2008 American Chemical Society.)

individual nanoparticle dimers into approximate distances. The latter requires precise knowledge of the distance-dependent optical response of pairs of noble metal nanoparticles. Quantitative calibration relationships for flexibly DNA-linked gold nanoparticles (gold plasmon rulers)[34,42] have been derived using DNA spacers of fixed length.[42] Silver nanoparticles are also very versatile labels in optical microscopy (Figure 12.12) and for plasmon coupling–based assays; silver nanoparticles have even some advantages over gold nanoparticles. Spherical silver nanoparticles have larger scattering cross sections and sharper plasmon resonances than gold nanoparticles of comparable size.[16,17,34] In addition, the spectral shift of the resonance wavelength with decreasing interparticle separation is larger for silver dimers than in pairs of gold nanoparticles.[28,69] Silver plasmon rulers can be assembled according to the procedures described in Section 12.3.2, and their distance-dependent spectral response was recently calibrated.[35] This calibration was achieved through correlation of single dimer spectroscopy with transmission electron microscopy (TEM).[56] The availability of both spectral and structural information for individual dimers enabled the correlation of the peak resonances with interparticle separations. In Figure 12.13a, the resonance energy in eV is plotted as a function of the ratio of center-to-center separation (*L*) divided by the average particle diameter (*D*) for approximately 30 dimers. The plot shows that the peak scattering wavelength continuously redshifts with decreasing interparticle separation until at separations below *L/D* = 1.05; the resonance energy does not continue to redshift but instead starts to blueshift. We ascribe this observation to the existence of different plasmon coupling regimes;[35] while the interactions for separations *L/D* > 1.05 are dominated by classic electromagnetic nanoparticle interactions, at very short separations *L/D* < 1.05, additional effects can become relevant such as direct charge transfer[70] or nonlocal effects that result in an effective screening[71] at points of high charge density. For applications in physiological buffers, the individual particles in the dimers need to be protected by a self-assembled monolayer of PEGs with typical thicknesses of 2–3 nm. This polymer layer prevents a very close approach of the particles in solution. Consequently, the "quantum plasmonic" range *L/D* < 1.05 is not relevant for most practical plasmon

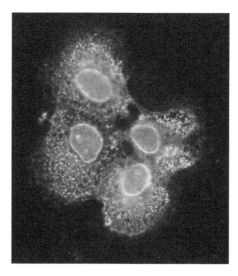

FIGURE 12.12   Dark field image of silver nanoparticle–labeled A431 cells (magnification: 96×). The silver NP labels on the surface create a vivid blue–green color. See for instance reference 89.

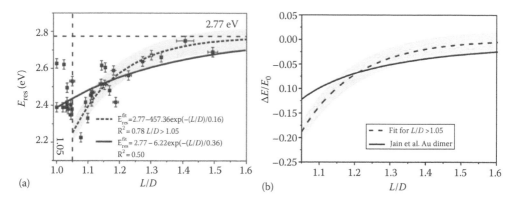

(a)

(b)

FIGURE 12.13   (a) Resonance energy (E) as function of the ratio of center-to-center separation divided by the particle diameter for silver plasmon rulers. (b) Shift in resonance energy (ΔE) normalized by the isolated monomer energy (E_0) for silver plasmon rulers (dashed line) and the universal scaling relationship (solid) derived for gold nanoparticle dimers.[38] (Adapted with permission from Yang, L., Wang, H., Yan, B., and Reinhard, B.M., *J. Phys. Chem. C,* 114, 4901–4908, 2010. Copyright 2010 American Chemical Society.)

ruler applications. In the classic electromagnetic regime with $L/D > 1.05$, the individual measurements are spread around a common trend—the resonance energy blueshifts with increasing interparticle separation—confirming that pairs of silver nanoparticles indeed act as plasmon rulers in this separation range. The spread of the data in Figure 12.13a is attributed to variations in the size and shapes of the nanoparticles used for the assembly of the dimers. The resonance energy of individual nanoparticles depends crucially on the particle geometry,[18,19] and differences in the resonance energies of the two particles within one dimer will affect the resonance of the coupled system. Further improvements of the monodispersity of noble metal nanoparticles preparations will improve the accuracy of the conversion of measured resonance wavelengths into distances in the future.

Figure 12.13b compares the experimental results for the silver plasmon rulers with the universal scaling relationship derived for gold particle dimers with well-defined separations.[38] The silver calibration curve is noticeably steeper than the gold curve, which means that comparable distance changes lead to a larger spectral shift in silver dimers. The achievable spatial resolution is therefore higher for silver than for gold dimers. Figure 12.13 shows that the resonance energy continuously decreases with increasing separation in the interval $L/D = [1.05-1.60]$. This indicates that silver particles with a diameter of $D = 30$ nm can be used to measure distances of up to 18 nm. Even longer distances become accessible with larger particles.

12.4.4 Characterizing Cancer Cell Surface Morphologies: Two-Dimensional Membrane Profiling through Silver Plasmon Ruler Tracking

The plasma membrane represents a two-dimensional diffusion system in which the spatial organization and dynamics of both lipids and proteins play important roles in regulating signal transduction.[1] The cell surface is compartmentalized,[72–76] and the nanostructured cell surface could be instrumental in providing confined regions with high local concentrations of cell surface moieties involved in signaling. Imaging methods that provide detailed information about the cell morphology with nanometer spatial resolution are therefore useful for improving our understanding of the role of the cell surface structure for regulating cancer-related cell signaling. The latter could offer new opportunities for cancer diagnostics and therapy.

We have demonstrated recently that plasmon rulers can provide detailed insight into the two-dimensional organization of mammalian plasma membranes.[46] The plasmon resonance in a coupled nanoparticle dimer contains information about the interparticle separation, whereas the polarization of the light scattered off an individual dimer yields information about the orientation of the long dimer axis. A plasmon ruler diffusing on the plasma membrane provides structural information on the length scale of the plasmon ruler: With decreasing size of the compartment, the spatial confinement of a plasmon ruler diffusing on a membrane is expected to increase, whereas the average interparticle separation and rotational mobility of the plasmon rulers are expected to decrease. HeLa cell surface compartments have been reported to be on the order of ~70 nm. These small sizes cannot be directly imaged in an optical microscope and require specialized tools for characterization. The potential value of the plasmon ruler technology for revealing structural details of these compartments was demonstrated with silver plasmon rulers assembled from spherical silver nanoparticles with an average diameter of approximately 30 nm and a DNA tether with an equilibrium separation in solution of approximately 14 nm.[46] In order to be sensitive to structural details of the confinement, the plasmon rulers applied in this work were chosen to be of similar size as the cell surface compartments. The plasmon rulers attached nonspecifically to isolated membranes, on which they performed a lateral diffusion. This motion was tracked through polarization-resolved PCM to determine R and P values as function of space and time. The measured R values were then converted into approximate interparticle separations (S) using the experimentally derived silver plasmon ruler calibration curve.[46]

Figure 12.14 summarizes the results for two different plasmon rulers (PR1 and PR2) tracked through polarization-resolved PCM. The figure shows the trajectories of the plasmon rulers together with the corresponding P, R, and S values as a function of time. PR1 (Figure 12.14a) is confined but exhibits a significantly higher degree of translational mobility than the more strongly confined PR2 (Figure 12.14b). The approximate average interparticle separation of PR1 is obtained as the average of all individual S measurements in the trajectory as $S_{av} \approx 13.4 \pm 2.8$ nm. Fluctuations in the five-point sliding average of S indicate systematic variations in the interparticle separation on short time scales. The recorded P values for PR1 show a rich dynamics and fluctuate between $P = -0.4$ and $P = +0.5$. Despite the relatively slow temporal resolution in these experiments of 5 Hz, the orientation dependence of the polarization is not averaged out. The observation of a large dynamic range in P for PR1 indicates that the lateral motion of this plasmon ruler on the cell surface is coupled to changes in the orientation of the long dimer axis that occur on the time scale of the temporal resolution of the experiment. The long plasmon ruler axis "hops" between different orientations on the surface.

PR2 shows a different behavior. Concurrent with a stronger confinement of the lateral mobility in Figure 12.14b, the rotational mobility of PR2 decreases significantly. The P value is almost constant except during the interval $t = 58-115$ s, when PR2 exhibits some rotational dynamics which is correlated with noticeable changes in the interparticle separation. Consistent with the interpretation of a strong confinement of PR2, the average interparticle separation is strongly decreased to $S_{av} \approx 9.7 \pm 1.3$ nm.

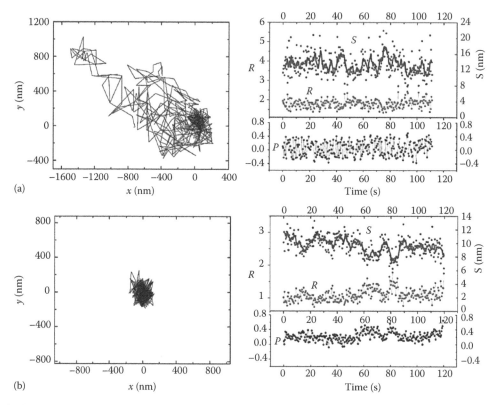

FIGURE 12.14   Polarization-resolved ratiometric tracking of two silver plasmon rulers (a) PR1 and (b) PR2. The panel on the left shows the ruler trajectory, the right panel contains the intensity ratios of the monitored wavelength channels (R), the approximate interparticle separation (S), and the reduced linear dichroism (P). The wavelength channels monitored were 430 and 470 nm in (a) and 450 and 490 nm in (b). (Adapted with permission from Rong, G.,Wang, H., and Reinhard, B.M., *Nano Lett.*, 10, 230–238, 2010. Copyright 2010 American Chemical Society.)

Since polarization-resolved PCM provides S and P values as function of space and time for individual plasmon rulers, further insight can be obtained through an analysis of the distribution of P and S on the cell surface. In the case of PR2, this analysis revealed a correlation of short S values with a specific orientation of the dimer axis in some areas of the confinement.[46] This observation indicates that PR2 needs to assume a specific spatial orientation to fit into this compartment. The cell surface compartments are anisotropic, and plasmon rulers enable now to measure the dimensions of individual compartments along different axes.

12.5 Discussion

Due to the distance-dependent electromagnetic interactions between individual noble metal nanoparticles, gold and silver nanoparticles provide new opportunities for measuring nanoscale distances and distance changes in conventional wide field microscopy. Some applications of the new technology with cancer relevance have been discussed in this chapter (e.g., resolving subdiffraction limit contacts on cellular surfaces and two-dimensional membrane profiling using plasmon rulers). The interest in noble metal nanoparticle–based imaging modalities is, however, only partly motivated by the gain in functionality that arises from the particles' ability to couple in the near field. The superb photophysical properties of noble metal nanoparticle labels are certainly additional beneficial aspects for biosensing and imaging applications. Gold and silver nanoparticles with diameters larger than 20 nm are efficient light scatterers that can be localized with high spatial precision in a short time,[77] provided that the scattering background can be eliminated. Since their signal is based on light scattering, noble metal nanoparticles provide stable signals without limitation in observation duration. Noble metal nanoparticles have, however, also disadvantages when compared with fluorescence dyes. The most obvious disadvantage is certainly the size of the applied probes; the nanoparticles commonly used as optical labels have diameters ≥20 nm, which are large in comparison with most conventional organic dyes. In addition, the selective labeling of specific proteins with nanoparticle tags in live cell microscopy remains significantly more difficult than in fluorescence microscopy, where genetically encoded fluorescent tags are established and routinely used. The availability of genetically encoded probes in combination with so-called fluorescent nanoscopies,[78] which provide deep subdiffraction limit resolution, such as stimulated emission depletion (STED) microscopy,[79,80] photoactivatable localization microscopy (PALM),[81] or stochastic optical reconstruction microscopy (STORM),[82] promises a new era of high-resolution fluorescent imaging. Although fluorescent nanoscopies can resolve deeply subdiffraction limit distances, the capabilities of these methods are still limited by the moderate photostabilities of fluorescent dyes. Consequently, it remains challenging to monitor fast dynamic interactions between individual molecules for extended periods of time. Plasmon coupling–based approaches have the potential to fill this gap, especially in applications where the groups of interest are amenable to an efficient labeling with nanoparticles. Dynamic cell surfaces processes fall in this category, since variable labeling strategies for live cell surface proteins, such as site-specific biotinylation procedures, have already been developed.[83]

In addition, interactions between nanoparticles and cells are of interest in their own rights.[84] Nanoparticles are finding important applications in cancer therapy[85] and drug delivery,[86] and a better understanding of the interactions between nanoparticles and living cells will contribute to improve the efficacies of these strategies. The use of noble metal nanoparticles as probes in PCM will in the future lead to a better understanding of the nano-bio interface.

12.6 Future Trends

Both PCM and plasmon rulers are emerging technologies and many aspects, including the interfacing of the nanoparticle probes with the target molecules, need to be further refined. We expect that most of these developments will occur in response to the needs of specific applications. A general aspect of the technology that needs to be addressed is the heterogeneity of the applied probes. For many applications of PCM, ideal spheres with constant size are the probes of choice, and we are optimistic that with improved synthetic control over nanoscale length scales, more homogenous building blocks for nanoscale devices will become available in the future. There is still substantial room for improvement of the temporal resolution in PCM and polarization-resolved PCM. In conventional particle tracking, microsecond temporal has been achieved in living cells,[87] and it is realistic to assume that similar temporal resolutions are accessible with PCM. In our group, we are currently working toward the development and application of high-speed PCM for a real-time monitoring of cell surface receptor clustering and for probing the dynamic structure and mechanical properties of nucleoprotein complexes.

ACKNOWLEDGMENTS

We acknowledge financial support from the National Institutes of Health through grants 5 R21 EB008822-02 and 1 R01 CA138509-01 to BMR and a Ruth L. Kirschstein National Service Award to LRS. We also acknowledge support from the National Science Foundation to BMR (CBET-0953121, CBET-0853798).

REFERENCES

1. Whitty, A. *Nature Chemical Biology* 2008, 4, 435–439.
2. Deniz, A. A.; Mukhopadhyay, S.; Lemke, E. A. Jr. *Soc. Interface* 2008, 5, 15–45.
3. Zhuang, X. W.; Bartley, L. E.; Babcock, H. P.; Russell, R.; Ha, T. J.; Herschlag, D.; Chu, S. *Science* 2000, 288(5473), 2048–2051.
4. Lakowicz, J. R. *Principles of Fluorescence Spectroscopy*, Vol. 2. Kluwer Academic: New York, 1999.

5. Roy, R.; Hohng, S.; Ha, T. *Nature Methods* 2008, 5(6), 507–516.
6. Churchman, L. S.; Okten, Z.; Rock, R. S.; Dawson, J. F.; Spudich, J. A. *Proceedings of the National Academy of Sciences of the United States of America* 2005, 102(5), 1419–1423.
7. Lacoste, T. D.; Michalet, X.; Pinaud, F.; Chemla, D. S.; Alivisatos, A. P.; Weiss, S. *Proceedings of the National Academy of Sciences of the United States of America* 2000, 97(17), 9461–9466.
8. van Oijen, A. M.; Kohler, J.; Schmidt, J.; Muller, M.; Brakenhoff, G. J. *Chemical Physics Letters* 1998, 292, 183–187.
9. Yun, C. S.; Javier, A.; Jennings, T.; Fisher, M.; Hira, S.; Peterson, S.; Hopkins, B.; Reich, N. O.; Strouse, G. F. *Journal of the American Chemical Society* 2005, 127(9), 3115–3119.
10. Michalet, X.; Pinaud, F. F.; Bentolila, L. A.; Tsay, J. M.; Doose, S.; Li, J. J.; Sundaresan, G.; Wu, A. M.; Gambhir, S. S.; Weiss, S. *Science* 2005, 307(5709), 538–544.
11. Hohng, S.; Ha, T. *ChemPhysChem* 2005, 6(5), 956–960.
12. Hohng, S.; Ha, T. *Journal of the American Chemical Society* 2004, 126, 1324–1325.
13. Fomenko, V.; Nesbitt, D. *Nano Letters* 2008, 8, 287–293.
14. Antelman, J.; Ebenstein, Y.; Dertinger, T.; Michalet, X.; Weiss, S. *Journal of Physical Chemistry C* 2009, 113, 11541–11545.
15. Kreibig, U.; Vollmer, M. *Optical Properties of Metal Clusters*. Springer: Berlin, Germany, 1995.
16. Yguerabide, J.; Yguerabide, E. E. *Analytical Biochemistry* 1998, 262(2), 157–176.
17. Yguerabide, J.; Yguerabide, E. E. *Analytical Biochemistry* 1998, 262(2), 137–156.
18. Kelly, K. L.; Coronado, E.; Zhao, L. L.; Schatz, G. C. *Journal of Physical Chemistry B* 2003, 107(3), 668–677.
19. Halas, N. *MRS Bulletin* 2005, 30(5), 362–367.
20. Felidj, N.; Laurent, G.; Aubard, J.; Levi, G.; Hohenau, A.; Krenn, J. R.; Aussenegg, F. R. *Journal of Chemical Physics* 2005, 123(22), 221103.
21. Krenn, J. R.; Dereux, A.; Weeber, J. C.; Bourillot, E.; Lacroute, Y.; Goudonnet, J. P.; Schider, G.; Gotschy, W.; Leitner, A.; Aussenegg, F. R.; Girard, C. *Physical Review Letters* 1999, 82(12), 2590–2593.
22. Lamprecht, B.; Schider, G.; Lechner, R. T.; Ditlbacher, H.; Krenn, J. R.; Leitner, A.; Aussenegg, F. R. *Physical Review Letters* 2000, 84(20), 4721–4724.
23. Salerno, M.; Krenn, J. R.; Hohenau, A.; Ditlbacher, H.; Schider, G.; Leitner, A.; Aussenegg, F. R. *Optics Communications* 2005, 248(4–6), 543–549.
24. Maier, S. A.; Brongersma, M. L.; Kik, P. G.; Atwater, H. A. *Physical Review B* 2002, 65(19), 193408.
25. Maier, S. A.; Brongersma, M. L.; Kik, P. G.; Meltzer, S.; Requicha, A. A. G.; Atwater, H. A. *Advanced Materials* 2001, 13(19), 1501–1505.
26. Li, K. R.; Li, X. T.; Stockman, M. I.; Bergman, D. J. *Physical Review B* 2005, 71(11), 115409.
27. Li, K. R.; Stockman, M. I.; Bergman, D. J. *Physical Review Letters* 2003, 91(22), 227402.
28. Gunnarsson, L.; Rindzevicius, T.; Prikulis, J.; Kasemo, B.; Kall, M.; Zou, S. L.; Schatz, G. C. *Journal of Physical Chemistry B* 2005, 109(3), 1079–1087.
29. Shalaev, M. *Physics Reports* 1996, 272(2–3), 61–137.
30. Shalaev, V. *Optical Properties of Nanostructured Random Media*. Springer: Berlin, Germany, pp. 1–158, 2002.
31. Shalaev, V. M.; Botet, R.; Tsai, D. P.; Kovacs, J.; Moskovits, M. *Physica A* 1994, 207(1–3), 197–207.
32. Yan, B.; Thubagere, A.; Premasiri, R.; Ziegler, L.; Dal Negro, L.; Reinhard, B. M. *ACS Nano* 2009, 3, 1190–1202.
33. Pinchuk, A. O.; Schatz, G. C. *Materials Science and Engineering B* 2008, 149, 251–258.
34. Sonnichsen, C.; Reinhard, B. M.; Liphardt, J.; Alivisatos, A. P. *Nature Biotechnology* 2005, 23(6), 741–745.
35. Yang, L.; Wang, H.; Reinhard, B. M. *The Journal of Physical Chemistry C* 2010, 114, 4901–4908.
36. Su, K. H.; Wei, Q. H.; Zhang, X.; Mock, J. J.; Smith, D. R.; Schultz, S. *Nano Letters* 2003, 3(8), 1087–1090.
37. Rechberger, W.; Hohenau, A.; Leitner, A.; Krenn, J. R.; Lamprecht, B.; Aussenegg, F. R. *Optics Communications* 2003, 220(1–3), 137–141.
38. Jain, P. K.; Huang, W. Y.; El-Sayed, M. A. *Nano Letters* 2007, 7(7), 2080–2088.
39. Nordlander, P.; Oubre, C.; Prodan, E.; Li, K.; Stockman, M. I. *Nano Letters* 2004, 4, 899–903.
40. Muskens, O. L.; Giannini, V.; Sanchez-Gil, J. A.; Rivas, J. G. *Optics Express* 2007, 15(26), 17736–17746.

41. Reinhard, B. M.; Sheikholeslami, S.; Mastroianni, A.; Alivisatos, A. P.; Liphardt, J. *Proceedings of the National Academy of Sciences of the United States of America* 2007, 104(8), 2667–2672.
42. Reinhard, B. M.; Siu, M.; Agarwal, H.; Alivisatos, A. P.; Liphardt, J. *Nano Letters* 2005, 5(11), 2246–2252.
43. Skewis, L. R.; Reinhard, B. M. *Nano Letters* 2008, 8(1), 214–220.
44. Wang, H.; Reinhard, B. M. *The Journal of Physical Chemistry C* 2009, 113, 11215–11222.
45. Jun, Y.-W.; Sheikholeslami, S.; Hostetter, D. R.; Tajon, C.; Craik, C. S.; Alivisatos, A. P. *Proceedings of the National Academy of Sciences* 2009, 42, 17735–17740.
46. Rong, G.; Wang, H.; Reinhard, B. M. *Nano Letters* 2010, 10, 230–238.
47. Rong, G.; Wang, H.; Skewis, L. R.; Reinhard, B. M. *Nano Letters* 2008, 8, 3386–3393.
48. Wang, H.; Rong, G.; Yan, B.; Yang, L.; Reinhard, B. M. *Nano Letters* 2011, 11, 498–504.
49. Cloyd, M. W.; Bigner, D. D. *American Journal of Pathology* 1977, 88, 29–52.
50. Iyer, S.; Gaikwad, R. M.; Subba-Rao, V.; Woodworth, C. D.; Sokolov, I. *Nature Nanotechnology* 2009, 4, 389–393.
51. Yan, B.; Reinhard, B. M. *The Journal of Physical Chemistry Letters* 2010, 1, 1595–1598.
52. Allen, T. D.; Iype, P. T.; Murphy Jr., M. J. *In Vitro* 1976, 12, 837–844.
53. Skewis, L. R.; Reinhard, B. M. *ACS Applied Materials and Interfaces* 2010, 2, 35–40.
54. Sonnichsen, C.; Franzl, T.; Wilk, T.; von Plessen, G.; Feldmann, J.; Wilson, O.; Mulvaney, P. *Physical Review Letters* 2002, 88(7), 077402.
55. Hu, M.; Novo, C.; Funston, A.; Wang, H. N.; Staleva, H.; Zou, S. L.; Mulvaney, P.; Xia, Y. N.; Hartland, G. V. *Journal of Materials Chemistry* 2008, 18(17), 1949–1960.
56. Yang, L.; Yan, B.; Reinhard, B. M. *Journal of Physical Chemistry C* 2008, 112, 15989–15996.
57. Schultz, S.; Smith, D. R.; Mock, J. J.; Schultz, D. A. *Proceedings of the National Academy of Sciences* 2000, 97(3), 996–1001.
58. Knight, M. W.; Fan, J.; Capasso, F.; Halas, N. J. *Optics Express* 2010, 18(3), 2579.
59. Aaron, J.; Nitin, N.; Travis, K.; Kumar, S.; Collier, T.; Park, S. Y.; Jose-Yacaman, M.; Coghlan, L.; Follen, M.; Richards-Kortum, R.; Sokolov, K. *Journal of Biomedical Optics* 2007, 12, 034007.
60. Aaron, J.; Travis, K.; Harrison, N.; Sokolov, K. *Nano Letters* 2009, 9, 3612–3618.
61. Kah, J. C. Y.; Olivo, M. C.; Lee, C. G. L.; Sheppard, C. J. R. *Molecular and Cellular Probes* 2008, 22, 14–23.
62. Kolb, H. C.; Finn, M. G.; Sharpless, K. B. *Angewandte Chemie International Edition* 2001, 40, 2004–2021.
63. Nicholson, R. I.; Gee, J. M. W.; Harper, M. E. *European Journal of Cancer* 2001, 37, S9–S15.
64. Sibilia, M.; Kroismayr, R.; Lichtenberger, B. M.; Natarajan, A.; Hecking, M.; Holcmann, M. *Differentiation* 2007, 75(9), 770–787.
65. Zimmermann, M.; Zouhaair, A.; Azria, D.; Ozsahin, M. *Radiation Oncology* 2006, 2, 1–11.
66. Schlessinger, J. *Cell* 2000, 103(2), 211–225.
67. Wei, C. Y.; Lu, C. Y.; Kim, Y.; Vanden Bout, D. *Journal of Fluorescence* 2007, 17, 797–804.
68. Sonnichsen, C.; Alivisatos, A. P. *Nano Letters* 2005, 5(2), 301–304.
69. Encina, E. R.; Coronado, E. A. *The Journal of Physical Chemistry C* 2010, 114, 3918–3923.
70. Zuluoga, J.; Prodan, E.; Nordlander, P. *Nano Letters* 2009, 9, 887–891.
71. de Abajo, F. J. G. *Journal of Physical Chemistry C* 2008, 112, 17983–17987.
72. Murase, K.; Fujiwara, T.; Umemura, Y.; Suzuki, K.; Iino, R.; Yamashita, H.; Saito, M.; Murakoshi, H.; Ritchie, K.; Kusumi, A. *Biophysical Journal* 2004, 86(6), 4075–4093.
73. Suzuki, K.; Ritchie, K.; Kajikawa, E.; Fujiwara, T.; Kusumi, A. *Biophysical Journal* 2005, 88(5), 3659–3680.
74. Kusumi, A.; Nakada, C.; Ritchie, K.; Murase, K.; Suzuki, K.; Murakoshi, H.; Kasai, R. S.; Kondo, J.; Fujiwara, T. *Annual Review of Biophysics and Biomolecular Structure* 2005, 34, 351–378.
75. Hugel, B.; Martinez, M.; Kunzelmann, C.; Blattler, T.; Aguzzi, A.; Freyssinet, J. M. *Cellular and Molecular Life Sciences* 2004, 61(23), 2998–3007.
76. Nagy, P.; Vereb, G.; Sebestyen, Z.; Horvath, G.; Lockett, S. J.; Damjanovich, S.; Park, J. W.; Jovin, T. M.; Szollosi, J. *Journal of Cell Science* 2002, 115(22), 4251–4262.
77. Thompson, R. E.; Larson, D. R.; Webb, W. W. *Biophysical Journal* 2002, 82(5), 2775–2783.
78. Hell, S. W. *Science* 2007, 316(5828), 1153–1158.
79. Hell, S. W.; Wichmann, J. *Optics Letters* 1994, 19(11), 780–782.

80. Westphal, V.; Rizzoli, S. O.; Lauterbach, M. A.; Kamin, D.; Jahn, R.; Hell, S. W. *Science* 2008, 320(5873), 246–249.
81. Betzig, E.; Patterson, G. H.; Sougrat, R.; Lindwasser, O. W.; Olenych, S.; Bonifacino, J. S.; Davidson, M. W.; Lippincott-Schwartz, J.; Hess, H. F. *Science* 2006, 313(5793), 1642–1645.
82. Rust, M. J.; Bates, M.; Zhuang, X. W. *Nature Methods* 2006, 3(10), 793–795.
83. Howarth, M.; Takao, K.; Hayashi, Y.; Ting, A. Y. *Proceedings of the National Academy of Sciences* 2005, 102(21), 7583–7588.
84. Thubagere, A.; Reinhard, B. M. *ACS Nano* 2010, 4(7), 3611–3622.
85. Hirsch, L. R.; Stafford, R. J.; Bankson, J. A.; Sershen, S. R.; Rivera, B.; Price, R. E.; Hazle, J. D.; Halas, N. J.; West, J. L. *Proceedings of the National Academy of Sciences of the United States of America* 2003, 100(23), 13549–13554.
86. De Jong, W. H.; Borm, P. J. A. *International Journal of Nanomedicine* 2008, 3, 133–149.
87. Nan, X. L.; Sims, P. A.; Xie, X. S. *ChemPhysChem* 2008, 9(5), 707–712.
88. Wang, J.; Boriskina, S.V.; Wang, H.; Reinhard, B.M. *ACS Nano* 2011, 5, 6619–6628.
89. Wang, H.; Rong, G.; Yan, B.; Yang, L; Reinhard, B.M. *Nano lett* 2011, 11, 498–504.

13

Cost-Effective Evaluation of Cervical Cancer Using Reflectance and Fluorescence Spectroscopy

Shabbir B. Bambot

CONTENTS

13.1 Introduction

The American Cancer Society estimates that there will be 12,710 new cases of invasive cervical cancer diagnosed in 2011 and about 4,290 women will die from the disease in the United States [1]. Worldwide, there are approximately 550,000 cases of cervical cancer diagnosed annually and approximately 300,000 deaths per year [1]. Like most cancers, cervical cancer is easily treated if diagnosed early, and the uterine cervix is one of the few cancer-afflicted organs, which can be accessed and visualized in a noninvasive manner. If detected in its precancerous state, treatment can lead to almost 100% survival in 5 years [1]. Moreover, the significant difference* in treatment cost between high-grade cervical precancer and stage 1 or higher cancer demonstrates a pressing economic need for early detection technologies. Estimates show the market potential for noninvasive cervical cancer detection to be at $1.25 billion annually in the United States and Europe.

13.1.1 Cervical Cancer Screening

An annual Pap test is currently the most widely used tool to screen women for neoplasia or cervical cancer. While its contribution to reducing patient mortality is widely acknowledged, it is prone to errors from low screening frequency, insufficient cell sampling, inadequate sample preparation, lack of exfoliation of abnormal cells, and technician reading error. The discrimination performance of this test is therefore limited, resulting in a trade-off between sensitivity and specificity as illustrated in a landmark meta-analysis conducted by Fahey [3]. Current practice [2] sets the sensitivity at 51% in order to achieve a specificity of 97%. Thus, Pap tests have been used as a means to "rule out" rather than "rule in" disease. One rationale behind this is to limit the large number of false positives that would inadvertently burden downstream health management systems. While this may be true, this also results in a delay of diagnosis at an annual cost of nearly $2 billion for treating advanced disease [4]. Alternatives to the traditional Pap test such as ThinPrep [5] and SurePath [6] have become popular with physicians absent clear evidence that they improve upon the traditional Pap test [7,8]. Using a different approach, the HPV test [9] used for ASCUS (Table 13.1) triage appears to be better than a repeat Pap test in finding patients with disease who are referred to colposcopy. Overall, however, it has a low specificity resulting in an increase in false positives [10]. Also, while the FDA has approved computerized aids to Pap test screening such as AutoPap and Papnet, based on higher sensitivities, the evidence regarding the impact of these technologies on the screening process is unclear [11–13]. These methods are being used primarily for rescreening of normal Pap smears for quality. Although these instruments may benefit laboratories through reduction of false-negative rates, increased sensitivity, and increased throughput, they are also expensive [14].

13.1.2 Cervical Cancer Diagnosis

Diagnosis of cervical precancers and cancers is done by colposcopy and colposcopy-directed biopsy followed by a histological examination of the biopsied tissue. Colposcopy is done following a suspicious Pap test or positive HPV test. Trained clinicians, who visually assess the cervix, localize likely disease and biopsy tissue, perform the colposcopy procedure. Since suspect areas are identified visually, colposcopy requires extensive training, experience, and a significant effort toward maintenance of skills. A pathologist, whose diagnosis is considered the gold standard, microscopically examines the biopsy specimens. The amount of tissue biopsied varies according to the extent of the assessed lesion, and in some cases, a large portion of the cervix is removed in what is known as a loop electrosurgical excision procedure

---

* The cost of treatment (in 1997 dollars) of high-grade cervical pre-cancer is $14,000 lesser than Stage I cancer and $37,000 lesser than Stage IV cancer [2].

TABLE 13.1

Acronyms of Terminology Used in Cytology and Pathology
Reporting for Cervical Precancers and Cancers

Cytology	
LSIL	Low-grade squamous intraepithelial lesion
HSIL	High-grade squamous intraepithelial lesion
ASC	Atypical squamous cells
ASC-H	Atypical squamous cells—favor high grade
ASCUS	Atypical squamous cells of undetermined significance
AGUS	Atypical glandular cells of undetermined significance
AGC	Atypical glandular cells
Pathology	
CIN1	Cervical intraepithelial neoplasia 1 (approximately 1/3 depth of epithelium from basement membrane having neoplastic cells)
CIN2	Cervical intraepithelial neoplasia 1 (approximately 2/3 depth of epithelium from basement membrane having neoplastic cells)
CIN3	Cervical intraepithelial neoplasia 1 (entire depth of epithelium from basement membrane having neoplastic cells)

Source: Apgar et al., *Am. Fam. Physician.*, 68(10), 1992, 2003; DeMay, M., *Practical Principles of Cytopathology*, Revised edition, American Society for Clinical Pathology Press, Chicago, IL, 2007.

(LEEP) or cone biopsy. The "colposcopic diagnosis" therefore involves a close interplay between the colposcopist's ability to correctly identify likely disease and the pathologist's ability to rule on it. As a result, colposcopic diagnosis has a limited performance. A meta-analysis of colposcopy summarizing the results of nine studies [15] lists the average sensitivity and specificity at 85% and 69% respectively for separating CIN1 and lower from CIN2/3 and higher. Further, the well-publicized ALTS (ASCUS LSIL triage study) trial demonstrated the inability of colposcopy alone to determine CIN3 for women with an ASC cytology result [16,17].

The key disadvantage of current practice is the low test specificity and sensitivity for both Pap and colposcopy. The result is that a large number of patients undergoing unnecessary biopsy and/or a large number of patients with cancer go untreated. Any improvement in sensitivity and specificity with new technology would significantly alleviate this situation. Another disadvantage of the current practice in cervical precancer management is the significant time delay in obtaining the results. A patient and care provider must wait 1–4 weeks for the results of the Pap test and pathology. Quite often, the colposcopy, biopsy, and histopathology sequence has to be repeated in order to localize and diagnose the disease definitively. Moreover, the 1–4 weeks required in obtaining a Pap test result or a histology evaluation results in increased patient anxiety and/or reduced patient commitment to seeking aggressive treatment. This is especially problematic in treating patients in developing countries and with indigent populations in the United States and other developed countries. Given that these are the same populations with the highest prevalence of cervical disease, a point of care approach in new technology will be a significant advantage.

Spectroscopy as a diagnostic method and our approach:

Quantitative optical spectroscopy has the potential to improve upon the current technology in the following four major areas:

1. The potential for a truly noninvasive test. Spectroscopy can obviate the need to remove patient tissue.
2. The potential for providing results at the point of care. Since the tissue in the patient is interpreted using an algorithm at the point of care, follow-up consultation with the test results in hand is made possible in the same visit.

3. The potential for improved detection and diagnosis. As outlined in Section 13.1.2.4, a number of investigators have shown promising results in improved discrimination using fluorescence and reflectance spectroscopy.

4. The potential for a test that can be performed by a "nonspecialist." The performance of the device is compared to colposcopy and biopsy/pathology. Given the subjective nature of colposcopy and the benefits it derives from experience and training, maintaining a uniform standard of performance is difficult. LuViva would remove or alleviate this "subjectivity" if used in an adjunctive or triage mode.

Our approach is to combine fluorescence and reflectance (multimodal) spectroscopy with visible and UV light. Both spectroscopic techniques are used simultaneously and in an imaging mode for interrogating the ectocervix. Since the natural variation in the spectroscopic signatures of normal tissue from patient-to-patient is relatively large, it is not possible to assign an absolute spectral intensity or signature to a disease state. Often, this natural variation is higher than the variation seen in the spectroscopic signatures going from normal to diseased tissue in the same patient. Rather all measurements must be normalized or baselined to "normal" tissue in the same patient, and it is this relative change that has diagnostic relevance. Given the inability to determine "a priori" the location of abnormal and normal tissue with certainty, the logical alternative is to measure the entire cervix.

13.1.2.1 Fluorescence Measurement

A fluorescence measurement is made by measuring the intensity of light emitted from the tissue at a wavelength red-shifted (longer) from that of light used to irradiate the tissue. Fluorescence measures biochemical changes, i.e., the earliest changes that occur in the course of normal cells becoming malignant. Among the natural fluorophores present in tissue are tyrosine, phenylalanine, and tryptophan, the metabolites NAD(H) and FAD(H), and structural proteins collagen and elastin. The key endogenous fluorophores present in tissue and their optimum excitation and emission wavelengths are listed in Table 13.2. The fluorescence from these molecules depends upon their physiochemical environment including pH, solvation, and oxidation state. The action of various proteases, secreted by tumor cells on structural proteins, causes fluorescent amino acids (tryptophan, phenylalanine, etc.) to be exposed to a different local environment (different solvation, viscosity, and hydrophobicity), thus altering their fluorescence.

13.1.2.2 Reflectance Measurement

This measurement is made by measuring the intensity of light returned from the tissue at the same wavelength as that used to irradiate the tissue. Reflectance measures the morphological changes associated with cancer progression. Although biochemical changes precede the morphological changes that occur as a result of the former, varying degrees of morphological change accompany the biological changes.

TABLE 13.2

Optimum Excitation and Emission Wavelengths for Primary Endogenous Tissue Fluorophores

Fluorophores	Excitation Wavelength Range (nm)	Emission Wavelength Peak (nm)
Tryptophan	280–300	340
NADH	340–400	460
FAD	360–600	550
Collagen	340–360	400
Elastin	330–400	405
Porphyrins	350–450	640

Morphological changes appear later in the course of tumor progression and are defined as any change in cell nuclei, cell size, cell appearance, cell arrangement, and the presence of nonnative cells. In addition, effects of the host response such as increased perfusion from angiogenesis result in an overall difference in tissue appearance. Morphological changes distort the fluorescence measurement by scattering and absorbing both the excitation and fluorescent light, thereby altering the true fluorescence signal [18,19]. Thus, it is difficult to make a fluorescence measurement that is truly independent of the effects of scattering and absorbance.

13.1.2.3 Multimodal Spectroscopy

The interactive nature of the information gathered from fluorescence and reflectance modes makes it necessary to use both modes to correct for interferences from one mode to the other. For example, by knowing the absorption and scattering at each site on the tissue, the corresponding error that these effects produce in the fluorescence yield can be corrected for. In addition, as explained earlier, the information content of each mode is partly exclusive with fluorescence being sensitive to earlier biochemical changes and reflectance being sensitive to later morphological changes. Thus, by combining the two modes, a better measurement is made. In order to gain this advantage, however, both measurements must be made on the same site at preferably the same time so as to ensure identical conditions.

13.1.2.4 Summary of Previous Work on Spectroscopic Diagnosis of Cancer

As an epithelial cancer, cervical cancer provides an ideal target for diagnosis using visible light spectroscopy. The path that light energy must travel to fully penetrate the cervical epithelium down to the basal layer is short (100 μm–1 mm). The light undergoes minimum absorption and scattering from nonspecific interactions and obtains the largest possible diagnostic information on its biochemical and morphological state.

Fluorescence and reflectance spectroscopy have been shown to be valuable in diagnosing epithelial cancers by several investigators. A number of studies appear in the literature showing the performance of either fluorescence or reflectance spectroscopy in discriminating between normal tissue and different epithelial cancer grades in vivo. These include cervical [20–23], colon [24–27], GI tract [28], skin [29], and lung [30] cancers. All of these studies with the exception of reference 30 use point measurements of areas that are either suspect or normal, based on heuristic determinations. Reference [30] uses the image to compute lesion area and border fragmentation. For cervical cancer, a discrimination of about 90/70 or 80/80 in sensitivity/specificity (discriminating between LSIL and HSIL) is routinely obtained. The measurements made by Rebecca Richards-Kortum and her associates [20–23] were made at discrete normal or abnormal points on the cervix based on the visual feedback of a colposcopist. Xillix Inc. used blue light–excited fluorescence spectroscopy combined with white light visualization to discriminate lung cancer from normal lung tissue [31]. The FDA approved the Xillix PMA application in 1997 despite the low specificity (42%), given the importance of the higher sensitivity (75%) compared with white light bronchoscopy. Xillix has since been acquired by Novadaq Technologies (Mississauga, Canada). Later, the FDA approved a spectroscopic device by Spectrascience Inc. (Minneapolis, MN) for detecting colon cancer. This system, using fluorescence only has shown a high sensitivity and specificity (97% and 75%) in discriminating between neoplastic and nonneoplastic colonic polyps [31]. More recently (in 2006), the FDA approved the MediSpectra (Lexington, MA) LUMA™ device as an adjunct to colposcopy in diagnosing cervical cancer. This device uses a combination of fluorescence and reflectance spectroscopy, making it the first FDA-approved spectroscopic device for cervical cancer diagnosis. MediSpectra was acquired by Spectrascience in 2007.

There is clearly significant promise and ongoing activity toward using optical spectroscopy to improve upon the current technology for cancer detection. Recent FDA approvals of devices using quantitative visible spectroscopy in diagnosing cancers encourage the possibility of it being able to show similar improvements with cervical cancer detection. Our own work in this area shows promise in the ability to improve upon the discrimination that can be obtained with current methods. Section 13.3 summarizes the results we have obtained in clinical measurements using our LuViva device. Commercial success, however, requires cost-effectiveness in addition to discrimination performance and FDA clearance. The costs of the two FDA-cleared devices listed earlier (from Xillix Inc. and Spectrascience Inc.) are estimated to be between $50,000 and $100,000 per unit. The significance and innovation of our approach is in producing a commercially viable, multimodal, hyperspectral instrument with a cost of goods target under $7,000, resulting in a retail cost of less than $15,000. Thus, based on extensive marketing data (Section 13.2.2), the LuViva device will be designed to demonstrate not only technical feasibility but also market viability.

13.2 Methods

13.2.1 Product Development Process

We adhere to a product development process modified from the well-known stage–gate [32] process to make it suitable for the development of medical diagnostic devices in compliance with design controls set forth by the U.S. Food and Drug Administration (FDA) in 21CFR part 820:30 [33] and ISO-13485:2003. An example of such a process is illustrated in Figure 13.1. The process arranges all development activity in a phased or staged manner where each stage is followed by a checkpoint in which a decision is made to either (a) continue as planned, (b) stop a project, or (c) modify the plan for business reasons. Our modifications to this approach allow for the very specific and unique activities related to medical devices, viz., conduct of clinical studies and the regulatory approval. Successful completion of each stage involves assembly and approval of a set of prescribed documents specific to the stage. Although the tasks contemplated in each stage must permit a go/no-go/alter decision at each gate, some of the tasks may extend into future stages as necessary.

13.2.2 Stage 1: Market, Technology, and Reimbursement Assessment

As part of our stage 1 diligence, we conducted studies to assess the market opportunity in the area of cervical cancer diagnosis and the suitability of optical spectroscopy in addressing this opportunity. An integral part of this was to determine the status of reimbursement in cervical cancer management. The information gathered from these studies has had a significant impact on our strategy for positioning the device as a tool to triage women who screen positive on Pap smears or other tests such as hybrid capture 2 HPV, where the LuViva test could assist in identifying women at risk for CIN2+ disease or higher. Table 13.3 summarizes the studies and how they have affected the program.

13.2.2.1 Human Factors Study to Determine Shape, Size, Weight, and Measurement Particulars for Preproduction Device

Prior to arriving at our current product concept, we conducted a human factors study in order to ascertain physician comfort/usability as well as patient comfort/tolerance of our proposed device at the following two clinical facilities:

- Institute for Women's Health at Jackson Memorial Hospital, University of Miami, Miami, FL
- Emory University School of Medicine, Atlanta, GA

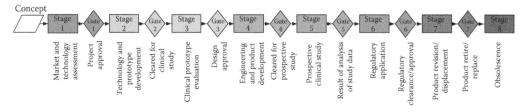

FIGURE 13.1   Example product development process using stage–gate method.

TABLE 13.3

Summary of Market Analyses in Terms of Impact upon Product Positioning and Design

Study/Date	Impact on Product Positioning and Device Design
Boston Healthcare Associates Positioning Strategy: Findings from Physician and Payer Research	Determined that ASCUS+ triage was optimal product position for initial launch, followed by expansion into primary screening 62% of physicians preferred a portable device 28% of physicians prefer a device that makes *no contact* with the cervix, 52% had no preference, and 8% preferred a device making *contact*
Boston Healthcare Associates Global Opportunity Assessment: Mexico, Brazil, Germany, France, and Italy	Improved our understanding of the international market and helped solidify the decision to target ASCUS+ triage first, followed by primary screening
Interviews with 15 physicians at ASCCP[a] meeting 10/2000	Confirmed some of our assumptions about the ASCUS triage market, and caused us to look more closely at whether or not an endocervical measurement capability would be required for the triage and primary screening markets
Telephone survey of 51 U.S. OB/GYNs 12/2000	Showed that accuracy and cost are the most important features. Test time is less critical. The measurement time with the preproduction device is intended to be 3 min Confirmed importance of endocervical probe for primary screening
Vincent McCabe Inc. Telephone survey of 100 U.S. OB/GYNs for ASCUS+ triage product	Confirmed marketability of an ASCUS+ triage product without an endocervical probe at initial launch Confirmed price point of $10,000–$19,000 for the instrument and $35 for the disposable. Showed that lower prices would lead to higher adoption (as expected) Confirmed that 4 min measurement sequence was acceptable and showed that difference between 3 and 5 min test time was not significant in terms of market acceptance Showed that immediate results, improved accuracy, and cervical analysis (localization) were the most important features
Reimbursement study	Showed that by adding colposcopic measurement capabilities (magnified image of the cervix), we would be able to obtain immediate reimbursement use colposcopy CPT code. This reinforced our decision to add a colposcopic imaging channel for ease-of-use reasons

[a] ASCCP, American Society of Colposcopy and Cervical Pathology. http://www.asccp.org/

In phase 1 of this study, 20 subjects of various ages ranging from 22 to 62 years were tested using a foam design model of the handheld unit (HHU, Figure 13.2) that could be weighted with three different weights (3, 4, and 5 lb). Ten different individuals (physicians and assistants) were asked to use the device. In phase 2 of the study, 20 subjects over the age of 18 were tested with a prototype HHU, weighing 4.5 lb, intended for the final device.

As expected, we found that the lower weights were easier for the physicians to handle and manipulate. Also, shorter time periods were easier on the physician with no physicians reporting fatigue in less than 2 min. More than half the subjects reported no discomfort, and all but one subject tolerated the probe. Image quality was good for 35/40 of the subjects, meaning that the cervix was in focus and the normal features of the cervix were clearly visible. We also found that greater than 85% of the measurements had user/subject movement of <2 mm. In conclusion, the study showed that a 3–5 lb instrument, with the shape shown in Figure 13.2, was acceptable for tests up to 5 min, and a user motion on the order of about 2 mm over a 5 min period is tolerable.

13.2.3 Stage 2: Technology and Prototype Development

In this stage, we developed a research prototype device for feasibility testing of our technology and approach. Our assessment of the clinical and market requirements as outlined in Section 13.1 pointed us to a number of key design requirements. For example, an accurate and comprehensive test requires

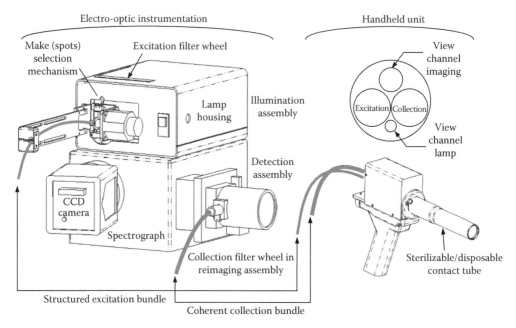

FIGURE 13.2 Simplified schematic of the research system showing key component details.

measuring the entire ectocervical tissue surface. In addition, our human factors study results highlighted the need for a compact and light easy-to-use patient interface and a fast (less than 5 min) test. Finally, our market research limited our product cost target to less than $20,000 in order to gain adoption. We designed our research device with these constraints in mind. In order to delay the cost of customizing the device until after feasibility was established, we selected off-the-shelf, scientific-grade components while making sure that future replacement with custom alternatives was possible.

Our research device, illustrated in Figure 13.2, consisted of two parts: (a) A handheld patient interface that is electrically and optically connected to (b) the electro-optic instrumentation located on a movable cart. The HHU looks like a hair dryer and has a removable snout called the "contact tube." The contact tube is designed, in conjunction with a vaginal speculum, for placement in the patient's vagina during the examination procedure. The contact tube is reusable after cleaning and steam sterilization and was made of a black-coated stainless steel material. Prior to subject measurement, the instrument is calibrated by making measurements on fluorescent and reflective calibration targets.

The cervix is illuminated with white light during probe placement and viewed through a video-imaging camera located in the HHU. This camera provides a "video view" of the cervix on a monitor and assists the physician in properly aligning and positioning the contact tube. The contact tube makes circumferential contact with the periphery of the cervix. The spectroscopic interrogation is done in a standoff manner on the area encircled by the contact tube. After the contact tube is satisfactorily positioned, the video camera is used to capture a still image of the cervix, which is digitally stored and used for later data analysis. The video lamp is then automatically turned off, and the spectroscopic measurement started. The device uses the line scan approach [34] to gather data from all interrogated points. In this method, a line or column of points is illuminated at any given time, and the returned radiation from the tissue is collected using a coherent fiber bundle. The column of light is transferred through the coherent fiber bundle and acts as a virtual slit at the entrance of the spectrograph used to spectrally resolve the light. Given the spectral resolution required and the light dispersion by the spectrograph, only one column can be measured at a given time. The system sequentially scans through the cervix, acquiring both fluorescence and reflectance spectra in a total time duration of 4 min.

Among the key device components are a 300 W short-arc Xe lamp (Perkin Elmer, Azusa, CA) with an integrated parabolic reflector that provided the illumination for spectroscopy; an imaging spectrograph SpectraPro SP-306I (Acton Research Corporation, Acton MA) with F#4 optics, a 300 mm focal length, and a 40 lines/mm grating; and a thermoelectrically cooled CCD camera NTE/CCD-SITE 512SB (Roper Scientific, Princeton, NJ) with a 512×512, square format, 24 µm pixel, back-illuminated detector. The coherent image transfer fiber bundle was built by Schott Fiber Optics (Elmsford, NY). This is a 2 m long bundle with 0.43 NA, 10 µm fibers=element arranged in a 6×6 mm square aperture in a coherent fashion to provide a one-to-one image transfer from the HHU to the spectrograph. For device calibration, we used a reusable reference target: 99% Spectralon (Labsphere Inc., North Sutton, NH). The device was controlled using a software/hardware package that consists of instrument control, graphical user interface, and data storage capabilities. The hardware portion is a compact PC with adequate ports and bays to accommodate the requisite interfaces and PCI cards, external controllers, and power supplies. The software (written in Microsoft Visual Basic 6.0), in addition to instrument control, also provides graphical feedback to the user showing images (video and spectroscopy) that were used to make real-time determinations of measurement adequacy. The measured data, which are stored by the program, were then downloaded onto recordable CDs.

13.2.4 Stage 3: Clinical Prototype Evaluation and Algorithm Training

We tested our research device at multiple clinical facilities in the United States using IRB-approved protocols. The list of clinical sites is provided in Table 13.4. The data gathered in this stage of our program helped us validate our approach and verify clinician and patient acceptance of human factors–related choices made in our device design. It also helped investigate factors that influence device discrimination performance such as spectral and spatial resolution, signal-to-noise ratio, the impact of confounders such as blood, mucus, and movement artifact as well as physiological factors such as age and menopausal state. The data were used to design and refine quality assurance procedures, calibration, data reduction, and preprocessing. Finally, we used the data for algorithm development and retrospective testing.

13.2.4.1 Clinical Safety Testing

We tested the device to establish safety for clinical use of our device prior to using in an investigational study. The device was appropriately labeled "Research device-Investigational use only" and similarly noted in the study protocol and instructions for use. Safety testing consisted of the following:

- Biocompatibility: We tested the contact tube which served as the patient interface for biocompatibility in accordance with ISO 10993. Specifically, the black coating on the stainless steel contact tube was tested for cytotoxicity, sensitization, and intracutaneous reactivity, given that the tube is in contact with a mucosal membrane during the measurement procedure.
- Sterilization: Given the reusable nature of the contact tube, we relied on the internal quality procedures of the clinical institution for cleaning and steam sterilization for reuse.
- Electrical safety: We tested the device for ground continuity and electrical leakage in accordance with the IEC60601 standard to assure that there is no risk of electrical shock prior to clinical use.

13.2.4.2 Clinical Procedure

Table 13.5 shows the subject measurement procedure used in our clinical study with the research device. We enrolled women who were referred to colposcopy per current American Society for Colposcopy and Cervical Pathology (ASCCP) guidelines [35] because of a suspicious Pap test result (ASCUS, ASC-H, HPV+, AGC) and/or follow-up for recent previous dysplasia. The overall time involved is approximately 30 min (including patient consent). The spectroscopy portion alone takes approximately 4 min

TABLE 13.4

Subject Enrollment for Algorithm Training

Clinical Site	Principal Investigator	Enrolled	Evaluable	Benign (Normal)	Dysplasia
Medical College of Georgia	Daron Ferris, MD	273	232	137	95
Emory University School of Medicine	Lisa Flowers, MD	45	21	4	17
University of Miami School of Medicine	Leo Twiggs, MD	175	131	46	85
St. Francis Hospital (Hartford)	M. Lashgari, MD	136	126	39	87
Total		629	510	228	284

to complete. Following the measurement procedure, the clinical team completed the case report forms (CRFs) to record patient details in an unidentifiable manner and per GCP guidelines [36] such as age, menopausal status, colposcopic findings, cytology, and, when available, pathology results.

13.2.4.2.1 Pathology QA

We established a consensus pathology procedure wherein a representative section of each biopsy taken from a subject was sent to both the clinical site pathologist and a QA 1 pathologist. Agreement between the two pathologists was recognized when both categorized the biopsy sample as

- Normal tissue (no pathologic findings) or benign changes (including metaplasia, inflammation, reactive atypia, HPV changes not associated with dysplasia, bacterial infections, and other benign changes)
- CIN1
- CIN2
- CIN3+ (CIN3, focal microinvasive carcinoma or invasive carcinoma)

If they disagreed, then a representative section of all biopsy samples from the subject were sent to a QA 2 pathologist. Two agreements on any category in the earlier text were considered a concordant or valid diagnosis. In the case of multiple concordant biopsies from a subject, the highest (most severe) concordant biopsy diagnosis was assigned to the subject. Three-way disagreements were considered discordant. If a discordant biopsy had a CIN2 or higher diagnosis by any pathologist, then the subject from whom the biopsy was taken was considered discordant. Our pathologists were

QA Pathologist 1. Dr. Edward Wilkinson. University of Florida Diagnostic Referral Laboratories, Gainesville, FL

QA Pathologist 2. Dr. Steven Raab. Allegheny General Hospital, Pittsburgh, PA

13.2.4.3 Subject Data Accrual

We enrolled women who were referred to colposcopy per current ASCCP guidelines [36] because of a suspicious Pap test result (ASCUS, ASC-H, HPV+, AGC) and/or follow-up for recent previous dysplasia. We enrolled subjects at the institutions listed in Table 13.5 using IRB-approved protocols. All subjects enrolled were referred to colposcopy because of an abnormal Pap result and in some cases because of a positive HPV test result or as for follow-up from previous disease.

A total of 648 consecutive women met inclusion criteria and consented to participate in the study. Data could not be collected from 19 subjects because they withdrew before the study could be completed. All of the remaining 629 study participants underwent colposcopic examination. Of these, 42 subjects lacked either a Pap test result or valid histopathology and 15 cases could not be analyzed because of a device or

TABLE 13.5

Subject Measurement Procedure

Patient orientation/consent

Pregnancy test, device calibration

Preanalytical preparation

- Insert speculum
- Clean cervix (by gently removing excess mucus) thoroughly with sterile water

LuViva measurements

Mount a fresh (or sterile) contact tube onto the research HHU

- Insert the HHU while viewing the video feedback to ascertain positioning and concentration on the cervix
- Record first image with video camera in research device
- Spectroscopy
- Record second image with video camera in research device
- Remove handheld device

Pap smear: The pap smear is read at/or contracted by the clinical site

Postanalytical

- Acetowhitening: Apply acetic acid on the cervix with cotton tipped swab
- Reinsert research HHU and record third image with video camera in research device
- Remove handheld device
- Discard contact tube and clean HHU per protocol

Colposcopy

- Identify biopsy sites, including a normal site within transformation zone
- Perform biopsies
- Mark biopsy sites and colposcopic impression on the post–acetic acid image

---

operator error. Further, 62 subjects had some type of measurement artifact that when analyzed separately were found to have varying impact upon performance. This left a remainder of 510 evaluable subjects.

In addition to the data accrual for algorithm training, we enrolled an additional 302 subjects at the University of Texas—Southwest Medical Center, Emory University (Grady Hospital) University, and Medical College of Georgia, University of Miami and Orange Coast Women's Center. Data from these subjects was used to help select and cross-validate the algorithm. Both research and preproduction device (see Section 13.2.5) versions were used in this study. This population included subjects tested only with the research device ($n = 141$), the preproduction device ($n = 115$), and both devices sequentially ($n = 46$).

13.2.5 Stage 4: Engineering and Product Development

Following successful completion of stage 3, we adopted a parallel path toward product development that included both stage 4 "Engineering and Product Development" and stage 5 "Prospective Clinical Study." The availability of a trained algorithm permitted starting the prospective study. Our power calculations based upon the variability seen in the algorithm training data found that we would require approximately 2000 subjects in order to meet the confidence limits on our algorithm performance targets in the prospective study. This implied a lengthy study period and consequently the need to start it early. Further, since the algorithm was trained on the research device, it became necessary to have data in the prospective study from the same device version in addition to data from the final preproduction version that would result from the engineering and product development effort in stage 4.

Figure 13.3a shows a block diagram of our research prototype. The base unit of this device (detailed in Section 13.2.3) contains the excitation and detection subsystems. The detection subsystem contains the spectrometer and camera, the cost and performance drivers of this device. They account for 65% of the overall device cost of the research prototype. Without a priori knowledge of the specific requirements of these two components, we chose a scientific-grade state-of-the-art camera and spectrograph for the research prototype placing cost and size at a lower priority and deferring any effort to reduce both to

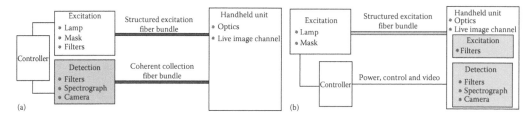

FIGURE 13.3   Transition from LuViva research (a) to preproduction (b) device. Components relocated are shown in gray.

when specific requirements were known. Clinical evaluation of the research device helped determine performance factors such as spectral resolution, device optical throughput, and signal-to-noise ratio. Knowing these allowed us to set minimum specifications for a custom detection subsystem. We designed and built a size- and cost-reduced integrated camera–spectrograph allowing us to move it into the HHU (Figure 13.3b). As a result of this, we will be able to dispense with the coherent imaging bundle and the associated cost since we no longer need an image transfer mechanism. An immediate advantage of doing this is that our system throughput will be increased twofold since the coherent bundle has a transmittance of 50%. This also helped reduce the cost share of the camera and spectrograph to 40% of the overall device cost. We designed custom excitation and collection optics to replace the catalog optics used in the research device. This helped improve optics performance, reduced aberration, and increased throughput by reducing vignetting.

13.2.5.1 Custom "Combined Camera–Spectrograph" Detection Module

This custom module has a concentric spectrograph design [17,37] using a convex aberration-corrected holographic grating and a concave mirror that focuses the spectrally dispersed light onto an integrated SONY ICX285AL sensor. The spectrograph has a numerical aperture of 0.12 and unity magnification (Figure 13.4).

13.2.5.2 Custom Excitation and Collection Optics

We designed custom optics using UV transmissive glasses for focusing light onto the cervix and for imaging reflectance and fluorescence light received from the cervical tissue. The optics have a ½ in. clear aperture and are placed alongside each other with a 5 mutual angle in confocal arrangement. The excitation optics has a 5× nominal magnification which allowed light from a 100 μm fiber to be imaged

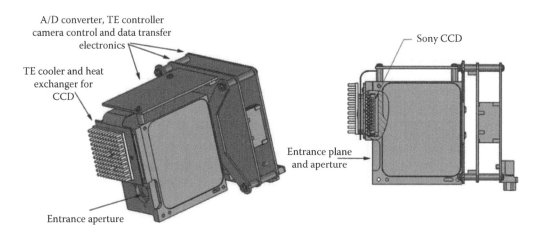

FIGURE 13.4   Custom combined camera–spectrograph module. Size 3.5×3.5×4.5 in.

as a 500 μm spot on the tissue. The collection optics was based on the Cooke triplet [39] design with a 5× demagnification allowing the 500 μm spots to be imaged back into 100 μm spots on the CCD. The optics has a depth of focus/field of ±5 mm. The custom design helped reduce aberrations and losses from vignetting due to the limited clear apertures resulting from space constraints in placing the optics in the HHU.

13.2.5.3 Custom Lamp Assembly

We designed a custom lamp assembly with mask (for scanning the spots on the cervix) using a 75 W short-arc, ozone-free, xenon lamp (L2194-11, Hamamatsu, Bridgewater, NJ); an elliptical reflector (f# 4.5, Photon Technologies, Trenton NJ); and a custom-built lamp housing. We chose this lamp for its longevity (2000 h) and stability (less than 5% drift in output) over its life. Although this lamp uses a fourth of the electrical wattage of the lamp used in the research system (a significant cost and power advantage), by virtue of its short, 1.0 mm, arc, it is able to couple about half the light energy per fiber compared to that measured in the research device. The lamp output is filtered by a WG295 filter to block all UV radiation below 295 nm (primarily UVC). This light is then reflected off a custom cold mirror that effectively removes all light above 800 nm from entering the excitation fiber optics. A software-actuated motorized safety shutter is placed between the lamp and the fiber optic bundle as shown in Figure 13.5. Although the lamp operates continuously, illumination is allowed into the system and through to the test subject only for the duration of the spectroscopic measurements and video collection by a safety shutter which is normally closed and designed to fail in this position. The lamp output beam is focused sequentially onto illumination fibers which in turn illuminate the spots measured on the cervix.

13.2.5.4 Disposable Contact Tube and Calibration Target

Our requirement for a disposable contact tube was driven by our business model as well as the need to ensure biosafety in our product. We chose a black CYCOLAC MG47MD-NA1000 plastic (SABIC, Selkirk, NY) as the material for this and had a multicavity tool built for rapid production molding of our contact tubes. These tubes were used in conjunction with a removable and disposable "paper" calibration

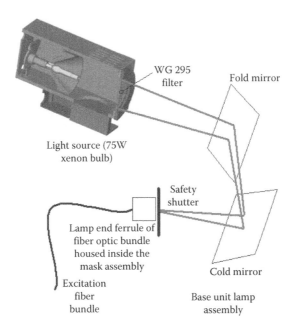

FIGURE 13.5 Lamp and mask assembly provides illumination for both spectroscopy and video imaging.

target. We selected and qualified this paper calibrant to replace the reusable Spectralon target used with the research device. The contact tube and calibrant were tested for biocompatibility as described in Section 13.2.4.1. The tube and calibrant assembly are intended for use clean albeit nonsterile and with reduced bioburden by gamma irradiation.

13.2.6 Stage 5: Prospective Pivotal Clinical Study

The results of our algorithm training and selection studies helped devise our pivotal study plan and to reach consensus with the FDA on the intended use of the LuViva as well as final claims.

13.2.6.1 Intended Use

The LuViva device is intended for use by a trained healthcare professional, (e.g., physician, nurse practitioner, physician's assistant) to test prior to colposcopy, women who have an abnormal or equivocal cytology result, a positive HPV test, or other risk factors per standard of care to identify those women with higher likelihood of high-grade dysplasia. LuViva is expected to provide clinical value if used as a triage step prior to colposcopy for the over 5.5 million expected tests per year in the United States.

13.2.6.2 Device Claims

Our primary claim is to demonstrate prospectively that the LuViva test, when positive, can be used to triage patients for earlier or more definitive care. When negative, the test result would allow safely returning the patient to routine screening. In order to support these claims, a sensitivity of approximately 93%–94% for CIN2+ cases and a specificity of 37%–39% are targeted, with lower confidence of 90% and 35%, respectively. In our intended use population, the prevalence of CIN2+ disease is expected to be approximately 20%. Therefore, the positive predictive value would be greater than 25% and the negative predictive value would be greater than 93%.

13.2.6.3 Sample Size Calculations for the Primary Claim

The pivotal study was conducted using both the research prototype and the preproduction device. All data gathered were quarantined with the exception of quality checks to ensure proper device functioning. The data quarantine is in accordance with a data quarantine plan previously submitted to the FDA. We estimated sample size for the pivotal study using NCSS PASS based on our expected sensitivity and specificity estimates.

The sensitivity of the LuViva for CIN2+ lesions in the ASCUS/LSIL Pap category is expected to be approximately 93%–94% with a lower bound of 90%. The minimum number of subjects with CIN2+ disease needed for this analysis is 165–213. (e.g., if a sensitivity of the LuViva is 93.5% with a lower bound of 90%, then 198 out of 213 subjects with CIN2+ disease would need to be identified as positive by the LuViva. The same lower bound of 90% can also be achieved with 94% sensitivity or 155 of 165 subjects with CIN2+ identified as positive by the LuViva.)

In both cases, this would achieve at least 80% power and a significance level (alpha) of 0.05. At 93.5% sensitivity, the specificity for subjects with benign cervices is expected to be 39% with a lower bound of 35%. The sample size required for this is 414 subjects. (Alternatively, if 1031 benign cases are enrolled, then specificity in the pivotal trial would need to be 37.5% to achieve the targeted lower bound of 35%.) In all cases, the sample sizes will be adequate to achieve 80% power with a significance level (alpha) of 0.05 (Table 13.6).

13.2.6.4 Sample Size Calculations for Equivalence and Repeatability Study

In addition to the main pivotal study, we conducted additional studies to evaluate device equivalence and repeatability. Given our use of two device versions in the pivotal study, the research device and the preproduction device, it was necessary to determine performance equivalence between the two.

TABLE 13.6

Estimated Accrual for the Pivotal Study

Subgroup	Estimated Prevalence of CIN2+	Target Sensitivity (Lower Bound)	Target Specificity (Lower Bound)	Number of CIN2+ Cases Required	Number of Benign Cases Required	Total Cases
ASC, LSIL	20%	93.5%–94% (90%)	37.5%–39% (35%)	165–213	414–1031	1600–1650

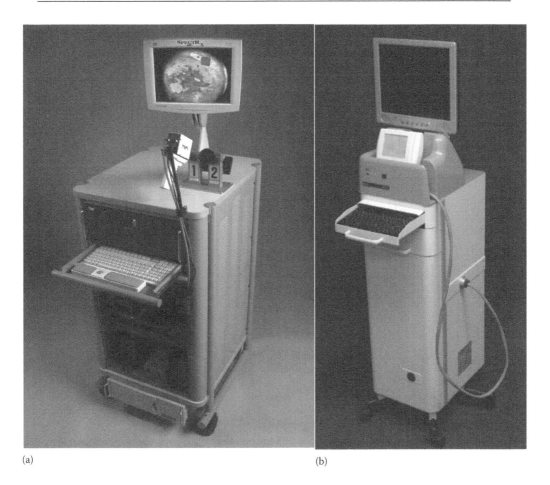

(a)   (b)

FIGURE 13.6 The CNDS research device (a) and preproduction device (b).

We had the same investigator measure the same subject using both device versions in this study. For the repeatability study, we made back-to-back measurements on the same subject with one copy of the preproduction device while changing users/operators. Both studies enrolled a total of 80 subjects each, and the data can be pooled with the main pivotal study to satisfy minimal accrual requirements.

13.2.6.5 Pivotal Study Data Accrual

We began our pivotal study in 2004 under an FDA-reviewed protocol and communicated the completion of subject enrollment to FDA on September 25, 2008. During this study period, we quarantined all data until study completion, with the exception of applying quality check routines to ensure data quality. A data quarantine plan was submitted to the FDA. The institutions involved in our study under

TABLE 13.7

Institutions That Participated in the Pivotal Study and the Number of Subjects Enrolled at Each

	Institution/PI	Enrollment
1	The University of Texas—Southwest Medical Center (Dr. Claudia Werner, PI)	234
2	Emory University School of Medicine, Atlanta, GA (Dr. Lisa Flowers, PI)	348
3	Medical College of Georgia, Augusta, GA (Dr. Daron Ferris, PI)	130
4	Institute for Women's Health at Jackson Memorial Hospital, University of Miami, Miami, FL (Dr. Leo Twiggs, PI)	313
5	Orange Coast/Saddleback Women's Medical Group, Laguna Hills, CA (Dr. Marc Winter, PI)	140
6	St. Francis Hospital, Hartford, CT (Dr. Manocher Lashgari, PI)	394
7	University of Arkansas for Medical Sciences, Little Rock, AK (Dr. Alexander Burnett, MD)	48
	Total	1607

TABLE 13.8

Demographics of Subjects Enrolled in the Pivotal Study

Age			Race			Ethnicity		
Category	Number	Percent	Category	Number	Percent	Category	Number	Percent
Median	27.00		African American	882	54.9	Hispanic	466	29.0
Range	16–84		White	702	43.7	Non-Hispanic	1141	71.0
Less than age 30	929	57.8	Asian American	16	1.0			
Age 30 and older	678	42.2	American Indian	3	0.2			
			Pacific Islander	4	0.2			

IRB-approved protocols and the numbers of subjects accrued are listed in Table 13.7 and the subject demographics in Table 13.8.

All CRFs were monitored and retrieved from all clinical sites, and double data entered into our clinical database. This includes all cytology and pathology results. No severe or unexpected adverse events occurred during this clinical trial.

One major change in protocol from that used in the algorithm training study was the addition of HPV testing. In addition, FDA has asked to obtain 2 year follow-up data where available on all women who participated in our study, given that 2 year follow-up was used in the ALTS trial [40] as the benchmark for definitive diagnosis.

13.3 Results

13.3.1 Algorithm Development

Data from the algorithm training study (Section 13.2.4) were used for algorithm development. The measurements were calibrated, normalized, reduced (image binning), and filtered for artifacts. This resulted in an intensity data matrix for each subject with spectral parameters in one dimension and the points measured in the other. We found, in general, that the partial least squares (PLS) models gave the best performance, especially with regard to its ability to maintain high sensitivity with the least shrinkage in performance upon cross-validation. Other approaches included: PLS analysis, classification and regression trees (CART), artificial neural nets (ANNs), support vector machines (SVMs), nonlinear CART, and back-propagation and radial basis ANNs. Per discussions with the FDA, algorithmic models were developed with and without Pap test results as an input parameter along with the spectroscopic measurements. We have since focused primarily upon using PLS as our method of choice in algorithm development. We have found that our algorithm performed poorly on those subjects that were found normal upon colposcopic examination but not biopsied to confirm diagnosis. (This is important considering that all subjects included in our study were either found to have a suspicious Pap result or were being followed up due to previous disease.) When we removed these subjects from our training data set of 510 subjects, a total of 451 subjects remained. All candidates for our final algorithm were trained on this subject set.

We identified a set of artifacts that can potentially reduce algorithm performance: excessive blood, mucus, specular light, movement artifact, and ambient light. We designed and applied algorithmic filters for these artifacts as part of our data preprocessing. Ambient light was found to have a significant effect on performance, and so ambient light detection monitoring has been added in our preproduction device.

13.3.1.1 Algorithm/Device Output

The clinically useful threshold for determining which women require treatment is widely accepted to be at CIN2. This is because lower pathology grades such as CIN1 have a high spontaneous regression to normality [40]. Our discussions with the FDA have confirmed this to be the appropriate threshold for a binary yes/no device output. Using the PLS approach, we built several algorithms. Each algorithmic selection was assigned a binary response variable, based on consensus histopathology,* at threshold CIN2. These selections varied from all spectral parameters (wavelengths) measured to a limited number of parameters depending upon the following considerations:

- They individually discriminate patients with CIN2+ disease from those with benign cervices.
- They provide, to varying degrees, independent information.
- They avoid spectral zones that may be relatively unstable (e.g., parts of the spectrum on a steep slope).
- They are consistent with known biological processes (e.g., blood absorption peaks, fluorescence peaks of target fluorophores) and previous reports in the literature.

The objective in limiting the number of parameters was to increase robustness of the algorithm. Further, we tested the algorithms both with and without the "day of study" Pap result as an additional coded parameter in the algorithm.

13.3.2 Final Algorithm Selection

We used the data from the algorithm selection study for this. Among the criteria used in this selection were the ability to achieve high sensitivity, uniform performance across data from different sites (and hence population demographics), and least performance shrinkage between training and selection sets (at the cost of overall performance). We reported our final algorithm selection to the FDA. The performance of the algorithm with and without the Pap test as well the performance of the Pap test by itself are shown in Figure 13.7 as receiver operating characteristic (ROC) curves, i.e., no curve fitting

---

* We used the same method as in the algorithm training study, Section 13.2.4.2.1, whereby final pathology is the consensus between the site pathologist and QA 1 pathologist both reviewing a representative slide. In the absence of consensus, a representative slide is sent to a third QA 2 pathologist, and agreement between any two of the three is chosen as the final diagnosis. In the absence of an agreement, the diagnosis is considered discordant. Approximately 3% of total enrolled cases had ambiguous histopathology results and therefore are not included in this analysis.

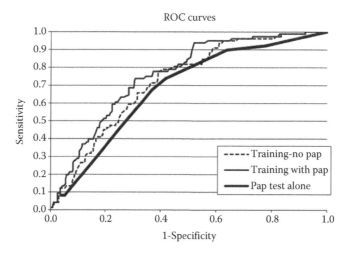

FIGURE 13.7   Raw ROC curves showing performance of algorithm on the 451 subject training data set. Curves show the algorithm performance with the Pap code input and without. Also shown is the performance of the Pap test by itself on the same subject cohort.

TABLE 13.9

Algorithm Performance Measured as AUC for the Training Set

	Training AUC
Algo 2.61 without Pap	0.715
Algo 2.61 with Pap (see Section 13.3.2)	0.755
Pap test by itself	0.675

or smoothing has been applied. The algorithm produces a numerical index for each subject which is compared with numerically coded consensus histopathology. Subjects with a consensus histopathology diagnosis of CIN2 or higher are considered positive, and those with a CIN1, benign, or normal diagnosis are considered negative. The area under the ROC curve (AUC) performance for the algorithm training is shown in Table 13.9.

13.3.3 Pivotal Study Results

We present the primary study results in this section. These results were presented at the American Society of Colposcopy and Cervical Pathology biennial meeting (Las Vegas) in May 2010. Of the 1607 women tested, all were used to evaluate safety and acceptability of the procedure while 160 were excluded from efficacy analysis for the following reasons: 54 women were used for training each of the 19 physicians and 7 nurses that conducted the tests, 33 had no or insufficient histopathology results and are pending follow-up, 37 had a three-way discordant histopathology result, 1 had an insufficient referral Pap result, 10 were excluded due to user errors, and the device malfunctioned in 25 cases and did not produce coherent spectroscopic data. A histopathology review procedure of the 1447 evaluable women involving two independent pathologists in addition to each site pathologist rendered diagnoses that included 278 women with cervical intraepithelial neoplasia CIN2+, 600 with CIN1, and 569 with no CIN (yield of positive biopsies = 19%). Table 13.10 summarizes the cytology and histology results for these subjects. The final diagnoses included data on 742 women who returned for follow-up visits within 2 years following the completion of our study and tested in accordance with the American Society for Colposcopy and Cervical Pathology (ASCCP) guidelines (please see Section 13.3.4).

TABLE 13.10

Number and Prevalence of Final QA Histopathology
as a Function of Reason for Referral to Colposcopy

Referral Cytology Result	Normal	CIN1	CIN2+	Total
AGC	6	2	1	9
AGUS	10	5	0	15
ASC-H	30	19	12	61
ASCUS	205	185	51	441
HSIL	8	24	85	117
LSIL	232	309	127	668
Negative	78	56	2	136
Total	**569**	**600**	**278**	**1447**

Excludes no or insufficient histopathology, three-way discordant histopathology ($n=74$).

13.3.3.1 Sensitivity and Specificity Estimates

The primary metrics used to evaluate efficacy of the LuViva technology are the AUC (area under the receiver operating characteristic "ROC" curve) and resulting sensitivity and specificity at a clinically important threshold as recommended by FDA medical reviewers. No adverse events were reported during or after the trial. When all 1447 evaluable cases are considered, sensitivity of LuViva was 91.0% (253/278) for women with CIN2+ lesions. Specificity for women with normal or benign histology was 38.8% (221/569), and specificity for women with CIN1 histology was 30.5% (183/600). Pooling the normal/benign cases with the CIN1 cases results in a specificity of 34.6% (404/1169).

In comparison, sensitivity and specificity of referral Pap were 72.5% and 46.1%, respectively. HSIL Pap as expected was found to be highly predictive of actual dysplasia (93.2% of HSIL cases were diagnosed histologically as CIN1 or worse and 72.6% were CIN2+). In contrast, all other referral cytology categories had prevalences below 66% for all dysplasia (CIN1+) in general and below 20% for CIN2+ in particular. To this point, Table 13.11 also shows the performance of LuViva when women with HSIL Pap cytology are excluded from analysis and that sensitivity, specificity, and predictive values do not change very much when women with HSIL are excluded from analysis. The sensitivity of LuViva when excluding HSIL Paps is 87% (168/193), and specificity pooling normal/benign with CIN1 cases is 35.5% (404/1137).

13.3.4 Follow-Up Study

The ALTS study [17] commissioned by the National Cancer Institute to evaluate the performance of alternative management strategies (including colposcopy) for ASCUS and LSIL Paps used 2 year

TABLE 13.11

Sensitivity/Specificity and Predictive Values (PPV and NPV) for PEP Arm Only, with and without HSIL Paps ($n=1447$)

All Sites and Devices	% Sensitivity CIN2+ (95% CI)	% Specificity Normal (95% CI)	% Specificity CIN1 (95% CI)	% PPV	% NPV
All Pap categories ($n=1447$)	91.0 (87.6–94.4)	38.8 (34.8–42.8)	30.5 (26.8–34.2)	25.8	93.9
All Pap categories except HSIL Paps ($n=1330$)	87.0 (82.3–91.7)	39.4 (35.4–43.4)	31.8 (28.0–35.6)	25.2	91.6

follow-up of patients for reducing verification error in the presence of disease. We conducted a similar follow-up evaluation of the subjects enrolled in the pivotal study. Approximately half (804) of the enrolled women returned for follow-up visits based on ASCCP guidelines as supervised by the PI at each site. A total of 742 of these women were evaluable, and 134 of these were diagnosed with CIN2+, including reclassification due to follow-up histology. If follow-up histology indicated that a subject had a more severe disease classification than the diagnosis made at the time of study enrollment, then the subject's final histology result was reclassified to that disease category (i.e., either CIN1 or CIN2+). A subject's final histology result was never reclassified to a less severe disease state based on 2 year follow-up data.

In all 31 cases initially diagnosed at the time of our study to be negative were reclassified as CIN2+ in follow-up. In 28 of these 31 cases, LuViva was positive for CIN2+, for a sensitivity of 90.3%. Table 13.12 lists these subjects by clinical site.

Of the 134 CIN2+ cases in this population, the standard of care detected 88 of these for a sensitivity of 65.7% while LuViva detected 121 of these for a sensitivity of 90.3% (Table 13.13). The 37.4% increase in sensitivity shown by LuViva over the standard of care is statistically significant using McNemar's test ($p < 0.0001$). Excluding HSIL cytology, the standard of care detected 48 of 83 CIN2+ (57.8%) while LuViva detected 70 of 83 CIN2+ (84.3%), an improvement of 45.8% ($p < 0.0001$). In addition, specificity for the group of subjects with follow-up data was slightly higher than in the pivotal study population as a whole. Specifically, for histologically normal women, LuViva specificity was 40.1% (112/279) and for women with CIN1, it was 35.9% (118/329). This most likely was due to the fact that without follow-up data, many of LuViva's true positive results would have been erroneously classified as false positives with a net effect of lower specificity.

TABLE 13.12

Follow-Up II: Subjects Reclassified as CIN2+ Based on up to 2 Year Follow-Up Histopathology

Clinical Site	Number of Subjects with CIN2+ at Follow-Up	Number Detected by Light Touch	LuViva Sensitivity (%)
The University of Texas—Southwest Medical Center	5	4	80.0
Emory University School of Medicine, Atlanta, GA	9	8	90.0
Institute for Women's Health at Jackson Memorial Hospital, University of Miami, Miami, FL	10	10	100.0
St. Francis Hospital, Hartford, CT	2	2	100.0
Medical College of Georgia, Augusta, GA	1	1	100.0
Orange Coast/Saddleback Women's Medical Group, Laguna Hills, CA	4	3	75.0
Total	31	28	90.3

TABLE 13.13

Sensitivity for Standard of Care and LuViva for 742 Subjects with Both Histopathology Review and up to 2 Year Follow-Up Data

Follow-Up Procedure	Standard of Care[a] (95% CI)	LuViva (95% CI)
Sensitivity	65.7% (88/134)	90.3% (121/134)

[a] Includes Pap cytology, HPV, and colposcopy.

To summarize the primary findings from the follow-up group analyses:

- Sensitivity for detection of CIN2+ of the current standard of care was 65.7% (88/134), which was similar to the sensitivity of the standard of care based on ALTS, 65% for CIN3.
- LuViva detected 90.3% of CIN2+ in the group of women with follow-up data, which was 37.4% more CIN2+ than the standard of care ($p < 0.0001$).

13.3.5 Equivalence and Repeatability

13.3.5.1 Equivalence Study

A total of 76 subjects were enrolled in the study, and data from 65 of these subjects were analyzable. The remaining subjects were excluded because they were intended for clinician training or due to absent cytology or pathology results. Table 13.14 shows the performance of our algorithm on these subjects.

Both devices detected 12 of the 14 CIN2+ cases. Overall agreement in algorithm output (positive or negative) was 72% with most of the disagreements occurring early in the study perhaps due to lack of clinician familiarity with the process of using two different devices on the same subject. Figure 13.8 shows a scatter plot and correlation coefficient between research device and preproduction device measurements on the same subject. As can be seen in the scatter plot, the correlation between the research device and preproduction devices was 0.755 ($p < 0.0001$). The small offset between the two devices, results in a difference in the number of cases identified as positive by the two devices ($p = 0.0339$) but is not enough to affect sensitivity at all or affect specificity in a clinically meaningful way.

13.3.5.2 Repeatability Study

We enrolled 87 subjects in this study, and data from 76 of these subjects were analyzable. The remaining subjects were excluded because they were intended for clinician training or due to absent cytology or pathology results. Table 13.15 shows the algorithm performance on these subjects.

Figure 13.9 shows a scatter plot and Pearson product correlation between the first and second measurements using the preproduction device on the same subject. As can be seen in the scatter plot, the correlation between the two successive tests using the same preproduction device was 0.762 ($p < 0.0001$).

The magnitude of the Pearson correlation coefficient calculated and shown in Figures 13.8 and 13.9 is indicative of the equivalence of the devices. We attribute the reason for the coefficient being less than 1.0 primarily to "clinical noise" which comes from the slight difference in the area of tissue measured between two measurements and how the cervix is presented to the device for measurement. The similar correlations of approximately 0.76 found between repeat measurements in both

TABLE 13.14

Performance of Algorithm on Equivalence Study Subjects

Disease Category	Sensitivity	Specificity		
	CIN2+	Normal	CIN1	Combined
Number tested	14	31	20	51
Number correct (research device)	12	13	10	23
Number correct (preproduction device)	12	7	6	13
Percent correct (research device)	85.7%	41.9%	50.0%	45.1%
Percent correct (preproduction device)	85.7%	22.6%	30.0%	25.5%

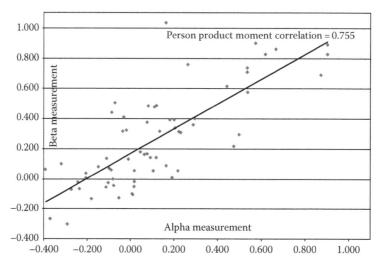

FIGURE 13.8   Scatter plot showing correlation between research device and preproduction device measurements on the same subject for 65 subjects.

TABLE 13.15

Performance of Algorithm 2.61 on Repeatability Study Subjects

Disease Category	Sensitivity	Specificity		
	CIN2+	Normal	CIN1	Combined
Number tested	8	28	40	68
Number correct (first test)	7	14	17	31
Number correct (second test)	7	14	15	29
Percent correct (first test)	87.5%	50.0%	42.5%	45.6%
Percent correct (second test)	87.5%	50.0%	37.5%	42.6%

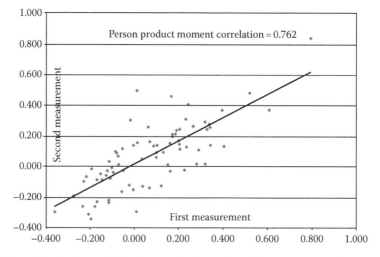

FIGURE 13.9   Scatter plot showing correlation between the first and second measurements on the same subject using the preproduction device for 76 subjects.

the repeatability and equivalence studies indicate that device-related differences are inconsequential when compared to the clinical noise. The offset seen in the equivalence study (Figure 13.8) is marginally statistically significant, is limited to specificity, and may have resulted from a few specific cases that differed due to clinical noise or sampling differences. Additional data from the preproduction device support this interpretation. For example, in the repeatability studies, the preproduction device sensitivity and specificity were very similar to that of the research device in the equivalence study.

The entire body of data therefore indicates that we can accept the null hypothesis that the research and preproduction devices are equivalent. In particular, the repeatability data demonstrate the robustness of the preproduction device and the algorithm used to derive the final output.

13.4　Light Touch in Clinical Use and Future Trends

Our proposed intended use (Section 13.2.6.1) for LuViva is to triage women directed to colposcopy in accordance with current ASCCP guidelines. The intended population is women who have an abnormal or equivocal cytology result, a positive HPV test, or other risk factors per standard of care in order to identify those women with higher likelihood of high-grade dysplasia. Our clinical studies have determined the high-grade dysplasia threshold to be at CIN2, i.e., LuViva will find women with CIN2 or higher grades of disease to be as positive. The abnormal or equivocal cytology classification includes women with an ASCUS, ASC-H, LSIL, AGUS, and AGC Pap result. It excludes women with an HSIL Pap result given the high prevalence of CIN2+ disease (67% in our prospective study) in this category. These women are referred directly to colposcopy.

Our results show that when LuViva is used by a trained healthcare professional, (e.g., physician, nurse practitioner, physician's assistant) to triage women scheduled for colposcopy (and excluding HSIL Paps), a total of 404 women can be ruled out of the need for colposcopy and biopsy although we miss 25 women with CIN2+ (193 − 168), a 16.2:1 ratio (Section 13.3.3.1). As a point of reference, the accepted ratio for adolescent women in favor of not conducting colposcopy/biopsy is about 6–10 avoided biopsies per case of missed CIN2+. However, LuViva detected 28 out of 31 CIN2+ missed at baseline and caught upon 2 year follow-up (Section 13.3.4), resulting in a net positive detection of CIN2+. The key benefits of using LuViva can be summarized as follows:

- Noninvasive, easy to use, and does not require a physician
- Immediate result
- Reduced morbidity
- Detects more disease (CIN2+) than the standard of care while saving 35% of women from unnecessary colposcopy and biopsy

There are several ramifications to these benefits, with increased healthcare efficiency and cost saving being the primary. In an age of drastic increases in healthcare demand both due to an aging population and laws requiring greater healthcare coverage combined with a projected and severe shortage of physicians [41], a solution can be found by increasing efficiency of care and reducing unnecessary care. LuViva provides both. Further, a positive output of LuViva can serve as a notice to the physician to exercise a higher degree of vigilance where needed on those patients that need it most. LuViva can also be used, given that it does not need a physician to perform the test, to increase and extend care in rural areas where approximately 40% of the population lives and which has historically remained underserved. Finally, the noninvasive nature of the test and immediate results provided by LuViva will help increase healthcare compliance among patients and reduce the number for patients who are lost to follow-up by allowing a physician to administer treatment in the same visit rather and wait to schedule an appointment at a future date.

REFERENCES

1. Cancer facts & figures, American Cancer Society, http://www.cancer.org (accessed July 21, 2011).
2. McCrory, D.C., Matchar, D.B., Bastian, L. et al. Evaluation of cervical cytology. Evidence Report No. 5. (Prepared by Duke University under Contract No. 290-97-0014.) AHCPR Publication No. 99-E010. Rockville, MD: Agency for Health Care Policy and Research. February 1999.
3. Fahey, M.T. et al. Meta-analysis of Pap test accuracy. *Am. J. Epidemiology.* 1995, 141(7), 680–689.
4. Companies exploit opportunities in women's healthcare as interest in 'gender specific' medicine increases. *Clinical World Medical Device and Diagnostic News*, No. 798, 03/02/98, p. 14.
5. Cytyc Corporation, Boxborough, MA. www.hologic.com/en/ (accessed January 9, 2012).
6. BD Tripath. http://www.bd.com//tripath//products//surepath//index.asp (accessed January 9, 2012).
7. Sawaya, G. Evidence-based medicine vs. liquid-based cytology. *Obst. Gynecol.* 2008, 111(1), 2–3.
8. Arbyn, M. et al. Liquid cytology compared with conventional cervical cytology. *Obst. Gynecol.* 2008, 111(1), 167–177.
9. Hybrid capture II HPV test. Digene Corp. Beltsville, MD. Cleared for US marketing in March 1999 by the US FDA.
10. ALTS Group. Human papilloma testing for triage of women with cytologic evidence of low-grade squamous intraepithelial lesions. Baseline evidence from a randomized trial. *J. Natl. Can. Inst.* 2000, 92(5), 397.
11. Despite the publicity, new Pap tests are no better than the old technology, say experts. *Clinical World Medical Device and Diagnostic News*, No. 820, 08/10/98, p. 13.
12. Specialists divided over computer Pap scanning system. *Am. Med. News.* 1996, 39(34), 36.
13. Rosenthal, D.L. Computerized scanning devices for Pap smear screening: Current status and critical review. *Clin. Lab. Med.* 1997, 17(2), 263–284.
14. American Society of Cytopathology. Cervical cytology practice guideline. *Diagn. Cytopathol.* 2001, 25, 3–24.
15. Mitchell, M.F., Schottenfeld, D., Tortolero-Luna, G., Cantor, S.B., and Richards-Kortum, R. Colposcopy for the diagnosis of squamous intraepithelial lesions: A meta-analysis. *Obstet. Gynecol.* 1998, 91(4), 626–631.
16. Solomon, D., Schiffman, M., and Tarone, R., ALTS Study group. Comparison of three management strategies for patients with atypical squamous cells of undetermined significance: Baseline results from a randomized trial. *J. Natl. Cancer Inst.* February 21, 2001, 93(4), 293–299.
17. The ASCUS-LSIL Triage Study (ALTS Group). Results of a randomized trial on the management of cytology interpretations of atypical squamous cells of undetermined significance. *Am. J. Obstet. Gyncol.* 2003, 188(6), 1383–1392.
18. Zhang, Q., Müller, M.G., Jun Wu, J., and Feld, M.S. Turbidity-free fluorescence spectroscopy of biological tissue. *Opt. Lett.* 2000, 25(19), 1451–1453.
19. Palmer, G.M. and Ramanujam, N. Monte-Carlo-based model for the extraction of intrinsic fluorescence from turbid media. *J. Biomed. Opt.* 2008, 13(2), 024017.
20. Ramanujam, N. et al. In vivo diagnosis of cervical intraepithelial neoplasia using 337 nm excited laser induced fluorescence. *PNAS*, 1994, 91, 10193–10197.
21. Ramanujam, N. et al. Development of a multivariate statistical algorithm to analyze human cervical tissue fluorescence spectra acquired in vivo. *Lasers Surg. Med.* 1996, 19, 46–62.
22. Ramanujam, N. et al. Spectroscopic diagnosis of cervical intraepithelial neoplasia (CIN) in vivo using laser induced fluorescence spectra at multiple excitation wavelengths. *Lasers Surg. Med.* 1996, 19, 63–74.
23. Ramanujam, N. et al. Cervical pre-cancer detection using a multivariate statistical algorithm based on laser induced fluorescence spectra at multiple excitation wavelengths. *Photochem. Photobiol.* 1996, 64(4), 720–735.
24. Kapadia, C.R. et al. Laser induced fluorescence spectroscopy in human colonic mucosa. *Gastroenterology.* 1990, 99(1), 150–157.
25. Marchesini, R. et al. Light induced fluorescence spectroscopy of adenomas, adenocarcinomas and non-neoplastic mucosa in human colon. *Photochem. Photobiol.* 1992, 14, 219–230.
26. Shoemacher, K.T. et al. Ultraviolet laser induced fluorescence of colonic polyps. *Gastroenterology.* 1992, 102, 1155–1160.

27. Shoemacher, K.T. et al. Ultraviolet laser induced fluorescence of colonic polyps. *Lasers Surg. Med.* 1992, 12, 63–78.
28. Cothren, R.M. et al. Gastrointestinal tissue diagnosis by laser induced fluorescence spectroscopy at endoscopy. *Gastrointest. Endosc.* 1990, 36(2), 105–111.
29. Tomatis, S. et al. Spectrophotometric imaging of cutaneous pigmented lesions: Discriminant analysis, optical properties and histological characteristics. *Photochem. Photobiol.* 1998, 42, 32–39.
30. Xillix Inc. Premarket approval memorandum. 1996. DHHS. FDA.
31. Kincade, K. Optical biopsy device nears commercial reality. *Laser Focus World.* March 2000, p. 59.
32. Stage-Gate is a registered trademark logo for Product Development Institute Inc. Stage-Gate was pioneered and developed by Dr. Robert G. Cooper and is a widely implemented and trusted product development process. http://www.prod-dev.com//stage-gate.php (accessed January 9, 2012).
33. Code of Federal Regulations Part 820:30 Quality System Regulations.
34. Marmo, J. Hyperspectral imager will view many colors of earth. *Laser Focus World.* August 1996, p. 85.
35. Wright Jr., T.C., Massad, L.S., Dunton, C.J. et al. 2006 Consensus guidelines for the management of women with abnormal cervical screening tests. *J. Low. Genit. Tract Dis.* 2007, 11(4), 201–222.
36. Clinical trial protocol and protocol amendments, Direct access to source data/Documents (International Conference of Harmonization ICH E6, Section 6.10).
37. Mertz, L. Concentric spectrographs. *Appl. Opt.* 1977, 16(12), 3122–3124.
38. Lobb, D.R. Theory of concentric designs for grating spectrometers. *Appl. Opt.* 1994, 33, 2648–2658.
39. Smith, W.J. *Modern Lens Design.* Academic Press.
40. Bansal, N. et al. Natural history of established low grade cervical intraepithelial (CIN 1) lesions. *Anticancer Res.* 2008, 28, 1763–1766.
41. Association of American Medical Colleges. https://www.aamc.org/initiatives/workforce (accessed March 17, 2012).
42. Apgar, B.S., Zoschnick, L., and Wright, T.C. The 2001 Bethesda system terminology. *Am. Fam. Physician.* November 2003, 68(10), 1992–1998.
43. DeMay, M. *Practical Principles of Cytopathology.* Revised edition. Chicago, IL: American Society for Clinical Pathology Press, 2007.

Part V

Optical Technologies for Cancer Detection and Diagnostics: Optical Imaging for Cancer Analysis

14

Location and Biomarker Characterization of Circulating Tumor Cells

H. Ben Hsieh, George Somlo, Xiaohe Liu, and Richard H. Bruce

CONTENTS

14.1 Introduction

Tumor cells observed in peripheral blood are believed to be the means for the spread of the disease to metastatic sites. These circulating tumor cells (CTCs) could provide up-to-date assessment of disease status. Peripheral blood is a particularly attractive body fluid for assessing the disease as its collection is a minimally invasive procedure that can be repeated throughout the course of disease progression and treatment. Consequently, the enumeration and characterization of CTCs are being explored for their role as prognostic and predictive diagnostic tools in cancer.

The concentration and change in concentration of CTCs have been shown to be independent prognostic indicators and can provide earlier assessment of treatment than traditional imaging [13]. High concentrations of CTCs are prognostic of poor disease-free survival and overall survival [10], and increases in CTC concentration during therapy are prognostic of poor response [13]. One technology platform, CellSearch, has FDA approval for enumeration of CTCs in breast, colorectal, and prostate cancer patients. The concentration of CTCs, however, does not provide information for predicting efficacy of a specific therapy.

The characteristics of cancer biomarkers in primary tissue or in metastases are essential in choosing therapies. Biomarkers such as human epidermal growth factor receptor 2 (HER2) and estrogen receptor (ER) are regularly used to prescribe receptor-targeted therapy for all stages of breast cancer. HER2 is a membrane receptor that gives higher aggressiveness in breast cancers, and its overexpression in 20% of breast cancer patients is associated with increased disease recurrence and worse prognosis. ER serves as a DNA-binding transcription factor that regulates gene expression. Persistent activation of downstream pathways is implicated in tumor formation. Expression of ER occurs in approximately 55% of breast cancer patients.

HER2 expression level is routinely tested in breast cancer patients as overexpression identifies patients who can be successfully treated with trastuzumab, a therapeutic antibody to the HER2 protein. About 35% of HER2-positive patients respond to trastuzumab alone, and when given in combination with chemotherapy, the response duration is 10–11 months for metastatic breast cancer (MBC). ER-positive tumors are amenable to antiestrogen therapy with an average duration of first line response of 10–12 months. No receptor-targeted treatment exists for patients who express none of these markers (plus progesterone receptor, triple negative breast cancer). While the response rate to chemotherapy is 50% for these triple negative patients, their prognosis is much worse with a response duration of only 4 months.

In MBC, choosing treatment options based on biomarkers from the primary tumor to predict treatment has limitations. Tumor biology may have changed in the time between primary tumor resection and onset of metastatic disease. Biopsy of distant metastases may not be possible without significant risk to the patient, and a single biopsy may not accurately represent the disease in patients who have metastases in multiple sites. Furthermore, given the invasive nature of the procedure, repeated biopsies to monitor changes during therapy are costly and pose added risk.

In MBC patients whose primary tumor has been removed, CTCs are presumed to originate from metastatic sites, and cancer biomarkers on CTCs are a promising source of additional information to guide therapy. For example, patients whose primary tumor is triple negative, but whose CTCs are HER2+ or ER+ may benefit from targeted therapy. Markers could also predict that a therapy will lack efficacy. For example, in lung cancer patients, platinum-based therapy has been much less effective for patients expressing ERCC1, a nuclear repair protein. The presence of ERCC1 in CTCs could provide evidence that platinum therapy would not be effective. Furthermore, a change in marker status could potentially indicate the onset of resistance to a targeted therapy.

CTCs are found in very low concentrations, and less than 1 CTC/mL of blood has clinical significance in advanced disease [1]. Even lower concentrations are observed in earlier stage cancer. Consequently, enrichment is needed for CTC detection. Intact CTCs possess heterogeneous phenotypes, varying by size, shape, and protein expression level. Any enrichment or capture method must comprehend this heterogeneity or risk loss of sensitivity.

Positive enrichment, the most widely practiced method, targets a distinguishing phenotypic characteristic of CTCs such as a protein or cell size that separates CTCs from the millions of surrounding leukocytes. The most often used protein target is EpCAM, an adhesion molecule found on the surface of epithelial cells such as cells from solid tumors but not normally on blood cells [2]. Membrane proteins are preferred targets for solution-based enrichment techniques because they can be accessed without rupturing the cell membrane, which can weaken the cell. However, EpCAM is known to be downregulated in CTCs [31], and expression levels were recently shown to be exceptionally low in some breast cancer cell subtypes that are prevalent in triple negative tumors [37]. For CTCs with low EpCAM levels, enrichment technologies that use EpCAM as a target suffer a loss of sensitivity. Enrichment based on CTC size is problematic because many CTCs are smaller than leukocytes.

Once enriched or separated from the huge population of leukocytes, CTCs are identified amid a population of non–tumor cells and other artifacts by morphology and the presence of positive and absence of negative markers. Typically positive markers consist of an epithelial cell protein such as a mixture of cytokeratin (CK) clones along with a nuclear label. To eliminate false positives from activated leukocytes that can express CK [17], CD45 is used as a negative marker.

High-quality imaging of well-preserved cells enables better visualization of these markers, which improves CTC identification and reduces false positives. High image fidelity benefits from a gentle labeling protocol and a planar substrate for optimal focusing. Some technologies inherently limit image quality. For example, some implementations of immunomagnetic enrichment distort cell morphology during sample preparation [2]. Microfluidic approaches using tall posts coated with EpCAM antibodies that capture the CTCs [21] are challenged by the need to quickly focus a high-magnification microscope on a cell that can be attached anywhere on the post. A microfluidic approach with a more planar design using herringbone pattern promises to enable higher-quality images [40].

With automated digital microscopy (ADM), samples can be prepared on a planar surface with minimal distortion of cell morphology, but the approach is challenged by long imaging times [19]. Microscopy using automated image analysis for recognition of rare cells has been demonstrated to be a highly sensitive method [3], but the minimum time to image 50 million nucleated cells was reported to be 12–16 h [3,4], which is prohibitively long for a clinical assay [45].

A major impediment to imaging approaches is the time and cost for sample preparation. Microfluidic approaches are being investigated to detect and characterize CTCs from whole blood samples [21,41]. However, these approaches rely on EpCAM capture of CTCs, and in a comparison of several CTC detection approaches, the microfluidic devices were shown to lack sensitivity to breast cancer cell lines that were low expressers of EpCAM and reported a high level of false positives [29].

Flow cytometry (FC) has a much higher scan rate with reported rates around 10^4 cells/s [30] but faces the major difficulty that without applying sophisticated cell-sorting technologies, the cells constituting the "positive" cells are lost and thus are not available for morphologic examination. Furthermore, system noise of FC makes reliable detection of objects below the frequency of $1:10^4$ a challenge. The one reported exception required preparation with eight different fluorescent tags and slower scan rates to achieve adequate specificity [11]. Laser scan cytometry is limited in field of view and thus has only moderate throughput for ultra rare events without pre-enrichment [6,26].

In this chapter, we will discuss the use of fiber array scanning technology (FAST) to locate CTCs on a planar substrate without the need of enrichment. The FAST scanner, described in more detail in Section 14.3, accurately locates CTCs for subsequent imaging in an ADM. The approach uses CK to locate CTCs and, like other approaches, as a marker for identification of CTCs. Cell labeling is done on a planar substrate to minimize distortion and optimize focus resulting in high-fidelity imaging.

14.2 Materials and Methods

14.2.1 Immunofluorescent Labeling

With the FAST system, the samples are processed on a large (64 cm^2) glass slide (Paul Marienfeld GmbH & Co. KG, Bad Mergentheim) that can support about 25 million nucleated blood cells (typically 5 mL of blood) without affecting CTC retention. The active surface of the slide has a poly-l-lysine-based coating [8]. Blood is drawn into preservative tube (cell-free DNA tube, Streck Laboratories, Omaha, NE) and processed within 24 h.

The processing involves lysing the erythrocytes and plating the leukocytes on the slide surface where they are permeabilized, fixed, and labeled (Figure 14.1). Erythrocytes are lysed with 5× volume of isotonic ammonium chloride buffer (155 mM NH$_4$Cl, 10 mM KHCO$_3$, 0.1 mM EDTA, pH 7.4) at room temperature for 5 min. After centrifugation, the remaining leukocyte pellet is washed once and resuspended in phosphate buffered saline (PBS). A 5 µL portion of cell suspension is used to assess the white blood cell concentration to determine the volume needed for optimal coverage on the slide. The cells are pipetted onto a slide where attachment occurs during incubation at 37°C for 40 min. The buffer cannot contain any protein or serum which may hinder charge-based adhesion. The attached cells are fixed with 2 mL of 2% paraformaldehyde (pH 7.2–7.4) for 10 min, rinsed with PBS, and then permeabilized in acetone for 5 min in −20°C. Cells are blocked with 20% human AB serum (Sigma H4522) for 30 min at 37°C.

The markers used for CTC identification include the epithelial cell marker, CK; a leukocyte marker for negative control, CD45; and a nuclear marker, DAPI. Primary antibodies include mouse anti-human CD45 (MCA87, AbD Serotec, Raleigh, NC) directly conjugated to Qdot 705 (Invitrogen custom conjugation) and a cocktail of mouse monoclonal anti-CK antibodies for CK classes 1, 4, 5, 6, 8, 10, 13, 18, and 19 (C2562, Sigma and RCK108, DAKO). These antibodies are incubated at 37°C for 1 h along with antibodies for tumor biomarkers. Subsequently, the slide is washed in PBS and incubated for an hour with a mixture of secondary antibodies including biotin goat anti-mouse IgG1 (A10519, Invitrogen) for staining CK and other secondary antibodies for staining tumor biomarkers. Streptavidin Alexa 555 conjugate (S-32355, Invitrogen) is incubated at 37°C for 30 min. Nuclear counterstaining with very dilute DAPI (0.01 µg/mL 4′,6-diamidino-2-phenylindole) (D-21490, Invitrogen) at room temperature for 10 min

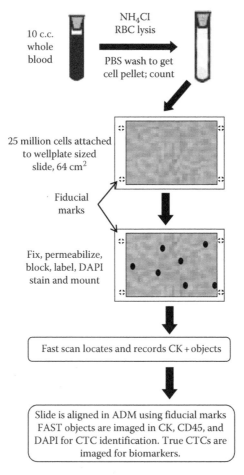

FIGURE 14.1 Sample preparation and imaging. Red blood cells from 10 mL of patient blood are lysed with 5× ammonium chloride buffer, and the pellets are washed by PBS buffer to remove debris and protein. Approximately 25 million cells are plated on a slide and allowed to attach. Cells are fixed and permeabilized. Blocking reagents are added and then cells are labeled with antibodies for five biomarkers. Slides are stained with DAPI, and LiveCell mounting media is applied under a coverslip. A FAST scan locates objects fluorescing with cytokeratin label. Fiduciary markers are used to align slide in ADM. ADM images three markers (CK, CD45, and DAPI) on CK+ objects for CTC identification. CTCs are imaged for ER, HER2, and ERCC, and biomarker expression levels are determined from emission intensities.

allows enough contrast for autofocus without interfering with other labels. The slide is mounted in an aqueous mounting medium (20 mM tris pH 8.0, 0.5% n-propyl gallate, and 90% glycerol), and the coverslip is sealed with nail polish.

14.2.2 Tumor Biomarker Labeling

The multiplex assay includes labeling of three tumor biomarkers along with the three labels needed for CTC identification. Secondary antibodies are employed to enhance the signals. In order to minimize cross-reactivity when employing multiple secondary antibodies, antibodies derived from different immunoglobulin isotype subclasses or different species are used. These antibodies are absorbed against the other isotype subclasses, the other immunoglobulin classes, or the serum of other species in the manufacturing process to minimize cross-reactivity. The addition of multiple markers was found to have no impact on the brightness of CK (data not shown).

Through choice of organic dyes and design of optical filters, fluorescent emissions in the visible to near-infrared spectrum (400–850 nm) can be comfortably divided into wavelength regions to accommodate

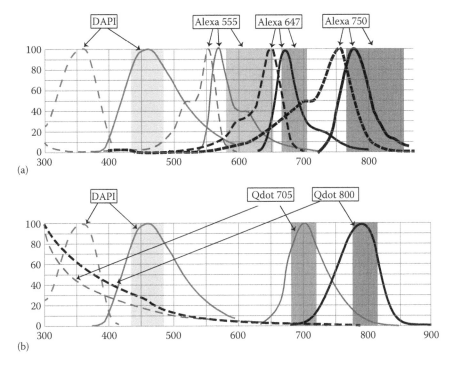

FIGURE 14.2 (a) Excitation (dotted lines) and emission (solid lines) curves for the four organic dyes (DAPI, Alexa 555, Alexa 647, and Alexa 750) used in this work. Overlap between labels with adjacent emission curves is small, and bleed-through between dyes is further minimized with the four emission notch filters in the Sedat Quad set. Shaded areas denote the transmission of the emission filters for the respective label as described in Table 14.1. (b) Excitation (dotted lines) and emission (solid lines) curves of the three dyes (DAPI, Qdot 705, and Qdot 800) excitable by UV light. The three emission notch filters minimize bleed-through between labels with adjacent emission curves. The excitation of the Qdots has minimal overlap with the excitation of the organic dyes shown in (a), and the organic dye excitation will not excite the Qdots. Shaded areas denote the transmission of the emission filters for the respective label as described in Table 14.1.

the four labels: DAPI, Alexa 555, Alexa 647, and Alexa 750. The centers of the emission in these labels are separated by 100 nm, and proper choice of emission filters insures minimal overlap eliminating the need for overlap subtraction or deconvolution (Figure 14.2). Because Alexa 555 is efficiently excited by the 488 laser used with the FAST scanner, it is used for labeling CK. Alexa 647 and Alexa 750 are small organic dyes and suitable for the intracellular proteins, ERCC1 and ER-α. Their emission wavelengths are sufficiently separated from the Alexa 555 emission wavelength to eliminate significant interference from CK emission in CTCs with high CK expression levels.

Because the excitation wavelengths of these organic dyes are close to the emission wavelength, inorganic labels that are efficiently excited with UV light having much shorter wavelength can be added without interfering with organic dyes even though their emissions overlap. Furthermore, the visible light that is used to excite the organic dyes does not excite the inorganic labels. We include two inorganic labels (Qdot 705 and Qdot 800) that emit in the near infrared for labeling of the membrane markers (CD45 and HER2). The emission wavelengths are sufficiently separated from the DAPI emission, which is also excited with UV, to insure negligible bleed through (Figure 14.2). Furthermore, Qdot 705 and Qdot 800 have emission wavelengths further from cellular autofluorescence minimizing background interference. The larger Qdot labels are best used for the membrane proteins, HER2 and CD45, to eliminate the need to diffuse through the membrane perforations.

Table 14.1 summarizes the six-label assay with nonoverlapping emission in the visible to near-infrared spectrum, which is accessible for fluorescent imaging with commercially available dyes. Possibly narrower bandwidth filtering could enable more channels; however, signal cross talk will increase, and subtraction or deconvolution may be necessary. Figure 14.2 shows the dye emission curves and

TABLE 14.1

Assay Labeling and Detection

Target	Label	Primary Ab	Secondary Ab	Tertiary Ab	Excitation Filter	Emission Filter	Dichroic Filter
Pan-CK	Alexa 555	Mouse anti-human IgG1	Biotinylated goat anti-mouse IgG1	Streptavidin, Alexa 555	555/28 (Chroma 86000v2)	617/73	Quad Chroma 86100bs
ERCC1	Alexa 647	Mouse anti-human IgG2b	Goat anti-mouse IgG2b	n/a	635/20 (Chroma 86000v2)	685/40	Quad Chroma 86100bs
ER-α	Alexa 750	Rabbit anti-human	Goat anti-rabbit	n/a	710/75 (Chroma 41009)	810/90	750LP
HER2	Qdot 800	Rat anti-human	Goat anti-rat	n/a	420/40 (Chroma 21021)	800/50	475 LP
CD45	Qdot 705	Mouse anti-human IgG2a	n/a	n/a	420/40 (Chroma 32015)	705/40	475 LP
Nucleus	DAPI	n/a	n/a	n/a	350/50 (Chroma 86000v2)	457/50	Quad Chroma 86100bs

Note: The assay uses three markers for CTC identification (CK, DAPI, and CD45) and three markers for characterization (ER-α, HER2, and ERCC1). The CK marker uses tertiary labeling to enable FAST detection of CTCs with low levels of CK expression. The ERCC1, εR-α, and HER2 markers use a secondary antibody to increase signal, while CD45 is conjugated to the label. The larger Qdot labels are used for membrane targets, and the smaller Alexa dyes are used to label intracellular markers. For microscopy, the excitation and emission filters are specified with center wavelength/width in nanometer. The dichroic filters are specified with the long pass (LP) cutoff and the part number for the quad notch filter (Sedat Quad).

emission filters which are optimized to capture emissions from a total of six channels with minimum overlap.

The assay for characterization of CTCs in breast cancer patients includes labels for HER2, ER-α, and ERCC1. The primary antibody for HER2 (rat anti-human, Serotec MCA1788) is labeled with a Qdot 800 conjugated goat anti-rat secondary antibody (custom conjugate, Invitrogen). The primary antibody for ER-α (rabbit polyclonal anti-human, Santa Cruz Biotech sc-543) is labeled with Alexa 750 tagged goat anti-rabbit secondary antibody (A-21039, Invitrogen). The primary antibody for ERCC1 (mouse monoclonal anti-human IgG2b, Santa Cruz Biotech sc-17809) is labeled with Alexa 647 tagged goat anti-mouse IgG2b secondary antibody (A-21242, Invitrogen), see Table 14.1.

14.3 Approach

14.3.1 Optical Enrichment

In the approach described here, CTCs are located with a separate scanning instrument using only the epithelial marker, CK, without subcellular resolution. Objects are subsequently imaged in an automated digital microscope at much higher resolution for the three markers used for CTC identification. The scanning instrument employs laser printing optics and collection through a wide field-of-view to rapidly locate CTCs [20]. This "optical enrichment" substantially reduces the number of exposures at high resolution. A key enabler of this approach is collection optics built with FAST that enables the wide field-of-view (Figure 14.3).

The scanning is high speed because no inertial changes are required in the optical system. A single 488 nm laser is directed across the substrate through reflections from the faceted surfaces of a spinning polygon mirror, as is done in laser printing. The laser reflects off one of the 10 spinning facets causing it to sweep 76 mm across the substrate at a speed of 10 m/s. At this rate, a 64 cm^2 substrate containing

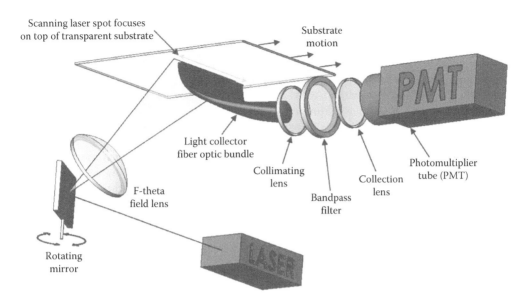

FIGURE 14.3 FAST optical system. The excitation laser is reflected from a rotating mirror through an F-theta lens to a focused spot (10 μm) that traverses the substrate with a constant speed (10 m/s). Emission is collected through the wide (76 mm) and narrow (1 mm) end of a bundle of optical fibers and is transmitted through a round (1 cm) aperture at the opposite end. The emitted light is collimated to ±10° for bandpass filtering. This figure shows a filtering leg that collects 580 nm light. A second leg normal to the optical axis of the first leg that collects 525 nm light is not shown. The ratio of the emitted light at these two wavelengths is used to eliminate autofluorescence. Printed with the permission of IEEE.

25 million nucleated blood cells can be scanned in 60 s [20]. A lens focuses the beam to a 10 μm spot size on the substrate surface.

While the scan velocity across the substrate is nonuniform, it is repeatable allowing the nonuniformities to be compensated. Here, a grading is used to create a rectilinear image at the focal plane enabling the scan time, which is measured in the controller, to be accurately translated into laser position on the substrate. The substrate is supported on a stage that is programmed to raster the scans with an 8 μm separation. The start of each scan is sensed by a custom start-of-scan circuit. The circuit uses a split detector to sense the movement of the laser beam across a start point during each scan and delivers an electronic pulse that triggers data acquisition. Objects are located to an accuracy of 40 μm by the FAST scan, which is well within the field of view of a 20× objective lens.

A bifurcated design (not shown) enables normal incidence of the excitation source, so the beam image is circular rather than elliptical. The bundle contains more than 40,000 fifty-micron diameter fibers and is placed in close proximity to the fluorescence to maximize capture of the emission. The numerical aperture of the glass bundle is 0.66. Fluorescence transmitted from the 1 cm² round end is collimated within 10° and passed through a dichroic mirror that splits the beam into two wavelength regions. Double cavity emission filters transmit fluorescence from the two wavelength regions (525 and 580 nm) to photomultiplier tubes. These filters have a contrast of 10^8 for incident light within 10°, which is adequate to block scattered laser light. The photomultiplier output is amplified and passed through low-pass filters before digitization in a high-speed A/D capture card.

14.3.2 Automated Digital Microscopy

Two automatic digital microscopes are used in this work, a Nikon (Melville, NY) TE2000U and a Nikon 80i. These fluorescent microscopes include motorized components (filter cube turret and objective changer) and a third-party motorized stage and motorized filter wheels. These components are controlled with custom image acquisition software through an application programming interface. A Nikon

20× Plan APO (NA = 0.6) microscope objective is used for automated image acquisition. Fluorescence illumination is achieved with an X-Cite 120 metal halide lamp (EXFO, Quebec, Canada) or Lambda LS Xenon arc lamp (Sutter Instrument, Novato, CA).

While the field-of-view diameter through the 20× objective is 1.1 mm, the area of the captured image size, however, is determined by the CCD imaging device in the camera. Digital images are acquired through a Retiga EXi Fast 1394 Mono Cooled digital camera (Qimaging, Burnaby, BC, Canada). The camera captures an area of 448 μm × 335 μm creating a 1392 × 1040 array of 6.45 μm square pixels on the imaging device. The camera is installed on a side port and receives 80% of the collected light.

The large size of the slide requires autofocus for each image. To avoid the extra time required by autofocusing using digital image analysis [6], a video-based (contrast) algorithm is used in hardware controller (MS-2000, Applied Scientific Imaging, Eugene, OR) which works in a way similar to what has been previously described [7,28]. This autofocus is implemented through a monochrome analog camera (CCD100, DAGE-MTI Inc., Michigan City, IN) that is installed on the trinocular head. Twenty percent of the signal from the DAPI excitation is used for focusing. The video images are fed to a MS-2000 controller for contrast analysis. A motorized drive moves the stage up to ±100 μm in each direction to determine the position of optimal contrast. Video images are acquired at 30 Hz enabling location of the focus position in less than 2 s.

Objects of interest located by the FAST cytometer are first imaged only with the CTC identification markers, CK, CD45, and DAPI, using a 20× objective. Autoexposure is used to obtain properly exposed images of the CK marker as its emission varies over several orders of magnitude. The gain on the CCD imager is set so that imaging of the 3 channels can be accomplished within 10 s enabling a slide with 150 objects to be imaged in less than 30 min. These images are ranked using analysis software and reviewed by trained personnel to identify CTCs.

CTCs are subsequently imaged for the three cancer biomarkers using a 20× objective. Each marker label is individually excited at the optimal excitation wavelength by passing the microscope lamp output through a notch filter centered at the excitation wavelength. Likewise emission from each marker is passed through a notch filter centered about the maximum emission wavelength of the marker's label. Dichroic bandpass filters enable excitation and emission optical paths to share focus/collection optics. The excitation and emission filters for each label are shown in Table 14.1. These filters are located on two filter wheels. For excitation, one wheel is placed after the lamp output, and for emission, a second wheel is placed in front of the CCD camera. The wheels are operated by a controller (Lamda 10-2, Sutter Instrument, Novalto, CA). Each wheel contains a Sedat Quad notch filter set, which has a filter each for DAPI, FITC, Alexa 555, and Alexa 647, and a filter for Cy7. In addition, the wheel for excitation contains a filter for the Qdots, and the emission filter wheel contains separate filters for Qdots 705 and Qdot 800 (Chroma Technology Corp., Rockingham, VT). A motorized cube turret contains dichroic filters that include a quad notch filter for the four emission wavelengths associated with Sedat Quad excitation, a long-pass filter for both Qdots, and a long-pass filter for Cy7. The notch filter provides additional blocking. Because the wheel position can be changed much more quickly than the turret (50 ms vs. 2 s), placing excitation and emission filters in wheels enables much faster multicolor imaging. The filters inside the wheels and the dichroic mirrors inside the turret are automatically changed during image acquisition using custom software.

Slides are placed in a custom slide holder installed on a motorized stage MS-2000 (Applied Scientific Instrumentation, Eugene, OR). Target locations predetermined by the FAST scan are translated into microscope coordinates using fiduciary marks predeposited on the slide prior to scanning (Figure 14.1). Custom software is used to position the stage, control the imaging, and acquire the images for the six labels. Images of each biomarker and CD45 are saved as four separate RGB images with the 3 biomarkers and CD45 displayed in green, CK in red, and DAPI in blue.

14.3.3 CTC and Biomarker Identification

Selecting only cells that originated from a tumor is essential to assay specificity. Including false positives in biomarker analysis will diminish the prognostic and predictive capacity. CK are the primary positive markers for identification of CTCs. While different types of epithelial cells express different CK isoforms, CK is universally expressed in epithelial cells. We use a mixture of antibodies to nine isoforms of CK (Sigma C-2562) for broad coverage [16]. Intact cells used for biomarker quantification have a nucleus labeled with DAPI that is surrounded by cytoplasm labeled with CK. A negative control (CD45) is used to insure against nonspecific binding and native fluorescence from granulocytes.

Tumor biomarker expression is quantified on each CTC, and an average expression level of all CTCs expressing the biomarker is used for computation of a sample score. A moderate expresser for each biomarker is used as a positive control and provides a metric for scoring CTCs that compensates for sample-to-sample variations in labeling intensity. Leukocytes from the patient's blood are used as a negative control for these biomarkers. The expression level for each biomarker is measured by averaging the intensity of the pixels in the region of the digital image where the marker expression is localized using a software tool that is integrated into the analysis and data acquisition software. However, the open source software tool, ImageJ, also has sufficient capability for this purpose. This methodology provides more consistent results than the subjective approach of evaluating expression by visual inspection especially for low expression levels [18].

A biomarker scoring methodology was adapted from tissue analysis that combines expression level and the percentage of expressing cells in the sample [15,23,27,35]. The marker expression level is scored relative to a moderate expressing cell line used as a positive control (Figure 14.4). CTCs with an expression level between the 16th and 84th quantiles of the positive control distribution are scored 2 while CTCs expressing higher levels are scored 3. CTC expression levels lower than the 16th quantile of the positive control line but higher than the 84th quantile of the negative control (leukocytes) are scored 1. Cells expressing lower than the 84th quantile of the negative control are scored a 0. Here, a quantile is 1% of the data points. The use of quantiles takes data skews into account, and the 16th and 84th quantiles were chosen to avoid outliers. The cell line controls are MDA-Mb-453 for HER2 [32,33,42], T-47D for

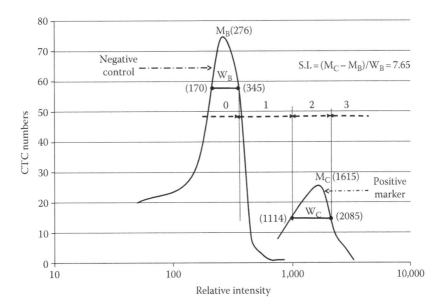

FIGURE 14.4 Histogram of ER expression intensity for positive and negative control cells of a typical patient sample. T47D cells, moderate expressers of ER, are used for positive control, and leukocytes are used for negative control. The separation index (SI) is used to quantify the staining quality and is defined as the difference between the median intensity of the positive control cells (M_C) minus the median intensity of the negative control cells (M_B) divided by the width of the distribution of the negative control cells between the 16th and 84th quantiles (W_B). The SI is 7.65 for these distributions. The controls are used in scoring the biomarker expression on the CTCs. A CTC is scored **0** when its intensity is less than the 68th quantile of the negative control cell distribution, **1** if it is between the 68th quantile of the negative control cell distribution and the 16th quantile of the positive control cell distribution, **2** if it is between the 16th quantile of the positive control cell distribution and the 84th quantile of the positive control cell distribution, and **3** if it is greater than the 84th quantile of the positive control distribution.

ER [39,44], and A-549 for ERCC1 [36]. The percent population is scored linearly on a 10-point scale with 0 for less than 10% of CTCs expressing and 10 for populations between 90% and 100%. The sample score is product of the expression average and the population scores.

14.4 Results

14.4.1 Sensitivity

The inherent system sensitivity was tested with a comparison to detection with an ADM. Test samples consisted of 1 mL of whole blood spiked with HT29 cells. Samples were first scanned by the FAST, and located objects were imaged at a 20× resolution by the ADM for identification. The samples were subsequently scanned at a much slower rate and lower numerical aperture by the ADM using a 4× magnification objective by stepping the microscope field-of-view (3.7 mm^2) across the sample. The HT29 cells have sufficient intensity to be easily detected by the 4× objective with a low numerical aperture (NA = 0.2). Image analysis located areas of fluorescence above the background, and these locations were subsequently imaged with the same 20× objective used to image the objects located by the FAST cytometer. The number of cells spiked in 1 mL of blood varied from 2 to 50 with a total of 238 cells in 13 samples. The same HT29 cells were detected by both the FAST scanner and the ADM (Figure 14.5).

14.4.2 Specificity

Because the FAST scanner does not have subcellular resolution and uses only one marker for CTC location, many CK-positive (CK+) objects are detected. While these objects are easily eliminated with high-resolution microscopy, the imaging time would be prohibitively long, so they need to be eliminated prior to microscopic imaging without loss of sensitivity. The three main sources of these objects are labeled cellular debris, dye aggregates, and autofluorescent particles. The vast majority of the CK+ detections originate from autofluorescing particles. These fluoresce broadly, and the fluorescence intensity diminishes as its wavelength shift increases.

A wavelength comparison technique is used to eliminate the effects of fluorescence from autofluorescing particles. The technique involves measuring emissions at two different wavelengths, one at the target emission wavelength (580 nm) and the other at 525 nm, a wavelength intermediate between the target emission wavelength and the laser excitation (488 nm). Because autofluorescence is typically more intense at wavelengths closer to the excitation, the ratio of the intermediate wavelength intensity

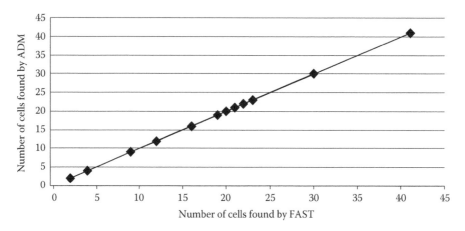

FIGURE 14.5 Number of HT29 cells detected by an ADM versus the number detected by the FAST cytometer. The cells were mixed in 1 mL of WBCs and scanned by each instrument. A total of 238 cells were scanned in 13 different samples. The FAST system found exactly the same number and same cells identified by ADM.

to the target wavelength intensity is greater than one, while for labeled objects, the ratio is less than one. A software filter used the ratio to eliminate the autofluorescing particles. CK+ objects originating from dye aggregates and cell fragments are reduced with software filtering for object size and brightness. With this filtering, 99.8% of the located objects are eliminated and need not be imaged. The typical specificity for a FAST scan is around 3×10^{-6} resulting in 150 unfiltered CK+ objects in a typical 10 mL sample containing 50 million leukocytes.

The FAST-identified objects are classified by visual inspection into several categories: (1) intact cells where the cytoplasm surrounds the nucleus, (2) large fragments where both cytoplasm and nucleus are present, (3) microfragments where only the nucleus is present, (4) objects that were CK+, CD45+ with a nucleus, and (5) objects that were only CK+. While objects belonging to the first three categories have been shown to correlate with outcome [9], objects in the fourth and fifth categories do not. Until it is known which of the first through third categories carry relevant tumor biomarker information, we only characterize intact cells.

14.4.3 Biomarker Characterization

Accurate quantification of biomarker expression benefits from high-fidelity imaging. Many biomarkers are localized in specific regions of the cell, and measuring label fluorescence only from these regions assures labeling is specific to the biomarker expression. With a 20× image, the localized regions are typically three times brighter than regions where the biomarker is not expressed. For example, with HER2, the membrane fluorescence is three times brighter than fluorescence in the cytoplasm or nuclear region (Figure 14.6). Likewise for ERCC1, fluorescence in the nuclear region is three times greater than in the cytoplasm. This localization contrast depends on many factors in addition to the specificity of the staining. For example, the depth and optimization of focus as well as the power of the imaging optics are relevant.

We use a separation index (SI) [5] to assess the specificity of the staining. The SI is the difference in the median expression level of the moderate expressing control and the median expression of the negative control divided by the width of the distribution of expression of the negative control (Figure 14.4). The index is degraded by either low staining levels of the biomarkers or high levels of nonspecific staining. In our assay, the SI is at least 4 for all of the biomarkers.

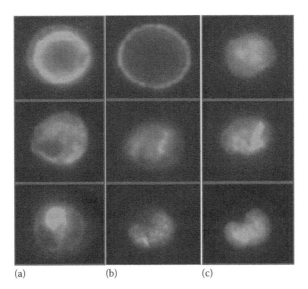

(a)   (b)   (c)

FIGURE 14.6 Images of tumor cells labeled for biomarkers. MDA-Mb-453 (top row) expresses HER2, T-47D (middle row) expresses ER, and A-549 (bottom row) expresses ERCC1. The first column (a) is an image of the CK expression, the second column (b) is an image of the biomarker, and the third column (c) is an image of the nucleus.

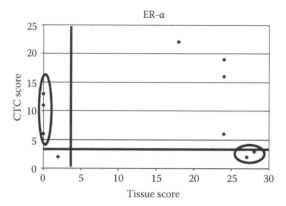

FIGURE 14.7   Immunofluorescence score for ER in CTCs versus IHC score of tumor tissue. Cutoff levels (marked by vertical and horizontal lines) are four for CTCs and three for tissue. The CTC cutoff was chosen to minimize the discordance.

14.4.4 Concordance and Discordance

The assay was used to characterize biomarker expression on CTCs in cancer patients with stage IV breast cancer. Patients with newly diagnosed MBC or with newly progressing tumors were included in the study. Primary tumor tissue from patients was characterized for all three markers for comparison to the CTC characteristics. Sample status for ER and HER2 was determined using conventional scoring for clinical pathology [12]. For HER2, the sample status was positive with protein expression score of 3 or a positive fluorescence in situ hybridization (FISH) status [12]. For ER, the sample status was positive if at least 10% of the cells expressed ER at a level 2 or 20% at a level 1 [27]. No clinically established scoring criteria are available for ERCC1 [24].

At least one CTC was detected in 13 of 14 patients who had HER2 primary tumors. The HER2 patients are actionable as a positive HER2 score on CTCs could mean that these patients could potentially benefit from HER2-targeted therapy. The CTC score was HER2+ in 23% (3 of 13) of these HER2 patients. The sample was considered HER2+ when 10% of the CTCs exhibited HER2+ expression at level 3. The CTC score was discordant for ER in 50% of the patients with 30% of the patients actionable (Figure 14.7). The cutoff score of 4 for CTCs was chosen because it minimizes the discordance level. For ERCC1, no correlation was observed between scores for tissue and scores for CTCs.

14.5 Discussion

Research results for predictive assays using CTCs are encouraging and are motivating development of the robust assays needed for clinical trials. Robust prediction is predicated on accurate measurement of biomarkers. Both nucleic acid and protein biomarkers have been measured in CTCs.

DNA mutations and RNA expression levels are generally measured from CTC-enriched samples, and the measurement is made on pooled not on individual CTCs [21,38]. These applications typically rely on PCR for measuring biomarker levels and multiple biomarkers are routinely evaluated. Microfluidic devices work well for enriching CTCs, enabling improved signal-to-noise for detection of genomic biomarkers. For example, microfluidic enrichment of CK+ objects has enabled improved genotyping of an EGFR mutation biomarker that is predictive of therapeutic response [21]. While analyzing a pooled sample for biomarkers can provide prognostic and predictive diagnostics, groupings of genotypic and phenotypic information, which would be available with individual cell analysis, are lost.

Protein expression levels are measured on individual CTCs by immunochemistry, as in tissue pathology. Measuring protein biomarker expression on CTCs fits well with current practice in breast cancer treatment, where protein biomarker expression measured with IHC is commonly used to predict efficacy of targeted therapy. Using a multiplex assay to measure multiple biomarkers in a single sample eliminates the need for separate samples for each marker and provides an opportunity to increase sample volume and therefore CTC number.

The concentration of CK+ objects reported in patient samples varies widely. This variation could depend on the detection technology but also on the evaluation criteria. A recent study using CellSearch evaluated several levels of stringency in CTC identification and showed that unfavorable patient outcome correlated with a wide range of CK+ objects such as nuclei with small amounts of associated CK-labeled regions and CK-labeled objects not associated with nuclei [9]. Only objects that were also positive for CD45 did not correlate with outcome. The number of these objects increased as the evaluation criteria were relaxed to include objects other than intact CTCs, as did the concentration of these objects observed in healthy donors. Other studies using different detection technologies have shown that high concentrations of CK positive objects in patient samples correlate with unfavorable outcome [21,25].

While the concentration of these CK positive objects correlates with unfavorable outcome, characterization of protein biomarkers on individual cells has typically been reported only on intact CTCs using CD45 as a negative marker to eliminate leukocyte false positives [14,22,47]. Protein biomarker expression for any classes of CK+ objects other than intact CTCs certainly needs validation.

Reports that identify CTCs through both positive and negative markers as well as morphology tend to report lower CTC concentrations. While the number of CTCs for any of the approaches has not approached the number of cells available in biopsy samples, increasing the sensitivity will benefit CTC characterization. Note those patients in greatest need of predictive diagnostics—those who are not responding to therapy—also have the highest CTC counts.

The sensitivity of the FAST scanner depends on the intensity of the CK labeling. CK expression levels vary considerably within a sample and between tumor types. CK expression in breast cancer patients is significantly higher than expression in prostate and lung cancer patients [34]. To detect CTCs at lower expression levels, brighter fluorescent labeling could be beneficial. The current assay employs a biotinylated secondary antibody and avidin Alexa 555 tertiary. Replacement with the protein dye, phycoerythrin, could increase the brightness by three- to sixfold.

Another opportunity for sensitivity improvement is through improving the negative marker. Without a robust negative marker, nonspecific binding of CK labels to leukocytes and autofluorescence from granulocytes could lead to false positives. Typically, in CTC identification, CD45 is the only negative marker. While CD45 is expressed strongly on lymphocytes, the distribution of expression levels is wide, and expression levels on monocytes and granulocytes are weaker. Consequently, some cells display only faint labeling. Sample permeabilization (necessary for CK staining) also diminishes the ability to stain for CD45. Additionally, with high-fidelity imaging, low-level CD45 transfer from lymphocytes to CTCs is observed [43] and can be indistinguishable from faint CD45 fluorescence especially on granulocytes and monocytes. Transferred CD45 can lead to the exclusion of some CTCs. Brighter labeling of monocytes and especially granulocytes with CD67 alleviates this problem as it results in sharper contrast between CD45-bearing CTCs and leukocytes that express low levels of CD45. Eosinophils are particularly troublesome, displaying low absolute staining intensity for CD45 and CD67 after permeabilization, but the highly positive charge of their abundant granules may be used to advantage [46]. Once permeabilized, eosinophils will stain brightly with many negatively charged dyes, allowing easy identification.

Because of its high-fidelity imaging, the FAST system is well suited for characterizing biomarkers on CTCs. Preliminary work has shown significant discordances in expression level of ER, HER2, and ERCC1 between CTCs and the primary tissue. For both tissue and the CTCs, specific cutoff scores are used to render a sample either positive or negative, and discordance levels are predicated on these cutoff scores. For ER, the cutoff score for CTCs is based on minimization of discordance levels (Figure 14.7). For HER2, cutoff scores are adopted from tissue scoring, with the exception that the score includes only cells with protein expression of level 3 and not level 2 cells that are FISH positive. Exclusion of these FISH-positive cells will result in lower scores. While such CTC cutoff scores provide an early estimation of discordance levels, the scores must be established through prospective clinical trials that show correlation between biomarker score and response to therapy.

Changes in CTC phenotypes were observed for some patients during the course of treatment for each of the three markers. While retrospective correlation of these CTC biomarker expression patterns to treatment response is ongoing, prospective clinical studies are needed to determine whether the CTC phenotypes are more predictive of therapy efficacy.

Breast cancer was chosen to first demonstrate the efficacy of protein biomarker phenotype on CTCs as a predictive diagnostic because of the effective use of biomarkers to prescribe targeted therapy for this disease. Additional work is also underway using the FAST scanner to analyze biomarkers in lung cancer, a disease where prognosis is much worse, and CTCs have been observed in blood samples from patients with prostate and colorectal cancer. The promise of a minimally invasive method to monitor the changing nature of cancer and to identify therapy to address these changes is motivating considerable effort.

14.6 Future Trends

CTC analysis promises to provide a valuable tool in both research and medical settings. CTC-based assays could predict therapeutic efficacy for each cancer patient, providing personalized care in a noninvasive manner. As CTCs are central to the metastatic process, their characterization could provide new insights into the metastatic process, potentially leading to new methods to arrest or prevent metastasis.

While preliminary data are encouraging, the medical value of CTCs as a surrogate for biopsy material has not yet been confirmed in the clinic. To move to the next step of clinical validation, CTC biomarker assays must be made robust and certified for use in clinical trials. Current work in the field to establish discordance levels for biomarkers between the primary tumor and CTCs will provide a basis for prospective trial design.

Investigations into CTC protein and genetic biomarkers have examined both pooled and individual CTCs. Pooled assays enable testing of biomarkers at higher concentrations than are present in serum and consequently increase the assay's effective sensitivity. Several PCR-based platforms are being used to investigate genetic biomarkers and early results are promising. Characterization of biomarkers on individual CTCs is more complex but offers greater resolution and opens the possibility of using well-established pathology techniques to determine patient status. Before these techniques may be employed, it must be determined whether CTC biomarkers have been significantly altered compared to the primary tumor and, if so, whether the markers retain predictive power. Identification of biomarkers robust to the biology of CTC creation needs investigation and validation in clinical trials. Another issue when characterizing individual CTCs is the susceptibility of the assay results to false positives. While several types of CK+ objects have been shown to have prognostic power, investigations to determine which CTCs have biomarkers with predictive power are incomplete. Note that a PCR assay on pooled cells is probably more susceptible to false positives than observing individual cells.

Some of these issues could be addressed with improved markers for CTC identification and new biomarkers for characterizing the disease. Two categories of markers are currently used for CTC identification. Epithelial markers such as CK and EpCAM distinguish epithelial cells from leukocytes but do not explicitly identify cells as tumor cells. Furthermore, CK is observed to be expressed in activated leukocytes, and expression levels of both markers vary due to the heterogeneity of tumor cells. Because the assays are used with patients undergoing powerful therapy regimens, rare cell phenotypes are also observed in leukocytes. Tumor markers such as HER2, mammaglobin, and CEA identify tumor cells directly but are not universally expressed in all tumors and consequently lack sensitivity. A universal tumor biomarker, particularly one that identifies the cancer type, would enable more robust assays. As more clinically relevant biomarkers are validated, multiplex assays capable of assessing them with a single blood sample will be needed. For protein assays, the imaging will need to be extended with a higher level of multiplexing.

Because CTCs concentrations are even lower in early stage cancer patients, most of the work to date was done with patients with advanced disease. For detecting the presence and providing prognostic capability in earlier stage disease, inclusion of additional CK+ objects such as macro- and microfragments could increase the number of cells that predict outcome and hence the sensitivity of a CTC-based assay. While additional cell types might not have useful protein or mRNA biomarkers for predicting therapy, their DNA could provide relevant information for predicting therapy.

REFERENCES

1. Allan, A. L. and M. Keeney. 2010. Circulating tumor cell analysis: Technical and statistical considerations for application to the clinic. *J Oncol* **2010**:426218.
2. Allard, W. J., J. Matera, M. C. Miller, M. Repollet, M. C. Connelly, C. Rao, A. G. Tibbe, J. W. Uhr, and L. W. Terstappen. 2004. Tumor cells circulate in the peripheral blood of all major carcinomas but not in healthy subjects or patients with nonmalignant diseases. *Clin Cancer Res* **10**:6897–6904.
3. Bajaj, S., J. B. Welsh, R. C. Leif, and J. H. Price. 2000. Ultra-rare-event detection performance of a custom scanning cytometer on a model preparation of fetal nRBCs. *Cytometry* **39**:285–294.
4. Bauer, K. D., J. de la Torre-Bueno, I. J. Diel, D. Hawes, W. J. Decker, C. Priddy, B. Bossy, S. Ludmann, K. Yamamoto, A. S. Masih, F. P. Espinoza, and D. S. Harrington. 2000. Reliable and sensitive analysis of occult bone marrow metastases using automated cellular imaging. *Clin Cancer Res* **6**:3552–3559.
5. Bigos, M. 2001. Separation Index: An easy-to-use metric for evaluation of different configurations on the same flow cytometer, *Current Protocols in Cytometry*. John Wiley & Sons, Inc., New York.
6. Bocsi, J., V. S. Varga, B. Molnár, F. Sipos, Z. Tulassay, and A. Tárnok. 2004. Scanning fluorescent microscopy analysis is applicable for absolute and relative cell frequency determinations. *Cytometry Part A* **61A**:1–8.
7. Bravo-Zanoguera, M., B. V. Massenbach, A. L. Kellner, and J. H. Price. 1998. High-performance autofocus circuit for biological microscopy. *Rev Sci Instrum* **63**.
8. Bross, K. J., G. A. Pangalis, C. G. Staatz, and K. G. Blume. 1978. Demonstration of cell surface antigens and their antibodies by the peroxidase-antiperoxidase method. *Transplantation* **25**:331–334.
9. Coumans, F. A., C. J. Doggen, G. Attard, J. S. de Bono, and L. W. Terstappen. 2010. All circulating EpCAM+CK+CD45- objects predict overall survival in castration-resistant prostate cancer. *Ann Oncol* **21**:1851–1857.
10. Cristofanilli, M., G. T. Budd, M. J. Ellis, A. Stopeck, J. Matera, M. C. Miller, J. M. Reuben, G. V. Doyle, W. J. Allard, L. W. M. M. Terstappen, and D. F. Hayes. 2004. Circulating tumor cells, disease progression, and survival in metastatic breast cancer. *N Engl J Med* **351**:781–791.
11. Gross, H. J., B. Verwer, D. Houck, R. A. Hoffman, and D. Recktenwald. 1995. Model study detecting breast cancer cells in peripheral blood mononuclear cells at frequencies as low as 10(−7). *Proc Natl Acad Sci USA* **92**:537–541.
12. Hatanaka, Y., K. Hashizume, Y. Kamihara, H. Itoh, H. Tsuda, R. Y. Osamura, and Y. Tani. 2001. Quantitative immunohistochemical evaluation of HER2/neu expression with HercepTestTM in breast carcinoma by image analysis. *Pathol Int* **51**:33–36.
13. Hayes, D. F., M. Cristofanilli, G. T. Budd, M. J. Ellis, A. Stopeck, M. C. Miller, J. Matera, W. J. Allard, G. V. Doyle, and L. W. Terstappen. 2006. Circulating tumor cells at each follow-up time point during therapy of metastatic breast cancer patients predict progression-free and overall survival. *Clin Cancer Res* **12**:4218–4224.
14. Hayes, D. F., T. M. Walker, B. Singh, E. S. Vitetta, J. W. Uhr, S. Gross, C. Rao, G. V. Doyle, and L. W. Terstappen. 2002. Monitoring expression of HER-2 on circulating epithelial cells in patients with advanced breast cancer. *Int J Oncol* **21**:1111–1117.
15. Hirsch, F. R., M. Varella-Garcia, F. Cappuzzo, J. McCoy, L. Bemis, A. C. Xavier, R. Dziadziuszko, P. Gumerlock, K. Chansky, H. West, A. F. Gazdar, L. Crino, D. R. Gandara, W. A. Franklin, and P. A. Bunn Jr. 2007. Combination of EGFR gene copy number and protein expression predicts outcome for advanced non-small-cell lung cancer patients treated with gefitinib. *Ann Oncol* **18**:752–760.
16. Hsieh, H. B., D. Marrinucci, K. Bethel, D. N. Curry, M. Humphrey, R. T. Krivacic, J. Kroener, L. Kroener, A. Ladanyi, N. Lazarus, P. Kuhn, R. H. Bruce, and J. Nieva. 2006. High speed detection of circulating tumor cells. *Biosens Bioelectron* **21**:1893–1899.
17. Jung, R., W. Krüger, S. Hosch, M. Holweg, N. Kröger, K. Gutensohn, C. Wagener, M. Neumaier, and A. Zander. 1998. Specificity of reverse transcriptase polymerase chain reaction assays designed for the detection of circulating cancer cells is influenced by cytokines in vivo and in vitro. *Br J Cancer* **78**:1194–1198.
18. Kersting, C., J. Packeisen, B. Leidinger, B. Brandt, R. von Wasielewski, W. Winkelmann, P. J. van Diest, G. Gosheger, and H. Buerger. 2006. Pitfalls in immunohistochemical assessment of EGFR expression in soft tissue sarcomas. *J Clin Pathol* **59**:585–590.

19. Kraeft, S. K., R. Sutherland, L. Gravelin, G. H. Hu, L. H. Ferland, P. Richardson, A. Elias, and L. B. Chen. 2000. Detection and analysis of cancer cells in blood and bone marrow using a rare event imaging system. *Clin Cancer Res* **6**:434–442.

20. Krivacic, R. T., A. Ladanyi, D. N. Curry, H. B. Hsieh, P. Kuhn, D. E. Bergsrud, J. F. Kepros, T. Barbera, M. Y. Ho, L. B. Chen, R. A. Lerner, and R. H. Bruce. 2004. A rare-cell detector for cancer. *Proc Natl Acad Sci USA* **101**:10501–10504.

21. Maheswaran, S., L. V. Sequist, S. Nagrath, L. Ulkus, B. Brannigan, C. V. Collura, E. Inserra, S. Diederichs, A. J. Iafrate, D. W. Bell, S. Digumarthy, A. Muzikansky, D. Irimia, J. Settleman, R. G. Tompkins, T. J. Lynch, M. Toner, and D. A. Haber. 2008. Detection of mutations in EGFR in circulating lung-cancer cells. *N Engl J Med* **359**:366–377.

22. Meng, S., D. Tripathy, E. P. Frenkel, S. Shete, E. Z. Naftalis, J. F. Huth, P. D. Beitsch, M. Leitch, S. Hoover, D. Euhus, B. Haley, L. Morrison, T. P. Fleming, D. Herlyn, L. W. Terstappen, T. Fehm, T. F. Tucker, N. Lane, J. Wang, and J. W. Uhr. 2004. Circulating tumor cells in patients with breast cancer dormancy. *Clin Cancer Res* **10**:8152–8162.

23. Olaussen, K. A., A. Dunant, P. Fouret, E. Brambilla, F. Andre, V. Haddad, E. Taranchon, M. Filipits, R. Pirker, H. H. Popper, R. Stahel, L. Sabatier, J. P. Pignon, T. Tursz, T. Le Chevalier, and J. C. Soria. 2006. DNA repair by ERCC1 in non-small-cell lung cancer and cisplatin-based adjuvant chemotherapy. *N Engl J Med* **355**:983–991.

24. Olaussen, K. A., P. Fouret, and G. Kroemer. 2007. ERCC1-specific immunostaining in non-small-cell lung cancer. *N Engl J Med* **357**:1559–1561.

25. Pachmann, K., O. Camara, A. Kavallaris, U. Schneider, S. Schunemann, and K. Hoffken. 2005. Quantification of the response of circulating epithelial cells to neodadjuvant treatment for breast cancer: A new tool for therapy monitoring. *Breast Cancer Res* **7**:R975–R979.

26. Pachmann, K., J. H. Clement, C. P. Schneider, B. Willen, O. Camara, U. Pachmann, and K. Hoffken. 2005. Standardized quantification of circulating peripheral tumor cells from lung and breast cancer. *Clin Chem Lab Med* **43**:617–627.

27. Phillips, T., G. Murray, K. Wakamiya, J. Askaa, D. Huang, R. Welcher, K. Pii, and D. C. Allred. 2007. Development of standard estrogen and progesterone receptor immunohistochemical assays for selection of patients for antihormonal therapy. *Appl Immunohistochem Mol Morphol* **15**:325–331.

28. Price, J. H. and D. A. Gough. 1994. Comparison of phase-contrast and fluorescence digital autofocus for scanning microscopy. *Cytometry* **16**:283–297.

29. Punnoose, E. A., S. K. Atwal, J. M. Spoerke, H. Savage, A. Pandita, R.-F. Yeh, A. Pirzkall, B. M. Fine, L. C. Amler, D. S. Chen, and M. R. Lackner. 2010. Molecular biomarker analyses using circulating tumor cells. *PLoS One* **5**:e12517.

30. Racila, E., D. Euhus, A. J. Weiss, C. Rao, J. McConnell, L. W. M. M. Terstappen, and J. W. Uhr. 1998. Detection and characterization of carcinoma cells in the blood. *Proc Natl Acad Sci USA* **95**:4589–4594.

31. Rao, C. G., D. Chianese, G. V. Doyle, M. C. Miller, T. Russell, R. A. Sanders Jr., and L. W. Terstappen. 2005. Expression of epithelial cell adhesion molecule in carcinoma cells present in blood and primary and metastatic tumors. *Int J Oncol* **27**:49–57.

32. Rhodes, A., D. Borthwick, R. Sykes, S. Al-Sam, and A. Paradiso. 2004. The use of cell line standards to reduce HER-2/neu assay variation in multiple European cancer centers and the potential of automated image analysis to provide for more accurate cut points for predicting clinical response to trastuzumab. *Am J Clin Pathol* **122**:51–60.

33. Rhodes, A., B. Jasani, J. Couturier, M. J. McKinley, J. M. Morgan, A. R. Dodson, H. Navabi, K. D. Miller, and A. J. Balaton. 2002. A formalin-fixed, paraffin-processed cell line standard for quality control of immunohistochemical assay of HER-2/neu expression in breast cancer. *Am J Clin Pathol* **117**:81–89.

34. Riethdorf, S., H. Wikman, and K. Pantel. 2008. Review: Biological relevance of disseminated tumor cells in cancer patients. *Int J Cancer* **123**:1991–2006.

35. Sauter, G., J. Lee, J. M. Bartlett, D. J. Slamon, and M. F. Press. 2009. Guidelines for human epidermal growth factor receptor 2 testing: Biologic and methodologic considerations. *J Clin Oncol* **27**:1323–1333.

36. Shimizu, J., Y. Horio, H. Osada, T. Hida, Y. Hasegawa, K. Shimokata, T. Takahashi, Y. Sekido, and Y. Yatabe. 2008. mRNA expression of RRM1, ERCC1 and ERCC2 is not associated with chemosensitivity to cisplatin, carboplatin and gemcitabine in human lung cancer cell lines. *Respirology* **13**:510–517.

37. Sieuwerts, A. M., J. Kraan, J. Bolt, P. van der Spoel, F. Elstrodt, M. Schutte, J. W. M. Martens, J.-W. Gratama, S. Sleijfer, and J. A. Foekens. 2009. Anti-epithelial cell adhesion molecule antibodies and the detection of circulating normal-like breast tumor cells. *J Natl Cancer Inst.* **101:**61–66.

38. Smirnov, D. A., B. W. Foulk, G. V. Doyle, M. C. Connelly, L. W. Terstappen, and S. M. O'Hara. 2006. Global gene expression profiling of circulating endothelial cells in patients with metastatic carcinomas. *Cancer Res* **66:**2918–2922.

39. Sommers, C. L., S. W. Byers, E. W. Thompson, J. A. Torri, and E. P. Gelmann. 1994. Differentiation state and invasiveness of human breast cancer cell lines. *Breast Cancer Res Treat* **31:**325–335.

40. Stott, S. L., C.-H. Hsu, D. I. Tsukrov, M. Yu, D. T. Miyamoto, B. A. Waltman, S. M. Rothenberg, A. M. Shah, M. E. Smas, G. K. Korir, F. P. Floyd, A. J. Gilman, J. B. Lord, D. Winokur, S. Springer, D. Irimia, S. Nagrath, L. V. Sequist, R. J. Lee, K. J. Isselbacher, S. Maheswaran, D. A. Haber, and M. Toner. 2010. Isolation of circulating tumor cells using a microvortex-generating herringbone-chip. *Proc Natl Acad Sci* **107:**18392–18397.

41. Stott, S. L., R. J. Lee, S. Nagrath, M. Yu, D. T. Miyamoto, L. Ulkus, E. J. Inserra, M. Ulman, S. Springer, Z. Nakamura, A. L. Moore, D. I. Tsukrov, M. E. Kempner, D. M. Dahl, C. L. Wu, A. J. Iafrate, M. R. Smith, R. G. Tompkins, L. V. Sequist, M. Toner, D. A. Haber, and S. Maheswaran. 2010. Isolation and characterization of circulating tumor cells from patients with localized and metastatic prostate cancer. *Sci Transl Med* **2:**25–23.

42. Szollosi, J., M. Balazs, B. G. Feuerstein, C. C. Benz, and F. M. Waldman. 1995. ERBB-2 (HER2/neu) gene copy number, p185HER-2 overexpression, and intratumor heterogeneity in human breast cancer. *Cancer Res* **55:**5400–5407.

43. Tabibzadeh, S. S., Q. F. Kong, and S. Kapur. 1994. Passive acquisition of leukocyte proteins is associated with changes in phosphorylation of cellular proteins and cell–cell adhesion properties. *Am J Pathol* **145:**930–940.

44. Thompson, E. W., S. Paik, N. Brunner, C. L. Sommers, G. Zugmaier, R. Clarke, T. B. Shima, J. Torri, S. Donahue, M. E. Lippman et al. 1992. Association of increased basement membrane invasiveness with absence of estrogen receptor and expression of vimentin in human breast cancer cell lines. *J Cell Physiol* **150:**534–544.

45. Tibbe, A. G., B. G. de Grooth, J. Greve, P. A. Liberti, G. J. Dolan, and L. W. Terstappen. 1999. Optical tracking and detection of immunomagnetically selected and aligned cells. *Nat Biotechnol* **17:**1210–1213.

46. Valnes, K. and P. Brandtzaeg. 1981. Selective inhibition of nonspecific eosinophil staining or identification of eosinophilic granulocytes by paired counterstaining in immunofluorescence studies. *J Histochem Cytochem* **29:**595–600.

47. Wulfing, P., J. Borchard, H. Buerger, S. Heidl, K. S. Zanker, L. Kiesel, and B. Brandt. 2006. HER2-positive circulating tumor cells indicate poor clinical outcome in stage I to III breast cancer patients. *Clin Cancer Res* **12:**1715–1720.

15

High-Resolution Microendoscopy for Cancer Imaging

Mark C. Pierce, Veronica Leautaud, Ann Gillenwater,
Sharmila Anandasabapathy, and Rebecca Richards-Kortum

CONTENTS

15.1 Introduction

Over the past decade, advances in medical imaging have revolutionized the field of cancer diagnosis and treatment. The use of computed tomography (CT), magnetic resonance imaging (MRI), and ultra-sound has become standard practice for visualizing anatomical structures within the body. Newer developments including positron emission tomography (PET) and single photon emission computed tomography (SPECT) are capable of identifying the functional and molecular aspects of disease and when combined with anatomic imaging can provide improved detection and localization of even small neoplastic lesions [6]. While these systems have shown remarkable capabilities in preclinical and clinical use [24,39,49,52,69], they have limited spatial resolution (typically ranging from 100 µm to 5 mm) and thus cannot visualize alterations in tumor morphology, function, and architecture at the cellular level. To overcome this limitation, current diagnostic approaches still rely heavily on optical microscopy of stained cytology or histology specimens obtained from fine needle aspiration or open biopsies. Specimen collection by these methods is often invasive, painful, expensive, and samples only a small fraction of the suspicious tissue site, contributing to the limited sensitivity of such techniques [10,35,65].

15.1.1 In Vivo Microscopy

Several optical imaging technologies have demonstrated the ability to acquire images at subcellular reso-lution from within intact, living tissues [14,38], raising the prospect of performing a noninvasive "optical biopsy." While histopathology remains the gold standard for diagnosis, several potential roles are emerg-ing for in vivo microscopy methods in cancer management, ranging from biopsy guidance to surgical margin delineation. These techniques include confocal microscopy in both reflectance and fluorescence modes, as well as nonlinear techniques including multiphoton, second-harmonic, and coherent anti-Stokes Raman scattering (CARS) microscopy [49]. Each of these methods has been demonstrated with laboratory-based platforms through imaging of ex vivo tissue specimens and in vivo animal models. Developments in fiber optics, microoptics, and microelectromechanical systems (MEMS) components have contributed to the translation of these benchtop systems to the clinical setting.

Confocal microscopy has become a standard technique in biology research, capable of imaging a thin "optical section" from within intact, thick tissues, providing real-time imaging of fluorescent or reflected light to depths of a few hundred microns, depending on the tissue type. An example which highlights the potential for confocal microscopy to improve patient care is its use for rapid intraoperative determination of margin status during Mohs surgery [47]. Confocal microscopy was used to generate large-scale (12 mm × 12 mm) histologic-quality images of resected tissue specimens in less than 9 min, compared to 20–45 min for standard frozen section processing. Diagnosis of basal cell carcinoma by these confocal images was shown to yield sensitivity and specificity of 96.6% and 89.2%, respectively, relative to histopathology [15]. Other studies using ex vivo tissue specimens also demonstrated the ability of confocal microscopy to image the characteristic morphological alterations in nuclear size, shape, and nuclear-to-cytoplasm ratio that pathologists use when evaluating stained tissue sections, motivating the development of confocal microscopy systems for in vivo imaging [8,71]. Clinically viable confocal systems have incorporated narrow-diameter optical fibers, miniaturized objective lenses, and compact beam scanning mechanisms to image at confined sites within the body [25,30,36,51,54,60]. Technical approaches can be broadly divided into two categories, based upon either [1] a single optical fiber for light delivery and collection, with a miniature scanning mechanism at the distal tip, or [2] a coherent fiber-optic bundle for light transmission, with beam translation by conventional scanners at the proximal end. In the latter case, the structure of the fiber-optic bundle is inherently imposed upon the sample structure, although image processing methods can be applied to diminish this effect [37]. The presence of this artifact is eliminated by use of single optical fiber methods, although the requirement of a robust, miniature mechanism for high-speed raster scanning at the distal tip can prove challenging and ultimately limit the minimum achievable probe diameter and bend radius.

Multiphoton microscopy is another technique that can acquire subcellular resolution images from within intact, living tissues. Fluorophores are excited through near-simultaneous absorption of two or more low-energy (near-infrared) photons at the focus of an ultrafast pulsed laser [7,75]. Due to the low probability of multiphoton absorption at any location other than the high-intensity beam focus, background fluorescence is eliminated, without the need for a conjugate pinhole in front of the detector. By using near-infrared wavelengths for excitation, multiphoton microscopy can acquire images to increased depths relative to confocal microscopy, often exceeding 500 μm [20,68]. In addition to multiphoton imaging of traditional fluorescent dyes emitting in the visible wavelength range, two-photon excitation of contrast agents with near-IR emission spectra can virtually eliminate the contribution of tissue autofluorescence [73]. The recent development and availability of optical fibers with tightly controlled dispersion characteristics has enabled remote delivery of the pico- and femtosecond pulses required for nonlinear excitation. The use of compact gradient index (GRIN) lenses alone or in combination with spherical lenses at the distal tip of these fibers has resulted in platforms compatible with in vivo clinical use [2,3,5,16,26,32]. Although the cost of a nonlinear microscopy system significantly exceeds that of a confocal platform (due to the expense of the ultrafast laser source), multiphoton imaging of internal organ sites has been demonstrated in animal studies [9,13] and human subjects [32], demonstrating the feasibility of fiber-optic nonlinear microscopy.

15.1.2 Molecular Imaging and Optical Contrast Agents

While second harmonic microscopy, CARS microscopy, and multiphoton autofluorescence imaging all use intrinsic tissue components as sources of contrast, most high-resolution in vivo imaging techniques use exogenous contrast agents. Current clinical confocal microendoscopes are primarily used in fluorescence mode, using nonspecific FDA-approved dyes such as fluorescein [51] and indocyanine green (ICG) [17]. Nuclear stains including acriflavine and acridine orange have been evaluated in animal [64] and human subject studies [51,61,62], though approval for routine clinical use has not yet been established.

In addition to these relatively mature dyes that enhance morphological features, several classes of optical contrast agent are under development to either label or respond to activity at the molecular level. Many of these agents are designed to report on overexpression of cancer related biomarkers and/or biochemical and physiological processes. Entire families of cancer biomarkers have been identified, including cellular factors involved in proliferation, differentiation, and metastasis [19,24,27,39,55,58,69]. Changes in molecular expression can be monitored via introduction of targeting antibodies, antibody fragments, or peptides, coupled with optically active reporters to generate a fluorescent or scattering signal. For example, labeling the epidermal growth factor receptor (EGFR) with either the EGF peptide [28] or EGFR antibody [23] conjugated to an organic dye has been demonstrated in confocal imaging of breast cancer xenografts and head and neck cancers. EGFR imaging was also demonstrated in an in vivo murine model of colorectal cancer and in ex vivo human specimens [18]. In a separate study, Hsiung et al. used phage display techniques to screen for a peptide specific to colon cancer, which was then conjugated to fluorescein [22]. The fluorescent contrast agent was shown to specifically highlight neoplastic regions under fiber bundle confocal microscopy of human subjects in vivo. Changes in metabolism that occur during cancer progression can also be imaged optically through the use of the fluorescent glucose analog 2-NBDG [72]. Conceptually similar in mechanism to the widely used radiolabeled tracer 2-[^{18}F] FDG used in PET imaging, elevated uptake of 2-NBDG in metabolically overactive cells has allowed nontransformed and cancer cells to be distinguished based on fluorescence intensity levels [44,45,56].

These contrast agents rely on the specificity of a targeting moiety to differentially highlight cancer biomarkers but are limited by employing an optical reporter that is always active. Any nonspecific uptake in normal tissue results in a reduction in the achievable tumor-to-background ratio. Activatable or "smart" probes are a class of contrast agents that remain in a dark state until interacting with the target species, minimizing the amount of background light. An early example of an enzyme-activated contrast agent was demonstrated by Weissleder et al., who conjugated multiple fluorescent Cy5.5 molecules in close proximity along a long synthetic copolymer backbone, remaining quenched until cleaved by MMPs [70]. A similar protease-cleavable contrast agent was used in an in vivo murine model of colonic adenocarcinoma, with cathepsin B activation enabling imaging of colonic neoplasms with a microcatheter [1]. Activatable contrast agents can also be released from a quenched state by cellular internalization following binding to targeted cell surface molecules [31]. A recent development in this area was noteworthy in that ICG was used as the fluorescent label, conjugated to monoclonal antibodies targeting overexpressed cell surface markers CD25, HER-1, and HER-2 [46]. Both the fluorescent reporter and the targeting components are already FDA approved, raising the potential for translation of these activatable probes to the clinical setting.

15.1.3 Optical Microendoscopy

Following demonstrations of in vivo confocal microscopy and two-photon microscopy by several research groups, commercial systems have reached the marketplace and several are currently involved in multicenter trials to establish clinical efficacy. Unfortunately, these systems tend to be expensive (confocal >$50k, multiphoton>$100k), requiring laser sources, scanning systems, and sensitive detectors. The comparatively simple approach of wide-field imaging through a coherent fiber bundle was reported in the 1990s by several groups exploring methods for neurophysiology imaging studies within an intact animal cortex [21,53]. These efforts employed a short length (few centimeters) of a rigid image guide to simply relay an image from the tissue site to the sample stage of a benchtop microscope or charge-coupled device (CCD) camera. During characterization of a fiber bundle–based scanning confocal microscope, Dubaj et al. compared the characteristics of the wide-field fiber bundle imaging approach to the point-scanning confocal method [12]. Using a filtered mercury arc lamp for excitation of fluorescein and a photographic camera for image capture, the authors qualitatively and quantitatively compared the degree of optical sectioning available with the wide-field imaging method to that of fiber bundle confocal microscopy. As expected from theory, the confocal system with the smallest pinhole diameter achieved the strongest optical sectioning strength, but imaging was still demonstrated in the wide-field configuration. Dromard et al. also demonstrated wide-field epifluorescence imaging through a coherent fiber bundle on a modified benchtop microscope, this time with a filtered halogen lamp source and scientific-grade cooled CCD camera [11]. These authors used fluorescein to nonspecifically stain corneocytes on the skin and demonstrated in vivo imaging of human skin in contact and noncontact (GRIN lens) modes.

We have extended these early demonstrations of wide-field, epifluorescence microendoscopy and developed a high-resolution microendoscope (HRME) which allows visualization of cellular morphology and tissue architecture in vivo. The HRME provides a platform to study cellular morphology, function, metabolism, and interaction with the tumor microenvironment, in situ and in real time [41,42,50]. The system is compact ($12'' \times 10'' \times 2.5''$), robust (no moving parts), cost-efficient (<$5k), and simple to use and modify. The primary components of the HRME system are described in the succeeding text, followed by a discussion of our clinical studies in the areas of head and neck, and gastrointestinal cancers.

15.2 Materials

A schematic diagram illustrating the key components of the HRME is shown in Figure 15.1a. Light from a high-power LED (Luxeon/CREE, $\lambda_0 \sim 455$ nm) is collected by an aspheric condenser lens, passed through an excitation filter (Semrock FF01-452/45), and reflected at a dichroic beam splitter (Chroma 485DCLP). The focal length and location of the condenser lens (Thorlabs ACL2520) are chosen such that an image of the LED chip fills the back aperture of the objective lens (Olympus Plan 10x/0.25), as in conventional Köhler illumination. This configuration ensures that the sample is uniformly illuminated, regardless of the structure of the LED chip. A coherent fiber-optic bundle (Sumitomo/Fujikura, see Table 15.1) is positioned at the working distance of the objective lens, where it delivers excitation light to, and collects fluorescence emission from the sample of interest. In the configuration shown in Figure 15.1a, an infinity-corrected objective lens is used in combination with a "tube" lens, to image the proximal face of the fiber bundle onto the image sensor (Point Grey Research 14S5M-C, 1384×1036 pixels, 6.45 µm sq., monochrome). The magnification provided by the objective plus tube lens combination should be chosen

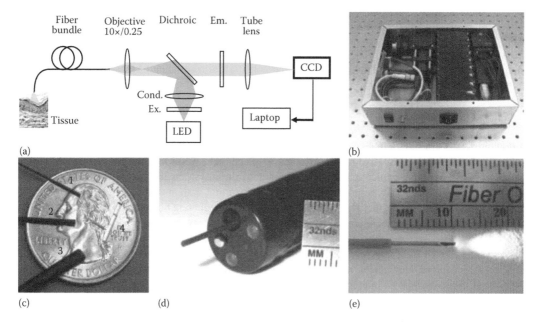

FIGURE 15.1 (a) Schematic diagram of the HRME system. Ex., fluorescence excitation filter; Em., emission filter; Cond., aspheric condenser lens. (b) Photograph of the HRME packaged in a portable battery-powered unit, measuring $10'' \times 8'' \times 2.5''$. (c) Three fiber-optic bundles used with the HRME, providing fields-of-view with diameters of 330, 720, and 1400 µm. A GRIN lens is shown alongside. (d) The 720 µm diameter fiber-optic bundle can be passed through the biopsy channel of standard endoscopes. (e) The 330 µm diameter fiber-optic bundle can be passed through the lumen of a 21 gauge biopsy needle.

TABLE 15.1

Summary of Fiber-Optic Bundle Parameters

Supplier	Part #	Image Diameter (mm)	Outer Diameter (mm)	Minimum Bend Radius (mm)	Fits within Needle (gauge)
Sumitomo	IGN-037/10	0.33	0.45	20	21
Sumitomo	IGN-08/30	0.72	0.96	40	17
Fujikura	FIGH-100–1500N	1.40	1.70	130	13

such that the image of each individual core of the fiber bundle spans at least two pixels on the sensor (to satisfy the Nyquist sampling criterion). The magnification of the system equals the focal length of the tube lens divided by the focal length of the objective lens. Once this magnification factor has been determined, the actual operating magnification can be increased until the image of the fiber bundle fills the image sensor area. The system shown in Figure 15.1 uses a 150 mm focal length tube lens (Thorlabs AC-254-150-A1) which provides a magnification factor of 8.3× when used with the Olympus 10× objective. This results in a sampling frequency of 2.8 pixels per core at the camera, with the image of the bundle face (720 μm imaging diameter) spanning the full 6 mm height of the image sensor. An emission filter (Semrock FF01-550/88) is placed in the optical path before the camera. The LED and filters listed earlier were chosen for imaging with the fluorescent dye proflavine, which exhibits peak absorption and emission wavelengths of 445 and 515 nm, respectively. The HRME can be configured to image with other fluorophores, provided the LED source and transmission spectra of the filter set (excitation filter, dichroic mirror, and emission filter) are chosen to match the fluorophore of interest. The simplicity of the system allows it to be packaged in a compact unit (10″×8″×2.5″) and powered by a single rechargeable battery (Figure 15.1b). A complete video demonstration of the assembly and use of the HRME can be found in [50].

15.3 Methods

As described earlier, the HRME can be easily configured for imaging tissues labeled with different fluorophores. The images presented here were acquired with the use of proflavine as the fluorescent contrast agent (0.01% w/v in PBS or sterile water). The solution was topically applied using a cotton swab in tissue imaging studies, or added to cells in culture by pipette for in vitro studies. Proflavine (3,6-diaminoacridine) (Sigma P2508) is a member of the acridine family of compounds and originally found widespread use as an antibacterial agent during the Second World War [66]. Proflavine and other acridines are still used today as clinical antimicrobials, for example, within triple dye, a topical solution applied to the umbilical cord stump in newborns [40]. The chemical structure of proflavine is similar to that of methylene blue and toluidine blue [67], both vital dyes used to aid visual diagnosis of neoplasia in several clinical fields. Under HRME imaging, proflavine provides fluorescent labeling of cell nuclei by interacting with nucleic acids.

A range of values for field-of-view and resolution can be accessed through the use of fiber bundles with varying diameter (Figure 15.1c), alone or in combination with miniature optical elements at the distal tip. We have evaluated the use of fiber bundles with 330, 720, and 1400 μm image diameters (Table 15.1); when used in direct contact with the sample, these figures directly correspond to the field-of-view. Miniature optical elements, such as GRIN lenses (Figure 15.1c), and compact objective lens systems [3,29,34,54] can be used to increase the spatial resolution of the system, albeit with a proportional reduction in field-of-view. When using a bare fiber bundle, spatial resolution is limited by the sampling frequency imposed by the discrete cores within the bundle. For example, the 720 μm diameter bundle, shown in Figure 15.1c, contains 30,000 individual fibers, each approximately 2.1 μm in diameter and separated from its neighbor by 4 μm (center-to-center distance). By using optical elements at the distal end of the bundle to project a demagnified image of the bundle onto the sample, the sampling frequency can be increased. When using the 2.5× GRIN lens shown alongside the bundles in Figure 15.1c, the fiber core spacing at the sample is reduced to 4 μm/2.5 = 1.6 μm, providing an increase in spatial resolution.

The fiber-optic bundles are flexible (see Table 15.1 for bend radii), and for relatively accessible sites such as the oral cavity and cervix, the fiber can simply be placed in light contact with the tissue by hand. The fiber bundle can also be passed through the accessory channel of standard endoscopes (Figure 15.1d) and within the lumen of needle biopsy instruments (Figure 15.1e). The small, medium, and large fibers shown in Figure 15.1c can be passed through needles of 21, 17, and 13 gauge, respectively.

15.4 Results

15.4.1 Field-of-View and Resolution

Figure 15.2 illustrates the spatial resolution and field-of-view achievable with the HRME using bare fiber bundles (without optical elements at the distal tip) in direct contact with the sample. Figure 15.2a through c corresponds to imaging of a standard U.S. Air Force (USAF) resolution target with fiber bundles of 330, 720, 1400 μm fields-of-view, respectively. The group 6, element 6 bars of the resolution target (4.4 μm wide), can be distinguished in each case. The lower panels (Figure 15.2d through f) present images of normal human oral mucosa, imaged in vivo immediately following topical application of proflavine (0.01% w/v), using each of the three fiber bundles described. Due to interaction between the dye and nucleic acids, epithelial cell nuclei appear as discrete bright dots, sparsely and evenly distributed throughout each image. Topical application of proflavine can be accomplished by brief contact (<5 s) with a cotton-tipped applicator, followed by imaging immediately afterward. In our experience, good images can be obtained with short (<100 ms) CCD exposure times, for up to around 8 min following initial application, with reapplication necessary beyond this time.

The improvement in spatial resolution that can be achieved with the use of distal optics is demonstrated in Figure 15.3. In Figure 15.3a, the resolution target is imaged with a bare 720 μm imaging diameter fiber bundle, with the limit of resolution at around group 6, element 6 (4.4 μm). With the addition

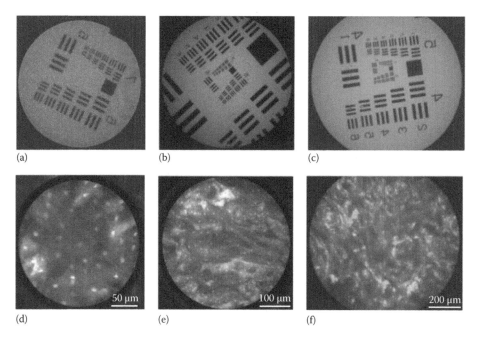

(a)　　　　　　　　(b)　　　　　　　　(c)

(d)　　　　　　　　(e)　　　　　　　　(f)

FIGURE 15.2 (a–c) HRME images of a standard USAF resolution target, using fiber-optic bundles with 330, 720, and 1400 μm field-of-view diameter, respectively. (d–f) HRME images of normal human oral mucosa, following topical application of proflavine solution (0.01% w/v). Images were acquired using the three fiber-optic bundles as shown in (a–c).

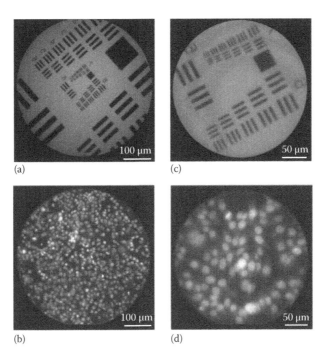

FIGURE 15.3 (a) HRME image of a USAF target through a bare fiber bundle with a 720 μm field-of-view. (b) HRME image of 1483 cells in culture following staining with proflavine (0.01% w/v) using the same fiber bundle as (a). (c) HRME image of a USAF target through the same fiber bundle as (a), following addition of a 2.5× GRIN lens to the distal tip. Spatial resolution is increased at the expense of field-of-view. (d) HRME image of 1483 cells in culture following staining with proflavine (0.01% w/v) using the same fiber bundle with GRIN lens attachment as (c).

of a 2.5× GRIN lens to the distal tip of the same bundle, all elements in group 7 are resolved, down to element 6 with a 2.2 μm line width (Figure 15.3c). The difference in resolution (and field-of-view) is also apparent in Figure 15.3b and d, where a monolayer of 1483 cells is imaged with the bare fiber (Figure 15.3b) and the fiber with the GRIN lens added (Figure 15.3d), following staining with proflavine. Labeling is again primarily confined to cell nuclei, but intracellular uptake is apparent at higher resolution in Figure 15.3d.

15.4.2 Ex Vivo Human Tissues

With a view toward clinical translation of high-resolution microendoscopy, we have carried out pilot studies on ex vivo tissue specimens, in collaboration with clinicians specializing in head and neck, and gastrointestinal cancers. The aim of these studies is to establish the capabilities of high-resolution microendoscopy techniques against a gold standard of histopathology.

Oral cavity: An example of HRME imaging of a surgical specimen containing an oral squamous carcinoma located in the buccal mucosa and retromolar trigone is shown in Figure 15.4 [42]. With the 720 μm diameter fiber bundle placed in contact with the specimen at a clinically normal region (Figure 15.4a), bright, sparsely distributed nuclei are visible in the HRME image (Figure 15.4b). In contrast, as the fiber probe was moved to the region clinically determined to contain the tumor (Figure 15.4d), nuclei qualitatively exhibited the characteristic increase in density and disorganization associated with cancer (Figure 15.4e). The corresponding histopathology sections are also presented, with the clinically normal appearing region (Figure 15.4c) exhibiting normal epithelium with some inflammation. Histopathology from the clinically abnormal region (Figure 15.4f) indicates the presence of squamous carcinoma cells extending to the tissue surface.

Upper and lower gastrointestinal tracts: Confocal microscopy has been shown to improve the sensitivity and specificity of detection of Barrett's esophagus over white light endoscopy with four-quadrant

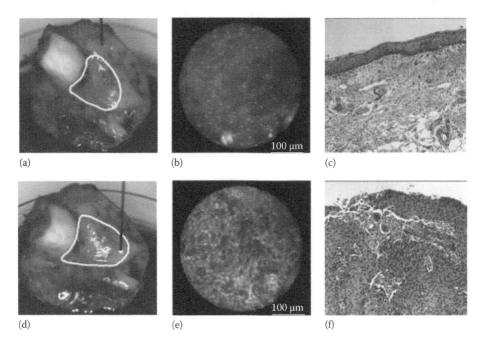

FIGURE 15.4 (a) Photograph of a surgical specimen resected from the buccal mucosa and retromolar trigone. The yellow line delineates the clinically abnormal region identified visually by the surgeon. The tip of the HRME probe is seen placed outside of the abnormal region. (From Muldoon, T.J. et al., *Opt. Express,* 15, 16413, 2007.) (b) HRME image from the specimen, with the fiber probe at the location shown in (a). (c) Histopathology (H&E stained) from the site imaged in (b). (d–f) Corresponding macroscopic photograph (d), HRME image (e), and pathology (f) at a site inside the abnormal region.

biopsy [48]. We sought to establish whether the HRME could also improve on current diagnostic techniques, without the level of cost associated with confocal microscopy. Patients undergoing surgical resection of the esophagus or colon were recruited to participate in a study on HRME imaging of ex vivo specimens. Patients gave written informed consent to participate in this study, which was approved by Institutional Review Boards at Rice University, the MD Anderson Cancer Center, and Mount Sinai Medical Center.

Immediately following surgical resection, each fresh tissue specimen was rinsed with saline and then photographed under white light illumination. Following application of proflavine (0.01% w/v) to the mucosal surface, the 720 µm diameter fiber bundle was placed in contact with grossly normal and abnormal areas, as identified by the study pathologist. In the esophagus, HRME images were evaluated qualitatively, noting nuclear size and density, nuclear-to-cytoplasmic ratio, glandular structure and organization, and fluorescence intensity [41,43]. Distinct patterns were observed for esophageal tissues with histopathologic diagnoses of [1] normal squamous epithelium, [2] Barrett's metaplasia/low-grade dysplasia, [3] Barrett's neoplasia (high-grade dysplasia [HGD] or intramucosal adenocarcinoma), and [4] esophageal adenocarcinoma (EAC).

Figure 15.5a shows an HRME image of normal squamous mucosa obtained from a site that appeared normal on white light endoscopy, with corresponding histopathology in Figure 15.5e. The squamous epithelium is distinguished by individual cell nuclei appearing as discrete bright dots, with polyhedral/polygonal cell shape, centrally situated nuclei, and well-defined cell membranes. Uniform spacing of the nuclei indicates the intact polarity that distinguishes benign squamous mucosa. This characteristic also helps in differentiating other, more randomly distributed cells (e.g., inflammatory cells) that might infiltrate the epithelium.

The HRME image in Figure 15.5b shows glandular epithelium characterized by a double-ring-like architecture with bright nuclei distributed uniformly along the basement membrane, consistent with

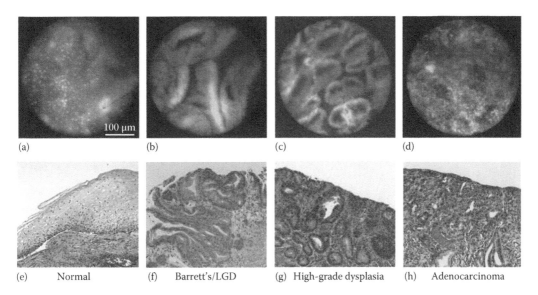

(a)　　　　　　　　(b)　　　　　　　　(c)　　　　　　　　(d)

(e)　　Normal　　　(f)　　Barrett's/LGD　　(g)　　High-grade dysplasia　　(h)　　Adenocarcinoma

FIGURE 15.5 HRME images from ex vivo specimens from the esophagus. (From Muldoon, T.J. et al., *J. Biomed. Opt.*, 15, 026027, 2010.) (a) Normal squamous epithelium, (b) Barrett's metaplasia and low-grade dysplasia, (c) HGD, (d) EAC. (e–h) Corresponding histopathology from the sites imaged in (a–d).

intact nuclear polarity. In addition, nuclei are uniformly distributed in the epithelium distant from the basement membrane, all features that are consistent with nondysplastic glandular mucosa. The histopathology section of the same specimen (Figure 15.5f) shows distinctive Barrett's mucosa, histologically defined as intestinal metaplasia. Intestinal metaplasia is characterized by the presence of goblet cells, easily identified by intracytoplasmic clear to light blue vacuoles. These features are consistent with Barrett's metaplasia.

HRME imaging of an EMR specimen with HGD is shown in Figure 15.5c, demonstrating a confluent proliferation of small glandular structures with variable size and shape, with occasional areas showing a gland-in-gland appearance. High nuclear density along with a large proportion of cells occupied almost entirely by nuclei result in high nuclear-to-cytoplasmic (N/C) ratio. These features are consistent with HGD. The H&E-stained section of the specimen (Figure 15.5g) shows distinctive Barrett's mucosa with HGD, consistent with the features observed in the HRME image. HGD is characterized by an architecturally complex arrangement of glands, with loss of nuclear polarity, nuclear overcrowding, and nuclei and mitotic figures often reaching up to the luminal surface.

Figure 15.5d illustrates the typical appearance of EAC under HRME imaging, showing an almost complete loss of glandular architecture with highly crowded nuclei and disorganization. These qualitative characteristics are also apparent in the corresponding histopathology (Figure 15.5h).

Figure 15.6 presents typical HRME images of ex vivo colon tissue with the corresponding histopathology sections [63]. Figure 15.6a shows normal colonic mucosa, characterized by evenly distributed, round tubular structures of consistent diameter with small, basally oriented fluorescent nuclei. The lamina propria appears less dense than the gland edges. Features observed in the HRME image correspond to the luminal openings of colonic glands in H&E sections from the same area (Figure 15.6e). Figure 15.6b shows a typical HRME image of colonic dysplasia. Glands appear elongated and neither circular nor evenly shaped, with nuclear crowding apparent in some areas. Figure 15.6c illustrates the appearance of severe dysplasia on HRME imaging, with cancer at the base of the mucosa. Glands are only visible infrequently, with crowded nuclei occupying the majority of the field-of-view. Figure 15.6d is an HRME image at the site of a grossly involved tumor, showing irregularly shaped tubular structures lined by enlarged fluorescent nuclei amid densely cellular tissue. Dysplastic glands characteristic of adenocarcinoma are observed surrounded by desmoplastic stromal reaction in the corresponding histology section (Figure 15.6h).

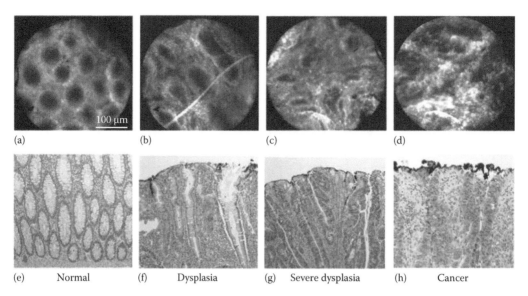

FIGURE 15.6   HRME images from ex vivo specimens from the colon. (From Thekkek, N. et al., *Gastrointest. Endosc.*, 877, 2012) (a) Normal colonic mucosa, (b) colonic dysplasia, (c) severe dysplasia, (d) cancer. (e–h) Corresponding histopathology from the sites imaged in (a–d).

15.4.3 Quantitative Image Analysis

In addition to establishing descriptive criteria for image interpretation (based upon features conventionally observed in histopathology slides), we are also developing quantitative image analysis criteria. By computationally extracting morphologic and texture-based parameters from HRME images, it may be possible to develop automated software to assist the clinician in their decision making, particularly in nonspecialized settings.

Morphological features are often qualitatively apparent in images; with proflavine staining, the predominant features are nuclear size, N/C ratio, nuclear crowding, and glandular architecture, as seen in Figures 15.5 and 15.6. In principle, these features can also be quantified by using image processing algorithms as has been demonstrated in confocal microscopy [8]. We can also compute a wide range of texture-based features from the raw pixel values that make up the image. These encompass familiar statistical measures such as mean intensity, standard deviation, and correlation of pixel values, as well as less visually apparent parameters such as those derived from frequency-domain analysis of an image. We analyzed a total of 128 HRME images obtained from six patients undergoing endoscopic mucosal resections and three patients receiving esophageal biopsies [43]. Fifty-nine features were evaluated, including first-order statistical features such as mean and variance of pixel values, textural features such as pixel value correlation over varying distances and frequency content, and morphological features of nuclear size and separation.

An algorithm was then developed to classify each image as nonneoplastic (Barrett's metaplasia and/or low-grade dysplasia) or neoplastic (HGD or EAC), based on the value(s) of one or more of the quantitative image features described earlier. The algorithm was based on two-class linear discriminant analysis (LDA), with sequential forward feature selection used to identify the best performing subset containing a maximum of 10 features to classify the images. Initially, the best performing single feature was identified, and then subsequent features were selected that gave maximum performance when combined with previously selected features. The algorithm was developed using fivefold cross-validation, whereby each measurement site was initially randomly assigned to one of five groups. Four-fifths of the data were then used to train the linear classification algorithm, and the remaining one-fifth of the data was used to test the algorithm. This cycle was repeated four additional times such that the algorithm was tested using data from each site. Performance was quantified by calculating the area under the curve (AUC) of the classifier.

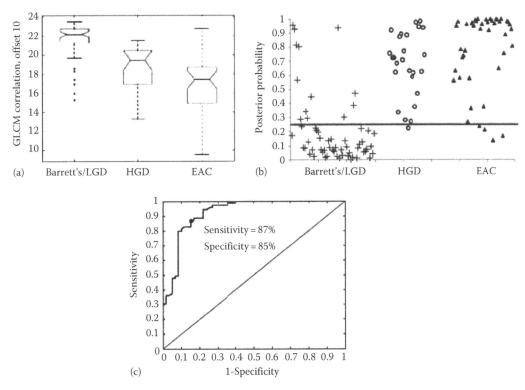

FIGURE 15.7 Quantitative HRME image analysis for ex vivo esophageal specimens. (From Muldoon, T.J. et al., *J. Biomed. Opt.,* 15, 026027, 2010.) (a) Box plot showing the calculated GLCM values for the three diagnostic categories of Barrett's and low-grade dysplasia/HGD/EAC. (b) Scatter plot of calculated posterior probabilities based on two-feature LDA, grouped by diagnostic category. (c) Receiver-operator characteristic (ROC) curve for the two-class, two-feature LDA algorithm.

The best classification performance was obtained by using a combination of two features; the gray-level co-occurrence matrix (GLCM) correlation value at an offset of 10 pixels, with the magnitude of the power spectrum within the sixth (frequency) partition. Figure 15.7a shows the GLCM values for this dataset in a box-and-whiskers plot for each diagnostic category. Figure 15.7b shows the posterior probability that each site is neoplastic, as calculated by the best performing two-feature classifier, with each measurement also grouped by pathology diagnosis. Figure 15.7c presents the ROC curve for this linear discriminant classifier, indicating sensitivity and specificity values of 87% and 85%, respectively, at the Q-point, and an AUC of 0.92.

We are currently following the same algorithm development process with HRME data acquired from head and neck specimens to determine the most diagnostically important image features in each case. We are also now applying the findings described earlier to in vivo images of esophageal tissues, to establish the diagnostic performance of the classifier in vivo.

15.5 Discussion

This chapter describes the development and application of a high-resolution fiber-optic microendoscope (HRME), based upon the simple approach of wide-field epifluorescence imaging through a coherent fiber bundle. Similar approaches have appeared previously in the literature [11,12,21,53], but we have taken advantage of technical developments in light sources, CCD cameras, and fiber optics to produce an instrument that is compact, robust, and entirely battery powered. We have begun to explore the capabilities of the system for applications in oncology through a series of studies with ex vivo and in vivo tissues of the oral cavity, esophagus, and colon. By correlating HRME images with histopathology from the same tissue site, we have assembled qualitative "image atlases" for the distinct tissue types found at these sites, including representative examples of normal, dysplastic, and neoplastic tissue. Data obtained in these studies have also enabled us to explore the potential for automated image analysis algorithms to provide feedback and guidance to the clinician.

As described earlier in this chapter, several in vivo microscopy techniques are under investigation for applications in clinical cancer management; the HRME has strengths and weaknesses relative to these. The availability of compact CCD cameras with good sensitivity and high-speed data transfer rates, combined with high-power LED sources, enabled development of a robust, compact, low-cost system with no moving parts. The imaging frame rate can be sustained at 14 frames per second, independent of field-of-view, and when the larger fiber-optic bundles are used with the HRME, the field-of-view can significantly exceed that of point-scanning microscope systems. A fundamental limitation of the HRME is its inability to collect light from only one single focal plane within tissue and allow the user to position this plane at different depths. Confocal and multiphoton methods can collect light from a narrow range of depths around the focal plane, although to actually translate the location of this plane in depth remains challenging in miniaturized probe-based systems. To enable the HRME to collect images from varying depths within tissue, we have adopted a simple needle-based approach, whereby the fiber bundle is positioned within the lumen of a hypodermic needle which is then advanced into the tissue, analogous to core needle biopsy or fine needle aspiration techniques. While the wide-field imaging approach used in the current HRME does not physically reject background light in the same sense as confocal and nonlinear microscopy techniques, we have found that contact imaging with proflavine can nevertheless still provide high-quality, real-time imaging of subcellular detail in epithelial tissues. The bright, nuclear targeting nature of proflavine contributes significantly to providing image contrast. In our experience, HRME image contrast in epithelial mucosa is considerably worse with nonspecific fluorescent agents such as fluorescein.

15.6 Future Directions

We have begun to evaluate the capabilities of the HRME for identifying cancer and precancerous lesions originating in the epithelia of the oral cavity, esophagus, colon, and cervix. These studies are in many ways analogous to current and previous studies on in vivo microscopy at large referral cancer centers. Our experience so far in early studies on ex vivo tissues has been encouraging [41,43], and we are currently evaluating whether sensitivity and specificity is maintained in vivo. The ultimate role of the HRME (and the broader range of in vivo microscopy techniques) remains to be defined, though one paradigm that has been suggested is the combination of high-resolution (microscopic) imaging with wide-field (macroscopic) techniques. Wide-field imaging in white light, fluorescence, or narrow-band reflectance can be used to survey an entire tissue surface at risk of disease with good diagnostic sensitivity but may be vulnerable to identifying false positive lesions due to inflammation and other benign conditions. A wide-field plus high-resolution combination could use the macroscopic mode to identify suspicious regions with high sensitivity, with the microscopic component used to inspect those regions at the cellular level, to eliminate false positives and improve specificity [4,33]. High-resolution optical imaging methods including HRME may also complement the macroscopic imaging capabilities of PET, CT, and MRI systems.

Characterizing and recognizing the appearance of inflammatory conditions under HRME is an important area that warrants further study, particularly if the goal is to use the platform in screening programs in low-risk populations. Distinguishing normal tissue from cancer is often not particularly challenging for expert clinicians; the real difficulty lies in identifying rare early precancerous lesions from among many common benign conditions. There is therefore a need to conduct studies in populations where disease prevalence is low, which will require large numbers of patients to be imaged and followed over time in order to establish sensitivity and specificity with narrow confidence limits. While the expense of optical imaging systems is generally low in comparison to radiological imaging systems, the ability to truly use a device as a point-of-care screening tool requires costs to be as low as possible. In this regard, the continual decline in costs and improvements in performance of consumer-grade optoelectronics offer significant opportunities for developers of optical imaging systems. The HRME was recently coupled to a $400 digital SLR camera, by replacing the standard camera lens with the same epifluorescence illumination and collection optics described in Section 15.2 to image the fiber bundle face onto the camera's image sensor [57]. Proflavine-stained cells could be visualized in real time on the camera's LCD screen, with images saved on a flash memory card, eliminating the requirement of a separate desktop or laptop PC for image acquisition and display.

While this chapter has focused on potential applications for the HRME in clinical cancer management, there is considerable potential for the system to be used as a lab research tool to study in vitro cell culture and in vivo animal models of disease and treatment. Zhong et al. used an HRME configured to image the 690–700 nm fluorescence emission from the photodynamic compound BPD-MA [74]. This enabled imaging of 3D cultured tumors, as well as in vivo micrometastases in an ovarian cancer mouse model. Significantly, the ability of HRME to acquire digital images in a minimally invasive fashion allowed the effect of the PDT drug to be quantified and treatment response monitored in a longitudinal study.

Our focus to date has been to develop a robust HRME platform for rapid translation to clinical studies. However, there are several technical developments that are currently under investigation by our group and others that will expand the capabilities of the current system. Modification of the HRME to allow imaging of scattered light will open up the possibility of imaging with contrast agents based upon metallic nanoparticles [59]. Such agents are the subject of considerable research effort, offering immunity to photobleaching and the ability to tune their peak scattering wavelength across visible and near-infrared wavelengths. However, HRME imaging in reflected light will require systems which can strongly suppress specular reflections arising within the system, particularly those generated at or near conjugate image planes, such as the proximal and distal faces of the fiber bundle. The HRME can also be configured to image samples labeled with multiple contrast agents, each generating scattered or fluorescent light at distinct wavelengths. Such an arrangement is made possible by the use of multiband filter sets, as in conventional fluorescence microscopy.

As described earlier in this chapter, the HRME probe can be passed through the lumen of a hypodermic needle and advanced into tissue to access subsurface locations. The size of the probe (and therefore the needle) is minimized when used without additional optics at the distal tip. However, the potential for integration of ultraminiature optics which only marginally increase the probe diameter has been demonstrated [29,34]. These optics can also be achromatized to accommodate HRME imaging with contrast agents spanning a wide range of visible and near-infrared wavelengths.

In summary, we have described the development and application of a fiber-optic microendoscope system for imaging with cellular resolution in tissues. The system is relatively inexpensive, robust, and straightforward to assemble and use. We have demonstrated potential application areas in oncology through collaborative work at large health-care centers and anticipate that future advances in telemedicine may also allow outreach of pathological and medical expertise to remote and underserved areas.

ACKNOWLEDGMENT

This research was funded by the National Institutes of Health, grant R01 EB007594.

REFERENCES

1. Alencar, H., M. A. Funovics, J. Figueiredo, H. Sawaya, R. Weissleder, and U. Mahmood. 2007. Colonic adenocarcinomas: Near-infrared microcatheter imaging of smart probes for early detection-study in mice. *Radiology* **244**:232–238.
2. Bao, H., J. Allen, R. Pattie, R. Vance, and M. Gu. 2008. Fast handheld two-photon fluorescence microendoscope with a 475 microm × 475 microm field of view for in vivo imaging. *Opt Lett* **33**:1333–1335.
3. Barretto, R. P., B. Messerschmidt, and M. J. Schnitzer. 2009. In vivo fluorescence imaging with high-resolution microlenses. *Nat Methods* **6**:511–512.
4. Bedard, N., M. Pierce, A. El-Naggar, S. Anandasabapathy, A. Gillenwater, and R. Richards-Kortum. 2010. Emerging roles for multimodal optical imaging in early cancer detection: A global challenge. *Technol Cancer Res Treat* **9**:211–217.

5. Bird, D. and M. Gu. 2003. Two-photon fluorescence endoscopy with a micro-optic scanning head. *Opt Lett* **28**:1552–1554.

6. Boyle, P. and B. Levin (ed.). 2008. World Cancer Report 2008, vol. WHO, Lyon, France.

7. Brown, E. B., R. B. Campbell, Y. Tsuzuki, L. Xu, P. Carmeliet, D. Fukumura, and R. K. Jain. 2001. In vivo measurement of gene expression, angiogenesis and physiological function in tumors using multiphoton laser scanning microscopy. *Nat Med* **7**:864–868.

8. Collier, T., M. Guillaud, M. Follen, A. Malpica, and R. Richards-Kortum. 2007. Real-time reflectance confocal microscopy: Comparison of two-dimensional images and three-dimensional image stacks for detection of cervical precancer. *J Biomed Opt* **12**:024021.

9. Deisseroth, K., G. Feng, A. K. Majewska, G. Miesenbock, A. Ting, and M. J. Schnitzer. 2006. Next-generation optical technologies for illuminating genetically targeted brain circuits. *J Neurosci* **26**:10380–10386.

10. Dellon, E. S. and N. J. Shaheen. 2005. Does screening for Barrett's esophagus and adenocarcinoma of the esophagus prolong survival? *J Clin Oncol* **23**:4478–4482.

11. Dromard, T., V. Ravaine, S. Ravaine, J. L. Leveque, and N. Sojic. 2007. Remote in vivo imaging of human skin corneocytes by means of an optical fiber bundle. *Rev Sci Instrum* **78**:053709.

12. Dubaj, V., A. Mazzolini, A. Wood, and M. Harris. 2002. Optic fibre bundle contact imaging probe employing a laser scanning confocal microscope. *J Microsc* **207**:108–117.

13. Engelbrecht, C. J., R. S. Johnston, E. J. Seibel, and F. Helmchen. 2008. Ultra-compact fiber-optic two-photon microscope for functional fluorescence imaging in vivo. *Opt Express* **16**:5556–5564.

14. Flusberg, B. A., E. D. Cocker, W. Piyawattanametha, J. C. Jung, E. L. Cheung, and M. J. Schnitzer. 2005. Fiber-optic fluorescence imaging. *Nat Methods* **2**:941–950.

15. Gareau, D. S., J. K. Karen, S. W. Dusza, M. Tudisco, K. S. Nehal, and M. Rajadhyaksha. 2009. Sensitivity and specificity for detecting basal cell carcinomas in Mohs excisions with confocal fluorescence mosaicing microscopy. *J Biomed Opt* **14**:034012.

16. Gobel, W., J. N. Kerr, A. Nimmerjahn, and F. Helmchen. 2004. Miniaturized two-photon microscope based on a flexible coherent fiber bundle and a gradient-index lens objective. *Opt Lett* **29**:2521–2523.

17. Goetz, M., I. Deris, M. Vieth, E. Murr, A. Hoffman, P. Delaney, P. R. Galle, M. F. Neurath, and R. Kiesslich. 2010. Near-infrared confocal imaging during mini-laparoscopy: A novel rigid endomicroscope with increased imaging plane depth. *J Hepatol* **53**:84–90.

18. Goetz, M., A. Ziebart, S. Foersch, M. Vieth, M. J. Waldner, P. Delaney, P. R. Galle, M. F. Neurath, and R. Kiesslich. 2010. In vivo molecular imaging of colorectal cancer with confocal endomicroscopy by targeting epidermal growth factor receptor. *Gastroenterology* **138**:435–446.

19. Hasina, R. and M. W. Lingen. 2004. Head and neck cancer: The pursuit of molecular diagnostic markers. *Semin Oncol* **31**:718–725.

20. Helmchen, F. and W. Denk. 2005. Deep tissue two-photon microscopy. *Nat Methods* **2**:932–940.

21. Hirano, M., Y. Yamashita, and A. Miyakawa. 1996. In vivo visualization of hippocampal cells and dynamics of Ca2+ concentration during anoxia: Feasibility of a fiber-optic plate microscope system for in vivo experiments. *Brain Res* **732**:61–68.

22. Hsiung, P. L., J. Hardy, S. Friedland, R. Soetikno, C. B. Du, A. P. Wu, P. Sahbaie, J. M. Crawford, A. W. Lowe, C. H. Contag, and T. D. Wang. 2008. Detection of colonic dysplasia in vivo using a targeted heptapeptide and confocal microendoscopy. *Nat Med* **14**:454–458.

23. Hsu, E. R., E. V. Anslyn, S. Dharmawardhane, R. Alizadeh-Naderi, J. S. Aaron, K. V. Sokolov, A. K. El-Naggar, A. M. Gillenwater, and R. R. Richards-Kortum. 2004. A far-red fluorescent contrast agent to image epidermal growth factor receptor expression. *Photochem Photobiol* **79**:272–279.

24. Jaffer, F. A. and R. Weissleder. 2005. Molecular imaging in the clinical arena. *JAMA* **293**:855–862.

25. Jean, F., G. Bourg-Heckly, and B. Viellerobe. 2007. Fibered confocal spectroscopy and multicolor imaging system for in vivo fluorescence analysis. *Opt Express* **15**:4008–4017.

26. Jung, W., S. Tang, D. T. McCormic, T. Xie, Y. C. Ahn, J. Su, I. V. Tomov, T. B. Krasieva, B. J. Tromberg, and Z. Chen. 2008. Miniaturized probe based on a microelectromechanical system mirror for multiphoton microscopy. *Opt Lett* **33**:1324–1326.

27. Katayama, A., N. Bandoh, K. Kishibe, M. Takahara, T. Ogino, S. Nonaka, and Y. Harabuchi. 2004. Expressions of matrix metalloproteinases in early-stage oral squamous cell carcinoma as predictive indicators for tumor metastases and prognosis. *Clin Cancer Res* **10**:634–640.

28. Ke, S., X. Wen, M. Gurfinkel, C. Charnsangavej, S. Wallace, E. M. Sevick-Muraca, and C. Li. 2003. Near-infrared optical imaging of epidermal growth factor receptor in breast cancer xenografts. *Cancer Res* **63**:7870–7875.

29. Kester, R. T., T. Christenson, R. R. Kortum, and T. S. Tkaczyk. 2009. Low cost, high performance, self-aligning miniature optical systems. *Appl Opt* **48**:3375–3384.

30. Kim, P., E. Chung, H. Yamashita, K. E. Hung, A. Mizoguchi, R. Kucherlapati, D. Fukumura, R. K. Jain, and S. H. Yun. 2010. In vivo wide-area cellular imaging by side-view endomicroscopy. *Nat Methods* **7**:303–305.

31. Kobayashi, H., M. Ogawa, R. Alford, P. L. Choyke, and Y. Urano. New strategies for fluorescent probe design in medical diagnostic imaging. *Chem Rev* **110**:2620–2640.

32. Konig, K., A. Ehlers, I. Riemann, S. Schenkl, R. Buckle, and M. Kaatz. 2007. Clinical two-photon micro-endoscopy. *Microsc Res Tech* **70**:398–402.

33. Lam, S., B. Standish, C. Baldwin, A. McWilliams, J. leRiche, A. Gazdar, A. I. Vitkin, V. Yang, N. Ikeda, and C. MacAulay. 2008. In vivo optical coherence tomography imaging of preinvasive bronchial lesions. *Clin Cancer Res* **14**:2006–2011.

34. Landau, S. M., C. Liang, R. T. Kester, T. S. Tkaczyk, and M. R. Descour. Design and evaluation of an ultra-slim objective for in-vivo deep optical biopsy. *Opt Express* **18**:4758–4775.

35. Lee, A., S. Krishnamurthy, A. Sahin, W. F. Symmans, K. Hunt, and N. Sneige. 2002. Intraoperative touch imprint of sentinel lymph nodes in breast carcinoma patients. *Cancer* **96**:225–231.

36. Lee, C. M., C. J. Engelbrecht, T. D. Soper, F. Helmchen, and E. J. Seibel. 2010. Scanning fiber endoscopy with highly flexible, 1 mm catheterscopes for wide-field, full-color imaging. *J Biophotonics* **3**:385–407.

37. Lin, K. Y., M. Maricevich, N. Bardeesy, R. Weissleder, and U. Mahmood. 2008. In vivo quantitative microvasculature phenotype imaging of healthy and malignant tissues using a fiber-optic confocal laser microprobe. *Transl Oncol* **1**:84–94.

38. MacAulay, C., P. Lane, and R. Richards-Kortum. 2004. In vivo pathology: Microendoscopy as a new endoscopic imaging modality. *Gastrointest Endosc Clin N Am* **14**:595–620, xi.

39. Massoud, T. F. and S. S. Gambhir. 2007. Integrating noninvasive molecular imaging into molecular medicine: An evolving paradigm. *Trends Mol Med* **13**:183–191.

40. McConnell, T. P., C. W. Lee, M. Couillard, and W. W. Sherrill. 2004. Trends in umbilical cord care: Scientific evidence for practice. *Newborn Inf Nursing Rev* **4**:211–222.

41. Muldoon, T. J., S. Anandasabapathy, D. Maru, and R. Richards-Kortum. 2008. High-resolution imaging in Barrett's esophagus: A novel, low-cost endoscopic microscope. *Gastrointest Endosc* **68**:737–744.

42. Muldoon, T. J., M. C. Pierce, D. L. Nida, M. D. Williams, A. Gillenwater, and R. Richards-Kortum. 2007. Subcellular-resolution molecular imaging within living tissue by fiber microendoscopy. *Opt Express* **15**:16413–16423.

43. Muldoon, T. J., N. Thekkek, D. Roblyer, D. Maru, N. Harpaz, J. Potack, S. Anandasabapathy, and R. Richards-Kortum. 2010. Evaluation of quantitative image analysis criteria for the high-resolution microendoscopic detection of neoplasia in Barrett's esophagus. *J Biomed Opt* **15**:026027.

44. Nitin, N., A. L. Carlson, T. Muldoon, A. K. El-Naggar, A. Gillenwater, and R. Richards-Kortum. 2009. Molecular imaging of glucose uptake in oral neoplasia following topical application of fluorescently labeled deoxy-glucose. *Int J Cancer* **124**:2634–2642.

45. O'Neil, R. G., L. Wu, and N. Mullani. 2005. Uptake of a fluorescent deoxyglucose analog (2-NBDG) in tumor cells. *Mol Imaging Biol* **7**:388–392.

46. Ogawa, M., N. Kosaka, P. L. Choyke, and H. Kobayashi. 2009. In vivo molecular imaging of cancer with a quenching near-infrared fluorescent probe using conjugates of monoclonal antibodies and indocyanine green. *Cancer Res* **69**:1268–1272.

47. Patel, Y. G., K. S. Nehal, I. Aranda, Y. Li, A. C. Halpern, and M. Rajadhyaksha. 2007. Confocal reflectance mosaicing of basal cell carcinomas in Mohs surgical skin excisions. *J Biomed Opt* **12**:034027.

48. Pech, O., T. Rabenstein, H. Manner, M. C. Petrone, J. Pohl, M. Vieth, M. Stolte, and C. Ell. 2008. Confocal laser endomicroscopy for in vivo diagnosis of early squamous cell carcinoma in the esophagus. *Clin Gastroenterol Hepatol* **6**:89–94.

49. Pierce, M. C., D. J. Javier, and R. Richards-Kortum. 2008. Optical contrast agents and imaging systems for detection and diagnosis of cancer. *Int J Cancer* **123**:1979–1990.

50. Pierce, M. C., D. Yu, and R. Richards-Kortum. 2011. High-resolution fiber optic microendoscopy for in situ cellular imaging. *J Vis Exp* http://www.jove.com/details.stp?id=2306

51. Polglase, A. L., W. J. McLaren, S. A. Skinner, R. Kiesslich, M. F. Neurath, and P. M. Delaney. 2005. A fluorescence confocal endomicroscope for in vivo microscopy of the upper- and the lower-GI tract. *Gastrointest Endosc* **62**:686–695.

52. Pysz, M. A., S. S. Gambhir, and J. K. Willmann. 2010. Molecular imaging: Current status and emerging strategies. *Clin Radiol* **65**:500–516.

53. Rector, D. and R. Harper. 1991. Imaging of hippocampal neural activity in freely behaving animals. *Behav Brain Res* **42**:143–149.

54. Rouse, A. R., A. Kano, J. A. Udovich, S. M. Kroto, and A. F. Gmitro. 2004. Design and demonstration of a miniature catheter for a confocal microendoscope. *Appl Opt* **43**:5763–5771.

55. Shah, N. G., T. I. Trivedi, R. A. Tankshali, J. A. Goswami, J. S. Shah, D. H. Jetly, T. P. Kobawala, K. C. Patel, S. N. Shukla, P. M. Shah, and R. J. Verma. 2007. Molecular alterations in oral carcinogenesis: Significant risk predictors in malignant transformation and tumor progression. *Int J Biol Markers* **22**:132–143.

56. Sheth, R. A., L. Josephson, and U. Mahmood. 2009. Evaluation and clinically relevant applications of a fluorescent imaging analog to fluorodeoxyglucose positron emission tomography. *J Biomed Opt* **14**:064014.

57. Shin, D., M. C. Pierce, A. M. Gillenwater, M. D. Williams, and R. R. Richards-Kortum. 2010. A fiber-optic fluorescence microscope using a consumer-grade digital camera for in vivo cellular imaging. *PLoS One* **5**:e11218.

58. Soni, S., J. Kaur, A. Kumar, N. Chakravarti, M. Mathur, S. Bahadur, N. K. Shukla, S. V. Deo, and R. Ralhan. 2005. Alterations of rb pathway components are frequent events in patients with oral epithelial dysplasia and predict clinical outcome in patients with squamous cell carcinoma. *Oncology* **68**:314–325.

59. Sun, J., C. Shu, B. Appiah, and R. Drezek. 2010. Needle-compatible single fiber bundle image guide reflectance endoscope. *J Biomed Opt* **15**:040502.

60. Sung, K. B., C. Liang, M. Descour, T. Collier, M. Follen, and R. Richards-Kortum. 2002. Fiber-optic confocal reflectance microscope with miniature objective for in vivo imaging of human tissues. *IEEE Trans Biomed Eng* **49**:1168–1172.

61. Tan, J., M. A. Quinn, J. M. Pyman, P. M. Delaney, and W. J. McLaren. 2009. Detection of cervical intraepithelial neoplasia in vivo using confocal endomicroscopy. *BJOG* **116**:1663–1670.

62. Tanbakuchi, A. A., J. A. Udovich, A. R. Rouse, K. D. Hatch, and A. F. Gmitro. 2010. In vivo imaging of ovarian tissue using a novel confocal microlaparoscope. *Am J Obstet Gynecol* **202**:90, e1–e9.

63. Thekkek, N., T. Muldoon, A. D. Polydorides, D. M. Maru, N. Harpaz, M. T. Harris, W. Hofstetter, S. P. Hiotis, S. A. Kim, A. J. Ky, S. Anandasabapathy, R. Richards-Kortum. 2012. Vital-dye enhanced fluorescence imaging of GI mucosa: metaplasia, neoplasia, inflammation. *Gastrointest Endosc* **75**:877–887.

64. Udovich, J. A., D. G. Besselsen, and A. F. Gmitro. 2009. Assessment of acridine orange and SYTO 16 for in vivo imaging of the peritoneal tissues in mice. *J Microsc* **234**:124–129.

65. van Sandick, J. W., J. J. van Lanschot, B. W. Kuiken, G. N. Tytgat, G. J. Offerhaus, and H. Obertop. 1998. Impact of endoscopic biopsy surveillance of Barrett's oesophagus on pathological stage and clinical outcome of Barrett's carcinoma. *Gut* **43**:216–222.

66. Wainwright, M. 2001. Acridine—A neglected antibacterial chromophore. *J Antimicrob Chemother* **47**:1–13.

67. Wainwright, M. 2003. Local treatment of viral disease using photodynamic therapy. *Int J Antimicrob Agents* **21**:510–520.

68. Wang, B. G., K. Konig, and K. J. Halbhuber. 2010. Two-photon microscopy of deep intravital tissues and its merits in clinical research. *J Microsc* **238**:1–20.

69. Weissleder, R. and M. J. Pittet. 2008. Imaging in the era of molecular oncology. *Nature* **452**:580–589.

70. Weissleder, R., C. H. Tung, U. Mahmood, and A. Bogdanov Jr. 1999. In vivo imaging of tumors with protease-activated near-infrared fluorescent probes. *Nat Biotechnol* **17**:375–378.

71. White, W. M., M. Baldassano, M. Rajadhyaksha, S. Gonzalez, G. J. Tearney, R. R. Anderson, and R. L. Fabian. 2004. Confocal reflectance imaging of head and neck surgical specimens. A comparison with histologic analysis. *Arch Otolaryngol Head Neck Surg* **130**:923–928.

72. Yamada, K., M. Saito, H. Matsuoka, and N. Inagaki. 2007. A real-time method of imaging glucose uptake in single, living mammalian cells. *Nat Protoc* **2**:753–762.

73. Yazdanfar, S., C. Joo, M. Y. Berezin, W. J. Akers, and S. Achilefu. 2010. Multiphoton microscopy with near infrared contrast agents. *J Biomed Opt* **15**:0305050.

74. Zhong, W., J. P. Celli, I. Rizvi, Z. Mai, B. Q. Spring, S. H. Yun, and T. Hasan. 2009. In vivo high-resolution fluorescence microendoscopy for ovarian cancer detection and treatment monitoring. *Br J Cancer* **101**:2015–2022.

75. Zipfel, W. R., R. M. Williams, and W. W. Webb. 2003. Nonlinear magic: Multiphoton microscopy in the biosciences. *Nat Biotechnol* **21**:1369–1377.

16

Lensless Fluorescent Imaging on a Chip: New Method toward High-Throughput Screening of Rare Cells

Ahmet F. Coskun, Ting-Wei Su, Ikbal Sencan, and Aydogan Ozcan

CONTENTS

16.1 Introduction

Optical imaging has become a powerful tool for biomedical research as well as clinical diagnostics. One of the most important features of optical imaging methods is that they can, in general, provide a decent spatial resolution without exposing the specimen to high-energy radiation. While such useful resolution is limited to small depths within the specimen (typically a few hundreds of microns), optical imaging methods are still widely used in biomedicine as they can also provide molecular specificity based on, e.g., absorption, fluorescence, or scattering processes. Among these contrast mechanisms, fluorescence deserves a special attention since it has been widely utilized not only for microscopy but also for various high-throughput screening applications including antibody microarrays [29] and biochemical assays [27], bringing specificity and sensitivity to the imaging platform within reasonable cost and ease of use [10,12,21]. In particular, fluorescent-based high-throughput separation and sorting of heterogeneous samples into purified cell populations has been quite valuable to study eukaryotic cells for enhancing our understanding of diseases and cell-based therapies [17].

Along the same lines, another high-throughput problem of significant interest is rapid detection and quantification of rare cell populations (e.g., circulating tumor cells—CTCs) within large sample volumes [16]. What makes this problem rather challenging is that these CTCs (that are detached from a primary tumor site) circulate within the bloodstream at extremely low densities, i.e., only a few cells per mL of whole blood, which implies less than one cell among, e.g., billion cells [19]. CTCs are considered to be one of the main reasons for metastasis [23] and also play a crucial role in tumor spread, starting from primary neoplasm and circulating in the peripheral blood, reaching distant locations within the body [13]. Despite the fact that surgery and therapy are quite successful to combat the disease in the primary site, metastasis is still far away from prognosis, i.e., the colonization of cancer cells in a distant organ has not yet been fully analyzed and understood, and it still remains as the main cause of ~90% of tumor-related patient deaths [13,25]. Furthermore, the release timing of CTCs from primary tumor sites to the bloodstream is not entirely known, and this necessitates periodic screening of CTCs over time [26].

To develop a complete solution to detect and quantify CTCs in whole blood samples, there are two major obstacles that need to be addressed: (1) specific, efficient, and sensitive isolation of CTCs from whole blood; and (2) rapid and accurate quantification of these isolated cells. Several methods have been developed so far for the first task, i.e., isolation of CTCs from whole blood. One separation method is done by using magnetic beads or ferrofluids that are coated with antibodies specific to cell-surface markers such as EpCAM [2,8,14,24,30,32]. Another isolation method is achieved based on the size and deformability differences between CTCs and hematological cells. Since CTCs are on average larger than leukocytes (>8 μm), filtration techniques can be used to separate CTCs from different cell types by using, e.g., membranes [31], channels [22], apertures [20], or even microcavities [15].

Recently, microfluidics-based lab-on-a-chip technologies have also been utilized to efficiently and specifically capture CTCs from whole blood samples [1,11,23,28]. To increase the sensitivity of cell isolation, surface chemistry–based large-area microfluidic chips were used such that CTCs were specifically captured over an active area of ~9 cm² [23]. While such microfluidic technologies offer powerful solutions for isolation and capture of CTCs at extremely low densities, the second challenge described earlier, namely, rapid quantification of these captured cells, still remains. The main reason is that conventional fluorescent microscopes that are used to analyze these large-area microfluidic devices do "not" provide sufficient throughput to rapidly image such large field-of-views (FOVs). Typical imaging FOV of a conventional microscope covers a few mm², which demands mechanical scanning and the capture of literally thousands of images that should be digitally stitched together to enumerate the captured CTCs. This is not only a slow process, taking hours, but also requires sophisticated and expensive optical hardware that needs to be fairly stable during the entire image acquisition.

To address this ultrawide FOV imaging challenge, we have recently introduced a high-throughput on-chip fluorescent imaging platform that can rapidly monitor fluorescent microobjects and labeled cells over an ultrawide FOV (e.g., >9 cm²) *without* the use of any lenses, thin-film filters, or mechanical scanners [6,7]. In this lens-free on-chip imaging technique (see Figure 16.1), the microfluidic chip of interest is loaded onto an optoelectronic sensor array (e.g., a charge-coupled device—CCD) where the

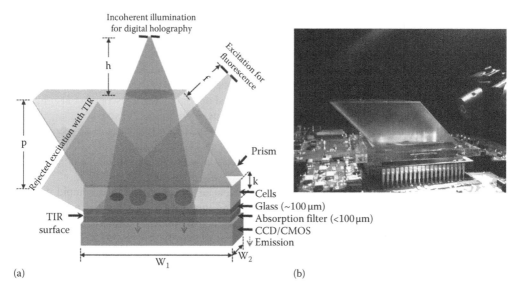

(a)   (b)

FIGURE 16.1 (a) The schematic diagram of on-chip lens-free fluorescent imaging platform is shown. Fluorescent excitation is achieved through the side facet of a rhomboid prism using an incoherent source (e.g., an LED or spectrally filtered xenon lamp). Holographic bright-field imaging of the same FOV can also be achieved through the top flat part of the same prism using an incoherent source. Dimensions: prism height, p (17 mm); imager area, $w_1 \times w_2$ (25×35 mm); depth of reservoir, k (10 μm–3 mm); distance of the vertical source, h (~5–10 cm); distance of the fluorescence excitation, f (~1–2 cm). (b) The experimental setup of the on-chip fluorescent imaging platform is shown. The incoherent fluorescent excitation is achieved through a fiber bundle that is coupled to a spectrally filtered xenon lamp (monochromator). The labeled cancer cells and the blood sample located within a microfluidic chip (dimensions: 25×35×0.3 mm) were excited through a prism interface, where a refractive index matching oil was used to assemble the chip and the prism. After rejection of the excitation light by TIR and a plastic color filter, only the fluorescent emission from the sample was recorded by the optoelectronic sensor array over an FOV of ~25×35 mm.

fluorescent cells are pumped (i.e., excited) from the side using an incoherent light source. After exciting the captured cells within the chip volume, this excitation light is then rejected using total internal reflection (TIR) process occurring at the bottom facet of the microchannel. Most of the fluorescent emission from these excited cells, however, does not obey TIR and can be detected by the optoelectronic sensor array within a detection numerical aperture (NA) of ~1. To filter out the weakly scattered excitation light and hence create a better dark-field background, we also employ an inexpensive absorption filter as illustrated in Figure 16.1a. Based on this lens-free detection principle, only the fluorescent emission of the labeled objects located on the chip can be recorded by an optoelectronic sensor array achieving an ultrawide FOV of, e.g., >9 cm^2. In this chapter, we will review the operation principles of this emerging biomedical imaging approach with a special emphasis on its unique application to high-throughput detection and quantification of rare cells within specially designed microfluidic channels. Among other applications, the same lensless on-chip fluorescent imaging platform can also be quite valuable for high-throughput imaging of, e.g., DNA or protein microarrays.

16.2 Materials

Here is a list of specific materials that we utilized to construct and test the performance of this lens-free on-chip fluorescent imaging platform:

- 2 μm and 10 μm fluorescent microspheres (excitation: 495/emission; 505 nm; FluoSpheres from Invitrogen)
- 20 μm and 30 μm polymer microspheres in water (Duke Scientific)
- Rhomboid prism (base: 35 × 25 mm; Height, 17 mm, Edmund Optics)
- Glass slide, 24 × 35 mm (Matsunami)
- Plastic-based absorption filter (Kodak Wratten color filter 12, <30 dB for <500 nm; <0.1 mm thick, Edmund Optics)
- CCD (KAI-11002, Kodak)
- Light-emitting diode (LED) (470 nm peak, 30 nm bandwidth, Luxeon Star)
- Monochromator (Newport)
- Calcein viability reagent (excitation: 495 nm/emission; 515 nm; LIVE viability assay; molecular probes).
- 1× PBS (Fisher Scientific)
- Fiber-optic faceplate (NA: 0.3; 6 μm pitch size, Incom, United States)
- PDMS elastomers part A and B (Sylgard® 184 from Dow Corning®)

16.3 Methods

In our lens-free on-chip fluorescent imaging modality (Figure 16.1), microfluidic chips that contain the fluorescent objects/cells are directly positioned onto an optoelectronic sensor array (e.g., a CCD chip). Since fluorescent emission is not directional, it diverges significantly over a distance of, e.g., 0.5–1 mm, and therefore, it is preferable to work with decapped sensors where the cover glass of the sensor chip is removed. This permits reduction of the vertical distance between the fluorescent microobjects and the active region of the sensor array to ≤0.5 mm, which significantly increases the detection signal-to-noise ratio (SNR) of lens-free fluorescent signatures on the chip. As shown in Figure 16.1, a spatially incoherent source, e.g., an LED or an appropriately filtered xenon lamp, can be used to excite the fluorescent microobjects within the microfluidic chip through the side facet of a prism, where the excitation light is entirely rejected by TIR process occurring at the bottom surface of the microfluidic chip. In conjunction with TIR, an inexpensive plastic-based absorption filter (see Figure 16.1b) is also used to eliminate the weakly scattered excitation light, creating the necessary dark-field background for fluorescent imaging.

As a result, only the fluorescent emission from the labeled specimen is detected over an imaging FOV of, e.g., >9 cm², without the use of any lenses or thin-film-based interference filters.

Note that a small fraction of the fluorescent emission from the target objects within the microchannel is also blocked by TIR; however, these fluorescent rays are extremely oblique, corresponding to an NA range of >1. Therefore, the detection NA of this lens-free fluorescent imaging system is around 1.0, implying that it is quite photon efficient compared to low NA optical systems. On the other hand, this large NA does not necessarily translate itself into a high spatial resolution, since a point source at the object plane would be widely spreading at the sensor plane, despite the fact that the detection occurs through a large NA. To control the extent of this spatial spreading and to engineer the point-spread function of our lens-free fluorescent imaging platform, passive optical components such as fiber-optic faceplates or tapers can also be used as illustrated in Figure 16.4. These planar optical elements are composed of two-dimensional arrays of fiber-optic waveguides that translate the intensity of the optical fields (fluorescent emission in our case) from one side to another. By placing such a fiber-optic faceplate between the fluorescent objects and the detector array, a better control over the spatial spreading of lens-free fluorescent signatures can be achieved without affecting the imaging FOV (see, e.g., Figure 16.5). Another important advantage of using a faceplate for on-chip lens-free fluorescent imaging is that it provides thermal isolation of the microfluidic chip and the cells from the sensor array.

In addition to fluorescent imaging, our on-chip microscopy platform also enables bright-field imaging of the same microchip through lens-free holographic microscopy as illustrated in Figures 16.2d and e. For this purpose, the same lens-free imaging platform can be coupled to a partially coherent light source (e.g., an LED) that is emanating from a relatively large aperture of ~0.1 mm to vertically illuminate the cells within the microfluidic chip (see Figure 16.1a). Each cell, depending on its size, morphology, refractive index, and subcellular features, interacts with this vertical illumination source and modifies the optical wave fronts passing through the cell volume. This transmitted optical wave, which carries the signature of the cells, then interferes with the background light (that is not affected by the presence of the cells) to create an interference pattern, i.e., a lens-free in-line hologram of the cells. This

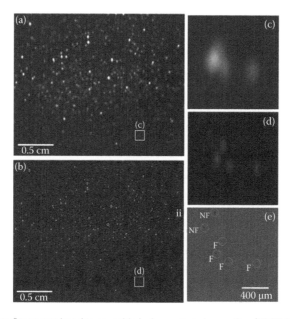

FIGURE 16.2 (a) Lensless fluorescent imaging on a chip is demonstrated over >8 cm² FOV for a mixture of 10 μm fluorescent particles and 20 μm nonfluorescent particles. (b) illustrates the digitally deconvolved (using the Lucy–Richardson algorithm) fluorescent image of the same FOV. (c) and (d) show a zoomed section of the raw fluorescent image and the deconvolution result of the same cropped FOV. Lensless holographic imaging of the same FOV is also shown in (e) which illustrates both nonfluorescent (NF) and fluorescent (F) particles, whereas the image in (c) only shows the fluorescent particles (F).

lens-free hologram can be rapidly processed to create bright-field images of the objects/cells within the microchip over the same FOV as the lens-free fluorescent image has. This dual-mode lens-free imaging capability over a large FOV is quite important for rare cell imaging in general since, e.g., specificity and sensitivity could potentially increase with multimodal analysis of the same cells captured within a given microchannel.

16.4 Results and Discussion

Initially, to validate the performance of our lens-free fluorescent on-chip imaging modality, we imaged a solution containing 10 μm fluorescent particles (495 nm excitation/505 emission) and 20–30 μm non-fluorescent particles. In this experiment, a sample solution of ~15–20 μL containing these particles was sandwiched between a microscope slide and the rhomboid prism. An LED at 470 nm is used for the excitation, and a long pass absorption filter (500 nm cutoff, 30 dB) was used to remove the randomly scattered excitation light that does not obey TIR. The raw imaging results of these initial experiments are summarized in Figure 16.2a, which demonstrates a fluorescent imaging FOV of >8 cm². Subsequently, using the same experimental platform, we also imaged fluorescently labeled white blood cells in ~400× diluted whole blood samples, where 0.5 μL of 4 mM calcein reagent (495 nm excitation/515 nm emission) was used for labeling. The green fluorescence was activated inside the cells after ~20–30 min of incubation. The blood sample was also mixed with 20 μm nonfluorescent particles serving as mechanical spacers to avoid cell damage due to pressure. The raw lens-free fluorescent image of this sample is illustrated in Figure 16.3a which, similar to Figure 16.2a, clearly shows the fluorescent signatures of the labeled white blood cells over an ultrawide FOV of >8 cm². Note that in this experiment, because platelets were also labeled, there is some increased background fluorescence as can be seen in Figure 16.3a.

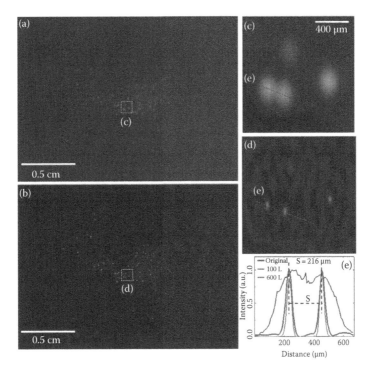

FIGURE 16.3 (a) illustrates lens-free fluorescent imaging of calcein-labeled white blood cells on a chip. Deconvolution results of the same FOV are also shown in (b). The zoomed raw fluorescent image (c) was cropped from (a), and (d) illustrates the deconvolved image of the same FOV as in (c). (e) illustrates the cross sections of the raw image in blue and the results of iterative deconvolution in green and red, for 100 and 600 L, respectively. The FWHM of the fluorescent spots is improved by ~5× from 191 to 38 μm after deconvolution.

In these initial lensless on-chip imaging results (Figures 16.2a and 16.3a), fluorescent emission from the particles/cells diverges significantly over a vertical distance of ~200 μm (between the sample plane and the sensor array), and therefore, the recorded spot size of each fluorescent object spreads over ~200–300 μm, resulting in a significant spatial overlap at the sensor plane (see, e.g., Figures 16.2c and 16.3c). While such a wide fluorescent point-spread function can still be useful for certain applications that demand high-throughput counting of, e.g., low-density cell solutions, a better spatial resolution would be important to widen the application domain of this lens-free on-chip platform.

For this end, we have initially implemented a digital reconstruction algorithm to improve the spatial resolution of this platform by ~fivefold, claiming a lens-free resolution of ~40–50 μm over a large FOV of, e.g., >8–9 cm² [3,7]. This initial approach involves a widely used deconvolution method, namely, the Lucy–Richardson algorithm, which iteratively converges to maximum-likelihood estimation (MLE) of fluorescent distribution at the object plane. For this purpose, the *measured* incoherent point-spread function (see, e.g., Figure 16.6a) of the lens-free system is used to compute the two-dimensional fluorescent pattern at the sensor plane (corresponding to an initial estimate of the fluorescent object distribution at the microfluidic chip) which is then compared to the measured lens-free fluorescent image to iteratively update the MLE. This iterative process of the Lucy–Richardson algorithm is terminated typically after a few hundred iterations (Itrs) before the emergence of noise amplification, which can otherwise introduce artifacts [7]. For an image size of, e.g., 4200-by-2700 pixels, computation of 100 Itrs using MATLAB® 2009 takes ~10 min on a dual-core 1.8 GHz Pentium CPU. Since this deconvolution operation is highly parallel, this computation time can be significantly reduced by, e.g., 40-fold using a graphics processing unit (GPU—e.g., NVIDIA GeForce GTX 285). As a result of this Lucy–Richardson deconvolution algorithm, the spatial resolution of our lens-free on-chip imaging platform was digitally improved ~fivefold without a trade-off in the imaging FOV (>8 cm²), enabling ~40–50 μm resolution (see, e.g., Figures 16.2 and 16.3).

To further increase the spatial resolution of our lens-free fluorescent platform, as briefly discussed in the previous section, we utilized an additional planar optical component (i.e., a fiber-optic faceplate—see Figure 16.4) that is placed between the fluorescent objects and the detector array [6]. One of the main functions of this passive optical component is to engineer the point-spread function of our lens-free

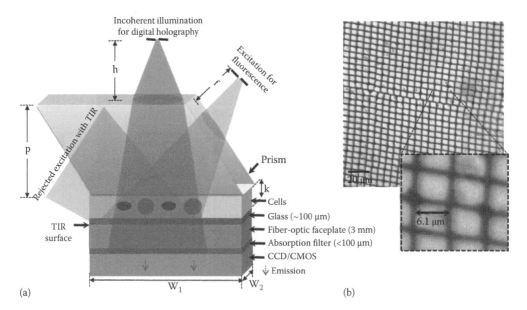

FIGURE 16.4 (a) An alternative lens-free on-chip fluorescent imaging geometry is illustrated. In this configuration, when compared to the setup shown in Figure 16.1, an additional component, i.e., a fiber-optic faceplate, is used to decrease the divergence of fluorescent signatures. (b) 10× microscope image of the faceplate that was used in our experiments is shown. The pitch size of the faceplate is 6.1 μm, and its NA is ~0.3.

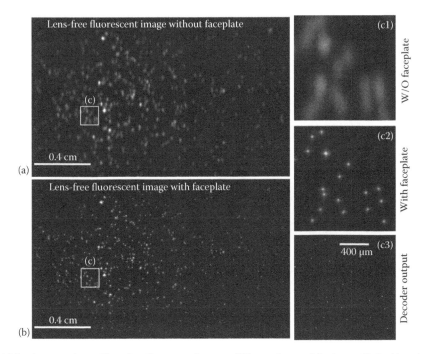

FIGURE 16.5 A comparison of lens-free fluorescent images of 10 μm microparticles is provided with and without the faceplate in (a) and (b), respectively. The FWHM of the fluorescent spots is decreased by ~fivefold from ~180 to ~36 μm with the use of a faceplate. (c1) and (c2) illustrate zoomed images cropped from (a) and (b), respectively. (c3) shows the compressive decoding results corresponding to the same zoomed FOV. As zoomed images (c1–c3) suggest, the resolution of the lens-free raw images can be improved *physically* by use of a fiber-optic faceplate and *digitally* by use of compressive decoding.

fluorescent imaging platform by converting the free space propagation modes of the fluorescent emission from the objects into guided modes of a two-dimensional fiber-optic waveguide array such that the spatial spreading of fluorescent intensity can be tuned to narrow our point-spread function and increase the digital SNR of the lens-free fluorescent image (see, e.g., Figure 16.5c1 and c2). Depending on the NA of the fiber-optic faceplate, coupling efficiency of the fluorescent emission into guided modes of the faceplate would vary, which creates another degree of freedom for engineering the spatial width and the SNR of the lens-free point-spread function of our platform.

In Figure 16.5, we illustrate the physical effect of the fiber-optic faceplate on our imaging platform, where the raw images of 10 μm fluorescent microobjects with and without a faceplate are compared against each other to demonstrate the decrease in the spatial divergence of the fluorescent spots, enabling a significantly narrower point-spread function together with an enhanced SNR. For instance, the spatial width (FWHM) of a single 10 μm fluorescent particle is now narrower by ~fivefold (36 vs. 180 μm) with the use of the fiber-optic faceplate.

While a fiber-optic faceplate provides the feasibility of engineering the point-spread function of our platform to achieve a better spatial resolution without sacrificing the imaging FOV, there is also room for further improving our resolution using more advanced computational tools that are better suited for lens-free fluorescent imaging. One emerging possibility toward this end is the use of compressive sampling or sensing theories [4,5,9] to decode the detected lens-free image. This new mathematical framework enables unique recovery of a signal from much fewer samples than normally required according to the sampling theorem. While such a recovery in general is not possible for any class of signals or objects, it makes an important difference especially for recovering or decoding of *sparse objects*, which can be represented or approximated with few coefficients in a given signal basis. Since fluorescent rare cell analysis on a chip, by definition, has a sparse object distribution at the microfluidic chip plane, its lens-free imaging could benefit from compressive decoding algorithms to achieve an even better spatial resolution.

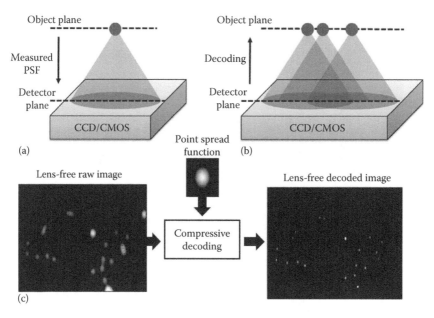

FIGURE 16.6 (a) Using isolated fluorescent particles, the lens-free point-spread function of a given object layer is physically measured. (b) Using the knowledge of this measured point-spread function, any arbitrary fluorescent object distribution at the chip plane can be decoded using a compressive sampling–based optimization algorithm. The procedure is also summarized in (c) where the raw lens-free fluorescent image is compressively decoded using the measured point-spread function of the platform to increase the spatial resolution of the fluorescent image.

In other words, for sparse fluorescent objects/cells that are captured within a large-area microfluidic channel, compressively decoded lens-free image of the microchip can exhibit a significantly larger space-bandwidth product than what the detector array physically has.

Figure 16.6c illustrates simplified schematics of this concept, where a compressive decoding algorithm processes the raw lens-free fluorescent image together with the measured point-spread function of the platform to create a sparse image of the fluorescent source distribution located at the object plane. In Figures 16.7 and 16.8, we further analyze the spatial resolution of this approach by reconstructing the fluorescent images of closely packed 10 and 2 μm fluorescent particles. Even for closely spaced particles shown in Figures 16.7 and 16.8, compressive decoding can resolve them apart, achieving ~10 μm spatial resolution over the entire active area of the optoelectronic sensor array [6]. The computation times for these decoded images vary between 0.1 and 0.5 min using an Intel Centrino Duo Core, 1 GHz PC. Considering the fact that, in these experiments, the physical pixel size at the CCD chip is ~9 μm and that a unit magnification is employed in our platform, a spatial resolution of ~10 μm indeed validates our previous statement that compressively decoded lens-free images of sparse objects exhibit a significantly larger space-bandwidth product than what the detector array physically has.

Furthermore, using the same optical setup shown in Figure 16.4, we also demonstrated lens-free imaging and compressive decoding of vertically stacked microchannels, where microparticles in a suspension were stacked together using thin glass slides creating a vertical separation of 50 and 100 μm (between each layer) as illustrated in Figures 16.9 and 16.10, respectively. Using this multilayer imaging configuration, the throughput of the lens-free fluorescent imaging platform can be increased by another factor of, e.g., 2–3. This requires compressive decoding of a three-dimensional fluorescent object distribution based on two-dimensional lens-free fluorescent images, as illustrated in Figures 16.9 and 16.10. In these decoding results, the lens-free point-spread functions corresponding to each vertical layer are separately measured, which are then used to compute the hypothetical lens-free fluorescent image at the detector plane for an initial estimate of the three-dimensional object distribution. This computed image is then compared with the measured lens-free fluorescent image for iterative modification of the three-dimensional fluorescent object estimation until convergence is achieved [6].

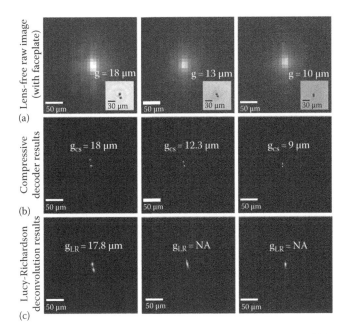

FIGURE 16.7 (a) The raw lens-free fluorescent images of various pairs of 10 μm particles which were all recorded using the setup shown in Figure 16.4. The inset images in (a) show the microscope comparisons of the same particles from which the center-to-center distances (i.e., "g") are calculated. (b) The compressive decoding results, where g_{cs} refers to the center-to-center distances of the resolved fluorescent particles. (c) The Lucy–Richardson deconvolution results, where g_{LR} refers to the center-to-center distances of the resolved fluorescent particles.

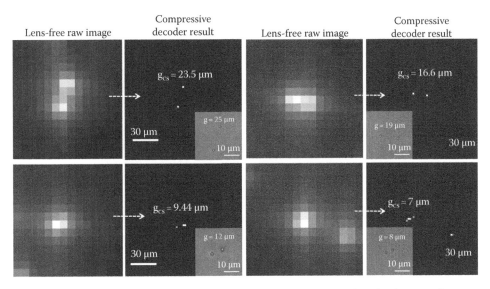

FIGURE 16.8 Same as in Figure 16.7, except for 2 μm fluorescent particles. Once again, using these experiments, system resolution of our platform is tested by resolving closely packed 2 μm particles, achieving a spatial resolution of ~10 μm using compressive decoding of the raw lens-free images. Note that the detector pixel size is 9 μm in these experiments.

This lens-free on-chip fluorescent imaging platform, with powerful digital algorithms behind it, could be very useful for rare cell analysis, high-throughput cytometry, and microarray research. Particularly, rare cell analysis might significantly benefit from wide-field on-chip imaging, since these cells occur at extremely low concentrations in whole blood, which makes them highly sparse on the object plane (a requirement for compressive decoding). Therefore, the presented lensless fluorescent on-chip imaging

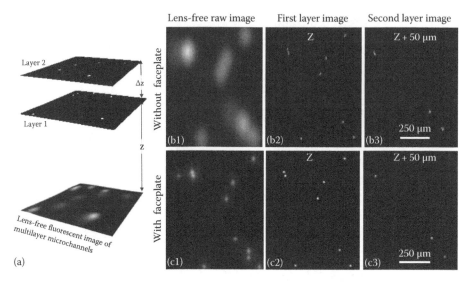

FIGURE 16.9 (a) illustrates the lens-free fluorescent imaging of two layers that are vertically separated by ~50 μm. (b1) and (c1) show a zoomed FOV of the raw lens-free fluorescent images without the faceplate and with the faceplate, respectively. Compressive decoding results of these raw fluorescent images for each vertical channel are shown in (b2–b3) and (c2–c3), respectively.

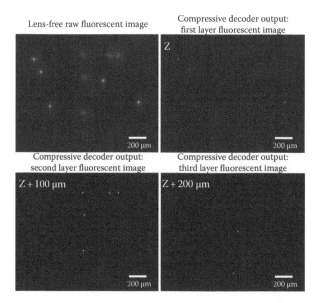

FIGURE 16.10 Similar to Figure 16.9, lens-free fluorescent imaging of three layers that are vertically separated by ~100 μm is shown. Compressive decoding results of each layer are also illustrated.

platform, achieving ~10 μm spatial resolution over an ultrawide FOV of >8 cm², would be quite valuable to rapidly characterize large-area microfluidic devices that can efficiently, sensitively, and specifically capture rare cells from whole blood samples.

To further explore the potentials of our lens-free on-chip imaging approach toward this application, we also imaged FITC (fluorescein isothiocyanate)-labeled breast cancer cells (MCF-7) over an FOV of ~8 cm². FITC labeling is specific to epithelial cells (MCF-7 cells are epithelial in origin), and in blood, there should be no epithelial cells unless they are CTCs. In Figure 16.11, we illustrate the full FOV lens-free fluorescent image of these cells, together with its compressive decoded image. In Figure 16.12, we also compare the lens-free fluorescent images of some of these labeled cancer cells against a conventional 10× fluorescent microscope image, further validating the imaging performance of our lens-free on-chip approach.

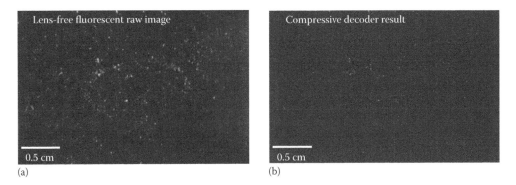

FIGURE 16.11 (a) Lens-free fluorescent on-chip imaging is demonstrated over >8 cm² FOV for FITC-labeled breast cancer cells (i.e., MCF-7). (b) illustrates compressively decoded fluorescent image of the same FOV.

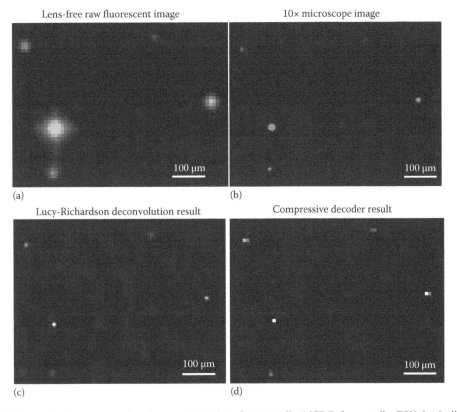

FIGURE 16.12 (a) illustrates lens-free fluorescent imaging of cancer cells (MCF-7) for a smaller FOV that is digitally cropped from Figure 16.11(a). The results of Lucy–Richardson deconvolution and compressive decoding for the same FOV are shown in (c) and (d), respectively. These deconvolved/decoded images shown in (c–d) are in good agreement with the 10× microscope objective image of the same FOV as shown in (b).

16.5 Conclusion and Future Trends

In this chapter, we reviewed a new high-throughput imaging modality for lens-free fluorescent imaging of labeled cells on a chip. In this wide-field imaging technique, fluorescent samples placed on a chip are excited through a prism interface, where the excitation light is filtered out by TIR after exciting the entire sample volume. The emitted fluorescent light from the specimen is then delivered to a wide FOV optoelectronic sensor array for lensless recording of fluorescent spots corresponding to the samples. A compressive sampling–based optimization algorithm can be used to rapidly reconstruct the sparse distribution of fluorescent sources to achieve ~10 μm spatial resolution over the entire active region of the sensor array, i.e., over an imaging FOV of >8 cm². Such a wide-field lensless fluorescent imaging platform could especially be significant for high-throughput imaging cytometry, rare cell analysis, as well as for microarray research.

The presented results in this chapter were obtained using a spatially invariant point-spread function, which can be considered to be at the heart of the limited resolution of this platform. To further improve the spatial resolution in lens-free on-chip fluorescent imaging without degrading the FOV, nanostructured substrates can be utilized as recently demonstrated by our group [18]. Such nanostructured surfaces can encode resolution information at the subpixel level through well-calibrated structures, and the far-field diffraction pattern of the fluorescent specimen on the chip, after being modulated by the substrate, can be decoded using, e.g., compressive sampling algorithms [18], to rapidly reconstruct a higher resolution image of the labeled specimen over a large FOV. Such novel imaging architectures that combine digital processing and nanotechnology would be highly valuable for further extending the limits of lens-free microscopy on a chip, which will especially impact the use of lab-on-a-chip devices for biomedical applications including detection and quantification of CTCs.

Another direction that we have not explored in this chapter toward achieving a better spatial resolution in lens-free on-chip fluorescent imaging is actually the use of *tapered* fiber-optic faceplates where the density of the fiber-optic waveguides on one end of the faceplate is significantly higher compared to the other facet. Such passive optical components can introduce some magnification into our lensless imaging platform to achieve a better resolution that is also at the subpixel level.

REFERENCES

1. Adams A. A., P. I. Okagbare, J. Feng, M. L. Hupert, D. Patterson, J. Gottert, R. L. McCarley, D. Nikitopoulos, M. C. Murphy, and S. A. Soper. 2008. Highly efficient circulating tumor cell isolation from whole blood and label-free enumeration using polymer-based microfluidics with an integrated conductivity sensor. *Journal of the American Chemical Society* **130**:8633–8641.
2. Allard W. J., J. Matera, M. C. Miller, M. Repollet, M. C. Connelly, C. Rao, A. G. J. Tibbe, J. W. Uhr, and L. W. M. M. Terstappen. 2004. Tumor cells circulate in the peripheral blood of all major carcinomas but not in healthy subjects or patients with nonmalignant diseases. *Clinical Cancer Research* **10**:6897–6904.
3. Biggs D. S. C. and M. Andrews. 1997. Acceleration of iterative image restoration algorithms. *Applied Optics* **36**:1766–1775.
4. Candès E. J., J. K. Romberg, and T. Tao. 2006. Stable signal recovery from incomplete and inaccurate measurements. *Communication on Pure and Applied Mathematics* **59**:1207–1223.
5. Candes E. and T. Tao. 2006. Near-optimal signal recovery from random projections: Universal encoding strategies? *IEEE Transactions on Information Theory* **52**:5406–5425.
6. Coskun A. F., I. Sencan, T. Su, and A. Ozcan. 2010. Lensless wide-field fluorescent imaging on a chip using compressive decoding of sparse objects. *Optics Express* **18**:10510–10523.
7. Coskun A. F., T. Su, and A. Ozcan. 2010. Wide field-of-view lens-free fluorescent imaging on a chip. *Lab on a Chip* **10**:824–827.
8. Deng G., M. Herrler, D. Burgess, E. Manna, D. Krag, and J. Burke. 2008. Enrichment with anti-cytokeratin alone or combined with anti-EpCAM antibodies significantly increases the sensitivity for circulating tumor cell detection in metastatic breast cancer patients. *Breast Cancer Research* **10**:R69.

9. Donoho D. 2006. Compressed sensing. *IEEE Transactions on Information Theory* **52**:1289–1306.
10. Fu A. Y., C. Spence, A. Scherer, F. H. Arnold, and S. R. Quake. 1999. A microfabricated fluorescence-activated cell sorter. *Nature Biotechnology* **17**:1109–1111.
11. Gleghorn J. P., E. D. Pratt, D. Denning, H. Liu, N. H. Bander, S. T. Tagawa, D. M. Nanus, P. A. Giannakakou, and B. J. Kirby. 2010. Capture of circulating tumor cells from whole blood of prostate cancer patients using geometrically enhanced differential immunocapture (GEDI) and a prostate-specific antibody. *Lab on a Chip* **10**:27.
12. Grepin C. and C. Pernelle. 2000. High-throughput screening. *Drug Discovery Today* **5**:212–214.
13. Gupta G. P. and J. Massagué. 2006. Cancer metastasis: Building a framework. *Cell* **127**:679–695.
14. Hardingham J. E., D. Kotasek, B. Farmer, R. N. Butler, J. Mi, R. E. Sage, and A. Dobrovic. 1993. Immunobead-PCR: A technique for the detection of circulating tumor cells using immunomagnetic beads and the polymerase chain reaction. *Cancer Research* **53**:3455–3458.
15. Hosokawa M., T. Hayata, Y. Fukuda, A. Arakaki, T. Yoshino, T. Tanaka, and T. Matsunaga. 2010. Size-selective microcavity array for rapid and efficient detection of circulating tumor cells. *Analytical Chemistry* **82**:6629–6635.
16. Hu X., P. H. Bessette, J. Qian, C. D. Meinhart, P. S. Daugherty, and H. T. Soh. 2005. Marker-specific sorting of rare cells using dielectrophoresis. *Proceedings of the National Academy of Sciences of the United States of America* **102**:15757–15761.
17. Ibrahim S. F. and G. van den Engh. 2003. High-speed cell sorting: Fundamentals and recent advances. *Current Opinion in Biotechnology* **14**:5–12.
18. Khademhosseinieh B., I. Sencan, G. Biener, T. Su, A. F. Coskun, D. Tseng, and A. Ozcan. 2010. Lens-free on-chip imaging using nanostructured surfaces. *Applied Physics Letters* **96**:171106.
19. Krivacic R. T., A. Ladanyi, D. N. Curry, H. B. Hsieh, P. Kuhn, D. E. Bergsrud, J. F. Kepros, T. Barbera, M. Y. Ho, L. B. Chen, R. A. Lerner, and R. H. Bruce. 2004. A rare-cell detector for cancer. *Proceedings of the National Academy of Sciences of the United States of America* **101**:10501–10504.
20. Kuo J. S., Y. Zhao, P. G. Schiro, L. Ng, D. S. W. Lim, J. P. Shelby, and D. T. Chiu. 2010. Deformability considerations in filtration of biological cells. *Lab on a Chip* **10**:837.
21. Malo N., J. A. Hanley, S. Cerquozzi, J. Pelletier, and R. Nadon. 2006. Statistical practice in high-throughput screening data analysis. *Nature Biotechnology* **24**:167–175.
22. Mohamed H., M. Murray, J. N. Turner, and M. Caggana. 2009. Isolation of tumor cells using size and deformation. *Journal of Chromatography A* **1216**:8289–8295.
23. Nagrath S., L. V. Sequist, S. Maheswaran, D. W. Bell, D. Irimia, L. Ulkus, M. R. Smith, E. L. Kwak, S. Digumarthy, A. Muzikansky, P. Ryan, U. J. Balis, R. G. Tompkins, D. A. Haber, and M. Toner. 2007. Isolation of rare circulating tumour cells in cancer patients by microchip technology. *Nature* **450**:1235–1239.
24. Riethdorf S., H. Fritsche, V. Müller, T. Rau, C. Schindlbeck, B. Rack, W. Janni, C. Coith, K. Beck, F. Jänicke, S. Jackson, T. Gornet, M. Cristofanilli, and K. Pantel. 2007. Detection of circulating tumor cells in peripheral blood of patients with metastatic breast cancer: A validation study of the CellSearch system. *Clinical Cancer Research* **13**:920–928.
25. Steeg P. S. 2006. Tumor metastasis: Mechanistic insights and clinical challenges. *Nature Medicine* **12**:895–904.
26. Stott S. L., R. J. Lee, S. Nagrath, M. Yu, D. T. Miyamoto, L. Ulkus, E. J. Inserra, M. Ulman, S. Springer, Z. Nakamura, A. L. Moore, D. I. Tsukrov, M. E. Kempner, D. M. Dahl, C. Wu, A. J. Iafrate, M. R. Smith, R. G. Tompkins, L. V. Sequist, M. Toner, D. A. Haber, and S. Maheswaran. 2010. Isolation and characterization of circulating tumor cells from patients with localized and metastatic prostate cancer. *Science Translational Medicine* **2**(25):25ra23.
27. Sundberg S. A. 2000. High-throughput and ultra-high-throughput screening: Solution- and cell-based approaches. *Current Opinion in Biotechnology* **11**:47–53.
28. Wang S., H. Wang, J. Jiao, K. Chen, G. Owens, K. Kamei, J. Sun, D. Sherman, C. Behrenbruch, H. Wu, and H. Tseng. 2009. Three-dimensional nanostructured substrates toward efficient capture of circulating tumor cells. *Angewandte Chemie International Edition* **48**:8970–8973.
29. de Wildt R. M. T., C. R. Mundy, B. D. Gorick, and I. M. Tomlinson. 2000. Antibody arrays for high-throughput screening of antibody-antigen interactions. *Nature Biotechnology* **18**:989–994.

30. Witzig T. E., B. Bossy, T. Kimlinger, P. C. Roche, J. N. Ingle, C. Grant, J. Donohue, V. J. Suman, D. Harrington, J. Torre-Bueno, and K. D. Bauer. 2002. Detection of circulating cytokeratin-positive cells in the blood of breast cancer patients using immunomagnetic enrichment and digital microscopy. *Clinical Cancer Research* **8**:1085–1091.
31. Zheng S., H. Lin, J. Liu, M. Balic, R. Datar, R. J. Cote, and Y. Tai. 2007. Membrane microfilter device for selective capture, electrolysis and genomic analysis of human circulating tumor cells. *Journal of Chromatography A* **1162**:154–161.
32. Zigeuner R. E., R. Riesenberg, H. Pohla, A. Hofstetter, and R. Oberneder. 2000. Immunomagnetic cell enrichment detects more disseminated cancer cells than immunocytochemistry in vitro. *Journal of Urology* **164**:1834–1837.

17

Multiphoton Luminescence from Gold Nanoparticles as a Potential Diagnostic Tool for Early Cancer Detection

Nicholas J. Durr, Marica B. Ericson, and Adela Ben-Yakar

CONTENTS

17.1 Introduction

Cancer is the second leading cause of death, after heart diseases [25]. It is well known that early detection of precancerous lesions can dramatically decrease morbidity and mortality [37]; thus, there is an urgent need for new diagnostic tools in order to aid in early cancer detection. More than 85% of all cancers begin as precancerous lesions that are confined to the superficial region of the epithelium [38]. However, of the currently used clinical imaging modalities, none have sufficient resolution and sensitivity to detect tumors less than a few cubic centimeters in volume [23]. The current gold standard for assessing suspicious cancerous lesions is histopathology of excised tissue biopsies. Generally the tissue biopsy is sectioned and stained before histopathological assessment. This is an invasive, labor intensive, and costly procedure. Furthermore, the accuracy of the pathological diagnosis depends on the subjective assessment of pathologists [58], which makes current tumor diagnostics ambiguous. Therefore, there is an urgent need for alternative techniques for early detection of cancer that can provide noninvasive, three-dimensional (3D), depth-resolved imaging with microscopic resolution, high sensitivity and specificity.

There are several techniques at the research stage that can meet the requirements for obtaining noninvasive imaging; however, their resolution and applicability are varying. For example, high-frequency ultrasound can image down to several millimeters deep in epithelial and connective tissue, but only provides a maximum resolution of approximately 100 μm [60]. Photoacoustic imaging is based on the photoacoustic effect in which laser pulses causing localized heating, leading to transient thermoelastic expansion and thus wideband (e.g., MHz) ultrasonic emission. The generated ultrasonic waves are then detected by ultrasonic transducers to form images. Compared to traditional ultrasound imaging, photo-acoustic imaging can provide slightly better resolution of about $50-100\,\mu m$ [78,79].

Optical techniques can provide substantially better resolution. For example, optical coherence tomography (OCT) can reach resolutions of $10\,\mu m$ in the lateral direction and submicron in the axial directions [16]. Other advantages of OCT include fast imaging and the capability of imaging deep into epithelial tissue—down to approximately $1\,mm$. OCT has found growing success in intravascular imaging over the last decade and is now being applied toward epithelial tissue imaging [24]. Confocal optical imaging technologies reject out-of-focus light to generate 3D images with diffraction-limited resolution, namely, with axial and lateral resolutions below 1 and $0.5\,\mu m$, respectively. Unfortunately, because the confocal pinhole nominally rejects all the scattered emission light, the confocal autofluorescence microscopy, using blue visible excitation light source, can generally probe intrinsic fluorophores in a very thin superficial layer of tissue that is less than $100\,\mu m$ thick. Therefore, confocal autofluorescence microscopy is limited for *in vivo* use as a deep tissue diagnostic tool. As an alternative, confocal reflectance microscopy (CRM) based on imaging of the back-scattering of near-infrared (NIR) light is preferentially applied for improving imaging depths to several hundreds of microns in tissue [59]. However, CRM can only provide structural information with limited diagnostic capabilities. To further improve imaging depth in high-resolution optical imaging while still obtaining both molecular and structural information, nonlinear optical imaging technologies can be applied [81].

Nonlinear microscopy based on multiphoton excitation of endogenous fluorophores, so-called autofluorescence, has emerged as a novel noninvasive diagnostic tool [29,39,47]. Two-photon microscopy (TPM) was first demonstrated by Webb and colleagues in 1990 [53] and has since become the method of choice for 3D imaging of biological tissue. The quadratic dependence of the two-photon excitation process confines the fluorescence emission to the focal volume, which means that a confocal pinhole is not necessary. Unlike in confocal fluorescence microscopy, all the generated fluorescence in TPM can be collected and assumed to originate from the focal volume. More importantly, TPM utilizes NIR light through two-photon absorption process to excite UV-visible absorbing fluorophores [13,47]. In general, the mean free path length is the largest in the NIR wavelength range between 700 and $1400\,nm$, where the scattering is reduced and the absorption of common biological molecules is minimized [36]. The reduced attenuation in the NIR wavelengths effectively makes the tissue more transparent. Consequently, this wavelength regime generally referred to as "the optical window" of biological tissue and is preferentially used when performing optical microscopy to probe deeper layers of biological tissue. Most recently, Ben-Yakar and colleagues have shown that it is possible to image two-photon induced autofluorescence in human oral tissue biopsies down to $370\,\mu m$, reaching the theoretical imaging depths of 3–5 mean free scattering lengths [18]. Even though tissue autofluorescence enables visualizing tissue morphology, cancer diagnostics based on TPM is complex mainly due to weak signal from epidermal layers [29,56]; thus, there is a need for strategies to improve tumor contrast. In addition, the weak signal of autofluorescence makes clinical use of TPM using miniaturized endoscopes very challenging.

An alternative strategy to relying on endogenous contrast is to introduce contrast agents into the sample. The ideal optical contrast agent should provide high contrast by being very bright and targeted to the specific tumor cells, in addition to having low toxicity. Fluorescent organic dyes suffer from problems with photobleaching. Quantum dots, while exhibiting bright fluorescence and photostability, are not suitable for clinical use due to toxicity. Instead, gold nanoparticles are attractive candidates as optical contrast agents because of their bright luminescence, as well as their biocompatibility [56]. Additionally, their optical properties can be tuned in the visible and NIR wavelengths [34], allowing us to take advantage of the optical window of biological tissues.

Since the first demonstration of gold nanoparticles as sensitive probes for Raman spectroscopy [35,51], single-molecule studies [22,64,76], and as contrast agents in the electron microscopy [28], the biomedical optics community has been investigating them as sources of optical contrast for biomedical imaging. The ability to control their optical properties by engineering the particle dimensions and geometry (e.g., nanospheres or nanorods) makes them an ideal source of contrast for a variety of optical imaging modalities. Their tunability is based on the localized surface plasmon resonance effect [21]. Surface plasmon resonance arises as a result of electrons collectively oscillating along the particle surface when

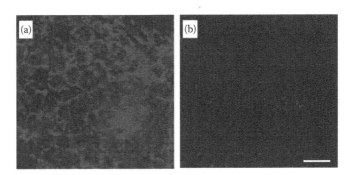

FIGURE 17.1 Laser scanning confocal reflectance images of precancerous (a) and normal fresh cervical ex vivo tissue (b) labeled with anti-EGFR gold conjugates obtained using 647 nm excitation wavelength. The same acquisition conditions have been applied to both images. The scale bar is 20 μm. (Adapted from Sokolov, K. et al., *Cancer Res.*, 63, 1999, 2003. With permission.)

interacting with the incoming electromagnetic radiation. This intense interaction leads the nanoparticles to have strong absorption, scattering, and/or luminescence properties that can be utilized in providing the desired contrast mechanism for various optical imaging modalities.

One of the first uses of gold nanoparticles in combination with optical imaging of tissues was demonstrated by Sokolov et al. [66]. They applied molecularly targeted gold bioconjugates with high scattering cross sections for providing tumor contrast in CRM of human biopsies (Figure 17.1). After incubating tissue slices (200 μm thick) of fresh cervical biopsies with gold nanospheres labeled with anti-EGFR (epidermal growth factor receptor) antibodies, they successfully demonstrated molecularly targeted contrast of tumor cells overexpressing EGFR. Figure 17.1a shows the elevated reflectance signal obtained in the precancerous tissue due to the presence of gold nanospheres, as compared to normal tissue shown in Figure 17.1b. In another study, El-Sayed et al. demonstrated the use of the scattering properties of gold nanoparticles for providing cellular contrast in optical dark field microscopy [20]. Furthermore, gold nanospheres, nanoshells, and nanorods have also been applied as contrast agents for OCT [8,55,65], which also relies on scattering from the nanoparticles for a signal source.

Another approach is to use the absorption properties of gold nanoparticles to provide contrast suitable in, for example, photoacoustic or optoacoustic imaging [2,12,46,75]. The photoacoustic technique is based on detecting the acoustic signal of the expanding media caused by local heating due to laser irradiation. Therefore, the strong absorption of the gold nanoparticles drastically enhances the localized heating, and, thus, the generated photoacoustic signal is increased. Figure 17.2 presents an example from the Emelianov and colleagues [46] where targeted gold nanoparticles significantly enhance the photoacoustic signal from an *in vitro* model for turbid tissue (a tissue phantom). The photoacoustic

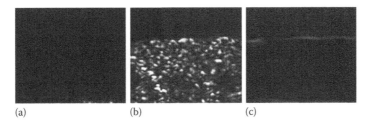

(a)   (b)   (c)

FIGURE 17.2 Photoacoustic images (λ = 680 nm) of tissue phantoms consisting of A431 epithelial skin carcinoma cells: (a) unlabeled control, (b) targeted phantom exposed to anti-EGFR conjugated gold nanoparticles (50 nm), and (c) phantom exposed to nontargeted gold nanoparticles (50 nm). The images measure 2 × 1.67 mm. The faint signal in the lower region of the images in (a) and (c) seems to be due to absorption of the laser by the plastic bottom of the Petri dish holding the phantom. (Adapted from Mallidi, S. et al., *Opt. Express*, 15, 6583, 2007. With permission.)

signal is nearly undetectable when no particles are present (Figure 17.2a), while the signal from cancer cells are clearly visible when labeled with gold nanoparticles functionalized with anti-EGFR (Figure 17.2b). Interestingly, they also demonstrate the possibility of using wavelength tuning, to preferentially generate signal from particles which are bound to cells. As shown by Figure 17.2c, the signal obtained from the tissue phantom exposed to nontargeted gold nanoparticles, i.e., particles without anti-EGFR, is very low, which is explained by a redshift in the absorption spectrum of the functionalized particles. In addition to photoacoustic imaging, the strong light absorption of gold nanoparticles can be used for photothermal therapy [19], which is an intriguing research field in itself and will not be covered in this chapter.

When it comes to providing contrast for nonlinear optical imaging, it is instead the bright photoluminescence from the gold nanoparticles becomes an attractive optical property to be utilized [6,9,15,17,33,49,73,77]. Gold nanorods are particularly appealing because their longitudinal plasmon modes are resonant in the NIR range, where the absorption of water and biological molecules is minimized [67]. The potential of gold nanorods for *in vivo* nonlinear optical imaging was first demonstrated by Cheng and colleagues [44,72]. They presented real-time noninvasive imaging of blood flow in a mouse ear using the multiphoton luminescence (MPL) from the nanorods. In another *in vivo* study, Tunnell and colleagues demonstrate the possibility of using MPL for enhancing the signal when performing tumor imaging using nonlinear microscopy. In this study, they administered gold nanoshells to murine tumors and compared the MPL images to autofluorescence signal [57]. These studies demonstrate the capability of gold nanoparticles as contrast agents in nonlinear optical microscopy; however, in order to provide tumor selectivity, the particles need to be targeted. Thus, our strategy is to adopt the tumor targeting protocol already implemented for other imaging modalities discussed earlier and combine with the strong MPL signal of gold nanorods, to provide tumor contrast. We here demonstrate the technology on experimental tissue phantoms and show that one to two orders of magnitude less excitation power are required when imaging labeled tumor cells compared to utilizing the autofluorescence, proving the strong brightness of gold nanorods when illuminated with femtosecond laser pulses at their resonance. In addition, we investigate the toxicity of the gold nanorods.

17.2 Materials and Methods

17.2.1 Preparation of Gold Nanospheres

We prepared gold nanospheres using the method introduced by Turkevich and colleagues in 1953 [71]. The method is based on reducing a solution consisting of gold chloride using sodium citrate which causes the auric ions to nucleate and form particles. The size distribution of the formed nanoparticles is dependent on the kinetics of the process. The diameter of the nanospheres was measured to be 50 nm with a linear absorption peak of 530 nm.

17.2.2 Preparation of CTAB-Coated Gold Nanorods

The synthesis of gold nanorods generally starts from small gold seeds (1.5–3 nm) [48,62], but their anisotropic growth is promoted by adding cationic surfactant (hexadecyl trimethyl ammonium bromide, CTAB), silver ion, and ascorbic acid to the solution. The CTAB and silver atoms prefer to be adsorbed on the sides of the rods rather than at the edges, which leads to the deposition of gold atoms only on the edges and, subsequently, the formation of the desired gold nanorods. The CTAB molecules also play an important role as a stabilizer for dispersion in an aqueous phase by forming a bilayer on the gold nanorod.

In this study, we synthesized CTAB-coated gold nanorods using this seed-mediated, surfactant-assisted growth method introduced earlier. Colloidal gold seeds (1.5 nm diameter) were first prepared by mixing aqueous solutions of CTAB (0.1 M, 9.75 mL) and hydrogen tetrachloroaurate(III) hydrate (0.01 M,

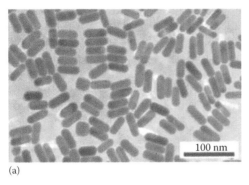

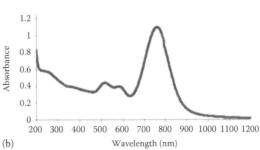

(a) (b)

FIGURE 17.3 Properties of CTAB-coated gold nanorods used as contrast agents. (a) TEM image of the produced gold nanorods indicates an average length and width of 48 ± 6 and 14 ± 2, respectively. (b) The corresponding absorbance spectrum of the gold nanorods in aqueous solution. The absorption peaks at 754 and 520 nm correspond to the longitudinal and latitudinal resonance, respectively.

250 μL). For nanorods to form, we found it necessary to use low-purity CTAB from Fluka (96% purity)* [17]. An aqueous solution of sodium borohydride (0.01 M, 600 μL) was added, and the colloidal gold seeds were injected into an aqueous "growth solution" of CTAB (0.1 M, 9.5 mL), silver nitrate (0.01 M, varying amounts of silver between 20 and 120 μL depending on desired nanorod aspect ratio), hydrogen tetrachloroaurate(III) hydrate (0.01 M, 500 μL), and ascorbic acid (0.1 M, 55 μL). Nanorods were purified by several cycles of centrifugation and resuspension in ultrapure water. They were then isolated in the precipitate, and excess CTAB was removed in the supernatant.

We characterized the physical dimensions of the formed gold nanorods with transmission electron microscopy (TEM) and measured the average aspect ratio of the nanorods as 3.4 ± 0.6. We air dried a 2 μL volume from the stock solution of each sample on a copper grid from Grid Tech© (Cu-400CN). Imaging was performed on Hitachi S-5500 TEM at 150 k magnification. A representative image is shown in Figure 17.3a. Figure 17.3b shows that the longitudinal plasmon mode of these nanorods was centered at 754 nm and the lateral resonance around 520 nm wavelengths.

17.2.3 Preparation of PEGylated Gold Nanorods

Given the known cytotoxicity of CTAB surfactant coating, we also explored labeling with a more biocompatible coating, PEG-coated gold nanorods. In this case, we purchased gold nanorods having the dimensions 39 ± 6 nm×9 ± 1 nm that were coated with neutravidin-terminated PEG [Ntherapy Gold Nanorods, Nanopartz]. The exact size the PEG used in the CTAB dialysis is a trade secret, but the manufacturer did specify that it is less than 5 kDa.

17.2.4 Molecular Targeting of Nanorods Using Antibodies

To obtain molecularly specific targeting of the CTAB-coated gold nanorods, we functionalized the nanorods by adsorbing anti-EGFR antibodies (clone 29.1, Sigma) to the surface of the nanorods. In this protocol, the positive surface potential of the gold nanorods was converted to a negative potential by coating the CTAB nanorods with polystyrene sulfonate (PSS). PSS (MW 14 kDa, 10 mg/mL in 1 mM NaCl

---

* The CTAB source was found to be critical for success in the gold nanorod synthesis. For nanorods to form using the published procedures, it was the necessary to use lower-purity CTAB. CTAB from both Fluka (96% purity) and MP Biomedical (98% purity) resulted in formation of gold nanorods; however, production of nanorods was not possible using CTAB obtained from the following manufacturers: Acros (99% purity), Aldrich (100% purity), and Sigma (99% purity). We speculate that this difference is due to degree of purity of the CTAB, but cannot fully understand the reason. However, reviewing the literature, it is clear that gold nanorod syntheses have only been reported using the CTAB of lower purity from either Fluka or MP Biomedical.

solution) was added to the nanorod suspension in a 1:10 volume ratio and allowed to react for 30 min. The particles were collected via centrifugation at 2000 g for 30 min, resuspended in NaCl (1 mM), and reacted with another aliquot of PSS solution. Following the second PSS incubation, the particles were washed twice in water and then resuspended in HEPES (40 mM, pH 7.4) for compatibility with the antibody solution. Anti-EGFR antibody was purified using 100 kD MW cut-off filters (Centricon) and resuspended in HEPES (40 mM, pH 7.4, 200 μg/mL). Antibody solution and nanorods were mixed at 1:1 volume ratio and allowed to interact for 45 min. Polyethylene glycol (PEG, MW 15 kD, 10 mg/mL in 1× PBS) was then added for stability, and the nanorods centrifuged to remove unbound antibodies. For the control, nanorods with unspecific PEG coating were used.

To conjugate the PEG-neutravidin-functionalized gold nanorods, we used biotin-labeled antibodies for EGFR (clone 111.6, MS-378-B0, Thermo Scientific). The density of the stock antibody solution was 200 μg/mL, and the size of an individual antibody is 145 kDa, which gives an approximate antibody density of 8.3×10^{14} antibodies/mL. We mixed the nanorods, antibody, and 40 mM HEPES solution at a 1:1:4 volume ratio, which results in an interaction of ~67 antibodies for every nanorod. The mixture was allowed to interact for 20 min and then centrifuged at 2000×g for 30 min to remove unbound antibodies. We performed a second washing step, resuspending the pellet in 2 mL of 40 mM HEPES, and centrifuging at 2000×G for 30 min to further remove unbound antibodies.

17.2.5 Tissue Models

In this work, the targeted gold nanorods were demonstrated as potential contrast agents using two model systems: (1) monolayer of cancer cells (either EGFR-overexpressing A431 human epithelial skin cancer cells, American Type Culture Collection, or MDA-MB-468 human epithelial breast cancer cells), and (2) 3D tissue phantoms. The 3D tissue phantoms were prepared from A431 skin cancer cells cultured in DMEM supplemented with 5% fetal bovine serum (FBS, Sigma) incubated at 37°C in a humidified 5% CO_2 incubator. Cells were harvested via trypsinization and resuspended in PBS at a concentration of 6×10^6 cells/mL. The cell suspension was mixed with either EGFR-targeted or PEG-coated nanorods in a 1:1 volume ratio and allowed to interact for 45 min. The cells were then spun down at 200 g for 5 min to remove unbound nanorods. The cells were resuspended in a buffered collagen solution at a concentration of 7.5×10^7 cells/mL. The collagen/cell mix was pipetted into a 500 μm well (Molecular Probes) and sealed with a coverslip for imaging.

17.2.6 Multiphoton Microscopy Imaging System

The MPL imaging in our studies is performed using two different home-built multiphoton microscopy systems, one inverted and the other upright. Both systems were connected to a Ti:sapphire, mode-locked laser oscillator (MaiTai, Spectra Physics, Newport). This source has a repetition rate of 80 MHz, an excitation wavelength, λ_x, tunable from 710 to 880 nm, and an average power, P_{ex}, of 0.6–1.1 W across the tuning range. The full width at half maximum (FWHM) of the spectral bandwidth, $\Delta\lambda$, was measured to be 7.5 nm. Using a time-bandwidth product of 0.44, a transform-limited pulse with this bandwidth would have an FWHM pulse duration, FWHM τ_p^{FWHM}, of 120 fs. We measured the $1/e^2$ diameter of the excitation beam to be approximately 1.5 mm in the vertical direction and 1.2 mm in the horizontal direction at the output of the laser oscillator. The slightly larger divergence of the smaller (horizontal) axis resulted in the beam shape being relatively circular at the position of the objective back aperture. We used two lenses in a Galileo-configuration beam expander with focal lengths of 30 and 75 mm for a 2.5× increase in beam size. Two sets of half-wave plate and polarizing beam cube are used to control the excitation power. In all experiments, the laser was operated at 760 nm.

We used the inverted microscope for single-cell-layer experiments, presented in this chapter. Figure 17.4 shows a schematic drawing and a photo of the inverted setup. The laser light is raster scanned into the back aperture of a high-NA oil immersion objective lenses (Zeiss 63×/1.4 and Olympus 40×/1.3) with a set of galvanometric scanning mirrors (6215H, Cambridge Technologies). Emitted

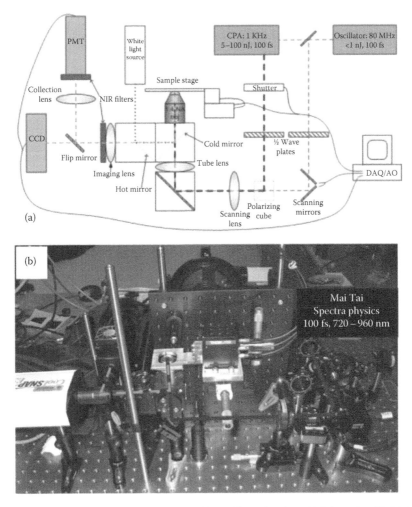

FIGURE 17.4 The inverted laser scanning microscope for nonlinear imaging. (a) Schematic with labeled parts. (b) Photograph of actual microscope. The lateral and axial resolutions are 320 and 625 nm, respectively, using the oil immersion, 1.4 NA, 63× Zeiss objective.

light is epi-collected, reflected by a cold mirror (400–700 nm reflectance, HT-1.00, CVI Laser), passed through a laser filter (BG-38, Schott, blocks > 700 nm), detected with a cooled GaAsP photomultiplier tube (PMT) (H7422-40, Hamamatsu), and assembled into an image in real time with a data acquisition card (6111E, National Instruments). Given the response of the PMT and optics used, the system collects emission light between 400 and 700 nm. The scan area at each image plane is variable to sizes up to 150 × 150 μm. The field of view can be scanned into a 512 × 512 pixel image at a rate of 1.5 frames per second. The sample can be moved in the axial direction with a piezoelectric actuator (P280, Physik Instrumente).

We used the upright microscope for phantom experiments, presented in this chapter. Figure 17.5 shows a photo and schematic drawing of the upright multiphoton microscope. The excitation path consists of a pair of scanning mirrors, two relay lenses, dichroic mirror, and a long working distance, large field-of-view objective lens (20×/0.95 water dipping objective, Olympus). The emission path uses non-descanned collection, large diameter optics, high-throughput filters, and a sensitive PMT to maximize visible-light sensitivity. A customized computer program controls and synchronizes image acquisition, mirror scanning, excitation power control, and sample position. The microscope is configured in an upright configuration—the sample is placed under the objective—to enable the use of a water dipping

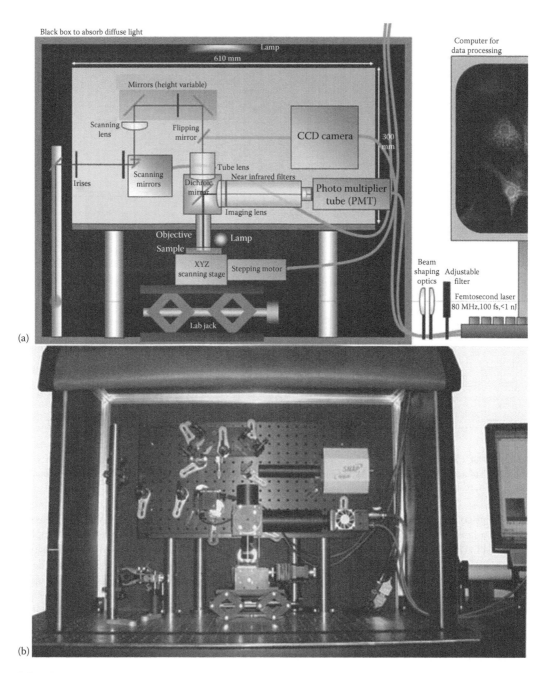

FIGURE 17.5 The upright laser scanning microscope for deep nonlinear imaging. (a) Schematic with labeled parts. (b) Photograph of actual microscope. The lateral and axial resolutions are 320 and 625 nm, respectively, using the water dipping, 0.95 NA, 20× Olympus objective.

objective and to facilitate future studies involving *in vivo* animal studies. Imaging is typically performed at 1×10^6 pixels per second to match the bandwidth of our preamplifier. This setup was used for deep imaging experiments with the tissue phantoms [18].

17.2.7 Toxicity Assay

We performed the toxicity test of the biocompatibility of the gold nanorods using an MTT assay (Promega CellTiter Aqueous One kit). Here, the proliferation of A468 cancer cells was assessed after incubation with either CTAB-coated or PEG-coated gold nanorods at different concentrations for 24 and 48 h. At each time point, the density of cells incubated with each solution was compared to the density of cells incubated in DMEM supplemented with 5%–10% FBS. Concentrations in the range 4–4000 pM were investigated, and around 100 pM was found sufficient for dense labeling of the cells.

17.3 Results

Here, we present the bright MPL properties of gold nanorods labeling cancer cells, as an attractive method to potentially improve tumor contrast for noninvasive diagnostics. First, we present data to demonstrate how the bright MPL significantly increases the signal obtained from nonlinear imaging of individual cancer cells labeled with molecularly targeted gold nanorods. We then present data validating that CTAB-produced gold nanorods can be made biocompatible, by capping the particles with PEG. Finally, we demonstrate the potential of MPL from molecularly targeted gold nanorods as a diagnostic tool in a tissue phantom consisting of cancer cells embedded in a collagen matrix, to simulate *in vivo* tissue.

To show the bright luminescence properties of gold nanorods, we present in Figure 17.6 a comparison between the properties of two-photon autofluorescence images of unlabeled cells and MPL images of nanorod-labeled human carcinoma skin cells (A431). Unlabeled cells show a relatively uniform distribution of two-photon-induced autofluorescence signal throughout cellular cytoplasm (Figure 17.6a), using a laser power of 9 mW. On the other hand, images of the cells, labeled with EGFR-targeting gold nanorods, show signal mainly in the periphery of the cells and interestingly can be acquired using much less laser excitation powers, as low as 0.14 mW (Figure 17.6b). At these low powers, the unlabeled cells cannot be visualized at all (Figure 17.6c). The bright rings observed in the cells labeled with functionalized gold nanorods are consistent with the expected distribution of

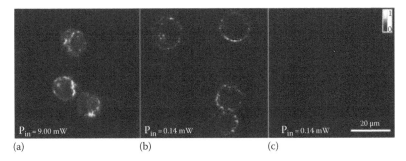

FIGURE 17.6 Multiphoton images of A431 cancer cells without and with labeling of EGFR-targeted gold nanorods (length = 48 nm, width = 14 nm) as imaged using the inverted microscope with an oil immersion 63×/1.4 objective lens and 760 nm excitation wavelength. (a) Two-photon autofluorescence image of unlabeled cells requires 9.00 mW of average laser excitation power. (b) MPL images of cells labeled with EGFR-targeted gold nanorods can be obtained using as little as 0.14 mW of average laser excitation power. The need for 64 times less power to generate the same signal level indicates that MPL from nanorods can be more than 4000 times brighter than two-photon autofluorescence from intrinsic fluorophores. (c) Two-photon autofluorescence image of unlabeled cells does not generate any signal at 0.14 mW of average laser excitation power used to obtain MPL images.

EGFR at the cellular membrane and have been reported also in confocal reflectance imaging using spherical gold nanoparticles [66]. This result therefore indicates successful labeling and confirms that the signal is coming directly from the multiphoton-induced luminescence of gold nanorods. There are also several discrete bright spots in the cytoplasm of nanorod-labeled cells that are indicative of endosomal uptake of labeled EGFRs inside cells [4,63]. Taken together, these results imply that the cells, labeled with functionalized gold nanorods, can be visualized using 64 times less power compared to autofluorescence to obtain similar signal levels, and prove the potential of the nanorods as bright contrast agents for tumor diagnostics.

To find the optimal excitation wavelength used for excitation of the gold nanorods, we tuned the laser excitation wavelength from 710 to 910 nm (data not presented) and found that 760 nm yielded the brightest MPL signal from the nanorods. This excitation wavelength corresponds to the longitudinal plasmon resonance frequency of the nanorods (Figure 17.3b) and also coincides with the optimal excitation wavelength of autofluorescence of the cells. The intrinsic fluorophores, such as NAD(P)H, flavins, retinol, and tryptophan, are mainly responsible for the autofluorescence as they have two-photon cross sections that increase with decreasing excitation wavelength from 1000 to 750 nm, and level off around 750 nm [80]. The similarity of the optimal excitation wavelengths for both the gold nanorods and cellular autofluorescence allows a comparison of the two imaging modalities under identical excitation conditions. As we showed earlier, MPL imaging of nanorod-labeled cells requires 64 times less power than the autofluorescence of unlabeled cells in order to achieve the similar collected intensity (Figure 17.6). Given the quadratic dependence of emission intensity on the incident power, this observation implies that, for equal excitation powers, MPL imaging of nanorod-labeled cancer cells can generate more than 4000 times larger emission signal than two-photon autofluorescence imaging of unlabeled cells.

CTAB is known to be cytotoxic at the concentrations used in nanorod labeling, raising concerns about the *in vivo* use of nanorods for molecular imaging [32,69,70]. To alleviate cytotoxicity concerns, CTAB can be exchanged with PEG, a much more biocompatible material, in a dialysis reaction [52]. In this study, we compare the biocompatibility of CTAB-coated gold nanorods and PEGylated gold nanorods in a cell viability assay (Figure 17.7). This comparison also incorporates gold nanospheres since they are synthesized without CTAB and are expected to show no significant cytotoxic effects. Indeed, the CTAB-coated nanorod samples exhibit significant cytotoxicity within 24 h at concentrations of 400 pM and above. For longer incubation times, above 48 h, even concentrations as low as 40 pM show cytotoxic effects. On the other hand, we observe that this toxicity problem can be eliminated by using PEGylated gold nanorods. As Figure 17.7 indicates, the PEGylated nanorods have very low cytotoxicity at all concentrations after 24 h of incubation. After 48 h exposure, only small effects on cell proliferation are observed and only at the highest concentration 4000 pM (data not shown). These results which are in agreement with other studies [61] emphasize the importance of taking toxicity into consideration in order to avoid unwanted toxic effects using gold nanorods as contrast agents. We can conclude that PEG coating should be preferentially used to avoid unspecific toxicity.

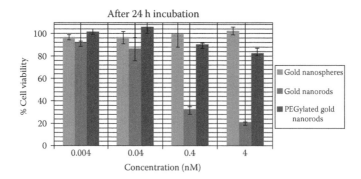

FIGURE 17.7 Cell viability of A468 cancer cells, exposed to various concentrations of gold nanospheres, CTAB-coated gold nanorods, and PEGylated gold nanorods for 24 h, is assessed using MTT assay. While the CTAB-coated nanorods exhibit strong cytotoxicity above 0.4 nM, the PEGylation is shown to drastically reduce cell death.

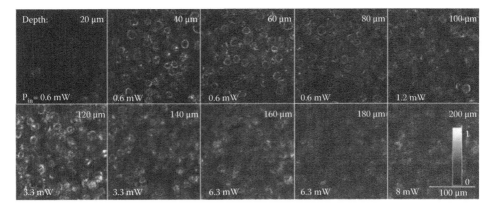

FIGURE 17.8 MPL images of a tissue phantom, consisting of cancer cells that are labeled with anti-EGFR-targeted nanorods as contrast agents and embedded in a collagen matrix. Imaging was performed using the upright microscope with a 20×/0.95 NA water dipping objective, allowing deep tissue imaging. The laser excitation power was increased to maintain constant emission intensity at each imaging depth.

To test the potential of the targeted gold nanorods not only to elevate the luminescence signal of a few cells but also luminescence of tissue, we used 3D cellular tissue phantom, mimicking epithelial tissue. The tissue phantom is constructed by human cancer cells (A431) that are labeled with targeted gold nanorods and embedded in a collagen matrix. Here, we used the upright microscope with long working distance, water dipping objective to facilitate deep tissue imaging because the inverted microscope limits imaging depths to a few tens of microns due to the spherical aberrations and short working distance of the oil immersion objective. Figure 17.8 shows the MPL signal of the tissue phantom obtained at different depths by performing optical sectioning. The increased PMT gain configuration used in this experiment allowed imaging at 10 times less power than the single-cell-layer experiments presented in Figure 17.6. To maintain a constant detected intensity throughout the phantoms, we increased the power when imaging deeper into the phantom. Even with the addition of gold nanoparticles or nanorods, the tissue extinction coefficient is dominated by the tissue scattering. Based on our calculations (unpublished data), the mean free path of a 780 nm photon in epithelial tissue only modestly decreases from 99 to 98 μm, when comparing the two different cases without or with the presence of nanorods. This approximate calculation is consistent with our experiments in which we observed no significant change in extinction coefficient in gold-nanorod-labeled tissue phantoms [17]. The ability to perform imaging down to a depth of 200 μm demonstrates the potential of gold nanorods to be used as contrast agents without inflicting on the maximum imaging depth achievable in biological tissue when performing TPM for *in vivo* imaging and tumor diagnostics.

17.4 Discussion

Here, we demonstrate the potential of gold nanorods as bright nonlinear plasmonic contrast agents based on MPL for molecularly targeted imaging of cancer cells. By functionalizing gold nanorods with anti-EGFR antibodies, we can molecularly target the cancer cells. The three orders of magnitude brighter signal of gold-nanorod-labeled cells as compared to the cell autofluorescence imply that we need much less laser excitation powers for *in vivo* imaging. This technique, therefore, has the potential to selectively diagnose tumor cells as the nanorod-targeted tumor cells will appear much brighter compared to the low-autofluorescence, unlabeled background cells, although this experiment still needs to be demonstrated.

Within the last couple of years, nonlinear microscopy has gained elevated attention because of its potential for becoming a noninvasive diagnostic tool [14,40,56]. While two-photon-induced autofluorescence of tissue may provide some contrast due to the presence of intrinsic fluorophores, such as

NAD(P)H, flavins, retinol, and tryptophan, it exhibits weak emission due to the small values of the two-photon absorption cross sections on the order of $10^{-4}-10^{-1}$ GM [80,81]. Second harmonic generation (SHG) imaging, being a nonlinear scattering process, can also provide contrast to native structures in biological tissues; however, it is restricted to fibrillar structures, such as collagen, axons, muscle filaments, and microtubule assemblies [11].

Exogenous fluorophores could potentially be used as alternative contrast agents to nonlinear imaging with only intrinsic fluorophores. Among the brightest of these fluorophores, some organic dyes have two-photon action cross sections on the order of 10–100 GM [3,74,81]. Also quantum dots are being considered as two-photon contrast agents, and extremely large two-photon cross sections of up to 50,000 GM have been reported [42]. However, problem with toxicity is the major obstacle of developing organic dyes or quantum dots to potential clinically applicable contrast agents. Instead, gold nanoparticles have been shown to have minimal cytotoxic effects and are believed to be relatively biocompatible [1,43]. We here demonstrate that PEG coating of CTAB-produced gold nanorods enhances their biocompatibility with cultured cancer cells.

Our measurements of gold nanoparticle brightness indicate that the effective two-photon absorption cross section of gold nanorods can be as high as 10^6 GM (unpublished data). Historically, the large multiphoton absorption cross-section values observed in metallic particles were somewhat unexpected. Instead, gold particles were known for their quenching of fluorescence rather than exhibiting fluorescence emission [41]. The lifetime of excited electrons in metals is extremely fast, on the order of tens to hundreds of femtoseconds [7,68], leading to very short emission lifetimes, on the order of 10^{-10} s [10]. Nonetheless, weak luminescence can be obtained. Single-photon-induced luminescence from bulk copper and gold was first reported in 1969 [50]. Later, it was found that roughened metal surfaces exhibited much higher luminescence efficiency than smooth surfaces [10]. In 2000, it was found that gold nanorods offered dramatically larger quantum yield than bulk gold, which was dubbed, the "lightning rod" effect [49]. The results presented here support the hypothesis that the presence of a surface plasmon resonance is important to the efficient generation of luminescence from gold nanorods.

As gold nanorods exhibit intense optical interaction properties when excited at their plasmon resonance, they can preferably be utilized to improve contrast for nonlinear optical technologies. Furthermore, the use of bright contrast agents instead of the dim endogenous fluorophores can enable imaging using clinically relevant nonlinear endoscopes that are less efficient [29,30]. Thus, the large brightness of gold nanoparticles can enable molecularly specific imaging in systems with poor collection efficiencies and/or limited excitation fluence, such as that typically found in nonlinear endoscopic probes [12–15].

17.5 Future Trends

To achieve molecularly specific nonlinear imaging of tissues with plasmonic contrast agents, several challenges remain. The biggest disadvantage of using gold nanoparticles as contrast agents is their large size. While some groups have had success in the topical application of gold nanoparticles by using permeation enhancers, such as chitosan [26], the delivery of gold nanoparticles, which are orders of magnitude larger than molecular fluorophores, remains a challenge. Therefore, more detailed studies exploring the topical delivery and biodistribution of gold particles are necessary to understand the true biocompatibility of this class of contrast agents. In addition, further studies on clearance of the compounds are important.

While the large size of gold nanoparticles makes topical delivery and clearance challenging, it can actually be advantageous when it comes to intravascular delivery and plasmonic photothermal treatment [31]. Because of the leaky vasculature phenomenon, their large size may allow preferentially delivering them to the regions of tumors [54]. Their large size may also be beneficial in preferentially heating the nanorods. There has been indeed important progress in using gold nanoparticles as therapeutic agents, based on the large absorption of gold nanoparticles at NIR wavelengths for targeted thermal therapy

[19,27,31,45,53,54]. Excitation of nanoparticles with ultrashort laser pulses opens the possibility for additional novel laser therapies [5]. Thus, a highly interesting future direction is the combination of diagnostic and therapeutic agents, in the so-called theragnostic contrast agent [57].

In conclusion, we here demonstrate the potential of the combination of nonlinear optical microscopy and molecularly targeted gold nanorods for high-resolution diagnostic imaging of cancer cells. By the use of nonlinear optical microscopy techniques based on femtosecond laser pulses operating in the optical window of tissue in the NIR wavelength region, we can perform deep tissue imaging as demonstrated by the 3D tissue phantoms. While endogenous fluorophores can visualize tissue morphology, enhanced tumor contrast is desirable, which we obtain using gold nanorods that exhibit bright MPL and can be molecularly targeted by conjugation to cancer-specific moieties such as antibodies. With some further refinement, the technique perhaps has its most important application for noninvasive multiphoton endoscopic diagnostics as improved tumor contrast and bright signal will here overcome the poor collection efficiencies with these types of miniaturized instruments.

ACKNOWLEDGMENTS

The authors acknowledge support from the National Science Foundation under grants BES-0548673, CBET-1014953, and Career Award, CBET-0846868, a grant from the Texas Ignition Fund by The University of Texas Board of Regents, and VINNOVA (2008-03414, Sweden).

REFERENCES

1. Abrams, M. J. and B. A. Murrer. 1993. Metal compounds in therapy and diagnosis. *Science* **261**:725.
2. Agarwal, A., S. W. Huang, M. O'Donnell, K. C. Day, M. Day, N. Kotov, and S. Ashkenazi. 2007. Targeted gold nanorod contrast agent for prostate cancer detection by photoacoustic imaging. *Journal of Applied Physics* **102**:064701.
3. Albota, M. A., C. Xu, and W. W. Webb. 1998. Two-photon fluorescence excitation cross sections of biomolecular probes from 690 to 960 nm. *Applied Optics* **37**:7352.
4. Barnes, C. J. and K. Rakesh. 2004. *Biology of the Epidermal Growth Factor Receptor Family*, Vol. 119. Springer, New York.
5. Ben-Yakar, A., D. S. Eversole, and O. Ekici. 2008. Spherical and anisotropic gold nanoparticles in plasmonic laser phototherapy of cancer. In C. Kumar (ed.), *Non-Magnetic Metallic Nanomaterials for Life Sciences of the 10 Volume Series on Nanomaterials for Life Sciences*. John Wiley & Sons, Weinheim, Germany, pp. 493–539.
6. Beversluis, M., A. Bouhelier, and L. Novotny. 2003. Continuum generation from single gold nanostructures through near-field mediated intraband transitions. *Physical Review B* **68**:115433.
7. Biagioni, P., M. Celebrano, M. Savoini, G. Grancini, D. Brida, S. Matefi-Tempfli, M. Matefi-Tempfli, L. Duo, B. Hecht, G. Cerullo, and M. Finazzi. 2009. Dependence of the two-photon photoluminescence yield of gold nanostructures on the laser pulse duration. *Physical Review B* **80**:045411.
8. Boppart, S. A., A. L. Oldenburg, C. Xu, and D. L. Marks. 2005. Optical probes and techniques for molecular contrast enhancement in coherence imaging. *Journal of Biomedical Optics* **10**:41208.
9. Bouhelier, A., M. R. Beversluis, and L. Novotny. 2003. Characterization of nanoplasmonic structures by locally excited photoluminescence. *Applied Physics Letters* **83**:5041.
10. Boyd, G. T., Z. H. Yu, and Y. R. Shen. 1986. Photoinduced luminescence from the noble metals and its enhancement on roughened surfaces. *Physical Review B, Condensed Matter* **33**:7923.
11. Brown, E., T. McKee, E. diTomaso, A. Pluen, B. Seed, Y. Boucher, and R. K. Jain. 2003. Dynamic imaging of collagen and its modulation in tumors in vivo using second-harmonic generation. *Nature Medicine* **9**:796.
12. Copland, J. A., M. Eghtedari, V. L. Popov, N. Kotov, N. Mamedova, M. Motamedi, and A. A. Oraevsky. 2004. Bioconjugated gold nanoparticles as a molecular based contrast agent: Implications for imaging of deep tumors using optoacoustic tomography. *Molecular Imaging and Biology: MIB: The Official Publication of the Academy of Molecular Imaging* **6**:341.

13. Denk, W., J. H. Strickler, and W. W. Webb. 1990. Two-photon laser scanning fluorescence microscopy. *Science* **248**:73.

14. Dimitrow, E., M. Ziemer, M. J. Koehler, J. Norgauer, K. Konig, P. Elsner, and M. Kaatz. 2009. Sensitivity and specificity of multiphoton laser tomography for in vivo and ex vivo diagnosis of malignant melanoma. *Journal of Investigative Dermatology* **129**:1752.

15. Drachev, V. P., E. N. Khaliullin, W. Kim, F. Alzoubi, S. G. Rautian, V. P. Safonov, R. L. Armstrong, and V. M. Shalaev. 2004. Quantum size effect in two-photon excited luminescence from silver nanoparticles. *Physical Review B* **69**:035318.

16. Drexler, W. 2004. Ultrahigh-resolution optical coherence tomography. *Journal of Biomedical Optics* **9**:47–74.

17. Durr, N. J., T. Larson, D. K. Smith, B. A. Korgel, K. Sokolov, and A. Ben-Yakar. 2007. Two-photon luminescence imaging of cancer cells using molecularly targeted gold nanorods. *Nano Letters* **7**:941.

18. Durr, N. J., C. T. Weisspfennig, B. A. Holfeld, and A. Ben-Yakar. 2011. Maximum imaging depth of two-photon autofluorescence microscopy in epithelial tissues. *Journal of Biomedical Optics* **16**:026008.

19. El-Sayed, I. H., X. Huang, and M. A. El-Sayed. 2006. Selective laser photo-thermal therapy of epithelial carcinoma using anti-EGFR antibody conjugated gold nanoparticles. *Cancer Letters* **239**:129.

20. El-Sayed, I. H., X. Huang, and M. A. El-Sayed. 2005. Surface plasmon resonance scattering and absorption of anti-EGFR antibody conjugated gold nanoparticles in cancer diagnostics: Applications in oral cancer. *Nano Letters* **5**:829.

21. El-Sayed, M. A. 2001. Some interesting properties of metals confined in time and nanometer space of different shapes. *Accounts of Chemical Research* **34**:257.

22. Elghanian, R., J. J. Storhoff, R. C. Mucic, R. L. Letsinger, and C. A. Mirkin. 1997. Selective colorimetric detection of polynucleotides based on the distance-dependent optical properties of gold nanoparticles. *Science* **277**:1078.

23. Frangioni, J. V. 2006. Translating in vivo diagnostics into clinical reality. *Nature Biotechnology* **24**:909.

24. Gambichler, T., G. Moussa, M. Sand, D. Sand, P. Altmeyer, and K. Hoffmann. 2005. Applications of optical coherence tomography in dermatology. *Journal of Dermatological Science* **40**:85–94.

25. Garcia, M., A. Jemal, E. M. Ward, M. M. Center, Y. Hao, R. I. Siegel, and M. J. Thun. 2007. Global cancer facts and figures 2007. *Global Cancer Facts & Figures 2007*. Atlanta, GA: American Cancer Society.

26. Ghosn, B., A. L. van de Ven, J. Tam, A. Gillenwater, K. V. Sokolov, R. Richards-Kortum, and K. Roy. 2010. Efficient mucosal delivery of optical contrast agents using imidazole-modified chitosan. *Journal of Biomedical Optics* **15**:015003.

27. Hirsch, L. R., R. J. Stafford, J. A. Bankson, S. R. Sershen, B. Rivera, R. E. Price, J. D. Hazle, N. J. Halas, and J. L. West. 2003. Nanoshell-mediated near-infrared thermal therapy of tumors under magnetic resonance guidance. *Proceedings of the National Academy of Sciences of the United States of America* **100**:13549.

28. Horisberger, M. and J. Rosset. 1977. Colloidal gold, a useful marker for transmission and scanning electron-microscopy. *Journal of Histochemistry & Cytochemistry* **25**:295–305.

29. Hoy, C., N. Durr, P. Y. Chen, D. K. Smith, T. Larson, W. Piyawattanametha, H. J. Ra, B. Korgel, K. Sokolov, O. Solgaard, and A. Ben-Yakar. 2008. Two-photon luminescence imaging using a MEMS-based miniaturized probe. *Conference on Lasers and Electro-Optics and Quantum Electronics and Laser Science Conference* **1–9**:951–952.

30. Hoy, C. L., N. J. Durr, P. Chen, W. Piyawattanametha, H. Ra, O. Solgaard, and A. Ben-Yakar. 2008. Miniaturized probe for femtosecond laser microsurgery and two-photon imaging. *Optics Express* **16**:9996.

31. Huang, X., I. H. El-Sayed, W. Qian, and M. A. El-Sayed. 2006. Cancer cell imaging and photothermal therapy in the near-infrared region by using gold nanorods. *Journal of the American Chemical Society* **128**:2115.

32. Huang, X., S. Neretina, and M. A. El-Sayed. 2009. Gold nanorods: From synthesis and properties to biological and biomedical applications. *Advanced Materials* **21**:4880.

33. Imura, K., T. Nagahara, and H. Okamoto. 2004. Plasmon mode imaging of single gold nanorods. *Journal of the American Chemical Society* **126**:12730–12731.

34. Jain, P. K., K. S. Lee, I. H. El-Sayed, and M. A. El-Sayed. 2006. Calculated absorption and scattering properties of gold nanoparticles of different size, shape, and composition: Applications in biological imaging and biomedicine. *Journal of Physical Chemistry B* **110**:7238.

35. Kneipp, K., A. S. Haka, H. Kneipp, K. Badizadegan, N. Yoshizawa, C. Boone, K. E. Shafer-Peltier, J. T. Motz, R. R. Dasari, and M. S. Feld. 2002. Surface-enhanced Raman spectroscopy in single living cells using gold nanoparticles. *Applied Spectroscopy* **56**:150.

36. Konig, K. 2000. Multiphoton microscopy in life sciences. *Journal of Microscopy-Oxford* **200**:83–104.

37. Kort, E. J., N. Paneth, and G. F. Vande Woude. 2009. The decline in U.S. cancer mortality in people born since 1925. *Cancer Research* **69**:6500.

38. Kumar, V., N. Fausto, and A. Abbas. 2004. *Robbins & Cotran Pathologic Basis of Disease*, 7th edn. Saunders, Philadelphia, PA.

39. König, K. and I. Riemann. 2003. High-resolution multiphoton tomography of human skin with subcellular spatial resolution and picosecond time resolution. *Journal of Biomedical Optics* **8**:432–439.

40. König, K., M. Speicher, R. Bückle, J. Reckfort, G. McKenzie, J. Welzel, M. J. Koehler, P. Elsner, and M. Kaatz. 2009. Clinical optical coherence tomography combined with multiphoton tomography of patients with skin diseases. *Journal of Biophotonics* **2**:389.

41. Lakowicz, J. R. 2006. *Principles of Fluorescence Spectroscopy*. Springer, New York.

42. Larson, D. R., W. R. Zipfel, R. M. Williams, S. W. Clark, M. P. Bruchez, F. W. Wise, and W. W. Webb. 2003. Water-soluble quantum dots for multiphoton fluorescence imaging in vivo. *Science* **300**:1434.

43. Lewinski, N., V. Colvin, and R. Drezek. 2008. Cytotoxicity of nanoparticles. *Small* **4**:26.

44. Liu, C.-L., M.-L. Ho, Y.-C. Chen, C.-C. Hsieh, Y.-C. Lin, Y.-H. Wang, M.-J. Yang, H.-S. Duan, B.-S. Chen, J.-F. Lee, J.-K. Hsiao, and P.-T. Chou. 2009. Thiol-functionalized gold nanodots: Two-photon absorption property and imaging in vitro. *The Journal of Physical Chemistry C* **113**:21082.

45. Loo, C., A. Lowery, N. Halas, J. West, and R. Drezek. 2005. Immunotargeted nanoshells for integrated cancer imaging and therapy. *Nano Letters* **5**:709.

46. Mallidi, S., T. Larson, J. Aaron, K. Sokolov, and S. Emelianov. 2007. Molecular specific optoacoustic imaging with plasmonic nanoparticles. *Optics Express* **15**:6583.

47. Masters, B. R., P. T. So, and E. Gratton. 1997. Multiphoton excitation fluorescence microscopy and spectroscopy of in vivo human skin. *Biophysical Journal* **72**:2405.

48. Mitamura, K. and T. Imae. 2009. Functionalization of gold nanorods toward their applications. *Plasmonics* **4**:23–30.

49. Mohamed, M. B., V. Volkov, S. Link, and M. A. El-Sayed. 2000. The 'lightning' gold nanorods: Fluorescence enhancement of over a million compared to the gold metal. *Chemical Physics Letters* **317**:517.

50. Mooradian, A. 1969. Photoluminescence of metals. *Physical Review Letters* **22**:185.

51. Nabiev, I. R., H. Morjani, and M. Manfait. 1991. Selective analysis of antitumor drug interaction with living cancer cells as probed by surface-enhanced Raman spectroscopy. *European Biophysics Journal* **19**:311–316.

52. Niidome, T., M. Yamagata, Y. Okamoto, Y. Akiyama, H. Takahashi, T. Kawano, Y. Katayama, and Y. Niidome. 2006. PEG-modified gold nanorods with a stealth character for in vivo applications. *Journal of Controlled Release* **114**:343.

53. Norman, R. S., J. W. Stone, A. Gole, C. J. Murphy, and T. L. Sabo-Attwood. 2008. Targeted photothermal lysis of the pathogenic bacteria, *Pseudomonas aeruginosa*, with gold nanorods. *Nano Letters* **8**:302.

54. O'Neal, D. P., L. R. Hirsch, N. J. Halas, J. D. Payne, and J. L. West. 2004. Photo-thermal tumor ablation in mice using near infrared-absorbing nanoparticles. *Cancer Letters* **209**:171.

55. Oldenburg, A. L., M. N. Hansen, D. A. Zweifel, A. Wei, and S. A. Boppart. 2006. Plasmon-resonant gold nanorods as low backscattering albedo contrast agents for optical coherence tomography. *Optics Express* **14**:6724.

56. Paoli, J., M. Smedh, and M. B. Ericson. 2009. Multiphoton laser scanning microscopy—A novel diagnostic method for superficial skin cancers. *Seminars in Cutaneous Medicine and Surgery* **28**:190–195.

57. Park, J., A. Estrada, K. Sharp, K. Sang, J. A. Schwartz, D. K. Smith, C. Coleman, J. D. Payne, B. A. Korgel, A. K. Dunn, and J. W. Tunnell. 2008. Two-photon-induced photoluminescence imaging of tumors using near-infrared excited gold nanoshells. *Optics Express* **16**:1590.

58. Raab, S. S., D. M. Grzybicki, J. E. Janosky, R. J. Zarbo, F. A. Meier, C. Jensen, and S. J. Geyer. 2005. Clinical impact and frequency of anatomic pathology errors in cancer diagnoses. *Cancer* **104**:2205–2213.

59. Rajadhyaksha, M., M. Grossman, D. Esterowitz, R. H. Webb, and R. R. Anderson. 1995. In vivo confocal scanning laser microscopy of human skin: Melanin provides strong contrast. *Journal of Investigative Dermatology* **104**:946.

60. Rallan, D. and C. C. Harland. 2003. Ultrasound in dermatology—Basic principles and applications. *Clinical and Experimental Dermatology* **28**:632–638.

61. Rayavarapu, R. G., W. Petersen, L. Hartsuiker, P. Chin, H. Janssen, F. W. B. van Leeuwen, C. Otto, S. Manohar, and T. G. van Leeuwen. 2010. In vitro toxicity studies of polymer-coated gold nanorods. *Nanotechnology* **21**:145101.

62. Sau, T. K. and C. J. Murphy. 2004. Seeded high yield synthesis of short Au nanorods in aqueous solution. *Langmuir* **20**:6414–6420.

63. Schlessinger, J. 2000. Cell signaling by receptor tyrosine kinases. *Cell* **103**:211–225.

64. Schultz, S., D. R. Smith, J. J. Mock, and D. A. Schultz. 2000. Single-target molecule detection with nonbleaching multicolor optical immunolabels. *Proceedings of the National Academy of Sciences of the United States of America* **97**:996.

65. Skala, M. C., M. J. Crow, A. Wax, and J. A. Izatt. 2008. Photothermal optical coherence tomography of epidermal growth factor receptor in live cells using immunotargeted gold nanospheres. *Nano Letters* **8**:3461.

66. Sokolov, K., M. Follen, J. Aaron, I. Pavlova, A. Malpica, R. Lotan, and R. Richards-Kortum. 2003. Real-time vital optical imaging of precancer using anti-epidermal growth factor receptor antibodies conjugated to gold nanoparticles. *Cancer Research* **63**:1999.

67. Sonnichsen, C. and A. P. Alivisatos. 2005. Gold nanorods as novel nonbleaching plasmon-based orientation sensors for polarized single-particle microscopy. *Nano Letters* **5**:301.

68. Sun, C. K., F. Vallee, L. H. Acioli, E. P. Ippen, and J. G. Fujimoto. 1994. Femtosecond-tunable measurement of electron thermalization in gold. *Physical Review B* **50**:15337.

69. Takahashi, H., Y. Niidome, T. Niidome, K. Kaneko, H. Kawasaki, and S. Yamada. 2006. Modification of gold nanorods using phospatidylcholine to reduce cytotoxicity. *Langmuir* **22**:2.

70. Tong, L., Q. Wei, A. Wei, and J.-X. Cheng. 2009. Gold nanorods as contrast agents for biological imaging: Optical properties, surface conjugation and photothermal effects. *Photochemistry and Photobiology* **85**:21.

71. Turkevich, J., P. C. Stevenson, and J. Hillier. 1953. The formation of colloidal gold. *The Journal of Physical Chemistry* **57**:670.

72. Wang, H., T. B. Huff, D. A. Zweifel, W. He, P. S. Low, A. Wei, and J. X. Cheng. 2005. In vitro and in vivo two-photon luminescence imaging of single gold nanorods. *Proceedings of the National Academy of Sciences* **102**:15752.

73. Wilcoxon, J. P., J. E. Martin, F. Parsapour, B. Wiedenman, and D. F. Kelley. 1998. Photoluminescence from nanosize gold clusters. *Journal of Chemical Physics* **108**:9137.

74. Xu, C. and W. W. Webb. 1996. Measurement of two-photon excitation cross sections of molecular fluorophores with data from 690 to 1050 nm. *Journal of the Optical Society of America B—Optical Physics* **13**:481.

75. Xu, M. H. and L. H. V. Wang. 2006. Photoacoustic imaging in biomedicine. *Review of Scientific Instruments* **77**:1–22.

76. Yasuda, R., H. Noji, M. Yoshida, K. Kinosita, and H. Itoh. 2001. Resolution of distinct rotational substeps by submillisecond kinetic analysis of F1-ATPase. *Nature* **410**:898.

77. Yelin, D., D. Oron, S. Thiberge, E. Moses, and Y. Silberberg. 2003. Multiphoton plasmon-resonance microscopy. *Optics Express* **11**:1385.

78. Zhang, E. Z., J. G. Laufer, R. B. Pedley, and P. C. Beard. 2009. In vivo high-resolution 3D photoacoustic imaging of superficial vascular anatomy. *Physics in Medicine and Biology* **54**:1035.

79. Zhang, H. F., K. Maslov, G. Stoica, and L. V. Wang. 2006. Functional photoacoustic microscopy for high-resolution and noninvasive in vivo imaging. *Nature Biotechnology* **24**:848.

80. Zipfel, W. R., R. M. Williams, R. Christie, A. Y. Nikitin, B. T. Hyman, and W. W. Webb. 2003. Live tissue intrinsic emission microscopy using multiphoton-excited native fluorescence and second harmonic generation. *Proceedings of the National Academy of Sciences* **100**:7075.

81. Zipfel, W. R., R. M. Williams, and W. W. Webb. 2003. Nonlinear magic: Multiphoton microscopy in the biosciences. *Nature Biotechnology* **21**:1368.

18

Early Detection of Oral Cancer Using Biooptical Imaging Technologies

Malini Olivo, Ramaswamy Bhuvaneswari, Kho Kiang Wei, Ivan Keogh, and Soo Khee Chee

CONTENTS

18.1 Introduction

Oral squamous cell carcinoma (OSCC) represents an important pathology of the upper digestive tract, being the sixth common cancer diagnosed around the world [40]. Almost 80% of oral cancer patients would have a 5-year survival rate if their disease had been detected early [28]. Although improvements have been achieved in surgical techniques, radiation therapy protocols, and chemotherapeutic regimes, oral cancer remains a lethal disease for over 50% of cases diagnosed annually, and it has one of the lowest survival rates of about 50%, within a 5-year period [8]. Despite improvements in diagnostic and therapeutic modalities and easy accessibility of the oral cavity, prognosis of patients with oral malignancies has remained poor. This is largely due to the fact that most cases of oral cancer are in advanced stages at the time of detection. Thus, early detection and diagnosis of neoplastic changes may be the best way to improve patient outcomes.

Current clinical diagnosis and surveillance of oral cancer typically involve the use of white light endoscopy and histopathology of biopsy samples. However, as oral tumors are mostly superficial, the lesion can easily go undetected even with white light endoscopy [19]. Also, performing invasive needle biopsies can cause psychological trauma and risk of infection to patients. Furthermore, conventional histopathological diagnosis is based on morphological and structural changes at the cellular or tissue level, which may not be obvious for early-stage tumors [42]. Given the difficulty of detecting oral cancer in early stages [27], any tool that provides rapid and accurate diagnostics for early oral cancer detection is highly desirable. Therefore, optical technologies that are simple to use, robust, portable, and of low cost are needed for effective clinical diagnosis.

Recent advancements in oral cancer research have led to the development of potentially useful diagnostic tools at the clinical and molecular level for early detection of oral cancer [36]. For example, several studies have demonstrated the ability of autofluorescence imaging to distinguish normal from premalignant and malignant oral tissue [3,25,30]. This promising approach uses the differences in native autofluorescence properties between normal and neoplastic tissue to visually detect abnormal oral mucosal areas. Living tissues contain endogenous fluorophores such as NADH (nicotinamide adenine dinucleotide), FAD (flavin adenine dinucleotide), and collagen and elastin cross-links that produce fluorescence after excitation with light of specific wavelengths. The images of the emitted fluorescence are visualized and recorded, producing a spatial depiction of areas with altered fluorescence [7]. Another useful method is fluorescence diagnosis (FD), which is a noninvasive imaging technique that uses photosensitizers such as 5-aminolevulinic acid (ALA) and hypericin as contrast agents to detect early oral cancer lesions with a superior level of sensitivity and specificity [32,46]. In addition to this, using novel endoscopic optical imaging systems such as laser confocal endomicroscopy [33] and optical coherence tomography (OCT), in vivo high-resolution imaging of oral epithelial tissues can be acquired for diagnostics [34].

During carcinogenesis in the oral cavity, structural and biochemical changes occur in both the epithelium and stroma, altering the optical and biological properties of dysplastic and cancerous tissue. These changes in molecular signatures can be detected in early oral cancers by using surface-enhanced Raman spectroscopic (SERS) imaging that uses specific antibody conjugated gold (Au) nanoparticles labeled with highly SERS-efficient reporter tags [12]. In this chapter, novel early oral cancer diagnostic methodologies such as FD and SERS will be discussed in detail.

18.1.1 Fluorescence Diagnosis

FD is of increasing interest in oral cancer detection as conventional white light endoscopy that is currently used for oral cancer detection may fail to detect small and flat mucosal neoplasms and thus are frequently overlooked during routine examination. Accurate detection and demarcation of early neoplasms followed by efficient treatment can significantly improve the survival rates of oral cancer patients. Though previous studies have evaluated the use of vital staining with Lugol's iodine and toluidine blue for improving the detection of early oral neoplasms, these methods are not yet clinically useful due to high false-positive or false-negative results [4,21]. FD using fluorescence endoscopy system is a technique to visualize the neoplastic lesions in a tumorous organ after topical or systemic application of a tumor-selective photosensitizer. Exact demarcation of tumor margins using this technique could contribute to optimum results in surgical excision and reconstruction.

Numerous photosensitizers are being investigated as fluorescent markers for in vivo detection and demarcation of tumors. One of the most promising photosensitizers for oral cancer diagnosis is 5-ALA. ALA is a precursor in the heme biosynthetic pathway of nucleated cells and is metabolized by certain endogenous enzymes to produce protoporphyrin (PPIX). PPIX preferentially accumulates in the tumor cells due to changes in the activity of two main enzymes, porphobilinogen deaminase (PBG) and ferrochelatase (FC) [22,37]. In the tumor cells, while the activity of PBG is increased, the activity of FC is decreased, resulting in the buildup of PPIX. Studies conducted with 5-ALA-induced PPIX fluorescence have shown a sensitivity of 95%–100% for oral cancer diagnosis. Labeling of mucosal lesions of the oral cavity with PPIX fluorescence seems to be a promising diagnostic procedure for neoplastic lesions of the oral cavity [17]. 5-ALA-induced PPIX fluorescence in the tissue is limited to the epithelium, and both normal and dysplastic epithelium showed PPIX fluorescence, suggesting the usefulness of PPIX fluorescence in the determination of superficial tumor margins [18]. In another study, Zheng et al. [46] explored the application of quantifying PPIX fluorescence images to improve the diagnostic specificity and detection of early oral lesions. The results demonstrated that the combination of quantified PPIX fluorescence images with the ratio diagnostic algorithms has the potential for noninvasive diagnosis of early oral cancers in vivo with high diagnostic accuracy. This technique can be a useful adjunct to pathological diagnosis for directing biopsies and assessing resection margins during oral surgery.

ALA-derived PPIX fluorescence spectroscopy has been employed for the detection of epithelial hyperkeratosis (EH) or epithelial dysplasia (ED) and lesions in oral submucous fibrosis (OSF) patients that could not be detected by autofluorescence spectroscopy. The results demonstrated that ALA-induced PPIX fluorescence spectroscopy could successfully identify the premalignant lesions on oral fibrotic mucosa [39]. FD with Photofrin has been successfully used to detect hyperplastic and malignant changes in oral tissue. A quantitative analysis of the fluorescence contrast between the neoplastic and healthy tissue established PPIX fluorescence as a sensitive, noninvasive technique for the early identification of malignant neoplasms in the oral cavity [1]. Another photosensitizer, hypericin, has shown excellent fluorescence diagnostic properties in oral cavity cancers [32]. Typical characteristics of hypericin include bright red fluorescence emission and exceptional photostability. It is a plant-based photosensitizer that belongs to the perylenequinone family and accumulates in abnormal cells including tumor cells [24]. Hypericin fluorescence diagnostic imaging as a technique can facilitate guided biopsies in the clinic, thereby reducing the number of biopsies taken. It can also provide visualization of tumor margins during surgical procedures and assist for same-day diagnosis in the clinic. Studies have reported that hypericin fluorescence can provide improved specificity and is subject to reduced photobleaching compared to 5-ALA [2]. Apart from the detection of macroscopic fluorescence in tumor tissue, studies are also under way to detect microscopic fluorescence using laser confocal endomicroscopy.

Though fluorescence imaging has long been the technique of choice to map the biochemical distributions in a living system, current cell biophysics research has generated great interest in SERS owing to its sensitivity and ease of use. This is because fluorescence signals emanated from both the intrinsic biomolecules as well as the exogenously introduced fluorophores may lack specificity due to overlapping signals between different fluorescent species. In addition, photobleaching that may occur as a result of the fluorophores undergoing exhaustive absorption/emission cycles is also a limiting factor in using the fluorescence technique. Therefore, in the next section, the use of SERS for oral cancer diagnosis will be discussed.

18.1.2 Surface-Enhanced Raman Spectroscopy Imaging

Raman spectroscopy (RS) is a laser-based technique that enables chemical characterization, provides information about molecular structure, and is therefore a potential tool for noninvasive tissue diagnosis. It could be considered as a complementary technique to biopsy, pathology, and clinical assays in many medical technologies. Using Fourier-transform (FT)-RS, Oliveira et al. [23] demonstrated the changes in the vibrational bands of normal, dysplastic (DE), and squamous cell carcinoma (SCC) tissues that seem to arise from the compositional, conformational, and structural changes of proteins. Another study has shown that principal components analysis (PCA) of the Raman spectral data can discriminate between normal, inflammatory, premalignant, and malignant conditions in oral tissue [20]. While RS offers molecular specificity and provides the ability to elucidate molecular structures, the small Raman cross section (10^{-26} cm^2)—compared to that of fluorescence which is on the order of 10^{-16} cm^2—of almost all biomolecules has limited the wide utility of RS imaging. Lengthy acquisition times and high excitation power are the major factors plaguing their general applications [9]. Fortunately, such a drawback of normal RS technique can be circumvented by the use of SERS. In SERS, analyte molecules of interest are first adsorbed on the surface of nanostructured metallic substrates, which could be in the form of colloids, nanowires, and aggregates. When these substrates are made to resonate with external optical fields, oscillating surface plasmons are produced (Figure 18.1) [5,11,31,38]. This brings about extremely intense local electric fields at certain sites, known as the SERS hot spots, on the metal surface, causing the adsorbed molecules there to experience tremendous excitation fields and, hence, the subsequent emission of strong Raman signals [13]. It has been shown that such a resonant response can greatly enhance the efficiency of Raman scattering, typically by up to 10^6–10^7-fold. Further improvement in the enhancement factor by a factor of 10^4–10^5 is also possible if the molecules are sandwiched in between particles within an aggregate, giving the possibility of single-molecule spectroscopy [13,16,31]. With such a high sensitivity and chemical specificity, SERS is beginning to gain popularity as a powerful optical tool for the quantitative multicomponent analysis of cells or tissues.

For instance, Huang et al. [10] demonstrated the applications of anti-EGFR conjugated Au nanorods for in vitro SERS spectroscopic analysis of cell surface on a malignant epithelial cell line, HSC 3 (human OSCC) cultured on a cover slip. Although not reported by Huang et al. in his study, it would be worthwhile to compare the SERS spectra from nonmalignant and malignant cells since the chemical composition of cell membrane is known to undergo alterations in cancerous cells. Qian et al. [26], on the other

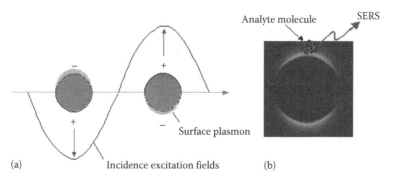

FIGURE 18.1   (a) SPR of metallic nanoparticles under oscillating electric fields. (b) An analyte molecule residing within the SPR hot spot and emitting enhanced Raman signals.

hand, demonstrated the use of SERS tags for in vivo targeting of human head-and-neck tumor (Tu686) in nude mice at the near-infrared region (785 nm). In order to achieve this, pegylated Au nanoparticles bearing malachite green SERS reporter and covalently attached ScFv antibody were injected systemically through tail vein into the tumor-bearing nude mice. Spectra derived from the tumor site indicated that the antibody conjugated Au nanoparticles were able to target the tumor in vivo.

In our earlier study, we demonstrated the use of Au nanoparticles in surface-enhanced Raman scattering to enhance the RS signal for the analysis of cancer-related chemical changes in saliva [12,14]. The aim was to develop colloidal Au nanoparticles as well as a self-assembled monolayer of these nanoparticles and exploit their ability to undergo coupled surface plasmon resonance (SPR) and set up strong electric fields when closely spaced for cancer diagnosis using both an imaging and chemical approach. Under the imaging approach, we demonstrated the use of these nanoparticles as an optical contrast agent for improving the optical contrast of clinically relevant biomarkers in epithelial cancer cells under reflectance-mode imaging in vitro. Therefore, we developed a simple and cost-effective method for preparing highly sensitive saliva assay based on SERS-active Au nanoparticle films and used to analyze biofluids spectroscopically; the SERS from the saliva biofilm interface seeks spectral features indicative of oral cancer. The use of saliva as a diagnostic fluid would offer huge advantages over previous sera-based counterparts in that saliva is easily accessible, is painlessly acquired, and presents lower risk of infection compared to serum.

18.2 Materials

Fluorescence diagnosis: (i) A fluorescence endoscope system (Karl Storz, Tuttlingen, Germany) was used to perform macroscopic fluorescence digital imaging. (ii) 5-ALA was purchased from Medac, Hamburg, Germany. (iii) Hypericin was purchased from Molecular Probes, United States. (iv) For ALA fluorescence imaging, patients ($n = 49$) received topical application of 5-ALA to the oral mucosa using 0.4% rinsing solution 3 h prior to endoscopy and underwent conventional white light endoscopy followed by fluorescence. (v) For hypericin fluorescence imaging, patients ($n = 23$) received 8 µM hypericin instillation solution for topical application. One hundred milliliters of hypericin solution was topically administered to each patient by oral rinsing over 30 min.

Surface-enhanced Raman spectroscopy: (i) The Raman experiments were carried out with a modified micro-Raman system, in which an Olympus microscope with a color closed circuit television (CCTV) system was coupled to a Spex 1704 spectrometer that was equipped with a liquid nitrogen–cooled charge-coupled device (CCD) detector. (ii) A 15 nm Au hydrosol (EMGC.15, particle concentration $= 1.4 \times 10^{12}$ particles per mL, %CV <10%) was purchased from British Biocell International and used as received. This colloid was stored at 4°C if not used. (iii) Ethanol was purchased from MERCK. (iv) Distilled water (18 MΩ) was obtained from Millipore Milli-Q water purification system. (v) SuperFrost Plus microscope glass slides were purchased from VWR Scientific. (vi) Fifteen-milliliter plastic containers for saliva collection and 1.5 mL centrifuge tubes were purchased from Eppendorf. (vii) Chemicals and reagents such as crystal violet, hydrogen tetrachloroaurate, and trisodium citrate were purchased from Sigma-Aldrich, silica gel from Chemicod, and anti-EGFR antibody from Santa Cruz Biotechnology Inc.

18.3 Methods

18.3.1 Fluorescence Diagnosis

The fluorescence endoscopy system consists of an illumination console, a fluorescence detection unit, a video displaying and recording unit, and a computing system for image acquisition, display, and processing (Figure 18.2). A 100 W xenon short arc lamp (D-Light AF system, Karl Storz, Tuttlingen, Germany) is used for both the white light illumination (WL mode) and fluorescence excitation when filtered by a band pass filter (375–440 nm). The excitation power of the violet-blue light at the endoscope tip is approximately 50 mW. Both illumination and observation of tissues of interest are achieved via a modified endoscope integrated with a long pass (LP) filter (cutoff wavelength at 470 nm). The observation LP filter in the eyepiece of the endoscope only transmits 8% of the backscattered excitation blue light with a peak at 450 nm, while it has a transmission of over 98% in the 470–800 nm range; thus, the green tissue autofluorescence and the red fluorescence can pass through the endoscope efficiently. The fluorescence and white light images were recorded by a sensitive color CCD video camera (Tricam SL-PDD, Karl Storz, Tuttlingen, Germany) connected to the modified endoscope, and an image frame integration time of 1/15 s was commonly used. Both the excitation light source and the CCD camera can be switched interchangeably between the fluorescence mode and white light mode. The RGB video outputs of the CCD camera can be grabbed simultaneously by a frame grabber (Matrox Genesis-LC, MD, United States), and the digitized image can be stored for image processing. The processed images were displayed on the computer screen in near real time for tissue diagnosis. Meanwhile, the RGB video signals were displayed on a video monitor in real time for tissue diagnosis and recorded by a video recorder for postreviewing of the entire examination process.

Data processing and analysis: The multicolor fluorescence images of tissues obtained were digitized and stored for image quantification. An image-processing program was developed to carry out image processing, such as contrast enhancement, threshold, segmentation, and quantification of the fluorescence parameters. The processing can extract the information such as perimeter and areas of lesions contained in the image acquired. The fluorescence intensities of the red and blue channels of the images

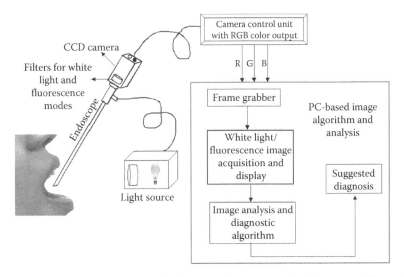

FIGURE 18.2 Schematic diagram of the setup showing an endoscopy system for fluorescence diagnostic imaging of lesions in the human oral cavity in a clinical setting.

can also be retrieved and quantified in near real time. The red channel registered fluorescence from the photosensitizer and the red tissue autofluorescence, and the blue channel showed the diffusely backscattered excitation light. It can be noted that the red-to-blue intensity ratio (I_R/I_B) of the same tissue site of the fluorescence image can effectively eliminate the influence of factors such as the illumination/observation geometries, the distance between the endoscope tip and the observed tissue, and the fluctuation of the excitation irradiance. Thus, this nondimensional intensity ratio I_R/I_B was selected as a diagnostic algorithm in this study.

Patient samples: Patients with clinically suspicious oral lesions or histologically proven malignancies of the oral cavity were enrolled for the study following informed consent. Topical application of the photosensitizers (5-ALA or hypericin) to the oral mucosa was performed via a rinsing solution. Patients continuously rinsed the oral cavity with the solution under supervision for 15–30 min. Fluorescence imaging was performed on the oral cavity after optimal incubation period of 1–2 h. Biopsies were taken from the tissue sites imaged to allow correlation with histology. Normal inflammation and hyperplasias were classified as benign, whereas severe dysplasias, carcinoma in situ (CIS), and invasive SCC were classified as malignant.

18.3.2 Surface-Enhanced Raman Spectroscopy

Preparation of SERS-active substrates: SERS-active Au particle arrays were fabricated by electrostatically immobilizing the negatively charged Au particles on positively charged glass slides (SuperFrost® Plus) as previously described [15]. Briefly, the microscope glass slides were rinsed thoroughly with 70% v/v aqueous solution of ethanol for 1 min, followed by rinsing with fresh distilled water four times to remove traces of ethanol. The slides were then dried, and a 150 µL droplet of 15 nm Au hydrosol (BBInternational, 1.4×10^{12} particles per mL) was applied to the surface of each of the slides. The slides were then dried in a desiccator for 24 h, followed by rinsing in distilled water, after which the slides were dried using a dust blower, concluding the preparation procedure.

Preparation of saliva samples: Saliva samples were collected from five normal healthy individuals and five oral cancer patients following their informed consent. This study was reviewed and approved by the Ethics Committee of the National Cancer Centre Singapore. Each subject was asked to rinse their mouth thoroughly with water for 30 min to remove any food particles in their oral environment before expelling saliva into a 15 mL plastic container (Nalgene, Eppendorf). About 1 mL of whole saliva was obtained each time. The specimen was then transferred into a 1.5 mL Eppendorf centrifuge tube and spun, using the Eppendorf centrifuge 5415C, at 14,000 rpm for 5 min to remove small particulates and exfoliated cells. The supernatant was then extracted and stored in a new centrifuge tube at −20°C until needed. Note that only samples with no sign of blood contamination were used for the experiment.

Raman measurements: The Raman experiments were carried out with a modified micro-Raman system, in which an Olympus microscope with a color CCTV system was coupled to a Spex 1704 spectrometer that was equipped with a liquid nitrogen–cooled CCD detector. In this modified system, the coupling optics was arranged in a box that was linked to the modified optical microscope by a mechanical arm and fixed to the Spex spectrometer. The laser light at 632.8 nm was introduced from the back of the box after passing through a plasma filter and was then directed into the modified microscope via a notch filter. This notch filter acted as a reflection mirror to the laser light, but to the signal returning from the sample, it was a very effective filter to the Rayleigh (laser) line. This filter prevented any backscattered laser from entering the spectrometer and, hence, from interfering with the Raman signals to be collected but allowed Raman signals to transmit with little attenuation. The returned signals then passed through a second notch filter, which was used to further improve the Rayleigh rejection. The Raman signal was then focused onto the entrance slit of the Spex spectrometer by a coated singlet focusing lens of 50 mm focal length. The inclusion of the CCTV system allowed both the laser beam and white light to be viewed directly from the monitor. A biological specimen could also be viewed directly from the monitor. The objective lens used in this study was from Olympus with magnification of 10× and NA of 0.25. The closely packed Au particle film was first coated with a drop (approximately 30 µL) of saliva. Immediately, a 633 nm laser was focused, via a 10× NA 0.25 objective, onto the sample-film interface. Laser intensity at sample is about 2 mW. Spectrum acquisition was then started immediately. The spectral resolution was 1 cm^{-1}, and an integration time of 10 s was used for all Raman measurements.

18.4 Results

18.4.1 Fluorescence Diagnosis

Forty-nine patients with various oral mucosal lesions were examined under ALA-induced PPIX fluorescence endoscopy. In the fluorescence images, suspicious lesions displayed bright reddish colors while normal surrounding regions exhibited blue color background (Figure 18.3). A total of 70 mucosal biopsies of the imaged tissues from different oral sites were taken for histopathological confirmation and were classified as normal, hyperplasias with chronic inflammation, severe dysplasia, CIS, and invasive SCC (Table 18.1).

For the assessment of diagnostic sensitivity and specificity for detecting oral cancers, histopathological results were regarded as the gold standard. Oral mucosa tissues were divided into three categories:

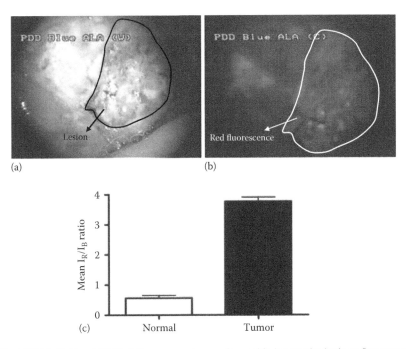

(a)   (b)

(c)   Normal   Tumor

FIGURE 18.3 (a) White light and (b) ALA fluorescence image of an oral lesion acquired using a fluorescence endoscopy system. (c) Bar charts showing the mean values of the red to blue intensity ratios in normal and tumor regions of the fluorescence image.

TABLE 18.1

Histopathological Classification of the 70 Biopsies

Histology	Classification of Biopsies
Normal	20
Inflammation and hyperplasia	10
Dysplasia (severe)	8
CIS	6
SCC	26

(a) benign (normal and hyperplastic tissues with inflammation), (b) intraepithelial neoplasia (dysplasia), and (c) cancer (CIS and SCC). The unpaired two-sided Student's t-test was performed on the fluorescence intensity ratio I_R/I_B to test if this diagnostic algorithm is statistically significant for differentiation between the two groups within each of the three categorical groupings.

By quantifying the multicolor ALA fluorescence images, it was observed that the red fluorescence intensity of the malignant tissues (dysplasia, CIS, SCC) is much stronger than that of benign (normal, inflammation, and hyperplasia) tissues, while the green fluorescence and the diffusely backscattered blue excitation light of malignant tissue are less than benign tissue. For oral cancer diagnosis, the red-to-green (I_R/I_G) and red-to-blue (I_R/I_B) intensity ratios were used as diagnostic algorithms to further enhance the contrast between malignant and benign tissue. Each ratio diagnostic algorithm (I_R/I_G or I_R/I_B) can be used to differentiate malignant tissue from benign tissue with equivalent high levels of sensitivity (90%–95%) and specificity (90%). The combined two ratio algorithms performed better than each diagnostic algorithm alone. The specificity improved from 90% to 97% while sensitivity remained at 95% in distinguishing malignant tissue from benign tissue using the combined diagnostic algorithms (Table 18.2) [17,45].

Next, we evaluated the use of hypericin fluorescence in oral cancer diagnosis. There were 32 lesions and 31 normal oral cavity sites imaged in patients after topical administration of hypericin. After biopsy, 13 lesions were histopathologically characterized as hyperplasia, 3 as cellular pleomorphic adenoma of the palate, 4 as dysplasia, and 12 as SCC. Therefore, a total of 63 lesion and normal sites from the tongue, buccal mucosa, palate, and gingiva were imaged using hypericin fluorescence endoscopy. The image parameters, hue, saturation, and intensity, as well as the ratiometric parameters, red to blue and red to green intensity ratios, were extracted and analyzed for their ability to distinguish between normal and lesional oral tissue. A representative white light and hypericin fluorescence image has been shown in Figure 18.3 a and b. The hypericin fluorescence images showed a progressive increase in the red to blue intensity ratios from normal to hyperplastic to SCC tissue. The fluorescence images showed a progressive increase in the red to blue intensity ratios from normal ($I_R/I_B = 0.3$) to hyperplastic ($I_R/I_B = 1.0$) to SCC ($I_R/I_B = 2.0$) tissue. Based on the results, the red to blue intensity ratio and the image hue could best distinguish between normal, hyperplastic, and SCC oral tissue. The red to green intensity ratio did reasonably well in distinguishing between normal and SCC tissue, but was not good for distinguishing between normal and hyperplastic tissue and between hyperplasia and SCC. The image intensity was good for distinguishing between normal and hyperplastic tissue and between normal and SCC tissue, but not for distinguishing between hyperplasia and SCC. For distinguishing between normal and hyperplastic oral tissue, the red to blue (I_R/I_B) ratio and image hue are good test parameters, offering 96% and 100% specificity, respectively, at the 100% sensitivity level. Likewise for distinguishing between normal and SCC tissue, the I_R/I_B ratio and hue are good tests, both offering 100% specificity at the 100% sensitivity level. For distinguishing between hyperplasia and SCC, the I_R/I_B ratio and hue offer the best results, a specificity of 90% and 80%, respectively, at the 92% sensitivity level. Overall, the I_R/I_B ratio and hue were the best test parameters, giving the best distinction between normal, hyperplastic, and SCC oral tissue. The red to green intensity ratio and image intensity serve reasonably well to distinguish between normal and hyperplastic and normal and SCC tissue, but distinguishes poorly between hyperplastic and SCC tissue [32]. Based on the earlier results, it can be seen that fluorescence imaging plays a crucial role in the staging of oral cancers.

TABLE 18.2

Sensitivity and Specificity Analysis Was Performed to Distinguish Malignant Tissue from Benign Tissues Using Ratio Diagnostic Algorithm of I_R/I_G and I_R/I_B, as well as the Combined I_R/I_G and I_R/I_B

	Sensitivity			Specificity		
Tissue Classification	I_R/I_G	I_R/I_B	I_R/I_G and I_R/I_B	I_R/I_G	I_R/I_B	I_R/I_G and I_R/I_B
Benign (Normal+inflammation/hyperplasia)+ Malignant (dysplasia + CIS + SCC)	90%	95%	95%	90%	90%	97%

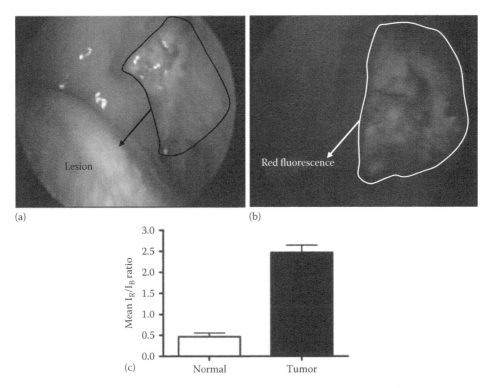

(a)

(b)

(c)

FIGURE 18.4 (a) White light and (b) hypericin fluorescence image of an oral lesion acquired using a fluorescence endoscopy system. (c) Bar charts showing the mean values of the red to blue intensity ratios in normal and tumor regions of the fluorescence image.

18.4.2 Surface-Enhanced Raman Spectroscopy of Human Saliva Samples

Figure 18.5 shows the typical AFM image of a SERS-active closely packed particle array. One could observe closely spaced but amorphously organized particles with an average particle-to-particle separation of 17 nm, which is comparable to the individual particle size (i.e., 15 nm). Thus, assuming a particle radius of 7.5 nm (=15 nm/2), this value can be translated into an interparticle spacing of just 2 nm. Note that this equilibrium distance is far too small to be explained by diffuse layer repulsion [6].

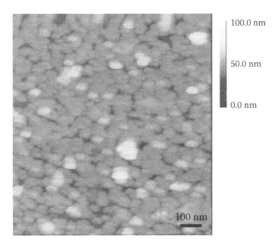

FIGURE 18.5 AFM mapping of a colloid film prepared from a drying 15 nm Au droplet. Scanned area = 1 μm × 1 μm.

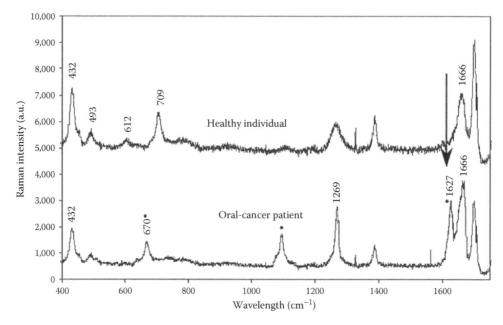

FIGURE 18.6 SERS spectra of raw human saliva samples. * highlights the differences in the spectra of oral cancer patients when compared with healthy individual.

Thus, the short particle spacing must be explained by the steric interaction of a stabilizing molecular layer adsorbed on the particle surface, in this case, the chloride ions. We have to note that the observation of such a closely packed metallic-particle array, formed by a drying suspension droplet, is rather surprising because this clearly disagrees with many of the previous claims that Au hydrosol are not stable enough to form 2D closely packed particle arrays without forming large aggregates [41,43] owing to their large Hamaker constant. Although particles can be made more stable with thiol- and amide-based protection layers to allow the formation of large, ordered 2D colloid lattices, and even 3D superlattices, [41–43] capping Au particles with a protection layer is not always desirable as far as SERS is concerned. This is because, firstly, the layer can increase the separation between the analyte molecules and the particle surface, thus compromising the SERS effect and, secondly, the adsorbed layer may introduce confounding SERS peaks to the final spectrum. We note that the later effect was not detected here, since no Au-Cl vibrations were observed during SERS measurements using our films.

To test the applicability of our films in analyzing fluidic biological samples, we used raw human saliva as the test sample. Saliva samples were collected from five normal healthy individuals and five oral cancer patients. Typical SERS spectra for these samples are shown in Figure 18.6. Tentative assignment of the major vibrational bands in the spectra based on existing knowledge is presented in Table 18.3. The upper spectrum is derived from normal saliva, and the lower spectrum, abnormal saliva (i.e., saliva from a patient). Striking differences between these spectra can be seen at 670, 1097, and 1627 cm⁻¹ (marked). Of these bands, special note must be given to that situated at 1627 cm⁻¹ (arrowed), which were observed in three of the five saliva samples collected from the patients.

18.5 Discussion

As cancer-associated biomarkers precede phenotypic manifestations of disease, there is a growing research interest in clinical molecular diagnostics for early cancer detection. Combining advances in biomedical optics and nanotechnology offers the opportunity to significantly impact future strategies toward the detection and diagnosis of cancer. In this section, we will discuss the usefulness of

TABLE 18.3

Assignment of Major Vibrational Bands in the SERS Spectra of Saliva Samples

Raman Shifts		Assignments
Normal		**Cancerous**
432	432	⎫
493	493	⎬ Skeletal bending
612	—	⎭
—	670	ν (C–S)
709	—	ν (C–S) met
—	1097	ν (C–N)
1114	—	ν (C–C)
1269	1269	δ (=CH) phospholipid
—	1627	Trp, Tyr, and Phe
1666–1700	1666–1700	Protein (amide I): α-helix, β-sheet, random coil

Notes: ν, stretching vibration; δ, bending vibration; Trp, tryptophan; Tyr, tyrosine; Phe, phenylalanine. "Skeletal bending" refers to vibrations in the secondary structure of the protein.

noninvasive imaging techniques in oral cancer diagnostics such as identifying early oral lesions, determining adequate resection margins, and evaluating tumor staging.

Quantifying fluorescence imaging using the fluorescence endoscopy system revealed that the red fluorescence intensity of malignant tissue is much stronger than that of benign tissue, while the diffusely backscattered blue light signals of malignant tissues are less than those of benign tissue. The red-to-blue intensity ratio (I_R/I_B) differentiation was achieved without requiring the comparison of the abnormal tissue to the surrounding normal tissue site within the same patient, and this misclassification caused by the incorrect selection of the control sites can be avoided. Staging of oral cancer is important for the early diagnosis of oral cancer and to provide timely and appropriate treatment. Because more than 90% of all oral cancers are SCCs, the vast majority of oral cancers will be diagnosed from lesions on the mucosal surfaces. Therefore, distinguishing normal from hyperplastic and SCC tissue can be effective in early cancer diagnosis. One notes that the fluorescence intensity differences between the benign tissue and the different stages of oral malignancies may involve the changes of morphological structures and fluorophores such as collagen, elastin, NADH, and flavins in tissues [29,44]. According to the tissue optical properties, the blue excitation light (380–450 nm) used in our study is able to penetrate into the tissue depth of up to 600 μm, inducing tissue fluorescence from several tissue layers in the oral cavity. The enhanced red fluorescence observed in malignant tissue is mainly attributed to the selective accumulation of ALA-induced PPIX in malignant lesions, and the decrease of the diffusely scattered blue excitation light may be due to thickening of epithelial tissue resulting in attenuation of excitation light and the increased hemoglobin absorption by a larger blood volume in malignant lesions.

Fluorescence imaging of the human oral cavity using hypericin has great potential as a complementary technique to standard white light examination and histopathology of oral cavity lesions. Red fluorescence was observed in lesions, indicating selective uptake of hypericin in lesions compared to normal tissue. Moreover, hyperplastic and SCC lesions exhibited different levels of fluorescence. The I_R/I_B ratio was observed to increase progressively from normal to hyperplastic to SCC oral tissue, as lesions tend to retain drugs for a longer period of time compared to normal tissue. At 2 h after topical application, most normal tissue cleared hypericin while lesions still exhibited fluorescence. SCC tissue is likely to be partially ulcerated and heavily inflamed. Highly vascular tissue is likely to be close to the tissue surface, resulting in brighter fluorescence compared to normal tissue. This indicates that the R/B ratio is a good image parameter for distinguishing between normal, hyperplastic, and SCC tissue. Therefore, fluorescence imaging is promising as a complementary technique to white light endoscopy of human oral cancers by facilitating guided biopsies in the clinic. This may reduce the number of biopsies taken, enable delineation of tumor margins during surgical procedures, and provide a means for same-day diagnosis in the clinic.

In our effort to develop noninvasive imaging for oral cancer diagnostics, we investigated the possibility of SERS to detect specific molecular markers in the saliva of oral cancer patients. Three Raman peaks that were noticed at 670, 1079, and $1627\,cm^{-1}$ are known to be the characteristics of saliva specimens collected from cancer patients. Of these bands, special note must be given to that situated at $1627\,cm^{-1}$, which were observed for three of the five saliva samples collected from the patients. This particular band was also observed in a previous Raman study of cancerous oral tissues [35]. As to whether there is any correlation in the Raman signals at $1627\,cm^{-1}$ between the saliva spectra and those derived from the tissues is not known at the moment. A comparison of SERS saliva spectra and the Raman spectra derived from matched oral tissues that is currently being performed in our lab would certainly provide an answer. We are also unable to evaluate the applicability of using the saliva spectrum for cancer staging at this stage, owing to insufficient samples. From further patient samples, we observed a sensitivity of the current technique in the range of around 70%, i.e., 70% of the "abnormal" saliva shows abnormal peaks. However, none of the normal samples show these indicative peaks. Currently, work is under way to develop a better SERS substrate consisting of periodic metallic nanostructures as opposed to the randomly distributed nanoparticles film that was employed in this study. This would improve signal reproducibility, thereby rendering quantitative analysis of the absolute signal intensity possible. However, the results of this study show the potential of these simple Au nanoparticle films to be used in routine clinical diagnosis of oral cancer.

A diagnostic technique capable of performing optical biopsy for noninvasive pathological diagnosis of cancer in situ and in real time should prove to be a powerful diagnostic modality in clinical medicine that will likely be a significant clinical advance with considerable impact on patient management.

18.6 Future Trends

Currently, using the fluorescence endoscopy system, the image analysis takes place off-line, and a digital diagnosis is determined only after the patient visit. Further development of a real-time image acquisition and analysis system interfaced to the fluorescence endoscope will allow acquired images to be analyzed on the spot within automatic or user-selected regions of interest to extract the image parameters. These image parameters can then be compared to predefined threshold levels to determine whether the tissue is normal, hyperplastic, or malignant. This will enable same-day diagnosis and cancer staging to be achieved in a clinical setting and provide a visual aid for demarcation of tumor margins, thus enhancing the clinical usefulness of fluorescence imaging.

To further improve SERS imaging, we will be using SERS tags rather than unfunctionalized NPs to perform SERS imaging on saliva samples to improve stability and reproducibility. Owing to the availability of a wide variety of RR molecules, each exhibiting nonoverlapping spectral signatures, the number of SERS tags that could be used simultaneously per sample would greatly exceed that of the fluorescent counterparts and is limited only by the resolution of the spectrometer. Coupled with the advances in computational technologies and improvements in the sensitivity of the CCD detector, it is likely that SERS tags will be the next generation of labelers superceding the fluorescent or chromophore counterparts in immunostaining of biological samples.

REFERENCES

1. Chang, C. J. and P. Wilder-Smith. 2005. Topical application of photofrin for photodynamic diagnosis of oral neoplasms. *Plast Reconstr Surg* **115:**1877–1886.
2. D'Hallewin, M. A., L. Bezdetnaya, and F. Guillemin. 2002. Fluorescence detection of bladder cancer: A review. *Eur Urol* **42:**417–425.
3. De Veld, D. C., M. J. Witjes, H. J. Sterenborg, and J. L. Roodenburg. 2005. The status of in vivo autofluorescence spectroscopy and imaging for oral oncology. *Oral Oncol* **41:**117–131.

4. Epstein, J. B., C. Scully, and J. Spinelli. 1992. Toluidine blue and Lugol's iodine application in the assessment of oral malignant disease and lesions at risk of malignancy. *J Oral Pathol Med* **21**:160–163.

5. Fleischmanna, M., P. J. Hendraa, and A. J. McQuillana. 1974. Raman spectra of pyridine adsorbed at a silver electrode. *Chem Phys Lett* **26**:163–166.

6. Giersig, M. and P. Mulvaney. 1993. Ordered two-dimensional gold colloid lattices by electrophoretic deposition. *J Phys Chem* **97**:6334–6336.

7. Gillenwater, A., V. Papadimitrakopoulou, and R. Richards-Kortum. 2006. Oral premalignancy: New methods of detection and treatment. *Curr Oncol Rep* **8**:146–154.

8. Greenlee, R. T., M. B. Hill-Harmon, T. Murray, and M. Thun. 2001. Cancer statistics, 2001. *CA Cancer J Clin* **51**:15–36.

9. Hu, Q., L. L. Tay, M. Noestheden, and J. P. Pezacki. 2007. Mammalian cell surface imaging with nitrile-functionalized nanoprobes: Biophysical characterization of aggregation and polarization anisotropy in SERS imaging. *J Am Chem Soc* **129**:14–15.

10. Huang, X., I. H. El-Sayed, W. Qian, and M. A. El-Sayed. 2007. Cancer cells assemble and align gold nanorods conjugated to antibodies to produce highly enhanced, sharp, and polarized surface Raman spectra: A potential cancer diagnostic marker. *Nano Lett* **7**:1591–1597.

11. Jeanmaire, D. L. and R. P. van Duyne. 1977. Surface Raman spectroelectrochemistry: Part I. Heterocyclic, aromatic, and aliphatic amines adsorbed on the anodized silver electrode. *J Electroanal Chem* **84**:1–20.

12. Kah, J. C., K. W. Kho, C. G. Lee, C. James, R. Sheppard, Z. X. Shen, K. C. Soo, and M. C. Olivo. 2007. Early diagnosis of oral cancer based on the surface plasmon resonance of gold nanoparticles. *Int J Nanomed* **2**:785–798.

13. Kerker, M. 1984. Electromagnetic model for surface-enhanced Raman-scattering (SERS) on metal colloids. *Acc Chem Res* **17**:271–277.

14. Kho, K. W., J. C. Y. Kah, C. G. L. Lee, C. J. R. Sheppard, Z. X. Shen, K. C. Soo, and M. Olivo. 2007. Applications of gold nanoparticles in the early detection of cancer. *J Mech Med Biol* **7**:19–35.

15. Kho, K. W., Z. X. Shen, H. C. Zeng, K. C. Soo, and M. Olivo. 2005. Deposition method for preparing SERS-active gold nanoparticle substrates. *Anal Chem* **77**:7462–7471.

16. Kneipp, J., H. Kneipp, B. Wittig, and K. Kneipp. 2009. Novel optical nanosensors for probing and imaging live cells. *Nanomedicine* **6**:214–226.

17. Leunig, A., M. Mehlmann, C. Betz, H. Stepp, S. Arbogast, G. Grevers, and R. Baumgartner. 2001. Fluorescence staining of oral cancer using a topical application of 5-aminolevulinic acid: Fluorescence microscopic studies. *J Photochem Photobiol B* **60**:44–49.

18. Leunig, A., K. Rick, H. Stepp, A. Goetz, R. Baumgartner, and J. Feyh. 1996. Photodynamic diagnosis of neoplasms of the mouth cavity after local administration of 5-aminolevulinic acid. *Laryngorhinootologie* **75**:459–464.

19. Lumerman, H., P. Freedman, and S. Kerpel. 1995. Oral epithelial dysplasia and the development of invasive squamous cell carcinoma. *Oral Surg Oral Med Oral Pathol Oral Radiol Endod* **79**:321–329.

20. Malini, R., K. Venkatakrishna, J. Kurien, K. M. Pai, L. Rao, V. B. Kartha, and C. M. Krishna. 2006. Discrimination of normal, inflammatory, premalignant, and malignant oral tissue: A Raman spectroscopy study. *Biopolymers* **81**:179–193.

21. Mashberg, A. 1980. Reevaluation of toluidine blue application as a diagnostic adjunct in the detection of asymptomatic oral squamous carcinoma: A continuing prospective study of oral cancer III. *Cancer* **46**:758–763.

22. Navone, N. M., C. F. Polo, R. M. Dinger, and A. M. Batlle. 1990. Heme regulation in mouse mammary carcinoma and liver of tumor bearing mice—I. Effect of allyl-isopropylacetamide and veronal on delta-aminolevulinate synthetase, cytochrome P-450 and cytochrome oxidase. *Int J Biochem* **22**:1005–1008.

23. Oliveira, A. P., R. A. Bitar, L. Silveira, R. A. Zangaro, and A. A. Martin. 2006. Near-infrared Raman spectroscopy for oral carcinoma diagnosis. *Photomed Laser Surg* **24**:348–353.

24. Olivo, M., H. Y. Du, and B. H. Bay. 2006. Hypericin lights up the way for the potential treatment of nasopharyngeal cancer by photodynamic therapy. *Curr Clin Pharmacol* **1**:217–222.

25. Poh, C. F., L. Zhang, D. W. Anderson, J. S. Durham, P. M. Williams, R. W. Priddy, K. W. Berean, S. Ng, O. L. Tseng, C. MacAulay, and M. P. Rosin. 2006. Fluorescence visualization detection of field alterations in tumor margins of oral cancer patients. *Clin Cancer Res* **12**:6716–6722.

26. Qian, X., X. H. Peng, D. O. Ansari, Q. Yin-Goen, G. Z. Chen, D. M. Shin, L. Yang, A. N. Young, M. D. Wang, and S. Nie. 2008. In vivo tumor targeting and spectroscopic detection with surface-enhanced Raman nanoparticle tags. *Nat Biotechnol* **26**:83–90.

27. Ram, S. and C. H. Siar. 2005. Chemiluminescence as a diagnostic aid in the detection of oral cancer and potentially malignant epithelial lesions. *Int J Oral Maxillofac Surg* **34**:521–527.

28. Rhodus, N. L. 2009. Oral cancer and precancer: Improving outcomes. *Compend Contin Educ Dent* **30**:486–488, 490–494, 496–498 passim; quiz 504, 520.

29. Richards-Kortum, R. and E. Sevick-Muraca. 1996. Quantitative optical spectroscopy for tissue diagnosis. *Annu Rev Phys Chem* **47**:555–606.

30. Roblyer, D., C. Kurachi, V. Stepanek, M. D. Williams, A. K. El-Naggar, J. J. Lee, A. M. Gillenwater, and R. Richards-Kortum. 2009. Objective detection and delineation of oral neoplasia using autofluorescence imaging. *Cancer Prev Res* **2**:423–431.

31. Scaffidi, J. P., M. K. Gregas, V. Seewaldt, and T. Vo-Dinh. 2009. SERS-based plasmonic nanobiosensing in single living cells. *Anal Bioanal Chem* **393**:1135–1141.

32. Thong, P. S., M. Olivo, W. W. Chin, R. Bhuvaneswari, K. Mancer, and K. C. Soo. 2009. Clinical application of fluorescence endoscopic imaging using hypericin for the diagnosis of human oral cavity lesions. *Br J Cancer* **101**:1580–1584.

33. Thong, P. S., M. Olivo, K. W. Kho, W. Zheng, K. Mancer, M. Harris, and K. C. Soo. 2007. Laser confocal endomicroscopy as a novel technique for fluorescence diagnostic imaging of the oral cavity. *J Biomed Opt* **12**:014007.

34. Tsai, M. T., C. K. Lee, H. C. Lee, H. M. Chen, C. P. Chiang, Y. M. Wang, and C. C. Yang. 2009. Differentiating oral lesions in different carcinogenesis stages with optical coherence tomography. *J Biomed Opt* **14**:044028.

35. Ullas, G., S. Sudhakar, and K. Nayak. 1999. Laser Raman spectroscopy: Some clinical applications. *Biomed Appl Lasers* **77**:101–107.

36. Upile, T., W. Jerjes, H. J. Sterenborg, A. K. El-Naggar, A. Sandison, M. J. Witjes, M. A. Biel, I. Bigio, B. J. Wong, A. Gillenwater, A. J. MacRobert, D. J. Robinson, C. S. Betz, H. Stepp, L. Bolotine, G. McKenzie, C. A. Mosse, H. Barr, Z. Chen, K. Berg, A. K. D'Cruz, N. Stone, C. Kendall, S. Fisher, A. Leunig, M. Olivo, R. Richards-Kortum, K. C. Soo, V. Bagnato, L. P. Choo-Smith, K. Svanberg, I. B. Tan, B. C. Wilson, H. Wolfsen, A. G. Yodh, and C. Hopper. 2009. Head & neck optical diagnostics: Vision of the future of surgery. *Head Neck Oncol* **1**:25.

37. Van Hillegersberg, R., J. W. Van den Berg, W. J. Kort, O. T. Terpstra, and J. H. Wilson. 1992. Selective accumulation of endogenously produced porphyrins in a liver metastasis model in rats. *Gastroenterology* **103**:647–651.

38. Vo-Dinh, T., M. Y. K. Hiromoto, G. M. Begun, and R. L. Moody. 1984. Surface-enhanced Raman spectroscopy for trace organic analysis. *Anal Chem* **56**:1667–1670.

39. Wang, C. Y., T. Tsai, C. P. Chiang, H. M. Chen, and C. T. Chen. 2009. Improved diagnosis of oral premalignant lesions in submucous fibrosis patients with 5-aminolevulinic acid induced PpIX fluorescence. *J Biomed Opt* **14**:044026.

40. Warnakulasuriya, S. 2009. Global epidemiology of oral and oropharyngeal cancer. *Oral Oncol* **45**: 309–316.

41. Weitz, D. A., M. Y. Lin, and C. J. Sandroff. 1985. Colloidal aggregation revisited: New insights based on fractal structure and surface-enhanced Raman scattering surface science **158**:147–164.

42. Wickline, S. A. and G. M. Lanza. 2002. Molecular imaging, targeted therapeutics, and nanoscience. *J Cell Biochem Suppl* **39**:90–97.

43. Zhao, S., S. Wang, and K. Kimura. 2004. The first example of ordered two-dimensional self-assembly of Au nanoparticles from stable hydrosol. *Langmuir* **20**:1977–1979.

44. Zheng, W., W. Lau, C. Cheng, K. C. Soo, and M. Olivo. 2003. Optimal excitation-emission wavelengths for autofluorescence diagnosis of bladder tumors. *Int J Cancer* **104**:477–481.

45. Zheng, W., M. Olivo, and K. C. Soo. 2004. The use of digitized endoscopic imaging of 5-ALA-induced PPIX fluorescence to detect and diagnose oral premalignant and malignant lesions in vivo. *Int J Cancer* **110**:295–300.

46. Zheng, W., K. C. Soo, R. Sivanandan, and M. Olivo. 2002. Detection of squamous cell carcinomas and pre-cancerous lesions in the oral cavity by quantification of 5-aminolevulinic acid induced fluorescence endoscopic images. *Lasers Surg Med* **31**:151–157.

19

Tactile Sensing and Tactile Imaging in Detection of Cancer

A. Sarvazyan, V. Egorov, and N. Sarvazyan

CONTENTS

19.1 Introduction

19.1.1 Background of Tactile Imaging Technology

Since Hippocrates, the human sense of touch has been the most prevalent and successful medical diagnostic technique. A great variety of diseases were diagnosed through tactile sensing including detection of malignant tumors. Hippocrates in 400 B.C. wrote as follows: "… Such swellings as are soft, free from pain, and yield to the finger, … and are less dangerous than the others. … then, as are painful, hard, and large, indicate danger of speedy death; but such as are soft, free of pain, and yield when pressed with the finger, are more chronic than these…" [28].

In ancient medicine, medical practitioners have used diagnostic methods based on assessment of mechanical properties of tissues for diagnosing and treating ailments without much help from other tests. During the last century, traditional physical examination techniques were becoming outmoded and were often considered of little clinical value in comparison with many other modern diagnostic technologies. Palpation is only briefly addressed in medical school, and few physicians have the tactile ability to detect subtle variations of tissue elasticity. Few physicians are willing to devote the necessary time to master the technique, despite the fact that the American Cancer Society guidelines suggest for women that clinical breast examination (CBE) be part of a periodic health examination [1] and the American Urological Association in their 2009 Best Practices Statement recommended that men who wish to be screened for prostate cancer should have both a prostate-specific antigen (PSA) test and a digital rectal examination (DRE) [2].

The development of modern diagnostic technologies has brought a decline in physical examination skills. But is this technique on the way to extinction? During the last two decades, several devices were developed which mimicked both the tactile sensors and the analysis part of traditional palpation techniques; many experimental and theoretical papers were published and patents filed. Numerous emerging technologies, such as various version of ultrasound and MR elasticity imaging, started to bring new life into the ancient technique of diagnostics based on assessment of tissue mechanical properties: *Le Roi Est Mort, Vive Le Roi!*

The task of developing tactile sensors fully mimicking the human fingertip and the mechanism of transduction of the tactile sensor stimuli into nerves and further analysis by brain is extremely difficult because Mother Nature created sophisticated and highly sensitive system for obtaining information through sense of touch. A tactile sensing system of a human uses sensory information derived from mechanoreceptors embedded in the skin to provide data from an area contacting with an object and from mechanoreceptors rooted in muscles, tendons, and joints to provide motion tracking data [37]. Each fingertip is equipped with about 2,000 tactile sensors [31] located inside two major layers of the skin (epidermis and dermis) and the underlying subcutaneous tissue. These layers have the four mechanoreceptor populations: Meissner corpuscles, Merkel cell neurite complexes, Pacinian corpuscles, and Ruffini endings. Primary functions of each mechanoreceptor population are found to be different. They provide texture perception, pattern and form detection, motion detection, stable precision grasp, and manipulation [80]. Average applied pressure differential sensitivity for human fingertips was found to be about 900 Pa [34]. The tactile spatial resolution is within the range from 0.8 to 1.6 mm [9,22,67]. The temporal resolution of human tactile sensing is about 0.05 s as measured between successive taps on the skin [21]. The tactile roughness sensitivity of human fingertip is typically about 0.2 mm [6], but may reach up to 0.02 mm [41].

However, manual palpation aiming at cancer detection, such as CBE or DRE, makes use of only a small fraction of plurality of the fingertip tactile system features. This makes the task of developing a diagnostic technology mimicking the sophisticated means implemented by Mother Nature in the human fingertip much more realistic.

19.1.2 Emergence of Tactile Imaging

The first description of a technical implementation related to tactile imaging (TI) was given in the late 1970s by Frei et al. [18,19] who proposed an instrument for breast palpation that used a plurality of spaced piezoelectric force sensors. The sensors were pressed against the breast tissue by a pressure member which applied a given periodic or steady stress to the tissue. A different principle for evaluating the pattern of pressure distribution over a compressed breast was proposed by Gentle [20] 8 years later. The pressure distribution was monitored optically by using the principle of frustrated total internal reflection to generate a brightness distribution (see Figure 19.1). Using this technique, simulated lumps in breast prostheses were detected down to a diameter of 6 mm. But the author was unable to obtain any quantitative data on lumps in a real breast. The failure has been explained by the insufficient sensitivity of the registration system and that "the load, that the volunteers could comfortably tolerate, was less than that used in the simulation." Then, Sabatini et al. built a robotic system for discriminating mechanical inhomogeneities in a soft tissue using a finger-like palpation device equipped with a fingertip piezoelectric polymer film tactile sensor [57].

TI as a modality of medical diagnostics based on reconstruction of tissue structure and elastic properties using mechanical sensors was introduced in the 1990s by Sarvazyan et al. in Artann Laboratories [58,63] and by Wellman et al. in Harvard University [76,77]. TI, which is also called "mechanical imaging" or "stress imaging," is most closely mimicking manual palpation because the TI probe with a pressure sensor array mounted on its tip acts similar to human fingers during clinical examination, slightly compressing soft tissue by the probe. In essence, TI "captures the sense of touch" and stores it permanently in digital format for analysis and comparison. Extensive laboratory studies on breast phantoms and excised prostates have shown that the computerized palpation is more sensitive than human finger [14,59,79].

The TI examination is performed through a set of manual compressions of the target tissue/organ by pressure sensor array mounted on a hand-held probe. The pressure response pattern shows spatial distribution of softer and harder areas of the palpated region, thus providing information on the presence,

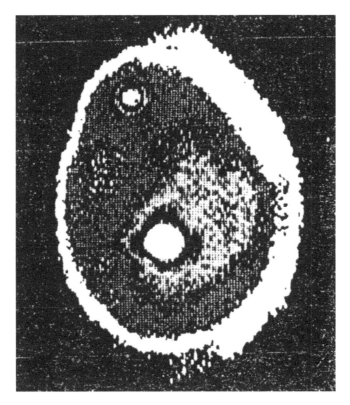

FIGURE 19.1 On one of the first elastographic images published in 1988, a stress pattern recorded on the surface of compressed breast phantom (rubber prosthesis filled with silicone rubber gel) containing two lumps (nylon balls of diameters 25 and 6 mm). (Reproduced from Gentle, C.R., *J. Biomed. Eng.,* 10, 124, 1988. With permission.)

dimensions, and location of hard inclusions. Stress profile resulting from the presence of a hard nodule strongly depends on the nodule shape, size, hardness, and depth.

During the last decade, TI was implemented in several devices for a variety of applications. These devices included the prostate mechanical imager (PMI) for 3-D prostate visualization highlighting prostate nodularity in terms of tissue elasticity [13,78], the breast mechanical imager for breast cancer detection [14], and the vaginal tactile imager for pelvic organ prolapse assessment [16].

19.1.3 Sensors for Tactile Imaging

A wide variety of technologies, optical, electromechanical, and ultrasonic, have been explored to address the tactile sensing problem in robotics and medicine [73]. Most of the sensors and transduction mechanisms tested in applications related to TI, such as capacitive, resistive, based on piezopolymers, fiber optics, conductive elastomers, and MEMS sensors, appeared to be far from optimal due to either insufficient sensitivity and reproducibility, excessive hysteresis, or fast aging [4,5,8,25,38,39,42,55,56].

The TI devices developed in the 1990s [44,76] employed Tekscan resistive pressure sensor array (Boston, MA) [55]. Tekscan sensor is based on changing electrical resistance under external pressure of electroconductive powder layer embedded between two flexible polyester sheets. The main drawback of Tekscan sensors limiting their applicability in TI applications is low reproducibility and fast aging.

The other type of sensor arrays which satisfied the requirements of TI and which is currently implemented in the TI systems appeared to be the capacitive tactile sensors developed by Pressure Profile Systems, Inc. (PPS), Los Angeles, CA [71]. The PPS sensor array is a capacitor grid produced by a set of orthogonal electrodes. The basic technical characteristics for PPS sensors are as follows: basic operational range is 0–60 kPa, noise level/sensitivity 0.06 kPa, reproducibility 0.8 kPa, and temperature sensitivity 0.05 kPa/°C. These characteristics fully correspond to the requirements of TI technology.

19.1.4 Potential of Tactile Imaging for Detecting Cancer

The very first clinical study on the application of TI technology for cancer detection was conducted on excised prostate glands in 1998 [44,45]. Seven radical prostatectomy and two cystoprostatectomy specimens were evaluated by a proof-of-concept TI prototype with a Tekscan sensor array. Excised prostates were placed with the posterior side facing down on 110 mm × 110 mm sensor array. Manual pressure was applied to the prostate anterior surface for 3–5 s to produce 300–500 tactile images of the gland. The prostates were further histopathologically analyzed for the presence of cancer. The results of the mechanical imaging and pathological analysis were closely correlated. Figure 19.2 provides an example illustrating the results of that study.

Mechanical properties of tissues are highly sensitive to the structural changes accompanying various physiological and pathological processes. A change in Young's modulus of tissue during the development of a tumor could reach thousands of percent [64,76]. Evaluation of tissue "hardness" (Young's or shear elasticity modulus) by various elasticity imaging techniques provides means for characterizing the tissue, differentiating normal and diseased conditions, and detecting tumors and other lesions [60]. Tumors or tissue blocked from its blood nutrients is stiffer than normal tissue. Further, benign and cancerous tumors have distinguishing elastic properties [35,70].

Over the last two decades, there has been significant development in different methods to visualize mechanical structure of tissues in terms of their elasticity properties. Every elasticity imaging method involves two common elements: the application of a force and the measurement of a mechanical response. The measurement method can be performed using differing physical principles including magnetic resonance imaging (MRI) [17,43,52,68] and ultrasound imaging [46,47,50,64,65]. TI is a branch of elasticity imaging; it differs from conventional ultrasonic and MR elasticity imaging in that it evaluates soft tissue mechanical structure using stress data rather than dynamic or static strain data.

The current surge of publications on ultrasonic and MR elasticity imaging covers practically all key human organs [61]. Researchers from numerous laboratories all over the world demonstrated the potential of elasticity imaging in cancer diagnostics and differentiating benign and malignant lesions. Table 19.1 illustrates the sensitivity and specificity of elasticity imaging in differentiating benign vs. malignant lesions using literature data on assessment of breast cancer by various elasticity imaging modalities: USE (ultrasound elastography), MRE (magnetic resonance elastography), and TI. These data show that elasticity imaging even in its least sophisticated version, like TI, has significant diagnostic potential comparable and exceeding that of conventional imaging techniques such as mammography, MRI, and ultrasound. Results of clinical studies indicate that TI has a potential not only to just detect tumors but also to distinguish between benign and malignant classes of fibroadenoma, cyst, fibrosis, ductal, lobular carcinoma, and other conditions [15].

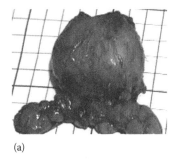
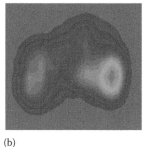
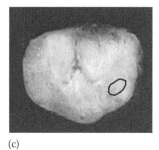

(a)   (b)   (c)

FIGURE 19.2 An excised prostate (a), its mechanical image (b), and corresponding pathology section (c) revealing a cancerous nodule (adenocarcinoma with Gleason score 4) at the location exactly corresponding to that on the mechanical image. (Adapted from Niemczyk, P. et al., *J. Urol.*, 160, 797, 1998. With permission.)

TABLE 19.1

Recent Clinical Data on Benign-Malignant Breast Lesion Differentiation by Elasticity Imaging

No.	Method	Number of Analyzed Lesions	Sensitivity (%)	Specificity (%)	Citation
1	USE	52 malignant/59 benign	86.5	89.8	Itoh et al. [29]
2	USE	49 malignant/59 benign	91.8	91.5	Thomas et al. [74]
3	MRE	38 malignant/30 benign	95.0	80.0	Sinkus et al. [69]
4	USE	50 malignant/48 benign	99.3	25.7	Burnside et al. [11]
5	USE	237 malignant/584 benign	97.5	48.0	Svensson et al. [72]
6	TI	32 malignant/147 benign	91.4	86.1	Egorov et al. [15]
7	SSI	82 malignant/110 benign	87.8	87.3	Cosgrove et al. [12]
8	USE	144 malignant/415 benign	92.4	91.1	Zhi et al. [82]
9	USE	61 malignant/127 benign	92.7	85.8	Raza et al. [54]

USE, ultrasound elastography; MRE, magnetic resonance elastography. TI, tactile imaging; SSI, supersonic shear imaging.

19.2 Materials and Methods

Several devices have been developed based on the TI technology for detection and 3-D visualization of cancer. General architecture of these devices has several common features and components, such as a probe with an array of tactile sensors, an electronic unit, and a touch screen laptop computer [13,14]. The configuration of the probe, structure of sensors, user interface, and data processing algorithms are specific for a particular application. Figures 19.3 and 19.4 show general view of the devices for breast (Medical Tactile, CA) and prostate TI (Artann Laboratories, NJ; ProUroCare Medical, MN). The probe for the tactile breast imager (TBI) has a pressure sensor array of 40 mm by 30 mm comprising 192 pressure sensors to acquire pressure patterns between the probe surface and the exterior skin layer of the breast during contact (Figure 19.3). Each pressure sensor has rectangular pressure sensing area of 2.5 mm by 2.0 mm (PPS, CA).

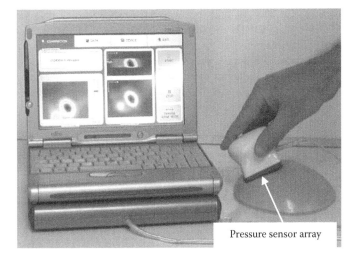

Pressure sensor array

FIGURE 19.3 General view of the TBI. The device comprises a probe with 2-D pressure sensor array, an electronic unit, and a laptop computer with touch screen capability. (Reproduced from Egorov, V. and Sarvazyan, A.P., *IEEE Trans. Med. Imaging,* 27, 1275, 2008. With permission.)

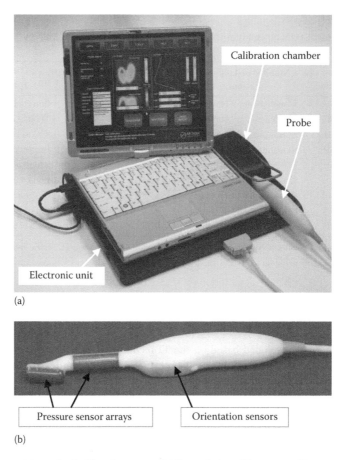

FIGURE 19.4 The prostate mechanical imaging system. (a) General view of the system; (b) transrectal probe.

The transrectal probe of the PMI is much more complex. It comprises two separate pressure sensor arrays and orientation sensors. The first pressure sensor array is installed on the probe head surface, and it is in contact with prostate through the rectal wall during the examination procedure. The second pressure sensor array is installed on the probe shaft surface for assessment of the pressure pattern in the sphincter area during the manipulation of the probe. The probe shaft and head pressure arrays differ by their geometry and sensor's size. The probe head pressure sensor array comprises 128 (16×8) pressure sensors covering an area of 40 mm by 16 mm while the shaft sensor array comprises 48 sensors (16×3), and the sensors have different dimensions (3.75 mm×2.5 mm). In contrast to the TBI probe, the PMI probe includes also an orientation system (InterSense, Inc., MA) mounted in the probe handle. The probe head pressure sensors are intended for acquisition prostate pressure patterns as well as for calculation of the possible prostate displacements during probe head pressing against the prostate. The probe shaft pressure sensors are capturing and tracking the sphincter position that allows real-time spatial visualization of the sphincter and prostate area to help an operator in finding the prostate and assist in probe manipulation. Another important function of the probe shaft sensor array is to provide quantitative information on the level of forces exerted by the operator on the sphincter. Displaying this information on the user interface helps the operator to avoid excessive stretching of patient's sphincter, which is one of the causes of patient's discomfort during examination. Calculated distance between the sphincter and prostate and the probe azimuth angle helps to compute left/right probe head displacement relative to a start reference line.

During the examination of the breast with TBI, the patient is placed in a position similar to that of a standard CBE with her breast in the supine position on a standard examination table. The examiner places a disposable sheath over the sensor head of the TBI and then applies a water soluble lubricating lotion to the sensor head or applied directly to the area of concern. The examination is recorded and stored by the TBI system in a digital format file. The duration of a typical lesion scan is approximately 1–2 min.

For the PMI examination, a patient is asked to bend over the examination table so as to form a 90° angle at the waist. The patient places his chest on a table, so that his weight is applied to the table surface in order to free leg muscles from tension. The rectum does not need to be evacuated prior to the examination. A lubricated probe is inserted into the rectum with the probe head sensor surface down until the prostate is visualized on the computer. The prostate scan is performed through a set of multiple compressions. The examiner is able to see in real time two orthogonal prostate cross sections with the relative location of the probe head pressure-sensitive area in both projections. The PMI scan takes 40–60 s on average, and the collected data are saved in a digital format.

Software of tactile imagers allows real-time visualization and 2-D/3-D computer-aided reconstruction of the examined tissue and detected abnormalities. An operator may look through various orthogonal slices of the examined tissue. The software also offers a standard range of data management features, such as data storage and retrieval and image printout. An additional set of image enhancement techniques are applied to improve the image, data acquisition, and lesion detection, as illustrated in Figure 19.5 [14,15]. Image enhancement techniques such as low-pass noise filtration, signal thresholding and 2-D interpolation, pixel neighborhood rating–based filtering, and 2-D interpolation are used to increase signal-to-noise ratio. Assessment of the motion of lesions relative to probe sensing surface is performed to identify the mobility of lesions, an important diagnostic feature. The 2-D image matching is performed to generate a compound image from a sequence of successive images obtained along the scanning trajectory over the detected lesion. Lastly, geometrical characteristics of the detected lesions, including lesion shape, edges, strain hardening, mobility as an ability to change shape, and position under applied stress are calculated.

An important feature of TI is the ability of 3-D reconstruction of internal structures using data of stress patterns on the surface of the examined tissue at different levels of compression (Figure 19.6). The input data for 3-D reconstruction comprise a continuous sequence of 2-D filtered images. The initial hypotheses enabling 3-D reconstruction are as follows: (a) the higher the compression force, the greater the representation of deeper structures in the imprint image and (b) the total pressure is proportional to the tissue deformation in the direction normal to the probe surface (Z-axis). The 3-D reconstruction starts with the formation of an initial (seed) 3-D structure by stacking the series of 2-D structure images along Z-axis during first tissue compression. Every 2-D imprint is further integrated by a parallel translation inside the 3-D structure image by a matching algorithm [14]. The final 3-D structure visualization

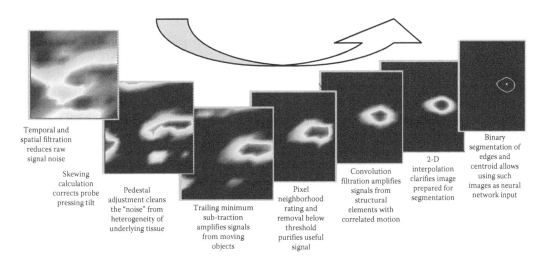

Temporal and spatial filtration reduces raw signal noise

Skewing calculation corrects probe pressing tilt

Pedestal adjustment cleans the "noise" from heterogeneity of underlying tissue

Trailing minimum sub-traction amplifies signals from moving objects

Pixel neighborhood rating and removal below threshold purifies useful signal

Convolution filtration amplifies signals from structural elements with correlated motion

2-D interpolation clarifies image prepared for segmentation

Binary segmentation of edges and centroid allows using such images as neural network input

FIGURE 19.5 A sequence of algorithms for isolating the lesion signal while rejecting artifacts.

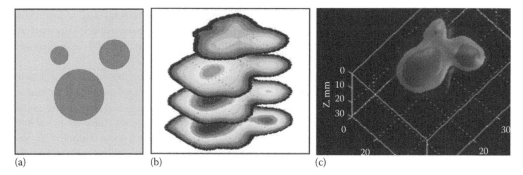

FIGURE 19.6 Illustration of an algorithm for 3-D image reconstruction. (a) Schematic top view of a test phantom (E=8 kPa) with three inclusions (E=125 kPa, D1=15 mm, D2=8 mm, and D3=5 mm) located at about 8 mm depth, (b) sequence of 2-D stress patterns obtained at different levels of compression, and (c) reconstructed 3-D image. See text for details.

is deployed by the computation of isosurfaces and 2-D image slices for the 3-D pressure field, which is related to the hardness distribution of the underlying structure. Figure 19.6 presents an example of the 3-D image reconstruction of a composite inclusion in the test phantom. Panel C in Figure 19.6 shows a composite inclusion visualized by means of three semitransparent isosurfaces (blue, green, and red). Each surface represents points of a constant value of pressure (10, 17, and 25 kPa, respectively) in 3-D space. The relative elasticity levels forming 2-D image slices (panel B in Figure 19.6) are represented by a color map. The colors from blue to red are selected to provide a visual demarcation between different levels of pressure response. The blue color corresponds to the lowest level, and red corresponds to the highest level of the pressure response. The ability of this approach to reproduce the underlying tissue structures was demonstrated on a variety of phantom models and clinical data [13,14,78].

19.3 Results

The results of clinical testing of various TBI and PMI prototypes are described in several publications [15,26,78,79]. Here we will show some representative results illustrating the potential of the TI technology in cancer detection.

19.3.1 Tactile Imaging of Breast

There were several clinical studies performed on diagnostic/screening potential of breast TI. In one clinical study that included 110 patients with a complaint of a breast mass, TI demonstrated detection of 94% of the breast mass, while CBE identified only 86% [32]. The positive predictive value for breast cancer using TI was 94%. Another study with 179 patients (147 benign and 32 malignant cases) was conducted at four clinical sites to evaluate the TBI capability for breast lesion characterization and differentiation [15]. Tissue biopsy data were used as a gold standard in tissue differentiation. Examples of actual tactile images of breast lesions are shown in Table 19.2. Five features of the detected lesions were calculated from the acquired pressure pattern data for breast lesion characterization: strain hardening (F1), loading curve average slope (F2), maximum pressure peak for the fixed total force applied to the probe (F3), lesion shape (F4), and lesion mobility (F5) [15]. All these features (F1–F5) plus a patient age (F6) were used as input data of a Bayesian classifier to calculate probability of lesion being benign P(b) and malignant P(m) for a given set of input features. Clinical examples of calculated features for detected lesions from the acquired pressure pattern data for breast lesion are presented in Table 19.2. The difference between P(b) and P(m) was used as a threshold parameter for the construction of the receiver operating characteristic (ROC) curve to analyze the ability of the TBI in differentiation of benign from malignant lesions. The area under the ROC curve (AUC) demonstrating the diagnostic accuracy in discrimination of benign

TABLE 19.2

Clinical Examples of Breast Lesion Imaging and Feature Calculations

Case number, patient ID	1, L13	2, S27	3, S24
Patient age (F6)	55	25	54
Lesion biopsy diagnosis	Benign (fibrosis)	Benign (fibroadenoma)	Malignant (ductal carcinoma)
Breast lesion image			
Calculated features			
Strain hardening (F1), rel. units	−5.4	22.9	20.9
Loading slope (F2), rel. units	6.1	14.4	17.7
Peak pressure (F3), kPa	4.7	18.0	32.4
Lesion shape (F4), %	128	117	106
Lesion mobility (F5), %	44	12	11
Classifier output, P(b)–P(m)	0.72	0.13	−0.51

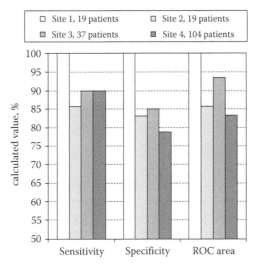

FIGURE 19.7 Differentiation for benign and malignant lesions with the use of Bayesian classifier evaluated for four clinical sites. (From Egorov, V. et al., *Breast Cancer Res. Treat.*, 118, 67, 2009.)

and malignant lesions was calculated separately for each clinical site. Figure 19.7 presents the calculated sensitivity, specificity, and AUC for each clinical site. We found that the sensitivity ranges from 85.7% to 100%, specificity from 78.7% to 100%, and AUC from 83.4% to 100%. An average sensitivity of 91.4% and a specificity of 86.8% with a standard deviation of ±6.1% were found. For clinical data combined from all sites, the AUC was equal to 86.1% with the 95% confidence interval (CI) from 80.3% to 90.9% while a significance level $P = 0.0001$ for the area of 50%; sensitivity was equal to 87.5% with the 95% CI from 71.0% to 96.4% ± 12% (95% CI) and specificity 84.4% with the 95% CI from 77.5% to 89.8% [15].

19.3.2 Tactile Imaging of Prostate

Capability of PMI technology to provide an objective image of the prostate and detect abnormalities was evaluated in a clinical setting. Figure 19.8 illustrates clinical examples of healthy and two prostate

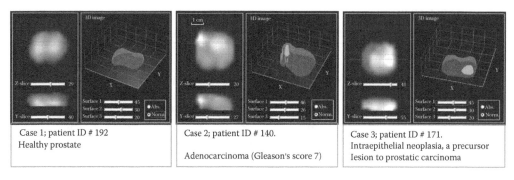

Case 1; patient ID # 192	Case 2; patient ID # 140.	Case 3; patient ID # 171.
Healthy prostate		Intraepithelial neoplasia, a precursor
	Adenocarcinoma (Gleason's score 7)	lesion to prostatic carcinoma

FIGURE 19.8   Examples of MI examination results. Left panel shows 2-D and 3-D mechanical images of a normal prostate, and two following panels show images of diseased prostates.

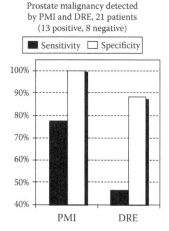

FIGURE 19.9   Malignancy detection by PMI vs. biopsy and DRE vs. biopsy.

pathology cases as were diagnosed by the transrectal ultrasound (TRUS)-guided biopsy. PMI 2-D coronal and transverse images (the top and bottom 2-D patterns, respectively) and a composite 3-D image of the prostate gland clearly visualize the prostate and confirm the presence of hard nodules [78].

The ROC analysis was conducted on the clinical study sample of 168 subjects to demonstrate the ability of the PMI to visualize nodules. The AUC was calculated to be 81%, with a 95% CI from 74% to 88%. The subgroup of the study was referred for further TRUS-guided biopsy testing as a result of patients having an elevated PSA level above 4.0 ng/mL, an abnormal DRE finding, or a combination of age or family prostate cancer history factors. For 13 patients in the study (PSA levels ranging from 1.0 to 26.7 ng/mL) with the presence of cancerous nodules confirmed by biopsy, PMI confirmed the biopsy results for 10 of the 13 cancer patients. The DRE identified only 6 of the 13. In the 8 cases (PSA levels ranging from 4.4 to 13.6 ng/mL) that were defined by the TRUS-guided biopsy as noncancerous, PMI depicted all 8 as normal images of the prostate, whereas the DRE detected 7 normal and 1 suspicious reading [78]. Figure 19.9 summarizes these findings and presents sensitivity and specificity for malignancy detection by PMI vs. biopsy and DRE vs. biopsy for a limited number of patients.

19.4 Discussion

Availability of health care is limited by its skyrocketing cost, and application of new advanced techniques frequently contributes to this escalation. Substantial disparity between industrialized countries and the rest of the world widens with more than 70% of all cancer deaths occurring in lower income

TABLE 19.3

Comparative Data for Breast Cancer Detection Effectiveness and Cost-Effectiveness

Screening/Diagnostic Technique	Sensitivity/ Specificity, %	Procedure Cost of Bilateral Exam, USD	Cost-Effectiveness, USD per Life Year Gained
CBE	56.5/93.7	—	522, India [10]
			31,900, Japan [48]
Mammography	73.7/94.3	112[a]	1,846, India [10]
			26,500–331,000 [75]
Ultrasound	Limited, see text	70[a]	—
MRI	87.7/92.8	1037[a]	55,420–130,695 [66]
Biopsy	96.6/100.0	2061[b]	2,250–77,500 [23,27]
TI	91.9[d]/88.9	5–50[c]	162[c]

Source: Adapted from Sarvazyan, A.P. et al., *Breast Cancer Basic Clin. Res.*, 1, 91, 2008. With permission.

[a] The U.S. average Medicare reimbursements in 2005.

[b] In average for one biopsy.

[c] Projections based on a physician's assistant performing the exam.

[d] Averaged for two clinical studies.

countries where availability of resources for diagnosis, prevention, and treatment of cancer is limited or nonexistent [3,36,81]. Introduction of cost-effective screening and diagnostic methods for cancer is especially important in the developing countries and lower income population all over the world having limited access to the sophisticated conventional devices.

Current methods of breast screening and diagnosis include breast self-examination (BSE), CBE, mammography, ultrasound, MRI, and biopsy. Table 19.3 presents a summary of breast cancer screening/diagnostic efficiency for various techniques, procedure cost, and cost-effectiveness numbers. Clinical results demonstrate that tactile elasticity imaging with a sensitivity of 91.9% and specificity of 88.9% has significant diagnostic potential comparable to that of conventional imaging techniques. Further, the ease-of-use, portability, and no-radiation position TI technology well for early detection, monitoring, and measurement of recurrence. In view of many countries with limited resources, effective yet less expensive modes of screening must be considered globally. The substantially lower TI procedure cost makes it, possibly, one method that has the potential to provide cost-effective breast cancer screening and diagnostics worldwide.

Another way of increasing the cost-effectiveness is enhancing diagnostic efficacy of screening that would lead to the higher rate of cancer detection. In the United States alone, more than one million breast biopsies are performed annually following mammography findings, and approximately 80% of these findings are benign [24,40]. To evaluate possible impact of TBI supplementing standard screening procedures (mammography alone or combination of mammography and conventional ultrasound) on the benign biopsy rate, one could apply TBI cancer sensitivity and specificity to the patient sample referred for the breast biopsy (20% malignant and 80% benign). The results signal that a 23% reduction of the benign biopsy is possible without missing cancer cases, while a 50% reduction of the benign biopsy with 4.6% missed cancer cases [15].

In case of the prostate cancer, frequently indolent and slow-progressing disease, adverse outcomes of overdiagnosis and overtreatment are weighted heavily against benefits of screening and early detection. In this context, active surveillance emerges as a new practical approach that provides for selective intervention on the individually defined need basis [33,49,51]. Introduction of a technology that could help discriminate between slow-growing and aggressive prostate tumors would be critical in guiding a curative treatment. PMI could augment PSA and ultrasound-guided biopsy by visual and quantitative assessment of changes in the mechanical properties of prostate tissue associated with disease progression. In addition, PMI imaging capability for visualization, recording, and tracking of the physical growth of the prostate nodule may significantly add to its diagnostic potential value. Another potential niche and clinical benefits of the use of PMI technology could be in expectant management. Such deferred treatment involves actively monitoring the course of the disease with the expectation to intervene if the cancer progresses or if symptoms become imminent [7,30].

The limitations of TI in detecting tissue abnormality are close to that of a highly skilled physician: detection limit of superficial nodules is about 2–4 mm in diameter and larger nodules of 10–15 mm in diameter could be detected at a depth of no more than 40 mm. Further progress in TI technology is expected mainly in expanding its fields of applications. We believe that any area of medical diagnostics, where the sense of touch is shown to provide useful information on the state of a tissue or an organ, could become a new field of application of TI. In any such new application, e.g., examination of lymph nodes or detecting an abnormality in thyroid, the geometry of the TI probe must be carefully redesigned to conform to a particular examination site.

19.5 Conclusion

There is a tremendous worldwide need for cost-effective means to detect cancer at its earliest stage and to monitor it progression through treatment. TI, a comparatively low-cost, no-radiation, and easy-to-use technology, electronically and quantitatively captures cancer-induced transformation of the mechanical properties of soft tissues. The TI applications, including the TBI for breast and PMI for prostate, have the potential to be positioned as an adjunct to conventional methods leading to enhanced screening, reduced negative biopsy rates, and improved cancer detection.

ACKNOWLEDGMENTS

This work was supported in part by the NCI NIH grants: R43 CA91392-01, R44 CA69175, R43 CA82620, R43 CA94444, R43 CA99094, and R44 CA82620-02.

REFERENCES

1. American Cancer Society. 2010. *Breast Cancer Facts & Figures 2009–2010*. Atlanta, GA: American Cancer Society, Inc.
2. American Urological Association. 2009. *Prostate-Specific Antigen Best Practice Statement: 2009 Update*. Linthicum, MD: Education and Research, Inc.
3. Anderson, B.O., R. Shyyan, A. Eniu, R.A. Smith, C.H. Yip, N.S. Bese, L.W. Chow, S. Masood, S.D. Ramsey, and R.W. Carlson. 2006. Breast cancer in limited-resource countries: An overview of the Breast Health Global Initiative 2005 guidelines. *Breast J.* **12**(Suppl.):S3–S15.
4. Beccai, L., S. Roccella, A. Arena, F. Valvoa, P. Valdastria, A. Menciassia, M.C. Carrozza, and P. Dario. 2005. Design and fabrication of a hybrid silicon three-axial force sensor for biomechanical applications. *Sens. Actuat. A Phys.* **120**:370–382.
5. Beebe, D.J., A.S. Hsieh, R.G. Radwin, and D.D. Denton. 1995. A silicon force sensor for robotics and medicine. *Sens. Actuat. A Phys.* **50**:55–65.
6. Bensmaïa, S.J. and M. Hollins. 2005. Pacinian representations of fine surface texture. *Percept. Psychophys.* **67**:842–854.
7. Bill-Axelson, A., L. Holmberg, M. Ruutu, M. Häggman, S.O. Andersson, S. Bratell, A. Spångberg, C. Busch, S. Nordling, H. Garmo, J. Palmgren, H.O. Adami, B.J. Norlén, J.E. Johansson, and Scandinavian Prostate Cancer Group Study No. 4. 2005. Radical prostatectomy versus watchful waiting in early prostate cancer. *New Eng. J. Med.* **352**:1977–1984.
8. Bloor, D., K. Donnely, K.J. Hands, P. Laughlin, and D. Lussey. 2005. A metal-polymer composite with unusual properties. *J. Phys. D Appl. Phys.* **38**:2851–2860.
9. van Boven, R.W., R.H. Hamilton, T. Kauffman, J.P. Keenan, and A. Pascual-Leone. 2000. Tactile spatial resolution in blind Braille readers. *Neurology* **54**:2230–2236.
10. Brown, M.L., S.J. Goldie, G. Draisma, J. Harford, and J. Lipscomb. 2006. Health service interventions for cancer control in developing countries. In D.T. Jamison, J.G. Breman et al. (eds.), *Disease Control Priorities in Developing Countries*, 2nd edn., Washington, DC: Oxford University Press, pp. 569–589.

11. Burnside, E.S., T.J. Hall, A.M. Sommer, G.K. Hesley, G.A. Sisney, W.E. Svensson, J.P. Fine, J. Jiang, and N.J. Hangiandreou. 2007. Differentiating benign from malignant solid breast masses with US strain imaging. *Radiology* **245**:401–410.

12. Cosgrove, D., C.J. Doré, C. Cohen-Bacrie, and J.P. Henry. 2010. Preliminary assessment of ShearWave(TM) elastography features in predicting breast lesion malignancy. *European Congress of Radiology*, Vienna, Austria. http://www.supersonicimagine.fr/fichiers/ts_etude/ecr2010_poster_c-0444_cosgrove.pdf

13. Egorov, V., S. Ayrapetyan, and A.P. Sarvazyan. 2006. Prostate mechanical imaging: 3-D image composition and feature calculations. *IEEE Trans. Med. Imaging* **25**:1329–1340.

14. Egorov, V. and A.P. Sarvazyan. 2008. Mechanical imaging of the breast. *IEEE Trans. Med. Imaging* **27**:1275–1287.

15. Egorov, V., T. Kearney, S.B. Pollak, C. Rohatgi, N. Sarvazyan, S. Airapetian, S. Browning, and A.P. Sarvazyan. 2009. Differentiation of benign and malignant breast lesions by mechanical imaging. *Breast Cancer Res. Treat.* **118**:67–80.

16. Egorov, V., H. van Raalte, and A.P. Sarvazyan. 2010. Vaginal tactile imaging. *IEEE Trans. Biomed. Eng.* **57**:1736–1744.

17. Fowlkes, J.B., S.Y. Emelianov, J.G. Pipe, A.R. Skovoroda, P.L. Carson, R.S. Adler, and A.P. Sarvazyan. 1995. Magnetic-resonance imaging techniques for detection of elasticity variation. *Med. Phys.* **22**:1771–1778.

18. Frei, E.H., B.D. Sollish, and S. Yerushalmi. March 1979. Instrument for viscoelastic measurement. U.S. patent 4144877.

19. Frei, E.H., B.D. Sollish, and S. Yerushalmi. February 1981. Instrument for viscoelastic measurement. U.S. patent 4250894.

20. Gentle, C.R. 1988. Mammobarography: A possible method of mass breast screening. *J. Biomed. Eng.* **10**:124–126.

21. Gescheider, G.A. 1974. Effects of signal probability on vibrotactile signal recognition. *Percept. Mot. Skills* **38**:15–23.

22. Grant, A.C., R. Fernandez, P. Shilian, E. Yanni, and M.A. Hill. 2006. Tactile spatial acuity differs between fingers: A study comparing two testing paradigms. *Percept. Psychophys.* **68**:1359–1362.

23. Groenewoud, J.H., R.M. Pijnappel, M.E. van den Akker-van Marle, E. Birnie, T. Buijs-van der Woude, W.P Mali, H.J. de Koning, and E. Buskens. 2004. Cost-effectiveness of stereotactic large-core needle biopsy for nonpalpable breast lesions compared to open-breast biopsy. *British J. Cancer* **90**:383–392.

24. Gur, D., L.P. Wallace, A.H. Klym, L.A. Hardesty, G.S. Abrams, R. Shah, and J.H. Sumkin. 2005. Trends in recall, biopsy, and positive biopsy rates for screening mammography in an academic practice. *Radiology* **235**:396–401.

25. Helsel, M., J.N. Zemel, and V. Dominko. 1988. An impedance tomographic tactile sensor. *Sens. Actuat. A Phys.* **14**:93–98.

26. Helvie, M.A., P.L. Carson, A.P. Sarvazyan, N. Thorson, V. Egorov, and M.A. Roubidoux. 2003. Mechanical imaging of the breast: A pilot trial. *Ultrasound Med. Biol.* **29**(Suppl.):S112.

27. Hillner, B.E. 2004. Cost and cost-effectiveness considerations. In J.R. Harris, M.E. Lipman, M. Mprrow, C.R. Osborne (eds.), *Diseases of the Breast*, 3rd edn., Philadelphia, PA: Lippincott Williams & Wilkins, pp. 1133–1142.

28. Hippocrates. 400 B.C. *The Book of Prognostics*. http://www.greektexts.com/library/Hippocrates/The_%20Book_Of_Prognostics/eng/4.html

29. Itoh, A., E. Ueno, E. Tohno, H. Kamma, H. Takahashi, T. Shiina, M. Yamakawa, and T. Matsumura. 2006. Breast disease: Clinical application of US elastography for diagnosis. *Radiology* **9**:341–350.

30. Johansson, J.E., O. Andrén, S.O. Andersson, P.W. Dickman, L. Holmberg, A. Magnuson, and H.O. Adami. 2004. Natural history of early, localized prostate cancer. *J. Amer. Med. Assoc.* **291**:2713–2719.

31. Johansson, R.S. and A.B. Vallbo. 1979. Tactile sensibility in the human hand: Relative and absolute densities of four types of mechanoreceptive units in glabrous skin. *J. Physiol.* **286**:283–300.

32. Kaufman, C.S., L. Jacobson, B. Bachman, and L. Kaufman. 2006. Digital documentation of the physical examination: Moving the clinical breast exam to the electronic medical record. *Am. J. Surg.* **192**:444–449.

33. Klotz, L. 2006. Active surveillance versus radical treatment for favorable risk localized prostate cancer. *Curr. Treat. Options Oncol.* **7**:355–362.

34. Kotani, K., S. Ito, T. Miura, and K. Horii. 2007. Evaluating tactile sensitivity adaptation by measuring the differential threshold of archers. *J. Physiol. Anthropol.* **26**:143–148.

35. Krouskop, T.A., T.M. Wheeler, F. Kallel, B.S. Garra, and T. Hall. 1998. Elastic moduli of breast and prostate tissues under compression. *Ultrason. Imaging* **20**:260–274.

36. Laxminarayan, R., J. Chow, and S.A. Shahid-Salles. 2006. Intervention cost-effectiveness: Overview of main messages. In D.T. Jamison, J.G. Breman et al. (eds.), *Disease Control Priorities in Developing Countries*, 2nd edn., Washington, DC: Oxford University Press, pp. 35–86.

37. Lederman, S.J. and R.L. Klatzky. 2009. Human haptics. In L.R. Squire (ed. in Chief), *Encyclopedia of Neuroscience*, vol. 5. San Diego, CA: Academic Press, pp.11–18.

38. Lee, H.K., S.I. Chang, and E. Yoon. 2006. A flexible polymer tactile sensor: Fabrication and modular expandability for large area deployment. *J. Microel. Systems* **15**:1681–1686.

39. Lee, M.H. and H.R. Nichols. 1999. Tactile sensing for mechatronics: A state of the art survey. *Mechatronics* **9**:1–31.

40. Liang, W., W.F. Lawrence, C.B. Burnett, Y.T. Hwang, M. Freedman, B.J. Trock, J.S. Mandelblatt, and M.E. Lippman. 2003. Acceptability of diagnostic tests for breast cancer. *Breast Cancer Res. Treat.* **79**:199–206.

41. Libouton, X., O. Barbier, L. Plaghki, and J.L. Thonnard. 2010. Tactile roughness discrimination threshold is unrelated to tactile spatial acuity. *Behav. Brain. Res.* **208**:473–478.

42. Mei, T., W.J. Li, Y. Ge, Y. Chen, L. Ni, and M.N. Chan. 2000. An integrated MEMS three-dimensional tactile sensor with large force range. *Sens. Actuat. A Phys.* **80**:155–162.

43. Multhupillai, R., D.J. Lomas, P.J. Rossman, J.F. Greenleaf, A. Manduca, and R.L. Ehman. 1995. Magnetic resonance elastography by direct visualization of propagating acoustic strain waves. *Science* **269**:1854–1857.

44. Niemczyk, P., A.P. Sarvazyan, A. Fila, P. Amenta, W.S. Ward, P. Javidian, K. Breslayer, and K.B. Cummings. 1996. Mechanical imaging, a new technology for prostate cancer detection. *Surg. Forum* **47**:823–825.

45. Niemczyk, P., K.B. Cummings, A.P. Sarvazyan, E. Bancila, W.S. Ward, and R.E. Weiss. 1998. Correlation of mechanical imaging and histopathology of radical prostatectomy specimens: A pilot study for detecting prostate cancer. *J. Urol.* **160**:797–801.

46. Nightingale, K., M.S. Soo, R. Nightingale, and G. Trahey. 2002. Acoustic radiation force impulse imaging: In vivo demonstration of clinical feasibility. *Ultrasound Med. Biol.* **28**:227–235.

47. Ophir, J., I. Cespedes, H. Ponnekanti, Y. Yazdi, and X. Li. 1991. Elastography: A quantitative method for imaging the elasticity of biological tissues. *Ultrason. Imaging* **13**:111–134.

48. Ohnuki, K., S. Kuriyama, N. Shoji, Y. Nishino, I. Tsuji, and N. Ohuchi. 2006. Cost-effectiveness analysis of screening modalities for breast cancer in Japan with special reference to women aged 40–49 years. *Cancer Sci.* **97**:1242–1247.

49. Parker, C. 2004. Active surveillance: Towards a new paradigm in the management of early prostate cancer. *Lancet Oncol.* **5**:101–106.

50. Parker, K.J., S.R. Huang, R.A. Musulin, and R.M. Lerner. 1990. Tissue response to mechanical vibrations for "sonoelasticity imaging." *Ultrasound Med. Biol.* **16**:241–246.

51. Patel, M.I., D.T. DeConcini, E. Lopez-Corona, M. Ohori, T. Wheeler, and P.T. Scardino. 2004. An analysis of men with clinically localized prostate cancer who deferred definitive therapy. *J. Urol.* **171**:1520–1524.

52. Plewes, D.B., I. Betty, S.N. Urchuk, and I. Soutar. 1995. Visualizing tissue compliance with MR imaging. *J. Mag. Res. Imaging* **5**:733–738.

53. Pressure mapping and force measurement. Sensor technology. http://www.tekscan.com/technology. html#1

54. Raza, S., A. Odulate, E.M. Ong, S. Chikarmane, and C.W. Harston. 2010. Using real-time tissue elastography for breast lesion evaluation: Our initial experience. *J. Ultrasound Med.* **29**:551–563.

55. Russell, R.A. 2002. A tactile sensor skin for measuring surface contours. In *Proceedings of IEEE Region 10 International Conference on Technology Enabling Tomorrow: Computers, Communications and Automation towards the 21st Century,* Melbourne, Victoria, Australia, pp. 262–266.

56. Russell, R.A. and S. Parkinson. 1993. Sensing surface shape by touch. In *Proceedings of IEEE International Conference on Robotics and Automation,* Atlanta, GA, pp. 423–428.

57. Sabatini, A.M., P. Dario, and M. Bergamasco. 1990. Interpretation of mechanical properties of soft tissues from tactile measurements. In *Lecture Notes in Control and Information Sciences*, vol. 139. *The First International Symposium on Experimental Robotics*, Springer-Verlag, London, U.K., pp. 452–462.

58. Sarvazyan, A.P. 1998. Mechanical imaging: A new technology for medical diagnostics. *Int. J. Med. Inf.* **49**:195–216.

59. Sarvazyan, A.P. 1998. Computerized palpation is more sensitive than human finger. In *Proceedings of the 12th International Symposium on Biomedical Measurements and Instrumentation*, Dubrovnik, Croatia, pp. 523–524.

60. Sarvazyan, A.P. 2001. Elastic properties of soft tissues. In M. Levy, H.E. Bass, R.R. Stern (eds.), *Handbook of Elastic Properties of Solids, Liquids and Gases,* vol. III, Chapter 5, New York: Academic Press, pp. 107–127.

61. Sarvazyan, A.P. 2006. Tissue viscoelasticity: Past and future, unexplored areas and brave projections. In *Proceedings of the 5th International Conference on the Ultrasonic Measurement and Imaging of Tissue Elasticity*, Snowbird, UT.

62. Sarvazyan, A.P., V. Egorov, J.S. Son, and C.S. Kaufman. 2008. Cost-effective screening for breast cancer worldwide: Current state and future directions. *Breast Cancer: Basic Clin. Res.* **1**:91–99.

63. Sarvazyan, A.P. and A.R. Skovoroda. June 1996. Method and apparatus for elasticity imaging. U.S. patent 5524636.

64. Sarvazyan, A.P., A.R. Skovoroda, S.Y. Emelianov, J.B. Fowlkes, J.G. Pipe, R.S. Adler, R.B. Buxton, and P.L. Carson. 1995. Biophysical bases of elasticity imaging. In J.P. Jones (ed.), *Acoustical Imaging*, vol. 21, New York: Plenum Press, pp. 223–240.

65. Sarvazyan, A.P., O.V. Rudenko, S.D. Swanson, J.B. Fowlkes, and S.Y. Emelianov. 1998. Shear wave elasticity imaging: A new ultrasonic technology of medical diagnostics. *Ultrasound Med. Biol.* **24**:1419–1435.

66. Saslow, D., C. Boetets, W. Burke, S. Harms, M.O. Leach, C.D. Lehman, E. Morris, E. Pisano, M. Schnall, S. Sener, R.A. Smith, E. Warner, M. Yaffe, K.S. Andrews, C.A. Russell, and American Cancer Society Breast Cancer Advisory Group. 2007. American Cancer Society guidelines for breast screening with MRI as an adjunct to mammography. *CA Cancer J. Clin.* **57**:75–89.

67. Sathian, K., A. Zangaladze, J. Green, J.L. Vitek, and M.R. DeLong. 1997. Tactile spatial acuity and roughness discrimination: Impairment due to aging and Parkinson's disease. *Neurology* **49**:168–177.

68. Sinkus, R., J. Lorenzen, D. Schrader, M. Lorenzen, M. Dargatz, and D. Holz. 2000. High-resolution tensor MR elastography for breast tumour detection. *Phys. Med. Biol.* **45**:1649–1664.

69. Sinkus, R., K. Siegmann, M. Tanter, T. Xydeas, and M. Fink. 2007. MR–elastography is capable of increasing the specificity of MR-mammography influence of rheology on the diagnostic gain. In *Proceedings of the 6th International Conference on the Ultrasonic Measurement and Imaging of Tissue Elasticity*, Snowbird, UT, p. 111.

70. Skovoroda, A.R., A.N. Klishko, D.A. Gusakian, E.I. Maevskii, V.D. Ermilova, G.A. Oranskaia, and A.P. Sarvazyan. 1995. Quantitative analysis of the mechanical characteristics of pathologically changed biological tissues. *Biophysics* **40**:1359–1364.

71. Son, J.S. and T. Parks. October 2008. Hybrid tactile sensor. U.S. patent 7430925.

72. Svensson, W.E., N. Zaman, N.K. Barrett, G. Ralleigh, K. Satchithananda, S. Comitis, V. Gada, and N.R. Wakeham. 2007. Breast elasticity imaging aids patient management in the one stop breast clinic. In *Proceedings of the 6th International Conference on the Ultrasonic Measurement and Imaging of Tissue Elasticity*, Santa Fe, NM, p. 128.

73. Tegin, J. and J. Wikander. 2005. Tactile sensing in intelligent robotic manipulation—A review. *Ind. Rob.* **32**:64–70.

74. Thomas, A., T. Fischer, H. Frey, R. Ohlinger, S. Grunwald, J.U. Blohmer, K.J. Winzer, S. Weber, G. Kristiansen, B. Ebert, and S. Kümmel. 2006. Real-time elastography—An advanced method of ultrasound: First results in 108 patients with breast lesions. *Ultrasound Obstet. Gynecol.* **28**:335–340.

75. Tosteson, A.N., N.K. Stout, D.G. Fryback, S. Acharyya, B.A. Herman, L.G. Hannah, E.D. Pisano, and DMIST Investigators. 2008. Cost-effectiveness of digital mammography breast cancer screening. *Ann. Int. Med.* **148**:1–10.

76. Wellman, P.S. 1999. Tactile imaging. PhD thesis. Harvard University's Division of Engineering and Applied Sciences, Cambridge, MA.

77. Wellman, P.S., E.P. Dalton, D. Krag, K.A. Kern, and R.D. Howe. 2001. Tactile imaging of breast masses: First clinical report. *Arch. Surg.* **136**:204–208.

78. Weiss, R., V. Egorov, S. Ayrapetyan, N. Sarvazyan, and A.P. Sarvazyan. 2008. Prostate mechanical imaging: A new method for prostate assessment. *Urology* **71**:425–429.

79. Weiss, R., V. Hartanto, M. Perrotti, K. Cummings, A. Bykanov, V. Egorov, and S.A. Sobolevsky. 2001. In vitro trial of the pilot prototype of the prostate mechanical imaging system. *Urology* **58**:1059–1163.

80. Wolfe, J.M., K.R. Kluender, D.M. Levi, L.M. Bartoshuk, R.S. Herz, R.L. Klatzky, and S.J. Lederman. 2008. Touch. In *Sensation and Perception*, Chapter 12, 2nd edn., Sunderland, MA: Sinauer, pp. 286–313.

81. World Health Organization. February 2006. Fact sheet N 297. http://www.who.int/cancer/en/index.html

82. Zhi, H., X.Y. Xiao, H.Y. Yang, B. Ou, Y.L. Wen, and B.M. Luo. 2010. Ultrasonic elastography in breast cancer diagnosis strain ratio vs 5-point scale. *Acad. Radiol.* **17**:1227–1233.

Part VI

Optical Technologies for Cancer Detection and Diagnostics: Fluorescence, Luminescence, Refractive Index Detection Technologies

20

Biomechanics-Based Microfluidic Biochip for the Effective Label-Free Isolation and Retrieval of Circulating Tumor Cells

Swee Jin Tan, Wan Teck Lim, Min-Han Tan, and Chwee Teck Lim

CONTENTS

20.1 Introduction

Cancer is a genetic disease that manifests as abnormal cell proliferation and growth consequently impairs normal body functions. In 2008 alone, 7.6 million people died from the disease according to the World Health Organization (WHO) [1]. Furthermore, the lack of telltale signs at the disease onset limits the ability for the eradication or early treatment of cancer [2]. There is a need for improved techniques of cancer detection and monitoring in order to reduce and alleviate the burden imposed on patients and health-care systems by cancer. Circulating tumor cells (CTCs) are cancerous cells that enter the blood circulation during hematogenous metastasis, and multiple studies have demonstrated an association between these cells and disease activity [3–5]. The documented evidence for the presence of CTCs dates back to over a century [6], which verifies the blood circulation as a means that allows transportation of tumor cells to other parts of the body. More recently, studies have demonstrated the presence of CTCs in patients with various metastatic carcinomas [7,8]. There have been numerous studies demonstrating that the number of CTCs in peripheral blood is an important prognostic indicator in metastatic breast, prostate, and colon cancers [9–12]. These studies have directly linked CTC count with disease progression, and there is emerging evidence for its use in prediction of treatment efficacy [13–16]. The isolation of CTCs from peripheral blood is more amenable to that of biopsies in terms of less invasiveness, patient acceptability, and increased frequency of tests to perform. Additionally, the isolation of CTCs provides a promising alternative source of tumor tissue for the detection and characterization of non-blood-related cancers. Thus, analyzing the blood specimens of patients which are routinely taken can be useful. The isolation, quantification, and study of these cells obtained from peripheral blood are thus of much interest.

Tumor cell dissemination through the circulatory system has been shown to be an inefficient process, with most tumor cells being eliminated before distant implantation [17–19]. As a result, CTCs are relatively rare in peripheral blood of cancer patients [20,21], and their detection presents a technical challenge [22]. Tumor cell count can be as low as one cancer cell to 1 mL of blood, the same volume containing approximately 4.8–5.4 billion erythrocytes, 7.4 million leukocytes, and 280 million thrombocytes [23]. Several different enrichment methodologies have been employed to isolate CTCs, most notably using biochemical means. Using antibodies targeted against epithelial specific antigens such as epithelial cell adhesion molecule (EpCAM), CTCs have been isolated using magnetic separation or flow-based assays [24,25]. Additional immunofluorescence staining or molecular-based techniques are utilized to confirm the presence of CTCs in the isolated cell population and distinguish them from hematopoietic cells. The affinity-based CTC separation is demonstrated in various clinical settings, but studies have established that the effectiveness is highly dependent on the specificity of the antibody used [26]. Alternative methodologies such as a direct visualization assay [27], fluorescent-activated cell sorter (FACS) [28], and fiber-optic array scanning technology (FAST) cytometer [29] have also been used to detect CTCs in blood samples. Complex procedures, tedious inspections, and long processing time are the limiting factors associated with most existing techniques. Furthermore, viability of the isolated cells is often lost with existing techniques, which require sample fixation. There is much to appreciate about the conditions of CTCs while in circulation [4,5], and having systems that are gentle to the tumor cells during enrichment will aid studies to be carried out on CTC subpopulations. This may provide valuable insights into the metastatic process which will influence therapeutic decisions. Thus, a sensitive enrichment method is crucial to aid in further examination of CTCs which can be clinically beneficial.

This chapter focuses on the development of a microfluidic biomechanics-based platform that enriches CTCs directly from the peripheral blood of cancer patients using the inherent biomechanical property differences in tumor and blood cells. Biological cells are small, typically in the range of several microns to tens of microns, and consequently are extremely hard to manipulate [30]. Traditional bench-top tools in cell and molecular biology are insufficient to address the needs for sensitive and accurate measurements of rare cell events such as the enumeration of CTCs. With microfluidic channels in the dimensions of a few to hundreds of micrometers, they are well suited for single-cell handling which is of comparable dimensions. The small size in microfluidic devices ensures laminar flow characteristics (low Reynolds number) which make the fluid flow predictable and aid in the precise control of cells in the enclosed environment. These methodologies also offer a high-throughput analysis using small sample volumes [31,32] and are suitable for the challenges involved in single-cell manipulations, and to analyze the molecular components such as DNA and RNA [33,34]. It also provides a means to integrate several devices with various functions to form a complete integrated laboratory-on-chip system [35]. Furthermore, the miniaturized platforms are cost-efficient, minimizing the volumes required for expensive reagents relative to similar conventional biological bench-top methods. The technology fundamentally introduces myriad possibilities to enhance and bring about new capabilities in a variety of analyses [36].

Using a microfluidic device, we aim to achieve an effective CTC separation from peripheral blood using distinctive physical differences between cancer cells [17,37] and blood constituents [38,39]. The system is label-free and is thus potentially able to capture a larger pool of tumor cells in an unbiased fashion as there is no initial selection as compared to affinity-based separation. Sample processing is user friendly, requires no laborious preparation of the sample, and only a single processing step is required for cell enrichment. As the microdevice is optically transparent, it is compatible on existing laboratory microscopes, and real-time imaging of the CTC isolation process can be recorded. In addition, the system was tested on a diverse range of cancer cell lines to show its versatility, given the genetic and phenotypic heterogeneity of cancer [40–42]. Our results have shown a consistent sensitivity to low counts of cancer cells in numerous tested samples. Using computational analyses, it was demonstrated that the shear forces due to the moving fluid were gentle to the isolated tumor cells, allowing retrieval of these viable cells for further study. Altogether, the system is attractive for applications in oncology research.

20.2 Materials

20.2.1 Microdevice Fabrication

Figure 20.1a highlights the fabrication steps involved in the manufacture of the biochip. The microfluidic biochip was produced using soft lithography [43]. Design of the microdevice was first printed on a photo mask (Infinite Graphics Inc., Minneapolis, MN). SU8-2025 (Microchem Corporation, Newton, MA), a negative photoresist, was spin coated onto an 8 in. silicon substrate to achieve the thickness of 18–20 μm. It then underwent an ultraviolet exposure through the photo mask where portions exposed became polymerized. Following that, the unexposed region was removed chemically. A final hard bake ensured better adhesion of the photoresist to the substrate. The designs on the substrate became the master molds (Figure 20.1b) for replica molding of the microdevice. Polydimethylsiloxane (PDMS, Sylgard 184, Dow Corning, Midland, MI) mixed according to manufacturer's recommendation (10:1) was degassed and poured over the mold. The mixture was then subjected to heat curing, and fluidic ports were punched on the patterned PDMS, as shown in Figure 20.1c, after removal from the mold. The PDMS block and a glass slide were then subjected to oxygen plasma treatment and bonded irreversibly. Prior to use, tubings were inserted directly into the fluidic ports to allow blood samples to be introduced.

20.2.2 Experimental Setup and Apparatus Preparation

The experimental setup (Figure 20.2) was integrated onto an inverted microscope (Leica Microsystems GmbH, Germany), permitting real-time imaging of the isolation process. Predetermined pressure differentials were used to drive the samples across the microdevice using compressed air from two large syringes to produce pressured lines. A program written in NI Labview (National Instruments, Austin, TX)

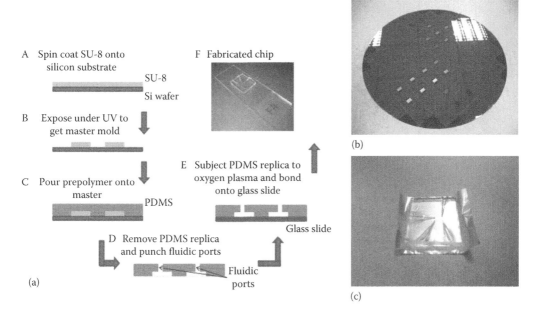

FIGURE 20.1 Major steps in the fabrication of the microdevice. (a) Device fabrication using soft lithography procedures. (b) Master mold on a precut 8 in. wafer. (c) Microdevice removed from the master mold.

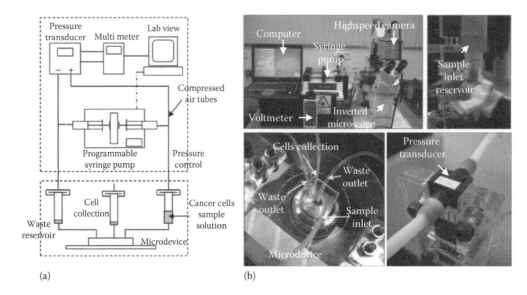

FIGURE 20.2   Experimental apparatus and setup. (a) Schematic of the entire setup showing the pressure control component and the microdevice system. (b) Entire CTCs isolation setup which is integrated onto an inverted microscope to allow real-time visualization and compactness to reduce dead volume. Microdevice is patent pending. (From Tan, S.J. and Lim, C.T., Microsystem for isolating viable circulating tumor cells, US Provisional Application No. 61/172, 250 [patent pending].)

controlled a programmable syringe pump (Harvard Apparatus, Holliston, MA). The differential pressure was measured by a pressure transducer (Honeywell, Morristown, NJ) to precisely control the pressure settings into the microdevice. This allowed semiautomation in the blood processing and also made the entire system easily portable without requiring external pressure sources.

For apparatus preparation prior to sample processing, the microfluidic system was first flushed with 5 mM EDTA (Sigma, St. Louis, MO) buffer through the sample inlet, as shown in Figure 20.2b. No other sample preparatory steps were required. Image capturing of the cell isolation process was taken using a high-speed camera (Photron, San Diego, CA).

20.2.3 Blood Collection

Informed consents from healthy volunteers and cancer patients were taken before blood extraction, as approved by the institutional review boards of participating institutions. Blood samples from healthy patients served as controls and were also used in spiking experiments for the characterization of the microdevice. Samples were stored in EDTA tubes (BD, Franklin Lakes, NJ) prior to use and discarded after the experiment. For cancer patients, 9–10 mL of blood was extracted each time with the first 0.5–1.0 mL of blood discarded to prevent false-positive responses. Two milliliters of blood is required for each test, and the excess amount drawn allowed other tests to be tested concurrently.

20.2.4 Cell Culture

The cell lines of human breast adenocarcinoma (MCF-7 and MDA-MB-231), gastric carcinomas (AGS and N87), hepatocellular (HepG2) adenocarcinomas, tongue squamous carcinoma (CAL27), and pharynx squamous carcinoma (FADU) were used to characterize the efficacy of the microfluidic chip. Culturing of cells was done in 25 cm^2 tissue culture flasks (Greiner Bio-One, Frickenhausen, Germany). AGS and N87 were maintained using RPMI 1640 (Sigma, St. Louis, MO), while the rest used Dulbecco's Modified Eagle's Medium (DMEM) (Sigma, St. Louis, MO). Both culture media were supplemented with 10% fetal bovine serum (FBS) (Hyclone, Logan, UT) and 1% penicillin G/streptomycin/amphotericin (Gibco, Carlsbad, CA).

The culturing media were continually changed every two days during the maintenance of the cancer cell lines. Upon cell confluence, they were subcultured. Cells were washed with 1× phosphate buffered saline (PBS) to remove traces of media in the culturing flask, and 2.5 mL of trypsin-EDTA (Gibco, Carlsbad, CA) was added which aided to release the cells from the culture flask surface. An additional 2.5 mL of fresh media was added into the mixture of suspended cells and spun down into a pellet using centrifugation at 1200 rpm for 5 min. The supernatant was removed, and fresh media of 1 mL were added to resuspend the cell pellet. The resuspended cells were then ready for experimentation.

20.2.5 Spiked Sample Preparation

For spiked sample preparation, cancer cell counting was done with a disposable hemocytometer (iN Cyto, Republic of Korea) and serial diluted to achieve the desired concentration of 100 cells/mL in 1× PBS. Cells were first grown to confluence and resuspended in culture media. A portion of the suspended cells were taken as control in the experiment and cultured normally. The rest was diluted to a concentration of approximately 100 cells/mL, and the sample solution was injected into the device for characterizing the isolation efficiency of cancer cells. After the cells were trapped in the microdevice, they were retrieved by reversing the pressure differential for cell viability experiments which directed the isolated cells to dislodge and flow towards the collection point. The collected solution was centrifuged at 1200 rpm for 5 min with the cell pellet resuspended later in culturing media DMEM. These were then reseeded in the T25 culture flask. Their proliferative rates were compared with normal cultures which acted as controls in the experiment over a 5-day period under normal culturing conditions.

The detection limit for the setup was ascertained with low cancer cell count in sample solutions. One to three cells were suspended in 1 mL of 1× PBS and passed through the device. The effectiveness is measured by the ability of the system to recover the cells. Isolation of individual cells from the suspended state was done by either manually pipetting or using the FACS (BD FACSAria II Cell Sorter, Franklin Lakes, NJ).

20.2.6 Immunofluorescence Staining to Identify CTCs

In order to ascertain the isolation purity, each of the cancer cell types was spiked into whole blood donated from healthy donors at approximately 100 cells/mL. It was further diluted with 5 mM EDTA in a 1:2 ratio to reduce the sample viscosity so that it could be processed easily. Isolated cells inside the microdevice were immunofluorescently stained to distinguish between cancer cells and hematopoietic cells. This allowed the visual examination of the isolation purity of cancer cells in the microdevice.

For the control experiment, the premixed sample of blood and cancer cells (200 μL) was incubated onto a 12 mm coverslip (polylysine-coated) for 30 min. The sample was then stained for EpCAM (1:50, Santa Cruz Biotechnology Inc., Santa Cruz, CA) to identify cancer cells, CD45 (1:50, Santa Cruz Biotechnology Inc., Santa Cruz, CA) for white blood cells (WBCs), and 4′,6-diamidino-2-phenylindole (DAPI) to permit nuclei visualization. For staining in the microdevice, a pressure differential of 5 kPa was used to induce fluid flow into the microdevice. The value of 5 kPa was chosen as it empirically best preserves the state of the isolated cells in the microdevice as compared to using higher-pressure differentials. Captured cells were first fixed by flowing 4% paraformaldehyde (PFA) for 30 min, permeabilized by 0.1% Triton X-100 in 1× PBS for 10 min, followed by washing with 1× PBS for 15 min and adding 10% goat serum to block out nonspecific bindings. To identify cancer cells, 0.2 mL of EpCAM antibodies were injected for 15 min, left to stand for another 45 min, and followed by washing with PBS. The procedures of antibody injection and PBS wash were repeated for the secondary antibody (1:500, goat anti-mouse Alexa Fluor 568, Invitrogen Corp., Carlsbad, CA). For the identification of WBCs, 0.2 mL of fluorescein isothiocyanate (FITC) conjugated with CD45 antibodies was injected for 15 min, left to stand for another 45 min, and followed by washing with PBS. Staining was completed by flowing DAPI for 15 min at 5 kPa followed by washing with PBS. For the identification of CTCs in metastatic lung cancer patients, anticytokeratin antibodies (CAM5.2, BD Biosciences, San Jose, CA) were used. A positive CTC count was defined to be cytokeratin positive, CD45 negative, and DAPI positive.

20.3 Methods

20.3.1 Biomechanical Approaches to Studying Cancer

Biomechanical properties of the blood cells allow for their passage through the body's microvasculature but not that of tumor cells [17,45,46]. The links between biomechanics and cancer have attracted considerable interests in recent years [47,48], where traditional engineering techniques are used to quantify, detect, and study the disease. Though most current research work on cancer emphasizes on the molecular, immunological, and pathological aspects, recent studies have suggested that the physical structure of diseased cells is a direct result of their anomalies and can be used as an indicator to detect them. Studying cancer from a biomechanical perspective can thus lead to a better understanding of the disease which occurs at the molecular level and can lead to changes at the cellular level.

Table 20.1 summarizes some of the investigations into the biophysical and mechanics of cancer cells. Technological advances have brought significant improvements in measurements at the micron and nanometer dimensions which make these studies possible [49]. For instance, using the atomic force microscope, Cross et al. [50] measured structural aberrations due to malignancies at the single-cell level and found distinctive variations in the cancer cell stiffness as compared to normal cells. The analyses are sensitive and evident that the mechanical phenotype can distinguish the diseased cells accurately in a nondestructive test. In another investigation, Remmerbach et al. [51] used the physical property of cells in an optical stretcher to process samples at a high throughput to aid cancer diagnosis. Other techniques such as the micropipette aspiration also showed hepatocellular carcinoma (HCC) cell lines having distinct elastic coefficients [52,53] and clearly support the fact that the genetic mutations during malignant transformation cascade downstream to affect the structural elements of the cell [50–55]. These physical changes extend also to hematopoietic cancers. Lam et al. [56] demonstrated

TABLE 20.1
Biomechanical Investigation of Cancer Cells and Their Link to the Disease States

Cancer Type	Biomechanical Assay	Parameter of Interest	Key Findings	References
Lung, breast, and pancreatic cancers	Atomic force microscopy (AFM)	Single-cell stiffness	1. Cellular structural changes due to malignancies at single cellular level 2. Cancer cells from pleural fluids isolated and found to be distinctively different from normal cells 3. Cells from different cancer types exhibit similar stiffness	[50,54]
Leukemia	AFM and microfluidics	Cell stiffness and deformability	1. Deformability of human myeloid was 18 times stiffer than lymphoid cells 2. Myeloid cells were six times stiffer than neutrophils 3. Effects of chemotherapy increase the leukemia cell stiffness by two orders of magnitude	[56,57]
Liver	Micropipette aspiration	Viscoelastic coefficient and cell adhesion force	1. Elastic coefficients were significantly higher for human HCC cells than normal hepatocytes 2. Adhesion studies on mouse hepatoma cells showed stronger bonds at G_1 phase than at S phase of the cell cycle	[52,53]
Oral SCC	Optical stretcher	Cell compliance	1. Mechanical phenotyping as an alternative for diagnostics for malignant change 2. Cancer cells were approximately 3.5 times more compliant than healthy donors	[51,55]

that the stiffness of leukemia cells increased by two orders of magnitude after chemotherapy, while Rosenbluth et al. [57] verified that human myeloid cells were much more rigid than neutrophils.

20.3.2 Physical Isolation of CTCs

The microdevice provides an efficient enrichment process using the physical differences of CTCs to blood constituents, and is analogous to the idea of CTCs being trapped in the lungs during circulation [17,18,46]. As mentioned earlier, the biomechanics of tumor cells differs significantly from blood constituents, making this an ideal parameter in separation which will also eliminate the need for any functional modifications. By controlling the flow conditions inside the microdevice, the highly deformable erythrocytes [38] and leukocytes [58] are removed as these cells traverse through without much difficulty. CTCs, which are generally larger and less deformable, are rapidly arrested. The CTC isolation efficiency can thus be enhanced with optimal placements of the isolation structures, while minimizing noise from the surrounding leukocytes in blood samples to ensure a high CTC isolation purity. In brief, the purity and efficiency of isolating CTCs from blood using the microdevice is directly related to the external pressure driving the sample fluid across the device as well as the design of the isolation structures that holds the cancer cells in place while essentially filtering out blood constituents.

Figure 20.3 shows the experimental setup and the design of the microdevice. During processing, blood specimens are pressure driven into the microdevice via control by a computer, achieving a semiautomated operation. Blood enters the microdevice via the sample inlet (Figure 20.3a) and passes through the cell isolation region and out from the waste collection at the bottom. A set of prefilters shown in Figure 20.3a positioned before the isolation traps also serve to prevent large clumps from clogging up the setup and linked to the waste outlet to effectively remove debris. Clogging prevention is important to ensure a feasible system which is a major issue normally associated with devices that attempt to separate and isolate cells through direct physical means [39,60,61]. In addition, the microdevice allows for recovery of cells in the isolation traps through the cell collection point, which can then be retrieved for downstream applications. A key advantage of using microfluidic devices lies in their laminar and uniform flow characteristics, making the profile predictable unlike commercial membrane filters. With a uniform array of 900 crescent-shaped cell isolation structures positioned in an optimal arrangement, the isolation efficiency is maximized.

The design of each trap follows a crescent shape with two gaps of 5 μm, as shown in Figure 20.3b and c, ensuring the exit of blood constituents. Multiple arrays of these crescent-shaped isolation traps

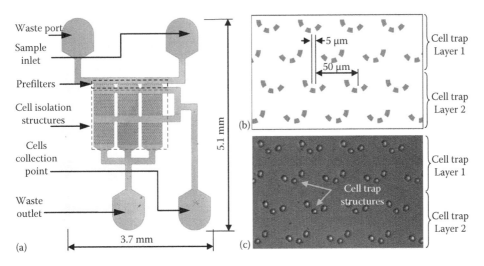

FIGURE 20.3 Design characteristics and view of fabricated device for CTCs trapping. (a) Microdevice layout. (b) Cadence schematic drawing of the cell isolation structures and the placement of the traps. (c) Corresponding fabricated PDMS structures. (From Tan, S.J. et al., Presented at the *International Conference on Biocomputation, Bioinformatics, and Biomedical Technologies, 2008, BIOTECHNO'08*, 2008, Bucharest, Romania.)

are created in the microchannel to process samples at a high throughput and resulting in a high purity of trapped cancer cells. Each trap is positioned with a pitch of 50 μm which effectively prevents cell buildup in any particular region. In addition, each row of isolation structures is offset by 25 μm to enhance the hydrodynamic efficiency (Figure 20.3b and c). The crescent shaped of the isolation traps which are alternated left and right favored the flow profile to prevent clogging and allow single-cell trapping in each structure. Incoming cells that try to enter an occupied trap are directed away due to lower fluidic resistance on the adjacent side. The design is compartmentalized into six different regions (Figure 20.3a), facilitating the maximal retrieval of isolated cells. This arrangement also allows the flow to be more uniform and permits future scaling up of the setup through the addition of more compartments. The scaling up of the setup will reduce sample processing time if larger volumes of blood are desired. During cell retrieval, the rounded inverted crescent-shaped structures provide a favorable path in the opposite flow direction to extract cells out from the cell collection point. It also minimizes physical interactions as the laminar streamlined curves around the structures to reduce possible mechanical damage during retrieval. Henceforth, using the microdevice requires no additional preparatory procedures for the blood specimens, and whole blood can be injected and processed in a single step.

20.4 Results and Discussion

20.4.1 Microdevice Efficiency and Isolation Purity

For characterizing the cell isolation efficiency, mixtures containing a low concentration of cancer cells (100 cells/mL) spiked in 1× PBS were injected into the microdevice at various pressure differentials. Small numbers of cancer cells in the sample solution mimic the rarity of CTCs in peripheral blood. Isolation purity can be determined with cancer cells spiked in blood samples from healthy volunteers. Spiking concentrations of 100 cancer cells/mL of blood which was further diluted in 5 mM EDTA buffer (1:2) were used in all the tests. EDTA buffer was used in place of PBS to prevent blood coagulation. Standard staining protocols were then followed in the identification of cancer cells from the isolated cell population in the microdevice.

Figure 20.4a shows the successful arrest of MCF-7 breast cancer cells and HT-29 colorectal adenocarcinoma in the cell traps. Consistent with the design hypothesis, mostly single cells were found in each of the traps. It was also noted that despite the diverse range of dimensions existed between different cell lines or even within the same cell line, the platform was able to retain most of the cancer cells in place. Also, the single or doublet cells arrested in each isolation trap facilitated cell counting with ease. For traps holding more than one cell, they were generally observed to be of smaller diameters as depicted in Figure 20.4a. This can be resolved by tuning the trap size to suit the dimensions of the tested sample. This aspect is advantageous for the system as controlling the size of the traps can potentially allow the microdevice to be used for a wide variety of cells and applications in single-cell studies. Furthermore, it is confirmed that the anticlogging mechanism is successful. From the time sequence images extracted from the high-speed camera (Figure 20.4b), it is clearly observed for cells that try to enter an occupied trap will be pushed downwards to engage an empty one. This mechanism allowed single-cell trapping to occur and prevented a local buildup of cells that might result in clogging.

By visually counting the number of trapped and escaped cancer cells, the efficiency of tumor cell capture can be determined. Efficiency is defined as a ratio of the trapped cells over the total number of cells that enters the microdevice. The main factor affecting cell isolation efficiency is the pressure applied as the input values directly alter the flow conditions. With a larger applied pressure differential, flow rates will increase leading to a proportionate increase of forces acting on the cells. A suboptimal pressure applied is likely to compromise efficiency of cancer cell capture. Although, higher flow rates meant lesser sample processing time, the larger shear forces on the cells are undesirable as they can damage the cells [65]. In this investigation, the selected pressure settings included 5, 10, and 15 kPa which are comparable to physiological conditions and verified using computational analyses [62]. For the isolation purity characterization, in order to differentiate between hematologic and cancer cells, immunofluorescence staining of the isolated cells was carried out. It is reported that the EpCAM

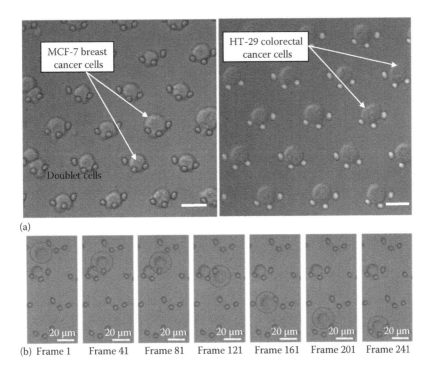

FIGURE 20.4 Cancer cells isolation in the microdevice which presented mostly single cell per trap. (a) MCF-7 (breast adenocarcinoma). Smaller cancer cells attracted doublet cell trapping but do not hinder the functionality of the device. HT-29 (colorectal adenocarcinoma) [62,63]. Scale bar represents 20 μm. (b) Time sequence images showing the capture of a cancer cell. The arrangement of the cell traps enables the capture of cells that circumvent occupied traps and prevents clogging in the microdevice. Images taken with a high speed camera at 1000fps. (From Tan S.J. et al., *Biosens Bioelectron.*, 26, 1701, 2010.)

is overexpressed in human carcinoma [66,67] which makes this an ideal marker to identify the cancer cells. All the selected cell lines in this study (MCF-7, MDA-MB-231, HT-29, FADU, CAL27, AGS, N87, HepG2, and HuH7) had been previously reported to be positive for EpCAM [67–73] and were confirmed in control experiments. In addition, DAPI was used to counterstain the cell nucleus to identify all peripheral blood mononucleated cells (PBMC). CD45, a transmembrane glycoprotein, is expressed among hematologic cells and was used to distinguish WBCs.

Using the computer to control and maintain the pressure differential in the system, the quantitative analyses from comparing cell isolation efficiencies and purity at various pressure settings are plotted in Figure 20.5. In the current study, nine dissimilar cell lines from six different cancer types were used to ascertain the microdevice performance. A total of 264 experimental runs were performed for the entire investigation, showing an approximate effective isolation rate of 80% for the tested samples, at an operating condition of 5 kPa. The Student's t-test verifies that isolation efficiencies at 5 kPa are significantly higher ($p < 0.01$) for all three samples than at higher-pressure inputs as illustrated in Figure 20.5a. A downward trend was observed as the pressure differential increased. The reduction in cell capture efficiency can be accounted for by the dislodging of the trapped cells due to increased hydrodynamic forces acting on these cells at higher-pressure differentials or the forcing of smaller cells through the gaps in the crescent traps. An average isolation efficiency of 82% was predicted from the results of all the different cell lines. Figure 20.5b shows the cancer cell capture purity in the experiments over the range of operating pressures applied. A mean isolation purity of 89% was obtained at the operating pressure of 5 kPa and showed insignificant variations at larger operating conditions. A Student's t-test performed on each individual cell line showed no significant differences in the mean isolation purity over the range of operating pressures at $p < 0.01$. The tumor cell purity in each of the tests was maintained over the entire pressure range. Cell isolation purity was calculated as the ratio of cancer cells to the total number of cells isolated from the blood mixture. A typical experimental run for immunofluorescence

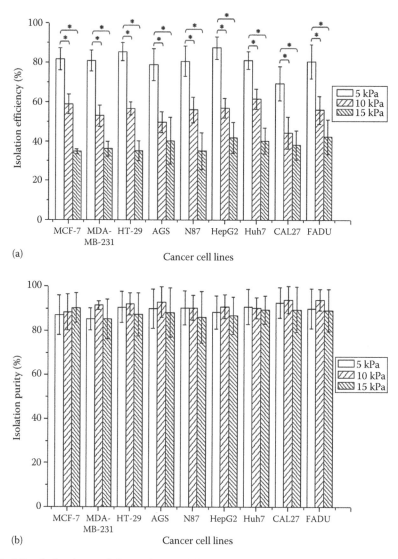

FIGURE 20.5 Microdevice characteristics. (a) Cancer cell isolation efficiency for various cancer cell lines in the microdevice over a range of operating pressure applied. Hypothesis testing was performed comparing the mean isolation efficiency at 5–10 and 15 kPa. * refers to $p < 0.01$ with a Student's t-test. (b) Cancer cell isolation purity for various cancer cell lines in the microdevice over a range of operating pressure applied. (From Tan, S.J. et al., *Biosens. Bioelectron.*, 26, 1701, 2010.)

on chip is illustrated in Figure 20.6. Given the minute volume occupied by the microdevice, it uses minimal amount of reagents to stain the cells which will save cost. Furthermore, the ability to perform immunofluorescence on chip reduces process complexity and ensures maximum yield of CTCs, as compared to retrieving and then to stain them in a separate procedure.

The insignificant change of capture purity over a wide range of applied pressure differentials and high mean capture purity obtained indicates that the physical properties of cancer cells are distinctively different from blood constituents. The results also highlighted that the microdevice was effective to remove blood constituents. The gap size of 5 μm in each of the crescent traps was sufficiently large enough to allow blood cells to deform through while effectively trapping the cancer cells. Other leading techniques to enrich cancer cells from peripheral blood have efficiency ranging from 20% to 90% [7,74,75]. However, there are also numerous restrictions and complex preparation procedures.

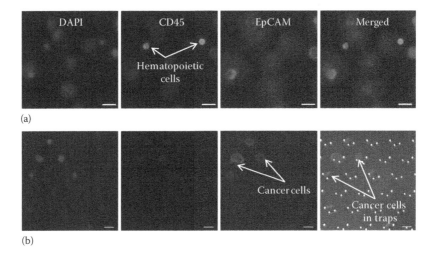

FIGURE 20.6 Isolation purity of tumor cells in a spiked sample using the microdevice. Immuno-fluorescence staining to detect cancer cells using DAPI to counterstain the cell nucleus, CD45 for hematopoietic cells and EpCAM to detect tumor cells. (a) Control was done with a mixture of blood and resuspended cancer cells. (b) Staining in the microdevice is to distinguish between the different cell types. Scale bar represents 20 μm. (From Tan S.J. et al., *Biomed Microdev.*, 11, 883, 2009.)

For example, there is limited purity when detecting low concentrations of CTCs [76] in peripheral blood, and these techniques also include various preparatory steps such as centrifugation, incubation, and functional modifications which can be tedious and time consuming. The proposed microdevice is comparable to other leading biochemical techniques in terms of cancer cell enumeration from blood and is done without any functional modifications. The results favor the lower input pressure of 5 kPa to effectively isolate cancer cells. The Student's *t*-test with a significance level of 0.01 confirmed that isolation efficiencies at 5 kPa were significantly higher than other operating pressures for all the samples. Single-cell trapping is also affirmed in our studies and that the system proves to be suitable to process large sample volume without clogging.

20.4.2 Versatility of the Separation Technique and Limits of Detection

A diverse range of cancer cell lines were used to determine the versatility of the technique to handle different cancers as the disease is genetically heterogeneous [40–42,77]. A total of nine different cancer cell lines from six different origins were utilized. Table 20.2 summarizes the tumor cell isolation efficiencies obtained at 5 kPa and the corresponding 95% confidence interval to gauge the degree of uncertainty around the mean estimates. Collectively, an average cell isolation efficiency of 82.0% [95% CI 80.3%, 83.7%] was obtained with the results acquired from all the cancer cell lines. This is indicative of an accurate platform using different cell lines. The sample standard deviation for the cell isolation efficiencies was approximately 2.6% which suggested that a rather small spread of data and indicates that the technique is reliable and versatile in isolating the different cancer cells. These made separation via size and deformability of cancer cells an attractive method with less bias relative to affinity-based separation.

The rarity of CTCs in the presence of a multitude of blood cells is the main technical challenge to identify them. An enrichment step is thus crucial, and a sensitive setup is desired to detect small numbers of cells in the mixture. Therefore, the detection limit of the system is of interest as it represents the sensitivity of the microfluidic chip in CTC detection. For these experiments, very low counts of cancer cells were added to 1× PBS, either with manual pipetting or FACS, and made to pass through the microdevice. One to three cells were used in each of the experiments to mimic the rarity in the specimen. All other experimental procedures are similar to that used in establishing the microdevice cell isolation efficiency as illustrated earlier. We define a positive ratio for the number of trials for which at least one of the total

TABLE 20.2

Maximum Isolation Efficiency and Corresponding Cell Capture Purity in the Microdevice Using Various Cancer Cell Lines

Cancer Cell Line	Cancer Type	Isolation Efficiency (%)	95% Confidence Interval	Positive Case Ratio[a]
MCF-7	Breast adenocarcinoma	81.5 ± 5.6	[78.7%, 84.3%]	1.00
MDA-MB-231	Breast adenocarcinoma	80.9 ± 5.3	[77.6%, 84.2%]	0.78
HT-29	Colorectal adenocarcinoma	85.4 ± 4.6	[82.7%, 88.1%]	0.80
AGS	Gastric adenocarcinoma	78.9 ± 8.0	[73.9%, 83.9%]	0.80
N87	Gastric carcinoma	80.7 ± 7.6	[76.0%, 85.4%]	0.60
HepG2	Hepatocellular adenocarcinoma	87.4 ± 5.8	[83.8%, 91.0%]	0.75
HuH7	Hepatocellular adenocarcinoma	81.1 ± 4.4	[78.4%, 83.8%]	1.00
CAL27	Tongue squamous carcinoma	81.6 ± 8.6	[76.3%, 86.9%]	0.68
FADU	Pharynx squamous carcinoma	80.6 ± 8.7	[75.2%, 86.0%]	0.90

[a] Isolation efficiency and positive ratio for low cancer cell count using spiked (one to three cells) samples with a sample size of 10 for each cell line tested.

number of cells spiked into PBS is retrieved, over the total number of cases (Table 20.2). A value closer to 1 denotes most of the trials were successful to recover the cells, while a lower value illustrates the converse situation. This provided a gauge of the detection limit for the microdevice with respect to each of the cell lines. A total of 90 experimental runs were performed for all the cell lines listed in Table 5.4. To sum up, the mean positive rate achieved from the investigation was approximately 0.81 from all the nine different cancer cell lines. The results indicate an average success rate for 8 out of every 10 tries in detecting low cell counts from the specimens using this microdevice. This is also coherent with the microdevice efficiency characterization in Section 20.4.1 which predicted an average isolation efficiency of 82%.

For the processing of samples at 5 kPa, the microdevice is capable of achieving a rate of 0.71 mL/h, with a high isolation efficiency needed for a sensitive and accurate diagnosis. Furthermore, the system is easily multiplexed to concurrently run several microdevices, which will reduce sample processing time or allow for simultaneous processing of different samples with minimal setup additions.

20.4.3 State of Retrieved Cancer Cells

The main aims of the system are to provide a direct means of detecting tumor cells from peripheral blood which will also keep cells viable so that downstream analyses can be supported. Furthermore, the conditions of cancer cells after isolation are of interest, as these tumor cells in circulation are likely to be responsible for the progression of the disease cancer. For this aspect, preserving the native state of the cells after isolation will help to determine their exact nature and allow a detailed study of CTC subpopulations such as in the investigation of cancer stem cells or drug-resistant subclones [42,78]. This will hopefully elucidate the metastatic process clearer to aid in diagnosis, prognosis, monitoring, and the search for new therapeutic targets.

Retrieval of the isolated cells in the microdevice can be achieved by reversing the flow conditions inside the microdevice, using the valve connections and the computer which controls the pressure lines. Isolated cells would then be dislodged, and the returned path did not hinder the disengagement of cells from the cell traps, and most of the events were instantaneous. Recovery of cells following the release from the traps was smooth as the inverted isolation structures directed the free floating cells to the adjacent sides and not restricting their movement. An average recovery rate of 92.2% was attained from all the cell lines, indicating a significant cell quantity that can be collected from the microdevice after the isolation process. The standard error for the mean cell recovery for this sampling was 92.2% ± 5.1%.

The integrity of the isolated cells was measured by the proliferation of the retrieved cells under normal culturing conditions. It is hypothesized that the cells initiating tumors have stem cell-like properties [79],

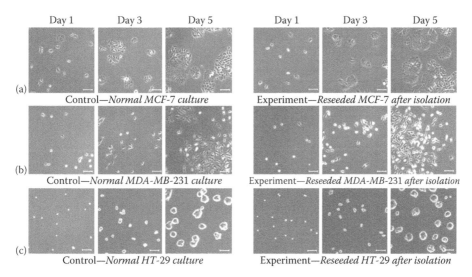

Day 1   Day 3   Day 5   Day 1   Day 3   Day 5

(a) Control—*Normal MCF-7 culture*   Experiment—*Reseeded MCF-7 after isolation*

(b) Control—*Normal MDA-MB*-231 *culture*   Experiment—*Reseeded MDA-MB*-231 *after isolation*

(c) Control—*Normal HT*-29 *culture*   Experiment—*Reseeded HT*-29 *after isolation*

FIGURE 20.7 Cell proliferation comparisons between normal cultures (control) and retrieved cells over a period of 5 days. No observable differences in cell proliferation rate for (a) MCF-7, (b) MDA-MB-231, and (c) HT-29 cancer cells. (Scale bar represents 100 μm.) (From Tan, S.J. et al., *Biomed. Microdev.*, 11, 883, 2009.)

and CTCs being circulated in blood are the potential targets. Isolating sufficient living CTCs has been challenging [4] mainly due to the harsh conditions that the CTCs were subjected to in the body before retrieval. Current leading techniques are not able to maintain the quality of these cells, and an optimal culturing condition needs to be further examined [80]. The proliferation and viability of the retrieved cells were compared with a normal culture from the same batch. Figure 20.7 illustrates an overview of a 5 day culture for MCF-7, MDA-MB-231, and HT-29. Similar conclusions were also obtained for the rest of the cancer cell lines that were tested. The proliferative rates of reseeded cells when compared to normal cultured cells ascertain that isolated cancer cells were not affected by the microdevice. Initial seeding of cells was approximately similar in all experiments and can be verified by observing the cultures at day 1. Cells were then allowed to grow for 5 days with a medium change every 2 days. Over the same duration, there were no observable differences in proliferation rate for all cell lines, as shown in Figure 20.7, when comparing against their respective normal cultures. The morphology of the retrieved cells and that of control experiment were also rather similar. For instance, reseeded MDA-MB-231 showed a spindle-shaped trait which was coherent with the control experiment, and reseeded HT29 proliferated in clusters which were also exhibited in the control experiments. This uniformity in cell behavior confirmed that the retrieved cells were unaffected after isolation in the microdevice, implying that processing in our system was appropriately gentle.

20.4.4 Clinical Sample Processing

With the system optimized, the platform was further evaluated using actual clinical specimens. Patients with metastatic lung cancer participated in the study after consents were taken as stipulated by the Institutional Review Board, and a summary of the data is presented in Table 20.3 for eight specimens. A linearity test is further conducted on the first five samples using different volumes from the same blood tube to determine the reliability of the tests. Successful negative controls with blood extracted from 10 healthy volunteers were observed, showing no cells being retained in the traps. Figure 20.8a illustrates the enumeration of the CTCs from the blood specimens with experimental splits of 1, 2, and 3 mL of whole blood being simultaneously processed. This determines the sensitivity of the platform as well as the variability for each test. The results were fitted with a linear model, and the goodness of fit was obtained to establish the linearity and reproducibility of isolating CTCs using the microdevice. With less than 3 mL of whole blood from each specimen, the adjusted R^2 values in this study group covered the

TABLE 20.3

CTC Isolation and Detection from Peripheral Blood of Metastatic Lung Cancer

Specimen Number	Cancer Origin	Volume Processed (mL)	Positive CTC Count	Leukocyte Count	Isolation Purity (%)
1	NSCLC	2	30	4	88.2
2	NSCLC	2	84	10	89.4
3	NSCLC	2	82	10	89.1
4	NSCLC	2	20	8	71.4
5	NSCLC	2	36	11	76.6
6	NSCLC	2	166	4	97.6
7	NSCLC	2	79	3	96.3
8	NSCLC	2	3	1	75.0

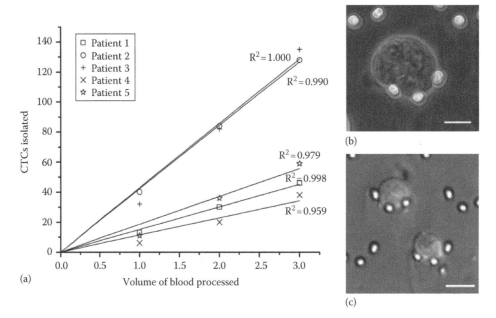

FIGURE 20.8 CTC isolation from peripheral blood of metastatic lung cancer patients. (a) Comparison of CTC enumeration with different volume of blood from the same specimen. (b) Isolated CTC directly after blood processing. (c) Immunofluorescence staining on chip that demonstrates the isolated cells being cytokeratin positive, CD45 negative, and DAPI positive. (Scale bar represents 10 μm.) (From Tan, S.J. et al., *Biosens. Bioelectron.*, 26, 1701, 2010.)

range of 0.959–1.000. This indicates the microdevice reproducibility to detect CTCs. The results also suggest a sensitive system and that blood volume of 1–2 mL is sufficient to gauge the concentration of tumor cells in whole blood. In contrast to other leading separation techniques which require larger volumes, this is advantageous as it reduces the patients' burden during blood extraction, and other tests can be concurrently performed with the remaining amount of blood. The yield and sensitivity are high due to the single processing step that minimizes tumor cell losses, which also simplifies sample processing procedures.

From Table 20.3, the results for processing 2 mL of blood were further analyzed. To sum up, a 100% detection rate was observed with the clinical blood samples from cancer patients. Negative responses from all healthy volunteers indicated the specificity of the system to detect CTCs in the peripheral blood of cancer patients. On the whole, the number of CTCs detected from eight independent samples ranges from 3 to 166 cells in 2 mL of blood. The distinct quantities of CTCs isolated from blood can be attributed to the dissimilar conditions of each of the participating patients, and trials to correlate with their disease states are currently ongoing. Figure 20.8b and c shows the isolated cells after blood processing and immunofluorescence staining. Real-time imaging of the process is possible and will facilitate the identification of the cells. Besides providing the CTC count for each of the samples, the noise associated with the system is also quantified by measuring the number of leukocytes after blood processing. Here, an average CTC isolation purity of 85.5% was achieved which implies an enrichment factor of approximately 5×10^9, assuming an initial tumor cell concentration of 1 CTC in 1 billion blood cells. The high purity obtained is beneficial for downstream applications as minimal noise from leukocytes implied lesser chances of false positives. It is evident that physical dissimilarities between CTCs and the blood constituents are significant in peripheral blood of cancer patients for an efficient tumor cell enrichment and separation.

20.5 Conclusions

Evolving techniques for isolating CTCs represent an unprecedented opportunity for clinical and biological insight into the nature of metastasis and cancer. We demonstrated a microfluidic biochip that can yield an accurate and sensitive measure of CTCs in peripheral blood. We also showed that effective isolation of cancer cells can be accomplished by utilizing their inherent physical differences of size and deformability from that of normal blood cells. Results showed that the microfluidic platform is versatile, able to handle diverse cancer cell lines with minimal noise and yet maintain cell viability and phenotype through a gentle and straightforward processing. More importantly, the system is label free, isolate cells in an unbiased fashion, and can achieve this sensitive detection in a single processing step. Given the simplicity of the procedures, multiplexing of the system to concurrently run numerous samples is also straightforward. Moving forward, the system is currently in the process of testing clinical specimens from cancer patients to determine disease correlation, as well as to study the epigenetics of the disease from the retrieved CTCs. Further efforts to improve upon the design of the microchip to allow greater flexibility in processing blood samples are currently being pursued. We anticipate its use in disease prognosis and therapeutic monitoring, thereby permitting personalization of therapy.

ACKNOWLEDGMENTS

The ARF funding provided by the National University of Singapore (R-265-000-309-112), and NMRC funding provided by the Ministry of Health, Singapore (NMRC 1225/2009), are gratefully acknowledged.

REFERENCES

1. Ferlay, J., H. Shin, F. Bray, D. Forman, C. Mathers, and D. M. Parkin. 2008. Posting date. Cancer incidence and mortality worldwide: IARC CancerBase No. 10. [Online.] International Agency for Research on Cancer: Lyon, France.
2. Chambers, A. F., A. C. Groom, and I. C. MacDonald. 2002. Dissemination and growth of cancer cells in metastatic sites. *Nat Rev Cancer* **2:**563–572.
3. Pantel, K. and S. Riethdorf. 2009. Pathology: Are circulating tumor cells predictive of overall survival? *Nat Rev Clin Oncol* **6:**190–191.
4. Pantel, K., R. H. Brakenhoff, and B. Brandt. 2008. Detection, clinical relevance and specific biological properties of disseminating tumour cells. *Nat Rev Cancer* **8:**329–340.
5. Pantel, K. and C. Alix-Panabieres. 2007. The clinical significance of circulating tumor cells. *Nat Clin Pract Oncol* **4:**62–63.
6. Ashworth, T. R. 1869. A case of cancer in which cells similar to those in the tumours were seen in the blood after death. *Aust Med J* **4:**146–147.
7. Allard, W. J., J. Matera, M. C. Miller, M. Repollet, M. C. Connelly, C. Rao, A. G. Tibbe, J. W. Uhr, and L. W. Terstappen. 2004. Tumor cells circulate in the peripheral blood of all major carcinomas but not in healthy subjects or patients with nonmalignant diseases. *Clin Cancer Res* **10:**6897–6904.

8. Steen, S., J. Nemunaitis, T. Fisher, and J. Kuhn. 2008. Circulating tumor cells in melanoma: A review of the literature and description of a novel technique. *Proc (Bayl Univ Med Cent)* **21:**127–132.

9. Slade, M. J. and R. C. Coombes. 2007. The clinical significance of disseminated tumor cells in breast cancer. *Nat Clin Pract Oncol* **4:**30–41.

10. Cristofanilli, M., D. F. Hayes, G. T. Budd, M. J. Ellis, A. Stopeck, J. M. Reuben, G. V. Doyle, J. Matera, W. J. Allard, M. C. Miller, H. A. Fritsche, G. N. Hortobagyi, and L. W. Terstappen. 2005. Circulating tumor cells: A novel prognostic factor for newly diagnosed metastatic breast cancer. *J Clin Oncol* **23:**1420–1430.

11. Helo, P., A. M. Cronin, D. C. Danila, S. Wenske, R. Gonzalez-Espinoza, A. Anand, M. Koscuiszka, R. M. Vaananen, K. Pettersson, F. K. Chun, T. Steuber, H. Huland, B. D. Guillonneau, J. A. Eastham, P. T. Scardino, M. Fleisher, H. I. Scher, and H. Lilja. 2009. Circulating prostate tumor cells detected by reverse transcription-PCR in men with localized or castration-refractory prostate cancer: Concordance with CellSearch assay and association with bone metastases and with survival. *Clin Chem* **55:**765–773.

12. de Bono, J. S., H. I. Scher, R. B. Montgomery, C. Parker, M. C. Miller, H. Tissing, G. V. Doyle, L. W. Terstappen, K. J. Pienta, and D. Raghavan. 2008. Circulating tumor cells predict survival benefit from treatment in metastatic castration-resistant prostate cancer. *Clin Cancer Res* **14:**6302–6309.

13. Serrano, M. J., P. Sanchez-Rovira, M. Delgado-Rodriguez, and J. J. Gaforio. 2009. Detection of circulating tumor cells in the context of treatment: Prognostic value in breast cancer. *Cancer Biol Ther* **8:**671–675.

14. Reuben, J. M., S. Krishnamurthy, W. Woodward, and M. Cristofanilli. 2008. The role of circulating tumor cells in breast cancer diagnosis and prediction of therapy response. *Expert Opin Med Diagn* **2:**339–348.

15. Cristofanilli, M., G. T. Budd, M. J. Ellis, A. Stopeck, J. Matera, M. C. Miller, J. M. Reuben, G. V. Doyle, W. J. Allard, L. W. Terstappen, and D. F. Hayes. 2004. Circulating tumor cells, disease progression, and survival in metastatic breast cancer. *N Engl J Med* **351:**781–791.

16. Nole, F., E. Munzone, L. Zorzino, I. Minchella, M. Salvatici, E. Botteri, M. Medici, E. Verri, L. Adamoli, N. Rotmensz, A. Goldhirsch, and M. T. Sandri. 2008. Variation of circulating tumor cell levels during treatment of metastatic breast cancer: Prognostic and therapeutic implications. *Ann Oncol* **19:**891–897.

17. Weiss, L. 1990. Metastatic inefficiency. *Adv Cancer Res* **54:**159–211.

18. Weiss, L. 1992. Biomechanical interactions of cancer cells with the microvasculature during hematogenous metastasis. *Cancer Metastasis Rev* **11:**227–235.

19. Brodland, D. G. and J. A. Zitelli. 1992. Mechanisms of metastasis. *J Am Acad Dermatol* **27:**1–8.

20. Zieglschmid, V., C. Hollmann, and O. Bocher. 2005. Detection of disseminated tumor cells in peripheral blood. *Crit Rev Clin Lab Sci* **42:**155–196.

21. Losanoff, J. E., W. Zhu, W. Qin, F. Mannello, and E. R. Sauter. 2008. Can mitochondrial DNA mutations in circulating white blood cells and serum be used to detect breast cancer? *Breast* **17:**540–542.

22. Pantel, K., R. J. Cote, and O. Fodstad. 1999. Detection and clinical importance of micrometastatic disease. *J Natl Cancer Inst* **91:**1113–1124.

23. Fournier, R. L. 1998. *Basic Transport Phenomena in Biomedical Engineering*. Taylor & Francis: Philadelphia, PA.

24. Riethdorf, S., H. Fritsche, V. Muller, T. Rau, C. Schindlbeck, B. Rack, W. Janni, C. Coith, K. Beck, F. Janicke, S. Jackson, T. Gornet, M. Cristofanilli, and K. Pantel. 2007. Detection of circulating tumor cells in peripheral blood of patients with metastatic breast cancer: A validation study of the CellSearch system. *Clin Cancer Res* **13:**920–928.

25. Nagrath, S., L. V. Sequist, S. Maheswaran, D. W. Bell, D. Irimia, L. Ulkus, M. R. Smith, E. L. Kwak, S. Digumarthy, A. Muzikansky, P. Ryan, U. J. Balis, R. G. Tompkins, D. A. Haber, and M. Toner. 2007. Isolation of rare circulating tumour cells in cancer patients by microchip technology. *Nature* **450:**1235–1239.

26. Sieuwerts, A. M., J. Kraan, J. Bolt, P. van der Spoel, F. Elstrodt, M. Schutte, J. W. Martens, J. W. Gratama, S. Sleijfer, and J. A. Foekens. 2009. Anti-epithelial cell adhesion molecule antibodies and the detection of circulating normal-like breast tumor cells. *J Natl Cancer Inst* **101:**61–66.

27. Kahn, H. J., A. Presta, L. Y. Yang, J. Blondal, M. Trudeau, L. Lickley, C. Holloway, D. R. McCready, D. Maclean, and A. Marks. 2004. Enumeration of circulating tumor cells in the blood of breast cancer patients after filtration enrichment: Correlation with disease stage. *Breast Cancer Res Treat* **86:**237–247.

28. Moreno, J. G., S. M. O'Hara, S. Gross, G. Doyle, H. Fritsche, L. G. Gomella, and L. W. Terstappen. 2001. Changes in circulating carcinoma cells in patients with metastatic prostate cancer correlate with disease status. *Urology* **58**:386–392.

29. Krivacic, R. T., A. Ladanyi, D. N. Curry, H. B. Hsieh, P. Kuhn, D. E. Bergsrud, J. F. Kepros, T. Barbera, M. Y. Ho, L. B. Chen, R. A. Lerner, and R. H. Bruce. 2004. A rare-cell detector for cancer. *Proc Natl Acad Sci USA* **101**:10501–10504.

30. Van Vliet, K. J., G. Bao, and S. Suresh. 2003. The biomechanics toolbox: Experimental approaches for living cells and biomolecules. *Acta Materialia* **51**:5881–5905.

31. Whitesides, G. M. 2006. The origins and the future of microfluidics. *Nature* **442**:368–373.

32. Bashir, R. 2004. BioMEMS: State-of-the-art in detection, opportunities and prospects. *Adv Drug Deliv Rev* **56**:1565–1586.

33. Hong, J. W. and S. R. Quake. 2003. Integrated nanoliter systems. *Nat Biotechnol* **21**:1179–1183.

34. Thorsen, T., S. J. Maerkl, and S. R. Quake. 2002. Microfluidic large-scale integration. *Science* **298**:580–584.

35. Melin, J. and S. R. Quake. 2007. Microfluidic large-scale integration: The evolution of design rules for biological automation. *Annu Rev Biophys Biomol Struct* **36**:213–231.

36. Sorger, P. K. 2008. Microfluidics closes in on point-of-care assays. *Nat Biotechnol* **26**:1345–1346.

37. Weiss, L. and D. S. Dimitrov. 1986. Mechanical aspects of the lungs as cancer cell-killing organs during hematogenous metastasis. *J Theor Biol* **121**:307–321.

38. Shelby, J. P., J. White, K. Ganesan, P. K. Rathod, and D. T. Chiu. 2003. A microfluidic model for single-cell capillary obstruction by Plasmodium falciparum-infected erythrocytes. *Proc Natl Acad Sci USA* **100**:14618–14622.

39. Mohamed, H., L. D. McCurdy, D. H. Szarowski, S. Duva, J. N. Turner, and M. Caggana. 2004. Development of a rare cell fractionation device: Application for cancer detection. *IEEE Trans Nanobioscience* **3**:251–256.

40. Shah, R. B., R. Mehra, A. M. Chinnaiyan, R. Shen, D. Ghosh, M. Zhou, G. R. MacVicar, S. Varambally, J. Harwood, T. A. Bismar, R. Kim, M. A. Rubin, and K. J. Pienta. 2004. Androgen-independent prostate cancer is a heterogeneous group of diseases: Lessons from a rapid autopsy program. *Cancer Res* **64**:9209–9216.

41. Braun, S., F. Hepp, H. L. Sommer, and K. Pantel. 1999. Tumor-antigen heterogeneity of disseminated breast cancer cells: Implications for immunotherapy of minimal residual disease. *Int J Cancer* **84**:1–5.

42. Reya, T., S. J. Morrison, M. F. Clarke, and I. L. Weissman. 2001. Stem cells, cancer, and cancer stem cells. *Nature* **414**:105–111.

43. McDonald, J. C., D. C. Duffy, J. R. Anderson, D. T. Chiu, H. Wu, O. J. Schueller, and G. M. Whitesides. 2000. Fabrication of microfluidic systems in poly(dimethylsiloxane). *Electrophoresis* **21**:27–40.

44. Tan, S. J. and C. T. Lim. Microsystem for isolating viable circulating tumor cells, US Provisional Application No. 61/172, 250 (patent pending).

45. Weiss, L. 1989. Biomechanical destruction of cancer cells in skeletal muscle: A rate-regulator for hematogenous metastasis. *Clin Exp Metastasis* **7**:483–491.

46. Weiss, L. and G. W. Schmid-Schonbein. 1989. Biomechanical interactions of cancer cells with the microvasculature during metastasis. *Cell Biophys* **14**:187–215.

47. Suresh, S. 2007. Biomechanics and biophysics of cancer cells. *Acta Biomater* **3**:413–438.

48. Lee, G. Y. H. and C. T. Lim. 2007. Biomechanics approaches to studying human diseases. *Trends Biotechnol* **25**:111–118.

49. Lim, C. T., E. H. Zhou, A. Li, S. R. K. Vedula, and H. X. Fu. 2006. Experimental techniques for single cell and single molecule biomechanics. *Mater Sci Eng: C* **26**:1278–1288.

50. Cross, S. E., Y.-S. Jin, J. Rao, and J. K. Gimzewski. 2007. Nanomechanical analysis of cells from cancer patients. *Nat Nano* **2**:780–783.

51. Remmerbach, T. W., F. Wottawah, J. Dietrich, B. Lincoln, C. Wittekind, and J. Guck. 2009. Oral cancer diagnosis by mechanical phenotyping. *Cancer Res* **69**:1728–1732.

52. Wu, Z. Z., G. Zhang, M. Long, H. B. Wang, G. B. Song, and S. X. Cai. 2000. Comparison of the viscoelastic properties of normal hepatocytes and hepatocellular carcinoma cells under cytoskeletal perturbation. *Biorheology* **37**:279–290.

53. Zhang, G., M. Long, Z. Z. Wu, and W. Q. Yu. 2002. Mechanical properties of hepatocellular carcinoma cells. *World J Gastroenterol* **8**:243–246.

54. Sarah, E. C. et al. 2008. AFM-based analysis of human metastatic cancer cells. *Nanotechnology* **19**:384003.

55. Guck, J., S. Schinkinger, B. Lincoln, F. Wottawah, S. Ebert, M. Romeyke, D. Lenz, H. M. Erickson, R. Ananthakrishnan, D. Mitchell, J. Kas, S. Ulvick, and C. Bilby. 2005. Optical deformability as an inherent cell marker for testing malignant transformation and metastatic competence. *Biophys J* **88**:3689–3698.

56. Lam, W. A., M. J. Rosenbluth, and D. A. Fletcher. 2007. Chemotherapy exposure increases leukemia cell stiffness. *Blood* **109**:3505–3508.

57. Rosenbluth, M. J., W. A. Lam, and D. A. Fletcher. 2006. Force microscopy of nonadherent cells: A comparison of leukemia cell deformability. *Biophys J* **90**:2994–3003.

58. Yap, B. and R. D. Kamm. 2005. Mechanical deformation of neutrophils into narrow channels induces pseudopod projection and changes in biomechanical properties. *J Appl Physiol* **98**:1930–1939.

59. Tan, S. J., L. Yobas, G. Y. H. Lee, C. N. Ong, and C. T. Lim. 2008. Microdevice for isolating viable circulating tumor cells. Presented at the *International Conference on Biocomputation, Bioinformatics, and Biomedical Technologies, 2008. BIOTECHNO'08*, Bucharest, Romania.

60. Di Carlo, D., L. Y. Wu, and L. P. Lee. 2006. Dynamic single cell culture array. *Lab Chip* **6**:1445–1449.

61. Pamme, N. 2007. Continuous flow separations in microfluidic devices. *Lab Chip* **7**:1644–1659.

62. Tan, S. J., L. Yobas, G. Y. Lee, C. N. Ong, and C. T. Lim. 2009. Microdevice for the isolation and enumeration of cancer cells from blood. *Biomed Microdev* **11**:883–892.

63. Tan, S. J., L. Yobas, G. Y. H. Lee, C. N. Ong, and C. T. Lim. 2009. Microdevice for trapping circulating tumor cells for cancer diagnostics. Presented at the *13th International Conference on Biomedical Engineering*, Singapore.

64. Tan, S. J., R. L. Lakshmi, P. Chen, W. T. Lim, L. Yobas, and C. T. Lim. 2010. Versatile label free biochip for the detection of circulating tumor cells from peripheral blood in cancer patients. *Biosens Bioelectron* **26**:1701–1705.

65. Chang, S. F., C. A. Chang, D. Y. Lee, P. L. Lee, Y. M. Yeh, C. R. Yeh, C. K. Cheng, S. Chien, and J. J. Chiu. 2008. Tumor cell cycle arrest induced by shear stress: Roles of integrins and Smad. *Proc Natl Acad Sci USA* **105**:3927–3932.

66. Baeuerle, P. A. and O. Gires. 2007. EpCAM (CD326) finding its role in cancer. *Br J Cancer* **96**:417–423.

67. Osta, W. A., Y. Chen, K. Mikhitarian, M. Mitas, M. Salem, Y. A. Hannun, D. J. Cole, and W. E. Gillanders. 2004. EpCAM is overexpressed in breast cancer and is a potential target for breast cancer gene therapy. *Cancer Res* **64**:5818–5824.

68. Flieger, D., A. S. Hoff, T. Sauerbruch, and I. G. Schmidt-Wolf. 2001. Influence of cytokines, monoclonal antibodies and chemotherapeutic drugs on epithelial cell adhesion molecule (EpCAM) and Lewis Y antigen expression. *Clin Exp Immunol* **123**:9–14.

69. Pauli, C., M. Münz, C. Kieu, B. Mack, P. Breinl, B. Wollenberg, S. Lang, R. Zeidler, and O. Gires. 2003. Tumor-specific glycosylation of the carcinoma-associated epithelial cell adhesion molecule EpCAM in head and neck carcinomas. *Cancer Lett* **193**:25–32.

70. Di Paolo, C., J. Willuda, S. Kubetzko, I. Lauffer, D. Tschudi, R. Waibel, A. Plückthun, R. A. Stahel, and U. Zangemeister-Wittke. 2003. A Recombinant immunotoxin derived from a humanized epithelial cell adhesion molecule-specific single-chain antibody fragment has potent and selective antitumor activity. *Clin Cancer Res* **9**:2837–2848.

71. Wenqi, D., W. Li, C. Shanshan, C. Bei, Z. Yafei, B. Feihu, L. Jie, and F. Daiming. 2009. EpCAM is overexpressed in gastric cancer and its downregulation suppresses proliferation of gastric cancer. *J Cancer Res Clin Oncol* **135**:1277–1285.

72. Joka, M., K. Pietsch, S. Paulick, R. Issels, K. Jauch, and B. Mayer. 2009. Heterogeneous expression of prognostic and predictive antigens in primary and metastatic gastric cancer. *ASCO Meeting Abstract* **27**:e22030.

73. Yamashita, T., A. Budhu, M. Forgues, and X. W. Wang. 2007. Activation of hepatic stem cell marker EpCAM by Wnt-{beta}-catenin signaling in hepatocellular carcinoma. *Cancer Res* **67**:10831–10839.

74. Balic, M., N. Dandachi, G. Hofmann, H. Samonigg, H. Loibner, A. Obwaller, A. van der Kooi, A. G. Tibbe, G. V. Doyle, L. W. Terstappen, and T. Bauernhofer. 2005. Comparison of two methods for enumerating circulating tumor cells in carcinoma patients. *Cytometry B Clin Cytom* **68**:25–30.

75. Lara, O., X. Tong, M. Zborowski, and J. J. Chalmers. 2004. Enrichment of rare cancer cells through depletion of normal cells using density and flow-through, immunomagnetic cell separation. *Exp Hematol* **32:**891–904.

76. Smirnov, D. A., D. R. Zweitig, B. W. Foulk, M. C. Miller, G. V. Doyle, K. J. Pienta, N. J. Meropol, L. M. Weiner, S. J. Cohen, J. G. Moreno, M. C. Connelly, L. W. Terstappen, and S. M. O'Hara. 2005. Global gene expression profiling of circulating tumor cells. *Cancer Res* **65:**4993–4997.

77. Sergeant, G., F. Penninckx, and B. Topal. 2008. Quantitative RT-PCR detection of colorectal tumor cells in peripheral blood-A systematic review. *J Surg Res* **150:**144–152.

78. Wicha, M. S. 2006. Cancer stem cells and metastasis: Lethal seeds. *Clin Cancer Res* **12:**5606–5607.

79. Marx, J. 2007. Molecular biology. Cancer's perpetual source? *Science* **317:**1029–1031.

80. Kaiser, J. 2010. Medicine. Cancer's circulation problem. *Science* **327:**1072–1074.

21

Sensitive Mesofluidic Immunosensor for Detection of Circulating Breast Cancer Cells onto Antibody-Coated Long Alkylsilane Self-Assembled Monolayers

François Breton and Phuong-Lan Tran

CONTENTS

21.1 Introduction

Breast cancer is the first leading cause of cancer death in women. Release of tumor cells into blood circulation may occur at early stage of the disease and may be responsible for its progression [1]. Metastasis in breast cancer patients leads to cancer-related death because early dissemination of tumor cells usually remains undetected at initial diagnosis on clinical, imaging, and biochemical examination. Thus, detection of circulating tumor cells (CTCs) in a blood sample becomes of potential value despite their rarity (median, ≤1 CTC/mL). CTCs have been successfully detected and isolated from blood in patients with metastatic carcinomas [2,3]. A simple blood test would allow the detection and analysis of CTCs to be frequently repeated. It would also help to noninvasively stage the disease at diagnosis as well as to monitor therapy and long-term patient management [4,5]. Several reports have shown that efficacy of treatment could be measured by the number of CTCs in blood [6–8]. CTCs provide easy access to patients' cancer cells for performing several molecular analyses. Studies on alterations in CTC oncogenes suggested heterogeneity among cancer cells from a single patient, for example, with respect to heterogeneity of *HER-2* and *uPA* gene copy number and expression in breast cancer [9,10]. Thus, identifying the CTC subpopulations contained in a blood sample could help to identify targets for treatment and exploring mechanisms that underlie metastases and drug sensitivity. Increasing the useful information on CTCs might help to tailor systemic therapies to the individual needs of a cancer patient.

During the past decades, most research on CTCs has been focused on the development of reliable methods for CTC enrichment and identification, which could overcome severe technical limitations [11–14]. The most currently used methods for CTC detection include quantitative immunomagnetic separation followed by immunocytochemistry detection [2,3,15,16] or RT-PCR to indicate the qualitative presence of CTCs in peripheral blood [17–19]. EpCAM is an epithelial cell adhesion molecule antigen that is overexpressed in breast cancer metastases [20]. Immunomagnetic beads coated with EpCAM antibody facilitate CTCs isolation. All cells that are magnetically labeled are identified by cytokeratin positivity, DAPI nuclear staining, and CD45 (leukocyte-specific antigen) negativity (Cell Search System, Veridex, Johnson & Johnson, Raritan, NJ). Other techniques used to isolate and enumerate CTCs in blood samples are based on cell size selection such as cell filtration [21] and flow cytometry [22,23]. Alternative methodologies addressing microfluidic procedures (i.e., at a micrometer scale) have been reported. CTC detection is performed on CTC-Chips made within a $967\,mm^2$ surface of 78,000 functionalized microposts (diameter×height: $100\,\mu m \times 100\,\mu m$) coated with anti-EpCAM antibody [24], in a biochip consisting of microchannels also coated with anti-EpCAM antibody [25], or in a label-free microdevice separating CTCs from blood constituents on biorheological property differences [26].

Most existing techniques are associated with protocols that allow capturing only cancer EpCAM-positive epithelial cell population in peripheral blood. This could be a limiting factor and leads to missing information on cancer EpCAM-negative cell subpopulations if more than one cancer cell subpopulations are present [27]. Therefore, an advanced tool for detection of various breast cancer cell subpopulations for "tailored" therapy should be investigated [28].

Faced up to challenge a highly sensitive diagnostic method for cancer disease, we developed a new mesofluidic (i.e., at a millimeter scale) immunosensor to immobilize breast cancer cells based on antigen-mediated adhesion of cells to specific antibody-binding surfaces, in a laminar flow field, using a parallel plate, millimeter scale, laminar flow chamber for fluid circulation [29,30]. The floor of the flow chamber can be achieved by surface chemical patterning for cell immobilization. Bioengineered surfaces have proved their performances in chip technology for immobilization of proteins including antibodies, DNA, or cells [31–34]. Self-assembled monolayers (SAMs) have drawn attention to devise such surfaces due to their promising surface properties [35]. Thus, surface chemical patterning can be functionalized by grafting long-chain organosilicon compounds on a Si/SiO_2 solid support to form dense organized SAMs (for a review, [36]). The quality of such functionalized surfaces is dependent on a wide variety of parameters including chain length, temperature, solvent, and reaction time. It has been shown that C_{22}-derivatized SiO_2 surfaces are readily applied to immobilize covalently oligonucleotides [37,38] or red blood cells [39]. Therefore, surface functionalization of the flow chamber floor is carried out by grafting long alkyl organosilicon chain, 21-aminohenicosyl trichlorosilane (AHTS), onto a standard microscopy glass slide to form dense organized SAMs. Control quality of the homogeneous AHTS SAMs displayed an excellent resistance to both physical constraints (temperature and shear stress) and chemical constraints (UV and hydrolysis), as well as preservation of the antibody functional activity [29,37,40]. Then tethering EpCAM antibody to AHTS SAMs provides selectivity and specificity of cell capture [29,41]. Accordingly, (1) design of the flow chamber at a millimeter scale, hence the so-called mesofluidic designation, enables to reduce the rheological phenomena not yet well-mastered in microfluidics; (2) optimization of the flow kinetics of the immunosensor for minimal shear forces, and maximal contact between cells and the active surface, provides high yield of captured MCF7 breast cancer cells spiked in background leukocytes [29] or metastatic breast cancer CTCs [30]; and (3) simultaneous combination of, in parallel, four independent parallel plate laminar flow chambers [30], onto which each independent surface is grafted with a specific monoclonal antibody, would allow patterning various antibodies for capturing breast cancer CTC subpopulations in a single blood sample.

21.2 Materials

1-Ethyl-3-(3-dimethylaminopropyl)carbodiimide hydrochloride (EDC) and (2-[morpholino]ethanesulfonic acid) (MES)-buffered saline, 0.09% NaCl, pH 4.7, were purchased from Perbio Science (Brebières, France), and human serum albumin (HSA) from Sigma-Aldrich (Saint Quentin Fallavier, France). Dulbecco's modified Eagle's medium (DMEM), Roswell Park Memorial Institute (RPMI), fetal calf serum (FCS), penicillin, streptomycin, and Hank's balanced salt solution (HBSS) were supplied by Gibco BRL (Invitrogen, France). The antihuman EpCAM antibody was purchased from R&D Systems (Lille, France), and Alexa-conjugated goat anti-rabbit IgG and Alexa-conjugated goat anti-mouse IgG from Molecular Probes (Invitrogen, France). ACCUSPIN tubes were purchased from Sigma-Aldrich (Saint Quentin Fallavier, France), and EDTA Vacutainer tubes from Becton Dickinson (Pont de Claix, France).

21.3 Methods

21.3.1 Glass Surface Silanization

Glass substrate and silane film preparation were performed as previously described [29,37,40]. Standard microscopy glass slides were first silanized under argon atmosphere with N-(21-trichlorosilanylhenicosyl) phthalimide (N-protected AHTS), then deprotection of the amino group of AHTS was performed as described [29,37,40]. Figure 21.1A displays both N-protected AHTS and AHTS after deprotection.

21.3.2 Physical Characterization of the Silane Films and Quality Control

Quality controls of the silanized surfaces were performed as described [29,40]. Fourier transform infrared (FTIR) spectroscopy was used as a quantitative quality control tool for routine monitoring of the buildup process of the films (Nicolet FT/IR Nexus 870 apparatus equipped with a mercury–cadmium telluride [MCT] detector and purged with dry air). Figure 21.1B shows the FTIR spectrum of long alkyl chains AHTS and displays both methylene antisymmetric ($\nu_{as}CH_2$) and symmetric ($\nu_s CH_2$) stretching modes near 2851 and 2923 cm^{-1}, respectively, which indicate that N-protected AHTS forms quasi-compact and

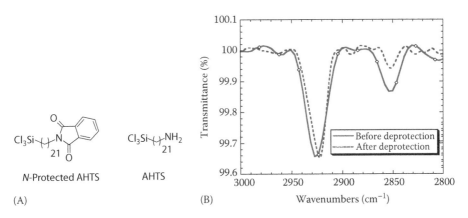

FIGURE 21.1 (A) Representation of N-(21-trichlorosilanylhenicosyl)-phthalimide (N-protected AHTS) and 21-aminohenicosyl trichlorosilane (AHTS). (From Navarre, S. et al., *Langmuir*, 17, 4844, 2001; From Bennetau, B., Bousbaa, J., and Choplin, F. CNRS FR Patent 0000695.) (B) Fourier transform infrared (FTIR) spectra for grafted long AHTS film on glass slide, respectively, before deprotection (straight line) and after deprotection (dotted line) of the amino group. (From *Biosens. Bioelectron.*, 24, Ehrhart, J.C., B. Bennetau, L. Renaud, J.P. Madrange, L. Thomas, J. Morisot, A. Brosseau, S. Allano, P. Tauc, and P.L. Tran., A new immunosensor for breast cancer cell detection using antibody-coated long alkylsilane self-assembled monolayers in a parallel plate flow chamber, 467–474, Copyright 2008 Elsevier.)

ordered monolayers. After deprotection (Figure 21.1B, dotted line), the FTIR spectrum shows a shift of the CH_2 vibration bands ($\nu_{as}CH_2$ and ν_sCH_2) to 2921 and 2850 cm^{-1}, respectively. This shift clearly indicates a better order in the alkyl chains when the protecting group is removed.

Modified surfaces by AHTS were studied for their wetting properties by contact angle measurements using a Krüss goniometer. Droplets (1.5–2 μL) of Nanopure water were placed randomly on the surface and contact angle values were determined within 1 min after droplet deposition. The value of contact angles was observed at $61.4° \pm 1.0°$, proving optimal spreading of water on AHTS surface.

The physical characterization and quality control of silane films were also performed by atomic force microscopy (AFM) using a modified Digital Instruments contact NII head and a NanoscopeEcontroller [40]. For the amplitude modulation mode, a lock-in amplifier (Perkin-Elmer 7280 DSP) produces the excitation of the cantilever piezoceramic and records the variations of amplitude and phase [40]. The frequency modulation data are recorded with a Nanosurf electronics including a phase lock loop. All the data were recorded using Si cantilevers (average stiffness of about 50 N/m), a resonance frequency around 150 kHz, and a quality factor of $Q \approx 400$ at 200 nm from the surface. The images were recorded with an "NCLW" tip and approach-retract curves with "Supersharp" tips [40].

21.3.3 Physical Characterization of Antibody-Coated Surfaces

Antibody immobilization on the AHTS-grafted glass surface was performed with the monoclonal anti-human EpCAM antibody in MES-buffered saline, 0.09% NaCl, pH 4.7 at a concentration of 200 μg/mL in the presence of EDC, overnight at 4°C. All slides were then washed in phosphate-buffered saline (PBS) and stored at 4°C under Ar atmosphere up to 1 month until use.

Quality controls were performed as described [29]. Antibody-coated AHTS slides were incubated with a secondary goat anti-mouse Alexa-conjugated IgG overnight at 4°C. After thorough washing in PBS, the slides were analyzed by AFM using the Explorer microscope (VEECO, Santa Barbara, CA) in tapping mode. A soft Si cantilever (AURORA NanoDevices Inc., Nanaimo, Canada) was used with a nominal force constant of ~40 N/m, a resonance frequency of 300 kHz, and a tip radius <15 nm. The epifluorescence images were obtained using a Leica SP2 inverted microscope (DMIRE2) and an oil immersion objective 40× (NA 1.25). Images were obtained with a charge-coupled device (CCD) camera ProgRes C14^{plus} (Jenoptik, Germany) that records colors. To select light, an N2.1 Leica cube filter was used (exciter BP 515–560, dichromatic mirror: 580, emitter LP 590).

21.3.4 Characteristics of Laminar Flow and Simulation in a Parallel Plate Flow Chamber

The physical characteristics of a parallel plate laminar flow chamber were previously described [29,41]. The parallel plate flow chamber was constructed by our mechanics facilities. It was made of a poly(methyl methacrylate) (PMMA) block bearing a milled cavity of $6 \times 16 \times 0.5$ mm^3 ($W \times L \times H$) surrounded by a toric gasket. The silanized glass slide was maintained against the PMMA block with a screwed steel plate. Figure 21.2A shows a picture of a parallel plate laminar flow chamber. The laminar flow was generated by a peristaltic pump (Wheaton Science, Millville, NJ). The chamber was set on the stage of an inverted NIKON microscope bearing a modified plate allowing easy positioning. Cell adhesion was observed by focusing the bottom surface of the chamber, and the data recorded with a SONY video camera, which was connected to the microscope (Figure 21.2B).

The simulation of fluid flow through the single parallel plate flow chamber was performed according to a model constructed with the geometry of the flow chamber and used the "incompressible Navier–Stokes" hypothesis of FEMLAB software 3.2 (Comsol, France) [29]. It showed the laminar regimen of flow, confirmed by the Reynolds number, Re, which can be expressed as a function of the flow rate Q: $Re = \rho \times Q/l \times \eta$ where ρ is the fluid density and η, the fluid viscosity. After a conversion of the variables using a practical unit system, we obtain $Re = 1.67 \times 10^{-5} \times Q/(l \times \eta)$, where Q is the flow rate (μL/min), l, the flow chamber width (mm), and η, the fluid viscosity (Pa s). Taking into account $l = 6$ mm and $\eta = 1.0 \times 10^{-3}$ Pa s for water at 20°C [42], $Re = 0.025$ when a flow rate of 90 μL/min was applied to the parallel plate flow chamber, and the outlet flow was considered as free under atmospheric pressure.

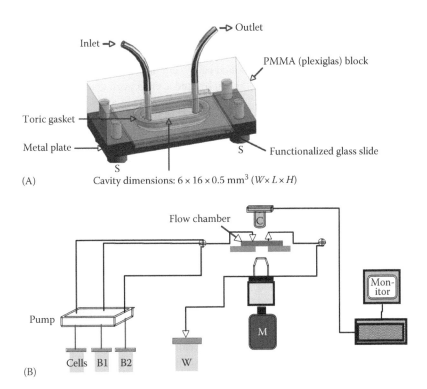

FIGURE 21.2 (A) A schematic representation of a flow chamber including a toric gasket. The silanized glass slide was maintained against the poly(methyl methacrylate) (PMMA) block with a screwed (S) steel plate. (B) Diagram of the system comprising a flow chamber set on the stage of a microscope (M) connected to a camera (C) and a desk computer, a peristaltic pump, and connectors plus various tubings. The containers B1 and B2 provide various buffers and W, a waste unit.

Re value clearly corresponded to a laminar regimen (the transition between laminar and turbulent regimen is around 2000).

Under a laminar regimen, the stream velocity in the chamber is determined by the "infinite plate" formula: $v_x(z) = 4v_{x,\max}/h^2 \times (z \times h - z^2)$, where $v_{x,\max}$ is the maximum velocity, h, the chamber depth, and z, the distance from the bottom wall of the chamber. $v_{x,\max}$ can be expressed as a function of the flow Q: $v_{x,\max} = 3Q/2h \times l$. Thus $v_x(z) = 6Q/l \times h^3 \times (z \times h - z^2)$. Close to the chamber floor wall, the stream velocity is approximately proportional to z: $v_x(z) = G \times z = 6Q/(l \times h^2) \times z$, where G is defined as the wall shear rate (s^{-1}), that is, the coefficient between the flow velocity and the distance from the wall. It can be calculated using a practical unit: $G = 0.10 \times Q/l \times h^2$, where Q is expressed as µL/min, and l and h in mm.

21.3.5 Isolation of MCF7 Breast Cancer Cells Spiked in Normal Blood Leukocytes

Four milliliters of fresh blood samples from healthy volunteers were drawn into heparin plus EDTA Vacutainer tubes. They were diluted into complete RPMI medium containing 10% FCS, 2 mM glutamine, 100 U/mL penicillin, and 100 µg/mL streptomycin. Isolation of nucleated leukocyte cells was performed onto ACCUSPIN tubes at 1600 g for 20 min at 15°C. They were washed with HBSS, and resuspended in HBSS and 0.3% HSA to a dilution of 10×10^6 cells/mL.

The human breast cancer cell line, MCF7, was grown in DMEM containing 10% FCS, 2 mM glutamine, 100 U/mL penicillin, and 100 µg/mL streptomycin, and cultured until subconfluence. Cells were then detached, washed with HBSS, and resuspended in HBSS and 0.3% HSA at various dilutions in 10×10^6 leukocytes/mL, as reported earlier. They were loaded onto AHTS-coated glass surface grafted with the monoclonal antihuman EpCAM antibody. The following dilutions of MCF7 cells were performed: 200–4000 MCF7 cells/mL.

Capture of MCF7 breast cancer cells spiked in buffer-containing leukocytes was performed at a flow rate of 90 μL/min [29]. Cells were driven into the flow chamber until the entire active surface was covered. Then 12 min cell incubation was performed at room temperature under stable conditions. At the end of the incubation period, nonspecifically bound cells were flushed with PBS by increasing the flow rate at 350 μL/min for 6 min. Following cell fixation by cold acetone, cell counting was performed with MATLAB® software [29].

21.3.6 Immunofluorescence Staining and Identification of MCF7 Cells by Fluorescence Microscopy

As reported [29,43], following cell fixation with 100 μL cold acetone, MCF7 cells were labeled by the primary polyclonal anti-pan-cytokeratin antibody followed by secondary anti-rabbit Alexa-conjugated IgG and leukocytes by either anti-CD45 allophycocyanin (APC), or anti-CD3, anti-CD20, and anti-CD35 and Alexa-conjugated anti-mouse IgG. Cell nuclei were labeled with DAPI (1 μg/mL) for 5 min.

According to [30], slides were analyzed on an inverted NIKON E50i (NIKON, France) microscope equipped with an automated stage (IMSTAR, France). Each slide was scanned automatically in a 1360 × 1024 pixels format using the programmable stage and capture software (version 6, imaging IMSTAR systems, France). Captured images at 10× magnification were carefully examined and the objects that met predetermined criteria were counted. Color, brightness, and morphological characteristics such as cell size, shape, and nuclear size were considered in identifying MCF7 tumor cells, excluding cell debris, nonspecific, and false positive cells. The criteria for their selection were as follows: cells that stained positive by the primary polyclonal anti-pan-cytokeratin antibody and secondary Alexa-conjugated anti-rabbit IgG (green fluorescence) and met the phenotypic morphological characteristics were scored; the same cells were stained by DAPI (blue stain) and negatively stained by either anti-CD45-APC, or anti-CD3, -CD20, and -CD35 and Alexa-conjugated anti-mouse IgG (no red fluorescence).

21.4 Results

21.4.1 Physical Characteristics of the Antibody-Coated AHTS Surface

The efficacy of breast cancer cell capture based on antigen–antibody recognition has to meet several criteria. The candidate antigen should be a membrane protein highly expressed in tumor cells, and antibody avidity for the tumor cells should be high. Therefore, to capture tumor cells efficiently, the antibody should be attached at an optimum titer on the functionalized surfaces.

As reported [29], the quality controls of a primary anti-EpCAM antibody and a secondary Alexa-conjugated anti-mouse IgG grafting onto AHTS-coated glass surfaces were followed by dynamic AFM and confocal microscopy. A systematic investigation was performed on AHTS monolayers before and after grafting the IgGs, using AFM in the tapping mode. Figure 21.3A displays an image of a negative control surface, while Figure 21.3B shows an AHTS surface grafted with the monoclonal antihuman EpCAM antibody. Features of ovoid protrusions observed above the silane monolayer had diameters from 40 to 70 nm and covered the surface irrespective of the topography. IgGs are large proteins of asymmetrical shape since they are composed of two heavy chains (MW 50 kDa) and two light chains (25 kDa) attached by disulfide bridges that form a Y-like structure. Large dimensions of the observed protrusions may correspond to the superimposition of the two grafted antibodies. These features appeared randomly dispersed on the surface and had no obvious tendency to form aggregates.

Figure 21.4A and B depicts immunofluorescent images of AHTS monolayer before and after grafting of Alexa-conjugated anti-mouse IgG. At the micrometer scale of fluorescence microscopy, fluorescent spots of high signal intensity of size <2 μm are observed. The surface showed very high sensitivity and good signal-to-background ratios. As already observed by AFM, the fluorescent spots were randomly dispersed on the substrate and had no obvious tendency to form aggregates [29].

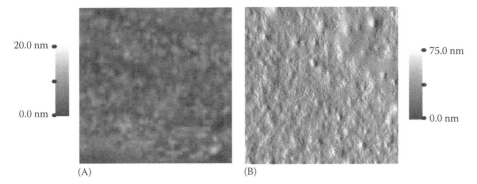

FIGURE 21.3 Tapping-mode AFM images show the topology of 21-aminohenicosyl trichlorosilane (AHTS) glass surface before (A) and (B) after grafting the primary antihuman EpCAM monoclonal antibody followed by a secondary Alexa-conjugated anti-mouse IgG. Features of ovoid protrusions of diameters from 40 to 70 nm, randomly dispersed, cover the silane monolayer surface. The image area is $2\,\mu m \times 2\,\mu m$.

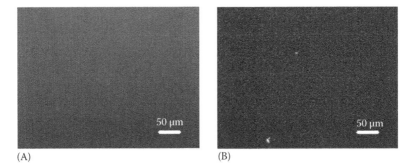

FIGURE 21.4 (A) Confocal microscopy picture of control AHTS glass slide without antibody grafting. (B) Confocal microscopy picture of control AHTS glass slide grafted with the primary antihuman EpCAM monoclonal antibody and a secondary Alexa-conjugated anti-mouse IgG. High-sensitive spots of fluorescence of size $<2\,\mu m$ are randomly dispersed. (Adapted from *Biosens. Bioelectron.*, 24, Ehrhart, J.C., B. Bennetau, L. Renaud, J.P. Madrange, L. Thomas, J. Morisot, A. Brosseau, S. Allano, P. Tauc, and P.L. Tran., A new immunosensor for breast cancer cell detection using antibody-coated long alkylsilane self-assembled monolayers in a parallel plate flow chamber, 467–474, Copyright 2008 Elsevier.)

21.4.2 Simulation of the Fluid Flow in a Single Parallel Plate Laminar Flow Chamber

Simulation of fluid flow in a parallel plate flow chamber using the "incompressible Navier–Stokes" protocol of Comsol software was performed at a flow rate of $90\,\mu L/min$ [29]. Figure 21.5 shows the calculated trajectories that would use cells flowing throughout the whole chamber including the area located on both sides of inlet and outlet of the flow, according to the postprocessing tools of Comsol software. We have repeated the same simulations in various flow rates from 10 to $1000\,\mu L/min$. All results led to similar figures because the flows were always laminar in this range. Figure 21.5 displays the output of a unique simulation, but all simulations led to the same figure. It would also indicate cell locations onto the chemically modified surface after their deposition. Thus the entire antibody-coated surface in a laminar flow chamber is actively exposed to cells.

21.4.3 Isolation of MCF7 Breast Cancer Cells Spiked in Background Normal Blood Leukocytes

To evaluate tumor cell capture under physiological conditions, we performed capture of unlabeled human breast cancer, MCF7, living cells spiked in leukocytes isolated from healthy blood samples.

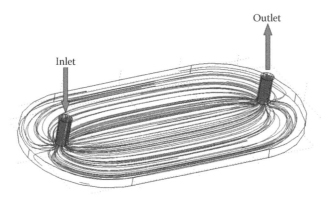

FIGURE 21.5 Stream visualization in the flow chamber at a flow rate of 90 µL/min. (From *Biosens. Bioelectron.*, 24, Ehrhart, J.C., B. Bennetau, L. Renaud, J.P. Madrange, L. Thomas, J. Morisot, A. Brosseau, S. Allano, P. Tauc, and P.L. Tran., A new immunosensor for breast cancer cell detection using antibody-coated long alkylsilane self-assembled monolayers in a parallel plate flow chamber, 467–474, Copyright 2008 Elsevier.)

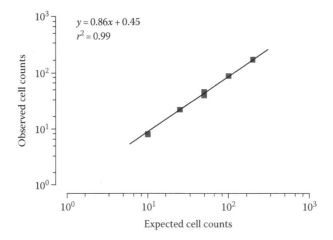

FIGURE 21.6 MCF7 cell capture efficiency in a parallel plate laminar flow chamber. The plot represents the number of MCF7 breast cancer cells spiked in 10×10^6 normal blood leukocytes/mL versus the number of MCF7 cells recovered.

Our procedure allows treating living cells instead of other methods that required permeabilized prelabeled cells. The chamber floor carried AHTS-coated surface grafted with the monoclonal antihuman EpCAM antibody [29]. A series of various cell suspensions was performed at concentrations ranging from 200 to 4000 tumor cells/mL of HBSS and 0.3% HSA containing 10×10^6 leukocytes. Immobilized cells were then submitted to staining of MCF7 cells and leukocytes and analysis for cell recovery yield. Figure 21.6 displays the regression analysis of capture efficiency of MCF7 cells per chamber in a competent volume of 48 µL, according to $y = 0.86x + 0.45$ ($r^2 = 0.99$). The recovery rate yielded from 70% to 85%. The large dead volume due to long tubing at the entrance of the chamber has been then surmounted by allowing 2 mL of cell suspension to load MCF7 cells onto four single parallel plate laminar flow chambers connected in parallel to a four-channel peristaltic pump. Recently, we have further drastically reduced the dead volume by designing a multiplex device [30,44]. Both reduction of dead volume and multiplexing cell detection surfaces allowed increasing recovery rate of MCF7 cells up to 92%, only 1.8% ± 1% of background leukocytes being observed on the capture surfaces (data not shown [30]). This depletion level of background cells demonstrates high capture specificity and sensitivity of our device and its functionalized surface.

21.4.4 Identification of MCF7 Breast Cancer Cells Spiked in Normal Blood Leukocytes

Following MCF7 cell capture, their identification on each capture surface consisted of staining with 4,6-diamino-2-phenylindole (DAPI) for DNA content, anti-pan-cytokeratin antibodies for epithelial tumor cells, and anti-CD45-APC for leukocytes (Figure 21.7A through D). Captured cells staining positive for cytokeratin and negative for leukocyte markers were recorded. The morphological characteristics exhibited by MCF7 cells were consistent with tumor cells, including large cellular size with high nuclear/cytoplasmic ratios (Figure 21.7D). The cellular viability of captured cells was assessed under transmission light using Trypan blue that attested integrity of the cell membrane.

21.5 Discussion

The development of this mesofluidic strategy with the design of a millimeter scale, parallel plate, laminar flow chamber aimed at reducing rheological phenomena not yet well mastered in microfluidics [29]. Recently we constructed a multiplex mesofluidic immunosensor purposed to highly reduce dead volume, thus allowing rapid access of cells to functionalized surfaces of the flow chambers, gentle immobilization of cells, and increasing cell capture yield [30,44].

Several key fundamental features of our detection system are as follows: (1) the versatility of the silanized surfaces, which can be used for any ligand–receptor interactions; (2) its ability to treat fresh and frozen samples that have undergone an initial step depleting red blood cells by the method of Ficoll. In a clinical resort, it will provide an easier way to handle patients' blood samples; and (3) our device works with living cells.

The proof of concept of the device needs testing under physiological conditions MCF7 cells spiked in background normal blood leukocytes for linearity, precision, and cell capture efficiency (regression

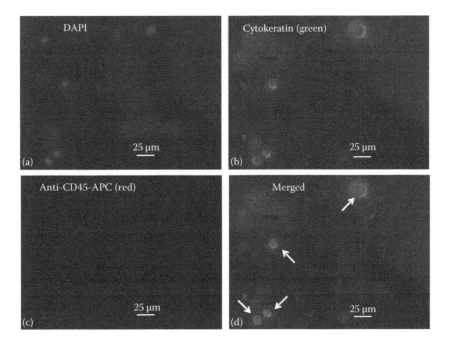

FIGURE 21.7 High magnification images of captured MCF7 cells spiked in normal blood leukocytes (1/500 dilution). (A) Nuclei were stained with DAPI. MCF7 cells were positively stained with the anti-pan-cytokeratin antibody and a secondary anti-rabbit Alexa-conjugated IgG. Leukocyte cells were stained with anti-CD45-APC antibody. Merged image (D) identifies four MCF7 cells stained positive for cytokeratin (B) and negative for leukocyte cell markers (C).

analysis). Spiking <10 cells in 1 mL of 10^7 background leukocytes [15] requires high dilution ratio that would achieve a 95% confidence interval of the spiking too large according to the Poisson distribution statistics ([2], Supplementary data). Therefore, we tested our procedure based on the dilution of cells according to the small 48 μL volume per parallel plate laminar flow chamber.

Once linearity and cell capture efficiency of the mesofluidic immunosensor are demonstrated, they have to be tested with clinical blood samples [30]. In analyses of clinical blood samples, the following issues of the competent volumes handled by the mesofluidic device and other fluidics-based systems can be noted. The CellSearch system (7.5 mL competent volume) showed a CTC detection in <60% of metastatic patients ($n = 964$; [2]). The microfluidics CTC-Chip (35.8 μL competent volume: microfluidic chamber volume = 97 μL; lost volume of the 78,000 microposts of 100 μm diameter × 100 μm height = 61.2 μL [24]; $97 - 61.2 = 35.8$ μL) showed a CTC detection in 99% of comparable metastatic patients ($n = 116$; [24]). Our new multiplex immunosensor (48 μL per chamber) proved enough accuracy to detect CTCs in 100% of metastatic breast cancer patients ($n = 21$; [30]) and 27% in localized breast cancer patients ($n = 11$; [30]). To summarize these observations, the suitable competent volume depends on the fluidic conditions. It is important to point out that for improving sensitivity several parallel plate flow laminar chambers can be connected in "parallel" to increase volume and that the current system can detect up to 20 cells/mL. Microfluidic and mesofluidic strategies have been developed to greatly enhance recovery yield and specificity of CTC capture.

Another important feature of the immunosensor is its active glass surface that provides an ideal transparent surface easy to read under fluorescence microscopy. On the same glass slide, cell capture surfaces can be then recovered and easily processed for further cellular, genetic, and biochemical analyses. In addition, we also demonstrated that amino-terminated long alkyl chain–coated surfaces are stable on long-term storage under Ar atmosphere. Once the antibodies were spotted on the surfaces and kept at 4°C under Ar, their activity was preserved for at least 1 month in cell capture experiments.

Time required for all steps of the automated detection system, from cell capture to cellular analysis, is about 2 h, which will reasonably allow incorporation of the medical device in clinical trials.

21.6 Future Trends

A key development of breast cancer CTC capture by immunosensors is the versatility of antibody grafting in order to detect CTC subpopulations. The ability to isolate, enumerate, and molecularly characterize CTC subpopulations could help improving the clinical management of breast cancer patients including early diagnosis, evaluation of tumor process, and monitoring therapy response.

Compared to existing tumor cell separation technologies, multiplexing several parallel plate laminar flow chambers in parallel is appealing to overcome technical limitations, with respect to detection of several subpopulations of rare circulating cells within a single blood sample. By using in parallel and simultaneously several parallel plate laminar flow chambers [30,44], it could be possible to pattern surfaces of this multiplex immunosensor with various types of antibodies or ligands capturing specific circulating cells in the peripheral blood, making easy the screening of cell subpopulations. The flexibility of the "smart surfaces" of our multiplex device makes them easily suitable to use simultaneously two co-expressed tumor markers instead of one on the cellular surface of breast cancer cells, thus highly increasing the accuracy of the detection as well as facilitating classification of CTC subpopulations [45]. This improvement would allow performing various protocols from cell immobilization on a standard microscopy functionalized glass slide to downstream analyses on the same stand, thus answering expressed medical needs, unmet by others technologies.

ACKNOWLEDGMENTS

Dr. Bernard Bennetau is thanked for his expertise in surface chemistry, Dr. Louis Renaud for his expertise in fluid mechanics, and Professor Jean-Claude Ehrhart for stimulating discussions and reading this chapter. This work was supported in part by contracts from Roche-Pharma, France (Prot. 031.0780.00), the Association pour la Recherche sur le Cancer (ARC #3675), the Direction de la Politique Industrielle

(CNRS #63875), and grants from the Agence Nationale de la Recherche (grant EMPB CAPCELL 2006-06), and the Institute Alembert (Ecole Normale Supérieure Cachan).

REFERENCES

1. Chambers, A.E., A.C. Groom, and I.C. MacDonald. 2002. Dissemination and growth of cancer cells in metastatic sites. *Nat. Rev. Cancer* 2:563–572.

2. Allard, W.J., J. Matera, M.C. Miller, M. Repollet, M.C. Connelly, C. Rao, A.G. Tibbe, J.W. Uhr, and L.W. Terstappen. 2004. Tumor cells circulate in the peripheral blood of all major carcinomas but not in healthy subjects or patients with nonmalignant diseases. *Clin. Cancer Res.* 10:6897–6904.

3. Cristofanilli, M., G.T. Budd, M.J. Ellis, A. Stopeck, J. Matera, M.C. Miller, J.M. Reuben, G.V. Doyle, W.J. Allard, L.W. Terstappen, and D.F. Hayes. 2004. Circulating tumor cells, disease progression, and survival in metastatic breast cancer. *N. Engl. J. Med.* 351:781–791.

4. Cristofanilli, M., D.F. Hayes, G.T. Budd, M.J. Ellis, A. Stopeck, J.M. Reuben, G.V. Doyle, J. Matera, W.J. Allard, M.C. Miller, H.A. Fritsche, G.N. Hortobagyi, and L.W. Terstappen. 2005. Circulating tumor cells: A novel prognostic factor for newly diagnosed metastatic breast cancer. *J. Clin. Oncol.* 23:1420–1430 [Erratum in *J. Clin. Oncol.* 23:4808 (2005)].

5. Hayes, D.F., M. Cristofanilli, G.T. Budd, M.J. Ellis, A. Stopeck, M.C. Miller, J. Matera, W.J. Allard, G.V. Doyle, and L.W. Terstappen. 2006. Circulating tumor cells at each follow-up time point during therapy of metastatic breast cancer patients predict progression-free and overall survival. *Clin. Cancer Res.* 12:4218–4224.

6. Liu, M.C, P.G. Shields, R.D. Warren, P. Cohen, M. Wilkinson, Y.L. Ottaviano, S.B. Rao, J. Eng-Wong, F. Seilleier-Moiseiwitsch, A.M. Noone, and C. Isaacs. 2009. Circulating tumor cells: A useful predictor of treatment efficacy in metastatic breast cancer. *J. Clin. Oncol.* 27:5153–5159.

7. Nolé, F., E. Munzone, L. Zorzino, I. Minchella, M. Salvatici, E. Botteri, M. Medici, E. Verri, L. Adamoli, N. Rotmensz, A. Goldhirsch, and M.T. Sandri. 2008. Variation of circulating tumor cell levels during treatment of metastatic breast cancer: Prognostic and therapeutic implications. *Ann. Oncol.* 19:891–897.

8. M. Tewes, B. Aktas, A. Welt, S. Mueller, S. Hauch, R. Kimmig, and S. Kasimir-Bauer. 2009. Molecular profiling and predictive value of circulating tumor cells in patients with metastatic breast cancer: An option for monitoring response to breast cancer related therapies. *Breast Cancer Res. Treat.* 115:581–590.

9. Cristofanilli, M. and J. Mendelsohn. 2006. Circulating tumor cells in breast cancer: Advanced tools for "tailored" therapy? *Proc. Natl Acad. Sci. USA* 103:17073–17074.

10. Meng, S., D. Tripathy, S. Shete, R. Ashfaq, B. Haley, S. Perkins, P. Beitsch, A. Khan, D. Euhus, C. Osborne, E. Frenkel, S. Hoover, M. Leitch, E. Clifford, E. Vitetta, L. Morrison, D. Herlyn, L.W. Terstappen, T. Fleming, T. Fehm, T. Tucker, N. Lane, J. Wang, and J. Uhr. 2004. HER-2 gene amplification can be acquired as breast cancer progresses. *Proc. Natl Acad. Sci. USA* 101:9393–9398.

11. Kahn, H.J., A. Presta, L.Y. Yang, J. Blondal, M. Trudeau, L. Lickley, C. Holloway, D.R. McCready, D. Maclean, and A. Marks. 2004. Enumeration of circulating tumor cells in the blood of breast cancer patients after filtration enrichment: Correlation with disease stage. *Breast Cancer Res. Treat.* 86:237–247.

12. Krivacic, R.T., A. Ladanyi, D.N. Curry, H.B. Hsieh, P. Kuhn, D.E. Bergsrud, J.F. Kepros, T. Barbera, M.Y. Ho, L.B. Chen, R.A. Lerner, and R.H. Bruce. 2004. A rare-cell detector for cancer. *Proc. Natl Acad. Sci. USA* 101:10501–10504.

13. Racila, E., D. Euhus, A.J. Weiss, C. Rao, J. McConnell, L.W. Terstappen, and J.W. Uhr. 1998. Detection and characterization of carcinoma cells in the blood. *Proc. Natl Acad. Sci. USA* 95:4589–4594.

14. Zieglschmid, V., C. Hollmann, and O. Böcher. 2005. Detection of disseminated tumor cells in peripheral blood. *Crit. Rev. Clin. Lab. Sci.* 42:155–196.

15. Riethdorf, S., H. Fritsche, V. Müller, T. Rau, C. Schindlbeck, B. Rack, W. Janni, C. Coith, K. Beck, F. Jänicke, S. Jackson, T. Gornet, M. Cristofanilli, and K. Pantel. 2007. Detection of circulating tumor cells in peripheral blood of patients with metastatic breast cancer: A validation study of the CellSearch system. *Clin. Cancer Res.* 13:920–928.

16. Talasaz, A.H., A.A. Powell, D.E. Huber, J.G. Berbee, K.-H. Roh, W. Yu, W. Xiao, M.M. Davis, R.F. Pease, M.N. Mindrins, S.S. Jeffrey, and R.W. Davis. 2009. Isolating highly enriched populations of circulating epithelial cells and other rare cells from blood using a magnetic sweeper device. *Proc. Natl Acad. Sci. USA* 106:3970–3975.

17. Aerts, J., W. Wynendaele, R. Paridaens, M.R. Christiaens, W. van den Bogaert, A.T. van Oosterom, and F. Vandekerckhove. 2001. A real-time quantitative reverse transcriptase polymerase chain reaction (RT-PCR) to detect breast carcinoma cells in peripheral blood. *Ann. Oncol.* 12:39–46.

18. Iakovlev, V.V., R.S. Goswami, J. Vecchiarelli, N.C. Arneson, and S.J. Done. 2008. Quantitative detection of circulating epithelial cells by Q-RT-PCR. *Breast Cancer Res. Treat.* 107:145–154.

19. Schröder, C.P., M.H. Ruiters, S. de Jong, A.T. Tiebosch, J. Wesseling, R. Veenstra, J. de Vries, H.J. Hoekstra, L.F. de Leij, and E.G. de Vries. 2003. Detection of micrometastatic breast cancer by means of real time quantitative RT-PCR and immunostaining in perioperative blood samples and sentinel nodes. *Int. J. Cancer* 106:611–618.

20. Osta, W.A., Y. Chen, K. Mikhitarian, M. Mitas, M. Salem, Y.A. Hannun, D.J. Cole, W.E. Gillanders. 2004. EpCAM is overexpressed in breast cancer and is a potential target for breast cancer gene therapy. *Cancer Res.* 64:5818–5824.

21. Wong, N.S., H.J. Kahn, L. Zhang, S. Oldfield, L.Y. Yang, A. Marks, and M.E. Trudeau. 2006. Prognostic significance of circulating tumor cells enumerated after filtration enrichment in early and metastatic breast cancer patients. *Breast Cancer Res. Treat.* 99:63–69.

22. Allan, A.L., S.A. Vantyghem, A.B. Tuck, A.F. Chambers, I.H. Chin-Yee, and M. Keeney. 2005. Detection and quantification of circulating tumor cells in mouse models of human breast cancer using immunomagnetic enrichment and multiparameter flow cytometry. *Cytometry A* 65:4–14.

23. Cruz, I., J. Ciudad, J.J. Cruz, M. Ramos, A. Gómez-Alonso, J.C. Adansa, C. Rodríguez, and A. Orfao. 2005. Evaluation of multiparameter flow cytometry for the detection of breast cancer tumor cells in blood samples. *Am. J. Clin. Pathol.* 123:66–74.

24. Nagrath, S., L.V. Sequist, S. Maheswaran, D.W. Bell, D. Irimia, L. Ulkus, M.R. Smith, E.L. Kwak, S. Digumarthy, A. Muzikansky, P. Ryan, U.J. Balis, R.G. Tompkins, D.A. Haber, and M. Toner. 2007. Isolation of rare circulating tumor cells in cancer patients by microchip technology. *Nature* 450:1235–1239.

25. Du, Z., K.H. Cheng, M.W. Vaughn, N.L. Collie, and L.S. Gollahon. 2007. Recognition and capture of breast cancer cells using an antibody-based platform in a microelectromechanical systems device. *Biomed. Microdevices* 9:35–42.

26. Tan, S.J., L. Yobas, G.Y. Lee, C.N. Ong, and C.T. Lim. 2009. Microdevice for the isolation and enumeration of cancer cells from blood. *Biomed. Microdevices* 11:883–892.

27. Sieuwerts, A.M., J. Kraan, J. Bolt, P. van der Spoel, F. Elstrodt, M. Schutte, J.W.M. Martens, J.-W. Gratama, S. Sleijfer, and J.A. Foekens. 2009. Anti-epithelial cell adhesion molecule antibodies and the detection of circulating normal-like breast tumor cells. *J. Natl Cancer Inst.* 101:61–66.

28. Cristofanilli, M. and S. Braun. 2010. Circulating tumor cells revisited. *JAMA* 303:1092–1093.

29. Ehrhart, J.C., B. Bennetau, L. Renaud, J.P. Madrange, L. Thomas, J. Morisot, A. Brosseau, S. Allano, P. Tauc, and P.L. Tran. 2008. A new immunosensor for breast cancer cell detection using antibody-coated long alkylsilane self-assembled monolayers in a parallel plate flow chamber. *Biosens. Bioelectron.* 24:467–474.

30. Breton, F., B. Bennetau, R. Lidereau, L. Thomas, G. Regnier, J.C. Ehrhart, P. Tauc, and P.L. Tran. 2011. A mesofluidic multiplex immunosensor for detection of circulating cytokeratin-positive cells in the blood of breast cancer patients. *Biomed. Microdevices* 13:1–9.

31. MacBeath, G. and S.L. Schreiber. 2000. Printing proteins as microarrays for high-throughput function determination. *Science* 289:1760–1763.

32. Chrisey, L.A., C.E. O'Ferrall, B.J. Spargo, C.S. Dulcey, and J.M. Calvert. 1996. Fabrication of patterned DNA surfaces. *Nucleic Acids Res.* 24:3040–3047.

33. Kusnezow, W., A. Jacob, A. Walijew, F. Diehl, and J.D. Hoheisel. 2003. Antibody microarrays: An evaluation of production parameters. *Proteomics* 3:254–264.

34. Chuang, H., P. Macuch, and M.B. Tabacco. 2001. Optical sensors for detection of bacteria. 1. General concepts and initial development. *Anal. Chem.* 73:462–466.

35. Senaratne, W., L. Andruzzi, and C.K. Ober. 2005. Self-assembled monolayers and polymer brushes in biotechnology: Current applications and future perspectives. *Biomacromolecules* 6:2427–2448 [Review].

36. Ulmann, A. 1996. Formation and structure of self-assembled monolayers. *Chem. Rev.* 96:1533–1554.

37. Bennetau, B., J. Bousbaa, and F. Choplin. CNRS. January 2000. Organosilicon compounds, preparation method and uses thereof. FR Patent 0000695.

38. Navarre, S., F. Choplin, J. Bousbaa, and B. Bennetau. 2001. Structural characterization of self-assembled monolayers of organosilanes chemically bonded onto silica wafers by dynamic force microscopy. *Langmuir* 17:4844–4850.

39. Deleris, G., S. Rubio-Albenque, B. Bennetau, B. Desbat, F. Buffiere, and J.-L. Chagnaud. 2007. Université Victor Segalen Bordeaux 2/CNRS. FR Patent 0754424.

40. Martin, P., S. Marsaudon, L. Thomas, B. Desbat, J.P. Aimé, and B. Bennetau. 2005. Liquid mechanical behavior of mixed monolayers of amino and alkyl silanes by atomic force microscopy. *Langmuir* 21:6934–6943.

41. Tran, P.L. and B. Bennetau. CNRS, ENS Cachan and Univ. Bordeaux 1. July 2004. Novel microfluidic system and method for capturing cells. FR Patent 0407722.

42. Richter, M., P. Woias, and D. Weiß. 1997. Microchannels for applications in liquid dosing and flow-rate measurement. *Sens. Actuators A Phys.* 62:480–483.

43. Tran, P.L., J. Weinbach, P. Opolon, G. Linares-Cruz, J.P. Reynes, A. Grégoire, E. Kremer, H. Durand, and M. Perricaudet. 1997. Prevention of bleomycin-induced pulmonary fibrosis after adenovirus-mediated transfer of the bacterial bleomycin resistance gene. *J. Clin. Invest.* 99:608–617.

44. Tran, P.L., G. Régnier, and F. Breton, CNRS-ENS Cachan. January 2009. Patent FR 09/00333.

45. Li, T., Q. Fan, T. Liu, X. Zhu, and G. Li. 2010. Detection of breast cancer cells specially and accurately by an electrochemical method. *Biosens. Bioelectron.* 25:2686–2689.

22

Micropatterned Biosensing Surfaces for Detection of Cell-Secreted Inflammatory Signals

Jun Yan and Alexander Revzin

CONTENTS

22.1 Introduction

Inflammation is an important process that involves immune cell activation and production of inflammatory signals. Inflammation is a natural and necessary response of the body to injury or infections and inflammatory signals play an important role in recruiting immune cells to the site of the injury to clear out pathogens. However, dysregulated inflammation can contribute to a number of pathologies including diabetes [27], liver fibrosis/cirrhosis [19], and cancer [8]. The connection between inflammation and cancer has drawn considerable attention recently [5,6]. While the causative link between inflammation and cancer is still tenuous, studies have suggested that treatment with anti-inflammatory drugs can significantly reduce cancer risk [4]. Other studies described functional impairment of T-cells in patients with malignant tumors [7]. Given the importance of inflammation in cancer formation and metastasis, the immune cells that produce inflammatory signals are likely to become either targets of anticancer therapies or diagnostic correlates of cancer progression. Importantly, histological analysis of the presence/absence of immune cells may not be sufficiently informative with regards to pro- or anti-inflammatory effects of these cells. We envision the need to monitor function of immune cells and are developing biosensors for cell function analysis.

Macrophages are immune cells that migrate from blood vessels and position themselves in various tissues. These cells are the first to respond to injury or infection happening in the tissue/organ and produce a battery of inflammatory molecules designed to destroy the invader and to recruit other immune cell types to the site [12]. Reactive oxygen species (ROS) and cytokines are some of the inflammatory molecules produced by activated macrophages. ROS is a collective term that refers to chemical species formed by incomplete reduction of oxygen and encompasses superoxide anion ($O_2{}^{\cdot-}$), hydrogen peroxide (H_2O_2), hydroxyl radical ($^\cdot OH$), and peroxynitrite ($ONOO^-$). H_2O_2 is the most stable of ROS types and serves as an important indicator of ROS production and inflammation [3].

Cytokines are small proteins secreted by the immune cells to help clear out the infection by recruiting more cells to the site of the injury and by inducing proliferation of immune cells. While macrophages are known to secrete a range of cytokines, tumor necrosis factor alpha (TNF-α) is one of the most important and abundant cytokines produced by these cells. TNF-α is an inflammatory cytokine that induces apoptosis and has been implicated in helping cancer cell survival [1]. It has been proposed that TNF-α secreted by the neighboring immune cells up-regulates expression of NF-κB in cancer cells, providing survival advantage to these cells [2,21]. Thus, detection and quantification of TNF-α production by the immune cells is an important area of research since TNF-α regulation represents a likely strategy for anticancer therapy development [2].

Given the importance of H_2O_2 and cytokines as markers of inflammation, a number of bioanalytical approaches have been developed for monitoring these molecules. The methods for detection of H_2O_2 include chemiluminescence [13], fluorescence [32], and electrochemistry [18]. These methods commonly rely on the use of horseradish peroxidase (HRP)—an enzyme that catalyzes breakdown of H_2O_2. One approach for monitoring this breakdown was described by Haugland and colleagues [32] who employed Amplex Red—a small molecule that becomes fluorescent upon HRP-catalyzed reduction of H_2O_2. In contrast to enzyme-based detection of H_2O_2, cytokines are most commonly detected using monoclonal antibodies (Abs) and sandwich immunoassays [17]. While robust and sensitive, traditional cytokine immunoassays are designed to analyze an amount of cytokine present in a given physiological sample (e.g., blood). These traditional assays reveal very little about cell type and cell-to-cell heterogeneity in cytokine secretion.

Over the past several years, our laboratory has been developing strategies for isolating specific cell types from a heterogeneous sample and detecting signaling molecules secreted by these cells [9,33–35]. We are particularly interested in designing cell–biosensor interface to allow placement of cells in the immediate vicinity of sensing elements. The proximity of cells and biosensors should translate into sensitive detection of cell-secreted analytes. One approach for designing cell–biosensor interface employed by us is poly(ethylene glycol) (PEG) hydrogel photolithography whereby non-fouling biomaterial (PEG) is micropatterned to define cell adhesive and nonadhesive regions on a substrate [23]. For example, creating PEG hydrogel microwells allowed organizing cells into high-density single-cell arrays on cell culture surfaces such as glass [24,25]. In this chapter, we describe how to convert non-fouling gel into a functional/sensing layer by incorporating sensing molecules inside and around hydrogel microstructures.

In one strategy, glass attachment sites inside PEG hydrogel microwells were modified with cytokine-specific Abs (e.g., anti-TNF-α) in order to capture cells and detect cell-secreted cytokine molecules in the same microwells. This approach allows establishing production of TNF-α by concrete individual macrophages and will enable cell-by-cell analysis of heterogeneity in cytokine release by these cells. Importantly, in addition to preventing cell attachment, PEG hydrogels provide an excellent matrix for entrapment of enzymes [11,22]. Therefore, hydrogel microstructures not only are useful for guiding cell attachment but also may contain enzymes (biorecognition molecules) and may serve as sensing elements [31]. To highlight this, we describe the development of HRP-carrying hydrogel microstructures that can be used for detection of H_2O_2 produced by macrophages. Finally, as a future application of micropatterned cell–biosensor surfaces, we describe PEG hydrogel microwells that may be used for detecting both TNF-α and H_2O_2 secreted by the same cells.

22.2 Materials and Methods

22.2.1 Materials

PEG diacrylate (PEG-DA, MW 575) (Catalog #437441), 2-hydroxy-2-methyl-propiophenone (photoinitiator) (Catalog #H55103), 99.9% toluene (Catalog #34866), hydrogen peroxide (35 wt% solution in water) (Catalog #349887), HRP (Catalog #P6782), phorbol 12-myristate 13-acetate (PMA) (Catalog #P1585), and lipopolysaccharide (LPS) (Catalog #L2630) were purchased from Sigma-Aldrich (St. Louis, MO). Amplex Red reagent (Catalog #A12222) and Alexa Fluor-546 Streptavidin (Catalog #S11225) were obtained from Invitrogen (Carlsbad, CA). 3-Acryloxypropyl trichlorosilane (Catalog #SIA0199.0) was purchased from Gelest, Inc. (Morrisville, PA). Dimethyl sulfoxide (DMSO) (Catalog #20688) was from Pierce (Rockford, IL). PBS (0.1 M, pH 7.4) without calcium and magnesium (Catalog #MT-21-031-CM) was obtained from Fisher Scientific and used to prepare aqueous solution. J774 Macrophages cell line (Catalog #HB-197) was purchased from American Type Culture Collection (ATCC, Manassas, VA). Anti-mouse TNF-α (Catalog #AF-410-NA), biotinylated anti-mouse TNF-α (biotinylated TNF-α) (Catalog #BAF410), and recombinant TNF-α (Catalog #410MT) were purchased from R&D Systems (Minneapolis, MN). Neutravidin fluorescein FITC conjugated (Catalog #31006) was from Pierce. Silicone gaskets were purchased from Grace Bio-Labs (Bend, OR). Chromium etchant (CR-4S) and gold etchant (Au-5) were from Cyantek Corporation (Fremont, CA). Positive photoresist (AZ 5214-E IR) and developer solution (AZ300 MIF) were bought from Mays Chemical (Indianapolis, IN). Poly-(dimethylsiloxane) (PDMS) and its curing agents (MSDS #01064291) were purchased from Dow Corning (Midland, MI). SU-8 (Catalog #Y 131269) was from MicroChem Corp (Newton, MA). All these chemicals were used as received without further purification.

22.2.2 Methods

22.2.2.1 Silane Modification of Glass Substrates to Ensure Attachment of Hydrogel Microstructures

Standard 3×1 in. glass slides were cleaned by immersion in 3:1 mixture of sulfuric acid and hydrogen peroxide. Right before silane modification, the clean glass substrates were exposed to O_2 plasma (YES-R3, San Jose, CA) for 3 min at 300 W. Then they were immediately placed into a glass dish containing toluene and moved into an N_2-filled glove bag. Two percentage (v/v) of 3-acryloxypropyl trichlorosilane was added to the dish inside the bag. Silane self-assembly was allowed to proceed for 1 h under N_2 atmosphere, after which the substrates were removed, rinsed in fresh toluene, and dried using N_2 gas. The substrates were then placed in an oven for 3 h at 100°C to cure the silane layer. This silanization procedure has been used by us previously for anchoring PEG hydrogel microstructures to glass substrates [23,25].

22.2.2.2 Fabricating Hydrogel Microstructures with Integrated Biorecognition Molecules

The layout of the micropattern was drafted using AutoCAD (Autodesk Inc.) and was then converted into a transparency-based photomask (CAD Art Services, Portland, OR). The pattern was later transferred from a transparency onto a chrome-coated soda lime plate (Nano Film Company, Westlake Villaget, CA) using standard photolithography and chrome-etching protocols. Briefly, positive resist (AZ 5214-E IR) was spin-coated at 800 rpm for 10 s followed by 4000 rpm for 30 s, resulting in formation of a 4 μm thick layer of photoresist. The photoresist-coated substrate was soft-baked on a hot plate (make/model) at 100°C for 105 s, then placed in contact with a photomask and exposed to 365 nm, 10 mW/cm² UV source for 55 s using Canon PLA-501F mask aligner. The substrate was then placed into a developer solution (AZ300 MIF) for 4 min. After the development step, the substrate was immersed in chrome-etching solution (1:1 v/v mixture of CR-4S etchant with H_2O) for 2 min.

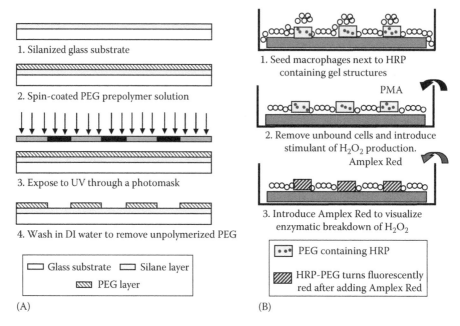

FIGURE 22.1 (A) A schematic of PEG hydrogel photolithography process. Photosensitive liquid prepolymer is cross-linked by UV exposure through a photomask, creating hydrogel microstructures. (B) HRP molecules (gray solid dots) can be entrapped inside hydrogel microstructures. Macrophages attach on glass next to hydrogel microstructures and are induced to secrete H_2O_2. Secreted ROS molecules are detected by change in fluorescence of hydrogel structures.

Figure 22.1 shows a step-by-step process for fabricating hydrogel microstructures on glass substrates. The first step in the process involved spin-coating photosensitive prepolymer solution onto the substrate using spinner from Headway Research Inc. (Garland, TX). This prepolymer was prepared by dissolving a photoinitiator—2% (v/v) (2-hydroxy-2-methyl-propiophenone)—in PEG-DA (MW 575). To fully mix the PEG-DA with the photoinitiator, the mixture was stirred for 15 min prior to use. When fabricating H_2O_2-sensing hydrogel structures, HRP solution (10 mg/mL in 1× PBS) was added to PEG prepolymer to make it 10% v/v. In cases when fluorogenic reagent Amplex Red was to be integrated into hydrogel structures, it was added to the HRP–PEG prepolymer solution to make it 1% v/v in prepolymer. After adding sensing molecules (HRP and Amplex Red) into the PEG prepolymer, this solution was spin-coated at 600 rpm for 5 s and then exposed to UV (60 mW/cm^2) from Omnicure 1000 (EXFO, Mississauga, Ontario, Canada) for 1.2 s. This resulted in free-radical polymerization/cross-linking of exposed regions of the prepolymer. After exposure, surfaces were immersed in distilled water for 2 min to remove the unpolymerized PEG and dried with N_2. This photopatterning process, described in Figure 22.1, resulted in physical entrapment of HRP and Amplex Red molecules inside the hydrogel microstructures.

In order to immobilize cytokine-sensing Abs, glass slides with micropatterned hydrogel structures were immersed in 0.2 mg/mL solution of anti-TNF-α, supplement with Tween 20 (0.005%) for 1 h. Ab molecules were physically adsorbed onto silanized glass regions but did not deposit on non-fouling PEG hydrogel structures.

22.2.2.3 Seeding Cells on Micropatterned Biosensing Surfaces

Mouse macrophage cells (J774A) were cultured at 37°C with 5% CO_2 in phenol red-free Dulbecco's modified Eagle's medium (DMEM) supplemented with 10% fetal bovine serum (FBS). The cells were grown in suspension culture in 50 mL bioreactor tubes (Techno Plastic Products) on a rolling apparatus (Stovall). The cells were passaged two times a week by centrifuging and resuspending in fresh culture media.

Figure 22.1B shows the scheme of seeding cells onto surfaces micropatterned with PEG hydrogels. Prior to cell seeding, glass slides with HRP-containing hydrogel micropatterns were diced into 0.5×0.5 in. pieces using a diamond scribing pen. These smaller glass pieces were placed into P35 Petri dishes. Fifty microliters of cell suspension at 10^6 cells/mL concentration was introduced into a Petri dish, allowing cells to sediment and interact with the micropatterned surface. Silicone gaskets (diameter: 6 mm; depth: 1 mm) were used to limit the cell suspension volume to 50 μL. After 30 min of incubation, the cell suspension was aspirated and replaced with fresh DMEM. In the process of cell seeding, macrophages were able to attach onto silane-modified glass regions, becoming localized next to but not on top of enzyme-carrying hydrogel microstructures. This particular macrophage cell line did not require adhesive ligands for attachment to silane-modified glass. When working with less "sticky" cells, we have precoated micropatterned surfaces with extracellular matrix (ECM) proteins (e.g., collagen (I) or fibronectin) to promote cell attachment to glass [14,25].

22.2.2.4 Detection of Macrophage-Secreted H_2O_2 Using Hydrogel Microstructures

As described in Section 22.2.2.2, Amplex Red, a reagent that fluoresces during HRP-catalyzed breakdown of H_2O_2, can be impregnated into PEG microstructures. However, we found this approach to be impractical for cell sensing because of oxidation/autofluorescence and leaching out of this reagent. Therefore, when detecting cell-secreted H_2O_2, Amplex Red added into cell culture media.

Prior to cell detection experiments, Amplex Red was dissolved in analytically pure DMSO to reach a concentration of 5 mM and was aliquoted and stored in a desiccant-loaded jar at -20°C. One aliquot was used per experiment for quality assurance due to the easy oxidation of Amplex Red. In order to induce secretion of H_2O_2, macrophages residing around HRP-containing PEG hydrogel structures were exposed for 1 h to a mitogen (PMA) dissolved in cell culture media. In order to visualize cellular production of H_2O_2 and its breakdown by HRP inside gel microstructures, 50 μM Amplex Red was added into the cell culture medium after cell activation (Figure 22.1B). As a result, HRP-carrying hydrogel structures were expected to become fluorescence with signal intensity corresponding to concentration of secreted H_2O_2. The appearance of fluorescence signal in the hydrogel microstructures was monitored using a fluorescence microscope as described later and was determined to occur rapidly (on the scale of minutes). In most of the experiments reported here, fluorescence images of sensing hydrogel structures were acquired 5 min after introducing Amplex Red into the Petri dish containing cells and hydrogel micropatterns. The calibration curves were constructed to convert fluorescence intensity signal into analyte concentration. In order to construct a calibration curve, HRP-containing hydrogel micropatterns without cells were exposed to H_2O_2 ranging in concentration from 0 to 20 μM and corresponding fluorescence signals were recorded as described later. All experiments were performed in triplicate ($n = 3$) for statistical significance. In order to account for variability in fluorescence signal from one experiment to the next—an artifact of autooxidation of Amplex Red under ambient conditions—the fluorescence response to 1 μM H_2O_2 was used to normalize other data points of the calibration curve. Therefore, an experiment for detecting oxidative burst from macrophages consisted of two components performed in parallel: (1) detection of H_2O_2 release from macrophages using sensing hydrogel microstructures, and (2) sensor response after exposure to a standard of 1 μM H_2O_2 and 50 μM of Amplex Red.

In order to determine concentration of a cell-secreted product, fluorescence signal recorded from the cell cultures was normalized (divided) by the fluorescence signal generated from 1 μM H_2O_2 reference point. This normalized fluorescence value was then compared with a calibration curve that was created using the same normalization technique. A Zeiss 200M epifluorescence microscope (Carl Zeiss MicroImaging, Inc. Thornwood, NY) equipped with an AxioCam MRm (charge-coupled device [CCD] monochrome, 1300×1030 pixels) was used in order to detect fluorescence signal from hydrogel microstructures. Image acquisition and fluorescence analysis were carried out using AxioVision software (Carl Zeiss MicroImaging, Inc. Thornwood, NY). In this work, fluorescence associated with Amplex Red reagent was detected using excitation 550 ± 25 nm/emission 605 ± 35 nm filter (550 ± 25 nm describes a band-pass filter with optimal excitation of 550 nm and bandwidth of 50 nm).

22.2.2.5 Integrating Hydrogel Micropatterns inside Microfluidic Devices

It is beneficial to integrate micropatterned biosensing surfaces inside a microfluidic device to permit easy reagent exchange and to perform experiments in a small volume. A PDMS-based microfluidic device was fabricated using standard soft-lithography approaches and is described in detail elsewhere [34]. Briefly, a transparency photomask (CAD Art Services, Bandon, OR) was generated based on AutoCAD drawing of the device. This photomask was then employed to micropattern SU-8 on a 4 in. silicon wafer in order to create a negative replica fluidic network. PDMS was mixed 10:1 with a curing agent, poured onto a Si wafer containing SU-8 features and cured for 12 h at 60°C. The elastomer with embedded channel architecture was released and inlet–outlet holes were punched with a blunt 16-gauge needle. This microfluidic device consisted of two flow chambers with width×length×height dimensions of 3×10×0.1 mm and a network of independently addressed auxiliary channels. The volume of each flow chamber was ~3 μL. The auxiliary channels were used to apply negative pressure (vacuum suction) to the PDMS mold and reversibly secure it on top of the micropatterned substrate.

Silicone tubing (1/32 in. I.D., Fisher) was used to connect a 10 mL syringe to the outlet through a metal insert cut from a 20-gauge needle. A blunt, shortened 20-gauge needle carrying a plastic hub was positioned at the inlet. The pressure-driven steady flow was generated by a precision syringe pump (Harvard Apparatus, Boston, MA).

22.2.2.6 Detection of Macrophage-Secreted TNF-α in Hydrogel Microwells

For analysis at single-cell level, arrays of hydrogel microwells with 30 μm in diameter were fabricated on glass. These microwells were made comparable to the size of single macrophages (~15–20 μm in diameter) to ensure capture of individual cells. For TNF-α detection, anti-TNF-α Ab molecules were physically adsorbed onto microwell arrays. The walls of microwells were composed of non-fouling PEG hydrogel and Ab molecules deposited selectively on the glass attachment sites at the bottom of the microwells. Macrophages express receptors for Fc domains on Ab molecules and therefore readily attach inside the microwells. These micropatterned surfaces were enclosed inside a PDMS microfluidic device. Prior to the introduction of cells, a PDMS device containing fluidic and vacuum channels was sterilized by UV for 30–45 min in a tissue culture hood. 0.5 mL of 1×10^6/mL of macrophages were concentrated by centrifugation and resuspension in phenol red-free DMEM media. The final concentration of the cells was ~20 million/mL. To remove air bubbles inside the microfluidic chamber, sterilized 1× PBS was introduced first and was followed by 50 μL of the concentrated cell solution injected at a flow rate of 20 μL/min. Flow was stopped for 15 to 20 min for cell attachment. Unbound cells were washed away by applying 1× PBS at a flow rate set at 50–100 μL/min for 5 min. The whole procedure was observed using a microscope (Nikon Inc., Melville, NY). After cell seeding, 1 μg/mL of LPS diluted in phenol red-free DMEM was introduced into the channel to stimulate macrophages to produce TNF-α. Once the LPS solution was introduced into the flow chamber, the flow was stopped and a surgical clamp was used to secure the outlet tubing and eliminate convective mixing. The sample was kept in a tissue culture incubator for 3 h.

After 3 h stimulation with LPS, micropatterned surfaces were taken out from the incubator for cytokine staining. First, the LPS solution was washed away with 1× PBS and flow channels were infused and incubated with biotinylated anti-TNF-α for 1 h followed by incubation with either streptavidin–Alexa 546 (red fluorescence) or neutravidin–FITC (green fluorescence) for 30 min. Cells were fixed with 4% PFA for 15 min and stained with DAPI for 5 min to visualize cell nuclei. Between each step, the sample was washed with 1× PBS for 5 min to remove the previous reagent. All staining and washing steps were performed inside a microfluidic device at room temperature. The fluorescently labeled cytokine was visualized and imaged with an A Zeiss 200 M epifluorescence microscope described in the preceding section.

To demonstrate detection of both H_2O_2 and TNF-α from the same micropatterned surface, HRP molecules were incorporated into the walls of hydrogel microwells as described earlier. The glass attachment pads of the microwells were modified with anti-TNF-α Ab molecules. Macrophages were seeded on these surfaces as stimulated with LPS as described before to commence TNF-α production. This protocol did not elicit sufficient production of H_2O_2 to cause a change in fluorescence of hydrogel microstructures. To demonstrate detection of two inflammatory markers from the same micropatterned surface, $1 \mu M$ of exogenous H_2O_2 and $50 \mu M$ of Amplex Red solution were added to the sample. Secreted TNF-α molecules were visualized with neutravidin–FITC (green signal) whereas H_2O_2 appeared as red fluorescence.

22.3 Results and Discussion

22.3.1 Non-Fouling Properties of Enzyme-Carrying Hydrogel Microstructures

To demonstrate sensing of H_2O_2, HRP/Amplex Red-carrying hydrogel microstructures were challenged with known concentration of this analyte. Figure 22.2A shows an array of hydrogel microstructures (30–500 μm diameter) fluorescing in response to incubation of $10 \mu M$ H_2O_2. This image highlights the possibility of fabricating sensors of various sizes to meet the need of different experiments. For example, as shown later in this chapter, the ability to control dimensions of hydrogel microstructures was used to capture individual cells inside the microwells and to detect cell-secreted TNF-α. As shown in Figure 22.2B, the HRP/Amplex Red molecules were uniformly distributed inside the hydrogel structures with uniform fluorescence signal emanating from different z-plane slices of these sensing structures.

Effective juxtaposing of small groups of cells with sensing PEG hydrogel micropatterns requires that enzyme-entrapping hydrogel structures remain non-fouling. To demonstrate that the incorporation of enzymes does not adversely affect non-fouling properties of PEG hydrogels, an HRP-containing PEG precursor solution was photopatterned on glass substrates and incubated with macrophages or fibroblasts. As shown in Figure 22.3A and B, macrophages and fibroblasts selectively attached on glass regions with limited or no adhesion observed on PEG domains that contained enzyme HRP. The introduction of exogenous $10 \mu M$ H_2O_2 into culture media resulted in the appearance of an optical (fluorescence) signal from sensing hydrogel microstructures. In addition, a multistep PEG fabrication process could be employed to micropattern enzyme-carrying as well as enzyme-free hydrogel structures on the same surface. Figure 22.3C demonstrates one example where polymer spin-coating and photopatterning were performed twice using two different prepolymer solutions to create sensing PEG structures integrated into a nonsensing PEG layer. The nonsensing PEG layer can serve as a negative control in metabolite detection experiments.

22.3.2 Characterization of H_2O_2-Sensing Hydrogel Microstructures

In our initial experiments, Amplex Red was entrapped inside hydrogel microstructures along with HRP. However, Amplex Red is irreversibly oxidized during the HRP-catalyzed breakdown of H_2O_2 [28];

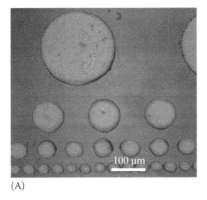
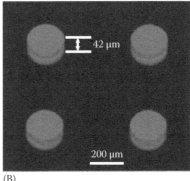

(A)   (B)

FIGURE 22.2   (A) HRP-containing PEG hydrogel micropattern challenged with $10 \mu M$ concentration of H_2O_2 in the presence of Amplex Red. (B) Z-stack images of hydrogel micropatterns underscore three-dimensionality of the structures and uniform distribution of HRP molecules inside the gel.

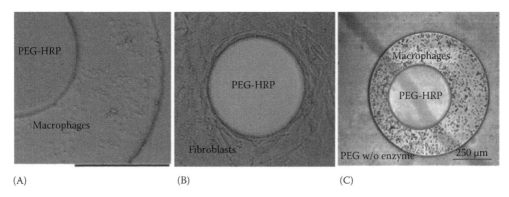

(A)   (B)   (C)

FIGURE 22.3 (A) SEM image of HRP-containing hydrogel micropattern after incubation with macrophages. The cells attached exclusively on exposed glass regions and did not bind to hydrogel microstructures. (B) Attachment of 3T3 fibroblasts around HRP-containing hydrogel microstructure 250 μm in diameter. Addition of 10 μM of H_2O_2 and 50 μM of Amplex Red resulted in appearance of fluorescence in the gel. This image highlights the dual role of hydrogel structures as sensing elements and non-fouling biomaterials. (C) Two sets of hydrogel microstructures fabricated in alignment on the same surface. HRP-containing cylindrical elements 500 μm in diameter were micropatterned inside an array of enzyme-free hydrogel wells 1 mm in diameter. After the addition of 10 μM of H_2O_2, the signal appeared only from HRP-containing gel structures. Both sets of microstructures were non-fouling and macrophages attachment was seen only on exposed glass regions.

therefore, the immobilized probe was expected to lose its ability to sense an analyte over extended periods of time. An alternative approach was to introduce soluble Amplex Red into cell culture media during the detection process. We noticed that the fluorescence intensity of Amplex Red changed over time even though the concentration of analyte was held constant (see Figure 22.4). This behavior was attributed to continuous production of red-fluorescent oxidant resorufin in the HRP-catalyzed reaction. Similar behavior of Amplex Red was reported previously [10]. We chose to record fluorescence intensity after 5 min of interaction between the sensing hydrogel structures and Amplex Red. This time frame was long enough to achieve ~40% of maximum signal strength [32]. To reduce the detection time, a 5 min time point was also frequently chosen by other groups in the study of Amplex Red [16,26].

The fluorescence signal was also concentration dependent. Figure 22.5A and B shows a difference in fluorescence signal after incubation of hydrogel sensors with 5 vs. 20 μM of H_2O_2 for 5 min. The calibration curve of [H_2O_2] vs. fluorescence intensity was constructed in order to quantify the amount of metabolite produced by macrophages. The fluorescence signal increased linearly over the range of 0–20 μM as shown in Figure 22.5C, indicating that the concentration-dependent response could be derived from the micropatterned PEG hydrogel structures. It should be noted that when constructing calibration curve

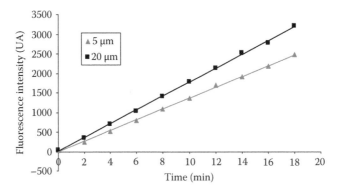

FIGURE 22.4 Time-dependent increase in fluorescence intensity of hydrogel microstructures in the presence of 5 or 20 μM H_2O_2 and 50 μM Amplex Red.

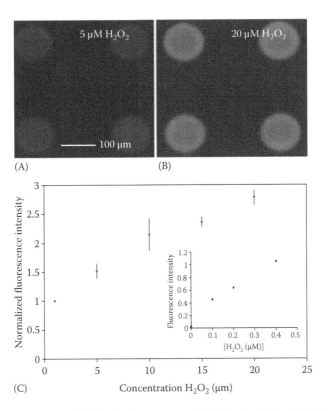

FIGURE 22.5 Fluorescence images of hydrogel microstructures after 5 min incubation with 5 µM H_2O_2 (A) and 20 µM H_2O_2 (B) in the presence of Amplex Red. (C) Calibration curve correlating fluorescence intensity of the hydrogel microstructures to concentration of H_2O_2. Because of experiment-to-experiment variability in optical signal, all fluorescence intensity values were divided (normalized) by the signal for 1 µM H_2O_2. Inset shows signal vs. analyte relationship for low concentrations of H_2O_2. When sensing for cell-secreted H_2O_2, images were collected 5 min after introduction of Amplex Red into cell culture medium.

shown in Figure 22.5C, absolute fluorescence intensity signals were normalized by the sensor signal due to 1 µM H_2O_2. We found this normalization procedure necessary in order to account for experiment-to-experiment differences in absolute fluorescence signals. When detecting endogenous H_2O_2 from macrophages, there was always a control experiment where the same hydrogel micropatterns without cells were exposed to 1 µM H_2O_2 providing a reference point. This allowed us to compare normalized signal generated from cell-secreted metabolite to the normalized calibration curve.

22.3.3 Detecting H_2O_2 Released from Activated Macrophages Using Sensing Hydrogel Microstructures

To investigate the possibility of detecting endogenous H_2O_2, macrophages were cultured on surfaces next to sensing hydrogel microstructures and were stimulated with a mitogen, PMA, for 1 h. This protocol was expected to result in oxidative burst and release of H_2O_2 from macrophages. Figure 22.6A and B shows representative bright field/fluorescence images of hydrogel microstructures after 1 h PMA stimulation and without stimulation. Analysis of fluorescence intensity from these images, presented in Figure 22.6C, shows that 1 h of PMA activation of macrophages led to a more than threefold increase in signal observed in hydrogel-sensing elements, compared to macrophages not exposed to this stimulant, and around twofold increase compared to cells with 3 h stimulation. While some fluorescence is observed in Figure 22.6B, this should be considered a background signal that is most likely due to autooxidation

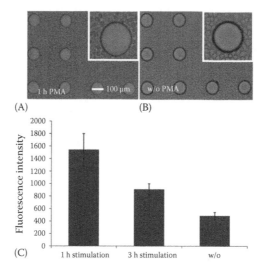

FIGURE 22.6   Detection of secreted H_2O_2 with PMA activation (A) and without (B) PMA activation. (C) Fluorescence intensity measurements of H_2O_2 signal in hydrogel microstructures after PMA activation for 1 and 3 h and without PMA.

of Amplex Red into fluorescent resorufin in the ambient environment. In addition, incubation of HRP-containing hydrogel micropatterns with Amplex Red but without cells resulted in fluorescence signals that were comparable to signals from unactivated macrophages (data not shown). These results highlight the connection between activation of macrophages and the appearance of H_2O_2 signal in the adjacent hydrogel biosensors. The fluorescence signal detected in HRP-containing hydrogel micropatterns was converted to an analyte concentration using the calibration curve presented in Figure 22.5C. The results of this experiment presented in Figure 22.6C point to 1 and 0.55 µM of H_2O_2 production in the cases of 1 and 3 h stimulation, respectively.

22.3.4 Creating Microdevices for Capture and Analysis of Single Macrophages

In the results described in preceding sections of this chapter, we demonstrated that sensing hydrogel microstructures were non-fouling and allowed placement of tens to hundreds of macrophages in defined locations on the glass surface. However, PEG hydrogel photolithography offers a precise control of bio-interfacial properties so that placement of single cells in desired locations of the surface is achievable [24]. We envision functional analysis at the single-cell level as a prerequisite to better understanding of heterogeneity of cancer cell and immune cell populations. Our laboratory is currently developing biosensors for detecting H_2O_2 at a single-cell level and has already demonstrated detection of cytokine release from single cells [35]. The strategy of detecting cell-secreted cytokines, shown in Figure 22.7A, involves capturing macrophages inside microwells that contain cytokine-specific Abs. Upon release, cytokines become bound next to the secreting cells and can be visualized using standard sandwich immunoassay approaches (see Figure 22.7A). This allows us to make a direct connection between individual cells and levels of secreted cytokines. Micropatterned sensing surfaces were integrated into microfluidic devices (see, e.g., Figure 22.7B) to permit ease of cell seeding, activation and to carry out immunoassays inside a small volume (~3 µL per channel).

Given that the size of macrophages used in the present study ranged from 15 to 20 µm diameter, we fabricated arrays of 30 µm diameter wells. These wells consisted of non-fouling hydrogel walls and silane-modified glass bottom that supported cell attachment (Figure 22.8A). As can be seen in Figure 22.8A, anti-TNF-α Ab immobilized onto the silanized glass bottom was visualized by sequential incubation of 500 ng/mL of recombinant TNF-α, anti-TNF-α–biotin, and neutravidin–FITC. Fluorescence emanated from glass attachment sites insides the microwells and no fluorescence was seen on non-fouling hydrogel

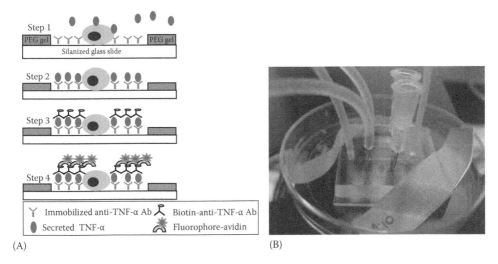

FIGURE 22.7   (A) Detection of TNF-α secreted by single macrophages. Macrophages are captured in microwells that are modified with anti-TNF-α Abs. Cell-secreted cytokine molecules become bound inside the microwells next to specific cells. The presence of TNF-α in microwells is detected using sandwich immunoassay. (B) Arrays of cell capture and cytokine detection microwells are incorporated into a PDMS microfluidic device. The device consists of two fluid flow chambers (needle hubs designate inlets of the channels) that are surrounded by a separate network of channels for vacuum suctioning PDMS to glass.

walls. Figure 22.8B seeding macrophages into 30 μm diameter PEG wells resulted in sequestration of single cells. As described by us in previous publications, this surface micropatterning/cell-seeding process allows forming large-scale single-cell arrays [24,25].

22.3.5 Detecting TNF-α from Single Macrophages

In order to detect the cytokine, anti-TNF-α Ab was immobilized at the bottom of hydrogel microwells by physical absorption. Macrophage attachment did not require precoating of the surfaces with adhesive ligands as these cells were able to attach onto silanized glass or Ab-containing surfaces. However, in case cell attachment needs to be promoted, microwells may be modified with ECM proteins or cell-specific Abs as described by us in several preceding reports [14,15]. To extend the strategy of capturing single cells and then detecting secreted proteins to other cell types, adhesion-promoting ECM proteins/Abs can be mixed with a detection Ab and co-immobilized inside microwells.

After seeding macrophages onto arrays of microwells, we proceeded to detect TNF-α released from single cells. The microdevice (hydrogel microwells integrated into PDMS microfluidic channels) was kept in the incubator for 3 h after stimulant LPS was introduced into the chamber. Stimulation was performed in tissue culture incubator (37°C, 5% CO_2) as the cells were found to be more responsive to stimulants under physiological conditions [20]. An identical vacuum system was set up inside the incubator to ensure the complete sealing of the PDMS device on top of the sample.

After LPS activation, microwell arrays with captured macrophages were stained with anti-TNF-α–biotin followed by streptavidin–Alexa 546. Staining was performed in situ, inside the microfluidic channels. Figure 22.8C and D shows representative bright field and fluorescence images of macrophages after activation with LPS. As seen from Figure 22.8C, macrophages had a "healthy" phenotype—cells were spread out inside the microwells. Importantly, fluorescence microscopy (Figure 22.8D) revealed a strong red fluorescence associated with anti-TNF-α–biotin/streptavidin–Alexa 546 staining. This fluorescence signal was due to TNF-α production by macrophages as it only appeared in the wells occupied by macrophages and only when macrophages were activated by a stimulant LPS. While not demonstrated here, a calibration curve of TNF-α concentration vs. fluorescence intensity may be constructed in order to quantify the amount of cytokine secreted by single cells [35].

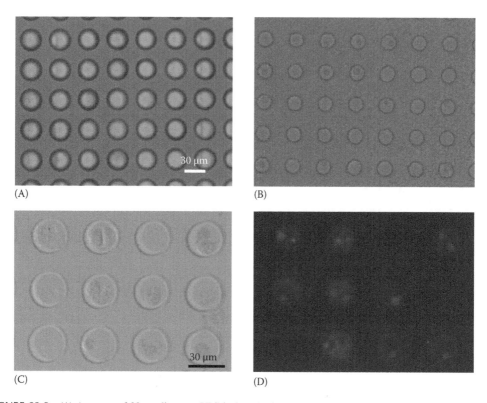

FIGURE 22.8 (A) An array of 30 μm diameter PEG hydrogel microwells. The deposition of anti-TNF-α Ab onto the silanized glass bottom was visualized by sequential incubation of 500 ng/mL of recombinant TNF-α, biotinylated anti-TNF-α, and neutravidin–FITC. (B) Single macrophages attaching inside microwells. Note that the gel surface is rough because of entrapped HRP molecules. (C) Higher magnification bright field image showing macrophages captured inside the PEG hydrogel microwells and stimulated with PMA for 3 h. (D) The same set of microwells stained with biotinylated anti-TNF-α Ab and streptavidin–Alexa 565 shows red fluorescence inside the wells. This fluorescence signal is due to cell-secreted TNF-α. Note that nuclei of cells are stained with DAPI (brighter gray color) and that presence of cytokine signal is associated with the presence of macrophages in the wells.

22.3.6 Integrating Biosensors for H₂O₂ and TNF-α Detection into the Same Platform

In this chapter, we have demonstrated two separate biosensing strategies based on PEG hydrogel microstructures: (1) entrapping enzyme molecules within the gel structures for H_2O_2 detection, and (2) using hydrogel micropatterns to adsorb anticytokine Abs onto glass regions inside the microwells. To demonstrate flexibility of hydrogel micropatterning, we fabricate arrays of PEG microwells where hydrogel walls contained entrapped HRP and the glass attachment sites contained anti-TNF-α Abs (see cartoon in Figure 22.9A for description of this biosensor). Therefore, the same hydrogel micropatterns were expected to detect both small ROS molecules and cytokines. In a proof of concept experiment, macrophages were seeded into sensing hydrogel microwells, stimulated with LPS for 3 h and stained for secreted TNF-α (neutravidin–FITC used for fluorescence labeling). As shown in Figure 22.9A, microwells with cells emitted green fluorescence due to secreted TNF-α. Figure 22.9A shows that the microwells contained macrophages (nuclei stained with blue dye DAPI) and that the cytokine signal was associated with specific cells. As shown by staining cell nuclei with DAPI, macrophages are captured almost exclusively inside the glass attachment sites of hydrogel microwells. To demonstrate that the same microwells may also be used for detecting H_2O_2, 1 μM of exogenous H_2O_2 and 50 μM of Amplex Red solution were added to the sample. As seen from Figure 22.9B, hydrogel microwells emitted red fluorescence in response to stimulation with H_2O_2. While this result stops short of detecting endogenous H_2O_2 in addition to TNF-α, it points out a future strategy for detecting two inflammatory markers secreted by single cells.

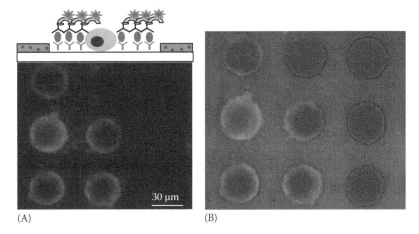

(A)　　　　　　　　　　　　　(B)

FIGURE 22.9 Merged bright field/fluorescence image showing integration of hydrogel-based H_2O_2 sensor with TNF-α sensor. (A) Macrophages were treated with LPS for 3 h and the sample was stained with biotinylated anti-TNF-α Ab and neutravidin–FITC. The green fluorescence indicates (bright gray color in the microwells) secreted TNF-α from cells. (B) The addition of 1 μM of H_2O_2 and Amplex Red resulted in appearance of fluorescence signal (seen in gray here) inside the walls of hydrogel wells.

22.4 Conclusions and Future Trends

This chapter describes fabrication of sensing hydrogel microstructures that can be used to both define cell attachment and detect cell-secreted molecules. Entrapment of HRP molecules inside the gel allowed detecting H_2O_2 release by macrophages cultured nearby. Creating hydrogel microwells and modifying glass attachment sites with anti-TNF-α allowed capturing single macrophages and detecting cytokine molecules secreted by these cells. Finally, we demonstrated micropatterned surfaces that can be used for detection of both H_2O_2 and TNF-α inflammatory signals produced by macrophages.

While we demonstrate the possibility of integrating biosensors for enzyme-based detection of a small molecule (H_2O_2) and antibody-based sensing of a cytokines (TNF-α) into the same micropatterned surface, the differences in function of biorecognition molecules would need to be reconciled in the future. Enzymes are constantly turning over, providing temporal information about substrate breakdown and signal generation. On the other hand, antibody-based immunoassays provide end-point measurements and reveal little about the change in cytokine signal over time. Replacing antibody molecules with aptamer-beacons [29,30] would allow real-time detection of cell-secreted cytokines or other proteins so that affinity sensors will function in a similar way as enzyme-based biosensors. In addition, the number of analytes detected from single cells will be expanded in the future.

The overarching goal of the studies described in this chapter is to enable local detection of cell function with novel micropatterned sensing surfaces. Analysis of cell function (i.e., what the cell produces) is different from standard approaches of phenotyping cells based on morphology or cell surface marker expression. The sensor-cell platform described in this chapter can be used to analyze two inflammatory signals, H_2O_2 and TNF-α, that play important roles in cancer formation and progression. The platform described here may be used for screening efficacy of anticancer drug therapies. In addition, we see the ability to detect and quantify cell-secreted products with single-cell resolution as an important step toward improved, more nuanced diagnosis of cancer and toward the development of personalized anticancer therapies.

REFERENCES

1. Balkwill, F., K. A. Charles, and A. Mantovani. 2005. Smoldering and polarized inflammation in the initiation and promotion of malignant disease. *Cancer Cell* 7:211–217.
2. Balkwill, F. and L. M. Coussens. 2004. Cancer: An inflammatory link. *Nature* 431:405–406.

3. D'Autréaux, B. and M. B. Toledano. 2007. ROS as signalling molecules: Mechanisms that generate specificity in ROS homeostasis. *Nat. Rev. Mol. Cell Biol.* 8:813–824.

4. Dannenberg, A. J. and K. Subbaramaiah. 2003. Targeting cyclooxygenase-2 in human neoplasia: Rationale and promise. *Cancer Cell* 4:431–436.

5. de Visser, K. E., A. Eichten, and L. M. Coussens. 2006. Paradoxical roles of the immune system during cancer development. *Nat. Rev. Cancer* 6:24–37.

6. DeNardo, D. G., M. Johansson, and L. M. Coussens. 2008. Immune cells as mediators of solid tumor metastasis. *Cancer Metastasis Rev.* 27:11–18.

7. Finke, J., S. Ferrone, A. Frey, A. Mufson, and A. Ochoa. 1999. Where have all the T cells gone? Mechanisms of immune evasion by tumors. *Immunol. Today* 20:158–160.

8. Halliwell, B. 2007. Oxidative stress and cancer: Have we moved forward? *Biochem. J.* 401:1–11.

9. Jones, C. N., J. Y. Lee, G. S. Stybayeva, M. A. Zern, and A. Revzin. 2008. Multifunctional protein microarrays for cultivation of cells and immunodetection of secreted cellular products. *Anal. Chem.* 80:6351–6357.

10. Kim, S. H., B. Kim, V. K. Yadavalli, and M. V. Pishko. 2005. Encapsulation of enzymes within polymer spheres to create optical nanosensors for oxidative stress. *Anal. Chem.* 77:6828–6833.

11. Koh, W. G. and M. Pishko. 2005. Immobilization of multi-enzyme microreactors inside microfluidic devices. *Sens. Actuators B Chem.* 106:335–342.

12. Kuby, J. 1998. *Immunology*, 3rd edn. W.H. Freeman & Co., New York.

13. Lee, D., V. Erigala, M. Dasari, J. Yu, R. Dickson, and N. Murthy. 2008. Detection of hydrogen peroxide with chemiluminescent micelles. *Int. J. Nanomed.* 3:471–476.

14. Lee, J. Y., S. S. Shah, J. Yan, M. C. Howland, A. N. Parikh, T. R. Pan, and A. Revzin. 2009. Integrating sensing hydrogel microstructures into micropatterned hepatocellular cocultures. *Langmuir* 25:3880–3886.

15. Lee, J. Y., C. Zimmer, G. Y. Liu, and A. Revzin. 2008. Use of photolithography to encode cell adhesive domains into protein microarrays. *Langmuir* 24:2232–2239.

16. Li, W. and W. Jin. 2006. Measurement of peroxidase activity in single neutrophils by combining catalyzed-enzyme reaction and epi-fluorescence microscopy. *Talanta* 70:251–256.

17. Li, Y., N. Nath, and W. M. Reichert. 2003. Parallel comparison of sandwich and direct label assay protocols on cytokine detection protein arrays. *Anal. Chem.* 75:5274.

18. Lindgren, A., T. Ruzgas, L. Gorton, E. Csoregi, G. B. Ardila, I. Y. Sakharov, and I. G. Gazaryan. 2000. Biosensors based on novel peroxidases with improved properties in direct and mediated electron transfer. *Biosens. Bioelectron.* 15:491–497.

19. Nussler, A., S. Konig, M. Ott, E. Sokal, B. Christ, W. Thasler, M. Brulport, G. Gabelein, W. Schormann, M. Schulze, E. Ellis, M. Kraemer, F. Nocken, W. Fleig, M. Manns, S. C. Strom, and J. G. Hengstler. 2006. Present status and perspectives of cell-based therapies for liver diseases. *J. Hepatol.* 45:144–159.

20. Pihel, K., E. Travis, R. Borges, and R. Wightman. 1996. Exocytotic release from individual granules exhibits similar properties at mast and chromaffin cells. *Biophys. J.* 71:1633–1640.

21. Pikarsky, E., R. M. Porat, I. Stein, R. Abramovitch, S. Amit, S. Kasem, E. Gutkovich-Pyest, S. Urieli-Shoval, E. Galun, and Y. Ben-Neriah. 2004. NF-kappa B functions as a tumour promoter in inflammation-associated cancer. *Nature* 431:461–466.

22. Quinn, C. A. P., R. E. Connor, and A. Heller. 1997. Biocompatible, glucose-permeable hydrogel for in situ coating of implantable biosensors. *Biomaterials* 18:1665–1670.

23. Revzin, A., R. J. Russell, V. K. Yadavalli, W.-G. Koh, C. Deister, D. D. Hile, M. B. Mellott, and M. V. Pishko. 2001. Fabrication of poly(ethylene glycol) hydrogel microstructures using photolithography. *Langmuir* 17:5440–5447.

24. Revzin, A., K. Sekine, A. Sin, R. G. Tompkins, and M. Toner. 2005. Development of a microfabricated cytometry platform for characterization and sorting of individual leukocytes. *Lab Chip* 5:30–37.

25. Revzin, A., R. G. Tompkins, and M. Toner. 2003. Surface engineering with poly(ethylene glycol) photolithography to create high-density cell arrays on glass. *Langmuir* 19:9855–9862.

26. Rupcich, N. and J. Brennan. 2003. Coupled enzyme reaction microarrays based on pin-printing of sol–gel derived biomaterials. *Anal. Chim. Acta* 500:3–12.

27. Simmons, R. 2006. Developmental origins of diabetes: The role of oxidative stress. *Free Radic. Biol. Med.* 40:917–922.

28. Towne, V., M. Will, B. Oswald, and Q. Zhao. 2004. Complexities in horseradish peroxidase-catalyzed oxidation of dihydroxyphenoxazine derivatives: Appropriate ranges for pH values and hydrogen peroxide concentrations in quantitative analysis. *Anal. Biochem.* 334:290–296.

29. Tuleuova, N., C. N. Jones, J. Yan, E. Ramanculov, Y. Yokobayashi, and A. Revzin. 2010. Development of an aptamer beacon for detection of interferon-gamma. *Anal. Chem.* 82:1851–1857.

30. Xiao, Y., A. A. Lubin, A. J. Heeger, and K. W. Plaxco. 2005. Label-free electronic detection of thrombin in blood serum by using an aptamer-based sensor. *Angew. Chem. Int. Ed.* 44:5456–5459.

31. Yan, J., Y. H. Sun, H. Zhu, L. Marcu, and A. Revzin. 2009. Enzyme-containing hydrogel micropatterns serving a dual purpose of cell sequestration and metabolite detection. *Biosens. Bioelectron.* 24:2604–2610.

32. Zhou, M., Z. Diwu, N. Panchuk-Voloshina, and R. P. Haugland. 1997. A stable nonfluorescent derivative of resorufin for the fluorometric determination of trace hydrogen peroxide: Applications in detecting the activity of phagocyte NADPH oxidase and other oxidases. *Anal. Biochem.* 253:162–168.

33. Zhu, H., M. Macal, M. D. George, S. Dandekar, and A. Revzin. 2008. A miniature cytometry platform for capture and characterization of T-lymphocytes from human blood. *Anal. Chim. Acta* 608:186–196.

34. Zhu, H., G. S. Stybayeva, M. Macal, M. D. George, S. Dandekar, and A. Revzin. 2008. A microdevice for multiplexed detection of T-cell secreted cytokines. *Lab Chip* 8:2197–2205.

35. Zhu, H., G. S. Stybayeva, J. Silangcruz, J. Yan, E. Ramanculov, S. Dandekar, M. D. George, and A. Revzin. 2009. Detecting cytokine release from single human T-cells. *Anal. Chem.* 81:8150–8156.

23

Quantum Dots Nanosensor Analysis of Tumor Cells

Lee-Jene Lai, Yi-Heui Hsieh, and Shih-Jen Liu

CONTENTS

23.1 Introduction

In biological science, nanotechnology is extensively adopted as a multidisciplinary approach. An optical biosensor is developed in materials physics and chemistry to sense and detect the state of a biological system and a living organism. The elementary functional units of biological system, including protein, DNA, and antibody, consist of complex nanoscale components. An integration of biomolecules with optical elements through biorecognition produces functional devices. A novel material of artificial nanoparticles of quantum dots (QDs) served as a sensor of hepatitis B virus DNA, DNA mutants, and *Escherichia coli* O157:H7 [1,2]. Nanoparticles of 2~6 nm attract considerable interest because of their dimensional similarities with biological macromolecules [3].

The principal difference between the macrocrystalline and nanocrystalline semiconductor is related to the size. With a nanocrystalline semiconductor is associated a large ratio of surface area to volume that is related to quantum confinement. Nanoparticles are smaller than the exciton Bohr radius of a bulk semiconductor. The effects of quantum confinement generate unique optical and electronic properties in semiconductor QDs. QDs received considerable attention for use in fluorescent biolabels such as proteins, pathogens, bacteria, DNA, and cells [4–9]. Relative to fluorophores such as organic dyes and fluorescent proteins, QDs have unique optical properties, including excellent resistance to photobleaching, a high quantum yield, broad excitation spectra, and a narrow emission line. These properties lead QDs to become highly promising alternatives to organic dyes in biological labeling to improve the sensitivity of molecular pathology and in vitro diagnosis [10–12]. Even so, the potential toxicity of

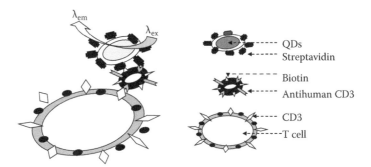

FIGURE 23.1 Concept of tumor cell coupling with quantum dots nanosensors. Biotinylated antihuman CD3 serves as a bridge to anchor the CD3 marker of T cell and the streptavidin of quantum dot. The fluorescent intensity at emission wavelength (λ_{em}) of 625 nm was measured for quantitative analysis of the tumor cell populations using an excitation wavelength (λ_{ex}) of 365 nm.

cadmium-containing QDs is still of great concern for imaging applications in vivo [13]. QDs fluorescence labeling has been so far performed in vitro for clinical diagnosis and prognosis.

We present here a quantitative method, based on a QD nanosensor, to detect rare tumor cells with large throughput. The human Jurkat lymphocyte (T cell) served as a model for circulating tumor cells (CTCs). CTCs are cells that have unfasten from a primary tumor and circulate in the bloodstream. CTCs may compose germs for following growth of additional tumors in different tissues. Red blood cells (RBC) behave the interference in the blood during the detection. QDs–streptavidin conjugation was applied for indirect fluorescent labeling based on the marker of CD3 of T cell through biotinylated antihuman CD3. A high affinity constant, about 10^{-15} M, between biotin and streptavidin [14] made their link a "nature bridge" between QD and antibodies, which self-assemble on a QD surface and represent the correlation between QD and biomolecules [15]. We labeled fluorescent QDs on T cells based on the characteristic of a high affinity between biotin and streptavidin. Figure 23.1 depicts the principle of T cell coupling with QDs as a fluorescence marker. Biotinylated antihuman CD3 represents a bridge between QD–streptavidin and T cells. The CD3 marker, which is on the membrane of a T cell, is conjugated with biotin antihuman CD3 through an antigen–antibody reaction, subsequently producing biotinylated T cell. The biotinylated T cell is again associated with the QD–streptavidin via biotin–streptavidin coupling, ultimately forming QD-labeled T cells (QD–T cell). Additionally, the number of T cells, which are linked with QDs, is evaluated with a system to measure the intensity of fluorescence with appropriate excitation and emission spectra. Moreover, the conditions for QDs fluorescence are optimized on controlling the pH and duration of photostimulation of the QD–T cell solution. The experimental results indicate that <15 min suffices to detect cells in a population, even when as few as 40 cells are present.

In this chapter, we focused on the number of cells or frequency of cells, not concentration of cells. The volume of developed system for detection of tumor cells is 200 μL. It means 40 rare specific tumor cells, under the interference of 10^6 mixed cells in total (frequency of 4×10^{-5}), could be concentrated to the volume of 200 μL for the detection. The detection time is just 15 min. Capable of detecting a specific tumor cell, the proposed QD nanosensor enables a simple, efficient, and sensitive count of CTCs in the early diagnosis, staging, and prognosis of cancer diseases.

23.2 Materials

23.2.1 Chemicals

Human T lymphocyte cells (Jurkat cell line, ATCC TIB-152) were obtained from American Type Culture Collection (ATCC, Manassas, VA). QD625–streptavidin conjugates (QD–stv) with a maximum emission wavelength of 625 nm were obtained from Invitrogen (Carlsbad, CA). Biotin antihuman CD3 (13-0038) was purchased from eBioscience, Inc. Ficoll-Paque PLUS reagent was obtained from GE Healthcare Amersham Biosciences (Uppsala, Sweden). Chemicals of Na_2HPO_4, KH_2PO_4,

$Na_2B_4O_7 \cdot 10H_2O$, H_3BO_4, boric acid, and fluorescein isothiocyanate–streptavidin (FITC–stv) were purchased from Sigma Chemical Co. (St. Louis, MO). Potassium chloride was purchased from Riedel-de Haën (Seelze, Germany). Ultrapure water (18 MΩ·cm), obtained using a Milli-Q purification system from Millipore Corporation (Billerica, MA), was used in preparation of all of the solutions. All chemicals were used as received without further purification.

23.2.2 Cell Culture

Jurkat cells (T cells) were cultured in RPMI-1640 (11875, Cibco, ATCC) supplemented with 25 mM HEPES SH30237.010, 10% (v/v) heat-inactivated fetal Clone III serum SH30109.03, from Hyclone Laboratories Inc. (Logan, Utah), 1% penicillin P0781, and 0.005 mM 2-ME M7522 from Sigma Chemical Co. (St. Louis MO) at 37°C and 5% CO_2. C1R cells (B cells) were maintained in Iscove's modified Dulbecco's medium (IMDM, 12200-028, Gibco) supplemented with 10% (v/v) heat-inactivated fetal Clone III serum (SH30109.03, Hyclone) and 1% penicillin (P0781, Sigma) at 37°C and 5% CO_2 in a humidified atmosphere. RBC were obtained with informed consent from healthy donors, and were separated with the Ficoll-Paque PLUS reagent.

23.3 Methods

23.3.1 Sample Preparation

T-lymphocyte cells, Jurkat cells, and RBC were prewashed twice in a phosphate-buffered saline (PBS) solution (0.1 M, pH 7.2). The populations of cells were counted with a hemocytometer before conjugation. Initially, 10^6 cells in total with a mixture of T cells and RBC were concentrated (to 100 μL) in PBS solution and reacted with biotin antihuman CD3 (0.5 mg/mL, 2 μL, 4°C) for 15 min. Next, mixed cells with biotinylated T cells were washed completely in a PBS solution and purified on centrifugation (2000 rpm, 10 min). Furthermore, the precipitated pellet of mixed cells was dissolved in a PBS solution (100 μL) and reacted with QD625–streptavidin (1 μM, 0.5 μL, 4°C) for 30 min. The mixed cells were then washed completely in a PBS solution and purified on centrifugation (2000 rpm, 10 min) to avoid nonspecific binding. Finally, the precipitated pellet of mixed cells with QD-conjugated T cells (QD–T cells) was dissolved in each required buffer and pH solution (200 μL) to determine the fluorescence.

23.3.2 Optical System

23.3.2.1 Fluorescence Microscopy Image System

The fluorescence images from QD–T cells (emission 625 nm) attached to the surface of the substrate were recorded with an inverted microscope (Eclipse series TE2000-U, Nikon Inc., Japan), a single Hg lamp, and a cooled, monochrome, charge-coupled device (CCD, Evolution QEi Digital Camera, Media Cybernetics Inc.) as detector. Filter sets for QD625 (Chroma Technology Corp., catalog no. 32110b; exciter, D405/90X; dichroic, 470DCXR; emitter, D625/40m) and for FITC (Nikon Inc., Japan, catalog no. B-2E/C; exciter, 465–495 nm; dichroic, 505 nm; emitter, 515–555 nm) were obtained from the indicated suppliers. The fluorescence images of QD–T cells were captured on exposure of each image for 1 s, but fluorescence images of FITC–T cells were captured with exposure of each image for 30 s. The color of the image was treated with software (Photoshop).

23.3.2.2 Fluorescence Spectroscopy

Figure 23.2 represents the system of tumor cells detection by fluorescent spectroscopy. Figure 23.2a displays the schematic diagram of system setup. The incident light from a Hg lamp was focused on the entrance slit of monochromator 1 (Mono-1) with Lens-1; a beam with a selected single wavelength of 365 nm from the exit slit of Mono-1 was focused onto the sample position with Lens-2; and the fluorescent emission from the sample was focused onto the entrance slit of monochromator 2 (Mono-2, SpectraPro™-275 Mono-2 from Acton Research Corporation) with Lens-3. A photomultiplier tube

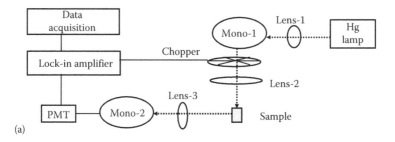

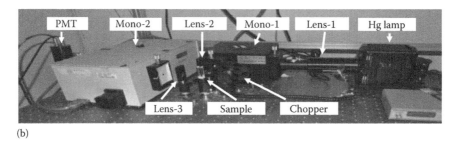

FIGURE 23.2 A system setup of tumor cell detection in fluorescence spectroscopy. (a) Schematic diagram of system setup. The dash line with arrow is the light path and direction. The solid line is the cable connection between two instruments. (b) A photograph of the optical elements with relative positions. The components of the fluorescent system are addressed as follows: Hg lamp, UV excitation light source; Lens-1, focusing the photons from UV light source to the entrance slit of Mono-1; Lens-2, focusing the photons with the wavelength of 365 nm from exit slit of Mono-1 to sample position; Lens-3, focusing the fluorescence that was emitted from the sample; Mono-1, with fixed grating to select the specific wavelength of 365 nm to be the excited UV light source; Mono-2, with rotated grating to obtain the fluorescence spectra from the sample; PMT, which houses a photocathode, several dynodes, and an anode. Incident photons strike the photocathode, producing a consequence of the photoelectric electrons. These electrons are multiplied by several dynodes by the process of secondary emission and collected at anode, resulting in a sharp current pulse. Chopper, with rotating metal disk, is similar to a motorcycle, allows the light source modulated by pass through or being blocked, and provides a synchronous known carrier wave to drive the reference signal for a lock-in amplifier. Lock-in amplifier, a type of amplifier, can extract a weak signal with a known carrier wave by chopper from an extremely noisy environment. It is also known as a phase-sensitive detector. Lock-in amplifier refers to the demodulation or rectification of an AC signal from the photodetector of PMT by a circuit with band-pass filter. The circuit is also controlled by a reference waveform which is derived from the chopper, causing the signal to be modulated. The AC signal from PMT is coherent (same frequency and phase) with the reference waveform; therefore, the other signals are rejected. The signals of the device are large and of high signal-to-noise ratio, and do not need further improvement.

(PMT, R955, Hamamatsu Photonics K. K., Japan) was adopted to collect the photons from the exit slit of Mono-2. The signal from the PMT was modulated at a reference frequency with a chopper positioned between Mono-1 and Lens-2. The modulated signal was transferred to a lock-in amplifier for data acquisition using a computer. The fluorescent emission was measured for samples at room temperature. Figure 23.2b shows a photograph of the optical elements with relative position.

23.4 Results

23.4.1 Photostability of QD–T Cells

The principal advantage of QDs over organic fluorescence dyes for the detection of tumor cells is the photostability. To demonstrate this property, we incubated T cells with biotin antihuman CD3 antibody, followed with either QD–streptavidin or FITC–streptavidin conjugation. Continuously illuminating the conjugated cells resulted in only a small loss of fluorescent signal when the cells were coupled with QDs even after continuous illumination for 10 min and exposure 1 s for each image as shown in Figure 23.3a.

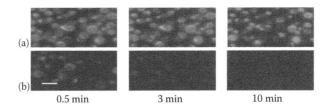

FIGURE 23.3 Comparison of the photostability of QD–T cells with FITC–T cells. Fluorescence images of T cells incubated with biotin anti-CD3, following incubation with either (a) QD–streptavidin or (b) FITC–streptavidin. Slides were continuously illuminated and images were captured at 0.5, 3, and 10 min. Each image of QD–T cells and FITC–T cells was obtained with an exposure for 1 and 30 s, respectively. Objective magnification is 40× and scale bar is 20 μm.

In contrast, a rapid loss of signal occurred when the cells were coupled with FITC; after illumination for 3 min, detection of the signal from the fluorescence image of the FITC–T cells was difficult, despite exposure for 30 s of each image as shown in Figure 23.3b. The results reveal that QD–T cells have greater photostability than FITC–T cells despite 10 min of continuous excitation. These results confirm the report by Wu et al. [16] that microtubules dimmed rapidly after irradiation for 60 s when stained with Alexa488, but microtubules were still bright after irradiation for 180 s when stained with QD608. Chan and Nie reported that QDs coupling to a protein applied to a bioassay exhibited 20 times the luminescent intensity of rhodamine [9]. Hahn et al. adopted QDs conjugated with streptavidin to assay *E. coli* and observed a sensitivity ~100 times that of a similar assay with FITC on excitation at 488 nm [10]. Our solution of T cells labeled with streptavidin-modified QDs yielded a signal ~400 times as bright as for cells labeled with an equal amount of FITC when the excitation photon flux at 365 and 488 nm was normalized for QDs and FITC, respectively. Semiconductor QDs are an attractive alternative to establish organic fluorescent markers for the biolabeling of living cells because of their higher photostability.

23.4.2 Effect of Biochemical Buffer on the Fluorescence of QD–T Cells

A biochemical buffer is an aqueous solution commonly used for a biological system. To test the effect of biochemical buffer on QD–T cells, we chose three biochemical buffers to investigate their fluorescent intensity. Figure 23.4a displays the fluorescence spectra of 10^5 QD–T cells in PBS, Tris, and borate buffer solutions at pH 11. The fluorescent intensity at 625 nm for QD–T cells in borate buffer is 1.12 and 1.3 times the intensity of these cells in Tris buffer and PBS, respectively. Figure 23.4b shows the fluorescent intensity at 625 nm of QD–T cells in PBS, Tris, and borate buffer, which were continuously irradiated for 10 min in the range of pH between 8 and 13. The fluorescence intensity of QD–T cells in these three buffers has the same property of enhancing with increasing pH, for pH between 8 and 11. The fluorescence intensity of QD–T cells in PBS and borate buffer continued to increase until pH 12, but declined to pH 13, whereas the fluorescent intensity of QD–T cells in Tris buffer decreased from pH 11 to 13. These experimental results demonstrate that QD–T cell in borate buffer solution emits fluorescence at 625 nm, which is superior to Tris buffer and PBS. These results resemble those reported by Boldt et al. [17]; their fluorescent stability of QDs was sensitive to the concentration of QDs, pH, and types of biochemical buffer, including Milli-Q H₂O, pH 5.0 citrate/NaOH, MES, 3-(*N*-morpholino)propanesulfonic acid (MOPS), piperazine-1,4-bis(2-ethanesulfonic acid) (PIPES), 500 mM Tris/HCl, 1 M Tris/HCl, pH 6.8 PBS, pH 8.0 PBS, pH 8.0 Tris, pH 9.0 B(OH)₃/KCl/NaOH, and pH 11.0 B(OH)₃/KCl/NaOH. We therefore chose the borate buffer solution for further investigation.

23.4.3 Quenching of the Fluorescence from QD–T Cells under Photoactivation

Although fluorescence of QD–T cells is more photostable than that of the FITC–T cells, the fluorescence of QD–T cells is slightly quenched on photoactivation. Figure 23.5 presents the quenching of fluorescence intensity at 625 nm under continuous illumination of QD–T cells in borate buffer solution at varied pH [18]. The quenching of fluorescence in a cell solution between pH 8 and 11 typically fluctuates during the first 10 min but becomes stable and slow. The emission intensity of QD–T cells improves with

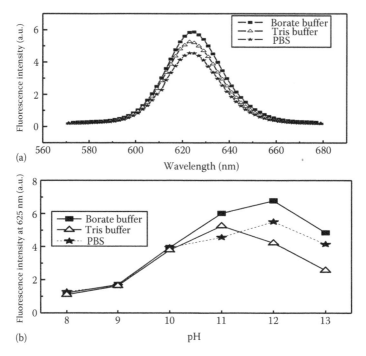

FIGURE 23.4 (a) Fluorescence spectra of 10^5 QD–T cells in solution of PBS, Tris, and borate buffer at pH 11 and (b) at varied pH after irradiation for 10 min, the intensity of fluorescence is recorded at wavelength of 625 nm. All buffer concentrations are 50 mM.

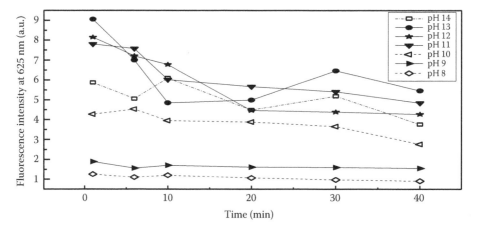

FIGURE 23.5 Fluorescent intensity as a function of duration of photoactivation of QD–T cells at varied pH. The fluorescent intensity is monitored at wavelength of 625 nm.

increasing the pH, but the quenching of fluorescence fluctuates extremely during continuous illumination for 40 min in cell solutions at pH 12 and 13. The observed results indicate that the fluorescence of QD–T cell depends on pH, consistent with the results of Cordero et al. [19]; in the latter work, the excitonic quantum yield of QD increased to 100 during the first 200 s of illumination in air and then steadily decreased to 30 after 5000 s. We accepted their explanation that the quenching of fluorescence is similar to that associated with the solution adsorbed on the QDs surface, in which condition light-induced modifications of surface states following the adsorption of water cause changes in fluorescence from the QDs. The structure of QDs from commercial products is inner shell of CdSe and outer shell of ZnS. Wang et al. [20] demonstrated that light induces a variation of QDs surface charge and the existence of oxygen in the aqueous solution produces a slow photocorrosion, resulting in the release of Cd^{2+} ions. Under these conditions, the formation of a Cd-OH hydroxide layer eliminates the surface defects, thus enhancing the fluorescent intensity. Our experimental results resemble those reported by Sato et al. [21]; their QDs surface under the alkaline conditions could remove the surface trap states to form Cd-OH bond, resulting in the fluorescence emission. The formation of Cd-OH is more easily achieved in a solution with a large pH. Sun et al. [22] observed that QD-labeled cells are sensitive to pH, but these authors reported that the fluorescence image of QD–ovarian cell is brighter at pH 7.0 than at pH 9.0 and 4.6.

For the application of QDs in cell quantification, two factors of photostability and intensity of the fluorescence should be considered. We therefore selected data measured under stable conditions, even though a slightly decreased sensitivity of detection was obtained. As a result, a period of 10 min of photo-activation of QD–T cells at pH 11 for a borate buffer solution was utilized before the cells were counted.

23.4.4 Specificity and Cross-Reaction of QD–T Cells

The biotin–streptavidin with a strong coupling reaction was employed to develop QD labels for optical immunoassay. The reagent of biotinylated antihuman CD3 was used in junction of T cell and QD–streptavidin. The specificity after conjugation was confirmed by the antigen–antibody recognition reaction that occurred between CD3 biomarker of T cell and biotinylated antihuman CD3, and the coupling reaction that occurred between biotinylated antihuman CD3 and QD–streptavidin. Therefore, QD–T cell was formed on incubating the T cell with biotin antihuman CD3, followed with QD–streptavidin at 4°C for 15 and 30 min, respectively. Figure 23.6 shows the evidence on the specificity of the T cell, when biotinylated antihuman CD3 and QD–streptavidin incubated with 10^3 and zero T cells in 9.99×10^5 and 10^6 B cells, respectively. An intense fluorescence was obviously observed in the target 10^3 T cells, but a weak fluorescent signal appeared with B cells (without CD3 biomarker), even though B cells numbered 10^6, indicating that QD–streptavidin successfully attached to the T cell membrane through the indirect specificity of antigen–antibody reaction. Moreover, control experiments involving incubating 10^5 T cells with only QD–streptavidin (without biotinylated antihuman CD3) and the self-fluorescence all resulted in dim fluorescent signals, confirming that the nonspecific adsorption of QD–streptavidin on T cell membrane is negligible.

The cross-reaction of QD–T cell was studied by selecting RBC, separated from whole blood, as the mixing cell. As shown in Figure 23.7, the cross-reaction was performed on mixing the T cell with RBC, when biotinylated antihuman CD3 and QD–streptavidin were incubated with 10^3 and zero T cells in 9.99×10^5 and 10^6 RBC, respectively. Although a weak fluorescent signal appeared with only 10^6 RBC, a significant fluorescence was clearly observed for the T cells in RBC. Moreover, control experiments were performed in which 10^5 T cells were incubated in 9×10^5 RBC, and 10^6 RBC with only QD–streptavidin

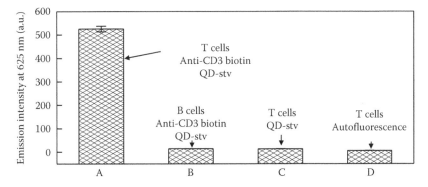

FIGURE 23.6 Specific binding test of the T cells. Columns A and B represented the fluorescent intensity for 10^3 T cells and 10^6 B cells, respectively, incubated with biotin antihuman CD3, followed with QD–streptavidin (stv). Column C described the fluorescent intensity with 10^5 T cells incubated only with QD–streptavidin. Column D depicted the self-fluorescence of 10^6 T cells under UV illumination. The fluorescent intensity at 625 nm was recorded for each sample.

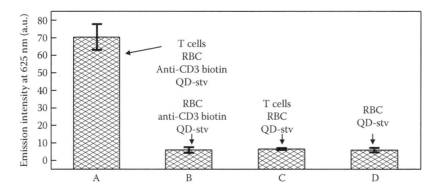

FIGURE 23.7 Cross-reaction test of T cells in RBC. Columns A and B represented the fluorescent intensity for 10^3 and zero T cells for 10^6 RBC mixed cells in total, respectively, incubated with biotin antihuman CD3, followed with QD–streptavidin (stv). Columns C and D referred to the fluorescent intensity with 10^5 and zero T cells in 10^6 RBC mixed cells in total incubated without biotin antihuman CD3 but with QD–streptavidin. The fluorescent intensity at 625 nm was recorded for each sample.

(without biotin antihuman CD3), all resulting in weak fluorescent signals. This finding confirms that the cross-reaction of QD–streptavidin on T cell by RBC is negligible. All these experimental results demonstrate that the QD–streptavidin coupled with biotinylated antihuman CD3 can function as a reliable sensor for the specific detection of T cells. Therefore, the QD-functionalized protein coupled with specific antibody can be applied for the detection of specific tumor cells.

23.4.5 Detection of T Cell in the Mixture of RBC

The populations of T cells in the mixture of RBC were detected on incubation with biotinylated antihuman CD3, followed by the use of a QD–streptavidin fluorescent probe. Figure 23.8a plots the fluorescent spectra of various populations of T cells in 10^6 RBC mixed cells in total. The results indicate that the fluorescent intensity enhances with increasing the populations of T cells in the mixed cells. Figure 23.8b plots the log–log calibration graph of the fluorescent intensity for the incubated QD–T cells in mixed cells. Data correlated with the blank 10^6 RBC were subtracted from the main experimental data. An approximately linear relationship with the T cell population was observed between 40 and 100,000. Notably, <15 min was required to detect the cells, including the period of photoactivation of the QD–T cells, even when the T cell population was as small as 40. A detection sensitivity as small as 40 rare specific cells was hence achieved to detect both the single line of T cells and the rare cells from the mixed cells.

In this chapter, we focused on the number of cells or frequency of cells, not on concentration of cells. The cells could be concentrated and formed the pellet by centrifugation; the pellet was then diluted to varied concentrations. The volume of developed system for detection of tumor cells is 200 μL. It means 40 rare specific tumor cells, under the interference of 10^6 mixed cells in total (frequency of 4×10^{-5}), could be concentrated to the volume of 200 μL for the detection. The detection time is just 15 min. It is unmeaning to look for 2 cells/mL, but the sample is only 0.1 mL. In this developed system, we assume that the concentration of the cells is 2 cells/mL, it might need 20 mL sample and concentrated to 200 μL for the detection. It is acceptable for the patient to collect 20 mL of blood for the detection.

23.5 Discussion

Evidence of one in four deaths in United States attributed to cancer, as well as 1,529,560 new cases and 569,490 related deaths in 2010 [23], indicates that this disease poses a major problem of public health worldwide. Early diagnosis and prognosis of cancer tumor are thus essential to improve cancer treatment. Detection and diagnosis of cancer typically depend on the changes of cells or tissues that are

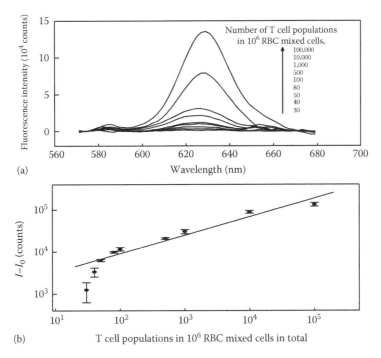

FIGURE 23.8 Relation of fluorescence intensity to the number of T cell populations in 10^6 RBC mixed cells in total. (a) Variation of fluorescent spectra of QD–T cell population in RBC mixture. (b) Log–log calibration plot of fluorescent intensity as a function of T cell populations in RBC mixture. I and I_0 represent the fluorescent intensities observed at 625 nm from biotin antihuman CD3, followed by QD–streptavidin reacted with T cell populations in 10^6 RBC mixed cells in total and 10^6 RBC, respectively. The error bar represents a standard deviation from at least three measurements.

detected by examination from a physician or an imaging method, but such approaches are ineffective to detect small changes in cells or tissues. Early diagnosis and prognosis of the tumor hence remain poor, even in specialized centers. For this purpose, the detection of CTCs is more effective than a physical examination or imaging technology in terms of achieving an early diagnosis [24,25]. CTCs in blood serve as an early marker for recurrence and relapse. As is generally assumed, a quantitative variation in their number reflects the chemotherapeutic sensitivity and metastatic growth activity of a tumor. However, CTCs are seldom found in the early stage, making detection difficult. Conventional methods of limiting dilution analysis (LDA) [26], enzyme-linked immunospot (ELISPOT) [27–29], flow cytometry [30,31], and polymerase chain reaction (PCR) [32,33] have been reported to be capable of detecting a few cells. Although capable of sensitive detecting of cells, LDA and ELISPOT methods require a few days for preparation and measurement. Methods of flow cytometry and PCR can detect a rare cell quickly, but the detection sensitivity is only around 0.001% [34]. Pantel et al. [35] suggested that although PCR can achieve a higher sensitivity of detection, false positive results might produce a nonlinear amplification curve. Although flow cytometry can also ensure highly sensitive detection, the number of cells must be increased by cell culture to an adequate number for detection. Conventional methods require a substantial duration for cell incubation and detection, subsequently making impossible the detection of a rare tumor cell in real time. Therefore, an efficient and sensitive method must be developed to detect tumor cells of small number.

Exploiting the advantages of QDs achieved the detection of rare tumor cell in our work. The ideal conditions of QD–T cell solution were optimized in boric buffer, at pH 11 and with irradiation for 10 min to obtain a stable and intense fluorescence. During the conjugation of QDs with T cell, the solution must be maintained in pH 7.4 PBS to conserve the activity of CD3 on the membrane, implying that the T cell is still alive in QD–T cells. During measurement of fluorescence, the solution of QD–T cells in PBS at

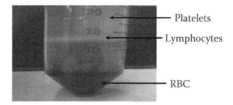

FIGURE 23.9 CTCs and RBC in blood, with different gradient density, were separated and formed different layers with Ficoll-Paque Plus reagent by centrifugation.

pH 7.4 is, however, replaced with a borate buffer at pH 11; there is no need for the activity of CD3 on the membrane to be conserved during this stage. The intensity of fluorescence is correlated with the number of QDs, and the number of QDs corresponds to the number of T cells.

In conclusion, the conjugation of QDs with T cells as a fluorescence probe can be exploited to ensure the rapid and sensitive detection of tumor cells. The parameters of the buffer solution, period of photo-activation, and pH affect the efficiency of the fluorescence of QD–T cells. Adequately controlling these parameters increases the optimal sensitivity. In terms of detecting the specific T cells from mixing with RBC, the fluorescent intensity varies linearly with the T cell population from 40 to 100,000 cells. Less than 15 min is required to detect the cells except 1 h of cell–QD conjugation. This technique can be applied in detection of CTCs in blood of 5 CTCs/mL, and 8 mL of blood is required for the CTCs detection. There is no problem for this technology to handle the volume of 8 mL. In the beginning, CTCs in blood volume of 8 mL added on top of adequate volume of Ficoll-Paque Plus reagent. After centrifugation, the solution in the tube formed different layers, due to the gradient density, as shown in Figure 23.9. CTCs are in the same layer of lymphocytes. After removing the first layer of platelet, the second layer of lymphocytes with CTCs is collected and labeled with QDs. Then, the CTCs can be concentrated to 200 μL for the fluorescent detection. The several advantages of this QD-based cell detection include ease of manipulation, small cost, rapid detection, and reliable sensitivity. Importantly, the proposed method can detect CTCs in the early diagnosis, staging, and prognosis of cancer diseases.

23.6 Future Trends

Optical detection remains the most widely used to detect biological binding events and to image in biological system. In the detection of tumor cells, most of the specific tumor cells are mixed with other cells. Rare specific tumor cells are generally difficult to detect in blood because RBC are present at a large concentration. In the future, Epstein–Barr virus (EBV)-specific memory T lymphocytes will be detected in the blood in this proposed method. The memory T lymphocytes become an interesting field in the immunology due to the potential for monitoring the vaccine efficacy. The population of memory T lymphocytes is rare in circulation, but memory T lymphocytes can expand quickly after the antigen engagement. EBV infection is associated with the development of many types of malignancies, including the Hodgkin's disease and nasopharyngeal carcinoma (NPC). However, the role of memory cytotoxic T lymphocytes for controlling the progress of NPC is still unclear. We plan to explore the possible application of QD-labeled biomolecules to detect EBV-specific cytotoxic T lymphocytes. To achieve a highly sensitive and rapid detection of rare tumor cells from a clinic, the separation of specific tumor cells from cell mixture is required by magnetic bead. The detecting sensitivity of tumor cells could be improved for using magnetic capturing to isolate the rare cells for the detection. We tried to use magnetic beads to isolate the cells and got a better sensitivity for the cell detection [36]. However, the cells loss would be a drawback for using magnetic capturing. The efficiency of the separation of specific tumor cells with commercial magnetic bead from the mixture cells is small, and involves loss of specific cells. Developing a highly efficient separation of magnetic nanoparticles and combining QD fluorescence to detect specific tumor cells from the blood are therefore the future trend.

REFERENCES

1. Wang X, Lou X, Wang Y, Guo Q, Fang Z, Zhong X, Mao H, Jin Q, Wu L, Zhao H, Zhao J. QDs–DNA nanosensor for the detection of hepatitis B virus DNA and the single-base mutants. *Biosens. Bioelectron.* 2010;25:1934–1940.

2. Liu YJ, Yao DJ, Chang HY, Liu CM, Chen C. Magnetic bead-based DNA detection with multi-layers quantum dots labeling for rapid detection of *Escherichia coli* O157:H7. *Biosens. Bioelectron.* 2008;24: 558–565.

3. Chan WCW, Maxwell DJ, Gao X, Bailey RE, Han M, Nie S. Luminescent quantum dots for multiplexed biological detection and imaging. *Curr. Opin. Biotechnol.* 2002;13:40–46.

4. Edgar R, McKinstry M, Hwang J, Oppenheim AB, Fekete RA, Giulian G, Merril C, Nagashima K, Adhya S. High-sensitivity bacterial detection using biotin-tagged phage and quantum-dot nanocomplexes. *Proc. Natl Acad. Sci. USA* 2006;103:4841–4845.

5. Chalmers NI, Palmer RJ, Jr, Du-Thumm L, Sullivan R, Shi W, Kolenbrander PE. Use of quantum dot luminescent probes to achieve single-cell resolution of human oral bacteria in biofilms. *Appl. Environ. Microbiol.* 2007;73:630–636.

6. Goldman ER, Anderson GP, Tran PT, Mattoussi H, Charles PT, Mauro JM. Conjugation of luminescent quantum dots with antibodies using an engineered adaptor protein to provide new reagents for fluoroimmunoassays. *Anal. Chem.* 2002;74:841–847.

7. Hahn MA, Keng PC, Krauss TD. Flow cytometric analysis to detect pathogens in bacterial cell mixtures using semiconductor quantum dots. *Anal. Chem.* 2008;80:864–872.

8. Dubertret B. Quantum dots–DNA detectives. *Nat. Mater.* 2005;4:797–798.

9. Chan WCW, Nie S. Quantum dot bioconjugates for ultrasensitive nonisotopic. *Science* 1998;281: 2016–2018.

10. Hahn MA, Tabb JS, Krauss TD. Detection of single bacterial pathogens with semiconductor quantum dots. *Anal. Chem.* 2005;77:4861–4869.

11. So MY, Xu C, Loening AM, Gambhir SS, Rao J. Self-illuminating quantum dot conjugates for in vivo imaging. *Nat. Biotechnol.* 2006;3:339–343.

12. Yezhelyev BMV, Al-Hajj A, Morris C, Marcus AI, Liu T, Lewis M, Cohen C, Zrazhevskiy P, Simons JW, Rogatko A, Nie S, Gao X, O'Regan RM. In situ molecular profiling of breast cancer biomarkers with multicolor quantum dots. *Adv. Mater.* 2007;19:3146–3151.

13. Derfus AM, Chan WCW, Bhatia SN. Probing the cytotoxicity of semiconductor quantum dots. *Nano Lett.* 2004;4:11–18.

14. Gonzílez M, Luis A, Bagatolli LA, Echabe I, Arrondo JLR, Argaraña CE, Cantor CR, Fidelio GD. Interaction of biotin with streptavidin. *J. Biol. Chem.* 1997;272:11288–11294.

15. Goldman ER, Balighian ED, Mattoussi H, Kuno MK, Mauro JM, Tran PT, Anderson GP. Avidin: A natural bridge for quantum dot–antibody conjugates. *J. Am. Chem. Soc.* 2002;124:6378–6382.

16. Wu X, Liu H, Liu J, Haley KN, Treadway JA, Larson JP, Ge N, Peale F, Bruchez MP. Immunofluorescent labeling of cancer marker Her2 and other cellular targets with semiconductor quantum dots. *Nat. Biotechnol.* 2003;21:41–46.

17. Boldt K, Bruns OT, Gaponik N, Eychmu1ller A. Comparative examination of the stability of semiconductor quantum dots in various biochemical buffers. *J. Phys. Chem. B* 2006;110:1959–1963.

18. Hsieh YH, Liu SJ, Chen HW, Lin YK, Liang KS, Lai LJ. Highly sensitive rare cell detection bases on quantum dot probe fluorescence analysis. *Anal. Bioanal. Chem.* 2010;396:1135–1141.

19. Cordero SR, Carson PJ, Estabrook RA, Strouse GF, Buratto SK. Photo-activated luminescence of CdSe quantum dot monolayers. *J. Phys. Chem. B* 2000;104:12137–12142.

20. Wang Y, Tang Z, Correa-Duarte MA, Pastoriza-Santos I, Giersig M, Kotov NA, Liz-Marzán LM. Mechanism of strong luminescence photoactivation of citrate-stabilized water-soluble nanoparticles with CdSe cores. *J. Phys. Chem. B* 2004;108:15461–15469.

21. Sato K, Kojima S, Hattori S, Chiba T, Ueda-Sarson K, Torimoto T, Tachibana Y, Kuwabata S. Controlling surface reactions of CdS nanocrystals: Photoluminescence activation, photoetching and photostability under light irradiation. *Nanotechnology* 2007;18:465702.

22. Sun YH, Liu YS, Vernier PT, Liang CH, Chong SY, Marcu L, Gundersen MA. Photostability and pH sensitivity of CdSe/ZnSe/ZnS quantum dots in living cells. *Nanotechnology* 2006;17:4469–4476.

23. Jemal A, Siegel R, Xu J, Ward E. Cancer Statistics, 2010. *CA Cancer J. Clin.* 2010; 60:277–300.
24. Budd GT, Cristofanilli M, Ellis MJ, Stopeck A, Borden E, Miller MC, Matera J, Repollet M, Doyle GV, Terstappen LWMM, Hayes DF. Circulating tumor cells versus imaging—Predicting overall survival in metastatic breast cancer. *Clin. Cancer Res.* 2006;12:6403–6409.
25. Sabbath KD, Schwartzberg LS. Circulating tumor cells: Ready for prime time. *Community Oncol.* 2008;5:516–524.
26. Zauderer M, Singer A. Limiting dilution analysis of primary cytotoxic T-cell precursors. *J. Immunol. Methods* 1997;208:85–90.
27. Pahar B, Li J, Rourke T, Miller CJ, McChesney MB. Detection of antigen-specific T cell interferon gamma expression by ELISPOT and cytokine flow cytometry assays in Rhesus macaques. *J. Immunol. Methods* 2003;282:103–115.
28. Alix-Panabières C, Brouillet JP, Fabbro M, Yssel H, Rousset T, Maudelonde T, Choquet-Kastylevsky G, Vendrell JP. Characterization and enumeration of cells secreting tumor markers in the peripheral blood of breast cancer patients. *J. Immunol. Methods* 2005;299:177–188.
29. Vlasselaers D, Schaupp L, Ingeborg van den Heuvel Ivd, Mader J, Bodenlenz M, Suppan M, Wouters P, Ellmerer M, Greet Van den Berghe GVd. Detection and characterization of putative metastatic precursor cells in cancer patients. *Clin. Chem.* 2007;53:537–539.
30. Daugherty PS, Iverson BL, Georgiou G. Flow cytometric screening of cell-based libraries. *J. Immunol. Methods* 2000;243:211–227.
31. Cruz I, Ciudad J, Cruz JJ, Ramos M, Gómez-Alonso A, Adansa JC, Rodríguez C, Orfao A. Evaluation of multiparameter flow cytometry for the detection of breast cancer tumor cells in blood samples. *Am. J. Clin. Pathol.* 2005;123:66–74.
32. Steurer M, Kern J, Zitt M, Amberger A, Bauer M, Gastl G, Untergasser G, Gunsilius E. Quantification of circulating endothelial and progenitor cells: Comparison of quantitative PCR and four-channel flow cytometry. *BMC Res. Notes* 2008;1:71.
33. Xenidis N, Perraki M, Kafousi M, Apostolaki S, Bolonaki I, Stathopoulou A, Kalbakis K, Androulakis N, Kouroussis C, Pallis T, Christophylakis C, Argyraki K, Lianidou ES, Stathopoulos S, Georgoulias V, Mavroudis D. Predictive and prognostic value of peripheral blood cytokeratin-19 mRNA-positive cells detected by real-time polymerase chain reaction in node-negative breast cancer patients. *J. Clin. Oncol.* 2006;24:3756–3762.
34. Neale G, Coustan-Smith E, Stow P, Pan Q, Chen X, Pui C-H, Campana D. Comparative analysis of flow cytometry and polymerase chain reaction for the detection of minimal residual disease in childhood acute lymphoblastic leukemia. *Leukemia* 2004;18:934–938.
35. Pantel K, Brakenhoff RH, Brandt B. Detection, clinical relevance and specific biological properties of disseminating tumour cells. *Nat. Rev. Cancer* 2008;8:329–340.
36. Hsieh YH, Lai LJ, Liu SJ, Liang KS. Rapid and sensitive detection of cancer cells by coupling with quantum dots and immunomagnetic separation at low concentrations. *Biosens. Bioelectron.* 2011;26:4249–4252.

24

Compact Discs Technology for Clinical Analysis of Drugs

Ángel Maquieira

CONTENTS

24.1 Introduction

Electrooptical devices derived from mass consuming electronics (compact discs [CD], cellular phones, TV and computer screens, photographic charge-coupled device [CCD] cameras, document scanners, etc.) possess an advanced technology and high potential [1] to be used as a basis for developing sensing systems such as bioelectromechanical systems (bioMEMs), cheap, reliable, small-sized, no maintenance, and low-energy consumption. One of the most promising is CD.

According to Austerberry [2], the estimated working numbers of CD and digital versatile disc (DVD) players surpass 600 million and there are more than 200 billion optical discs in circulation. This widely used optoelectronic hardware would be addressed to obtain chemical information at very competitive conditions, the main reason being to exploit new applications of CD technology.

The idea of performing assays on circular plastic surfaces has been around for many years following two different courses. Rigid polymeric supports with disc shape and size similar to CDs have been used as centrifugal microfluidic platforms with channels and reservoirs developing several process steps such as filtering, mixing, and quantitative analysis [3–5].

The second strategy exploits the operation principle and elements of audio–video disc technology to perform the assays on its surface and read the chemical results. In this way, there is a huge potential in the development of analytical tools to fit on CD and read by standard hardware. Thus, the CD/DVD-based technology is very well established, robust, and affordable; and both discs and drives are everywhere, this ubiquity being a key advantage. The cost of a CD is about 10 to 20 cents, and players run in the tens to hundreds of dollars and compare quite favorably to other support, detector, or hardware device (glass, silicon, gold, etc. and fluorometers, chip scanners, and microscopes) whose cost runs in the tens of thousands of dollars. Besides, standard disc materials (polycarbonate and polymethylmethacrylate) have demonstrated good properties for passive adsorption and covalent immobilization of coating probes (proteins and nucleic acids) in high-density scales [6–9]. In addition, there is much potential for developing the analysis recording also the results on the same disc, offering confidential data that would assist, for example, in the clinical field. This combination is promising to perform simultaneously multianalyte microarray-based analyses, especially for screening purposes, addressing the requirements of point-of-care or in situ analysis demands [10].

Microarray technology offers high sensitivity using small amounts of bioreceptors, arranged in minimum spot size on a planar surface, in agreement with Ekins' theory [11]. In brief, this pioneering author demonstrated that there is a positive correlation between increasing signal-to-noise ratios and reducing diameters of spots down to a theoretical limit of 3.5 µm. Therefore, the development of microarrays of bioreceptors on CD has considerable analytical potential.

In an original paper, Kido et al. [12] reported (Figure 24.1) the prospective of using CD as microarraying platform for the quantitative analysis. For the first time, it was demonstrated the possibility to determine three pesticide residues simultaneously by immunoassay at micrograms per liter level.

Also, Gordon [13] claimed the use of CD hardware for conducting bioanalysis by using low-reflectivity discs and modifying the optical system of the drive, incorporating one or two optical detectors. In this context, Alexander et al. [14] developed DNA microarrays for genomic analysis by modifying both disc and reader. The determination of low abundant compounds on low-reflectivity discs in a high-density format with microgram per liter sensitivity using a modified CD drive detector [15] has been recently reported as well.

Within the label-free domain, CD-based technology has been adapted for the detection of molecular recognition events by an error determination routine [16]. The optical pickup head as separate detection unit outside CD/DVD drive has been used for several applications [17–20]. However, the use of unmodified optical disc reader has significant advantages for analysis over the systems based on CD technology. First, the use of standard disc players; second, the use of standard discs, especially DVD or Blu-ray; and finally, the original analytical protocols and reagents are applicable to this sensing mode.

FIGURE 24.1 First published experimental results demonstrating the viability of the CD technology to determine simultaneously several substances. In the picture, final competitive immunoassay results detecting hydroxyatrazine (HA), carbaryl (C), and molinate (M), individually and mixed. O is the control. (From Kido, H. et al., *Anal. Chim. Acta*, 411, 1, 2000.)

So far, some progress has been made in this idea and the advantage of using standard CDs and DVDs is winning followers. In this sense, Potyrailo et al. [21] report the determination of cations in water on standard DVD by attaching on it films (3×4 mm) containing colorimetric reagents. Further, other approaches have been performed in array format for bioanalysis on raw disc surfaces, and the effect of the type of discs used for this application has also been studied. Briefly, author of this chapter, using DVD disc as support, has shown the potential of both disc and driver to detect multiplex results for several residues of cholinesterase inhibitors and sulfonamide residues. In other approach, the same technology has been applied [22] to quantify a pool of microcystin toxins directly in water without sample treatment.

On the other hand, Kumar and coworkers [23,24] developed a strategy based on the use of DVDs as spinning-disc interferometric biosensor, utilizing optical interference of reflected light from a multilayered structure, to monitor, in a label-free format, biointeractions (nucleic acid hybridization, RNA–protein, RNA–small molecules, and antigen–antibody) on the modified surface. DVD driver measures the reflected light intensity reaching a very good sensitivity for bioanalytical assays.

In brief, the analytical capabilities of the CD technologies working with CD and DVD using unmodified disc players to read the results have been demonstrated. Also, additional detectors have been installed in commercial drivers to read the chemical information developed on discs. Applications are distributed in three areas. Determination of inorganic anions and cations using reactive films, immunoanalytical determination of low and high molecular mass analytes, and determination of nucleic acid targets, working with microarray heterogeneous assay formats, mainly employing tracers, but recently interferometric techniques have been addressed to develop label-free assays on disc.

24.2 CD Technology

A CD is an optical storage medium with digital data recorded on its surface. The disc player is a device that reproduces the recorded information, being designed for audio, video, and computer applications.

The CD bases were established from the 40′ (twentieth century), but the technology development started later, when practical advances in digital electronics, laser optics, took place. Also, methods of modulating the audio signals in optical form by digital means and large-scale integration (LSI) technologies development, allowing processing a huge amount of data stored on the disc, were established.

A CD (Figure 24.2) is basically a polycarbonate disc substrate with pits and lands molded onto its surface and lands, covered by a reflective metallic layer (aluminum, gold, etc.) protected with a lacquer film. The standard disc mostly used is 12.065 cm in diameter, 1.2 mm thick, and has a center hole of 15 mm; but smaller discs are in the market also with different presentation such as business card, and so on.

Data are stored in a spiral (track) of about 2 billion small shallow pits on the inner polycarbonate surface. The pits are the coded data and carry the information. The disc driver has a laser emitting at 780, 650, or 405 nm wavelengths, depending if CD, DVD, or Blu-ray is read and 3.2, 1.48, or 0.64 µm spot diameter focused on the data side of the disc (Figure 24.3).

Small scratches on the surface of CD do not directly influence the data reading, because the big spot diameter size on the disc surface (0.8 mm) integrates the signal over the large area, making the system much less sensitive to dirt and disc scratches, even with wavy discs. Traditional CD drives employ a single laser beam directed to one track of the data continuous spiral.

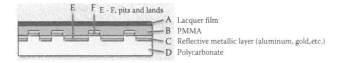

FIGURE 24.2 Scheme of a compact disk component layers (not in scale).

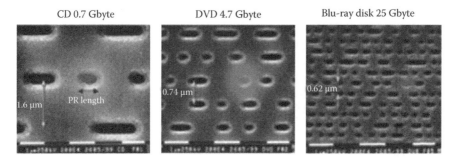

CD 0.7 Gbyte   DVD 4.7 Gbyte   Blu-ray disk 25 Gbyte

FIGURE 24.3 Comparison between the pit dimensions and laser beam size of the CD, DVD, and Blu-ray disks. (From http://www.circuitstoday.com/blu-ray-technology-working)

Instead of illuminating a single track on the surface of a CD-ROM or DVD-ROM discs, the technology illuminates multiple tracks, detects bits simultaneously, and reads them in parallel, and can be used without changes to the CD or DVD disc standards or basic drive design. Multibeam technology at constant linear velocity (CLV) enables optical drives to read and transfer data from the disc at a constant speed, which corresponds to the drive's true spin X-rating. Increase in the data rates due to the multibeam approach allows reducing disc spin rates, decreasing the speed vibration and noise. The TrueX Multibeam method uses a conventional laser beam sent through a diffraction grating which splits the beam into seven evenly spaced beams. These beams illuminate seven different tracks simultaneously (Figure 24.4). Currently the commercial disc drivers use multi-beam design.

Improvements on data storage capacity create new CD technologies such as DVD and Blu-ray, evolutions of optical disc that uses denser recording techniques in addition to layering and two-sided

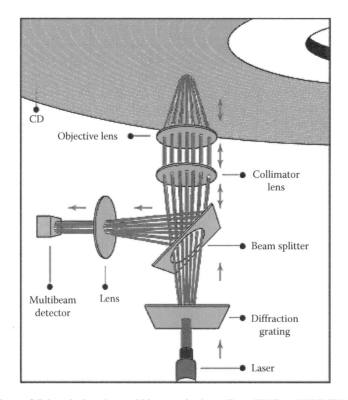

FIGURE 24.4 Scheme of disk reader based on multi-beam technology. (From 52X True-X IDE CDROM, Zen Research.)

manufacturing to achieve very large disc capacities. DVD and Blu-ray drives are also able to read CD-ROMs because they incorporate both reading systems (405, 650, and 780 nm wavelengths). These CD types are marketed with different specifications such as ROM, RAM, Audio, R and RW, and so on.

24.3 Use of CD as Analytical Systems

The strategies based on CD technologies, developing analytical tools to perform biochemical assays, are discussed.

Chemical and biochemical analysis can be performed on standard CD and DVD CD systems. Currently, optical disc players are available with three lasers operating at 405, 650, and 780 nm, making these systems very interesting for multiwavelength detection. Figure 24.5a shows a readout prototype based on a commercial disc player adapted to read analytical assay results developed on any type of standard CD. Figure 24.5b shows detector output signals.

24.3.1 Assay Developments on CD

Using unmodified standard CD and drivers as analytical tools, heterogeneous assays in high-density microarray format are the most interesting, due to the large active surface of the discs, possibility to immobilize a huge collection of probes, develop multiplexing formats, work in microscale saving significant quantities of reagents, reducing the analysis time between 10 and 100 min.

To set up a heterogeneous assay format on disc, we must consider the disc surface composition, probe to be immobilized, printing mode and buffers, washing solutions, and markers to display the final analytical

(a)

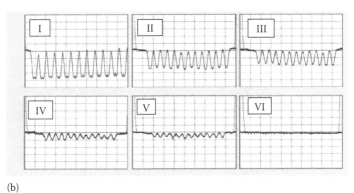

(b)

FIGURE 24.5 (a) Prototype of the multiwavelength disk driver adapted to read analytical results developed on CD, DVD, or Blu-ray disks. (b) Graphical records of signal acquisition when scanning a DVD. The displayed signal is inversely proportional to analyte concentration. Panels I–VI: 0, 0.06, 0.10, 0.25, 1.0, and 4.0 µg atrazine L^{-1}.

result on disc surface (see Section 24.3.3). Probes such as proteins are directly adsorbed on both polymers (polycarbonate and PMMA) composing the CDs, due to its hydrophobicity. Protein probes dissolved in printing buffer (mainly PBS 0.1 M, pH 7.5 or 50 mM carbonate buffer, pH 9.5) are directly and effectively adsorbed on the CD polymer surfaces. Table 24.1 shows the influence of printing buffer composition on sensitivity.

In the case of oligonucleotides, the direct adsorption on raw polycarbonate or PMMA does not work properly, but precoating the surface with streptavidin that immobilizes biotinylated single chain nucleic acid fragments is straightforward. Covalent bond of oligos on derivatized disc polymers is also an immobilization practical way [25] but considering the hybridization reached yields, it is better to add a spacer arm between the disc surface and the probe. The same strategy has important advantages to immobilize directly, on polymeric supports, small molecular mass targets such as drugs. In this case, aminated polycarbonate allows attaching carboxylated drugs to develop heterogeneous indirect immunoassay on disc [26]. Analytical performances (sensitivity and selectivity) reached by covalent linking are better than the classical adsorption strategy of protein conjugate probes.

24.3.1.1 Experimental

- Low-reflectivity CD (L-CD) were from Media Corp. (Tau-Yuan Shien, China). The disc surface was gold metalized to reflect at 780 nm 30% of the incident light.
- Bulk DVD-R discs were purchased from MPO Iberica (Madrid, Spain).

TABLE 24.1

Effect of Printing Buffer Composition on Sensitivity and Signal-to-Noise Ratio (SNR) in a Chlorpyrifos Competitive Immunoassay

Variable	IC50 (µg/L)	SNR
Tween 20 (%)[a]		
0.010	0.96	12.41
0.025	4.87	8.51
0.050	4.00	5.49
0.100	2.49	2.42
pH[b]		
4.6	3.17	10.57
6.0	1.57	24.01
7.2	0.96	12.41
9.6	1.33	4.20
PBS (mM)[c]		
0	13.77	6.23
1	2.54	8.96
5	3.19	6.32
10	0.96	12.41
20	2.57	10.78
45	2.81	10.30

[a] PBS-T 10 mM, pH 7.2, competition time 5 min.
[b] Sodium acetate pH 4.6, MES pH 6.0, phosphate pH 7.2, and sodium carbonate pH 9.6 buffers at 10 mM with NaCl at 150 mM, 0.01% (v/v) Tween 20, competition time 5 min.
[c] PBS-T pH 7.2, 0.01% (v/v) Tween 20, competition time 5 min.
[d] PBS-T 10 mM, pH 7.2, 0.01% (v/v) Tween 20.

TABLE 24.2

Comparison on Reagent Savings: Microarray vs. ELISA Plate

Format	Assay Volume	Reagent Saving (Fold)	Probe Cost Per Assay (U.S.$)	Cost Saving (Fold)
96-well plate	100–200 μL	—	1.0–2.0	—
384-well plate	20–50 μL	2–10	0.20–0.50	2–10
1536-well plate	2–10 μL	10–100	0.05–0.10	10–40
Microarray	1–10 nL	>1000	0.001–0.005	>200

24.3.2 Microarray Printing

Depositing probes on disc surface, printing one or several microarrays is simple, using manual or robotic devices. Commercial tip contact manual arrayers have been successfully used. Also, inkjet and piezo-electric office printers give good results at very low cost and good printing performances including a high throughput capacity [27]. However, it is difficult to adapt the standard paper printers to microarraying probe solutions because information about the drivers is not open to the users.

Contact and noncontact robotic printers have demonstrated to be the best devices for microarray probes on discs, especially because they can print any type of array reaching a high versatility and working capacity. The volume of solution necessary to print a disc with ~3000 dots distributed along is nearly minimum if compared with that employed to develop the same assay on ELISA plate format (Table 24.2).

The dot size after printed and dried on disc surface is related to the dispensed probe solution volume and composition. Thus, if the printed surface is polycarbonate and is given its hydrophobicity, the average of resulting dot diameters are about 400–500 μm for protein or oligonucleotides. Only reducing the dispensed volumes it is possible to arrive to 100 μm dot sizes, reaching diameters down to 10 μm if nano-printing devices are used. Considering the large active surface (94 cm^2) of the standard discs and a dot size of 100 μm and 100 μm pitch, 300,000 spots could be printed on a 12 cm diameter disc. This colossal working capacity indicates the potential of these supports. Nevertheless, 500 μm dot diameter is a good compromise between the spot size, printing density, and the reached sensitivity.

24.3.2.1 Experimental

For manual printing, a disc arrayer from V & P Scientific, Inc. (San Diego, CA) is used, matched with four rows of five pins (20 in total); each one carries 6 nL and leave a 3 nL drop on the CD, producing 500 μm diameter dots. This size allows arraying 320 spots with a horizontal and vertical pitch of 1.125 μm, and eight arrays fit on a disc (2560 spots/disc).

For automatic application of the spots in microarray or another layout, a noncontact printing device is used (AD 1500 BioDot, Irvine, CA), delivering 20 nL probe volume. Also, the working temperature and relative humidity are controlled (25°C and 90%, respectively) because these parameters have dramatic effects on microarray printing quality. Design of the microarray is carried out using Microsoft Office package software.

24.3.3 Staining Strategies

The chemical and biochemical assays to be performed using the CD technology are developed in heterogeneous format, based on immobilizing a probe on a surface to trap, by chemical reaction or molecular recognition (protein–protein, nucleic acids interaction, etc.), the targets. In that case, the assay must yield a detectable product (dye or precipitate) at the disc player wavelength. However, many biochemical assays do not exhibit a detectable post-reaction change by the in use techniques, requiring markers. Consequently, enzymes or gold particles yielding a precipitate from a soluble, are used as markers. All the products to be detected must absorb, disperse, or reflect the laser beam of the disc player.

1. First approach use substrates that are enzymatically transformed in nonsoluble products which absorb and partially reflect the light emitted by the laser of the CD reader. The enzymatic conversion of an organic substrate to a dark-blue insoluble reaction product is widely used as a detection method in immunohistochemistry. 3,3′,5,5′-Tetramethylbenzidine (TMB) and 3,3′-diaminobenzidine (DAB) peroxidase substrates and 5-bromo-4-chloro-3′-indolyl phosphate (BCIP)/nitro blue tetrazolium chloride (NBT) alkaline phosphatase substrate have been studied as developers. The best results in terms of sensitivity were obtained with peroxidase/TMB. The use of TMB leads to an assay with a sensitivity at least eightfold of that obtained with alkaline phosphates BCIP/NBT. TMB yields reproducible analytical results with variation coefficients in the range 1.4%–11.2%.

2. Second approach on tracing and developing the assay results, as, for example, in immunoassay, was based on the use of nanogold-labeled immunoglobulins (Figure 24.6). Under the tested conditions, adding a silver solution as a developer, the gold nanoparticles (optimum: 40 nm diameter) act as a nucleation point around which a solid reaction product is formed (silver enhancement process), resulting in a highly sensitive method with good SNR values, and slightly better than that obtained with TMB. Silver precipitates are stable for months and block light at a broad spectrum of wavelengths, allowing the use of a wide variety of laser diodes and photodetectors. Applying this tracing method to nucleic acids analysis is not straightforward because of the difficulty to directly mark oligos with gold. However, it is possible to exploit the gold–silver staining method using biotinylated targets and gold-marked streptavidin.

24.3.3.1 Experimental

24.3.3.1.1 Staining Strategies

On the basis of the use of enzymatic or metallic particle, marked tracers and substrates yielding nonsoluble products, which absorb the light emitted by the laser of the CD reader, are used.

24.3.3.1.2 Reagents

Nanogold-labeled goat anti-rabbit immunoglobulins (GAR-Au, 40 nm diameter of gold particles) and gold enhance kit are from Nanoprobes, Inc. (Yaphank, NY). Silver enhancer solution, DAB, TMB liquid substrate, BCIP *p*-toluidine salt with NBT purple liquid substrate (BCIP/NBT), and horseradish peroxidase (HRP)-labeled goat anti-rabbit immunoglobulins (GAR-HRP) are from Sigma-Aldrich (Madrid, Spain). Alkaline phosphatase–labeled goat anti-rabbit serum is provided by EMD Biosciences, Inc. (San Diego, CA).

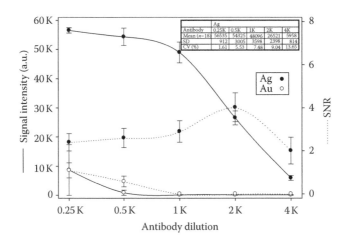

FIGURE 24.6 Signal intensity (solid lines) and signal-to-noise ratio (SNR) (dash lines) of tracing immunoreactions using gold and silver amplification approaches.

24.3.3.1.3 Enzymatic Approach

The enzymatic substrate, TMB for HRP, is used.

24.3.3.1.4 Protocol of Microimmunoassays on Disc

The coating conjugates, dissolved in carbonate buffer, are passively adsorbed on the polycarbonate side of the disk, dispensing 30 µL evenly distributed. Then, the coated disc is set into a CD box for 16 h at 4°C. After, the disc surface is thoroughly washed by immersing in deionized water for 1 min and air dried. Specific antibody solutions in PB with or without analyte are dispensed in a 384-well plate (20 µL/well) and incubated for 15 min. Then, the solutions are arrayed on CD. After 10 min incubation, it is washed with PBS-T buffer for 1 min and rinsed with deionized water. Next, GAR-HRP (1:100 in PBS-T) dispensed onto disc surface areas makes sure of an even distribution. After 10 min at room temperature, the disc is washed and dried.

For displaying the immunoreaction, the disc is incubated with TMB solution developing in ten minutes a dark-blue deposit, stopping the reaction by washing with deionized water. The disc is then ready to be read by the CD player.

24.3.3.1.5 Nanogold-Labeled Tracers Approach

Seven hundred and fifty microliters of nanogold-labeled goat anti-rabbit immunoglobulin is dispensed on to the disc and the solution evenly distributed using a 12 cm diameter 0.6 mm thick dummy plastic surface. After 8 min at room temperature, the cover surface is removed, and the disc is washed and dried as before. The immunoreaction is developed by homogeneously dispensing 750 µL of silver enhancer solution on disc, and the reaction is stopped by washing the disc with water after 8 min.

24.3.4 Analytical Signal Collection

When the analytical results developed on the disc surface are read by the CD player, an image is displayed. One interesting point is the method to obtain this final image of results without saturating the computer memory and in the minimum time. For this, the developed software activates the reading driver only in the analytical zones. Selecting the specific areas on disc is allocated using trigger signals. A trigger is a physical indication, printed, glued, or marked on the disc surface, quote in to read it. Thus, the optical changes in the disc surface are scanned by the laser. This saves a lot of memory, obtaining a file of the only needed information.

Trigger marks are not necessarily used to read the disc. Thus, a trigger-free detection mode implemented in the reading control software has been developed [28], simplifying the disc layout of the chemical assay and saving physical space on the same disc.

On the other hand, Figure 24.7 shows the raw image obtained after reading a section of the disc. Software also corrects the aberration making a proportional image (Cartesian coordinate) that is used to obtain the final signal value of each microarray dot.

24.3.4.1 Experimental

24.3.4.1.1 CD Detection System

Any commercial optical CD/DVD standard drive can be used. In our case, we utilize an LG Electronics (Englewood Cliffs, NJ, model GSA-H42N) holding two lasers, 650 and 780 nm, to alternatively read CD and DVD.

To read semitransparent discs, standard drive is modified with a planar photodiode SLSD-71N6, Silonex, Montreal, Canada (25.4 mm long and 5.04 mm width, spectral sensitivity 0.55 A/W at 940 nm, spectral range 400–1100 nm, acceptance half-angle of 60°), located 1 mm up disc surface, to detect the transmitted laser light (λ 780) and convert it into an analog electrical signal.

Analytical areas are located by a reflective photosensor (EE-SY125, Omron) that includes an infrared light-emitting diode (LED) of 950 nm and a phototransistor, with a sensing distance ranging from 0.5 to 2 mm. This photosensor locates the different reflectivities between the sensing object and the disc. For

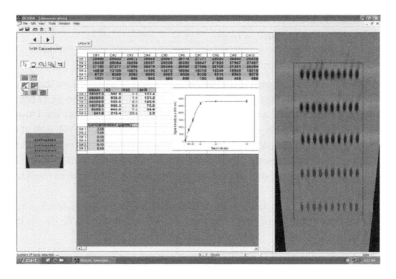

FIGURE 24.7   Raw data (center) and image (right side) displayed from the disk driver.

that, the analytical areas are marked in the outer rim of the disc by trigger footprints of 3.5 cm. Because the unmarked perimeter presents higher reflectivity, the photosensor detects the marked areas providing a trigger signal to the data acquisition board in order to start capturing data exclusively from those zones.

24.3.4.1.2 Electronic Board

A custom-built electronic board (DAB) incorporates the planar photodiode and the photosensor. The board is twofold, amplification of the analog signal, and detection of the analytical areas. Both functions are carried out at the same time; the CD drive performs its original function of reading and writing data. In this way, during the data acquisition process only the signals from the detection areas are digitized by the data acquisition board (DAQ), stored in the computer, and deconvoluted into an image to further quantification.

Custom software is written in Visual C++ (BioDisc). The software provides control to the CD/DVD driver to scan the surface of the disc, to modulate the rotation velocity to a specified spatial resolution, and to write data in the CD or DVD. Also, BioDisk software provides control to the data acquisition board to configure the sampling frequency according to the disc rotation speed and desired angular resolution. For scanning completely the surface of the discs, software simulates the writing process of a 700 MByte or 3.8 GB size file for CD and DVD, respectively.

Because of the spatial difference between samples taken horizontally (each 13 μm) and vertically (each 1.6 μm), a graphical adjustment is done to display a proportional $x-y$ image. This software allows the image to be exported in a gray-scale code to a compressed tif format or bitmap. Then, the images is processed with Photoshop 7.0 (Adobe Systems Inc., San Jose, CA) to map the lightest and darkest pixels into black and white before quantifying with GenePix software (Axon Inst., Union City, CA). Signal intensities of each spot are calculated by background subtraction.

24.3.5 Detection Approaches

Detecting the analytical signal developed on disc surface by the player, means measuring the attenuation of the incident laser beam strength, working at different wavelengths (780 nm; 650 nm). Thus, either the original laser beam is attenuated, dispersed, and/or reflected back to the pickup; or the same beam crosses the disc till reaching an extra detection system, added to the standard disc player (Figure 24.8). In the first approach, high-reflectivity discs (CD or DVD) are employed. In the second one, L-CDs are selected. In all cases, polycarbonate is the standard material but other polymers such as polystyrene, PMMA, etc., have been applied to modify the disc surface successfully [29].

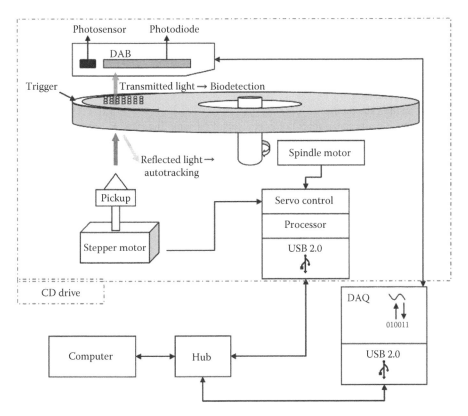

FIGURE 24.8 Schematic representation showing the detection system used to read low- or high-reflectivity CD and DVD. The set of servo systems (spindle and stepper motors) of the CD drive allows disk rotation and laser scanning. The transmitted light through the low-reflectivity disk is transformed by the photodiode into an analog electrical signal (RF signal). At the same time, the photosensor detects the trigger footprints, starting the data collection. For high-reflectivity CD and DVD, the laser incident light on disk comes back to the pickup. The data acquisition board (DAQ) digitizes the analog signals and transfers them to the computer for processing. The amplification/detection board (DAB) is integrated into the CD drive unit, controlling the planar photodiode and the photosensor.

During the disc reading, the laser that follows the spiral track hits the immunoreaction product. As a consequence of the interaction, a variation of the back-reflected light to the photodiode transducer is produced, the analog-acquired signals being different from those detected in the absence reaction product (background).

The optical density of the spots is correlated well with the analyte concentration. The photodiode transducer of the pickup head converts the reflected light into an analog electrical signal (bandwidth of 10 MHz) which is collected and amplified by a custom-built electronic board. The data acquisition board provides two 16-bit analog-to-digital converters, an input voltage range of ±10 mV and up to a 2 MHz sampling rate.

Signals coming from triggered areas are processed for digitization, stored in the computer (5 MB size file), and deconvoluted into an image for further quantification.

The CD/DVD drive is controlled by software running on a Windows-based computer and connected to the PC through a USB2.0 universal serial bus interface. The collected data of each detection zone are represented in an ordered, rectangle-shaped array, being stored in independent, uncompressed binary format files, and displayed them in a graphical image. The software allows for the exporting of the image in a gray-scale code to a compressed tif format or bitmap for further quantification. Calibration curves are mathematically analyzed.

For the reliable quantification of the spots, a 2 MHz data acquisition rate was proved to be adequate. According to the following equation, given a spot of 500 μm in diameter and a scan speed of 21 m/s (16×), an accurate identification of each spot is obtained when the number of the data samples (M) taken on a CD and DVD is 11,795 and 9502, respectively:

$$M = \frac{2}{\Delta x} \sum_{n=-R/\Delta y}^{R/\Delta y} \sqrt{R^2 - (n * \Delta y)^2} \quad \Delta x = \frac{v_L}{F_m}$$

where
- R is the radius of the spot
- Δy the track pitch of the media
- Δx is the distance between samples taken at a specified sampling rate (F_m) and a linear velocity of the disc (v_L)

A lower linear velocity would increase the angular resolution and identify smaller spots. However, the maximum spatial resolution is limited to 1.6 μm in CD and 0.74 μm in DVD by the media track pitch.

Figure 24.9 shows a disc player prototype and the graphical signals acquired scanning a DVD.

24.3.5.1 Experimental

Table 24.3 shows the characteristics of the different CD types used as analytical platforms.

24.4 Applications: Experimental Results

In this section, several practical applications based on heterogeneous immunoassay and hybridization complementary oligonucleotides in high-density microarray format are presented, paying attention on clinic and research cancer interest topics. Table 24.4 resumes the applications based on CD technology developed by our research group.

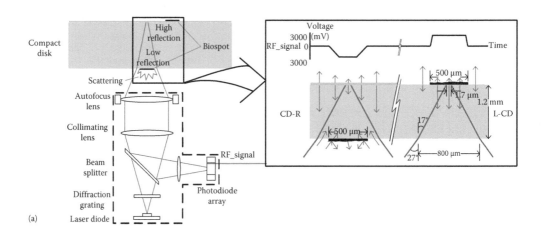

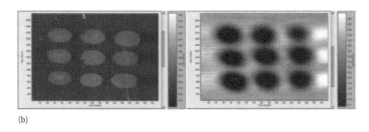

FIGURE 24.9 (a) Detail of the reading strategies on down side standard CD-R and top side L-CD. (b) Results displayed by the both reading modes: L-CD (left) and CD-R (right).

TABLE 24.3

Characteristics of Different Compact Disk Types Used as Analytical Platforms

Disk Type	Acronym	Laser Wavelength (nm)	Active Surface	Disk Reading Device	Reading Mode	Reading Speed	Software[a]
Standard compact disk	CD (R−)	780	PC[b]	Pickup	Reflection	24×; 36× till 52×	Writing simulation
Semitransparent compact disk	CD	780	PC, PS,[c] PMMA,[d] gold	Auxiliary photodetector	Transmission reflection	4×	Writing simulation
Standard digital versatile disk	DVD (R−)	650	PC	Pickup	Reflection	From 4× to 12×	Writing simulation
Blu-ray	Blu-ray R	405	Lactone composite	Pickup	Reflection	2× till 4×	Disk record using the Nero disk-speed 4 software. After, the disk is read to display an image
Light-scribe	Transparent CD or DVD	780 or 650	PC	Pickup or an auxiliary photodetector	Reflection or transmission	1× (~500 rpm)	Burning recording or rewritable simulation

[a] Biodisk (own software).
[b] Polycarbonate.
[c] Polystyrene.
[d] Polymethylmethacrylate.

TABLE 24.4

Resume of the Applications Based on Compact Disk Technology Developed by our Research Group*

Assay	Analyte/s	Sample Nature	Analytical Technique	Disk Type	LD	References
Single nucleotide polymorphisms (SNPs)	• Plum pox virus	Vegetal extract	DNA hybridization	CD	• 5 nM	[8]
Pesticides	• 2,4,5-TP	Natural waters	IEC	CD	• 0.02 µg/L	[16]
	• Chlorpyrifos				• 0.33 µg/L	
	• Metolachlor				• 0.54 µg/L	
Cancer biomarker	• Alfa-fetoprotein	Human serum	IES	DVD	• 8.0 µg/L	[26]
Multiplex pesticide and antibiotics	• Atrazine	Natural waters	IEC	DVD	• 0.06 µg/L	[33]
	• Chlorpyrifos				• 0.25 µg/L	
	• Metolachlor				• 0.37 µg/L	
	• Sulfathiazole				• 0.16 µg/L	
	• Tetracyclin				• 0.10 µg/L	
Flu	• Aviar virus	Saliva	IES	DVD	• 5.0 µg/L	UD
Marine toxins	• Microcystins	Natural waters	IEC	DVD	• 0.65 µg/L	UD
	• Domoic acid				• 1.50 µg/L	
Pathogens	• *Salmonella* spp.	Food extract	PCR products/ hybridization	DVD	• 30 fM	UD
Genetically modified organisms	• Soybean	Food extract	PCR products/ hybridization	DVD	• 100 nM	UD

LD, Limit of detection; IEC, competitive immunoassay (adsorption); IES, sandwich immunoassay (adsorption); UD, unpublished data.

24.4.1 Immunoassay

Three representative immunoassay applications are described as proof of concept, highlighting key points in each case. The first one shows the capabilities of using low-reflectance discs. The second one present a multiplexed assay on DVD and the third one treats on the determination of α-fetoprotein cancer biomarker.

24.4.1.1 *Competitive Microimmunoassays on Standard CD to Determine Chlorpyrifos, an Inhibitor of Acetyl Cholinesterase*

According to a previous characterization study [30] by ELISA, polyclonal antibodies (C2-II) and coating conjugates (ovalbumin [OVA]-C5) were used to set up a high sensitivity method to determine an organophosphorus compound inhibiting the acetyl cholinesterase, chlorpyrifos, and frequently determined in blood serum. Following, the development of a microimmunoanalysis to this compound performed on both sides (down and top) of L-CD is presented.

Using a manual CD arrayer from V & P Scientific, Inc. (San Diego, CA), four rows of five pins (20 in total) were matched, each one carrying 6 nL and leaving a 3 nL drop on the CD surface producing 500 µm diameter spots.

24.4.1.1.1 *Down Side Disc Approach*

Sensing on the polycarbonate face of low-reflectivity discs, different competition times (5, 10, 15, 30, 45, and 60 min) were tested to study the effect on absolute signal intensity and sensitivity. It was observed that long competition times reduced assay sensitivity (higher IC_{50} values) while the absolute signal intensity was similar for all incubation tested times. The dose–response curve obtained after

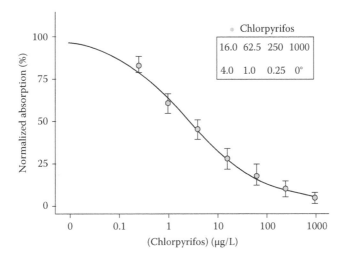

FIGURE 24.10 Chlorpyrifos dose–response curve.

10 min incubation is shown in Figure 24.10. As analyte concentration increased, the gray-scale intensity changed from black (100% normalized signal corresponding to 0 μg/L) to white (2.5% at 1000 μg/L) (frame inset in Figure 24.10 depicts the experimental intensity change). The resulting silver deposit from the highest analyte concentration was minimal (<5%), indicating a small degree of nonspecific response. The limit of detection (IC_{10}) was 0.39 μg/L and the sensitivity reached (IC_{50}) was 2.51 μg/L. Assays performed on four independent discs (eight calibration curves per disc) showed an average variation of 13.8%, indicating a suitable reproducibility between microarrays in different discs. The dose–response curves and the limits of detection were similar and independent of the processed disc. Also, the intra-discs variations (CV) average 8%. Analogous results are achieved by ELISA plate format using the same immunoreagents.

24.4.1.1.2 Top Side

To sense on this side of the disc, a 0.6 mm thick polycarbonate film was assembled onto the L-CD to carry out the assays. The microimmunoassay protocol was the same as described earlier. At this working side, the original incident diameter of the laser beam (around 728 μm entering the polycarbonate) is focused to 13 μm, the resolution being limited by the scan speed and the RF bandwidth. The resolution could be easily improved up to 1.7 μm by reading the disc at $1x$ but it would imply a decrease on the scan speed to 1.3 m/s taking 60 min to scan the whole disc.

Although the sensitivity of the assays was not improved by reading the disc at lower speed, a better precision of the readings was achieved, increasing the reproducibility of the signals. In addition, the system resolution was good enough to detect 500 μm diameter spots. Also, the sensitivity of the microimmunoassay arrays, in all cases, was equivalent to that obtained on the down side of the disc. The limit of detection (IC_{10}) was 0.33 μg/L and the sensitivity (IC_{50}) 1.81 μg/L. Also the coefficient of variations intra-discs are similar to that reached on down disc side, while the inter-disc variation was 13%.

In terms of reagents consumption, an assay performed on disc consumes <2% coating conjugate of that required on plate (100 μL/well) per sample. Also, working with L-CDs, <0.05% specific antibody and 2.64% labeled secondary antibody of that used per assay on plate are required, so there is an important immunoreagent saving. Moreover, the assays can be developed with a well established and inexpensive silver staining process. It is worth mentioning that an important time saving on critical steps, compared to the optimized 96-well plate ELISA format, is reached.

Related to reproducibility, similar results have been reported microarraying on microscopic glass slide support for small molecules. Thus, an average chip variance between 11% and 14% was shown [31]. Other authors also reported comparable reproducibility (9%–12%) and sensitivity [32] for pesticide residues quantification. In both cases, fluorescent detection was used.

24.4.1.2 Multiplexed Microimmunoassays on DVD

In this practical application, determination of five drug targets, using DVD technology for simultaneous and high-sensitivity multi-analysis of several compounds from different chemical families, is presented.

The calibration was accomplished in a practical and simple mode to achieve a powerful multi-analyte methodology. As proof of concept, very different analytes (pesticides—atrazine, chlorpyrifos, and metolachlor—and antibiotics—sulfathiazole and tetracycline) were chosen, the selection criteria being twofold. First is the need for highly sensitive, effective, and simple screening techniques to determine chemical residues in different scenarios. Second, the current standard protocol for their quantification is off-site laboratory analysis by SPE-LC tandem MS^2 and SPE-GC-MS for antibiotics and pesticides, respectively.

For microarraying, DVD-R discs were conditioned and the coating protein conjugates printed diluted in printing buffer, dispensing 50 nL/spot. The array layout consists of 10 columns of 10 dots each; five of them correspond to single-target systems; three are positive controls; one is the negative control; and the final column is the internal standard. A total of eight arrays (800 spots) were printed on the polycarbonate surface of the disc separated from each other by 45° (Figure 24.11). In this configuration, spots are 500 μm in diameter with a track pitch (center to center distance) of 1.5 mm, achieving an array density of 1.0 spot/mm². The working temperature (25°C) and relative humidity (90%) are controlled.

The microimmunoassays on DVDs are based on an indirect competitive format. After the coating conjugate solutions are arrayed in columns, internal standard OVA-BT at 0.2 mg/L and BSA (5 mg/L), KLH (1 mg/L), and 1/4000 dilution of nonimmunized rabbit sera (RIgG) are arrayed as positive controls for the first and second immunoreaction steps, respectively. OVA solution (5 mg/L) is included as negative control. After 16 h at 4°C, the disc is thoroughly washed with PBS-T, rinsed with deionized water, and dried by centrifugation.

For multiplexed assays, 0.9 mL deionized water or the same quantity of water sample was conditioned with 0.1 mL TSAB, and mixed then with 7.5 μL of the polyclonal antibody solution cocktail (1/1000 dilution for R12, C2-II, R48, pBT, and KOTC3-III and 1/2000 for S3-I). After, the resulting solution is dispensed onto the disc and covered with a 12 cm diameter 0.6 mm thick dummy plastic surface. After 5 min incubation at room temperature, the cover is removed, and the disc is washed with PBS-T buffer and rinsed with deionized water. Next, 1.0 mL gold-labeled secondary antibody solution (1/50 in PBS-T)

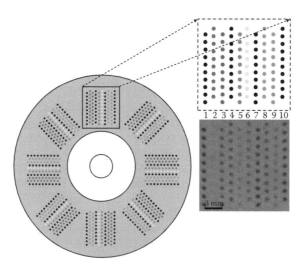

FIGURE 24.11 DVD layout scheme. The disk holds eight arrays of 10 columns of 10 spots each. From left to right, the columns correspond to (1) BSA, (2) OVA, (3) OVA–atrazine, (4) OVA–biotin, (5) OVA–chlorpyrifos, (6) OVA–metolachlor, (7) KLH, (8) OVA–sulfathiazole, (9) OVA–tetracycline, and (10) rabbit serum.

is dispensed onto the disc, and it is evenly distributed using a DPS. After 8 min incubation at room temperature, the disc is washed and dried as before. The immunoreaction is developed by homogenously dispensing 1.0 mL of silver enhancer solution onto the disc, and the reaction is stopped by washing the disc with water after 8 min.

For the CD scanning and data acquisition, two disc reading strategies are compared: one acquiring attenuated analog signals and the other being based on the analysis of reading errors. First approach shows high sensitivity and shorter reading disc time. However, using the analysis of reading errors allows using the standard disc player without adding any supplementary hardware of software [33].

To fully scan the DVD surface (5 min at 16× speed), the software simulates the writing process. During the disc scanning, only signals coming from selected areas are processed for digitization. Software identifies spots with SNR ≥ 2 to calculate the mean signal intensity by averaging data points from a circular area of 150 µm in diameter. For sample analysis, the measured signal intensity is related to that of the internal standard response. Inhibition curves are mathematically analyzed by fitting experimental results to a sigmoidal four-parameter logistic equation. Figure 24.12 shows detailed images displayed after disc reading.

For comparison purposes, simultaneous analysis of natural water samples for five analytes at trace levels are performed by an all-in-one reaction using the multiplexed DVD-based microimmunoassay. The analytical procedure needs only sample mix with conditioning buffer in order to adjust pH and ionic strength. The performance of the multiplexed DVD microimmunoassay method is evaluated through the determination of spiked water samples at three different levels. Results are shown in Table 24.5, being in good agreement with those obtained through reference method (SPE-GC/MS). Recovery range analyte-by-analyte is very good. Also, the obtained results of the nonvolatile targets

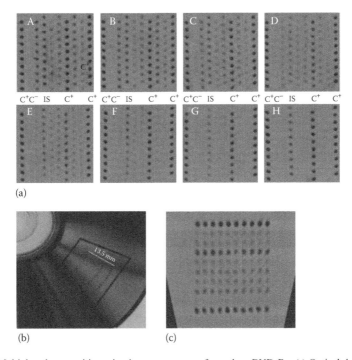

FIGURE 24.12 Multiplexed competitive microimmunoassay performed on DVD-Rs. (a) Optical density images of the developed arrays. Columns identification for all the displayed microarrays: columns 1, 7, and 10 are positive controls (C+), column 2 is the negative control (C–), and column 4 is the internal standard (IS). The other columns are the analytical results displayed by mixtures containing all the analytes (atrazine, column 3; chlorpyriphos, column 5; metolachlor, column 6; sulfathiazole, column 8 and tetracycline, column 9) at 0 (A), 0.06 (B), 0.25 (C), 1.0 (D), 4.0 (E), 16.0 (F), 64.0 (G), and 256 (H) µg of mixed analyte/L, concentrations. (b) Photograph of a DVD showing the microarray (framed) after the assay development with the systems printed in rows from the center of the disk toward the rim. (c) Optical density image of the framed area obtained after reading the DVD shown in (b). At first glance, the spots are lightly unfocused due to optical resolution of the DVD drive at the polycarbonate surface.

TABLE 24.5

Determination of Chemical Residues in Natural Water Samples by CD Immunoanalytical Multiplexed and SPE-GC-MS Techniques

| Analyte Added | Analyte Found (µg/L) | | | | | | | |
| | DVD | | | | | SPE-GC-MS | | |
	ATZ	CLP	MTL	STZ	TC	ATZ	CLP	MTL
Sample 1								
0	<LD	<LD	<LD	<LD	<LD	<LD	<LD	<LD
Level #1	0.25 ± 0.02	2.92 ± 0.31	1.35 ± 0.28	3.13 ± 0.62	8.63 ± 0.67	0.23 ± 0.13	4.21 ± 0.4	1.69 ± 0.19
MR	101 ± 12	79 ± 5	95 ± 7	85 ±16	86 ± 2	88 ± 9	104 ± 2	104 ± 8
Sample 2								
0	<LD	<LD	<LD	<LD	<LD	<LD	<LD	<LD
Level #1	0.25 ± 0.04	4.08 ± 0.47	1.49 ± 0.22	2.41 ± 0.22	14.58 ± 1.16	0.28 ± 0.15	4.50 ± 0.23	1.68 ± 0.17
MR	102 ± 14	96 ± 6	100 ± 21	92 ± 12	116 ± 33	110 ± 10	113 ± 1	112 ± 6
Sample 3								
0	<LD	<LD	<LD	<LD	<LD	<LD	<LD	<LD
Level #1	0.26 ± 0.02	4.08 ± 0.58	1.71 ± 0.12	2.66 ± 0.46	11.78 ± 1.26	0.29 ± 0.12	3.44 ± 0.11	1.63 ± 0.13
MR	100 ± 12	113 ± 12	105 ± 9	100 ± 15	112 ± 21	111 ± 4	92 ± 7	109 ± 3
Sample 4								
0	<LD	<LD	<LD	<LD	<LD	<LD	<LD	<LD
Level #1	0.28 ± 0.03	4.63 ± 0.58	1.93 ± 0.32	2.77 ± 0.37	11.47 ± 2.00	0.23 ± 0.11	4.24 ± 0.15	1.33 ± 0.12
MR	102 ± 11	116 ± 17	117 ± 19	108 ± 16	111 ± 3	100 ± 7	106 ± 1	91 ± 16

ATZ, Atrazine; CLP, chlorpyrifos; MTL, metolachlor; STZ, sulfathiazole; TC, tetracycline; MR, mean recovery; Level #1: Atrazine: 0.25 µg/L. Chlorpyrifos: 4.0 µg/L. Metolachlor: 1.5 µg/L. Sulfathiazole: 3.0 µg/L. Tetracycline: 10.0 µg/L.

(antibiotics STZ and TC) spiked water samples, by CD and LC-MS^2, are concordant. It is important to note that in the case of the multiplexed DVD assays, water analysis is performed directly, diluting only the samples 10-fold. These results confirmed the applicability of microimmunoassays on DVDs for direct multiplexed screening of drug residues at very low level and without sample preconcentration (Figure 24.13).

24.4.1.3 Quantitative Immune Determination of the α-Fetoprotein Cancer Marker

As other view of the CD applications, a simple and effective immunoanalytical sandwich method to quantify the protein tumor marker α-fetoprotein is presented. Briefly, the method was developed using the previously explained L-CD working technique, but in this case the disc was spin-coated with a polystyrene (PS) film [34].

Spin-coating of a polystyrene film onto the gold layer improves the antibody immobilization efficiency, reaching the best results, in terms of film transparency, smoothness, and assay sensitivity, by spin-coating a 3% (w/v) polystyrene solution (Mw: 250,000 Da) in Dowanol, resulting in a uniform film averaging 60 nm thick. Thus, immobilization of the capture antibody was carried out by passive adsorption on to PS-CDs, as it was supplied by the manufacturer. The best results were obtained after incubation times of 30 and 60 min for α-fetoprotein (AFP) and detector antibody (1:1 dilution), respectively.

The determination of AFP was based on a sandwich immunoassay format, using commercial immunoreagents from Abbot Laboratories (Abbott Park, IL). The capture antibody was arrayed on to PS-CD and incubated for 16 h at 4°C. The discs were then thoroughly washed with PBS-T and deionized water.

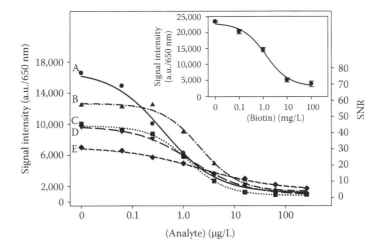

FIGURE 24.13 Standard calibration curves of competitive multiplexed microimmunoassays on DVDs for (A) atrazine, (B) metolachlor, (C) chlorpyrifos, (D) sulfathiazole, and (E) tetracycline. Error bars are eliminated to avoid overlaps with symbols. The frame inset depicts the calibration graph for biotin used as internal standard [33].

Then, 50 μL antigen at AFP concentrations ranging from 15 to 350 μg/L in sodium phosphate-buffered saline, containing 5% fetal calf serum, were dispensed on to the disc and incubated for 30 min at room temperature. Afterward, the discs were washed with Tris-T buffer, pH 8.0 for 1 min and rinsed with deionized water. Next, 750 μL of the secondary antibody labeled with alkaline phosphatase in Tris buffer solution was dispensed on to the disc and covered with a dummy plastic plate. After 1 h at room temperature, 750 μL of BCIP/NBT developing solution was dispensed over the disc and distributed as described. The immunoreaction was stopped with water after 20 min, and purple deposits occurred as spots, the discs thus being ready for reading.

Figure 24.14 depicts the calibration curve plot of gray-scale intensity obtained as a function of the AFP concentration. The framed inset shows a schematic diagram of the sandwich immunoassay. Representative gray-scale images as a function of the AFP concentration in the samples are shown in the panels.

Performances of the assay are really good because dynamic range covers two orders of magnitude with a detection limit of 8 μg/L and takes 135 min, making this methodology suitable for massively detection of AFP below normal human fluid basal levels.

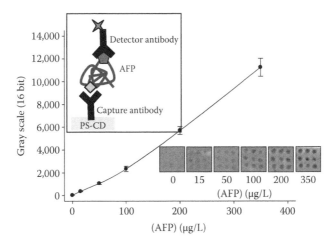

FIGURE 24.14 Calibration plot for the determination of AFP on a compact disk (PS-CD) [34].

24.4.2 Single Nucleotide Polymorphisms Detection

The covalent attachment of oligonucleotide probes, on the contrary to adsorption, offers the advantage of a robust resulting surface while retaining its chemical integrity over an extended period of time. Besides, covalent interactions allow the performance of highly homogeneous and reproducible surfaces as self-assembled monolayers.

The chemical derivatization of PC, the main CD constituent, to be applied on discs for microarray-based assays can be achieved by means of different chemical or physical processes. The problem is that most of the employed methods for the chemical surface modification affect their optical (transparency) and mechanical (resistance, hardness, roughness, etc.) properties, not being applicable to polymers without perturbing the original performance.

From the several approaches of covalent modification of PC discs carried out in our laboratory, nitration–amination of the PC disc surface has offered the best results. Based on this advantage, we developed an assay for single nucleotide polymorphisms (SNPs) discrimination on polycarbonate disc surface. A CD player was also tested as detection system to demonstrate the feasibility of the developed approach.

Aminated oligos were immobilized on aminated disc using glutaraldehyde as cross-linker reagent. For this, NH_2-functionalized PC disc surface was immersed in aqueous glutaraldehyde (5%, 1× PBS, pH 7) and stirred for 2 h at room temperature. After washing, 3′ NH_2-ended oligonucleotide (SYM22) was dissolved in carbonate buffers (1×, 10×) pH 9.6, at 0.1, 0.5, 1, and 5 µM concentrations. Due to the big sample spotting CD surface, eight different zones of 3×3 microarrays of each solution were printed onto the PC disc surface and incubated in a wet, dark chamber for 1 h at 42°C or 4 h at room temperature. Finally, discs were thoroughly rinsed with PBS-T and water and air dried.

To demonstrate the usefulness of the method, hybridization was performed employing serial dilutions of the targeted oligonucleotide SYM 25 (0.2–0.001 µM). Afterward, the CD was treated with gold-conjugated streptavidin for 20 min, washed, and developed with silver enhancer reagent for 12 min. A black precipitate was detected by the CD player.

Working with different concentrations of probe (from 0.1 to 20 µM), times, and temperatures, black spots were obtained until 2.5 nM of target for all the probe concentrations (from 0.5 to 100 nM). Also, hybridization positive spots were obtained for all the target concentrations, being visible even to the naked eye [25]. The performed controls did not develop black spots. The optimal values were found for 5 µM of probe concentration, being possible to detect hybridization with concentrations as low as 0.1 µM probe. Results are shown in Figure 24.15.

Covalently probe immobilization procedure results slightly better than that obtained on PC CD surfaces by adsorption immobilization of probes and target hybridization reading with a fluorescence

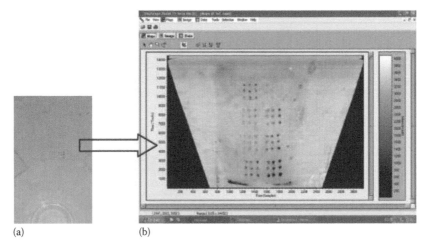

(a) (b)

FIGURE 24.15 (a) Microarray spots developed on the PC disk after the chemical treatment and the hybridization assays, and (b) computer screen image displayed by the CD reader.

TABLE 24.6

Nucleotide Sequence of Probes and Targets Used

Name	Sequence (5′ to 3′)	5′ End	3′ End
SYM8	AATGCTAGCTAATCAATCGGG	Cy5	None
SYM22	$(T)_{15}$CCCGATTGATTAGCTAGCATT	None	C_7-NH_2
SYM23	TTACGATCGAATTAGTTAGCCC$(T)_{15}$	Cy5	C_7-NH
SYM25	AATGCTAGCTAATCAATCGGG	Biotin	None
SYM26	AATGCTAGATAATCAATCGGG	Biotin	None

microarray scanner [7]. In this case, 25 nM of target was detected by employing probe concentrations between 0.5 and 1.0 µM. This demonstrated the feasibility of this approach, which provides results as good as or even better than those from the standard techniques, using fluorescent tags.

The application of the methodology to discriminate SNP was attempted. Two oligonucleotide strands differing both in one mismatch: SYM25, the complementary strand of the immobilized probe on the PC disc (SYM22), and SYM26 differs from SYM25 in the base number nine (see Table 24.6).

At the best assay conditions, SNPs resulted at 40°C, for 1 h and 25% of formamide in the hybridization solution, being possible to discriminate between SYM25 and SYM26 at concentrations of 2.5 nM using probe concentrations ranging from 0.1 to 5 µM. These results show the usefulness of the aimed goals, which proves the possibilities of discs as suitable analytical platforms and CD players as effective and competitive detection systems, also in the genomic field.

24.5 Materials

24.5.1 Chemicals

24.5.1.1 Printing Buffer

Printing buffer is 0.1 M sodium carbonate buffer, pH 9.6 containing 1% glycerol (v/v)

24.5.1.2 Analysis Buffer

Analysis buffer is 10 mM sodium phosphate buffer 0.15 M NaCl, 0.005% Tween 20, pH 7.4. Disc washing solution was Milli-Q water, conductimetry of 18.22 Ω/cm. All the liquid reagents must be filtered through a 0.22 µm pore size disc before use.

24.5.1.3 Immunoreagents

Coating conjugates are synthesized and purified in house or purchased. Polyclonal antibiotin antibody (PAb-BT) is from Abcam (Cambridge, United Kingdom). Biotin, biotin *N*-succinimidyl ester (used to prepare OVA–biotin conjugate; OVA-BT), gold-labeled goat anti-rabbit and anti-mouse immunoglobulins (GAR-Au and GAM-Au, respectively), and silver enhancer solutions (A and B) are from Sigma-Aldrich (Madrid, Spain).

24.5.2 Microarray Printing

Discs are first conditioned by gentle ethanol washing, water rinsing, and dried by centrifugation. The coating reagents, diluted in printing buffer, is dispensed in a 384-well plate (25 µL/well) and transferred to the disc (25 nL) with a microprinter. The disc layout consisted in several zones (30° to 45° separation) with microarrays (at a distance ranging about 9 mm, corresponding to the space between nozzles in standard multichannel pipettes) of four subarrays each, arranged in blocks of 2 × 2 spots, printed on the polycarbonate surface under controlled temperature and relative humidity (25°C and 80%, respectively).

Within each microarray, the blocks correspond to internal standard, targeted compounds, and positive controls, respectively, are printed. Resulting spots are between 300 and 500 μm diameter with a track pitch (center to center distance) of 0.9 mm.

24.5.3 Microimmunoassay Protocol

The working principle of the sensor arrays on disc is based on an indirect competitive microimmunoassay format. Competition occurs between free analyte (sample) and immobilized conjugate for specific antibody. Also, BSA- and OVA-BT at few mg/L are used as internal calibration, and nonimmunized mouse and rabbit sera at 1/1000 and 1/4000 dilution, respectively, as positive controls are arrayed on the polycarbonate surface of the discs. After 16 h at 4°C, discs are thoroughly washed with PBS-T, rinsed with deionized water, dried by handshaking, and stored at 4°C until use.

Related to the specific antibodies volumes, depending on the title of the sera, 10 μL of antibodies of the different IgGs cocktail solution (in average 1/5000 dilution) in PBS-T with and without analyte is dispensed onto the array. After 8 min incubation at room temperature, the disc is washed with deionized water. Next, 500 μL of gold-labeled secondary cocktail antibody solution in PBS-T are dispensed onto the disc and evenly distributed. After 8–15 min, disc is washed and dried. The immunoreaction is developed by dispensing 1.0 mL of silver enhancer solution onto the disc and distributed as before. Finally, the reaction is stopped after 8 min by washing the disc with water. Total assay takes between 30 and 150 min, depending on the assay format and characteristics of the immunoreagents.

24.6 Future Prospects

The advantage of the CD technology as analytical tool is threefold. First, the surface of a single standard disc (94 cm^2) can hold thousands of dots, so thousands of replicates can be made simultaneously in a sample or a huge number of analytes can be quantified in the same sample, including the calibration standards and controls. This makes the technology truly scalable, thinking not only in the working capability and performances, but simplicity, robustness, and price must also be considered to establish the authentic scalability of this technology.

Second, commercial discs have demonstrated good properties for passive adsorption and covalent immobilization of coating probes (proteins and nucleic acids) in high-density scales to develop method of application in general analysis, but with important advantages in the biomolecular areas.

Third, microarraying developments on optical discs are attractive because of their primarily cost-efficiency, minimal reagent consumption, and the ability to both conduct assays and record and/or read data from the same disc support, using standard drives. The interest in using this hardware (disc and readers) arises from their low cost, ubiquity, portability, reliability, and easy operation.

Future has several challenges, some easy to accomplish and others needing much time and effort. First, development approaches based on new disc and driver such as Blu-ray would allow reaching best performances. Thus, this technology works at lower wavelength (405 nm), achieving better sensitivities. Also, reduction of the dot size by using nanoimprinting devices must allow the reduction of the detection limits by increasing the printing density and the number of studied targets. This is very interesting for ohmic applications.

Other possibilities to improve and enlarge the CD analytical applications are the development of label-free assays. Working in massive analysis of proteins, this method is really interesting because it do not need target labeling and report postdevelopers.

For that, two developments are very promising if significant changes of the detection system (disc driver) are not required. One is the accomplishment of surface plasmon resonance on disc. The feasibility of this principle has been demonstrated on disc, but many research and development effort must be carried out to set up an SPR equipment and applications, based on CD technology.

The second future development is the incorporation of interferometry as detecting principle in CD supports and players. This approach is more reliable than the SPR and several research groups are working on it. Advantages of interferometric detection is the compatibility with label-free format assays, feasibility of making interferometric structures on CDs, and the adaptation of disc drivers to read interferometric signals.

From the application point of view, future trends must be focused on the development of high-density microarraying protocols addressed to a pool of targets potentially present in the same sample at different levels. To reach this goal, CD analytical methodologies are very competitive, because currently determining several substances such as proteins, drugs, and inorganic anions needs to split each sample into several, spending much time and requires expensive laboratory equipment and specialized staff. Also, developing screening applications used out of the laboratory, for example, for monitoring patients at home or water sources in field, is really necessary and affordable. For this, analytical CD technology is well established, being an exceptional tool to accomplish these goals with affordable costs.

ACKNOWLEDGMENTS

Part of the data of this chapter were from research funded through FEDER project CTQ2007-64735; CTQ2010-15943 (MICINN, Spain) and GVA Prometeo grant 2010/008. The author thanks S. Morais and M.A. Gonzalez for helping in the preparation of the data and revising the manuscript.

REFERENCES

1. Filippini, D., A. Alimelli, C. DiNatale, R. Paolesse, A. D'Amico, I. Lundström. 2006. Chemical sensing with familiar devices. *Angew. Chem. Int. Ed.* 45:3800–3803.
2. Philips celebrates 25th anniversary of the compact disk. 2007. http://news.ecoustics.com/bbs/messages/10381/379607.html
3. Dittrich P.S., A. Manz. 2006. Lab-on-a-chip: Microfluidics in drug discovery. *Rev. Drug Discov.* 5:210–218.
4. Honda, N., U. Lindberg, P. Andersson, S. Hoffmann, H. Takei. 2005. Simultaneous multiple immunoassays in a compact disk–shaped microfluidic device based on centrifugal force. *Clin. Chem.* 51(10):1955–1961.
5. Madou, M., J. Zoval, G. Jia, H. Kido, J. Kim, N. Kim. 2006. Lab on a CD. *Annu. Rev. Biomed. Eng.* 8:601–628.
6. Mira, D., R. Llorente, S. Morais, R. Puchades, A. Maquieira, J. Marti. 2004. High-throughput screening of surface-enhanced fluorescence on industrial standard digital recording media. *Proc. SPIE 2004* 5617, 364–373.
7. Morais, S., R. Marco-Molés, R. Puchades, A. Maquieira. 2006. DNA microarraying on compact disk surfaces. Application to the analysis of single nucleotide polymorphisms in Plum pox virus. *Chem. Commun.* 22:2368–2370.
8. Bañuls, M.J., V. González-Pedro, R. Puchades, A. Maquieira. 2007. PMMA isocyanate-modified digital disc as a support for oligonucleotide-based assays. *Bioconjug. Chem.* 18(5):1408–1414.
9. Li, Y.C., Z. Wang, L.M.L. Ou, H.Z. Yu. 2007. DNA detection on plastic: Surface activation protocol to convert polycarbonate substrates to biochip platforms. *Anal. Chem.* 79:426–433.
10. Ligler, F.S., J.S. Erickson. 2006. Bioengineering: Diagnosis on disc. *Nature* 440:159–160.
11. Ekins, R.P. 1989. Multi-analyte immunoassay. *J. Pharm. Biomed. Anal.* 7:155–168.
12. Kido, H., A. Maquieira, B.D. Hammock. 2000. Disc-based immunoassay microarrays. *Anal. Chim. Acta* 411:1–11.
13. Gordon, J.F. 1999. Apparatus and method for carrying out analysis of samples. U.S. Patent 5,892,577.
14. Alexandre, I., Y. Houbion, J. Collet, S. Hamels, J.L. Demarteau, J. Gala, J. Remacle. 2002. Compact disc with both numeric and genomic information as DNA microarray platform. *Biotechniques* 33:435–439.
15. Morais, S., J. Carrascosa, D. Mira, R. Puchades, A. Maquieira. 2007. Microimmunoanalysis on standard compact discs to determine low abundant compounds. *Anal. Chem.* 79:7628–7635.
16. Najmabadi, P., J.J. La Clair, M.D. Burkart. 2007. Affinity analyses on moldable optical polycarbonate. *Org. Biomol. Chem.* 5:214–222.

17. Quercioli, F., B. Tiribilli, C. Ascoli, P. Baschieri, C. Frediani. 1999. Monitoring of an atomic force microscope cantilever with a compact disk pickup. *Rev. Sci. Instrum.* 70:3620–3624.
18. Chu, C.L., C.H. Lin. 2005. Development of an optical accelerometer with a DVD pick-up head. *Meas. Sci. Technol.* 16:2498–2502.
19. Lange, S.A., G. Roth, S. Wittemann, T. Lacoste, A. Vetter, S. Grässle, S. Kopta, M. Kolleck, B. Breitinger, M. Wick, J.K.H. Hörber, S. Dübel, A. Bernard. 2006. Measuring biomolecular binding events with a compact disc player. *Angew. Chem. Int. Ed.* 45:270–273.
20. Perraut, F., A. Lagrange, A. Pouteau, O. Pdeyssonneaux, P. Puget, G. McGall, L. Menou, R. Gonzalez, P. Lebeye, F. Ginot. 2002. A new generation of scanners for DNA chips. *Biosens. Bioelectron.* 17:803–813.
21. Potyrailo, R., W.G. Morris, A.M. Leach, T.M. Sivavec, M.B. Wisnudel, S. Boyette. 2006. Analog signal acquisition from computer optical disk drives for quantitative chemical sensing. *Anal. Chem.* 78:5893–5899.
22. Morais, S., J. Tamarit-López, R. Puchades, A. Maquieira. 2010. Determination of microcystins in river waters using microsensor arrays on disk. *Environ. Sci. Technol.* 44:9024–9029.
23. Subash, C.B., G.K. Awazu, J. Tominaga, K.R. Kumar. 2008. Monitoring biomolecular interactions on a digital versatile disk: A bioDVD Platform Technology. *ACS Nano* 2(9):1885–1895.
24. Subash, C.B., G.K. Awazu, P. Fons, J. Tominaga. 2009. A sensitive multilayered structure suitable for biosensing on the bioDVD platform. *Anal. Chem.* 81:4963–4970.
25. Bañuls, M.J., F. García-Piñón, R. Puchades, A. Maquieira. 2008. Chemical derivatization of compact disc polycarbonate surfaces for SNPs detection. *Bioconjug. Chem.* 19:665–672.
26. Tamarit-López, J., S. Morais, M.-J. Bañuls, R. Puchades, A. Maquieira. 2010. Development of hapten linked microimmunoassays on polycarbonate disks format. *Anal. Chem.* 82:1954–1963.
27. Allain, L.R., D.N. Stratis-Cullum, T. Vo-Dinh. 2004. Investigation of microfabrication of biological sample arrays using piezoelectric and bubble-jet printing technologies. *Anal. Chim. Acta* 518:77–85.
28. Potyrailo, R., W.G. Morris, A.M. Leach, L. Hassib, K. Krishnan, C. Surman, R. Wroczynski, S. Boyette, C. Xiao, P. Shrikhande, A. Agree, T. Cecconie. 2007. Theory and practice of ubiquitous quantitative chemical analysis using conventional computer optical disk drives. *Appl. Opt.* 46(28):7007–7017.
29. Tamarit-López, J., S. Morais, R. Puchades, A. Maquieira. 2008. Use of polystyrene spin-coated compact discs for microimmunoassaying. *Anal. Chim. Acta* 609:120–130.
30. Brun, E.M., M. Garces-Garcia, R. Puchades, A. Maquieira. 2005. Highly sensitive enzyme-linked immunosorbent assay for chlorpyrifos. *J. Agric. Food Chem.* 53:9352–9360.
31. Du, H., Y. Lu, W. Yang, M. Wu, J. Wang, S. Zhao, M. Pan, J. Cheng. 2004. Preparation of steroid antibodies and parallel detection of multianabolic steroid abuse with conjugated hapten microarray. *Anal. Chem.* 76:6166–6171.
32. Belleville, E., M. Dufva, J. Aamand, L. Bruun, L. Clausen, C.B.V.J. Christensen. 2004. Quantitative microarray pesticide analysis. *Immunol. Methods* 286:219–229.
33. Morais, S., L. Tortajada-Genaro, T. Arnandis-Chover, R. Puchades, A. Maquieira. 2009. Multiplexed micro-immunoassays on digital versatile disk (DVD). *Anal. Chem.* 81(14):5646–5654.
34. Morais, S., J. Tamarit-López, J. Carrascosa, R. Puchades, A. Maquieira. 2008. Analytical prospective of compact disc technology to immunosensing. *Anal. Bioanal. Chem.* 391:2837–2844.

25

Colorimetric Multiplexed Immunoassay for Sequential Detection of Tumor Markers

Jing Wang and Genxi Li

CONTENTS

25.1 Introduction

Cancer is a leading cause of death worldwide, and the number of global cancer deaths is projected to increase 45% from 2007 to 2030 (from 7.9 million to 11.5 million deaths). In some less developed continents, such as South America and Asia, already more than half of all cancer cases occur.[1] According to the World Health Organization (WHO), early detection of cancer greatly increases the chances for successful treatment.[2] So the early screening and accurate diagnosis have been considered as one of the most important treatments against cancers.

There are some factors that should be taken into account for a successful sensor of cancer detection: sensitivity, specificity, and acceptability. Sensitivity means that the cancer screening technique has the ability to identify people who have the disease; specificity means that it can identify people who do not have the disease, and this approach also need to be accepted and affordable by patients. The common cancer screening techniques include biopsy and endoscopy, which show the presence, location, and size of an abnormal mass, such as x-ray, ultrasonography, computed tomography (CT) scan, magnetic resonance imaging (MRI), and positron emission tomography (PET).[3–5] These methods are particularly powerful in the detection of solid tumors, but some of them have limitation in confirming the cause and stage of cancer development especially when the cancer is still at the very early stage. Another embarrassing situation for these "high-technology" facilities is that some developing countries lack the infrastructure and resources to use them appropriately or to achieve adequate coverage of the population.[6] Therefore, we need more efficient and affordable screening strategies that are specific and reliable for the diagnosis of cancers, and are possible to have sufficient numbers of personnel to perform.

As is known, both cancer cells and some other noncancer cells produce tumor markers (TMs) that are also called tumor-related antigens, as a response to the presence of tumor or certain benign conditions.[7] Not only TMs express in different types of cancers, single kind of cancer also expresses multiple

TABLE 25.1

List of Tumor Markers, Their Normal Levels in Human, and Potential Related Diseases When They Overexpress

Tumor Markers (TMs)	Normal Level	Related Diseases
AFP (α_1-fetoprotein)	<20 ng/mL	Ovarian/testicular embryonal carcinoma; primary hepatoma; nonseminomatous germ cell tumors
β-hCG (β subunit of human chorionic gonadotropin)	<20 unit/mL	Gestational trophoblastic neoplasia; hydatidiform mole; testicular embryonal carcinoma; choriocarcinoma
β_2-Microglobulin	1.1–2.4 µg/mL	Multiple myeloma; some lymphomas
CA 15-3	<20 unit/mL	Metastatic breast cancer
CA 19-9	<37 unit/mL	Colorectal/pancreatic cancers
CA 125	<35 unit/mL	Ovarian cancer
CA 27-29	<38 unit/mL	Metastatic breast cancer
CEA (carcinoembryonic antigen)	<2.5 ng/mL	Colon/liver/breast/pancreas/bladder/ovary/cervix cancers; cirrhosis; ulcerative colitis
Chromogranin A	<125 ng/mL	Carcinoid; other neuroendocrine tumors
Estrogen receptor/progesterone receptor	<6 fmol/mg	Breast cancer
HER-2/neu	<15 ng/mL	Breast cancer
PSA (prostate-specific antigen)	<4 ng/mL	Prostate cancer
Thyroglobulin	<2 ng/mL	Thyroid disorders
TA-90 (melanoma-associated antigen)	Still being studied	Breast/colon/lung cancers; melanomas; soft-tissue sarcomas

species of TMs simultaneously (Table 25.1).[8–17] And with the different stages of cancer development, the expression level of TMs may alter. Due to the significant role of TMs in cancer occurrence, growth, and metastasis, various kinds of TM-based detection techniques have been developed for screening, diagnosis, and evaluation of cancer development and therapy effects as well.[18] Compared with imaging facilities that are more promising in the screening of solid tumor, TM-targeted immunoassays that are based on the high affinity between TMs and their specific antibodies have superiority in the diagnosis of early stage cancers.

An abnormal TM level may suggest cancer; however, this alone is usually not enough to diagnose cancer with high accuracy. Therefore, multiplexed TM immunoassays that are able to detect two or even more kinds of TMs simultaneously/sequentially have been received more and more attention.[19–22] Compared with single-analyte assays, they possess the advantages of shortened analysis time, improved detection efficiency, decreased sample volume, and reduced costs.[23] But these assays have high requirement for multiple signal output, which means each target TM needs an identified output signal without overlaying with other targets, and these signal outputs would be available in easily understandable form. During recent years, colorimetry, fluorescence, and chemiluminescence have been wildly used in the multiplexed immunoassays. These assays use various kinds of visible chemical products and organic dyes to specifically discriminate different TM targets, so that they achieve not only fast and qualitative observation using naked eyes, but also quantitative and accurate measurements.[24–26] Nevertheless, some sensors based on fluorescence and chemiluminescence still suffer from overlapping spectral features, fast photobleaching, the need for multiple excitation sources, and also instrumentation for signal readout. In comparison, colorimetric methods show distinct and last-for-long color observed by naked eyes, so there would be no dependence on expensive apparatus. Also because of this advantage, they have great value in the promotion of disposable and affordable test kits and indicator papers for cancer diagnosis and monitoring.

The conventional colorimetric assay for TM detection is enzyme-linked immunosorbent assay (ELISA),[27] a biochemical technique used mainly in immunology to detect the presence of antigens in the sample (Figure 25.1). The commonly used ELISA is called "sandwich" ELISA, which involves capture antibody and detection antibody with high specificity with the TM antigen, as well as one

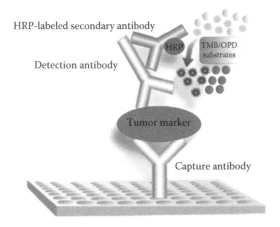

FIGURE 25.1 Working principle of colorimetric enzyme-linked immunosorbent assay (ELISA).

secondary antibody against the detection antibody. In ELISA, the capture antibody is immobilized on a solid support (usually a polystyrene microtiter plate), and then the unpurified sample with unknown amount of TM flows through the surface. Once the antigen is captured, the detection antibody is added to form a "sandwich" complex. Since the secondary antibody is labeled with enzymes (horseradish peroxidase [HRP]/alkaline phosphatase) or chromogenic reporters, adding enzymatic substrates would produce an observable color change that indicates the presence of TM. By using ELISA microplate reader, the concentration of TM can also be quantified. For multi-analyte detection, sequential ELISA can be used by repeating the cycles earlier.[28] During first cycle, in brief, limited volume of sample solution such as plasma containing multiple TM target proteins would be flushed through the ELISA kit against one TM target. After the initial assay, the plasma sample is collected and reused to measure the second TM. After several consecutive tests, multiple species of TMs in the sample can be detected. According to the performance of some commercially available ELISA kits, the detection limit of colorimetric ELISA is variable from 80 pg/mL to 100 ng/mL based on different substrates.[29] In order to improve the sensitivity of ELISA to detect TMs in early stage of cancer, chemiluminescent and chemifluorescent ELISA kits have also been developed by changing the enzyme substrates. Considering the price and ease-of-use, however, colorimetric ELISA is still the most frequently used immunoassay for TM detection.

In recent years, with the fast development of nanotechnology, a large number of new nanomaterials have been synthesized and put into biological and medical applications.[30] These materials have excellent biocompatibility and unique physical properties which make them visible or have distinct light emission at different excitation wavelengths.[31] They have been widely used in many colorimetric biosensing platforms as signal transducers and/or signal amplifiers, which create more opportunities for efficient colorimetric TM detection systems. In this chapter, we will introduce an ultrasensitive colorimetric immunoassay for multiplexed TM detection based on nanotechnology.[32] The liver cancer–associated TMs, carcinoembryonic antigen (CEA), and α-fetoprotein (AFP) have been chosen as the model analytes. This multi-analyte immunoassay utilizes two species of nanomaterials: streptavidin-conjugated paramagnetic particles (SMPPs) and gold nanoparticles (AuNPs). While magnetic nanoparticles conjugated with antibodies are used for antigen collection and separation, AuNPs loaded with HRP-labeled antibodies are for fast recognition of TMs using naked eyes and sensitive detection using UV–visible spectroscopy. Under optimal conditions, CEA and AFP antigens can be detected clearly by naked eyes. And the detection limits from spectrometric measurements are as low as 20 pg/mL. This proposed method shows significant clinical values for potential applications in TM-based cancer screening and convenient point-of-care testing. We will discuss in detail the experimental procedure, the performance of this nanomaterial-based multi-analyte immunoassay, and as well as the comparison with chemiluminescent and fluorescent assays.

25.2 Materials

Mouse monoclonal biotinylated CEA and AFP antibodies (biotin-*anti*-CEA/AFP), HRP-labeled goat polyclonal CEA and AFP antibodies (HRP-*anti*-CEA/AFP), human CEA and AFP antigens were all available from Sigma-Aldrich and Thermo Scientific Pierce-Antibodies. The activities of HRP conjugated with the antibodies were ~80% of their original activities, about 144 units/mg. Streptavidin MagneSphere® paramagnetic particles (ca. 1.0 μm in diameter, 1 mg/mL) were purchased from Promega Co. (Catalog No. MD1471).* 3,3',5,5'-Tetramethylbenzidine (TMB), *o*-phenylenediamine (OPD), and bovine serum albumin (BSA) were obtained from Sigma-Aldrich. Hydrogen tetrachloroaurate (III) trihydrate ($HAuCl_4 \cdot 3H_2O$, 99.9%) and trisodium citrate were obtained from Sigma-Aldrich (Catalog No. 27988-77-8, S4641). Serum samples were prepared by dissolving CEA and AFP antigens in standard human serum, instead of buffer. All other reagents were of analytical grade and used without further purification.

The buffer that was used for the antigen–antibody binding procedure was 10 mM phosphate-buffered saline (PBS), pH 7.4. For the HRP-based catalytic reactions, the substrate solution of TMB was 10 mM PBS, containing 0.2 mM TMB and 0.1 mM H_2O_2, pH 6.2; and the substrate solution of OPD was 20 mM phosphate–citrate buffer, containing 7.4 mM OPD and 5.9 mM H_2O_2, pH 5.0. Both of the substrate solutions were freshly prepared before use. All the solutions were prepared by double-distilled water, which was purified with a Milli-Q purification system (Millipore Co., Billerica, MA) to a specific resistance of 18 MΩ cm.

25.3 Methods

25.3.1 Apparatus

UV–visible spectroscopy was performed using a UV-2450 UV–visible spectrophotometer (Shimadzu Co., Kyoto, Japan). Transmission electron micrographs (TEM) were taken using a Jeol JEM-2000EX (Jeol Ltd., Tokyo, Japan). All the incubation processes during the preparation of antibody-loaded nanoparticles and the immunoreactions were performed in ThermoShaker (Boxun Industry Co., Shanghai, China). Magnetic separation was carried out on MagneSphere Technology Magnetic Separation Stand (Promega Co., Madison, WI).

25.3.2 Preparation of AuNPs Loaded with HRP-Labeled Antibodies

Here AuNPs were synthesized by reducing $HAuCl_4$ with citrate·3Na.† In brief, a 100 mL aqueous solution of 0.01% (w/v) $HAuCl_4$ was added into a round-bottom flask and stirred to boil. Then 3.5 mL 1% trisodium citrate was added rapidly into the boiling solution, the color of which became wine red from colorless after boiling for another 15 min with vigorous stirring. The size of the nanoparticles was 12.5 ± 2.3 nm, determined by TEM (Figure 25.2A). The concentration of AuNPs was 2.3 nM, which was calculated from the quantity of starting material ($HAuCl_4$) and the size of AuNPs at the wavelength of 520 nm.

AuNPs loaded with HRP-labeled antibodies were prepared as follows. Twenty microliters of HRP-*anti*-CEA or HRP-*anti*-AFP solution (1 mg/mL) was added into 5 mL of the previously prepared AuNPs solution, and the pH was adjusted to 8.5 by adding 1 M NaOH solution. Then, the mixed solution was incubated for 30 min at 37°C with gentle shaking and concentrated to 1 mL by centrifuging (13,000 rpm, 30 min, 4°C). After that, 100 μL of 10% (w/v) BSA solution used as nonspecific blocker was added

---

* Other magnetic microparticles functionalized with groups or proteins are also commercial. For example, Dynabeads® magnetic microparticles modified with many kinds of groups and proteins (size 2.8 μm) can be purchased from Invitrogen Corporation. Amino-functionalized superparamagnetic microparticles (size 1–2 μm) can be obtained from Polysciences, Inc., Warrington, PA (Catalog No. 18879).

† Monodispersed gold colloid nanoparticles with size from 2 to 250 nm are also commercial from Ted Pella Inc., Redding, CA (AuNPs Catalog No. 15701-15714).

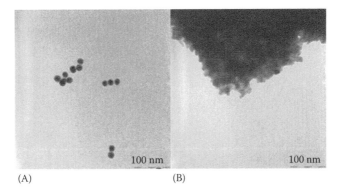

FIGURE 25.2   TEM images of (A) 13 nm prepared AuNPs loaded with HRP-anti-CEA. (B) Antigen–antibody sandwich mixture (SMPPs conjugated with biotin-*anti*-CEA/CEA antigen/AuNPs loaded with HRP-*anti*-CEA).

into the aforementioned solution to passivate the surface of AuNPs. After a 30 min duration, the mixed solution was then centrifuged (13,000 rpm, 30 min, 4°C). The supernatant was subsequently decanted and rinsed with 1 mL of washing buffer (10 mM PBS, pH 7.4). The centrifuging/rinsing procedure was repeated three times to remove the unbound antibodies and BSA. It has to be mentioned that the residual AuNPs loaded with HRP-labeled antibodies, which were not combined with antigens, could be retrieved and reused for the next application after the detection procedure.

25.3.3 Preparation of SMPPs Conjugated with Biotinylated Antibodies

SMPPs were first washed for three times with 0.5× SSC buffer (75 mM NaCl, 7.5 mM citrate·3Na, pH 7.0) prior to use. Then they were diluted to 0.1 mg/mL with PBS, pH 7.4. After that, a solution of biotin-*anti*-CEA or biotin-*anti*-AFP was added into the SMPP solution at a ratio of 100 μg protein to 1 mg SMPPs. The mixed solution was incubated for 20 min at room temperature (ca. 25°C). Finally, the antibody-conjugated SMPPs were thoroughly washed for three times with pH 7.4 PBS.

25.3.4 Optimization of Experimental Conditions

The salt-induced AuNP aggregation test was employed to determine the appropriate concentrations of HRP-labeled antibodies used for modification. NaCl was able to cause the aggregation of AuNPs and the shift of their absorbance peak from 520 to 580 nm. When proteins adsorbed on the surface of AuNPs, however, they could be prevented from aggregation in the presence of salts. So the concentration of antibodies that stabilizes the AuNPs was determined by measuring the difference between absorbances at 520 and 580 nm. From the results in Figure 25.3A, it could be observed that the minimum concentrations of HRP-labeled antibody needed to keep the stability of AuNPs were 4 μg/mL.

Moreover, due to the excellent biocompatibility of AuNPs, neutral or weakly negative-charged proteins could adsorb on the surface of AuNPs easily and stably through coordinate-covalent bond.[33] So the pH value of AuNP solution was also optimized. Figure 25.3B indicated that when the pH value of AuNP solution was 8.5, the neutral HRP-labeled antibody (isoelectric point = 8.6)[34] adsorbed onto the surface of AuNPs, preventing their aggregation induced by NaCl.

To quantify the number of antibodies loaded on each AuNP, standard spectrometric assay was performed. From the standard curves in Figure 25.3C and D, we calculated the concentrations of HRP-labeled antibodies modified on the surfaces of AuNPs and estimated that each AuNP carried with three HRP-*anti*-CEA or six HRP-*anti*-AFP molecules. As a rough comparison, one 13 nm AuNP could, as a maximum, carry around 13 densely packed HRP molecules (ca. 6.0 nm in diameter) on its surface.[35] Although antibody has larger molecular weight compared with HRP molecules, a protein monolayer still formed on the surface of AuNPs, leaving sufficient space for the subsequent combination between target antigens and antibodies.

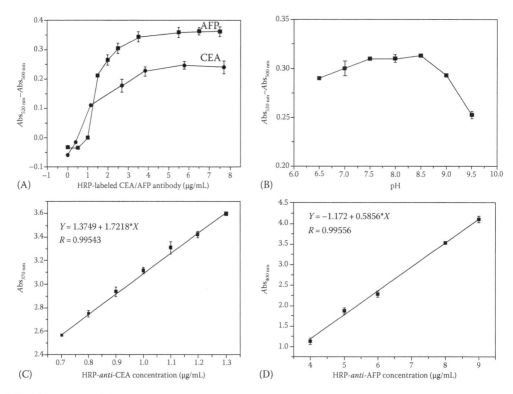

FIGURE 25.3 (A) Salt-induced AuNP aggregation test to determine the minimum concentrations of HRP-labeled CEA and AFP antibody to be used for the modification of AuNPs. When 30 μL of HRP-anti-CEA/AFP is added into 200 μL of AuNP solution to get the final protein concentration of 4 μg/mL, the antibodies can form a monolayer on the surface of AuNPs, preventing their aggregation induced by salts. (B) Optimization of pH value for the modification of HRP-labeled antibodies onto AuNPs. (C) and (D) are standard curves between the concentrations of HRP-labeled CEA and AFP antibody in the solution and the absorbance values of catalytic products at the wavelength of 370 and 490 nm. Data were collected from three independent sets of experiments.

Some other experimental conditions also influenced the sensitivity and efficiency of the immunoassay. First, to simplify the analytical process, both the incubation procedures and catalytic reactions were performed at room temperature. Second, the immunoassay speed is usually controlled by the mass transport of immunoreagents and kinetics of immunoreactions, so the incubation condition is an important element that affected the assay efficiency. From Figure 25.4A, we also optimized the incubation time to be 20 min, much shorter than 1–3 h at 37°C for ELISA approach. Besides, the kinetic behaviors of the two reactions catalyzed by HRP molecules were examined. Figure 25.4B indicated that the absorbance peak values of catalytic products tended to maximum in 10 min, which was used in the further study as reaction time.

25.3.5 Colorimetric Reaction of TMB and OPD

Both TMB and OPD are chromogenic reagents utilized for HRP-catalyzed oxidation. TMB was first dissolved into ethanol or dimethyl sulfoxide (DMSO), and then added 10 mM PBS, pH 6.2 to make a 0.2 mM solution.* With the catalysis of HRP, TMB was oxidized into a radical form of two TMB molecules, which had a characteristic absorbance peak at the wavelength of 370 nm. OPD can be dissolved in water, and with the catalysis of HRP, it yields a yellow-orange reaction product having an

---

* TMB·HCl is a hydrochloride form of TMB that is readily soluble in water, which can also be used as substrate of HRP catalysis.

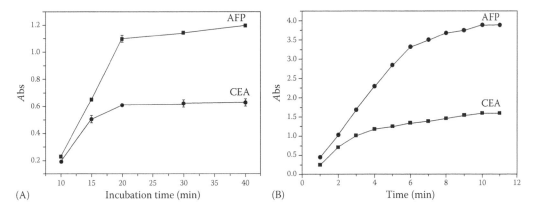

FIGURE 25.4   (A) Optimization of incubation time for the immunoreactions between CEA and AFP antigens and the antibodies loaded on the surface of SMPPs and AuNPs. (B) Kinetic curves of the reactions catalyzed by immune complex SMPP/biotin-anti-CEA/CEA/HRP-anti-CEA/AuNP and SMPP/biotin-anti-AFP/AFP/HRP-anti-AFP/AuNP at the antigen concentration of 200 ng/mL. Data were collected from three independent sets of experiments.

absorbance maximum of 492 nm. Furthermore, OPD is considered hazardous (potential carcinogens), it always needs to be handled with care. Right now, the substrate combo containing TMB/OPD substrate tablet and H_2O_2 buffer is commercially available.

25.4 Results

Figure 25.5 illustrates this nanomaterial-based colorimetric multi-analyte immunoassay for the sequential detection of CEA and AFP. In a typical experiment, 100 μL of serum sample solution containing different concentrations of CEA, AFP antigens, and other nonspecific proteins was added into a 1.5 mL microcentrifuge tube. Then, 200 μL of biotin-*anti*-CEA-conjugated SMPPs were added into the tube and incubated for 20 min with gentle shaking. After that, the mixed solution was magnetically separated. CEA antigen captured by biotin-*anti*-CEA immobilized on the surface of SMPPs was collected, and the supernatant was collected as sequential AFP sample solution. The CEA antigen was suspended in 200 μL buffer after being rinsed for three times. Subsequently, 100 μL of AuNPs solution loaded with HRP-*anti*-CEA was added into the CEA antigen solution and incubated with gentle shaking for another 20 min, during which HRP-*anti*-CEA loaded on the surface of AuNPs capture the CEA antigen conjugated on the surface of SMPPs (see Figure 25.2B). Each AuNP is loaded with at least 3~6 HRP molecules that are able to catalyze colorimetric reaction. When TMB and H_2O_2 were added into the solution containing immune complex of SMPP/CEA antigen/AuNPs, it turned from colorless into bluish green. The detection of AFP antigen in the serum sample solution was carried out by repeating the previous procedure, using biotin-*anti*-AFP-conjugated SMPPs and AuNPs loaded with HRP-*anti*-AFP instead. When OPD and H_2O_2 were added, its color changed into orange-yellow. Table 25.2 is the flowchart of the experimental procedure, showing that the sequential detection for two species of TMs can be accomplished in 94 min, less than colorimetric ELISA that requires about 1–3 h to wash and block microplates that support capture antibodies and target TMs. This detection period of 94 min is also shortened compared with those obtained in most of the reported literatures,[19,20,36,37] though it is longer than the time needed in a few detection methods.[38,39]

This multiplexed immunoassay is simple and easily operated. The presence of CEA and AFP can be directly observed and discriminated by naked eyes (Figure 25.6). And the specificity of this colorimetric detection approach was also demonstrated by testing the cross-reactivity between different antigens and their noncognate antibodies. Even if nonspecific antibodies were added into the system, there was no color change, showing that cross-reactivity can be negligible. Meanwhile, in order to test the amount of antigen more accurately, UV–visible spectroscopy was also performed. The absorbance at 370 and

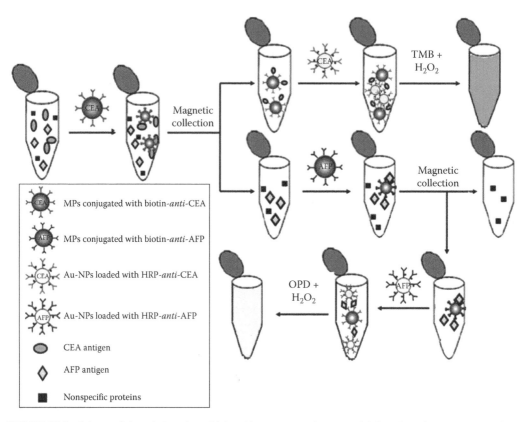

FIGURE 25.5 Scheme of the colorimetric multiplexed immunoassay for sequential detection of tumor markers, CEA and AFP.

TABLE 25.2

Flowchart of Colorimetric Multiplexed Immunoassay for the Sequential Detection of Tumor Markers CEA and AFP

Step No.	Step	Starting Time (min)
1	Addition of SMPPs conjugated with biotin-*anti*-CEA into the contaminated sample solutions containing CEA, AFP antigens, and nonspecific proteins	00:00
2	Magnetic separation	20:00
3	Addition of AuNPs loaded with HRP-*anti*-CEA into SMPPs and magnetic separation	21:00
4	Addition of substrate TMB and H_2O_2 to start catalytic reaction	42:00
5	Addition of SMPPs conjugated with biotin-*anti*-AFP into the residual solution after step (II)	52:00
6	Magnetic separation	72:00
7	Addition of AuNPs loaded with HRP-*anti*-AFP into SMPPs and magnetic separation	73:00
8	Addition of substrate OPD and H_2O_2 to start catalytic reaction	94:00

490 nm were, respectively, used for the detection of CEA and AFP (Figures 25.7 and 25.8). The detection limits could be as low as 0.02 ng/mL ($\approx$0.1 pM) for detection of both the two antigens with the relative standard deviation of 2.20% and 1.35%, respectively. Furthermore, the linear detection ranges were separately 2.0–40 ng/mL for CEA and 5.0–35 ng/mL for AFP (Figures 25.7B and 25.8B). Since the cut-off values of these two TMs in clinical diagnosis are 5.0 and 20.0 ng/mL, respectively (Table 25.1), both of the sensitivity and linear ranges of this multi-analyte immunoassay are satisfactory for clinical applications.

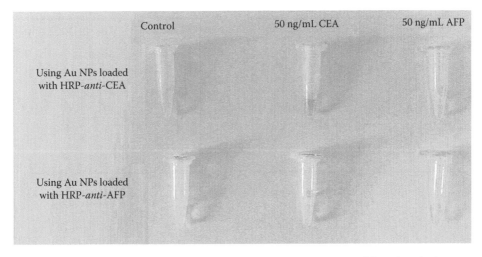

FIGURE 25.6 Colorimetric detection of CEA and AFP antigens. Only in the presence of the antigen that is cognate with the antibodies loaded on the surfaces of SMPPs and AuNPs can the immune complex be formed and the characteristic blue or yellow color of the catalytic reaction product be observed. No color can be observed either in the absence of the target antigen (control) or with the addition of nontarget antigen.

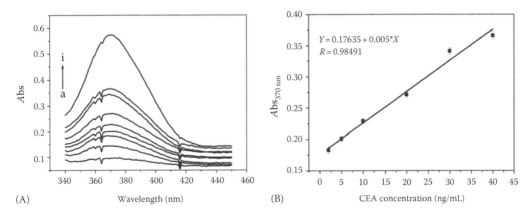

FIGURE 25.7 (A) UV–visible spectra of the test solution in the presence of (a) 0, (b) 0.02, (c) 2, (d) 5, (e) 10, (f) 20, (g) 30, (h) 40, (i) 50 ng/mL CEA antigen. (B) Linear relationship between CEA concentration and absorbance of TMB oxidized product. Data are collected from at least three independent sets of experiments.

The reproducibility of this multiplexed immunoassay was assessed by measuring both of the intra- and inter-assay coefficients of variation (CVs). The intra-assay CV is the difference among three measurements of the same sample incubated in a vessel, and the inter-assay CV is the difference among the measurements of three samples incubated in three different vessels. The error bars in Figures 25.7B and 25.8B were the inter-assay CV of three independent measurements at each concentration. When testing serum samples containing 30 ng/mL CEA and 30 ng/mL AFP antigens, the intra- and inter-assay CVs for the detection of 30 ng/mL CEA were 2.20% and 2.74%, and the intra- and inter-assay CVs for the detection of 30 ng/mL AFP were 0.94% and 4.09%. So we believe that the precision and reproducibility of this proposed method are acceptable.

However, most clinical applications do not require the sensing of pure TM targets; instead, it necessitates the detection of multiple analytes in a complex environment that contains large quantity of proteins and other components in a much higher concentration than the targets. Many species of TMs including CEA and AFP exist in serum that is composed of antibodies, clotting factors, mineral, vitamin transport proteins, and hormones as well. In order to evaluate the performance of this multiplexed immunoassay

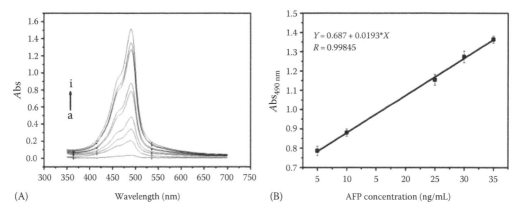

FIGURE 25.8 (A) UV–visible spectra of the test solution in the presence of (a) 0, (b) 0.02, (c) 1, (d) 2, (e) 5, (f) 10, (g) 30, (h) 40, (i) 100 ng/mL AFP antigen. (B) Linear relationship between AFP concentration and absorbance of OPD-oxidized product. Data are collected from at least three independent sets of experiments.

TABLE 25.3

Immunoassay Results of Contaminated Serum Samples Using the Proposed Method and the Comparison with Given Antigen Concentrations in Serum Samples

Sample No.	Proposed Method	Standard	Relative Error (%)
CEA (ng/mL)			
1	5.25	5	5.00
2	25.50	25	2.00
3	43.25	40	8.13
AFP (ng/mL)			
1	4.77	5	4.60
2	22.12	20	10.60
3	30.67	30	2.23

in clinical applications, we also measured serum samples containing CEA and AFP targets. The serum samples were prepared by dissolving different amounts of CEA and AFP antigens into undiluted human serum. The comparison between the assay results obtained by using this proposed method and the given concentrations in these contaminated samples is shown in Table 25.3. The relative errors were <10.60%, and the detection sensitivity for both of two TMs is as low as 5 ng/mL. The satisfactory performance of this nanomaterial-based colorimetric approach in the presence of excess serum makes it feasible in clinical detection of TMs and early diagnosis of cancers.

25.5 Discussion

The working principles between traditional colorimetric ELISA and this nanomaterial-based immunoassay have some similarities, for example, both of them utilize "sandwich" structure containing target antigens and their antibodies. However, compared with ELISA, the participation of two species of nanomaterials has demonstrated advantages exceeding ELISA. First of all, while ELISA uses multi-well plates as solid support, both SMPPs and AuNPs in this immunoassay are assembled with multiple antibodies on their surfaces. Due to their small granule diameters and large surface areas, nanomaterials produced a short diffusion distance between immunoreagents and so that accelerated the mass transport. As a result, both of the incubation time and catalytic reaction time are shortened, improving detection efficiency. In ELISA, furthermore, to prevent nonspecific adsorption of protein onto multi-well plates, they have to be blocked the remaining protein-binding sites in wells. Usually the blocking takes at least 1–2 h at room temperature or overnight at 4°C. This time-consuming step can be compensated by using SMPPs that achieve fast separation of antigens and nontarget proteins. Another advantage of nanomaterials is both of SMPPs and AuNPs are loaded with multiple antibodies, which lead to double signal amplification. The SMPPs collect target TMs and concentrate them onto their surfaces; AuNPs loaded with multiple HRP-labeled antibodies work not only as recognition components, but also as a "probe" of signal output. The number of HRP in the "sandwich" structure is always excess than the target TMs, so that the detection performance of this immunoassay is much prior to ELISA. Moreover, not only the detection target of this nanomaterial-based detection system is limited in TM-related immunoassay, but instead, it can also be easily extended to the detection of various analytes, such as proteins, DNA, heavy metal irons, some other small molecules, and even cells by only replacing the components on the surface of nanoparticles.

The ultimate goal of cancer screening using TM is to detect and diagnose cancers at their earliest stage by blood or urine tests, which could prevent the deaths of millions of people. Unfortunately, we have to mention that the TM detection alone may not be enough for accurate diagnosis of cancers. On the one hand, not all cancers have high levels of specific TMs; on the other hand, even when the levels of these markers are high, it does not always mean that cancer is present. For example, according to U.S. National Cancer Institute,[40] prostate-specific antigen (PSA) levels are often used to screen men for prostate cancer, but it is not yet known that elevated PSA levels are caused by prostate cancer or some other benign conditions, and the statistics show some men with elevated PSA levels turn out not to have prostate cancer. One more thing is it is still not clear if the benefits of PSA screening outweigh the risks of following diagnostic tests and cancer treatments. Right now, the measurements of TMs are usually combined with other tests. In clinical applications, TM levels are measured before patient treatment to help doctors plan appropriate therapy. The increase and decrease of TM levels also reflect the extent of the disease development, indicate the response of cancer to therapy, or monitor the recurrence after the treatment. Considering these situations, the performance of cancer screening sensors that are based on detecting TM level need to be further improved in the future: (1) Develop sensors that can detect three or even more species of TMs to elevate the precision of cancer diagnosis, such as epithelial ovarian tumor and lung cancer.[41] This also requires more feasible signal readout ways that are able to simplify the detection procedure of cancer sensors. To resolve it, nanomaterials that conjugate multiple kinds of biomolecules such as enzymes and antibodies can be used as general signal output platform by adding corresponding enzyme substrates into the system when detecting different TM targets; and (2) Improve the sensitivity and specificity of colorimetric sensors to make them monitor slight change of TM expression level. Scientists have already been working on discovering new TMs and also some other genome- and proteome-based biomarkers having high specificity with cancers.

25.6 Conclusion

In summary, we present a nanomaterial-based colorimetric multi-analyte immunoassay for sequential detection of TMs CEA and AFP. In this detection system, AuNPs loaded with multiple HRP-labeled antibodies are used for signal amplification, and SMPPs for fast magnetic separation and target concentration. Although its working principle has some similarities with colorimetric ELISA, the participation of nanomaterials makes this immunoassay having the capability of detecting CEA and AFP antigens with significant sensitivity and specificity. Furthermore, this proposed method also shows acceptable accuracy and reproducibility, which make it a promising approach to perform the diagnosis and detection of cancers in clinical applications.

REFERENCES

1. World Health Organization. Cancer. http://www.who.int/mediacentre/factsheets/fs297/en/index.html (accessed February 18, 2011).

2. World Health Organization. Screening and early detection of cancer. http://www.who.int/cancer/detection/en/index.html (accessed February 18, 2011).

3. Leichner, P. K., Koral, K. F., Jaszczak, R. J., Green, A. J., Chen, G. T. Y., Roeske, J. C. An overview of imaging techniques and physical aspects of treatment planning in radioimmunotherapy, *Medical Physics*, 1993, 20, 569–577.

4. Sieren, J. C., Ohno, Y., Koyama, H., Sugimura, K., McLennan, G. Recent technological and application developments in computed tomography and magnetic resonance imaging for improved pulmonary nodule detection and lung cancer staging, *Journal of Magnetic Resonance Imaging*, 2010, 32, 1353–1369.

5. Lynch, T. B. *PET/CT in Clinical Practice*, Springer-Verlag London Limited, London, U.K., 2007, pp. 1–16.

6. Christopher Kwesi, O. W., Olufunmalayo, I. O., Carla, I. F. *Breast Cancer in Women of African Descent*, Springer, Berlin, Germany, 2006, pp. 74–76.

7. Kumar, S., Mohan, A., Guleria, R. Biomarkers in cancer screening, research and detection: Present and future: A review, *Biomarkers*, 2006, 11, 385–405.

8. Safi, F., Kohler, I., Rottinger, E., Beger, H. The value of the tumor marker CA15-3 in diagnosing and monitoring breast cancer, *Cancer*, 1991, 68, 574–582.

9. Harris, L., Fritsche, H., Mennel, R., Norton, L., Ravdin, P., Taube, S., Somerfield, M. R., Hayes, D. F., Bast, R. C. Update of recommendations for the use of tumor markers in breast cancer, *Journal of Clinical Oncology*, 2007, 25, 5287–5312.

10. Chen, D. S., Sung, J. L. Serum alpha-fetoprotein in hepatocellular carcinoma, *Cancer*, 1977, 40, 779–783.

11. Jacobsen, G. K., Jacobsen, M. Alpha-fetoprotein (AFP) and human chorionic gonadotropin (HCG) in testicular germ cell tumors: A prospective immunohistochemical study, *Acta Pathologica, Microbiologica et Immunologica Scandinavica A*, 1983, 91, 165–176.

12. Benchimol, S., Fuks, A., Jothy, S., Beauchemin, N., Shirota, K., Stanners, C. P. Carcinoembryonic antigen, a human tumor marker, functions as an intercellular adhesion molecule, *Cell*, 1989, 57, 327–334.

13. Stamey, T. A., Yang, N., Hay, A. R., McNeal, J. E., Freiha, F. S., Redwine, E. Prostate-specific antigen as a serum marker for adenocarcinoma of the prostate, *The New England Journal of Medicine*, 1987, 317, 909–916.

14. Wong, W. W., Vijayakumar, S., Weichselbaum, R. R. Prognostic indicators in node-negative early stage breast cancer, *American Journal of Medicine*, 1992, 92, 539–548.

15. Mccann, A. H., Dervan, P. A., Oregan, M., Codd, M. B., Gullick, W. J., Tobin, B. M. J., Carney, D. N. Prognostic significance of c-Erbb-2 and estrogen-receptor status in human breast cancer, *Cancer Research*, 1991, 51, 3296–3303.

16. Modlin, I. M., Gustafsson, B. I., Moss, S. F., Pavel, M., Tsolakis, A. V., Kidd, M. Chromogranin A—Biological function and clinical utility in neuro endocrine tumor disease, *Annals of Surgical Oncology*, 2010, 17, 2427–2443.

17. Robbins, J., Merino, M. J., Boice, J. D., Ron, E., Ain, K. B., Alexander, H. R., Norton, J. A., Reynolds, J. Thyroid cancer—A lethal endocrine neoplasm, *Annals of Internal Medicine*, 1991, 115, 133–147.

18. Bates, S. E. Clinical-applications of serum tumor-markers, *Annals of Internal Medicine*, 1991, 115, 623–638.

19. Wilson, M. S., Nie, W. Multiplex measurement of seven tumor markers using an electrochemical protein chip, *Analytical Chemistry*, 2006, 78, 6476–6483.

20. Wu, J., Yan, Y., Yan, F., Ju, H. Electric field-driven strategy for multiplexed detection of protein biomarkers using a disposable reagentless electrochemical immunosensor array, *Analytical Chemistry*, 2008, 80, 6072–6077.

21. Gkalpakiotis, S., Arenberger, P., Kremen, J., Arenbergerova, M. Quantitative detection of melanoma-associated antigens by multimarker real-time RT-PCR for molecular staging: Results of a 5 years study, *Experimental Dermatology*, 2010, 19, 994–999.

22. Donach, M., Yu, Y. H., Artioli, G., Banna, G., Feng, W. W., Bast, R. C., Zhang, Z., Nicoletto, M. O. Combined use of biomarkers for detection of ovarian cancer in high-risk women, *Tumor Biology*, 2010, 31, 209–215.

23. Kricka, L. J. Multianalyte testing, *Clinical Chemistry*, 1992, 38, 327–328.

24. Stoeva, S. I., Lee, J. S., Smith, J. E., Rosen, S. T., Mirkin, C. A. Multiplexed detection of protein cancer markers with biobarcoded nanoparticle probes, *Journal of American Chemical Society*, 2006, 128, 8378–8379.

25. Pei, X. P., Chen, B. A., Li, L., Gao, F., Jiang, Z. Multiplex tumor marker detection with new chemiluminescent immunoassay based on silica colloidal crystal beads, *Analyst*, 2010, 135, 177–181.

26. Bacher, J. W., Flanagan, L. A., Smalley, R. L., Nassif, N. A., Burgart, L. J., Halberg, R. B., Megid, W. M. A., Thibodeau, S. N. Development of a fluorescent multiplex assay for detection of MSI-high tumors, *Disease Markers*, 2004, 20, 237–250.

27. Voller, A., Bartlett, A., Bidwell, D. E. Enzyme immunoassays with special reference to ELISA techniques, *Journal of Clinical Pathology*, 1978, 31, 507–520.

28. Osuchowski, M. F., Siddiqui, J., Copeland, S., Remick, D. G. Sequential ELISA to profile multiple cytokines from small volumes, *Journal of Immunological Methods*, 2005, 302, 172–181.

29. Selection Guide: Enzyme Substrates for ELISA. *Colorimetric ELISA Substrates*, Thermo Scientific, Rockford, IL. http://www.piercenet.com/browse.cfm?fldID=EEB28337-5056-8A76-4E7C-1FA2CA04F788#ELISAColorimetric (accessed February 18, 2011).

30. Ferrari, M. Cancer nanotechnology: Opportunities and challenges, *Nature Reviews Cancer*, 2005, 5, 161–171.

31. West, J. L., Halas, N. J. Engineered nanomaterials for biophotonics applications: Improving sensing, imaging, and therapeutics, *Annual Review of Biomedical Engineering*, 2003, 5, 285–292.

32. Wang, J., Cao, Y., Xu, Y., Li, G. Colorimetric multiplexed immunoassay for sequential detection of tumor markers, *Biosensors and Bioelectronics*, 2009, 25, 532–536.

33. Chen, Y. H., Tsai, C. Y., Huang, P. Y., Chang, M. Y., Cheng, P. C., Chou, C. H., Chen, D. H., Wang, C. R., Shiau, A. L., Wu, C. L. Methotrexate conjugated to gold nanoparticles inhibits tumor growth in a syngeneic lung tumor model, *Molecular Pharmaceutics*, 2007, 4, 713–722.

34. Hamilton, R. G. *The Human IgG Subclasses*, Calbiochem-Novabiochem Corporation, La Jolla, CA, 2001, pp. 12–15.

35. Green, A. J., Johnson, C. J., Adamson, K. L., Begent, R. H. Mathematical model of antibody targeting: Important parameters defined using clinical data, *Physics in Medicine and Biology*, 2001, 46, 1679–1693.

36. Knecht, B. G., Strasser, A., Dietrich, R., Martlbauer, E., Niessner, R., Weller, M. G. Automated microarray system for the simultaneous detection of antibiotics in milk, *Analytical Chemistry*, 2004, 75, 646–654.

37. Wang, K. Y., Chuang, S. A., Lin, P. C., Huang, L. S., Chen, S. H., Ouarda, S., Pan, W. H., Lee, P. Y., Lin, C. C., Chen, Y. J. Multiplexed immunoassay: Quantitation and profiling of serum biomarkers using magnetic nanoprobes and MALDI-TOF MS, *Analytical Chemistry*, 2008, 80, 6159–6167.

38. Fu, Z., Yang, Z., Tang, J., Liu, H., Yan, F., Ju, H. Channel and substrate zone two-dimensional resolution for chemiluminescent multiplex immunoassay, *Analytical Chemistry*, 2007, 79, 7376–7382.

39. Liu, H., Fu, Z., Yang, Z., Yan, F., Ju, H. Sampling-resolution strategy for one-way multiplexed immunoassay with sequential chemiluminescent detection, *Analytical Chemistry*, 2008, 80, 5654–5659.

40. U.S. National Cancer Institute. Prostate-Specific Antigen (PSA) Test. http://www.cancer.gov/cancertopics/factsheet/detection/PSA (accessed February 18, 2011).

41. Duffy, M. J., Crown, J. A personalized approach to cancer treatment: How biomarkers can help, *Clinical Chemistry*, 2008, 54, 1770–1779.

26

Molecular Pincers for Detecting Cancer Markers

Ewa Heyduk and Tomasz Heyduk

CONTENTS

26.1 Introduction

Immunodetection techniques, for example, enzyme-linked immunosorbent assay (ELISA) [1], have been a workhorse of detection and quantitation of disease biomarkers, including cancer markers. While ELISA is a well-established and robust technique, there continues to be an interest in developing alternative biomarker detection techniques that would be more suitable for practical clinical applications. For example, it becomes increasingly clear that in order to achieve acceptable sensitivity and specificity, molecular diagnostics assays will have to employ a panel of biomarkers rather then relying on a single biomarker. Classical ELISA assays do not allow easy multiplexing. Therefore, efforts to develop multiplexed ELISA designs or to develop alternative multiplexing-friendly assay formats (e.g., protein arrays [2]) have been implemented to overcome this ELISA shortcoming. Simplification of biomarker detection is another area of alternative assay development. ELISA is a relatively lengthy process involving multiple steps of reagent addition and washing. Development of greatly simplified biomarker detection technologies could reduce the cost of the assays and could allow point-of-care deployment of the assays.

The interest of our laboratory has been in developing new detection methodologies for biomedical targets that could provide answers to some of the challenges of classical assays described earlier. The methodology that could be particularly useful for detecting cancer biomarkers is the molecular pincer assay [3–8] illustrated in Figure 26.1. Target recognition elements (DNA, aptamers, antibodies, or peptides) (A and B in Figure 26.1) labeled with short complementary signaling oligonucleotides (d and e in Figure 26.1) via nanometer-scale flexible linkers (c in Figure 26.1) are the critical components of molecular pincer assay. The length of these signaling oligonucleotides is designed such that very little association between the oligonucleotides occurs in the absence of the target. When the target

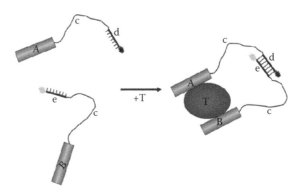

FIGURE 26.1 Design of molecular pincer assay. (Reprinted from *Biophys. Chem.*, 151, Heyduk, T., Practical biophysics: Sensors for rapid detection of biological targets utilizing target-induced oligonucleotide annealing, 91–96, Copyright 2010, with permission from Elsevier.)

(T in Figure 26.1) is present in a sample, the recognition elements form a complex with the target bringing the signaling oligonucleotides to close proximity (increasing greatly they local concentration), which triggers the annealing of the signaling oligonucleotides (Figure 26.1). Very long nanometer-sized linkers that are used for attaching signaling oligonucleotides assure that large targets could be accommodated by the design shown in Figure 26.1. Fluorescence probes that are incorporated into the signaling oligonucleotides are brought to close proximity as a result of target-induced annealing of signaling oligonucleotides. This target-induced proximity between fluorescence probes can be utilized to produce a measurable signal reporting the presence of the target by employing fluorescence resonance energy transfer (FRET) [9]. FRET is a process where excitation energy of one fluorophore (donor) can be transferred to another fluorophore (acceptor) in a radiationless process involving resonance between energy levels of excited donor and acceptor. For the FRET to occur, the emission spectrum of the donor has to overlap with the excitation spectrum of the acceptor. FRET efficiency exhibits steep dependency on the distance between the donor and acceptor (typical range of FRET is <100 Å). To practically implement FRET detection in a sensor design illustrated in Figure 26.1, the sample would be illuminated with the light matching excitation of the donor fluorophore and the fluorescence emission will be measured at the wavelength of the acceptor emission. In the absence of the target, such setup should produce low fluorescence corresponding to the light scattering background and background fluorescence due to direct excitation of the acceptor and spillover of donor emission to acceptor channel. In the presence of the target, large acceptor fluorescence intensity increase with the excitation of the donor should be observed. Since detecting of FRET involves a simple fluorescence intensity measurement, common nonspecialized instrumentation (e.g., fluorescence plate readers) could be used to read FRET signals produced by molecular pincers.

In this chapter, we describe the detailed experimental protocol that we have implemented for preparing molecular pincer assay for a cancer marker, carcinoembryonic antigen (CEA) protein. CEA was discovered in 1967 as an antigen present in fetal colon as well as in colon adenocarcinoma but absent in healthy adult colon [10,11]. Subsequent studies demonstrated the presence of CEA in serum [12] and elevated levels of CEA have been found in colon, breast, pancreatic, and lung cancers [13]. CEA is a ~200 kDa glycoprotein belonging to immunoglobin superfamily. Its normal functions are not entirely clear but in vitro data suggest the role in cell adhesion. Such role would be consistent with a proposed role of CEA in cancer invasion and metastasis [14]. The most useful applications of CEA as a biomarker are in gastrointestinal cancers, especially colon cancer. Despite improvements in colon cancer detection due to implementation of colonoscopy, there continues to be a great interest in developing noninvasive clinically useful biomarkers. Although CEA lacks sufficient sensitivity and specificity for the use in general cancer screening of healthy subjects, it can be applied for assessing prognosis [14]. Studies have demonstrated that high preoperative concentrations of CEA are predictive of worse outcomes compared to low marker concentration. Thus, CEA determinations could be used to identify patients who might benefit from aggressive adjuvant chemotherapy. Postoperative CEA determinations could be used for early detection of cancer recurrence. Another promising use of CEA is detection of liver metastasis and in assessing prognosis after hepatectomy.

26.2 Materials

N-Hydroxysuccinimide (NHS)-PEO$_8$-maleimide cross-linker (cat # 22108) and Traut's reagent (cat # 26101) were from Thermo-Fisher Scientific (Waltham, MA). Monoclonal antihuman CEA antibodies (clone SP-1 (cat # 5909) and clone SPTN-5 (cat # 5910)) and CEA antigen (cat # J62000126) were purchased from BioSpacific (Emeryville, CA). Superdex™ 200 10/300 GL FPLC size exclusion column (cat # 17-5175-01) was from GE Healthcare (Piscataway, NJ) and HPLC PRP-1 (cat # 79246) column was from Hamilton (Reno, NV). The oligonucleotides were prepared by standard automated oligonucleotide synthesis and were obtained from IDT (Coralville, IA) or from W.M. Keck facility at Yale University (New Haven, CT). The following oligonucleotides were prepared (names are given in parentheses): 5′amino-TA GGA GAG AGA GAG AGG A (A1), 5′amino-TAG GTG CTC GAC GCT GAC (A2), 5′ fluorescein-GCT CAT TGT CAG CGT CGA GCA CCT A (A3), 5′ Cy5-ATG AGC TTC CTC TCT CTC TCT CCA T (A4). Fluorescently labeled oligonucleotides were purified on PRP-1 column using reversed-phase HPLC. 5′-Amino-labeled oligonucleotides that were used for antibody modification were ethanol-precipitated and were used without further purification.

26.3 Methods

26.3.1 Modification of Antibodies with Signaling Oligonucleotides

Antibodies can be labeled with fluorophore-modified oligonucleotides either directly (as schematically illustrated in Figure 26.1) or indirectly by first attaching an unlabeled oligonucleotide to the antibody followed by annealing of a complementary fluorophore-labeled oligonucleotide. We found the latter approach more robust, less expensive (since smaller amounts of fluorophore-labeled oligonucleotides are used), and more flexible. For example, different fluorescent labels can be attached to the antibody by a simple hybridization reaction. Additionally, we found that the indirect labeling strategy usually produced better FRET signals. Thus, the antibody-labeling procedure describes five steps that are required for the indirect labeling strategy.

26.3.1.1 Modification of 5′-Amino Oligonucleotide with NHS-PEO$_8$-Maleimide

The first step involves preparation of a thiol-reactive oligonucleotide that is subsequently used to react with thiolated antibody. Two hundred microliters of 5′-amine containing A1 or A2 oligonucleotides at ~250 μM in 20 mM NaH$_2$PO$_4$ (pH 7.4), 150 mM NaCl, and 2.5 mM EDTA buffer (conjugation buffer) were mixed with 5 μL of ~250 mM of a NHS-PEO$_8$-maleimide freshly dissolved in DMF. The reaction mixtures were incubated for 1.5 h at room temperature. Oligonucleotide was purified from the excess of the cross-linker by ethanol precipitation in the presence of 1 mg/mL of glycogen. Precipitated oligonucleotides were dried in SpeedVac and were stored at −20°C until they were used for antibody modification. The fraction of the oligonucleotides that were successfully conjugated with the linker can be estimated by native 10% polyacrylamide gel electrophoresis. Addition of the cross-linker to the oligonucleotide results in a changed (retarded) mobility of the oligonucleotide. Typically, ~50%–75% of the oligonucleotide can be modified with the cross-linker.

Modification of the oligonucleotide with the cross-linker in most cases proceeds with no problems. 5′-Amino oligonucleotide should be ethanol-precipitated before using it in the reaction with the cross-linker because otherwise occasionally poor yields of modified oligonucleotide are observed. In place of ethanol precipitation, the 5′-amino oligonucleotide can also be run first on reversed-phase HPLC column, which in some instances resulted in better outcomes of reaction with the cross-linker compared to ethanol precipitation.

26.3.1.2 Thiolation of Antibodies with Traut's Reagent

Seventy-five microliters of SPTN-5 antibody (5 mg/mL) was desalted on Zeba™ spin column equilibrated in conjugation buffer. Seven microliters of 14 mM Traut's reagent dissolved in DMF were added and a sample was incubated for 1.5 h at room temperature. Excess of Traut's reagent was removed by running the sample through Zeba spin column equilibrated in conjugation buffer. Three hundred and seventy-five microliters of SP-1 antibody at 1 mg/mL was concentrated to ~85 µL SpeedVac and was desalted on Zeba spin column equilibrated in conjugation buffer. The reaction with Traut's reagent was performed as described for SPTN-5 antibody.

Thiolation of antibodies is usually quite reproducible and we did not find reasons to change the reaction conditions described earlier. The buffers in which commercial antibodies are sold can vary significantly and care should be exercised to make sure that none of the buffer components interfere with the thiolation reaction. For example, antibodies are often shipped in buffers that contain bovine serum albumin (BSA) or gelatin. Both of these components would also be thiolated and this would cause serious problems in assay performance. Such interfering substances need to be removed before the reaction with Traut's reagent using, for example, protein A chromatography. Thiolation and subsequent oligonucleotide addition work best if the concentration of the antibody is in the range of few milligrams per milliliter. Diluted antibody solutions can be concentrated using Microcon concentrators. We also often concentrate antibody solution by first exchanging the buffer (on Zeba spin column) to appropriately diluted (e.g., 5× if 5× concentration will be performed) reaction buffer and reducing the volume of such solution on SpeedVac.

26.3.1.3 Conjugation of Antibodies with Unlabeled Oligonucleotides

Freshly prepared thiolated antibodies were reacted with a 15–20 molar excess of cross-linker-conjugated oligonucleotide by adding antibody solutions prepared as described in Section 26.3.1.2 to appropriate amount of dried oligonucleotide–cross-linker conjugate (A1 and A2 oligonucleotides for SPTN-5 and SP-1 antibodies, respectively). Reaction mixtures were incubated for 4 h at room temperature followed by an overnight incubation at 4°C.

26.3.1.4 Purification of Unlabeled Oligonucleotide–Antibody Conjugate

Oligonucleotide–antibody conjugates were purified from the excess of the unreacted oligonucleotides by size exclusion FPLC chromatography on 10/300 GL Superdex 200 column equilibrated with 10-fold diluted 20 mM Tris (pH 8.0), 100 mM NaCl, 10 mM EDTA buffer. The column was run at 0.5 mL/min and 0.5 mL fractions were collected. Fractions containing modified antibodies were pooled and concentrated 10-fold in a SpeedVac. Concentration of labeled antibodies and degree of antibody labeling were estimated from UV absorbance spectra by fitting the observed ratio of absorbance at 260 and 280 nm to a linear combination of the values for unmodified antibody and free oligonucleotide. The observed ratios of oligonucleotide to antibody were 5 and 2.7 for SPTN-5 and SP-1 antibodies, respectively. Modified antibodies were stored at 4°.

A good method to check the antibodies labeled with oligonucleotides is to do a titration of a known concentration of fluorochrome-labeled A3 (or A4, whichever is complementary to the oligonucleotide attached to the antibody) with increasing concentrations of labeled antibody (Figure 26.2). The samples are then run on native 7.5% polyacrylamide gel and the gel is imaged by fluorescence. Fluorescent oligonucleotide exhibits much retarded mobility in the gel when annealed with oligonucleotide-labeled antibody allowing measurement of the amount of fluorescent oligonucleotide that could be effectively bound to the oligonucleotide-labeled antibody (Figure 26.2).

26.3.1.5 Annealing of Fluorophore-Labeled Oligonucleotide with Oligonucleotide–Antibody Conjugate

Purified oligonucleotide–antibody conjugates were hybridized with corresponding complementary fluorophore-labeled oligonucleotides (A4 and A3 for SPTN-5 and SP-1 antibodies, respectively) by

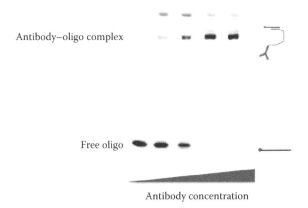

Antibody–oligo complex

Free oligo

Antibody concentration

FIGURE 26.2 Testing labeling of the antibody with the oligonucleotide by native polyacrylamide gel electrophoresis. Constant amount of fluorescein-labeled oligonucleotide complementary to oligonucleotide attached to the antibody was titrated with increasing concentrations of labeled antibody. Higher mobility band is free fluorescein-labeled oligonucleotide whereas slow mobility band corresponds to the fluorescein-labeled oligonucleotide annealed to the labeled antibody.

incubating 100 nM antibody solution with the oligonucleotides added at 1:1 molar ratio for 30 min at room temperature in 20 mM Tris (pH 8.0), 100 mM NaCl, and 10 mM EDTA buffer.

26.3.2 Molecular Pincer Assay

CEA assays were performed in 20 mM Tris (pH 8.0), 100 mM NaCl, 10 mM EDTA, and 0.2 mg/mL BSA in 384-well low-volume black microplates (cat # 3820; Corning, Lowell, MA) using Analyst AD (LJL Biosystems, Sunnyvale, CA). The final assay volume was 20 μL.

Ten microliters of the labeled antibody mix prepared (from 100 nM stocks described in 5.1.5) at 2× of desired final assay concentration were added to the samples containing varied amounts of CEA diluted to 10 μL with the assay buffer. The samples were gently mixed and were incubated at room temperature for 30 min followed by readout of the fluorescence. Two readouts were obtained for each well: one with the excitation at 485 nm and emission at 520 nm (donor signal) and one with the excitation at 485 nm and emission at 670 nm (sensitized acceptor signal). Background signals measured with a sample-containing assay buffer only were subtracted from all measured fluorescence intensities and ratio of sensitized acceptor and donor signal were calculated for each sample. The results were plotted as FRET signal expressed as the ratio of the signal observed at a given CEA concentration to a signal observed for the antibody mix alone.

26.4 Results

Figure 26.3 summarizes the results obtained with molecular pincer assay for CEA. Large CEA-dependent FRET signal was observed that changed essentially linearly with the concentration of CEA up to a specific CEA concentration where a sharp break in a response curve was observed. Further increase in CEA concentration resulted in a decrease of FRET signal. Such behavior of the assay suggested essentially stoichiometric complex formation between CEA and pincer assay components indicating high affinity-binding constants of the antibodies. Decrease of FRET signal at high CEA concentration is a result of the dissociation of ternary antibody–CEA complex to two binary antibody–CEA complexes driven by high concentration of CEA. Changing the concentrations of the antibodies in the assay mix preserved the shape of the response curve but shifted the curve such that the inflection point moved to a lower CEA concentration at lower antibody concentrations. Changing the concentrations of the labeled antibodies in the assay mix can be thus used to modify the range of CEA concentrations where the linear response of the assay is observed. Additionally, lowering the antibody concentrations enhances the response to low

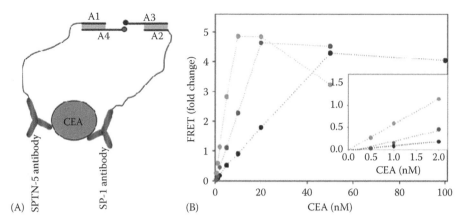

FIGURE 26.3   Molecular pincer assay for CEA. (A) Design of CEA assay. (B) FRET signal at different concentrations of CEA. Concentrations of labeled antibodies were 20 nM (black symbols), 10 nM (dark gray symbols), and 5 nM (light gray symbols).

CEA concentrations (Figure 26.3B inset). For example, the limit of detection at 20 nM antibody concentration was ~1 nM whereas at 2 nM antibody concentration the limit of detection improved to ~100 pM. This is because at low antibody concentration, a higher fraction of the antibodies will be in the complex with the target producing relatively higher FRET signal compared to the background signal from unbound antibodies. This strategy of increasing response of the assay to low target concentrations has two limitations. Decreasing labeled antibody concentrations to the values low enough that fluorescence signals become small and noisy will be counterproductive. Also, if antibody concentrations are lowered below the dissociation constant of antibody–target complex, inefficient binding of antibody to the target will compromise the performance of the assay.

26.5 Discussion

Successful application of molecular pincer design to the detection of CEA illustrates applicability of molecular pincers for preparing assays for cancer biomarkers. This further supports the general applicability of molecular pincer sensor design since our previous studies have demonstrated the applicability of molecular pincers for detecting a variety of other targets [3–8]. As an example, Figure 26.4A and B shows the application of molecular pincers for detecting protein targets (thrombin, troponin, and C-reactive protein) related to cardiovascular disease. For each of these targets, detection limits similar to those obtained with CEA (tens of pM) were observed [4]. While molecular pincers for detecting CEA were prepared using antibodies as the target recognition elements of the pincers, other target-specific binders could also be used. For example, DNA aptamers were used to prepare molecular pincer for detecting thrombin [3] (Figure 26.4A) or double-stranded DNA fragment containing binding site for the target was used to prepare molecular pincer for p53 protein [4] (Figure 26.4C). Results obtained with molecular pincers were shown to exhibit outstanding correlation with well-established commonly used techniques such as ELISA or radioimmunoassay (RIA) [7].

Molecular pincer sensor concept illustrated in Figure 26.1 can be extended to include designs that go beyond signaling resulting from target co-binding of two binders. Two published variants of molecular pincers illustrate this point [5,6]. The first example is molecular pincer design for detecting intact cells utilizing cell surface target [6] (Figure 26.5A). In this design, target recognition elements labeled with the oligonucleotides are brought to proximity by binding to cell surface epitopes that are located on cell surface with the density compatible with the length of flexible linkers used to attach oligonucleotides to the antibody. This design was tested using pathogenic bacteria as the target and produced robust homogeneous mix-and-read assay (Figure 26.5A) capable of detecting ~300 cells. The second example is

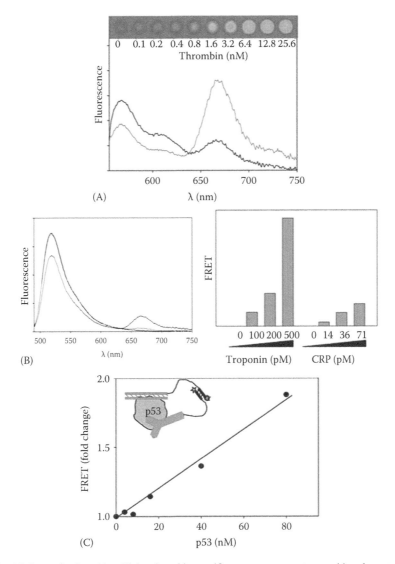

FIGURE 26.4 (A) Sensor for thrombin utilizing thrombin-specific aptamers as target recognition elements. Fluorescence spectra of Cy3- and Cy5-labeled sensor mix in the absence (black) and presence (gray) of thrombin. FRET fluorescence image of microplate wells containing sensor and indicated concentrations of thrombin is also shown. (B) Sensors for cardiac troponin and C-reactive protein (CRP) utilizing target-specific antibodies as target recognition elements. Left panel: Fluorescence spectra of fluorescein and Cy5-labeled sensor mix in the absence (black) and presence (gray) of troponin. Right panel: FRET signals observed at low target concentrations. (C) Sensor for p53 utilizing dsDNA and antibody as target recognition elements. (Reprinted from *Biophys. Chem.*, 151, Heyduk, T., Practical biophysics: Sensors for rapid detection of biological targets utilizing target-induced oligonucleotide annealing, 91–96, Copyright 2010, with permission from Elsevier.)

molecular pincer design for detecting antibodies [5] (Figure 26.5B). In this version of the assay, epitope peptide recognized by the antibody is used as target recognition element of molecular pincer. The peptide is conjugated with fluorochrome-labeled oligonucleotides via a flexible linker. When such conjugate binds to a bivalent antibody, the oligonucleotides are brought to proximity resulting in their annealing (Figure 26.5B) and producing FRET signal in a manner analogous to original molecular pincer design illustrated in Figure 26.1. Molecular pincers for detecting antibodies were shown to work in serum samples and allow rapid detection of antibodies with pM limit of detection. We are currently working on strategies for utilizing molecular pincers for detecting cancer-specific autoantibodies.

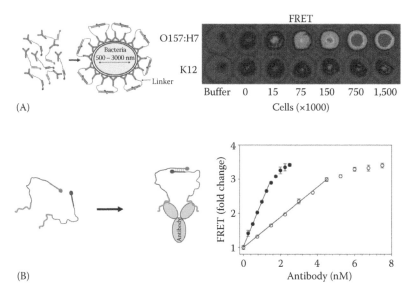

FIGURE 26.5 (A) Left panel: Molecular pincer sensor for detecting *Escherichia coli* O157:H7. Right panel: Fluorescence FRET images of 96-well microplate containing indicated amounts of *E. coli* O157:H7 cells or control cells (*E. coli* K12). (Reprinted from *Anal. Biochem.*, 396, Heyduk, E. and Heyduk, T., Fluorescent homogeneous immunosensors for detecting pathogenic bacteria, 298–303, Copyright 2010, with permission from Elsevier.) (B) Left panel: Design of antigen peptide–based sensors for detecting antibodies. Right panel: Response of the antibody sensor to indicated concentrations of the antibody measured at two concentrations of sensor components (filled and open symbols, respectively). (Reprinted with permission from Tian, L. and Heyduk, T., Antigen peptide-based immunosensors for rapid detection of antibodies and antigens, *Anal. Chem.*, 81, 5218–5225. Copyright 2009 American Chemical Society.)

The advantages of molecular pincers include simple mix-and-read assay design that is derived from the homogeneous nature of the assay. Simplicity of the assay and its speed make molecular pincers good candidates for deployment in point-of-care setting. Current limit of detection for CEA (~100 pM) is similar to what we observed with other targets [3–7]. Limits of detection in pM range will be sufficient for some markers and too high for others. Sensitivity of detection is currently the most important limitation of molecular pincer assays.

26.6 Future Trends

Practical applicability of molecular pincer sensors will be to a large extent determined by the lowest amounts of target molecules that could be detected by the sensors. Thus, one important future trend will be to develop strategies to improve detection sensitivity of the sensors without compromising their beneficial characteristics derived from their homogeneous nature. These improvements could be achieved by using fluorescence probes that would produce much larger signal changes in response to the target compared to "classical" fluorescence probes. Lanthanide chelate probes together with gated time-resolved FRET signal measurements are one of the alternative probes that could provide desired signal enhancement [15,16]. Additionally, the use of time-resolved fluorescence measurements will greatly reduce fluorescence background issues that are associated with determinations of the biomarkers in serum.

Another technological development that we expect to happen in the future is a large-scale multiplexing of molecular pincer assay. Low-level multiplexing would be currently possible through the use of multicolor fluorescence detection. Higher multiplexing could be achievable by modifying molecular pincer assay such that sensors immobilized on solid support could perform the detection. This will allow microarray-like multiplexing where single color fluorescence detection is coupled with using the position of the sensor on the solid support to indicate the identity of the biomarker.

REFERENCES

1. Reen, D. J. (1994) Enzyme-linked immunosorbent assay (ELISA), *Methods Mol Biol* 32, 461–466.
2. Haab, B. B., Dunham, M. J., and Brown, P. O. (2001) Protein microarrays for highly parallel detection and quantitation of specific proteins and antibodies in complex solutions, *Genome Biol* 2, RESEARCH0004.
3. Heyduk, E. and Heyduk, T. (2005) Nucleic acid-based fluorescence sensors for detecting proteins, *Anal Chem* 77, 1147–1156.
4. Heyduk, E., Dummit, B., Chang, Y. H., and Heyduk, T. (2008) Molecular pincers: Antibody-based homogeneous protein sensors, *Anal Chem* 80, 5152–5159.
5. Tian, L. and Heyduk, T. (2009) Antigen peptide-based immunosensors for rapid detection of antibodies and antigens, *Anal Chem* 81, 5218–5225.
6. Heyduk, E. and Heyduk, T. (2010) Fluorescent homogeneous immunosensors for detecting pathogenic bacteria, *Anal Biochem* 396, 298–303.
7. Heyduk, E., Moxley, M. M., Salvatori, A., Corbett, J. A., and Heyduk, T. (2010) Homogeneous insulin and C-peptide sensors for rapid assessment of insulin and C-peptide secretion by the islets, *Diabetes* 59, 2360–2365.
8. Heyduk, T. (2010) Practical biophysics: Sensors for rapid detection of biological targets utilizing target-induced oligonucleotide annealing, *Biophys Chem* 151, 91–95.
9. Selvin, P. R. (1995) Fluorescence resonance energy transfer, *Methods Enzymol* 246, 300–334.
10. Gold, P. and Freedman, S. O. (1965) Specific carcinoembryonic antigens of the human digestive system, *J Exp Med* 122, 467–481.
11. Gold, P. and Freedman, S. O. (1965) Demonstration of tumor-specific antigens in human colonic carcinomata by immunological tolerance and absorption techniques, *J Exp Med* 121, 439–462.
12. Thomson, D. M., Krupey, J., Freedman, S. O., and Gold, P. (1969) The radioimmunoassay of circulating carcinoembryonic antigen of the human digestive system, *Proc Natl Acad Sci USA* 64, 161–167.
13. Chevinsky, A. H. (1991) CEA in tumors of other than colorectal origin, *Semin Surg Oncol* 7, 162–166.
14. Duffy, M. J. (2001) Carcinoembryonic antigen as a marker for colorectal cancer: Is it clinically useful? *Clin Chem* 47, 624–630.
15. Selvin, P. R. (2002) Principles and biophysical applications of lanthanide-based probes, *Annu Rev Biophys Biomol Struct* 31, 275–302.
16. Selvin, P. R. and Hearst, J. E. (1994) Luminescence energy transfer using a terbium chelate: Improvements on fluorescence energy transfer, *Proc Natl Acad Sci USA* 91, 10024–10028.

27

Fluorescent Nanoparticles for Ovarian Cancer Imaging

Xu Hun, Liang Tiao, and Zhujun Zhang

CONTENTS

27.1 Introduction

According to the American Cancer Society's annual report, about 565,650 Americans are expected to die of cancer, more than 1500 people a day. Cancer is the second most common cause of death in the United States, exceeded only by heart disease. In the United States, cancer accounts for one of every four deaths. About 1,437,180 new cancer cases are expected to be diagnosed in 2008 [1].

In addition, ovarian cancer is the leading cause of death from gynecological cancer in the world [2–6]. It is the fifth leading cause of cancer death for women after lung, breast, colorectal, and pancreatic cancers. In 2007, 22,430 new cases of ovarian cancer were diagnosed, accounting for ~3% of all the cancers in women and about 15,280 women will die this year because of the disease in the United States [7].

Because there are few warning signs or symptoms, and malignant cells can escape from the ovarian capsule and disseminate throughout the peritoneal cavity, early detection of ovarian cancer is very difficult [8–13]. Early cancer diagnosis, in combination with the precise cancer therapies, could eventually save millions of lives. Over the last 70 years, despite tremendous advances in our understanding of the molecular and cellular processes of cancer, there has been no change in the age-adjusted mortality due to cancer [14]. In order to further reduce the morbidity and mortality due to cancer, the diagnosis of cancer at the early stage is extremely challenging and has been an active research area these days.

The diagnosis of cancer has been greatly improved with the recent developments in nanobiotechnology. Nanobiotechnology, defined as biomedical applications of nanosized systems, is a rapidly developing area within nanotechnology. Nanomaterials, which measure 1–1000 nm, allow unique interaction with biological systems at the molecular level. They can also facilitate important advances in detection, diagnosis, and treatment of human cancers and have led to a new discipline of nano-oncology [15,16]. Nanoparticles (NPs) are being actively developed for tumor imaging in vivo, biomolecular profiling of cancer biomarkers, and targeted drug delivery [17].

Fluorescent NPs can be used for profiling of tumor biomarkers and for detection of multiple genes and matrix RNA with fluorescent in situ hybridization. And fluorescent NPs, such as polymer fluorescent NPs (PFNPs), quantum dots (QDs), and up-conversion NPs enable highly sensitive optical imaging of cancer at cellular and animal level. In this chapter, we first review the types of fluorescent NPs for ovarian cancer imaging. Then a novel kind of PFNPs was synthesized and its application for ovarian cancer imaging with fluorescence microscopic imaging technology was presented.

27.1.1 QDs Imaging

Fluorescent QDs are inorganic fluorescent nanocrystals that overcome many limitations of dyes and provide a useful alternative for studies that require long-term and multicolor imaging of cellular and molecular interactions [18,19]. QDs have broad potential for use in various applications in the research, management, and treatment of cancer [20–24].

Monodisperse CHP–QD hybrid NPs were prepared using simple mixing as reported by Urara Hasegawa. They found that CHPNH$_2$–QD NPs were delivered into human cells more efficiently than QDs complexed with cationic liposome (Figure 27.1) (cholesterol-bearing pullulan [CHP], CHP modified with amino groups [CHPNH$_2$]). The CHPNH$_2$–QD complexes were uniformly internalized into cells without being aggregated. Thus, CHPNH$_2$ nanogel has high potential for use in long-term live-cell

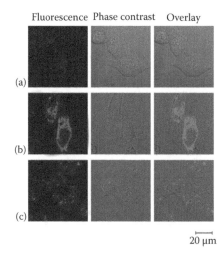

FIGURE 27.1 CLSFM images of HeLa cells labeled with (a) QD, (b) CHPNH$_2$ (15)–QD, and (c) liposome–QD. HeLa cells were cultured in DMEM containing 1 nM QD, CHPNH$_2$ (15)–QD hybrid nanoparticles, and liposome–QD complexes for 3 h. The concentration of QD was 1 nM. After cells were washed with PBS and treated with fresh DMEM, they were observed by CLSFM. (From Hasegawa, U. et al., *Biochem. Biophys. Res. Commun.*, 331, 917–921, 2005.)

imaging. The interaction of QDs with cells was successfully controlled by the amino-group content of the CHPNH₂ nanogel. In particular, neutral CHP nanogels suppressed the internalization of QD. Such stealth-type nanocarriers that do not interact with cells or tissue are important for in vivo imaging to target specific tissue by the intravenous injection of QD. These internalization characteristics of CHP and CHPNH₂ nanogels may be used to develop carriers for not only QDs but also drugs and other medical diagnostic products.

Yang et al. [25] prepared a novel imaging agents, folate receptor (FR)-targeted liposomes encapsulating hydrophilic CdTe QDs, and they also use them as luminescence probes for live-cell imaging. Hydrophilic CdTe QDs were directly synthesized in the water phase, and FR-targeted QD liposomes were prepared by hydrating the lipid thin film with CdTe suspension. Formulations were characterized by UV–visible and fluorescent measurements, liposomal particle size, and zeta potential. The targeting and imaging ability of FR-targeted liposomes were investigated against the human uterine cervix cancer cell line (HeLa). Furthermore, the cytotoxicity of QD liposomes was evaluated by HeLa cells incubated with FR-targeted QD liposomes, nontargeted QD liposomes, and free QDs. The results showed that FR-targeted QD liposomes were spherically shaped with high fluorescence yield, excellent photochemical stability, good cancer targeting, and minimal cytotoxicity. The average size of FR-targeted fluorescence liposomes was ~105 nm, and their size distribution was rather narrow. After storage at 4°C for 11 months, QD liposomes maintained similar size and did not show any leakage of QDs. FR-targeted CdTe QD liposomes, which can target tumor cells via FR-mediated endocytosis, would become an attractive probe for tumor cell or tissue imaging for a long-time monitoring [25].

Zdobnova et al. also described a construct based on fluorescent CdSe/CdS nanocrystals and genetic engineering recombinant mini-antibodies, intended for detecting cells overexpressing the HER-2/neu oncomarker [26].

Sun et al. reported a reversible photoinduced fluorescence enhancement (photoactivation) of endocytosed mercaptoacetic acid–capped CdSe QDs (MAA QDs) and the pH dependence of MAA QD photoluminescence (PL) in SKOV-3 human ovarian cancer cells. They found that the fluorescence emission of MAA QDs taken up directly by SKOV-3 cells without the need for extra capping ligands or permeabilization steps remains bright and stable for at least 14 days. These intracellular fluorescent nanocrystals do not colocalize with low-pH lysosomes, and the emission of the MAA QDs in fixed cell preparations is quenched by acidic buffer, suggesting that a low-pH environment in cellular vesicles quenches QD fluorescence. Photoactivation of intracellular MAA QD luminescence is dependent on the excitation energy and is related to the metabolic activity of the cells. These active interactions between cells and nanocrystals demonstrate the potential of MAA QDs as intracellular environmental sensors [27].

Pan et al. prepared water-soluble QDs conjugated with epidermal growth factor (EGF) using 1-ethyl-3-(3-dimethylamino propyl)-carbodiimide (EDAC) as a cross-linking reagent. The properties of the QDs and QDs–EGF conjugates were characterized by capillary electrophoresis (CE), UV–visible absorption, and PL spectra. The QDs–EGF conjugates recognized the ovarian cancer cells; SKOV-3 was also presented [28].

27.1.2 PFNPs Imaging

Fluorescent organic NPs have received considerable attractive research interactions, because of the high flexibility in material synthesis and ease for NP generation. However, investigations are in initial stage research; very few research groups were working in this area [29–32]. As far as the application is concerned, Nakanishi and coworkers [33–36] have demonstrated that perylene and phthalocyanine NPs showed size-dependent fluorescent properties from those of bulk samples [37–41].

To develop HER-2-targeted drug-encapsulated NPs, Alexis et al. conjugated the anti-HER-2 affibody to the thiol-reactive maleimide group of the poly(D,L-lactic acid)–poly(ethylene glycol)–maleimide (PLA–PEG–Mal) copolymer of the previously formed NPs through a stable thioether bond and evaluated the targeting specificity and efficacy using fluorescence microscopy (Figures 27.2 and 27.3). They have demonstrated that the NP–affibody bioconjugates are specifically and efficiently internalized to HER-2-positive cancer cells (i.e., ovarian, breast, and pancreatic cancers), thereby providing a promising way to deliver chemotherapeutic drugs in a selective manner [42].

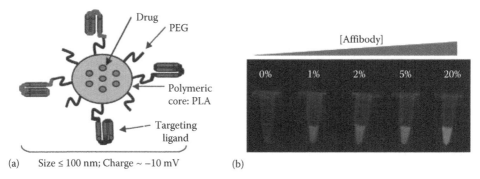

FIGURE 27.2 (a) Schematic diagram of the PLA–PEG–Mal NP–affibody bioconjugates with encapsulated drug. The hydrophilic poly(ethylene glycol) (PEG) chains on the surface limit protein absorption at the hydrophobic polymeric surface to form "stealth" nanoparticles. (b) Fluorescence images of fluorescent affibody (Alexa fluor 532, white) conjugated to NPs. After washing the NP–affibody bioconjugates, the fluorescence signal increases with an increased amount of fluorescent affibody (0% → 20% polymer/affibody molar ratio) at the NP surface, confirming the efficiency of the chemical conjugation.

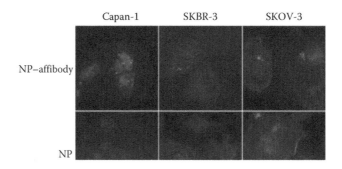

FIGURE 27.3 Fluorescence microscopy of NP–affibody bioconjugates incubated with HER-2-positive cell lines. Capan-1 cells, SKBR-3 cells, and SKOV-3 cells were grown on chamber slides and incubated in OptiMEM medium supplemented with 50 mg NBD fluorescent dye encapsulated into PLA–PEG–affibody bioconjugate NPs (shown as white dots) with (upper row) or without affibody (lower row) for 2 h prior to imaging with fluorescence microscopy (60× magnification). Cell nuclei and the actin cytoskeleton are stained blue (4′,6-diamidino-2-phenylindole) and red (Alexa fluor phalloidin 488), respectively. They show up here as light colored nuclei and darker cytosol. The deconvolved fluorescence images represent the mid-cross-section of the cells after wash (3× magnification), permeabilization, and staining steps.

Nobs et al. explored two approaches to target biodegradable PLA NPs to tumor cells. Anti-HER-2 mAbs (trastuzumab, Herceptin) and anti-CD20 mAbs (rituximab, Mabthera) were used as targeting ligands. Two cell lines were used: SKOV-3 human ovarian cancer cells, expressing HER-2 antigen, and Daudi lymphoma cells, expressing CD20 antigen. On each cell line, the antibody directed against the nonexpressed antigen served as isotype-matched irrelevant control immunoglobulin. In the direct approach, NPs exposing mAbs at their surface were incubated with the two tumor cells. Cell interaction of fluorescence-labeled NPs was measured by flow cytometry [43].

A water-soluble fluorescent hyperbranched conjugated polyelectrolyte (HCPE) with a unique double-layered architecture synthesized via the combination of alkyne polycyclotrimerization and alkyne–azide "click" reaction for live-cell imaging was reported by Pu et al. The resulted organic nanospheres possess high quantum yield (30% in buffer), good solution stability, and low cytotoxicity. Using breast cancer cell MCF-7 as an example, these core–shell nanospheres are internalized efficiently by the cells and accumulated in the cytoplasm to give bright fluorescence (Figure 27.4). Photostability study reveals that these nanospheres are significantly more stable than commercial dyes, such as fluorescein, rhodamine,

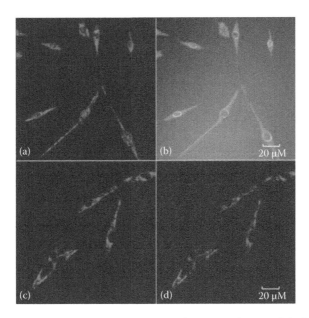

FIGURE 27.4 Confocal laser scanning microscopy (CLSM): (a) fluorescence image and (b) fluorescence–transmission overlapped image of MCF-7 cells co-stained by P2 and P1; time-resolved CLSM fluorescence images of MCF cells stained by P2 under laser scanning for (c) 0 min and (d) 15 min. P1, hyperbranched poly[9,9-bis((60-*N*,*N*,*N*-trimethylammonium)-hexyl)-2,7-fluorene-*co*-phenylene dibromide]. P2, core–shell hyperbranched conjugated polyelectrolyte (HCPE).

and Cy5, which demonstrates the great potential of organic polymeric nanomaterials in long-term clinical applications [44].

27.1.3 Organic–Inorganic Hybrid NPs Imaging

In order to create more robust emitters, researchers have developed hybrid organic–inorganic NPs from organic dye molecules and amorphous silica. As a matrix material for fluorescent probes, silica provides a chemically and mechanically stable vehicle, which can protect the encapsulated dye molecules from external perturbations, while exposing a biocompatible and easily functionalized surface to the environment and in some cases enhancing the photophysical properties of the encapsulated dyes [45]. The versatility of organic–inorganic silica as a host material for fluorescent dyes is used for many applications, specifically in the fields of nanobiotechnology and the life sciences.

The organic–inorganic silica NPs are usually synthesized by two main methods namely the Stober method and the microemulsion method. The size varies from 2 to 200 nm and can be controlled. The NPs produce light of high intensity due to the large number of dye molecules within each particle and they are quite photostable. The photostability is mainly due to the polymer coating that prevents the penetration of oxygen, thereby reducing the bleaching [46]. Many of these NPs exhibit good biocompatibility and water solubility and universal bioconjugation strategies can be used for attaching biomolecules to them. The versatile silica chemistry is used for bioconjugation through functional groups such as thiol, amino, and carboxyl groups. Interactions between avidin and biotin are also employed [47].

Huang et al. synthesized fluorescent silica NPs doped with an organic dye rhodamine B isothiocyanate (TRITC) with water-in-oil microemulsion method and detected the ovarian carcinoma cells SKOV-3 based on the prepared NPs [48]. The results demonstrated that the fluorescent NPs as labels in tumor cell showed a great improvement in sensitivity, selectivity, and multiplexing capacity (Figures 27.5 and 27.6)

Except organic–inorganic silica NPs were used in the ovarian cancer imaging, other kinds of organic–inorganic hybrid NPs were prepared with the researchers. Shi and their colleagues have designed dendrimer-entrapped gold NP (GNP) systems that are stable, water-soluble, and biocompatible. The researchers start out by synthesizing GNPs within amine-terminated dendrimers. Next, dye molecules

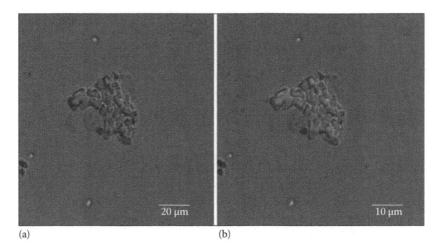

(a)   (b)

FIGURE 27.5 (a) Optical image and (b) fluorescent image of antibody nanoparticles culture with SKOV-3 cells.

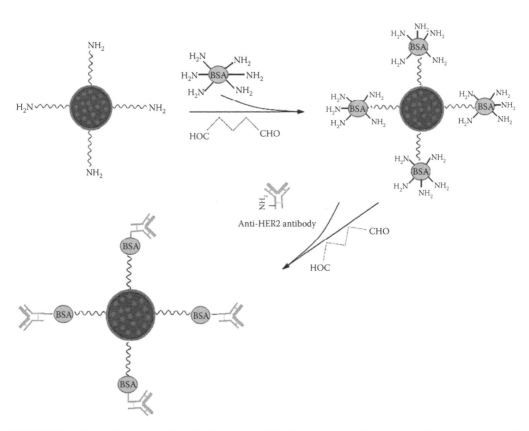

FIGURE 27.6 Schematic representation of antibody immobilization process onto functionalized fluorescent core–shell nanoparticles.

and a targeting molecule, folic acid, are attached to the ends of the dendrimers. Many cancer cells, including those implicated in cancers of the ovary, kidney, uterus, testis, brain, colon, and lungs, tend to overexpress folic acid receptors. Owing to the folic acid attachments grafted onto the dendrimer NPs, the dendrimer NPs are seen to latch onto the cancer cells via these folic acid receptors. Since the dendrimer NPs are also equipped with dye molecules, the high concentrations of NPs accumulated in the cancer cells can be imaged by confocal microscopy, and indeed diseased cells can be easily identified apart from healthy cells [49].

27.1.4 Up-Converting NPs Imaging

Up-converting NPs show great promise for use in biological imaging techniques [50–57]. These NPs are capable of emitting visible and near-infrared (NIR) light when excited with NIR radiation. Unlike two-photon absorption in fluorescent dyes and QDs, the absorption proceeds through real intermediate energy levels which increase the efficiency of the process by orders of magnitude over conventional two-photon absorption. The improved efficiency allows for the use of lower excitation densities and thus relatively inexpensive laser diodes as excitation sources as opposed to costly ultrafast pulsed lasers that are employed in traditional two-photon techniques. These NPs thus combine all the advantages of traditional two-photon imaging such as absence of photodamage, low autofluorescence, high detection sensitivity, and high penetration depth of the excitation light at low power densities (1–100 W/cm^2). Also, up-converting NPs may have several additional advantages over traditional fluorophores used in two-photon imaging. They possess nonoverlapping absorption and emission bands which lead to the ability to multiplex. The NIR up-conversion emission of Tm^{3+} (i.e., 800 nm emission) is in a portion of the electromagnetic spectrum where human tissue is relatively transparent [55]. Recent studies have also suggested that the up-converting NaYF$_4$ NPs have low cytotoxicity [50,51,55,56,58].

Boyer et al. present a technique for the replacement of oleate with a PEG-phosphate ligand as an efficient method for the generation of water-dispersible NaYF$_4$ NPs [59]. The PEG-phosphate ligands are shown to exchange with the original oleate ligands on the surface of the NPs, resulting in water-dispersible NPs. The up-conversion intensity of the NPs in aqueous environments was found to be severely quenched when compared to the original NPs in organic solvents. This is attributed to an increase in the multiphonon relaxations of the lanthanide-excited state in aqueous environments due to high-energy vibrational modes of water molecules. This problem could be overcome partially by the synthesis of core–shell NPs which demonstrated improved photophysical properties in water over the original core NPs. The PEG-phosphate-coated up-converting NPs were then used to image a line of ovarian cancer cells (CaOV3) to demonstrate their promise in biological application (Figure 27.7).

In addition to the familiar fluorescent NPs that were used in the ovarian cancer imaging, other kinds of NPs, such as magnetic NPs and GNPs, were also reported by the researchers.

27.1.5 Magnetic NPs Imaging

Magnetic resonance imaging (MRI), or nuclear MRI (NMRI), is primarily a medical imaging technique most commonly used in radiology to visualize detailed internal structure and limited function of the body. MRI provides much greater contrast between the different soft tissues of the body than computed tomography (CT) does, making it especially useful in neurological (brain), musculoskeletal, cardiovascular, and oncological (cancer) imaging. Unlike CT, it uses no ionizing radiation, but uses a powerful magnetic field to align the nuclear magnetization of (usually) hydrogen atoms in water in the body. Radio frequency (RF) fields are used to systematically alter the alignment of this magnetization, causing the hydrogen nuclei to produce a rotating magnetic field detectable by the scanner. This signal can be manipulated by additional magnetic fields to build up enough information to construct an image of the body [60]. Recently, magnetic NPs were used as the MRI agents.

MRI provides a noninvasive means for tumor detection with excellent soft-tissue contrast and anatomic resolution, without radiation exposure. Wang and coworkers investigate the diagnostic performance of new formulations of FR-targeted MR contrast agents (P866 or P1048) in an ovarian tumor model with MRI (Figure 27.8). They focused their studies on ovarian cancer as a representative tumor with high levels of FR overexpression in order to evaluate the feasibility of FR-targeted MR imaging [61,62].

Xu et al. report a novel approach for the preparation of a biotinylated dendrimer-based MRI magnetic NPs agent 5, in which a unique disulfide bond in the core of the Gd(III)-1B4M-DTPA-chelated G2 PAMAM dendrimer was reduced and then attached to a maleimide-functionalized biotin. The new

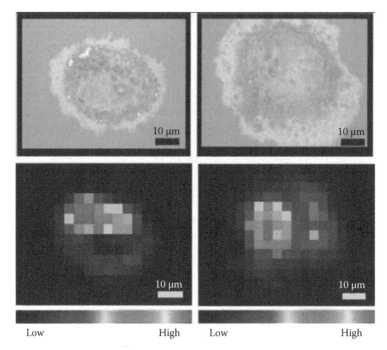

FIGURE 27.7 Up-conversion imaging of ovarian cancer cells, CaOV3 incubated with PEG(2000)-phosphate-coated NaYF$_4$: Tm^{3+} 0.5 mol% and Yb^{3+} 30 mol% NPs using confocal microscopy. The top row consists of bright field images of the cells, while the bottom images were collected using 800 nm up-conversion emission from Tm^{3+} under 980 nm laser diode excitation.

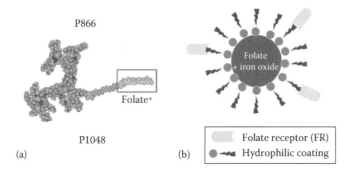

FIGURE 27.8 FR-targeted MR contrast agents: (a) gadolinium P866 and (b) iron oxide P1048. P866 is a high-relaxivity dimeric gadolinium chelate conjugated to folate. P1048 consists of an iron oxide core conjugated to folate.

MRI agent 5 features a well-defined dendron structure and a unique biotin functionality. Immobilization of up to four copies of biotinylated dendrimer 5 to fluorescently labeled avidin yields a supramolecular avidin–biotin–dendrimer–Gd(III) complex. Validation of the complex in mice bearing ovarian tumors demonstrates that the avidin–biotin–dendrimer-targeting system efficiently targets and delivers sufficient amounts of chelated Gd(III) and fluorophores (e.g., Rhodamine green) to ovarian tumors to produce visible changes in the tumors by both MRI and optical imaging, respectively. Thus, the avidin–biotin–dendrimer complex may be used as a tumor-targeted probe for dual-modality MRI and fluorescence imaging [63,64].

27.1.6 GNPs Imaging

To date, GNPs are finding increased applications in certain biological fields, such as DNA hybridization detection, immunoassay, single particle tracking, drug delivery, and cancer diagnostics and therapy

[65–74]. Among these applications, optical signal is mostly based on the fact that GNPs resonantly scatter light upon excitation of the surface plasmon. Recently, He et al. developed a new method for imaging of cells using GNPs as fluorescent probes. They first synthesized colloidal GNPs with sizes varying between 16 and 55 nm by the citrate reduction of chloroauric acid and then investigated the optical properties of the as-prepared GNPs by various optical techniques including UV–visible absorption spectroscopy, fluorescence spectroscopy, fluorescence correlation spectroscopy (FCS), and fluorescence microscopy. On the basis of this excellent anti-photobleaching of GNPs and easy photobleaching of cellular autofluorescence, they developed a new method for imaging of cells using GNPs as fluorescent probes. After photobleaching of cellular autofluorescence by illumination of strong light, they clearly observed GNPs in living HeLa cells, which entered cells via the endocytic pathway during cell differentiation and proliferation processes. Furthermore, they conjugated GNPs to anti-EGF receptor (anti-EGFR) antibodies and successfully used the anti-EGFR/GNPs conjugates as targeted probes for fluorescent imaging of cancer cells [74].

In this study, a new type of PFNPs has been developed with precipitation polymerization method. This kind of PFNPs shows the advantages of excellent photostability, low leakage, and uniform particles' size. And the PFNPs were successfully modified with anti-Her-2 monoclonal antibody. This kind of fluorescence probe based on the anti-Her-2 monoclonal antibody–conjugated PFNPs has been used to detect ovarian cancer cells with fluorescence microscopic imaging technology.

27.2 Experimental

27.2.1 Materials

Methacrylic acid (MAA), acetonitrile, KH_2PO_4, and Na_2HPO_4 were purchased from Xi'an Chemical Reagent Company (Xi'an, China). Butyl rhodamine B (BTRB) and azobisisobutyronitrile (AIBN) were obtained from Shanghai Chemical Plant (Shanghai, China) and used as supplied. Trimethylolpropane trimethacrylate (Trim) was from Tokyo Kasei Kogyo Co., Ltd. (Tokyo, Japan). *N*-Hydroxysuccinimide (NHS) and EDAC were obtained from Acros (Morris Plains, NJ). Anti-Her-2 monoclonal antibody was obtained from Beijing Biosynthesis Biotechnology Co., Ltd (Beijing, China). RPMI-1640 medium was purchased from HyClone Biochemical Product Co., Ltd. (Beijing, China). Unless otherwise stated, all chemicals and reagents used in this study were of analytical grade quality.

27.2.2 Synthesis of PFNPs

PFNPs were synthesized by using precipitation polymerization method described in the following. The dye BTRB, MAA (0.8 mmol), acetonitrile (40 mL), and Trim (0.48 mmol) were placed into a 100 mL glass tube and the mixture was sonicated for about 10 min. Subsequently, AIBN (3 mg) was added. Then the mixture was purged with N_2 for 10 min. The glass tube was sealed and thermostated at 58°C to start the polymerization process. After 12 h, the production was collected by centrifugation, successively washed three times with methanol and once with acetonitrile. The obtained NPs were air-dried. The synthesized PFNPs were characterized by spectrofluorophotometer for fluorescence intensity and by transmission electron microscope (TEM; Hitachi H700) for the morphology. The size and size distribution of PFNPs dispersed. in water were measured by particle size analyzer (PSA, BI-90Plus, Brookhaven Instruments Corp, Holtsville, NY).

27.2.3 Test of Photostability

To investigate whether dye BTRB molecules is photostable when embedded in polymer NPs, the intensity of the fluorescence vs. time was monitored [75,76]. Measurements were performed in solution of PFNPs or free dye. The solutions were continuously illuminated by the 150 W xenon lamp at their optimal excitation wavelengths for 60 min using spectrofluorophotometer (Shimadzu RF-540). And the fluorescence intensity was acquired every 2 min over a 60 min period. A leaching test was performed as described by Wang and coworkers [77] with 1 mg/mL NPs solution, monitoring the fluorescence of solution of PFNPs over 24 h.

27.2.4 Covalent Immobilization of the Anti-Her-2 Monoclonal Antibody onto PFNPs Surface

The anti-Her-2 monoclonal antibody was directly immobilized onto the PFNPs with well-established carbodiimide method. The immobilization protocols were the following: (1) 2 mg of PFNPs was dispersed into 1.0 mL phosphate-buffered saline (PBS); (2) 10 mg EDAC and 1 mg NHS in 1.0 mL water were added and the mixture was allowed to react for 15 min at room temperature with continuous mixing; (3) after the NPs were washed twice with PBS, 2 mg anti-Her-2 monoclonal antibody dissolved into 0.5 mL PBS was added and the mixture was allowed to react at room temperature for 4 h with constant mixing; and (4) the product, namely, anti-Her-2 monoclonal antibody–conjugated PFNPs, was washed with PBS several times and kept at 4°C in PBS.

27.2.5 Cell Culture

Human ovarian cancer cells (SKOV-3) and MRC-5 normal cell were obtained from Type Culture Collection of Chinese Academy of Sciences (Shanghai, China). SKOV-3 cell and MRC-5 cells were routinely maintained in RPMI-1640 medium containing 10% fetal serum bovine (FBS) at 37°C in a humidified 5% CO_2–95% air atmosphere. To performance cell imaging, the cells were first plated on poly-L-lysine-coated six-well plastic dishes for 24 h before microscopic observation. And then the anti-Her-2 monoclonal antibody–conjugated PFNPs-dispersed culture media was added. Incubated for another fixed time, the cells were washed several times with PBS solution (pH 7.4) to remove nonspecifically adsorbed anti-Her-2 monoclonal antibody–conjugated PFNPs. The plastic dishes with cells were mounted over the microscope stage for fluorescence microscopic observation. During observation, an incubator was mounted on the stage for the maintenance of proper temperature and humidity.

27.2.6 Fluorescence Microscopic Imaging

Fluorescence microscopic imaging was performed in Olympus inverted microscope system (Olympus, Model IX70, Tokyo, Japan) with a 100 W high-pressure mercury lamp (Olympus, Model BH2-RFL-T3, Tokyo, Japan) used as the light source. The six-well plastic dishes used mentioned earlier was mounted over the microscope stage. The excitation light which comes from high-pressure mercury was introduced through the inverted microscope objective from underneath the chamber. The fluorescence image of the cell is collected by a microscope objective. A charge-coupled device (CCD) camera (Pixera, model PVC100C, Los Gatos, CA) interfaced with a Pentium computer was employed for the acquisition of imaging of the cell.

27.3 Results and Discussion

27.3.1 Synthesis of PFNPs

In the present study, PFNPs were prepared by utilizing a precipitation polymerization technique. The technique is based on the precipitation of polymeric spherical particles as they reach their solubility limit in the system due to their increasing molecular weight or as a result of cross-linking. PFNPs were synthesized by using MAA as hydrophilic monomer, Trim as cross-linker, and BTRB as fluorophore. When radical initiator, AIBN, was added and the mixture was heated, the polymerization reaction was initiated by free radicals. The fluorescent dye was embedded into the three-dimensional network of the polymer when the polymer was produced. With this method, the PFNPs can be prepared easily. This technique has the advantage that the PFNPs can be prepared under mild conditions using inexpensive reagents and simple procedures. The resulting PFNPs have advantages of uniform size, good dispersivity, and easily being labeled which makes them popular for various research and clinical applications.

27.3.2 PFNPs–Anti-Her-2 Monoclonal Antibody Conjugation

After the polymerization reaction of MAA and Trim, some carboxyl groups have been introduced to the NP's surface. These carboxyl groups made the modification and the bioconjugation of the NPs easier. PFNPs–anti-Her-2 monoclonal antibody conjugation were developed using the coupling strategy between anti-Her-2 monoclonal antibody and NPs using EDAC as the coupling agent. Figure 27.9 depicts the fabrication of a PFNPs–anti-Her-2 monoclonal antibody conjugation and how it captures a cancer cell. The Her-2 receptor is a glycoprotein that is found to be expressed in ovarian cancer cells. With such a unique labeling method, thousands of BTRB could be labeled with the Her-2 monoclonal antibody and targeted one cell, which led to the strong fluorescence signal and finally resulted in the increased sensitivity. In this study, we use the anti-Her-2 monoclonal antibody–conjugated PFNPs to target the ovarian cancer cells which express the Her-2 receptor.

27.3.3 Characterization PFNPs

The preparation of NPs yielded spherical beads as seen in the TEM micrograph in Figure 27.10a. And the PFNPs conjugated with anti-Her-2 monoclonal antibody were also characterized by using TEM. It can be seen from Figure 27.10b that the PFNPs were also kept well-dispersed after they were conjugated with anti-Her-2 monoclonal antibody. The size distribution of PFNPs was measured by PSA. The typical diameter of the PFNPs is about 105 nm, with a rather narrow particle size distribution. Figure 27.10c shows a typical size distribution profile of PFNPs. PSA studies showed that the PFNPs obtained had values ranging from 91 to 132 nm. The narrow particle size distribution and good dispersivity of the PFNPs would make them convenient for various research and clinical applications.

The used precipitation polymerization technique allows NPs to be prepared without the use of stabilizing surfactants. Further conjugation of the particles to biomolecules is easier because there is no need to wash the particles off surfactant molecules, which often requires multiple washing steps when inverse microemulsion techniques are used to prepare NPs.

In addition, spectrofluorometric measurements were used to characterize the NPs. The excitation and emission spectra of fluorescent dye and PFNPs were measured. Free fluorescent dye exhibited an emission at 584 nm when excited at 540 nm in aqueous solution. However, the emission maximum of the PFNPs shifted 6 nm to the longer wavelength compared with the free dye (Figure 27.11), indicating that the spectral characterization of the fluorescent dye did not change to a great extent when it was embedded inside the polymer NPs.

27.3.4 Leaching Test

As always, leaching of fluorescent dye for NPs is a critical factor. The experimental result showed fluorescence intensity of the PFNPs only decreased 5% after 24 h continuous washing (Figure 27.12). The leaching speed of this kind of PFNPs was obviously lower than that of other dye-doped silica NPs in the issued papers [77]. The hydrogen bond interaction between the dye molecule and the polymer molecule may be contributed to the low leaching of dye from the polymer NPs. And the other reason maybe was that the structure of the polymer is more compact and the pore in the molecule of the polymer is much smaller than that in silica NPs. This was because the diameter of the pore in the silica NPs is about

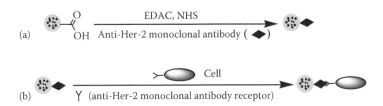

FIGURE 27.9 Schematic representation of (a) the fabrication of a PFNPs–anti-Her-2 monoclonal antibody conjugation and (b) its capturing of a cancer cell.

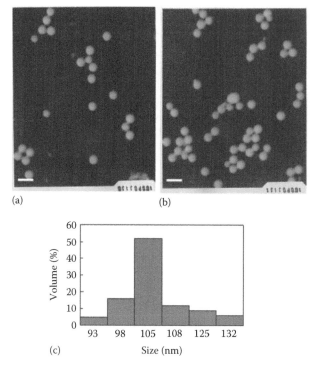

(a)   (b)

(c)

FIGURE 27.10   TEM image of (a) PFNPs and (b) after their conjugation to anti-Her-2 monoclonal antibody. (c) The particle size distribution profiles of PFNPs. Scale bars are 100 nm for (a) and (b).

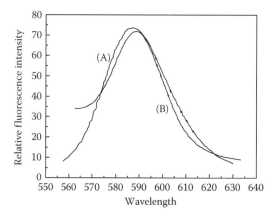

FIGURE 27.11   Fluorescence emission spectra of (A) free dye and (B) PFNPs in aqueous solution.

2.27 nm [78]. We also found that the speed of leaching is very fast during the first hours in aqueous solution, and then the speed of leaching slows. One possibility is that free dye that was not washed away during the synthesis procedure may have been attached to the outer surface of the polymer NPs. Carefully washing would eliminate the phenomenon in future experiment.

27.3.5 Photostability

Photobleaching is known to be dependent on solvent interactions and is thought to occur as a bimolecular reaction between fluorescent and, for example, dissolved oxygen [79,80]. Figure 27.13 shows the

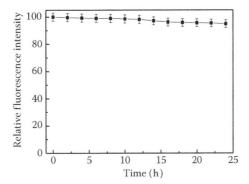

FIGURE 27.12 Leaching stability of PFNPs.

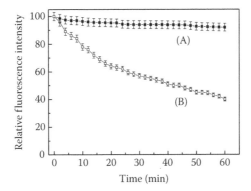

FIGURE 27.13 Photostability of (A) PFNPs and (B) free dye in aqueous solutions.

photobleaching behavior of the PFNPs and free fluorescent dye. The results showed that there was no markedly photobleaching for the NPs over a long period of continuous excitation with a 150 W xenon lamp at optimal excitation wavelength of PFNPs. The PFNPs show significantly less photobleaching than free fluorescent dye. And this result comes with the experimental phenomenon of Tan and coworkers [81] suggesting that the polymer NPs matrix surrounding free fluorescent dye molecules acted as a barrier to protect BTRB from the surrounding environment.

27.3.6 Application of the PFNPs in Ovarian Cancer Imaging

Immunolabeling of cell surface markers is commonly used in cell biology, immunology, and clinical laboratories. Her-2 receptor is a surface marker of ovarian cancer (SKOV-3). We tested the ability to label Her-2 receptor–positive ovarian cancer by using anti-Her-2 monoclonal antibody–conjugated PFNPs. When incubated in the presence of NPs, the cells grow at a normal rate and the NPs target the cells. Using fluorescence microscopy, the imaging of cells is obtained (Figure 27.14a). It was found that the PFNPs with anti-Herr-2 monoclonal antibody could capture SKOV-3 cell. For anti-Her-2 monoclonal antibody–free PFNPs, there was no obvious interaction between the PFNPs and SKOV-3 cells. Therefore, the cell can be seen in the bright field (Figure 27.14b) but no fluorescence imaging was observed in the fluorescent field (Figure 27.14c). Control experiments showed that the anti-Her-2 monoclonal antibody–free PFNPs were incubated with MRC-5 cells for 30 min under the same conditions; no capture occurred (Figure 27.14d). All of these results indicate that the anti-Her-2 monoclonal antibody–conjugated PFNPs can capture the SKOV-3 cells through the specific recognition interaction between anti-Her-2 monoclonal antibody and Her-2 receptor.

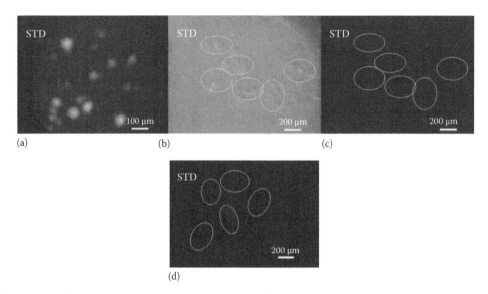

FIGURE 27.14   Fluorescence microscopic images of SKOV-3 cells (a) after incubation with anti-Her-2 monoclonal antibody–conjugated fluorescent nanoparticle and (c) SKOV-3 cells with anti-Her-2 monoclonal antibody–free PFNPs. (b) Images of (c) in the bright field. (d) Fluorescence microscopic images of MRC-5 cells after incubation with anti-Her-2 monoclonal antibody–conjugated fluorescent nanoparticle.

27.4 Conclusions

In conclusion, we have developed a new kind of highly fluorescent, extremely photostable, and easily biomodificatory PFNPs. The proposed PFNPs were easily synthesized with the precipitation polymerization method. The application of the proposed PFNPs was tested by their targeting the ovarian cancer with fluorescence microscopic imaging when they were conjugated with anti-Her-2 monoclonal antibody. The experimental results demonstrate that the anti-Her-2 monoclonal antibody–conjugated PFNPs can effectively recognize ovarian cancer cells and exhibited good sensitivity and exceptional photostability, which would provide a novel way for the diagnosis and curative effect observation of ovarian cancer cells. The proposed PFNPs will be extensively used in biochemical analysis, biomolecular interaction studies, and clinical research for their excellent fluorescence properties and easy preparation and modification [82].

27.5 Future Trends

Cancer is one of the major causes of mortality and morbidity in the world and the incidence of cancer continues to increase. The important issues in improving treatment regimens are (1) development of advanced imaging technologies for early diagnosis and to track therapeutic responses; (2) utilization of targeting moieties to specifically and efficiently deliver drugs to tumor tissues; and (3) discovery in cancer biology and genetics leading to new knowledge for cancer treatment. Furthermore, more studies are needed to explore new methods that improve the biocompatibility of NPs should be developed to reduce the undesirable side effects.

ACKNOWLEDGMENTS

This research was supported by the National Natural Science Foundation of China (21005045); the Scientific and Technical Development Project of Qingdao (09-1-3-45-jch); and the Scientific Research Startup Foundation of Qingdao University of Science and Technology for Talents.

REFERENCES

1. American Cancer Society. *Cancer Facts & Figures*. Atlanta, GA: American Cancer Society, 2008, pp. 2–3, http://www.cancer.org/Research/CancerFactsFigures/index (accessed 3 Feb, 2012).
2. Indraccolo, S., V. Tisato, S. Agata, L. Moserle, S. Ferrari, M. Callegaro, L. Persano, M. Palma, M. Scaini, and G. Esposito. 2006. Establishment and characterization of xenografts and cancer cell cultures derived from BRCA1$^{-/-}$ epithelial ovarian cancers. *Eur. J. Cancer* 10:1475–1483.
3. Simon, I., Y. Liu, K. L. Krall, N. Urban, R. L. Wolfert, N. W. Kim, and M. W. McIntosh. 2007. Evaluation of the novel serum markers B7-H4, Spondin 2, and DcR3 for diagnosis and early detection of ovarian cancer. *Gynecol. Oncol.* 106:112–118.
4. Edwards, B. K., M. L. Brown, P. A. Wingo, H. L. Howe, E. Ward, and L. A. G. Ries. 2005. Annual report to the nation on the status of cancer, 1975–2002, featuring population-based trends in cancer treatment. *J. Natl Cancer Inst.* 97:1407–1427.
5. Mi, R. R. and H. Ni. 2003. Mdm2 sensitizes a human ovarian cancer cell line. *Gynecol. Oncol.* 90:238–244.
6. Wang, W. W., D. Das, S. A. McQuarrie, and M. R. Suresh. 2007. Design of a bifunctional fusion protein for ovarian cancer drug delivery: Single-chain anti-CA125 core-streptavidin fusion protein. *Eur. J. Pharm. Biopharm.* 65:398–405.
7. American Cancer Society. *Cancer Facts & Figures*. Atlanta, GA: American Cancer Society, 2007, http://www.cancer.org
8. Zhang, X. Y., J. Feng, X. Ye, Y. Yao, P. Zhou, and X. X Chen. 2006. Development of an immunocytokine, IL-2-183B2scFv, for targeted immunotherapy of ovarian cancer. *Gynecol. Oncol.* 103: 848–852.
9. Roby, K. F., C. C. Taylor, J. P. Sweetwood, Y. Cheng, J. L. Pace, and O. Tawfik. 2000. Development of a syngeneic mouse model for events related to ovarian cancer. *Carcinogenesis* 21: 585–591.
10. American Cancer Society. *Learn About Cancer*. Atlanta, GA: American Cancer Society, http://www.cancer.org/Cancer/OvarianCancer/index (accessed 3 Feb, 2012).
11. Giles, J. R., H. L. Shivaprasad, and P. A. Johnson. 2004. Ovarian tumor expression of an oviductal protein in the hen: A model for human serous ovarian adenocarcinoma. *Gynecol. Oncol.* 95:530–533.
12. Rosen, D. G., L. Wang, N. Atkinson, Y. Yu, K. H. Lu, and E. P. Diamandis. 2005. Potential markers that complement expression of CA125 in epithelial ovarian cancer. *Gynecol. Oncol.* 99: 267–277.
13. Rodriguez Burford, C., M. N. Barnes, W. Berry, E. E. Partridge, and W. F. Grizzle. 2001. Immunohistochemical expression of molecular markers in an avian model: A potential model for preclinical evaluation of agents for ovarian cancer chemoprevention. *Gynecol. Oncol.* 81:373–379.
14. American Cancer Society. *Cancer Facts & Figures*. Atlanta, GA: American Cancer Society, 2005, http://www.cancer.org
15. Jain, K. K. 2005. Nanotechnology in clinical laboratory diagnostics. *Clin. Chim. Acta* 358:37–54.
16. Ferrari, M. 2005. Cancer nanotechnology: Opportunities and challenges. *Nat. Rev. Cancer* 5:161–171.
17. Yezhelyev, M. V., X. H. Gao, Y. Xing, A. Al Hajj, S. M. Nie, and R. M. O'Regan. 2006. Emerging use of nanoparticles in diagnosis and treatment of breast cancer. *Lancet Oncol.* 7:657–667.
18. Jaiswal, J. K., H. Mattoussi, J. M. Mauro, and S. M. Simon. 2003. Long-term multiple color imaging of live cells using quantum dot bioconjugates. *Nat. Biotechnol.* 21:47–51.
19. Wu, X. Y., H. J. Liu, J. Q. Liu, K. N. Haley, J. A. Treadway, J. P. Larson, N. F. Ge, F. Peale, and M. P. Bruchez. 2002. Immunofluorescent labeling of cancer marker HER2 and other cellular targets with semiconductor nanocrystals. *Nat. Biotechnol.* 21:41–46.
20. Bruchez, M., M. Moronne, P. Gin, S. Weiss, and A. P. Alivisatos. 1998. Semiconductor nanocrystals as fluorescent biological labels. *Science* 281:2013–2016.
21. Seydel, C. 2003. Quantum dots get wet. *Science* 300: 80–81.
22. Jaiswal, J. K., E. R. Goldman, H. Mattoussi, and S. M. Simon. 2004. Use of quantum dots for live cell imaging. *Nat. Methods* 1:73–78.
23. Michalet, X., F. F. Pinaud, L. A. Bentolila, J. M. Tsay, S. Doose, J. J. Li, G. Sundaresan, A. M. Wu, S. S. Gambhir, and S. Weiss. 2005. Quantum dots for live cells, in vivo imaging, and diagnostics. *Science* 307:538–544.
24. Hasegawa, U., S. M. Nomura, S. C. Kaul, T. Hirano, and K. Akiyoshi. 2005. Nanogel-quantum dot hybrid nanoparticles for live cell imaging. *Biochem. Biophys. Res. Commun.* 331:917–921.
25. Yang, C., N. Ding, Y. B. Xu, X. L. Qu, J. Zhang, C. Y. Zhao, L. Hong, Y. Lu, and G. Y. Xiang. 2009. Folate receptor-targeted quantum dot liposomes as fluorescence probes. *J. Drug Target.* 17:502–511.

26. Zdobnova, T. A., S. G. Dorofeev, P. N. Tananaev, V. P. Zlomanov, O. A. Stremovskiy, E. N. Lebedenko, I. V. Balalaeva, S. M. Deyev, and R. V. Petrov. 2010. Imaging of human ovarian cancer SKOV-3 cells by quantum dot bioconjugates. *Doklady Biochem. Biophys.* 430:41–44.

27. Sun, Y. H., Y. S. Liu, P. T. Vernier, C. H. Liang, S. Y. Chong, L. Marcu, and M. A. Gundersen. 2006. Photostability and pH sensitivity of CdSe/ZnSe/ZnS quantum dots in living cells. *Nanotechnology* 17:4469–4476.

28. Pan, M., M. Q. Chu, Y. Sun, and Z. Q. Ding. 2006. *Shanghai J. Biomed. Eng.* 26(2005):198.

29. Ding, L., D. A. M. Egbe, and F. E. Karasz. 2004. Photophysical and optoelectronic properties of green-emitting alkoxy-substituted PE/PV hybrid conjugated polymers. *Macromolecules* 37:6124–6131.

30. Uthirakumar, P., E. K. Suh, C. H. Hong, and Y. S. Lee. 2005. Synthesis and characterization of polyesters containing fluorescein dye units. *Polymer* 46:4640–4646.

31. Gao, C., J. Hou, D. Yan, and Z. Wang. 2005. Preparation and characterization of fluorescent hyperbranched polyether. *React. Funct. Polym.* 58:65–72.

32. Morawetz, H. 1988. Studies of synthetic polymers by nonradiative energy transfer. *Science* 240: 172–176.

33. Kasai, H., Y. Yoshikawa, T. Seko, S. Okada, H. Oikawa, H. Matsuda, A. Watanabe, O. Ito, H. Toyotama, and H. Nakanishi. 1997. Optical properties of perylene microcrystals. *Mol. Cryst. Liq. Cryst.* 294:173–176.

34. Kasai, H., H. Kamatani, S. Okada, H. Oikawa, H. Matsuda, and H. Nakanishi. 1996. Size-dependent colors and luminescences of organic microcrystals. *Jpn. J. Appl. Phys.* 35:L221–L223.

35. Kasai, H., H. Kamatani, Y. Yoshikawa, S. Okada, H. Oikawa, A. Watanabe, O. Ito, and H. Nakanishi. 1997. Crystal size dependence of emission from perylene microcrystals. *Chem. Lett.* 26:1181–1182.

36. Komai, Y., H. Kasai, H. Hirakoso, Y. Hakuta, S. Okada, H. Oikawa, T. Adschiri, H. Inomata, K. Arai, and H. Nakanishi. 1998. Size and form control of titanylphthalocyanine microcrystals by supercritical fluid crystallization method. *Mol. Cryst. Liq. Cryst.* 322:167–172.

37. (a) Fu, H. B. and J. N. Yao. 2001. Size effect on the optical properties of organic nanoparticles. *J. Am. Chem. Soc.* 123:1434–1439. (b) Levitus, M., K. Schmieder, H. Ricks, K. D. Shimizu, U. H. F. Bunz, and M. A. Garcia-Garibay. 2001. Steps to demarcate the effects of chromophore aggregation and planarization in poly(phenyleneethynylene)s. 1: Rotationally interrupted conjugation in the excited states of 1,4-bis(phenylethynyl)benzene. *J. Am. Chem. Soc.* 123:4259–4265.

38. Uthirakumar, P., C. H. Hong, E. K. Suh, and Y. S. Lee. 2006. Preparation and photophysical properties of organic fluorescent polymers and their nanoparticles. *J. Appl. Polym. Sci.* 120:5344–5350.

39. Deans, R., J. Kim, M. R. Machacek, and T. M. Swager. 2000. A poly(*p*-phenyleneethynylene) with a highly emissive aggregated phase. *J. Am. Chem. Soc.* 122:8565–8566.

40. Luo, J., Z. Xie, J. W. Y. Lam, L. Cheng, H. Chen, C. Qiu, H. S. Kwok, X. Zhan, Y. Liu, D. Zhu, and B. Z. Tang. 2001. Aggregation-induced emission of 1-methyl-1,2,3,4,5-pentaphenylsilole. *Chem. Commun.* 381:1740–1741.

41. Miteva, T., L. Palmer, L. Kloppenburg, D. Neher, and U. H. F. Bunz. 2000. Interplay of thermochromicity and liquid crystalline behavior in poly(*p*-phenyleneethynylene)s: Pi–pi interactions or planarization of the conjugated backbone. *Macromolecules* 33:652–654.

42. Alexis, F., P. Basto, E. Levy-Nissenbaum, A. F. Radovic-Moreno, L. F. Zhang, E. Pridgen, A. Z. Wang, S. L. Marein, K. Westerhof, L. K. Molnar, and O. C. Farokhzad. 2008. HER-2-targeted nanoparticle–affibody bioconjugates for cancer therapy. *ChemMedChem* 3:1839–1843.

43. Nobs, L., F. Buchegger, R. Gurny, and E. Allémann. 2006. Biodegradable nanoparticles for direct or two-step tumor immunotargeting. *Bioconjug. Chem.* 17:139–145.

44. Pu, K. Y., K. Li, J. B. Shi, and B. Liu. 2009. Fluorescent single-molecular core–shell nanospheres of hyperbranched conjugated polyelectrolyte for live-cell imaging. *Chem. Mater.* 21:3816–3822.

45. (a) Ow, H., D. Larson, M. Srivastava, B. Baird, W. Webb, and U. Wiesner. 2005. Bright and stable core–shell fluorescent silica nanoparticles. *Nano. Lett.* 5:113–117. (b) Burns, A., H. Ow, and U. Wiesner. 2006. Fluorescent core–shell silica nanoparticles: Towards "Lab on a Particle" architectures for nanobiotechnology. *Chem. Soc. Rev.* 35:1028–1042.

46. Zhou, X. C. and J. Z. Zhou. 2004. Improving the signal sensitivity and photostability of DNA hybridization on microarrays by using dye-doped core–shell silica nanoparticles. *Anal. Chem.* 76:5302–5312.

47. Tapec, R., X. J. J. Zhao, and W. H. Tan. 2002. Development of organic dye-doped silica nanoparticles for bioanalysis and biosensors. *J. Nanosci. Nanotechnol.* 2:405–409.

48. Huang, S. S., R. N. Li, Y. X. Qu, and J. S. Jia Liu. 2009. Fluorescent biological label for recognizing human ovarian tumor cells based on fluorescent nanoparticles. *J. Fluoresc.* 19:1095–1101.

49. Shi, X. Y. 2007. Dendrimer-entrapped gold nanoparticles as a platform. *Small* 3:1245–1252.

50. Abdul Jalil, R. and Y. Zhang. 2008. Biocompatibility of silica coated NaYF4 upconversion fluorescent nanocrystals. *Biomaterials* 29:4122–4128.

51. Chatterjee, D. K., A. J. Rufaihah, and Y. Zhang. 2008. Upconversion fluorescence imaging of cells and small animals using lanthanide doped nanocrystals. *Biomaterials* 29:937–943.

52. Hu, H., L. Xiong, J. Zhou, F. Li, T. Cao, and C. Huang. 2008. Multimodal-luminescence core–shell nanocomposites for targeted imaging of tumor cells. *Chem. Eur. J.* 15:3577–3584.

53. Jiang, S., Y. Zhang, K. M. Lim, E. K. W. Sim, and L. Ye. 2009. NIR-to-visible upconversion nanoparticles for fluorescent labeling and targeted delivery of siRNA. *Nanotechnology* 20:155101–155109.

54. Kumar, R., M. Nyk, T. Y. Ohulchanskyy, C. A. Flask, and P. N. Prasad. 2009. Combined optical and MR bioimaging using rare earth ion doped NaYF4 nanocrystals. *Adv. Funct. Mater.* 1:853–859.

55. Nyk, M., R. Kumar, T. Y. Ohulchanskyy, E. J. Bergey, and P. N. Prasad. 2008. High contrast in vitro and in vivo photoluminescence bioimaging using near infrared to near infrared up-conversion in Tm^{3+} and Yb^{3+} doped fluoride nanophosphors. *Nano Lett.* 8:3834–3838.

56. Shan, J., J. Chen, J. Meng, J. Collins, W. Soboyejo, J. S. Friedberg, and Y. J. Ju. 2008. Biofunctionalization, cytotoxicity, and cell uptake of lanthanide doped hydrophobically ligated NaYF4 upconversion nanophosphors. *Appl. Phys.* 104:094308, doi: 10.1063/1.3008028 (7 pages).

57. Yu, M., F. Li, Z. Chen, H. Hu, C. Zhan, H. Yang, and C. Huang. 2009. Laser scanning up-conversion luminescence microscopy for imaging cells labeled with rare-earth nanophosphors. *Anal. Chem.* 81:930–935.

58. Wang, F., D. K. Chatterjee, Z. Li, Y. Zhang, X. Fan, and M. Wang. 2006. Synthesis of polyethylenimine/NaYF4 nanoparticles with upconversion fluorescence. *Nanotechnology* 17:5786–5791.

59. Boyer, J. C., M. P. Manseau, J. I. Murray, and F. C. J. M. van Veggel. 2010. Surface modification of upconverting NaYF4 nanoparticles with PEG-phosphate ligands for NIR (800 nm) biolabeling within the biological window. *Langmuir* 26:1157–1164.

60. Squire, L. F. and R. A. Novelline. 1997. *Squire's Fundamentals of Radiology*, 5th ed. Cambridge, MA: Harvard University Press.

61. Toffoli, G., C. Cernigoi, A. Russo, A. Gallo, M. Bagnoli, and M. Boiocchi. 1997. Overexpression of folate binding protein in ovarian cancers. *Int. J. Cancer* 74:193–198.

62. Wang, Z. J., S. Boddington, M. Wendland, R. Meier, C. Corot, and H. Daldrup-Link. 2008. MR imaging of ovarian tumors using folate-receptor-targeted contrast agents. *Pediatr. Radiol.* 38:529–537.

63. Xu, H., C. A. S. Regino, Y. Koyama, Y. Hama, A. J. Gunn, M. Bernardo, H. Kobayashi, P. L. Choyke, and M. W. Brechbiel. 2007. Preparation and preliminary evaluation of a biotin-targeted, lectin-targeted dendrimer-based probe for dual-modality magnetic resonance and fluorescence imaging. *Bioconjug. Chem.* 18:1474–1482.

64. Schultz, S., D. R. Smith, J. J. Mock, and D. A. Schultz. 2000. Single target molecule detection with nonbleaching multicolor optical immunolabels. *Proc. Natl. Acad. Sci. U.S.A.* 97:996–1001.

65. Sonnichsen, C., B. M. Reinhard, J. Liphardt, and A. P. Alivisatos. 2005. A molecular ruler based on plasmon coupling of single gold and silver nanoparticles. *Nat. Biotechnol.* 23:741–745.

66. Paciotti, G. F., L. Myer, D. Weinreich, D. Goia, N. Pavel, R. E. McLaughlin, and L. Tamarkin. 2004. Colloidal gold: A novel nanoparticle vector for tumor directed drug delivery. *Drug Deliv.* 11:169–183.

67. Kusumi, A., Y. Sako, and M. Yamamoto. 1993. Confined lateral diffusion of membrane receptors as studied by single particle tracking (nanovid microscopy). Effects of calcium-induced differentiation in cultured epithelial cells. *Biophys. J.* 65:2021–2040.

68. Meier, J., C. Vannier, A. Serge, A. Triller, and D. Choquet. 2001. Fast and reversible trapping of surface glycine receptors by gephyrin. *Nat. Neurosci.* 4:253–260.

69. Borgdorff, A. J. and D. Choquet. 2002. Regulation of AMPA receptor lateral movements. *Nature* 417:649–653.

70. Sokolov, K., M. Follen, J. Aaron, I. Pavlova, A. Malpica, R. Lotan, and R. Richards-Kortum. 2003. Real-time vital optical imaging of precancer using anti-epidermal growth factor receptor antibodies conjugated to gold nanoparticles. *Cancer Res.* 63:1999–2004.

71. El-Sayed, I. H., X. H. Huang, and M. A. El-Sayed. 2005. Surface plasmon resonance scattering and absorption of anti-EGFR antibody conjugated gold nanoparticles in cancer diagnostics: Applications in oral cancer. *Nano Lett.* 5:829–834.

72. Huang, X. H., I. H. El-Sayed, W. Qian, and M. A. El-Sayed. 2006. Cancer cell imaging and photothermal therapy in the near-infrared region by using gold nanorods. *J. Am. Chem. Soc.* 128:2115–2120.
73. Kumar, S., N. Harrison, R. Richards-Kortum, and K. Sokolov. 2007. Plasmonic nanosensors for imaging intracellular biomarkers in live cells. *Nano Lett.* 7:1338–1343.
74. He, H., C. Xie, and J. C. Ren. 2008. Nonbleaching fluorescence of gold nanoparticles and its applications in cancer cell imaging. *Anal. Chem.* 80:5951–5957.
75. Zhang, C. Y., H. Ma, S. M. Nie, Y. Ding, L. Jin, and D. Y. Chen. 2000. Quantum dot-labeled trichosanthin. *Analyst* 125:1029–1031.
76. Lian, W., S. A. Litherland, H. Badrane, W. H. Tan, D. H. Wu, H. V. Baker, P. A. Gulig, D. V. Lime, and S. G. Jin. 2004. Ultrasensitive detection of biomolecules with fluorescent dye-doped nanoparticles. *Anal. Biochem.* 334:135–144.
77. Duan, J. H., K. M. Wang, W. H. Tan, X. X. He, C. M. He, B. Liu, D. Li, S. S. Huang, X. H. Yang, and Y. Y. Mo. 2003. A study of a novel organic fluorescent core–shell nanoparticle. *Chem. J. Chin. Univ.* 24:255–259.
78. Zhang, L. H. and S. J. Dong. 2006. Electrogenerated chemiluminescence sensors using Ru(bpy)32+ doped in silica nanoparticles. *Anal. Chem.* 78:5119–5123.
79. Song, L., E. J. Hennink, I. T. Young, and H. J. Tanke. 1995. Photobleaching kinetics of fluorescein in quantitative fluorescence microscopy. *Biophys. J.* 68:2588–2600.
80. Soper, S. A., H. L. Nutter, R. A. Keller, L. M. Davis, and E. B. Shera. 1993. The photophysical constants of several fluorescent dyes pertaining to ultrasensitive fluorescence spectroscopy. *Photochem. Photobiol.* 57:972–977.
81. Bagwe, R. P., C. Y. Yang, L. R. Hilliard, and W. H. Tan. 2004. Optimization of dye-doped silica nanoparticles prepared using a reverse microemulsion method. *Langmuir* 20:8336–8342.
82. Hun, X., Z. J. Zhang, and L. Tiao. 2008. Anti-Her-2 monoclonal antibody conjugated polymer fluorescent nanoparticles probe for ovarian cancer imaging. *Anal. Chim. Acta* 625:201–206.

28

Detection of Cancer-Associated Autoantibodies as Biosensors of Disease by Tumor Antigen Microarrays

Steven P. Dudas, Madhumita Chatterjee, Wei Chen, and Michael A. Tainsky

CONTENTS

28.1 Introduction: Background and Hypothesis of Autoantibodies as Biosensors of Disease Conditions

Autologous cellular components generally referred to as tumor-associated antigens (TAAs) have been recognized and evaluated as markers of disease states for decades [1,2]. The application of cancer-associated autoantibodies to the understanding of human cancer has been ongoing for a similar period of time and has provided insight into molecular mechanisms of human carcinogenesis. The majority of the autologous cellular components that present reactive epitopes to the immune system have been found to be intracellular proteins [3]. Often the proteins representing these epitopes function in cell cycle regulation, DNA repair and replication, and RNA processing. We and others are exploiting our understanding of the immune response to evaluating panels of cancer-associated autoantibodies as measurable circulating molecular biosensors in blood to develop diagnostic tests for the presymptomatic detection of cancer.

28.1.1 Basics of the Autoimmune Response to Disease

The principal function of the immune system is to protect the body from infectious organisms and the toxic substances they may produce. The immune system can also be divided into (1) the innate immune system which refers to a limited set of recognition receptors that are germ line encoded and conserved by natural selection, and (2) the acquired or adaptive immune system which responds with a virtually limitless number of possible recognition receptors to neoantigens. Each of these systems contains both humoral and cellular components. The acquired or adaptive immune system is composed of the humoral immune response (the aspect of immunity that is mediated by secreted antibodies produced in the cells of the B-lymphocyte lineage) and the cell-mediated immunity (which involves T lymphocytes that present antigen-binding receptors on their surface and directly kill foreign organisms and cells). All associated parts of the immune system are known to be actively involved in the body's own response to cancer. The experimental approach described in this chapter is designed to evaluate the qualitative profile of autoantibodies in sera and as such it is an evaluation of the cancer-induced adaptive humoral immune response. The later sections briefly cover the general aspects of autoimmunity and the physiological aspects of the generation of autoantibodies in disease.

28.1.1.1 Neoantigens

The central premise for this work is that the immunogenicity of certain aberrantly expressed cellular components from human tumors drives an adaptive immune response mediated by lymphocytes. Neoantigens for the context of this topic refer to those novel immunogens that arise through multiple mechanisms resulting in alterations of macromolecules of the body's own origin and may be related to the disease progression. The focus of our approach is to characterize the humoral immune responses (i.e., circulating autoantibodies) that are indicative of the presence of a neoplasm. Neoantigens represent a class of TAAs that are not normally present in normal tissues (except the testis) and elicit an adaptive immune response. Some general categories of reported mechanisms of neoantigens in cancer patients include mutations, novel posttranslational modifications of proteins, and aberrant expression.

28.1.1.1.1 Mutations

A missense mutation (a type of nonsynonymous mutation) is a point mutation in which change in a single nucleotide results in a codon for a different amino acid. The altered peptide sequence may then present a new immunogenic epitope and elicit an adaptive immune response. This class of neoantigen is best represented by studies with autoantibodies to p53 that have been reported in a number of tissue types [4–7] and a predominant number of p53 autoantibodies have been associated with the missense mutations in that gene.

A frameshift mutation is a mutation that is the result of an insertion or deletion in the DNA sequence where the number of nucleotides involved is not evenly divisible by 3. Due to the triplet nature of the genetic code, the insertion or deletion can change the reading frame, resulting in a completely different peptide sequence being translated after the point of the mutation. The neoantigen could arise within this novel peptide sequence. A number of reports of immunological responses against frameshift mutations have been documented in colorectal cancer patients [8–11].

Chromosomal translocations are genetic aberrations which result from rearrangements between segments of nonhomologous chromosomes. Gene fusion may occur when a translocation joins two separated genes and can result in the formation of chimeric proteins which can present as a neoantigen in cancer patients. For example, the chimeric Bcr–Abl fusion protein produced by the translocation, t(9;22), is the hallmark of Philadelphia (Ph) chromosome-positive leukemia [12]. Presentation of Bcr–Abl-derived peptides by chronic myelogenous leukemia (CML)-derived dendritic cells (DC) have been shown to activate fusion peptide-specific CD^{4+} T lymphocytes in an HLA class II-restricted manner [13]. A humoral immune response to the Bcr–Abl fusion protein was demonstrated in both Ph-positive and Ph-negative leukemias, as well as in healthy individuals [14]. The conclusion of that study suggested that Bcr–Abl fusion protein autoantibodies have a greater recognition to the Abl portion than the Bcr part.

28.1.1.1.2 Posttranslational Modifications of Proteins

Alteration in the posttranslational modifications on proteins is a well-documented protein variation occurring during malignant cell transformation. The best studied examples of this aberrant cellular modification in cancer deal with alter structure related to mucin-type O-glycosylation found on mucins and other O-glycoproteins [15]. There are two major forms of protein glycosylation, N- and O-glycosylation. Mucin-type O-glycosylation is initiated by a large family of UDP-GalNAc:polypeptide α-N-acetylgalactosaminyltransferases (ppGalNAc Ts) that transfer GalNAc from UDP-GalNAc to the Ser and Thr residues of polypeptide acceptors. One of the most extensively studied mucins in cancer is MUC1. Extracellular MUC1 is overexpressed and aberrantly glycosylated in many cancers and autoantibodies generated to aberrant O-glycoforms of MUC1 are being evaluated as diagnostic biomarkers for cancer [16]. The MUC1 mucin is heavily O-glycosylated in a large 20-amino-acid tandem repeat region. Aberrantly glycosylated MUC1 has been detected in the serum of late-stage cancer patients [17–19]. Cancer-associated IgG autoantibodies to several O-glycopeptide epitopes were identified in MUC1, whereas IgG antibodies to peptide epitopes were not detected [16].

28.1.1.1.3 Aberrant Expression

Cancer/testis (CT) antigens are protein antigens with normal expression restricted to adult testicular germ cells, but they are aberrantly activated and expressed in a proportion of various human cancers. This class of antigens have been found to elicit spontaneous humoral and cell-mediated immune responses in cancer patients [20]. More than 100 CT antigen genes have been reported, with approximately 30 being members of multigene families on the X chromosome, so-called CT-X genes (reviewed in [21]). These proteins are aberrantly expressed during tumorigenesis and are among a class of neoantigens that are presently undergoing studies toward diagnostic and therapeutic uses in cancer treatment.

Alternative splicing of pre-messenger RNA is a key molecular mechanism resulting in the formation of different mRNA variants that encode different polypeptides endowed with different biological functions and thus could have a profound impact on tumor development and progression [22,23]. Both mutations in *cis*-acting splicing elements and changes in expression pattern of effector molecules involved in splicing regulatory network affect the splicing profile of many cancer-related genes. Bourdon et al.

showed that different isoforms of p53 that originate through alternative splicing are expressed in a tissue-dependent manner and that their expression pattern is altered in human breast tumors [24]. Line et al. identified an alternatively spliced isoform known as transforming acidic coiled-coil protein 1 (TACC1) in gastric cancer using SEREX technology [25].

28.1.1.2 Self-Antigens

28.1.1.2.1 Overexpression

Endogenous self-antigens have been shown to exhibit immunogenicity when they are overexpressed during tumorigenesis [26]. Although T-cells encounter most of the self-antigens due to promiscuous gene expression by medullary thymic epithelial cells [27] during their maturation in thymus and get tolerized, growing evidence still supports the presence of self-reactive T-cells in the T-cell repertoire. Studies have shown that only self-antigen determinants expressed in a physiological normal manner are tolerized by the immune system. However, the speculation is that self-antigens may have "subdominant" or "cryptic" determinants that exist at a suboptimal threshold in native molecules and are therefore inefficiently presented to lymphocytes [28]. When these self-antigens are overexpressed in cancer, the cryptic determinants are eventually presented to lymphocytes in a costimulatory environment thus eliciting immunological responses [29].

28.1.1.2.2 Immune Tolerance

Autoimmune diseases are characterized by the activation of the body's immune responses directed against its own tissues due to breakdown of immune tolerance to specific self-antigens causing prolonged inflammation and subsequent tissue destruction. For example, autoimmunity can be associated with the production of increased concentration of antibodies to ribonucleoproteins, double-stranded DNA, phospholipids, and other cellular constituents as observed in patients diagnosed with systemic lupus erythematosus (SLE) which provoke proinflammatory responses [30]. The autoantibodies most frequently detected in RA patients are directed toward the rheumatoid factor and citrullinated peptides (anti-CCP) [31,32]. The humoral immune response to TAAs elicited in cancer patients exhibit many similarities to the immune response in classical autoimmune diseases [33]. There may be a significant overlap in the autoantibody repertoire observed between these two conditions. This aspect must be taken into account in identifying useful cancer-specific autoantibody biomarkers. Therefore, studies in which cancer-specific autoantigens are being sought, benign non-neoplastic disease cases, must be included in the experimental design to recognize and eliminate overlapping immunoreactive antigens for a truly high specificity diagnostic test.

28.1.1.3 Natural Autoantibody Repertoire in Tissue Homeostasis (Inflammation and Apoptosis)

28.1.1.3.1 Natural Antibodies

Natural antibodies (NAbs), the humoral part of innate immunity, provide a nonredundant role in the first line of defense against pathogens, but are also believed to provide important homeostatic housekeeping functions against self-antigens. Innate immunity utilizes a limited set of receptors called pattern recognition receptors (PRRs) that are germ line encoded, conserved by natural selection, and bind to pathogen-associated molecular patterns (PAMPs) common on multiple structures. NAbs produced by B lymphocytes are typically of the IgM class. Subgroups of NAbs have been demonstrated to target oxidation-specific epitopes such as oxidized LDL or oxidized phospholipids and may aid the neutralization of proinflammatory response and mediate atheroprotection. One hypothesized function of NAbs in homeostasis is the recognition of stress-induced self-antigens such as oxidation-specific epitopes on apoptotic cells. These NAbs might represent a part of the body's natural mechanism of clearance of such dying cells [34,35].

28.1.1.3.2 Heat Shock Proteins Autoantibodies

Heat shock proteins (HSPs) are stress-inducible proteins that act as chaperones in the correct folding of both nascent and denatured proteins, promoting the degradation of unrecoverable denatured proteins. HSPs maintain cellular homeostasis and protect against cell death through a mechanism called thermotolerance. However, if cells are subjected to severe stress, HSPs promote apoptosis. In innate immunity, extracellular HSPs induce various proinflammatory cytokines. In acquired immunity, they interact with antigenic polypeptides and assist in antigen presentation. HSPs are released from cells as a response to stress and trauma thereby inducing an immunological "danger signal" and anti-HSP autoantibodies are frequently found in patients with autoimmune diseases and inflammatory disorders. Studies have suggested that these autoantibodies themselves can modulate the "danger signal" induced by the extracellular HSPs [36]. These studies suggest that circulatory levels of HSP proteins and their corresponding autoantibodies function in a tightly regulated choreographed manner to maintain cellular homeostasis. In atherosclerosis, ELISA studies have detected soluble HSP90 in 42% and anti-HSP90 antibodies in 46% of the patients' sera evaluated. The conclusion showed that with atherosclerosis HSP90 overexpression induces an immune response and implicates HSP90 as a possible target autoantigen in the pathogenesis of carotid atherosclerosis [37]. Anti-HSP70 antibodies are significantly elevated in myasthenia gravis, an autoimmune disorder where patients develop autoantibodies toward skeletal muscle proteins (e.g., acetylcholine receptor and muscle-specific kinase), causing weakness in striated muscles. Anti-HSP60 antibodies were found in significant numbers of early stage breast cancer and DCIS patients, compared to healthy subjects. This was supported by immunohistochemistry (IHC) studies which found that HSP60 expression gradually increases from normal through DCIS to invasive tissues [38].

These data support the idea that repertoires of natural autoantibodies might be useful measurable biosensors of inflammatory diseases, but one could conceive they lack significant specificity to be considered useful due to the spectrum of diseases in which they would be present.

28.1.2 Significance

Early detection of any cancer is generally accepted as the best approach to reduce cancer mortality rates [39]. The objective of this work is to identify molecular biosensors capable of discriminating the "healthy" state from the "disease" state, preferentially in the asymptomatic phase of the cancer, and therefore may improve the patient's outcome. Many TAAs that have been identified to date are biomarkers of disease state yet few if any have been found to be valid diagnostic markers. To meet a criterion for clinical applications, a set for diagnostic markers must perform with adequate levels of specificity and sensitivity to discriminate the disease state on an individual level. The literature of this field represents an enormous amount of effort to identify thousands of TAAs, yet most have not been evaluated for diagnostic efficacy in large, controlled cohorts alone or in combination with other markers. The emergent concept in biomarker research is that panels of biomarkers will be required to produce assays with sufficient sensitivity and specificity for the early detection of cancer in patients. Our experimental approach, described later, was designed to identify panels of complementing biomarkers from its conception. Results from this discovery phase analysis will be applied to the development of a clinically relevant platform for routine clinical testing.

28.2 Materials

28.2.1 Construction of Phage Display cDNA Library

1. T7Select®10-3 *OrientExpress*™ cDNA Cloning System, Random Primer (Novagen–EMB Chemical Inc., San Diego, CA: cat. no. 70580, cDNA manual, TB247; T7Select System manual, TB178)

28.2.2 Printing of Antigen Microarrays

1. Nitrocellulose Film Slides
   a. Whatman-GE Healthcare: FAST Slides—1 pad, bar-coded, 20×51 mm, 20/pk cat. no. 10484182
   b. Grace Bio-Labs: ONCYTE® NOVA 1–20 mm $\times$ 51 mm NC pad per slide; frosted end glass $25 \times 75 \times 1$ mm 20/pk cat. no. 505170
2. Cartesian ProSys 5510A: (Note: This is an obsolete model no longer in production yet has been an invaluable workhorse to our scientific endeavors. Comparable systems are available at multiple manufactures)
3. Stealth Pins: Arrayit Corporation cat. no. SMP3B

28.2.3 Control Spots (Positive and Negative)

1. EBV protein (EBNA-1) Recombinant Epstein Barr Virus Nuclear Antigen Type 1 cat. no. R57523

28.2.4 Immunoassays

1. Immunoassay Chamber: Clontech—Glass Hybridization Chamber—cat. no. 634706
2. BSA: Sigma—Albumin from bovine serum, lyophilized powder, protease, essentially free, $\geq 98\%$ (agarose gel electrophoresis) cat. no. A3294
3. Nonfat dry milk (whatever is on sale at the local grocery)
4. AlexaFluor 647 goat antihuman IgG (H+L): Invitrogen Molecular Probes cat. no. A21445
5. AlexaFluor 532 goat anti-mouse IgG (H+L): Invitrogen Molecular Probes cat. no. A111002
6. T7-tag monoclonal antibodies: Novagen–EMB Chemical Inc., San Diego, CA: cat. no. 69522-4
7. 1% BSA–PBS: Dissolve 5 g of BSA in 500 mL PBS
8. 0.1% PBST: 1 mL Tween 20 to 1 L of PBS

(*Note*: All solutions that are applied to the microarrays are filtered and, where possible, ultrafiltered to avoid nonspecific spots on the microarrays.)

28.2.5 Scanning

1. Microarray Scanner: Genepix Autoloader 4200AL, Molecular Devices, MDS Analytical Technologies, Sunnyvale, CA
2. Image Analyzer Software: IMAGENE®, BioDiscovery, Inc., El Segundo, CA

28.2.6 Data Analysis

1. R—This free software is available at http://www.r-project.org/index.html

28.3 Methodology: Phage Display Antigen Microarrays

The objective of this autoantibody approach is to accurately identify peptide epitope sequences that are recognized by the cancer-specific humoral immune response via an undirected screening of phage display libraries. Our approach is a high throughput version of the SEREX method [40]. The goal is to isolate phage clones that express proteins which bind to antibodies present in patient sera of the cancer being studied. As part of our experimental design, we subtract out phage clones immunoreactive to antibodies in sera from healthy controls or patients with benign inflammatory conditions. Phage display antigen microarrays are employed as a high-throughput approach allowing for the simultaneous screening of thousands of antigens as monoclonal cDNA phage clones with only a minimal amount of sera. Advanced microarray scanning technology is used to analyze the results. In this process, cDNA libraries are prepared using mRNA from cancer cell lines or fresh tumor specimens using T7 phage display cloning technology in which each library contains approximately 10^7 original random-primed clones. Using a random-primed cDNA library permits representation of N- and C-terminal epitopes in the library. Using cDNA clones affords the ability to rapidly determine TAA identity through DNA sequencing of the selected immunoreactive clones. This approach developed in our laboratory has been employed for the identification of biomarkers for early detection studies in ovarian [41], head and neck [42], prostate [43], and lung cancers [44,45].

28.3.1 Phage Display cDNA Library

28.3.1.1 Construction of cDNA Libraries

Custom cDNA libraries were made in our laboratory from mRNA preparations from tissue as well as cell lines. In addition, for some studies, commercially available cDNA libraries were used. The OrientExpress Random Primer System is used to achieve orientation-specific cloning between *Eco*RI and *Hind*III sites. Briefly, first and second strand cDNA synthesis are sequentially carried out in the presence of 5-methyl dCTP. After second strand synthesis, the cDNA is treated with T4 DNA polymerase to blunt the ends. Addition of *Eco*RI/*Hind*III directional linker d(GCTTGAATTCAAGC) at the d(A)n:d(T)n end created a *Hind*III site d(*A*AGCTT) in which the two underlined bases are derived from cDNA. The two dTs are provided on the 5′ end of each first strand by the *Hind*III random primer d(TTNNNNNN). In order to differentiate cDNAs from different sources, we have introduced bar-coding sequences as triplet nucleotide bases at two different locations of the *Eco*RI/*Hind*III directional linker in such a way that the insertion of triplet nucleotides does not introduce stop codons and at the same time maintains the palindromic sequence of the linker. The translation of the triplet nucleotides in the correct reading frame results in the addition of an extra amino acid in the linker region and thus the sequence results of each bar-coded linker present in T7 phage clone indicates the source of the cDNA used for the construction of the T7 phage cDNA library. Excess linkers and small cDNAs (<300 bp) are removed by a gel filtration step. Digestion of cDNAs with both *Hind*III and *Eco*RI thus yield cDNA molecules ready for directional insertion into *Eco*RI/*Hind*III vector T7Select 10-3b arms. After vector ligation and packaging using T7 packaging extracts, the phage is plated to determine the library titer. About 50 phage clones are randomly picked up and PCR is performed with T7 forward primer (TCTTCGCCCAGAAGCTGCAG) and T7 reverse primer (CCTCCTTTCAGCAAAAAACCCC), to determine the insert sizes. The insert size range is usually found to be between 300 bp and 2.0 kb. The coding capacity of the vector system is as much as 1200 amino acids.

28.3.1.2 Amplification of Packaged Libraries by Liquid Culture Method

Ten milliliters of LB/carbenicillin medium is inoculated with a single colony of BLT5615 from a freshly streaked plate. The mixture is shaken at 37°C overnight. Ten milliliters of the overnight culture is added to 90 mL of LB/carbenicillin medium and is allowed to grow until OD_{600} reaches 0.4–0.5. Next, IPTG (1 mM), M9 salts (1×), and glucose (0.4%) are added and the cells are allowed to grow for 20 min. An appropriate volume of culture is infected with phage library at MOI of 0.001–0.01 (100–1000 cells for each pfu). The infected bacterial culture is incubated with shaking at 37°C for 1–2 h until lysis is observed. Glycerol (0.02%) and PMSF (0.02 M) are added to the cell lysate to block proteolysis of the capsid fusion proteins. The phage lysate is centrifuged at 8000g for 10 min. The supernatant is collected and is stored at 4°C. The lysate is titered by plaque assay under standard conditions. The libraries are stored at −80°C after purification by polyethylene glycol precipitation and ultracentrifugation through a stepwise CsCl gradient (please refer to T7 Select System Manual TB178 for detail procedure).

28.3.2 Subtractive Biopanning

A high-throughput technology for tumor antigen cloning has been developed by our group ("subtractive biopanning") to isolate cancer antigens from within the human proteome [46,47]. Subtractive biopanning involves immunoscreening of T7 phage tumor-derived cDNA libraries using a two-step process, starting with serum IgGs pooled from different normal healthy individuals. The subtractive biopanning with sera from normal healthy individuals reduces nontumor/common antigens that bind to IgGs in normal sera. Next, serum IgGs from cancer patients are used to capture antigen expressing phage clones and enriching the cDNA library for potential candidate tumor antigens for subsequent immunoscreening on protein arrays. In attempts to also capture antigens more relevant to early cancer detection, we have biopanned with sera from individuals having precancerous or early cancerous conditions, such as high-grade dysplasia of the colon or female patients with ductal carcinoma in situ. A discovery cohort of equivalent sera types is used to perform subsequent immunoscreening on protein arrays. Through this approach, we are isolating potential panels of phage clones that can be used for the development of early detection assays (Figure 28.1).

28.3.2.1 Selection of T7 cDNA Phage Displayed Libraries with Human Sera

28.3.2.1.1 Subtractive Biopanning with Healthy Control Sera

Protein G plus agarose beads (25 μL) are washed twice with 1× PBS and blocked with 1% BSA at 4°C for 1 h. The beads are then incubated at 4°C for 1 h with 250 μL of a 1:20 dilution pooled sera from 20 healthy individuals. After 3 h of incubation, beads are washed three times with 1× PBS and then incubated with ~10^{10} phage particles from a freshly amplified tumor-derived phage library. After incubation, the mixture is centrifuged at 850g for 2 min to remove phage not specifically bound to the beads and the supernatant (phage library) is collected for positive immunoselection in the next step.

28.3.2.1.2 Immunoselection: Biopanning with Cancer Patient–Derived Sera

Protein G plus agarose beads are washed two times with 1× PBS and blocked with 1% BSA at 4°C for 1 h. The beads are then incubated at 4°C for 3 h with 250 μL of serum from individual cancer patients at a 1:20 dilution. After this incubation, the beads are washed three times with 1× PBS and are incubated with phage library supernatant from above at 4°C for overnight for immunoselection (shorter times of incubation have not proven successful using model antibody systems). After incubation, the mixture is

---

1. T7 phage cDNA libraries are constructed from tumors obtained from cancer patients or cancer cell lines.

2. T7 phage cDNA libraries are immunoscreened with pooled serum samples obtained from normal individuals

3. For enrichment of candidate TAAs, the T7 phage cDNA libraries (free of normal antigens) are next immunoscreened with patient's serum

4. Normally 3200 phage clones are individually picked from BP3 or BP4 and are printed on first-generation antigen microarrays

5. Serological screening is performed with sera obtained from cancer patients, healthy or benign individuals

6. Clone selection is performed by statistical analysis (*T*-test, ROC or Wilcoxon rank sum method)

7. The reduced numbers of significant clones are printed on second-generation microarrays

---

FIGURE 28.1 Flowchart identifying cancer biomarkers using subtractive biopanning in combination with serological screening on antigen microarrays.

centrifuged at 850*g* for 2 min and supernatant is discarded. Beads are washed three times with 1× PBS. To elute the bound phage, 1% SDS (freshly prepared) is added to the washed beads and the mixture is incubated at room temperature for 10 min. The bound phage is removed from the beads by centrifugation at 6000*g* for 8 min. Eluted phage (termed as Biopanning 1, BP1) is transferred to liquid culture for amplification (50 µL of elution to 5 mL culture). Four rounds of affinity immunoselection are carried out with amplified phage obtained after each biopanning. The number of biopanning cycles generally determines the extent of the enrichment for phage that binds to the sera of patients with ovarian cancer. This strategy allows for one cycle of biopanning to be performed in a single day.

Assessment of the biopanning procedure is done on the enriched phage sets by PCR to determine clone diversity. Plaque assays are performed on biopanning cycles 3 and 4, BP3 and BP4. Aliquots from each biopanning are diluted to enable the isolation of individual clones (~300 clones per standard 15 cm plate). Randomly picked phages (20–40 individual plaques) are amplified and PCR, using T7 phage–specific primers, is performed to determine variation in insert sizes (typical range 300–2000 base pairs). From this rough evaluation, we typically find one or two pairs of clones with apparent equivalent size inserts. A diversity of 80%–90% is standard by this level of evaluation (lesser levels of diversity [i.e., <60%] would indicate over-biopanning). The biopanning cycles exhibiting the greatest diversity in insert size are then selected for subsequent work.

Aliquots from each biopanning are diluted, plated, and unique individual clones are plaque-picked and amplified. Initially, a minimum of 192 clones (two 96-well plates) are isolated from each biopanning per patient. This number may vary from experiment to experiment due to the number of selective patients utilized for biopanning. A typical experiment starts with a range of 3000–5000 individual clones. After amplification, 15 µL aliquots of phage lysates are pipetted into 384-well plates using a BioMek liquid-handling robot. These plates are the source plates for printing of the microarrays. The 384-well configuration is required to fit the microarrayer 32-pin printhead size.

28.3.3 Printing of Antigen Microarrays

28.3.3.1 Reference Spots (Positive and Negative Signals)

We employ a series of known proteins, phage clone lysates, and diluted human sera as positive or negative signal reference spots. The objective of reference spots are twofold: (1) indicate the performance of the assay as a technical qualitative indicator, and (2) produce a background basal level value to compare the possible true positive signaling.

28.3.3.1.1 Positive Control Spots

To ensure the technical performance of our assay, we routinely print known proteins that will produce a signal in the red channel. The Epstein Barr virus protein (EBNA-1) has proven to be immunoreactive in >95% of the sera we have analyzed. The red signal is generated from binding human anti-EBNA-1 IgGs to the protein and made detectable by the labeled secondary, AlexaFluor 647-labeled goat antihuman immunoglobulin G (IgG). The viral protein EBNA-1 is printed at a concentration of 10 µg/mL. An intense positive signal in the red channel can be produced by spotting a series of diluted human sera, though this does not represent a truly immunoreactive signal but rather an indication of the presence and binding of the labeled secondary antibody to IgGs in the spotted sera. Alternatively, we have spotted both protein G and protein A on arrays which generally gives intensive strong fluorescent signals and serves a similar function as the spotted sera. These spots function as positive red channel signals useful during the calibration of the scanner parameters (see Section 28.3.5).

28.3.3.1.2 Negative Control Spots

Multiple individual amplifications of the empty cloning vector in phage are printed at multiple locations on the antigen microarray. Three variants of the empty vectors are used. Each version translates slightly different small peptide tails on the phage capsid protein at the cloning site. We utilize the manufacturer's original vectors, 10.3b and 2.1a, as well as a third independent variant isolated in this lab which has a stop codon immediately after the *Eco*RI insertion site. All three variants have consistently exhibited low immunoreactivity to human IgGs over multiple studies involving thousands of sera. Each empty vector variant is printed 20–60 times per microarray. A strong green channel signal and a weak red channel signal are produced in all variants.

28.3.3.1.3 Negative Spots

In most printing schemes, a series of negative spots are printed as a quality control for contamination during printing. Negative spots consist of either an aliquot of the media from a bacterial culture, unused media, or an empty well with no fluid. These spots should not produce either a red or green signal. A faint signal is occasionally seen with the bacteria/media. A strong positive signal in either channel is an indication that cross-contamination may have occurred during the printing process.

28.3.3.2 Printing

All printings of antigen microarrays have been conducted on a Cartesian ProSys 5510A microarrayer fitted with a 32-pin printhead. We used SMP3B Arrayit Stealth Pins. The microarrayer platform is capable of handling 80 nitrocellulose slides per run. The arrays are printed at ambient temperature and 50% relative humidity within the printing chamber. Each sample in the printing source plate is spotted as five replicates on each slide. Due to the general viscosity of the sample fluid and hydrophilic nature of the nitrocellulose material, we limit dwell time of the pin contact with the surface matrix to 20 ms. After printing, slides are stored dry in their original containers at room temperature until immunoassays. Typically, the microarray slides are used within 4 weeks of printing as longer storage times can introduce variability with an experiment.

28.3.4 Immunoassays

28.3.4.1 Prewash

The microarray slides are prewashed two times (4 min per wash) in PBS using a tabletop shaker (at a moderate speed).

28.3.4.2 Blocking

We recommend a blocking procedure for the nitrocellulose slides. Through empirical observation, we found that the performance characteristics of different blocking agent vary with the different manufacturers' slides. A generally acceptable blocking solution is 1% BSA–PBST. A comparable blocking buffer we have also used is 4% milk in PBST. The blocking treatment is performed for 1 h at room temperature using a tabletop shaker.

28.3.4.3 Immunoassay Blocking to Absorb Antibacterial Human Antibodies

Because each slide/assay chamber requires 2 mL of solution, a stock solution is prepared in 1% BSA–PBS–0.1% Tween 20, plus an appropriate amount of *Escherichia coli* bacterial lysate to obtain a final concentration of 60 μg/mL. To a 2 mL aliquot of 1% BSA–PBST–lysate solution, the mouse anti-T7-tag antibody is added to yield a concentration of 0.15 μg/mL. To each individual tube, 6.6 μL of human serum sample (1:300 dilution) is added. The mixture is incubated in 5 mL tubes for 1.5 h at room temperature on a rocking platform to allow the bacteria lysate to react with any *E. coli* relative human antibodies. This step has been found necessary to reduce nonspecific binding of human serum antibodies to bacterial proteins present in phage clone lysates spotted on antigen microarrays.

28.3.4.4 Primary Immunoassay

The slide from the pretreatment vessel is transferred to an immunoassay chamber and the 2 mL of primary antibody solution is pipetted carefully into the immunoassay chamber. Any air bubbles that may form upon the addition of the primary antibody solution are dislodged by gentle tapping. The immuno-assay chamber is shaken for 1 h at room temperature using the tabletop shaker. After immunoassay, the slide is removed from the immunoassay chamber and the slide is washed three times (4 min each) with 0.1% PBST using the tabletop shaker.

28.3.4.5 Secondary Immunoassay

The secondary labeled antibodies, AlexaFluor 532 (green) antibody anti-mouse IgG (2 mg/mL) at a dilution of 1:40,000 and AlexaFluor 647 (red) antibody antihuman IgG (2 mg/mL) at a dilution of 1:2000, are added to 1% BSA–PBS–0.1% Tween 20 (prepared just prior to use and kept in the dark). After transferring the slides to a clean immunoassay chamber, 2 mL of the fluorescent secondary antibody solutions are added to each chamber and shaken for 1 h at room temperature using the tabletop shaker. The chambers are covered with aluminum foil to avoid photobleaching of the fluorescent secondary antibodies. After 1 h, the microarrays are removed from the chambers and washed three times (4 min each) with 0.1% PBST, followed by two washes of the slide (4 min each) with PBS. The chambers remain covered with aluminum foil during all washes. An optional wash in deionized H_2O can be preformed to decrease background levels if necessary.

28.3.4.6 Drying Microarrays

Before storage, the excess liquid is wicked off from slides with an absorbent Kimwipe and then the slide is placed into a 50 mL plastic conical centrifuge tube stuffed with a cushion of Kimwipes in the bottom. The slides are centrifuged for 3 min at 1200 rpm in a tabletop centrifuge. Any residual liquid can be removed by using an aerosol (pressurized) air to completely dry it. Dried slides are stored at room temperature in the dark until scanning.

28.3.5 Scanning

The objective of selecting the proper scanner settings is to achieve a functional working range that is able to handle variation in background and signal levels characteristic to individual serum/plasma samples. Not all samples produce an equal background or signal due to inherent variations. This is a profound issue when dealing with convenience sets where the need to employ multiple institutional sources to obtain sufficient numbers of body fluids for a study is an unavoidable fact. One needs to be cognizant of this fact and knows there may be inconsistency in sample handling. Occasionally, sera have extremely high backgrounds and therefore the settings will have to accommodate a range of conditions. In general, we try to ensure we have a good background-to-foreground ratio. For the red channel, we look for a signal at least 1:2 to 1:3 over the background and a ratio of around a 1:5 for the green channel. This tends to result in a color–color background ratio near 1:1, although the green background typically tends to be slightly higher. These scanning conditions also result in a general ratio of green to red spots of something close to ~5:1. Whatever serum/slide set used to setup the PMTs and power settings, it must be very representative of the typical middle range signals seen for a series of slides. We suggest that one should make initial PMT/power settings and then follow the results for a group of four or five different sera/slides to ensure that the signals stay within a general range. We routinely scan the arrays with an Axon GenePix 4200AL scanner using the following settings: 635 nm (Cy5) channel: PMT = 380 V; power = 10% and 532 nm (Cy3) channel: PMT = 370 V; power = 10%.

On a pre-scan of the slide, the control spots can be used as a guide for setting the scanner. The control spots EBNA-1 generally give a fluorescent signal of 10,000–40,000 fluorescence units (FU). If the EBNA-1 spots are >50,000 FU, this is an indication that the scanner settings are too high; the slides should be rescanned using lower settings. The positive control spots (spotted sera, protein G, and protein A) generally give a fluorescent signal in the range of 50,000 FU at a printed concentration of 0.1 µg/µL. This high level is typical and is an indication that the scanned slides have sufficiently high titer of secondary antibodies. Higher FU readings would be an indication of high scanner settings. Counts in the green channel for any particular spot are in the 6,000–10,000 count range. Counts in the red channel have a

FIGURE 28.2 Analyses of antigen microarrays: two fluorescent dye system. The above scan is a representative section of an antigen microarray interrogated with human colorectal serum. Potential TAA clones are robotically spotted on ONCYTE® NOVA nitrocellulose-coated glass slides. Two separate phage cultures of each clone are printed in quintuplicate on arrays (the five spots of each culture are printed vertically in a single column). The array was interrogated with a 1:300 dilution of human serum and a 1:7000 dilution of mouse anti-T7-tag monoclonal antibody. AlexaFluor 647 (red fluorescent dye)-labeled antihuman IgG antibody is used to detect immunoreactivity of human IgGs to antigen spots. An AlexaFluor 532 (green fluorescent dye)-labeled goat anti-mouse IgG is applied to detect the T7 phage allowing for the normalization of the amount of each antigen spotted. This second antibody pair provides internal normalization for each spot. Log2 dye ratios are determined for each spot using global median normalized data and then replicate combining.

greater variability representing the positive signals and spots can range from just above background to twice that of a middle level green FU for a individual case. Slides should be scanned within 24 h after processing. We typically scan slides the following day. Longitudinal analyses of the stability of the fluorescent signal have determined that there is little variation over the first 5–6 days after processing. Figure 28.2 shows a representative scan from a processed antigen microarray immunoassayed with a human colon cancer serum.

28.3.6 Quality Control and Normalization

In this section, we discuss our approach for image quality assessment and normalization. All the computer applications used to analyze the data generated from the image analysis were obtained as R packages [48]. R is a free language and environment for statistical computing that is available on the Web.

28.3.6.1 Data Structure

Each microarray was scanned on a GenePix scanner followed by image analysis using the IMAGENE software package. This software generates quantitative intensity value and a qualitative detection call for each antigen spot on the microarray. Two raw outputs text files (one for each fluorescent channel) are generated for each microarray. Text files are read into the R environment as RGList objects using the R package limma [49]. One component of the image analysis program outputs is the spot quality weights, which provides information on the quality of each spot. A zero weight indicates that the spot is unreliable and should be ignored. The weights are stored in the RGList object (RG$weights) and used automatically by normalization functions in the limma package. RGList objects also have other components such as gene ID (RG$genes$"Gene ID") and printer layout (RG$printer). The corresponding position on the microarray for each gene is stored in RG$genes.

28.3.6.2 Quality Assessment

A widely used quality assessment tool for two-color microarray is the *MA*-plot [50], which plots the log–intensity ratios of the red to green channels ($M = \log_2(R/G)$) versus the mean log–intensity of the two ($A = \log_2 \sqrt{RG}$). The *MA*-plot function in the limma package produces a *MA*-plot on unnormalized microarray data. It is essentially the same as the Tukey mean difference plot [51]. Under the assumption

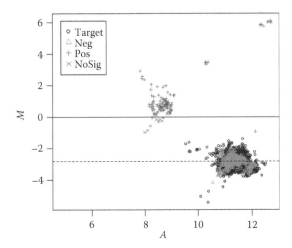

FIGURE 28.3 *MA*-plot: The vertical axis, labeled as *M*, is the log–intensity ratio ($M = \log_2(R/G)$). The horizontal axis, labeled as *A*, is the mean log–intensity of the red and green ($A = \log_2 \sqrt{RG}$). The dashed line is the median log–intensity ratio of all the spots. The black "o" spots represent the clones that are of interest. The green "Δ" and red "+" spots are negative and positive control clones, respectively. The blue "×" spots are a third type of quality control, negative spots, which consist of media or empty well with no fluid. No large variation at either in low-intensity end (small *A* value) or high-intensity end (large *A* value) is seen. No presence of truncation at the high-intensity end due to scanner saturation is seen.

that most antigens are not differentially reactive, most points in a *MA*-plot should fall along a horizontal line. In practice, *MA*-plot reveals systematic dependence of the ratios on fluorescent intensity with grouping by spot type (i.e., positive and negative control spots, negative spots, and antigen clones spots; see Figure 28.3). Diversity patterns (scattering) of spot groups suggest low quality of individual arrays. A detailed discussion and illustration of some of the commonly observed features of *MA*-plots can be found in Cui et al. [52]. Once the positive control spot and negative spots are used for quality control, they were removed before subsequent data preprocessing. The negative control spots (empty vector) are retained during normalization.

28.3.6.3 Data Normalization

Normalization is a common post hoc procedure to adjust spot intensities to account for the systematic experimental variations such as labeling, print-tip, and plate effects. Various packages for microarray normalization are available in the Bioconductor project library located at the R homepage. There are certain trade-offs to different normalization methods to consider for particular sets of array data: more complex normalization procedures tend to remove more technical variations (the undesirable variations) but may also remove or affect the biological signals (the desirable variations) in comparison to simple procedures [53,54]. From our experience with the data generated from protein microarrays, simpler procedures result in the optimal normalization of the data and are used with our studies. All procedures are found in the R package, limma [49].

The initial normalization step is background correction. We employ a straightforward method, "minimum," that replaces any zero or negative background subtracted intensity with half the minimum of the positive corrected intensities for that microarray. This backgroundCorrect function available in R can be reviewed (as well as other options) by seeing the help file of backgroundCorrect function. Next, global normalization is done using the R function "normalizeWithinArrays" and method "median." The method "median" subtracts the weighted median from the *M*-values for each microarray. Location bias is intended to be corrected using this within microarray global normalization. Application and options of the commonly used normalization procedures within the "normalizeWithinArrays" function are described in Smyth and Speed [55], including choices of tuning parameters. The replicates of each marker are combined using the median values after normalization. The between microarray normalization is performed only if the scale bias is fairly large after examining the data on a case-to-case basis. In studies where this is apparent, we have applied the function "normalizeBetweenArrays" and method "quantile" from the limma package.

28.4 Results

The objective of this approach is to isolate selective panels of immunoreactivity antigen markers capable of distinguishing a cancer patient serum. Our initial library pools of clones consist of a diversity on the order of 10^7 different phage antigens. From our four rounds of biopanning, we enriched the pool for potential candidate TAAs and have a reduction of the diversity of clones by a factor of 10^5. These phage display cDNA clones are robotically spotted in five replicates on nitrocellulose membranes and interrogated with human serum samples to identify immunoreactivity profiles using our two fluorescent dye system as described earlier. Candidate TAAs that exhibit immunoreactivity and the potential ability to be informative markers are re-assayed on a secondary microarray series to reconfirm reactivity and evaluated for their performance characteristics with sera not previously used to identify these clones. In this manner, panels of selected clones are generated with the ability to distinguish the healthy control sera (or relevant benign sera) from the cancer cases. Table 28.1 summarizes our results in the reduction of clones in our studies to the selected clone panels.

As a follow-up, we routinely determine the gene identity of selected antigen clones by DNA sequencing of the phage clones. This approach has proven to be very productive as a discovery platform for the identification of novel TAAs as well as re-isolating and confirming the identity of previously described TAAs in the literature. In the OVCA studies, 4 of the 65 antigens were identified as NASP, RCAS1, NBS1, and eIF5A [2]. These genes which had commercially available monoclonal antibodies were assessed by IHC. NASP and RCAS1 proteins were more frequently expressed in ovarian cancer tissues than with normal ovarian tissue and serous cystadenomas [56]. In order to determine diagnostic value of biomarkers, markers were first evaluated one at a time for their ability to discriminate between healthy ovaries and serous cystadenoma from ovarian cancer. We found that NASP ($p < 0.001$) and RCAS1 ($p = 0.01$) were significantly more likely to be positive in tissues from women with OVCA. None of these proteins expression levels were prognostic for survival. Together, these results suggest that induction of a humoral immune responses in OVCA patients to some TAAs is the result of overexpression [56]. From our CRC study, we also sequenced analyses of the top performing selected clones. Of interest in this gene list was that a number of clones had sequence similarities to clones listed in the SEREX database. These findings confirm this platform's ability to identify and reconfirm known TAA/cancer associations, but also the ability to identify novel genes and novel gene/cancer associations.

TABLE 28.1

Clone Reduction: Summary of the Reduction in the Number of Clones in Two Studies to Selected Antigen Panels

	Discovery Set		Number of Clones by Biopanning	Secondary Clone Set	Selected Clones for Classification	Performance Characteristics		
Study	Cancer	Controls				Sensitivity	Specificity	Accuracy
Ovarian [41]	32	25	480	N/A	65	0.55	0.98	0.76
Colorectal (Dudas et al., in preparation)	37	78	3800	487	48	0.84	0.40	0.65

Notes: Selection of clones for classification was done by using Wilcoxon rank sum test one-sided at $\alpha = 0.05$ level without adjustment for multiple testing. Performance characteristics were conducted in the R environment using the RandomForest package and using sample sets independent of the discovery sets.

28.5 Discussion

Two scientific approaches that are most directly comparable to our phage display method are natural protein antigen microarrays and nucleic acid programmable protein array (NAPPA). Natural protein antigen microarrays are printed arrays of natural proteins isolated from human cancer cell lines retain the naturally occurring native conformations and posttranslational modifications of each protein. Of the methodologies reviewed, this approach exploits the unique ability to present the native autologous cellular antigens as a target antigen for autoantibody studies. NAPPA generates proteins used as candidate target TAAs on the microarray by printing cDNA expression vectors directly onto glass slides and then transcribing and translating target proteins in situ with T7 RNA polymerase and rabbit reticulocyte lysate [57]. Both of these protein microarray methods have certain advantages for their use in studies to determine panels of diagnostic markers for the detection of cancer.

28.5.1 Natural Protein Antigen Microarrays

The initial phase of this approach is to conduct multidimensional cellular fractionation to separate proteins from cell lines before they are printed on microarrays. By way of example for this technique, two recent studies that applied this approach to the detection of autoantibodies in either pre-diagnostic lung cancer sera [58] or sera from pancreatic cancer patients [59]. This approach has also been used in other studies as well [60–62]. In both the lung and pancreatic cancer studies, candidate target TAAs were fractionated from lysates of human cancer cell line and the identities of the proteins in useful fractions determined using mass spectroscopic approaches. Protein fractions were printed in replicates onto nitrocellulose slides to identify candidate immunoreactive antigens. In the lung study, fractions were printed without further treatment, whereas in the pancreatic study, the fractions were further treated with a cyanogen bromide digestion before printing to expose additional epitopes. In lung, a series of 85 lung cancer sera and 85 matched controls were applied to individual microarray immunoassays. For the pancreatic studies, a series of 10 normal controls, 10 chronic pancreatitis, and 10 pancreatic cancer cases were used. Detection of autoantibodies was performed using an indirect immunofluorescence with AlexaFluor 647-labeled antihuman immunoglobulin G (IgG). In both cases, spotted fractions that exhibited high immunoreactivity to cancer sera were analyzed by mass spectrometry for protein identification.

A significant outcome of this proteomic approach in both studies was their ability to identify novel TAAs for each cancer. In presymptomatic lung cancer sera, this group identified the novel lung cancer antigen, LAMR1. This group also presented data that supported previous studies on the occurrence of annexin 1 and 14-3-3 theta autoantibodies in lung cancer sera. For the pancreatic cancer study, three novel pancreatic TAAs were found, in which two were mitochondrial proteins and the third protein was identified as ubiquitin-conjugating enzyme E2 variant 1. These results demonstrated the ability of this approach to reconfirm earlier data and also identify additional TAAs that can be used in diagnostic studies. Though this approach does demonstrate some clear advantages, there are some inherent disadvantages limiting its ability to be directly translated into reproducible large-scale studies to develop a clinical test. Initially, the source of the target TAAs comes from extensive cellular fractionation of cellular lysate which is technically difficult and labor-intensive. This approach has advantages in discovery phase studies, but may not be applicable to large-scale production of the epitopes for diagnostic testing as they are frequently complex carbohydrates. Also, using a single cell line as a source of proteins limits the available repertoire of possible TAAs to those expressed in that cell line which may not be sufficiently representative to be informative of very heterogeneous cancers. Identification of immunoreactive TAAs requires subsequent steps via MS. Validation of antigens would require additional steps in the production of the TAA as a recombinant protein and development of an ELISA test. In either case, this step may result in the loss of the natural protein epitope recognition and therefore the TAA's recognition as a diagnostic antigen.

28.5.2 Nucleic Acid Programmable Protein Array

A significant advantage of this methodology is the ability to construct custom microarrays with known target TAAs suited for the study of specific cancer types [63]. Custom microarrays, as in the NAPPA approach, were constructed with proven breast cancer TAAs. This is an advantage in one sense, but also a limitation in another. Possible target TAAs are limited to available cDNA clones that can be expressed in this system and also to known proteins in standard open reading frames. An advantage is that because each spot is known, those proteins that show immunoreactivity can be easily and quickly be advanced to the next phase of analysis. Another aspect of this approach is that full-length proteins are analyzed on the microarrays. This advantage allows for the presentation of multiple epitopic regions for any particular protein in the assay. A limitation in this system is it does not have the ability to present the proteins in the native form with natural cellular-directed posttranslational modifications and conformations.

28.6 Future Trends

Preliminary results in the development of serological diagnostic assays for early detection of cancer have demonstrated that the basic experimental tools to accomplish this goal exist. In the future, autoantibody patterns against tumor-specific proteins may achieve high specificity and sensitivity to diagnose disease in screening populations. Development of highly accurate reliable assays using this technology must be integrated into clinically applicable strategies in patient care. Autoantibody signatures in cancer patients' sera have shown the ability to discriminate various cancers from healthy patients and those with related benign diseases even in these early studies. The quest for accurate panels of biomarkers requires the need for robust assays that are both reliable and reproducible that can be taken to the next phase of large-scale validation studies. Understanding the nature of the autoimmune response in cancer has helped to establish the tools that will permit detecting tumor-specific autoantibodies for cancer diagnosis.

28.6.1 Development of Clinical Diagnostic Platforms

Emerging technologies with the ability to do multiplexing analyses of large numbers of candidate TAAs could lead to the development of clinically applicable diagnostic tests. This research is still in an initial discovery phase that has not yet evolved into large-scale validation studies or clinical trials. Further development is required to distinguish the reproducibility of each approach of their early results. Many of the antigen/antibody reactions seen during the discovery phase may not result in the development of serological assays for cancer detection. Appropriate study designs must be implemented for these findings to have future utility. Therefore, it is advisable to use a discovery platform that is as technically similar to the anticipated clinical diagnostics platform as possible. The lack of effectiveness (sensitivity or specificity) of the discovery platform may translate into a loss of overall accuracy for a panel of analytes in any proposed clinical laboratory platform. Two arising technical platforms that could be potentially adapted to our approach to develop a clinically feasible platform are multiplex planar array ELISAs and multiplex microsphere assays (xMAP Technology).

28.6.1.1 Multiplex ELISA

Multiplex ELISA assays have been performed for high-throughput screening in cancer [64]. The advantages of this multiplex ELISA technology are the ease of automation and the feasibility of measuring the autoantibody titers in patient serum. A high-throughput version of multiplex ELISA approach, reverse capture antibody microarray, is based on the dual-antibody sandwich immunoassay of enzyme-linked immunosorbent assay in which tumor or healthy tissue–derived native antigens are allowed to bind monoclonal antibodies that are spotted on the array surface. The antigen–antibody arrays are next incubated with fluorescently labeled IgG from cancer and control samples. Another set of microarrays incubation is carried on simultaneously by swapping the dyes (used to label cancer and control IgGs) to include the dye effects. The reverse capture antibody array platform uses native proteins as the target and the autoantibody respond to native proteins on the array making it superior to other protein microarrays that use recombinant proteins.

28.6.1.2 Multiplex Microsphere Assays: xMAP Technology

In xMAP technology, molecular reactions take place on the surface of microspheres containing two fluorophore dyes of differing concentrations generating distinct microsphere signatures. Using this method, 100 different microspheres can be created to capture different analytes allowing multiplexing analyses. Established assays utilize capture antibodies covalently bound to the surface of the beads to immobilize the analytes of interest from body fluids. After a washing step to remove unbound materials, detector antibodies are reacted with the beads, followed by addition of an R-phycoerythrin (RPE) conjugate that labels the immune complexes on the beads. xMAP assays have been applied to demonstrate high levels of various circulating analytes in cancer patients with the advantage of low serum volume requirements [65–68]. In the adaptation of this technology for our autoantibody application, we are directly binding the cloned antigen or full-length proteins to the microsphere. The primary antibodies are the serum-derived autoantibodies (IgGs) and a phycoerythrin-labeled goat antihuman IgG monoclonal antibody that is used for detection. The adaptation of this technology has the potentiality of making significant advancements in this field due to its ability to produce robust, reproducible epitomic profiles from patient sera and is a readily translatable platform for clinical implementation.

REFERENCES

1. Gold, P. and S. O. Freedman. 1965. Demonstration of tumor-specific antigens in human colonic carcinomata by immunological tolerance and absorption techniques. *J. Exp. Med.* 121:439–462.
2. Gold, P. and S. O. Freedman. 1965. Specific carcinoembryonic antigens of the human digestive system. *J. Exp. Med.* 122:467–481.
3. Chen, Y. T., M. J. Scanlan, U. Sahin, O. Tureci, A. O. Gure, S. Tsang, B. Williamson, E. Stockert, M. Pfreundschuh, and L. J. Old. 1997. A testicular antigen aberrantly expressed in human cancers detected by autologous antibody screening. *Proc. Natl. Acad. Sci. U.S.A.* 94:1914–1918.
4. Winter, S. F., J. D. Minna, B. E. Johnson, T. Takahashi, A. F. Gazdar, and D. P. Carbone. 1992. Development of antibodies against p53 in lung cancer patients appears to be dependent on the type of p53 mutation. *Cancer Res.* 52:4168–4174.
5. Ralhan, R., S. Arora, T. K. Chattopadhyay, N. K. Shukla, and M. Mathur. 2000. Circulating p53 antibodies, p53 gene mutational profile and product accumulation in esophageal squamous-cell carcinoma in India. *Int. J. Cancer* 85:791–795.
6. Soussi, T. 2000. p53 antibodies in the sera of patients with various types of cancer: A review. *Cancer Res.* 60:1777–1788.
7. Jezersek, B., Z. Rudolf, and S. Novakovic. 2001. The circulating auto-antibodies to p53 protein in the follow-up of lymphoma patients. *Oncol. Rep.* 8:77–81.
8. Line, A., Z. Slucka, A. Stengrevics, K. Silina, G. Li, and R. C. Rees. 2002. Characterisation of tumour-associated antigens in colon cancer. *Cancer Immunol. Immunother.* 51:574–582.
9. Reuschenbach, M., D. M. von Knebel, and N. Wentzensen. 2009. A systematic review of humoral immune responses against tumor antigens. *Cancer Immunol. Immunother.* 58:1535–1544.
10. Schwitalle, Y., M. Kloor, S. Eiermann, M. Linnebacher, P. Kienle, H. P. Knaebel, M. Tariverdian, A. Benner, and D. M. von Knebel. 2008. Immune response against frameshift-induced neopeptides in HNPCC patients and healthy HNPCC mutation carriers. *Gastroenterology* 134:988–997.
11. Ishikawa, T., T. Fujita, Y. Suzuki, S. Okabe, Y. Yuasa, T. Iwai, and Y. Kawakami. 2003. Tumor-specific immunological recognition of frameshift-mutated peptides in colon cancer with microsatellite instability. *Cancer Res.* 63:5564–5572.
12. Rowley, J. D. 1973. Chromosomal patterns in myelocytic leukemia. *N. Engl. J. Med.* 289:220–221.
13. Yasukawa, M., H. Ohminami, K. Kojima, T. Hato, A. Hasegawa, T. Takahashi, H. Hirai, and S. Fujita. 2001. HLA class II-restricted antigen presentation of endogenous Bcr–Abl fusion protein by chronic myelogenous leukemia-derived dendritic cells to CD4(+) T lymphocytes. *Blood* 98:1498–1505.
14. Talpaz, M., X. Qiu, K. Cheng, J. E. Cortes, H. Kantarjian, and R. Kurzrock. 2000. Autoantibodies to Abl and Bcr proteins. *Leukemia* 14:1661–1666.
15. Tarp, M. A. and H. Clausen. 2008. Mucin-type O-glycosylation and its potential use in drug and vaccine development. *Biochim. Biophys. Acta* 1780:546–563.

16. Wandall, H. H., O. Blixt, M. A. Tarp, J. W. Pedersen, E. P. Bennett, U. Mandel, G. Ragupathi, P. O. Livingston, M. A. Hollingsworth, J. Taylor-Papadimitriou, J. Burchell, and H. Clausen. 2010. Cancer biomarkers defined by autoantibody signatures to aberrant O-glycopeptide epitopes. *Cancer Res.* 70:1306–1313.

17. Burchell, J., D. Wang, and J. Taylor-Papadimitriou. 1984. Detection of the tumour-associated antigens recognized by the monoclonal antibodies HMFG-1 and 2 in serum from patients with breast cancer. *Int. J. Cancer* 34:763–768.

18. Taylor-Papadimitriou, J., J. M. Burchell, T. Plunkett, R. Graham, I. Correa, D. Miles, and M. Smith. 2002. MUC1 and the immunobiology of cancer. *J. Mammary Gland Biol. Neoplasia* 7:209–221.

19. Bon, G. G., G. J. van Kamp, R. A. Verstraeten, S. Mensdorff-Pouilly, J. Hilgers, and P. Kenemans. 1999. Quantification of MUC1 in breast cancer patients. A method comparison study. *Eur. J. Obstet. Gynecol. Reprod. Biol.* 83:67–75.

20. Simpson, A. J., O. L. Caballero, A. Jungbluth, Y. T. Chen, and L. J. Old. 2005. Cancer/testis antigens, gametogenesis and cancer. *Nat. Rev. Cancer* 5:615–625.

21. Caballero, O. L. and Y. T. Chen. 2009. Cancer/testis (CT) antigens: Potential targets for immunotherapy. *Cancer Sci.* 100:2014–2021.

22. Srebrow, A. and A. R. Kornblihtt. 2006. The connection between splicing and cancer. *J. Cell Sci.* 119:2635–2641.

23. Wang, E. T., R. Sandberg, S. Luo, I. Khrebtukova, L. Zhang, C. Mayr, S. F. Kingsmore, G. P. Schroth, and C. B. Burge. 2008. Alternative isoform regulation in human tissue transcriptomes. *Nature* 456:470–476.

24. Bourdon, J. C., K. Fernandes, F. Murray-Zmijewski, G. Liu, A. Diot, D. P. Xirodimas, M. K. Saville, and D. P. Lane. 2005. p53 isoforms can regulate p53 transcriptional activity. *Genes Dev.* 19:2122–2137.

25. Line, A., Z. Slucka, A. Stengrevics, G. Li, and R. C. Rees. 2002. Altered splicing pattern of TACC1 mRNA in gastric cancer. *Cancer Genet. Cytogenet.* 139:78–83.

26. Disis, M. L., K. L. Knutson, K. Schiffman, K. Rinn, and D. G. McNeel. 2000. Pre-existent immunity to the HER-2/neu oncogenic protein in patients with HER-2/neu overexpressing breast and ovarian cancer. *Breast Cancer Res. Treat.* 62:245–252.

27. Kyewski, B. and J. Derbinski. 2004. Self-representation in the thymus: An extended view. *Nat. Rev. Immunol.* 4:688–698.

28. Cibotti, R., J. M. Kanellopoulos, J. P. Cabaniols, O. Halle-Panenko, K. Kosmatopoulos, E. Sercarz, and P. Kourilsky. 1992. Tolerance to a self-protein involves its immunodominant but does not involve its subdominant determinants. *Proc. Natl Acad. Sci. U.S.A.* 89:416–420.

29. Nanda, N. K. and E. E. Sercarz. 1995. Induction of anti-self-immunity to cure cancer. *Cell* 82:13–17.

30. Arbuckle, M. R., M. T. McClain, M. V. Rubertone, R. H. Scofield, G. J. Dennis, J. A. James, and J. B. Harley. 2003. Development of autoantibodies before the clinical onset of systemic lupus erythematosus. *N. Engl. J. Med.* 349:1526–1533.

31. Schellekens, G. A., H. Visser, B. A. de Jong, F. H. van den Hoogen, J. M. Hazes, F. C. Breedveld, and W. J. van Venrooij. 2000. The diagnostic properties of rheumatoid arthritis antibodies recognizing a cyclic citrullinated peptide. *Arthritis Rheum.* 43:155–163.

32. Vittecoq, O., S. Pouplin, K. Krzanowska, F. Jouen-Beades, J. F. Menard, A. Gayet, A. Daragon, F. Tron, and X. Le Loët. 2003. Rheumatoid factor is the strongest predictor of radiological progression of rheumatoid arthritis in a three-year prospective study in community-recruited patients. *Rheumatology (Oxford)* 42:939–946.

33. Lu, H., V. Goodell, and M. L. Disis. 2008. Humoral immunity directed against tumor-associated antigens as potential biomarkers for the early diagnosis of cancer. *J. Proteome Res.* 7:1388–1394.

34. Chou, M. Y., L. Fogelstrand, K. Hartvigsen, L. F. Hansen, D. Woelkers, P. X. Shaw, J. Choi, T. Perkmann, F. Backhed, Y. I. Miller, S. Horkko, M. Corr, J. L. Witztum, and C. J. Binder. 2009. Oxidation-specific epitopes are dominant targets of innate natural antibodies in mice and humans. *J. Clin. Invest.* 119:1335–1349.

35. Chou, M. Y., K. Hartvigsen, L. F. Hansen, L. Fogelstrand, P. X. Shaw, A. Boullier, C. J. Binder, and J. L. Witztum. 2008. Oxidation-specific epitopes are important targets of innate immunity. *J. Intern. Med.* 263:479–488.

36. Yokota, S. and N. Fujii. 2010. Immunomodulatory activity of extracellular heat shock proteins and their autoantibodies. *Microbiol. Immunol.* 54:299–307.

37. Businaro, R., E. Profumo, A. Tagliani, B. Buttari, S. Leone, G. D'Amati, F. Ippoliti, M. Leopizzi, D. D'Arcangelo, R. Capoano, L. Fumagalli, B. Salvati, and R. Rigano. 2009. Heat-shock protein 90: A novel autoantigen in human carotid atherosclerosis. *Atherosclerosis* 207:74–83.

38. Desmetz, C., F. Bibeau, F. Boissiere, V. Bellet, P. Rouanet, T. Maudelonde, A. Mange, and J. Solassol. 2008. Proteomics-based identification of HSP60 as a tumor-associated antigen in early stage breast cancer and ductal carcinoma in situ. *J. Proteome Res.* 7:3830–3837.

39. Etzioni, R., N. Urban, S. Ramsey, M. McIntosh, S. Schwartz, B. Reid, J. Radich, G. Anderson, and L. Hartwell. 2003. The case for early detection. *Nat. Rev. Cancer* 3:243–252.

40. Scanlan, M. J., Y. T. Chen, B. Williamson, A. O. Gure, E. Stockert, J. D. Gordan, O. Tureci, U. Sahin, M. Pfreundschuh, and L. J. Old. 1998. Characterization of human colon cancer antigens recognized by autologous antibodies. *Int. J. Cancer* 76:652–658.

41. Chatterjee, M., S. Mohapatra, A. Ionan, G. Bawa, R. Ali-Fehmi, X. Wang, J. Nowak, B. Ye, F. A. Nahhas, K. Lu, S. S. Witkin, D. Fishman, A. Munkarah, R. Morris, N. K. Levin, N. N. Shirley, G. Tromp, J. Abrams, S. Draghici, and M. A. Tainsky. 2006. Diagnostic markers of ovarian cancer by high-throughput antigen cloning and detection on arrays. *Cancer Res.* 66:1181–1190.

42. Lin, H. S., H. S. Talwar, A. L. Tarca, A. Ionan, M. Chatterjee, B. Ye, J. Wojciechowski, S. Mohapatra, M. D. Basson, G. H. Yoo, B. Peshek, F. Lonardo, C. J. Pan, A. J. Folbe, S. Draghici, J. Abrams, and M. A. Tainsky. 2007. Autoantibody approach for serum-based detection of head and neck cancer. *Cancer Epidemiol. Biomarkers Prev.* 16:2396–2405.

43. Wang, X., J. Yu, A. Sreekumar, S. Varambally, R. Shen, D. Giacherio, R. Mehra, J. E. Montie, K. J. Pienta, M. G. Sanda, P. W. Kantoff, M. A. Rubin, J. T. Wei, D. Ghosh, and A. M. Chinnaiyan. 2005. Autoantibody signatures in prostate cancer. *N. Engl. J. Med.* 353:1224–1235.

44. Zhong, L., S. P. Coe, A. J. Stromberg, N. H. Khattar, J. R. Jett, and E. A. Hirschowitz. 2006. Profiling tumor-associated antibodies for early detection of non-small cell lung cancer. *J. Thorac. Oncol.* 1:513–519.

45. Zhong, L., G. E. Hidalgo, A. J. Stromberg, N. H. Khattar, J. R. Jett, and E. A. Hirschowitz. 2005. Using protein microarray as a diagnostic assay for non-small cell lung cancer. *Am. J. Respir. Crit. Care Med.* 172:1308–1314.

46. Chatterjee, M., A. Ionan, S. Draghici, and M. A. Tainsky. 2006. Epitomics: Global profiling of immune response to disease using protein microarrays. *OMICS* 10:499–506.

47. Draghici, S., M. Chatterjee, and M. A. Tainsky. 2005. Epitomics: Serum screening for the early detection of cancer on microarrays using complex panels of tumor antigens. *Exp. Rev. Mol. Diagn.* 5:735–743.

48. R Development Core Team. 2011. R: A language and environment for statistical computing. R Foundation for statistical computing. Vienna, Austria, http://www.R-project.org

49. Smyth, G. K. 2005. Limma: Linear models for microarray data. In: Gentleman, R., Carey, V., Dudoit, S., Irizarry, R., Huber, W. (eds.), *Bioinformatics and Computational Biology Solutions using R and Bioconductor*. Springer, New York, pp. 397–420.

50. Dudoit, S. and J. Fridlyand. 2002. A prediction-based resampling method for estimating the number of clusters in a dataset. *Genome Biol.* 3:RESEARCH0036.

51. Bland, J. M. and D. G. Altman. 1986. Statistical methods for assessing agreement between two methods of clinical measurement. *Lancet* 1:307–310.

52. Cui, X., M. K. Kerr, and G. A. Churchill. 2003. Transformations for cDNA microarray data. *Stat. Appl. Genet. Mol. Biol.* 2:Article 4.

53. Yang, Y. H., S. Dudoit, P. Luu, D. M. Lin, V. Peng, J. Ngai, and T. P. Speed. 2002. Normalization for cDNA microarray data: A robust composite method addressing single and multiple slide systematic variation. *Nucleic Acids Res.* 30:e15.

54. Yang, Y. H. and A. C. Paquet. 2005. Preprocessing two-color spotted arrays. In Gentleman, R., Carey, V. J., Huber, W., and Irizarry, R. A. (eds.), *Bioinformatics and Computational Biology Solutions using R and Bioconductor*. Springer Science+Business Media, New York, pp. 49–69.

55. Smyth, G. K. and T. Speed. 2003. Normalization of cDNA microarray data. *Methods* 31:265–273.

56. Ali-Fehmi, R., M. Chatterjee, A. Ionan, N. K. Levin, H. Arabi, S. Bandyopadhyay, J. P. Shah, C. S. Bryant, S. M. Hewitt, M. G. O'Rand, O. M. Alekseev, R. Morris, A. Munkarah, J. Abrams, and M. A. Tainsky. 2010. Analysis of the expression of human tumor antigens in ovarian cancer tissues. *Cancer Biomarkers* 6:33–48.

57. Ramachandran, N., E. Hainsworth, B. Bhullar, S. Eisenstein, B. Rosen, A. Y. Lau, J. C. Walter, and J. LaBaer. 2004. Self-assembling protein microarrays. *Science* 305:86–90.

58. Qiu, J., G. Choi, L. Li, H. Wang, S. J. Pitteri, S. R. Pereira-Faca, A. L. Krasnoselsky, T. W. Randolph, G. S. Omenn, C. Edelstein, M. J. Barnett, M. D. Thornquist, G. E. Goodman, D. E. Brenner, Z. Feng, and S. M. Hanash. 2008. Occurrence of autoantibodies to annexin I, 14-3-3 theta and LAMR1 in prediagnostic lung cancer sera. *J. Clin. Oncol.* 26:5060–5066.

59. Patwa, T. H., C. Li, L. M. Poisson, H. Y. Kim, M. Pal, D. Ghosh, D. M. Simeone, and D. M. Lubman. 2009. The identification of phosphoglycerate kinase-1 and histone H4 autoantibodies in pancreatic cancer patient serum using a natural protein microarray. *Electrophoresis* 30:2215–2226.

60. Nam, M. J., J. Madoz-Gurpide, H. Wang, P. Lescure, C. E. Schmalbach, R. Zhao, D. E. Misek, R. Kuick, D. E. Brenner, and S. M. Hanash. 2003. Molecular profiling of the immune response in colon cancer using protein microarrays: Occurrence of autoantibodies to ubiquitin C-terminal hydrolase L3. *Proteomics* 3:2108–2115.

61. Madoz-Gurpide, J., R. Kuick, H. Wang, D. E. Misek, and S. M. Hanash. 2008. Integral protein microarrays for the identification of lung cancer antigens in sera that induce a humoral immune response. *Mol. Cell Proteomics* 7:268–281.

62. Bouwman, K., J. Qiu, H. Zhou, M. Schotanus, L. A. Mangold, R. Vogt, E. Erlandson, J. Trenkle, A. W. Partin, D. Misek, G. S. Omenn, B. B. Haab, and S. Hanash. 2003. Microarrays of tumor cell derived proteins uncover a distinct pattern of prostate cancer serum immunoreactivity. *Proteomics* 3:2200–2207.

63. Anderson, K. S., N. Ramachandran, J. Wong, J. V. Raphael, E. Hainsworth, G. Demirkan, D. Cramer, D. Aronzon, F. S. Hodi, L. Harris, T. Logvinenko, and J. LaBaer. 2008. Application of protein microarrays for multiplexed detection of antibodies to tumor antigens in breast cancer. *J. Proteome Res.* 7:1490–1499.

64. Tang, L., J. Yang, S. K. Ng, N. Rodriguez, P. W. Choi, A. Vitonis, K. Wang, G. J. McLachlan, R. J. Caiazzo, Jr., B. C. Liu, W. R. Welch, D. W. Cramer, R. S. Berkowitz, and S. W. Ng. 2010. Autoantibody profiling to identify biomarkers of key pathogenic pathways in mucinous ovarian cancer. *Eur. J. Cancer* 46:170–179.

65. Gorelik, E., D. P. Landsittel, A. M. Marrangoni, F. Modugno, L. Velikokhatnaya, M. T. Winans, W. L. Bigbee, R. B. Herberman, and A. E. Lokshin. 2005. Multiplexed immunobead-based cytokine profiling for early detection of ovarian cancer. *Cancer Epidemiol. Biomarkers Prev.* 14:981–987.

66. Mor, G., I. Visintin, Y. Lai, H. Zhao, D. C. Ward, P. Bray-War, P. Schwartz, T. Rutherford, and L. Yue. 2005. Serum protein markers for early detection of ovarian cancer. *Proc. Natl Acad. Sci. USA* 102:7677–7682.

67. Arellano-Garcia, M. E., S. Hu, J. Wang, B. Henson, H. Zhou, D. Chia, and D. T. Wong. 2008. Multiplexed immunobead-based assay for detection of oral cancer protein biomarkers in saliva. *Oral Dis.* 14:705–712.

68. Linkov, F., A. Lisovich, Z. Yurkovetsky, A. Marrangoni, L. Velikokhatnaya, B. Nolen, M. Winans, W. Bigbee, J. Siegfried, A. Lokshin, and R. L. Ferris. 2007. Early detection of head and neck cancer: Development of a novel screening tool using multiplexed immunobead-based biomarker profiling. *Cancer Epidemiol. Biomarkers Prev.* 16:102–107.

Part VII

Optical Technologies for Cancer Detection and Diagnostics: Photoacoustic for Cancer Analysis

29

Detecting Circulating Melanoma Cells in Blood Using Photoacoustic Flowmetry

John A. Viator, Benjamin S. Goldschmidt, and Kyle D. Rood

CONTENTS

29.1 Introduction

Circulating tumor cells (CTCs) are those cells that separate from tumors and spread through the blood and lymphatic systems to create secondary tumors [1–3]. This process is known as metastasis and constitutes the most dangerous phase of cancer. Detection of CTCs can indicate the onset of metastasis or relapse and the absence of CTCs may indicate remission of disease. Thus, detection of CTCs is an important, emerging area of cancer research and many methods are being investigated for optimal detection. Researchers have used RT-PCR, fluorescence flow cytometry, immunomagnetic separation, and microfluidic devices to detect and separate CTCs [4–7]. Each of these technologies brings advantages and unique challenges and limitations, though they all suffer from long preparation and testing time, lack of sensitivity, or the need for dedicated technicians to evaluate results. For instance, in immunomagnetic separation, CTCs are captured on a magnetic array after they have been tagged with ferrous beads via antibody-specific attachment. After the cells have been captured, a cytologist must view the magnetic array to positively identify captured cells.

The use of light in medical diagnostics has increased over the last decade in research and clinical arenas, particularly in oncology. These techniques include optical coherence tomography for microscopic imaging, imaging with nanoparticle fluorescence enhancement, Raman spectroscopy, and hybrid techniques such as those that combine microfluidics and optical sensing [8–13]. Of the optical methods, photoacoustic generation and detection brings the advantages of optical contrast and its associated selectivity and acoustic signal propagation and its robust sensitivity [14].

Photoacoustics is a process in which optical energy is converted into high-frequency ultrasonic pulses. Photoacoustic waves can be generated by thermoelastic expansion, radiation pressure, cavitation, or ablation [15]. Thermoelastic expansion is the mechanism that we exploit and describe in this chapter. Thermoelastic expansion arises when laser energy is taken up by a light-absorbing material. The laser energy is immediately converted into heat, which causes subsequent heating and with it, a volume expansion of the irradiated area on a nanosecond timescale. Thus, this rapid expansion results in an outward propagating pressure wave that is, in fact, a sound wave. If conditions are right, this sound wave can be analyzed to determine optical properties of the absorbing medium. Certainly the sound wave can be studied to triangulate the position of the source, allowing for detection and localization of the light absorber. With a sophisticated array of acoustic sensors, photoacoustic imaging is possible, similar to traditional ultrasonic imaging.

The difference between photoacoustics and traditional ultrasound is that (a) photoacoustics produces a broad range of ultrasonic frequencies, rather than a single harmonic used in ultrasound, and (b) photoacoustics induces the object of interest to become an active source of sound. This is in contrast to ultrasound, in which sound waves are directed into a medium and material interfaces produce reflections that are reconstructed for imaging.

While photoacoustics produces active sources of sound in targets of interest, the fact that the sound waves are usually dominated by high frequencies, often as high as 100 MHz, the signal is attenuated within a centimeter or so. Thus, photoacoustics is ideal for cutaneous imaging or other applications where whole-body or organ imaging is not needed.

Photoacoustics has been used for vascular imaging and depth profiling of tissue due to these inherent advantages. In such cases, hemoglobin in blood provides optical contrast over tissue with no optical absorption, such as collagen. In a similar fashion, pigments such as melanin can provide optical contrast for photoacoustic generation and detection. We have detected melanoma cells in blood using a photoacoustic flowmeter we developed [16,17] and have used it to sense single melanoma cells suspended among white blood cells [18,19]. Zharov et al. have performed some in vivo studies in which they detected melanoma cells in a rabbit model in addition to other cancer cells tagged with gold nanoparticles [20,21].

We propose to use photoacoustics as an in vitro diagnostic test in which blood samples from patients at risk for metastatic melanoma are tested for the presence of melanoma cells. This testing paradigm would entail taking a blood sample at weekly or monthly intervals and processing them in the photoacoustic flowmeter where any melanoma cells present would be detected and counted by monitoring photoacoustic events.

We have detected melanoma cells spiked in human blood samples simulating circulating melanoma cells (CMCs). Most melanomas are highly melanotic, with estimates of amelanotic melanoma being <5% [22] or 1.8%–8.1%, though this latter figure includes partially pigmented melanoma [23]. Thus, the great majority of melanomas contain native light absorbers that can be exploited using photoacoustic generation and detection. As noted earlier, a photoacoustic effect occurs when the optical energy of a photon is transduced into a mechanical disturbance, resulting in an acoustic wave [24–29].

In this chapter, we present a sensitive and specific photoacoustic system for label-free detection of CMCs in blood. The apparatus is shown in Figure 29.1, where a laser irradiates the mononuclear cell layer (MNCL) obtained from separating a blood sample from a patient. The mononuclear cells separate in a standard centrifugation procedure, since these cells are less dense than the iron-laden red blood cells and are denser than the plasma layer. CMCs, if present, will reside in the MNCL due to similar density to white blood cells. Any melanoma cells within the flow chamber will selectively absorb laser light, subsequently producing photoacoustic waves that signal their presence in the cell suspension. Melanoma cells can be counted under flow by enumerating the transient photoacoustic waves detected. We have successfully detected single melanoma cells in blood using this technology.

29.2 Materials and Methods

29.2.1 Flow Chamber

The system utilized a polyvinylidene difluoride (PVDF) piezoelectric transducer for acoustic sensing. The PVDF film (K-Tech, Albuquerque, New Mexico) was 25 μm in thickness and was ~1 cm^2. It was coupled to two electrodes that were linked to the detection electronics for amplification and storage on the oscilloscope. Laser light was delivered at an angle of 90° from the transducer element. A single cell is

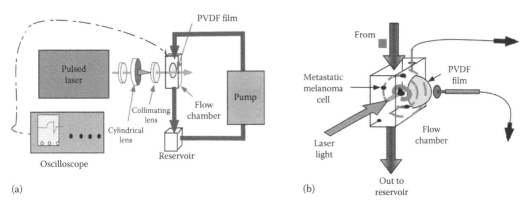

FIGURE 29.1 (a) The photoacoustic system for detection of circulating melanoma cells consists of a circulation system, including a flow chamber, and a photoacoustic generation system, consisting of a tunable laser source, an acoustic sensor, and an oscilloscope. (b) A rendering of the flow chamber showing flow of cellular material downward with laser radiation exciting melanoma cells, producing acoustic pulses sensed by the PVDF acoustic sensor.

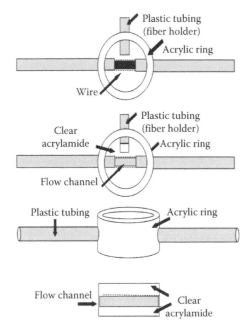

FIGURE 29.2 Schematic of the procedure for fabricating acrylamide flow chambers.

considered an optical point source and therefore emits photoacoustic waves radially in all directions. The flow chamber was held together with an acrylic ring that had three holes drilled at 90° from each other as shown in Figure 29.2. The ring was filled with clear acrylamide but must create a mold for constructing the flow channel and the entrance for the optical fiber. Masterflex tubing (Cole Parmer, Vernon Hills, IL) was fed through the holes and a 1 mm diameter wire to match the tubing's inner diameter was suspended through the two opposing holes and a wire was fed through the third hole to prevent acrylamide from entering the tubing hole. Once the acrylic ring was prepared, Parafilm was stretched across the bottom of the ring and clear acrylamide was poured into it, gelling around the tubing and wire. The acrylamide was made from 10 mL of 20% acrylamide solution (Sigma Aldrich, St. Louis, MO), 0.04 g of ammonium persulfate (Sigma Aldrich), and 20 μL of TEMED (Sigma Aldrich). After adding the TEMED, the mixture was poured immediately before gelling. After gelling, the wires were removed and the tubing that

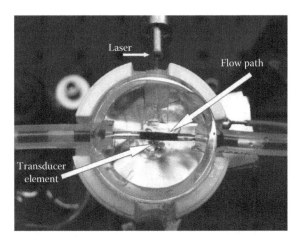

FIGURE 29.3 View from above the detection chamber, where the laser fires from the side and the transducer senses the photoacoustic wave from below.

was used for the fiber was pulled out ~4 mm. The acrylamide was used to acoustically match saline as the acoustic waves propagated from melanoma cells to the sensor. Thus, this procedure creates a flow path for the MNCL suspension while providing a right angle port for insertion of an optical fiber for laser light delivery. This procedure was used to create the photoacoustic flow chamber (Figure 29.3).

29.2.2 Cell Preparation

An HS 936 melanoma cell line was cultured for use in photoacoustic experiments. The cells were fixed in ethanol and resuspended in phosphate-buffered saline (PBS) (Sigma Aldrich). Approximately 15 min before experiments, the cells were diluted with a PBS and 2% Tween 20 solution (Sigma Aldrich) to the desired cell concentration. Tween 20 was used to promote flow through the microfluidic system. The cells were then counted manually using a hemocytometer.

In order to simulate the clinical situation, in which a patient's blood would be tested for CMCs, we prepared to spike whole blood samples with these cultured melanoma cells. Approximately 10 mL of whole blood was drawn from healthy human volunteers and added to a centrifuge tube that contained Histopaque 1077 (Sigma Aldrich). Histopaque is a sugar solution that provides a liquid component with a distinct density, thus separating different layers of blood after centrifugation. In this case, Histopaque 1077 is a sugar solution with a density of $1.077 \, g/cm^3$. This density provides separation between the red blood cells that settle to the bottom of the centrifuge tube and the MNCL, allowing easier extraction of the white blood cells without contaminating with red blood cells.

The tube was centrifuged at 3000 rpm for 10 min, after which the white blood cell layer and the contents near it were added to a thinner centrifuge tube. This tube was then centrifuged for 3 min and the white blood cell layer was removed. For white blood cells spiked with melanoma, cultured melanoma cells are added to the whole blood before the centrifugation process and the separation procedure remains the same thereafter. The white blood cells were resuspended in 10 mL of PBS.

The resulting cell concentrations pumped through the flowmeter with a syringe pump (Kent Scientific, Torrington, CT) had ~1000 white blood cells per microliter. The melanoma cells were irradiated at a concentration of 10, 25, 100, and 800 cells per microliter with laser pulse energies of 6 mJ at 10 Hz. These concentrations were prepared using serial dilutions. In order to account for dilution error and partial recovery in the white blood cell collection, the cells were counted after irradiation using a hemocytometer and averaged 13, 26, 57, and 798 cells per microliter. The detection volume defined by the laser beam was ~1 μL.

The flow rate was 1 mL/min; thus, a 10 mL sample took 10 min to process. We anticipate that this timeframe will be typical of a clinical test.

29.2.3 Detection Electronics

The output of the piezoelectric sensor was sent to an amplifier (BR-640A, Ritec, Inc., Warwick, RI) with a gain of 32 dB. This amplified signal was displayed on a 200 MHz oscilloscope with an input impedance of 1 MΩ (TDS 2024, Tektronix, Wilsonville, OR).

29.3 Results

29.3.1 Developing Photoacoustic Flowmetry System

The photoacoustic flowmeter in Figure 29.1 consists of a pulsed laser system, a flow chamber, an acoustic sensor, and detection electronics. It is a system that pumps a white blood cell suspension through a detection chamber wherein optically absorbing light particles, such as CMCs, will produce photoacoustic waves. White blood cells, since they lack color, merely scatter laser light and do not produce photoacoustic waves. The laser system is a tunable Nd:YAG pumped optical parametric oscillator (OPO) capable of delivering laser light with a pulse duration of 5 ns from 410 to 2400 nm with pulse energies in the range of 5–25 mJ (Surelite Continuum, Santa Clara, CA). In these experiments, we used 532 nm light as this green wavelength is appropriate for absorption by melanin. We used a plano-convex cylindrical lens with a focal length of 20 mm (Thorlabs, Newton, NJ) to convert the OPO beam to a circular spot. The circular spot was focused using a plano-convex lens with a focal length of 25 mm (Thorlabs). The laser light is delivered via a 1000 µm optical fiber (Thorlabs) to a custom-made flow chamber. The flow chamber is a polyacrylamide tube that is 1 mm in diameter and incorporates an acoustic sensor. Polyacrylamide was used as it acoustically matches the cell suspension and therefore allows acoustic energy to be sensed unimpeded. The laser spot was ~1 mm in diameter and diverged from the optical fiber at ~22°. Further details on the flow chamber are given later.

29.3.2 Photoacoustic Detection of Melanoma

Figure 29.4 shows representative waveforms from photoacoustic generation in the flowmeter for a control as well as various concentrations of melanoma cells. We show 10, 25, 100, and 800 melanoma cells per microliter. Separation of the white blood cells from whole blood recovered about half of the melanoma cells; thus, actual concentrations detected are approximately half the concentration indicated.

29.4 Discussion

Photoacoustics has been used in numerous research areas for biomedical diagnosis, including small animal imaging, cutaneous imaging, and trace gas detection. The ability to discriminate light-absorbing materials from distinctly absorbing or non-absorbing surroundings gives photoacoustics an advantage over other diagnostic methods. Additionally, an acoustic signal propagates in acoustically homogeneous, but optically turbid media, such as human tissue, with little degradation. These same advantages make photoacoustic flowmetry an attractive technique for detecting small particles under flow in human body fluids.

Photoacoustic flowmetry provides a means for detecting pathological analytes in body fluids. In the case described in this chapter, it is used to detect pigmented melanoma cells in blood. The optical contrast of melanoma cells suspended among millions of white blood cells enriched from whole blood allows specific and sensitive detection [18,19]. Figure 29.4 shows increasing signal strength with higher melanoma cell concentration. The control, consisting of white blood cells, shows some deviations from 0 to about 4.8–5.8 µs. This noise consisted of random, transient spikes that changed from pulse to pulse. All signals in this time window in the cases with melanoma cells were persistent and lasted about 2 s, corresponding to the passage of the melanoma cells through the laser beam and the signal averaging on the oscilloscope. While increasing concentration increased the signal-to-noise ratio, the effect was not

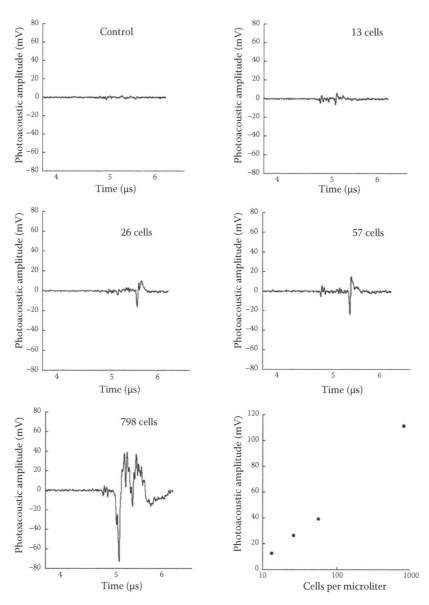

FIGURE 29.4 Representative photoacoustic waveforms for control, with no melanoma cells, 13, 26, 57, and 798 melanoma cells. On the bottom right is a graph showing the photoacoustic signal strength as a function of cell concentration.

linear. Considering that the horizontal scale is logarithmic, the signal strength of 800 cells should have been ~400 mV.

The dilution curve shown in the bottom right of Figure 29.4 indicates that a detection threshold is fewer than five cells in the irradiation volume. However, signal extraction can discriminate stationary noise from photoacoustic waves generated in cells; thus, the threshold is postulated to be single cell detection. Additionally, laser fluence and electronic amplification can be increased to ensure robust detection of single melanoma cells. However, testing for CMCs in human patients introduces many factors that have high biological variability, including melanin content in each CMC, thus affecting sensitivity of our instrument. However, we used a cell line that weakly expresses melanin under the culture conditions we used. Thus, our testing paradigm was performed in the challenging case of a weakly pigmented cancer cell. In fact, our preliminary work with human patients suggests that CMCs may be highly pigmented as

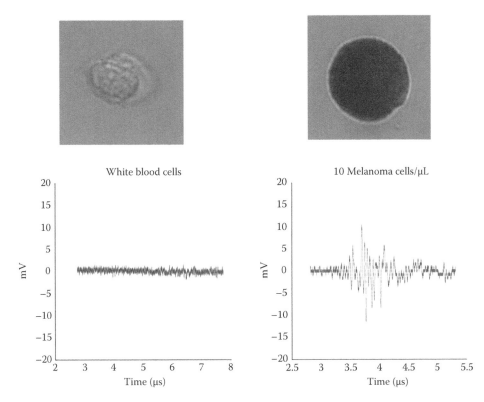

FIGURE 29.5 The photoacoustic wave obtained from irradiating the pigmented cell captured after testing in the photo-acoustic apparatus.

shown in Figure 29.5. This figure shows a photoacoustic waveform from testing the blood of a metastatic melanoma patient and the captured cell, along with a control sample of a white blood cell that does not generate a photoacoustic signal.

A closer inspection of the waveform for 10 cells per microliter shows distinct events that may constitute individual cells distributed in the flow chamber. Cells more distant from the piezoelectric sensor show up later in time, while closer cells show up earlier. This behavior is also evident in the higher cell concentrations, though larger waveforms are almost certainly a product of aggregations of melanoma cells.

29.5 Future Trends

Photoacoustic flowmetry can be used to detect CMCs in human blood samples. We are investigating several methods to extend this work and to increase its accuracy.

Since individual pigmented cells are detected, there is no limit to the volume of blood that can be tested. The limiting factor is the time to process the given volume and the volume of blood obtained from a patient. We anticipate blood draws of 10–20 mL. In processing the samples, the MNCLs will be resuspended in a similar volume, thus with a typical flow rate of 1 mL/min, the test will take about 10 min. Including blood separation, the entire procedure will take about an hour.

We are working to use photoacoustic flowmetry for detection of non-pigmented CTCs by adding exogenous chromophores. Using antibody tagging, this method can be extended to non-pigmented CTCs by specifically attaching exogenous chromophores. These chromophores, which may be dyed microspheres, absorbing nanoparticles, or any other light-absorbing contrast agent, will provide the target for photoacoustic generation. Figure 29.6 shows epigallocatechin gallate (EGCG)–conjugated gold nanoparticles internalized within prostate cancer cells from the PC-3 cell line. This concentration

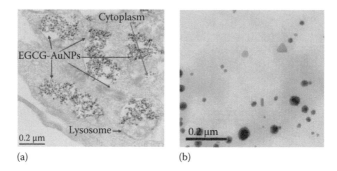

(a)   (b)

FIGURE 29.6   (a) Gold nanoparticle uptake is shown in a prostate cancer cell in this electron micrograph. (b) EGCG gold nanoparticles shown in suspension.

of nanoparticles provides optical contrast great enough for single cell detection in photoacoustic flow-metry (data not shown).

Thus, any cell that can be identified by specific monoclonal antibodies may allow detection using pho-toacoustic flowmetry, providing that the antibodies can be selectively targeted and attached to exogenous optical absorbers. Such antibody attachment is standard practice for fluorophores. The photoacoustic technology would merely replace dyed microspheres for fluorophores, for example.

Additionally, we are developing a microfluidic system for capturing CTCs. The system is shown con-ceptually in Figure 29.7. In this setup, photoacoustic signals generated in the active area of the flow chamber will trigger a suction pump downstream after an appropriate delay to account for the cell to travel to the suction volume. In clinical implementation, the CTC will be captured along with thousands of white blood cells. The suction volume, ~10 μL, will be diluted to about 1 mL and run through the system again. Upon capture the second time, the concentration and hence the number of captured white blood cells will be reduced by a factor of 100. This process can be repeated until the CTC is isolated.

This system can be adapted to test for other pathological analytes in human body fluids. Some appli-cations may include detection of bacteria for analysis of septicemia, and detection of malaria parasites in blood by targeting hemozoin present in infected red blood cells and ingested by white blood cells.

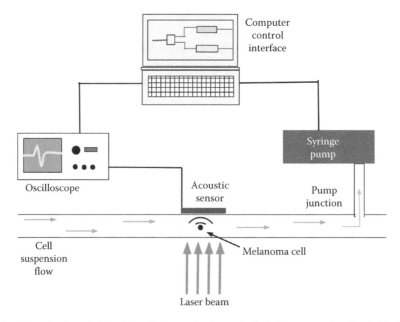

FIGURE 29.7   Future implementation of the photoacoustic flowmeter includes a pump to extract detected melanoma cells.

Furthermore, this method may be used for forensic testing of small samples of body fluids contaminated by blood or other optically absorbing particles.

The ability to isolate CMCs from patients provides a unique opportunity to understand the biological properties of these cells that determine their ability to form metastases. CMCs have never been isolated from patients in quantities that have allowed their biological properties to be characterized. For example, we do not know if the CTCs are proliferating while they travel around in the bloodstream and, if so, what growth factors, cytokines, and hormones facilitate their survival and stimulate cell proliferation. Nor do we know that the CMCs are mesenchymal as predicted by the hypothesis that they have undergone the epithelial–mesenchymal transition. These basic questions have never been answered during the 40 year long war on cancer, simply because we have not been able to purify CTCs from the blood of a cancer patient. Capturing CTCs will tell us much about the basic biological properties of circulating cancer cells and will provide critical knowledge toward our long-term goal, which is the development of therapeutic strategies that will kill CTCs and prevent the formation of secondary metastatic tumors.

REFERENCES

1. A. Gilbey, D. Burnett, R. Coleman, and I. Holen, The detection of circulating breast cancer cells in blood, *J. Clin. Pathol.*, (57), 903–911, 2004.
2. R.T. Krivacic, A. Ladanyi, D.N. Curry, H.B. Hsieh, P. Kuhn, D.E. Bergsrudt, J.F. Keprost, T. Barbera, M.Y. Ho, L.B. Chen, R.A. Lerner, and R.H. Bruce, A rare-cell detector for cancer, *Proc. Natl. Acad. Sci. USA*, 101(29), 10501–10504, 2004.
3. W. Check, Circulating tumor cell tests-Boon or bust? *Cap Today*, (19), pp. 1,76–1,86, 2005.
4. R. Loberg, Y. Fridman, B. Pienta, E. Keller, L. McCauley, R. Taichman, and K. Pienta, Detection and isolation of circulating tumor cells in urologic cancers: A review, *Neoplasia*, (6), 302–309, 2004.
5. A. Ring, L. Zabaglo, M. Ormerod, I. Smith, and M. Dowsett, Detection of circulating epithelial cells in the blood of patients with breast cancer: Comparison of three techniques, *Br. J. Cancer*, (92), 906–912, 2005.
6. D. Pellegrino, C. Bellina, G. Manca, G. Boni, M. Grosso, D. Volterrani, I. Desideri, F. Bianchi, A. Bottoni, V. Ciliberti, G. Salimbeni, D. Gandini, M. Castagna, V. Zucchi, and A. Romanini, Detection of melanoma cells in peripheral blood and sentinel lymph nodes by RT-PCR analysis: A comparative study with immunochemistry, *Tumori*, 86(4), 336–338, 2000.
7. S. Nagrath, L. Sequist, S. Maheswaran, D. Bell, D. Irimia, L. Ulkus, M. Smith, E. Kwak, S. Digumarthy, A. Muzikansky, P. Ryan, U. Balias, R. Tompkins, S. Haber, and M. Toner, Isolation of rare circulating tumor cells in cancer patients by microchip technology, *Nature*, 450(7173), 1235–1239, 2007.
8. S. Jiang, M. Gnanasammandhan, and Y. Zhang, Optical imaging-guided cancer therapy with fluorescent nanoparticles, *J. R. Soc. Interface*, 7(42), 3, 2010.
9. J. Holmes, OCT technology development: Where are we now? A commercial perspective, *J. Biophotonics*, 2(6–7), 347–352, 2009.
10. S. Kumar and R. Richards-Kortum, Optical molecular imaging agents for cancer diagnostics and therapeutics, *Nanomedicine*, 1(1), 23–30, 2006.
11. S. Annieke Nijssen, C. Tom, J. Peter, and J. Gerwin, Towards oncological application of Raman spectroscopy, *J. Biophotonics*, 2(1–2), 29–36, 2009.
12. F. Myers and L. Lee, Innovations in optical microfluidic technologies for point-of-care diagnostics, *Lab Chip*, 8(12), 2015–2031, 2008.
13. E. Cauberg, D. de Bruin, D. Faber, T. van Leeuwen, J. de la Rosette, and T. de Reijke, A new generation of optical diagnostics for bladder cancer: Technology, diagnostic accuracy, and future applications, *Eur. Urol.*, 56(2), 287–297, 2009.
14. M. Xu and L. Wang, Photoacoustic imaging in biomedicine, *Rev. Sci. Instrum.*, 77, 041101, 2006.
15. M. Sigrist, Laser generation of acoustic waves in liquids and gases, *J. Appl. Phys.*, 60, R83–R121, 1986.
16. R. Weight, P. Dale, C. Caldwell, A. Lisle, and J. Viator, Photoacoustic detection of metastatic melanoma cells in the human circulatory system, *Opt. Lett.*, (31), 2998–3000, 2006.
17. S. Holan and J. Viator, Automated wavelet denoising of photoacoustic signals for circulating melanoma cell detection and burn image reconstruction, *Phys. Med. Biol.*, 53, N227–N236, 2008.

18. G. Gutiérrez-Juárez, S. Gupta, R. Weight, L. Polo-Parada, C. Papagiorgio, J. Bunch, and J. Viator, Optical photoacoustic detection of circulating melanoma cells in vitro, *Int. J. Thermophys.*, 31, 784–792, 2010.

19. G. Gutierrez-Juarez, S. Gupta, M. Al-Shaer, L. Polo-Parada, P. Dale, C. Papageorgio, and J. Viator, Detection of melanoma cells in vitro using an optical detector of photoacoustic waves, *Lasers Surg. Med.*, 42(3), 274–281, 2010.

20. V. Zharov, E. Galanzha, E. Shashkov, N. Khlebtsov, and V. Tuchin, In vivo photoacoustic flow cytometry for monitoring of circulating single cancer cells and contrast agents, *Opt. Lett.*, 31(24), 3623–3625, 2006.

21. V. Zharov, E. Galanzha, E. Shashkov, J. Kim, N. Khlebtsov, and V. Tuchin, Photoacoustic flow cytometry: Principle and application for real-time detection of circulating single nanoparticles, pathogens, and contrast dyes in vivo, *J. Biomed. Opt.*, 12, 051503, 2007.

22. S. Swetter, Malignant melanoma. http://www.emedicine.com/DERM/topic257.htm, (accessed November 10, 2011).

23. E. Wain, C. Stefanato, and R. Barlow, A clinicopathological surprise: Amelanotic malignant melanoma, *Clin. Exp. Dermatol.*, 33, 365–366, 2008.

24. J. A. Viator, S. L. Jacques, and S. A. Prahl, Depth profiling of absorbing soft materials using photoacoustic methods, *J. Select. Top. Quantum Electron.*, 5, 989–996, 1999.

25. C. Hoelen, A. Dekker, and F. de Mul, Detection of photoacoustic transients originating from microstructures in optically diffuse media such as biological tissue, *IEEE Trans. Ultrason. Ferroelectr. Freq. Control*, 48, 37–47, 2001.

26. X. Wang, Y. Pang, G. Ku, G. Stoica, and L. Wang, Three-dimensional laser-induced photoacoustic tomography of mouse brain with the skin and skull intact, *Opt. Lett.*, 28, 1739–1741, 2003.

27. J. Viator, B. Choi, M. Ambrose, J. Spanier, and J. Nelson, In vivo port wine stain depth determination using a photoacoustic probe, *Appl. Opt.*, 42, 3215–3224, 2003.

28. R. Kolkman, E. Hondebrink, W. Steenbergen, T. van Leeuwen, and F. de Mul, Photoacoustic imaging of blood vessels with a double-ring sensor featuring a narrow angular aperture, *J. Biomed. Opt.*, 9, 1327–1335, 2004.

29. G. Ku, X. Wang, G. Stoica, and L. Wang, Multiple-bandwidth photoacoustic tomography, *Phys. Med. Biol.*, 49, 1329–1338, 2004.

Part VIII

Electrochemical Biosensors

30

Self-Contained Enzymatic Microassay Biochip for Cancer Detection

Jianwei Mo, JrHung Tsai, and Brian J. Sullivan

CONTENTS

30.1 Introduction

This chapter describes the development of a microelectromechanical system (MEMS)-based microchip (lab-on-a-chip [LOC], BioMEMS chip) for the point-of-care measurement of prostatic acid phosphatase (PAP) activity in a blood sample obtained by a small hollow microneedle integrated into the chip. The chip is fabricated from silicon using MEMS processing technology. The chip is designed to be a single-use disposable (similar to a blood glucose test strip) and to be used in conjunction with a small handheld instrument.

The incidence of new cases of prostate cancer in the United States is estimated at 217,730 for 2010, of which 32,050 men will succumb. Only lung cancer has a higher mortality rate among all other forms of malignancies [1]. Although 1 man in 6 will be diagnosed with prostate cancer during his lifetime, only 1 man in 34 will die of it. Often prostate cancer grows slowly, and "watchful waiting" is the chosen course of intervention [2]. However, there are aggressive forms that require early diagnosis and treatment before metastasis to the bone or other organs occurs. There is a clear need for highly specific and sensitive biomarkers for early diagnostics and treatment.

In a reivew on prostatic acid phosphatase, Helms et al. [3] cite earlier work by Gutman in which he observed increased acid phosphatase (ACP) activity both at the site of Osteoplastic metastasis and in the serum of 11 out of 15 patients with metastasis carcinoma. In benign prostatic hypertrophy and in non-prostatic diseases, in contrast, the activity of serum ACP was rarely, and only mildly, elevated [4]. The activity of ACP derived from the human prostate has been observed to be inhibited upon the addition of L-(+)-tartrate and serum levels of this form have been considered useful in the diagnosis and management of prostatic diseases [4,5]. Elevated serum ACP activity is not unique to prostatic diseases, also being present in Paget's, Gaucher's, and bone diseases among others; however, most of these are of tartrate-resistant forms [4–6].

ACPs (EC 3.1.3.2) represent a heterogeneous group of enzymes containing many isoenzymes, each specific to one particular type of tissue [4,6]. Isoenzyme 2 of ACP is the most active in the human prostate, and is believed to be responsible for the tartrate-inhibited activity in serum. Li et al. [7] discovered that four of five isoenzymes of ACP separated from leukocytes were also inhibited by tartrate; therefore, tartrate sensitivity is not a specific criterion for PAP as a cancer marker. There has been considerable debate as to the best inhibitors and substrates to specifically assay PAP [3,4,6,7] with no resolution. With the discovery of the isoenzymes of ACP, it has been suggested that a serum isoenzyme pattern might be more valuable in pinpointing a particular tissue(s), and that the pattern would show abnormalities in the face of normal total serum enzyme [4,8]. Since PAP does not possess adequate sensitivity for detection of early stage cancer, the more sensitive assay for prostate-specific antigen (PSA) has become the standard for screening and early detection, but not without controversy [9,10].

New biomarkers are being sought for early detection of prostate cancer and cancers in general and include enzymes [8], mass spectroscopic proteomic patterns [11,12], serum amyloid A [13], human tissue kallikreins, of which PSA is a member, serum selenium [14], genetic markers, and microarray technology [15].

Traditional enzyme activity measurements, for diagnostic or therapy monitoring, are performed in a clinical laboratory by highly skilled operators using sensitive and often delicate instrumentation [5]. The ability to perform health status monitoring at the point of care, in an automated way requiring minimal training, has advantages in such areas as home health monitoring, emergency situations, and at remote or isolated locations. An ideal point-of-care device consists of a self-contained disposable microchip used in conjunction with a small portable device that can actuate the microchip to obtain the sample, perform the analysis, and report the results quickly. Results can be stored within the device and sent by telemetry to physicians or to a central location at desired intervals.

With the publication of the seminal paper, "Silicon as a Mechanical Material," Kurt E. Petersen ushered in the field of MEMS of which LOC is a subset [16]. MEMS employs much of the technology developed for the semiconductor industry, which is capable of creating miniaturized intricate structures reproducibly in high volumes and at low unit cost. Common commercial MEMS devices are miniature accelerometers used to trigger deployment of automobile airbags. The central premise of LOC is to reduce the functionalities of a clinical or analytical laboratory onto a single disposable chip not much larger and often smaller than the U.S. quarter. The past two decades have witnessed tremendous growth in the field and have spawned several companies commercializing in LOC technology. In creating an analytical microchip system, miniaturized versions of common laboratory procedures are required. These include sample acquisition and metering, separation, mixing with reagents and buffers, and detection of the desired analyte. To this end, a large collection of miniaturized fluidic components has been developed including microneedles, microvalves, stop junctions, microfluidic networks, on-chip reagent reservoirs and waste storage, microfluidic mixers, and the analyte detection system [17,18]. With the inclusion of a biosensor, the chip becomes an analytic BioMEMS device. Biosensors are analytical devices that combine a biological agent with a physicochemical detection component. The biological component of a biosensor can include enzymes, proteins and antibodies, cell receptors, nucleic acids, or other biologically derived materials. The physicochemical component (transducer) of a biosensor transforms the interaction between the analyte and the biological element into a signal that can be more easily measured and quantified. Common biosensor transducers include optical techniques such as optical absorption, fluorescence, and surface plasmon resonance, piezoelectric mass sensors, and electrochemical methods. The blood glucose meter, used by diabetics to control their disease, is an example of a commercial biosensor, which uses an enzyme specific for glucose as the biological component. The conversion of glucose by the enzyme is monitored using an electrochemical transducer where the resulting current is usually a measure of the blood glucose concentration, although some glucose biosensors are affected by other blood constituents.

Although LOC devices have been fabricated from a variety of materials such as glass, quartz, and plastics, the focus of this chapter will be on silicon because this is the technology being developed in the authors' laboratory and because it offers greater fabrication flexibility than other materials. Single crystal silicon wafers, used in the semiconductor industry, are of very high purity and of reasonable cost. The single crystal nature allows the fabrication of small strong hollow bore microneedles, precise etching of intricate channels, and other features of micrometer dimensions. In addition, established batch-processing technology allows high volume production of identical chips at low unit cost through economies of scale. A brief tutorial on MEMS technology will be presented in Section 30.3.

Enzyme activities are commonly performed assays [5], because the increase or decrease in the activity of a particular enzyme can indicate organ damage due to injury or disease. Enzyme activity is defined as that amount of enzyme that catalyzes the conversion of 1 μmol of substrate per minute and is designated as 1 unit (U). The conditions under which the assay is performed must be tightly specified: temperature is usually taken as 25°C while the other parameters, such as pH, substrate concentration, ionic strength to name a few, are those that produce maximum rate of substrate conversion. Optical techniques are frequently employed to follow the buildup of the product resulting from the enzymatic cleavage of the substrate. Calorimetric, fluorometric, chemiluminescence, and light scattering are methods that have been applied to enzyme assays [19–22]. Much of the early work in detecting PAP activity was to find substrates that were highly specific to PAP and whose enzymatic product was strongly adsorbing at optical wavelengths. Although optical detection methods are sensitive and straightforward to perform in the laboratory, miniaturizing them to microchip dimensions becomes difficult. However, electrochemical methods are quite amenable to miniaturization and are simpler to implement because the only interfaces to the instrument are electrical contacts, obviating the need for light sources, lenses, optical filters, often-delicate optical detectors, and the problem of optical shielding to prevent ambient stray light from entering the detector reducing signal to noise. In support of electrochemical enzyme assays, it should be pointed out that all blood glucose test strips currently on the market employ this type of detection and are produced in quantities of billions by roll-to-roll printing.

In developing an electrochemical enzyme activity assay, selection of the enzymatic substrate is critical in that the substrate must be stable for long periods of time, be rapidly turned over by the enzyme, and the product must be electroactive with a low oxidation potential relative to the reference electrode, and have reasonably stable oxidized and/or reduced forms. The last property allows for increased sensitivity through electrochemical amplification, which will be discussed in Section 30.3.

Elevated levels of serum PAP cannot be taken as the sole criterion for prostate cancer; however, it appears to have significant prognostic value, which will be discussed further in Section 30.4. The BioMEMS enzyme assay chip discussed here can be expanded to include other enzymes and/or analytes.

30.2 Materials

30.2.1 Microchip Fabrication

1. Silicon wafers (primary grade) were supplied by Silicon Quest International (Santa Clara, CA) Web identification SQ12388. Wafers are 100 mm in diameter with a wafer-to-wafer thickness range from 500 to 550 μm. All wafers used in fabrication were P-Type (boron-doped) with <1–0–0> crystal orientation. Wafers come in lots of 25 in a cassette suitable for Class-100 clean room environments, and commercial processing equipment.

2. Pyrex cover wafers are supplied by University Wafer (Boston, MA, catalog # 7740 DSP) and are 100±0.3 mm in diameter with a thickness of 500±25 μm. Surface roughness is <1.5 nm (Ra) with a scratch–dig of 60–40. Wafers come in lots of 25 per cassette suitable for Class-100 clean room environments.

3. Bonding between the Pyrex cover and the silicon wafer was accomplished using DYMAX Ultra Light-Weld 3069 UV cured adhesive. Curing was performed with a DYMAX BlueWave 50 AS UV Curing Light Source.

4. Thermally activated valves integrated within the microfluidic chip were fabricated using paraffin with a melting point <100°C supplied by Sigma-Aldrich, St. Louis, MO.

5. Gold and titanium sputtering targets were purchased from Process Materials (Pleasanton, CA). Both targets were 3 in. in diameter 0.125 in. thick and of 99.99% purity.

6. Double-sided biocompatible adhesive film was supplied by Tyco Plastics and Adhesives, Charlotte, NC.

7. Other chemicals and reagents used are of analytical grade purchased from Sigma-Aldrich.

30.2.2 Enzyme Activity Assay

1. Human orthophosphoric-monoester phosphohydrolase (PAP, isoenzyme 2 EC 3.1.3.2) was obtained from Sigma-Aldrich. *Para*-aminophenyl phosphate (APP), enzyme substrate, was obtained from Universal Sensors (Cork, Ireland).

2. Buffers were purchased in preweighed envelopes from Sigma-Aldrich. Buffers were made in Type 1 ultrapure water produced by a Barnstead NANOpure DIamond purification system.

30.3 Methods

30.3.1 Electrochemistry

Electrochemistry deals fundamentally with the process of electron transfer between an electrode surface and electroactive substances in solution adjacent to this surface. The electrode surface can be, for example, a metal (e.g., gold, platinum), carbon, a semiconductor, conducting polymers, or composite materials, whereas the electroactive substances consist of ions, inorganic, or organic molecules that can be either oxidized (species loses electrons by transfer to electrode) or reduced (species gains electrons

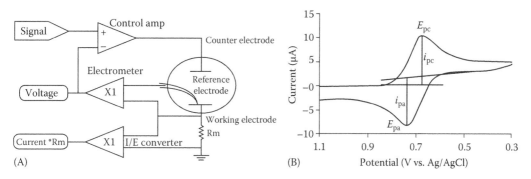

FIGURE 30.1 A schematic of an electrochemical measurement system is shown in (A) consisting of a potentiostat, a three-electrode cell, an amplifier, and data acquisition. (B) is a drawing depicting the cyclic voltammogram of a reversible redox couple for the case of a single electron transfer.

by transfer from electrode) under given conditions (e.g., appropriate applied potential, electrode surface, and composition of the electrolyte solution). Oxidation and reduction always occur in pairs; when one species is oxidized, another is reduced. A species that can be reversibly cycled between its oxidized and reduced form is of importance in electrochemistry and is known as a redox (reduction–oxidation) couple. Electron transfer causes a current to flow in an electrochemical cell, which requires a minimum of two electrodes to form a closed circuit, one where oxidation occurs (the anode) and the other where reduction takes place (the cathode). To make meaningful electrochemical measurements, all applied potentials must be set relative to a stable fixed potential produced by the reference electrode. A reference electrode has a stable and well-established equilibrium electrode potential resulting from a stable redox couple where each of the participants is at constant concentration. A common reference is produced from the silver (Ag) silver chloride (AgCl) redox couple designated as Ag/AgCl. The Ag/AgCl reference electrode [23] is robust and easy to fabricate and used exclusively in our laboratory. A three-electrode scheme is the most commonly employed arrangement used in making electrochemical measurements and is comprised of a "working electrode" (where the reaction being studied occurs), the "counter" electrode (this completes the electrochemical circuit), and the reference electrode (the potential against which all other potentials are measured). Figure 30.1A is a schematic of a typical three-electrode electrochemical circuit. Electrochemical measurements are normally performed using a potentiostat to control the three-electrode electrochemical cell and to obtain data. This instrument can be used to either precisely hold the working electrode at a given potential, or can sweep the potential in time. In addition to poising the working electrode at the desired potential, the instrument also measures the resulting current between the working and counter electrodes. The resulting current from an electroactive species is directly related to its concentration in the solution.

Electrochemistry is a rich subject and a full overview of the field is beyond the scope of this chapter; interested readers are directed to consult the extensive literature on this subject. Electroanalytical chemistry is a subset that answers questions such as what is the concentration of a particular species in a sample, what is the rate of oxidation of an electroactive species, what electrical potential is required to convert one substance into another? Two common formats for displaying experimental data are current (i) against applied potential (E) (voltammetry) and current (i) against time (t) (chronoamperometry). Most electroanalytical techniques are variations on these two data formats. Cyclic voltammetry is a potentiodynamic method that is routinely performed in our laboratory because it provides considerable information about electrochemical systems under study in a rapid manner. In this method, the potential of the working electrode is scanned linearly from a lower limit to an upper limit at a constant rate and then back again as shown in Figure 30.1B as the resulting current is recorded. During the forward sweep, current peaks will appear for any species that can be oxidized or reduced (dependent on initial scan direction) in the applied potential window. If the species are reversible, redox couples current peaks, corresponding to either reduction or oxidation, will be observed on the return sweep. Figure 30.1B shows the current vs. potential of a reversible single electron transfer reaction. Reversible couples will display a near unity ratio ($i_{pa}/i_{pc} \approx 1$) of the peak oxidation current (i_{pa}) to peak reduction current (i_{pc}). Asymmetry in this ratio indicates that the system under study is semi-reversible or that other kinetic reactions are taking place. Cyclic voltammetry is a powerful investigatory tool that can be used to elucidate details regarding kinetic and diffusional properties, thermodynamic and equilibrium properties of both single and multiple component systems; the interested reader is directed to the excellent text and reference book by Bard and Faulkner [24].

30.3.1.1 Electrochemical Amplification

Enzymes are biological catalysts and as such are not consumed in the reaction with their substrates; this is why they are useful as signal amplifiers in biosensors. Enzymes are often employed in immunoassay to report the presence of a particular analyte. As an example, enzymes are frequently conjugated to the secondary antibody in a sandwich assay format where the product from the enzymatic cleavage of the substrate can be easily detected by optical, electrochemical, or other techniques. As the substrate is turned over, more and more product is generated resulting in an increased signal level. A gain factor of close to a million can be achieved through the buildup of product. High analyte specificity of the antibodies coupled with the high signal gain from the enzymatic turnover makes sandwich immunoassays popular in low-level detection.

When the enzymatic product forms a relatively stable redox couple, an additional signal gain factor can be obtained through electrochemical amplification. The principle behind electrochemical amplification is rather straightforward. Consider two very closely spaced electrodes in a solution containing a stable single electron transfer redox couple where Ox is the oxidized form and Re is the reduced form. Now poise the anode to the oxidation potential of Re and the cathode to the reduction potential of Ox (remember all potentials are measured relative to a reference electrode potential). At the anode, an electron will be transferred from Re to the anode, generating a current and the oxidized form Ox. If the spacing between the electrodes is sufficiently small, Ox will diffuse directly to the cathode, rather than into the bulk solution, where it is reduced back to Re with the generation of current. As the number of oxidation–reduction cycles increases so too does the signal current and therefore amplification. This can be taken further by considering an array of interdigitated electrodes where now the gain is related to the total number of electrode pairs. The principle behind electrochemical amplification is illustrated in Figure 30.2.

The electrochemical amplification gain factor is dependent upon the dimensions of the gap between the anodes and cathodes. For coplanar interdigitated (as shown in Figure 30.2) electrodes (IDA) [25,26], the width of the anode and cathode is on the order of several micrometers (micrometer is one millionth of a meter) as is the gap between them, whereas the combined thickness of the adhesion and electrode layer is on the order of 120 nm (nanometer is one billionth of a meter). At these dimensions, fabricated

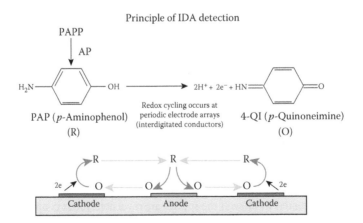

FIGURE 30.2 Principle of electrochemical amplification. Electroactive species are cycled between the anode and cathode increasing signal current during each cycle.

electrode structures are often delicate, suffering from both limited photolithographic resolution and poor adhesion. Moving to a bilayer electrode design results in more robust structures with a submicrometer gap between the anode and cathode allowing for a greater gain factor and simpler fabrication, which is important for single-use disposable BioMEMS chips.

30.3.2 Electrochemical Measurements

All electrochemical experiments were performed with a CH Instruments electrochemical analyzer (CHI 832B, Bee Cave, TX). The analyzer can perform a number of standard electrochemical techniques via computer control, running software supplied with the instrument. Initial experiments were performed in small (5 mL) glass beakers. Test electrode structures were fabricated on 100 mm diameter silicon wafers using MEMS processing methods. The thin gold working and counter electrodes were sputtered from a high purity gold target onto a previously sputtered titanium adhesion layer. Each electrode structure was 1 mm by 1 mm. Figure 30.3 shows the wafer after fabrication of the array of electrode structures, but before singulation. Each wafer contains 100 structures; individual units were singulated using a dicing saw, stored in Fluoroware containers, and cleaned before use. All potentials were measured against a Ag/AgCl reference electrode at room temperature if not specifically stated otherwise.

It has been shown that the product from the enzymatic hydrolysis of 4-APP, 4-aminophenol (4-AP), can be oxidized to quinonimine, which can then be reduced back to 4-AP. This reversibility can be utilized to accomplish electrochemical amplification, as discussed in Section 30.3.1.1.

30.3.3 Enzyme Activity Scheme

PAP (EC 3.1.3.2: isoenzyme 2) has been shown to hydrolyze the substrate 4-APP, resulting in electrochemically active product 4-AP and phosphate [29]. At the anode, 4-AP is oxidized to quinonimine at a potential of 0.40 V, resulting in a change in oxidation current against time (current/time) that is proportional to the enzyme activity.

30.3.3.1 Enzyme Activity Measurements

Initial activity measurements were conducted in small (5 mL) beakers placed in a glass jacket that could be kept at a constant temperature by circulating water from a heat bath (Neslab RTE-111) maintained at 25°C. A working and counter electrode structure, as described in Section 30.3.1 (see Figure 30.1) was employed with an external Ag/AgCl reference electrode. The beaker contained a continuously stirred

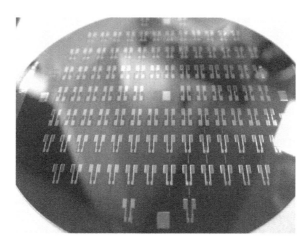

FIGURE 30.3 Silicon wafer with an array of deposited gold microelectrode structures. MEMS fabrication affords multiple copies of a device lowering cost through economies of scale; individual devices are singulated by dicing.

solution of substrate (4-APP) in a pH 4.8 acetate buffer and at a given concentration to insure enzyme saturation kinetics for the particular enzyme of study. A volume of known enzyme activity in acetate buffer was added; the working electrode was poised to +0.4 V. Current vs. time was recorded to obtain the enzymatic progress curve.

30.3.3.2 Optimized Assay Conditions

ACP exhibit optimal activity below pH 7.0, with PAP showing peak activity in the pH 5–6 range. The substrate 4-APP and the product 4-AP must also be stable at this pH range. However, it has been reported [27–29] that alkaline media are beneficial for the electrochemical behavior of 4-AP, but will destroy the catalytic activity of PAP. Ellenberger et al. [30] observed electrochemical activity of 4-AP over range of pH 6.8–7.6 but no activity above pH 9.0. Cyclic voltammograms were run with 4-AP in buffers ranging from pH 4.8 to 7.4 and it was observed that both 4-AP and its oxidation product quinonimine were stable at pH 4.8. Enzyme activity assays were initially run using wheat germ ACP in a pH 4.8 acetate buffer at 25°C over the range of 0.0–100 U/L using 4-APP as the substrate. Acceptable progress curve (current vs. time) was obtained indicating that both 4-APP and 4-AP are stable and function at this pH. Once satisfactory assay conditions were determined, further studies were performed using human PAP.

30.3.3.3 Calibration of PAP Assay

The calibration of the enzymatic activity assay followed the procedure outlined in Section 30.3.3.1, but using human PAP. PAP activities covered the range 0.0–100 U/L, although the physiologically relevant range is 0–20 U/L.

30.3.3.4 On-Chip Mixing

Mixing of solutions in the clinical laboratory is often accomplished by a number of mechanical means that include shaking, stirring, and vortex mixing. All these methods produce turbulent flow resulting in a homogeneous mixture. However, creating turbulent flow in microfluidic networks, used in BioMEMS devices, is challenging as scaling into the micrometer domain causes fluids to mix solely by diffusion. Although diffusional mixing can be rapid in the nanoscale, it becomes slow in the micro- and macroscale (micrometers to millimeters). The internal length of the chip shown in Figure 30.13 is 4 mm and the time for a molecule of serum albumin (molecular weight 66 kDa) to travel this length by diffusion is about 15 h. Mixing in microfluidic devices can be generally categorized into active and passive methods. Active mixing is defined as any technique that requires an additional source of energy to induce mixing (e.g., magnetic stirring), whereas passive mixing requires no such additional source. However, passive mixing does require that the fluid be in motion. Numerous reviews in microfluidic mixing have been published and the interested reader is directed to consult these [31–34].

The PAP assay chip described here is intended to be a single-use disposable, and therefore any mixing strategy employed must be simple, reliable, and inexpensive. Passive mixing requires a pumping mechanism; on-chip pumping increases the chip complexity and fabrication steps thus increasing cost, whereas the off-chip approach requires that the pump be external and that a reliable seal be repeatedly established between it and the chip. In the chips described in this chapter, there is no active flow once the sample and substrate have been loaded into the chip. Magnetically activated mechanical mixing is an attractive means of achieving homogeneous mixing in this chip because it is straightforward and simple. Attempts at using commercially available miniaturized magnetic stir bars proved not to be fruitful so alternative magnetic structures were explored. The oscillatory motion of a small paramagnetic disk along the long axis of the chip proved to be effective in creating a homogeneous mixture in a rapid manner. This disk was designed to fit within the fluid chamber of the chip as shown in Figure 30.13. The disk was fabricated from permalloy (a paramagnetic alloy comprised of iron and nickel) through an electroplating processing using a mold made from photoresist. A paramagnetic material will be attracted to an applied magnetic field yet will not be magnetized; thus the attraction ceases when the field is removed.

30.3.3.5 Measurement of PAP Activity in Blood

Measurements of PAP activity in whole blood were performed in a 1.0 μL BioMEMS assay chip shown in Figure 30.13. The chip is capable of both accurate volume metering and active mixing. Volumetric metering was accomplished by first halting the capillary-driven flow, at a precise location, by means of an on-chip component, the stop-flow junction. After the sample and reagent were introduced into the chip, and flow ceased at the desired location, thermally activated paraffin microvalves sealed the input ports. The sample and substrate were mixed by driving a small paramagnetic disk (as shown in Figure 30.12) in an oscillatory manner by mechanically moving an external magnet along the long axis of the BioMEMS chip. The sequence of events of filling, metering, and mixing is shown in Figure 30.13. Details on the design and fabrication of these two microfluidic components will be given in Sections 30.3.4 and 30.3.5.

Blood samples were obtained from a local blood center that sells surplus blood for research purposes where it is tested for pathogens and stripped of all means of identifying the donors. Before the assays were performed, a volume of blood was spiked with a solution of PAP in acetate buffer (pH 4.8), resulting in blood sample with a PAP activity of 10 U/L. The substrate solution contained 4 mM of 4-APP in acetate buffer (pH 4.8). The assays were conducted as follows: (1) 0.5 μL of the spiked blood sample was drawn into the BioMEMS chip by capillary action and stopped at the junction, and the valve at the sample input port was sealed; (2) 0.5 μL of substrate solution was drawn into the opposite side of the chip and again stopped at the junction, and the substrate input port valve was sealed; and (3) the mixing disk was driven by the external magnet for 30 s resulting in a homogeneous mixture, then the working electrode was poised at 0.4 V (vs. Ag/AgCl), and the enzymatic progress curve was recorded as current vs. time, the slope of which was proportional to the enzymatic activity. The assay was repetitively performed with the same spiked blood sample in six different BioMEMS assay chips. Results will be given in Section 30.4.1.4.

30.3.4 Design and Fabrication of the BioMEMS Chip

30.3.4.1 Microelectromechanical Systems

MEMS is the technology of the design and fabrication of very small mechanical devices that can be driven by electricity, although this is not the only means of power that can be employed. MEMS devices generally range in size from 20 μm to several millimeters, whereas individual components within the device have dimensions between 1 and 100 μm. Richard Feynman, in his famous 1959 lecture, "There's Plenty of Room at the Bottom" [35], predicted that MEMS devices were achievable well before the technology for their fabrication was available. Only with the adaptation and modification of the high-precision bulk processing techniques developed for the semiconductor industry was it practical to create commercial MEMS devices: accelerometers to trigger automobile airbags and micromirror arrays used in optical projectors are two popular examples. Materials and methods used in the fabrication of MEMS devices are vast and an extensive review is beyond the scope of this chapter. Those interested can consult the literature, a good overview being given in the text and reference book by Madou [36]. Before discussing the details of the methods employed in the fabrication of our microchips, the steps involved in going from concept to finished microchip will be summarized. Figure 30.4 shows the design flow. Initial concepts are translated into computer models where their mechanical and microfluidic behavior can be simulated for design optimization (Figure 30.4A). A series of photolithography masks are drawn (Figure 30.4B) which will be used in the fabrication of the various layers of the device. Electronic files of each of the masks are then sent to a vendor where a set of chrome on quartz photolithographic masks is fabricated (Figure 30.4C).

Figure 30.5 shows the general microchip fabrication process flow. Fabrication starts with a commercial off-the-shelf silicon wafer (Figure 30.5A); for ease of handling, 4 in. diameter wafers are routinely used in research and development. For commercial scale manufacturing, however, 6, 8, or 12 in. wafers are available and would be used. The wafer is cleaned and prepared for photolithographic transfer of the desired pattern (Figure 30.5B) followed by multiple processing steps resulting in a wafer containing a large number of identical devices (Figure 30.5C). Finally, the individual devices are separated ready for

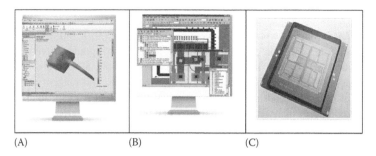

FIGURE 30.4 Microchip design process flow. (A) Computer aided design (CAD) software is used for device design. (B) The design is transferred to a set of digital files for the device layers. (C) A set of photolithography masks specific to the design is generated.

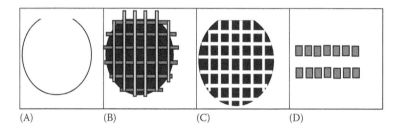

FIGURE 30.5 MEMS fabrication flow. (A) Off the shelf silicon wafer, ready for processing. (B) Devices are fabricated using the photomasks and multiple processing steps. (C) Many devices are produced on one wafer. (D) The devices are separated, ready for use, or packaging.

use or packaging (Figure 30.5D). This type of bulk manufacture allows the economies of scale, so successful at bringing sophisticated and powerful electronic devices and computer to the mass market, to be brought to the medical device and diagnostics market.

30.3.4.2 Design Considerations

The design of a BioMEMS chip suitable for performing on-board enzymatic activity measurement was based on consideration of both the type and the minimum number of components necessary to make an accurate measurement. To perform an assay as described earlier, the following components are required: (1) microneedle for obtaining blood sample, (2) metering structure for both sample and reagent volumes, (3) microvalves, (4) storage reservoirs, (5) electrode structures, and (6) magnetic mixing device. A conceptual drawing of a chip is shown in Figure 30.6.

Several potential chip designs were conceived and the final design refined through mathematical analysis and computer simulations. Microfluidics was modeled using the Hagen–Poiseuille flow [37] and the surface energy properties of the materials used in fabrication. Flow is laminar (low Reynolds number regime where viscous force prevails) [38] and mixing occurs through diffusion if no external agitation is exerted. Diffusional mixing is too slow for a practical device of this size and active mixing was installed.

30.3.4.3 Microfluidic Component Design

30.3.4.3.1 Microneedle

Blood samples are obtained through a hollow bore microneedle integrated into the BioMEMS chip. The single crystal nature of the silicon wafer, which provides strength and flexibility, and the bulk mechanical properties of single crystal silicon have been considered in the computer modeling and simulation of prospective designs. Silicon fails by brittle fracture. Designs that minimize stress-causing fracture resulting from an applied lateral force are desired. Finite element analysis, a computer modeling and

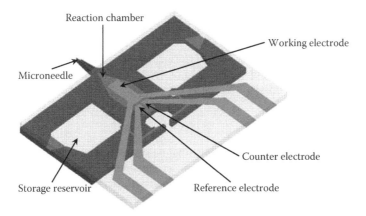

FIGURE 30.6 BioMEMS chip with integrated on-chip storage reservoir, electrodes, microneedle, and reaction chamber.

simulation software routine, using the bulk mechanical properties of single crystal silicon was employed to generate stress maps along the length of the microneedle for different designs under a variety of imposed lateral forces. Designs that minimize stress were chosen for fabrication and experimental evaluation. Typical dimensions are 2–3 mm in length, 175 μm in thickness, and 200 μm in width at widest part. These microneedles are roughly the size of a human hair resulting in painless insertion for blood sampling.

30.3.4.3.2 Volumetric Metering

Precise volumetric metering is accomplished using a flow stop junction in conjunction with on-chip microvalves. When a capillary-driven flow reaches an abrupt change in the geometry of the flow channel, the flow will cease and will require an applied force for it to proceed [39]. An abrupt change in geometry can be accomplished by etching a trench perpendicular to the flow channel, resulting in a discontinuity, and by modifying surface energy at selective positions to form a pressure barrier preventing flow. Figure 30.7 is a drawing of a stop junction showing pertinent parameters.

The pressure (P) at the stop junction can be estimated with the following equations (Equation 30.1) derived from total surface energy U and liquid volume V as $P = -dU/dV$. Equation 30.1 gives the pressure as of function of the parameters [39]:

$$P = \frac{2\gamma \cos\theta_1}{h}; \quad \alpha = \frac{\pi}{2} - \theta_1$$

$$P = \frac{(2\gamma/h)\left\{-\cos\theta_2\left[1+\cot^2(\alpha+\theta_2)\right]+\left((-1+\alpha\cot(\alpha+\theta_2)+\alpha\cot\alpha)/(\sin(\alpha+\theta_2)\sin\alpha)\right)\right\}}{-\left[1+\cot^2(\alpha+\theta_2)\right]+\left((-1+\cot(\alpha+\theta_2)(\alpha-\sin\alpha\cos\alpha)+\alpha\cot\alpha)/(\sin^2(\alpha+\theta_2)\sin^2\alpha)\right)};$$

$$\frac{\pi}{4} - \frac{\theta_1+\theta_2}{2} \le \alpha \le \frac{\pi}{2} - \theta_1 \tag{30.1}$$

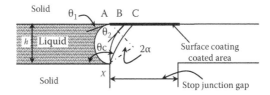

FIGURE 30.7 Schematic drawing of the cross-sectional view of the stop junction showing the parameters used in the modeling calculations.

Equation 30.1 represents the capillary pressure at the stop junction as a function of the parameters as follows: θ_1 and θ_2 are the contact angles of the liquid-surface interface at point A and at the liquid-modified surface at point B respectively, 2α is the meniscus arc angle, γ is the surface tension, and h is the microchannel depth. Calculations were performed for the case of a rectangular microchannel 50 μm deep and 100 μm wide, initial energy (U_0) of 1×10^{-9} J, surface tension (γ) of 0.0728 N/m, and contact angle (θ_1) of 30° and where the contact angle of the modified region was varied from 30° to 80°.

30.3.4.3.3 Microfluidic Microvalves

Building simple, efficient multiuse valves for microfluidic systems has proven to be difficult and much work has been done in this area [40]. Many of the designs involve three-dimensional structures and flexible membranes, resulting in complex fabrication steps. Because the devices discussed here are to be used only once, simple limited-use microvalves are desired for both ease of fabrication and minimum cost. Simple microvalves based on thermally activated paraffin were developed. The principle is rather straightforward, the paraffin is heated and melts expanding into the channel at the desired location and then solidifies forming a plug. Structures for holding the paraffin and the microheater were fabricated using MEMS process techniques that will be discussed in Section 30.3.4. Fabricated microvalves and their behavior will be presented in the results in Section 30.4.2.3.

30.3.4.3.4 Storage Reservoirs

A fully integrated self-contained assay chip requires on-chip reservoirs, which can be adequately sealed to prevent fluid evaporation during storage (and thus permit reasonable shelf life). Figure 30.8 is a conceptual design of a storage subsystem responsive to these issues. On the left, the drawing shows the configuration during storage. On the right, the drawing shows how the fluid is transferred into the interconnection microchannel.

Silicon (which has a native oxide coating), silicon oxide, glass, and other epitaxial layers used in standard silicon processes are suitable materials for fabrication of long-term on-chip storage reagent fluid reservoirs. Further, these materials can be readily formed into membranes. In the conceptual design shown, silicon is used for the primary structure of the storage chamber to provide longer shelf life. A polymer layer was used as an extensible membrane to drive the fluid out of the storage chamber. The polymer was selected so that when it is depressed beyond its elastic deformation limit the membrane will not bounce back to its original shape and thus draw the fluid back into the storage chamber. Potential actuation mechanisms to deform the polymer layer could be human power, off-chip motors, and some

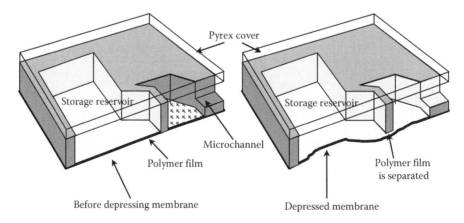

FIGURE 30.8 Schematic drawing of releasing liquid reagent out of the storage reservoir. When the release button is pressed, the pressure in the reservoir is increased. As the pressure increases beyond the bonding strength of the adhesive on the polymer film, the stored liquid reagent is forced out of the reservoir.

on-chip actuators such as electrostatic actuators [41], piezoelectric actuators [42], and thermopneumatic actuators [43].

30.3.4.3.5 Magnetic Mixing Disk

Magnetic mixing disks were designed to fit within the test chamber of the assay chip. Disks were fabricated from permalloy (a paramagnetic alloy comprised of iron and nickel) by an electroplating technique using a mold formed from photoresist [44].

30.3.5 Fabrication of BioMEMS Chips

Fabrication of MEMS devices shares many common processing techniques developed for the semiconductor industry such as photolithography, thin film deposition of materials by a variety of well-developed methods, wafer bonding, chemical etching, deep reactive ion etching, and ion doping as examples [36]. However, since this is not a review, only the methods used in fabricating the BioMEMS enzyme assay chip will be presented and discussed.

30.3.5.1 Photolithography

Designs for the photomasks were drawn in L-Edit (Tanner Research, Monrovia, CA), a commercial mask drawing software program. The generated files were sent to Image Technology (Palo Alto, CA) where sets of chromium on quartz photomasks were manufactured. Wafers were prepared; positive photoresist was applied by spin-coating; and the wafers baked. Contact photolithography was performed on a Karl Suss mask aligner. After exposure, the wafers were placed in photoresist developer where the exposed areas of the photoresist are removed. The wafers are now ready to be etched.

30.3.5.2 Deep Reactive Ion Etching

Microfluidic channels and other components are etched into silicon wafers by deep reactive etching (DRIE). In this work, silicon etching was accomplished by a Alcatel 601E deep reactive ion etcher employing the patented Bosch Process (licensed from Robert Bosch GmbH) and equipped with an electrostatic chuck. The Bosch process, also known as pulsed or time-multiplexed etching, alternates repeatedly between an etching phase and a passivation phase allowing for the creation of vertical structures with high aspect ratios. When operating properly, the etch rate is constant and depth of etch is proportional to time. Etching of the microfluidic network is performed at the wafer level, where each wafer can contain hundreds of individual chips. In the final etch step, the wafer is attached to a carrier wafer using photoresist and the individual chips singulated by completely etching through the wafer (punch out). Individual chips are released by ashing the photoresist in a high-temperature oven.

30.3.5.3 Electrode Fabrication

The pattern of the electrode structure was transferred onto the Pyrex or silicon wafer through the photolithographic process after which the wafer was cleaned and placed into the vacuum chamber of the sputterer. First, a thin adhesion layer of titanium was deposited after which a thicker layer of gold was applied. The reference electrode was fabricated by electroplating a layer of silver at the desired location after which it was partially chloridized through an electrochemical process [23]. The gold acts as both the working and counter electrodes.

30.3.6 Assembly of Complete BioMEMS Assay Chip

At present, individual assay chips are assembled by hand using custom designed tools, alignment jigs, and micromanipulators. The assembly starts with aligning and clamping the glass cover and the silicon chip together. Ultraviolet (UV)-activated adhesive is applied and drawn in between the glass cover and the silicon chip by capillary force. Once the glue fills all the bonding area, UV light is shined through the glass cover for 60 s to solidify the UV adhesive. Adhesive film for the reservoir cover is laminated over the reservoir. Reagents are drawn into the storage reservoirs by capillary force from the filling opening. After reagent filling, the filling opening is sealed with UV adhesive. The final assembly is the placement of the paraffin that will form the microvalves. The paraffin beads are packed into the filling holes and the filling holes are sealed with UV adhesive.

30.4 Results

30.4.1 Enzymatic Activity Assay

30.4.1.1 ACP Activity: From Wheat Germ

Initial assay development was performed using ACP derived from wheat germ, using the protocol in Section 30.3.3.2. Figure 30.9A presents the results for the assay run in pH 4.8 acetate buffer at room temperature. Linear progress curves were observed for ACP activities over the range of 0.0–100 U/L as seen in the insert.

When the substrate concentration is much greater than the Michaelis constant, the Michaelis–Menten [45] equation reduces to a rate equation linear in enzyme concentration as shown in Equation 30.2.:

$$d[P] = V_S = k_3[E]dt \tag{30.2}$$

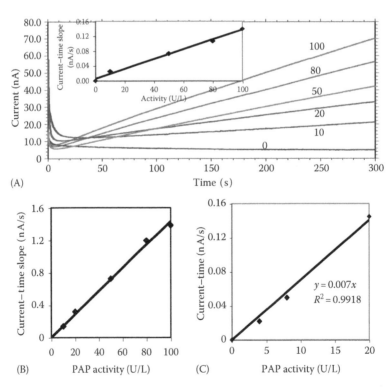

FIGURE 30.9 Results of electrochemical measurements of enzymatic activity. (A) Initial progress curves for wheat germ acid phosphatase for activities of 0, 10, 20, 50, 80, and 100 U/L run in pH 4.8 acetate buffer containing 1 mM 4-aminophenyl phosphate potential poised at 0.4 V sedimented Ag/AgCl. Inset: Current–time slope (nA/s) as function of activity shows linear behavior. (B) Initial slope vs. activity of human prostatic acid phosphatase run under same conditions as the earlier ones. (C) Initial slope vs. activity of prostatic acid phosphatase ran in BioMEMS chip for the physiologically relevant range of 0–20 U/L.

Equation 30.2 illustrates enzyme kinetics under substrate saturation conditions. Slope is proportional to enzyme activity.

In Equation 30.2, $d[P]$ is the change in product concentration, V_S is maximum velocity for the enzyme, k_3 is the turnover number (1/s), and $[E]$ is the enzyme concentration. The signal current vs. time is linear and the slope is proportional to the activity. Activity vs. slope is illustrated in Figure 30.9B, which clearly shows linear behavior as predicted for Michaelis–Menten kinetics at substrate saturation conditions. Human PAP is a dimeric glycoprotein, and there is evidence that it follows cooperative kinetics and is better represented by the Hill rate equation, yet under substrate saturation conditions linear behavior should be observed [46].

30.4.1.2 Activity for Human Prostatic Phosphatase

Progress curves were run as described earlier. Figure 30.9B shows the corresponding calibration plot for PAP in pH 4.8 acetate buffer over the 0–100 U/L range with the substrate 4-APP. This assay also shows linear behavior under substrate saturation conditions.

30.4.1.3 PAP Activity in BioMEMS Assay Chip

The PAP activity assay was performed in the 1 μL assay chip shown in Figure 30.13 using the protocol outlined in Section 30.3.3.5 using spiked buffer instead of blood. Linear behavior was observed over the physiologically relevant range of 0–20 U/L as seen in Figure 30.9C.

30.4.1.4 Activity of PAP in Whole Blood

Figure 30.10 gives the measurements of PAP activity in spiked blood (protocol in Section 30.3.3.5) performed in six separate 1.0 μL BioMEMS assay chips (as shown in Figure 30.12); error bars represent one standard deviation. The mean is 10.2 U/L in agreement with the spiking level of 10 U/L and with a standard deviation of ±1.4, which represents a coefficient of variation of 14%. The level of agreement demonstrates that enzymatic activity can be reproducibly determined within miniature BioMEMS chips. Refinements in both the assay and the manufacturing of the BioMEMS chips will increase accuracy and precision.

As outlined in the introduction and amplified in a recent review by Taira et al. [47], PAP is a rather poor marker for early stage screening and diagnostics of persistent disease. In contrast to its inadequacies in staging and posttreatment monitoring, an elevated PAP has consistently been identified as a

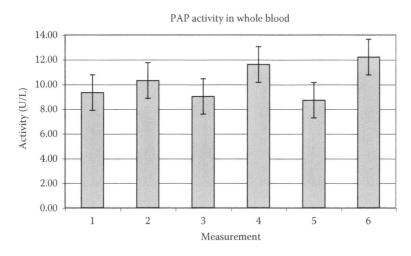

FIGURE 30.10 Measurement of PAP activity in spiked blood.

significant prognostic factor [48–50]. Investigators at Walter Reed studied 295 higher-risk prostatectomy patients and found that those with elevated pre-prostatectomy PAP had only 39% 4 year biochemical disease–free survival vs. 79% for those with normal PAP levels. Similar results have been observed in higher-risk patients after brachytherapy. Dattoli et al. report that pretreatment PAP is a stronger predictor of biochemical failure than PSA or Gleason score after palladium (Pd)-103 brachytherapy in 161 higher-risk patients [51]. The previous conclusion has been recently reinforced by the findings of Dattoli and colleagues that elevated PAP is the single strongest independent predictor of prostate cancer–specific death [52].

There is significant evidence suggesting that PAP activity should be reintroduced into the clinical arsenal, and that the BioMEMS assay system presented in this chapter offers a rapid and convenient means for performing this assay. Additionally, multiple assays can be integrated into the chip to provide additional information supporting a particular diagnosis or the efficacy of a chosen course of therapy by providing physicians with real-time status of their patients' response.

30.4.2 BioMEMS Assay Chip

30.4.2.1 Microneedle

Blood sampling microneedles are integral to the BioMEMS chip and were fabricated using methods developed in the author's laboratory and discussed earlier. They are strong to dependably puncture the skin (human and animal) and can reliably draw blood into its integrated chip chamber. The properties of the silicon hollow bore microneedles are demonstrated in Figure 30.11.

Blood sampling is accomplished with a mechanical actuator, which provides proper alignment and delivers precisely determined velocity and force to drive the microneedle into the skin. Pain perception trials were conducted on 65 subjects roughly half of who were diabetics. The cohort contained men,

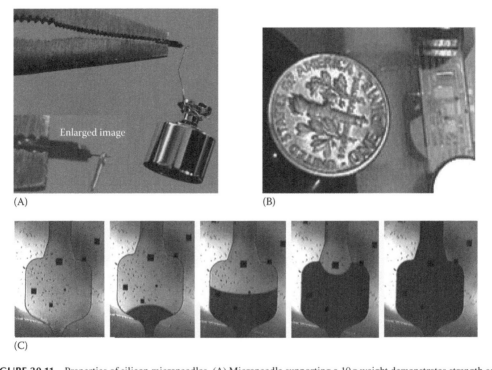

(A)   (B)

(C)

FIGURE 30.11 Properties of silicon microneedles. (A) Microneedle supporting a 10 g weight demonstrates strength and flexibility. (B) BioMEMS chip in comparison with time, demonstrating minimization. (C) Filling sequence of integrated sampling chamber through integrated hollow bore microneedle captured with infrared video camera. Total filling time of cavity is ~1.5 s.

women, and children with ages ranging from 12 to 75 years old. Most subjects reported no pain when compared to that from a standard metal lancet.

30.4.2.2 Volumetric Metering

Figure 30.12 shows the results of the calculations based on Equation 30.1. The graph on the left is that of surface energy vs. liquid volume for various values of contact angle θ_2 in the coated area and that on the right is liquid pressure vs. liquid volume for some values of θ_2 at stop junction.

As the contact angle (θ_2) between the liquid and the surface modified section (coated area, see Figure 30.12A) is increased, the surface energy at point A also increases forming a potential barrier preventing fluid from flowing beyond the junction region. This increase in surface energy manifests as a pressure opposing flow as seen in the graph on the right (Figure 30.12B). These results provide a qualitative explanation of how the stop junction stops the flow, and quantitative estimations of this pressure barrier for various values of the contact angle θ_2 (note at $\theta_2 = 80°$, the pressure is negative and opposing the flow). This is demonstrated in Figure 30.13 where the fluid is seen advancing until it reaches the junction where it abruptly stops. Sealing the valves at the base of the microneedle and reagent input port provides precise volume metering of the sample and reagent.

30.4.2.3 Microvalves

Thermally activated microvalves in the microfluidic network design consist of a small alcove to contain the thermally activated paraffin. A resistive microheater located below the alcove activates the paraffin. Electrical current provides localized heating at the desired location. The microvalves are designed to be open prior to use. The paraffin storage is next to the detection chamber. On top of the paraffin storage, there is a resistive microheater. To close the detection chamber, the heater is activated causing the

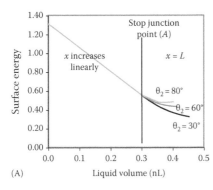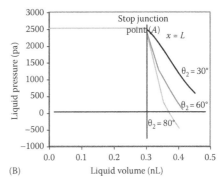

FIGURE 30.12 The estimated pressure at the stop junction based on Equation 30.1. (A) Surface energy vs. liquid volume at various contact angles θ_2: 30°, 60°, and 80°. (B) Liquid pressure vs. liquid volume at various contact angles θ_2: 30°, 60°, and 80°.

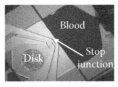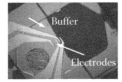

500 nL Blood filled   500 nL Buffer solution filled   Complete on-chip mixing achieved in under 30 s

FIGURE 30.13 Stop junction in action and mixing. The left photograph shows that blood is stopped at the stop junction. The middle photograph shows the buffer fills up the second half of the chamber and meets with the blood. The right hand photograph shows that the blood is fully mixed with the buffer after moving the paramagnetic disk in the chamber.

FIGURE 30.14   Photographs in series show the microvalves in action. The left photograph shows the chamber filled with water. The middle photograph shows melted paraffin flowing into the microchannel forming seal between channel and reaction chamber. The right photograph shows that paraffin has sealed reaction chamber.

paraffin to flow out of the storage and eventually blocking the outlet of the detection chamber as shown in the series of photographs in Figure 30.14.

30.4.2.4 Electrochemical Amplification

4-AP and quinonimine form a reversible redox couple allowing cycling between the anode and cathode required for electrochemical amplification. Quinonimine also undergoes irreversible hydrolysis to quinine. If this hydrolysis rate is significant compared to its diffusion rate to the cathode, it will decrease the amplification gain factor (less amplification). Cyclic voltammograms were run in pH 4.8 acetate buffer in the absence and presence of 1 mM 4-APP and 100 U/L ACP for 0, 100, and 300 s incubation time after mixing the enzyme and substrate together. The signal from the 300 s incubation was significantly higher than that from the 100 s incubation demonstrating that there is no significant hydrolysis of quinonimine during the time that the assay is to be conducted. Figure 30.15 shows an example of the amplification achieved with the bilayer electrode structure. Although electrochemical amplification was not employed for the measurement of PAP activity, it was presented here to demonstrate that additional gain can be obtained if higher sensitivity is required.

30.4.2.5 On-Board Mixing

Homogeneous mixing by the on-chip scheme was established by comparing assay results obtained in this manner to assay results where the sample and substrate are first mixed externally by conventional means then introduced into the chip. Since the kinetics are first order in PAP activity and it is the slope that is being measured, the slight time delay between on-chip mixing and external mixing will not significantly alter the results. For the externally mixed case, 0.75 μL of 2 U/L of PAP in a pH 4.8 buffer was mixed with an equal volume of 4 mM 4-APP in a pH 4.8 acetate buffer in an Expender tube. The solution mixture was mixed for 30 s after which it was introduced into the microchip where the buildup of product

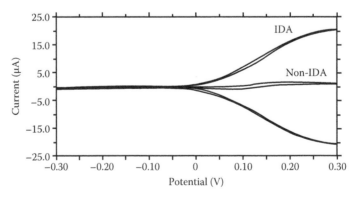

FIGURE 30.15   The cyclic volammagrams of 1 mM 4-aminophenol in pH 4.8 acetate buffer performed with a conventional gold electrode (non-IDA) and with the bilayer microelectrode array (IDA).

was monitored electrochemically. For on-chip mixing, sample and substrate solutions identical to those used previously were separately introduced into the chip. Each solution is drawn into the chip until capillary flow ceases at the stop junction. Once both solutions have filled their respective space, input ports are sealed and a small magnet is brought to close proximity to the chip and rapidly moved in the oscillatory motion described earlier. After 30 s, mixing was stopped and the measurements were made. There was substantial agreement between the results obtained by the two mixing methods to conclude that on-chip mixing produced homogeneously mixed solutions rapidly and reliably.

30.4.2.6 Storage Reservoirs

Storage reservoirs were fabricated by first etching through the wafer defining the volume. A biocompatible polymer film was then attached creating the bottom of the reservoir. The chip was flipped over, revealing the detection chamber and microfluidic network, and the glass top was then applied using UV adhesive, resulting in an empty cavity with a flexible membrane on the bottom. The reservoir is filled with a desired fluid for long-term storage through a small port, after which the port is sealed. The membrane seals the reservoir from the microfluidic network unless significant pressure is applied to peel the membrane away from the post between the reservoir and the network allowing the fluid stored in the reservoir to flow out into the detection chamber. Figure 30.16 shows photographs of the storage reservoir before and after transfer of the fluid to the detection chamber. In the photograph on the left, the blue dye solution is confined within the reservoir. In the right-hand photograph, the seal has been breached, allowing the stored fluid to flow into the chamber.

30.4.2.7 Integration and Handheld Instrument

Integrated BioMEMS assay chips designed and fabricated for a particular assay may employ a variety of microfluidic components previously developed. Although the microchips are self-contained, in that they include all necessary reagents and components to perform a particular assay, they need to be used in conjunction with a handheld instrument to reliably obtain the sample, perform the assay, and deliver the results. For this, a small handheld instrument has been developed and constructed. The unit has a precision actuation mechanism, under microprocessor control, that propels the microchip at the proper velocity and with the proper force to puncture skin and obtain sample. Additionally, the instrument contains electronics and software that perform the electrochemical assay, analyze the data, and display the results. In use, the disposable microchip fits into a socket within the instrument where electrical connections are made. Electrochemical assays have the distinct advantage for use in small portable devices in which the electrical circuitry is well established, and are robust and very amenable to miniaturization. In addition, a large range of assays and assay formats can be performed with a single instrument, because different protocols/algorithms can be stored within memory in advance and called forth for a specific assay protocol and algorithm when the coded disposable is inserted. Prototype instruments have been designed and constructed using standard engineering practices; thus, details will not be provided here.

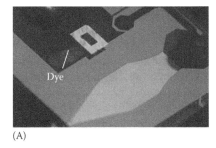
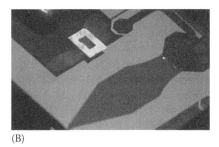

(A)   (B)

FIGURE 30.16 In (A), the storage reservoir is filled with water containing a blue dye and sealed. In (B), pressure has been applied to the membrane rupturing seal between reservoir and microchannel allowing fluid to flow into reaction chamber.

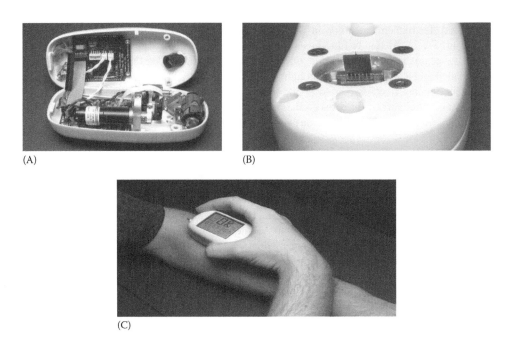

(A)   (B)

(C)

FIGURE 30.17   Handheld instrument with disposable BioMEMS assay chip. (A) is of the interior of the instrument show-ing the compact electrical and mechanical components. (B) shows the PAP activity assay BioMEMS chips mounted into socket. (C) shows how the instrument would be used.

Figure 30.17 shows a series of photographs of a prototype device. Figure 30.17A displays the interior of the instrument showing the compact electrical and mechanical components. Figure 30.17B shows the PAP activity assay BioMEMS mounted into the instrument. Figure 30.17C is a photograph of how the instrument would be used.

30.5 Discussion

This chapter presents one application of the powerful platform BioMEMS technology, that being the measurement of enzymatic activity of PAP in a small blood sample obtained with a silicon microsam-pling and assay chip. Enzyme activity is measured electrochemically by following the buildup of the enzymatic product 4-AP. The assay was run under substrate saturation kinetics, where the initial slope of the progress curve is directly proportional to enzyme activity. As discussed earlier, electrochemical detection methods are far easier to miniaturize than conventional optical methods. In addition, a new bilayer electrode structure was described which can be used for electrochemical amplification to provide higher sensitivity. The electrode structure is both easier to fabricate and more robust than conventional interdigitated electrode arrays.

The ability to measure enzyme activities, or other health status parameters, in a small blood sample obtained by an integrated microsampling and assay chip in conjunction with a small portable instrument offers advantages over conventional laboratory methods. A physician may make measurements at the point of care for monitoring response to treatment without delay. Additionally, multiple assays can be integrated within the BioMEMS chip so that several analytes can be measured simultaneously in a single blood sample. Additional assays are not limited to activity measurements; immunoassays and other can-cer marker assays can be included. Figure 30.18 is a concept drawing of a multiple analyte BioMEMS chip showing three separate assays on a single chip.

Clearly the measurement of increased activity of a single enzyme, such as PAP, is insufficient to confirm or reject a diagnosis, but it can be part of a panel to follow trends during "watchful waiting"

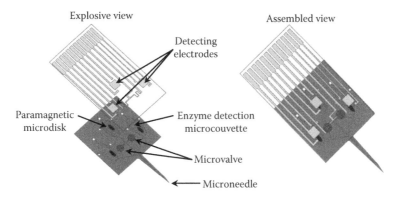

FIGURE 30.18   Concept drawing of a three-chambered enzymatic assay BioMEMS chip. The drawing on the left is an exploded view showing the internal structure of the chip, although there will be on-chip reservoirs they are not shown. The drawing on the left is a view of the assembled chip. Assays are not confined to enzymatic activity only; other types of assays can be included immunoassays being given as an example.

or response to therapy. Methods employing HPLC or mass spectrometry do not lend themselves, at the present, to miniaturization. However, BioMEMS sampling devices can be used to interface with these instruments.

30.6 Future Trends

As healthcare costs escalate, there is a push to contain these costs through a variety of means. Having patients monitor their health status, parameters at their present location with the results encrypted and transmitted to the physician via a wired or wireless network can dramatically reduce the burden. Only when results are abnormal will the patient need to go to the office of clinic. BioMEMS chips, as described in this chapter, offer easy-to-use self-testing with painlessness (no finger stick with separate lancet) so that patient compliance is substantially improved. In addition, cost-effective BioMEMS chips can be readily manufactured with commercially available facilities and industrially mature processes to allow the chips to be disposable. Outside of the personal monitoring area, BioMEMS chips have application in emergency situations where first responders need to perform rapid triage. Additionally, miniaturized BioMEMS chips will bring healthcare to remote and impoverished areas of the world by offering modern health monitoring that can be administered by personnel requiring minimal training.

Critical analytes, such as glucose, lactate, alcohol, total hemoglobin, and other health markers in whole blood, have been measured with BioMEMS assay chips developed in the authors' laboratory. Self-contained immunoassay BioMEMS chips have also been developed. Targets include insulin and protein markers of nerve agent exposure. A BioMEMS assay system has been developed to detect trace amount of heavy metals (e.g., Pub, Cod, Cu, Zn) in whole blood as part of a study correlating environmental exposure to genetic alterations.

ACKNOWLEDGMENTS

The authors would like to thank NCI for their funding of this work through grant 5 R44CA103335-03 and the superb assistance of Mr. Steve Jones of Kumetrix.

REFERENCES

1. National Cancer Institute, September 2008, *What You Need to Know About™ Prostate Cancer,* Publication number P035, NIH number 08-1576, http://www.cancer.gov/cancertopics/wyntk/prostate/
2. Prostate Cancer Institute, posting 2011, What is Prostate Cancer? http://www.prostate-cancer-institute.com/

3. Helms, S. R., M. G. Britain, T. G. Prêt low, and J. I. Kleinberg. 1977. "Prostatic acid phosphatase?" A comparison of acid phosphatase activities in epithelial cells, granulocytes, monocots, lymphocytes, and platelets purified by velocity sedimentation in kinetic gradients of Ficoll in tissue culture medium. *Am. J. Pathol.* 100: 529–538.

4. Yam, L. T. 1971. Clinical significance of human acid phosphatase: A review. *Am. J. Med.* 56: 604–616.

5. Moss, D. W. and A. R. Henderson. 1999. Clinical enzymology, in C. A. Burtis and E. R. Ashwood (eds.), *Tietz Textbook of Clinical Chemistry*, 3rd edn. W. B. Saunders Company, Philadelphia, PA, pp. 711–715.

6. Lam, K. W., O. Li, C. Y. Li, and L. T. Lam. 1973. Biochemical properties of human prostatic acid phosphatase. *Clin. Chem.* 19: 483–487.

7. Li, C., L. T. Yam, and K. W. Lam. 1970. Studies on acid phosphatase isoenzymes of human leukocytes: Demonstration of isoenzyme-cell specificity. *J. Histochem. Cytochem.* 18: 901–910.

8. Stefannini, M. 1985. Enzymes, isozymes, and enzyme variants in the diagnosis of cancer: A short review. *Cancer* 55: 1931–1936.

9. National Cancer Institute, reviewed 3/18/2009, Prostate-Specific Antigen (PSA) Test: Fact Sheet, http://www.cancer.gov/cancertopics/factsheet/Detection/psa

10. Carter B. H., Winter 2005, Interview, When the PSA test cannot do it all: New biomarkers to fill the gaps, Prostate Cancer Discovery, publication of the James Buchanan Brady Urological Institute John Hopkins Medical Institutions, http://urology.jhu.edu/newsletter/prostate_cancer710.php

11. Bañez, L. L., P. Prasanna, L. Sun, A. Ali, Z. Zou, B.-L. Adam, D. G. McLeod, J. W. Moul, and S. Srivastava. 2003. Diagnostic potential of serum proteomic patterns in prostate cancer. *J. Urol.* 170: 442–446.

12. Diamandis, E. P. 2003. Point proteomic patterns in biological fluids: Do they represent the future of cancer diagnostics? *Clin. Chem.* 48: 1272–1278.

13. Le, L., K. Chi, S. Tyldesley, S. Flibotte, D. L. Diamond, M. A. Kuzyk, and M. D. Sadar. 2005. Identification of serum amyloid A as a biomarker to distinguish prostate cancer patients with bone lesions. *Clin. Chem.* 51: 695–707.

14. Nomura, A. M. Y., J. Lee, G. N. Stemmermann, and G. F. Combs, Jr. 2000. Serum selenium and subsequent risk of prostate cancer. *Cancer Epidemiol. Biomarkers Prev.* 9: 883–887.

15. Foley, R., D. Hollywood, and M. Lawer. 2004. Molecular pathology of prostate cancer: The key to identifying new biomarkers of disease. *Endocr. Related Cancer* 11: 477–488.

16. Petersen, K. E. 1982. Silicon as a mechanical material. *Proc. IEEE* 70: 420–457.

17. Saliterman, S. S. 2006. *Fundamentals of BioMEMS and Medical Microdevices,* Vol: PM153. SPIE Press Book, Bellingham, WA.

18. Urban, G. (ed.). 2006. *BioMEMS (Microsystems)*, Vol. 16. Springer, Part of Springer Science + Business Media, New York.

19. Todd, M. R. and J. Gomez. 2001. Enzyme kinetics determined using calorimetry: A general assay for enzyme activity. *Anal. Biochem.* 296: 179–187.

20. Palamakumbura, A. H. and P. C. Trackman. 2002. A fluorometric assay for detection of lysyl oxidase enzyme activity in biological samples. *Anal. Biochem.* 300: 245–251.

21. Buxton, R. C., B. Edwards, R. R. Juo, J. C. Voyta, M. Tisdale, and R. C. Bethell. 2000. Development of a sensitive chemiluminescent neuraminidase assay for the determination of influenza virus susceptibility to zanamivir. *Anal. Biochem.* 280: 291–300.

22. Weetall, H. H. and A. K. Gaigalas. 1993. A method for the assay of hydrolytic enzymes using dynamic light scattering. *Appl. Biochem. Biotechnol.* 41: 139–144.

23. Janz, G. J. 1961. Silver–silver halide electrodes, in D. J. G. Ives and G. J. Janz (eds.), *Reference Electrodes Theory and Practice*. Academic Press, New York and London, pp. 179–226.

24. Bard, A. J. and L. R. Faulkner. 2001. *Electrochemical Methods Fundamentals and Applications*, 2nd edn. John Wiley & Sons, Inc., New York.

25. Iwasaki, Y. and M. Morita. 1995. Electrochemical measurements with interdigitated array microelectrodes. *Curr. Separations* 14: 2–8.

26. Aoki, K., M. Morita, O. Niwa, and H. Tabei. 1988. Quantitative analysis of reversible diffusion-controlled currents of redox soluble species at interdigitated array electrodes under steady-state conditions. *J. Electroanal. Chem.* 265: 269–282.

27. Duan, C. and M. E. Meyerhoff. 1994. Separation-free sandwich enzyme immunoassays using micro porous gold electrodes and self-assembled monolayer/immobilized capture antibodies. *Anal. Chem.* 66: 1369–1377.

28. Meyer off, M. E., C. Duane, and M. Mussel. 1995. Novel no separation sandwich-type electrochemical enzyme immunoassay system for detecting marker proteins in undiluted blood. *Clin. Chem.* 41: 1378–1384.

29. Nina, O., Y. Up, H. B. Halsall, and W. R. Heineman.1993. Small volume voltammetric detection of 4-aminophenol with interdigitated array electrodes and its application to electrochemical enzyme immunoassay. *Anal. Chem.* 65: 1559–1563.

30. Ellenberger, M., M. Pastiche, and R. Hintsche. 1994. Interdigitated array microelectrodes for the determination of enzyme activities. *Analysis* 119: 1245–1249.

31. Hessel, V., H. Löwe, and F. Schöfeld. 2005. Micromixers—A review on passive and active mixing principles. *Chem. Eng. Sci.* 60: 2479–2501.

32. Fiorini, G. S. and D. T. Chiu. 2005. Disposable microfluidic devices: Fabrication, function, and application. *BioTechniques* 38: 429–446.

33. Mansur, E. A., Y. E. Mingxing, W. Yundong, and D. A. I. Youyuan. 2008. A state-of-the-art review of mixing in microfluidic mixers. *Chin. J. Chem. Eng.* 16: 503–516.

34. Agarwal, A. K., J. Atencia, D. J. Beebe, and H. Jiang. 2004. Magnetically-driven temperature-controlled microfluidic actuators, in *Proceedings of the 1st International Workshop on Networked Sensing System.* Tokyo, Japan, pp. 51–55.

35. Feynman, R. P. 1959. There's plenty of room at the bottom. http://www.its.caltech.edu/~feynman/plenty.html

36. Madou, M. 1997. *Fundamentals of Microfabrication.* CRC Press, New York.

37. Bird, R. B., W. E. Stewart, and E. N. Lightfoot. 1962. *Transport Phenomena*, 2nd edn. John Wiley & Sons, New York.

38. Brody, J. P., P. Yager, R. E. Goldstein, and R. H. Austin. 1996. Biotechnology at low Reynolds numbers. *Biophys. J.* 71: 3430–3441.

39. Man, P. F., C. H. Mastrangelo, M. A. Burns, and D. T. Burke. 1998. Microfabricated capillarity-driven stop valve and sample injector, in *MEMS 98'.* Heidelberg, Germany, pp. 45–50.

40. Oh, K. W. and C. H. Ahn. 2006. A review of microvalves. *J. Micromech. Microeng.* 16: R13–R39.

41. Zengerle, R., J. Ulsich, S. Kluge, M. Richter, and A. Richter. 1995. A bi-directional silicon micropump. *Sens. Actuators* A50: 81–86.

42. Maillefer, D., H. van Lintel, G. Ray-Mermet, and R. Hirschi,1999, A high-performance silicon micropump for implantable drug delivery system, *Proc. Of 12th IEEE MEMS 1999 Technical Digest,* Orlando, Fl, USA, 1/17-21: 541–546.

43. Van De Pol, F. C. M., H. T. G. Van Lintel, M. Elwenspoek, and J. H. J. Fluitman. 1990. A thermopneumatic micropump based on micro-engineering techniques. *Sens. Actuators* A21–A23: 198–202.

44. Quemper, J.-M., S. Nicolas, J. P. Gilles, J. P. Grandchamp, A. Bosseboeuf, T. Bourouina, and E. Dufour-Gergam. 1999. Permalloy electroplating through photoresist molds. *Sens. Actuators* 74: 1–4.

45. Price, N. C. and L. Stevens. 1999. An introduction to enzyme kinetics, in *Fundamentals of Enzymology: The Cell and Molecular Biology of Catalytic Protein*, 3rd edn. Oxford University Press, Published in United States by Oxford University Press Inc., New York.

46. Luchter-Wasylewska, E. 2001. Cooperative kinetics of human prostatic acid phosphatase. *Biochem. Biophys. Acta* 1548: 257–264.

47. Taira, A., G. Merrick, K. Wallner, and M. Dattoli. 2007. Reviving the acid phosphatase test for prostate cancer. *Oncology* 21: 1–12.

48. Johnson, D. E., G. R. Prout, and W. W. Scott, J. D. Schmidt, and R. P. Gibbons. 1976. Clinical significance of serum acid phosphatase in prostatic cancer. *Urology* 8: 123–126.

49. Berry, W. R., J. Laszlo, E. Cox, A. Walker, and D. Paulson. 1976. Prognostic factors in metastatic and hormonally unresponsive carcinoma of the prostate. *Cancer* 44: 763–775.

50. Ganem, E. J. 1956. The prognostic significance of an elevated serum acid phosphatase level in advanced prostatic carcinoma. *J. Urol.* 76: 179–181.

51. Dattoli, M., K. Wallner, L. True, J. Cash, and R. Sorace. 2003. Long-term outcomes after treatment with external beam radiation therapy and palladium 103 for patients with higher risk prostate carcinoma: Influence of prostatic acid phosphatase. *Cancer* 97: 979–998.

52. Fang, L. C., M. Dattoli, K. Wallner, L. True, J. Cash, and R. Sorace. 2007. Prostatic acid phosphatase predicts cancer-specific survival after palladium-103 brachytherapy for prostate cancer. *Brachytherapy* 6: 79–80.

31

Electrochemical Protein Chip for Tumor Marker Analysis

Michael S. Wilson

CONTENTS

31.1 Introduction

31.1.1 Tumor Marker Immunoassays and Electrochemical Immunosensors

Tumor markers (TMs) are substances that are often present in abnormally high concentrations in the blood, urine, or tissue of patients with certain types of cancer. They include proteins, hormones, enzymes, and other molecular species [8,23]. The measurement of the concentrations of TMs can be useful in the early detection of cancer, differentiating benign from malignant conditions, evaluating the extent of the disease, monitoring the response of the tumor to therapy, and predicting recurrence [21]. Most TMs have limited specificity, and so the measurement of a single TM is usually not sufficient to diagnose cancer. However, most cancers have more than one marker associated with their incidence, and in many cases the diagnostic value of TMs can be significantly increased if panels of markers are measured [1,3,13,20,29]. In addition to tumor-specific markers, the inclusion of tumor-associated markers and growth-related TMs into screening panels may also be of benefit. The presence of these substances and the order of their appearance may be of major significance in cancer screening.

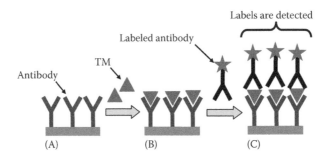

FIGURE 31.1 Basic TM immunoassay process. (A) Antibodies are immobilized on a solid support, typically the bottom of a polystyrene well in a 96-well plate; (B) the sample containing TM is added to the well and incubated. The TMs bind to the antibodies and are captured in the well. Excess sample is then removed; (C) a labeled antibody is added to the well and binds to the captured TM. The label is a molecule that can be easily detected, and is often an enzyme or fluorescent dye. Excess labeled antibody is removed by washing and the label that is bound to the well is detected and quantified using absorbance, fluorescence, or other method. In the case of an enzyme label, a substrate is added that is transformed by the enzyme into a detectable product (e.g., may be colored).

Proteins are one of the most promising classes of TMs for cancer diagnostics and therapy monitoring. For example, elevated levels of the serum proteins carcinoembryonic antigen (CEA), carbohydrate antigen (CA) 15-3, CA 19-9, and CA 125 may be associated with several types of cancer [21]. The concentration of a protein TM or other large biomolecule in a biofluid can be measured using an immunoassay. A basic immunoassay is shown in Figure 31.1, although many variations exist. In immunoassay, highly selective antibodies are used to bind to the target TM in the test sample. Detection of this binding event through the use of a second, labeled antibody enables the TM to be quantified, usually by absorbance or other optical measurements. The most common immunoassays use enzyme labels, and are called enzyme-linked immunosorbent assays (ELISA).

Electrochemical methods can also be used as the detection scheme in TM immunoassays, although these methods are less well-established than fluorescence or absorbance, especially in commercial products. Electrochemical detection is attractive for diagnostic test development because it offers excellent analytical sensitivity, and the required instrumentation is typically inexpensive and can be readily miniaturized for use in point-of-care (POC) applications. In electrochemical immunoassays, the TMs are usually quantified by measuring the electrical current generated at an electrode by the electrochemical oxidation or reduction of an electroactive antibody label or enzyme-generated product (amperometric sensor). Oxidation and reduction of a molecule involve the loss and gain of electrons, respectively, and an electrode can accept or supply these electrons if the appropriate oxidation or reduction electrical potential is applied. This flow of electrons through the electrode constitutes an electrical current that can be measured.

Electrochemical detection in immunoassays falls into two broad categories. In the first, an external electrode is used to make the assay measurement, and this can be used when the assay uses an enzyme-generated detection product. Here, the antibodies can be immobilized on any type of usable surface (such as the standard polystyrene well) and the electrode may be simply inserted into the assay solution during the final step, where the detection product is then oxidized or reduced (Figure 31.2A). In the second category, the antibodies (or other capture molecules) are immobilized on the electrode itself and the entire assay is conducted on the electrode surface. These systems are called electrochemical immunosensors (EISs, Figure 31.2B). In addition to redox currents, EISs also have other detection pathways available to them, such as the measurement of changes in electrical potential or impedance of the underlying electrode caused by the binding of the analyte to the immobilized capture molecules. A detailed discussion of the use of EISs for single-analyte TM measurement can be found in several recent reviews [18,31,41].

31.1.2 Multi-Analyte TM Analysis and Electrochemical Protein Chips

The increased diagnostic value provided by multi-TM measurements means that there is a need for the development of immunoassays that can easily measure panels of TMs at the same time, preferably using

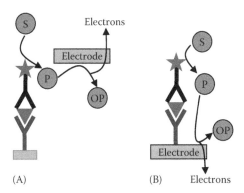

FIGURE 31.2 (A) Electrochemical detection using an external electrode. In the final step of the immunoassay, enzyme substrate S is added and is converted by the enzyme label to the product P. An electrode is inserted into the solution and an oxidizing electrical potential is applied. The product P is electrochemically oxidized to the product OP, with the lost electrons being accepted by the electrode. (B) Detection using an EIS. In the EIS, the electrode acts as the immobilization matrix upon which the assay is performed, as well as being the detection electrode where the electrochemical signal is measured.

a single sample and single assay. These types of assays, which are called simultaneous multi-analyte immunoassays, have the advantage that TMs can be measured using less sample, reagents, and technical operations, compared to when panels of TMs are measured by performing multiple single-analyte tests in parallel. This is important because sample amounts may be limited, particularly in patients weakened by disease or treatment. Simultaneous multi-analyte analysis also makes the test process more efficient, as well as increasing the throughput, and reducing the labor and cost-per-test.

One way of achieving multi-analyte immunoassay analysis is to use protein chips (also called protein microarrays). These chips consist of glass slides or other substrates that contain multiple types of immobilized antibodies. Each antibody is immobilized as a spot in a separate region, with each spot being capable of binding to a different target analyte. To perform an assay, the sample and reagents are added to the entire chip (or at least to the region containing the immobilized antibodies) and therefore come into contact with all of the different capture antibodies at the same time. In this way, multiple assays are performed simultaneously. The vast majority of assays performed on commercial and noncommercial chips use fluorescently labeled antibodies as the detection scheme, although mass spectrometry, chemiluminescence, and other optical detection methods are also used. These types of protein arrays have been used in cancer research for measuring protein abundances in a diverse range of sample types, including biofluids such as serum, cell culture supernatant, and interstitial fluid, as well as measuring cellular protein levels in cells from resected tissue and cell culture, and measuring the levels of cell surface proteins on whole cells [2,9,10,38].

Protein chips can also use electrochemical detection to measure multiple analytes [5]. These types of protein chips offer the potential benefits of good sensitivity, low-cost, and simplified instrumentation. One way of achieving multi-analyte testing is to combine multiple electrochemical immunoassays into a single test, with each analyte being detected by a different label. This method enables easy identification of the specific signal for each assay. Antibodies with different enzymes [16], metal ions [11], and nanoparticles [19] have been used in this type of assay, mostly in non-chip-based formats. For example, Liu et al. described an immunoassay in which the capture antibodies were attached to magnetic beads that were dispersed in the test sample, and the detection antibodies for each analyte were labeled with a different inorganic nanoparticle, for example, ZnS, CdS, or PbS [19]. At the end of the immunoassay, the beads were separated from the solution using a magnet, and any attached nanoparticles were then dissolved. The metal ions from the dissolved nanoparticles were specifically detected by a single electrode using a technique called stripping voltammetry. Here, each metal ion gave a distinct voltammetric peak with a position that reflected its identity and with a size that reflected its concentration (and therefore, that of the corresponding analyte). Multi-label approaches have several limitations, however, primarily that the degree of multiplexing is limited by the number of different labels available. In addition, the preparation of antibodies with multiple labels may be complicated and tedious, and the assay may become overly complex if different assay environments are needed for optimal performance of each label (e.g., different pH).

An alternative method for performing electrochemical-based multi-analyte assays is to use a single-label or label-free approach to detect all of the analytes. The main consideration here is how to discriminate between different detection events, i.e., how to identify the specific assay signal for each analyte. One method of achieving this is to use a chip-based sensor that contains an array of multiple EISs, with each EIS performing an immunoassay for a different analyte and having an individually measurable signal (addressable). Several groups have used a range of approaches to develop chips using this type of assay. These include chips with detection schemes that measure the oxidation current of enzyme-generated molecules [15,22,39]; the reduction of the electroactive area of the EIS caused by the enzyme-mediated precipitation of insoluble products [14]; the hindrance, caused by analyte binding, of the direct electrical communication between the electrode and an enzyme label [40]; or the changes in electrode conductance following protein binding [43]. Many of these chips have been used for the analysis of TMs, and have been discussed in a recent review [4].

Our group has developed several array-based biosensors, with a focus on sensors that can quantitatively measure panels of analytes with accuracy and precision comparable to established single-analyte absorbance-based ELISA [33,36,37]. These sensors measure analyte concentrations using a single-label simultaneous electrochemical multi-analyte immunoassay (SEMI). In one example, which has applications in cancer diagnostics, we developed a chip-based amperometric SEMI sensor that measured a panel of seven well-known TMs. The panel was selected to include markers associated with ovarian, colorectal, lung, and other types of cancers.

The electrochemical nature of the SEMI sensors brings significant flexibility to the system. On one hand, they can be used with standard ELISA protocols where sample addition, reagent additions, and washes are performed in multiple steps in a sample well using a pipette. Alternatively, by miniaturization and incorporation of cassette-based microfluidics, the chips can be developed into automated, portable, low-power POC devices. In this chapter, we describe the development of the SEMI sensor platform in detail, with a focus on multi-analyte TM measurement. We highlight the advantages and disadvantages of the SEMI sensor, describe the role that electrochemical-based protein chips can play in the field of multiplexed assays, and discuss the future prospects for electrochemical biosensors in cancer diagnostics.

31.2 Apparatus and Materials

Electrochemical measurements on the biosensor were performed in a Faraday cage using a custom-built PC-controlled eight-channel potentiostat (InnerSea Technologies, Bedford, MA). However, any commercial multichannel potentiostat that provides independent potential control for each electrode can be used (e.g., model 1000A, CH Instruments, Inc., Austin, TX). The measurements were performed using a three-electrode format consisting of a working electrode (the EIS), a counter electrode, and a reference electrode. The potentiostat was connected to the biosensor's contact pads using a 10-point PCB edge connector (Digi-Key, Thief River Falls, MN). Electrode cleaning and IrOx growth were performed using a commercial Ag/AgCl reference electrode (Microelectrodes, Inc., Bedford, NH) and platinum mesh (Sigma-Aldrich Chemical Company, Milwaukee, WI) as the counter electrode. The assay measurements were performed using the biosensor's on-board Ag/AgCl reference electrode.

Materials for fabricating the biosensor were purchased from the following vendors: Corning 1737 glass slides were from S. I. Howard Glass Co. (Worcester, MA); UltraMet Sonic detergent was from Buehler (Lake Bluff, IL); Shipley 1818 photoresist and CD-30 developer were from MicroChem Corporation (Newton, MA); CYTOP solution for spin-coating was prepared by mixing CTL-809M CYTOP and CT-SOLV.180 solvent from Bellex International (Wilmington, DE) in a ratio of 1:10 w/w; polyacetal for fabricating the sample well was from McMaster Carr (Princeton, NJ); adhesive gaskets for attaching the sample well were from Grace Bio Labs (Bend, OR). The photomask patterns were designed in-house using TurboCad (IMSI, Novato, CA), and the photomasks were produced by Advance Reproductions Corp. (North Andover, MA).

Materials for performing the SEMI were purchased from the following vendors: CEA, human chorionic gonadotropin (hCG-β), CA 19-9, alpha-fetoprotein (AFP), anti-CA 19-9 (clone M8073021), anti-CA 125 (clone M002203), anti-CA 15-3 (clone M8071022), and anti-hCG-β (clone M94139.7) were from Fitzgerald Industries International, Inc. (Concord, MA). CA 125 and CA 15-3 were from BioProcessing, Inc. (Scarborough, ME). CA 15-3, CA 19-9, and CA 125 were supplied with concentrations stated in U/mL. Ferritin and anti-ferritin (clone 090-12710) were from BioDesign International (Saco, ME). Goat anti-CEA was from BioSpacific (Emeryville, CA). Goat anti-AFP was from R&D Systems (Minneapolis, MN). AP-labeled donkey anti-goat IgG (AP-anti-GIgG) and AP-labeled donkey anti-mouse IgG (AP-anti-MIgG) were from Jackson ImmunoResearch (West Grove, PA).

Hydroquinone diphosphate (HQDP) was prepared as described later [34]. All other chemicals and reagents were purchased from Sigma-Aldrich Chemical Company and used as received. Water was purified using a Millipore MilliQ A10 system (Bedford, MA). Reaction buffer consisted of 100 mM Tris–HCl, 200 mM KCl, 10 mM $MgCl_2$, pH 9.0. Phosphate-buffered saline (PBS) consisted of 100 mM PBS (125 mM NaCl), pH 7.2. Blocking buffer consisted of Seablock (Pierce, Rockford, IL; 4× dilution with water).

31.3 Methods

31.3.1 SEMI Biosensor Design

We have developed two chip-based biosensor platforms for performing SEMI, the first-generation chip (SEMI-01) and a miniaturized second-generation chip (SEMI-02), shown in Figure 31.3. This chapter will focus on the use of SEMI-01, although the assay principles and fabrication process are the same for SEMI-02.

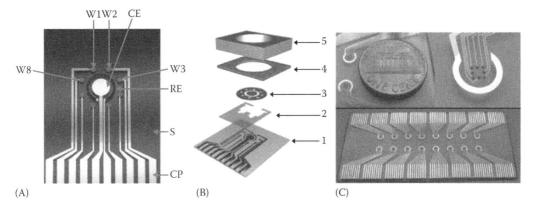

FIGURE 31.3 (A) Photograph of the first-generation SEMI biosensor (SEMI-01) showing the arrangement of EIS electrodes (W1–W8) on the glass slide (S). Electrodes W4–W7 continue in sequence between those marked as W3 and W8. Each EIS was 1 mm in diameter and was located 2.5 mm from its neighbor (edge to edge). The glass slide also contained a counter electrode (CE, 4 mm diameter), which was used to complete the electrical cell and provided a path for current flow during the assay measurement step, an Ag/AgCl reference electrode (RE), which was used to control the potential applied to the EIS, and electrical contact pads (CP) for connecting the chip to the potentiostat. The sample well is not shown. (B) Schematic of the SEMI-01 sensor showing the glass slide containing the electrodes (1) and other additional components. The other components included an insulation layer (2) that covered the leads connecting the contact pads to the EIS, a layer of hydrophobic polymer layer (3) that was deposited around the EIS to facilitate deposition of the antibody solution on the W1–W8 during the immobilization step (formation of the EIS), an adhesive silicone gasket (4), and a plastic well (5) for containing the test sample. The gasket was used to attach the sample well to the biosensor and provided a watertight seal. The enclosure formed by sample well contained all of the EIS, and so multiple analytes could be measured simultaneously using a single sample and by performing a single assay. (C) Miniaturized second-generation biosensor (SEMI-02), with smaller EIS (250 μm diameter and 300 μm separation). Here, the iridium counter electrode was also used as the reference electrode. The glass slide contained 16 separate SEMI-02 biosensors for the analysis of up to 16 samples at once (50 μL each) using a multi-well sample module (image not shown) and a multichannel pipette.

The SEMI-01 biosensor consisted of a glass slide containing a patterned array of seven EISs (the thin-film electrodes W1–W7) and one background sensor (electrode W8). Each EIS measured a different analyte using an amperometric electrochemical immunoassay, and the control electrode measured the background signal from nonspecific binding of reagents and sample during the assays. The EISs were prepared by immobilizing a different antibody or other binding protein onto the surface of each electrode. The array of EISs was located in the same sample well, and so all of the EISs were exposed to the same sample and reagents at the same time in the assay. Thus, by using mixtures of detection antibodies, one for each target, multiple analytes could be measured simultaneously. The signals generated by all of the analytes were read at the same time during the assay by connecting the chip to a multichannel potentiostat that made simultaneous, but independent electrochemical measurements on each EIS.

The electrodes W1–W8 on the biosensor were made from iridium and contained a surface coating of iridium oxide, denoted "IrOx" [35]. The oxide coating is electronically conducting and thus provides both a path for electron flow during electrochemical measurement and a surface for immobilizing antibodies or other proteins. Importantly, the IrOx is porous, has a high surface area, and provides a hydrous three-dimensional environment for immobilized proteins, with a high degree of transport for water, protons, and other ions. Proteins are susceptible to denaturing (unfolding with potential loss of binding ability) when they are immobilized on two-dimensional substrates [12]. We anticipate that the use of IrOx, or similar surfaces, will address this problem by providing stabilizing electrostatic interactions, and in general, we have observed excellent stability for proteins immobilized on IrOx surfaces.

We have used several different immunoassay formats to measure groups of analytes using the SEMI-01 sensor, including the "antibody-down" sandwich immunoassay and the competitive "antigen-down" immunoassay format described in Figures 31.4 and 31.5, respectively. Both of these assays use the same detection scheme, where the concentration of each analyte is determined by measuring the current resulting from the electrochemical oxidation of HQ at the corresponding EIS. The HQ is generated by the enzymatic hydrolysis of the substrate HQDP at each electrode containing bound AP. A crucial design element that makes multi-analyte analysis possible is that the spacing between the EISs is large enough that individual oxidation current measurements can be made without interference due to cross talk resulting from the diffusion of HQ from a neighboring EIS. This is facilitated because HQ is only generated in close proximity to the electrode surface, and so most of the HQ is efficiently electrochemically oxidized and is unavailable for diffusion (the diffusion coefficient of HQ is $\sim 1 \times 10^{-5}$ cm^2/s) [6]. Furthermore, radial diffusion of the HQ between the EISs is hindered because each EIS is surrounded by a layer of hydrophobic polymer that is several micrometers thick and thus is slightly recessed in the chip.

The main differences between the two assay formats are that antibodies are immobilized on the EISs in the sandwich assay and the size of the oxidation current measured on each EIS is proportional to the concentration of that analyte in the original sample, whereas for the competitive assay, a purified form of the analyte is immobilized on the EISs, and the oxidation current measured on each EIS is inversely proportional to the concentration of that analyte in the original sample. Typically, the different formats are not mixed on the same array, and one format may be preferred over the other, depending on the application. For example, we have used the sandwich assay to develop a SEMI for the TMs AFP and CEA [33], and another variation of this format to develop a SEMI sensor to measure immunoglobulins (IgGs) from four different species. The latter assay could be used to detect trace contaminations in transgenic products [36].

For single-analyte immunoassays, sandwich-type assays are preferred when possible. This is because they are generally easier to develop than competitive assays, and are typically more sensitive. A competitive assay is most useful when the target analyte is too small to allow the binding of two antibodies in a sandwich format, or when only one specific antibody is available. For higher degrees of multiplexing in assays, however, we prefer to use the competitive antigen-down format, even for larger analytes. For example, we used this assay system to develop the biosensor for measuring seven TMs [37], denoted SEMI-01-TM07, that is the focus of this chapter. The SEMI-01-TM07 assay was performed by following the steps outlined in Figure 31.5, but mixing antibodies toward all seven TMs with the sample, and using a mixture of AP-labeled anti-M IgG and AP-labeled anti-G IgG as the detection antibodies (all of the anti-TM antibodies used in the SEMI were derived from either mouse or goat hosts).

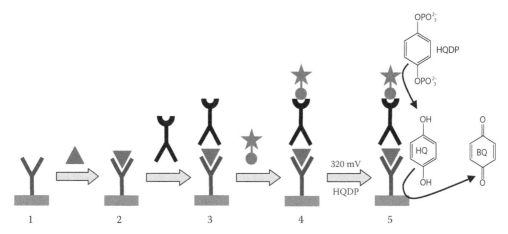

FIGURE 31.4 Sandwich immunoassay steps. (1) Consider a single EIS on the array containing an immobilized anti-TM antibody (Y); (2) sample containing TM (▲) is added to the sample well. During incubation, the TMs become bound to the anti-TM on the electrode. The amount of TM that binds to the electrode is proportional to the TM concentration in the test sample. The sample is then removed; (3) a biotin-labeled anti-TM is then added and incubated with the EIS. During incubation, this antibody binds to a different region of the TM, forming a sandwich complex. Excess antibody solution is then removed; (4) alkaline phosphatase (AP)-labeled streptavidin is then added, and the streptavidin binds to biotin-labeled antibody on the electrode (streptavidin is a protein that binds very strongly to biotin). The excess AP-labeled streptavidin is then removed; and (5) an oxidizing potential of 320 mV is applied to the EIS using a potentiostat, and the enzyme substrate hydroquinone diphosphate (HQDP) is added. AP attached to the electrode through streptavidin–biotin binding hydrolyzes the HQDP to hydroquinone (HQ), which is then electrochemically oxidized to benzoquinone (BQ) by the electrode. The oxidation of the HQ produces an electrical current which is measured (HQ transfers electrons to the EIS electrode during oxidation). The amount of HQ generated, and therefore the size of the oxidation current, is proportional to the concentration of TM in the original sample.

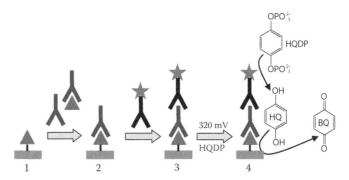

FIGURE 31.5 Competitive immunoassay steps. (1) Consider a single EIS on the array, say electrode W1 performing a TM immunoassay. Electrode W1 is converted from a plain electrode to an EIS by immobilizing a TM (▲) on the electrode surface; (2) a limited quantity of anti-TM is then added to the test sample containing the TM, and following an incubation period, the mixture is added to the sample well containing the EIS. The TM that is immobilized on the electrode (in step 1) competes with the TM present in the sample for binding to the limited amount of anti-TM. Anti-TM that is not bound to TM present in the sample is free to bind to TM immobilized on the electrode; (3) the sample is then removed and AP-labeled anti-IgG antibody is added, which binds to the anti-TM on the electrode. Anti-IgG antibodies (secondary antibodies) are antibodies that bind to other antibodies, specifically those with the IgG structure. All of the anti-TM antibodies used in these assays were of the IgG type (these are the typical "Y-shaped" monomeric antibodies). Following binding, excess AP-anti-IgG is removed; and (4) finally, the bound AP-anti-IgG is quantified by electrochemically measuring AP-generated HQ (from HQDP) by the same process as described in step 5 of Figure 31.4. However, in this assay, the amount of AP on the electrode is inversely proportional to the amount of TM in the original test sample, and so the EIS current response is inversely proportional to the concentration of the TM.

When compared to the sandwich assay format, the competitive assay has the significant advantage of requiring only one (unlabeled) specific antibody per analyte, and only one or two "off-the-shelf" AP-labeled anti-IgGs for detection, regardless of the degree of multiplexing. By comparison, a typical antibody-down sandwich assay requires at least two anti-TM antibodies per analyte, each reacting with a different region (epitope) of the TM, and usually will require preparation of a specific labeled antibody for each analyte (labeled anti-IgG can usually not be used here because both the capture and detection antibodies are typically monoclonal, and therefore both derived from mouse). This can represent a significant labor and cost burden when many antibodies are needed, as well as presenting challenges for storing and maintaining large numbers of labeled antibodies, many with labels that often lack long-term stability. This is true for almost any sandwich-based multi-analyte assay, not only the SEMI described here. Obviously, the advantages of the competitive assay increase with the degree of multiplexing. The primary limitation of the competitive assay is that very pure samples of the target analytes are required for immobilization on the electrodes, and this can sometimes present challenges. For example, the purity of commercially available protein TMs is variable, and some preparations may contain nontarget TMs in significant quantities (>20% w/w). However, the increasing need for protein-based diagnostic tests, refinements in recombinant production methods, and improved protein isolation and purification techniques mean that highly purified proteins are increasingly available.

In the next section, we describe the fabrication of the biosensor in detail, followed by a discussion of the processes used to develop the SEMI-01-TM07 assay, highlighting the factors used to refine and optimize the assay performance.

31.3.2 Detailed Fabrication and Experimental Procedures

The experimental details described later are for the fabrication of the SEMI-01 biosensor and the assays performed using this sensor. The SEMI-02 sensor was fabricated by identical methods, but using different photomasks. Immunoassay protocols were also the same for SEMI-02 as for SEMI-01, except that SEMI-02 used smaller sample and reagent volumes of 50 µL.

31.3.2.1 Fabrication of the SEMI-01-TM07 Biosensor

The SEMI-01-TM07 biosensor was fabricated using photolithographic methods. The fabrication was performed in a clean room to ensure a clean surface at each stage of the fabrication procedure, which was critical for good adhesion of the layered materials. For convenience, 12 SEMI-01 biosensors, each consisting of seven EISs, a control electrode, and a counter and reference electrode, were fabricated on each glass slide (Corning 1737 glass, 50 mm × 110 mm × 1 mm), and the slide could be used in this form with a multichannel pipette to perform assays on 12 samples at once. Here, a multi-well sample module was attached to the glass slide, with each well containing a SEMI biosensor.

In the first step, the glass slide was thoroughly cleaned with detergent (UltraMet Sonic), then rinsed with water, acetone, and isopropanol, and finally dried at 110°C for 45 min. Next, the glass slide was coated with photoresist (Shipley 1818), using a spin-coater (2500 rpm, 30 s) and then soft-baked at 90°C for 45 min. A photomask (the top half of mask A, Figure 31.6) was then positioned over the slide, as close as possible without contacting, using a photolithography instrument (Model 87330, Oriel Corp., CT). The mask contained the pattern to be transferred to the photoresist, which was the pattern for the electrodes W1–W8, counter electrode, connecting leads, and contact pads. The slide was then exposed to UV light from above (400 nm wavelength, 56 mW/cm^2, 5.4 s). The etched regions of the photomask were clear and allowed UV light to pass through. Here, the UV light changed the chemical structure of the photoresist, making it soluble in the developer solution (a KOH-based solution). The non-etched regions of the mask were dark and did not allow UV light to pass through. Following UV exposure, the mask was removed, and the slide was immersed in developer solution (CD-30, 1 min with shaking), and then rinsed with water, and dried with a stream of air. During the developing stage, the UV-exposed regions of photoresist dissolved in the developer solution, leaving behind a pattern of photoresist that was the negative of the desired electrode pattern. The slide was then exposed to UV again, for 5.4 s, this time without

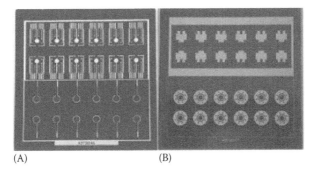

FIGURE 31.6   Photomasks used for fabricating the SEMI-01 sensor. (A) Each glass mask (5 in. × 5 in.) contained the pattern for two different steps in the fabrication process, with one pattern being on the top half of the plate, and the other pattern on the bottom half. (B) The plates contained the pattern for fabricating 12 separate biosensors at the same time on a single, large glass slide (50 mm × 110 mm × 1 mm). (A) Top half is the liftoff mask for the main electrodes and bottom half is for the Ag/Agcl electrode. (B) Top half is for the hard-baked photoresist isolation layer and bottom half is for the fluorocarbon layer.

the photomask. The purpose of the second exposure was to increase the solubility of the photoresist in acetone and make the subsequent metal "liftoff" step easier (described later).

In the next step, the slide was coated with a thin film of iridium using a sputtering process. To perform the metallization, the slide was cleaned in oxygen plasma for 5 min (Harrick PCD-3XG plasma cleaner, Pleasantville, NY) and then placed in a DC magnetron sputtering chamber (Ibex-2000, Ion Tech Inc., CO), the chamber evacuated (2×10^{-6} Torr) and filled with argon (10 mTorr Ar plasma pressure). The chamber contained both titanium and iridium targets, which were separately controlled. The slide was first coated with ~25 nm of titanium (100 W, 9 min), which acted as an adhesion layer, followed by ~100 nm of iridium (84 W, 13 min). Next, the substrate was placed in an acetone bath and sonicated for up to 1 h. Here, the remaining photoresist dissolved, lifting off any metal that was deposited on top, and leaving behind the patterned iridium electrodes (W1–W8), counter electrode, and electrical connections.

Either a commercial reference electrode or an on-chip Ag/AgCl reference electrode can be used for the SEMI. To fabricate the on-chip reference electrode, a second round of photolithography was performed using a procedure similar to that described earlier, except this time a different photomask was used to pattern the photoresist (Figure 31.6, bottom half of mask A). Here, silver was deposited by sputtering, and the adhesion layer consisted of titanium, followed by gold (using three separate metal targets in the sputtering chamber). The silver electrode was converted to AgCl at a later stage (described later).

To create electrodes with isolated, defined areas for immobilizing proteins (the circular electrodes W1–W8), the iridium leads running from the contact pads to the electrodes were insulated by covering them with hardened photoresist. This was achieved by spin-coating another layer of photoresist (spin-coated at 2500 rpm, 30 s, ~2 μm thickness) and patterning it using the photomask shown in Figure 31.6 (top half of the mask B). The photoresist was then hardened to a permanent layer by hard-baking at 180°C for 60 min. A final layer of photoresist was then spin-coated, soft-baked, and patterned using another photomask (Figure 31.6, bottom half of mask B); and a hydrophobic fluorocarbon polymer (CYTOP) was deposited by spin-coating (deposition: 1000 rpm, 30 s, 3–4 μm thickness) and then cured at 150°C for 60 min. Liftoff using sonication in acetone left a layer of CYTOP around the working electrodes.

The Ag/AgCl reference electrode was prepared at this stage by galvanically coupling the Ag electrode to a Pt mesh electrode in a stirred solution of KCl (0.1 M, 100 mL) and adding AgNO$_3$ (50 mg). The electrode was removed from the solution after 30 s, rinsed with water, and air-dried. In the buffer in which the immunoassay measurements were made (reaction buffer), the electrode had a stable potential of 80–85 mV, as compared to the commercial Ag/AgCl reference electrode. The increase in potential was likely due to the reduced Cl$^-$ content of the reaction buffer compared to the saturated KCl solutions used inside the commercial electrode.

A layer of IrOx was then grown on the surface of the iridium electrodes (W1–W8) by electrochemically activating them in PBS. For convenience, the electrodes were electrically joined as a common working electrode by placing adhesive conductive tape across the contact pads (except the pad connecting the reference electrode). The electrodes W1–W8 were immersed in PBS (500 mL), and a platinum mesh counter electrode (2 cm × 2 cm) and commercial Ag/AgCl reference electrode were used to complete the electrical cell. The electrodes were electrochemically cleaned by applying 3 V for 3 min, and then activated by applying 0.1 Hz square wave potential pulses with limits of −600 mV and +800 mV for 80 min using a potentiostat (with stirring). The electrodes were then washed with water, followed by isopropanol, and dried using a stream of air. If needed, the growth of IrOx was monitored by periodically performing cyclic voltammograms between −600 mV and +800 mV (50 mV/s scan rate) using a PC-controlled potentiostat and integrating the area under the curve. This area corresponded to the cathodic charge storage capacity of the oxide. To ensure complete electrode coverage, the electrodes were activated until the IrOx had a cathodic charge storage capacity of ~20 mC/cm², which corresponded to ~100 nm thickness [26]. The oxide growth was very reproducible, and different electrodes with the same activation time had charge storage capacities with <1% coefficient of variation (CV, $n = 16$).

The EISs were prepared by immobilizing the TMs CEA, AFP, hCG-β, CA 19-9, CA 125, CA 15-3, and ferritin on the surface of the electrodes W1–W7 on the SEMI-01 array (one TM per electrode). The TMs were immobilized through adsorption, by depositing drops of solution on the electrodes using a micropipette (0.5 μL in PBS) followed by incubation at 4°C overnight (if desired, 30% glycerol (v/v) can be included in the spotting solution to prevent drying). TM concentrations in the spotting solutions were 0.08 mg/mL (AFP), 0.14 mg/mL (ferritin), 0.2 mg/mL (CEA), 0.67 mg/mL (hCG-β), 14 kU/mL (CA 15-3), 42 kU/mL (CA 125), and 83 kU/mL (CA 19-9). The deposition solutions were contained within the area of each electrode during the immobilization step because the area between the electrodes was coated with the hydrophobic fluoropolymer (CYTOP) [17]. Electrode W8 was spotted with blocking buffer (Seablock) and was used as a control electrode. Finally, a plastic well (15 mm diameter), which was machined from a polyacetal block, was attached to the slide using an adhesive silicone gasket. This well provided a chamber for loading sample solution and performing assays. Excess unbound TM was removed from the sensors by washing with blocking buffer (3 × 1 mL) and then filling the well with blocking buffer (1 mL) and incubating for 20 min.

31.3.2.2 Preparation of HQDP

HQDP was used in these assays because, unlike most commercial AP substrates (e.g., phenyl phosphate), it does not foul the electrodes, and therefore provides significantly enhanced electrochemical signals. The structure of HQDP is shown in Figure 31.5. HQDP was prepared from hydroquinone (10 g, 0.091 mol) by adding KCl (500 mg) and POCl₃ (100 mL, 1 mol) and refluxing the mixture in an argon atmosphere for 18 h [34]. The excess POCl₃ was then removed under vacuum to leave the bis(phosphorodichloridate) of hydroquinone as an oily residue, which was then hydrolyzed to the corresponding diphosphate by spreading the residue in a Petri dish and incubating for 72 h in a partially evacuated desiccator containing aqueous sodium hydroxide. The residue was then dissolved in water and the pH of the solution was adjusted to 8.5 by the addition of cyclohexylamine. Acetone was then added to initiate crystallization of HQDP as the cyclohexylamine salt (71% yield).

31.3.2.3 SEMI-01-TM07 Assay Procedure

The simultaneous assay for the seven TMs was performed as follows. The 250 μL samples (typically blocking buffer) containing AFP, CEA, and hCG-β (0–200 ng/mL), ferritin (0–400 ng/mL), CA 125, CA 15-3, and CA 19-9 (0–200 U/mL) were mixed with samples of blocking buffer (250 μL) containing anti-AFP (600 ng/mL), anti-ferritin (140 ng/mL), anti-CEA (30 ng/mL), anti-hCG-β (25 ng/mL), anti-CA 125 (30 ng/mL), anti-CA 15-3 (200 ng/mL), and anti-CA 19-9 (50 ng/mL). The samples were incubated for 60 min with brief mixing at 15 min intervals using a micropipette. This gave the TMs in the sample time to react with the anti-TMs before the sample was exposed to the TMs immobilized on the electrodes, increasing the assay sensitivity. The assays had reduced sensitivities when this incubation time was <60 min, and it is likely that longer incubation times would increase the assay sensitivity (not investigated).

The mixtures (500 μL) were then added to the sensor and incubated for 2 h with gently shaking. The sensors were washed with blocking buffer (3 × 1 mL) and incubated with a mixture of AP-anti-GIgG (5 μg/mL) and AP-anti-MIgG (7.5 μg/mL) in blocking buffer (600 μL, 2 h with shaking). Here, the AP-anti-M IgG bound to any electrodes containing anti-CA 125, anti-CA 15-3, anti-CA 19-9, anti-ferritin, or anti-hCG-β, since mouse monoclonal anti-TM antibodies were used for these TMs in the assay, and the AP-anti-G IgG bound to any anti-AFP and anti-CEA on the electrodes, since polyclonal goat anti-TM antibodies were used for these TMs in the assay.

After washing with blocking buffer (5 × 1 mL) and reaction buffer (1 mL), the well was filled with reaction buffer (1 mL), and a potential of 320 mV was applied to the working electrodes (400 mV if an external commercial Ag/AgCl reference electrode was used). The initial transient signal due to capacitive charging of the IrOx was allowed to subside to a steady background response (0–1 μA/cm²), and then the enzyme substrate HQDP was added (1 mL, 2 mM final concentration). The solution was immediately mixed using a pipette by withdrawing and expelling 1 mL of solution four times, and the current response of electrodes W1–W8 was recorded simultaneously as a function of time. Measurements were taken at the initial transient maximum amperometric response (non–steady state), which was typically reached 0.5–2 min following addition of HQDP. Measurements were corrected for nonspecific background by subtracting the current response of the control electrode (W8).

31.3.3 Development and Optimization of the SEMI-01-TM07 Assay

31.3.3.1 Reagent Validation

One of the most important steps in developing the multi-analyte immunoassay was reagent validation. The use of highly specific monoclonal antibodies for most of the assays, rather than polyclonal antibodies, significantly reduced the likelihood for cross reaction with nontarget analytes. To validate the reagents, arrays containing all seven immobilized TMs were first incubated with individual anti-TM antibodies. Several different antibody clones for each TM were evaluated and the assay responses were compared (200 ng/mL anti-TM). This enabled identification of the best antibodies for each assay, these being the antibodies that produced the highest assay signal (for a fixed concentration) with no (or minimal) cross-reactivity with non-cognate TMs. The control electrode was used to differentiate between signals from nonspecific binding and those from cross-reactivity. Cross-reactivity was identified when an EIS produced a higher response than the control electrode when a sample absent of the TM specific for that EIS (but containing other TMs) was used. As expected, monoclonal antibodies gave the best performance and specificity for most of the TM assays, although better performance was achieved in two systems using polyclonals (AFP and CEA assays).

31.3.3.2 Optimization of the Immobilization of TMs on the EISs

Once the preferred reagents were identified, the assays were optimized by adjusting several parameters. First, saturation of the electrode with immobilized capturing element (in this case TM) was preferred in this type of assay. This was to ensure that the TM did not limit the amount of antibody binding in the assay, which would adversely affect the assay sensitivity and range. The amounts of TM immobilized on the IrOx electrodes were maximized by adjusting the concentration of each TM in the spotting solution until the assay signals reached a maximum (anti-TM concentrations were 200 ng/mL, 2 h incubation, and AP-anti-MIgG and AP-anti-GIgG were 10 μg/mL each, 2 h incubation). We found that the EISs could be saturated using spotting solutions with concentrations of 0.08 (AFP), 0.14 (ferritin), and 0.20 mg/mL (CEA), and 83 (CA 19-9) and 42 kU/mL (CA 125). The current responses of the EISs at the saturation points were 38 (AFP), 44.6 (ferritin), 63.05 (CEA), 22.7 (CA 19-9), and 26.8 μA/cm² (CA 125). HCG-β and CA 15-3 were used at non-saturating concentrations of 0.67 mg/mL and 14 kU/mL, respectively, as these were the concentrations of the supplied commercial TM samples. Under these conditions, these EISs had current responses of 59.7 and 92.5 μA/cm², respectively.

31.3.3.3 Optimization of the AP-Anti-IgG Incubation Step

Next, the AP-anti-IgG incubation step was optimized by varying the AP-anti-MIgG and AP-anti-GIgG concentrations (2.5–10 µg/mL) and incubation time (1–3 h) to maximize the assay signal (anti-TM concentrations were 200 ng/mL, 2 h incubation). Again, these reagents should be used under saturating conditions and not limit the assay signal. Here, we found that 7.5 µg/mL of AP-anti-MIgG and 5 µg/mL of AP-anti-GIgG were sufficient to reach saturation with a 2 h incubation time.

31.3.3.4 Optimization of the Anti-TM Concentrations

We wanted to develop assays that were sensitive over the 0–100 ng/mL or U/mL concentration range (0–200 ng/mL for ferritin), since this represented the typical normal and moderately elevated serum levels for the TMs. The most critical parameters in the SEMI development were the concentrations of the anti-TM antibodies added to the test sample. Because of the competitive nature of the assay, these reagents did not want to be at saturating concentrations (unlike for a noncompetitive assay). To determine the anti-TM saturation points, the response of each EIS was measured following incubation with anti-TMs with a range of concentrations (0–200 ng/mL, 2 h; 7.5 µg/mL of AP-anti-MIgG; and 5 µg/mL of AP-anti-GIgG, 2 h). With the exception of the CA 15-3 electrode, the sensors were almost saturated at antibody concentrations of 100 ng/mL (Figure 31.7A). The next optimization experiment was to evaluate the effect of the time that the anti-TMs were incubated on the biosensor (0.5–3 h). For convenience, 2 h was chosen as the preferred incubation time, even though it did not represent the optimum time for all samples.

The anti-TM saturation point data provided a guide for the range of anti-TM concentrations to be used in the next series of optimization experiments. Here, dose–response curves were generated for samples

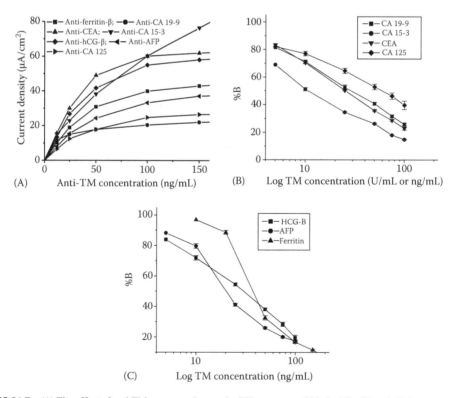

FIGURE 31.7 (A) The effect of anti-TM concentration on the EIS responses (500 µL, 2 h). (B) and (C) Inter-assay dose–response curves of the SEMI-01-TM07 assay. Data points and error bars represent the mean and ±1 SD of five measurements, respectively. All electrochemical measurements were recorded at 320 mV versus Ag/AgCl, in 100 mM Tris, 200 mM KCl, 10 mM MgCl$_2$, 2 mM HQDP, pH 9.0.

containing TMs (0–200 ng/mL for ferritin, 0–100 ng/mL (or units/mL) for the other TMs) using several different anti-TM concentrations (the anti-TM concentration was fixed for each dose–response curve). The aim of these experiments was to generate dose–response curves with the best slope (sensitivity) over the concentration range of interest. As expected, the range of the assays could be readily controlled by adjusting the anti-TM concentration. Increasing the anti-TM concentration increased the range of the assay, usually at the expense of assay sensitivity.

The assay responses for samples containing 100 ng/mL (or units/mL) TMs (S_{100}) were normalized by calculating the percentage of antibody binding ($\%B_{100}$) using the relationship, $\%B_{100} = (S_{100} - S_0)/(S_{max} - S_0) \times 100$, where S_{max} is the current response for samples containing zero TM, and S_0 is the background response measured on the control electrode (W8). The dose–response curves had the following characteristics, represented as $\%B_{100}$ and antibody concentration, respectively: AFP (8.2%–18.6%, 75–300 ng/mL); ferritin (6.5%–60.3%, 50–200 ng/mL); CEA (20.0%–59.3%, 12.5–100 ng/mL); hCG-β (18.1%–98.1%, 12.5–100 ng/mL); CA 15-3 (7.7%–26.6%, 75–300 ng/mL); CA 125 (38.4%–89.7%, 12.5–100 ng/mL); and CA 19-9 (19.4%–41.7%, 25–100 ng/mL). Based on these results, the antibody concentrations chosen for the SEMI were 300 (anti-AFP), 70 (anti-ferritin), 15 (anti-CEA), 12.5 (anti-hCG-β), 100 (anti-CA 15-3), 15 (anti-CA 125), and 25 ng/mL (anti-CA 19-9).

31.4 Results

31.4.1 Electrochemical Performance

For the competitive SEMI, the largest electrochemical current responses (S_{max}) were obtained using samples containing zero TMs. Because of this, high precision for repeated S_{max} measurements (at least for normalized values) was critical for good sensitivity at low TM concentrations (the detection limit is dependant on the standard deviation of S_{max}). The mean current responses of the EISs at S_{max} were 35.4 (AFP), 32.3 (ferritin), 17.1 (CEA), 12.5 (hCG-β), 61.1 (CA 15-3), 7.0 (CA 125), and 14.5 μA/cm² (CA 19-9). The S_{max} values had excellent precision, which was determined by calculating the CVs of the current signals. Intra-assay CVs for S_{max} were 1.4% (AFP), 1.5% (ferritin), 0.7% (CEA), 2.0% (hCG-β), 1.5% (CA 15-3), 1.7% (CA 125), and 0.7% (CA 19-9, $n = 6$); inter-assay CVs were 1.4% (AFP), 3.0% (ferritin), 1.2% (CEA), 2.4% (hCG-β), 1.6% (CA 15-3), 1.7% (CA 125), and 1.3% (CA 19-9, $n = 9$). The sensor also exhibited precise current responses over the entire range of TM concentrations measured, with inter-assay CVs of <3.9% (AFP), <5.6% (ferritin), <4.9% (CEA), <8.4% (hCG-β), <3.7% (CA 15-3), <7.5% (CA 125), and <3.4% (CA 19-9, $n = 5$).

31.4.2 Analytical Performance

The use of the SEMI-01-TM07 sensor for measuring panels of TMs has been rigorously demonstrated for the TMs AFP, ferritin, CEA, hCG-β, CA 15-3, CA 125, and CA 19-9. For convenience, samples consisting of salmon serum (Seablock) spiked with varying concentrations of the human TMs were used in the preliminary validation. The biosensor successfully detected all seven TMs with good specificity, with each EIS giving a current response that was inversely proportional only to the concentration of the TM specific for that sensor. To take into account the potentially large variations in absolute assay responses between different TMs in the SEMI, and also to provide signal normalization, the dose–response curves were plotted as percent binding ($\%B$) versus concentration, using the relationship: $\%B = (S_x - S_0)/(S_{max} - S_0) \times 100$, where S_x was the assay response for samples containing x ng/mL (or U/mL) TM, and S_{max} and S_0 had their previous definitions. Using this method, the EIS dose–response data were typical of those exhibited by standard competitive assays, being linear when plotted on log–linear or log–log scales (Figure 31.7B and C).

The analytical performance of the SEMI was demonstrated by measuring the concentration of each TM in samples containing "unknown mixtures" of the seven TMs. Here, a six-point SEMI calibration curve was obtained and used to quantify the TMs (made from Seablock samples with known TM concentrations). The concentration measurements made by the SEMI had excellent precision (intra-assay

TABLE 31.1

Performance of the SEMI Using Samples of Salmon Serum Spiked with TMs

	Sample 1				Sample 2			
	Intra-assay[a]		Inter-assay[b]		Intra-assay[a]		Inter-assay[b]	
Analyte	Mean (ng/mL)	CV (%)	Mean (ng/mL)	CV (%)	Mean (ng/mL)	CV (%)	Mean (ng/mL)	CV (%)
Ferritin	116.6	4.5	115.0	5.7	35.4	0.8	35.6	2.2
CEA	24.5	1.2	24.5	3.7	8.1	3.7	7.3	5.5
hCG-β	23.2	4.3	23.1	7.8	9.7	4.1	9.9	8.1
CA 15-3[c]	47.3	1.7	46.6	3.4	20.8	2.9	22.7	5.3
CA 125[c]	42.3	6.6	42.5	6.8	16.0	12.5	15.7	15.9
CA 19-9[c]	45.3	3.3	45.2	3.5	18.7	3.7	19.1	5.2
AFP	25.5	1.2	25.7	1.9	7.9	2.5	8.1	3.7

[a] $n=6$.
[b] $n=9$.
[c] U/mL.

CVs of 0.8%–4.5%, and inter-assay CVs of 1.9%–8.1%), as shown in Table 31.1, for example. This is comparable in performance to commercial single-analyte ELISAs [42]. The low slope of the CA 125 dose–response curve resulted in less-precise concentration determinations for this TM, with intra-assay and inter-assay CVs of 6.6%–12.5% and 6.8%–15.9%, respectively. For the most part, however, the level of precision of the CA 125 assay fell within the acceptable range for immunoassays ($\leq$10% and $\leq$20% for intra-assay and inter-assay, respectively) [25]. The SEMI had good accuracy over the entire TM concentration range, with plots of measured concentration versus expected concentration being linear. The graphs had slopes of 0.967 ($r^2=0.9981$, AFP); 1.012 ($r^2=0.9985$, ferritin); 0.964 ($r^2=0.9978$, CEA); 1.086 ($r^2=0.9906$, hCG-β); 1.008 ($r^2=0.9991$, CA 15-3); 1.062 ($r^2=0.9921$, CA 125); and 1.006 ($r^2=0.9981$, CA 19-9). The detection limits of the sensor, calculated as three times the standard deviation of S_{max}, were 1.4 (AFP), 7.0 (ferritin), 1.2 (CEA), and 1.8 ng/mL (hCG-β), and 0.7 (CA 15-3), 1.2 (CA 125), and 1.0 U/mL (CA 19-9). It should be noted that these limits do not represent the detection threshold for the EIS technology, but rather the detection limit for the assay when performed using these specific anti-TM concentrations and conditions. The detection limit could likely be lowered by reducing the anti-TM concentrations (at the expense of assay range).

Finally, the SEMI has been validated using human serum–based control samples (Liquid QC Tumor Marker Control, Cliniqa, Fallbrook, CA). These controls are supplied with TM concentrations that have been determined by several laboratories using commercial single-analyte clinical analyzers. For comparison, we analyzed the samples using the SEMI, again using a six-point calibration curve to quantify the TMs. The SEMI sensor measured the TMs with good precision (intra-assay CVs of 2.1%–7.1%), and the SEMI results had excellent correlation with the results obtained with the single-analyte clinical analyzers (within 8% agreement, Table 31.2). These results demonstrated that the SEMI could accurately and quantitatively measure multiple TMs in human serum samples. We are currently performing larger-scale validations of the SEMI-01-TM07 and SEMI-02 sensors using clinical specimen sets (serum) from cancer patients. This is a blind study and will provide extensive analytical and clinical sensitivity data for the use of the SEMI sensor in cancer diagnostics.

31.5 Discussion

Electrochemical protein chips represent a promising platform for performing multi-analyte immunoassays in cancer diagnostics. However, the development of these types of chips for TM measurement or any other application has been very limited to date. CombiMatrix Corp. (Mukilteo, WA), which is one of the companies to use electrochemical detection in commercial microarrays, appears focused on

TABLE 31.2

Performance of the SEMI Compared to Single-Analyte Analyzers Using Samples of Human Serum

	Roche Elecsys 1010	Beckman Coulter Access		SEMI
Analyte	Mean (ng/mL)	Mean (ng/mL)	Mean[a] (ng/mL)	CV (%)
Ferritin	120.3	114.3	128.6	4.2
CEA	7.6	—	8.4	7.1
hCG-β	33.7	32.6	34.5	5.8
CA 15-3[b]	17.8	—	18.9	2.1
CA 125[b]	60.3	—	60.1	6.8
CA 19-9[b]	30.7	—	31.3	4.8
AFP	57.3	65.3	59.5	3.2

[a] $n = 5$.
[b] U/mL.

DNA-based chips, with no apparent commercial development of antibody-based immunoassay arrays [5]. In addition, a handful of non-electrochemical protein chips that quantitatively measure a small number of cancer-related proteins have been developed academically and commercially [27,28,32], as well as larger fluorescence-based or mass spectrometry–based protein-profiling type arrays [9,10].

There are many different needs for multi-analyte analysis, and it is unlikely that one detection method will be suitable for all applications. For example, highly multiplexed arrays are most useful for performing qualitative or semiquantitative profiling experiments. Here, the relative abundance, or mere presence of a protein, is determined. Accurate quantitative data are difficult to obtain because of the significant challenges posed by the management and control of reagents when the number of immunoassays is high. For example, each antibody needs to be characterized for performance and specificity by itself and in combination with the other reagents. For large numbers of reagents, this can present difficult quality control problems. Furthermore, the system may need to be reexamined every time a new batch of reagent is introduced. Additional difficulties may arise from the different shelf lives, stability and binding affinities of the reagents, antibody cross-reactivity, and analyte concentrations that may vary over wide ranges. Finally, the sensitivity of multi-analyte assays may be lower than comparable single-analyte tests because of higher background signals. This arises because the total concentrations of antibodies is likely higher in the multi-analyte test (a mixture of multiple antibodies are used), leading to potentially higher degrees of nonspecific antibody binding. Because of these factors, multi-analyte assays for providing accurate quantitative data need to be of significantly lower assay density, with a limit estimated at 30–50 analytes [10].

Optical-based detection methods are usually used in high-density array-based assays since they are amenable to making hundreds or thousands of measurements. However, quantitative challenges remain, even in lower-multiplexed assays, from accuracy issues that can arise from nonuniform rates of photobleaching and quenching of optical labels, irregular protein spot drying (ring formation) during array fabrication, and interference from fluorescing or absorbing species in biological samples. These issues, together with the expense and sophistication of the required equipment, mean that optical methods are often not preferred for lower-density quantitative multiplexed assays [7].

On the other hand, electrochemical detection methods are well suited for use these types of tests. EISs are attractive because they directly give an electronic signal than can be easily and accurately quantified using instrumentation that is simple and low cost. They are typically robust, economical to mass produce, and can achieve excellent detection limits [24]. The instrumentation can be easily miniaturized to circuit board level with low-power requirements, facilitating the development of disposable devices and methodologies for small sample amounts [30]. EISs are not affected by specular reflection, scattering, quenching, or irregular protein spot drying, which aids accuracy in quantification, although interference from electroactive species found in biofluids can occur.

The SEMI sensor described in this chapter measured seven TMs with accuracy and precision comparable to traditional single-analyte ELISA, while significantly increasing the assay efficiency and throughput, and reducing the cost-per-test and sample amounts. The SEMI sensor provides a convenient method for measuring a wide range of cancer biomarkers. The main criteria for inclusion of biomarkers in the assay are the availability of highly specific antibodies and very pure samples of the marker. If samples of sufficient purity are not available, an alternative approach may be taken in which the electrochemical chip is derivatized with antibodies. The number of TMs measured in this type of sensor could be increased by using additional electrodes, although the issues associated with reagent management remain a consideration. The sensor could also be miniaturized by using even smaller electrodes, further reducing sample and reagent volumes. The primary limiting factor for miniaturization is the ability to accurately deposit protein spots onto the electrodes for immobilization, although it is likely that robotic instruments could be used for this purpose.

The SEMI provides significant advantages over performing multiple parallel single-analyte immunoassays, although some challenges still remain in achieving optimized application. This is because these types of assays place compromises on assay conditions that may represent a deviation from optimal conditions used for single-analyte assays, and this may adversely affect the assay performance. For example, the SEMI uses a single sample dilution and common incubation time to measure all analytes. Nevertheless, many of these issues can be addressed, for example, by choosing analytes that have similar physiological concentration ranges or by using appropriate combinations of low- and high-affinity antibodies to control the dynamic range of the assays, and in general, the advantages of the multi-analyte test outweigh the limitations.

31.6 Future Prospects

As new TMs are discovered and applications for existing TMs are refined and expanded, the need for multi-analyte TM analysis will undoubtedly increase. In particular, highly accurate quantitative tests for carefully selected panels of TMs are likely to be critical in the early detection of cancer. Electrochemical-based protein chips are an attractive platform for this type of application, bringing excellent analytical sensitivity that can be readily packaged in small, low-power formats. One of the main appeals of electrochemical protein chips is the simplicity of the required instrumentation compared to many other assay detection methods, which makes low-cost, widespread POC screening for cancer a real possibility. These strengths are embodied by the glucose test, an electrochemical-based assay which is the most successful commercial biosensor in history. It can be envisioned that sensors like the ones described in this chapter could be developed into similar types of devices.

The versatility of EISs means that they can be integrated into most existing immunoassaying platforms, making them adaptable to the testing environment. For example, we have recently integrated EIS arrays into the lateral flow immunoassay platform, the familiar test format used in many store-bought at-home tests. We are also developing sample flow-through modules for use with the SEMI-01 and SEMI-02 chips which may lead to lab-on-a-chip devices with significantly reduced assay times. The unique features of electrochemistry therefore provide a wide range of options for assay development, and electrochemical-based biosensors will likely have a strong future in cancer diagnostics.

REFERENCES

1. Amonkar, S. D., G. P. Bertenshaw, T. H. Chen, K. J. Bergstrom, J. Zhao, P. Seshaiah, P. Yip, and B. C. Mansfield. 2009. Development and preliminary evaluation of a multivariate index assay for ovarian cancer. *PLoS One* 4:e4599.
2. Borrebaeck, C. A. 2006. Antibody microarray-based oncoproteomics. *Exp. Opin. Biol. Ther.* 6:833–838.
3. Carpelan-Holmström, M., J. Louhimoa, U. H. Stenmanb, H. Alfthanb, H. Järvinena, and C. Haglunda. 2004. Estimating the probability of cancer with several tumor markers in patients with colorectal disease. *Oncology* 66:296–302.

4. Chen, H., C. Jiang, C. Yu., S. Zhang, B. Liu, and J. Kong. 2009. Protein chips and nanomaterials for application in tumor marker immunoassays. *Biosens. Bioelectron.* 24:3399–3411.

5. Dill, K., D. D. Montgomery, A. L. Ghindilis, K. R. Schwarzkopf, S. R. Ragsdale, and A. V. Oleinikov. 2004. Immunoassays based on electrochemical detection using microelectrode arrays. *Biosens. Bioelectron.* 20:736–742.

6. Ding, Y., L. Zhou, H. B. Halsall, and W. R. Heineman. 1999. Feasibility studies of simultaneous multi-analyte amperometric immunoassay based on spatial resolution. *J. Pharm. Biomed. Anal.* 19:153–161.

7. Fryer, R. M., J. Randall, T. Yoshida, L. L. Hsiao, J. Blumenstock, K. E. Jensen, T. Dimofte, R. V. Jensen, and S. R. Gullans. 2002. Global analysis of gene expression: Methods, interpretation, and pitfalls. *Exp. Nephrol.* 10:64–74.

8. Gregory, J. J. Jr. and J. L. Finlay. 1999. Alpha-fetoprotein and beta-human chorionic gonadotropin: Their clinical significance as tumour markers. *Drugs* 57:463–467.

9. Haab, B. 2003. Methods and applications of antibody microarrays in cancer research. *Proteomics* 3:2116–2122.

10. Haab, B. B. 2005. Antibody arrays in cancer research. *Mol. Cell. Proteomics* 4:377–383.

11. Hayes, F. J., H. B. Halsall, and W. R. Heineman. 1994. Simultaneous immunoassay using electrochemical detection of metal ion labels. *Anal. Chem.* 66:1860–1865.

12. Holtz, B., Y. Wang, X.-Y. Zhu, and A. Guo. 2007. Denaturing and refolding of protein molecules on surfaces. *Proteomics* 7:1771–1774.

13. Kim, K., I. Visintin, A. B. Alvero, and G. Mor. 2009. Development and validation of a protein based signature for the detection of ovarian cancer. *Clin. Lab. Med.* 29:47–55.

14. Ko, J. S., H. C. Yoon, H. Yang, H. B. Pyo, K. H. Chung, S. J. Kim, and Y. T. Kim. 2003. A polymer-based micro-fluidic device for immunosensing biochips. *Lab Chip* 3:106–113.

15. Kojima, K., A. Hiratsuka, H. Suzuki, K. Yano, K. Ikebukuro, and I. Karube. 2003. Electrochemical protein chip with arrayed immunosensors with antibodies immobilized in a plasma-polymerized film. *Anal. Chem.* 75:1116–1122.

16. Kricka, L. J. 1992. Multianalyte testing. *Clin. Chem.* 38:327–328.

17. Lee, C. S., S. H. Lee, S. S. Park, Y. K. Kim, and B. G. Kim. 2003. Protein patterning on silicon-based surface using background hydrophobic thin film. *Biosens. Bioelectron.* 18:437–444.

18. Lin, J. and H. Ju. 2005. Electrochemical and chemiluminescent immunosensors for tumor markers. *Biosens. Bioelectron.* 20:1461–1470.

19. Liu, G., J. Wang, J. Kim, and M. R. Jan. 2004. Electrochemical coding for multiplexed immunoassays of proteins. *Anal. Chem.* 76:7126–7130.

20. Louhimo, J., P. Finne, H. Alfthan, U. H. Stenman, and C. Haglund. 2002. Combination of hCGb, CA 19-9 and CEA with logistic regression improves accuracy in gastrointestinal malignancies. *Anticancer Res.* 22:1759–1764.

21. Maruvada, P., W. Wang, P. D. Wagner, and S. Srivastava. 2005. Biomarkers in molecular medicine: Cancer detection and diagnosis. *Biotechniques* (Suppl. 1):9–15.

22. Meyerhoff, M. E., C. Duan, and M. Meusel. 1995. Novel nonseparation sandwich-type electrochemical enzyme immunoassay. *Clin. Chem.* 41:1378–1384.

23. Niklinski, J. and M. Furman. 1995. Clinical tumour markers in lung cancer. *Eur. J. Cancer Prev.* 4:129–138.

24. Nyholm, L. 2005. Electrochemical techniques for lab-on-a-chip applications. *Analyst* 130:599–605.

25. O'Connell, M. A., B. A. Belanger, and P. D. Haaland. 1993. Calibration and assay development using the four-parameter logistic model. *Chemom. Intell. Lab. Syst.* 20:97–114.

26. Rauh, R. D. and S. F. Cogan. 1993. Design model for electrochromic windows and application to the WO_3/IrO_2 system. *J. Electrochem. Soc.* 140:378–386.

27. Song, S., B. Li, L. Wang, H. Wu, J. Hu, M. Li, and C. Fan. 2007. A cancer protein microarray platform using antibody fragments and its clinical applications. *Mol. BioSyst.* 3:151–158.

28. Sun, Z., X. Fu, L. Zhang, X. Yang, F. Liu, and G. Hu. 2004. A protein chip system for parallel analysis of multi-tumor markers and its application in cancer detection. *Anticancer Res.* 24:1159–1165.

29. Tsao, K.-C., J.-H. Hong, T.-L. Wu, P.-Y. Chang, C.-F. Sun, and J. T. Wu. 2007. Elevation of CA 19-9 and chromogranin A, in addition to CA 125, are detectable in benign tumors in leiomyomas and endometriosis. *J. Clin. Lab. Anal.* 21:193–196.

30. Wang, J. 2002. Portable electrochemical systems. *Trends Anal. Chem.* 21:226–232.

31. Wang, J. 2006. Electrochemical biosensors: Towards point-of-care cancer diagnostics. *Biosens. Bioelectron.* 21:1887–1892.

32. Weissenstein, U., M. J. Schneider, M. Pawlak, J. Cicenas, S. Eppenberger-Castori, P. Oroszlan, S. Ehret, A. Geurts-Moespot, F. C. Sweep, and U. Eppenberger. 2006. Protein chip based miniaturized assay for the simultaneous quantitative monitoring of cancer biomarkers in tissue extracts. *Proteomics* 6:1427–1436.

33. Wilson, M. S. 2005. Electrochemical immunosensors for the simultaneous detection of two tumor markers. *Anal. Chem.* 77:1496–1502.

34. Wilson, M. S. and R. D. Rauh. 2004. Hydroquinone diphosphate: An alkaline phosphatase substrate that does not produce electrode fouling in electrochemical immunoassays. *Biosens. Bioelectron.* 20:276–283.

35. Wilson, M. S. and R. D. Rauh. 2004. Novel amperometric immunosensors based on iridium oxide matrices. *Biosens. Bioelectron.* 19:693–699.

36. Wilson, M. S. and W. Nie. 2006. Electrochemical multianalyte immunoassays using an array-based sensor. *Anal. Chem.* 78:2507–2513.

37. Wilson, M. S. and W. Nie. 2006. Multiplex measurement of seven tumor markers using an electrochemical protein chip. *Anal. Chem.* 78:6476–6483.

38. Wong, S. C., C. M. Chan, B. B. Ma, M. Y. Lam, G. C. Choi, T. C. Au, A. S. Chan, and A. T. Chan. 2009. Advanced proteomic technologies for cancer biomarker discovery. *Exp. Rev. Proteomics* 6:123–134.

39. Wu, J., F. Yan, J. Tang, C. Zhai, and H. Ju. 2007. A disposable multianalyte electrochemical immunosensor array for automated simultaneous determination of tumor markers. *Clin. Chem.* 53:1495–1502.

40. Wu, J., F. Yan, X. Zhang, Y. Tan, J. Tang, and H. Ju. 2008. Disposable reagentless electrochemical immunosensor array based on a biopolymer/sol–gel membrane for simultaneous measurement of several tumor markers. *Clin. Chem.* 54:1481–1488.

41. Wu, J., Z. F. Fu, F. Yan, and H. X. Ju. 2007. Biomedical and clinical applications of immunoassays and immunosensors for tumor markers. *Trends Anal. Chem.* 26:679–688.

42. Wu, T. L., Y. C. Sun, P. Y. Chang, K. C. Tsao, C. F. Sun, and J. T. Wu. 2003. Establishment of ELISA on 384-well microplate for AFP, CEA, CA 19-9, CA 15-3, CA 125, and PSA-ACT: Higher sensitivity and lower reagent cost. *J. Clin. Lab. Anal.* 17:241–246.

43. Zheng, G., F. Patolsky, Y. Cui, W. U. Wang, and C. M. Lieber. 2005. Multiplexed electrical detection of cancer markers with nanowire sensor arrays. *Nat. Biotechnol.* 10:1294–1301.

32

Characterization of Cancer Cells Using Electrical Impedance Spectroscopy*

Dorielle Price, Abdur Rub Abdur Rahman, and Shekhar Bhansali

CONTENTS

32.1 Introduction

The magnitude of cancer-related deaths is a result of late diagnosis and the fact that cancer is challenging to treat, due to the highly heterogeneous nature of the tumor. All cancer cells do not have equal malignant or invasive potential and thus need to be uniquely treated [33]. In order to characterize and treat individual cells based on their malignant potential, it is important to have a measurement technique with enhanced spatial resolution and increased sensitivity, ideally at the single cell level. This requires the study of individual or small groups of cells that make up the entire tissue mass. Toward this end, research has been performed to investigate methods to enhance cancer cell detection and provide fundamental information about

---

* Portions of this chapter were reprinted, with permission, from Refs [23,28].

cancer cell characteristics, through the design and optimization of a whole-cell biosensor and development of an automated cell culture monitoring system. Application of the optimized system can advance drug discovery and ultimately lead to implementation of personalized healthcare, where an individual's cells can be subjected to potential drugs to monitor response to therapy, thus leading to personalized medicine.

32.1.1 Morphology and Impedance Characteristics of Cancer Cells

Cellular morphology is one of the most important parameters in cancer biology [13]. In general, tumor masses are less organized and structured than normal tissues. There are various degrees of abnormality of cells within tumors. Slightly abnormal tumors can contain only an excessive amount of cells, whereas in more abnormal tumors, the cells take on an irregular appearance. These cytological changes may include variability in the nuclear size, increased mitotic activity, and lack of cytoplasmic features [35], all of which can be detected with impedance measurements [1,6,10,17]. When cells attach and spread onto the electrode surface, the measured impedance increases because the cell membranes act as insulators and block current flow. Similarly, micromotion of the cells alters the measured impedance, as movement and changes of the structural features alter or restrict the path of the applied electrical signal. These structural changes in individual cells have a major impact on the tumor mass as a whole. To distinguish these behavioral characteristics, a space and time-resolved measurement system is needed.

Beginning in the 1920s with Frick and Morse, significant capacitive differences have been found between malignant breast tumors and normal tissues [37]. Due to increased cellular water and salt content, altered membrane permeability, and changed packing density, malignant tumors typically exhibited a lower electrical impedance [37]. Han et al. demonstrated that cancer cells at different stages had unique impedance signatures. Since the cell membrane of cancer cells are more permeable, a decrease in capacitance resulted, relative to the stage of cancer [8]. A study by Beetner et al. demonstrated differences between basal cell carcinoma (BCC), benign lesions, and normal skin using electrical impedance measurements. They noted that BCC had a larger nuclei and smaller intercellular spacing, causing a difference in measured impedance [3].

32.1.2 Impedance Spectroscopy Theory

Electrical impedance spectroscopy (EIS) can be used to characterize biological cells over a wide range of frequencies. Impedance measurements of adherent cells have been shown to detect changes in cell morphology on the order of nanometers [16], providing sensitivity much greater than that obtained through optical inspection.

Figure 32.1 shows a plot of the frequency dependence of relative permittivity and specific conductivity of biological tissue. This plot shows three regions, originally defined by Schwan [31], as alpha (α),

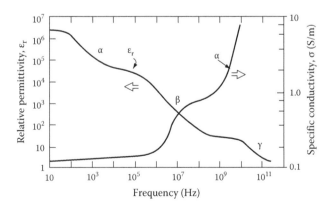

FIGURE 32.1 Frequency dependence of the relative permittivity and specific conductivity of complex biological tissue. (Reprinted from Springer Science+Business Media: *Applied Bioelectricity: From Electrical Stimulation to Electropathology*, 1st edn., 1998, Reilly, J.P., Antoni, H., Chilbert, M.A., and Sweeney, J.D.)

beta (β), and gamma (γ) dispersions, at low, mid, and high frequencies, respectively. As measurement frequency increases, tissue conductivity increases and permittivity decreases. Each of these dispersions is caused by a physical process. The alpha dispersion, which occurs below 10 kHz, is due to ionic diffusion in layers of the cell culture medium [21]. Within the beta dispersion region, between 1 and 100 MHz, the electric field begins to penetrate the cell interior. According to Tamura et al., the beta dispersion is dependent on the charging properties of the cell membranes, along with intra- and extracellular fluids [32]. Above 10 GHz (gamma dispersion), it has been shown that measured quantities tend to vary over frequency in proportion to the water content of the cells [30].

EIS of cell cultures has the following advantages: (a) small, ~10 mV, applied AC signals for nondestructive, noninvasive measurements within a pseudo-linear region [19], (b) ability to provide quantitative information about cell morphology, motility, attachment and spreading, mitosis, and apoptosis, and (c) sensitive measurements in a label-free and mediator-free environment. In addition to the ease of incorporation into a portable system, the potential has to be cost-efficient and miniaturized, and the ability to automate long-term measurements and data analysis.

Impedance data can be used to differentiate between cells without the need for fluorescent markers and other probe molecules or particles, which can adversely affect the cells. This is a major advantage of IS, thus eliminating the need for pre- and postprocessing of cells and providing rapid, quantitative analysis.

32.1.2.1 Electric Cell–Substrate Impedance Sensing

An established bioimpedance measurement technique for cell monolayers is known as electric cell–substrate impedance sensing (ECIS). ECIS was developed by Giaever and Keese in 1991 to study the electrical properties of adherent cells. A small, 250 μm diameter, gold working electrode (WE) and a larger counter electrode (CE) are used in the two-electrode system. A small AC electrical signal is passed between the two electrodes and the resulting impedance is measured. When cells attach and spread onto the surface of the electrodes, the measured impedance increases because the cell membranes act as insulators and block current flow. The cells are not adversely affected by the small signal, thus making the measurement noninvasive and nondestructive and facilitating real-time, live cell recordings. A photograph of a commercial ECIS-8W1E device is shown in Figure 32.2. ECIS can be used to monitor attachment and spreading, invasion, barrier function, toxicology, proliferation, motility, cell growth, among other applications. Monitoring of these basic cell functions, along with toxicology studies, can be used to characterize cancer cells and potential drugs.

The commercial ECIS measurement system uses mathematical modeling to fit and extract the parameters of the cell–electrode system. The data are analyzed as a function of frequency to obtain morphological information including barrier resistance (R_b), cell membrane capacitance (C_m), and the restraint of current flow beneath the cells (α). Details on the mathematical model can be found in [6,14,15].

Another modeling technique is equivalent circuit modeling. System parameters such as solution resistance (R_s), double-layer capacitance (C_{dl}), cell membrane capacitance (C_{cell}), and cell resistance (R_{cell}) are obtained using this technique. Parameters can be extracted using complex nonlinear least squares

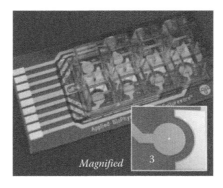

FIGURE 32.2 Photograph of commercial ECIS device.

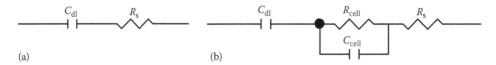

FIGURE 32.3　Equivalent circuits for systems (a) without cells and (b) with adherent cells.

(CNLS) fitting. Different circuits can be used to fit a single dataset; therefore, it is necessary to derive a circuit with the least number of physically relevant components in order to obtain an accurate model of the system. The equivalent circuits for electrode systems with and without cells are shown in Figure 32.3.

32.1.3 Requirements and Challenges of Microelectrode Impedance Biosensors

In a controlled, static environment, in which cells are not flowing through, cell characterization can provide fundamental information about cells' long-term behavior, functional characteristics, and growth cycle. This fundamental information can prove to be invaluable in applications such as cancer research, toxicology and cytopathic studies, drug screenings, wound healing, and personalized therapy. Development of a practical biosensor for cancer research requires (a) sensor design optimization [22,27], (b) measurement setup with reduced noise and enhanced spatial resolution [28], and (c) sensitive detection and differentiation of cancer cells.

A primary hurdle in achieving spatial resolution, using distributed impedance mapping, is the additional hardware requirement of switching between electrodes without introducing measurement parasitics. Additionally, data collection, processing, analysis, and visualization are difficult to implement and integrate for large multidimensional datasets, especially in a real-time environment. IS data are predominantly interpreted by parameter extraction via equivalent circuit modeling or mathematical modeling. Traditional spectroscopic measurements involve few datasets which are fit to models of choice to extract parameters. A multiple electrode system performing impedance spectroscopy (IS) in real-time generates large datasets, where each electrode datum has to be individually fit to a model for parameter extraction. This task is cumbersome unless automated and hence is omitted in most commercial systems, where a single frequency data point is collected as a function of time. Even if data are recorded for the entire frequency spectrum, it is up to the user to process each dataset and extract the relevant parameters. Hence, there is a need for a system that can perform automated data analysis and parameter extraction for multielectrode measurements and present each of these parameters as a function of culture monitoring time so as to allow multidimensional analysis and interpretation of results. A key challenge in the implementation of automated parameterization of evolving cell cultures is that the model representing the system continuously changes with changes in cell morphology.

32.2 Materials

- ECIS-8W1E electrode arrays (Applied Biophysics,Troy, NY)
- Ovarian cancer cells, OvCa429 (from Samuel Mok, Harvard Medical School, Boston, MA)
- 100 mm Pyrex 7740 DSP 700 μm (University Wafers, South Boston, MA)
- NR1-3000PY photoresist (Futurrex, Franklin, NJ)
- TG-25E RTU gold electroplating solution (Technic Inc., Providence, RI)
- Cloning cylinders (Fisher Scientific 09-552-22)
- Cell culture medium (ATCC, Manassas,VA)
- M199 and MCDB 105 cell culture medium (Sigma, St. Louis, MO)
- Fetal calf serum (Sigma)
- Glutamine (Sigma)
- Penicillin (Sigma)
- Streptomycin (Sigma)
- Trypsin (Sigma)
- Hank's balanced salt solution (HBSS) (Mediatech, Inc., Herndon, VA)

32.3 Methods

The following sections describe research that was performed by the BioMEMS and Microsystems group at the University of South Florida. The goals of the research were to (a) extend and characterize the frequency range of traditional ECIS measurements [23], and (b) enhance spatial resolution of bioimpedance measurements using a multielectrode array to "map" the impedance distribution of cell cultures [25].

32.3.1 High-Frequency Impedance Characterization of Ovarian Cancer Epithelial Cell

32.3.1.1 Cell Culture Procedure

Ovarian cancer (OvCa429) cells were grown in M199 and MCDB 105 (1:1) supplemented with 10% fetal calf serum, 2 mM L-glutamine, 100 U/mL penicillin, and 100 μg/mL streptomycin under 5% CO_2 and 37°C high-humidity atmosphere. Cell culture medium was changed every 2–3 days, depending on the rate of cell growth, and cells were subcultured at 80% confluence using 5× trypsin solution. Forty-eight hours after inoculating cells into electrode wells at 8×10^4 cm^{-2}, confluent cell layers were formed and the normal medium was replaced by HBSS without phenol red. Cells were seeded and measured in a commercially available electrode array, Applied Biophysics' ECIS-8W1E.

32.3.1.2 Equivalent Circuit Model

Figure 32.4 is a schematic of the equivalent circuit model of a partially coated metal electrode in a solution. R_{exp} and CPE_{coat} are the resistance and capacitance (constant phase element [CPE]) of the exposed and coated areas of the WE, respectively. CPE_{dl} is the double-layer capacitance of the exposed portion of the electrode. In most practical electrochemical situations, capacitive effects are represented by CPEs [9]. The impedance of a CPE is given by the expression $1/(A(j\omega)^n)$, where A is the magnitude of the element and $\omega = 2\pi f$ is the angular frequency, where f is the frequency in hertz. The parameter n is such that if $n = 1$, the impedance of a CPE is that of an ideal capacitor, and when $n = 0$, the CPE is a pure resistor. It has been stressed that any attempt to characterize the interfacial double layer exhibiting

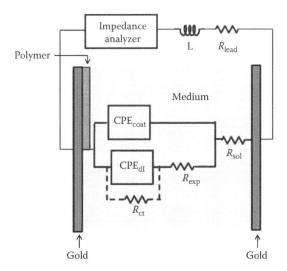

FIGURE 32.4 Illustration of equivalent circuit components of a partially coated metal electrode. R_{sol} is the solution resistance, CPE_{coat} is the CPE representing coating effect, CPE_{dl} is the CPE representing double-layer effect, R_{exp} is the spreading resistance through the area of the gold WE directly in contact with the electrolyte, and R_{ct} is the charge transfer resistance, applicable only in the case of a corroding electrode or in the presence of a redox species.

a CPE behavior by electrical capacitance is misleading and that such a simplifying assumption causes a modeling error and influences all parameters of the system [36].

32.3.1.3 Impedance Measurements on ECIS-8W1E

Measurements were performed with an Agilent 4294A Impedance Analyzer in connection with a Cascade Summit 11000 probe station. A small 10 mV signal was applied to the system between 100 Hz and 10 MHz. The impedance data were modeled using equivalent circuit modeling through the LEVM, a CNLS simulation and fitting software [18], which uses the Levenberg–Marquardt algorithm [20] for fitting and optimization.

Baseline measurements were performed with HBSS. The ECIS culture wells were filled with HBSS and allowed to equilibrate overnight before impedance measurements were conducted. Subsequently, cell measurements were performed with OvCa429 cells, which were allowed to grow to confluence across the WE and CE. Impedance data were collected at room temperature.

The cell layer can be conceptualized as a blanket covering throughout the surface area of the electrodes. Consequently, it is expected that the impedance of the cell layer should be in series with that of the electrode–electrolyte interface. Since the cells form close contact with the electrode, the interfacial impedance parameters will also be altered. The cell layers are usually represented by a parallel resistor–CPE combination [34]. The CPE of the cell layer represents the averaged effect of the membrane capacitance over the area of the electrodes, whereas the resistance is due to cell–cell junction constricting the flow of ions and the intracellular resistance.

32.3.2 CellMap: Multielectrode Impedance Spectroscopy Measurements

Multielectrode systems offer several advantages over two-electrode systems (i.e., ECIS), such as (a) spatial physiological sampling of in vitro cell systems, (b) multipoint sampling for statistical correlation, (c) in pathological cell systems with distinguished growth patterns from the normal cells [5], spatial sampling will provide cell density distribution for heterogeneous cell populations such as cancer cells which is not possible using conventional two-electrode systems, and (d) using spatially distributed electrodes, it is possible to determine the velocity of cell migration [12]. An impedimetric sensor system has been demonstrated that can map the cell distribution in the culture space by recording the impedance at various sites in the culture space using a radial electrode array and IS. This system facilitates multidimensional (two-dimensional [2D] space and time) characterization of the cell cultures [25]. The multielectrode array–based IS system maps the impedance distribution of the cell culture system, facilitating continuous monitoring of the adhesion, motility, distribution, and confluence of the cells. The CellMap system was used for the study of cell adhesion, proliferation, detachment, and cell–drug interaction of OvCa429 ovarian cancer cells.

32.3.2.1 Electrode Design and Fabrication

An eight-electrode array was used in monitoring cell behavior by cell–substrate sensing. This arrangement has the advantage of simplified channel compensation, which is a procedure used to account for fixture and lead parasitics during impedance measurements. Each electrode of the multielectrode array is 5000 μm in length and 50 μm in width. Out of the total length of the electrode, ~2500 μm is in contact with the electrolyte inside the cell culture cylinder.

The electrode array fabrication process is described later and in [26]. A thin (30 nm) film of chromium (Cr) was deposited, followed by a 100 nm thick film of gold (Au) by thermal evaporation. The chromium layer serves as an adhesion promoter between the glass surface and gold. Next, photolithography steps were performed using 3000PY negative photoresist. The resist was patterned on top of the Cr/Au layer using a bright-field mask with electrode patterns. This was followed by a timed, wet-etching of the Cr/Au layer. The resulting pattern was a circular array of eight electrodes. Each one of the eight electrodes serves as an independent WE during multielectrode scanning. Several electrode

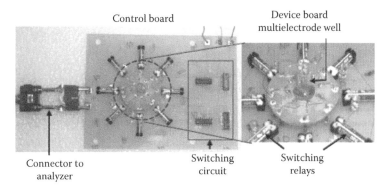

FIGURE 32.5 Photographs of the control and device boards. The carrier board containing the multielectrode array, fitted with a culture well and CE, plugs into the control board which contains the switching electronics for multielectrode scanning. The control board has two terminals which facilitate bipolar impedance measurements.

arrays were fabricated on a 2 in. diameter glass wafer. The glass wafer was then diced to yield the individual electrode arrays.

A CE of much larger surface area than the WE was manufactured by electroplating gold on a group of five-soldered brass cylinders of 1 mm diameter and 0.5 cm length each. The thickness of electroplated gold on the brass cylinders was ∼5 μm. This created a large CE surface area in comparison with the WE area. Such an arrangement of CE and WE minimizes the impedance contribution of the CE to the overall impedance of the bipolar system and thus simplifies the data analysis and parameterization procedures. The devices assembled with CE were then solder-mounted on to a carrier board manufactured using the printed circuited board (PCB) technology. The carrier board has the same electrode arrangement as the device but of a slightly larger footprint, thus allowing easier handling and connectivity to the control board. Connector pins soldered to the carrier board corresponded to the connector pin receptacles soldered to the control board.

In the final step of the assembly process, a glass-cloning cylinder of 500 μL capacity was fitted to the device to serve as a cell culture well. The assembled device modules (carrier board) being separate from the control board allowed easy interchange of devices between experiments. Figure 32.5 shows the assembled carrier board with devices and the control board.

Upon inoculation of the cell suspension on the microfabricated device, the cells will adhere to the glass substrate and proliferate. Since the Au electrodes are fabricated on the surface of the glass substrate, encroachment of the cells on the Au surface causes a change in the impedance between the WE and CE. The WE was switched sequentially to record impedances of all eight electrodes against the common CE. To facilitate this switching, an electronic switching module was developed which automatically senses the completion of an impedance scan and switches the WE to the next sequential electrode. A set of eight such impedance scans during a time instance constitutes a measurement frame. Several such "frames" are recorded at user-defined time intervals for in vitro cell culture monitoring.

32.3.2.2 Impedance Measurements on Multielectrode Devices

Impedance was measured using the frequency response analyzer module (FRA2) of the Autolab PGSTAT 30 potentiostat/galvanostat. The instrumentation was set up to acquire data automatically over a period of time using the programming capability of the FRA in PGSTAT. The impedance was recorded in the frequency range between 25 Hz and 1 MHz. A 10 mV amplitude signal was used as the excitation potential.

Impedance measurements were recorded on two separate devices (control and variable) beginning at ∼80% confluence of the OvCa429 culture. After 5 h of recording measurements at an interval of ∼1 h, 20 μL of trypsin was added to the variable cell culture medium. Trypsin was not added to the control. The impedance was recorded under room conditions and the devices were placed back in the incubator after each measurement.

32.4 Results

32.4.1 High-Frequency Impedance Characterization of OvCa429 on ECIS-8W1E Electrodes

Figure 32.6 is the Bode diagram of device 3 (D3) of ECIS-8W1E with OvCa429 cells and with medium only. The high-frequency region displays similar characteristics in HBSS and OvCa429 systems, as is evident from the impedance magnitude and phase angle beyond ~1 MHz. This implies that this section of the impedance spectra is dominated by the coating impedance. Hence, it is attributed to the polymer-covered portions of the gold WE. The impedance magnitudes and phase angle of cells and medium are considerably different at frequencies below 1 MHz. The addition of cell layers has not only contributed two additional impedance elements to the circuit, but also modified the other parameters such as double-layer capacitance and pore resistance. The presence of cells has considerably altered the interfacial space charge distribution of the electrode–HBSS system. This is because the cells form an intimate and adherent layer to the substrate, thereby altering the equilibrium charge density distribution that extends a few micrometers into the bulk of the electrolyte. The cell-layer impedance appears as a single slope. This characteristic is due to the overlap of time constants of the RC elements in the equivalent circuit.

The generally accepted value of cell membrane capacitance is $1\,\mu F/cm^2$ [7]. Cell membranes are composed of phospholipid bilayers that act as insulators, except for ion channel openings that conduct transcellular currents. Unlike the membrane capacitance, cell-layer capacitance indicates the capacitance of a layer of interconnected cells. Cell layers consist of transcellular junctions, folded cell walls, and in some cases, multiple cell layers that lead to a capacitance value other than that of a flat sheet cell wall capacitance of $1\,\mu F/cm^2$. Since the cell wall capacitance is in series with the coating and double-layer capacitance in the coated regions, the smallest of the capacitances, namely, the coating capacitance, dominates the impedance. Consequently, the effect of the membrane capacitance will be visible in the frequencies higher than those at which the double-layer capacitance appears. OvCa429 ovarian cancer epithelial cell layer was characterized in terms of electrical equivalent circuit elements. The trans-layer resistance and capacitance were determined to be $152 \pm 59\,\Omega\cdot cm^2$ and $8.5 \pm 2.4\,\mu F/cm^2$, respectively. Equivalent circuit models developed for device containing HBSS and OvCa429 cells indicate that the polymer-coated regions contribute to the impedance spectrum above 1 MHz.

32.4.2 CellMap: Multielectrode Impedance Spectroscopy Measurements

The previous section described the results of a single WE–CE system. This is a simplified system, in which impedance data are obtained from a single location within the cell culture. The following sections

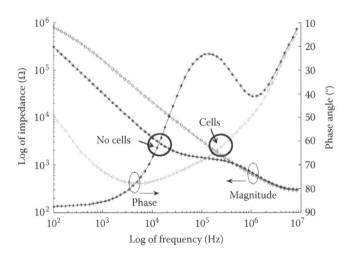

FIGURE 32.6 Bode diagram of device 3 of ECIS-8W1E with and without OvCa429 cell layer.

describe the results of a multielectrode array (eight WEs) with a common CE. This design allows one to measure the impedance distribution of a cell culture, in multiple locations. An automated data analysis program is used to accommodate the large datasets acquired from multifrequency, multielectrode measurements.

32.4.2.1 Monitoring the Adhesion and Proliferation of OvCa429

Figure 32.7 shows the changes in the phase angle of impedance as a result of cell adhesion and proliferation for one electrode of the multielectrode array. With the progression of culture time, a second phase minimum is seen to develop in the frequency range between ~10 and 300 kHz. This feature is absent in the control, from which it can be inferred that it is due to the contribution of cells' adhesion to the substrate.

32.4.2.2 Effect of Trypsin on Confluent Cell Culture

In order to affirm the utility of impedance mapping for tracking physiological changes, the effect of trypsin on the confluent cell culture was studied. OvCa429 cells (10^6 cells/cm^2) were seeded onto two separate electrode arrays and incubated under the same set of conditions until 80% confluence (~40 h post-inoculation). One served as a control and the other was used to monitor the effect of trypsin on the confluent cell culture, by adding 20 μL of trypsin to the cell culture medium. Trypsin was added after 5 h of monitoring the impedances of the confluent control and variant (trypsinization) cultures. After addition of trypsin to the OvCa429 culture, the phase angle shape displayed significant change and transition toward the shape which is representative of the absence of cells. It was observed that trypsin caused detachment of the cells from the substrate as expected. Further monitoring of control and trypsinized culture did not indicate significant changes.

32.4.2.3 Parameterization of Impedance Data to Monitor the Effect of Trypsin

The impedance data of control and trypsinized cell cultures were parameterized to monitor changes in specific components of the electrochemical system using the equivalent circuit modeling. Figure 32.8a and c are the extracted cell-layer parameters, namely, cell-layer resistance and cell-layer CPE magnitude of control. Figure 32.8b and d are the aforementioned parameters for the trypsinized cell culture.

The addition of trypsin at 5 h into the observation period causes a decrease in the cell-layer resistance and CPE followed by the elimination of these parameters from the model representing the data. Hence,

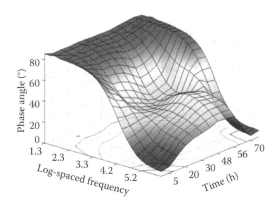

FIGURE 32.7 Changes in the phase angle of electrode designated 1 in the multielectrode array with the progression of culture time. A second phase minimum is seen to develop as a consequence of cell adhesion and proliferation. The phase angle is restored back to the pre-culture level upon cell detachment. The sign of the phase angle has been reversed for graphing convenience.

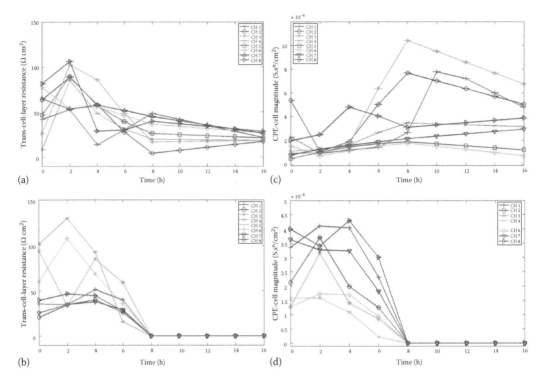

FIGURE 32.8 (a) Cell-layer resistance of control, (b) cell-layer resistance of the trypsinized culture (trypsin added after 5 h of monitoring), (c) cell-layer CPE magnitude of control, and (d) cell-layer CPE magnitude of the trypsinized culture. The vertical lines indicate the time instance at which trypsin was added to the cell culture. The cell-layer parameters disappear from the electrochemical model upon trypsinization, indicating cell detachment.

this could be a suitable way to study cytotoxicity. A similar trend is seen for the cell-layer CPE magnitude and the power factor (data not shown).

The electrode-to-electrode variability of the parameters for the eight electrodes emphasizes the ability of the CellMap system in resolving cellular behavior in the culture subspace. The fact that the confluent culture is not stagnant but continuous motion [16] also contributes toward the variability of these parameters in time and space. In addition, the electrode arrays used in this study were manufactured and assembled separately which also introduce a small device-to-device variability of parameters.

The penetration depth of the AC signal increases with frequency. Consequently, the information available from spectroscopy data is dependent on the frequency range of investigation. In the electrode–cells–electrolyte system, the lower frequencies reflect the cell–substrate interactions (interfacial), whereas the higher frequencies reflect the cell-layer (morphological) properties.

Cell adhesion and motility are well represented at the lower frequency range by way of alterations in double-layer parameters, whereas the cell-layer capacitance and tight-junction resistance are reflected at higher frequencies by the addition of cell-layer parameters. The spreading resistance parameter represents the opposition to the flow of current from the working to the CE. Since the field lines spread in planar electrodes, rather than remaining confined to the geometry of the electrodes (e.g., ideal parallel plate electrodes), it is referred to as spreading resistance. In the cell culture well, the spreading resistance depends on the available ionic charges as well as physical obstruction to the flow of these charges. The available ionic charge depends on the concentration of ionic species at any given point, which is dependent upon the rate of consumption and metabolism by the cells. The physical obstruction to the flow of the ions stems from the close approach of anchorage-dependent cells to the substrate, whereby the ions have to circumvent the cells at low frequencies to travel to the CE. Thus, the spreading resistance is an indicator of cell metabolism and cell–substrate interaction. Similarly, the double-layer CPE is also a function of the two factors mentioned earlier, although it is likely (based on the cell–substrate separation) to be influenced more by the change in the diffusion layer capacitance due to the presence of the cell membrane.

The resistance and the CPE parameters representing the cell layers display a clear correlation between cell-layer formation, confluence, and detachment. The cell junction resistance is dependent upon the cell–cell gap also referred to as the gap junctional resistance. The higher this parameter, the closer the cells are. It is highest at confluence which is consistent with optical observations. Beyond confluence, the layer begins to detach due to nonsupply of fresh nutrients and upon trypsinization. The cell layer–related parameters are not required in the fitting of data when the cells are suspended and when detached by trypsinization, as evident from the shape of the phase angle curves.

Physiological processes, both normal and affected (chemical stimulus), have been successfully visualized and quantified by spatiotemporal impedance mapping. This technique provides information on cellular behavior as well as morphology. It has the potential for remote monitoring of cell cultures in long-term studies in conjunction with or as an alternative to optical monitoring. Impedance mapping can potentially be more representative of the real system than optical techniques due to its label-free nature. Labels, besides involving additional sample preparation and processing, may interfere with the system under investigation and hence distort the analysis. IS on the other hand has proven to be noninvasive (small signal) and sensitive, with the ability to resolve microstructural domains in dielectric and polymeric materials. The impedance mapping system presented in this work provides a noninvasive, visual, and quantitative method for cellular monitoring with a wide application potential in areas such as drug discovery and cancer research. This system is scalable to higher numbers of electrodes for enhanced spatial resolution.

32.5 Discussion

Generally, the methods employed to study cell–substrate interactions can be broadly classified as either label and reporter based or label free. Label and reporter-based technologies require elaborate sample preparation, pre- and posttreatment. Label-free technologies are preferred due to their noninvasiveness and noninterference with biological functions [24].

The most widely used method for assessing and quantifying cell adhesion and spreading is to seed the cells onto the surface and allow the cells to adhere and spread for a length of time. The unbound cells are then washed away in a buffer solution. The attached cells are labeled with a fluorescent reagent and viewed in an inverted or confocal microscope. Alternatively, the cells can be stained and counted using light microscopy. The cells can also be pre-labeled with a fluorescent dye. After washing the unbound cells, the bound cells are quantified using a plate reader. An additional method for assessing the role of adhesion proteins is to coat the surface with antibodies or peptides that are specific for various receptors and then seed the cells that are expressing the appropriate receptors. The interaction of the receptor on the cell surface with the antibody or peptide can then be detected using one of the aforementioned methods.

The assays described earlier are known as endpoint assays. They provide one-time information about the cell system at the end of an observation point [2]. They require pre-labeling, post-labeling, fixation, and permeabilization. Additionally, the peptides, dyes, or proteins introduced for signal transduction may interfere with the biological function and add background noise to the data. Alternatively, electrical impedance–based techniques offer several advantages such as (a) noninvasive/nondestructive, (b) label-free, (c) real-time, and (d) dynamic monitoring of cell culture systems. Impedance-based cell–substrate sensing is fast emerging as a viable and reliable technique for investigating cell–substrate interactions. A major contribution of the aforementioned multielectrode impedance research is that both cell–cell and cell–substrate interactions are recorded and analyzed with spatiotemporal resolution. This has led to the development of an automated cell culture monitoring "CellMap" system. The cell-growth mapping of this system is able to resolve nonuniform cell distributions in the culture, which may be useful in differentiating normal and pathological cells, continuously and label-free.

32.6 Future Trends

Accurate and reliable measure of various cellular/tumor responses subjected to medicinal drugs has been a major indicator of drug efficacy. In particular, investigation of nonspecific (untargeted) cell interaction with the remedial components has been in great demand and is crucial within clinical diagnostics, drug discovery, and innovative medicine. Future trends are focusing on characterizing and comparing the drug response of three-dimensional (3D) tumor spheroids versus the more popularly studied 2D cell cultures. Most tissue consists of distinct 3D spatial arrangement of cells, in close contact and communication with each other. In particular, 3D architecture and communication with extracellular matrix and stromal cells are essential to reproduce in vivo like behavior in in vitro systems. Various studies have shown that 3D spheroidal cell cultures demonstrate significantly more in vivo like behavior than 2D cultures [4,11]. Notwithstanding these developments, significant challenges still exist in using these models reliably in the laboratory or clinical setting.

Drug development, drug delivery, and personalized cancer treatment have been hampered by the poor predictability of current chemosensitivity and resistivity assays. The importance of new ex vivo assays encompassing both the tumor and its microenvironment has been increasingly recognized. Animal models, while invaluable for drug discovery, are highly complex and are not predictive of human effects due to organism-level variations and unknown systemic interactions. 2D, adherent monocultures under static incubation environments, while relatively easy and in practice today, are phenotypically very different and do not reproduce clinical features such as drug resistance and clonal dominance.

REFERENCES

1. Arndt, S., J. Seebach, K. Psathaki, H.-J. Galla, and J. Wegener. 2004. Bioelectrical impedance assay to monitor changes in cell shape during apoptosis. *Biosens. Bioelectron.* 19:583–594.
2. Atienza, J. M., J. Zhu, X. Wang, X. Xu, and Y. Abassi. 2005. Dynamic monitoring of cell adhesion and spreading on microelectronic sensor arrays. *J. Biomol. Screen.* 10:795–805.
3. Beetner, D. G., S. Kapoor, S. Manjunath, X. Zhou, and W. V. Stoecker. 2003. Differentiation among basal cell carcinoma, benign lesions, and normal skin using electric impedance. *IEEE Trans. Biomed. Eng.* 50:1020–1025.
4. Carlsson, J. and T. Nederman. 1989. Tumour spheroid technology in cancer therapy research. *Eur. J. Cancer Clin. Oncol.* 25:1127–1133.
5. Evan, G. I. and K. H. Vousden. 2001. Proliferation, cell cycle and apoptosis in cancer. *Nature* 411:342–348.
6. Giaever, I. and C. R. Keese. 1991. Micromotion of mammalian cells measured electrically. *Proc. Natl Acad. Sci. U.S.A.* 88:7896–7900.
7. Grimnes, S. and O. G. Martinsen. 2000. *Bioimpedance and Bioelectricity Basics.* Academic Press, San Diego, CA.
8. Han, A., L. Yang, and A. B. Frazier. 2007. Quantification of the heterogeneity in breast cancer cell lines using whole-cell impedance spectroscopy. *Clin. Cancer Res.* 13:139–143.
9. Huang, V. M.-W., V. Vivier, I. Frateur, M. E. Orazem, and B. Tribollet. 2007. The global and local impedance response of a blocking disk electrode with local constant-phase-element behavior. *J. Electrochem. Soc.* 154:C89–C98.
10. Keese, C. R., K. Bhawe, J. Wegener, and I. Giaever. 2002. Real-time impedance assay to follow the invasive activities of metastatic cells in culture. *BioTechniques* 33:842–850.
11. Kunz-Schughart, L. A., J. P. Freyer, F. Hofstaedter, and R. Ebner. 2004. The use of 3-D cultures for high-throughput screening: The multicellular spheroid model. *J. Biomol. Screen.* 9:273–285.
12. Linderholm, P., T. Braschler, J. Vannod, Y. Barrandon, M. Brouard, and P. Renaud. 2006. Two-dimensional impedance imaging of cell migration and epithelial stratification. *Lab Chip* 6:1155–1162.
13. Liu, Q., J. Yu, L. Xiao, J. C. O. Tang, Y. Zhang, P. Wang, and M. Yang. 2009. Impedance studies of bio-behavior and chemosensitivity of cancer cells by micro-electrode arrays. *Biosens. Bioelectron.* 24:1305–1310.
14. Lo, C.-M. and J. Ferrier. 1998. Impedance analysis of fibroblastic cell layers measured by electric cell–substrate impedance sensing. *Phys. Rev. E* 57:6982.

15. Lo, C.-M., C. Keese, and I. Giaever. 1995. Impedance analysis of MDCK cells measured by electric cell–substrate impedance sensing. *Biophys. J.* 69:2800–2807.

16. Lo, C.-M., C. R. Keese, and I. Giaever. 1993. Monitoring motion of confluent cells in tissue culture. *Exp. Cell Res.* 204:102–109.

17. Luong, J. H. T., M. Habibi-Rezaei, J. Meghrous, C. Xiao, K. B. Male, and A. Kamen. 2001. Monitoring motility, spreading, and mortality of adherent insect cells using an impedance sensor. *Anal. Chem.* 73:1844–1848.

18. Macdonald, J. R. 2005. Impedance spectroscopy: Models, data fitting, and analysis. *Solid State Ionics* 176:1961–1969.

19. McAdams, E. T., A. Lackermeier, J. A. McLaughlin, D. Macken, and J. Jossinet. 1995. The linear and non-linear electrical properties of the electrode–electrolyte interface. *Biosens. Bioelectron.* 10:67–74.

20. Mor, J. J. 1977. The Levenberg–Marquardt algorithm: Implementation and theory. *Lecture Notes Math* 630:105–116.

21. Polk, C. and E. Postow. 1996. *Handbook of Biological Effects of Electromagnetic Fields*, 2nd edn. CRC Press, Boca Raton, FL.

22. Price, D., A. Rahman, and S. Bhansali. 2009. Design rule for optimization of microelectrodes used in electric cell–substrate impedance sensing (ECIS). *Biosens. Bioelectron.* 24:2071–2076.

23. Rahman, A., L. Chun-Min, and S. Bhansali. 2009. A detailed model for high-frequency impedance characterization of ovarian cancer epithelial cell layer using ECIS electrodes. *IEEE Trans. Biomed. Eng.* 56:485–492.

24. Rahman, A. R. A. 2007. Cellmap: An automated multielectrode array cell culture analysis system based on electrochemical impedance spectroscopy. Dissertation. University of South Florida, Tampa, FL.

25. Rahman, A. R. A. and S. Bhansali. 2007. Data acquisition and impedance mapping using a multielectrode bioimpedance spectroscopy system. In *13th International Conference on Electrical Bioimpedance and the 8th Conference on Electrical Impedance Tomography*, pp. 221–224, August 29–September 2, 2007, Graz, Austria.

26. Rahman, A. R. A., C.-M. Lo, and S. Bhansali. 2006. A micro-electrode array biosensor for impedance spectroscopy of human umbilical vein endothelial cells. *Sens. Actuat. B Chem.* 118:115–120.

27. Rahman, A. R. A., D. T. Price, and S. Bhansali. 2007. Effect of electrode geometry on the impedance evaluation of tissue and cell culture. *Sens. Actuators B* 127:89–96.

28. Rahman, A. R. A., J. Register, G. Vuppala, and S. Bhansali. 2008. Cell culture monitoring by impedance mapping using a multielectrode scanning impedance spectroscopy system (CellMap). *Physiol. Meas.* 29:S227.

29. Reilly, J. P., H. Antoni, M. A. Chilbert, and J. D. Sweeney. 1998. *Applied Bioelectricity: From Electrical Stimulation to Electropathology*, 1st edn. Springer, New York.

30. Schepps, J. and K. Foster. 1980. The UHF and microwave dielectric properties of normal and tumour tissues: Variation in dielectric properties with tissue water content. *Phys. Med. Biol.* 25:1149–1159.

31. Schwan, H. P. 1963. Electrical characteristics of tissues. *Biophysik* 1:198–208.

32. Tamura, T., M. Tenhunen, T. Lahtinen, T. Repo, and H. P. Schwan. 1994. Modelling of the dielectric properties of normal and irradiated skin. *Phys. Med. Biol.* 39:927.

33. Vassilopoulos, A., R. Wang, C. Petrovas, D. Ambrozak, R. Koup, and C. Deng. 2008. Identification and characterization of cancer initiating cells from BRCA1 related mammary tumors using markers for normal mammary stem cells. *Int. J. Biol. Sci.* 4:133–142.

34. Wegener, J., S. Zink, P. Rösen, and H.-J. Galla. 1999. Use of electrochemical impedance measurements to monitor β-adrenergic stimulation of bovine aortic endothelial cells. *Pflüg. Arch. Eur. J. Physiol.* 437:925–934.

35. Weinberg, R. A. 2007. *The Biology of Cancer*. Garland Science. Taylor & Francis Group, New York.

36. Zoltowski, P. 1998. On the electrical capacitance of interfaces exhibiting constant phase element behaviour. *J. Electroanal. Chem.* 443:149–154.

37. Zou, Y. and Z. Guo. 2003. A review of electrical impedance techniques for breast cancer detection. *Med. Eng. Phys.* 25:79–90.

33

Electrochemical Immunosensor for Detection of Proteic Cancer Markers

Alex Fragoso and Ciara K. O'Sullivan

CONTENTS

33.1 Introduction

Biosensors are self-contained integrated devices capable of providing specific quantitative or semiquantitative analytical information using a biological recognition element (biochemical receptor), which is retained in direct spatial contact with transduction element (usually physical, chemical, or electrical) responsible for detecting the biological reaction and converting it into a signal which can be processed in response to the concentration or level of either a single analyte or a group of analytes [1].

    Electrochemical transducers are one of the oldest and most commonly used in immunosensors, with the sensing principle being based on the electrical properties of the electrode or solution that is affected

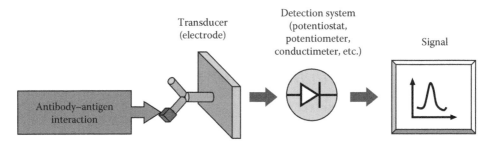

FIGURE 33.1 Block diagram showing the main components on an electrochemical immunosensor.

by the antigen–antibody interaction. These changes are detected at the surface of electrodes which, in the case of immunosensors, are modified with antibodies or antigens that provide the required specificity to the molecular recognition event that generates the signal (Figure 33.1) [2]. Electrochemical immunosensors can determine the levels of antigen by measuring the change of electric potential (potentiometric biosensors), conductance (conductimetric biosensors), current (amperometric biosensors), or impedance (impedimetric biosensors) caused by the immunoreaction [3,4].

Amperometric immunosensors measure the concentration-dependent current generated by an electrochemical reaction at constant voltage following immunocomplex formation. Most proteins, however, are not intrinsically electroactive and cannot be directly detected amperometrically. Therefore, electrochemically active labels (enzymes) are incorporated to catalyze redox reactions that facilitate the production of electroactive species, which then can be determined electrochemically. Alkaline phosphatase and horseradish peroxidase are by far the most employed enzymatic labels in immunosensors. In the most common assay format, a sandwich is formed between the target and two antibodies that bind to different regions of the target, one immobilized on the surface (capture) and the other enzyme-labeled (detecting). This yields enhanced selectivity, compensating the disadvantage of having an indirect sensing system. Other advantages of amperometric detection systems are their excellent sensitivity, low interferences from matrix components, and simple instrumentation [3].

In the electrochemical impedance spectroscopy (EIS) technique, an alternating current of small sinusoidal amplitude is forced to pass an interface. This technique is very sensitive to surface changes since the attachment of molecules to the surface of an electrode produces an increased hindrance to electron transfer, resulting in an increase of the impedance of the system that can be easily detected [5]. In the field of biosensors, EIS has been widely used for the characterization of surface modifications occurring, for example, during the immobilization of biomolecules on the transducer surface as well as for the development of impedimetric biosensors exploiting the biorecognition of target molecules by the attached probes [6,7]. This technique has the advantage of label-free detection, although its sensitivity is lower compared to amperometry and requires a relatively more complex instrumentation due to the need of frequency response analyzers.

The most common electrode arrangement used in amperometric and impedimetric biosensors is the three-electrode setup (Figure 33.2), which comprises a working electrode (WE), a reference electrode (RE), and a counter or auxiliary electrode (CE). The WE is in contact with the analyte and facilitate the charge transfer to and from the analyte and can be made of several materials such as noble metals, glassy carbon, semiconductors, etc. The RE is a high impedance half cell with a known potential (i.e., Ag/AgCl) that has the function to control the WE potential and through which no current passes during the measurement. Finally, the CE is fabricated of an inert metal (usually platinum) and serves to close the electrochemical circuit by receiving the current generated at the WE [8].

Electrochemical immunosensors are not affected by sample turbidity, quenching, or interferences from light-absorbing or light-emitting compounds commonly found in biological samples, as is the case with optical immunosensors [9]. Hence, they combine the high specificity of traditional immunoassay methods with the low detection limits and low cost of electrochemical instrumentation, and also exhibit other advantages such as ease of use, multiplexing ability, rapid response, and miniaturization, allowing point-of-care measurements [10].

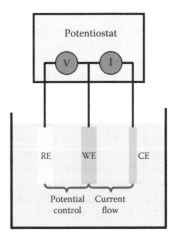

FIGURE 33.2 Schematic representation of a three-electrode cell for electrochemical detection. V, electrometer; I, current-to-voltage converter; WE, working electrode counter; CE, counter or auxiliary electrode; RE, reference electrode.

The modification of noble metals with thiols to form various types of self-assembled monolayers (SAMs) is one of the most exploited strategies for transducer activation in electrochemical biosensors. SAMs are ordered molecular arrays formed by the spontaneous adsorption of an active molecule on a solid surface, often thiol molecule, which forms a covalent bond with gold by chemisorption through the sulfur group [11]. The thiol molecules can consist of a spacer chain usually terminated with a functional group that can be modified to obtain a reactive surface. In biosensors, this strategy is very often used to immobilize the biorecognition element (typically an enzyme, antibody, or nucleic acid sequence) on the transducer surface since it offers the possibility of controlling the orientation, distribution, and spacing of the sensing element [12]. For example, long-chain SAMs modified with polyethylene glycol backbones have been used in optical biosensors with surface plasmon resonance detection due to their stability and effectiveness in reducing/eliminating nonspecific interactions [13]. Besides these interfacial properties, in electrochemical biosensors it is also essential to have a surface with permeability to electron transfer in order to avoid a possible blocking of the electrochemical response. This requirement can be fulfilled using either short-chain SAMs, which have the disadvantage of being relatively instable, or alkanethiol mixtures with different chain lengths, which results in a relatively low biomolecule immobilization capacity and negatively influences the performance of the sensor [14].

To date, cancer detection at an early stage is the only tool to initiate appropriate treatment, and potentially avoid a fatal outcome. Certain proteins such as prostate-specific antigen (PSA) and carcinoembryonic antigen (CEA) can signal the presence or recurrence of cancer. CEA is a glycoprotein of 200 kDa that has been extensively studied as a tumor marker for clinical diagnosis and is routinely used as part of an individual's annual medical checkup in many countries. It is present in about 95% of all colon tumors [15] and 50% of breast tumors [16] and is also associated with ovarian carcinoma [17], lung cancer [18], and others. Therefore, it can either be part of a panel of cancer markers for different cancers or, more importantly, can also be used as an independent prognostic factor [19,20]. The normal CEA levels in healthy adults lie in the range 3–5 ng/mL, although some benign diseases can increase these levels up to 10 ng/mL. When the CEA level is abnormally high before therapy, it is expected to fall to normal following successful surgery or other treatment to remove the tumor. A rising CEA level indicates progression or recurrence of the cancer. In addition, levels >20 ng/mL before therapy are commonly associated with cancer in metastatic state [19,20]. Several strategies have been developed for the detection of this important marker in the past such as enzyme-linked immunoassay [21] and fluoroimmunoassay [22].

Detection of CEA by surface plasmon resonance biosensor modified with mixed SAMs of ethylene glycol–terminated alkanethiols in spiked serum samples achieved a detection limit of 25 ng/mL [23]. Mercaptoundecanoic acid/mercaptoundecanol [24] and phenyl boronic acid–terminated alkanethiol monolayers [25] have also been exploited in the SPR detection of CEA. In addition, anti-CEA antibodies have been attached to thiourea [26] and cystamine monolayers [27] and used for the electrochemical detection of CEA in standard solutions with detection limits of 0.01 and 0.2 ng/mL, respectively, although no application to real samples of these systems was reported.

PSA is a 30 kDa single-chain glycoprotein expressed predominantly by human prostate [28]. It is usually present in the blood at low concentrations (<4 ng/mL) in healthy males and because of the correlation of PSA concentration with tumor volume and tissue specificity, its use as a tumor marker for prostate cancer has flourished over the past decade [29]. PSA is not prostate-specific and a number of studies have shown that it is present in some female hormonally regulated tissues, principally the breast and its secretions [30]. These findings indicate that PSA may have potential diagnostic utility in breast cancer [31]. High levels of PSA have been reported in nipple aspirate fluid (NAF) and recent studies have shown that the concentration of PSA in NAF is inversely related to breast cancer risk, indicating that NAF PSA may represent a clinical tool for breast cancer risk assessment [32].

As in the case of CEA, enzyme-linked immunoassays of different types have been developed for PSA [33], including membrane strip tests [34] and modified ELISA tests in which the enzymatic product is detected voltammetrically [35]. Recent progress in the biosensor technology has allowed the development of new strategies for PSA detection [36]. Flow-injection electrochemical detection has been applied to the detection of PSA on a thin-layer electrochemical cell with glassy carbon WEs using an anti-PSA–alkaline phosphatase conjugate [37]. Screen-printed sensors have also been used on the detection of PSA [38] as well as labeless formats based on impedimetric [39] and capacitive [40,41] detection.

In this chapter, we exemplify the advantages of electrochemical biosensor technology for the detection of PSA and CEA based on SAMs of dithiols **1** and **2** (Figure 33.3). These molecules are derived from 3,5-dihydroxybenzyl alcohol and contain two identical alkylthiol substituents attached to the phenyl ring through phenolate bridges that provide two attachment points on the metallic surface [42]. The benzylic core in **1** is modified with a carboxyl group onto which several types of biomolecules can be attached via amide bonds, while **2** has a terminal hydroxyl group. The aromatic core and terminal group are linked together by a polyethylene glycol chain that is used to prevent nonspecific interactions. Using these dithiols, we present an example of optimization of the specific response of the surface to avoid a possible intermolecular steric hindrance caused by a nonoptimal separation of the immobilized recognition units [43]. For this, the EIS technique was used by measuring the impedance changes exerted by the recognition of PSA on anti-PSA-modified surfaces having different molar compositions of dithiols **1** and **2**.

The electrochemical immunosensors for PSA and CEA are based on the covalent immobilization of a monoclonal capture antibody on the dithiol-modified surface. After antigen incubation, a reporter peroxidase–antibody conjugate is used and the product of the enzymatic reaction is detected amperometrically (Figure 33.4). Results on the multiplexed detection of CEA and PSA and on the detection of CEA in serum samples obtained from colon cancer patients are also presented to demonstrate the versatility of the developed amperometric immunosensor [43,44].

For these purposes, a microsystem platform integrating electrochemical detection with fluidic operations has been developed [45]. The microfluidic platform was realized by high-precision milling of polymer sheets and features two well-distinguishable sections: a detection zone incorporating the electrode

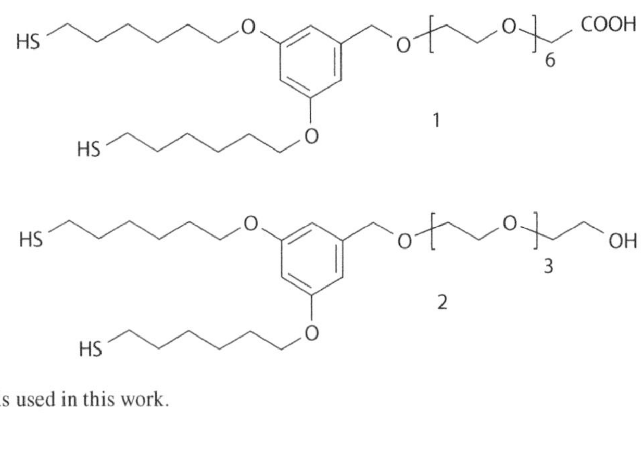

FIGURE 33.3 Dithiols used in this work.

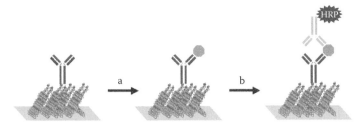

FIGURE 33.4 Detection principle: (a) antigen, and (b) HRP–antibody reporter conjugate.

arrays and the fluid storage part. The detection area is divided into separate microfluidic chambers addressing selected electrodes for the measurement of samples and calibrators and integrates an array of 16 WEs fabricated using the sputtering technique [46]. The fluidic storage part of the platform consists of five reservoirs to store the reagents and sample, which are interfaced by septa. These reservoirs have the appropriate volume to run a single assay per cartridge and are manually filled. The liquids from the reservoirs are actuated by applying a positive air pressure (i.e., via a programmable syringe pump) through the septa and are driven to the detection zone via two turning valves.

33.2 Materials

32.2.1 Reagents

Dithiols **1** (22-(3,5-bis((6-mercaptohexyl)oxy)phenyl)-3,6,9,12,15,18,21-heptaoxadocosanoic acid, catalog number: SPT0014A) and **2** (13-(3,5-bis((6-mercaptohexyl)oxy)phenyl)-3,6,9,12-tetraoxatridecanol, catalog number: SPT0013) were purchased from SensoPath Technologies (Bozeman, MT). Stock solutions (1 mM) were prepared in absolute ethanol, purged with argon, and kept at −20°C when not in use.

Potassium hexacyanoferrate (catalog number: 455946), 1-ethyl-3-(3-dimethylaminopropyl) carbodiimide hydrochloride (EDC, catalog number: E6383), and *N*-hydroxysuccinimide (NHS, catalog number: 56480) were purchased from Sigma-Aldrich España (Madrid, Spain).

CEA (catalog number: P364-4) and PSA (catalog number: P117-1) were purchased from Scipac Ltd. (Kent, United Kingdom).

Monoclonal anti-CEA antibodies (capture: catalog number: 12-140-10; detecting: catalog number: 12-140-1) and anti-PSA (capture: catalog number: PSA66; detecting: catalog number: PSA10) were obtained from Fujirebio Diagnostics AB (Göteborg, Sweden).

33.2.2 Materials for the Fabrication of Electrode Arrays and Microfluidic Chip

Borofloat glass wafers were purchased from Siegert Consulting e.K. (Aachen, Germany). Liftoff resin AZ® 5214E (catalog number: SXR081505) and protection lacquer AZ® 6632 (catalog number: SXR081570) were purchased from MicroChemicals GmbH (Ulm, Germany). Negative photoresist Nano™ SU-8 (catalog number: Y111022) was from Micro Resist Technology GmbH (Berlin, Germany). Polycarbonate sheets (catalog number: 10-0669-0000-03) were obtained from microfluidic ChipShop GmbH (Jena, Germany).

33.2.3 Serum Samples Collection for CEA Detection

Serum samples from colon cancer patients who underwent surgery at Robert Roessle Hospital (Berlin, Germany) with complete tumor removal (R0) were evaluated. All patients were free of distant metastasis (M0) at surgery. One third of the cases were diagnosed as colon adenocarcinoma, as sigma adenocarcinoma, or as mucinous colon carcinomas. A trained nurse carried out collection of serum samples on the surgical ward or the acute day ward. Serum was collected in specific cuvettes (about 5 mL), stored at 4°C when not processed, and centrifuged at 3000 rpm for 10 min. Thereafter, the supernatant was collected and stored at −80°C. Patient specimens were collected after informed consent and only used after the Ethics Committee approval of Charité, Berlin. The CEA levels of these samples were determined using a commercial ELISA kit (CanAg CEA EIA) from Fujirebio Diagnostics AB (Göteborg, Sweden) and using the developed immunosensor.

33.3 Methods

33.3.1 Fabrication of Electrode Arrays [46]

The lithographic process employed to fabricate the electrode arrays was achieved as follows (Figure 33.5a): a circular borofloat glass wafer ($\phi = 12$ cm) was fully covered with 5 µm of LOR5214 liftoff resist, later UV patterned to reveal the RE areas and corresponding connections. Ten nanometers of titanium, used as a glass/metal adhesion promoter, were sputtered followed by a 150 nm thick silver layer. The patterned substrate was fully coated a second time with the same liftoff resist and UV patterned to reveal WE and CE areas and their corresponding connections, and finally coated with a 10 nm thick titanium layer followed by a 150 nm thick gold layer. The remaining photoresist was finally removed with acetone, revealing the planar gold and silver patterns. Since the connection lines between the electrodes and their connections points have to cross the fluidic channels, an insulation of the chip was required. Thus, an SU-8 negative photoresist was applied to the whole surface of the chip and was only partially removed at the electrode and connection areas. This ensured that only electrode would be later exposed to the sample fluids. In order to extend storage lifetime by reducing contamination of the electrode surfaces, the patterned wafers were coated with a 5 µm protection layer of AZ 6632, which can be easily removed by rinsing the arrays with acetone, isopropanol, and water immediately prior to use.

The resulting electrode arrays consisted of 16 gold WEs (dimensions: 1×1 mm^2) in a 4×4 arrangement. Each WE is placed between a silver pseudo reference (0.2×1 mm^2) and a gold CE of the same size (Figure 33.5b).

33.3.2 Fabrication and Assembly of the Electrochemical Microfluidic Cell [45]

The microfluidic cell was fabricated in a polycarbonate substrate and diced to have the format of a microscopic slide ($75.5 \times 25.5 \times 5$ mm). The channels (1 mm depth), reservoirs, and valve holes were realized by high-precision milling and sealed with a layer of a pressure-sensitive adhesive. Valve bodies were fabricated by injection molding and integrated on the chip by means of a metallic holder.

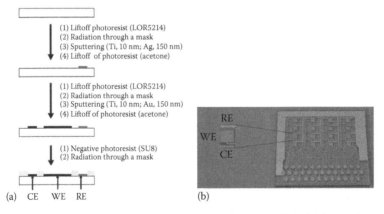

(a)   CE   WE   RE

(b)

FIGURE 33.5 (a) Fabrication process for the electrode array. (b) Photograph of the fabricated electrochemical array comprising 16 square gold working electrodes (WE), and common gold counter (CE) and silver reference (RE) electrodes.

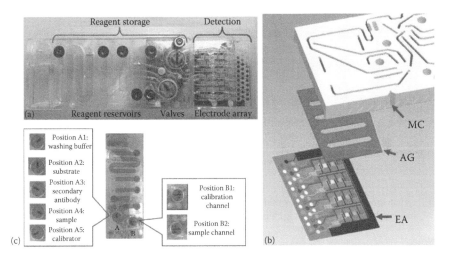

FIGURE 33.6 (a) Photograph of the microfluidic platform. (b) Strategy for the integration of electrode arrays (EA) in the microfluidic chip (MC) using a double-sided adhesive gasket (AG). (c) Valve positions in the microfluidic platform.

The microfluidic cell contains two well-distinguishable sections: a detection zone incorporating the electrode arrays and a fluid storage part (Figure 33.6a). The electrode array was integrated on the fluidic chip using a double-sided medical grade adhesive foil of 50 µm thickness previously laser machined to generate four microchannel structures of 1 mm width (Figure 33.6b) and is connected via pogopin connectors to an external multichannel potentiostat.

The fluidic storage part of the platform consists of five reservoirs to store the reagents and sample, which are interfaced by rubber septa plugged into the injection ports. These reservoirs have the appropriate volume to run a single assay per cartridge and are manually filled. A waste reservoir is placed at the top left corner of the chip to collect the liquids from the different reservoirs during the assay. The liquids from the reservoirs are actuated by applying a positive air pressure (i.e., via a programmable syringe pump) through the septa and are driven to the detection zone via two turning valves (Figure 33.6c).

33.3.3 Instrumentation

33.3.3.1 Microfluidic Setup

The main components of the microfluidic setup were purchased from Tecan Group Ltd., Switzerland. The five reservoirs on the cartridge were connected to a Cavro® Smart Valve 9-port valve. This valve was then connected to a Cavro® XLP 6000 syringe pump and interfaced to a personal computer via the Cavro® Fusion Software. A dedicated program was written on this software to allow the automated control of the fluidic driving of the reagents over the microfluidic cartridge. A photograph of the complete setup is shown in Figure 33.7.

33.3.3.2 Electrochemical Instrumentation

Electrochemical measurements were performed on a PC-controlled PGSTAT12 Autolab potentiostat (EcoChemie, The Netherlands) with an in-built frequency response analyzer FRA2 module.

The amperometric measurements were carried out by applying a potential of −0.20 V between WE and RE. The background current response was first recorded in PBS followed by injection of a mixture of 1 mM hydroquinone/1 mM H_2O_2 in PBS pH 6 and recording of the developed current at the same potential.

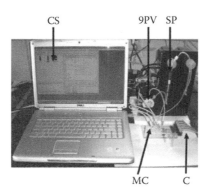

FIGURE 33.7   Photograph of the immunosensor setup. SP: Syringe pump (Cavro® XLP 6000 pump), 9PV: 9-port valve (Cavro® Smart Valve), MC: microfluidic chip, CS: control software (Cavro® Fusion Software), and C: connector to potentiostat.

The electrochemical impedance spectra (EIS) were recorded in 1 mM $Fe(CN)_6^{3-/4-}$ in 0.1 M KCl by applying a bias potential of +0.220 V at the WE. A sinusoidal perturbation of ±5 mV was superimposed to the bias potential and the impedance of the circuit was recorded for 25 frequency values in the interval of 10 kHz to 0.1 Hz. The impedance data were represented in the complex impedance plot (Nyquist plot), which plots the imaginary impedance component (Z'') against the real impedance component (Z') at each excitation frequency. The electrochemical parameters were obtained from simulation using the Frequency Response Analyzer software provided with the Autolab potentiostats by the manufacturer (EcoChemie, The Netherlands). For this, the raw data (Z'' vs. Z') were fitted to an equivalent circuit (Randles circuit), which comprises the uncompensated resistance of the electrolyte (R_s) in series with the capacitance of the dielectric layer (C_{dl}) and the charge transfer resistance (R_{ct}) of the redox probe present in the electrochemical cell (in this case, $Fe(CN)_6^{3-/4-}$), which are connected in parallel [5].

33.3.4 General Procedure for Immobilization of Capture Antibodies [43,44]

The WEs were spotted with 5 μL of a stock solution of **1** for 3 h. The carboxyl groups of the SAM were sequentially modified with an aqueous mixture of EDC (0.2 M) and NHS (50 mM) for 30 min followed by rinsing with ethanol. Capture antibodies (anti-CEA or anti-PSA) were covalently immobilized on the NHS-activated SAM by spotting with 5 μL of a 0.5 mg/mL solution in 10 mM acetate buffer pH 5.0 for 1 h at 37°C. The remaining NHS-active ester sites were blocked with 1.0 M ethanolamine pH 8.5 for 30 min at 37°C.

33.3.5 Optimization of Surface Composition [43]

Ethanolic solutions having different **1 + 2** molar compositions (100:0, 50:50, 25:75, and 5:95) were prepared to have a 1 mM total dithiol concentration. Clean gold electrodes were immersed in these solutions for 3 h and washed with ethanol. PSA66 monoclonal antibody was then immobilized on these SAMs as described earlier followed by blocking with ethanolamine. The electrodes were then exposed to a 1 μg/mL solution of PSA for 5 min, followed by washing with 0.1 M Tris buffer pH 7.6 (containing 0.1% v/v Tween® 20) to remove nonspecifically bounded antigen. EIS was then recorded in 1 mM $Fe(CN)_6^{3-/4-}$ in 0.1 M KCl as described in Section 33.3.1.

33.3.6 Amperometric Detection of CEA and PSA in Buffered and Real Samples [44]

CEA and PSA antigens were diluted in phosphate buffer pH 7.2 to the desired concentration. Undiluted serum samples containing CEA were used as received. The analytes (buffered or in serum) were applied to the surface of antibody-modified electrodes and left to incubate for 2 min to allow the recognition of the target protein by the immobilized receptors. After rinsing with PBS, a 1 µg/mL aliquot of HRP-labeled reporter antibody was added and left to incubate to form a sandwich immunocomplex for 2 min.

33.3.7 Multiplexed Detection of CEA and PSA

In this case, the electrode array was modified with anti-CEA and anti-PSA on alternating electrodes as described earlier. The electrodes were then incubated with a mixture of CEA (50 ng/mL) and PSA (20 ng/mL) in PBS for 2 min, followed by a mixture of HRP-labeled reporter anti-CEA and anti-PSA (1 µg/mL each). Detection was carried out by first recording the background response at −0.2 V in PBS followed by injection of a mixture of 1 mM hydroquinone/1 mM H_2O_2 in PBS pH 6.

33.4 Results

33.4.1 Optimization of Surface Composition

Nonoptimal separation of the immobilized recognition units may cause intermolecular steric hindrance in the antigen recognition step resulting in poor performance of the immunosensor. A common method to optimize the spacing of capture antibodies is to "dilute" a COOH-terminated thiol with another thiol carrying a nonreactive terminal group such as OH over a shorter alkyl chain. This strategy was followed for **1** and **2** using anti-PSA as capture antibody [47,48].

Gold surfaces were modified with SAMs of different **1**:**2** molar compositions (100:0, 50:50, 25:75, and 5:95) onto which anti-PSA was covalently immobilized. Impedance measurements revealed that there is an increase in the PSA66 immobilization level with increasing proportions of **1** (Figure 33.8) as expected for an increased number of attachment points of the surface. More interestingly, the response of each system to the presence of 1 µg/mL of antigen is markedly higher for the 100% composition with respect to those with lower proportion of **1**. In this SAM, the total impedance after antigen incubation increased by a factor of 3 with respect to the antibody-modified surface. This value is comparable with the 2.2-fold increase observed at the SAM with 5% of **1**. However, the monolayer of 100% of **1** exhibited the highest absolute impedance variation, indicating optimal antigen recognition for this composition.

This observation differs from previous reports in which mixed SAMs with low carboxylate group content exhibited the maximum sensing performance [47,48]. Therefore, we can postulate that the structure of a pure monolayer of **1**, in which the attachment points (COOH groups) are more separated from each other than in the case of a SAM of aliphatic chain monothiols, is sufficient to allow an optimal spacing

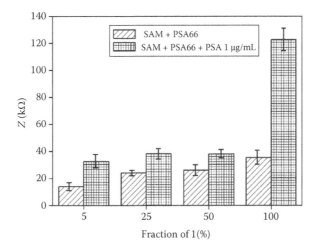

FIGURE 33.8 Total impedance (at 0.118 Hz) for different monolayer compositions in the absence and in the presence of 1 µg/mL PSA antigen.

and orientation of the linked antibodies for maximal binding. This represents an advantage of **1** over previous approaches since one-component SAMs are easier to prepare and their structure is more reproducible and controllable than mixed SAMs.

33.4.2 Detection

33.4.2.1 Amperometric Detection of CEA and PSA in Buffered Solutions

Figure 33.9 shows two examples of application in the detection of CEA and PSA using the developed SAM-based immunosensor. The current variations showed a linear relationship with the concentration of CEA over the range of 0–200 ng/mL with a sensitivity of 3.8 nA·mL/ng and a detection limit of 0.2 ng/mL. In the case of PSA, the linear range was from 0 to 75 ng/mL with a sensitivity of 35 nA·mL/ng and a detection limit of 2 ng/mL. The values of the limit of detection obtained were well below the commonly accepted concentration thresholds for CEA (5 ng/mL) and PSA (10 ng/mL) used in clinical diagnosis, highlighting the effectiveness of the developed immunosensors in the detection of these important markers [19,29].

33.4.2.2 Amperometric Detection of CEA in Serum Samples

The developed immunosensor was then applied to the detection of CEA in serum samples from cancer patients. Samples were taken and divided into two aliquots, one for ELISA and the other for immunosensor detection. The results obtained from both methods are shown in Table 33.1. As can be seen from

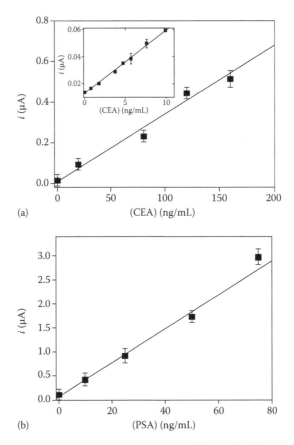

FIGURE 33.9 Calibration plots for the amperometric detection of (a) CEA and (b) PSA in buffered solutions.

TABLE 33.1

CEA Levels (in ng/mL, Duplicate
Measurements) Measured in Serum
Samples with the Developed
Immunosensor and ELISA

Sample	Immunosensor	ELISA
1	4.4 ± 1.7	3.6 ± 0.7
2	5.8 ± 1.1	6.9 ± 0.4
3	6.8 ± 1.4	5.9 ± 0.6
4	7.7 ± 0.7	8.9 ± 1.1
5	20.1 ± 2.9	22.5 ± 1.1
6	43.6 ± 2.7	41.5 ± 3.3

Figure 33.10a, the slope of the immunosensor vs. ELISA plot was 1.04 (from a theoretical value of 1), indicating an excellent correlation between the CEA levels obtained by both methods.

The results were also analyzed by the Bland–Altman method [49], which plots the difference between ELISA and immunosensor results (the bias) against the average of the two measurements (Figure 33.10b). If one method is sometimes higher, and sometimes the other method is higher, the average of the differences will be close to zero. If it is not close to zero, this indicates that the two assay methods are producing different results. The standard deviation (SD) of the differences between the two assay methods is used to calculate the limits of agreement, computed as the mean bias ±1.96 times its SD.

As can be seen from Figure 33.10b, the average of the differences shows a small bias of −0.14 ng/mL toward the immunosensor results, although this value is very close to zero and smaller than the limit of detection obtained with the electrochemical immunosensor, which means that both methods essentially give the same results. In addition, all measurements are within the area defined by the calculated limits of agreement. Therefore, there is an excellent correlation between the CEA values obtained with the electrochemical immunosensor and a commercial immunoassay, which validates the results obtained.

33.4.2.3 Multiplexed Electrochemical Detection of CEA and PSA

In cancer, the multifactorial nature of oncogenesis and the heterogeneity in oncogenic pathways make it unlikely that a single biomarker will detect all cancers with high specificity and sensitivity. Multiplexed analysis of panels of markers has thus the potential to provide information about patient prognosis and response to therapy that cannot be obtained by analysis of single markers in the panel alone [50]. Hence, quantitative identification of biomarkers in a mixture without separation and at clinically relevant concentrations is a crucial requirement for the development of more effective and simpler diagnostic devices.

To test the possibility of simultaneous detection of several protein markers, an electrode array was modified with anti-CEA and anti-PSA in alternating electrode spots via standard EDC/NHS chemistry on a SAM of **1**. An antigen mixture composed of CEA and PSA was applied to the channels and amperometric detection was carried out simultaneously using a multiplex potentiostat.

Figure 33.11a shows the response obtained in eight channels measured, and Figure 33.11b shows the average responses obtained for CEA, PSA, and the blank signal. As shown, the signal obtained with the respective electrodes is very reproducible, with an RSD lower than 6% for both CEA and PSA, demonstrating the possibility to differentiate both signals and detect CEA and PSA simultaneously.

33.5 Discussion

Due to their specificity, sensitivity, miniaturization, and low cost, immunosensors offer great opportunities for numerous decentralized clinical applications such as home self-testing or doctor's office analysis. The most common transduction mechanisms employed in immunosensors are electrochemical,

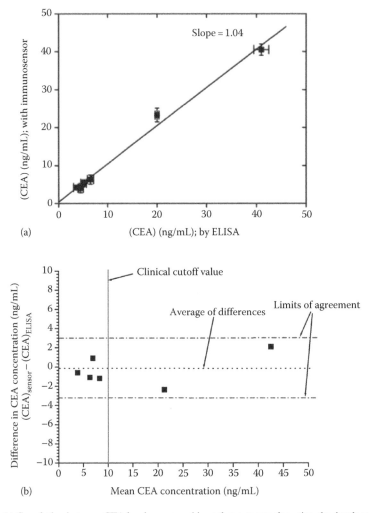

FIGURE 33.10 (a) Correlation between CEA levels measured in real serum samples using the developed immunosensor and ELISA results. (b) Bland–Altman plot comparing the results of the electrochemical measurements with the reference ELISA kit.

optical, and piezoelectric. From the instrumentation point of view, optical transduction requires a light source and several optical components to generate a light beam that is detected by a photodetector. These elements may be costly and suffer from stability problems and may not be amenable to miniaturization, although the advent of integrated optic technologies could allow the development of miniaturized compact devices. On the other hand, piezoelectric immunosensors are based on quartz crystals that resonate on application of an external alternating electric field and detect mass changes on the crystal surface. The frequency responses are influenced by many factors, such as viscosity, dielectric constant, electrode morphology, density, and ionic strength, requiring a careful evaluation of operating conditions and requiring relatively long times to establish a stable baseline and steady-state response. Compared with the other transduction technologies, electrochemical immunosensors have traditionally received a great deal of attention as they present a simple, inexpensive, and sensitive platform for clinical diagnostics.

The results described in this chapter demonstrate the versatility of electrochemical immunosensors and its applicability in cancer diagnostics. As we have shown, dithiol-modified surfaces can be prepared in relatively short periods of time as compared with other alkanethiol SAMs, which in many cases require at least an overnight period of interaction. These SAMs are also stable, permeable to electron

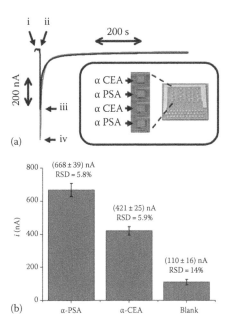

FIGURE 33.11 (a) Typical amperometric responses obtained in the simultaneous detection of CEA (black trace) and PSA (gray trace); i: buffer injection, ii: hydroquinone/H_2O_2 mixture injection (a blank measurement is taken between points i and ii); iii and iv: current peaks for PSA and CEA, respectively. Inset: electrode array modification map. (b) Average current values obtained for CEA, PSA, and the blank signal in eight channels. RSD: residual standard deviation.

transfer, possess low nonspecific-binding properties, and provide an adequate spacing of the biorecognition molecules. The low detection limits obtained in the amperometric detection of CEA and PSA are well below the commonly accepted concentration thresholds used in clinical diagnosis and the immunosensor is also able to simultaneously detect CEA and PSA in mixtures of antigens. The developed immunosensor was successfully applied to the detection of CEA in serum samples obtained from colon cancer patients and the results show an excellent correlation with those obtained using a commercial ELISA kit. The flexibility of the electrochemical approach is also demonstrated by the possibility to develop a multiplexed detection platform.

The possibility to integrate an electrochemical detection scheme with microfluidic functions for the development of automated assays has also been demonstrated. In terms of prospective commercial applications of this or similar devices, it is interesting to note that the cost of production of one prototype as that described in [45], which at the prototype level is about 25€, can be vastly reduced to less than 1€ per chip by the implementation of mass production techniques such as injection molding (to manufacture the microfluidic chip) and screen-printing (for the fabrication of electrode arrays).

33.6 Future Trends

33.6.1 Novel Surface Functionalization Tools

Surface engineering relies on the modification of materials at the molecular scale with the aim of providing an interface of enhanced performance and improved physical and (bio)chemical properties. This is especially relevant for the construction of biosensors where the challenge is not only to engineer the surface of the device such that the biorecognition molecule tolerates the "foreign" object but also to correctly orientate the surface functional groups in order to control the communication between the device and its bioenvironment, while taking into account other important biosensor operational parameters such as sensitivity and stability [51–53]. As we have seen in this chapter, the functionalization of noble metals with SAMs of thiolated compounds is a route to modify the electrode surfaces with functional organic molecules allowing a control of the microenvironment of the interface and the distance and/or orientation of the attached biomolecule.

On the other hand, the last decade has seen an outbreak of new methods for surface modification based on the fabrication of highly organized molecular systems with the aid of supramolecular interactions [54]. This technology has opened up new prospects for the control of molecular interactions and the design of novel functional materials and devices and offers an unconventional route for the reversible immobilization of different biomolecules based on interfacial, and often multivalent, interactions of noncovalent nature such as hydrogen bonds, host–guest, hydrophobic and electrostatic interactions [55,56].

33.6.2 Toward Integrated Immunodetection Systems

The determination of diagnostic markers generally still requires specialized instrumentation and/or laboratory facilities, including specialized personnel, not only for sample treatment and laboratory detection but also for blood sampling itself. Routine diagnostics systems are still expensive, and furthermore require a high number of manual steps, with the available systems being more semi-automated than fully automated, and highly time-consuming. This is due to the fact that most immunosensor systems are based on heterogeneous assay formats involving the separation of bound from unbound material from the sensor surface and thus requiring multiple assay steps and lengthy washes [57].

Recent advances in the fabrication of microfluidic platforms initiated during the late 1990s have facilitated the realization of micro total analysis systems as they enable the miniaturization, integration, and automation of (bio)chemical assays [58,59]. These platforms combine a series of fluidic unit operations (transport, metering, mixing, incubation, detection, etc.) within a device with interconnected elements, preferably in a single packaged support manufactured using cheap polymeric materials [60]. The integration of miniaturized fluidic handling and delivery systems with chemical and biochemical sensors provides applied scientists with powerful tools for in-field measurements away from central laboratories. Furthermore, the associated electronics used to drive the electrochemical detection and signal processing can also be easily miniaturized and integrated onto the same platform by carefully designing application-specific integrated circuits [61]. Assay automation greatly reduces user intervention and testing costs by incorporating full assays in laboratory workstations or other integrated systems usually in a modular manner, containing elements for sample and reagent treatment, washing, calibration, and signal measurement [62,63]. In this way, the quality of the assays is enhanced, resulting in improved precision, sensitivity, and reproducibility [64,65].

Figure 33.12 shows an example of different integration stages of an electrochemical immunosensor. The classical three-electrode cell connected to an external commercial potentiostat (Figure 33.12a) can be first integrated into arrays of electrodes attached to a platform having very basic fluidic features and manual reagent injection (Figure 33.12b). This enables the realization of multiplexed determinations and may be useful in the first stages of biosensor development. Further evolutions of the integration strategy involve the introduction of automated fluidic handling (i.e., via a programmable syringe pump) combined with microsystems now integrating fluidic storage elements such as reservoirs and valves (Figure 33.12c). These elements can finally be housed into an instrument also integrating a dedicated potentiostat (i.e., fabricated on a PCB board) and a computer or other control element that enables to run a completely automated assay. This instrument may also contain communication elements in order to transmit the results to a hospital database or doctor's office.

It is therefore evident that there are several advantages of integrating the microfluidic operations required to execute on-chip electrochemical immunosensor detection in a single platform. These include a reduction in assay time, low sample volumes, and low reagent consumption, the possibility of high-throughput assays. Advances in microfabrication and packaging will necessarily need to meet with the specifications of stable reagents for storage with the objective of achieving fully disposable and integrated detection platforms. Hence, the prospect of achieving commercial low-cost disposable platforms amenable for point-of-care applications can become a reality in a near future.

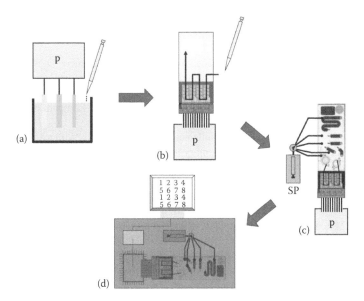

FIGURE 33.12 Possible integration steps for electrochemical biosensors: (a) standard three-electrode cell with external potentiostat, (b) electrode array integrated into a polymer chip with manual fluidic manipulation, (c) automated system integrating reagent storage and detection, and (d) fully integrated desktop instrument.

ACKNOWLEDGMENTS

The authors thank the EU SmartHEALTH project (FP6-2004-IST-NMP-2-016817) for financial support and Fujirebio Diagnostics AB for providing the monoclonal antibodies. A. F. thanks Ministerio de Ciencia e Innovacion, Spain, for a "Ramón y Cajal" Research Professorship.

REFERENCES

1. Theâvenot, D. R., K. Toth, R. A. Durst, G. S. Wilson. 1999. Electrochemical biosensors: Recommended definitions and classification. *Pure Appl. Chem.* 71:2333–2348.
2. North, J. 1985. Immunosensors: Antibody-based biosensors. *Trends Biotechnol.* 3:180–186.
3. Luppa, P. B., L. J. Sokoll, D. W. Chan. 2001. Immunosensors: Principles and applications to clinical chemistry. *Clin. Chim. Acta* 314:1–26.
4. Tothill, I. E. 2009. Biosensors for cancer markers diagnosis. *Semin. Cell Dev. Biol.* 20:55–62.
5. Katz, E., I. Willner. 2003. Probing biomolecular interactions at conductive and semiconductive surfaces by impedance spectroscopy: Routes to impedimetric immunosensors, DNA-sensors, and enzyme biosensors. *Electroanalysis* 15:913–947.
6. Lisdat, F., D. Schäfer. 2008. The use of electrochemical impedance spectroscopy for biosensing. *Anal. Bioanal. Chem.* 391:1555–1567.
7. Prodromidis, M. I. 2010. Impedimetric immunosensors—A review. *Electrochim. Acta.* 55:4227–4233.
8. Bard, A. J., L. R. Faulkner. 2001. *Electrochemical Methods. Fundamentals and Applications*, 2nd Edition. Wiley, Hoboken, NJ, Chapter 1.
9. Skládal, P. 1997. Advances in electrochemical immunosensors. *Electroanalysis* 9:737–745.
10. Wang, J. 2006. Electrochemical biosensors: Towards point-of-care cancer diagnostics. *Biosens. Bioelectron.* 21:1887–1892.
11. Ulman, A. 1996. Formation and structure of self-assembled monolayers. *Chem. Rev.* 96:1533–1554.
12. Gooding, J. J., D. B. Hibbert. 1999. The application of alkanethiol self-assembled monolayers to enzyme electrodes. *Trends Anal. Chem.* 18:525–533.
13. Frederix, F., K. Bonroy, G. Reekmans, W. Laureyn, A. Campitelli, M. Abramov, W. Dehaen, G. Maes. 2004. Reduced nonspecific adsorption on covalently immobilized protein surfaces using poly(ethylene oxide) containing blocking agents. *J. Biochem. Biophys. Methods* 58:67–74.

14. Bonroy, K., F. Frederix, G. Reekmans, E. Dewolf, R. de Palma, G. Borghs, P. Declerck, B. Goddeeris. 2006. Comparison of random and oriented immobilisation of antibody fragments on mixed self-assembled monolayers. *J. Immunol. Methods* 312:167–175.

15. Duffy, M. J., A. van Dalen, C. Haglund, L. Hansson, R. Klapdor, R. Lamerz, O. Nilsson, C. Sturgeon, O. Topolcan, 2003. Clinical utility of biochemical markers in colorectal cancer: European Group on Tumour Markers (EGTM) guidelines. *Eur. J. Cancer* 39:718–727.

16. Bremer, K., S. Micus, G. Bremer. 1995. CEA, CA 15-3 and MCA: Comparative clinical relevance in breast cancer. *Eur. J. Cancer* 31:S262–S262.

17. Engelen, M. J. A., H. W. A. de Bruijn, H. Hollema, K. A. Koor, P. H. B. Willemse, J. G. Aalders, A. G. J. van der Zee. 2000. Serum CA 125, carcinoembryonic antigen, and CA 19–9 as tumor markers in border-line ovarian tumors. *Gynecol. Oncol.* 78:16–20.

18. Hernandez, L., A. Espasa, C. Fernandez, A. Candela, C. Martin, S. Romero. 2002. CEA and CA 549 in serum and pleural fluid of patients with pleural effusion. *Lung Cancer* 36:83–89.

19. Molina, R., J. Jo, X. Fililla, G. Zanon, J. Pahisa, M. Munoz, B. Farrus, M. Latre, L. Escriche, J. Estape. 1998. c-erbB-2 oncoprotein, CEA, and CA15-3 in patients with breast cancer: Prognostic value. *Breast Cancer Res. Treat.* 51:109–119.

20. Macdonald, J. S. 1999. Carcinoembryonic antigen. Screening: Pros and cons. *Semin. Oncol.* 26:556–560.

21. De la Rosa, A. M., K. Minoru.1995. Trapping method of antibodies on surfaces of polymerizing discs for enzyme immunoassay. *Anal. Chim. Acta* 312:85–94.

22. Yuan, J., G. Wang, K. Majima, K. Matsumoto. 2001. Synthesis of a terbium fluorescent chelate and its application to time-resolved fluoroimmunoassay. *Anal. Chem.* 73:1869–1876.

23. Su, F., C. Xu, M. Taya, K. Murayama, Y. Shinohara, S. I. Nishimura. 2008. Detection of carcinoembry-onic antigens using a surface plasmon resonance biosensor. *Sensors* 8:4282–4295.

24. Zhang, B., X. Zhang, H. H. Yan, S. J. Xu, D. H. Tang, W. L. Fu. 2007. A novel multi-array immuno-assay device for tumor markers based on insert-plug model of piezoelectric immunosensor. *Biosens. Bioelectron.* 23:19–25.

25. Zhang, X., Y. Wu, Y. Tu, S. Liu. 2008. A reusable electrochemical immunosensor for carcinoembryonic antigen via molecular recognition of glycoprotein antibody by phenylboronic acid self-assembly layer on gold. *Analyst* 133:485–492.

26. Limbut, W., P. Kanatharana, B. Mattiasson, P. Asawatreratanakul, P. A. Thavarungkul. 2006. A reusable capacitive immunosensor for carcinoembryonic antigen (CEA) detection using thiourea modified gold electrode. *Anal. Chim. Acta* 561:55–61.

27. Dai, Z., J. Chen, F. Yan, H. Ju. 2005. Electrochemical sensor for immunoassay of carcinoembryonic antigen based on thionine monolayer modified gold electrode. *Cancer Detect. Prev.* 29:233–240.

28. Stenman, U. H., J. Leinonen, W. Zhang, P. Finne. 1999. Prostate-specific antigen. *Semin. Cancer Biol.* 9:83–93.

29. Oesterling, J. E. 1991. Prostate specific antigen: A critical assessment of the most useful tumor marker for adenocarcinoma of the prostate. *J. Urol.* 145:907–923.

30. Mannello, F., G. Gazzanelli. 2001. Prostate-specific antigen (PSA/hK3): A further player in the field of breast cancer diagnostics? *Breast Cancer Res.* 3:238–243.

31. Romppanen, J., R. Keskikuru, V. Kataja, M. Eskelinen, V.-M. Kosma, K. Savolainen, M. Uusitupa, I. Mononen. 1999. Measurement of prostate-specific antigen in detection of benign or malignant breast disease in women. *Br. J. Cancer* 79:1583–1587.

32. Sauter, E. R., G. Klein, C. Wagner-Mann, E. P. Diamandis. 2004. Prostate-specific antigen expression in nipple aspirate fluid is associated with advanced breast cancer. *Cancer Detect. Prev.* 28:27–31.

33. Johnson, E. D., T. M. Kotowski. 1993. Detection of prostate-specific antigen by ELISA. *J. Forensic Sci.* 38:250–258.

34. Jung, K., J. Zachow, M. Lein, B. Brux, P. Sinha, S. Lenk, D. Schnorr, S. A. Loening. 1999. Rapid detec-tion of elevated prostate-specific antigen levels in blood: Performance of various membrane strip tests compared. *Urology* 53:155–160.

35. Zhang, S., P. Dua, F. Lia. 2007. Detection of prostate specific antigen with 3,4-diaminobenzoic acid (DBA)–H_2O_2–HRP voltammetric enzyme-linked immunoassay system. *Talanta* 72:1487–1493.

36. Healy, A., C. J. Hayes, P. Leonard, L. McKenna, R. O'Kennedy. 2007. Biosensor developments: Application to prostate-specific antigen detection. *Trends Biotechnol.* 25:125–131.

37. Chen, S.-F., Y. Xu, M. P. Ip. 1997. Electrochemical enzyme immunoassay for serum prostate-specific antigen at low concentrations. *Clin. Chem.* 43:1459–1461.
38. Sarkar, P., P. S. Pal, D. Ghosh, S. J. Setford, I. E. Tothill. 2002. Amperometric biosensors for detection of the prostate cancer marker (PSA). *Int. J. Pharm.* 238:1–9.
39. Fernández-Sánchez, C., A. M. Gallardo-Soto, K. Rawson, O. Nilsson, C. J. McNeil. 2004. Quantitative impedimetric immunosensor for free and total prostate specific antigen based on a lateral flow assay format. *Electrochem. Commun.* 6:138–143.
40. Fernández-Sánchez, C., C. J. McNeil, K. Rawson, O. Nilsson. 2004. Disposable noncompetitive immunosensor for free and total prostate-specific antigen based on capacitance measurement. *Anal. Chem.* 76:5649–5656.
41. Briman, M., E. Artukovic, L. Zhang, D. Chi, L. Goodglick, G. Gruner. 2007. Direct electronic detection of prostate-specific antigen in serum. *Small* 3:758–762.
42. Spangler, C. W., B. D. Spangler, E. S. Tarter, Z. Suo. 2004. Design and synthesis of dendritic tethers for the immobilization of antibodies for the detection of class A bioterror pathogens. *Polym. Prepr.* 45:524–525.
43. Fragoso, A., N. Laboria, D. Latta, C. K. O'Sullivan. 2008. Electron permeable self-assembled monolayers of dithiolated aromatic scaffolds on gold for biosensor applications. *Anal. Chem.* 80:2556–2563.
44. Laboria, N., A. Fragoso, W. Kemmner, D. Latta, O. Nilsson, M. L. Botero, K. Drese, C. K. O'Sullivan. 2010. Amperometric immunosensor for carcinoembryonic antigen in colon cancer samples based on monolayers of dendritic bipodal scaffolds. *Anal. Chem.* 82:1712–1719.
45. Fragoso, A., D. Latta, N. Laboria, F. von Germar, T. E. Hansen-Hagge, W. Kemmner, C. Gärtner, R. Klemm, K. S. Drese, C. K. O'Sullivan. 2011. Integrated microfludic platform for the electrochemical detection of breast cancer markers in patient serum samples. *Lab Chip.* 11:625–631.
46. Henry, O. Y. F., A. Fragoso, V. Beni, N. Laboria, J. A. Acero Sánchez, D. Latta, F. von Germar, C. K. O'Sullivan. 2009. Design and testing of a packaged microfluidic cell for the multiplexed electrochemical detection of cancer markers. *Electrophoresis* 30:3398–3405.
47. Frederix, F., K. Bonroy, W. Laureyn, G. Reekmans, A. Campitelli, W. Dehaen, G. Maes. 2003. Enhanced performance of an affinity biosensor interface based on mixed self-assembled monolayers of thiols on gold. *Langmuir* 19:4351–4357.
48. Huang, L., G. Reekmans, D. Saerens, J. M. Friedt, F. Frederix, L. Francis, S. Muyldermans, A. Campitelli, C. van Hoof. 2005. Prostate-specific antigen immunosensing based on mixed self-assembled monolayers, camel antibodies and colloidal gold enhanced sandwich assays. *Biosens. Bioelectron.* 21:483–490.
49. Bland J. M., D. G. Altman. 1986. Statistical methods for assessing agreement between two methods of clinical measurement. *Lancet* i:307–310.
50. Zheng, G., F. Patolsky, Y. Cui, W. U. Wang, C. M. Lieber. 2005. Multiplexed electrical detection of cancer markers with nanowire sensor arrays. *Nat. Biotechnol.* 23:1294–1301.
51. Zhavnerko, G., K. Ha. 2004. Biosensor applications: Surface engineering. In *Dekker Encyclopedia of Nanoscience and Nanotechnology.* Schwarz, J. A., C. I. Contescu, K. Putyera (Eds.). Marcel Dekker Inc., New York. pp. 351–360.
52. Leca-Bouvier, B., L. J. Blum. 2005. Biosensors for protein detection: A review. *Anal. Lett.* 38:1491–1517.
53. Knopf, G. K., A. Bassi (Eds.). 2007. *Smart Biosensor Technology.* CRC Press, Boca Raton, FL.
54. Geckeler, K. E. (Ed.). 2003. *Advanced Macromolecular and Supramolecular Materials and Processes.* Kluwer, Dordrecht, the Netherlands.
55. Palmer, L. C., Y. S. Velichko, M. O. Cruz, S. I. Stupp. 2007. Supramolecular self-assembly codes for functional structures. *Phil. Trans. R. Soc.* 365:1417–1433.
56. Crespo-Biel, O., B. J. Ravoo, D. N. Reinhoudt, J. Huskens. 2006. Noncovalent nanoarchitectures on surfaces: From 2D to 3D nanostructures. *J. Mater. Chem.* 16:3997–4021.
57. Marquette, C. A., L. J. Blum. 2006. State of the art and recent advances in immunoanalytical systems. *Biosens. Bioelectron.* 21:1424–1433.
58. West, J., M. Becker, S. Tombrink, A. Manz. 2008. Micro total analysis systems: Latest achievements. *Anal. Chem.* 80:4403–4419.
59. Tudos, A. J., G. A. J. Besselink, R. B. M. Schasfoort. 2001. Trends in miniaturized total analysis systems for point-of-care testing in clinical chemistry. *Lab Chip* 1:83–95.
60. Becker, H., C. Gärtner. 2008. Polymer microfabrication technologies for microfluidic systems. *Anal. Bioanal. Chem.* 390:89–111.

61. Polsky, R., J. C. Harper, D. R. Wheeler, S. M. Brozik. 2008. Multifunctional electrode arrays: Towards a universal detection platform. *Electroanalysis* 20:671–679.
62. Truchaud, A., T. Le Neel, H. Brochard, S. Malvaux, M. Moyon, M. Cazaubiel. 1997. New tools for laboratory design and management. *Clin. Chem.* 43:1709–1715.
63. Orsulak, P. J. 2000. Stand-alone automated solutions can enhance laboratory operations. *Clin. Chem.* 46:778–783.
64. Sokoll, L. J., Chan, D. W. 2005. Choosing an automated immunoassay system. In *The Immunoassay Handbook*. Wild, D. (Ed.). Elsevier, Oxford, U.K., Chapter 20.
65. Blow, N. 2008. Lab automation: Tales along the road to automation. *Nat. Methods* 5:109–112.

34

Electrochemical Biosensors for Measurement of Genetic Biomarkers of Cancer

Robert Henkens and Celia Bonaventura

CONTENTS

34.1 Introduction

This chapter describes an electrochemical biosensor system, the eSystem, for measurement of cancer gene expression and explains its potential use in research on cancer drugs, in the diagnosis and treatment of cancer, how the biosensors are created, how they function, and how they can be applied in point-of-care settings. The aim is to describe research and development of the eSystem by Alderon Biosciences (Beaufort, NC) and potential use of its disposable biosensors for the measurement of cancer gene expression. Because the eSystem is small, portable, and robust, it can be used in point-of-care settings and play a key role in the further development and application of truly personalized medicine. The eSystem characteristics, further described in this chapter, thus show that this state-of-the-art electrochemical technology is poised for use in cancer research and could play a significant role in cancer drug development and in the improved early detection and treatment of cancers.

Progress on understanding the complexities of cancer has led to the discovery of a number of biomarkers whose levels are related to specific types and stages of cancer onset and progression [12,22,36]. The eSystem and its disposable biosensors can be used singly or in 8- or 96-sensor arrays to measure the expression of genetic biomarkers of cancer and follow their changes during cancer therapy. The idea of using biomarkers to detect disease and improve treatment goes back to the very beginnings of medical treatment. Originally, the term biomarker referred to such biological indicators as abnormal body temperature, blood pressure, and heart rate that signaled an imbalance in the body. Established cancer biomarkers now include genes, mRNA, microRNA, proteins, and metabolites. Oncologists are making important use of cancer biomarkers in drug development and patient care [8,26]. One example of a bio-marker finding increasing use in oncology is the level of circulating tumor cells (CTCs) present in the blood of cancer patients. At the moment, CTC determinations are used in the development of anticancer drugs as an objective and cancer diagnostics and direct measurement of the response of a cancer to therapeutic treatment.

Probably the most famous biomarker in current cancer diagnostics and drug development history is the *HER-2* gene and receptor, discovered in the mid-1980s. Between 20% and 30% of breast cancer patients show an overexpression of the HER-2 receptor on their cancer cells. Although the presence of this bio-marker indicates a higher risk of adverse outcomes, it also gave clinicians a new target for development of novel therapies.

DNA microarrays have been used to discover cancer biomarkers. However, DNA microarrays have had, up to now, fairly limited penetration into the field of molecular cancer diagnostics [6]. Better tools for gene expression measurement are needed, especially for use in point-of-care settings and personalized medicine. Tools that more effectively quantify the levels of genes expressed at varied stages of cancer development are required to make the development and use of personalized cancer drugs and therapies a reality. Improved biosensors for detection of genetic biomarkers of cancer offer exciting possibilities for this kind of clinical testing [31,32,40]. The use of modern molecular biosensors for analysis of cancer biomarkers has immediate application in defining cancer subtypes and in research on obtaining new drugs to more effectively deal with the disease [2,4,5,17,21]. Biosensor profiling of cancer biomarker levels could also facilitate cancer research and in particular aid in the identification of targets for new drugs and vaccines.

This chapter describes the eSystem developed by Alderon Biosciences. The eSystem is an innovative biosensor product that can be used for rapid, low-cost, quantitative measurement of cancer gene expression and biomarker detection in point-of-care settings, with advantages compared to more traditional molecular methods such as quantitative PCR. As described in this chapter, highly sensitive measurements of gene expression based on probe-based assays with the electrochemical sensors of the eSystem have been proven possible. Recent advances in the eSystem incorporate new biosensor-based plates and array readers that require very small volumes of samples and reagents. The screen-printed electrodes (SPE technology) used in the eSystem gave an important boost to the amperometric detection in biosensors by providing low-cost, disposable, more reliable and stable, and easy-to-prepare supports for immobilization of biological compounds. Most of the recent scientific papers dealing with biosensors based on amperometric detection report on the development of new enhanced sensing layers and components immobilized on SPEs or on improved assay methods [11,14,15,25,33,39,42,43].

At the moment, relatively few portable electrochemical devices based on SPEs have been developed and are available on the market for research, diagnostic, and monitoring applications. Among these are the products offered by Alderon Biosciences, Dropsens, and Palm Instruments. These systems provide the SPE amperometric readout and output signal processing needed for the use of disposable biosensors. The eSystem assays can be conducted on small, inexpensive, robust, point-of-care instruments. Our aim in this chapter is to describe our research and that of others on the eSystem and its potential application in the battle against cancer.

In brief, better tools for cancer gene expression analysis can result in improved cancer drug discovery and development, cancer prognosis, early diagnosis, and therapy. While PCR-based methods have been extensively used in research laboratories, they are generally unsuited for use in point-of-care settings or physician's offices where individualized treatments could result from more facile assays of cancer gene expression. Compared to quantitative PCR, the eSystem described in this chapter is more robust and portable, much easier to use, has comparable sensitivity (less than pM/assay), faster processing time, and very low cost. The eSystem can show the levels of genes expressed at varied stages of cancer development, allowing for monitoring of therapy and for development and the use of personalized cancer drugs and therapies. The advanced electrochemical technology of the eSystem can be further developed for cancer research and diagnostics, and then extended to other human health products. Markets for biosensors will likely grow rapidly for diagnosis and therapeutic aspects of each major disease [16].

34.2 Materials and Methods

34.2.1 eSystem

The biosensor assays reported here were done with an integrated electrochemical measurement system, the eSystem, whose major components are small, portable readers and disposable SPEs that can be used singly or in arrays. The evolution of the eSystem is described in prior publications and patents describing Alderon's biosensor research [7,10,18–20,37,44–49,51] and the research of others using the Alderon system [11,14,15,25,33,39,42,43]. Probe-based assays done with the eSystem provide high specificity and avoid major obstacles to the use of conventional gene-probe methods. The eSystem assays can be used to monitor and compare PCR products without the use of gels or other ancillary techniques [1] and they can be used without PCR or other forms of target amplification. The assays can be done with complex, unpurified mixtures, and generally have sufficient sensitivity that they can be preformed without the need for any target amplification steps such as PCR [45,46].

Alderon's eSystem technology uses screen-printed, disposable electrodes on test strips of single or eight sensors or in wells of 96-sensor plates. Small and inexpensive electrochemical monitors read the target-specific currents generated by the biosensors. The eSystem has been used to detect specific DNA or RNA sequences in bacteria, viruses, and human genes [7,18,20,37,44–49].

Although outside the scope in this chapter, Alderon's eSystem technology can be used to detect protein and small molecule targets via electrochemical immunoassays [1,27,28], as well as the DNA/RNA targets described here. In electrochemical immunoassays, the target protein is captured and detected via enzyme-labeled antibodies. Greater sensitivity is generally obtainable when an enzyme-linked immunosorbent assay (ELISA) is adapted to run on the eSystem, since the electrochemical detection is appreciably more sensitive than the traditional optical detection [1,14,43].

Alderon's eSystem is a universal hardware platform, and broad-based software operating system, and disposable screen-printed biosensors with target-specific coatings capable of quick and accurate determination of levels of specific nucleic acid targets. As demonstrated in the following sections, the distinctive characteristics of the eSystem can greatly facilitate studies of nucleic acid biomarkers of cancer.

34.2.2 Biosensors of the eSystem

Figure 34.1 shows a single sensor of the eSystem. The disposable sensor is comprised of a three-electrode system. Each electrode has a different function. The silver electrode provides an electrochemical reference during the measurement. A small carbon electrode serves as a counter electrode. A second, larger, carbon electrode serves as the working electrode and provides the detection surface. The eSystem electrodes are prepared by screen-printing three electrodes with conductive silver ink and then screen-printing carbon ink over the counter and working electrode. For the DNA/RNA detection applications described in this chapter, the carbon working electrode is completed by applying a coating of Neutravidin or by the controlled volumetric deposition and evaporation of a known volume of colloidal gold sol/poly(ester-sulfonic acid) polymer mixture on the electrode surface. The colloidal gold used for nanogold coatings was generated by addition of a solution of 1% aqueous sodium citrate to a rapidly boiling and stirred solution of gold trichloride and refluxing the solution for 30 min. The final concentrations (wt%) were 0.01% $HAuCl_4$ and 0.03% sodium citrate. Colloidal gold sols with particle diameters ranging from 100 to 1000 Å can also be purchased from BBI International, United Kingdom. Research has shown that the nanogold-modified electrodes act as microelectrode arrays. Evidence for this is based on the fact that, for the same geometric area, the stripping detection signal of the nanogold-coated surface is greater than the stripping detection signal for a planar electrode.

Figure 34.1, showing the single-sensor strip, also provides a schematic representation of the current flow that allows for electrochemical detection on the sensor of a target–enzyme hybrid using a horseradish peroxidase label (HRP) and an electrochemical mediator (Med). The enzyme label captured at the sensor surface on the target–enzyme hybrid is used to generate detectable electroactive products at the

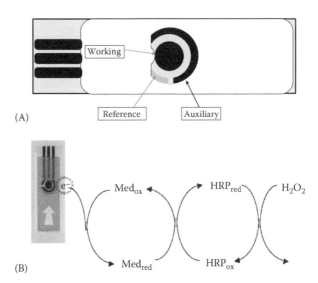

(A)

(B)

FIGURE 34.1 Single-sensor components are shown in (A) and current flow is shown in (B). The sensor components (connectors and electrode components) are screen-printed as described in the text. Current flow is shown for electrochemical detection of an immobilized enzyme-labeled target using a horseradish peroxidase (HRP) enzyme and an electrochemical mediator (Med). The enzyme captured at the surface of the working electrode is used to generate detectable electroactive products at the sensor surface. Catalytic reduction of the electroactive products gives rise to a target-dependent current that is proportional to the amount of target–enzyme complex captured at the sensor surface.

sensor surface. The electroactive products give rise to a target-dependent current that is proportional to the amount of target–enzyme complex captured at the sensor surface.

The biosensors of the eSystem for detection of cancer gene targets can employ a range of coatings and nanoscale chemistries, but all involve probe-based assays using oligonucleotides with sequences that are complementary to the target sequence. The probe-based assays confer high specificity in immobilization of targets on the electrode surface and electrochemical detection of the immobilized target. The eSystem uses amperometric biosensors in which the electrical current that is generated as the result of an applied potential is proportional to the amount of target immobilized on the electrode surface.

The disposable eSystem sensors are generated by highly reproducible screen-printing methods. As a consequence, they have low (<5%) relative standard deviations for sensor-to-sensor readout. This is a major advance, since biosensors have traditionally been plagued with problems of sensor-to-sensor variability. The disposable biosensors of the eSystem can be used singly or in 8- or 96-sensor arrays. A screen-printed sheet of multiple sensors is shown in Figure 34.2A and B shows an individual eight-sensor array. Figure 34.3 shows the use of a standard eight-tip micropipette to apply samples to an eight-sensor array. The small variations between batches of sensor strips can be accommodated by the use of a touch-key provided with each batch of sensors, where internal calibration is achieved by touching the key to the calibration button on the right side of the eSystem reader.

The 8-sensor and 96-sensor versions of the eSystem are designed with similar principles of operation. Each of the multiple biosensor elements on an 8-sensor test strip or 96-sensor ePlate is separately addressable. When used with enzyme-enhanced signal transduction methods as described in Section 34.2.3, the eSystem allows users to make sensitive and selective simultaneous measurement of up to 96 separate tests, some of which will be positive and negative controls and calibration standards that confirm system performance. As further discussed in Section 34.2.5, the readout of current from the sensors that quantifies DNA or RNA sequences is very rapid, with results for a 96-sensor array obtained in <2 s after pushing the start button.

For probe-based assays, the working electrodes of the biosensors are modified with oligonucleotides whose sequences are complementary to a cancer gene target sequence. These oligos function as capture probes in eSystem assays as shown in Figure 34.4. The amount of target captured is electrochemically

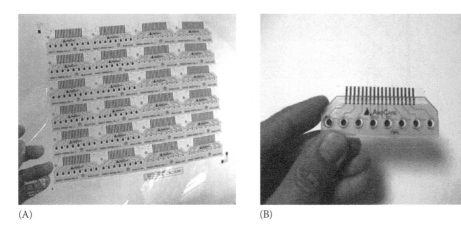

(A) (B)

FIGURE 34.2 Screen-printed elements for eight-sensor eSystems (A) and an individual eight-sensor array (B) for use in electrochemical determinations of cancer biomarker levels.

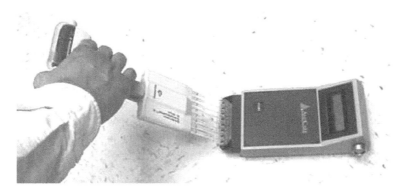

FIGURE 34.3 Use of a standard micropipette to add samples to an eight-sensor array for analysis by an eSystem reader.

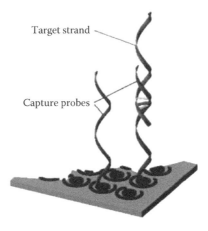

FIGURE 34.4 Schematic of probe-based capture of nucleic acid targets on an eSystem biosensor array. Nucleic acid (DNA or RNA) targets are captured on the working electrodes of eSystem biosensor arrays via hybridization to oligomers with complementary target sequences that serve as capture probes.

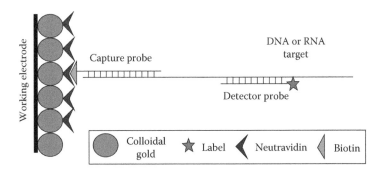

FIGURE 34.5 Schematic representation of a two-probe sandwich assay for electrochemical detection of nucleic acid targets. Nucleic acid (DNA or RNA) targets are captured on the working electrodes of a biosensor via hybridization to oligomers with complementary sequences. A detection probe, also complementary to the target sequence, completes the "sandwich." A fluorescein label on the detection probe allows for binding by the target/probe sandwich of an electroactive enzyme (HRP) that is linked to a fluorescein antibody. A current proportional to the amount of immobilized target is generated and detected when the enzyme's substrate is added. In this two-probe sandwich assay, hybridization of the target to both capture and detection oligomers is required for enzyme-enhanced current generation.

detected using an electrochemical probe–based assay as shown in Figure 34.5. The assay steps involved are shown schematically in Figure 34.6.

Nucleic acid targets captured on the sensor surfaces are enzyme-labeled (directly in single-probe assays or with complementary detection oligomers that carry enzyme labels in two-probe sandwich assays), and detected via electrochemical measurements of the enzyme catalytic reaction using sensitive yet simple and inexpensive readers (see Section 34.2.5). Since the target nucleic acids can be captured with high specificity by probe-based methods, the assays can be done even with targets in complex mixtures, greatly simplifying assay procedures.

The probe design is an important part of configuring a sensitive and selective eSystem assay. Similar considerations come into play as in design of primers for PCR amplification. Since many nucleic targets have been quantified by PCR methods, their validated primer sequences are commonly available. These can be used as a first step in design of capture and detection probes for eSystem assays. Software programs, such as ArrayOligoSelector, can be used as a design tool (see http://arrayoligosel.source-forge.net). Such programs optimize probe designs for expressed genes on the basis of uniqueness in the genome, sequence complexity, lack of self-binding, and GC content. Key factors that enter into probe design for eSystem assays are (a) avoidance of probe sequences that can lead to hybridization between the capture and detection probes, and (b) selection of probe lengths that provide reasonable specificity in light of anticipated variations in target sequence. For example, if a target has a lot of variability, the use of longer probes are desirable to permit hybridization even when some mismatch occurs. For our work with cancer biomarkers, that typically have conserved sequences, oligomeric probes of 20–40 bases have performed well for both capture and detection of targets. Peptide–nucleic acid probes (PNA probes) can be used for applications when very tight binding is desired with little or no tolerance for base mismatches.

34.2.3 Sensor Size and Sample Volume Considerations

Electrochemical target detection is fundamentally a surface-level event, taking place via electron exchange on the biosensor surface. Much of the electroactive product generated by enzyme labels on the captured targets is lost into solution when assays are conducted with 100–200 μL volumes typically used in optical microplate assays. We find that large improvements of sensitivity of cancer gene target detection were possible when small-volume samples (≤1 μL) are used in the eSystem assays. These low-volume protocols also greatly conserve sample and reagents.

Even greater improvements in sensitivity were demonstrated when small-volume assay protocols were used with sensors of reduced size. The eSystem biosensors in current use are typically about 3.05 mm

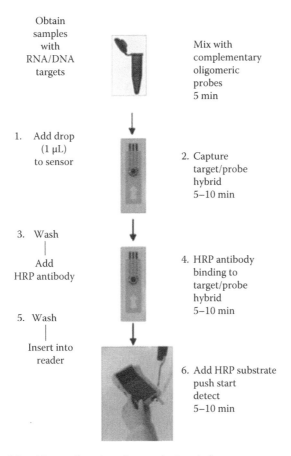

Obtain samples with RNA/DNA targets

Mix with complementary oligomeric probes 5 min

1. Add drop (1 μL) to sensor

2. Capture target/probe hybrid 5–10 min

3. Wash

Add HRP antibody

4. HRP antibody binding to target/probe hybrid 5–10 min

5. Wash

Insert into reader

6. Add HRP substrate push start detect 5–10 min

FIGURE 34.6 Steps of nucleic acid target detection using a probe-based eSystem assay.

(0.120″) in diameter. Biosensors with diameters half this size give a three- to fivefold improvement in signal to noise at low target levels (see Section 34.2.3).

An advantage of smaller-size electrodes has to do with reducing the charging current. Charging current is the major component of the background current of the electrochemical signal. In general, smaller electrodes generate smaller charging currents relative to target-dependent Faradaic currents. The use of smaller sensors allows the use of shorter monitoring pulses and thereby leads to increased target-specific signals. Additionally, for a fixed amount of target, the signal-to-background ratio increases with decreasing size of the working electrode, resulting in the ability to accurately measure targets at lower concentrations.

34.2.4 Assay Steps in Enzyme-Enhanced Target Detection

Numerous approaches to electrochemical detection have been developed over a period of >10 years, including direct electrochemistry of the nucleic acid bases [34,38] and various label-based detection methods [13,23,24,50]. The high sensitivity and specificity of eSystem assays is based on the use of amperometric probe-based detection methods with enzyme labels.

The steps required for enzyme-enhanced detection of nucleic acid targets using the eSystem approach are shown schematically in Figure 34.6. In these eSystem assays, nucleic acid targets are first immobilized on the sensors via interaction with oligomeric capture probes. The immobilized targets are subsequently labeled with enzymes, either directly or via enzyme-labeled detection probes. For some cancer gene expression applications (as described in Section 34.3) a single probe for capture of the target provides sufficient specificity and the target itself carries the enzyme label. For example, a cDNA replicates of mRNA from a tissue sample can be made to carry fluorescein, allowing for the use of an anti-fluorescein antibody complexed with an enzyme to provide for direct addition of enzyme labels to the target. This is just one of many ways a cancer biomarker target can be made to directly carry enzyme labels. Indirect modes of applying one or more enzyme labels to a target in two-probe assays that confer more specificity to the target detection process are described in Section 34.2.3 and illustrated in Figure 34.5.

Electrochemical signals that result from electroactive product formation by the enzyme labels can be quantitatively measured using Alderon's amperometric readers (see Section 34.2.5). Quantitation is achieved by measurement of the catalytic oxidation current of the enzyme reaction as shown in Figure 34.1, which is monitored by the reader after addition of enzyme substrate mixture and the application of potential pulses. Although many enzyme labels are possible, our work has primarily involved the use of HRP or alkaline phosphatase (AP) as enzyme labels whose products can be electrochemically detected after substrate addition.

We typically use HRP as an enzyme label. Many complexes of HRP appropriate for target labeling are commercially available, making it possible to tailor assays to the needs of a given assay and target type. We have often used HRP bound to an anti-fluorescein antibody (Thermo Scientific, Pittsburgh, PA: PA1-2680) that selectively binds to fluorescein-labeled targets. Alternatively, we have used HRP bound to avidin that binds to biotin-labeled targets or HRP. The electrochemical detection of HRP activity was done with K-Blue Max Substrate (Neogen Corp. Lansing, MI: catalog no. 304176), a commercially available mixture of 3,3′,5,5′-tetramethylbenzidine (TMB) as substrate and hydrogen peroxide (H_2O_2) as the co-substrate. The TMB serves as an electrochemical mediator for electron flow, as shown in Figure 34.1. The substrate/mediator solution is added to sensors prepared for electrochemical measurement as described earlier. The electrochemical measurement of the catalytic activity of the HRP label gives us an amplification factor of 10^3 to 10^4, compared with measurements involving non-catalytic labels. This high magnitude of amplification provides a strong justification for the extra layer of assay complexity that results from using an enzyme reagent.

34.2.5 Indirect Enzyme Labeling of Targets in Two-Probe "Sandwich" Assays

High specificity in target detection can be achieved using two-probe approaches that indirectly label the target nucleic acids through hybridization with an enzyme-labeled detector probe. Two separate hybridization events are required for current generation in these assays. The first hybridization is involved in target capture and the other is between the target and an enzyme-labeled detection oligomer (a detector probe) that also has regions of complementarity to the target. The requirement for two hybridization events to occur can be useful when one needs to discriminate between several target species. Figure 34.5 is a schematic representation of a classical two-probe "sandwich" assay on a biosensor of the eSystem.

34.2.6 Blockers for Reduction of Nonspecific Current Generation

The high sensitivity of the eSystem for target-specific current detection is made possible by avoiding or blocking the binding of enzyme labels to the sensor surface in the absence of the target. This is possible due to the high inherent specificity of the nucleic acid hybridizations involved in probe–target interactions. To block any residual nonspecific binding, we use protein-based blockers (typically casein or a purified form of bovine serum albumin). These blockers are selected on a case-by-case basis. Very high ratios of signals obtained when targets are present to signals for samples without targets are thereby achieved, even at very low target levels.

34.2.7 Readers of the eSystem

The eSystem readers are multiplexed electrochemical sensing instruments designed to deliver fast, accurate, and precise measurements of nucleic acids and other analytes. The instruments are designed to measure current generated at single sensors or to provide a test platform supporting electrochemical detection in 8-sensor or 96-sensor arrays. The instruments can be controlled and monitored via two separate data interfaces: user direct or remote USB-linked computer control. To comply with the restrictive requirements of high-speed data acquisition and data transfer, the system was designed around the ATMEL AVR microprocessor that uses "Flash EEPROM" for the firmware storage. This technology allows the processor to be programmed multiple times using the USB link and a host computer. Accompanying software was written for use in a Windows PC environment. The software is able to configure and run the instrument and receive and analyze data.

The eSystem meters are, in essence, modified potentiostats whose functions are controlled by the eSystem software. Single sensors or arrays of sensors (in 8- or 96-sensor arrays) are separately addressable by the eSystem readers, and current readings are completed in <1 s. Readers for single-sensor strips and eight-sensor arrays are capable of battery operation. All are capable of operation (complete target-dependent current detection and readout) without computer linkage. Data obtained can be uploaded via USB linkage to computers for storage and further analysis. The amount of target captured on the individual biosensors (singly or in array formats) is determined after insertion into a biosensor reader. The meters allow for rapid and quantitative detection of the labeled targets that have been captured on each biosensor surface.

For detection of HRP-labeled targets, the meter applies a pre-programmed pattern of intermittent pulses of −100 mV potential (vs. Ag/AgCl reference electrode) and measures current due to the electroreduction of H_2O_2 in a catalytic cycle involving the enzyme HRP, H_2O_2, and a mediator (Med) as shown in Figure 34.1. For HRP detection, the mediator of choice is TMB, which serves as a co-substrate for HRP in the oxidative process. When the calibration mode is active, the meter converts the difference in current to the corresponding concentration of the target analyte using calibration data (which differs to some degree from batch to batch of sensors). The calibration data are uploaded from a calibration button that is triggered by a calibration memory key. The instrument can be controlled and monitored via two separate data interfaces: user direct or remote computer control.

The meters use Alderon's patented mode of intermittent pulse amperometry (IPA) for target detection, with advantages of target detection sensitivity and stability of current generation compared to direct or continuous amperometric methods [20,45,46]. IPA measurement involves a series of millisecond pulses of the same potential applied to the sensing (working) electrode, separated by longer periods when the electrode is disconnected from the potentiostat circuit. As shown in Figure 34.7, current signals, which are measured during the last 100 μs of each pulse, are significantly larger than those measured by conventional DC amperometry (DCA). This is due to reduction of the effect of concentration depletion created by continuously applied potential in DCA. In comparison to differential pulse amperometry (DPA), IPA offers a better control of currents measured for one form of a redox couple in the presence of the other form. Via this IPA detection method, used by all types of eSystem readers, the current that is

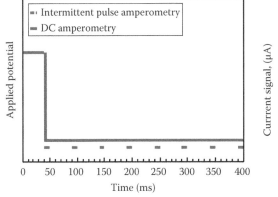

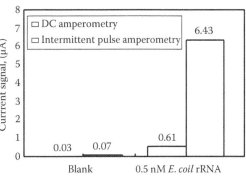

FIGURE 34.7 (A) Typical potential waveforms used in DC amperometry and intermittent pulse amperometry (IPA). (B) Comparison of target detection sensitivity using DC amperometry and IPA in probe-based detection of rRNA.

generated after enzyme substrate addition is measured, typically over a period of 1–5 min. The current level is directly proportional to the amount of HRP-labeled target captured on the sensor surface and, thus, to the amount of target captured on the biosensor surface.

34.3 Results

In this section we summarize some representative results on the measurement of cancer gene expression using the eSystem biosensors, meters, and assay protocols described in Section 34.2. Human cell lines U-937 (histiocytic lymphoma) and K-562 (chronic myelogenous leukemia) from the American Type Culture Collection (ATCC; catalog nos. CRL-1593.2 and CCL-243, respectively) were obtained and cultured for use in these studies. The cells were differentially treated to promote differential expression of cancer genes. Large batches of total RNA and mRNA from cancer cell lines were prepared as stocks for use in experiments. The mRNA quality was assured by RT-PCR amplification and gel analysis. For some studies, the isolated mRNA was converted to cDNA using either oligo(dT)18 or random hexamer primers using SuperScript II Reverse Transcriptase (Invitrogen, Carlsbad, CA). Control cDNA templates were synthesized by PCR amplification of human total cDNA using gene-specific primers equipped with a T7 RNA promoter (5′-TAATACGACTCACTATAGGGAGA-3′) at the 5′ end of the forward primer. Control cDNAs included a ~1623 nt transcript of the housekeeping gene TATA box–binding protein (TBP; accession number NM_001101) and a ~324 nt transcript of the *BRCA1* gene (accession number AY273801). Some cDNAs were labeled with fluorescein (Fl) using the Label-IT Fluorescein Labeling Kit (Madison, WI). The labeled and unlabeled cDNAs were used in subsequent gene expression assays with the eSystem biosensors and readers.

The eSystem was used in studies to detect differences in specific mRNAs resulting from differences in treatment of cancer cell cultures. Interleukin levels are known to be greatly enhanced in cultured cells chemically stressed with PMA compared to normal culture conditions. The mRNA isolated from both culture conditions was converted into cDNA-FL and examined in probe-based assays for levels of the target sequences. The eSystem assays were done using cDNA-specific capture probes and equivalent amounts of cDNA derived from the total RNA extracts from both stressed (induced) and nonstressed cells. Relatively low signals for interleukin were observed compared to the housekeeping gene *TBP* in the untreated cells. Consistent with interleukin induction in the treated cells, the eSystem assay gave between 2- and 10-fold higher signals for the stressed cells. Table 34.1 shows representative electrochemical data illustrating these results.

Table 34.2 presents representative data showing that a single hybridization event (that of a target-specific capture-oligo interacting with a singly labeled cDNA sample) provides adequate specificity for eSystem gene expression analysis. For these studies, we made fluorescein-labeled cDNA from a purified TBP mRNA standard. As shown in Table 34.2, single-probe eSystem assays done with a noncomplementary capture probe (one specific for an *Escherichia coli* sequence) generated negligible signals (equal to background) for the TBP cDNA and for cDNA generated from human liver cancer

TABLE 34.1

Electrochemical Detection of Elevated Levels of Interleukin in Chemically Stressed Cultured Cancer Cells

Capture Probe	Blank (No cDNA) (μA)	cDNA from Nonstressed Cells (μA)	cDNA from Stressed Cells (μA)
Interleukin	0.001	3.7	7.4
TBP	0.001	5.2	5.0

Notes: Probe-based eSystem assays, using capture probes complementary to the sequence of the cancer biomarker interleukin or of a housekeeping gene (*TBP*), were carried out to detect levels of these markers in equivalent amounts of the fluoresceinated cDNA that was generated from cultured cancer cells. The results shown illustrate the ability of the eSystem assays to detect a little as a twofold stress-induced elevation of interleukin level in the single probe–based assays. In other results (data not shown), stressed cells showed as high as 10-fold elevation in interleukin levels. The same samples, when probed for the housekeeping gene *TBP*, showed little or no difference between stressed and unstressed cells.

TABLE 34.2

Specificity in Electrochemical Detection of cDNA Targets Using Single-Probe Assays

Capture Probe	Blank (No cDNA) (μA)	Purified TBP cDNA (μA)	Human cDNA (μA)
TBP	0.01	3.7	0.3
E. coli	0.01	0.01	0.01

Notes: Equal amounts of labeled (fluoresceinated) cDNA of purified human TBP or the labeled cDNA from cultured human liver cancer cells were analyzed in a probe-based eSystem assay with an *E. coli*-specific capture probe (5′-biotin TCAATGAGCAAAGGTATTAACTTT-ACTCCCTTCCT-3′) or with a human TBP-specific capture probe TBPcap1(−). As shown, the human cDNA samples gave no signals above background when probed for the *E. coli* sequence, while the human samples probed for the TBP sequence gave detectable signals of the expected magnitude.

samples. Using equivalent amounts of cDNA, the eSystem assays done with the TBP probe generated a large signal (3.7 μA) for the purified target (Pure TBP cDNA) and a much lower signal (0.3 μA) for the multicomponent human cDNA sample derived from mRNA of the cultured liver cancer cells. This performance is, as expected, based on the much lower amount of TBP in the multicomponent human cDNA sample. Taken together, these measurements showed the ability of a single-probe eSystem assay to specifically detect the TBP target sequence in the total RNA extracted from cultured liver cancer cells.

Selective and effective target capture and electrochemical detection with single-probe assays as described earlier was demonstrated for cancer genes *BRCA1*, *BRCA2*, *ApoA1*, and two housekeeping genes (*TBP* and *GPDH*). In each case, the electrochemical signal generated showed a linear relationship to the target level. Multiple experiments done with varied types of targets added to complex matrices verified that non-target RNA did not alter the target-dependent signal response. This finding was substantiated with a wide range of targets in complex sample matrices.

We made extensive use of ApoA1 as a model system for our investigations of the utility of the eSystem assays for cancer gene detection. ApoA1 RNA is a moderately abundant message, with about 780 mRNA molecules/cell in cultured hepatic carcinoma cells. We generated purified ApoA1 RNA by in vitro transcription of a DNA clone for use as a purified cancer gene analyte. Figure 34.8 shows a titration of this purified cancer gene biomarker in a two-probe eSystem assay. The detector probe carried a single fluorescein label, which was used in combination with HRP bound to an anti-fluorescein antibody for

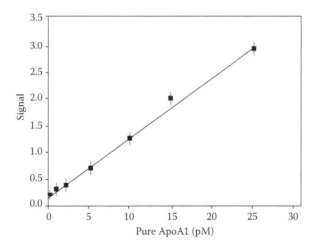

FIGURE 34.8 Detection of varied levels of a purified ApoA1 RNA sequence in two-probe sandwich assays as shown in Figure 34.5. The purified analyte was prepared by in vitro transcription of a DNA clone of ApoA1 RNA.

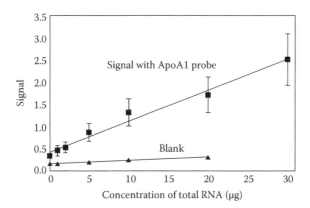

FIGURE 34.9   Detection of ApoA1 mRNA in total RNA extracted from cultured liver cancer cells using two-probe sandwich assays as shown in Figure 34.5. The results shown are the average response of nine experiments done over the course of 2 months and represent three different total RNA extractions.

enzyme-enhanced signal detection. The results demonstrate the ability of the eSystem assays to detect nucleic acid cancer biomarker targets at low levels without PCR amplification.

In our "standard volume" assays, the purified ApoA1 target was detectable in a 40 μL sample at target concentration above 2.5 pM, which corresponds to a detection limit of 6×10^7 molecules. This sensitivity is adequate for direct (unamplified) detection of moderately abundant mRNA in small tissue samples and is sufficient for many types of cancer biomarker assays. Good linearity of detection of the purified target was observed over the range examined (from no target up to 1000 pM). Studies like those shown in Figure 34.8 provided a calibration curve for estimating ApoA1 mRNA levels in unpurified samples.

Probe-based assays for the presence of ApoA1 were done with the target in a crude sample of unpurified (total) RNA from a small sample (0.8×10^4 cells) of cultured hepatic carcinoma cells. The results obtained are shown in Figure 34.9. The cancer biomarker could be detected (less than threefold ratio of signal to background) in 1 μg of the total RNA extracted from the very small sample of cancer cells. Based on the calibration curve obtained in studies like those shown in Figure 34.8, the average abundance of ApoA1 mRNA was estimated to be 0.19% of the total RNA. This value agrees closely with values reported in the literature for this message.

Data were also obtained on ApoA1 detection with small-volume protocols and new half-diameter sensor ePlates. Four orders of magnitude better detection, with detection limits of 5×10^3 molecules (instead of 6×10^7 molecules), were demonstrated with small-area sensors and small-volume protocols, showing the feasibility of highly sensitive eSystem detection of even rare gene expression targets without PCR amplification. The "small-volume" modification of the standard assay requires that ≤1 μL of a sample be applied to the working electrode, rather than the 40–100 μL normally used. Gene targets were detected in <0.1 ng of total RNA (instead of 1 μg) with the small-volume assays and reduced size electrodes.

Probe-based assays of the cancer biomarker BCRA1 gave similar results. Small volumes of sample application, together with electrodes of reduced electrode size, resulted in dramatic improvements of sensitivity of nucleic acid target detection. Figure 34.10 shows linear log–log detection of a synthetic BRCA1 RNA target with the small-volume protocol and reduced size sensors over a large range of target concentrations. *The eSystem assays done with half-diameter sensors and small-volume protocols can reliably detect gene expression targets at levels below 10^4 molecules. Even rarely expressed genes (one to three copies per cell) can be detected at this level in extracts from small (10^4 cells) tissue samples.*

34.4   Discussion

The eSystem and its hardware and software components as described in the foregoing sections can be applied to many areas of molecular cancer research and diagnostics. The technology should find a range

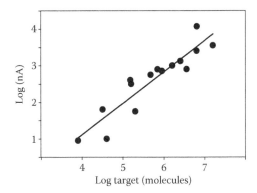

FIGURE 34.10 Linearity and high sensitivity of eSystem target detection with reduced diameter sensors and small-volume protocols. This log–log plot shows the linear relationship between electrochemical signals (in nA) vs. target concentration (in molecules) over a large range. Targets could be detected in samples containing <10⁴ molecules. The data shown were for detection of a synthetic BRCA1 target in probe-based eSystem assays with reduced diameter sensors and a small-volume assay protocol.

of applications in the development of new and improved therapies, treatments, and vaccines for cancers. Simultaneous quantification of multiple targets or multiple samples can be accomplished in the 8-sensor strip and 96-sensor plate eSystem formats. The high sensitivity and low sample volumes required for the electrochemical assays can support the detection and quantification in as few as 10^4 cells. With this level of sensitivity, eSystem assays can be used for direct (PCR-independent) detection and quantification of cancer gene biomarkers and other nucleic acid targets.

Suffering associated with cancer and huge costs for medical care could be reduced through earlier detection and more effective treatment. The primary cause of poor survival of patients with most types of cancers is that the cancers are detected after they have metastasized to distant sites [3,26,41]. Better tools for cancer biomarker detection could improve the prospects for survival of many patients. Moreover, there is a need for more personalized treatment. Consequently, there is a need to develop better ways to detect cancers at an early stage of development and to find appropriate ways to diagnose and treat individual cancers at varying stages of development.

The discovery and validation of cancer biomarkers for clinical diagnostics and patient care is expanding at an ever-increasing rate. It is unlikely that any single cancer biomarker will be sufficient by itself for early cancer detection, cancer risk assessment, or cancer progress determination. Thus, there is a need to be able to cheaply and easily measure multiple cancer biomarkers. Cost-effective, easy-to-accomplish, and accurate measurement of multiple cancer biomarkers could significantly improve cancer diagnosis, risk management, and care of cancer patients. Accurate measurement of levels of multiple cancer biomarkers with a low-cost method is the most promising approach to respond to this need. The measurement of multiple cancer biomarkers with innovative new tools such as Alderon's eSystem can increase patient survival by facilitating early-stage cancer detection, as well as aiding in other aspects of cancer patient care.

Point-of-care diagnostic methods that use probes to detect DNA or RNA targets are still in a relatively rudimentary stage of development. This is especially the case in the development of rapid, accurate, and inexpensive methods. The molecular testing industry is currently stuck in the high-investment, high-throughput approach and essentially ignores the potentially large but fragmented diagnostic market. The traditional industry leaders are providing products for a small selection of high-profit tests and leaving the majority of testing opportunities to small players or, for the most part, as unmet needs. Alderon Biosciences believes that this situation creates an opportunity for its small, low-cost instruments to take the lead in the point-of-care molecular diagnostics market. This untapped market has been estimated to be four times larger than existing diagnostic markets for some applications.

Alderon's eSystem and disposable biosensors can accurately measure cancer biomarkers. Research in this area, briefly described in Sections 34.2 and 34.3, builds on prior work in which we have shown that the electrochemical signals generated in probe-based eSystem assays are proportional to the target sequence, so that the amount of the target sequence in crude mixtures can be quantitatively determined [7,20,37,49]. The relatively inexpensive and rapid determinations of levels of multiple biomarkers of cancer made possible by the eSystem could aid in early cancer diagnosis and better therapy. The Alderon's eSystem technology can be implemented with very low-cost, rugged instruments (U.S. Patent Number: 5,873,990) and implemented using proven probe-based methods for RNA/DNA detection (U.S. Patent Number 6,391,558).

The eSystem platform has significant potential for speedily studying cancer biomarkers that today require expensive and cumbersome tests. The quick, easy, and accurate determination of levels of multiple cancer-related biomarkers can be used in many aspects of cancer research and diagnosis.

The lower cost and ease-of-use of the eSystem assays compared to more expensive and difficult bProbe or real-time PCR and RT-PCR assays should make them attractive to many researchers and in many diagnostic applications. The eSystem will meet the growing demand for simpler, more accessible, and widely applicable tools for molecular assays to measure clinical samples for cancer biomarkers in the early diagnosis of cancers. The eSystem can lead to assays with the appropriate analytical accuracy, reproducibility, sensitivity, specificity, and dynamic range for direct detection of multiple known mRNA cancer markers in clinical samples and cell cultures. Results with eSystem assays have shown:

- Accurate detection of low copy number (<10 copies/cell) mRNAs in 1 µL samples
- Sensitive detection at biomarker levels below 0.5 attomoles mRNA
- Selective detection (low copy number, <10 copies/cell) mRNA in total RNA
- Simultaneous detection and quantification of eight mRNA targets in total RNA
- Low variability for identical assays, with coefficient of variation of <10%
- Dynamic range >1000-fold for mRNA detection
- Quantitative results in <3 h total assay time

In conclusion, the eSystem is a low-cost, high-performance instrument for measuring nucleic acid biomarkers of cancer. The performance of the eSystem for cancer diagnostics and cancer research is equivalent to real-time RT-PCR for speed, sample size, and reliability. The assay cost and complexity is far less. The eSystem approach to cancer gene expression analysis is anticipated to gain widespread use in cancer research and clinical cancer diagnostics.

34.5 Future Trends

The use of biosensors for the measurement of cancer biomarkers is expected to play a large and increasing role in the early detection and treatment of cancer and in the further development of personalized point-of-care medicine. The use of biomarkers is clearly changing the traditional paradigms of cancer diagnosis, treatment, and drug development. The rapidly growing use of biomarkers in drug development will also have major clinical benefits that lie in the screening, diagnosing, or monitoring of the state of diseases or in assessing therapeutic responses. The expanding number of cancer genome sequencing projects will lead to the identification of novel molecular pathways of carcinogenesis, new classification of heterogeneous disease entities, and the elucidation of molecular mechanisms of tumor evolution and identification of more biomarkers of cancer development. Cancer biomarker assays will grow to have many more applications in cancer diagnostics and research. In oncology, biomarker assays are being used to better understand the biological mechanisms underlying oncogenesis and to discover new targets and new drugs. Biomarkers will also be found to estimate cancer risk and cancer recurrence, as well as to predict good outcome vs. poor outcome for specific cancer therapies in specific individuals. These advances will lead to truly individualized treatments for personalized medicine, to deliver the right treatment to the right person at the right time [6,9,12,22,29,30,35]. The use of biomarkers in both diagnostics and drug discovery and development thus shows much promise. Newly discovered biomarkers and better ways to measure them will likely change both the development and use of therapeutics and will become integrated into drug development and clinical trials as well as into diagnostics and the trend toward personalized medicine. Advances in cancer biomarker assays will allow for better stratification of patients with specific forms of cancer in clinical trials, and find use in evaluating new treatments or new therapeutic targets for individual patients [36], supporting the trend toward point-of-care testing and personalized medicine.

This chapter has brought out applications and challenges of electrochemical biosensors for next-generation cancer diagnostics. Notably, point-of-care diagnostic systems hold promise for the rapid move toward more personalized medicine based on an individual(s) genetic makeup. For this purpose, it is important that the biosensors be small, rugged, and very simple to use. Next-generation biosensors for gene expression assays such as those required for point-of-care applications will allow for lower cost and more rapid analyses. New tools, such as the eSystem described here, are expected to have increasing market acceptance as clinics and hospitals gain experience in molecular diagnostics and benefit from the use of new tools that make biomarker assays practical. The low cost and ease-of-use anticipated for these tools will eventually gain them a major foothold in the marketplace, but it is realistic to predict that significant returns on the investment needed for product manufacture and commercialization will be several years down the road. Potential research uses are much more immediate, because researchers are typically eager to bring new tools into their studies if they can do so at reasonable cost. In particular, the lower cost and ease-of-use of the eSystem compared to sophisticated and expensive real-time PCR and RT-PCR should make it attractive to many researchers. A steadily growing market for research-level electrochemical molecular detection systems is anticipated and is expected to lead to expanded applications in the clinical diagnostics arena.

ACKNOWLEDGMENTS

The work described in this chapter was supported by Grant Numbers R44 CA105622 from the National Cancer Institute and R43AI073266 from the National Institute of Allergy and Infectious Diseases. The content is solely the responsibility of the authors and does not necessarily represent the official views of the National Cancer Institute or the National Institute of Allergy and Infectious Diseases or the National Institutes of Health.

REFERENCES

1. Aitichoua, M., R. Henkens, M. Afroz, R. G. Sultana, M. Ulrich, and S. Ibrahim. 2004. Detection of *Staphylococcus aureus* enterotoxin A and B genes with PCR-EIA and hand-held electrochemical sensors. *Mol. Cell. Probes* 18:373–377.
2. Antoniou, A. 2003. Average risks of breast and ovarian cancer associated with mutations in BRCA1 or BRCA2 detected in case series unselected for family history: A combined analysis of 22 studies. *Am. J. Hum. Genet.* 72:1117–1130.
3. Aslakson, C. and F. Miller. 1992. Selective events in the metastatic process defined by analysis of the sequential dissemination of subpopulations of a mouse mammary tumor. *Cancer Res.* 52:1399–1405.
4. Baldassarre, G., S. Battista, B. Belletti, S. Thakur, F. Pentimalli, F. Trapasso, M. Fedele, G. Pierantoni, C. M. Croce, and A. Fusco. 2003. Negative regulation of *BRCA1* gene expression by HMGA1 proteins accounts for the reduced BRCA1 protein levels in sporadic breast carcinoma. *Mol. Cell. Biol.* 23:2225–2238.
5. Bergamaschi, A., Y. H. Kim, P. Wang, T. Sørlie, T. Hernandez-Boussard, P. E. Lonning, R. Tibshirani, A. Børresen-Dale, and J. R. Pollack. 2006. Distinct patterns of DNA copy number alteration are associated with different clinico-pathological features and gene-expression subtypes of breast cancer. *Genes Chrom. Cancer* 45(11):1033–1040.
6. Butte, A. 2002. The use and analysis of microarray data. *Nat. Rev. Drug Discov.* 1:951–960.
7. Castillo, S. A., R. W. Henkens, V. Kazantseva, N. Naser, J. P. O'Daly, T. N. Stewart, S. E. Wegner, M. Wojciechowski, H. Zhang, and D. M. Thompson. 1997. Electrochemical assays for nucleic acid detection of disease causing agents. *Fifth Annual Advances in Nucleic Acid Amplification and Detection*, June 1997, San Francisco, CA, pp. 55–64.

8. Chau, C. H., O. Rixe, H. McLeod, and W. D. Figg. 2008. Validation of analytic methods for biomarkers used in drug development. *Clin. Cancer Res.* 14(19):5967–5976.

9. Collins, M. L., B. Irvine, D. Tyner, E. Fine, C. Zayati, C. Chang, T. Horn, D. Ahle, J. Detmer, L. P. Shen, J. Kolberg, S. Bushnell, M. S. Urdea, and D. Ho. 1997. A branched DNA signal amplification assay for quantification of nucleic acid targets below 100 molecules/ml. *Nucleic Acids Res.* 25(15):2979–2984.

10. Crumbliss, A. L., J. P. O'Daly, S. C. Perine, J. Stonehuerner, K. R. Tubergen, J. Zhao, and R. W. Henkens. 1992. Colloidal gold as a biocompatible immobilization matrix suitable for the fabrication of enzyme electrodes by electrodeposition. *Biotech. Bioeng.* 40:483–492.

11. Díaz-González, M., A. Escosura-Muñiz, M. B. González-García, and A. Costa-García. 2008. DNA hybridization biosensors using polylysine modified SPCEs. *Electrochim. Acta* 53(10):3635–3642.

12. Dobbe, E., K. Gurney, S. Kiekow, J. S. Lafferty, and J. M. Kolesar. 2008. Gene-expression assays: New tools to individualize treatment of early-stage breast cancer. *Am. J. Health Syst. Pharm.* 65(1):23–28.

13. Drummond, T. G., M. G. Hill, and J. K. Barton. 2003. Electrochemical DNA sensors. *Nat. Biotechnol.* 21:1192–1199.

14. Du, D., Z. Zou, Y. Shin, J. Wang, H. Wu, M. H. Engelhard, J. Liu, I. A. Aksay, and Y. Lin. 2010. Sensitive immunosensor for cancer biomarker based on dual signal amplification strategy of graphene sheets and multienzyme functionalized carbon nanospheres. *Anal. Chem.* 82(7):2989–2995.

15. Fanjul-Bolado, P., M. B. González-García, and A. Costa-García. Multichannel electrochemical detection system for quantitative monitoring of PCR amplification. *Clin. Chem.* 45:1690–1693.

16. Genomics World Markets. 2008. TriMark Publications, LLC. p. 16–41.

17. Gutmann, D. H., N. M. Hedrick, and J. Li. 2002. Comparative gene expression profile analysis of neurofibromatosis 1-associated and sporadic pilocytic astrocytomas. *Cancer Res.* 62:2085–2091.

18. Henkens, R. W., J. P. O'Daly, S. C. Perine, J. Stonehuerner, K. R. Tubergen, J. Zhao, and A. L. Crumbliss. 1991. Electrochemistry of colloidal gold supported oxidase enzymes. *J. Inorg. Biochem.* 43:120–134.

19. Henkens, R. W., B. Kitchell, J. P. O'Daly, S. C. Perine, and A. L. Crumbliss. 1987. Bioactive electrodes using metalloproteins attached to colloidal gold. *Recl. Trav. Chim. Pays. Bas.* 106:298–318.

20. Henkens, R., C. Bonaventura, V. Kasantseva, M. Moreno, J. O'Daly, R. Sundseth, S. Wegner, and M. Wojciechowski. 2000. Use of DNA technologies in diagnostics, Section 2. In G. G. Kennedy and T. B. Sutton, Eds. *Emerging Technologies for Integrated Pest Management: Concepts, Research, and Implementation.* APS Press, St. Paul, MN, pp. 52–66.

21. Huang, E., S. H. Cheng, H. Dressman, J. Pittman, M. H. Tsou, C. F. Hornig, A. Bild, E. S. Iversen, M. Liao, C. M. Chen, M. West, J. R. Nevins, and A. T. Huang. 2003. Gene expression predictors of breast cancer outcomes. *Lancet* 361(9369):1590–1596.

22. Innocenti, F. and R. L. Schilsky. 2009. Translating the cancer genome into clinically useful tools and strategies. *Dis. Models Mech.* 2:426–429.

23. Kricka, L. J. 1993. Labelling and detection of nucleic acid. *Mol. Diagn.* 2:26–40.

24. Lacroix, M., N. Zammatteo, J. Remacle, and G. Leclercq. 2002. A low-density DNA microarray for analysis of markers in breast cancer. *Int. J. Biol. Markers* 17(1):5–23.

25. Lin, Y. Y., J. Wang, G. Liu, H. Wu, C. M. Wai, and Y. Lin. 2008. A nanoparticle label immuno-chromatographic electrochemical biosensor for rapid and sensitive detection of prostate-specific antigen. *Biosens. Bioelectron.* 23(11):1659–1665.

26. Marx, J. 2000. DNA arrays reveal cancer in its many forms. *Science* 289:1670–1672.

27. O'Daly, J. P. and R. W. Henkens. 1995. Applications of electrochemistry to enzyme-linked immunoassay. *Rev. Pest. Toxicol.* 3:45–60.

28. O'Daly, J. P., J. Zhao, P. A. Brown, and R. W. Henkens. 1992. Electrochemical enzyme immunoassay for detection of toxic substances. *Enzyme Microb. Technol.* 14:299–302.

29. Perez, E. A., L. Pusztai, and M. Van de Vijver. 2004. Improving patient care through molecular diagnostics. *Semin. Oncol.* 31(110):14–20.

30. Pusztai, L., F. W. Symmans, and G. N. Hortobagyi. 2005. Development of pharmacogenomic markers to select preoperative chemotherapy for breast cancer. *Breast Cancer* 12:73–85.

31. Rasooly, A. and J. Jacobson. 2006. Development of biosensors for cancer clinical testing. *Biosens. Bioelectron.* 21(10):1851–1858.

32. Rasooly, A. and K. E. Herold (Eds.). 2009. *Biosensors and Biodetection: Methods and Protocols (Volumes 1 and 2). Methods in Molecular Biology.* Humana Press, Totowa, NJ.

33. Rivas, G. A., M. L. Pedano, and N. F. Ferreyra. 2005. Electrochemical biosensors for sequence-specific DNA detection. *Anal. Lett.* 38(15):2653–2703.

34. Ropp, P. A. and H. H. Thorp. 1999. Site-selective electron transfer from purines to electrocatalysts: Voltammetric detection of a biologically relevant deletion in hybridized DNA duplexes. *Chem. Biol.* 6(9):599–605.

35. Simon, R. 2005. Roadmap for developing and validating therapeutically relevant genomic classifiers. *J. Clin. Oncol.* 23:7332–7341.

36. Sotiriou, C. and L. Pusztai. 2009. Gene-expression signatures in breast cancer. *N. Engl. J. Med.* 360:790–800.

37. Sundseth, R., D. M. Thompson, G. Taylor, V. Kasensteva, A. Kasensteva, H. Zhang, M. Wojciechowski, N. Naser, M. A. Wojciechowski, C. Herfst, S. Castillo, J. O'Daly, S. Wegner, and R. Henkens. 1998. Electrochemical detection and quantitation of DNA and RNA. *Proceedings of Cambridge Healthtech Institute DNA and RNA Diagnostics Meeting*, May 1998, Washington, DC, pp. 236–245.

38. Thorp, H. H. 2003. Reagentless detection of DNA sequences on chemically modified electrodes. *Trends Biotechnol.* 21(12):522–524.

39. Tibuzzi, A., G. Pezzotti, T. Lacvecchia, G. Rea, and M. Teresa. 2008. A portable light excitation equipped bio-amperometer for electrogenic biomaterials to support the technical development of most biosensors. *Sensors Transducers J.* 88(2):9–20.

40. Tothill, I. E. 2009. Biosensors for cancer markers diagnosis. *Semin. Cell Dev. Biol.* 20(1):55–62.

41. van't Veer, L. J., H. Dai, M. J. van de Vijver, Y. D. He, A. A. Hart, M. Mao, H. L. Peterse, K. van der Kooy, M. J. Marton, A. T. Witteveen, G. J. Schreiber, M. Kerkhoven, C. Roberts, P. S. Linsley, R. Bernards, and S. H. Friend. 2002. Gene expression profiling predicts clinical outcome of breast cancer. *Nature* 415(6871):530–536.

42. Wang, J. 2006. Electrochemical biosensors: Towards point-of-care cancer diagnostics. *Biosens. Bioelectron.* 21(10):1887–1892.

43. Wang, J., G. Liu, H. Wu, and Y. Lin. 2007. Quantum-dot-based electrochemical immunoassay for high-throughput screening of the prostate-specific antigen. *Small* 4(1):82–86.

44. Wegner, S. E., M. A. Harpold, T. M. McCaffrey, S. Morris, N. Naser, J. P. O'Daly, M. Wojciechowski, J. Zhao, and R. W. Henkens. 1995. Inventors of "Colloidal Gold Electrosensor Measuring Device," Assigned to AndCare, Inc., Filed on September 30, 1994, Patent Number: 5,468,366.

45. Wojciechowski, M., F. Ebeling, N. Naser, R. Sundseth, D. M. Thompson, and R. Henkens. 1998. Intermittent pulse amperometry—A faster and more sensitive method for detection and quantitation of nucleic acids. *50th SE Regional Meeting*, American Chemical Society, November 4–7. Research Triangle Park, NC, pp. 43–56.

46. Wojciechowski, M., R. Sundseth, M. Moreno, and R. W. Henkens. 1999. Sub-attomol detection of nucleic acids using disposable microarray sensors and intermittent pulse amperometry. *218th National Meeting of American Chemical Society*, August 22–26. New Orleans, LA, pp. 74–85.

47. Wojciechowski, M., R. Sundseth, M. Moreno, and R. W. Henkens. 1999. Intermittent Pulse Amperometry. *218th National Meeting of American Chemical Society*, August 22–26. New Orleans, LA, pp. 192–203.

48. Zhang, H., N. Naser, M. Wojciechowski, T. N. Stewart, L. Shafer, J. P. O'Daly, J. Zhao, R. W. Henkens, J. Stonehuerner, and A. L. Crumbliss. 1992. Direct electron transfer at horseradish peroxidase/colloidal gold modified electrodes. *J. Electroanal. Chem.* 327:109–119.

49. Zhang, H., D. Thompson, R. Sundseth, N. Naser, J. P. O'Daly, S. Wegner, R. W. Henkens, and M. Wojciechowski. 1997. Disposable sensor-based pulse amperometric detection of pathogens and DNA mutations. *Gordon Research Conference on Electrochemistry*, January 18–23, Ventura, CA.

50. Zhang, S., J. Xia, and X. Li. 2008. Electrochemical biosensor for detection of adenosine based on structure-switching aptamer and amplification with reporter probe on DNA modified Au nanoparticles. *Anal. Chem.* 80(22):8382–8388.

51. Zhao, J., R. W. Henkens, J. G. Stonehuerner, J. P. O'Daly, and A. L. Crumbliss. 1992. Direct electron transfer at horseradish peroxidase/colloidal gold modified electrodes. *J. Electroanal. Chem.* 327:109–119.

35

Microimpedance Measurements for Cellular Transformation and Cancer Treatments

Chang Kyoung Choi, Giljun Park, and Tim E. Sparer

CONTENTS

35.1 Introduction

One out of four deaths in the United States is due to some form of cancer [1]. According to the American Cancer Society, approximately 570,000 Americans will die due to cancer in 2010, which means more than 1,500 people are dying per day from this disease. Cancer is the second leading cause of death after heart disease in the United States [2]. The National Cancer Institute (NCI) defines cancer as a disease "in which abnormal cells divide without control and can invade nearby tissues. Cancer cells can also spread to other parts of the body through the blood and lymph systems" [3]. In terms of a single cell, mutations in cell cycle control genes [4], growth factors [5], transcription factors [6], and degradation pathways [7] induce cellular transformation. Immortalization, growth in reduced serum, anchorage-independent growth, loss of contact inhibition, and tumor formation in nude mice are general characteristics of transformed cells [8–10]. Full cellular transformation is an indicator of tumorigenic potential.

There are many experimental approaches to characterize transformed cells *in vitro*. Foci formation assays address the loss of contact inhibition and soft agar growth indicates anchorage-independent growth. Molecular biological techniques (e.g., siRNA treatments, phosphoprotein screens) are also utilized to uncover the molecular mechanisms of tumorigenesis. These conventional approaches generally require days and/or weeks and are limited to a fixed time point. Because these assays are terminal, their ability to examine and explain the dynamic processes of tumorigenesis is limited.

Electrical impedance measurements employ a novel biosensor setup designed to continuously measure cellular physiological changes such as attachment, spreading, motility, and growth. In 1984 and subsequently commercialized in the early 1990s, Drs. Ivar Giaever and Charles R. Keese developed the first impedance sensing systems by applying a weak $1\,\mu A$ current to a cell culture system and measuring the voltage across the microelectrode. This had no detrimental effects on the cells and showed that an increase in the resistance/reactance was an indicator of either greater cellular mass (i.e., more cells), coverage of the electrode (i.e., spread), and/or greater attachment to the surface. These features are common with cellular transformation. This chapter reviews several examples that use an impedance measurement system for assessing cellular transformation and potential cancer treatments.

Small microelectrodes were first applied to the movement of fibroblastic cell lines to measure cellular impedances with a large counter electrode [11]. Since then, impedance measurements have allowed for quantitative real-time estimates of cellular proliferation and physiological alterations during oncogenesis. Recently, microelectrical impedance measurements were used to measure dynamic changes in foci formation in transformed NIH3T3 cells, a mouse fibroblast line, overexpressing a constitutively active chemokine receptor [12]. We have also employed microimpedance measurements to human colorectal adenocarcinoma cells, HCT-116, as a measure of their cell–cell and cell–substrate adhesive properties. This technique can be used to assess different treatment protocols and cell signaling pathway usage [13–15].

Dynamic noninvasive real-time measurements using an ITO-based biosensor allows the acquisition of data on *in vitro* growing cells without interrupting or disturbing sample growth. Optoelectrical measurements are a powerful tool for quantitatively examining cellular dynamics. This technique is capable of monitoring cellular proliferation, apoptosis, morphology, cell–cell and cell–matrix attachment, and foci formation. By streamlining the analytical process, the ITO-based biosensor has immense potential for the assessment of clinical biopsies and with additional modifications could be used to assess the susceptibility of a patient's cancer to different anticancer drugs.

In this chapter, we discuss several aspects and applications of microimpedance measurements. In the first section, we discuss the setup and description of both optical and electrical impedance measurements. This is followed by four examples of how these techniques can be applied to the study of cancerous cells:

- Impedance measurements of cellular transformation
- Tolfenamic acid (TA) induction of apoptosis in colorectal cancer cells
- Green tea catechins repression of NUDT6 in human colorectal cancer cells
- Cyclooxygenase (COX) and lipoxygenase (LOX) effects on cellular adhesion

35.2 Optical and Electrical Measurements

Because of the complexity of tumor growth and metastasis, there is an increasing need for time-dependent methods to investigate cellular physiological changes. Optically, thin ITO electrodes are used for an optoelectrical biosensor to simultaneously take optical images while obtaining microimpedance measurements. This ITO-based biosensor provides a wealth of data without interrupting or disturbing sample growth. The optoelectrical ITO biosensor is recyclable, robust, resistant to ethanol sterilization, noninvasive, and provides real-time measurements [16–19]. This technique provides a wealth of data on the characteristics of a given cancer and the efficacy of potential treatments.

35.2.1 Materials

35.2.1.1 Impedance Measurement Equipment

- SR830 Lock-in Amplifier (Stanford Research Systems, Sunnyvale, CA)
- SCXI-1331 Switch (National Instruments, Austin, TX)
- $1 \times 1 M\Omega$ and $1 \times 10 M\Omega$ resistors
- 1×10 pF Capacitor
- LCR meter
- Indium tin oxide electrode
- LABVIEW software
- Data Acquisition PCI card (DAQ-card)

35.2.1.2 Microscopy Equipment

- Olympus model IX-71 microscope
- Optical components for imaging techniques:
  - Phase contrast microscopy (PCM)
  - Differential interference contrast microscopy (DICM)
  - Interference reflection contrast microscopy (IRCM)
  - Bright field microscopy (BFM)
- Halogen lamp

35.2.1.3 Indium Tin Oxide Electrode Fabrication

The fabrication schematic and electrode arrays of ITO are shown in Figures 35.1 and 35.2. To fabricate an ITO biosensor array, six 100 nm ITO film electrodes were first deposited on a glass slide ($1'' \times 3''$) using an AJA ATC2000 RF magnetron sputtering system. Five of the electrodes are used as working electrodes, while the sixth is used as a common counter electrode. The sputtering target consists of 90 wt% In_2O_3 and 10 wt% SnO_2. During sputter deposition, a flow rate of 25 sccm of Ar–H_2 (5% H_2) is introduced into the chamber to produce a 3 mTorr pressure. The ITO sputtering power was 200 W and the film was deposited for 15 min at 400°C. Subsequent to the ITO thin-film sputtering, photoresist (955CM-2.1) was spin-coated for 50 s at 3000 rpm, resulting in a nominal 1.5 μm thickness. After spin-coating, the photoresist was soft-baked at 90°C for 45 s, then exposed to a 365 nm wavelength for

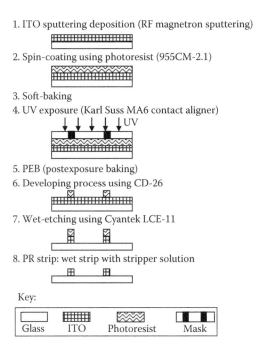

1. ITO sputtering deposition (RF magnetron sputtering)

2. Spin-coating using photoresist (955CM-2.1)

3. Soft-baking
4. UV exposure (Karl Suss MA6 contact aligner)

5. PEB (postexposure baking)
6. Developing process using CD-26

7. Wet-etching using Cyantek LCE-11

8. PR strip: wet strip with stripper solution

Key:

| Glass | ITO | Photoresist | Mask |

FIGURE 35.1 Fabrication schematic of an ITO thin-film microelectrode on a slide glass. A 90% In_2O_3/10% SnO_2 100 nm layer was sputter-coated onto a slide glass. Using standard photolithography methods, an array of five ITO electrodes and a single counter electrode was generated. (Reprinted with permission from Choi, C.K. et al., *Sensors*, 8, 3257, 2008.)

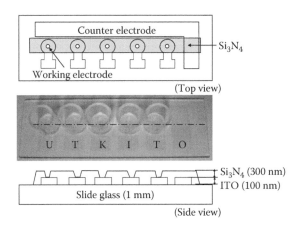

FIGURE 35.2 Indium tin oxide–silicon nitride microelectrode array. Each electrode makes an electrical contact with the common 2 cm^2 counter electrode via a 400 µL well. (Reprinted with permission from Choi, C.K. et al., *Biosens. Bioelectron.*, 22, 2585, 2007.)

1.5 s using a Karl Suss MA6 contact lithography system. Postexposure, it was baked at 120°C for 45 s and developed in CD-26 developer for 70 s. Finally, the ITO was wet-etched with Cyantek LCE-11 etchant at 40°C. Stripper solution was used to remove the photoresist at 70°C. On the top of the patterned ITO electrode, RFmagnetron sputtering in argon–hydrogen (95% Ar–5% H_2) and nitrogen gases, an insulating 300 nm silicon nitride film was deposited. The sputter deposition condition of the silicon nitride film was 100 W RF power, 5 mTorr pressure, 25 sccm Ar–H_2, and a 25 sccm N_2 gas flow at 300°C. A similar photolithography process, as described for the ITO film pattern, was used to produce the silicon nitride layer. The silicon nitride insulating wells on the array were patterned with reactive ion etching (RIE) with sulfur hexafluoride/oxide (SF_6/O_2) chemistry [16,18].

35.2.2 General Microelectrical Biosensing Method

This electrical impedance method measures the frequency-dependent resistance and reactance of a cell-covered thin-film electrode as a function of time (Figure 35.3). The impedance measurement consists of lock-in amplifier, data acquisition board, computer, and electrode. A data acquisition and analysis system was implemented using LabVIEW. The lock-in amplifier provides 1 µA, which is obtained by a reference voltage source of AC 1 Vp-p reference signal via a series 1 MΩ resistor to the electrode array. A National Instruments SCXI-1331 switch made successive connections between the five working electrodes and the counter electrode of each array. By providing a constant current of 1 µA and using the culture medium as the electrolyte, time-dependent and frequency-dependent voltages between the working electrode and the counter electrode are measured by the amplifier and converted to resistance and reactance. The source voltage generator resistance was 50 Ω. An SR830 lock-in amplifier measured the electrode voltage. The input impedance of the lock-in amplifier was equivalent to a parallel combination of 10 MΩ resistor and 10 pF capacitor. Direct measurements of the cable parasitic capacitances were made using an LCR meter and incorporated into a circuit model to estimate the impedance based on the lock-in voltage measurements.

Preliminary naked scans were performed to optimize the sensitivity at each frequency. The preliminary naked scan checked for any debris on the ITO–Si_3N_4 electrodes as well as electrode defects. A 16 s naked scan, sampled at a rate of 32 Hz, was performed at each frequency. Using the optical imaging system, images of all the working electrodes were taken just after the naked scans. One well of each of the electrode arrays was filled with medium alone as a control. Data were acquired at a rate of 32 Hz for 1 s using a 30 ms filter time constant and 12 dB/decade roll off. These same measurements were performed repeatedly, followed by the capture of cell microscopic images for each working electrode to confirm cellular attachment [16].

35.2.3 Results and Discussion

Microimpedance measures the voltage across the microelectrode while applying an approximately constant 1 µA current through it. Microimpedance consists of resistance and reactance as a function of the cellular morphology. These measurements, analyzed with the corresponding microscopic imagery, provide a more complete picture of its cellular physiology (Figure 35.4). Because microimpedance measurements are the sum of multiple different cellular parameters, additional visual assessment of cell coverage, relative packing density, and any cellular layering is necessary. Microscopic visualization compliments the quantitative measurement of cellular attachment, growth, and proliferation using the microimpedance system.

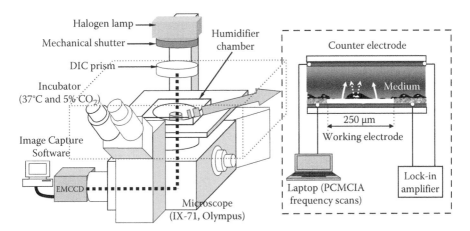

FIGURE 35.3 Schematic of integrated optical and electrical impedance measurement system. An environmental chamber maintains a humidified, 37°C, and 5% CO_2 atmosphere for dynamic long time-lapse measurements. The measured impedance is a function of the current flow under the cells, between the cells, and the capacitively coupled current through the cell membranes. (Reprinted with permission from Choi, C.K. et al., *J. Biomed. Opt.*, 12, 64028, 2007.)

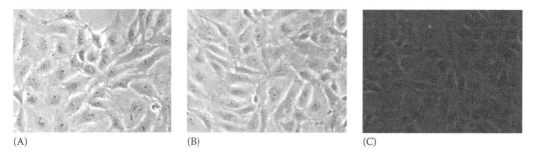

(A)   (B)   (C)

FIGURE 35.4 Images of endothelial cells on (A) a culture dish, (B) a 100 nm thick ITO layer on glass, and (C) a 47.5 nm gold film on 2.5 nm titanium-coated slide glass. Each image size is 308 m × 231 m. (Reprinted with permission from Choi, C.K. et al., *Biosens. Bioelectron.*, 22, 2585, 2007.)

This means that optoelectrical sensing has incredible potential for assessing the cellular physiology of cancerous cells. The ability to accurately and dynamically monitor cellular physiology throughout an experimental period allows for an unprecedented in-depth examination of the cellular activity of cancerous cells. The primary attributes of cancerous cells are increased cellular proliferation and foci formation, which are the largest contributors to resistance and reactance of a sample. This makes the ITO electrode an ideal biosensor for cancer cell analysis. The noninvasive nature of bioelectric sensing microimpedance measurements allows the monitoring of interior changes in a cell through an external means to track morphological changes with microscopy. Optoelectrical biosensors may prove to be the quickest and most effective means for determining effective cancer treatments or, conversely, the carcinogenic potential of a specific mutation or chemical.

Simultaneous and dynamic optoelectrical biosensing using the ITO electrode has been utilized in a number of studies to monitor cell morphology, cell–cell adhesion, cell–matrix attachment, cytotoxicity, apoptosis, proliferation, and cell growth (Figures 35.5 and 35.6) [17,18]. The ITO electrode is particularly well suited for this dual approach because of its unique combination of transparency and electrically

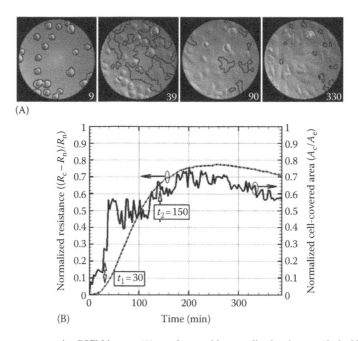

FIGURE 35.5 Four representative DICM images (A) are shown with normalized resistance (dashed lines) and normalized cell-covered area (solid lines) versus time for the subconfluence condition of ECs (B). (Reprinted with permission from Choi, C.K. et al., *J. Biomed. Opt.*, 12, 64028, 2007.)

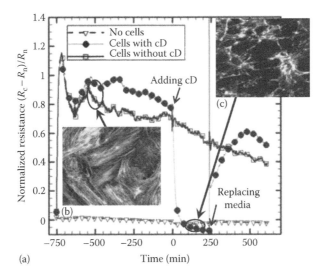

(a)

FIGURE 35.6 Microimpedance measurements of the cytotoxic effects of the actin inhibitor, cytochalasin D, on endothelial cells. (a) Normalized resistance response. Data were obtained every 1.2 s. However, in order to distinguish lines, only 20 data points are shown. Stained actin filaments in confocal images before (b) and after (c) the addition of cytochalasin D. (Reprinted with permission from Choi, C.K. et al., *Sensors*, 8, 3257, 2008.)

conductive properties. Traditionally, similar assays were performed using gold electrodes, which is the industry's standard electrode, but images taken at the same light intensity are clearly visible through the ITO electrode while obscured through the gold electrode (Figure 35.4) [16,18].

35.3 Impedance Measurements of Cellular Transformation

Cellular transformation is the first step in cancer development and has two distinctive characteristics: (1) proliferation in reduced serum, and (2) loss of cellular contact inhibition. Electrical impedance measurements were used to distinguish between normal cells and cells transformed with a constitutively active chemokine receptor, CXCR2. CXCR2, a member of the G-protein-coupled receptor (GPCR) family, is normally involved in cellular activation and migration, but a single amino acid substitution leads to constitutive activity. The substitution of a valine (V) for an aspartic acid (D), called D143V_CXCR2, in the second intracellular loop of CXCR2 leads to continual signaling and cellular transformation [20].

In this section, NIH3T3 cells were transformed with D143V_CXCR2 and growth in reduced serum was measured using a hemocytometer, the traditional method of counting cells. Foci formation of the stable transfectants was measured using established biological assays and compared to data from the microimpedance measurements. This shows that impedance measurements provide a quick and reliable way of measuring cellular transformation and could be used to distinguish cancerous cells from noncancerous cells [12].

35.3.1 Materials

35.3.1.1 Cell Lines and Growth Media

NIH3T3 mouse fibroblasts (American Type Culture Collection [ATCC], Manassas, VA), were grown in Dulbecco's modified Eagle's medium (DMEM) containing 10% fetal calf serum (FCS) (Hyclone, Logan, UT).

35.3.1.2 Reagents

- Transfection reagent: Lipofectamine 2000™ (Invitrogen, Carlsbad, CA)
- CXCR2 antagonist (SB225002 [Calbiochem, La Jolla, CA])

- Takara LA taq™ polymerase (Takara Bio, Madison, WI)
- MAX Efficiency® DH5α™ Competent Cells (Invitrogen)
- G418 Sulfate (Cellgro, Manassas, VA)
- Crystal violet solution (25% ethanol, 1% formaldehyde, 0.125% NaCl, 0.25% crystal violet)
- Trypan blue stain (0.4%, Lonza, Allendale, NJ)
- Trypsin–EDTA solution (Invitrogen)
- Phosphate-buffered saline (Hyclone)

35.3.1.3 Construction of the Expression Vector and Stable Transfectants

NIH3T3 cultured in DMEM containing 10% FCS (Hyclone) were transfected with pRc/CMV (Invitrogen) expressing either WT_CXCR2 or D143V_CXCR2 with the Lipofectamine 2000 (Invitrogen) according to the manufacturer's instructions. Stable transfectants were selected with 800 µg/mL of G418 sulfate. To generate D143V_CXCR2, the WT_CXCR2 pRc/CMV expression plasmid was mutagenized using the complementary primers (D143V_Forward: 5'-GCA TCA GTG TGG TTC GTT ACC TGG CCA TTG TCC ATG C-3'; D143V_Reverse: 5'-GCC AGG TAA CGA ACC ACA CTG ATG CAG GCC AGT AGC-3') and a modified QuickChange® method (Stratagene, Santa Clara, CA). Briefly, Takara LA taq polymerase (Takara Bio) was used for the mutagenesis. *Dpn*I treatment was followed by PCR amplification (98°C for 3 min, 35 cycles of 98°C for 30 s, 55°C for 30 s, and 1 cycle of 72°C for 10 min, and a final extension of 72°C for 20 min). This reaction, cleaned up with a QIAquick PCR purification kit (QIAGEN, Valencia, CA), was transformed into MAX Efficiency DH5α Competent Cells (Invitrogen). Mock PCR setup without polymerase and/or without primers provided negative controls.

35.3.1.4 Cellular Proliferation Assay

Stable transfectants expressing WT_CXCR2, D143V_CXCR2, or untransfected NIH3T3 cells were plated at 1×10^5 cells and grown in reduced (2%) serum at 37°C and 5% CO_2. At the time points indicated, cells were trypsinized, stained with trypan blue, and counted using a hemocytometer. Each time point was measured in triplicate.

To determine cellular proliferation using impedance measurements, a total of 3×10^4 transfectants and untransfected NIH3T3 cells were inoculated in 400 µL normal growth medium onto the electrode. One well remained untreated to provide a negative control.

35.3.1.5 Foci Formation Assay

One hundred stably transfected NIH3T3 cells were seeded on top of 2×10^5 untransfected NIH3T3 cells in normal growth medium. The growth medium was exchanged every 2 or 3 days. After 5–7 days, the cells started to form colonies. After 10–14 days of initial seeding, the cells were fixed with 70% ethanol and stained with a crystal violet solution. To enumerate the number of colonies, the wells were photographed and analyzed using *ImageJ* version 1.43 [21].

Using the impedance apparatus, electrodes were inoculated with 400 µL of normal growth medium containing 6×10^4 untransfected NIH3T3 cells to produce a cellular base layer to provide growth factors. A total of 3×10^4 transfectants or untransfected controls were seeded on the NIH3T3 layer at the saturated growth time point (~26 h) in 100 µL of normal growth medium in the presence of 1 mM CXCR2 inhibitor (SB225002). This was added in order to minimize background growth of WT_CXCR2 due to chemokines in the serum. Impedance measurements were acquired for another 61 h to produce a second set of scans.

35.3.2 Results and Discussion

Cellular transformation is the first step in cancer development. The initial transition from a normal cell to a transformed cell may identify specific mutations that lead to cellular transformation. Identification

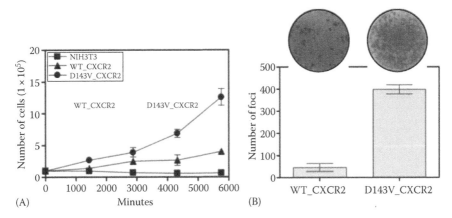

FIGURE 35.7 Traditional biological measurements of cellular proliferation (A) and foci formation (B). (Reprinted with permission from Park, G. et al., *Cell Biol. Int.*, 33, 429, 2009.)

of transforming mutations could allow the development of "personalized" treatments for specific mutations in a cancer. Electrical impedance measurements, particularly electronic cell–substrate impedance sensing (ECIS), have proven to be an effective method for detecting the precursors of cellular transformation and subsequent tumorigenesis [12]. This technique has been utilized to identify transformed cells expressing constitutively active chemokine receptors, D143V_CXCR2. CXCR2 is a GPCR normally involved in the activation and migration of cells but becomes constitutively active with a single amino acid substitution (D to V at position 143). Overexpression of this mutation leads to increased proliferation and loss of contact inhibition (Figure 35.7), which was reflected in the impedance measurements (Figure 35.8).

In this study, the rate of proliferation and foci formation were analyzed in both CXCR2-transfected and CXCR2-untransfected NIH3T3 mouse fibroblasts. Preliminary scans were made of the "naked" electrode in cell growth media. Then, three groups of cells were seeded onto their respective electrodes: WT_CXCR2 and D143V_CXCR2 transfectants, and untransfected NIH3T3 cells. Impedance and reactance measurements were made continuously over the course of the experiment (about 2 weeks). At regular time intervals throughout the experiment, the cells were trypsinized, stained, and counted.

The results for both cellular proliferation and foci formation align with the coincident impedance measurements. Figure 35.7 shows traditional biological measurements of cellular proliferation (A) and foci formation (B). As shown in Figure 35.8, the resistance (Figure 35.8A) and reactance (Figure 35.8B), which were normalized to the naked electrode scans, follow the same distribution as the number of cells counted using conventional techniques (Figure 35.7A). The normalized resistance (Figure 35.8C) and reactance (Figure 35.8D) of the D143V_CXCR2-treated electrodes greatly eclipse that of both the untransfected control and the weakly active WT_CXCR2 transfectants, which is in agreement with the physical enumeration of foci (Figure 35.7B).

35.4 TA Induction of Apoptosis in Colorectal Cancer Cells

Electrical impedance was used to investigate the role of nonsteroidal anti-inflammatory drug (NSAID) and a green tea catechin, TA, on colorectal cancer cells. In order to dissect the pathway induced upon TA treatment, colorectal cancer cells' stable transfectants overexpressing intermediates of this pathway were generated and impedance measurements taken. TA binds to the epithelial-specific ETS-1 (ESE-1) protein inducing early growth response-1 (EGR-1) expression and the apoptotic pathway. As seen by its affect on impedance measurements, ESE-1/EGR-1 plays an important role in TA-induced apoptosis [13]. This example shows that impedance measurements can be used to dissect the pathways involved in anticancer treatments.

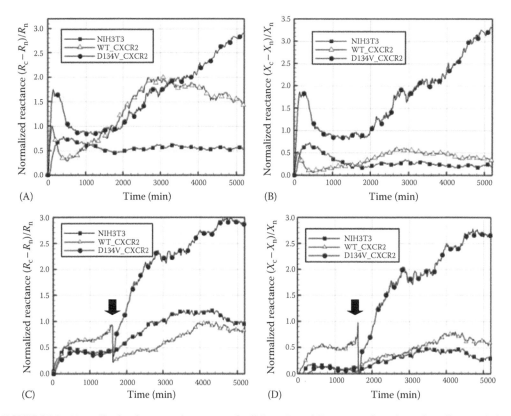

FIGURE 35.8 Normalized resistance and reactance of cell lines. R_c and X_c represent impedance for cell-covered electrodes and R_n and X_n represent naked scans. (A) and (B) measurements are from transfectants grown in reduced serum. (C) and (D) are scans following foci formation growth experimental setup. Arrow indicates the addition of transfectants on top of the untransfected monolayers. (Reprinted with permission from Park, G. et al., *Cell Biol. Int.*, 33, 429, 2009.)

35.4.1 Materials

35.4.1.1 Cell Lines and Growth Media

A human colorectal cancer cell line, HCT-116, was used. HCT-116 cell line purchased from ATCC was maintained in McCoy's 5A medium.

35.4.1.2 Reagents

- TA (Cayman Chemical, Ann Arbor, MI)
- CellTiter96 Aqueous One Solution Cell Proliferation Assay (Promega, Madison, WI)
- Lipofectamine (Invitrogen)
- EGR-1 expression plasmid

35.4.1.3 Cellular Proliferation Assay

Cells were incubated at 37°C in 5% CO_2 until wells were between 70% and 80% confluent. Cellular proliferation assays were performed using CellTiter96 Aqueous One Solution Cell Proliferation Assay (Promega). Cells were seeded at 2000 per well in 96-well tissue culture plates in quadruplicate and treated with 0, 1, 5, 10, 20, 30, 50 µM TA. At 0, 24, and 48 h after treatment, 20 µL CellTiter96 Aqueous One solution was added to each well, after which the plate was incubated at 37°C for 1 h. Absorbance

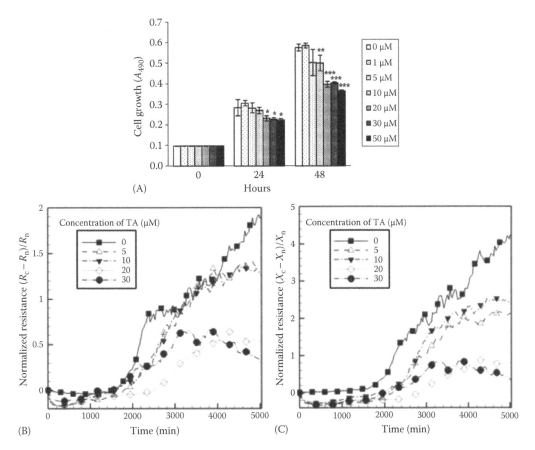

FIGURE 35.9 Treatment of colorectal cell lines with TA diminishes cellular growth. (A) HCT-116 growth using traditional methods. (B) Normalized resistance, $(R_c - R_n)/R_n$, and (C) reactance, $(X_c - X_n)/X_n$, in relation to TA concentrations. The terms R and X represent the resistance and reactance, respectively, and the subscripts c and n indicate cell-covered and naked scans, respectively. Measurements were performed simultaneously using the same batch of HCT-116 cells. The representative time-dependent normalized resistances and reactances were scanned at 5.62 and 100 kHz. Data shown in the figure were chosen at approximately every 30 min. (Reprinted with permission from Lee, S.H. et al., *Mol. Cancer Ther.*, 7, 3739, 2008, PMCID: 2643071.)

at 490 nm was measured using a 96-well plate reader. Figure 35.9A shows cell growth using the CellTiter96 assay. Values are expressed as mean ± SD of three independent experiments. *, $P < 0.05$; **, $P < 0.01$; ***, $P < 0.001$ versus dimethyl sulfoxide (DMSO)-treated cells at each time point.

35.4.1.4 EGR-1 Transfections

For assessing the role of TA and EGR-1 in apoptosis of HCT-116, HCT-116 cells were transfected with an EGR-1 expression plasmid [22] using Lipofectamine (Invitrogen) according to the manufacturer's instructions.

35.4.1.5 Impedance Measurements

Cells were treated with 0, 5, 10, 20, and 30 μM of TA and impedance was measured continuously. Data points are only shown every 8 h for simplicity. The experimental settings for these readings are similar to the ones described earlier. For assays with transfectants overexpressing EGR-1, transfectants were treated with 30 μM of TA after transfection and during microimpedance measurements.

35.4.2 Results and Discussion

NSAIDs have been shown to prevent tumorigenesis of cancers including colorectal cancers [13]. Although their primary mode of action is the inhibition of COX, other mechanisms are also at work. In particular, TA is an NSAID that exhibits anticancer activities including the inhibition of cell growth and the induction of apoptosis. TA induces translocation of ESE-1 transcription factor into the nucleus and subsequent upregulation of EGR-1 expression, which is linked to tumor suppression through apoptosis.

TA treatment of HCT-116 cells was monitored both bioelectronically and optically. Both methods confirmed the anticancer function of TA-induced EGR-1 overexpression [13]. Figure 35.9 shows the normalized resistance (B) and reactance (C) of the TA-treated samples in relation to samples treated with vehicle (DMSO). The data show a dose-dependent decrease in resistance and reactance in samples treated with the TA compared to those treated with DMSO, thus demonstrating that TA has anticancer activities that could be detected using impedance measurements.

For dissecting the role of TA on colorectal cancers, HCT-116 cells were transfected with an EGR-1 overexpression plasmid. Empty vector control transfectants were used as a negative control. Transfectants were treated with TA (30 μM) or vehicle (DMSO) and impedance was measured over time. TA treatment of HCT-116 with the empty vector had decreased growth matching the results in Figure 35.9. Interestingly, TA treatment of cells overexpressing EGR-1 had an even further reduction in growth as indicated with decreased resistance/reactance (Figure 35.10). These findings demonstrate that

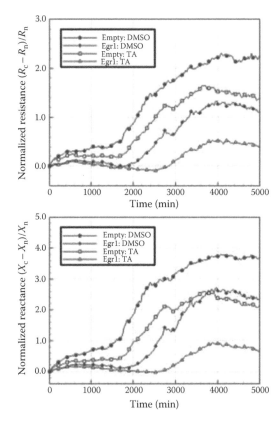

FIGURE 35.10 Treatment of colorectal cell lines with TA is enhanced with overexpressing EGR-1. HCT-116 cells were transfected with either empty plasmids (empty) or EGR-1 overexpression plasmids (EGR-1). Transfectants were treated with either TA (30 μM) or vehicle (DMSO) and impedance measurements were taken. The representative time-dependent normalized resistances and reactances shown here were scanned at 5.62 and 100 kHz, respectively. Filled symbols represent cell measurements with DMSO and open symbols represent cells treated with 30 μM of TA. For the sake of clarity, symbols are selectively marked. (Reprinted with permission from Lee, S.H. et al., *Mol. Cancer Ther.*, 7, 3739, 2008, PMCID: 2643071.)

impedance measurements can be used for assessing drug treatments for cancer and their mechanisms of action.

35.5 Green Tea Catechins Repression of NUDT6 in Human Colorectal Cancer Cells

Catechins found in green tea, specifically epigallocatechin-3-gallate (EGCG), were shown to repress nudix-type motif 6 (NUDT6) expression, thereby lowering the rate of tumorigenesis in human colorectal cancer cells [15]. Microimpedance measurements confirm that NUDT6-expressing lines had significantly larger resistance and reactance, indicating higher rates of growth and proliferation. The effect of green tea catechins on NUDT6 suppression and NUDT6's biological activity was characterized using impedance measurements [15].

35.5.1 Materials

35.5.1.1 Cell Lines

A human colorectal cancer cell line, HCT-116, was used. HCT-116 cell line purchased from ATCC was maintained in McCoy's 5A medium.

35.5.1.2 Reagents

- Lipofectamine (Invitrogen)
- NUDT6 expression vector (pcDNA3.1/V5-His-TOPO/NUDT6) or CONTROL vector (pcDNA3.1/V5-His-TOPO/CONTROL)
- Trizol (Invitrogen)
- Master Cycler Gradient (Eppendorf, Hauppauge, NY)
- PCR Primers
- Anti-V5 tag antibody (Invitrogen)
- G418 (Stratagene)
- *p*-Iodonitrotetrazolium violet solution (Sigma, St. Louis, MO)
- CellTiter96 Aqueous One Solution Cell Proliferation Assay (Promega)

35.5.1.3 NUDT6 Expression Constructs

The full-length human NUDT6 cDNA (915 bp) was isolated using reverse transcriptase PCR (see the following) from HCT-116 cells using forward (5′-ggacgaattaagcggcgtggaga-3′) and reverse (5′-atcaattcctttcatagtttttat-3′) primers designed from the reported human NUDT6 cDNA sequence (GenBank no. NM_007083). Amplified PCR products were cloned into the expression vector, pcDNA3.1/V5-His-TOPO vector (Invitrogen). NUDT6 cloned in the correct 5′–3′ orientation was named pcDNA 3.1/NUDT6 while cloning in the reverse orientation, the negative control, was named pcDNA 3.1/CONTROL.

35.5.1.4 RNA Isolation and RT-PCR

RNA extraction was performed using Trizol (Invitrogen) and cDNA was prepared with an i-script synthesis kit (Bio- Rad Laboratories, Hercules, CA). RT-PCR was carried out as previously described [23]. The PCR primers used were as follows: human NUDT6, 5′-catcctccaaagccgattta-3′ (forward) and 5′-aacttctcgaaccgctgtgt-3′ (reverse). The thermal cycle settings used on a Master Cycler Gradient (Eppendorf) were as follows: 94°C for 2 min as initial denaturation, 94°C for 30 s, 55°C for 30 s, and 72°C for 1 min. Amplification products (25–30 cycles) were analyzed on a 1.2% agarose gel and bands were visualized using ethidium bromide [15].

35.5.1.5 NUDT6-Expressing Stable Cell Lines

HCT-116 cells were plated in 6 cm plates and transfected with either NUDT6 expression vector (pcDNA3.1/V5-His-TOPO/NUDT6) or CONTROL vector (pcDNA3.1/V5-His-TOPO/CONTROL) using Lipofectamine (Invitrogen) in accordance with the manufacturer's protocol. After 24 h, cells were transferred to a 10 cm plate with G418 (500 µg/mL) (Stratagene). Selection with G418 was carried out for 3 weeks and then western blot analysis using an antibody against the V5 tag (Invitrogen) was used to confirm expression in the stable cell lines [15].

35.5.1.6 Soft Agar Growth

Soft agar assays were established as previously described [24]. After growth, colonies were visualized by staining with 0.5 mL of *p*-iodonitrotetrazolium violet (Sigma). Colonies were counted using the Multi Gauge Program (Fujifilm Co., Tokyo, Japan) in accordance with the manufacturer's instructions.

35.5.1.7 Impedance Measurements

In order to examine the effect of NUDT6 on cellular growth, 2.4×10^4 NUDT6 and control stable transfectants were inoculated in 400 µL of McCoy5A onto the microelectrodes. Impedance measurements were recorded over 3.5 days.

35.5.2 Results and Discussion

To confirm the tumor-promoting behavior of NUDT6, NUDT6-expressing cells or control cells were seeded and their growth and apoptosis rates were monitored. Stable cell lines expressing NUDT6 had higher proliferation and lower apoptosis rates in comparison to control cells (Figure 35.11). In order to examine cellular growth (Figure 35.11A), control and NUDT6 stable cell lines were plated onto a 96-well plate. Cell growth was measured using the CellTiter96 Aqueous One Solution Cell Proliferation Assay. Values are expressed as the mean ± SD of six replicates. The NUDT6 cell lines showed markedly higher absorbance as an indication of growth rates compared to the control cells. Soft agar growth (i.e., anchorage-independent growth) is used as an indicator of the degree of transformation of a cancerous cell. The soft agar assay shows a significant increase in colony formation in cell lines expressing NUDT6 (Figure 35.11B). These results are representative of three different experiments. Bars represent the mean ± SD. Normalized resistance and reactance measurements (Figure 35.11C) of the stable HCT-116 cell lines support the findings using the proliferation kit and the soft agar assay. The subscripts "c" and "n" indicate cell-covered and naked scans, respectively. For the sake of clarity, symbols represent 20 data intervals. The NUDT6-expressing lines had significantly larger resistance and reactance values, indicating higher rates of growth and proliferation. These data support the hypothesis that NUDT6 expression promotes transformation/increased growth rate in colorectal cell lines.

In this study, epicatechin-3-gallate (EGCG) catechins found in green teas act as NUDT6 down-regulators by affecting mRNA stability. In fact, there was a 4.44-fold decrease in NUDT6 gene expression in samples treated with the EGCG versus untreated. The conclusion of this study is that NUDT6 decreases apoptosis and increases cellular proliferation, which can be repressed with the green tea catechin EGCG [15].

35.6 COX and LOX Effects on Cell Adhesion Detected Using Impedance Measurements

Expression of COX and LOX has been linked to many pathophysiological phenotypes, including cell adhesion [25]. In order to dynamically investigate the effect of COX and LOX on cell adhesion, stable

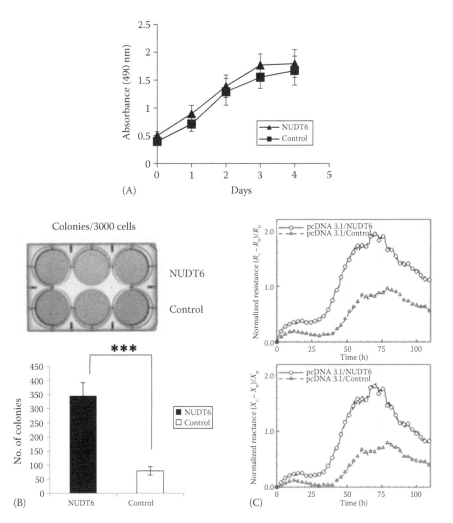

FIGURE 35.11 Characterization of the biological activity of NUDT6 in human colorectal cancer cells. (A) Traditional proliferation assay measuring growth in NUDT6-transfected cells versus nontransfectants. (B) Overexpressing NUDT6 cell lines form colonies in soft agar. (C) Impedance measurements show that NUDT6-overexpressing cell lines have increased growth rate/cell density as seen in traditional proliferation assay. *** Indicates significant difference ($P < 0.001$). (Reprinted with permission from Sukhthankar, M. et al., *J. Nutr. Biochem.*, 21, 98, 2010.)

human colorectal cancer cell lines ectopically expressing COX-1, COX-2, and 15LOX-1 were generated. The optical and electrical data from the impedance measurement show that both COX- and LOX-expressing cells have strong cell–cell and cell–substrate adhesions. Microscopic image analysis was coupled with cellular impedance data giving reliable cell adhesion measurements [14].

35.6.1 Materials

35.6.1.1 Cell Lines

HCT-116 cell lines purchased from ATCC were maintained in McCoy's 5A medium with 10% fetal bovine serum, 100 µg/mL penicillin, and 100 µg/mL streptomycin. The cells were grown at 37°C under 5% CO_2. HCT-116 transfectants overexpressing COX-1, COX-2, or 15LOX-1 were previously generated [26,27].

35.6.1.2 Reagents

- HCT-116 overexpressing 15LOX-1, COX-1, or COX-2
- McCoy's 5A media (Media Tech, Manassas, VA)
- Fetal bovine serum (Hyclone)

35.6.1.3 Impedance Measurement and Microscopic Imaging

HCT-116 cells were used as the "empty" controls, while stable overexpressing 15LOX-1, COX-1, and COX-2 cell lines were tested in the optoelectrical setup. Four hundred microliters of cells in media (10^5 cells/mL) were inoculated into each well. Optical and electrical data were continuously gathered using the electrodes coupled with both IRCM and DICM. Between these techniques, comprehensive data on cellular adhesion were gathered over 4½ days. The resistance and reactance values were normalized to "naked" scans performed prior to the experiment.

35.6.2 Results and Discussion

COX-1 and COX-2 generate prostaglandins from arachidonic acids. These enzymes have been implicated in tumor development and are the targets of NSAIDs with anticancer effects. COX-1 is constitutively expressed in many cell types while COX-2 is mostly expressed in transformed cell lines. Reports have shown that both COX-dependent and -independent methods are involved in carcinogenesis [28,29]. Several of their primary metabolites regulate receptor pathways that lead to increased transduction events increasing adhesion to the extracellular matrix (ECM) [25]. An impedance system was employed in tandem with microscopic techniques to monitor cell–cell and cell–substrate adhesions of COX- and 15LOX-1-overexpressing cancerous cells. These measurements supply information on cellular adhesion, COX pathways, and cancerous activity. Cellular adhesion has been extensively studied using only static assays prior to the application of the impedance system. The dynamic nature of the cell, especially its cellular interaction with other cells and the ECM, requires time-dependent systems to accurately and actively investigate the role of internal signaling with cellular communications. The impedance of the culture is a function of such intra- and extracellular interactions as well as growth rates. Given the complexity of the COX responses, microimpedance measurements were used to investigate the role of COX expression on colorectal tumor lines.

LOXs are lipid-peroxidizing enzymes that oxygenate arachidonic acid. Overexpression of the LOX, 15LOX-1, is associated with inhibition of colorectal tumor cell growth [27]. HCT-116 cell lines, which do not express COX-1, COX-2, or 15LOX1, were transfected with either an empty vector or plasmids overexpressing COX-1, COX-2, or 15LOX1. These lines were used in optoelectrical impedance assays to assess differences in cellular physiology of cancerous cells overexpressing these proteins.

The electrical characteristics of the different cell lines did not diverge until 2 days into the experiment after which the untransfected HCT-116 and 15LOX-1 cell cultures both plateaued (Figure 35.12). This agrees with the overall inhibition of growth due to 15LOX-1 while the COX-1 and COX-2 transfectants displayed continued growth/adherence. The optical analysis showed an even more complex phenotype (Figure 35.13). Both the COX- and LOX-overexpressing cell lines showed strong cell–cell and cell–substrate adhesions. The pixel gray levels (PGL) of the IRCM images were analyzed. Darker PGLs (i.e., more intense) indicate tighter cellular adhesions. Figures representing the gray scale intensity and the number of pixels at those intensities are shown for Days 3 and 4.5. These data show that the COX- and 15LOX-1-overexpressing cells had higher numbers of contacts at these time points indicating an alteration of the signaling pathways in the cancerous cells.

35.7 Concluding Remarks

In each of these cases, the optoelectrical biosensing approach was used to confirm and quantify results obtained through traditional methods. Impedance sensing confirmed the increased rate of proliferation

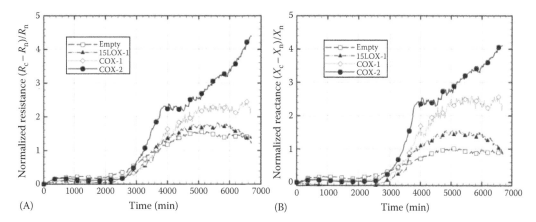

FIGURE 35.12 Normalized resistance $(R_c - R_n)/R_n$, recorded at 5.62 kHz, and normalized reactance $(X_c - X_n)/X_n$, recorded at 56.2 kHz, for empty vector, 15LOX-1-, COX-1-, and COX-2-transfected HCT-116 cells. The terms R and X represent the resistance and reactance, respectively, and the subscripts c and n indicate cell-covered and naked scans, respectively. (Reprinted with permission from Choi, C.K. et al., *Biochem. Biophys. Res. Commun.*, 391, 1385, 2010, PMCID: 2814420.)

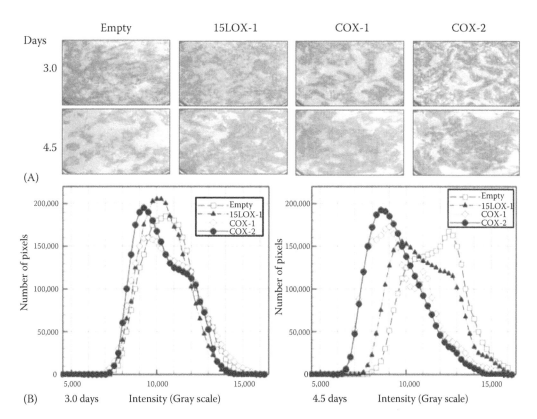

FIGURE 35.13 Interference reflection contrast microscopy (IRCM) (A) at days 3 and 4.5 after inoculating cells onto the electrode. The corresponding distribution of pixel gray levels (PGLs) representing cellular contact characteristics between cells and substrates, focal contacts, and close contacts (B). Digital imaging processing of IRCM images was employed to measure cell–substrate distance. The level of PGL, whose range is from 0 to 16,384, corresponds to the cell–substrate gap distance. At 3 days after inoculation, cells were confluent. Smaller PGL values represent the smaller gaps between cell and substrate. The absence of pixels indicates an area devoid of cells. Darker PGL values mean smaller gaps between the substrate and the cell (i.e., focal contacts). (Reprinted with permission from Choi, C.K. et al., *Biochem. Biophys. Res. Commun.*, 391, 1385, 2010, PMCID: 2814420.)

and foci formation of constitutively active CXCR2 transfectants. This shows that electrical impedance can be used to assess cellular transformation induced following overexpression of the constitutively active chemokine receptor. If this were modified for high-throughput analysis, impedance measurements could be used to assess whether *any* mutation results in cellular transformation.

Microimpedance measurements can also be utilized to assess treatments and important signaling pathways involved in tumor development. In the TA experiments, a combination of absorbance measurements and impedance readings separately confirmed that TA treatments decrease cellular growth. Using overexpressors of the important signaling molecules, impedance measurements confirmed the role of EGR-1 in inducing apoptosis. The biosensor was also able to contribute to the investigation of how a green tea catechin reduces NUDT6 expression resulting in decreased proliferation. The increased proliferation of NUDT6-expressing cells was confirmed using a cellular proliferation/absorbance assay demonstrating its role in transformation. Using impedance measurements and both IRCM and DICM, differences in adhesion and growth of 15LOX-1-, COX-1-, or COX-2-overexpressing colorectal cell lines were demonstrated. This showed the importance and versatility of optoelectrical measurements for assessing the role of these targets of COX inhibitors might have in tumorigenesis. In the near future, microimpedance will serve as a fast, reliable, and relatively easy method for the analysis of cellular activities. Impedance measurements can be used to determine whether a given biopsy contains transformed cells, to examine drug treatment regimes on an individual's cancer, or to dissect the cellular pathways involved in a particular cancer. These results have shown that optoelectrical impedance measurements are particularly well suited to many aspects of cancer research, a discipline always in search of more efficient techniques for discovering and treating cancers.

REFERENCES

1. Jemal A, Siegel R, Ward E, Murray T, Xu J, Smigal C, Thun MJ. Cancer statistics. *CA Cancer J. Clin.* 2006;56(2):106–130.
2. Cancer Facts & Figures 2010. 2010 (updated 2010; cited July 14, 2010). Available from: http://www. cancer.org/docroot/stt/stt_0.asp
3. Dictionary of Cancer Terms. 2010 [updated 2010; cited July 14, 2010]. Available from: http://www. cancer.gov/dictionary/?CdrID=45333
4. Cordon-Cardo C. Mutations of cell cycle regulators. Biological and clinical implications for human neoplasia. *Am. J. Pathol.* 1995;147(3):545–560.
5. Inoue A, Nukiwa T. Gene mutations in lung cancer: Promising predictive factors for the success of molecular therapy. *PLoS Med.* 2005;2(1):e13.
6. Latchman DS. Transcription-factor mutations and disease. *N. Engl. J. Med.* 1996;334(1):28–33.
7. Guardavaccaro D, Pagano M. Oncogenic aberrations of cullin-dependent ubiquitin ligases. *Oncogene* 2004;23(11):2037–2049.
8. McGlennen RC, Ghai J, Ostrow RS, LaBresh K, Schneider JF, Faras AJ. Cellular transformation by a unique isolate of human papillomavirus type 11. *Cancer Res.* 1992;52(21):5872–5878.
9. Bais C, Santomasso B, Coso O, Arvanitakis L, Raaka EG, Gutkind JS, Asch AS, Cesarman E, Gershengorn MC, Mesri EA. G-protein-coupled receptor of Kaposi's sarcoma-associated herpesvirus is a viral oncogene and angiogenesis activator. *Nature* 1998;391(6662):86–89.
10. Alberts B, Wilson JH, Hunt T. *Molecular Biology of the Cell*, 5th ed. New York: Garland Science, 2008.
11. Giaever I, Keese CR. Monitoring fibroblast behavior in tissue-culture with an applied electric-field. *Proc. Natl Acad. Sci. U.S.A. Biol. Sci.* 1984;81(12):3761–3764.
12. Park G, Choi CK, English AE, Sparer TE. Electrical impedance measurements predict cellular transformation. *Cell Biol. Int.* 2009;33(3):429–433.
13. Lee SH, Bahn JH, Choi CK, Whitlock NC, English AE, Safe S, Baek SJ. ESE-1/EGR-1 pathway plays a role in tolfenamic acid-induced apoptosis in colorectal cancer cells. *Mol. Cancer Ther.* 2008;7(12):3739–3750. PMCID: 2643071.
14. Choi CK, Sukhthankar M, Kim CH, Lee SH, English A, Kihm KD, Baek SJ. Cell adhesion property affected by cyclooxygenase and lipoxygenase: Opto-electric approach. *Biochem. Biophys. Res. Commun.* 2010;391(3):1385–1389. PMCID: 2814420.

15. Sukhthankar M, Choi CK, English A, Kim JS, Baek SJ. A potential proliferative gene, NUDT6, is down-regulated by green tea catechins at the posttranscriptional level. *J. Nutr. Biochem.* 2010;21(2):98–106.

16. Choi CK, English AE, Jun S, Kihm KD, Rack PD. An endothelial cell compatible biosensor fabricated using optically thin indium tin oxide silicon nitride electrodes. *Biosens. Bioelectron.* 2007;22(11):2585–2590.

17. Choi CK, English AE, Kihm KD, Margraves CH. Simultaneous dynamic optical and electrical properties of endothelial cell attachment on indium tin oxide bioelectrodes. *J. Biomed. Opt.* 2007;12(6):64028.

18. Choi CK, Margraves CH, Jun SI, English AE, Rack PD, Kihm KD. Opto-electric cellular biosensor using optically transparent indium tin oxide (ITO) electrodes. *Sensors* 2008;8:3257–3270.

19. Choi CK, Kihm KD, English AE. Optoelectric biosensor using indium–tin-oxide electrodes. *Opt. Lett.* 2007;32(11):1405–1407.

20. Burger M, Burger JA, Hoch RC, Oades Z, Takamori H, Schraufstatter IU. Point mutation causing constitutive signaling of CXCR2 leads to transforming activity similar to Kaposi's sarcoma herpesvirus-G protein-coupled receptor. *J. Immunol.* 1999;163(4):2017–2022.

21. Rasband, W.S., *ImageJ*, U.S. National Institutes of Health, Bethesda, MD, 1997–2009, http://rsb.info.nih.gov/ij/

22. Baek SJ, Wilson LC, Hsi LC, Eling TE. Troglitazone, a peroxisome proliferator-activated receptor γ (PPARγ) ligand, selectively induces the early growth response-1 gene independently of PPARγ. *J. Biol. Chem.* 2003;278(8):5845–5853.

23. Cho K-N, Sukhthankar M, Lee S-H, Yoon J-H, Baek SJ. Green tea catechin (−)-epicatechin gallate induces tumour suppressor protein ATF3 via EGR-1 activation. *Eur. J. Cancer* 2007;43(16):2404–2412.

24. Baek SJ, Kim K-S, Nixon JB, Wilson LC, Eling TE. Cyclooxygenase inhibitors regulate the expression of a TGF-β superfamily member that has proapoptotic and antitumorigenic activities. *Mol. Pharmacol.* 2001;59(4):901–908.

25. Tsujii M, DuBois RN. Alterations in cellular adhesion and apoptosis in epithelial cells overexpressing prostaglandin endoperoxide synthase 2. *Cell* 1995;83(3):493–501.

26. Baek SJ, Wilson LC, Lee C-H, Eling TE. Dual function of nonsteroidal anti-inflammatory drugs (NSAIDs): Inhibition of cyclooxygenase and induction of NSAID-activated gene. *J. Pharmacol. Exp. Ther.* 2002;301(3):1126–1131.

27. Kim J-S, Baek SJ, Bottone FG, Sali T, Eling TE. Overexpression of 15-lipoxygenase-1 induces growth arrest through phosphorylation of p53 in human colorectal cancer cells. *Mol. Cancer Res.* 2005;3(9):511–517.

28. Greenhough A, Smartt HJM, Moore AE, Roberts HR, Williams AC, Paraskeva C, Kaidi A. The COX-2/PGE$_2$ pathway: Key roles in the hallmarks of cancer and adaptation to the tumour microenvironment. *Carcinogenesis* 2009;30(3):377–386.

29. Garcia MC, Ray DM, Lackford B, Rubino M, Olden K, Roberts JD. Arachidonic acid stimulates cell adhesion through a novel p38 MAPK-RhoA signaling pathway that involves heat shock protein 27. *J. Biol. Chem.* 2009;284(31):20936–20945.

36

Multiplexible Electrochemical Sensor for Salivary Cancer Biomarker Detection

Fang Wei, Wei Liao, and David T.W. Wong

CONTENTS

36.1 Introduction

Biomarker detection is rapidly emerging as an important clinical tool, especially in the battle against cancer. Cancer is a complex disease with time-dependent dynamics, and presymptomatic indicators are useful in early detection, accurate diagnosis, and developing therapeutic strategies. Such biomarkers provide a "signature" of the health state and are found in biological fluids in blood or urine, and recently in saliva. Saliva is easily accessible in a noninvasive manner and can be collected with less patient discomfort relative to blood. In addition, the levels of interfering material (cells, DNA, RNA, and proteins) and inhibitory substances are lower and less complex in saliva than in blood. As we know, molecular markers from tumors are released into the surrounding environment and enter into blood and other body fluids. Recent results have shown that both cell-free mRNAs and proteins in saliva present diagnostic value for oral cancer and other systemic diseases. Therefore, a highly sensitive and accurate assay for salivary mRNA/protein biomarkers would make saliva a valuable diagnostic fluid and offer the potential for identification of disease in a noninvasive and specific manner [1–3].

One of the requirements for point-of-care (POC) cancer detection is identification of a small amount of target molecules in a mixture. Detection of a low-number-count target necessitates high sensitivity together with high specificity because of the complexity of any mixture. However, current processes require a compromise between specificity and sensitivity. Electrochemical (EC) sensors, initially developed to detect biomolecules in a laboratory setting, have recently been extensively applied for on-site biosensing and detection, especially in medical and clinical diagnostics [4–7]. While offering simplicity in operation and sample manipulation, the contemporary EC biosensor also provides highly sensitive and specific measurements for a broad spectrum of biomolecules. The sample size required for current EC sensors is small, ranging from several microliters to hundreds of nanoliters, which includes the sample pretreatment reagents. Additionally, the detection time is relatively fast, varying from a few minutes to tens of seconds. However, the most important feature of EC sensors is their potential to be easily transformed from a laboratory-based instrument to a POC device. Because of these advantages, EC biosensing for cancer detection is becoming a promising area of research and development [7–9].

One issue related to biosensing is increasing the signal-to-noise ratio. To achieve this goal, steric hindrance (SH) is applied to EC sensors for analysis of biomarkers in body fluids. SH suppresses nonspecific signals and generates a signal-on amplification process for target detection. The stem-loop configuration brings the reporter end of the probe into close proximity with the surface and makes it unavailable for binding with the mediator. Target binding opens the hairpin structure of the probe, and the mediator can then bind to the accessible reporter. This signal-on process is characterized by a low basal signal, a strong positive readout, and a large dynamic range [9–12].

Nanomaterials have shown emerging potential in EC cancer diagnostics and may serve to address some of the signal-to-noise ratio issues because of their enhanced electrical properties compared to traditional EC sensor materials. These nanomaterials additionally offer improved biocompatibility and additional binding sites [13–18]. For sensors detecting immobilized biomarkers, the interface between the surface and the fluid medium plays an important role in determining the levels of signal and noise in the EC detection process. When proteins are directly immobilized on the metal electrode, their denaturation by surface–protein interaction results in low activity and a low signal level. A conducting polymer–based interface, such as a nanomaterial provides, can prevent the protein conformation change and alleviate this problem [19–23]. A new type of nanomaterial, the DNA dendrimer, has been introduced into the interfacial film on the sensor surface to further improve the sensor performance. The DNA dendrimer is a nanoscale dendrite constructed of short DNA sequences that can be easily incorporated into the abiotic conducting polymer matrix and that is biocompatible with most biospecies [24].

Because of the complexity of biological systems, especially for human diseases, a single biomarker alone is not effective enough for accurate diagnosis. Medical decisions based on a single biomarker usually have a high possibility of relying on false-positive or false-negative findings. Accurate diagnosis often requires multiplex detection of endogenous biomarkers. The typical accuracy for salivary biomarkers ranges from 0.65 to 0.85; however, this range is still far from what is needed in the real clinical situation. Recently, using a combination of multiple biomarkers rather than a single biomarker has been shown to result in improved accuracy [25–27], indicating the importance of multiplexing assays for biomarkers. In addition, the combination of multiple biomarkers does not necessarily need to be limited to biomarkers of single type, that is, proteomic or genomic. The biomarkers in the combination could also include nucleic acids, proteins, and small molecules. Multiplex detection of different types of biomarkers may prove to be essential for accurate clinical diagnosis; however, because of difficulties in measuring the low levels of proteins/RNA/small molecules under the same conditions at the same time, no technology has yet been reported addressing this type of multiplexing mode.

36.2 Materials

All chemicals, buffers, and biological reagents were bought from commercial manufacturers and used without further modification (Table 36.1).

36.2.1 EC Sensors and Reader

The EC sensor is an array of 16 bare gold electrode chips (GeneFluidics, CA). Each unit of the array has a working electrode, a counter electrode, and a reference electrode. The species are immobilized on the working electrode (Figure 36.1A). EC current is measured between the working electrode and counter electrode under the potential between the working electrode and the reference electrode. The potential profile could be a constant value, a linear sweep, or a cyclic square wave. Sixteen arrays of plastic wells separate each three-electrode set, which avoid the cross contamination between different sensors. A conducting polymer was deposited on the working electrodes as the supporting film. The 16-channel EC reader (GeneFluidics) controls the electrical field applied onto the 16-array sensors and read out the amperometric current both simultaneously (Figure 36.1B).

All the electrical potentials in the following steps are referred to the gold reference electrode, which was determined to be +218 mV versus SCE by measuring cyclic voltammetric curves of 0.1 mM $[Fe(CN)_6]^{3-/4-}$. For the experiments, solutions were loaded onto the whole area of the three-electrode region including the working, counter, and reference electrodes, which was confined and separated by the 16 arrays of plastic wells. After each step, the EC sensors were rinsed with ultrapure water (18.3 MΩ·cm) and then dried under pure N_2.

36.2.2 SH Design for mRNA Detection in Saliva

For the low copy number of mRNA in saliva, SH [9] is applied in EC sensors to increase the signal-to-noise ratio. SH suppresses the unspecific signals and generates a signal-on amplification process for target detection [28]. The stem-loop configuration brings the reporter end of the probe into close proximity with the surface and makes it unavailable for binding with the mediator. Target binding opens the hairpin structure of the probe, and the mediator can then bind to the accessible reporter. This signal-on process

TABLE 36.1

Reagents for the Electrochemical Sensors of Salivary Biomarkers

Chemicals	
Pyrrole, 98%	Sigma
KCl, 3 M	Mettler-Toledo
Buffers	
1× PBS, pH 7.5	Invitrogen
Casein/PBS blocker	Pierce
Ultrapure water, RNase free	Invitrogen
Biological reagents	
Streptavidin DNA dendrimer	Genisphere
Human IL-8 mAb-Biotin	Pierce
IL-8 mRNA hairpin probe	MWG
Human IL-8 protein standard	Pierce
Human IL-8 mAb-HRP	Abazyme
Anti-FITC-HRP	Roche
Tetramethylbenzidine substrate, low activity	Neogen

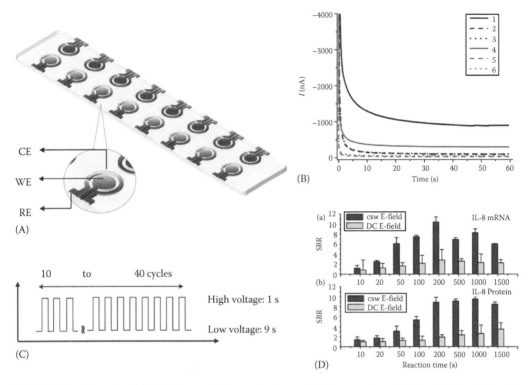

FIGURE 36.1 Illustration of (A) the 16-array electrochemical sensor. (B) Amperometric detection of IL-8 mRNA probe with (1) 5 nM IL-8 IVT RNA, (2) 5 nM S100A8 IVT RNA, and (3) blank control; and IL-8 protein probe with (4) 12.5 ng/mL IL-8 protein standard, (5) 12.5 ng/mL IL-1β protein standard, and (6) blank control. (C) The cyclic square-wave electrical field applied on each sensor electrode. For each sensor, three electrodes are included as the working electrode (WE), counter electrode (CE), and reference electrode (RE) and (D) signal-to-background ratio with different reaction time under csw E-field and DC E-field-assisted biosensing with (a) 5 nM IL-8 IVT RNA and (b) 12.5 ng/mL IL-8 protein standard. Mean value and standard deviation are illustrated with triplet experiment.

is characterized by a low basal signal, a strong positive readout, and a large dynamic range. The hairpin sequence for IL-8 mRNA was 5′-<u>GAG GGT TGC</u> *TCA GCC CTC TTC AAA AAC TTC TCC ACA ACC CTC*-3′, based on the calculation of M-fold [29,30].

36.2.3 Saliva Collection

A well-defined and standardized protocol was used for saliva collection, storage, and processing of all samples under exactly the same conditions. Unstimulated whole saliva samples were collected with prior mouth rinsing with water [31]. The donors were asked to abstain from eating, drinking, smoking, or using oral hygiene products for at least 1 h before collection. The samples, once collected, were centrifuged at 2600*g* for 15 min at 4°C to remove debris and cells. The supernatant was then removed, and protease inhibitors were added to the collected samples to ensure preservation of protein integrity. The sample aliquots were stored at −80°C to ensure prompt and adequate inhibition of salivary RNA/protein degradation.

36.2.4 Patient Biospecimens

Oral cancer and control participant saliva samples were obtained from the Gujarat Cancer & Research Institute (GCRI) under IRB approval from the respective institutions (GCRI and University of California, Los Angeles, CA). Saliva collection, processing, and stabilization were performed as described in Section 36.2.2 [32]. Each saliva sample was measured in multiplexing mode. For each sample, both EC and PCR/ELISA were performed separately. The volume of saliva requested for the EC sensor is 50 μL.

36.3 Methods

36.3.1 Cyclic Square-Wave Electrical Field–Enhanced Surface Reaction

A cyclic square-wave electrical field (csw E-field) was applied for the electropolymerization and surface-recognition processes, which provides more effective and versatile way to control the assay (Figure 36.1C). With the csw E-field, both the hybridization and protein binding are finished on the same chip within minutes, while previously these processes have to be done separately and the incubation time varies from 1 to 24 h. The positive potential in the csw E-field help accumulate the molecules onto the working electrode, while the negative potential removes the weak nonspecific binding which generates high specificity. The flapping between positive and negative potentials also provides good mixing during the incubation, which accelerates the binding process as well.

Each cycle of square wave consisted of 9 s at low voltage and 1 s at high voltage. For EC polymerization, the low voltage was +350 mV and the high voltage was +950 mV. For the surface recognition, the low voltage was −200 mV and the high voltage was +300 mV. In total, 20 cycles of square waves were applied for each surface reaction. The whole process lasted for 200 s. The number of cycles of csw E-field was chosen based on the experimental results for both mRNA and protein. From the studies in total reaction time, there exists an optimized condition for the csw E-field by spiking the standard in vitro translated (IVT) RNA and protein standard into saliva (Figure 36.1D). The signal-to-background ratio (SBR) increases at the very beginning and reaches to the highest SBR of about 10. Long time of incubation does not contribute to further improvement in SBR. Detections under DC E-field are also studied. Without the negative potential to remove the nonspecific binding, DC E-field only has the low SBR around 3.0 for RNA. In addition, DC E-field could not generate good mixing for protein binding. The highest SBR is only around 3.5 for IL-8 protein binding. With the current saliva sample and EC sensor setup, the optimized condition for both RNA and protein detection in saliva is 20 cycles of 9 s at −300 mV and 1 s at +200 mV (200 s total). However, for each specific clinical sample (serum, urine, saliva, etc.), because of the different properties of targeting molecule and supporting media, the csw E-field condition needs to be optimized independently.

36.3.2 EC Polymerization

The interface between the surface and the fluid medium in biosensors plays an important role in determining the levels of signal and noise in the EC detection process. When protein and RNA is directly immobilized on the metal electrode, denaturation by surface–biomolecule interaction results in low activity and low signal level. We introduce the DNA dendrimer into the interfacial polymer film on the sensor surface to further improve the sensor's biocompatibility. DNA dendrimer is a nanoscale dendrite constructed of short DNA sequences, which can be easily incorporated into the abiotic conducting polymer matrix and is biocompatible to most biospecies.

In this work, DNA dendrimer and polypyrrole (DDPpy) form the bio/abiotic interface on EC sensors. Streptavidin-modified dendrimer nanoparticles were introduced into the polymer matrix. The streptavidin dendrimer (Genisphere, PA) with a diameter of 70–90 nm is labeled with 2–4 streptavidins for each unit. The basic unit of the dendrimer is the oligonucleotide, which is heavily negatively charged and allows the incorporation of the dendrimer into the polymer matrix.

For electropolymerization, the dendrimer was diluted together with pyrrole (Sigma, MO) in 1× PBS (pH 7.5, Invitrogen, CA) in a volume ratio of 1:200. Potassium chloride was added at a final concentration of 300 mM to achieve high ionic strength. The final concentration of pyrrole was 10 mM. After loading of the mixture onto the gold electrode, a csw E-field was applied for electropolymerization. Each square wave consisted of 9 s at a potential of +350 mV and 1 s at +950 mV, and 20 cycles of square waves were applied. The whole process lasted for 200 s.

After the polymerization, the electrode was rinsed with ultrapure water (18.3 MΩ·cm) and then dried under pure N_2. The effects of DDPpy thicknesses on signal-to-background level were carefully studied. Under the optimized condition, the thickness of the polymer film was measured in triplet mode using a profilometer (Dektak 6 Surface Profile Measuring System, Veeco, NY) with a value of 51.5 ± 3.0 nm.

36.3.3 Probe Immobilization—RNA and Protein

A total of 50 μL of 10 nM biotin and fluorescein dual-labeled hairpin probe for IL-8 mRNA (MWG, AL) in 1× PBS (pH 7.5, Invitrogen), and 50 μL of 0.005 mg/mL biotinylated human IL-8 monoclonal antibody (Pierce/Thermo Fisher Scientific, IL) in 1× PBS were loaded separately onto different gold electrodes on the same chip array. After loading, the csw E-field was applied again for probe immobilization, with 20 cycles of 9 s at −300 mV and 1 s at +200 mV. The whole process lasted 200 s (Figure 36.2).

36.3.4 Sample Incubation

A total of 50 μL of saliva supernatant was loaded onto the probe-coated electrode. Regarding electrodes coated with probes for IL-8 mRNA and IL-8 protein, the same saliva supernatant was loaded. For the

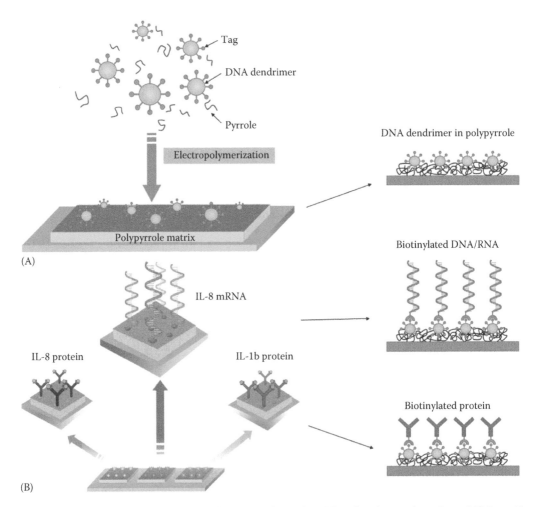

FIGURE 36.2 Schematic process of (A) DNA dendrimer polypyrrole (DDPpy)-based sensor formation and (B) its application in the DDPpy-directed multiplexing immobilization of biomolecules, including IL-8 mRNA, IL-8 protein, and IL-1β protein.

concentration calibration process, IVT IL-8 RNA and standard IL-8 protein (Pierce/Thermo Fisher Scientific, IL) were spiked into the saliva supernatant first and then loaded onto the electrodes. For the specificity control, S100A8 RNA and IL-1β protein, both biomarkers in saliva, were loaded onto the electrodes. Details regarding preparation of IVT IL-8 RNA/IVT S100A8 RNA have been published previously [9]. The protein standards for human IL-8 and IL-1β were both obtained from Pierce. In the spiking process, the delivery solution for IVT RNA was RNase-free distilled water (Invitrogen). The delivery solution for the protein standards was 1× PBS (pH 7.5). After saliva loading, a csw E-field was applied again for saliva incubation, with 20 cycles of 9 s at −300 mV and 1 s at +200 mV. The whole process lasted for 200 s.

36.3.5 Reporter Incubation

A mixture of a 1:1000 dilution of 150 U/mL anti-fluorescein horseradish peroxidase (HRP) (Roche, IN) and a 1:100 dilution of original HRP-conjugated human IL-8 monoclonal antibody (Pierce/Thermo Fisher Scientific, IL) in blocker casein/PBS (Pierce/Thermo Fisher Scientific, IL) was prepared. A total of 50 μL of the mixture was loaded onto the saliva-incubated electrodes separately. After saliva loading, a csw E-field was applied again for reporter binding with 30 cycles of 9 s at −300 mV and 1 s at +200 mV. The whole process lasted for 300 s.

36.3.6 Signal Readout

The 3,3′,5,5′-tetramethylbenzidine (TMB/H_2O_2, low activity) substrate (Neogen, FL) was loaded, and amperometric detection was carried out by application of −200 mV potential to each electrode unit, followed by a parallel amperometric signal readout after 60 s of equilibration. TMB acts as a mediator and is reduced at −200 mV; the reduced TMB in turn reduces the oxidized form of HRP. HRP then reduces H_2O_2 to two H_2O molecules, and the HRP is oxidized (Figure 36.3).

36.3.7 Statistical Analysis

The area under the curve (AUC) is based on construction of a receiver operating characteristic (ROC) curve that plots the sensitivity (*y*-axis in our analysis) versus one minus the specificity (*x*-axis in our analysis). The AUC value is computed by numerical integration of the ROC curve. The typical range for this value is from 0.5, which indicates that the biomarker has no diagnostic utility (i.e., that the biomarker is no better than a coin flip), to 1, which would indicate perfect diagnostic accuracy. The cutoff point corresponding to maximum (sensitivity + specificity) − 100% was found and used as the optimal cutoff in the logistic regression model. The cutoff points for IL-8 mRNA and protein were separately determined by identifying the cutoff that yields the maximum sum of the sensitivity plus specificity. The ROC curves

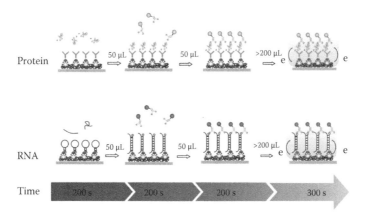

FIGURE 36.3 Schematic illustration of the multiplexing detection of protein and RNA with electrochemical sensors. The detection is carried out with cocktail reagents. Total reaction time is also given at the bottom.

were constructed using the ROC package in R (version 2.70). This function constructs the ROC curve from the continuous marker values by first ordering the marker values and then computing the sensitivity and specificity for each possible cut-point of the markers. For a combination of the two, the cutoff point corresponding to the maximum of sensitivity and specificity was found and used as the optimal cutoff in the logistic regression model.

36.4 Results

36.4.1 Calibration of the EC Sensor for Oral Cancer Salivary Biomarker Detection

The EC sensors proved to have high sensitivity and specificity for the oral cancer salivary biomarkers. The specificity was investigated by comparing the signals from perfect match targets (IL-8 mRNA/IL-8 protein) and interferents (S100A8 RNA/IL-1β protein) in saliva. For both mRNA and protein, non-specific targets exhibited only signal just above the background level. These results indicate the good specificity of the EC sensors.

The sensitivity of the EC sensors for the two salivary biomarkers was investigated by spiking the standard IVT RNA and protein into saliva under multiplexing mode. The sensitivity was 3.9 fM for mRNA and 7.4 pg/mL for protein with a cutoff at two standard deviations based on the linear regression of the calibration curve. The linearity values for the two species were $R^2 = 0.98$ and 1, respectively. These results show the high sensitivity of this multiplex salivary biomarker detection. In addition, the dynamic range covered about four orders of magnitude (IL-8 mRNA: 5 fM to 50 pM; IL-8 protein: 10 to 12,500 pg/mL). This wide dynamic range indicates a powerful tool for variant saliva samples (Figure 36.4).

36.4.2 Multiplexed Assay of Salivary Protein and mRNA Biomarkers by EC

The application of EC sensors in salivary diagnostics for oral cancer has been demonstrated in clinical saliva samples. In total, 56 saliva samples including 28 oral cancer samples and 28 control samples from India were measured using the EC sensor. Overall, each biomarker showed higher levels in the cancer samples than in the control samples (Figure 36.5A and B). Statistical analysis showed a significant difference for both IL-8 mRNA ($W = 810$, $P = 6.637 \times 10^{-9}$) and IL-8 protein ($W = 820$, $P = 1.863 \times 10^{-9}$), estimated using the Wilcoxon signed-rank test. For the sample size requirements, at $\alpha = 0.05$, $\beta = 0.2$, the sample size required was 13 for each group. Here α refers to the Type I error, which means false positive; β refers to the Type II error, which means the false negative [33]. Therefore, a sample size of 28

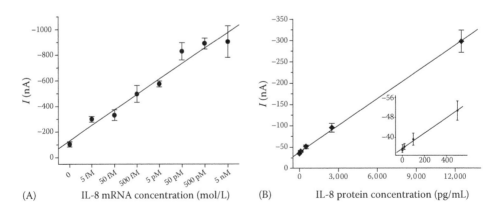

(A)   IL-8 mRNA concentration (mol/L)   (B)   IL-8 protein concentration (pg/mL)

FIGURE 36.4 Sensitivity of an electrochemical sensor for multiplexing detection of salivary biomarkers under a cyclic square-wave electrical field with linear fit. (A) IL-8 mRNA ($R^2 = 0.98$). (B) IL-8 protein ($R^2 = 1$). The sensor response from 0 to 600 pg/mL of IL-8 protein is listed in the inlets. Mean value and standard deviation are both illustrated with triplicate experiments.

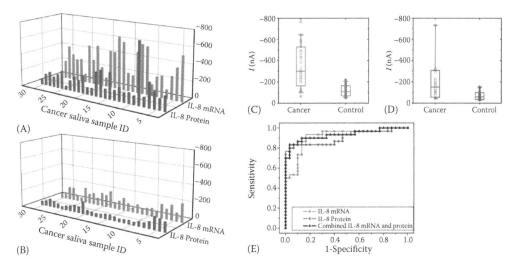

FIGURE 36.5 Multiplex detection of IL-8 mRNA and IL-8 protein with an electrochemical (EC) sensor of 56 saliva samples. (A) Results from cancer samples and (B) results from control samples and statistical analysis for clinical saliva samples tested by EC sensors. Boxes indicating standard deviation and mean value of (C) IL-8 mRNA and (D) IL-8 protein are shown. Statistical analysis showed a significant difference for both IL-8 mRNA ($P = 6.637 \times 10^{-9}$) and IL-8 protein ($P = 1.863 \times 10^{-9}$). (E) Receiver operating characteristic (ROC) analysis on a ROC curve with individual IL-8 mRNA with an area under the curve (AUC) value of 0.90, IL-8 protein with an AUC of 0.91, and combinational IL-8 mRNA + protein with an AUC of 0.93.

provided sufficient power. Despite the wide spread in EC values for the markers in the oral squamous cell carcinoma (OSCC) group, the box plots in Figure 36.5C and D show very little overlap between the two groups. For example, for IL-8 mRNA, 75% of the OSCC cases had lower EC values than the minimum EC value for the controls.

An equal number of age- and sex-matched subjects with comparable chewing, smoking, and alcohol histories were selected as a control group. Among the two subject groups, there were no significant differences in terms of mean age: OSCC patients, 46 ± 12.8 years; normal subjects, 44.9 ± 11.9 years (Wilcoxon rank sum test, $P > 0.8$); gender ($P = 1$); chewing history ($P > 0.08$); smoking history ($P > 0.27$); or alcohol drinking history ($P > 0.06$) (Table 36.2).

Figure 36.5E illustrates the results of the ROC analysis of the EC sensor for both IL-8 mRNA and IL-8 protein. The AUC is 0.90 for IL-8 mRNA and 0.91 for IL-8 protein. From the results measured with the same saliva samples by PCR and ELISA, the AUC values were 0.91 and 0.87, respectively. Logistical regression showed that the combined IL-8 mRNA and protein analysis increased the

TABLE 36.2

Statistical Evaluation of Correlation between Oral Cancer and Control Subjects

		Normal Group [28]	OSCC Group [28]	P Value
Age (years)	Mean	44.9	46.0	0.83
	Standard deviation	11.9	12.8	
Habits/factors	Chewing (yes)	28	22	0.08
	Smoking (yes)	7	12	0.27
	Alcohol (yes)	1	7	0.06
Gender	Female	4	5	1
	Male	26	25	
Tumor stage	T (1, 2, 3, 4)%	N/A	(14, 61,14,11)	N/A
	N (0, 1, 2, 3)%	N/A	(29, 46, 25, 0)	N/A
	M (0, 1)%	N/A	(100, 0)	N/A

AUC to 0.93. The improvement in ROC indicates that the combined IL-8 mRNA and protein further increased AUC. This clinical test suggests that the EC sensor is an accurate method for measuring salivary biomarkers, one that is as good as conventional PCR and ELISA in identification of oral cancer using saliva-based screening.

36.5 Discussion

EC sensors offer several advantages over other detection methods, making them attractive for saliva-based cancer diagnostics. They offer some advantages compared to PCR/ELISA in a multiplexing assay for salivary diagnostics, as well as comparable sensitivity and accuracy. In addition to being highly specific and sensitive, EC sensors are extremely efficient, in terms of fast detection times and low power consumption. There are several key features that EC sensors offer for salivary cancer diagnostics, which are discussed later.

36.5.1 Detection Speed

The detection of biomarkers in saliva involves several processes, including probe/target recognition, sample manipulation, and signal readout. Using EC sensors, probe/target recognition can be completed within minutes, and in some cases, within several seconds. The entire detection process, including sample preparation, sample delivery, and signal readout, can be completed in several minutes. Such rapid detection is ideal for POC clinical diagnostics.

36.5.2 Sensitivity

With appropriate electrical field profiles, EC sensors can achieve sensitivities to concentrations in the fM range, and in some cases, aM for short oligonucleotides, and pg/mL to fg/mL for protein in saliva. Such sensitivities can be achieved without sample amplification (i.e., PCR), a major advantage for POC systems. Especially in this work, the sensitivity was 3.9 fM for mRNA and 7.4 pg/mL for protein with a cutoff at two standard deviations. Additionally, the total amount of sample required for detection is on the order of several microliters.

36.5.3 Specificity

The stringency of EC-controlled recognition allows for high specificity, even in clinical samples in multiplex mode. Usually, the protocol for protein and mRNA analysis must be optimized separately to achieve high accuracy. However, the optimized conditions for measuring protein and mRNA biomarkers are often incompatible. With EC sensors, instead of optimizing with pH, ionic strength, and temperature, a csw E-field is applied to optimize the assay. Compared to other conditions, the csw E-field has several advantages. First, it is effective in improving binding efficiency for both mRNA and protein. For mRNA, the positive E-field in the cycles helps the strand hybridize, and the negative E-field removes the nonspecific binding. By optimizing the parameters in the csw E-field, high SBR can be achieved (Figure 36.1B and D). Meanwhile, the csw E-field generates mass transportation in the solution, which results in good mixing in a short time. This effective mixing also facilitates the protein binding. Second, the csw E-field can be easily controlled in multiplexing mode. The profile of the csw E-field onto each electrode is versatile, including the high and low voltage, the intervals of each cycle, and the total number of cycles. Because the E-field is highly localized, using microelectromechanical systems (MEMs) technology, the EC sensor could be easily expanded to high-throughput multiplexing mode with a high-density array of microelectrodes. The final advantage is that the csw E-field will not add a heavy load to the biological system, avoiding the damage caused by a DC electrical field. The high electrical field is applied for as little as 1 s in each cycle. During the remaining period, the biological system is retained at a very safe E-field (about −200 mV) so that damage from a high-voltage E-field could be eliminated.

36.5.4 Convenience

With contributions from MEMS and nanotechnology, EC sensors can be integrated into portable platforms, enabling POC monitoring and on-site biosensing [34]. Regarding the assay procedure, most obstacles originate from sample pretreatment and handling. Sample pretreatments include storage, transportation, amplification, and lysis processes. These complicated procedures, when stabilizing the sample and amplifying signal, introduce several uncertainties into a saliva sample. Usually, a quality control with traditional detection methods should be carried out before measurement of the saliva. With EC sensors, no pretreatment is required for the saliva supernatant. The total measurement takes around 10 min from the saliva sample loading, which means the saliva could be measured on-site, right after sample collection. This rapidity removes a major hindrance from the sample pretreatment part of the process.

Sample loading includes delivery, mix, and incubation. Especially for the sandwich assay, multiple steps are involved in the loading processes, including detection sample loading, reporter loading, and final readout reagent loading. These multiple steps can lead to technical problems in multiplexing modes. Increased throughput results in a nonlinear increase in the complexity of device design by integrating each individual sample loading process. Thus, simplifying the procedure will remove the uncertainty of the device and increase assay reliability. With the EC sensors, cocktail reagents are applied during the reporter loading process. Reporters for protein and mRNA, respectively, are premixed in the same buffer and loaded onto the device. During this step, only one cocktail reagent is loaded, and no separate control of sample loading is required.

36.5.5 Multiplexing

A single biomarker is not sufficient for high-specificity detection in clinical samples. Combining several biomarkers can greatly improve detection accuracy, making multiplex detection for DNA diagnostics important [4]. Because the electrical fields that the electrodes generate are highly localized, they can be employed for simultaneous detection within a miniaturized platform in which the detection condition for each DNA target can be optimized separately.

Based on these advantages, the EC system is promising for multiplexing detection. By overcoming key hurdles in developing POC detection systems for salivary biomarkers, EC sensors will become more prominent clinical diagnostic tools for detecting a broad spectrum of diseases and conditions using multiplex detection of RNA and proteins in salivary diagnostics.

36.6 Future Trends

EC sensors hold great potential for molecular diagnosis in cancer research, offering the high sensitivity and specificity essential for low-concentration biomarker detection. With the multiplexing detection mode, accuracy for diagnosis has been improved remarkably. Future trends for EC sensors will involve combination with nanotechnologies, specifically fabrication techniques [35,36] and new nanomaterials [16,37,38]. In particular, the detection sensitivity will be enhanced through highly specific molecular recognition (via appropriately designed targets and probes), improved EC signal generation, transduction, and amplification, and enhanced electrical conductivity for minimized background noise. Additionally, EC sensors are extremely efficient in terms of rapid detection speed, low power consumption, and electrode multifunctionality (i.e., sample manipulation, polymer electropolymerization, and DNA detection). Contributions from microfluidics and MEMS fabrication will allow EC sensors to be integrated with relevant sample handling and fluidic processes on a portable diagnostic platform, which will enable rapid, multiplexed, and high-throughput analysis.

ACKNOWLEDGMENTS

Supported by NIDCR/NIH grant U01DE017790, the Felix and Mildred Yip Endowed Professorship and the Barnes Family Research Funds.

REFERENCES

1. Mandel, I.D. (1993) Salivary diagnosis—More than a lick and a promise. *J. Am. Dent. Assoc.*, 124, 85–87.
2. Mandel, I.D. (1990) The diagnostic uses of saliva. *J. Oral Pathol. Med.*, 19, 119–125.
3. Hu, W.H., Li, C.M., Cui, X.Q., Dong, H., and Zhou, Q. (2007) In situ studies of protein adsorptions on poly(pyrrole-*co*-pyrrole propylic acid) film by electrochemical surface plasmon resonance. *Langmuir*, 23, 2761–2767.
4. Wei, F., Patel, P., Liao, W., Chaudhry, K., Zhang, L., Arellano-Garcia, M., Hu, S., Elashoff, D., Zhou, H., Shukla, S. et al. (2009) Electrochemical sensor for multiplex biomarkers detection. *Clin. Cancer Res.*, 15, 4446–4452.
5. Wei, F., Lillehoj, P.B., and Ho, C.M. (2010) DNA diagnostics: Nanotechnology-enhanced electrochemical detection of nucleic acids. *Pediatr. Res.*, 67, 458–468.
6. Ricci, F., Bonham, A.J., Mason, A.C., Reich, N.O., and Plaxco, K.W. (2009) Reagentless, electrochemical approach for the specific detection of double- and single-stranded DNA binding proteins. *Anal. Chem.*, 81, 1608–1614.
7. Ahmed, M.U., Hossain, M.M., and Tamiya, E. (2008) Electrochemical biosensors for medical and food applications. *Electroanalysis*, 20, 616–626.
8. Zhang, J., Wan, Y., Wang, L.H., Song, S.P., and Fan, C.H. (2007) The electrochemical DNA biosensor. *Progr. Chem.*, 19, 1576–1584.
9. Wei, F., Wang, J.H., Liao, W., Zimmermann, B.G., Wong, D.T., and Ho, C.M. (2008) Electrochemical detection of low-copy number salivary RNA based on specific signal amplification with a hairpin probe. *Nucleic Acids Res.*, 36(11), e65.
10. Wei, F., Qu, P., Zhai, L., Chen, C.L., Wang, H.F., and Zhao, X.S. (2006) Electric potential induced dissociation of hybridized DNA with hairpin motif immobilized on silicon surface. *Langmuir*, 22, 6280–6285.
11. Fan, Y., Chen, X.T., Trigg, A.D., Tung, C.H., Kong, J.M., and Gao, Z.Q. (2007) Detection of microRNAs using target-guided formation of conducting polymer nanowires in nanogaps. *J. Am. Chem. Soc.*, 129, 5437–5443.
12. Fan, C.H., Plaxco, K.W., and Heeger, A.J. (2003) Electrochemical interrogation of conformational changes as a reagentless method for the sequence-specific detection of DNA. *Proc. Natl. Acad. Sci. USA*, 100, 9134–9137.
13. Yang, Y.H., Wang, Z.J., Yang, M.H., Li, J.S., Zheng, F., Shen, G.L., and Yu, R.Q. (2007) Electrical detection of deoxyribonucleic acid hybridization based on carbon-nanotubes/nano zirconium dioxide/chitosan-modified electrodes. *Anal. Chim. Acta*, 584, 268–274.
14. Xu, K., Huang, J.R., Ye, Z.Z., Ying, Y.B., and Li, Y.B. (2009) Recent development of nano-materials used in DNA biosensors. *Sensors*, 9, 5534–5557.
15. Wang, Z.D. and Lu, Y. (2009) Functional DNA directed assembly of nanomaterials for biosensing. *J. Mater. Chem.*, 19, 1788–1798.
16. Pandey, P., Datta, M., and Malhotra, B.D. (2008) Prospects of nanomaterials in biosensors. *Anal. Lett.*, 41, 159–209.
17. Mao, X. and Liu, G.D. (2008) Nanomaterial based electrochemical DNA biosensors and bioassays. *J. Biomed. Nanotechnol.*, 4, 419–431.
18. Castaneda, M.T., Merkoci, A., Pumera, M., and Alegret, S. (2007) Electrochemical genosensors for biomedical applications based on gold nanoparticles. *Biosens. Bioelectron.*, 22, 1961–1967.
19. Ramanavicius, A., Ramanaviciene, A., and Malinauskas, A. (2006) Electrochemical sensors based on conducting polymer–polypyrrole. *Electrochim. Acta*, 51, 6025–6037.
20. Liao, W. and Cui, X.T. (2007) Reagentless aptamer based impedance biosensor for monitoring a neuroinflammatory cytokine PDGF. *Biosens. Bioelectron.*, 23, 218–224.
21. Gerard, M., Chaubey, A., and Malhotra, B.D. (2002) Application of conducting polymers to biosensors. *Biosens. Bioelectron.*, 17, 345–359.
22. Cosnier, S. (2007) Recent advances in biological sensors based on electrogenerated polymers: A review. *Anal. Lett.*, 40, 1260–1279.
23. Cosnier, S. (2005) Affinity biosensors based on electropolymerized films. *Electroanalysis*, 17, 1701–1715.
24. Wei, F., Liao, W., Xu, Z., Yang, Y., Wong, D.T., and Ho, C.M. (2009) Bio/abiotic interface constructed from nanoscale DNA dendrimer and conducting polymer for ultrasensitive biomolecular diagnosis. *Small*, 5, 1784–1790.

25. Oikonomopoulou, K., Li, L., Zheng, Y., Simon, I., Wolfert, R.L., Valik, D., Nekulova, M., Simickova, M., Frgala, T., and Diamandis, E.P. (2008) Prediction of ovarian cancer prognosis and response to chemotherapy by a serum-based multiparametric biomarker panel. *Br. J. Cancer*, 99, 1103–1113.

26. Li, Y., St. John, M.A.R., Zhou, X.F., Kim, Y., Sinha, U., Jordan, R.C.K., Eisele, D., Abemayor, E., Elashoff, D., Park, N.H., and Wong, D.T. (2004) Salivary transcriptome diagnostics for oral cancer detection. *Clin. Cancer Res.*, 10, 8442–8450.

27. Kozak, K.R., Su, F., Whitelegge, J.P., Faull, K., Reddy, S., and Farias-Eisner, R. (2005) Characterization of serum biomarkers for detection of early stage ovarian cancer. *Proteomics*, 5, 4589–4596.

28. Huang, T.J., Liu, M.S., Knight, L.D., Grody, W.W., Miller, J.F., and Ho, C.M. (2002) An electrochemical detection scheme for identification of single nucleotide polymorphisms using hairpin-forming probes. *Nucleic Acids Res.*, 30, e55.

29. SantaLucia, J. (1998) A unified view of polymer, dumbbell, and oligonucleotide DNA nearest-neighbor thermodynamics. *Proc. Natl. Acad. Sci. USA*, 95, 1460–1465.

30. Zuker, M. (2003) Mfold web server for nucleic acid folding and hybridization prediction. *Nucleic Acids Res.*, 31, 3406–3415.

31. Navazesh, M. (1993) Methods for collecting saliva. *Ann. N.Y. Acad. Sci.*, 694, 72–77.

32. Hu, S., Wang, J.H., Meijer, J., Leong, S., Xie, Y.M., Yu, T.W., Zhou, H., Henry, S., Vissink, A., Pijpe, J., Kallenberg, C., Elashoff, D., Loo, J.A., and Wong, D.T. (2007) Salivary proteomic and genomic biomarkers for primary Sjogren's syndrome. *Arthritis Rheum.*, 56, 3588–3600.

33. Allchin, D. (2001) Error types. *Perspect. Sci.*, 9, 38–59.

34. Gau, V. and Wong, D. (2007) Oral fluid nanosensor test (OFNASET) with advanced electrochemical-based molecular analysis platform. *Ann. N.Y. Acad. Sci.*, 1098, 401–410.

35. Soper, S.A., Brown, K., Ellington, A., Frazier, B., Garcia-Manero, G., Gau, V., Gutman, S.I., Hayes, D.F., Korte, B., Landers, J.L., Larson, D., Ligler, F., Majumdar, A., Mascini, M., Nolte, D., Rosenzweig, Z., Wang, J., and Wilson, D. (2006) Point-of-care biosensor systems for cancer diagnostics/prognostics. *Biosens. Bioelectron.*, 21, 1932–1942.

36. Liao, J.C., Mastali, M., Li, Y., Gau, V., Suchard, M.A., Babbitt, J., Gornbein, J., Landaw, E.M., McCabe, E.R.B., Churchill, B.M., and Haake, D.A. (2007) Development of an advanced electrochemical DNA biosensor for bacterial pathogen detection. *J. Mol. Diagn.*, 9, 158–168.

37. Radwan, S.H. and Azzazy, H.M.E. (2009) Gold nanoparticles for molecular diagnostics. *Exp. Rev. Mol. Diagn.*, 9, 511–524.

38. Lord, H. and Kelley, S.O. (2009) Nanomaterials for ultrasensitive electrochemical nucleic acids biosensing. *J. Mater. Chem.*, 19, 3127–3134.

37

Microelectrode Array Analysis of Prostate Cancer

Frank Davis, Andrew C. Barton, and Séamus P.J. Higson

CONTENTS

37.1 Introduction

Prostate cancer is a disease most frequently encountered in men over 50 and second only to lung cancer for the number of male deaths in the United States [1] (estimated 33720 deaths in 2011) and United Kingdom [2] (10,382 deaths in 2009). Early detection of the cancer enhances patient recovery, especially if the disease can be detected at the organ-confined stage. A rapid diagnosis of prostate cancer in its early stages could greatly reduce the severity of treatment needed and enhance survival rates. One approach to this could be via the development of methods for detection of prostate-specific antigen (PSA).

The growth of cancer cells can lead to the presence of overexpressed proteins related to these cells and these have been employed as biomarkers for cancer diagnosis. PSA is a 34 kDa glycoprotein manufactured almost exclusively by the prostate gland which is responsible for the liquefaction of seminal fluid. PSA is present in the serum of male patients, usually at levels below 4 ng/mL in serum [3]. However the presence of localized and metastatic cancerous prostate cells often (but not always) leads to elevated PSA levels, as indeed can other prostate disorders [4]. A blood test to measure PSA is the most effective test currently available for the early detection of prostate cancer but currently requires sending the patients' sample off to a laboratory with the associated time and financial costs. There is in this context an obvious market for an inexpensive, rapid reliable diagnostic test for the rapid determination of both free PSA and total PSA.

One potential solution to this would be via the use of an immunosensor. Immunoassays were first established in 1959, exploiting the specificity of antibodies and leading to the development of the widely used radioimmunoassay [5]. Rather than having a laboratory-based "wet" assay though, it could be much more useful to have the necessary components immobilized on a sensor chip. Within this work, we will describe how we and others have utilized this approach. Antibodies to PSA can be immobilized on a number of substrates by a variety of methods. These can then be interrogated by electrochemical, optical, or mass-sensitive techniques after exposure to sample solutions and the binding of the antigen PSA detected. This is the principle of the immunosensor.

One of the earliest immunosensors was fabricated by immobilization of antibodies to human serum albumin (HSA) into polypyrrole films grown electrochemically onto a platinum wire [6]. The resultant electrode was found to respond specifically to the antigen. Much of the development of electrochemical immunosensors is detailed in several reviews [7,8]. The availability of antibodies to PSA has led to the construction of immunosensors for PSA, some of which will be given here. A range of electrochemical immunosensors for PSA have already been developed as detailed in a recent review [3]. One of the earliest involved immobilizing anti-PSA at the surface of a gold-coated membrane and then exposing it to the sample. A second enzyme-labeled antibody was then attached to captured PSA and used to electrochemically assess the level of PSA binding to the membrane [9]. Other workers also used a sandwich-type assay with horseradish peroxidase–labeled antibodies at the surface of screen-printed carbon electrodes to create disposable PSA sensors [10]. Nanogold-modified electrodes and alkaline phosphatase enzymes have also been utilized [11] to create screen-printed electrodes capable of simultaneous detection of free and total PSA.

The methods detailed earlier required the use of secondary labeled antibodies. Label-free techniques which do not require these moieties are usually much simpler and less expensive to use and accordingly much recent activity has concentrated on these types of immunosensors. PSA antibodies could be immobilized at the surface of a pH-sensitive polymer [12,13] with the binding of the antigen leading to localized increases in pH. This then led to degradation of the polymer which could be detected via AC impedance, allowing quantification of PSA with a detection limit of 3 ng/mL. Very recently, the use of carbon nanotubes and field-effect transistors has also been reported with good sensitivity [14] and these sensors also have the ability to be used in human serum [15].

Other methods have also been used for the detection of PSA, such as piezoresistive microcantilever sensors containing immobilized anti-PSA [16], which were incorporated within microfluidic flow cells to give devices capable of detecting 10 ng/mL PSA. Varying the geometry of these cantilevers led to the development of immunosensors capable of detecting PSA between 0.2 ng/mL and 60 μg/mL in solutions containing human-specific albumin (HSA) and human plasminogen [17]. Optical methods have also proved successful [18] and attaching anti-PSA to gold surfaces allowed direct detection of 10 ng/mL PSA by surface plasmon resonance. Colloidal gold [19] could be used to amplify the response, allowing detection limits to be lowered to 0.15 ng/mL. Other workers have also demonstrated the use of gold nanoparticles to enhance surface plasmon resonance sensitivity by a factor of 100 [20].

Our group has attempted over the last few years to develop a range of label-free electrochemical immunosensors with ultralow detection limits and high specificities. Initial work utilized the simple entrapment of antibodies in electrochemically grown polypyrrole films and shown that up to 2–3 μg antibodies for bovine serum albumin (BSA) and digoxin may be successfully entrapped in the polymer film when deposited on a platinum electrode of 3.14 mm^2 working area [21]. AC impedance was later chosen as the method of interrogation of these systems [22] since it is extremely sensitive to the nature of the electrode–solution interface and does not require chemical modification of the samples or use of expensive labels.

Due to polypyrrole being somewhat unstable, easy to overoxidize, and moreover being difficult to modify chemically, later work utilized polyaniline-coated screen-printed carbon electrodes as substrates for antibody immobilization. One major problem furthermore with simple entrapment is that antibodies are located randomly within the film and in random orientations. This can be problematical especially for large antigens which may not be able to reach the majority of the entrapped antibodies because of the physical constraints associated with the polymer matrix. It was decided that it could be much more effective to selectively immobilize antibodies at the surface of the conductive polymer only where they can be easily accessed by solvated antigens. Therefore, studies were made using an avidin–biotin affinity protocol to construct immunosensors for the cardiac drug digoxin. Comparison between affinity

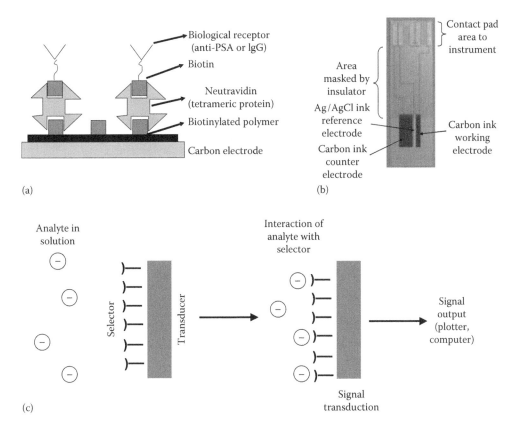

FIGURE 37.1 (a) Schematic of antibody modified electrodes, (b) commercial screen-printed electrodes used within this work, and (c) schematic of the principle of electrochemical biosensing.

and entrapment-based immunosensors demonstrated a 1000-fold lowering of the detection limit for the affinity-based sensors [23]. This protocol was further used for the construction of immunosensors for ciprofloxacin [24,25] and myelin basic protein—a marker for stroke and multiple sclerosis [26]. A schematic of the structure of these films is shown in Figure 37.1a.

It is known that microelectrodes often display superior qualities to similar electrodes on a macroscopic scale. For example, microelectrodes often display enhanced sensitivity, are stir independent in their response characteristics due to their hemispherical diffusion profiles, and can be used in high-resistive media. Unfortunately microelectrodes often display much smaller signals than their larger brethren, although this can be addressed by construction of an array of microelectrodes on a common substrate where the contribution of each microelectrode is summed to give a detectable signal. Several methods such as photolithography have been used to construct such arrays although they can be time-consuming and expensive. Our group has pioneered the development of method of constructing microarrays of conductive polymers using sonochemical ablation of insulating polymer films [27]. A schematic of this system is shown in Figure 37.2. The first step of this process is to take a conductive surface such as gold or carbon and then electrochemically deposit an insulating layer onto this surface. We have utilized commercial screen-printed electrodes (Figure 37.1b) as the basis of this sensor and electrochemically coated the working electrodes with an insulating polymer of 1,2-diaminobenzene. Sonochemical ablation is then used to ablate holes in this insulating material, leading to formation of an array of up to 120,000 pores/cm^2. It has been previously shown that ultrasound (in the kHz range) passing through water causes thermal agitation and localized hotspots of up to a few thousand K, thereby giving rise to the formation of superheated vapor bubbles which are then cooled by the solvent at ambient temperature. The bubbles then asymmetrically implode with the ejection of microjets of solvent at speeds of up

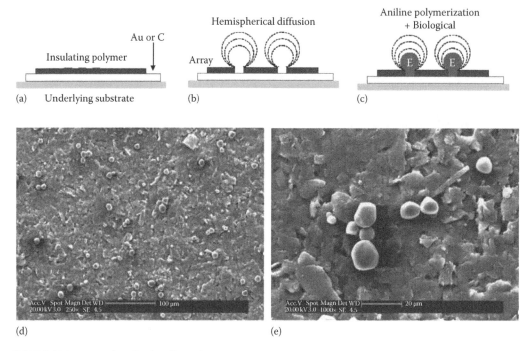

FIGURE 37.2 Formation of polyaniline microarrays: (a) deposition of insulating layer, (b) sonochemical formation of pores, (c) polymerization of aniline, (d) SEM images (250× magnification) of polyaniline protrusions, some of which are ringed for easy viewing, and (e) SEM images (1000× magnification) of polyaniline protrusions. SEM images were taken from directly above the carbon working electrodes.

to several hundred meters per second. These microjets can cause the shattering of hard brittle solids whereas soft solid surfaces such as thin films of insulating polymers may be ablated by such jets [27]. The pores are randomly arranged and vary between 0.1 and 3 μm in diameter. This gives us an array of pores all allowing electrical contact between the base electrode and the environment. These pores cannot be addressed individually using this construction method; instead, the electrochemical responses at each pore are summed over the electrode as a whole.

These arrays could be chemically modified to give sensors for chlorine [28] or alternatively further electropolymerization reactions could be utilized to grow conducting polyaniline protrusions. SEM images (Figure 37.2d and e) show a number of protrusions with sizes typically in the region of 8–15 μm. We have utilized this method to grow microarrays containing entrapped enzymes for the amperometric detection of glucose [27], alcohol [29], and a range of organophosphate pesticides [30] with extreme low limits of detection (10^{-17} M). We have also compared immunosensors containing anti-BSA with the antibody entrapped in either a planar film or a polyaniline microarray and demonstrated much lowered detection limits for the microarrays [23]. Figure 37.1c shows the principle of electrochemical biosensing. A specific biological receptor for an analyte (in our case an antibody but enzymes, aptamers, and whole cells have been utilized within this technology) is immobilized at the surface of a transducer. This transducer can be an optical chip, a mass-sensitive device such as a quartz crystal microbalance or in the case of electrochemical sensors, an electrode. Specific adsorption of the analyte leads to a change in the electrochemical properties of the adsorbed film which can be assessed by changes in its AC impedance.

Once the proof of principle for these arrays was established, we turned our attention to the development of microarrays which could serve for the detection of cancer markers such as PSA. Within this work we will describe, compare and contrast our successful immobilization of PSA using two different methods, either via random entrapment within the polyaniline matrix or alternatively specific immobilization via biotin–avidin interactions. One major drawback to the use of these sensors in real media is that nonspecific adsorptions can occur and lead to false-positive results and accordingly we will describe how we attempted to circumvent this by using nonspecific control electrodes.

37.2 Materials

Commercial screen-printed carbon electrodes (Figure 37.1b) containing carbon working and counter electrodes and an Ag/AgCl reference electrode were obtained from Microarray Ltd. (Manchester, United Kingdom). The surface area of the working electrode was $0.2178\,cm^2$. The conductive tracks were coated with an insulating layer to ensure any electrochemical reactions take place only at the designated electrodes. Sodium dihydrogen orthophosphate, disodium hydrogen orthophosphate, sodium chloride, sodium acetate, acetic acid, sodium perchlorate, and hydrochloric acid were obtained from BDH (Poole, Dorset, United Kingdom) and were of analytical grade. Aniline, polyclonal human anti-IgG (AIgG) (cat. no. I5260), biotin 3-sulfo-*N*-hydroxysuccinimide (cat. no. B5161), the biotinylation kit (cat. no. BK101), HSA, BSA, potassium ferrocyanide, and potassium ferricyanide were obtained from Sigma-Aldrich (Gillingham, Dorset, United Kingdom). PSA and its monoclonal antibody APSA (cat. no. 333-01), stroke marker proteins NSE and S-100[β] and CA125 (tumor marker protein, cat. no. 210-05) all with sodium azide preservative, were supplied by Canag Diagnostics Ltd. (now Fujirebio Diagnostics Inc., Malvern, PA). All water used was obtained from a Purelab UHQ Deioniser (Elga, High Wycombe, United Kingdom). Neutravidin (cat. no. PN31000) was purchased from Fisher Ltd. (Loughborough, United Kingdom).

For antibody biotinylation, the procedure outlined in the BK101 kit was followed (see manufacturers' instructions for details). Antibodies against PSA were reconstituted in pH 7.4 phosphate buffer. Biotinylated antibodies were kept frozen in aliquots of $200\,\mu L$ at a concentration of $1\,mg/mL$ until required. Aniline buffer (pH 1–2) was prepared containing $0.5\,mol/L$ KCl, $0.3\,mol/L$ HCl, and $0.2\,mol/L$ aniline. Phosphate-buffered saline (PBS, pH 7.4) was prepared comprising $52.8\,mmol/L$ disodium hydrogen orthophosphate 12-hydrate, $13\,mmol/L$ sodium dihydrogen orthophosphate 1-hydrate, and $5.1\,mmol/L$ sodium chloride. Acetate buffer (pH 4.0) was prepared comprising $0.4\,mol/L$ sodium acetate, $0.4\,mol/L$ acetic acid, and $0.4\,mol/L$ sodium perchlorate.

37.3 Methods

37.3.1 Formation of Microarrays

The initial step is to deposit the insulating layer of polydiaminobenzene onto the commercial screen-printed carbon electrode. A $5\,mmol/L$ solution of 1,2-diaminobenzene in pH 7.4 phosphate buffer was thoroughly purged with N_2 for 20 min in a sealed cell to remove oxygen. The screen-printed electrode was submerged in the solution and an initial 1 s blast of ultrasound was also applied to displace surface-trapped air bubbles. The working electrode potential was then scanned from 0 to +1000 mV (vs. Ag/AgCl) and back to the starting potential for 20 sweeps at a scan rate of 50 mV/s to form an insulating film. At low potentials (0 mV) there is no reaction; however, as the potential increases, the diaminobenzene is oxidized to form a radical cation. This species then reacts with another molecule of diaminobenzene forming a dimer. As the electrochemical procedure continues, higher and higher molecular weight oligomer and polymers form at the electrode surface. Since these conjugated polymers often have very limited solubility, this causes deposition of the polymer on the electrode surface. Insulation of the electrode could be confirmed by loss of electrochemical activity (no peaks in the cyclic voltammagram). The deposition and study of these films have been previously reported [27]. The advantages of this method are mainly due to the fact that as the film is deposited, the electrode becomes progressively more and more insulated. This renders the process self-limiting in that eventually the electrode is completely insulated, no more current can flow, and deposition of polymer stops. This means that insulating films of thicknesses of ~50 nm can be reproducibly deposited using this comparatively simple method.

Sonication experiments were performed using a custom-built 2 kW, 25 kHz ultrasound tank with internal dimensions of $750 \times 750 \times 600$ mm (working volume $750 \times 750 \times 500$ mm) (Ultrawave Ltd., Cardiff, United Kingdom). Ultrasound was applied at a frequency of 25 KHz for 10 s duration. Previous work has shown clear formation of microelectrode arrays using this process [28].

37.3.1.1 Construction of Antibody Microarrays via Entrapment Protocols

A 0.2 M aniline hydrochloride solution was prepared in a pH 4.0 acetate buffer (the solution pH fell to ~2.6 upon addition of monomer) and the nonbiotinylated antibody for PSA incorporated into the buffered monomer solution immediately prior to polymerization at a concentration of 0.5 mg/mL. The pH of the monomer solution increases upon addition of antibody but was required to stay below pH 4.0 so that the conductive "emeraldine" form of polyaniline was deposited at the working electrode. The potentiodynamic electrodeposition of polyaniline into the microelectrode array was achieved electrochemically by sequentially cycling the working electrode potential from -200 to $+800$ mV (vs. Ag/AgCl) and back to the starting potential at a scan rate of 50 mV/s. A linear sweep from -200 to $+800$ mV (50 mV/s) was performed at the end of cyclic voltammetry to leave the polyaniline in its protonated emeraldine salt form. Samples were then washed in pH 7.4 buffer and then treated with HSA solution (10^{-6} mol/L in PBS, 1 h) to minimize nonspecific interactions, rinsed with water, and stored dry at 4°C until required.

37.3.1.2 Construction of Antibody Arrays via an Affinity Protocol

Polyaniline was deposited electrochemically from aniline buffer, pH 1–2 into the microelectrode array again by sequentially cycling the working electrode potential as described earlier. The polyaniline microarray was then modified by binding of antibodies. Thirty microliters of biotin–sulfo-NHS (10 mg/mL in water) was placed on the microarray electrode surface for 24 h to chemically attach biotin groups to the polymer. The sensors were rinsed with copious water and 30 μL of neutravidin (10 μg/mL in water) was placed on the working electrode for 1 h, followed by further rinsing in water. Then, 30 μL biotinylated antibody (1 mg/mL in water, 1 h) was added, followed by rinsing. Finally, nonspecific interactions were blocked by HSA (10^{-6} mol/L in PBS, 1 h).

37.3.2 Determination of Antigen Concentration

AC impedance measurements were performed using an ACM Auto AC DSP frequency response analyzer. Immunosensor electrodes were immersed in pH 7.4 phosphate buffer, that is, containing no antigen, as a baseline trace. This buffer solution did, however, contain a 50:50 mixture of $[Fe(CN)_6]^{3-/4-}$, at a concentration of 10 mmol/L as redox mediator so as to perform Faradaic impedance spectroscopy. Following immobilization of antibody, impedance analyses were performed from 1 to 10,000 Hz (± 5 mV amplitude perturbation) with the potential of the electrochemical cell being offset to the formal potential of the redox probe ($+0.12$ V vs. Ag/AgCl identified via amperometric cyclic voltammetry). The sensors were then exposed to a range of concentrations from 1 to 300 ng or pg/mL PSA in pH 7.4 phosphate buffer containing $[Fe(CN)_6]^{3-/4-}$ for a period of 30 min. After exposure they are rinsed with pH 7.4 buffer, the AC impedance is measured again, and the change in impedance is obtained.

37.3.3 Regeneration Studies

The corrected calibration profiles for five APSA-immobilized sensors and five matched nonspecific AIgG-immobilized sensors made by either entrapment or affinity protocols were interrogated over the full range of PSA antigen concentrations used earlier. A mean "corrected" electron transfer change for each concentration was obtained with a standard deviation for the five pairs.

To split the antibody–antigen complex and regenerate the sensors, samples were dipped three times into 0.1 M HCl for 1 min and rinsed with PBS between each dip. A final rinse in 50 mL PBS produced PSA-free sensors which could then be used for fresh analysis. After the regeneration treatment, the same experimental procedure was again followed to assess the recovery of the initial response in terms of "corrected" electron transfer change.

37.4 Results

37.4.1 Impedance Profiles of the Electrodes

Nyquist curves were obtained for all the samples measured, this profile plots the Z' (real) component of the impedance against the Z'' (imaginary) component for 100 data points obtained at frequencies between 1 and 10,000 Hz. What happens is that the impedance equipment sequentially measures the impedance using an AC waveform at 100 different frequencies and assesses both the real and imaginary components of the impedance. The real component (Z') is where the change in impedance is in phase with the AC waveform and corresponds to the resistive component of the impedance; the imaginary component (Z'') is 180° out of phase with the AC waveform and corresponds to the capacitance of the electrode surface. One of the strengths of AC impedance as an interrogation technique is the ability to vary the frequency of the applied voltage. From the variable responses obtained at different frequencies, it is possible to gain an idea of the relative charge transfer resistive (real) and capacitive (imaginary) components of the electrode–solution interface. The Nyquist curve plots the real component (x-axis) against the imaginary component (y-axis) for each individual frequency.

In all cases, the Nyquist curves display a semicircular response where Z' component of the impedance increases steadily with decreasing frequency. This is because at lower frequencies the redox probe becomes exhausted at the surface and more must diffuse in. This limits the number of electrons available for transfer and increases the impedance. However, the behavior of the Z'' component is more complex, increasing as frequency decreases until a maximum is reached before falling as the frequency approaches 1 Hz. Within these types of systems, there are generally two rate-limiting steps for the electrochemical processes occurring at the electrode surface. During the measurement, there is an exchange of electrons between the electrode and the redox probe in solution. This process can be limited by either the diffusion of the redox probe and the electron transfer between electrode and redox probe. The electron transfer kinetics and diffusional characteristics can be extracted from the curves. The Nyquist curves shown within this system are semicircular in shape which is indicative of a surface-modified electrode system where the electron transfer is slow and the impedance is controlled by the interfacial electron transfer [31]. Systems where redox probe diffusion is the rate-limiting factor tend to give approximately straight lines with both Z' and Z'' increasing steadily with decreasing frequency.

A typical impedance profile is shown in Figure 37.3a which is for an entrapped APSA sensor exposed to concentrations between 0 and 10 ng/mL PSA for a PANI-modified, carbon microelectrode array sensor assembly with APSA immobilized via an entrapment method. As can be seen, there are noticeable increases in both the Z' and Z'' components of the impedance spectrum with increased antigen concentrations. This is probably due to increased insulation of the electrode surface due to the binding of the protein hindering electron transfer. Figure 37.3b shows similar plots for electrodes with APSA immobilized via an affinity protocol. We found that these types of sensors responded to much lower concentrations of PSA than the entrapped sensors, as expected from earlier work. In fact, noticeable responses were obtained from pg/mL concentrations of the antigen.

If the semicircle is extrapolated to where it crosses the x-axis, there are two values at which $Z''=0$. The difference between these two values is a measurement of the electron transfer resistance between the electrode and the redox couple [31]. These can be determined manually or by using dedicated software supplied with the ACM. The relative electron transfer resistance changes from the baseline response at each concentration are plotted in Figure 37.4a and b. As can be seen, immunosensors with entrapped APSA gave linear responses from 1 to 200 ng/mL PSA whereas affinity-based immunosensors were much more sensitive (linear response from 1 to 100 pg/mL PSA, above which the sensor reaches saturation and a plateau in response becomes apparent, Figure 37.4b). Limits of detection (three times the standard deviation of the baseline value) were 1 ng/mL (entrapped APSA) or 1 pg/mL (affinity-bound APSA). Both types of sensors eventually reached a plateau indicating saturation of the immunosensor.

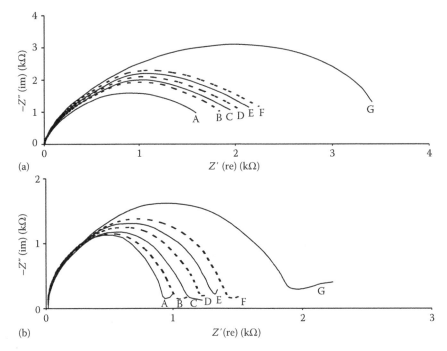

FIGURE 37.3   Specific binding of PSA for (a) entrapped APSA, Nyquist plot, concentrations of PSA: (A) 0, (B) 1, (C) 2, (D) 3, (E) 4, (F) 5, and (G) 10 ng/mL and (b) affinity-immobilized APSA, Nyquist plot, concentrations of PSA: (A) 0, (B) 1, (C) 2, (D) 3, (E) 4, (F) 5, and (G) 10 pg/mL.

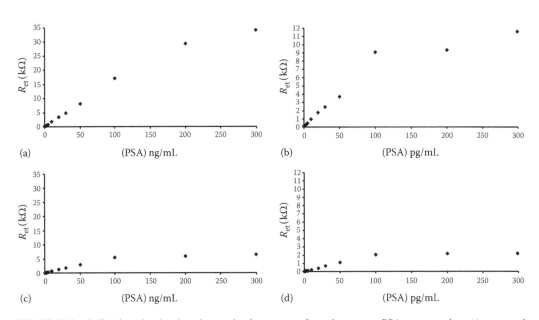

FIGURE 37.4   Calibration plot showing changes in electron transfer resistance vs. PSA concentration: (a) entrapped APSA (0–300 ng/mL), (b) affinity-immobilized APSA (0–300 pg/mL), (c) entrapped AIgG (0–300 ng/mL), and (d) affinity-immobilized AIgG (0–300 pg/mL).

As mentioned earlier, nonspecific interactions can lead to inaccurate results. Therefore, as control electrodes, similar immunosensors were fabricated but with APSA replaced by a nonspecific IgG antibody (AIgG). These electrodes were measured as were the specific electrodes. Nyquist plots for these systems (data not shown) demonstrate that while binding of PSA to the nonspecific immunosensors occurs, responses are much smaller (about 20% of the specific response) than for APSA-modified sensors (Figure 37.4c and d). To remove the effects of nonspecific binding, changes in electron transfer resistance for these electrodes were subtracted from those of the specific electrodes. This allowed the production of corrected calibration plots for all these sensors by subtracting nonspecific responses from the specific responses over the entire analytical concentration range. These are presented in Figure 37.5a for groups of 10 APSA and 10 AIgG sensors of each format, interrogated over their active concentration range. This allows an assessment of the reproducibility of these systems, shown by the error bars which are the standard deviations obtained for the 10 matched sensor pairs from the mean "corrected" values.

From Figure 37.5a, it can be seen that entrapped immunosensors give a linear response in the range 1–200 ng/mL PSA. Figure 37.5b shows an expanded view from 1 to 10 ng/mL PSA and shows linear behavior with $R^2 = 0.998$. Similar behavior could be obtained for affinity-immobilized immunosensors except with much higher sensitivity. These sensors displayed a linear range from 1 to 100 pg/mL PSA (Figure 37.5c) with $R^2 = 0.997$ between 1 and 10 pg/mL PSA (Figure 37.5d). Above the linear ranges, only minimal further changes in impedance are observed again indicating saturation of the sensors, perhaps due to all available antibodies having bound an antigen. From the error bars of the "corrected" responses obtained from the standard deviation for 10 matched sensor pairings, discrimination of concentrations at 1 ng/mL (entrapped) or 5 pg/mL (affinity) PSA can be obtained. The limit of detection of the APSA immunosensor (three times the standard deviation of the baseline value) is 1 ng/mL (entrapped) or 1 pg/mL (affinity).

Probably, the most interesting result is the much higher sensitivity and lower limit of detection of sensors produced by affinity immobilization in comparison to those produced via entrapment. Figure 37.6 shows a schematic of these two systems and gives an explanation for this much higher sensitivity for the systems fabricated using the affinity protocol. During the electrochemical growth of the polyaniline during the entrapment procedure, antibodies are incorporated at random within the polyaniline protrusions and with random orientation. This means that potentially most of the antibody could be located within

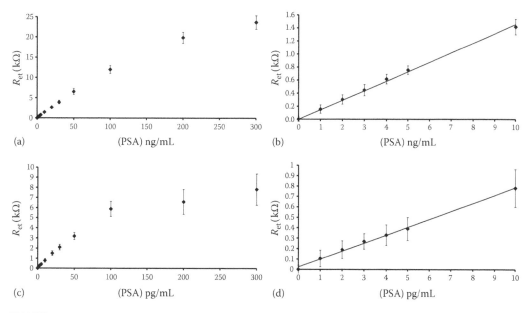

FIGURE 37.5 Corrected calibration plot showing changes in electron transfer resistance vs. PSA concentration: (a) entrapped APSA 0–300 ng/mL PSA, (b) 0–10 ng/mL PSA, (c) affinity-immobilized APSA 0–300 pg/mL PSA, and (d) 0–10 pg/mL PSA.

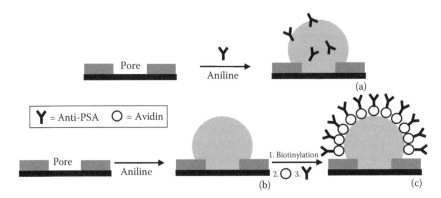

FIGURE 37.6 Comparison of antibody microelectrodes obtained by the entrapment and affinity protocols: (a) simultaneous deposition of anti-PSA and aniline to give polyaniline protrusions containing a low density of antibody in random locations and orientations, (b) deposition of aniline to give unsubstituted polyaniline protrusions, and (c) modification of protrusions with biotinylation agent, avidin, and then biotinylated antibody to give polyaniline protrusions containing a high density of antibody in specific locations and orientations.

the body of the protrusion and only a small fraction immobilized at the surface. Since both antibody and antigen are relatively large molecules, this means that only the small fraction of antibodies immobilized at the surface can bind the PSA. The affinity protocol leads however to a specific immobilization of the antibodies at the surface of the polyaniline protrusions, thereby leading to a much higher local density. This means the effective concentration of the antibody on the electrodes is much higher, thereby allowing easy access of antigen to the antibody and a relatively much less hindered reaction. Therefore, much smaller concentrations of antigen will lead to measurable effects.

A second possible factor is that within the entrapment procedure deposition of polyaniline must be carried out at a pH of 4.0 to maintain the polymers' conductive state and there may be as a result some denaturing of the antibody.

37.4.2 Regeneration Studies

Attempts were made to find whether it was possible to regenerate the immunosensors by washing them in 0.1 M HCl. The change in electron transfer resistance of the affinity sensor after treatment of 100 pg/mL PSA is shown in Figure 37.7a. After one regeneration, negligible changes in "corrected" electron transfer response are observed. Both specific and nonspecific sensors exhibited reproducible behavior. However, a further regeneration significantly lowers the response to ~85% of the initial response with further regeneration attempts causing significant loss of sensor response. The initial reversibility after one regeneration attempt shows that the avidin–biotin interactions are not disrupted by the acid wash and that the antibody–antigen complex disassociates to give free APSA on the sensor. This is limited, however, with further washing reducing sensitivity—possibly due to antibody denaturation. Further work using different washing procedures and methods of antibody stabilization could enhance sensor reversibility. Immunosensor binding is rarely reversible and is a highly desirable property of these systems.

Entrapped APSA sensors were also treated using the same washing protocol; however, regeneration proved impossible since the first acid wash removed all activity and the sensor proved impossible to regenerate. We postulated earlier that only antibodies located at the surface of the protrusion are active in these systems and these could be leached out during an acid wash.

37.4.3 Stability of the Immunosensors

Another desirable quality of any immunosensor system is stability. Therefore, batches of APSA (specific) and AIgG (nonspecific) immunosensors fabricated using the affinity and entrapment protocols were refrigerated at 4°C in sealed, sterilized vessels. Every 2 weeks, a set of three sensors of each type were removed from storage and interrogated as described earlier.

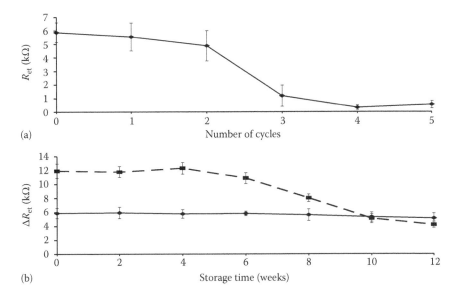

FIGURE 37.7 (a) Effect of regeneration on the mean change in electron transfer resistance at 100 pg/mL PSA for affinity-immobilized immunosensors and (b) effect of storage on the mean change in electron transfer resistance at 100 pg/mL PSA for affinity (solid) and 100 ng/mL for entrapped (dashed) APSA immunosensors.

Figure 37.7b shows the mean "corrected" electron transfer resistances for the sensors obtained at 100 pg or ng/mL PSA for each time interval. Only minimal changes occur for the affinity-based sensor, indicating that up to 12 weeks storage had minimal effects on activity, suggesting a satisfactory shelf-life. The entrapped sensors were stable up to 6 weeks. After that the response dropped—with the responses after 12 weeks being a third of those for fresh sensors.

37.4.4 Effect of Interferents

Interference from other materials in biological samples is a problem for many systems. To attempt to assess this problem, the response of a series of 10 affinity-based sensors to PSA were evaluated over the full range of concentration described earlier. However besides PSA, the antigen solutions contained five other proteins, each at a concentration of 10 ng/mL, namely, BSA, HSA, CA125 (an ovarian cancer marker), and the stroke marker proteins neuron-specific enolase and S-100[β]. Variations of mean "corrected" electron transfer changes were <5% for all concentrations and standard deviations compared to those obtained in the absence of interferents. The presence of large excesses of BSA, HSA, CA125, neuron-specific enolase, and S-100[β] therefore has only minimal effects. Similar behavior was observed for the AIgG sensors (nonspecific). These results indicate that the HSA blocking step during fabrication minimizes nonspecific binding almost totally. These results suggest that sensing of a target antigen in biological samples such as serum is a plausible proposal.

37.5 Discussion and Conclusions

We describe two fabrication techniques and one experimental interrogation protocol as a viable approach toward sensing of PSA without enzymatic labeling, thus demonstrating the generic nature of the technology. The affinity-based PSA immunosensors described have good linear response from 1 to 100 pg/mL with minimal nonspecific responses to other proteins. Sensors of this type can also be stored for at least 12 weeks and have the potential for regeneration.

The immunosensors formulated by an entrapment technique were found to be much less sensitive, although they still have limits of detection as low as 1 ng/mL which is within the range of clinical significance and can be stored for 6 weeks.

Many of the other biosensors for PSA mentioned in the introduction have ng/mL sensitivities but none of them approach the 1 pg/mL limit of detection of the affinity-based microelectrode array immunosensor. In clinical applications, ELISA is the preferred method of testing for PSA. Currently, ELISA testing kits such as those marketed by Fujirebio Diagnostics Inc. permit the quantification of levels of PSA in human serum from approximately 1–150 ng/mL. The extreme sensitivity of our sensors means that samples can be diluted before use, thereby lowering the amount of sample required and also minimizing the effects of any possible interferents by diluting them.

The low quantification levels and limits of detection for PSA have illustrated the potential use of the affinity-based microelectrode array immunosensor assemblies for the labeless detection of PSA. Sensors fabricated using an entrapment method, although less sensitive were found to be capable of determining PSA levels within the range of clinical significance. The lack of effect of interfering proteins is also promising for real applications. The generic nature of this technology has allowed the construction of immunosensors for not only PSA [32] but also the stroke marker proteins NSE [33] and S-100[β] [34]. The next step in our research will be to apply these techniques to clinical samples such as serum or blood. Once its application to clinical samples has been fully assessed, this technology will provide a sensitive, accurate, and relatively inexpensive method for PSA testing which could be applied as a point-of-care approach such as in a general practitioners surgery.

ACKNOWLEDGMENTS

This work, including funding for ACB, has been supported by the European Community QLRT-2001-02583 (SMILE) and NMP2-CT-2003-505485 (ELISHA) Framework VI contracts.

REFERENCES

1. http://seer.cancer.gov/statfacts/html/prost.html (accessed 27 Feb 2012).
2. Office for National Statistics. 2009. *Mortality Statistics: Deaths registered in 2009, England and Wales*, National Statistics: London, U.K.; General Register Office for Scotland. 2010. *Deaths Time Series Data, Deaths in Scotland in 2009*. Registrar General Annual Report 2010. Northern Ireland Statistics and Research Agency: Belfast, Northern Ireland.
3. Wu, J., Z. Fu, F. Yan, and F. H. Ju. 2007. Biomedical and clinical applications of immunoassays and immunosensors for tumor markers. *Trends Anal. Chem.* 26: 679–688.
4. Thompson, I., D. K. Pauler, P. J. Goodman, C. M. Tangen, M. S. Lucia, H. L. Parnes, L. M. Minasian, L. G. Ford, S. M. Lippman, E. D. Crawford, J. J. Crowley, and C. A. Coltman. 2004. Prevalence of prostate cancer among men with a prostate-specific antigen level ≤4.0 ng per millilitre. *N. Engl. J. Med.* 350: 2239–2246.
5. Yalow, R. S. and S. A. Berson. 1959. Assay of plasma insulin in human subjects by immunological methods. *Nature* 184: 1648–1649.
6. John, R., M. Spencer, G. G. Wallace, and M. R. Smyth. 1991. Development of a polypyrrole-based human serum albumin sensor. *Anal. Chim. Acta* 24: 381–385.
7. Cosnier, S. 2005. Affinity biosensors based on electropolymerized films. *Electroanalysis* 17: 1701–1715.
8. Diaz-Gonzalez, M., M. B. Gonzalez-Garcia, and A. Costa-Garcia. 2005. Recent advances in electrochemical enzyme immunoassays. *Electroanalysis* 17: 1901–1918.
9. Meyerhoff, M. E., C. Duan, and M. Meusel. 1995. Novel nonseparation sandwich-type electrochemical enzyme immunoassay system for detecting marker proteins in undiluted blood. *Clin. Chem.* 41: 1378–1394.
10. Sarkar, P., P. S. Pal, D. Ghosh, S. J. Setford, and I. E. Tothill. 2002. Amperometric biosensors for detection of the prostate cancer marker (PSA). *Int. J. Pharma.* 238: 1–9.
11. Escamilla-Gómez, V., D. D. Hernández-Santos, M. B. González-García, J. M. Pingarrón-Carrazón, and A. Costa-García. 2009. Simultaneous detection of free and total prostate specific antigen on a screen-printed electrochemical dual sensor. *Biosens. Bioelectron.* 24: 2678–2683.
12. Fernandez-Sanchez, C., C. J. McNeil, K. Rawson, and O. Nilsson. 2004. Disposable noncompetitive immunosensor for free and total prostate-specific antigen based on capacitance measurement. *Anal. Chem.* 76: 5649–5656.

13. Fernandez-Sanchez, C., A. M. Gallardo-Soto, K. Rawson, O. Nilsson, and C. J. McNeil. 2004. Quantitative impedimetric immunosensor for free and total prostate specific antigen based on a lateral flow assay format. *Electrochem. Commun.* 6: 138–143.

14. Palaniappana, A., W. H. Goha, J. N. Tey, I. P. M. Wijaya, S. M. Moochhala, B. Liedberg, and S. G. Mhaisalkar. 2010. Aligned carbon nanotubes on quartz substrate for liquid gated biosensing. *Biosens. Bioelectron.* 25: 1989–1993.

15. Kim, A., C. S. Ah, C. W. Park, J.-H. Yang, T. Kim, C. G. Ahn, S. H. Park, and G. Y. Sung. 2010. Direct label-free electrical immunodetection in human serum using a flow-through-apparatus approach with integrated field-effect transistors. *Biosens. Bioelectron.* 25: 1767–1773.

16. Wee, K. W., G. H. Kang, J. Park, J. Y. Kang, D. S. Yoon, J. H. Park, and T. S. Kim. 2005. Novel electrical detection of label-free disease marker proteins using piezoresistive self-sensing micro-cantilevers. *Biosens. Bioelectron.* 20: 1932–1938.

17. Wu, G. H., R. H. Datar, K. M. Hansen, T. Thundat, R. J. Cote, and A. Majumdar. 2001. Bioassay of prostate-specific antigen (PSA) using microcantilevers. *Nat. Biotechnol.* 19: 856–860.

18. Huang, L., G. Reekmans, D. Saerens, J. M. Freidt, F. Frederix, L. Francis, S. Muyldermans, A. Campitelli, and C. Van Hoof. 2005. Prostate-specific antigen immunosensing based on mixed self-assembled monolayers, camel antibodies and colloidal gold enhanced sandwich assays. *Biosens. Bioelectron.* 21: 483–490.

19. Besselink, G. A. J., R. P. H. Kooyman, P. J. H. J. van Os, G. H. M. Engbers, and R. B. M. Schasfoort. 2004. Signal amplification on planar and gel-type sensor surfaces in surface plasmon resonance-based detection of prostate-specific antigen. *Anal. Biochem.* 333: 165–173.

20. Jung, J., K. Na, J. Lee, K.-W. Kim, J. Hyun. 2009. Enhanced surface plasmon resonance by Au nanoparticles immobilized on a dielectric SiO_2 layer on a gold surface. *Anal. Chim. Acta* 651: 91–97.

21. Grant, S., F. Davis, J. A. Pritchard, K. A. Law, S. P. J. Higson, and T. D. Gibson. 2003. Labeless and reversible immunosensor assay based upon an electrochemical current-transient protocol. *Anal. Chim. Acta* 495: 21–32.

22. Grant, S., F. Davis, K. A. Law, A. C. Barton, S. D. Collyer, S. P. J. Higson, and T. D. Gibson. 2005. A reagentless immunosensor for the detection of BSA at platinum electrodes by an AC impedance protocol. *Anal. Chim. Acta* 537: 163–168.

23. Barton, A. C., S. D. Collyer, F. Davis, G.-Z. Garifallou, G. Tsekenis, E. Tully, R. O'Kennedy, T. D. Gibson, P. A. Millner, and S. P. J. Higson. 2009. Labeless AC impedimetric antibody based sensors with pg ml^{-1} sensitivities for point-of-care biomedical applications. *Biosens. Bioelectron.* 24: 1090–1095.

24. Garifallou, G.-Z., G. Tsekenis, F. Davis, P. A. Millner, D. G. Pinacho, F. Sanchez-Baeza, M.-P. Marco, T. D. Gibson, and S. P. J. Higson. 2007. Labeless immunosensor assay for fluoroquinolone antibiotics based upon an AC impedance protocol. *Anal. Lett.* 40: 1412–1442.

25. Tsekenis, G., G.-Z. Garifallou, F. Davis, P. A. Millner, D. G. Pinacho, F. Sanchez-Baeza, M.-P. Marco, T. D. Gibson, and S. P. J. Higson. 2008. Detection of fluoroquinolone antibiotics in milk via a labeless immunoassay based upon an alternating current impedance protocol. *Anal. Chem.* 80: 9233–9239.

26. Tsekenis, G., G. Garifallou, F. Davis, P. A. Millner, T. D. Gibson, and S. P. J. Higson. 2008. Labeless immunosensor assay for myelin basic protein based upon an AC impedance protocol. *Anal. Chem.* 20: 2058–2062.

27. Barton, A. C., S. D. Collyer, F. Davis, D. D. Gornall, K. A. Law, E. C. D. Lawrence, D. W. Mills, S. Myler, J. A Pritchard, M. Thompson, and S. P. J. Higson. 2004. Sonochemically fabricated microelectrode arrays for biosensors offering widespread applicability. Part I. *Biosens. Bioelectron.* 20: 328–337.

28. Davis, F., S. D. Collyer, D. D. Gornall, K. A. Law, D. W. Mills, and S. P. J. Higson. 2007. New techniques in monitoring water pollution—Development of sonochemically fabricated microarrays for the determination of pollutants. *Chim. Oggi/Chem. Today* 25: 28–31.

29. Myler, S., S. D. Collyer, F. Davis, D. D. Gornall, and S. P. J. Higson. 2005. Sonochemically fabricated microelectrode arrays for biosensors—Part III. AC impedimetric study of aerobic and anaerobic response of alcohol oxidase within polyaniline. *Biosens. Bioelectron.* 21: 666–671.

30. Davis, F., K. A. Law, N. A. Chaniotakis, D. Fournier, T. D. Gibson, P. A. Millner, J. Marty, M. A. Sheehan, V. I. Ogurtsov, G. Johnson, J. Griffiths, A. P. F. Turner, and S. P. J. Higson. 2007. Chapter 15 Ultrasensitive determination of pesticides via cholinesterase-based sensors for environmental analysis. *Comp. Anal. Chem.* 49: 311–330.

31. Katz, E. and I. Willner. 2003. Probing biomolecular interactions at conductive and semiconductive surfaces by impedance spectroscopy: Routes to impedimetric immunosensors, DNA-sensors, and enzyme biosensors. *Electroanalysis* 15: 913–947.

32. Barton, A. C., F. Davis, and S. P. J. Higson. 2008. Labeless immunosensor assay for prostate specific antigen with picogram per milliliter limits of detection based upon an AC impedance protocol. *Anal. Chem.* 80: 6198–6205.

33. Barton, A. C., F. Davis, and S. P. J. Higson. 2008. Labeless immunosensor assay for the stroke marker protein neuron specific enolase based upon an alternating current impedance protocol. *Anal. Chem.* 80: 9411–9416.

34. Barton, A. C., F. Davis, and S. P. J. Higson. 2010. Labeless immunosensor assay for the stroke marker protein S-100[β] based upon an AC impedance protocol. *Anal. Lett.* 43: 2160–2170.

38

Graphene-Based Electrochemical Immunosensor for the Detection of Cancer Biomarker

Minghui Yang, Alireza Javadi, and Shaoqin Gong

CONTENTS

38.1 Introduction

Early detection of cancer biomarkers plays an important role in early diagnosis and treatment of cancers [1–3]. Prostate-specific antigen (PSA) is a protein produced by cells of the prostate gland [4]. The PSA test measures the level of PSA in the blood. It is normal for men to have a low level of PSA in their blood; however, prostate cancer can lead to a high level of serum PSA. The serum PSA concentration ranges from 1 to 4 ng/mL and 4 to 10 ng/mL for normal person and cancer patient, respectively [5]. Hence, developing rapid, simple, and sensitive detection methods to accurately quantify the serum PSA concentration in clinic can significantly facilitate the early diagnosis of prostate cancer. Due to the specific binding of antibody to its antigen, immunosensors based on antibody–antigen interaction are one of the most widely used analytical techniques in quantitative detection of cancer biomarkers [6–8]. Immunosensors are biosensors based on specific antigen–antibody interactions. The transducers used in the immunosensors can detect the immunochemical reactions either directly or indirectly. Some of the common transducers used in immunosensors include optical, electrochemical, thermometric, piezoelectric, magnetic, or micromechanical. In recent years, various methods, such as enzyme-linked immunosorbent assays (ELISA) [9], fluorescence immunoassay [10], chemiluminescent assay [11], and time-resolved immunofluorometric assay [12], have been developed for PSA detection. Compared to these methods, electrochemical immunosensors have gained considerable attention over the past decades due to their high sensitivity, short analysis time, low cost, and ease of miniaturization [13,14]. Typically, the electrochemical immunosensors have a sandwich-type structure. Specifically, the primary antibody (Ab$_1$) is first immobilized onto a solid surface. Then, the specific antigen binds to the Ab$_1$ site. Finally,

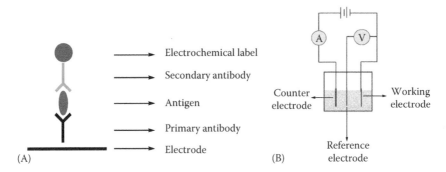

FIGURE 38.1 Schematic demonstration of the electrochemical sandwich immunosensor (A) and the three-electrode setting (B).

labeled secondary antibody (Ab$_2$) binds to the antigen [15–17]. The signal of the immunoassay is mainly from the label, which is in accordance with the antigen concentration. Figure 38.1A shows the principle of electrochemical immunosensor and Figure 38.1B is a demonstration of typical three-electrode electrochemical setting.

There are several electrochemical methods used to characterize the immunosensors, for example, cyclic voltammetry, square-wave voltammetry, and amperometry. Cyclic voltammetry is generally used to study the electrochemical properties of an analyte in solution [1]. In cyclic voltammetry, the electrode potential scans linearly within a given range. The scan produces current peaks for any analytes that can be reduced or oxidized at the electrode surface within the range of the potential scanned and the data is then plotted as current (*i*) vs. potential (*E*). While cyclic voltammetry is for general study of the redox properties of analytes, square-wave voltammetry is used to precisely detect the concentration of analytes, which can lower the detection limits to nanomolar concentrations [15]. Amperometry is directly used to study the electrochemical current response of an analyte on the electrode. The measurements from the electrode originate from electroactive species released into the buffer [8].

Enzymes (e.g., horseradish peroxidase [HRP]) and quantum dot (QD) are among the most popular labels for electrochemical immunosensors [18,19]. Yang et al. reported an immunosensor for the detection of human IgG using carbon nanotube for the immobilization of antibody and HRP-modified mesoporous silica nanoparticles (MSN) as label. The catalytic current of HRP toward H$_2$O$_2$ was used as signal of the immunosensor [18]. Ho et al. reported the detection of protein marker carcinoembryonic antigen (CEA) using carbon nanoparticles for the immobilization of antibody and CdS QD as label. The presence of the carbon nanoparticles improved the performance of the electrochemical reaction of the substrate and increased the sensitivity of the immunosensor for the diagnosis of CEA [19]. However, for HRP, direct electrical communication between the redox sites of enzymes and the electrode is usually not possible because the redox sites of the enzymes are located deep inside the protein shell [20,21]. Therefore, redox mediators are required to facilitate the electron transfer between enzyme and electrode. Indeed, in order to achieve high sensitivity, the amount of mediator, as well as that of enzyme, needs to be sufficiently high. Various methods have been investigated to introduce mediators into the immunosensors, including directly adding mediators into the buffer solution or separately immobilizing mediators onto the electrode surface [22,23]. QD is a semiconductor. To use QD as label, after the assemble of the immunosensor, QDs immobilized onto electrode are dissolved in an acid solution to form the Cd^{2+} ions. The broad use of QD as electrochemical label is due to the strong current generated when the Cd^{2+} in the QD is oxidized on the electrode. For QD-based label, various approaches have been investigated to increase the loading content of QD in the label, such as incorporating QD into polymer micelle [24], or immobilizing QD onto the surface of silica nanoparticles [25].

Recently, graphene sheet (GS), a flat monolayer of carbon atoms tightly packed into a two-dimensional honeycomb lattice, has stimulated a vast amount of research due to its fascinating properties [26–28]. It has a large specific surface area, good electrical conductivity, and excellent mechanical properties. Therefore, GS offers many potential applications in areas such as biosensors, energy-storage materials, and functional polymer nanocomposites. For electrochemical biosensors, GS's large specific surface area allows the immobilization/loading of a large amount of biomolecules and/or sensing materials while its good electrical conductivity can promote electron transfer [29,30], both of which will improve the sensitivity of the biosensor.

In this study, the electrochemical immunosensors were fabricated using GS both for the immobilization of primary antibody (Ab$_1$) and as the secondary antibody (Ab$_2$) label. Two different kinds of Ab$_2$ GS-based labels were prepared: (1) Mediator thionine (TH) was adsorbed onto GS, and the amino group of TH was then used to conjugate enzyme HRP and Ab$_2$. The resulting GS-TH-HRP-Ab$_2$ nanoparticles were used as label for the immunosensor; and (2) GS was functionalized with both CdS QD and Ab$_2$ and the resulting QD-GS-Ab$_2$ nanoparticles were also used as label for the immunosensor. Using these labels, sensitive immunosensors for the detection of PSA have been fabricated which displayed a wide range of linear response and low detection limit.

38.2 Materials

PSA and monoclonal anti-PSA antibody were purchased from Millipore. Graphite, HRP, sodium sulfide, cadmium chloride, mercaptoundecanoic acid (MUA), and 1-pyrenebutanoic acid succinimidyl ester (PBSE) were obtained from Fisher Scientific. TH and glutaraldehyde were obtained from Acros. Phosphate-buffered saline (PBS, 0.1 M, pH 7.4) was used as electrolyte for all electrochemistry measurements. All other reagents were of analytical grade and deionized water was used throughout the study. All electrochemical measurements were performed on a VersaStat 3 electrochemical workstation (Princeton Applied Research, Oak Ridge, TN). A conventional three-electrode system was used for all electrochemical measurements: a glassy carbon (GC) electrode as the working electrode, an Ag/AgCl electrode as the reference electrode, and a platinum wire electrode as the counter electrode.

38.3 Methods

38.3.1 Preparation of GS

GS was prepared from graphite oxide (GO) through a thermal exfoliation method [27]. First, GO powders were produced from graphite using a modified Hummer's method [28]. In a typical experiment, 5 g of graphite was oxidized by reacting them with 100 mL of concentrated H$_2$SO$_4$ under stirring for 12 h. Next, while immersing the reaction vessel in an ice bath, 30 g of KMnO$_4$ was added slowly. After the addition of KMnO$_4$, the solution was stirred at 100°C for another 12 h to fully oxidize graphite to GO. The GO was then thoroughly washed and dried. The resulting GO was then exfoliated thermally in a quartz tube. To do so, an argon inlet was first inserted through the tube. The quartz tube was fluxed with argon for 10 min and then quickly inserted into a furnace preheated to 300°C and held in the furnace for about 7 min.

38.3.2 Preparation of GS-TH-HRP-Ab$_2$ Label

As shown in Figure 38.2A, typically, the as-prepared GS (1 mg) was dispersed in 1 mL of PBS at pH 7.4. This dispersion was then mixed with 1 mL of 1 mM TH solution. The resulting mixture was set on a shaker at 37°C for 48 h and then centrifuged at 14,000 rpm. After discarding the supernatant, the GS again was dispersed in 1 mL of PBS. One mL of 2.5% glutaraldehyde solution was added into this GS buffer solution and stirred for 1 h. Glutaraldehyde was used as the cross-linking agent to conjugate enzyme HRP and Ab$_2$ onto TH. Then, 1.0 mg of HRP and 0.01 mg of anti-PSA Ab$_2$ were added into the solution. The mixture was allowed to react at room temperature under stirring for 24 h, followed by centrifuge. The resulting GS-TH-HRP-Ab$_2$ label was washed with PBS (pH 7.4) and then re-dispersed in 1 mL of buffer and stored at 4°C before use.

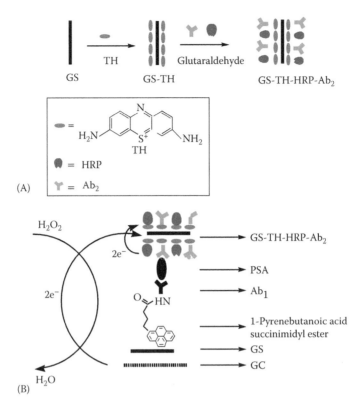

FIGURE 38.2 Schematic representation of the preparation of the GS-TH-HRP-Ab₂ labels (A) and immunosensor (B).

38.3.3 Preparation of GS-QD-Ab₂ Label

To prepare CdS QD–functionalized GS, first, CdCl₂ solution (10 mM in water) was added dropwise into the MUA solution (10 mM in water pH 9.0). After the mixture was stirred for 30 min, 10 mg of GS was added into 10 mL of the mixture and stirred for 24 h at room temperature. After stirring, the solution was centrifuged. After discarding the supernatant, the resulting Cd²⁺ functionalized GS were re-dispersed in PBS. Subsequently, Na₂S solution (1 mM) was slowly added into the aforementioned solution. The mixture was stirred for another 1 h and then centrifuged. After discarding the supernatant, the QD-GS again were dispersed in PBS (1 mg/mL). Ethyl(dimethylaminopropyl)carbodiimide (EDC) and N-hydroxysuccinimide (NHS) (400 mM) were added into this QD-GS buffer solution and stirred for 1 h. Then, 0.01 mg of anti-PSA Ab₂ was added into the solution. The mixture was allowed to react at room temperature under stirring for 24 h, followed by centrifuge. The resulting GS-QD-Ab₂ labels were washed with PBS (pH 7.4) and then re-dispersed in buffer and stored at 4°C before use.

38.3.4 Fabrication of the Immunosensor

Figure 38.2B shows the fabrication procedure of the immunosensor using GS-TH-HRP-Ab₂ as label. Primary anti-PSA antibody (Ab₁) was immobilized onto the surfaces of GS through PBSE based on an amidation reaction between the succinimidyl ester of PBSE and the available amine groups of the Ab₁. First, PBSE was dissolved in water to reach a concentration of 0.5 mg/mL, and then into 1 mL of such solution, 1 mg of GS was added. This mixture was allowed to react for 1 h. The resulting GS-PBSE conjugates were stored at 4°C before use.

GC electrode (3 mm diameter) was polished repeatedly using alumina powder and then thoroughly cleaned before use. Onto the electrode, 5 μL of GS-PBSE buffer solution was added. After the GS-PBSE-coated electrode was dried and washed with buffer, it was incubated in 50 μg/mL of Ab₁ solution for 1 h.

Subsequently, after washing, it was incubated in 1 wt% BSA solution for 1 h to eliminate nonspecific binding between the antigen and the electrode surface. Thereafter, PSA buffer solution with a varying concentration was added onto the electrode surface and incubated for 1 h at room temperature. After that, the electrode was washed extensively to remove unbounded PSA molecules. Finally, the prepared nanolabel (i.e., GS-TH-HRP-Ab$_2$) buffer solution was dropped onto the electrode surface and incubated for 1 h. After washing, the electrode was ready for electrochemical measurement.

Immunosensor using QD-GS-Ab$_2$ as label was prepared following the same procedures described earlier except the GS-TH-HRP-Ab$_2$ label was replaced with QD-GS-Ab$_2$.

38.3.5 Electrochemical Measurement Procedures

A conventional three-electrode system was used for all electrochemical measurements: a GC electrode as the working electrode, an Ag/AgCl electrode as the reference electrode, and a platinum wire electrode as the counter electrode. For immunosensors based on GS-TH-HRP-Ab$_2$ as label, amperometric measurement of the immunosensor was performed at a detection potential of −0.2 V. After the background current was stabilized, 0.5 mM H$_2$O$_2$ was added into the buffer and the current change was recorded. For the immunosensor based on QD-GS-Ab$_2$ as labels, before the electrochemical measurements were conducted, the immunosensor was immersed in PBS containing 0.1 M HNO$_3$ and 10 mg/mL Hg^{2+} to release the cadmium ion from the QD anchored onto the GS-based label. The electrochemical stripping detection was performed with pretreatment at +0.6 V for 1 min, electrodeposition at −1.4 V for 1 min, and stripping from −0.3 V to −1.3 V using square-wave voltammetry (potential steps, 5 mV; frequency, 25 Hz; amplitude, 25 mV). All the data obtained were then imported into Microcal Origin for further analysis.

38.4 Results and Discussion

38.4.1 Immunosensor Using GS-TH-HRP-Ab$_2$ as Label

In this chapter, the immunosensor using GS-TH-HRP-Ab$_2$ as label was discussed in more detail [29]. Figure 38.3 shows the transmission electron microscopic (TEM) image of the as-synthesized GS, which had a wrinkled paper-like structure. The large surface area of the GS was used to increase the loading content of TH, HRP, and Ab$_2$. After TH was adsorbed onto the GS surface through π−π stacking, the large number of the TH amino groups was then used to conjugate enzyme HRP and Ab$_2$ (GS-TH-HRP-Ab$_2$) using glutaraldehyde as the cross-linking agent. To immobilize Ab$_1$, PBSE was adsorbed onto GS through π−π stacking, which was then used to conjugate Ab$_1$ via an amidation reaction between the succinimidyl ester group of PBSE and the amino group of Ab$_1$ [30]. PSA was selected as a model analyte in this work.

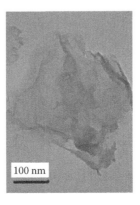

100 nm

FIGURE 38.3   TEM image of the graphene sheet.

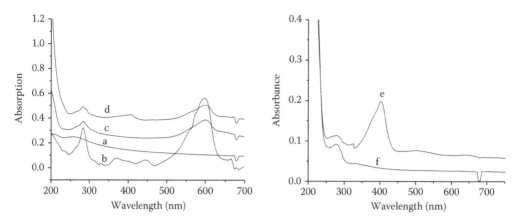

FIGURE 38.4   UV–vis spectra of (a) GS, (b) TH solution, (c) GS-TH, (d) GS-TH-HRP-Ab$_2$, (e) HRP, and (f) Ab$_2$.

The GS-TH-HRP-Ab$_2$ label was characterized with UV–visible (vis) spectroscopy. As shown in Figure 38.4, no absorption peak was observed for the bare GS (curve a). For pure TH solution, there were two major peaks (at approximately 290 and 590 nm) and two smaller peaks (at approximately 370 and 460 nm) (curve b). After TH was adsorbed onto the GS, two absorption peaks, around 290 and 600 nm, were shown (curve c). Changes in the TH UV absorption peaks after adsorption can be attributed to the interaction between GS and TH molecules. The amount of TH adsorbed onto GS, determined by UV–vis spectroscopy, was about 13.1 wt%.

After HRP was conjugated onto the GS, an additional peak appeared at 420 nm (curve d), which can be ascribed to the HRP attached onto the GS. Curve e displays the adsorption spectrum of free HRP in solution. According to the UV–vis spectra analysis, both TH and HRP were successfully conjugated onto GS. No distinct adsorption peak from the antibody was found on the spectrum of GS-TH-HRP-Ab$_2$, which may be attributed to (1) the relatively weak adsorption peak from antibody itself (282 nm, curve f), and (2) peak overlap between the antibody and TH (around 290 nm).

The performance of GS-TH-HRP-Ab$_2$ toward the detection of H$_2$O$_2$ was investigated first because the sensitivity of the GS-TH-HRP-Ab$_2$ label toward H$_2$O$_2$ determines the sensitivity of the immunosensor. The electrochemical signal of the immunosensor mainly results from the HRP-catalyzed H$_2$O$_2$ reduction in the presence of TH. For comparison purposes, three kinds of labels were used to prepare the electrode: GS-HRP-Ab$_2$, GS-TH-Ab$_2$, and GS-TH-HRP-Ab$_2$. To prepare the electrodes, 5 μL of each nanolabel solution (1 mg/mL) was dropped onto the surface of the electrode. After the electrodes were dried, the cyclic voltammograms of the prepared electrodes with and without the presence of H$_2$O$_2$ were measured.

Cyclic voltammetry is generally used to study the electrochemical properties of modified electrode, which could reveal the redox reactions on the electrode surface. As shown in Figure 38.5A, no peak was observed for the GS-HRP-Ab$_2$ label–modified electrode in PBS, and there was no obvious current change after the addition of 1 mM H$_2$O$_2$. For the GS-TH-Ab$_2$ (Figure 38.5B)- and GS-TH-HRP-Ab$_2$ (Figure 38.5C)-coated electrodes in PBS, the redox peaks of mediator TH appeared at −140 and −170 mV, respectively. The strong redox peaks indicated that TH was adsorbed onto GS successfully. However, for the GS-TH-Ab$_2$-modified electrode, there was only a very small catalytic current after the addition of 1 mM of H$_2$O$_2$, indicating that the catalytic efficiency of TH to H$_2$O$_2$ was quite low. On the contrary, a dramatic increase of the reduction current (current change of about 40 μA) and a sharp decrease of the oxidation current (current change of 5 μA) were observed for GS-TH-HRP-Ab$_2$-modified electrode. The high sensitivity observed with the GS-TH-HRP-Ab$_2$-modified electrode may be attributed to several factors: (1) a large amount of HRP and TH was immobilized onto the GS; (2) the immobilized HRP retained high enzymatic catalytic activity; and (3) GS exhibiting good conductivity and electron-transfer ability provided an efficient electrical conducting network, which promoted electrons transfer among the electrode, the redox center of HRP, mediator TH, and H$_2$O$_2$ [31,32].

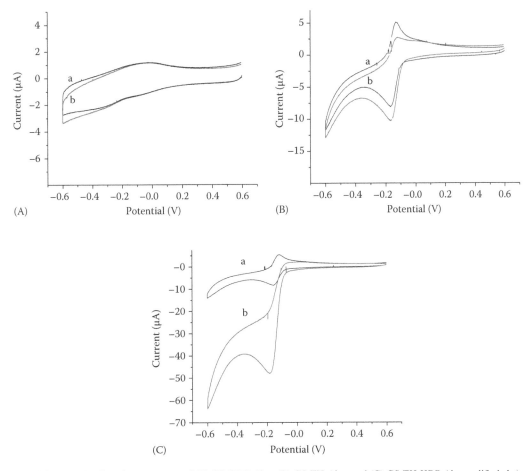

FIGURE 38.5   Cyclic voltammograms of (A) GS-HRP-Ab$_2$-, (B) GS-TH-Ab$_2$-, and (C) GS-TH-HRP-Ab$_2$-modified electrode in PBS before (a) and after (b) the addition of 1 mM H$_2$O$_2$. Scan rate: 100 mV/s.

Since the high sensitivity of GS-TH-HRP-Ab$_2$ toward the detection of H$_2$O$_2$ was proved, immunosensors for detecting PSA using GS-TH-HRP-Ab$_2$ as label were built and characterized. The Ab$_1$ was conjugated onto GS via PBSE, which was first adsorbed onto GS through π–π stacking. The high surface-to-volume ratio of GS can substantially increase the loading of Ab$_1$ [33]. As controls, immunosensors using GS-TH-Ab$_2$ and GS-HRP-Ab$_2$ as labels were also prepared and characterized. Amperometry was used here as it can directly observe the response of the immunosensor to the substrate, H$_2$O$_2$. The amperometric responses of the immunosensors to 1 ng/mL of PSA are shown in Figure 38.6A. The current change before and after the addition of H$_2$O$_2$ was considered as the response of the immunosensor. As expected, the immunosensor using GS-TH-HRP-Ab$_2$ as label displayed the highest current change, which was about 500 times higher than those using GS-TH-Ab$_2$ and GS-HRP-Ab$_2$ as labels. GS-TH-HRP-Ab$_2$ was used as label to prepare immunosensors for detecting different concentrations of PSA. The relationship between the current response and PSA concentration is shown in Figure 38.6B. The catalytic current increased linearly with the PSA concentration within the range of 0.002 to 10 ng/mL with a detection limit of 1 pg/mL (*S/N* = 3). The serum PSA concentration ranges from 1 to 4 ng/mL and 4 to 10 ng/mL for normal person and cancer patient, respectively, which fall in the linear range of this immunosensor. The low detection limit may be attributed to two factors: (1) GS can substantially increase the loading of Ab$_1$ due to its high surface-to-volume ratio, and (2) as discussed earlier, the high catalytic activity of GS-TH-HRP-Ab$_2$ toward H$_2$O$_2$ increased the sensitivity of the immunosensor and lowered the detection limit.

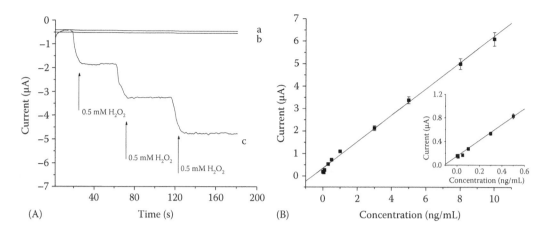

FIGURE 38.6 (A) Amperometric response of the immunosensor for detecting 1 ng/mL PSA with different label at −0.2 V vs. Ag/AgCl toward successive addition of 0.5 mM H_2O_2: (a) GS-HRP-Ab_2, (b) GS-TH-Ab_2, and (c) GS-TH-HRP-Ab_2. (B) Calibration curve of the immunosensor toward different concentrations of PSA. Inset: Amplification of the calibration curve in the low PSA concentration range. Error bar=RSD ($n=5$).

To clarify the advantage of using GS for the preparation of immunosensor, the performance of the immunosensor using GO instead of GS for the immobilization of Ab_1 and as label was tested. The sensitivity of the immunosensor based on GS was about 100 times higher than that based on GO, and the immunosensor based on GO had a much narrower linear range (0.2–2 ng/mL).

To evaluate the reproducibility of the immunosensors, a series of five electrodes were prepared for detecting 1 ng/mL PSA. The relative standard deviation (RSD) of the measurements for the five electrodes was 7.5%, suggesting that the precision and reproducibility of the immunosensor were quite good. The stability of the GS-TH-HRP-Ab_2 label was also investigated. When not in use, the label was stored in PBS at 4°C. After 3 weeks, the response of the immunosensor using GS-TH-HRP-Ab_2 as label retained about 90.3% of initial value. The good long-term stability may be ascribed to two factors: (1) TH were stably adsorbed onto GS through π–π stacking, and (2) Ab_2 and HRP were covalently immobilized onto the GS surface.

To investigate the selectivity of the immunosensors, the amperometric response of the immunosensor toward human IgG, BSA, vitamin C, and glucose were studied. One nanogram per milliliter of PSA solution containing 100 ng/mL of interfering substances was tested by the immunosensor and the current variation due to the interfering substances was <8% of the current measured without the presence of interfering agents, indicating the selectivity of the immunosensor was good.

In an effort to evaluate the reusability of the immunosensor, 0.2 M glycine–HCl solution was used to break the antibody–antigen linkage. After detecting 1 ng/mL of PSA, the immunosensor was dipped into the glycine–HCl solution for 30 s and used to detect PSA again. The immunosensor retained 90.6% of the initial value after five regeneration cycles and a RSD of 7.4% was obtained. The reason for the good regeneration ability may be due to the good stability of Ab_1 on the electrode surface through covalent binding. After 10 regeneration cycles, the response of the immunosensor decreased to about 70% of the initial value.

To investigate the utility of this immunosensor for real biological sample analysis, various concentrations of PSA were added into the normal serum and then analyzed. As shown in Table 38.1, the percent of PSA detected by the immunosensor from these serum samples ranged from 93.0% to 113.1% of the PSA amount added to the serum samples, which is acceptable.

The performance of the immunosensor was further tested for the detection of PSA in five patients' serum samples. The patients are from 30 to 45 years old; some are normal and some with prostate cancer. The patients' serum samples were analyzed using both the GS-based immunosensor done in our lab and the standard ELISA done by the hospital. The PSA contents determined by the immunosensor agreed well with the ELISA measurement. As shown in Figure 38.7, the plot of the PSA contents obtained by these two methods gave a straight line with a correlation coefficient of 0.993.

TABLE 38.1

Recovery of PSA from Serum Samples (ng/mL)

Sample No.	1	2	3	4	5
Amount of PSA added to the serum	1.0	2.0	3.0	4.0	5.0
Amount of PSA recorded by the sensor	1.06 ± 0.03	1.86 ± 0.05	3.13 ± 0.09	4.36 ± 0.10	5.66 ± 0.14
Percent of detection (%)	106.0	93.0	104.3	109.0	113.1

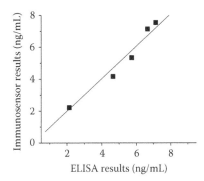

FIGURE 38.7 Comparison of the PSA concentrations in patient serum samples determined with the new GS-based immunosensor and the ELISA method.

38.4.2 Immunosensor Using QD-GS-Ab$_2$ as Label

For the immunosensor using QD-GS-Ab$_2$ as label, the signal of the immunosensor was originated from the oxidation of the Cd^{2+} ion in the CdS QD [34]. Figure 38.8 shows the TEM images of the GS and QD-GS. The GS exhibited a transparent, paper-like structure. For QD-GS, a large number of QD, with a diameter of ~5 nm, were shown on the surface of the GS. As demonstrated previously, nanomaterials with good electrical conductivity, such as carbon nanotubes and carbon nanoparticles, could effectively enhance the detection sensitivity of electroactive species [19,31]. Here, the effect of GS on the detection sensitivity of Cd^{2+} was investigated. Figure 38.9A compares the square-wave voltammograms (SWV) response of the same amount of CdS QD on bare and GS-modified electrode. Square-wave voltammetry is used as it can significantly lower the detection limits. Both electrodes produced one well-defined peak at about −0.8 V, which was due to the oxidation of Cd^{2+} ions. However, as expected, the GS-modified electrode enhanced the peak current about three times compared to the bare electrode.

Figure 38.9B shows the SWV response of the immunosensor to different concentrations of PSA. The peak current, attributed to the oxidation of Cd^{2+}, increased with the PSA concentration linearly in the range of 0.005–10 ng/mL with a detection limit of 3 pg/mL (Figure 38.9C). The low detection limit may be attributed to three factors: (1) a large number of QDs were immobilized onto the GS surface; (2) a large

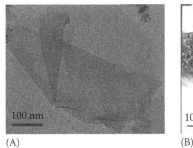

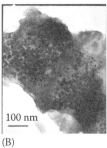

(A) (B)

FIGURE 38.8 TEM images of GS (A) and QD-functionalized GS (B).

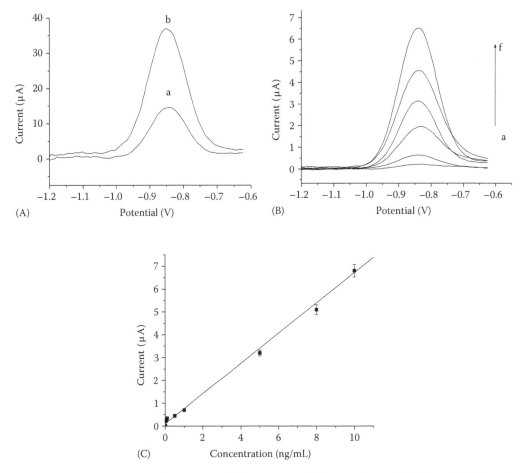

FIGURE 38.9 (A) Square-wave voltammograms (SWV) response of QD on (a) bare and (b) GS-modified electrode. (B) SWV response of the immunosensor to different concentrations of PSA: (a) 0.05, (b) 0.5, (c) 2, (d) 5, (e) 8, and (f) 10 ng/mL. (C) Calibration curve of the immunosensor at different concentrations of PSA. Error bar = RSD ($n = 5$).

amount of Ab_1 was immobilized to the GS-modified electrode; and (3) GS's good conductivity enhanced the detection sensitivity of Cd^{2+}.

The detection limit of our immunosensors was better than those of previously reported PSA immunosensors using label based on just HRP (1 ng/mL) [35], carbon nanotube–HRP (4 pg/mL) [36], or alkaline phosphatase–encapsulated liposome (7 pg/mL) [37]. Compared with several other types of immunosensors based on fluorescence, chemiluminescence, surface plasmon resonance, or quartz crystal microbalance, electrochemical immunosensors offer a number of significant advantages, including high sensitivity, low cost, and ease of miniaturization.

Both types of immunosensors reported earlier utilized functionalized GS as electrochemical labels. One of it used HRP-modified GS and the other used QD-modified GS. Both types of immunosensors exhibited high sensitivity; however, one drawback of HRP is that it can be easily denatured. Compared to HRP, QDs are more stable, but the intrinsic toxicity of QDs prevented them from widespread applications.

38.5 Future Trends

To increase the sensitivity of the electrochemical immunosensors, there are two basic strategies. The first one is to increase the amount of primary antibody captured onto the electrode surface and the second one is to optimize the signal amplification techniques. Various nanomaterials with a large surface-to-volume ratio and good electrical conductivity can be used to enhance the loading levels of antibodies and sensing materials, thereby achieving a lower detection limit and higher sensitivity. Future trends for the development of immunosensors include simplifying the immunosensor preparation process, decreasing the cost, and shortening the assay time. Recently, lateral flow test strip, also called dry-reagent strip biosensor, has become a powerful tool for protein analysis and has attracted increasing attention [38,39]. In addition, electrochemical immunosensors can be assembled into microfluidic chips to achieve high-throughput detection for cancer-screening applications. The microfluidic chips can be applied to the simultaneous detection of different kinds of cancer markers, or for the simultaneous detection of different concentrations of the same cancer marker.

Acronyms

Ab	Antibody
EDC	Ethyl(dimethylaminopropyl)carbodiimide
ELISA	Enzyme-linked immunosorbent assays
GC	Glassy carbon
GO	Graphite oxide
GS	Graphene sheet
HRP	Horseradish peroxidase
MUA	Mercaptoundecanoic acid
NHS	N-Hydroxysuccinimide
PBS	Phosphate-buffered saline
PBSE	1-pyrenebutanoic acid succinimidyl ester
PSA	Prostate specific antigen
QD	Quantum dot
TEM	Transmission electron microscopy
TH	Thionine

REFERENCES

1. Wu, Y., Zheng, J.W., Li, Z., Zhao, Y.R., Zhang, Y., 2009. A novel reagentless amperometric immunosensor based on gold nanoparticles/TMB/Nafion-modified electrode. *Biosens. Bioelectron.* 24: 1389–1393.
2. Bi, S., Zhou, H., Zhang, S.S., 2009. Multilayers enzyme-coated carbon nanotubes as biolabel for ultrasensitive chemiluminescence immunoassay of cancer biomarker. *Biosens. Bioelectron.* 24: 2961–2966.
3. Rasooly, A., Jacobson, J., 2006. Development of biosensors for cancer clinical testing. *Biosens. Bioelectron.* 21: 1851–1858.
4. LeBeau, A.M., Singh, P., Singh, J.T., Denmeade, S.R., 2009. Prostate-specific antigen is a "chymotrypsin-like" serine protease with unique P1 substrate specificity. *Biochemistry* 48: 3490–3496.
5. Lilja, H., Ulmert, D., Vickers, A.J., 2008. Prostate-specific antigen and prostate cancer: Prediction, detection and monitoring. *Nat. Rev. Cancer* 8: 268–278.
6. Liu, J.H., Ju, H.X., 2005. Electrochemical and chemiluminescent immunosensors for tumor markers. *Biosens. Bioelectron.* 20: 1461–1470.
7. Huang, K.J., Niu, D.J., Xie, W.Z., Wang, W., 2010. A disposable electrochemical immunosensor for carcinoembryonic antigen based on nano-Au/multi-walled carbon nanotubes-chitosans nanocomposite film modified glassy carbon electrode. *Anal. Chim. Acta* 659: 102–108.
8. Rusling, J.F., Sotzing, G., Papadimitrakopoulosa, F., 2009. Designing nanomaterial-enhanced electrochemical immunosensors for cancer biomarker proteins. *Bioelectrochemistry* 76: 189–194.
9. Acevedo, B., Perera, Y., Ruiz, M., Rojas, G., Benitez, J., Ayala, M., Gavilondo, J., 2002. Development and validation of a quantitative ELISA for the measurement of PSA concentration. *Clin. Chim. Acta* 317: 55–63.
10. Yu, F., Persson, B., Knoll, W., 2004. Surface plasmon fluorescence immunoassay of free prostate-specific antigen in human plasma at the femtomolar level. *Anal. Chem.* 76: 6765–6770.

11. Seto, Y., Iba, T., Abe, K., 2001. Development of ultra-high sensitivity bioluminescent enzyme immunoassay for prostate-specific antigen (PSA) using firefly luciferase. *Luminescence* 16:285–290.

12. Ye, Z., Tan, M., Wang, G., Yuan, J., 2004. Preparation, characterization, and time-resolved fluorometric application of silica-coated terbium(III) fluorescent nanoparticles. *Anal. Chem.* 76: 513.

13. Liu, S.Q., Zhang, X.T., Wu, Y.F., Tu, Y.F., Lin, H., 2008. Prostate-specific antigen detection by using a reusable amperometric immunosensor based on reversible binding and leasing of HRP-anti-PSA from phenylboronic acid modified electrode. *Clin. Chim. Acta* 395: 51–56.

14. Kim, D.J., Lee, N.E., Park, J.S., Park, I.J., Kim, J.G., Cho, H.J., 2010. Organic electrochemical transistor based immunosensor for prostate specific antigen (PSA) detection using gold nanoparticles for signal amplification. *Biosens. Bioelectron.* 25: 2477–2482.

15. Cui, R.J., Pan, H.C., Zhu, J.J., Chen, H.Y., 2007. Versatile immunosensor using CdTe quantum dots as electrochemical and fluorescent labels. *Anal. Chem.* 79: 8494–8501.

16. Fan, A.P., Lau, C., Lu, J.Z., 2005. Magnetic bead-based chemiluminescent metal immunoassay with a colloidal gold label. *Anal. Chem.* 77: 3238–3242.

17. Papkovsky, D.P., O'Riordan, T.C., Guilbault, G.G., 1999. An immunosensor based on the glucose oxidase label and optical oxygen detection. *Anal. Chem.* 71: 1568–1573.

18. Yang, M.H., Li, H., Javadi, A., Gong, S.Q., 2010. Multifunctional mesoporous silica nanoparticles as labels for the preparation of ultrasensitive electrochemical immunosensors. *Biomaterials* 31: 3281–3286.

19. Ho, J.A., Lin, Y.C., Wang, L.S., Hwang, K.C., Chou, P.T., 2009. Carbon nanoparticle-enhanced immunoelectrochemical detection for protein tumor marker with cadmium sulfide biotracers. *Anal. Chem.* 81: 1340–1346.

20. Ruan, C.M., Yang, F., Lei, C.H., Deng, J.Q., 1998. Thionine covalently tethered to multilayer horseradish peroxidase in a self-assembled monolayer as an electron-transfer mediator. *Anal. Chem.* 70: 1721–1725.

21. Lei, C., Deng, J., 1996. Hydrogen peroxide sensor based on coimmobilized methylene green and horseradish peroxidase in the same montmorillonite-modified bovine serum albumin–glutaraldehyde matrix on a glassy carbon electrode surface. *Anal. Chem.* 68: 3344–3349.

22. Tang, D.P., Ren, J.J., 2008. In situ amplified electrochemical immunoassay for carcinoembryonic antigen using horseradish peroxidase-encapsulated nanogold hollow microspheres as labels. *Anal. Chem.* 80: 8064–8070.

23. Mani, V., Chikkaveeraiah, B.V., Patel, V., Gutkind, J.S., Rusling, J.F., 2009. Ultrasensitive immunosensor for cancer biomarker proteins using gold nanoparticle film electrodes and multienzyme-particle amplification. *ACS Nano.* 3: 585–594.

24. Xiang, Y., Zhang, Y.Y., Chang, Y., Chai, Y.Q., Wang, J., Yuan, R., 2010. Reverse-micelle synthesis of electrochemically encoded quantum dot barcodes: Application to electronic coding of a cancer marker. *Anal. Chem.* 82: 1138–1141.

25. Qian, J., Zhang, C.Y., Cao, X.D., Liu, S.Q., 2010. Versatile immunosensor using a quantum dot coated silica nanosphere as a label for signal amplification. *Anal. Chem.* 82: 6422–6429.

26. Aleiner, I.L., Efetov, K.B., 2006. Effect of disorder on transport in graphene. *Phys. Rev. Lett.* 97: 236801.

27. McAllister, M.J., Li, J.L., Adamson, D.H., Schniepp, H.C., Abdala, A.A., Liu, J., Herrera-Alonso, M., Milius, D.L., Car, R., Prud'homme, R.K., Aksay, I.A., 2007. Single sheet functionalized graphene by oxidation and thermal expansion of graphite. *Chem. Mater.* 19: 4396–4404.

28. Liu, Z., Robinson, J.T., Sun, X.M., Dai, H.J., 2008. PEGylated nanographene oxide for delivery of water-insoluble cancer drugs. *J. Am. Chem. Soc.* 130: 10876–10877.

29. Yang, M.H., Javadi, A., Li, H., Gong, S.Q., 2010. Ultrasensitive immunosensor for the detection of cancer biomarker based on graphene sheet. *Biosens. Bioelectron.* 26: 560–565.

30. Shao, N., Wickstrom, E., Panchapakesan, B., 2008. Nanotube–antibody biosensor arrays for the detection of circulating breast cancer cells. *Nanotechnology* 19: 465101.

31. Zhang, M.G., Gorski, W., 2005. Electrochemical sensing platform based on the carbon nanotubes/redox mediators-biopolymer system. *J. Am. Chem. Soc.* 127: 2058–2059.

32. Yang, M.H., Jiang, J.H., Yang, Y.H., Chen, X.H., Shen, G.L., Yu, R.Q., 2006. Carbon nanotube/cobalt hexacyanoferrate nanoparticle–biopolymer system for the fabrication of biosensors. *Biosens. Bioelectron.* 21: 1791–1797.

33. Du, D., Zou, Z.X., Shin, Y.S., Wang, J., Wu, H., Englehard, M.H., Liu, J., Aksay, I.A., Lin, Y., 2010. Sensitive immunosensor for cancer biomarker based on dual signal amplification strategy of graphene sheets and multienzyme functionalized carbon nanospheres. *Anal. Chem.* 82: 2989–2995.

34. Hansen, J.A., Wang, J., Kawde, A.N., Xiang, Y., Gothelf, K.V., Collins, G., 2006. Quantum-dot/aptamer-based ultrasensitive multi-analyte electrochemical biosensor. *J. Am. Chem. Soc.* 128: 2228–2229.

35. Chikkaveeraiah, B.V., Bhirde, A., Malhotra, R., Patel, V., Gutkind, J.S., Rusling, J.F., 2009. Single-wall carbon nanotube forest arrays for immunoelectrochemical measurement of four protein biomarkers for prostate cancer. *Anal. Chem.* 81: 9129–9136.

36. Yu, X., Munge, B., Patel, V., Jensen, G., Bhirde, A., Gong, J., Kim, S.N., Gillespie, J., Gutkind, J.S., Papadimitrakopoulos, F., Rusling, J.F., 2006. Carbon nanotube amplification strategies for highly sensitive immunodetection of cancer biomarkers. *J. Am. Chem. Soc.* 128: 11199–11205.

37. Qu, B., Guo, L., Chu, X., Wu, D.H., Shen, G.L., Yu, R.Q., 2010. An electrochemical immunosensor based on enzyme-encapsulated liposomes and biocatalytic metal deposition. *Anal. Chim. Acta* 663:147–152.

38. Liu, G.D., Lin, Y.Y., Wang, J., Wu, H., Wai, C.M., Lin, Y.H., 2007. Disposable electrochemical immunosensor diagnosis device based on nanoparticle probe and immunochromatographic strip. *Anal. Chem.* 79: 7644–7653.

39. Xu, H., Mao, X., Zeng, Q.X., Wang, S.F., Liu, G.D., 2009. Aptamer-functionalized gold nanoparticles as probes in a dry-reagent strip biosensor for protein analysis. *Anal. Chem.* 81: 669–675.

39

Label-Free Electrochemical Sensing of DNA Hybridization for Cancer Analysis

Venkataraman Dharuman and Jong Hoon Hahn

CONTENTS

39.1 Introduction

39.1.1 Electrochemical DNA Biosensors

Electrochemical sensing of DNA hybridization using low-density DNA microarrays attracts tremendous interest in various fields including disease diagnosis and environmental and warfare agent detection. The electrochemical method is simple, sensitive, cost-effective, compatibility with mass fabrication of sensor elements, and miniaturized portable sensor devices for onsite evaluation of samples compared to the conventional fluorescence and radioactivity-based assays. A typical electrochemical DNA sensor consists of a solid electrode (Figure 39.1A), an electrochemical analyzer (Figure 39.1B), and surface-confined DNA probes (called hereafter as recognition layer) (Figure 39.1A). The hybridization between the pre-immobilized HS-single-stranded DNA (ssDNA) (Figure 39.1C) and target DNA probes is transduced into an electrical signal (Figure 39.1D). Numerous papers have been published on various electrochemical DNA sensors and reviewed recently [1–18]. In-depth discussion is out of scope of this chapter and limited to cancer diagnosis by DNA hybridization using gold transducer and electrochemical technique. Cancer caused by mutations in somatic cells and other genetic diseases are caused solely by mutations in germ line [19]. Predisposition of inherited genetic mutations results in a specific type of cancer. Depending

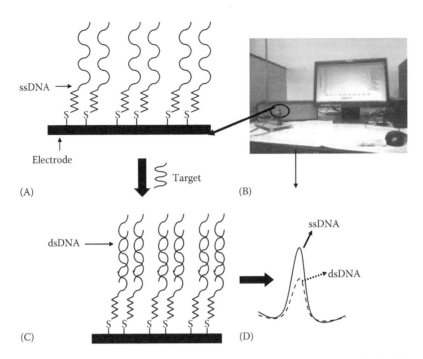

FIGURE 39.1   (A) Transducer carrying pre-immobilized single-stranded DNA. (B) Electrochemical instrument from CH Instruments Inc., USA, connected to a computer. (C) Hybridization of complementary target with the preimmobilized single stranded DNA on the transducer. (D) Behavior electrical signal before and after target hybridization with the capture DNA probes.

on the type of cancer, minimum 3–20 mutations required for the proliferation of particular cancer cell. Identification of sensitive and specific biomarker for the early diagnosis of cancer is an area with immense potential benefits for the patients. Most of the reports by electrochemical method focused on detection of breast cancer. The detection of breast cancer genes *BRCA1* and *BRCA2* was made by direct oxidation of guanine base [15], indirectly by catalytic oxidation of guanine using redox intercalator [16], and metal complex $[Ru(NH_3)_6]^{3+}$ [17], enzymatic detection (biotin labeling on the target probe coupled with avidine labeling of alkaline phosphatase) of 1000 HeLa cancer cells [18] and electrochemical immunoassays [20–23]. Sandwich enzyme immunoassays using array immunochips immobilized with different antibodies were reported for the simultaneous amperometric measurements of the two tumor markers: carcinoembryonic antigen (CEA) and α-fetoprotein (AFP) [24]. One-dimensional semiconductors, conducting polymer wires, and carbon nanotubes functionalized with DNA, PNA, antibodies [25–28], etc., were reported toward the development of miniaturized sensors for point-of-care testing systems.

Cytokeratin 20 (CK20) belongs to the epithelial subgroup of the intermediate filament family. Because of lack of immunological cross-reactivity, with other CKs, CK20 has become important tool for detecting and identifying metastatic cancer cells by immunohistochemistry and by PCR analysis [29,30]. CK20 expression is mostly used in diagnosing colorectal carcinomas, ovarian, and pancreatic—biliary tumor as cancer biomarker. CK20 is co-expressed with CK7 in several types of other cancers such as bladder cancer, urothelial, pancreatic adenocarcinoma, cholangiocarcinoma, lung adenocarcinoma, thyroid carcinoma, prostate carcinoma, gastrointestinal cancer, etc. [31,32], and hence, the combination of CK7 and CK20 became de facto standard against which other markers are judged. Other CKs, CK8, CK18, and CK19, have also been reported to co-occur with CK20 in human pancreas.

Screening the anticancer compounds for activity also profiles the cells for their drug sensitivity and provides a unique perspective on the molecular pharmacology of cancer and enhances profiling of cells at the molecular level. Recently, cell profiling at the level of DNA, mRNA, and proteins using cDNA microarrays has been reported using in vitro-cultured cell which are not full representative of in vivo cells but carries the signatures of RNA and proteins originating tissues [30]. Therefore, tumor genesis is thought to typically occur through series of multiple genetic mutations or hits with some tumor-associated alterations that affect molecular diagnosis. Molecular diagnosis involves the detection of clinically useful genetic genes at DNA, mRNA, or protein levels.

Brain cancer is the most deadly and intractable diseases, diagnosed after a patient exhibits symptoms such as nausea, dizziness, uncharacteristic behavior changes, or paralysis. Treatment is elusive since it spreads quickly and difficult to treat with surgery and radiation therapy. Early diagnosis of brain tumor being done by computed tomography (CT) scan, magnetic resonance imaging (MRI), angiogram by surgery, radiation, and chemotherapy. Other techniques such as intrinsic dielectric permittivity change, active microwave imaging technique [33], and thermal infrared imaging [34,35] have been used for early detection of breast cancer which tends to be more aggressive in young than in older women [36–39]. The electrochemical detection of brain tumor requires pretreatments for isolation of DNA or proteins from tissue or cells before subjected to diagnosis. This will be done by developing the micro-total analytical systems (µ-TAS) [40–42] integrated with fluidic channels for sample injection, separation of bioanalytes and preconcentration, electronic sensors for detection and quantification of separated bioanalytes. Before doing this, it is worthwhile to understand the genosensor behavior using short-chain tumor sequence and to optimize the conditions required for real sample analysis.

The authors have used synthetic brain tumor CK20 sequences [43–47] for demonstrating the applicability of electrochemical DNA hybridization sensing technology (shown in Figure 39.1). It is understood from our and other groups that the hybridization discrimination (signal difference between the ssDNA- and double-stranded DNA (dsDNA)-modified surfaces) effect of the electrochemical sensor is controlled mainly by the structural control of DNA [48,49], probe length, geometry [50], orientation distribution of DNA probes [51–54], transduction technique [47,55,56], and HS-ssDNA density [56]. Collectively, these studies indicate that the selectivity and sensitivity are affected mainly by the presence of nonspecific adsorption of DNA, protein, antibodies in a given sample on to the electrode surface and should be prevented for cancer diagnosis and hence in-depth basic and applied researches are encouraged in this field. Here, we constructed different types of mixed monolayers to discourage the nonspecific adsorption of DNA onto gold and to improve the sensitivity and selectivity by restricting the charge transport only through dsDNA and not by direct reactions at the electrode surface. The advantage of charge transport through dsDNA for the detection of point mutations and single nucleotide polymorphism (SNP) using short-chain synthetic oligonucleotides is used (this will be explained in later sections) [57]. Different electrochemical techniques such as cyclic voltammetry (CV), impedance, and chrono (time) techniques were used to understand the role of surface chemistry on discrimination of single-stranded (noncomplementary DNA reaction with the capture probe), double-stranded (complementary reaction with the capture probe), and double strands with SNP (point mutation). This chapter deals with the principles of different electrochemical techniques briefly and their application in detection of CK20 sequence on Au electrode and detailed information are available in specialized books.

39.2 Materials and Methods

39.2.1 Materials

$K_3[Fe(CN)_6]$, H_2SO_4, H_2O_2, NaCl, NaH_2PO_4, NaOH, KOH, KCl, sodium dodecyl sulfate (SDS), 3,6-diaminoacridine hydrochloride (proflavine, PF), methylene blue (MB), $[Ru(NH_3)_6]^{3+}$, 6-mercapto-1-hexanol (MCH), 3-mercaptopropane (MP), and 3-mercaptopropanoic acid (MPA) were all obtained from Sigma-Aldrich (St. Louis, MO). Deionized water (DI) was prepared from Milli-Q purifying system (Millipore, Milford, MA). Gold wires of 99.99% purity grade and 1 mm diameter, chromium metal were purchased from Sigma-Aldrich. Silicon wafer was purchased from LG Siltron, Korea.

The 0.01 M phosphate buffer containing 120 mM NaCl and 2.7 mM KCl (pH adjusted to 7.4 using NaOH) was used for all electrochemical measurements. Sodium saline citrate (SSC) buffer at pH 8.0 was also obtained from Sigma-Aldrich and used for DNA hybridizations. Ten millimolar Tris buffer of pH 7.0 (adjusted with 1 M HCl) was used in the HS-ssDNA density evaluation studies.

Short-chain 27 mer synthetic oligonucleotides were all synthesized by MWG Biotech (Ebersberg, Germany) with HPLC purification. The used synthetic brain tumor sequences are as follows:

- Capture probe: 5′-HS-(CH$_2$)$_6$-CGA T CTG TTT TAT GT<u>A</u> GGG TTA GGT CA-3′ (I)
- Complementary target to I: 5′-TG ACC TAA CCC TAC ATA AAA CAG-3′ (II)
- Noncomplementary target to II: 5′-TAC CAT TCT CAT CTC TGA AAA CTT CCG-3′ (III)
- Single base mismatch target to I: 5′-TGA CCT AAC CC<u>C</u> ACA TAA AAC AG-3′ (IV)

39.2.2 Electrochemical Methods

39.2.2.1 Cyclic Voltammetry

Generally, a three-electrode system consisting of Au working, Ag/AgCl reference, and Pt wire counter electrodes is used in most of the electrochemical measurements (Figure 39.1) (Potentiostat from CH Instruments Inc., Austin, TX and electrochemical cell shown, for example, in Figure 39.1B. All the techniques used in this study are integrated in an instrument and the required technique can be selected by the user. Similar Potentiostat are available from various companies). CV is a technique in which potential is applied between working and reference electrodes and the output current is monitored. The potential is scanned from the initial to the final potential and reversed back to the initial potential completing a cycle, represented by saw tooth form (Figure 39.2A). The scanning potential window is selected based on the solution pH and redox probe under study. The plot of current vs. potential is known as voltammogram (Figure 39.2B). Conventionally, two types of current measurements are made: (1) non-Faradaic (Figure 39.2B, dark dashed line), and (2) Faradaic currents (Figure 39.2B, solid and dotted lines). The non-Faradaic current occurs by double-layer charging in the absence of any electroactive species in solution and gives information on the capacitance of the electrode because the double layer behaves as parallel-plate capacitor. The Faradaic current measured in presence of redox probe and used under conditions where capacitive current contribution is very small. The Faradaic current depends on two things: (1) the kinetics of electron transfer, and (2) the rate at which electroactive species diffuses to the electrode surface. For redox probes [Fe(CN)$_6$]$^{3-/4-}$, the kinetics of electron transfer are reasonably fast, the concentrations of [Fe(CN)$_6$]$^{3-}$ and [Fe(CN)$_6$]$^{4-}$ can be described by the Nernst equation:

$$E = E^0 - 0.0592 \log\left(\frac{\left[\text{Fe}\left(\text{CN}\right)_6\right]^{4-}}{\left[\text{Fe}\left(\text{CN}\right)_6\right]^{3-}}\right)$$

where

E is the applied potential
E^0 is the formal potential

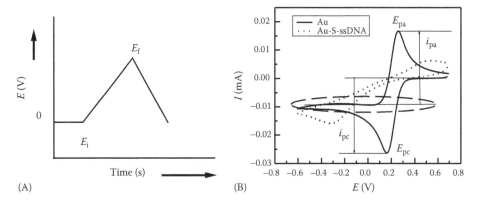

FIGURE 39.2 (A) Saw tooth representation of potential scanning from initial potential (E_i) to final potential (E_f) and reverting back. (B) Typical cyclic voltammetric curves for Au in phosphate buffer without and with [Fe(CN)$_6$]$^{3-}$ before and after immobilization of single-stranded DNA onto the Au surface. Dark dashed line: Double-layer charging current in pure buffer without redox probe. Solid line: In presence of electroactive redox probe [Fe(CN)6]$^{3-}$ in buffer. Dotted line: In presence of [Fe(CN)$_6$]$^{3-}$ for single-stranded DNA immobilized on Au transducer.

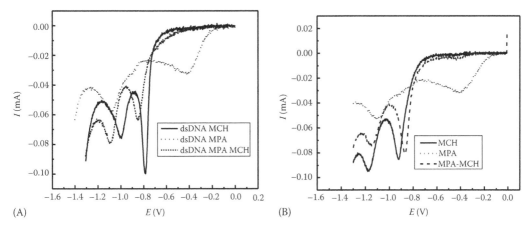

FIGURE 39.3 (A) Thiol-reductive voltammogram of binary HS-dsDNA/MCH (solid line), HS-dsDNA/MPA (dotted line), and ternary HS-dsDNA/MPA/MCH (dashed line) layers made in 0.5 M KOH at a scan rate of 100 mV/s after impedance measurements. (B) Thiol-reductive voltammogram of MCH (solid line), MPA (dotted line), and MPA/MCH (dashed line) monolayers made in 0.5 M KOH at a scan rate 100 mV/s after impedance measurements.

Figure 39.2B shows the voltammetric behavior for the $[Fe(CN)_6]^{3-/4-}$ redox probe in phosphate buffer. In the figure, E_{pa} and E_{pc} are anodic and cathodic peak potentials, respectively, and the corresponding peak currents are represented as i_{pa} and i_{pc}. Variation of these parameters for given redox probe decides the nature of electrochemical reaction such as reversible, quasireversible, and irreversible.

In this work, CVs were recorded between the potential windows 700 to −600 mV at a scan rate 50 mV/s in PBS buffer of pH 7.4 for DNA-sensing studies. Thiol reduction was made in 0.5 M KOH at a scan rate 100 mV/s between 0 to −1400 mV (Figure 39.3A and B). In this, the thiol functional group attached through covalent bond on the gold electrode is reduced by applying negative potential. From the integration of this peak, the total number of thiol molecules Γ_0 immobilized on to the Au surface is obtained from the formula $Q = nFA\Gamma_0$, where Q is charge (C/cm²), n is the number of electrons exchanged during the reaction, F is Faraday constant, and A is electrode (area/cm²).

39.2.2.2 Chronoamperometry and Chronocoulometry

In these techniques, a constant voltage is applied to the electrode–film interface and the change in current is monitored as a function of time. The resulting i–t curves are called amperogram (Figure 39.4), and the integration of the same gives the total charge passed through the sensors. Since the charge is integral current with respect to time, charge for chronocoulometry (CC) is simply obtained by integrating the chronoamperometry (CA) curves. Hence, in this study, CA was made at −350 mV for 10 s. CA technique was used for measuring the amount of ssDNA present on the transducer surface using

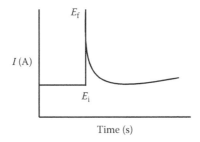

FIGURE 39.4 Typical chronoamperometric curve obtained by applying the final potential E_f at faster scan rate (1 V/s) from the initial potential E_i. The final potential E_f is determined from CV peak potentials $((E_{pa} + E_{pc})/2)$.

[Ru(NH$_3$)$_6$]$^{3+}$ redox probe [58,59]. The measuring principle is based on charge compensation method. That is, the positively charged [Ru(NH$_3$)$_6$]$^{3+}$ easily reacts with negatively charged phosphate backbone of HS-ssDNA. The formal potential $(E_{pa} + E_{pc})/2$ of the surface-confined [Ru(NH$_3$)$_6$]$^{3+}$ at the DNA-modified gold electrodes is more negative than that of the diffused redox cations in solution [60]. The observed formal potential for the surface-confined [Ru(NH$_3$)$_6$]$^{3+}$ was −335 ± 10 mV. Therefore, the CA measurements were made in the blank TBE buffer before and after incubation of the sensor surfaces in 100 μM [Ru(NH$_3$)$_6$]$^{3+}$ solution by applying −350 mV potential pulses and recording the current decay for 10 s. The incubation period of 10 min was sufficient to saturate all the phosphate backbone at this concentration. Since the measurement was made in pure Tris buffer after [Ru(NH$_3$)$_6$]$^{3+}$ incubation, the resulting signal must be from the surface-confined [Ru(NH$_3$)$_6$]$^{3+}$. The effect of different reduction potentials, viz., −300, −400, and 450 mV, on the CC charge was tested and an insignificant change in the total charge was noticed. The following Cottrell expression was used to calculate the DNA density, Γ_{DNA} [59,60]:

$$Q = \left(\frac{2nFAD_0^{1/2}C_0}{\pi^{1/2}} \right) t^{1/2} + Q_{dl} + nFA\Gamma_0$$

where

    n is the number of electrons exchanged
    F is the Faraday constant, 96,485 (C/equiv)
    A is the geometric area of electrode (cm^2)
    D_0 is the diffusion coefficient (cm^2/s)
    C_0 is the bulk concentration (mol/cm^2)
    Q_{dl} is the capacitive charge (C)
    $nFA\Gamma_0$ is the charge from the reduction of Γ_0 (mol/cm^2) of adsorbed [Ru(NH$_3$)$_6$]$^{3+}$

The surface DNA density, Γ_{DNA}, is calculated from the relation $\Gamma_{DNA} = \Gamma_0 \, (z/m) \, (N_A)$, where z is the charge of the redox molecule, m is the number of bases in the probe DNA, and N_A is the Avogadro's number. From the slope and intercept of plot of Q vs. $t^{1/2}$ (Figure 39.5A and B), one can calculate the Γ_{DNA} present on the sensor surface.

39.2.2.3 Impedance Technique

Impedance technique is a very versatile electrochemical tool regularly used to characterize the intrinsic electrical properties of electrode–film interface. The basis of this technique is the analysis of the impedance (resistance of alternating current) of the system in the subject to the applied frequency and exciting signal. Exciting signal involves imposition of a small amplitude AC sinusoidal voltage on the DC potential (equilibrium or formal potential of the redox reaction under investigation) at different frequencies across the device.

The total impedance is related to current and potential by Ohm's law

$$Z(t) = \frac{E(t)}{I(t)}, \quad Z(\omega) = \frac{E(\omega)}{I(\omega)}$$

The frequency-dependent potential and current are given as

$$E(t) = E_0 \cos(\omega t), \; \omega = 2\pi f$$

$$I(t) = I_0 \cos(\omega t - \phi)$$

$$Z(t) = \frac{E_0 \cos(\omega t)}{I_0 \cos(\omega t - \phi)} \quad \text{or } Z_0 \frac{\cos(\omega t)}{\cos(\omega t - \phi)}$$

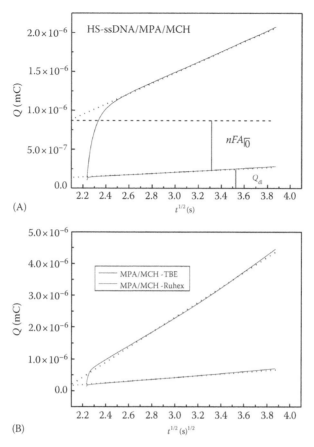

FIGURE 39.5 Q vs. $t^{1/2}$ plots for evaluation of HS-ssDNA density following the charge compensation of phosphodiester backbone of DNA strand in presence of $100\,\mu M$ $[Ru(NH_3)_6]^{3+}$ using the chronocoulometry at an applied potential of $-350\,mV$. Buffer used is $10\,mM$ Tris buffer at pH 7.0. (A) Q vs. $t^{1/2}$ for HS-ssDNA/MPA/MCH surface obtained in blank TBE buffer after incubation in $100\,\mu M$ $[Ru(NH_3)]^{3+}$. (B) Q vs. $t^{1/2}$ for MPA/MCH surface without HS-ssDNA obtained in blank TBE buffer after incubation in $100\,\mu M$ $[Ru(NH_3)]^{3+}$.

In complex form,

$$E(t) = E_0 \exp(i\omega t)$$

$$I(t) = I_0 \exp(i\omega t - i\phi)$$

$$Z(t) = Z_0 \exp(i\phi) = Z_0(\cos\phi + i\sin\phi)$$

From the last equation, the frequency-dependent impedance is measured in terms of real and imaginary impedance. The phase angle between the applied voltage and current is measured and plotted in different forms. The electrical resistance and capacitance are sensitive indicators for changes of surface properties and hence the DNA hybridization can be monitored and frequently used for monitoring biomolecular affinity reactions.

Impedance measurements in this work were made using a Solartron Instruments (Farnborough, Hampshire, England) model SI 1255 high-frequency response analyzer connected to an EG&G PAR (Gaitherburg, MD) model 273 potentiostat/galvanostat. Sine wave AC potential of $\pm5\,mV$ in the frequency range between $100\,kHz$ and $1\,Hz$ was applied over the DC potential redox potential $350\,mV$. Resulting impedance and phase shifts are presented in three different forms: Nyquist (Figure 39.6A), Bode (Figure 39.6B), and phase angle (Figure 39.6C). The R_{CT} and capacitance values were obtained by simulation with ZsimpWin program (EG&G PAR) using an appropriate equivalent circuit (Figure 39.6D)

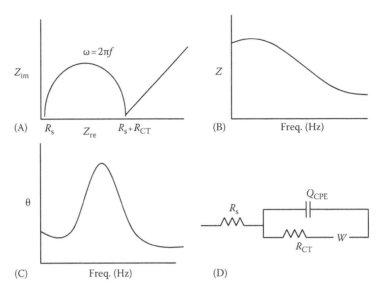

FIGURE 39.6 Impedance plots: (A) Imaginary impedance vs. real impedance (Nyquist plot). (B) Total impedance variation with frequency (Bode plot). (C) Phase angle θ variation with frequency. (D) Equivalent circuit used to model the electrode–film interface.

which mimics the electronic behavior of molecules at the electrode–film interface. Understanding this would help for further development of sensors with improved performance [55]. Other instruments from CH Instruments, Inc., USA, Autolab from Netherland, IVUM from Netherland, Gamry Potentiostats are also used for the aforementioned purpose.

39.2.3 Immobilization of HS-ssDNA, Mixed Monolayer Formation, and Target Hybridization

Most of the experiments were made using Au wire electrode of 1 mm diameter. Electrodes were cleaned in piranha ($1:3\ H_2O_2:H_2SO_4$) solution for 2–12 h followed by thorough rinsing with distilled water. The electrodes were polished using Al_2O_3 powders of 5.0, 1.0, and 0.05 μm dimensions, sonicated in DI water, and potentially pretreated in 1 M H_2SO_4 between −300 and 1500 mV at a scan rate 100 mV/s for constant current behavior. The target recognition layer on Au was constructed as follows. Seven microliters of 5 μM HS-ssDNA (I) in 1 M NaCl is dropped on Au surface and allowed to stand for 2.30 h without drying. Longer immobilization does not have significant effect on the monolayer quality in presence of 1 M NaCl [56] and DNA density was evaluated following the procedure suggested by Tarlov et al. [59,60]. After the extensive rinsing with the blank buffer, the absorptions of MCH and/or MPA were made sequentially by immersing the HS-ssDNA-modified surface in 500 μL of 5 mM solutions of the individual diluents in PBS buffer for 1 h. The surface is now allowed to hybridize with 7 μL of 5 μM (used in most of the experiments except in target concentration variation) of either complementary (II) or noncomplementary (III) targets in the SSC buffer (4× SSC) hybridization buffer for 2 h. The Au surface state is assessed by electrochemical techniques (CV, CC/CA, and impedance) intermittently in presence of 25 mM $[Fe(CN)_6]^{3-}$ in PBS buffer at pH 7.4. Hybridization discrimination in the absence and presence of dsDNA-specific intercalator was made following the charge compensation method using SDS surfactant and intercalators [61] (Figure 39.7). Experimental details for mixed monolayer construction and hybridization detection based on charge compensation principle and charge transport through dsDNA are presented in scheme. For experiments on deposited Au on Si surfaces, similar experiment protocol for the immobilization and hybridization was followed. Interdigitated Au electrodes embedded on SiO_2 were fabricated using standard silicon technology and photolithography [62]. Only results are taken here to explain the microelectrode behavior in presence of immobilized DNA layers.

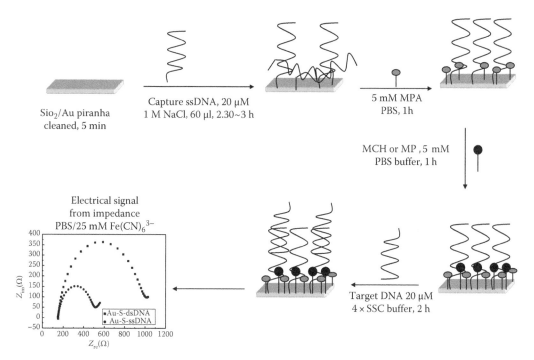

FIGURE 39.7 Thiol-capped DNA immobilization on gold, ternary mixed monolayer construction, and signal transduction in presence of $[Fe(CN)_6]^{3-}$.

39.3 Results and Discussion

39.3.1 Discrimination of ssDNA and dsDNA Surfaces by CV and Impedance in Presence of $[Fe(CN)_6]^{3-}$

Thiol-capped DNAs form self-assembled monolayer on gold due to high affinity of thiol toward gold. CV and impedance are used to understand the behavior of microscopic nature of thiol self-assembled DNA-modified Au electrodes [53,55,63–68]. Both techniques measure changes in electrochemical parameters (redox peak currents and potentials by CV, capacitance, and resistance by impedance) during the biomolecular interactions. Behavior of these thiol-DNA monolayers is strongly dependent on the monolayer density, arrangements, monolayer defects (pinholes, collapsed sites, grain boundaries), and anionic nature of the surface-confined DNA base sequences. Although the use of non-Faradaic method is simple, the results are not reliable. Hence, monitoring DNA hybridization by Faradaic processes in presence of redox probes $[Ru(bpy)_3]^{3+}$, $[Ru(NH_3)_6]^{3+}$, $[Fe(CN)_6]^{3-}$, and intercalators like MB is suggested to increase the sensitivity and reliability. The use of the negatively charged redox probe $[Fe(CN)_6]^{3-}$ in DNA hybridization sensing is most preferable due to the fact that it is repelled by the negatively charged DNA.

The CV behaviors of the bare Au (curve a), HS-ssDNA immobilized Au (curve b), and complementary target hybridized (II, see Section 39.2) (curve c) Au surfaces are shown in Figure 39.8. The immobilization of HS-ssDNA decreased the peak current of bare Au to 1.16×10^{-6} A (curve b) from 3.91×10^{-6} A and the peak potentials are difficult to measure owing to their diffused nature (curves b and c). The decrease is correlated to two factors: (1) diffusion restriction interaction of $[Fe(CN)_6]^{3-}$ directly on gold surface, and (2) repulsion between negatively charged DNA and $[Fe(CN)_6]^{3-}$. Upon hybridization, the Faradaic current for $[Fe(CN)_6]^{3-}$ decreased further due to doubly increased negative charge density and molecular crowding at electrode–film interface. The results show that the CV is able to discriminate the ssDNA (curve b) and dsDNA (curve c) in presence of $[Fe(CN)_6]^{3-}$, but change in the signal magnitude between the single-stranded (named as unhybridized) and double-stranded (named as hybridized) surface is insignificant for practical purpose. The results are reproduced for the bare Au

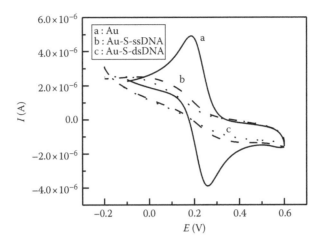

FIGURE 39.8 Detection of DNA hybridization using sensor by CV method. Curves: (a) Unmodified Au; (b) HS-ssDNA modified Au; and (c) HS-dsDNA in 0.01 M PBS buffer, pH 7.4. Both capture and target probe concentration used is 5 μM. Voltammograms were recorded in PBS buffer containing 25 mM $[Fe(CN)_6]^{3-}$.

and the HS-ssDNA immobilization, but it is difficult to reproduce the exact trend between HS-ssDNA and HS-dsDNA. This is related to several factors such as electrode roughness, HS-ssDNA arrangement, layer quality, and density and transduction technique as well. The smaller difference between single- and double-stranded surfaces in CV is also due to high charging current contribution to the total sensor signal. When the difference in current/potential between the unhybridized and hybridized surface is too insignificant, electrochemical impedance technique is used to obtain large discrimination effect by measuring impedance, Z, which is related to current inversely by Ohm's law, $Z = \Delta E/\Delta I$, as presented in Section 39.2. This technique is being utilized to develop label-free biosensors [46,47,53,69,70] but has greater challenges [55].

The observed impedance changes in presence of $[Fe(CN)_6]^{3-}$ for ssDNA and dsDNA along with unmodified electrodes are presented (Figure 39.9). Impedance data are modeled using the simple Randel's equivalent circuit $R_s(Q_{CPE}(R_{CT}W))$, in which R_s is solution resistance, R_{CT} charge transfer resistance, Q_{CPE} constant phase element (CPE) accounting the electrical double layer at the electrode–film interface, and W Warburg's

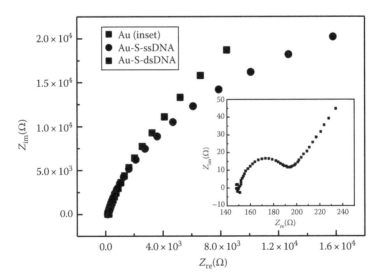

FIGURE 39.9 Impedance behavior measured in the frequency range 100 kHz to 1 Hz at applied potential 350 mV and AC potential amplitude 5 mV for the same surfaces under similar conditions.

TABLE 39.1

Impedance Parameters for Different Monolayers on Au Surface Obtained Using $R_s(Q_{CPE} (R_{CT}W))$

Surfaces/Circuit Fit Parameters	R_s (Ω)	Q_{dl} (F/cm²)	n	R_{CT} (Ω)	W	χ^2
HS-dsDNA	27.88	4.45×10^{-5}	0.87	1.28×10^4	8.5×10^{-5}	1.44×10^{-4}
HS-ssDNA	27.14	2.68×10^{-5}	0.86	7105	1.4×10^{-4}	1.49×10^{-4}
HS-dsDNA/MCH	28.01	5.78×10^{-6}	0.97	107.9	7.99×10^{-3}	1.75×10^{-3}
HS-ssDNA/MCH	29.69	6.74×10^{-6}	0.97	98.2	1.54×10^{-2}	7.71×10^{-5}
HS-dsDNA/MPA	29.13	4.34×10^{-5}	0.89	88.35	1.46×10^{-2}	1.25×10^{-4}
HS-ssDNA/MPA	27.09	3.88×10^{-5}	0.90	66.73	2.09×10^{-2}	1.14×10^{-4}
HS-dsDNA/MPA/ MCH	29.9	1.83×10^{-5}	0.94	734.3	8.35×10^{-3}	2.07×10^{-4}
HS-ssDNA/MPA/ MCH	30.9	9.41×10^{-5}	0.95	489.9	8.30×10^{-3}	1.42×10^{-4}
MCH	36.75	7.10×10^{-6}	0.95	3548	1.58×10^{-5}	1.76×10^{-4}
MPA	28.99	2.26×10^{-5}	0.91	37.24	2.8×10^{-2}	5.57×10^{-4}
MPA/MCH	28.11	1.15×10^{-5}	0.95	377.3	1.35×10^{-2}	1.73×10^{-4}

diffusion element. The CPE is defined as $1/(Aj\omega)^n$, where A is analogous to a capacitance, ω is the frequency expressed in rad/s, and n is the CPE exponent that takes the values $0.5 < n < 1$. When $n = 1$, Q corresponds to a perfect capacitor and $n = 0.5$ indicates a Warburg element; n for surface capacitance is typically 0.85 and 0.98 [67,68]. This introduces a sub-90° phase shift, or equivalently a frequency-dependent resistance in addition to a pure capacitor. The Kramers–Kronig data validation test showed that our data fits are of good quality with $\chi^2 \leq 10^{-4}$ (Table 39.1), and the parameters are well within 30% error limit. Since the target hybridization increases the negative charge density around the electrode–film interface, an increase in the R_{CT} (1.288×10^4 Ω) for the HS-dsDNA surface is obvious. However, we observed both increased and decreased (1757 Ω) R_{CT} values for the complementary hybridized surfaces and near constant value for the noncomplementary unhybridized surfaces (Table 39.1). Possibly, the decrease in impedance following the hybridization could occur either from "ion-gating" effect [71] or from the loosely bound DNAs or large capacitance contribution (varies from 20 to 50 $\mu C/cm^2$ between substrates) from the unmodified surfaces that access the $[Fe(CN)_6]^{3-}$. However, the behavior is not reproducible consistently. This could be due to the disordered arrangement, orientation, improper control of capture probe density, and nonspecific interaction of DNA on Au surface. These observations suggest the fact that irrespective of the techniques used for probing DNA hybridization, the reproducibility seems to be questionable. Hence, immobilization of thiolated DNA onto the gold surface in a controlled manner is essential for the production of reliable sensors. To overcome these difficulties, mixed monolayer consisting of thiolated ssDNA and MCH by two-step sequential adsorption method has been widely employed [48–54]. It is believed that the negatively charged OH^- head group in MCH repels the negatively charged ssDNA and keeps the ssDNA in upright position and enhances the hybridization efficiency. Recently, it has been proved that neither its posttreatment nor its co-deposition could control the sensor property consistently [53,60,72,73]. One possible reason is that MCH's pK_a is >12, at physiological pH 7.4 it might not provide high negative charge and remove all weakly or nonspecifically bound DNAs on the Au surface. Hence, we used thiol molecules having both less and high pK_a with same chain length ($n = 3$) and different head group functionalities such as MPA ($pK_a = 5.2$ [74,75]) and MP ($pK_a > 12$). These results suggest two important aspects: (1) Compared to the conventional MCH diluent, the MPA removes effectively the weakly bound HS-ssDNAs on the surface indicated by the increased peak currents and decreased ΔE_p (attainment of high reversibility characteristic features) for $[Fe(CN)_6]^{3-}$ (Figure 39.10, curve e); and (2) In presence of MP, decreased peak current and increased potential differences are observed [45] (Figure 39.10, curve d) than the HS-ssDNA/MCH binary layer. That is, the $[Fe(CN)_6]^{3-}$ has restricted movement toward electrode surface since the MP acted as a defect back filler by site competition insertion at defect sites without affecting the Au–S-ssDNA layer arrangement. In contrast, the MCH and MPA diluents create larger defects by orienting DNA and allow free access of

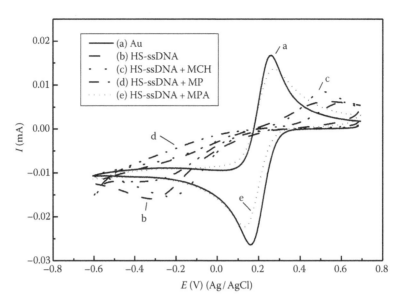

FIGURE 39.10 Effect of different diluents on the HS-ssDNA-covered Au surface. Curves: (a) Unmodified Au; (b) HS-ssDNA-modified Au; (c) HS-ssDNA+MCH; (d) HS-ssDNA+MP; and (e) HS-ssDNA+MPA. The solution concentrations of the monolayer components used are HS-ssDNA 5 μM in 1 M NaCl, MCH=MP=MPA=5 mM in 0.01 M PBS buffer, pH 7.4. Voltammograms were recorded in PBS buffer containing 25 mM [Fe(CN)$_6$]$^{3-}$. (Republished from *Sens. Actuators B*, 127, Dharuman, V. and Hahn, J.H., Effect of short chain alkane diluents on the label free electrochemical DNA hybridization discrimination at the HS-ssDNA/diluent binary mixed monolayer in presence of cationic intercalators, 536–544, Copyright 2007, from Elsevier.)

[Fe(CN)$_6$]$^{3-}$ that increases the current signal. Next effect of target hybridization on these surfaces is made. After target hybridization, the HS-ssDNA/MPA layer showed an insignificant discrimination indicating the presence of larger molecular defects (Figure 39.11A). In contrast, both the HS-ssDNA/MP and HS-ssDNA/MCH binary layers showed significant discrimination effect between the single-stranded and double-stranded surfaces. Comparison to the HS-ssDNA/MCH binary layer, the HS-ssDNA/MP layer showed 75% increased discrimination from CV technique. This is attributed to the fact that MP being a neutral molecule (pK_a > 12) it could easily diffuse through the molecular defects and occupies the active sites. This reduces the surface defects further and increases the surface passivation and improves the electrostatic blockade as well as discrimination effect. Reconsidering the system HS-ssDNA/MPA that showed nearly zero discrimination of hybridized and unhybridized surfaces, further blocking of pinholes using other diluent by back filling method may enhance the discrimination effect. Hence, we used MCH as the back filler for HS-ssDNA/MPA and formed a new layer, HS-ssDNA/MPA/MCH. Now the surface carries HS-ssDNA, MPA, and MCH, and named as ternary layer. The insertion of MCH as pinhole back filler is confirmed by decreased peak current and increased ΔE_p for the HS-ssDNA/MPA/MCH ternary layer (Figure 39.12, curve d), compared to the binary layer HS-ssDNA/MPA (Figure 39.12, curve c). Impedance also showed higher R_{CT} for the ternary layer (Figure 39.13, curve f) than that of both binary layer HS-ssDNA/MPA (Figure 39.13, curve e) and HS-ssDNA/MCH (Figure 39.12, curve a and Table 39.1). The discrimination effect after hybridization with complementary and noncomplementary targets has been verified again by both CV and impedance techniques. This ternary layer HS-ssDNA/MPA/MCH showed 60%–80% discrimination effect between the hybridized and unhybridized surfaces [43,44]. The ΔR_{CT} ($R_{CThyb} - R_{CTun-hyb}$) values for HS-ssDNA/MPA/MCH (489.9 Ω) has increased by four times compared with the individual binary layer HS-ssDNA/MCH (98.2 Ω) and HS-ssDNA/MPA (66.7 Ω) (Figure 39.13).

39.3.2 Evaluation of HS-ssDNA Density in Presence of [Ru(NH$_3$)$_6$]$^{3+}$

Molecular displacement is quite obvious in the construction of mixed monolayer. Therefore, to asses the number of HS-ssDNA available for the target reaction, chronoamperometric experiment has been

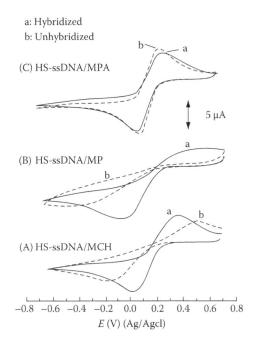

FIGURE 39.11 Hybridization discriminations at binary mixed monolayers constructed on Au surfaces after intercalation with proflavine. (A) HS-ssDNA/MCH. (B) HS-ssDNA/MP. (C) HS-ssDNA/MPA. Voltammograms recorded at 50 mV/s. Other conditions are similar to those in Figure 39.10. (Republished from *Sens. Actuators B*, 127, Dharuman, V. and Hahn, J.H., Effect of short chain alkane diluents on the label free electrochemical DNA hybridization discrimination at the HS-ssDNA/diluent binary mixed monolayer in presence of cationic intercalators, 536–544, Copyright 2007, from Elsevier.)

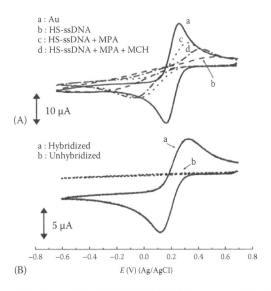

FIGURE 39.12 (A) Behavior of the ternary HS-ssDNA/MPA/MCH interface on gold in PBS buffer of pH 7.4 + 25 mM $[Fe(CN)_6]^{3-}$ at a scan rate 50 mV/s. Curve a: Bare Au. Curve b: HS-ssDNA. Curve c: HS-ssDNA + MPA. Curve d: HS-ssDNA + MPA + MCH. (B) DNA hybridization discrimination at the ternary HS-ssDNA/MPA/MCH interface after the interaction of electro-inactive PF intercalator in presence of $[Fe(CN)_6]^{3-}$. Curve a: Hybridized surface. Curve b: Unhybridized surface. The PF charge compensates the negative charge on the dsDNA effectively and hence higher current for dsDNA. But at the unhybridized ssDNA, the incoming $[Fe(CN)_6]^{3-}$ is repelled and no current flow. (Republished from *Biosens. Bioelectron.*, 23, Dharuman, V. and Hahn, J.H., Label free electrochemical DNA hybridization discrimination effects at the binary and ternary mixed monolayers of single stranded DNA/diluent/s in presence of cationic intercalators, 1250–1258, Copyright 2008, from Elsevier.)

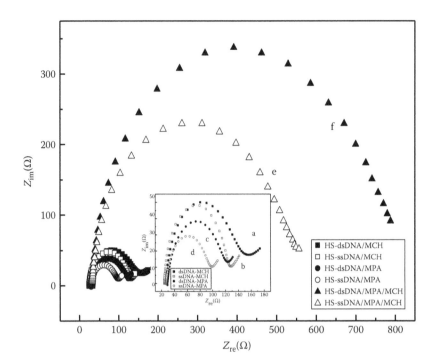

FIGURE 39.13 Nyquist behavior of binary HS-ssDNA/MCH (squares), HS-ssDNA/MPA (circles), and HS-ssDNA/MPA/MCH layers (triangles) for the complementary (closed symbols)- and noncomplementary (open symbols)-hybridized surfaces measured in presence of 25 mM [Fe(CN)6]3$^-$ in phosphate buffer, pH 7.4. Solid line: $Rs(Q_{CPE}\,(RW))$ circuit fit. Curve a: HS-ssDNA/MCH. Curve b: HS-dsDNA/MCH. Curve c: HS-ssDNA/MPA. Curve d: HS-dsDNA/MPA. Curve e: HS-ssDNA/MPA/MCH. Curve f. HS-dsDNA/MPA/MCH. (Republished from *Biosens. Bioelectron.*, 25, Dharuman, V. et al., Ternary mixed monolayers for simultaneous DNA orientation control and surface passivation for label free DNA hybridization electrochemical sensing, 2129–2134, Copyright 2010, from Elsevier.)

made (see Section 39.2), before and after interaction of [Ru(NH$_3$)$_6$]$^{3+}$ on monolayer-immobilized surface. Intercepts of Q vs. $t^{1/2}$ plots for the gold substrates modified with MCH, MPA, and MPA/MCH are nearly the same in the absence and presence of the [Ru(NH$_3$)$_6$]$^{3+}$, confirming the no adsorption of the metal complex on the diluent's head group. However, the ternary layer bearing HS-ssDNA showed distinctive intercepts due to the effective charge compensation between the anionic phosphodiester and cationic [Ru(NH$_3$)$_6$]$^{3+}$ (Figure 39.14). The calculated surface densities of HS-ssDNA and dsDNA without mixed monolayer formation are $1.6 \pm 5 \times 10^{13}$ and $2.9 \pm 7 \times 10^{13}$ molecules/cm^2, respectively, which are in agreement with previous reports based on ^{32}P radiometric [49,76] and electrostatic [43] methods. The MPA and MCH treatments to form the binary mixed monolayers reduce the surface DNA probe density to 10^{12} molecules/cm^2, indicating the removal of nonspecific DNAs by both the diluents. The probe density in the ternary layer also falls in the same order (10^{12} molecules/cm^2), which reconfirms the fact that the MCH acts only as the back filler and occupies the defect sites through site competition. Comparison of total thiol layer density (2.2×10^{14} molecules/cm^2, Figure 39.3A) obtained by thiol reduction in KOH with the HS-ssDNA density ($5.9 \pm 2 \times 10^{12}$ molecules/cm^2) from [Ru(NH$_3$)$_6$]$^{3+}$ measurements reveals that 3.7×10^2 molecules of diluents/cm^2 decorate the electrode surface. This presents higher electrostatic blockade for the direct access of [Fe(CN)$_6$]$^{3-}$ and enhances the R_{CT}. The double peaks seen in Figure 39.3 for all systems indicate the fact that the molecules are in two different phases on gold surface with different desorption energy requirement.

Achieving the orientation-controlled structure of the HS-ssDNA is the prime requirement in the mixed monolayer approach. This effect is further probed long with reduction (RS-Au(I) + e$^-$ $\leftrightarrow$ Au(0) + RS$^-$) voltammetry. The substrate–thiol bond, the lateral thiol-to-thiol interaction, the distribution of the adsorbate thiols generally determine the structure and electrochemical properties of

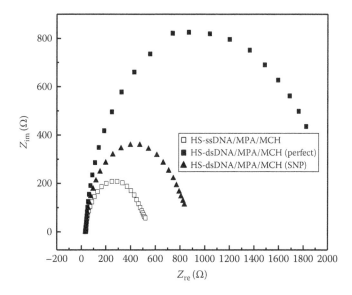

FIGURE 39.14 Point mutation detection in the cancer sequence used in the presence of 25 mM $[Fe(CN)_6]^{3-}$ at the ternary layer constructed HS-ssDNA/MPA/MCH using impedance spectroscopy. 25, Dharuman, V. et al., Ternary mixed monolayers for simultaneous DNA orientation control and surface passivation for label free DNA hybridization electrochemical sensing, 2129–2134, Copyright 2010, from Elsevier.

the monolayer. Therefore, the discrimination effects between the binary HS-ssDNA/MCH and ternary HS-ssDNA/MPA/MCH layers could be attributed to different orientations of the HS-ssDNA probes and different molecular crowding and defects. That is, the ssDNA strands in the ternary layer HS-ssDNA/MPA/MCH may well be separated from each other. This enhances the freedom for target coiling and leads to the increased hybridization efficiency. Further, high negative environment surrounding the HS-ssDNA provided by the MPA and MCH controls this layer property and repels $[Fe(CN)_6]^{3-}$. The observed charge from the area integration of the P1 and P2 peaks, 70.9 μC/cm^2 (Figure 39.3), is closer to the ideal amount of the $(\sqrt{3} \times \sqrt{3})R30°$ structure of the alkane thiol monolayers. This shows the perpendicular arrangement of DNA and diluent molecules with respect to the flat gold surface, and the thiol layer is tilted 30° away from 90°. That is, the total thiol surface concentration of 3.7×10^{-9} mol/cm^2 is arranged in the $(\sqrt{3} \times \& \sqrt{3})R30°$ configuration. The desorption peak potentials for the HS-ssDNA and MCH diluent are located nearly at the same potential, viz., ~ −1.2 (±0.1) V for all the three HS-ssDNA/MPA/MCH, HS-ssDNA/MCH, and MPA/MCH monolayers. In contrast, the positive shift of P1 peak potential for MPA in presence of HS-ssDNA from that in presence of the MCH diluent indicates high segregation of hydrophilic MPA and hydrophobic HS-ssDNA domains. Therefore, it is plausible that the MCH molecules occupy near the HS-ssDNA domain and reduce the defect size while forming the HS-ssDNA/MCH and HS-ssDNA/MPA/MCH interfaces. In addition, the more negative peak potential shift of P1 following the insertion of MCH into the MPA defects indicates high interaction of these two diluents to form a closely packed structure through bending supported by the FT-IR data.

Detection of mutations and damaged DNA bases is important for the early genetic disease diagnosis. Impedance spectroscopy is capable of detecting the SNP or point mutation. The effective control of layer orientation and surface passivation is further supported from the decreased R_{CT} of 789 Ω for the single base–mismatched dsDNA, mutation present in the middle of the sequence, compared with the perfect dsDNA that showed R_{CT} of 1791 Ω (Figure 39.8). The presence of point mutation in the middle of the sensor decreases the R_{CT} two times because of the fact that less number of dsDNA only formed than that of the number of perfect dsDNA and hence, less negative repulsion force exerted toward $[Fe(CN)_6]^{3-}$ (Figure 39.14). But R_{CT} is higher than the ssDNA. It seems that the structure of dsDNA formed with point mutations induces structural changes that might have significant effect on layer defects as well. However, this hypothesis needs to be verified by other experimental techniques.

39.3.3 Effect of Intercalation on the Electrical Signal at the Ternary HS-ssDNA/MPA/MCH Layer

Monitoring the charge transport through dsDNA offers an alternative approach to the point mutation detection with increased sensitivity and selectivity. It has been demonstrated that the combination of dsDNA-specific electroactive intercalator MB with $[Fe(CN)_6]^{3-}$ provides a sensitive method for the easy detection of point mutations [57,77–79]. The principle of detection is shown in Figure 39.15.

In this method, the electroactive MB intercalates at the top of the dsDNA. When it receives electron through dsDNA, it undergoes reduction reaction to a product leucomethylene blue (LB). Since $[Fe(CN)_6]^{3-}$ could not penetrate through the monolayer as the Au surface passivated by multilayer construction, the $[Fe(CN)_6]^{3-}$ interacts with the LB. In these processes, LB oxidized to MB by $[Fe(CN)_6]^{3-}$ and gives $[Fe(CN)_6]^{4-}$. This redox cycle continues and enhances the signal intensity and detection sensitivity. Therefore, here, we evaluated the influence of surface chemistry on the charge transport rate through dsDNA by using both electro-inactive PF [61] dsDNA-specific intercalator followed by electroactive MB [78–80], respectively, on the same surface. Presence of single point mutation at the center of dsDNA efficiently prevents the charge transport through dsDNA indicated by 60% decrease in the R_{CT} was observed for the perfect duplex in presence of the electrochemically active MB, than in presence of the inactive PF (Figure 39.16). Therefore, the involvement of MB in the mediated charge transport through π-stack of dsDNA and solution phase electrocatalysis of $[Fe(CN)_6]^{3-} \leftrightarrow [Fe(CN)_6]^{4-}$ is well controlled using this multi-monolayer construction. This is further supported by the minimized nonspecific intercalators adsorption onto the unhybridized HS-ssDNA, negative head groups of these diluents [81–83], and negligible access of MB through pinholes. Although impedance technique is quite sensitive to the layer composition, the technique is quite complicated and difficult to study the fast charge transport process due to the competitive charge compensation and intercalation process. In addition, impedance showed significant discrimination effect only in the low-frequency region where the data instability deteriorates the sensor during frequent dehybridization and rehybridization processes for constructing calibration curves. On the other hand, chorono techniques are demonstrated to be more sensitive to the surface-confined species as it retains information on the electrolysis of the species that occurs essentially simultaneously with the potential step [77–80]. Hence, we used CC technique for the construction of calibration curves of the sensor. We used both PF and MB for intercalation one after the other, and measured the charge transport in presence of $[Fe(CN)_6]^{3-}$ by applying a potential pulse of $-350\,mV$. The sensor surface after treatment with PF intercalator is incubated first in 5% SDS to remove the cationic intercalators present on the dsDNA [77,78,84] and negatively charged diluents [83] through the micelle sequestration, and followed by dehybridization in 0.5 M KOH solution for 20 min. The constancy of the $[Fe(CN)_6]^{3-}$ signal for the dehybridized ssDNA and unhybridized surface (data not shown) confirms the effective denaturation of the dsDNA strands and the high integrity of the HS-ssDNA/MPA/MCH layer on blocking the $[Fe(CN)_6]^{3-}$ redox reaction. The surfaces are now rehybridized and then intercalated

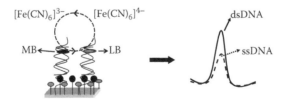

FIGURE 39.15 Scheme of hybridization discrimination in presence of $[Fe(CN)_6]^{3-}$ and electroactive double-stranded DNA-specific intercalator methylene blue (MB). MB receives electron through dsDNA and reduced to leucomethylene blue (LB). The electron released by LB is received by the solution phase $[Fe(CN)_6]^{3-}$ which then reduced to $[Fe(CN)_6]^{4-}$. This cycle continues and hence increased current or charge for the perfect complementary dsDNA and not for unhybridized ssDNA.

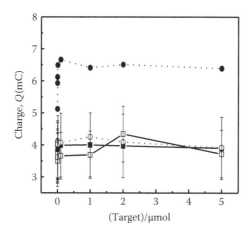

FIGURE 39.16 Effect of target concentration on the HS-ssDNA/MPA/MCH on chronocoulometry charge observed for the hybridized (closed symbols) and unhybridized (open symbols) measured by applying potential width −350 mV in phosphate buffer containing 25 mM [Fe(CN)$_6$]$^{3-}$ after intercalation with electrochemically inactive PF (squares) and active MB (circles) at the HS-ssDNA/MPA/MCH ternary layers. This figure clearly suggests the efficient DNA orientation and control of charge transfer through double-stranded DNA. (Republished from *Biosens. Bioelectron.*, 25, Dharuman, V., Chang, B.Y., Park, S.M., and Hahn, J.H., Ternary mixed monolayers for simultaneous DNA orientation control and surface passivation for label free DNA hybridization electrochemical sensing, 2129–2134, Copyright 2010, from Elsevier.)

with MB. The [Fe(CN)$_6$]$^{3-}$ reaction was interrogated. Figure 39.15 compares the discrimination effects in presence of MB and PF intercalators. In presence of the MB, the sensor signal intensity increases by 50% compared to the presence of PF confirming the fact that PF only charge compensates and not involved in the intercalator-assisted charge transport. Similar results were observed on the wire electrodes [39]. Comparison of both the binary HS-ssDNA/MCH and ternary HS-ssDNA/MPA/MCH layers revealed the discrimination efficiency of 1.60 in presence of MB intercalator for the ternary layer and insignificant discrimination efficiency for the binary layer. It may be noted that similar effect has been noticed by impedance technique. Good specificity and selectively were noticed, especially at the lower concentrations, and Langmuir type behavior was clearly visible in pM to nM range in presence of MB intercalators (Figure 39.16). Here, the ternary layer showed discrimination efficiency of 1.76 in presence of MB intercalator and nearly zero efficiency in presence of PF intercalator. The lowest concentration detected was 10 pM. The near constant signal for the [Fe(CN)$_6$]$^{3-}$ for the hybridized and unhybridized surfaces of the ternary recognition layer after treatment with electrochemically inactive PF reveals the presence of electrostatic blockade for the [Fe(CN)$_6$]$^{3-}$ diffusion by the surface passivation at this ternary monolayer. It is to note that the detection sensitivity is quite impressive with its high reproducibility; however, brain tumor CK20 sequences used are synthesized by oligonucleotide synthesizer. This attempt showed elegant approach for the construction of probe sequences in an orientation-controlled manner on the Au surfaces for surface-based molecular diagnosis sensors. The use of longer sequences like cDNA and PCR product detection may still have larger problems to solve. Hence, further research is underway to use different surface chemistry and to analyze real samples. Therefore, further improvement is needed to compete with the existing highly sensitive enzyme-based sensors.

39.4 Conclusions

The electrochemical methods such as CV, impedance, and CA techniques have been demonstrated to be useful to detect DNA hybridization taking brain tumor CK20 sequence as a model system. For the practical applications of label-free sensors, the sensor behavior should have more repeatability and reliability which are most essential parameters determining the scope of the sensors. In this chapter, we described how to overcome these problems by the control of Au-DNA-nonspecific interactions,

HS-ssDNA orientation and density, unwanted signal contribution from the non-Faradaic and Faradaic currents using MPA and MCH diluents by stepwise self-assembling to form HS-ssDNA/MPA/MCH monolayer and MCH is inserted into the local defects. Efficient control of microenvironment around the HS-ssDNA and higher target hybridization efficiency reveled by the higher R_{CT} values by EIS and higher discrimination effect by CC than the binary interfaces. This multicomponent ternary monolayer strategy holds greater promise in achieving the higher discrimination efficiency, a critical requirement for detecting fewer molecular targets present in the nanoliter biosamples. Further, these types of model surfaces may find applications in area of biological and technological interests, including biocompatibility studies of artificial biomaterials where the chemical composition, distribution, and mobility of diluents on the surface are known to have greater influence on the adsorption characteristics of the biomolecules. The method showed the presence of HS-ssDNA density of $5.9 \pm 2 \times 10^{12}$ molecules/cm^2 similar to densities reported based on the radiolabeling methods, confirming the fact that MPA diluent is not damaging the Au–S bonds, rather it removes only the nonspecifically bound DNAs on Au at shorter incubation times ≤ 1 h. Hence, the third SAM component, MCH, occupies the free active Au sites and thus enhances the electrostatic blockade for $[Fe(CN)_6]^{3-}$ diffusion. Charge transport studies in presence of PF and MB have confirmed the better control of the long-range charge transport through the perfect duplex than the case of the direct diffusion of $[Fe(CN)_6]^{3-}$ through layer defects. The use of longer-chain cDNA and detection of long-chain M13 phage DNA and polymerase chain reaction (PCR), which may expected to pose greater challenges in reproducibility, products are under way for clinical diagnosis applications of our method.

ACKNOWLEDGMENTS

Financial supports to Dr. Jong Hoon Hahn from Basic Science Research Program through the National Research Foundation of Korea (NRF) funded by the Ministry of Education, Science and Technology (Grant Number: 2010-0024306) are gratefully acknowledged. Financial supports to Dr. Venkataraman Dharuman from University Grants Commission ((UGC) SR(36-1/2008)), New Delhi, and by Alagappa University, Karaikudi, India are gratefully acknowledged.

REFERENCES

1. Palecek, E. 2002. Past, present and future of nucleic acids electrochemistry. *Talanta* 56:809–819.
2. Brett, A. M. O., M. Vivan, I. R. Fernandes, and J. A. Piedade. 2002. Electrochemical detection of in situ adriamycin oxidative damage to DNA. *Talanta* 56:959–970.
3. Fojta, M. 2002. Electrochemical sensors for DNA interactions and damage. *Electroanalysis* 14:1449–1463.
4. Brett, A. M. O., S. H. P. Serrano, T. A. Macedo, D. Raimundo, M. H. Marques, and M. A. La-Scalea. 1996. Electrochemical determination of carboplatin in serum using DNA-modified glassy carbon electrode. *Electroanalysis* 8:992–997.
5. Brett, A. M. O., T. R. A. Macedo, D. Raimundo, M. H. Marques, and S. H. P. Serrano. 1998. Voltammetric behaviour of mitoxantrone at a DNA-biosensor. *Biosens. Bioelectron.* 13:861–867.
6. Rusling, J. F. 2004. Sensors for toxicity of chemicals and oxidative stress based on electrochemical catalytic DNA oxidation. *Biosens. Bioelectron.* 20:1022–1028.
7. Palecek, E. and M. Fojta. 2001. Detecting DNA hybridization and damage. *Anal. Chem.* 73:74A–83A.
8. Palecek, E., M. Fojta, M. Tomschik, and J. Wang. 1998. Electrochemical biosensor for DNA hybridization and DNA damage. *Biosens. Bioelectron.* 13:621–628.
9. Lucarelli, F., G. Marrazza, A. P. F. Turner, and M. Mascini. 2004. Carbon and gold electrodes as electrochemical transducers for DNA hybridisation sensors. *Biosens. Bioelectron.* 19:515–530.
10. Drummond, T. G., M. G. Hill, and J. K. Barton. 2003. Electrochemical DNA sensors. *Nat. Biotechnol.* 21:1192–1199.
11. Wang, J. 2006. Electrochemical biosensors: Towards point-of-care cancer diagnostics. *Biosens. Bioelectron.* 21:1887–1892.
12. Mikkelsen, S. R. 1996. Electrochemical biosensors for DNA sequence detection. *Electroanalysis* 8:15–22.
13. Gooding, J. J. 2002. Electrochemical DNA hybridization biosensors. *Electroanalysis* 14:1149–1156.

14. (a) Zhao, Y. D., D. W. Pang, Z. L. Wang, J. K. Cheng, and Y. P. Qi. 1997. DNA modified electrodes Part 2. Electrochemical characterization of gold electrodes modified with DNA. *J. Electroanal. Chem.* 431:203–209. (b) Zhou, Y. and Z. Guo. 2003. A review of electrical impedance techniques for breast cancer detection. *Med. Eng. Phys.* 25:79–90. (c) Tiwari, A. and S. Gong. 2009. Electrochemical detection of a breast cancer susceptible gene using cDNA immobilized chitosan-*co*-polyaniline electrode. *Talanta* 77:1217–1222.

15. Wang, J. and A. N. Kawde. 2000. Pencil-based renewable biosensor for label-free electrochemical detection of DNA hybridization. *Anal. Chim. Acta* 431:219–225.

16. Tansil, N. C., F. Xie, H. Xie, and Z. Gao. 2005. An ultrasensitive nucleic acid biosensor based on the catalytic oxidation of guanine by a novel redox threading intercalator. *Chem. Commun.* (8):1064–1066.

17. Armistead, P. M. and H. H. Thorp. 2002. Electrochemical detection of gene expression in tumor samples: Overexpression of Rak nuclear tyrosine kinase. *Bioconjug. Chem.* 13:172–176.

18. Pavlov, V., I. Willner, A. Dishon, and M. Kotler. 2004. Amplified detection of telomerase activity using electrochemical and quartz crystal microbalance measurements. *Biosens. Bioelectron.* 20:1011–1021.

19. Voet, D., J. G. Voet, and C. W. Pratt. 2006. Chapter 3, Regulation of eukaryotic gene expression. *Fundamentals of Biochemistry: Life at the Molecular Level*, 2nd ed. John Wiley, New York, pp. 1054–1060.

20. Ronkainen-Matsuno, N. J., J. H. Thomas, H. B. Halsall, and W. R. Heineman. 2002. Electrochemical immunoassay moving into the fast lane. *Trends Anal. Chem.* 21:213–225.

21. Lin, J. and H. Ju. 2005. Electrochemical and chemiluminescent immunosensors for tumor markers. *Biosens. Bioelectron.* 20:1461–1470.

22. Dai, Z., F. Yan, J. Chen, and H. Ju. 2003. Reagentless amperometric immunosensors based on direct electrochemistry of horseradish peroxidase for determination of carcinoma antigen-125. *Anal. Chem.* 75:5429–5434.

23. He, Z., N. Gao, and W. Jin. 2003. Determination of tumor marker CA125 by capillary electrophoretic enzyme immunoassay with electrochemical detection. *Anal. Chim. Acta* 497:75–81.

24. Wiesner, A. 2004. Detection of tumor markers with ProteinChip technology. *Curr. Pharm. Biotechnol.* 5:45–67.

25. Hahm, J. and C. M. Lieber. 2004. Direct ultrasensitive electrical detection of DNA and DNA sequence variations using nanowire nanosensors. *Nano Lett.* 4:51–54.

26. Melosh, N. A., A. Boukai, F. Diana, B. Gerardot, A. Badolato, P. M. Petroff, and J. R. Heath. 2003. Ultrahigh-density nanowire lattices and circuits. *Science* 300:112–115.

27. Ramanathan, K., M. A. Bangar, M. Yun, W. Chen, N. V. Myung, and A. Mulchandani. 2005. Bioaffinity sensing using biologically functionalized conducting-polymer nanowire. *J. Am. Chem. Soc.* 127:496–497.

28. (a) Katz, E. and I. Willner. 2004. Biomolecule-functionalized carbon nanotubes: Applications in nanobioelectronics. *Chem. Phys. Chem.* 5:1085–1104. (b) Bindu, G., S. J. Abraham, A. Lonappan, V. Thomas, C. K. Aanandan, and K. T. Mathew. 2006. Active microwave imaging for breast cancer detection. *Prog. Electromagn. Res.* 58:149–169.

29. Wildi, S., J. Kleeff, H. Maruyama, C. A. Maurer, H. Friess, M. W. Buchler, A. D. Landur, and M. Korc. 1999. Characterization of cytokeratin 20 expression in pancreatic and colorectal cancer. *Clin. Cancer Res.* 5:2840–2847.

30. Nishizukua, S., T. Chen, F. G. Gwadry, J. Alexander, S. M. Major, U. Scherf, W. C. Reinhold, M. Waltham, L. Charboneau, L. Young, K. J. Bussey, S. Kim, S. Lababidi, J. K. Lee, S. Pittaluga, D. A. Scudiero, E. A. Sausville, P. J. Munson, E. F. Petricoin, 3rd, L. A. Liotta, S. M. Hewitt, M. Raffeld, and J. N. Weinstein. 2003. Diagnostic markers that distinguish colon and ovarian adenocarcinomas: Identification by genomic, proteomic and tissue array profiling. *Cancer Res.* 65:5243–5250 and back references therein.

31. Varadhachary, G. R., J. L. Abbruzzese, and R. Lenzi. 2004. Diagnostic strategies for unknown primary cancer. *Cancer* 100:1776–1785.

32. Futamura, M., Y. Takagi, H. Koumura, H. Kida, H. Tanemura, K. Shimokawa, and S. Saaji. 1998. Spread of colorectal cancer micrometastases in regional lymph nodes by reverse polymerase transcriptase chain reactions for carcinoembryonic antigen cytokeratin 20. *J. Surg. Oncol.* 68:34–40.

33. Jones, B. F. 1998. A reappraisal of the use of infrared thermal image analysis in medicine. *IEEE Trans. Med. Imaging* 17:1019–1027.

34. Keyserlingk, J. R., P. D. Ahlgren, E. Yu, N. Belliveau, and M. Yassa. 2000. Functional infrared imaging of the breast. *IEEE Eng. Med. Biol. Mag.* 19:30–41.

35. Peer, P. G., A. L. Verbeek, M. Mravunac, J. H. Hendriks, and R. Holland. 1996. Prognosis of younger and older patients with early breast cancer. *Br. J. Cancer* 73:382–385.

36. Fisher, C. J., M. K. Egan, P. Smith, K. Wicks, R. R. Millis, and I. S. Fentiman. 1997. Histopathology of breast cancer in relation to age. *Br. J. Cancer* 75:593–596.

37. Yildirim, E., T. Dalgic, and U. Berberoglu. 2000. Prognostic significance of young age in breast cancer. *J. Surg. Oncol.* 74:267–272.

38. Xiong, Q., V. Valero, V. Kau, S. W. Kau, S. Taylor, T. L. Smith, A. U. Buzdar, G. N. Hortobagyi, and R. L. Theriault. 2001. Female patients with breast carcinoma age 30 years and younger have a poor prognosis: The M.D. Anderson Cancer Center experience. *Cancer* 92:2523–2528.

39. Zou, Y. and Z. Guo. 2003. A review of electrical impedance techniques for breast cancer detection. *Med. Eng. Phys.* 25:79–90.

40. Saadi, W., S.-J. Wang, F. Lin, and N. L. Jeon. 2006. A parallel gradient microfludic chamber for quantitative analysis of breast cancer cell chemotaxis. *Biomed. Microdev.* 8:109–118.

41. Abhyankar, V. V., M. W. Toepke, C. L. Cortesio, M. A. Lokuta, A. Huttenlocher, and D. J. Beebe. 2008. A platform for assessing chemotactic migration within a spatiotemporally defined 3D microenvironment. *Lab Chip* 8:1507–1515.

42. Sung, K. E., N. Yang, C. Pehlke, P. J. Keely, K. W. Eliceiri, A. Friedl, and D. J. Beebe. 2011. Transition to invasion in breast cancer: A microfluidic in vitro model enables examination of spatial and temporal effects. *Integr. Biol.* 3(4):439–450. DOI: 10.1039/c0ib00063a.

43. Dharuman, V. and J. H. Hahn. 2008. Label free electrochemical DNA hybridization discrimination effects at the binary and ternary mixed monolayers of single stranded DNA/diluent/s in presence of cationic intercalators. *Biosens. Bioelectron.* 23:1250–1258.

44. Dharuman, V., B. Y. Chang, S. M. Park, and J. H. Hahn. 2010. Ternary mixed monolayers for simultaneous DNA orientation control and surface passivation for label free DNA hybridization electrochemical sensing. *Biosens. Bioelectron.* 25:2129–2134.

45. Dharuman, V. and J. H. Hahn. 2007. Effect of short chain alkane diluents on the label free electrochemical DNA hybridization discrimination at the HS-ssDNA/diluent binary mixed monolayer in presence of cationic intercalators. *Sens. Actuators B* 127:536–544.

46. Dharuman, V., T. Grunwald, E. Nebling, J. Albers, L. Blohm, and R. Hintsche. 2005. Label free impedance detection of oligonucleotide hybridization on interdigitated ultramicro electrodes using electrochemical redox probes. *Biosens. Bioelectron.* 21:645–654.

47. Dharuman, V., E. Nebling, T. Grunwald, J. Albers, L. Blohm, E. Eiholz, R. Worl, and R. Hintsche. 2006. DNA hybridization detection on electrical microarrays using coulostatic technique. *Biosens. Bioelectron.* 22:744–751.

48. Levicky, R., T. M. Herne, M. J. Tarlov, and S. K. Satija. 1998. Using self-assembly to control the structure of DNA monolayers on gold: A neutron reflectivity study. *J. Am. Chem. Soc.* 120:9787–9792.

49. Steel, A. B., R. L. Levicky, T. M. Herne, and M. J. Tarlov. 2000. Immobilization of nucleic acids at solid surfaces: Effect of oligonucleotide length on layer assembly. *Biophys. J.* 79:975–981.

50. Lubin, A. A., B. V. S. Hunt, R. J. White, and K. W. Plaxco. 2009. Effect of probe length, probe geometry, and redox tag placement on the performance of the electrochemical E-DNA sensors. *Anal. Chem.* 81:2150–2158.

51. Steel, A. B., T. M. Herne, R. Levicky, and M. J. Tarlov. 1998. Electrochemical quantification of DNA immobilized on gold. *Anal. Chem.* 70:4670–4677.

52. Satjapipat, M., R. Sanedrin, and F. Zhou. 2001. Selective desorption of alkanethiols in mixed self-assembled monolayers for subsequent oligonucleotide attachment and DNA hybridization. *Langmuir* 17:7637–7644.

53. Keighley, S. D., P. Estrela, P. Li, and P. Migliorato. 2008. Optimization of label-free DNA detection with electrochemical impedance spectroscopy using PNA probes. *Biosens. Bioelectron.* 24:906–911.

54. Huang, E., M. Satjapipat, S. Han, and F. Zhou. 2001. Surface structure and coverage of an oligonucleotide probe tethered onto a gold substrate and its hybridization efficiency for a polynucleotide target. *Langmuir* 17:1215–1224.

55. Daniels, J. S. and N. Pourmand. 2007. Label-free impedance biosensors: Opportunities and challenges. *Electroanalysis* 19:1239–1257.

56. (a) Peterlinz, K. A. and R. M. Georgiadis. 1997. Observation of hybridization and dehybridization of thiol-tethered DNA using two-color surface plasmon resonance spectroscopy. *J. Am. Chem. Soc.* 119:3401–3402. (b) Peterson, A. W., R. J. Heaton, and R. M. Georgiadis. 2001. The effect of surface probe density on DNA hybridization. *Nucleic Acids Res.* 29:5163–5168.

57. Kelley, S. O., N. M. Jackson, M. G. Hill, and J. K. Barton. 1999. Long-range electron transfer through DNA films. *Angew. Chem. Int. Ed.* 38:941–945.

58. Petrovykh, D. Y., H. Kimura-Suda, L. J. Whitman, and M. J. Tarlov. 2003. Quantitative analysis and characterization of DNA immobilized on gold. *J. Am. Chem. Soc.* 125:5219–5226.

59. Herne, T. M. and M. J. Tarlov. 1997. Characterization of DNA probes immobilized on gold surfaces. *J. Am. Chem. Soc.* 119:8916–8920.

60. Lao, R., S. Song, H. Wu, L. Wang, Z. Zhang, L. He, and C. Fan. 2005. Electrochemical interrogation of DNA monolayers on gold surfaces. *Anal. Chem.* 77:6475–6480.

61. Park, N. and J. H. Hahn. 2004. Electrochemical sensing of DNA hybridization based on duplex-specific charge compensation. *Anal. Chem.* 76:900–906.

62. Hintsche, R., M. Paeschke, U. Wollenberger, U. Schankenberg, B. Wagner, and T. Lisec. 1994. Microelectrode arrays and application to biosensing devices. *Biosens. Bioelectron.* 9:697–705.

63. Finklea, H. O. 1996. A. J. Bard (Ed.), *Electroanalytical Chemistry*, Vol. 19. Marcel Dekker, New York, pp. 109.

64. Bard, A. J. and L. R. Faulkner, 2001. *Electrochemical Methods: Fundamentals and Applications*, 2nd ed. Wiley, New York, pp. 156–260.

65. Krysinski, P., M. R. Moncelli, and F. A. Tadini-Bouninsegni. 2000. Voltammetric study of monolayers and bilayers self-assembled on metal electrodes. *Electrochim. Acta* 45:1885–1892.

66. Diao, P., M. Guo, and R. Tong. 2001. Characterization of defects in the formation process of self-assembled thiol monolayers by electrochemical impedance spectroscopy. *J. Electroanal. Chem.* 495:98–105.

67. Protsailo, L. V. and W. R. Fawcett. 2000. Studies of electron transfer through self-assembled monolayers using impedance spectroscopy. *Electrochim. Acta* 45:3497–3505.

68. Sadkowski, A., A. J. Motheo, and R. S. Neves. 1998. Characterization of Au(111) and Au(210) in aqueous solution interface by electrochemical admittance spectroscopy. *J. Electroanal. Chem.* 455:107–119.

69. Ma, K. S., H. Zhou, J. Zoval, and M. Madou. 2006. DNA hybridization detection by label free versus impedance amplifying label with impedance spectroscopy. *Sens. Actuators Chem. B* 114:58–64.

70. Park, J. Y. and S. M. Park. 2009. DNA hybridization sensors based on electrochemical impedance spectroscopy as a detection tool. *Sensors* 9:9513–9532 and back references therein.

71. Gooding, J. J., A. Chou, F. J. Mearns, E. L. S. Wong, and K. L. Jericho. 2003. The ion gating effect: Using a change in flexibility to allow label free electrochemical detection of DNA hybridization. *Chem. Commun.* 15:1938–1939.

72. Arinaga, K., U. Rant, M. Tornow, S. Fujita, G. Abstreiter, and N. Yokoyama. 2006. The role of surface charging during the coadsorption of mercaptohexanol to DNA layers on gold: Direct observation of desorption and layer reorientation. *Langmuir* 22:5560–5562.

73. Wong, E. L. S. and J. J. Gooding. 2006. Charge transfer through DNA: A selective electrochemical DNA biosensor. *Anal. Chem.* 78:2138–2144 and other references therein.

74. Kim, K. and J. Kwak. 2001. Faradaic impedance titration of pure 3-mercapto propionic acid and ethane thiol mixed monolayers on gold. *J. Electroanal. Chem.* 512:83–91.

75. Zhao, J., L. Luo, X. Yang, E. Wang, and S. Dong. 1999. Determination of surface pK_a of SAM using electrochemical titration method. *Electroanalysis* 11:1108–1111.

76. Gong, P., C. Y. Lee, L. J. Gamble, D. G. Castner, and D. W. Grainger. 2006. Hybridization behavior of mixed DNA/alkylthiol monolayers on gold: Characterization by surface plasmon resonance and ^{32}P radiometric assay. *Anal. Chem.* 78:3326–3334.

77. Boon, E. M., D. M. Ceres, T. G. Drummond, M. G. Hill, and J. K. Barton. 2000. Mutation detection by electrocatalysis at DNA-modified electrodes. *Nat. Biotechnol.* 18:1096–1100.

78. Kelley, S. O., J. K. Barton, N. M. Jackson, and M. G. Hill. 1997. Electrochemistry of methylene blue bound to a DNA-modified electrode. *Bioconjug. Chem.* 8:31–37.

79. Kelley, S. O., E. M. Boon, J. K. Barton, N. M. Jackson, and M. G. Hill. 1999. Single-base mismatch detection based on charge transduction through DNA. *Nucleic Acids Res.* 27:4830–4837.

80. Liu, T. and J. K. Barton. 2005. DNA electrochemistry through the base pairs not the sugar–phosphate backbone. *J. Am. Chem. Soc.* 127:10160–10161.

81. Vericat, C., F. R. Lenicov, S. Tanco, G. Andreasen, M. E. Vela, and R. C. Salvarezza. 2002. Building complex two-dimensional structures: Methylene blue on self-assembled monolayer–covered Au(111). *J. Phys. Chem. B* 106:9114–9121.

82. Benitez, G., C. Vericat, S. Tanco, F. R. Lenicov, M. F. Castez, M. F. Vela, and R. C. Salvarezza. 2004. Role of surface heterogeneity and molecular interactions in the charge-transfer process through self-assembled thiolate monolayers on Au(111). *Langmuir* 20:5030–5037.

83. Yau, H. C. M., H. L. Chan, S. F. Sui, and M. Yang. 2002. Integrity and redox properties of homogeneous and heterogeneous DNA films on gold surface probed by cyclic voltammetry. *Thin Solid Films* 413:218–223.

84. Westerlund, F., L. M. Wilhelmsson, B. Nordén, and P. Lincoln. 2003. Micelle-sequestered dissociation of cationic DNA-intercalated drugs: Unexpected surfactant-induced rate enhancement. *J. Am. Chem. Soc.* 125:3773–3779.

40

Electrochemical Biosensor for Detection of Chronic Myelogenous Leukemia and Acute Promyelocytic Leukemia

Yuanzhong Chen, Xinhua Lin, Ailin Liu, and Kun Wang

CONTENTS

40.1 Introduction

Leukemia is a kind of malignant blood disease characterized by the accumulation or infiltration of undifferentiated hematopoietic precursors or "blasts" in tissues, especially in blood and bone marrow. Clinically, leukemia is divided into two major categories, acute and chronic. Furthermore, the disease is subdivided into lymphocytic and myelogenous leukemia according to which lineage blood cells are involved. Myelogenous leukemia, contains acute myelogenous leukemia (AML) and chronic myelogenous leukemia (CML), while the acute promyelocytic leukemia (APL) is a subtype of AML.

CML is a clonal myeloproliferative disorder, resulting from the neoplastic transformation of the primitive hemopoietic stem cell [1–4].The chimeric oncogene breakpoint cluster region (BCR) gene and the cellular *abl* gene (BCR/ABL) exist in almost all CML patients [5–7]. The BCR/ABL chimerical fusion gene encodes a cytoplasmic hybrid protein with an important role in pathogenesis of CML [8–10]. There are many types of BCR/ABL genes, and Type b3a2, as one of the most common mutation types, has been often studied. The detection of BCR/ABL gene will afford an early diagnosis and monitor of the disease, which in turn improves the facility of detecting minimal residual leukemia cells in the CML patients, especially after the bone marrow transplantation (BMT).

APL which is a subtype of AML comprising 5%–15% of all AMLs is characterized by selective expansion of immature myeloid precursors that are blocked at the promyelocytic stage [11,12]. In APL, the reciprocal translocation t(15;17) (q22;21), which involves the retinoic acid receptor alpha (RARα) and promyelocytic leukemia (PML) genes, results in the expression of the PML/RARα fusion protein [13,14]. This single gene rearrangement plays an important role in leukemogenesis through antagonizing retinoic acid signaling and the regulatory pathways mediated by APL [15]. Thus, the availability of genotype-specific therapy for PML/RARα APL requires a precise diagnosis of the disease.

Recent monitoring methods for clinical diagnosis and prognosis of the CML fusion gene and PML/RARα fusion gene in APL include chromosome analysis, southern blot, polymerase chain reaction (PCR), fluorescence in situ hybridization (FISH), and so on; but all of these methods have some limitations, such as long assay time, insensitivity, poor precision, complexity, and high cost. Thus, it is important to develop and investigate novel detecting methods for CML gene and PML/RARα fusion gene with improved properties.

In recent years, an impressive number of studies focused on developing DNA electrochemical biosensors for the rapid and inexpensive diagnosis of genetic diseases have appeared [16–19]. These typical DNA electrochemical biosensors combine single-stranded DNA (ssDNA) probes on different electrodes with electrochemical transducers or other methods to test the hybridization between the DNA probes and their complementary DNA fragments. So the preparation of effective DNA-modified electrodes is very important for developing electrochemical DNA biosensors. Several DNA probes have been developed for the construction of DNA biosensors, including single-stranded cDNA, peptide nucleic acid (PNA), locked nucleic acid (LNA), and hairpin DNA. Each different kind of DNA probe affects the hybridization efficiency, which plays an important role in the sensitivity and specificity of the DNA electrochemical biosensor.

LNA probes have been developed for detection of BCR/ABL fusion gene in CML. As for LNA probe, it has many advantages including low toxicity, reduction of the conformational flexibility of the ribose, enhanced triplex formation, nuclease resistance, and increasing the phosphate backbone's local organization. LNA bases can increase the hybridization affinity between the capture probe and the target, which may allow single-base discrimination. In this chapter, we will introduce LNA as the capture probe and develop electrochemical DNA biosensors for BCR/ABL fusion gene detection in CML and PML/RARα fusion gene in APL [20–22].

40.2 Materials

40.2.1 Reagents

40.2.1.1 Reagents for CML Fusion Gene

The synthetic oligonucleotides were purchased from TaKaRa Biotechnology Co., Ltd. (Dalian, China). LNA-modified probe (18-base sequence S_1)-5'-NH_2-A^LGA GTT CA^LA AAG CCC^L TTC -3' (L: 2'-O,4'-C-methylene-(D-ribofuranosyl)nucleotides LNA); complementary (18-base sequence S_2)-5'-GAA GGG CTT TTG AAC TCT-3'; single-base mismatch (18-base sequence S_3)-5'-GAA GGG C<u>A</u>T TTG AAC TCT-3'; noncomplementary (18-base sequence S_4)-5'-ACG TGG TCC CCA GCT CTC-3'. Their base sequences are as follows: immobilized LNA-modified capture probe (22-base sequence S_5)-5'-$\underline{C}^L$GG CCA GTA GC^LA TCT GAC $\underline{TTT}^L$ G-TTT TTT TTT T-SH-3' (L: 2'-O,4'-C-methylene-(D-ribofuranosyl) nucleotides LNA); immobilized LNA-modified reporter probe (22-base sequence S_6)-5'-Biotin-GC $\underline{A}^L$GA $\underline{G}^L$TT $\underline{C}^L$AA $\underline{A}^L$AG $\underline{C}^L$CC $\underline{T}^L$TC $\underline{A}^L$G-3'; complementary (44-base sequence S_7)-5'-CA AAG TCA GAT GCT ACT GGC CGC TGA AGG GCT TTT GAA CTC TGC-3'; single-base mismatch (44-base sequence S_8)-5'-CA AAG TCA GAT GCT ACT GGC CGC TGA AGG GCT CTT GAA CTC TGC-3'; noncomplementary (44-base sequence S_9)-5'-AG GGT GAT CGG TAG GTA CCG TCA ACG TCT AGA AGC ATG GAG GCT-3'.

Tris-(hydroxymethyl)aminomethane was from Sinopharm Chemical Reagent Co., Ltd. (Beijing, China). Ethylenediaminetetraacetic acid (EDTA), Tris(2-carboxyethyl)phosphine hydrochloride (TCEP), and mercaptohexanol (MCH) were purchased from Sigma-Aldrich (St. Louis, MO). 3,3',5,5'-Tetramethylbenzidine (TMB) substrate (Neogen K-blue low activity substrate) was from Neogen (Lexington, KY). Avidin–horseradish peroxidase (HRP) was from Beijing Biosynthesis Biotechnology Co., Ltd. (Shanghai, China). Methylene blue (MB) was purchased from The Third Agents Factory of Shanghai (Shanghai, China). Stock solutions of MB (1 mM) were prepared with deionized water. PCl_5 and 4-aminobenzene sulfonic acid (4-ABSA) were purchased from Sinopharm Chemical Reagent Co., Ltd (Beijing, China). The buffer solutions involved in the study are as follows: Buffer for hybridization measurement was TMB substrate. Hybridization buffer was 1 M NaCl and 10 mM PBS buffer (pH 7.4). DNA immobilization buffer: 10 mM Tris–HCl, 1 mM EDTA, 10 mM TCEP (pH 7.4), and 1 M NaCl. Washing buffer was 0.1 M NaCl and 10 mM PB buffer (pH 7.4). Enzyme diluent was 0.01 M PBS buffer with 1% BSA (pH 7.4). All solutions were prepared with Milli-Q water (18 MΩ·cm resistivity) from a Millipore system.

40.2.1.2 Reagents for PML/RARα Fusion Gene

The synthetic oligonucleotides were obtained from TaKaRa Biotechnology Co., Ltd. Both probe and target sequences were shown in Table 40.1.

Streptavidin–HRP was supplied from Roche Diagnostics (Mannheim, Germany). The buffer solutions involved in the study were as follows: Buffer for hybridization measurement was TMB substrate. Hybridization buffer was 10 mM PBS buffer containing 0.5 M NaCl (pH 7.4). DNA immobilization buffer: 10 mM Tris–HCl containing 1 mM EDTA, 10 mM TCEP, and 1 M NaCl (pH 7.4). Washing buffer was 10 mM PBS buffer containing 0.1 M NaCl (pH 7.4). Enzyme diluent was 0.01 M PBS buffer with 1% BSA (pH 7.4). All solutions were prepared with Milli-Q water (18 MΩ·cm resistivity) from a Millipore system. All of the other chemicals were of analytical reagent grade and used without further purification.

40.2.2 Apparatus

Electrochemical measurements were performed using a CHI 660C Electrochemical Workstation (CH Instrument, Austin, TX). The electrochemical system consisted of gold electrode (GE, 2 mm in diameter, CH Instruments) as a working electrode, a platinum wire as the auxiliary electrode, and the reference electrode (Ag/AgCl). All potentials mentioned in this chapter refer to Ag/AgCl.

TABLE 40.1

All Synthetic Oligonucleotides Used in Experiment (L: 2′-O,4′-C-Methylene-(D-Ribofuranosyl) Nucleotides LNA)

Synthetic Oligonucleotides	Sequence
Immobilized LNA-modified capture probe (S_1)	5′-C_6-SH-TTTTTTTTTT CTTCAG^{L}AAC TGC^{L}TGCTCT^{L}GGGTCT^{L}CAATGG^L-3′
Immobilized LNA-modified reporter probe (S_2)	5′-C^LTG CCTC^{L}CCCGGC^{L}GCCACT^{L}GGCCAC^{L}GTGGT-biotin-3′
DNA capture probe (S_3)	5′-C_6-SH-TTTTTTTTTTCTTCAGAACTGCTGCTCTGGGTCTCAATGG -3′;
DNA reporter probe (S_4)	5′-CTGCCTCCCCGGCGCCACTGGCCACGTGGT-biotin-3′;
Complementary (S_5)	5′-ACCACGTGGCCAGTGGCGCCGGGGAGGCAGCCATTGAGACCCAGAGCAG CAGTTCTGAAG-3′
Single-base mismatch (S_6)	5′-ACCACGTGGCCAGT<u>C</u>GCGCCGGGGAGGCAGCCATTGAGACCCAGAGCAG CAGTTCTGAAG-3′
Three-base mismatch (S_7)	5′-ACCACGTGGCCA<u>A</u>TGGCGCCGGGGAGGCA<u>A</u>CCATTGAGACCC<u>T</u>GAGCAG CAGTTCTGAAG-3′
Five-base mismatch (S_8)	5′-ACCACGTGGCCA<u>A</u>TGGCGCC<u>T</u>GGGAGGCA<u>A</u>CCATTGAGACCC<u>T</u>GAGCAG C<u>T</u>GTTCTGAAG-3′
Ten-base mismatch (S_9)	5′-A<u>G</u>CACT<u>T</u>GGCCA<u>A</u>TGGCGCC<u>T</u>GGGAGGCA<u>A</u>CC<u>T</u>TTG<u>T</u>GACCC<u>T</u>GAGCAG C<u>T</u>GT<u>G</u>CTGAAG-3′
PML partial sequence (S_{10})	5′-ACCACGTGGCCAGTGGCGCCGGGGAGGCAGAGGAACGCGTTGTGGTGAT CAGCAGCTCGG-3′
RARα partial sequence (S_{11})	5′-TTAGTGGATATAGCACACCATCCCCAGCCACCATTGAGACCCAGAGCAGCA GTTCTGAAG-3′

40.3 Methods

40.3.1 Procedure for Fabrication of DNA Biosensor Using MB as Indicator

40.3.1.1 Preparation of the Biosensor Surface and Its Modification with DNA

First, the electrochemical modification of the clean GCE was carried out in 0.025 M KH_2PO_4 and Na_2HPO_4 solution containing 20 mM 4-ABSA by cyclic voltammetry (CV) scanning between +0.5 and +1.40 V (versus SCE) for four cycles with scan rate of 10 mV/s. Second, the capture probe was covalently attached on the 4-ABSA monolayer–modified GCE through the free amines of DNA bases based on the acyl chloride cross-linking reaction. The terminal sulfonic acid groups of the 4-ABSA/GCE were activated by immersing this modified electrode in acetone solution containing 40 mM PCl_5 for 0.5 h [23]. The linker/4-ABSA/GCE was rinsed with 20 mM Tris–HCl buffer (pH 7.00) to wash off the excess PCl_5. Five microliters solution containing 0.1 nM LNA probe was then pipetted onto the modified GCE and air-dried to dryness. Thus, a probe-modified GCE was obtained. The electrode surface was then washed with water to remove the unbound oligonucleotides. Finally, the hybridization was performed by pipetting 5 μL of different concentrations of target DNA (18-mer sequence S_2) onto the probe-modified GCE for 30 min at 45°C to obtain DNA hybrid-modified GCE. The electrode surface was then washed thoroughly with water to remove the unbound oligonucleotides. The same protocol was applied to probe-modified GCEs for hybridization of probe (18-mer sequence S_1) with one-base mismatch (18-mer sequence S_3) and also with noncomplementary sequences (18-mer sequence S_4). Details of the LNA electrochemical probe and target DNA sequences are shown in Table 40.2.

40.3.1.2 Intercalation of MB and Voltammetric Transduction

MB was accumulated onto the surface of hybrid-modified GCE by immersing the electrode into stirring 20 mM Tris–HCl buffer solution (pH 7.0) containing 20 μM MB and NaCl for 5 min without applying any

TABLE 40.2

Details of the LNA Electrochemical Probe and Target DNA Sequence LNA and Target DNA Sequence

LNA-modified probe (S_1)	5′-NH$_2$-A^{L}GA GTT CA^{L}A AAG CCC^L TTC-3′
	(L: 2′-O,4′-C-methylene-(D-ribofuranosyl)nucleotides LNA)
Target (S_2)	5′-GAA GGG CTT TTG AAC TCT-3′
Single-base mismatch (S_3)	5′-GAA GGG CAT TTG AAC TCT-3′
Noncomplementary (S_4)	5′-ACG TGG TCC CCA GCT CTC-3′

potential. After accumulation of MB, the electrode was rinsed with 20 mM Tris–HCl buffer (pH 7.0) in ultrasonic bath for 10 s to remove the nonspecifically bound MB. MB was intercalated into the DNA to form DNA/MB system on the probe electrode after hybridization. Differential pulse voltammetry (DPV) was then scanned from +0.40 to −0.50 V with amplitude of 5 mV.

40.3.2 Procedure for Detection of CML Fusion Gene with Sandwich-Type Electrochemical Enzyme-Based LNA-Modified DNA Biosensor (EC-ELDB)

40.3.2.1 Formation of the Self-Assembled Monolayers at Gold Electrode Surfaces

Gold electrodes were first polished using 1.0, 0.3, and 0.05 μM Al$_2$O$_3$ suspension, respectively, and then extensively rinsed with ethanol and Milli-Q water in an ultrasonic bath. Then electrodes were electrochemically cleaned in fresh 0.5 M H$_2$SO$_4$ solution. The electrochemically cleaned gold electrodes were then rinsed with a great amount of Milli-Q water. After being dried with nitrogen, electrodes were immediately used for DNA immobilization. The fresh electrodes were incubated in the immobilization buffer which contained capture probes modified with thiolate at appropriate concentrations for 2 h at room temperature. After that, the SH-DNA-modified electrodes were treated with either 1 mM MCH for 1 h to obtain mixed SH-DNA/MCH-modified electrodes.

40.3.2.2 Hybridization with Target DNA or PCR-Amplified Real Samples

The synthetic targets or the PCR amplicons were first mixed with the biotinylated reporter probe (100 nM) in the hybridization buffer and heated to 90°C for 10 min and then cooled in the ice bath before use. The sensor surface modified with capture probes was incubated in the hybridization buffer which contained the biotinylated reporter probes and target DNA at 50°C for 45 min. The sensor was rinsed with the washing buffer and then incubated with 3 μL of avidin–HRP (0.5 U/mL) for 25 min at room temperature. The sandwich formation was formed and subjected to electrochemical measurement.

40.3.2.3 Electrochemical Measurement

The electrochemical signal of the enzymatically produced TMB substrate was measured by CV and amperometric *i–t* curve detection. CV was carried out at a scan rate of 100 mV/s. Amperometric detection was performed with a fixed potential of 100 mV and the steady state was usually reached and recorded within 100 s.

40.3.3 Procedure for the Detection of PML/RARα Fusion Gene with EC-ELDB

40.3.3.1 Self-Assembled Monolayers at Gold Electrode Surfaces

The procedures for creating self-assembled monolayers (SAMs) on the gold electrode surfaces are very similar to those described in Section 40.3.2.1.

40.3.3.2 Hybridization with Target DNA and Detection

The synthetic targets were first mixed with the biotinylated reporter probes (100 nM) in the hybridization buffer and heated to 90°C for 10 min and then cooled in the ice bath before use. The sensor surface–modified capture probes were incubated in the hybridization buffer which contained the biotinylated reporter probes and target DNA at 63°C for 1 h. The sensor was rinsed with the washing buffer and then incubated with 3 μL of streptavidin–HRP (0.5 U/mL) for 15 min at room temperature. After removing any remaining HRP adsorbed nonspecifically, the "sandwich" structure was formed and subjected to electrochemical measurement. The detection method is similar to that described in Section 40.3.2.3.

40.4 Results

40.4.1 Specificity and Sensitivity of DNA Biosensor Using MB as Indicator

40.4.1.1 Thermostability of the LNA and DNA Capture Probe

Melting temperature (T_m) is the temperature at which, under a given set of conditions, double-stranded DNA (dsDNA) is changed (50%) to ssDNA. The T_m is a physical property of nucleic acids that gives information about the stability of duplexes in a specified environment. T_m values are useful in a variety of fields ranging from practical assay design in molecular biology to theoretical biophysics. The T_m for the hybridization of LNA with their complementary DNA was examined to confirm their potential for selective recognition of complementary sequences (see Figure 40.1 and Table 40.3). As seen in Figure 40.1 and Table 40.3, the T_m value of the LNA capture probe (62.8°C) was greatly higher than that of its corresponding DNA probes (48.3°C). Comparing with the analogous DNA–DNA hybrids, T_m value for LNA binding to DNA was increased 14.5°C, which is equivalent to 4.8°C per LNA base. It could also be seen from Table 40.3 that the T_m value for LNA binding to single base–mismatched DNA was decreased to 38.3°C. The T_m of oligonucleotides can be changed by the incorporation of LNA residues into the oligo sequence, thus making possible to discriminate fully complementary and single mismatch sequences better.

Theoretically, the optimum hybridization temperature was about 20°C lower than T_m. As seen in Table 40.3, the T_m value of the LNA capture probe binding to complementary DNA was 62.8°C, while for single base–mismatched DNA was decreased to 38.3°C. Therefore, the theoretical hybridization temperature was 42.8°C and 18.3°C for the complementary DNA and single base–mismatched DNA, respectively. The earlier experiments indicated that when the selected hybridization temperature was farther away from the theoretical hybridization temperature, the speed of hybridization would be slower. On the other hand, higher hybridization temperature would accelerate the denaturation of dsDNA, resulting in the decrease of the absolute hybridization number. Accordingly, 45°C was selected as hybridization temperature. At this temperature, hybridization would occur only for the complementary DNA strand

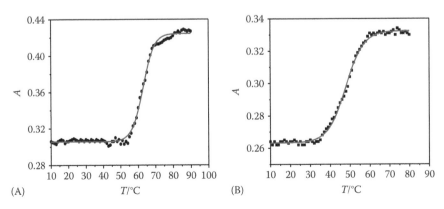

FIGURE 40.1 Melting curve of LNA–DNA (A) and DNA–DNA (B) duplex.

TABLE 40.3

Thermostability of the LNA and DNA Capture Probe Sequences against Complementary and Single Base–Mismatched DNA Targets

Capture Probe	Sequence	T_m, °C		ΔT_m, °C
		DNA Target (5′-GAAGGGCTT TTGAACTCT-3′)	m-DNA Target (5′-GAAGGGC<u>A</u>TT TGAACTCT-3′)	
DNA	5′-AGAGTTCAAA AGCCCTTC-3′	48.3	41.1	7.2
LNA	5′-<u>A</u>^{L}GAGTTCA^{L}A AAGCCC^{L}TTC-3′	62.8	38.3	24.5

but not for the single base–mismatched DNA strand. Therefore, the hybridization specificity was dramatically increased.

40.4.1.2 Electrochemical Responses of MB-Binded LNA-Modified Probe

As we know, MB interacts in a different way with ssDNA and dsDNA [24,25]. Here, MB was used as an electrochemical indicator to study hybridization between LNA and the target DNA. The CV responses of 2.0 μM MB at an LNA-modified electrode are shown in Figure 40.2. The plot of cathodic peak current (I) versus scan rate (v) is linear (see Figure 40.2A, inset), indicating that MB was strongly bound to the LNA-modified surface. MB could bind specifically to the guanine bases and readily intercalate into dsDNA as well. Different binding modes of MB with LNA-modified ssDNA and dsDNA resulted in variation in electrochemical responses of LNA probe. Figure 40.2B shows the CV signals at the LNA/4-ABSA/GCE before and after hybridization with 18-mer complementary target DNA in Tris–HCl buffer solution containing 20 μM MB. The voltammetric reduction signals of MB at LNA/4-ABSA/GCE decreased after hybridizing with complementary target DNA. Thus, the decrease of current signal after DNA hybridization was due to the inaccessibility of the guanine bases in dsDNA.

40.4.1.3 Hybridization Specificity of LNA Probe

Detection of target DNA was also monitored by measuring the change of peak current of MB on the LNA probe electrode with DPV. Typically, there is a small signal of MB on a 4-ABSA/GCE obtained as

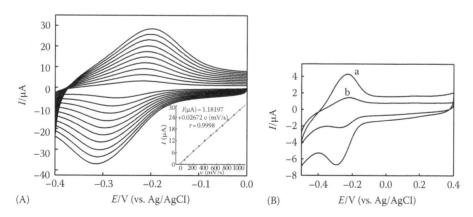

FIGURE 40.2 (A) Cyclic voltammograms after incubation of 2.0 μM MB in 20 mM Tris–HCl buffer (pH 7.0) at the LNA probe–modified GCE with increasing scan rate from inner to outer: 50, 100, 200, 300, 400, 500, 600, 700, 800, 900, 1000 mV/s. Inset shows plot of reduction peak current versus scan rate. (B) Cyclic voltammograms after incubation of 20 μM MB in 20 mM Tris–HCl buffer (pH 7.0) at the probe-modified GCE before (a) and after (b) hybridization with complementary target.

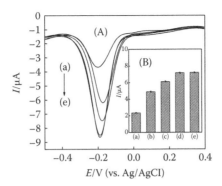

FIGURE 40.3 (A) Differential pulse voltammetry of 20 μM MB at a 4-ABSA/GCE (a) and at the redox indicator for (e) an LNA probe–modified GCE electrode and after hybridization with the (d) noncomplementary sequence (c) one-base mismatch sequence and (b) complementary target sequence. (B) The bar graph of the peak current of MB when the LNA probe hybridized with different gene fragments. Error bars=±relative standard deviation of three independent experiments. Buffer solution: 20 mM Tris–HCl buffer (pH 7.0) containing 20 mM NaCl.

shown in Figure 40.3 (curve a), which is due to the few MB molecular adsorbed on the 4-ABSA-modified GCE surface. The selectivity of this assay was investigated by using the LNA/4-ABSA/GCE as the capture probe to hybridize with various DNA sequences (complementary oligonucleotide S_2, one-base mismatch oligonucleotide S_3, and noncomplementary oligonucleotide sequence S_4). Figure 40.3 shows the DPV signal of MB at capture probe before hybridization (see Figure 40.3, curve e), and after hybridization with S_2 (see Figure 40.3, curve b), S_3 (see Figure 40.3, curve c), and S_4 (see Figure 40.3, curve d). It is clear that Figure 40.3 (curve e) shows the highest peak current of MB on the LNA/4-ABSA/GCE before hybridization, and after hybridization with complementary DNA the peak current of MB is dramatically decreased (see Figure 40.3, curve b). It can also be known that the LNA capture probe has high hybridization specificity, and it can easily discriminate the complementary from single-base mismatch target DNA. In the presence of oligonucleotide containing a single-base mismatch, significantly increased voltammetric signal can be observed (see Figure 40.3, curves b and c), which indicates that the complete hybridization is not accomplished due to the base mismatch. In addition, as expected, no significant difference of peak current can be observed for the LNA-modified ssDNA GCE and its hybridization with noncomplementary sequence (see Figure 40.3, curves d and e), since no successful hybridization occurs due to the sequence mismatch between the LNA-modified ssDNA and the noncomplementary sequence.

The sensitivity of this electrochemical biosensor for the target DNA was investigated by varying the target oligonucleotide concentration according to the procedure described in Section 40.3. The different current value obtained in the DPV response of MB after hybridization of probe with target was recorded with three repetitive measurements. The current response at about −0.21 V decreased in proportion to the amount of the target sequence used. Electrochemical responses of the complementary target with different concentration can also be quantitatively analyzed. Under the constant MB, the response of DNA hybridization between LNA probe and increasing concentration levels of complementary target was displayed in Figure 40.4A. The response for the reduction of MB after hybridization with the target DNA decreased with target concentration up to 50 pM and then tended to keep constant, which indicated that all the immobilized probes on the electrode were involved in hybridization. The formation duplex caused the decrease of the MB signal and the interaction of guanine base with MB was prevented.

The average current response shows excellent correlation with the amount of the complementary oligonucleotides in the range of $1.0 \times 10^{-12} \sim 1.1 \times 10^{-11}$ M (see Figure 40.4B). The regression equation is

$$I \, (\mu A) = -0.0444 \, C_{dsDNA} (pM) + 10.234 \; R = 0.9936$$

A detection limit of 9.4×10^{-13} M for the target DNA can be estimated using 3σ (where σ is the standard deviation of the blank solution, $n=8$). The reproducibility of the biosensor for detection of 7.0×10^{-12} M target DNA is 7.26% ($n=8$).

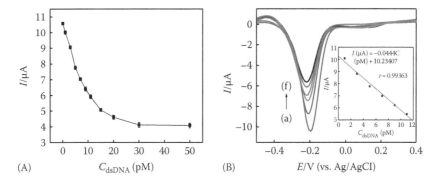

FIGURE 40.4 (A) Response of the capture probe with increasing concentration of target oligonucleotides. (B) Differential pulse voltammograms of MB accumulated on the LNA after its hybridization with different concentration of the target sequence. Target concentration ($\times 10^{-12}$ M): (a) 1.0, (b) 3.0, (c) 5.0, (d) 7.0, (e) 9.0, and (f) 11.0. Inset shows the plot of the peak current of MB as a function of the target concentration. Error bars = ±relative standard deviation of three independent experiments.

40.4.2 Specificity and Sensitivity Study of Sandwich-Type EC-ELDB

Electrochemical DNA biosensors were prepared using a solid electrode with immobilized short ssDNA probes and redox-active hybridization indicators. Most short ssDNA probes present poor specificity. In order to improve the specificity of the hybridization between the DNA probes and their complementary DNA fragments, the selectivity of DNA probes still needs to be enhanced. As a novel oligonucleotide analog, LNA has received more attention in recent years according to its promising properties. LNAs posses extraordinarily high affinities for complementary sequences and hybridize with very high affinity to perfectly complementary targets. Besides, typical intercalator-based DNA sensors often suffer from high background signals that are associated with nonspecific binding of intercalators to unhybridized ssDNA. In order to overcome this problem, the sandwich-type strategy was proposed and popularly applied. The typical "sandwich" DNA biosensor involves a pair of DNA probes (capture and redox-labeled reporter probes) that hybridize with the target DNA sequence. In the presence of target DNA, the capture probe modified on the electrode serves as a link to bring both the target DNA and the reporter probe to the proximity of the electrode surface. The formation of capture probe/target/reporter probe was formed. The electrochemical current signal was measured by the reporter probe–labeled enzyme. Through dual hybridization processes, the signal-to-noise ratio can be significantly improved.

40.4.2.1 Thermostability of the LNA and DNA Capture Probe

Experiments indicated that the hybridization temperature acted as a very important parameter. As a physical property of nucleic acids, T_m can give information about the stability of duplexes in a specified environment to confirm nucleic acids potential for selective recognition of complementary sequences. In order to improve the efficiency of hybridization, the values of T_m for the hybridization of LNA-modified probes with complementary DNA sequences were measured by Varian Cary 100 UV–vis spectrophotometer (see Figure 40.5A and B). The T_m of the LNA-modified capture probe (66.7°C) was higher than that of its corresponding DNA probes (44.1°C). The differential value of T_m between them reached to 22.6°C; T_m value of the LNA-modified reporter probe (76.3°C) was greatly higher than that of its corresponding DNA probe (52.9°C). Compared with the analogous DNA/DNA hybrids, T_m value for LNA/DNA duplex was increased 23.4°C. These results demonstrated that the increase of T_m was equivalent to 4.6°C per LNA base substitution, which were close to the preceding study [20]. Accordingly, the hybridization temperature was further investigated in the range from 55°C to 65°C by amperometric measurement (Tables 40.4 and 40.5). According to experiment, we found that relatively lower temperature led to poor efficiency of hybridization. On the other hand, with the increase of hybridization temperature, the denaturation of LNA/DNA would accelerate, and the absolute hybridization number would decrease.

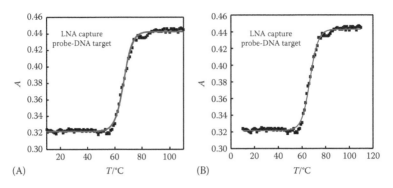

(A)   (B)

FIGURE 40.5 (A) Melting cure of LNA capture probe sequences against complementary DNA targets. (B) Melting cure of LNA reporter probe sequences against complementary DNA targets.

TABLE 40.4

Thermostability of LNA and DNA Capture Probe Sequences against Complementary DNA Targets

Complementary DNA	Sequence	Tm, °C		ΔTm, °C
		LNA Capture Probe (5′-CTTCAGLAACT GCLTGCTCTLGGGTC TLCAATGGL-3′)	**DNA Capture Probe (5′-CTTCAGAACT GCTGCTCTGGGTC TCAATGG-3′)**	
DNA	5′-CCATTGAGACCCAGA GCAGCAGTTCTGAAG-3′	66.7	44.1	22.6

TABLE 40.5

Thermostability of LNA and DNA Reporter Probe Sequences against Complementary DNA Targets

Complementary DNA	Sequence	Tm, °C		ΔTm, °C
		LNA Reporter Probe (5′-CLTGCCTCLCCCG GCLGCCACTLGGCCA CLGTGGT-3′)	**DNA Reporter Probe (5′-CTGCCTCCCCG GCGCCACTGGC CACGTGGT-3′)**	
DNA	5′-ACCACGTGGCCAGTG GCGCCGGGGAGGCAG-3′	76.3	52.9	23.4

Thus, 63°C was selected throughout our experiments to ensure relatively higher hybridization efficiency and improve the capability of LNA/DNA recognition.

40.4.2.2 Electrochemical Responses of the Enzyme-Based LNA-Modified Biosensor

In order to confirm the catalytic nature of HRP signals, we compared with the total electrochemical signal changes of bare electrode in different substrate solutions. From Figure 40.6, we can see there are two pairs of redox peaks at the surface of bare electrode (Figure 40.6, curve b). When HRP was added into TMB, the prominent catalytic reduction peak was found in CV of bare electrode (Figure 40.6, curve d). This reflected that with the catalysis of HRP, TMB was oxidized into a colored compound by H_2O_2, and thousands of reduction reactions of H_2O_2 could be efficiently catalyzed, leading to significantly amplified electrochemical current signals [26]. The different responses of the capture DNA (ssDNA) or dsDNA-modified biosensors were studied. The different DNA-modified biosensors were compared under the same condition. The avidin–HRP conjugate was used as the model protein since its adsorption could be electrochemically monitored. The CV responses of different DNA-modified electrodes are also shown in Figure 40.6. Comparing two curves (Figure 40.6, curves a and c), we can found a prominent catalytic

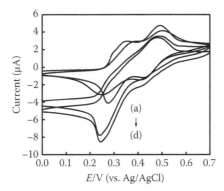

FIGURE 40.6 Cyclic voltammograms of bare electrode in TMB substrate (b), bare electrode in TMB substrate containing 3 μL 0.5 U/mL avidin–HRP (d), dsDNA-modified electrode in TMB substrate (a), and ssDNA-modified electrode in TMB substrate (c). The base solution for hybridization measurement was TMB substrate, and the potential scanning ranged from 0.0 to 0.7 V; the scanning rate was 100 mV/s; the sample interval was 0.001 V; and the quiet time was 2 s.

reduction peak appeared in CV of dsDNA-modified biosensor (Figure 40.6, curve c). That is because along with hybridization, the sandwich formation was formed, and avidin–HRP can conjugate with the biotin modified at an end of reporter probe via affinity bond. And the catalytic reduction of H_2O_2 with the redox reaction of TMB at the electrode surface happened [27,28]. The result indicates that avidin–HRP can induce the catalytic reduction of H_2O_2 with the redox reaction of TMB at the electrode surface and the enzyme-based LNA-modified biosensor can discriminated ssDNA and dsDNA very well.

40.4.2.3 Hybridization Specificity of the Enzyme-Based LNA-Modified Biosensor to the CML Fusion Gene

Detection of target DNA was also monitored by amperometric technique, which provides a rapid and direct measure of the current associated with the HRP-catalyzed electrochemical process. As we all know, HRP does not directly exchange electrons with the electrode due to the fact that its redox site is shielded within insulating peptide backbones [29]. So as an electron shuttle, a small redox molecule, TMB, was employed and it can diffuse in and out of the redox site of the macromolecule, thus coupling the catalytic reduction of H_2O_2 with the redox reaction of TMB at the electrode surface [27,28]. The potential was held at the catalytic reduction potential for H_2O_2 (100 mV). A decay curve for current (I) versus time (t) was observed instantly after the onset of the potential, which rapidly became steady-state current within 100 s (Figure 40.7). The selectivity of this assay was investigated by using the LNA-modified probes as the probe to hybridize with various DNA sequences (complementary oligonucleotide S_7, one-base mismatch oligonucleotide S_8, and noncomplementary oligonucleotide sequence S_9). It is clear that Figure 40.7 shows a large amperometric signal (1520 nA, Figure 40.7, curve a) obtained after hybridization with complementary DNA, which is more than an order of magnitude compare with background amperometric current (90 nA, Figure 40.7, curve e). It indicates that LNA-modified probes have complete hybridization with complementary target DNA. It also can be known that the LNA-modified probes have high hybridization specificity; they can easily discriminate the complementary from single-base mismatch target DNA. In the presence of oligonucleotide containing a single-base mismatch, significantly decreasing amperometric signal can be observed (see Figure 40.7, curve b), which indicates that the complete hybridization is not accomplished due to the base mismatch. In addition, as expected, no significant difference of currents can be observed for capture probe and the hybridization with noncomplementary sequence of LNA probes (see Figure 40.7, curves c and d), since no successful hybridization occurs due to the sequence mismatch between the capture probe and the noncomplementary sequence.

Electrochemical responses of the complementary target with different concentration can also be quantitatively analyzed. The sensor was challenged with the CML fusion gene sequence of a series of concentration across the range of 10 fM to 10 nM, spanning a response region of 7 orders of magnitude.

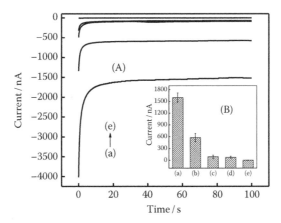

FIGURE 40.7   (A) Current–time curves of background (e), a capture probe–modified GE (d) and after hybridization with the noncomplementary sequence (c), one-base mismatch sequence (b), and complementary target sequence (a). (B) The bar graph of the amperometric currents when the EC-ELDB hybridized with different gene fragments. Error bars show the standard deviations of measurements taken from at least three independent experiments. The base solution for hybridization measurement was TMB substrate.

The amperometric signal was found to be a nonlinear logarithmic function related to the target concentration (Figure 40.8B), and the detection limit was experimentally found to be <10 fM (Figure 40.8A).

40.4.2.4 Specificity and Sensitivity of Sandwich-Mode Electrochemical LNA Biosensor to the PML/RARα Fusion Gene

The specificity of this assay was studied by using the LNA probes as the probe to hybridize with various PML/RARα fusion gene sequences (complementary S_5, one-base mismatch S_6, three-base mismatch S_7, five-base mismatch S_8, ten-base mismatch S_9, PML partial sequence S_{10}, and RARα partial sequence S_{11}). Figure 40.9A presented a large amperometric signal (1786 nA, curve a) obtained after hybridization with complementary DNA, which was more than an order of magnitude compared with background amperometric current (125 nA, curve h). It indicated that LNA probes showed the excellent affinity for complementary target. The amperometric signal obtained after hybridization with oligonucleotide containing a single-base mismatch decreased significantly (1042 nA, curve b), which indicated that the complete

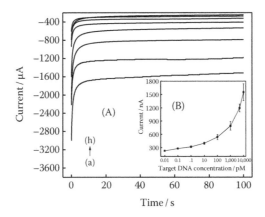

FIGURE 40.8   (A) Amperometric measurements for EC-ELDB hybridized with synthetic target DNA at a series of concentrations (from (a) to (h): 10 nM, 5 nM, 1 nM, 100 pM, 10 pM, 1 pM, 100 fM, and 10 fM). (B) Logarithmic plot of currents versus target DNA concentration. Error bars show the standard deviations of measurements taken from at least three independent experiments. The base solution for hybridization measurement was TMB substrate.

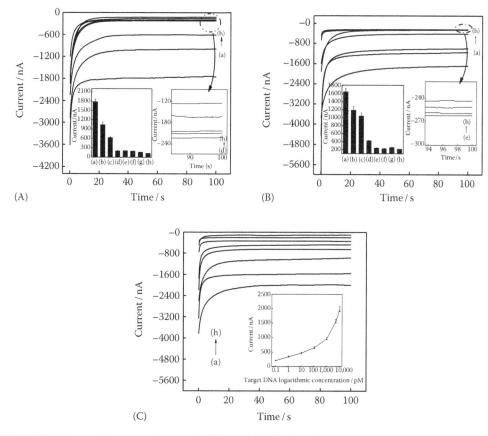

FIGURE 40.9 Current–time curves in respective different hybridizations with LNA probe (A) and DNA probe (B) after hybridization with S_5 (a), S_6 (b), S_7 (c), S_8 (d), S_9 (e), S_{10} (f), S_{11} (g), and the capture probe sequence (h). Insets show the bar graphs of the peak currents when the biosensor hybridized with different gene fragments. (C) Amperometric measurements for sandwich-mode electrochemical LNA biosensor hybridized with synthetic target DNA at a series of concentrations (from (a) to (h): 10 nM, 5 nM, 1 nM, 100 pM, 10 pM, 1 pM, 0.1 pM, and 0 pM). Inset shows logarithmic plot of currents versus target DNA concentration. Error bars show the standard deviations of measurements taken from at least three independent experiments. The base solution for hybridization measurement was TMB substrate.

hybridization was not accomplished due to the base mismatch. In addition, as expected, with the increasing of mismatch base number, the values of amperometric signal became smaller. When the number of mismatch base was ≥5, there were no significant differences of currents for capture probe and the hybridization with mismatch sequences of LNA probes (see Figure 40.9A, curve h versus curves d–g), since no successful hybridization occurred due to the mismatch bases between the capture probe and these mismatch sequences. Comparison between Figure 40.9A and B, we can find the main distinguish was that the differential value of current signals obtained for LNA probes hybridizing, respectively, with complementary sequence and single-base mismatch sequence was two times larger than that obtained for DNA probes. This result indicated that LNA probe exhibited an enhanced discriminatory power for the complementary and single-base mismatch target sequence, outperforming the DNA probe.

Electrochemical responses of the complementary target with different concentration in simple hybridization system were also quantitatively analyzed. The sensor was challenged with the target DNA of a series of concentration across the range of 100 fM to 10 nM, spanning a response region of 6 orders of magnitude. The amperometric signal was found to be a nonlinear logarithmic function related to the target concentration (Figure 40.9C), and the detection limit was 74 fM with a linear response range of 0.1 to 10 pM for synthetic PML/RARα fusion gene in APL. The detection limit of 74 fM for the target DNA can be estimated using 3σ (where σ is the standard deviation of the blank solution, $n = 5$).

The reproducibility of the biosensor for detection of 5.0 pM target DNA is 9.50% ($n = 5$). It was worthwhile to point out that this novel biosensor achieved even higher gain than the previously reported intercalator-based LNA-modified DNA sensors.

40.4.3 Study on Specificity and Sensitivity of Sandwich-Mode Electrochemical LNA Biosensor in Mixed Hybridization System to the PML/RARα Fusion Gene in APL

Via the homologous comparison analyzed by DNAMAN software, we found that the sequence in the human genome with the highest match with the 30 bp PML partial sequence in S_5 still exhibits 10 mismatch bases. At the same time, the similarity of the 30 bp RARα partial sequence in S_5 was also analyzed, and the highest match between this 30 bp sequence with the human genome was 24 bp. When the number of mismatch bases was ≥ 5, the biosensor possessed good nucleic acid recognition. We decided to design a simulation test in the mixed hybridization solution containing different kinds of mismatch sequences (S_8, S_9, S_{10}, and S_{11}) as interference background.

In order to evaluate the performance of the electrochemical DNA sensor under interference conditions, we compared its performance in the mixed hybridization solutions containing equal amounts of complementary and mismatch targets. The mismatch targets included S_8, S_9, S_{10}, and S_{11}. Figure 40.10A shows different interference backgrounds: (a) S_8, (b) $S_8 + S_9$, (c) $S_8 + S_9 + S_{10}$, and (d) $S_8 + S_9 + S_{10} + S_{11}$. The effect of increasing mismatch types on specificity of the biosensor was investigated. It was found that the biosensor still maintained high selectivity to discriminate complementary sequence from the interference background, even from the mixed hybridization solution containing four different kinds of mismatch sequences as interference background.

In addition, the effect of different interference background concentrations on specificity of the biosensor was also investigated. The composition of mixed hybridization is shown in Table 40.6.

The findings were as follows: when the range of total interference concentration was 0.4–40 nM, the values of amperometric signal from the hybridization with complementary sequences remained basically unchanged and kept a large difference value with the current values from interference background. However, when the total concentration of interference background was up to 400 nM, the signal obtained from the hybridization with complementary sequences significantly decreased (Figure 40.10B).

According to these results, the sensitivity of the DNA biosensor was further investigated under the mixed hybridization solution containing four kinds of mismatch sequences in 40 nM as interference background. Under the optimal condition of the hybridization, while detecting the complementary DNA standard concentration range of PML/RARα fusion gene from 0.05 nM to 5.0 nM, there was a good linear relationship between the current signal and the logarithmic function of complementary DNA concentration, the lower limit was 9.6 pM, which can be estimated using 3σ (where σ is the standard deviation of the blank solution, $n = 5$). The reproducibility of the biosensor for detection of 0.5 nM target DNA is 7.91% ($n = 5$) (Figure 40.10C). It could be seen that under a certain interference background, the electrochemical DNA biosensor still possessed good molecular recognition and can hybridize with completely complementary sequences well. It provided a good basis of further detection research in practical samples.

40.5 Discussion

The interactions between specific DNA probe and target play an important role on the overall performance of the DNA biosensors. Most of typical short ssDNA probes designed for the sensors present poor stability or specificity. Therefore, sensitive and selective DNA probes still need to be developed for the assay of the hybridization events between the DNA probes and their complementary DNA fragments. In this chapter, we applied LNA as the probe to develop electrochemical biosensor to test the genes related to CML and APL. LNAs have many advantages, including enhanced triplex formation, remarkable antisense activity, none detectable toxicity in vivo and nuclease resistance, and so on. LNA bases can be interspersed with DNA bases, allowing binding affinity to be tailored for particular applications.

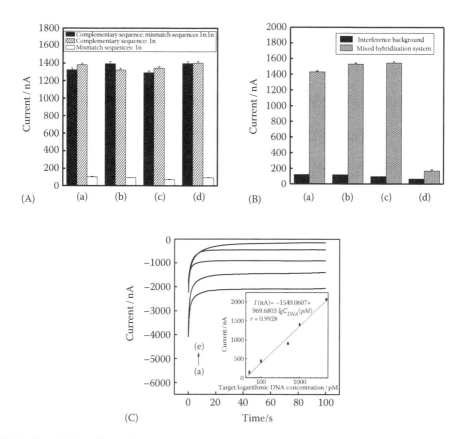

FIGURE 40.10 (A) The effect of increasing mismatch types on specificity of sandwich-mode electrochemical LNA biosensor. Different interference background: (a) S_8, (b) $S_8 + S_9$, (c) $S_8 + S_9 + S_{10}$, and (d) $S_8 + S_9 + S_{10} + S_{11}$. Different color bar charts show mixed hybridization solution containing 1 nM S_5 and 1 nM mismatch interference background (black), hybridization solution containing 1 nM S_5 (dense), hybridization solution containing 1 nM mismatch interference background (none). (B) The effect of interference background concentration on specificity of sandwich-mode electrochemical LNA biosensor. Different color bar charts showed interference background (black), mixed hybridization solution containing 1 nM S_5 (none). Different interference background concentrations (from (a) to (d) were 0.4, 4, 40, and 400 nM, respectively. (C) Current–time curves of hybridization with different concentrations of the target sequence in mixed hybridization solution containing four kinds of mismatch sequences in 40 nM as interference background. Target concentration (from (a) to (e): 5 nM, 1 nM, 500 pM, 100 pM, and 50 pM. Inset shows the logarithmic plot of currents versus target DNA concentration (from 0.05 to 5 nM). Error bars show the standard deviations of measurements taken from at least three independent experiments. The base solution for hybridization measurement was TMB substrate.

TABLE 40.6

The Compositions of Mixed Hybridization Systems

Mixed Hybridization	S8 (nM)	S9 (nM)	S10 (nM)	S11 (nM)	S5 (nM)
(a)	0.1	0.1	0.1	0.1	1
(b)	1	1	1	1	1
(c)	10	10	10	10	1
(d)	100	100	100	100	1

The ability of LNA bases to confer dramatic increases in hybridization affinity suggests that they have great potential for optimizing nucleic acid recognition. Due to the high affinity of the LNA molecules, very short probes can be used and show an extraordinary specificity to discriminate the targets that differ by a single base. In addition, we should recognize that the sensitivity of typical electrochemical DNA biosensors based on LNA probe is still not sufficient. More research on LNA-based electrochemical DNA biosensors is needed to achieve the sensitivity provided by PCR technology.

40.6 Future Trends

In future work, the LNA-modified DNA electrochemical biosensors should be combined with nanotechnology to further increase sensitivity. LNA-based sensors have the potential to contribute significantly to the early diagnosis and prognosis monitoring of CML and APL. Furthermore, the further development of LNA-modified DNA electrochemical biosensors is also promising for other diseases detection.

REFERENCES

1. Fialkow, P. J., Gartler, S. M., Yoshida, A. (1967). Clonal origin of chronic myelocytic leukemia in man. *Proc. Natl Acad. Sci. USA* 58: 1468–1471.
2. Champlin, R. E., Golde, D. W. (1985). Chronic myelogenous leukemia: Recent advances. *Blood* 65: 1039–1047.
3. Kantarjian, H. M., Deisseroth, A., Kurzrock, R. (1993). Chronic myelogenous leukemia: A concise update. *Blood* 82: 691–703.
4. Lin, X. H., Wu, P., Chen, W., Zhang, Y. F., Xia, X. H. (2007). Electrochemical DNA biosensor for the detection of short DNA species of chronic myelogenous leukemia by using methylene blue. *Talanta* 72: 468–472.
5. Butturini, A., Ralph, B. (1996). BCR/ABL and leukemia. *Leuk. Res.* 20: 523–529.
6. Jorge, E. C., Moshe, T., Hagop, K. (1996). Chronic myelogenous leukemia: A review. *Am. J. Med.* 100: 555–570.
7. Yoo, S. J., Eul-Ju, S., Lee, J. H. (2006). A complex, four-way variant t(15;17) in acute promyelocytic leukemia. *Cancer Genet. Cytogenet.* 167: 168–171.
8. Daley, G. Q., Van, E. R. A., Baltimore, D. (1991). Blast crisis in a murine model of chronic myelogenous leukemia. *Proc. Natl Acad. Sci. USA* 88: 11335–11338.
9. Druker, B. J., Lydon, N. B. (2000). Lessons learned from the development of an Abl tyrosine kinase inhibitor for chronic myelogenous leukemia. *J. Clin. Invest.* 105: 3–7.
10. Patricia, M. G., Maria, T. C., Pedro, N. A., Irene. B. L. (2003). Amplification of the BCR/ABL fusion gene clustered on a masked Philadelphia chromosome in a patient with myeloblastic crisis of chronic myelocytic leukemia. *Cancer Genet. Cytogenet.* 143: 140–144.
11. Franco, M., Giuseppe, A., Francesco, L. C. (2002). Advances in the understanding and management of acute promyelocytic leukemia. *Rev. Clin. Exp. Hematol.* 6: 60–71.
12. Raymond, P. W., Hugues, D. T., Zhen-Yi, W., Laurent, D. (1993). Medical progress: Acute promyelocytic leukemia. *N. Engl. J. Med.* 329: 177–189.
13. Borrow, J., Goddard A. D., Sheer, D., Solomon, E. (1990). Molecular analysis of acute promyelocytic leukemia breakpoint cluster region on chromosome 17. *Science* 249: 1577–1580.
14. Rowley, J. D., Golomb, H. M., Dougherty, C. (1977). 15/17 translocation, a consistent chromosomal change in acute promyelocytic leukaemia. *Lancet* 1: 549–550.
15. Andres, S., Ari, M., Arthur, Z., Jonathan, D. L. (2003). Molecular pathogenesis of acute promyelocytic leukaemia and APL variants. *Best Pract. Res. Clin. Haematol.* 16: 387–408.
16. Mickelsen, S. R. (1996). Electrochemical biosensors for DNA sequence detection. *Electroanalysis* 8: 15–19.
17. Wang, J., Jiang, M., Nilson, T. W., Getts, R. C. (1998). Dendritic nucleic acid probes for DNA biosensors. *J. Am. Chem. Soc.* 120: 8281–8282.
18. Zhang, S., Wright, G., Yang, Y. (2000). Materials and techniques for electrochemical biosensor design and construction. *Biosens. Bioelectron.* 15: 273–282.

19. Wei, F., Sun, B., Liao, W., Ouyang, J. H., Zhao, X. S. (2003). Achieving differentiation of single-base mutations through hairpin oligonucleotide and electric potential control. *Biosens. Bioelectron.* 18: 1149–1155.

20. Chen, J. H., Zhang, J., Wang, K., Lin, X. H., Huang, L. Y., Chen. G. N. (2008). An electrochemical biosensor for detection of BCR/ABL fusion gene using locked nucleic acids on 4-aminobenzene sulfonic acid modified glassy carbon electrode. *Anal. Chem.* 80: 8028–8034.

21. Wang, K., Chen, J. H., Chen, J., Liu, A. L., Li, G. W., Luo, H. B., Lin, X. H., Chen, Y. Z. (2009). A sandwich-type electrochemical biosensor for detection of BCR/ABL fusion gene using locked nucleic acids on gold electrode. *Electroanalysis* 21: 1159–1166.

22. Wang, K., Sun, Z. L., Feng, M. J., Liu, A. L., Yang, S. Y., Lin, X. H., Chen. Y. Z. (2011). Design of a sandwich-mode amperometric biosensor for detection of PML/RARα fusion gene using locked nucleic acids on gold electrode. *Biosens. Bioelectron.* 26: 2870–2876.

23. Zhuang, H. S., Huang, J. L., Chen, G. N. (2004). Synthesis of a new biacridine and its use as the chemiluminescent probe for immunoassay of carcinoembryonic antigen. *Anal. Chim. Acta* 512: 347–353.

24. Erdem, A., Kerman, K., Meric, B., Akarca, U. S., Ozsoz, M. (2000). Novel hybridization indicator methylene blue for the electrochemical detection of short DNA sequences related to the hepatitis B virus. *Anal. Chim. Acta* 422: 139–149.

25. Kerman, K., Ozkan, D., Kara, P., Meric, B., Gooding, J. J., Ozsoz, M. (2002). Voltammetric determination of DNA hybridization using methylene blue and self-assembled alkanethiol monolayer on gold electrodes. *Anal. Chim. Acta* 462: 39–47.

26. Volpe, G., Compagnone, D., Draisci, R., Palleschi, G. (1998). 3,3′,5,5′-Tetramethylbenzidine as electrochemical substrate for horseradish peroxidase based enzyme immunoassays. A comparative study. *Analyst* 123: 1303–1307.

27. Fanjul-Bolado, P., Gonzalez-Garcia, M. B., Costa-Garcia. A. (2005). Amperometric detection in TMB/HRP-based assays. *Anal. Bioanal. Chem.* 382: 297–302.

28. Das, A., Hecht, M. H. (2007). Peroxidase activity of de novo heme proteins immobilized on electrodes. *J. Inorg. Biochem.* 101: 1820–1826.

29. Liu, G., Wan, Y., Gau, V., Zhang, J., Wang, L. H., Song, S. P., Fan, C. H. (2008). An enzyme-based E-DNA sensor for sequence-specific detection of femtomolar DNA targets. *J. Am. Chem. Soc.* 130: 6820–6825.

Part IX

Electronic and Magnetic Technologies for Cancer Analysis

41

Nanowire Transistor–Based DNA Methylation Detection

Wusi C. Maki, Gary K. Maki, and Niranka Mishra

CONTENTS

41.1 Introduction

Early cancer diagnostics is one of the most important issues in health care. Early diagnostics directly influences the choice and effectiveness of therapy treatment which impacts cancer patient survival. To improve diagnostics in the earlier stage of tumor development, research is focused on two important factors: (a) discovery of new cancer biomarkers which report early events in tumor initiation, and (b) development of new molecular detection technologies which utilize small samples and are ultrasensitive and low cost. Current cancer diagnosis depends on a variety of factors including illness symptoms, imaging, blood test for protein markers, and biopsy pathological microscopic analysis. The limited sensitivity associated with these methods can result in the delay of diagnosis and treatment. Numerous new biomarkers for cancer are being discovered [1–3]. Among these new discoveries, epigenetic alteration such as methylated cytosine in the regions of high CpG content in gene promoters is one of the earlier events that occur in tumor development [4–7]. These high CpG content regions, referred to as CpG islands, are considered an important factor that influences cell growth, differentiation, proliferation, and apoptosis through mechanisms of transcription or posttranscriptional regulation of gene expression [8]. Abnormal DNA methylation at specific gene transcription sites can result in epigenetic silencing of tumor suppressor genes that prevent tumor formation or play a critical role in DNA repair [9,10]. Both are important factors of tumor initiation. In normal cells, most promoter-associated CpG islands are unmethylated, but in cancer cells, methylated CpG islands are often observed. A high level of hypermethylation in the promoter region of tumor suppressor genes was identified within sets of tumors. Thus, detection of DNA methylation in the promoter region of tumor suppressor genes can be an important assay in early cancer diagnosis. In addition, detection of DNA methylation in diagnosed cancer patients may provide important information about the subtype of cancer and the effectiveness of potential treatments. This information will allow oncologists the ability to better understand each individual case and develop strategies for personal treatment.

The detection and quantification of DNA methylation in the gene promoter region is a challenge. No simple and sensitive method has been reported to address such a challenge. Conventional methods for DNA methylation detection use methylation-specific and/or methylation-sensitive restriction enzymes for restriction landmark analysis. Several advanced methods were developed for DNA methylation detection, including bisulfite sequencing, methylation-specific PCR, MethyLight, and CpG island microarray [11–14]. Bisulfite treatment is the most popular and powerful approach used in many detection platforms. It changes the DNA sequence through a reaction with sodium bisulfate which converts unmethylated cytosine to uracil. This reaction allows identification of DNA methylation through sequence-specific primer extension or PCR amplification. However, the reaction is time-consuming and incomplete conversion is often a problem giving rise to potential false positives. Although MethyLight (real-time PCR) avoids the time-consuming aspect of polyacrylamide gel electrophoresis (PAGE) analysis, the lack of accuracy due to incomplete conversion and nonspecific primer extension limits its application in clinical laboratories. Development of new technologies for DNA methylation detection is urgently needed.

Nanotechnology, applied to biomolecular detection, has demonstrated that nanowire field effect transistor (nano-FET) devices and nanocantilever biosensors have the promising aspects of being label free, ultrasensitive, and able to reside on a simple chip-based detection platform [15,16]. Numerous researches on nano-FET biosensor has been conducted in the last decade, based on polysilicon, monocrystal, and carbon nanotubes. The proof-of-concept in electronic biodetection has been demonstrated in the detection of DNA, protein, and virus [17,18]. Fabrication of nano-FET devices has been intensively studied [19,20]; the work reported herein mostly used polysilicon-based devices. The fundamental concept associated with a nano-FET device is that being small, on the same order as the biomolecular structure being detected, it can be electronically affected when charged biomolecules are near the device surface. Since the detection of biomolecule is often on the order of tens of nanometers, devices with feature sizes much larger than 100 nm lack the sensitivity to discern the presence of charged biomolecules.

Most nanodevices have a semiconductor residing between two conductor pads (often gold). The goal is to design a semiconductor that is affected by the presence of a charged biomolecule that attaches to the semiconductor surface. The effect is usually measured in terms of change in resistance or

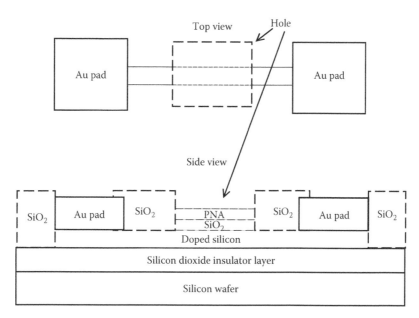

FIGURE 41.1 A diagram of nano-FET biosensor structure. Silicon wafer with a doped device layer is used to fabricate nano-FET. Gold pads are connected to each end of the nanowire. A thin layer of native silicon dioxide (1–2 nm) is formed on the nanowire surface. An additional silicon dioxide layer is deposited to cover a part of the gold pad that protects the metal pads from liquid sample. A window (hole) is open above the sensing surface which allows PNA immobilized on the nanowire.

impedance (ohms). The semiconductor characteristics, through doping, must be designed to respond to charged biomolecules on the device surface. A third terminal is usually attached to the body of the silicon die to provide a voltage called a back bias. The resulting device is not a traditional transistor. There is no traditional gate; however, the presence of a charged biomolecule on the semiconductor surface represents a gate signal. In DNA detection, the electronic charge (negative) on the DNA backbone acts analogously to a voltage on the gate electrode (region between the gold pads) to either increase or decrease the resistance of the device, depending on the semiconductor type. Figure 41.1 is a diagram of nano-FET.

Such devices are very sensitive to noise. Any charged particle on the device surface can impact the resistance, so care must be taken to insure only the desired charged molecules are present. Low-valued "gate" signals produce small current changes in the device which present challenges for the device design and measurement. Unanticipated current leakage paths must be avoided through the fabrication process. Measurement circuitry and conductive paths must be adequate to provide proper signal-to-noise values. Alligator clips with wrapped twisted pair wires, for example, should be avoided. All of these issues are solvable using state-of-the-art fabrication and electronic detection techniques.

A unique problem in the electronic detection of nucleic acid is negative charge in the sugar-phosphate backbone of DNA and RNA which exhibit electrostatic repulsion between complementary strands. Peptide nucleic acid (PNA) is a structural mimic of nucleic acid, in which the sugar-phosphate backbone is replaced by an achiral, nonionic pseudopeptide backbone composed of *N*-(2-aminoethyl)-glycine unit. Therefore, the backbone of PNA is electrically neutral, which avoids electrostatic repulsion between PNA and the complementary DNA, and decreases nonspecific binding. In addition, using PNAs as capture molecules on the sensing surface greatly reduces the noise background compared with using negatively charged oligonucleotide as capture probes. With the feature of higher annealing temperature of PNA, mismatch tolerance of PNA:DNA hybridization is lower than that of DNA:DNA hybridization. Moreover, the polypeptide backbone of PNA resists nuclease digestion. Therefore, PNAs are commonly used as capture molecules in the nano-FET biosensors.

The major challenge using nano-FET biosensors for DNA methylation detection is that both methylated and nonmethylated DNA has the same amount of electronic charge or molecular mass; hence, both have the same detection characteristics. Methods described in this section address this challenge and report the first use of a nano-FET-based electronic device to determine the methylation status in the promoter of tumor suppressor genes. This method does not require bisulfate treatment and a PCR amplification procedure. It features a simple, ultrasensitive procedure using a potentially low-cost device that could have a great impact on early cancer detection and treatment.

41.2 Materials

All modified and unmodified oligonucleotides were custom made by Integrated DNA Technology Inc. Streptavidin (SA) magnetic beads and desthiobiotin were purchased from Invitrogen. Anti-5-methylcytosine was obtained from Aviva System Biology. All restriction enzymes and DNA modification enzymes were obtained from New England Biolab. Nano-Thinker (11-mercaptoundecanoic acid [MUA]), 1-ethyl-3-(3-(dimethylamino)propyl) carbodiimide (EDAC), 25 mM *N*-hydroxysuccinimide (NHS), Sephadex G-25, dithiothreitol (DTT), and 3-aminopropyltriethoxysilane (APTES) were purchased from Sigma. DNA array blocking buffer was obtained from ArrayIt and alkaline phosphatase detector blocking solution was obtained from KPL. Bovine serum albumin (BSA), succinimidyl-4-(*N*-maleimidomethyl) cyclohexane-1-carboxylate (sulfo-SMCC), and Hellmanex were purchased from Fisher Scientific. AutoGlow 450 was purchased from Michigan Diagnostics LLC. PNAs were custom-synthesized by Bio-Synthesis. Silicon wafers were purchased from Silicon Quest. Silicon-on-insulator wafers were purchased from Ultrasil. All chemicals used in device fabrication were supplied by Cornell Nano-fabrication Facility.

Two major instruments were used in electronic test. Ref 600 Electrochemical System was acquired from Gamry Instrument Inc. and source meter models 2601 (single channel) and 2602 (dual channel) were obtained from Keithley Instrument Inc.

41.3 Methods

41.3.1 Design and Fabrication of Nano-FET Devices

There are two ways to fabricate a nanowire transistor: top down or bottom up. Both strategies are reported in the biosensor fabrication literature. Various protocols for devices fabrication are available in published papers [19,20]. In this project, a conventional semiconductor industry bottom up approach and photolithography technique was used to fabricate the devices. Nanowire semiconductor material, size, and doping level have a major effect on a nano-FET's overall behavior. Nanowire devices were designed, fabricated, and tested with various sizes and doping levels during the course of the project. Two types of nano-FET devices were fabricated in our previous work.

1. Polysilicon nanowire transistor: A 30–50 nm of a p+ type polysilicon film was deposited on 4 in. p-type silicon (300–500 μm thick)/silicon dioxide (50–200 nm thick) wafers. N_2/B_2H_6 (Boron p-type) gas was used to create a semiconductor film. Three doping levels, 2×10^{12}, 2×10^{16}, and 2×10^{20} cm^3, were used to investigate semiconductor properties of the nanowire.

2. Monocrystal nanowire transistor: Commercially available silicon in insulator wafers was used in the fabrication of nano-FETs. The doping level of monocrystal silicon was 2×10^{16} cm^3.

These transistors were ~70 nm thick, 50 nm wide, and several microns long. To prevent gold pad shorting from the test sample liquid, a thin layer of polyimide or silicon dioxide was deposited on the sensing area part of the contact gold pads. In both devices, the nanowire was placed under the gold pads. All thickness measurements used a profilometer, FilMetrices, and JOEL scanning electron micrograph. A profilometer is a useful metrology tool used to measure surface characteristics in the micron or finer scale. It is an ideal tool for film thickness measurements, wear scar measurements, step height measurements in semiconductor fabrication, etch depth, and surface roughness. The unit is a stylus-based surface profiler, with a motorized XY stage with variable speed, capable of 1 Å resolution, 1024 μm vertical dynamic range, 60 mm scan length, 1 µm/s to 25 mm/s scan speed, and 150–600× image magnification. The F50-EXR FilMetrices system for measurements of thickness, refractive index (*n*), and dispersion (*k*), as well as wafer mapping of any of these properties from 30 nm up to 250 µm. The F40 system is attached to a microscope for film thickness measurement from 20 nm to 20 µm with a measurement spot size ranging from 100 to 10 µm square. The use of the microscope allows you to make measurements on specific features of your pattern devices. Obtaining a good contact between the gold pads and semiconductor is critical, especially with sub-100 nm structures.

A nanowire was designed with a branch shape at both ends to greatly improved device performance and reduced variations from device to device.

41.3.2 Nano-FET Characterization

To characterize electronic properties of nano-FET devices, AC voltages from 10 to 250 mV were applied through two electrodes linked with semiconductor nanowires under varying DC voltages of 0–2.5 V. EIS Bode and Nyquist plots were recorded from 0.01 Hz to 100 kHz frequency range in a grounded Faraday cage. The spectra were analyzed with Gamry's Echem software. In general, a Bode or Nyquist plot measures the impedance values associated with resistance component, nonfrequency-dependent, and the frequency-dependent component of the nanodevice. Frequency-dependent components are attributed to capacitor elements associated with the device.

Electrical characterization of fabricated nanowire devices was also performed using Keithley Instruments models 2601 single-channel and 2602 dual-channel source meters, configured as a three-channel system. The source meters were controlled using Keithley Instruments LabTracer 2 software. First, the nanotransistors were tested to determine functionality and to characterize the devices. Second, validity of transistors was determined by running repeated characterization tests over a period of time. Poorly fabricated devices often fail after minutes of operation. Figure 41.2 shows the experimental setting.

41.3.3 Antibody Immobilization on the Sensing Surface

Plasma cleaning and sonication were used for cleaning the sensing surface. A self-assembled monolayer (SAM) of Nano-Thinker (MUA) was placed on the polyimide-coated nanowire surface by incubation for 16 h in a wet chamber. To remove excess MUA molecules, devices were sonicated in 100% ethanol for 5 min. To activate the carbolic terminal, an aqueous solution of 100 mM EDAC and 25 mM NHS (1:1 v/v) was placed on its surface and incubated for 1 h at room temperature (RT). The surface was then washed with PBS buffer (pH 7.2) and incubated with 0.5 µL of 10 µg/mL protein G in a wet chamber for 45 min, and then rewashed with PBS buffer. A monoclonal anti-5′-methylcytosine antibody solution was added to the surface and incubated for 30 min. To prevent nonspecific binding, the modified surface was blocked

FIGURE 41.2 A photograph of experimental setting for electronic detection. Electronic detection was performed using a probe station in a black box which protects nano-FET from light effect. The metal probes, which were connected to Kethely source meter, contacted to the gold pads of nano-FET for measuring changes of impedance.

using a blocking buffer containing 1 mM PEG, 1 mM glycine, and 0.1% BSA. An electrochemical imped-ance spectrum (EIP) was generated to verify the presence of SAM, protein G, and antibodies after each step.

41.3.4 Immobilization of Oligonucleotide and PNA on Sensing Surface

41.3.4.1 Sensing Surface Cleaning

Silicon chips with a thin layer of silicon dioxide were used to investigate cleaning and PNA immobiliza-tion. Two methods were evaluated. In method 1, silicon chips were rinsed with ethanol and sonicated in a 2% Hellmanex solution for 5 min. After rinsing with ddH$_2$O, chips were sonicated in ddH$_2$O for another 3 min. In method 2, silicon chips were cleaned using plasma. Further cleaning in both methods was performed with an acid solution of HCl:Meth (1:1) for 30 min, rinsed with ddH$_2$O, and dried in argon gas.

To activate the silicon dioxide surface, chips were immersed in 5% APTES ethanol solution and silanized at 50°C for 3 h. Alternatively, 300 μL APTES was applied on a set of nanodevice surfaces in an argon gas chamber for 1 h. After rinsing with 1% acetic acid and water, nanowire devices were dried in argon gas and stored in desiccators.

41.3.4.2 Immobilization of Oligonucleotide and PNA

Oligonucleotide was used to investigate the immobilization method. This oligonucleotide contains a protected thiol group at 5′-end for covalent linking to the surface, and biotin at 3′-end for chemilu-minescence detection. A bifunctional linker sulfo-SMCC was used to link the oligonucleotide on the APTS-activated surface. Four milligrams per milliliter sulfo-SMCC was prepared in a 100 mM NaHCO$_2$ buffer, pH 9.0. A hundred microliters of sulfo-SMCC solution was applied on a set of nanodevice sur-faces and incubated in a sealed moisture chamber for 1 h. After washing with ddH$_2$O, devices were dried in argon gas. Since a free SH group is required for immobilization, oligonucleotide deprotection was carried out in the presence of 50 mM DTT for 15 min in RT. The sample was then purified through a Sephadex G-25 spin column to remove DTT molecules. A drop of purified sample was applied on a sulfo-SMCC-modified surface and incubated in a sealed moisture chamber overnight. Unbound mol-ecules were washed away with a TE buffer and ddH$_2$O. After drying in argon gas, these modified chips can be dried and stored at 4°C for several months. The same method was used in the immobilization of PNA on the sensing surface. A cysteine residue was designed at the N-terminal of PNA, which provides thiol group for covalent linkage.

41.3.5 Characterization of Immobilized Biomolecules

Chemiluminescence detection method was used for characterization of immobilized oligonucleotides on the sensing surface. To prevent nonspecific binding, nanowire devices were blocked using a block-ing buffer containing 1 mM PEG and 1% BAS for 3 h. SA and alkaline phosphatase conjugate (SA-AP, Invitrogen) was diluted 1:1000 in a TBS buffer. Devices were incubated in SA-AP solution for 30 min at RT. After washing with a TBS buffer, chemiluminescence detection was performed in the presence of substrate AutoGlow 450 (Michigan Diagnostics). Images were taken using a charge-coupled device (CCD) camera (E-260, Apogee Instrument).

To determine the presence of PNA on the sensing surface, biotin-labeled oligonucleotides with complementary sequences were used as the reporter. Hybridization was performed at 33°C for 3 h. Unbound oligonucleotides were washed away. The devices were then blocked and incubated with SA-AP conjugate as described earlier. Chemiluminescence signal was detected and recorded using a CCD camera. Hybridization specificity was investigated through optimizing hybridization conditions. Three PNAs were immobilized on a set of six devices. Three oligonucleotides with complementary sequences and one oligonucleotide with an unrelated sequence (negative control) were used in the studies.

41.3.6 Direct Electronic Detection of DNA Methylation

An overview of the direct detection procedure is depicted in Figure 41.3. A desthiobiotin-labeled sequence-specific probe is used to capture the target DNA. Captured DNA molecules are bound to nitro-SA magnetic beads through biotin and SA interaction. After removal of unbound DNA, a small volume of D-biotin, a binding competitor, is used to reverse the binding between desthiobiotin and nitro-SA which results in target DNA release from the magnetic beads. The concentrated methylated or non-methylated target DNA molecules are then introduced to the sensing surface of the nano-FET device. Methylated DNA molecules are captured by monoclonal anti-5-methylcytosine antibodies immobilized on the surface. Nonmethylated DNA is washed away. Electronic charge in the DNA phosphate backbone on the device surface induces a change of impedance in the nanowire and generates a detectable electronic signal. This antibody-modified device can be used for detecting any methylated DNA in a simple and sensitive way.

41.3.6.1 Target DNA Generation and Methylation Modification

The promoter of the p16^{INK} gene was chosen as the target in the investigation of the direct detection method. P16^{INK} is a tumor suppressor gene. Its protein product inhibits cell cycle progression. Methylation of CpG islands in the p16^{INK} promoter and the first exon regions inactivates transcription of the p16^{INK} gene. This epigenetic alteration has been confirmed in several cancers [21]. Particularly, CpG island methylation is significantly associated with lung cancer [22].

To obtain a target DNA fragment, 100 base pair DNA representing the promoter and the first exon of p16^{INK} gene was generated through oligonucleotide synthesis and PCR amplification. Sequences of the full-length target DNA fragment, the top and bottom strand of oligonucleotides, and the capture probe are listed in Table 41.1.

A 90 bp fragment of p16 gene promoter was chosen as the target DNA molecule. The sequences of top (anticodon) and bottom (codon) strands with 15 bases overlap are partial fragments of the target DNA molecule. The desthiobiotin-labeled capture probe has the sequence complementary to the top strand at its 3′-end.

To generate a full length of the target DNA fragment, two strands of oligonucleotides with 15 bp overlap were used. The 3′-end of top strand contains a complementary sequence with the 3′-end of bottom strand. That allows hybridization between two strands. Full-length target DNA fragment was generated

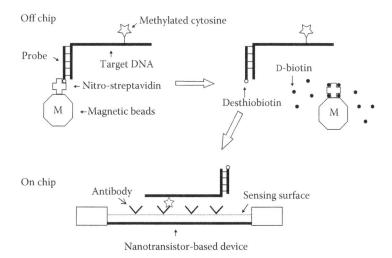

FIGURE 41.3 An overview of direct electronic detection of DNA methylation. There are two processes in the direct detection of methylated DNA: the "off-chip" process involves capturing, concentrating, and purifying target DNA and the "on-chip" process involves electronic detection of captured methylated DNA. A reversible binding approach is used to capture target DNA on magnetic beads and release target DNA in the presence of a competitive molecule.

TABLE 41.1

Sequences of the Target Area of p16 Promoter and Probe

The target region of p16 promoter DNA	5'-ACCAGAGGGTGGGGCGGACCGCGTGCGCTCGGCGGCTGCGGAGAGGGGGAGAGC AGGCAGCGGGCGGCGGGGAGCAGCATGGAGCCGGCGGCGGGGAGCA-3'
Top strand of the target DNA	5'-ACCAGAGGGTGGGGCGGACCGCGTGCGCTCGGCGGCTGCGGAGAGGGGGAGAG CAGGCAGC-3'
Bottom strand of the target DNA	5'-TGCTCCCCGCCGCCGGCTCCATGCTGCTCCCCGCCGCCCGCTGCCTGCTCTCCC-3'
Desthiobiotin-labeled capture probe	5'-Desthiobiotin-TGCTCCCCGCCGCCGGCTCCA-3'

using strand extension method in the presence of dNTP and a DNA polymerase. Hybridization of the top and bottom strand of oligonucleotides was carried out in 2× SSC solution at 1 μM concentration. One cycle of 95°C for 10 min, 4°C for 3 min, and 50°C for 60 min was performed in a thermal cycler. Strand extension was carried out in the presence of 200 μM dNTP and 5 units of Klenow fragment of DNA polymerase exo⁻ at 37°C for 30 min. Full-length target DNA fragment was verified using PAGE analysis. Probe extended full-length target DNA fragments were purified using a PCR purification kit (Qiagen). Purified DNA fragment were, then, used as the target DNA molecules in the detection experiments.

DNA methylation reaction was performed using the methylation modification kit following manufacturing product instructions from New England Biolab. Methylation status of the target DNA fragment was verified by methylation-sensitive restriction enzyme *Aci*I digestion and gel electrophoresis.

41.3.6.2 Electronic Detection of Methylated p16^{INK} Promoter

Two picomoles methylated or nonmethylated p16^{INK} DNA fragments were used as target DNA in the initial test samples. Twenty picomoles of the desthiobiotin-labeled probe were hybridized with target DNA in a hybridization solution containing 10 mM MOPS, 4% PEG, and 0.05% Tween 20. Hybridization was carried out at 95°C for 10 min, 58°C for 60 min, and 22°C for 5 min. After hybridization, 10 μL of SA magnetic beads in 40 μL 2× binding buffer were added to the sample and incubated at RT for 60 min. After washing away unbound DNA, bound DNA molecules were eluted from the beads in the presence of 0.5 mM D-biotin. The eluted sample was diluted from 1:10 to 1:1000,000 in a TE buffer. These diluted samples were used in the electronic detection on nano-FET devices. Individual samples, from low concentration to high concentration, were added to the sensing surface. TE buffer without DNA was used as blank control. Nonmethylated p16^{INK} fragment was used as negative control. Electronic detection was performed by EIP.

41.3.7 Indirect Electronic Detection of DNA Methylation

The indirect method is designed to detect multiple target DNAs on a universal detection platform. The principle of indirect electronic detection is depicted in Figure 41.4. Two probes are designed for each target DNA. A 3'-biotin-labeled probe with a complementary sequence of the target DNA is designed as the capture probe. The other is a signal probe which contains two parts of the sequence: part 1 is complementary with a CpG island region of the target DNA, and part 2 is a target unrelated sequence which is used to generate the electronic signal on a nano-FET. A deoxyuridine residue is designed in between the two parts which creates an enzymatic cleavage site for releasing part 2 of the probe. This released oligonucleotide is the signal molecule. Sequence-specific hybridization occurs between probes and the target DNA region with and without methylation. Target DNA are captured on SA magnetic beads through a biotin–SA interaction. After anti-5'-methylcytosine antibodies bind to methylated cytosine, *Escherichia coli* exonuclease III, a 3'-end dsDNA-specific nuclease, is used to degrade signal probe in

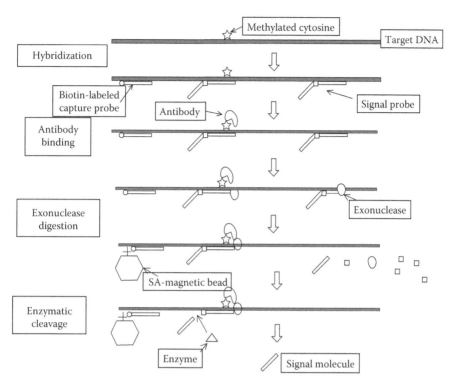

FIGURE 41.4 An overview of indirect electronic detection of DNA methylation. Indirect electronic detection involves a molecular signal transduction process. Two probes are used to capture target DNA and generate signal molecules in the presence of methylated DNA. Anti-methylcytosine antibody is used for binding methylated cytosine which blocks the exonuclease digestion of signal probe. Signal molecule is then generated through enzymatic cleavage of signal probe. The sequence of signal molecule is target-independent. Therefore, it serves as a universal signal which can be detected on a universal platform.

the hybrid of nonmethylated DNA. This enzymatic reaction was blocked in the presence of the antibody which binds to methylated CpG island. Degraded signal probes from nonmethylated DNA were washed away. Uracil-DNA glycosylase (UDG) and *E. coli* endonuclease IV are then used to cleave at the site of deoxyuridine and release part 2 signal probe from methylated DNA. These oligonucleotide signal molecules are detected on nano-FET devices.

41.3.7.1 Target Selection and Probe Design

Three promoters were selected as the target DNAs in the study: p16[INK], MGMT, and DAPK genes, in which CpG islands are highly methylated in tumors. These target DNA fragments were custom-synthesized and PCR-amplified. Purified PCR products were then methylated by using a CpG methylase, *Msss*I as described previously. The methylation status of target DNA fragments was verified using methylation-sensitive restriction enzymatic digestion and PAGE analysis. Nonmethylated DNA fragments were used as negative control. Three sets of probes were designed and synthesized. Each set contains a biotin-labeled capture probe and a signal probe. The sequences of targets and probes are listed in Table 41.2.

The target regions of p16, MGMT, and DAPK promoters were chosen in the CpG island rich area. The sequences of three 90 bp DNA fragments are listed in Table 41.2. Biotin-labeled capture probes are complementary to the underline sequences. Signal probes contain two parts: part 1 is complementary to the bold sequences, and part 2 sequence is target-independent and listed in the table. PNA molecules having complementary sequences to part 2 signal probes are also listed in the table.

TABLE 41.2

Sequences of the Target DNAs, Signal Probes, and PNA

The target region of 16 promoter	5'-CGG**ACCGCGTGCGCTCGGC**GGCTGCGGAGAGGGGG AGAGCAGGCAGCGGGCGGCGGGGAGCAGCATGGAGC CGGCGGCGGGGAGCAGCAT-3'
The target region of MGMT promoter	5'-**CCG**ACCCGGTCGGGCGGGAACACCCGCCCCGCCC GGGCTCCGCCCCAGCTCCGCCCCCGCGCGCCCCGGCCCC GCCCCCGCGCGCTCTC-3'
The target region of DAPK promoter	5'-TCCCTAGCTGTGTTCCCGCCGCCGCCCCGGC**TAGTC TCCGGCGCTGGCGC**CTATGGTCGGCCTCCGACAGCG CTCCGGAGGGACCGGGGG-3'
Part 2 of Signal probe 1	5'-AGAGGAAGATAA-3'
Part 2 of Signal probe 2	5'-ATAACAAGTGAA-3'
Part 2 of Signal probe 3	5'-ATAATAGTGGTG-3'
PNA-1	Ac-Cys-TTATCTTCCTCT-CONH$_2$
PNA-2	Ac-Cys-TTCACTTGTTAT-CONH$_2$
PNA-3	Ac-Cys-CACCACTATTAT-CONH$_2$

41.3.7.2 Enzymatic Cleavage and Release of Signal Molecule

To investigate the feasibility of a universal detection platform, signal probes were labeled with AF-488 fluorescent dye at their 5'-end. Two tests were performed: antibody blocking exonuclease digestion and enzymatic cleavage of signal probe in hybrid. Exonuclease III was used to digest methylated DNA with and without antibody presence. Two enzymes were used to cleave the part 2 signal probe at dU residue from DNA hybrids. UDG removes uridine from deoxyuridine residue and generates an abasic site (AP site). This AP site is further processed by endonuclease IV which incises the phosphodiester bond adjacent to the AP site. The conditions of enzymatic reaction were performed following manufacturing instructions.

41.3.7.3 Detection of Methylated CpG Islands in the Promoter of p16^{INK}, MGMT, and DARK Genes

A 2 pmol methylated target DNA fragment, 50 pmol capture probe, and 80 pmol signal probe of each gene were used in the hybridization. The hybridization solution contained 10 mM MOPS, 4% PEG, and 0.05% Tween 20. Hybridization was carried out at 95°C for 15 min, 63°C for 90 min, and 22°C for 5 min. After hybridization, 15 μL of SA magnetic beads were added to the sample and incubated at RT for 60 min. After washing and removing unbound DNA and probes, anti-5'-methylcytosine antibody was added. The antibody concentration was 10 μg/mL. After incubation for 30 min at RT, excessive antibodies were washed away. Exonuclease digestion was carried out at 37°C for 15 min and followed a washing step. The signal molecules were released from methylated DNA in a 10 μL buffer containing two enzymes at 37°C for 30 min. This sample was further diluted from 1:10 to 1:1,000,000 for electronic detection on nano-FET.

41.4 Results

41.4.1 Device Characterization and Surface Modification

Figure 41.5 has images of the nanowire devices used in the study. Device characterization results are shown in Figure 41.4. Transistor curves in Figure 41.6A show a spread of the curve families ISD (VSD) that indicates the device is sensitive to changes in the applied electric field. Transistor curves determine drain/source (S/D) current variations relative to gate voltage and S/D voltage. The source and drain are the pads that attach to the end points of the nanowire. It is typical to use a backgate, a terminal attached to the back of the silicon, that serves as the gate to measure the voltage/current variations. An

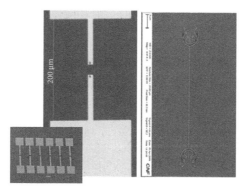

FIGURE 41.5 Microscopic images of nanowire and nano-FET devices. The image of a set of six devices is presented at the bottom corner on the left. The larger image on the left is a nano-FET device. White color is the area of gold pads. The black dots at the end of two gold pads are the contact of nanowire and gold. The large image on the right is a nanowire with branch-shaped contact area at each end.

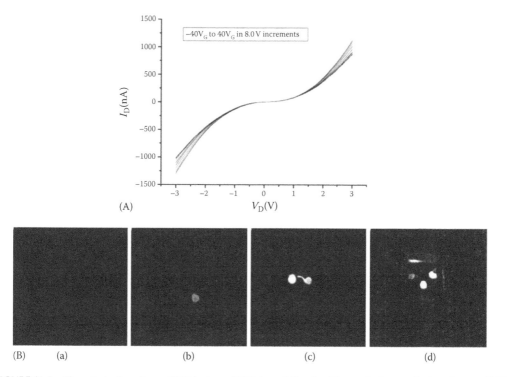

FIGURE 41.6 Characterization of nano-FET device and PNA immobilization. Electronic characterization of a nano-FET device is shown in (A). Characterization of PNA immobilization on silicon chips is shown in (B). Three PNA were immobilized on each chip. Biotin-labeled oligonucleotides were used to report PNA immobilized on chip using a chemiluminescence detection method. From left to right: (a) hybridization with no complementary oligonucleotide; (b) hybridization with one complementary oligonucleotide; (c) hybridization with two complementary oligonucleotides; and (d) hybridization with three complementary oligonucleotides.

Electrometer/Source (Keithley 6517A Electrometer) was used to generate standard transistor curves. Device transconductance was also measured.

PNA immobilization on the chip was confirmed using biotin-labeled oligonucleotides and chemiluminescence detection. Three PNAs were immobilized on chips. After hybridization with oligonucleotides, the results of chemiluminescence detection were recorded using a CCD camera. Figure 41.6B shows (a) buffer only, (b) buffer containing one oligonucleotide, (c) buffer containing two oligonucleotides, and (d) buffer containing three oligonucleotides. The presence of all three PNAs on chips was confirmed and specific hybridization was demonstrated.

41.4.2 Direct Electronic Detection of DNA Methylation

A serial dilution of methylated p16[INK] DNA fragment was used in the proof-of-concept experiments. The 0.25 µL samples containing 2.5×10^{-14} to 2.5×10^{-20} moles of methylated DNA molecules were applied to the sensing surface with immobilized anti-5′-methylcytosine antibodies. The results depicted in Figure 41.5 were obtained using a small feature size (28 nm) for nanowire devices. Figure 41.7 shows total impedance in terms of Z_{real} and Z_{mag}. Z_{real} and Z_{mag} refer to the impedance components that comprise the total impedance. Z_{real} is the component that measures "resistance" associated with a zero frequency signal. Z_{mag} is the impedance component that measures "resistance" associated with reactive components such as capacitance; this component is frequency-sensitive. Figure 41.7A shows that a significant decrease of impedance (20 MΩ) was detected in the sample containing 2.5×10^{-19} moles of methylated DNA. However, no significant change in impedance was observed in the sample containing 2.5×10^{-20} moles of methylated DNA. This detection sensitivity was not observed in the nano-FET with a relatively large feature size (100 nm). The electronic impedance spectra showed a linear correlation between DNA concentration and electronic signal.

To demonstrate the feasibility of this method, samples containing methylated and nonmethylated p16[INK] DNA fragments were hybridized with a capture probe. After captured on magnetic beads, the DNA fragment was released and detected on a nano-FET device (50 nm feature size). Impedance changes of 0.5 and 1.2 MΩ were detected in the samples containing 2.5×10^{-18} moles (Figure 41.7C) and 2.5×10^{-16} moles (Figure 41.7D) of methylated DNA in the initial sample. No significant change in impedance was observed in a sample containing 2.5×10^{-14} moles of nonmethylated DNA as the initial

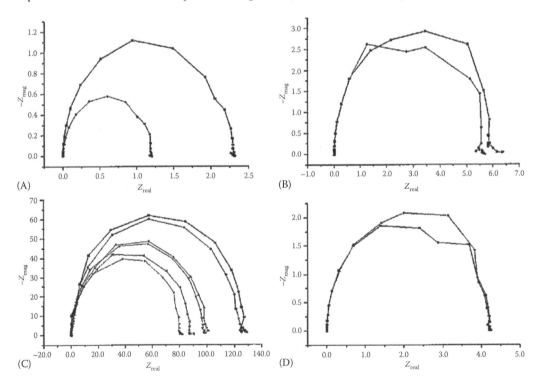

FIGURE 41.7 Direct electronic detection of DNA methylation. Electrochemical impedance spectra of detection sensitivity are shown in (A). The test samples contained 2.5×10^{-14} to 2.5×10^{-20} moles of methylated DNA fragment. Electronic impedance spectra of p16 promoter methylation detection are shown in (B) through (D). The initial samples in the "off-chip" processes contained (B) 2.5×10^{-14} moles of nonmethylated p16 promoter fragment; (C) 2.5×10^{-18} moles of methylated p16 promoter fragment; and (D) 2.5×10^{-16} moles of methylated p16 promoter fragment.

material (Figure 41.7B). The ratio of no-false-positive detection versus maximum detection sensitivity on 50 nm devices is 10^4. Nano-FETs with a feature size of 28 nm were destroyed by excessive current while testing samples containing 2.5×10^{-13} moles of DNA. Devices with feature size of 70–100 nm detected 2.5×10^{-12} moles of DNA in the test sample without being destroyed by excessive current. However, the detection sensitivity using these relatively large devices decreased for samples containing 2.5×10^{-16} moles and less of DNA. These observations indicate that the size of nanowire is directly related to the detection sensitivity.

41.4.3 Indirect Electronic Detection of DNA Methylation

41.4.3.1 Enzymatic Cleavage of Signal Probe

Enzymatic degradation of part 1 of the signal probe and release signal molecule was investigated using a fluorescent-labeled signal probe. Results from the proof-of-concept experiments are shown in Figure 41.8. In the absence of anti-5′-methylcytosine antibody, the enzymatic reaction resulted in signal molecules released from both nonmethylated DNA (Figure 41.8A) and methylated DNA (Figure 41.8B). In the presence of the antibody, the enzymatic reaction was not affected in nonmethylated DNA, since no antibody was bound to the DNA (Figure 41.8C). However, this reaction was blocked in methylated DNA, as the antibody binds to methylated cytosine (Figure 41.8D). Therefore, this antibody protection against enzymatic reaction can be used to distinguish methylated and nonmethylated DNA.

41.4.3.2 Electronic Detection of Signal Molecules

Detection of signal molecules on nano-FET devices was carried out using AC electronic impedance measurements and an EIP. These signal molecules are part 2 of signal probes which contains complementary

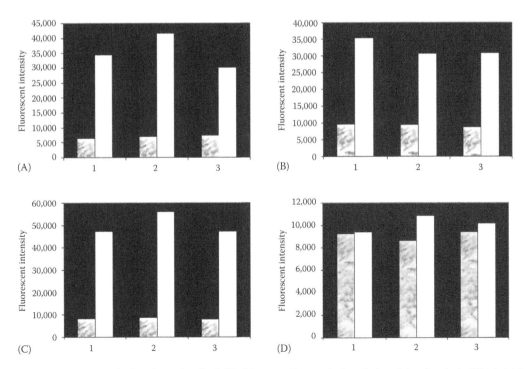

FIGURE 41.8 Enzymatic digestion and antibody blocking tests. Enzymatic degradation of signal probe in DNA hybrid was performed under the conditions of with and without the presence of anti-methylcytosine antibody. Fluorescent intensity of released signal molecule was measured. Gray columns are samples without enzyme digestion. White columns are samples with enzyme digestion. (A) Nonmethylated p16 promoter hybrid; (B) methylated p16 promoter hybrid; (C) nonmethylated p16 promoter hybrid in the presence of antibody; and (D) methylated p16 hybrid in the presence of antibody.

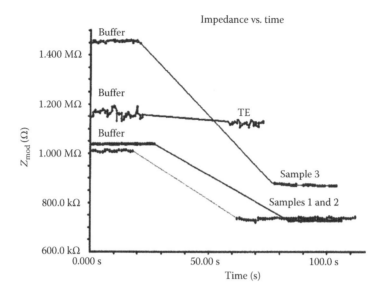

FIGURE 41.9 Electronic detection of signal molecules. Three signal molecules and control were detected on nano-FET devices: Sample 1 containing the TE buffer only, Sample 2 containing 1pM signal-1 molecules, Sample 2 containing 1pM signal-2 molecules and Sample 3 containing 1pM signal-3 molecules.

sequences with three target promoters and three PNAs (see Table 41.2). Signal molecules captured on the sensing surface through hybridization with PNAs and generated detectable electronic signal. Figure 41.9 shows the detection of three signal molecules. The relatively flat curve is TE buffer. The curves of samples 1, 2, and 3 show a significant decrease in impedance in the presence of three samples containing 1 pM of signal molecules. Initially a small change in impedance was observed when the TE buffer was applied on the nano-FET because of salt in the buffer, since applying organic solvent trimethylamine acetate (TTAA) did not change impedance (data not shown). The detection of the three signal molecules also varied as different devices were used in the tests. Some fabricated nanostructures had no nanowires between the gold pads. These devices were used as a negative control in the tests. No response was observed when DC voltages or charged molecules were applied to the sensing surface of these devices.

41.4.4 Detection Sensitivity

The detection limitation was also investigated. A set of samples with concentrations of 200 fM, 200 pM, and 200 nM signal molecules was sequentially applied from low to high concentrations on the same nano-FET device which avoids the detection variation using different devices. The EIP was used to analyze the detection sensitivity. In Figure 41.10, the curve of TE buffer is from a sample containing buffer only which presents a detection baseline. The other three curves are samples containing 200 fM, 200 pM, and 200 nM signal molecules. The results show a 200 fM signal molecule concentration can be detected. This is close to the 100 fM result reported by Hahm and Lieber [18].

41.5 Discussion

A major disadvantage of using bisulfate treatment method in the detection of methylated DNA is uncompleted conversion which reduces the accuracy in the detection. Even though PCR-based detection methods are highly sensitive, false-positive results from the amplification process are commonly observed. In addition, the quantity of the starting material is still a critical issue, as a purification process is required after bisulfate treatment which may result in the loss of the sample when a small amount of starting material is used. Electronic detection of DNA methylation demonstrated in this study reveals great promise

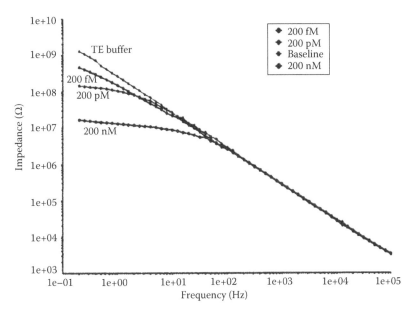

FIGURE 41.10 Electronic impedance measurement of detection sensitivity. Electrochemical impedance spectrum was used to measure the changes resulting from the molecular charge of signal molecules in a series concentrations. Four samples were sequentially applied onto a nano-FET device. The TE curve is a sample containing buffer only. Signal-1 molecules at the concentration of 200 fM, 200 pM, and 200 nM were detected.

for simple detection of very small amounts of methylated DNA. It has several advantages over existing DNA methylation detection technologies.

41.5.1 Label-Free and Ultrasensitive

Since DNA carries a negative charge on its backbone, electronic detection of that molecular charge on a sensing surface is possible. Therefore, no label is required for detection. The ultrasensitivity of electronic detection eliminates the need for PCR amplification which is required for most current DNA methylation detection methods. The feature size of the nanowire is one of the most important factors affecting detection sensitivity. For FET-based biosensors to be ultrasensitive, they must have nanoscale dimensions to detect the relatively weak effects of biomolecules which are at the nanoscale also. The nano-FET devices used in this work were linear wires of semiconductor material with diameters varying from 28 to 100 nm. The electrical responses to small electronic charge on the sensing surface varied with wire size. Nanowires of smaller size resulted in higher detection sensitivity. Detection sensitivity with nano-FETs (28–50 nm) was achieved at the 200 fM level, which is 10^4 times higher than that of fluorescent detection and is similar to real-time PCR.

41.5.2 Improved Detection Specificity

Specificity is another important issue in biodetection. Most DNA methylation detection methods require PCR amplification and must consider false-positive problems. In electronic detection, a nonspecific binding could result in a false-positive signal due to the ultrasensitive capability of the nano-FET device. Therefore, it is critical to eliminate nonspecific DNA molecule binding on the sensing surface. Blocking reagents were used to the block sensing surface after antibody, oligonucleotide, or PNA immobilization. In addition, by choosing a good monoclonal antibody and using PNA as capture molecules on the sensing surface, nonspecific binding and mismatch hybridization can be reduced. Importantly, nano-FET devices did not respond to nonmethylated DNA with identical sequences. Only molecules directly binding to the sensing surface within the Debye length distance (10 nm in this case) affected the electronic properties of the nano-FET. Nonmethylated DNA may float in solution at a distance greater than 10 nm from the surface. This important feature makes electronic detection more accurate.

41.5.3 Simple Procedure without Bisulfate Treatment

Current methods for DNA methylation detection are usually complicated, particularly when a bisulfite treatment is involved. The simplicity of the method presented herein is based on the detection of an electrical charge. Therefore, no sequence conversion is required. A monoclonal antibody is used to identify methylated DNA as the direct detection method. Antibody protection against enzymatic digestion is used to distinguish methylated and nonmethylated DNA as in indirect detection method. This is a simple and straightforward approach.

41.5.4 Universal and Low-Cost Devices

Biosensors used in the electronic detection are semiconductor devices. With advanced semiconductor nanotechnology, the fabrication capability for a feature size of 32 nm has been recently achieved. Therefore, mass production of nano-FET devices has the potential to be low cost. In addition, the methods presented herein describe a universal detection platform, which is target-independent and largely eliminates the modification of the sensing surface for each target molecule. The same device can be used to detect any methylated DNA without further modification, which also reduces the cost of device production.

Because nano-FET biosensor for DNA methylation detection is in the development stage, there are some issues that need to be addressed.

41.5.4.1 Nano-FET Surface

A major difference between a normal semiconductor device and a nano-FET device is the surface. In the normal device, the semiconductor material is embedded under a thick insulator layer for protection from moisture and other environmental effects. However, in nano-FET biosensors, the semiconductor nanowire must be on or near the surface in near proximity to the signal biomolecules which are in a liquid solution. Creating an ultrathin insulator layer on the nanowire to protect it from the liquid is critical. Currently, it is difficult to control the thickness of an insulator layer within few nanometers.

41.5.4.2 pH and Salt Concentration in the Samples

A biological sample is usually in a liquid solution. The pH and salt concentration in the solution directly affect electronic detection. A buffer with neutral pH is recommended. However, it is difficult to perform biological reactions in a solution without salt. High ionic strength solutions enhance hybridization efficiency. However, high ionic strength also reduces the Debye length. The Debye length is the distance for detectable molecular charge and the sensing surface. It is determined by the ionic strength of the solution. Therefore, the ionic strength of the solution can significantly affect the detection sensitivity due to the limitation of the distance between charged molecules and the sensing surface. A low ionic strength solution should be used in nano-FET-based biodetection. Therefore, there is tension between reasonable ionization to enhance hybridization efficiency and low ionization to enhance sensitivity.

41.6 Future Trends

41.6.1 Where the Technology Is Going

Advancements in nanoelectronic detection of epigenetic alteration biomarkers, such as ultrasensitive detection of methylated CpG islands in the tumor suppressor gene promoters, will provide oncologists with extremely valuable tools to identify molecular mechanisms involved in tumor initiation. With such knowledge, specific treatments can be designed to treat the underlying mechanisms involved in tumor initiation. Treatments can be specific and targeted for each individual. In addition, treatment therapies can be evaluated by monitoring the treatment's effect on observable cancer biomarkers. When hypermethylation is detected in the tumor suppressor gene promoters, potential therapies can be developed to demethylate cytosine in the specific gene promoters, and thereby restore the expression of tumor suppressor genes. The effectiveness of the treatments can be followed through continued biomarker monitoring.

Electronic biosensors make it possible to utilize the commercial semiconductor infrastructure for mass production of these devices. There are two subsystems associated with the proposed system. The first subsystem is the nanoelectronics for biomolecular detection as described herein. The second is a "microelectronic" subsystem which consist of circuits that determine the presence of cancer biomarkers by identifying the change in nanowire impedance, translation of that analog signal to a digital signal, digital signal processing to enable reliable biomarker identification, and finally a microprocessor which performs on-chip analysis and input/output off-chip communication. An on-chip computer would allow operation of the "intelligent" cancer biosensor with the ease of a simple cell phone. These subsystems can be manufactured in a modern semiconductor foundry and placed on a single die. High volume production could produce devices that cost on the order of tens of dollars and are designed for a throw-away market.

41.6.2 Future Applications

Detection of epigenetic alteration in the promoter of tumor suppressor genes is a promising diagnostic approach for early cancer diagnosis and treatment. Nano-FET-based electronic detection of DNA methylation is simple, label free, ultrasensitive, and potentially low cost. Future technology applications will not be limited to diagnostic and routine screening of epigenetic alterations of DNA molecules in clinic laboratories. With sensing surface changes, a nano-FET-based system can be used to detect various biomolecules in clinical samples.

ACKNOWLEDGMENTS

This work was supported in part by National Cancer Institute Contracts HHSN261200800064C and HHSN261200900052C, and NASA grant NNX06AB17G.

REFERENCES

1. Whitfield, M. L., L. K. George, G. D. Grant, and C. M. Perou. 2006. Common markers of proliferation. *Nat Rev Cancer* 6: 99–106.
2. Parrish, M. "Genomics Considerations for Successful Biomarker Development". Assay Development, Covance Genomics Laboratory. Biomarker World Congress. 2010. http://www.biomarkerworldcongress. com
3. Bast, R.C., Jr. 2010. Biomarkers for ovarian cancer: New technologies and targets to address persistently unmet need. *Cancer Biomarkers*. 8(4–5): 161–166. Cancer Biomarkers. ISSN: 1574-0153, Volumes 7–8; 12 issues.
4. Jones, P. A. and S. B. Baylin. 2007. The epigenomics of cancer. *Cell* 128: 683–692.
5. Wilson, A. S., B. E. Power, and P. L. Molloy. 2007. DNA hypomethylation and human diseases. *Biochim Biophys Acta Rev Cancer* 177: 138–162.
6. Baylin, S. B. and J. E. Ohm. 2006. Epigenetic gene silencing in cancer—A mechanism for early oncogenic pathway addiction? *Nat Rev Cancer* 6: 107–116.
7. Esteller, A. 2004. Aberrant DNA methylation as a cancer-inducing mechanism. *Annu Rev Pharmacol Toxicol* 45: 629–656.
8. Baylin, S. B. and J. G. Hermam. 2000. DNA hypermethylation in tumorigenesis: Epigenetics joins genetics. *Trends Genet* 16: 168–174.
9. Merlo, A., J. G. Herman, L. Mao, D. J. Lee, E. Gabrielson, P. C. Burger, S. B. Baylin, and D. Sidransky. 1995. 5′ CpG island methylation is associated with transcriptional silencing of the tumour suppressor *p16/CDKN2/MTS1* in human cancers. *Nat Med* 1: 686–692.

10. Wood, R. D., M. Mitchell, J. Sgouros, and T. Lindahl. 2001. Human DNA repair genes. *Science* 291: 1284–1289.

11. SeqWrite DNA Technology Service. http://www.seqwright.com/researchservices/methylationanalysis.html

12. EpigenDx Solutions for Genomic and Epigenomic Research. http://www.epigendx.com/Methylation. html?gclid=CJOct6LmjqICFRdSagodDisJUQ

13. Cottrell, S. E., J. Distler, N. S. Goodman, S. H. Mooney, A. Kluth, A. Olek, I. Schwope, R. Tetzner, H. Ziebarth, and K. Berlin. 2004. A real-time PCR assay for DNA-methylation using methylation-specific blockers. *Nucleic Acids Res* 32: e10.

14. Eads, C. A., K. D. Danenberg, K. Kawakami, L. B. Saltz, C. Blake, D. Shibata, P. V. Danenberg, and P. W. Laird. 2000. MethyLight: A high-throughput assay to measure DNA methylation. *Nucleic Acids Res* 28: E32.

15. Patolsky, F., B. P. Timko, G. Zheng, and C. M. Lieber. 2007. Nanowire-based nanoelectronic devices in the life sciences. *MRS Bull* 32: 142–149.

16. Patolsky, F., G. Zheng, and C. M. Lieber. 2006. Fabrication of silicon nanowire devices for ultra-sensitive, label free, real time detection of biological and chemical species. *Nat Prot* 1: 1711–1724.

17. Kim, D.-S. 2006. An extended gate FET-based biosensor integrated with a Si microfluidic channel for detection of protein complexes. *Sens Actuat B Chem* 117: 488–494.

18. Hahm, J. and C. M. Lieber. 2004. Direct ultrasensitive electrical detection of DNA and DNA sequence variations using nanowire nanosensors. *Nano Lett* 4: 51–54.

19. Stern, E., J. F. Klemic, D. A. Routenberg, P. N. Wyrembak, D. B. Turner-Evans, A. D. Hamilton, D. A. LaVan, T. M. Fahmy, and M. A. Reed. 2007. Label-free immunodetection with CMOS-compatible semiconducting nanowires. *Nature* 455: 519–522.

20. Barbaro, M., A. Bonfiglio, L. Raffo, A. Alessandrini, P. Facci, and I. Barák. 2006. Fully electronic DNA hybridization detection by a standard CMOS biochip. *Sens. Actuators B Chem* 118: 41–46.

21. Liggett, W. H., Jr. and D. Sidransky. 1998. Role of the p16 tumor suppressor gene in cancer. *J Clin Oncol* 16: 1197–1206.

22. Belinsky, S. A., K. J. Nikula, W. A. Palmisano, R. Michels, G. Saccomanno, E. Gabrielson, S. B. Baylin, and J. G. Herman. 1998. Aberrant methylation of *p16*[INK4a] is an early event in lung cancer and a potential biomarker for early diagnosis. *Proc Natl Acad Sci USA* 95: 11891–11896.

42

Cancer Cell Detection and Molecular Profiling Using Diagnostic Magnetic Resonance

Cesar M. Castro, Hakho Lee, and Ralph Weissleder

CONTENTS

42.1 Introduction

Despite advances in our understanding of solid tumor biology and genetics, conventional pathology continues to dictate much of today's routine cancer treatments. However, clinicians and investigators are now beginning to recognize that molecular profiling of tumors could help prune ineffective therapies and potentially streamline clinical trial subject selection, critical ingredients for achieving personalized medicine [1,13,14]. Practical considerations, including the risks and costs associated with serial core biopsies or surgical resections, often limit the amount of tissue sampled. Yet conventional methods for tumor characterization and molecular profiling, such as immunohistochemistry, flow cytometry, or western blot analyses, require tumor sample sizes that exceed the yield of other less invasive sampling techniques such as fine needle aspirates (FNA). Despite promising advances in nanotechnology-based biosensing strategies, there remains an outstanding need for clinically useful and practical sensing platforms that have high sensitivity/specificity, require minimal sample preparation, and provide robust measurements—key driving factors in biodiagnostic development [4].

Nuclear magnetic resonance (NMR) not only allows measurements in turbid samples, but also forms the basis for clinical imaging systems. Moreover, NMR can exploit the negligible magnetic susceptibilities of most biological objects. NMR is a powerful analytical method, enabling nondestructive, structure-specific identification of chemical and biological species. It is based on the thermal equilibrium property of magnetic nuclei. When nuclei with nonzero spin (e.g., protons in water) are placed in a static magnetic field, a slightly larger portion of nuclei will occupy a lower-energy state where their spins are aligned in parallel to the external fields (Figure 42.1A). By radiating the samples with electromagnetic pulses, these nuclei can absorb energy to populate the excited state. Subsequently, the excited nuclei emit energy to the environment, and the spins *relax* back to their initial states. The diagnostic magnetic resonance (DMR) assay exploits the changes in this relaxation process, caused by magnetic nanoparticles (MNPs) in water. When placed in static magnetic fields for NMR detection, MNPs produce local dipole fields with strong spatial dependence. Such field inhomogeneities accelerate the relaxation processes, leading to faster return to the thermal equilibrium.

Innovations in bioconjugation chemistries [2,3], nanotechnology [7], bio-microelectromechanical systems (MEMS), and innovative microfabrication techniques have recently catalyzed our development of a unique magnetic resonance technique termed DMR [10]. This technology couples novel labeling chemistries with MNPs to be able to sense targets of interest within minimally processed samples, in a multiplexed, point-of-care, and ex vivo manner.

Both MR imaging and NMR spectroscopy can sense magnetic resonance through spin–spin relaxation time (T_2) measurements. Advantages of these well-established techniques include their capability for strong magnetic field generation (1–11 T), their sophisticated data acquisition schemes, and fast measurements. The realities of high operating costs, cumbersome equipment, and large sample volume requirements (hundreds of microliters), however, restrict their widespread use. Whilst benchtop NMR systems operate at lower frequencies (100 kHz–50 MHz) and fields (<1 T), they still require large sample volumes and are not conducive to larger-scale clinical sample measurements. To overcome the limitations of available detectors, we have developed a DMR device, capable of performing multiplexed measurements on smaller sample volumes. Since miniaturization of the sensing components lessen the need for spatial homogeneity of the field, inexpensive and portable permanent magnets can be used to generate polarizing magnetic fields (0.1–0.5 T). In contrast to benchtop relaxometers, the DMR system directly measures molecular targets and cells in a multiplex manner, needing only microliter volumes. The performance of our handheld DMR system was initially characterized using avidin–biotin interactions, in which it achieved detection limits of approximately 1×10^{-12} M (~5 attomoles; mouse IgG to Tag peptide). This detection limit is better than the detection limits achieved by ELISA [8]. We have continued to refine the features of the DMR device. Recent iterations weigh <100 g (including NMR probes, radiofrequency generators, power) and are over 60 times more sensitive than conventional NMR, with sample volume requirements as low as 1 μL [11]. The DMR "hardware" system consists of four major components (Figure 42.2):

1. The micro-NMR chip containing microcoils to excite samples and to receive subsequent signals
2. A microfluidic network for sample handling
3. On-board NMR electronics
4. A small permanent magnet

We continue to improve the signal-to-noise ratio (SNR) by devising NMR probes with improved filling factors (>1) and lower electrical noise. The latest iteration probe consists of a solenoidal microcoil embedded within a microfluidic structure [11]. The coils are first wound around polyethylene tubes and this is followed by their immersion in a polymer (polydimethylsiloxane). After polymer curing, the tubes are retracted to open up the fluidic channels. The entire bore of the solenoid is then filled with sample, which results in significant enhancement in NMR signal levels. The advantages of this system over previously used lithographically patterned and metal-plated planar coils include homogeneous radiofrequency magnetic fields and less electrical resistance [9,10]. Additional systems have recently been used for clinical trials [15,16].

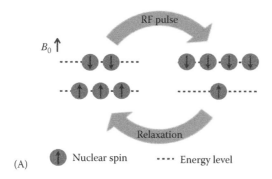

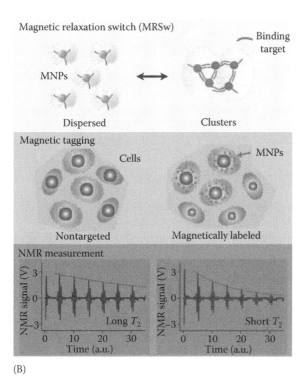

(B)

FIGURE 42.1 (A) NMR principle. When nuclei with nonzero spin (e.g., protons in water) are placed in a static magnetic field, a slightly larger portion of nuclei will occupy a lower-energy state where their spins are in parallel to the external fields (left). Radiofrequency (RF) pulses can cause the population inversion to place more nuclei into the higher energy state. Subsequently, those excited nuclei radiate energy and relax back to the thermal equilibrium state. (B) DMR assay configurations. *Magnetic relaxation switches* (MRSw): Functionalized MNPs can be induced to switch from dispersed to clustered states by analytes. Unlike evenly dispersed MNPs, aggregates of nanoparticles (self-assembled magnetic clusters) are more efficient at enhancing the transverse relaxation and hence faster NMR signal decays. *Magnetic tagging for cell detection*: To detect larger biological objects such as bacteria and mammalian cells, the DMR assay employs the "tagging" mode, wherein target-specific MNPs are used to label the detection targets. Samples containing magnetically tagged cells have lower T_2 or higher relaxation rate against background.

42.1.1 Microcoils

Microcoils are used to excite samples and to receive subsequent NMR signals. Our first DMR prototype was designed to perform eight-multiplexed measurements in 5–10 µL sample volumes. A 2×4 planar microcoil array was fabricated on a glass substrate using standard microfabrication technology (Figure 42.2b). Lately, we changed our design to solenoidal coils, as such geometry provides higher SNR by producing more homogeneous radiofrequency magnetic fields for sample excitation [11]. We have

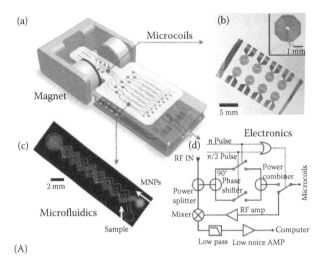

(A)

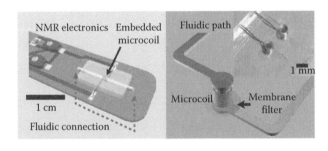

(B)

FIGURE 42.2 (A) DMR system schematic. (a) This particular DMR system consists of an array of microcoils for NMR measurements, microfluidic networks for sample handling and mixing, embedded NMR electronics, and a permanent magnet for polarizing magnetic field generation. (b) Micrograph of a microcoil array. The microcoil (inset) generates RF magnetic fields to excite samples and receives the resulting NMR signal. (c) Example of a microfluidic network. Effective mixing between magnetic nanoparticles and the target analytes is achieved by generating chaotic advection through the meandering channels. (d) Schematic of the NMR electronics. The circuit is designed to perform T_1 and T_2 measurements via inversion-recovery and CPMG pulse sequences, respectively. (B) (a) Solenoidal coils can provide higher signal-to-noise ratio (SNR) through more homogeneous radiofrequency magnetic fields for sample excitation. SNR is further increased by integrating the coil with a microfluidic network which facilitates handling of small sample volumes and ensures sample confinement to the most sensitive microcoil region. (b) A membrane filter can size-selectively retain cells, and remove smaller contaminants, notably unbound MNPs. This is effective for concentrating cells from large sample volumes, and allows on-chip washing steps.

further increased SNR by integrating the coil with a microfluidic channel. The coils were first wound around polyethylene tubes and subsequently immersed into a polymer (PDMS). Following polymer curing, the tubes were retracted to open up fluidic channels. The entire bore of the solenoid thus can be filled with samples to achieve maximal filling factor (~1), the fraction of the coil volume occupied by the samples. Due to the larger cross-sectional area of winding wires, the solenoidal coils also had less electrical resistance than lithographically patterned coils. With these advantages, the sample volume for DMR detection could be reduced by a factor of ~10 (to 1 μL) compared to the previous devices (~10 μL).

42.1.2 Microfluidics

Microfluidics represents a convergence between several highly interdisciplinary scientific fields. The microfluidic system within the DMR device provides additional magnetic separation and concentration of targets from biological samples [12]. In particular, the microfluidic network enables the handling and distribution of small sample volumes, the mixing of varying flow streams, and also ensures the confinement of samples to the most sensitive region of a given microcoil. Furthermore, a membrane filter can be included at the outlet of the latest DMR's solenoidal microcoils to size-selectively retain large species such as cells, and to remove smaller contaminants such as unbound MNPs. This is effective for concentrating cells from large sample volumes, and allows on-chip washing steps.

42.1.3 NMR Electronics

Custom-made NMR electronics is designed to provide pulsed NMR measurements. The electronics generates versatile RF (radiofrequency) pulse sequences to measure the longitudinal (T_1) and transverse (T_2) relaxation times, process raw NMR signals (amplification, frequency conversion, filtering) for acquisition by a computer, and handle multiplexed operation of an array of coils. Figure 42.2d shows the blueprint of NMR electronics for such operations. In our first prototype, the NMR electronics was constructed as a tabletop system using discrete RF components (e.g., AD9830 for RF generation and AD604 for NMR signal amplification; Analog Device) and off-the-shelf RF components (e.g., ZAD-1 mixer, ZMSC-2 power splitter, and ZYSWA-2 RF switch; Mini-Circuits). Following the successful demonstration of the first prototype, we integrated the entire NMR electronics on a CMOS IC chip. The packaged IC had a small footprint ($\sim 1\,cm^2$) and could be placed close to NMR microcoils to minimize electrical noise and environmental interferences. The overall size of the NMR system was considerably reduced; our latest prototype enables true portability for bedside operation (Figure 42.2A).

42.1.4 Signal Acquisition

DMR systems can measure both the longitudinal (T_1) and transverse (T_2) relaxation times by generating different pulse sequences. As the main sensing mechanism, however, we focused on measuring T_2 times, as MNPs are more efficient in changing the transverse relaxation of water protons. To accurately measure T_2 values, we used Carr-Purcell-Meiboom-Gill (CPMG) spin echo pulse sequences that can compensate for the inhomogeneity of the polarizing magnetic field by portable magnets. Figure 42.3 shows a typical signal pattern from CMPG measurements. After the first excitation by the 90° pulse, the precessing spins (in water) are continually refocused by the 180° pulses. The refocusing causes the signal to arise from an otherwise noisy state (hence the name "spin echo") before the signal fades out again. To measure T_2, the envelopes of spin echoes are collected and fitted to an exponentially decaying function (Figure 42.3, dashed line, and Table 42.1).

For typical DMR measurements, we used the following parameters: echo time (TE): 4 ms; repetition time (TR): 6 s; the number of 180° pulses: 500; the number of scans: 9. These values are chosen to match with those of our commercial relaxometer (Minispec mq20; Bruker) so as to enable the direct comparison of system performance. The 90° and 180° pulse widths were determined for individual microcoils.

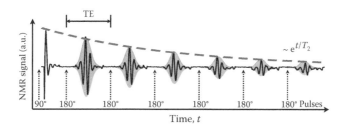

FIGURE 42.3 To accurately measure T_2 values, we use Carr-Purcell-Meiboom-Gill (CPMG) spin echo pulse sequences that can compensate for the inhomogeneity of the polarizing magnetic field by portable magnets. A typical signal pattern from CMPG measurements is shown here. After the first excitation by the 90° pulse, the precessing spins (in water) are continually refocused by the 180° pulses. The refocusing causes the signal to arise from an otherwise noisy state (hence the name "spin echo") before the signal dies out again. To measure T_2, the envelopes of spin echoes are collected and fitted to an exponentially decaying function (dashed line).

TABLE 42.1

List of Parameters for T_1 and T_2 Measurements

Parameter		T_1 Measurement	T_2 Measurement
Pulse sequence		Inversion recovery	CPMG
		Pulse separation: 5–1000 ms	Echo time: 4 ms
Pulse width	90°	41	41
(μs)	180°	82	82
Data points per scan		26	36–500 (number of 180° pulses)
Number of scans		16	64
Repetition time (ms)		2000	1000–6000

42.1.4.1 DMR Assay

The DMR or assay exploits two major components:

1. MNPs
2. Antibodies or other affinity ligands

42.1.4.1.1 Magnetic Nanoparticles

Ideal MNPs used for DMR sensing require a strong magnetic moment to induce pronounced T_2 changes, whilst at the same time exhibiting superparamagnetic behavior (i.e., no permanent remanence). They also require unique coatings in order to avoid spontaneous magnetic aggregation. Essentially, the higher the relaxivity, the fewer the number of nanoparticles needed to produce a detectable signal. Cross-linked iron oxide (CLIO) nanoparticles have driven the majority of our DMR biosensing due to their favorable in vivo and bioconjugation properties [5]. CLIOs comprise a superparamagnetic iron oxide core, usually ferromagnetic magnetite (Fe_3O_4), and are coated with dextran to form stabilizing cross-links, as well as with ammonia to provide primary amine group functionality (amino-CLIO). These nanoparticles have an average hydrodynamic diameter of 25–40 nm, and each particle has between 40 and 80 amines available for conjugation of biomolecules. Small nanoparticles are advantageous for cell-labeling applications since they can pack more closely. This maximizes surface coverage, unlike the micron-sized particles often used for cell separation.

Depending on the size of detection targets, the DMR assay can have two different configurations: magnetic relaxation switches or magnetic tagging for cell detection (Figure 42.1B).

42.1.4.1.2 Different Assay Configurations

Functionalized MNPs can be induced to switch from dispersed to clustered states by analytes. Unlike evenly dispersed MNPs, aggregates of nanoparticles (self-assembled magnetic clusters) are more efficient at enhancing the transverse relaxation and hence faster NMR signal decays. This unique phenomenon, known as MRSw, has been employed to detect small molecular targets with sizes less than or comparable to that of MNPs. Note that MRSw assays are performed without removing excess unbound MNPs.

42.1.4.1.3 Magnetic Tagging for Cell Detection

To detect larger biological objects such as bacteria and mammalian cells, the DMR assay employs the "tagging" mode, wherein target-specific MNPs are used to label the detection targets. As this assay mode is relevant to cancer detection, we explain it in more detail.

Samples containing magnetically tagged cells have lower T_2 or higher relaxation rate against background. The degree of R_2 changes can be formulated as

$$R_2 = R_W + r_2 \cdot \frac{N_P}{V}, \tag{42.1}$$

where

R_W is the relaxation rate of the background (usually water)

V is the NMR detection volume

N_P is the total number of MNPs in V

r_2 is the intrinsic capability of MNPs to induce T_2 changes, defined as transverse relaxivity

If a targeted cell has n MNPs and the total number of cells is N_C ($= N_P/n$), the change of R_2 ($\Delta R_2 = R_2 - R_W$) due to MNPs is

$$\Delta R_2 = r_2 \cdot \frac{N_P}{V} = \frac{n \cdot r_2}{V} N_C = r_2^{\text{cell}} \cdot \frac{N_C}{V}, \tag{42.2}$$

where r_2^{cell} ($= n \cdot r_2$) is defined as the *cellular relaxivity* (transverse relaxivity per given cell concentration). As r_2^{cell} is directly proportional to n, it is indicative of the abundance of a target biomarker, and DMR thus attains the capacity for quantitative molecular profiling on cells.

Equation 42.2 shows two major ways to achieve large ΔR_2 changes and thereby high detection sensitivity. One strategy is to increase the number of MNPs (n) loaded on target cells, which led to the development of novel MNP-labeling platform, bio-orthogonal nanoparticle detection (BOND) as described later. Another way is to decrease the NMR detection volume (V) so as to increase the effective cell concentration (N_C/V) inside the NMR detection region. This is the most important motivation for implementing a miniaturized NMR system.

42.1.4.2 Bio-Orthogonal Nanoparticle Detection II (BOND-2)

Numerous assay configurations have been previously investigated. Driven by the need for enhanced efficiency of nanoparticle loading (Figure 42.4), we have recently developed a novel, catalyst-free

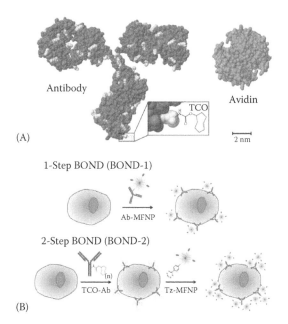

FIGURE 42.4 (A) Shown are comparative sizes of a representative antibody (mouse IgG2a), available lysine residues for TCO modification (in lighter shade), and a TCO tag and avidin protein for comparison. (B) Application of a bio-orthogonal nanoparticle detection (BOND) for one-step (direct) and two-step targeting of nanoparticles to cells (BOND-2). The latter demonstrates improved nanoparticle loading.

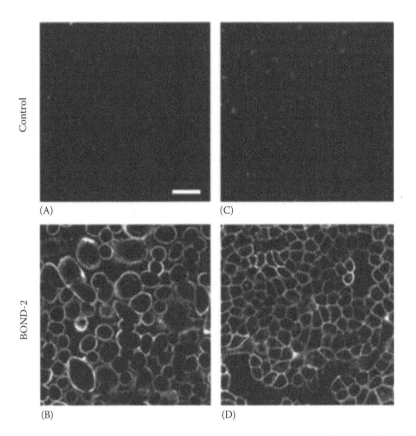

FIGURE 42.5 Confocal microscopy images of similarly labeled live cells. Control: nonbinding, TCO-modified control antibody (clone MOPC-21). HER2 (A, B) and EpCAM (C, D). Scale bar, 50 mm (A).

two-step universal cell-labeling approach. Coined BOND-2, the reactions rely on the bio-orthogonal tetrazine (Tz) and *trans*-cyclooctene (TCO) cycloaddition. This method is both rapid and chemoselective and effectively targets nanoparticles to live cells. Antibodies against any cancer marker of interest are initially modified with TCO and then used as scaffolds to later couple Tz-modified nanoparticles to live cells. In Figure 42.5, we performed confocal microscopy to confirm the efficacy of nanoparticle targeting to live cancer cells using BOND-2. Cells (SK-BR-3 and HCT 116) known to maintain elevated levels of HER2 and EpCAM, respectively, were incubated with TCO antibody (i.e., anti-HER2 or anti-EpCAM) followed by Tz-magnetic fluorescent nanoparticles (MFNPs; as described in Section 42.2). A strong fluorescence signal was detected at the cell membranes in all instances except for TCO-modified control. Prior work has demonstrated that BOND-2 markedly improves the targeting of cells, as compared to the widely used avidin–biotin interaction, which is typically employed for in vitro applications [6]. We have also demonstrated that Tz's small size results in effective interactions with TCO on antibody surfaces, without encountering steric hindrance or influencing nearby TCO sites. The small-molecule BOND-2 chemistry allows nanoparticles to pack more densely onto the antibody scaffolds, thus yielding greater signal amplification and significant improvements in detection thresholds to near single-cell levels (Table 42.2). Using the cancer marker HER2 as an example, BOND-2 outperformed direct antibody–nanoparticle conjugation and avidin–biotin approaches with improvements in detection threshold reaching fivefold.

Our exquisitely sensitive and versatile DMR platform offers portable, quantitative, and multiplexed measurements of cancer markers. Coupled with its rapid, operator-independent readouts, DMR could catalyze widespread adoption of real-time bedside molecular diagnostics, potentially redefining the standard-of-care during diagnostic workup.

TABLE 42.2

Comparison of Nanoparticle Targeting Strategies

Comparison of Nanoparticle Targeting Strategies	BOND-2	Avidin/Biotin	Direct Conjugation
Antibody modification	*Trans*-cyclooctene	Biotin	Various (thiol, amine, cycloaddition)
Nanoparticle–antibody linkage	Covalent	Noncovalent	Covalent
Antibody valency potential	~30 to 50	~30 to 50	NA
Nanoparticle modification	Tetrazine	Avidin	Various (thiol, amine, cycloaddition)
Nanoparticle valency (N_v)	84	8	2–3
Kinetic reaction rate (monovalent, k_{on}, $M^{-1}s^{-1}$)	6×10^3	1×10^7	10^5–10^6
Net nanoparticle reaction potential ($N_v x\, k_{on}$, $M^{-1}s^{-1}$)	5×10^5	8×10^7	~10^5–10^6
Relative labeling efficacy	15	5	1

42.2 Materials and Methods

42.2.1 BOND-2 Assay

42.2.1.1 Antibody Modifications

Monoclonal antibodies are modified with (*E*)-cyclooct-4-enyl 2,5-dioxopyrrolidin-1-yl carbonate (TCO-NHS) and synthesized as previously reported [2]. For each case, 0.5 mg of antibody is buffer-exchanged into phosphate-buffered saline (PBS) (pH 8.0) using 2 mL Zeba desalting columns (Thermo Fisher). TCO-NHS is then reacted in 10% dimethylformamide (DMF) for 3 h at room temperature. The relevant antibodies of interest and controls (e.g., anti-MUC-1, anti-EGFR, and anti-CD45) are then reacted with TCO-NHS. Samples are purified by buffer exchange into PBS using 2 mL Zeba columns, and concentrations are determined by absorbance measurement.

42.2.1.2 Tz Modification of MFNPs

Amino-CLIO is modified with 2,5-dioxopyrrolidin-1-yl 5-(4-(1,2,4,5-tetrazin-3-yl)benzylamino)-5-oxopentanoate (Tz-NHS) to create Tz-CLIO. The reaction is performed using 500 equivalents of Tz-NHS relative to amino-CLIO, and proceeds in PBS containing 5% DMF for 3 h at room temperature. Excess Tz-NHS is removed with Sephadex G-50.

42.2.2 Flow Cytometry Experiments

Cells are grown in culture flasks to 90% confluency, released using 0.05% trypsin/0.53 mM ethylenediaminetetraacetic acid (EDTA), and washed twice by centrifuging at 300g for 5 min, aspirating the supernatant, and resuspending with PBS containing 2% fetal bovine serum (PBS+). Cells (250,000/sample) are then labeled with TCO-modified monoclonal antibodies (10 µg/mL) in 0.1 mL PBS+ for 10 min at room temperature. Following centrifugation and aspiration of the antibody solution, cells are directly resuspended with Tz-MFNPs (0.2–200 nM), incubated for 30 min at room temperature, and washed twice by centrifugation with ice-cold PBS+. Control samples lack the antibodies.

Once the labeling procedures are complete, VT680 fluorescence on CLIOs is assessed using a flow cytometer (e.g., LSRII, Becton Dickinson, San Diego, CA), and mean fluorescence intensity is determined using FlowJo software (Tree Star Inc., Ashland, OR). All measurements are performed in triplicate and the data are presented as the mean ± standard error.

42.2.3 Measurement of Biomarker Expression Levels on Cells

When large numbers of cells are available, actual expression levels of the tumor markers of interest within the cell lines examined can be determined by antibody staining and flow cytometry. Cells are prepared as described earlier for the cell-labeling studies and for DMR measurements. Cells (10^6 per sample) are incubated in 0.1 mL PBS+ containing 1 µg/mL primary antibody for 15 min at room temperature. Following a centrifugal wash with ice-cold PBS+, appropriate fluorescein-conjugated secondary antibodies are added at 1 µg/mL and incubated for 30 min on ice. Cells are then washed twice with ice-cold PBS+ and analyzed on a flow cytometer (e.g., LSRII). Correlative DMR measurements for calibration can be performed on a smaller number of cells, as described earlier.

42.2.4 FNA Protocols

FNA sampling experiments in xenograft models (e.g., murine) [11] demonstrate the utility of DMR measurements. When the tumors reach ~1 cm in size, they are aspirated with 22G needles. Following the aspiration, the needle is washed in cell dissociation solution (Cellstripper; Mediatech) to dislodge the cells. Four needle washes per tumor are pooled and then equally divided into five 1 mL aliquots. Four aliquots are incubated with unmodified or target-specific MNPs (e.g., HER2, EGFR, and EpCAM) under identical conditions (5 min at 37°C). Excess MNPs are subsequently removed via triple washes (with centrifugation at 200*g* for 5 min), and cell pellets are resuspended in PBS (1 mL). As a control, the fifth aliquot is prepared in a similar manner as the other four, but without incubation with the MNPs. The analytical accuracy of our DMR assay can be benchmarked against gold standards in molecular analysis, notably flow cytometry and western blot analysis, by measuring tumor marker expression in cell lines (e.g., HER2 in breast cancer cells).

42.3 Results

42.3.1 Cancer Detection in Xenograft Models

We extended our inquiries from cell lines to xenograft mouse models. Here, the detection limit with the most recent DMR iterations approached the single-cell level (~2 cells in 1 µL detection volume). This detection limit far surpassed the sensitivities achieved with earlier DMR iterations (~10^3 cells in 10 µL samples), and even those of current clinical standards (e.g., cytology and histology). The R_2 changes ($\Delta R_2^{HER2}/\Delta R_2^{\theta}$) from DMR (requiring ~$10^3$ cells) demonstrated excellent correlation ($R^2 > 98\%$) with both flow cytometry (requiring >10^5 cells) and western blot analysis (requiring >10^7 cells). Importantly, the DMR sensor not only required fewer cells than both approaches but also achieved faster time to readouts. For perspective, specimen processing of about 30–50 samples for flow cytometry including LS column separation can take 3–4 h and minutes to record events. The other accepted standard, western blots, takes much longer principally due to specimen preparation which, combined with measurement time, can be achieved in 1–2 days. DMR processing time and readout can be achieved in ~1 h.

42.3.2 Cancer Cell Profiling

For a given sample, DMR's multiplexing capabilities and exquisite sensitivity allow for considerably broader cell profiling, in comparison to conventional standards, where only limited markers can be profiled before the original sample is depleted. To be clinically useful, cellular assays should not only be sensitive but also have the capacity for multiparameter analysis. The latter is all the more important considering that (1) biological samples from bodily fluids or tissues are inherently complex in composition and variable in cell numbers, and (2) there is no single ubiquitous biomarker, especially for cancer detection.

42.3.3 Protocol for Analytical Measurements

We have established an assay protocol that reports the target cell density as well as the expression level of a specific biomarker. With such improvements, DMR could achieve the same level of analytical capability as flow cytometry and western blotting, the gold standards in cellular analysis.

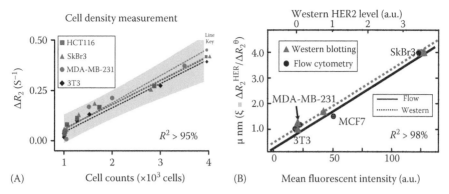

FIGURE 42.6   To independently measure cell numbers by DMR, we exploited a previously characterized phenomenon of low-grade phagocytosis of nontargeted MNPs by tumor cells. When mammalian cells were incubated (15 min at 37°C) with unmodified MNPs (MNP-θ), linear and cell-number-dependent R_2 changes (ΔR_2^θ) were observed (A). Of note, this method was extensively verified by comparing DMR measurements to other standard methods, flow cytometry, and western blotting). (B) A proof-of-concept example in which we measured the expression level of HER2 in breast cancer cell lines. The measured ξ^{HER2} ($= \Delta R_2^{HER2}/\Delta R_2^\theta$) from DMR (requiring ~$10^3$ cells) showed good agreement ($R^2 > 98\%$) with both flow cytometry (requiring ~10^5 cells) and western blotting (requiring ~10^7 cells), validating the analytical capability of DMR. Note that DMR detection was much faster (~15 min) and performed using >10^2 times fewer cells.

To independently measure cell numbers by DMR, we exploited a previously characterized phenom-enon of low-grade phagocytosis of nontargeted MNPs by tumor cells. When mammalian cells were incu-bated (15 min at 37°C) with unmodified MNPs (MNP-θ), linear and cell-number-dependent R_2 changes (ΔR_2^θ) were observed (Figure 42.6A). Interestingly, these changes were similar across a wide variety of cell types. The results were fitted to Equation 42.2; $\Delta R_2^\theta = r_2^\theta \cdot n_C$, where r_2^θ is the cellular relaxivity for MNP-θ and n_C is the cell concentration (N_C/V). The cellular relaxivities (r_2^θ) were statistically identical ($p > 0.99$) among different tumor cell lines, suggesting that the method may provide a universal measure for estimating n_C.

Using Equation 42.2 and the cell density information mentioned earlier, we defined the expression level (ξ) of a select marker as

$$\xi^+ = \frac{\Delta R_2^+}{\Delta R_2^\theta} = \frac{r_2^+ \cdot n_C}{r_2^\theta \cdot n_C} = \frac{r_2^+}{r_2^\theta}, \tag{42.3}$$

where ΔR_2^+ and r_2^+ are R_2 changes and the cellular relaxivity, respectively, with a marker-specific MNP. In this normalized form, ξ now reports the *cellular* expression level of a targeted marker, providing a way to molecularly profile target cells regardless of the cell numbers in a sample. The method was extensively verified by comparing DMR measurements to other standard methods (i.e., flow cytometry and western blotting). Figure 42.6B shows a proof-of-concept example in which we measured the expression level of HER2 in breast cancer cell lines. The measured ξ^{HER2} ($= \Delta R_2^{HER2}/\Delta R_2^\theta$) from DMR (requiring ~$10^3$ cells) showed good agreement ($R^2 > 98\%$) with both flow cytometry (requiring ~10^5 cells) and western blotting (requiring ~10^7 cells), validating the analytical capability of DMR. Note that DMR detection was much faster (~15 min) and performed using >10^2 times fewer cells.

42.3.4 Multiplexed Profiling

NMR is intrinsically a single-parameter detection that measures the abundance of one marker at a time. The development of DMR overcomes this limitation by enabling sensitive measurements in small vol-umes (~1 μL) of samples. A parent specimen thus can be divided into a large number of aliquots and each interrogated for a different molecular marker. Such a capacity has become even more practical by the development of the BOND (especially BOND-2) technology that allows modular detection (same MNPs for readout, different antibodies for targeting).

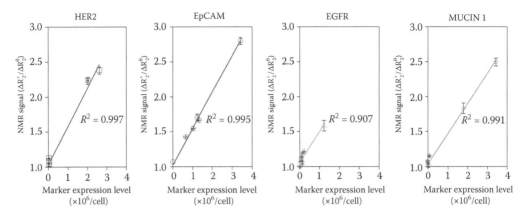

FIGURE 42.7   DMR tumor signals in cell lines showed excellent correlation with measured marker expression levels, as determined independently by flow cytometry.

Figure 42.7 shows an example of a multiplexed assay using DMR and BOND-2. We profiled scant samples (~1000 cells) from human tumor cell lines (A431, A549, NCI-H1650, HCT116, SK-BR-3, and SK-OV-3) with a panel of cancer markers (HER2, EpCAM, EGFR, and Mucin 1) [6]. NMR signal levels for HER2, EpCAM, EGFR, and Mucin 1 detection showed excellent correlation with flow cytometry measurements previously determined on larger sample amounts due to inherent detection limitations. Of note, the detection limit with DMR was ~100-fold better in comparison to flow cytometry. The use of BOND-2 to drive nanoparticle targeting is feasible in scant cell populations and its detection threshold can exceed even accepted standards such as flow cytometry.

Working under the hypothesis that multiplex analyses beyond one marker improves cancer detection in a xenograft model [11], our marker assays were found to yield false-negative rates of 57%–72% for single targets, 28% for two targets, and down to negligible values when investigating three targets (Figure 42.8B). We were thus able to generate a malignancy signature for each of our samples (Figure 42.8A). Moreover, we found that the DMR assay could be expanded beyond three markers to profile

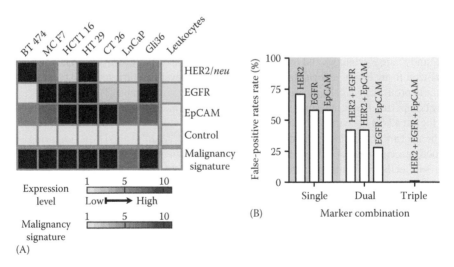

FIGURE 42.8   (A) A panel of xenograft tumors underwent FNA and profiled by DMR. The R_2 changes (ΔR_2) for three cancer makers (HER2, EGFR, and EpCAM) were divided by the corresponding ΔR_2 with unmodified nanoparticles to report the cellular expression levels. The maximum ratio among the three markers was used to determine the malignancy of a given aspirate. With multiple-marker targeting, the accuracy for correctly diagnosing cells as malignant could be considerably enhanced. (B) Bar graph representation of false-positive rates.

more complex cancer signatures. Given that well-controlled experimental conditions, favored by many competing nanosensors, routinely overestimate real life and practical performance characteristics, we sought to extend DMR into the human clinical arena.

We recently completed a clinical study at the Massachusetts General Hospital (Boston, MA) in which 70 patients with suspicious intra-abdominal lesions, who had undergone FNA and core biopsies for diagnostic purposes, were enrolled [15]. We demonstrated the feasibility of (1) processing complex human samples for DMR using BOND-2, and (2) applying our DMR device to derive measurements on materials of variable yield (a problem inherent to biopsy sampling techniques). Furthermore, we successfully interrogated a total of 10 intra- and extracellular cancer markers in these samples (EpCAM, HER2, Muc-1, Vimentin, Ki-67, B7-H3, EGFR, CK18, p53, and CD45). A priori selection of these protein markers was based on current practice (e.g., EpCAM and CK18 for epithelial cancer detection) or on reports of clinically relevant overexpression. For example, increased HER2 expression in subpopulations of breast and gastric cancers has been associated with particularly aggressive biological behavior. At the same time, the tumor cells' reliance on such growth-promoting cancer proteins serve as an Achilles Heel evident by the demonstrated clinical benefit of targeted therapies against HER2. Similarly, EGFR overexpression and/or mutations have been found to confer aberrant proliferative properties to cancer cells. Indeed, therapeutics against EGFR has gained traction over the past few years. We have used DMR to identify a four-panel marker set, with a >95% diagnostic accuracy in the clinical setting [15]. Beyond the quantitative data generated, we also encountered several situations in which conventional pathology, the clinically accepted standard and metric of diagnostics, was either falsely positive or falsely negative, when compared to DMR findings. Thus, having identified such lapses with conventional diagnostic measurements, we anticipate that the enhanced accuracy rendered by DMR could result in tangible treatment additions and/or modifications to improve outcomes.

The portable, point-of-care aspect of DMR, together with its quantitative readouts, makes it an attractive diagnostic device for use in larger prospective treatment or screening clinical trials. In particular, developing real-time readouts of cancer-related pathway inhibition would be extremely useful, especially since incorporating the myriad of existing and emerging targeted therapies into large clinical trials has proven extremely inefficient. A priori evidence of tumor response to particular treatment strategies could streamline patient eligibility and filter out ineffective agents. For this reason, we are currently expanding the findings from our pilot data, which have demonstrated that protein levels of phospho-S6RP/S6RP (a downstream readout of the PI3K-Akt and MAP kinase pathways) can be detected with high sensitivity. Thus far, in preliminary studies, we have shown successful target inhibition ($ED_{50} = 0.9\,nM$) in $\sim$100 rapamycin-treated and rapamycin-untreated SK-BR-3 cells, each assessed using DMR analyses. These and other related data suggest that DMR could be a valuable device for drug efficacy testing in scant cell populations harvested from FNA, and potentially from other sources (e.g., blood or bone marrow samples).

42.4 Discussion

In the current era of chemotherapy and targeted treatments for epithelial cancers, improved serial detection and profiling of patients' tumors could help modify or maintain treatment. Likewise, more sensitive profiling methods could be used to prune ineffective therapies from the myriad of drug combinations found in many existing or nascent clinical trials, and/or potentially redirect patients toward more effective strategies. Furthermore, by having the ability to identify malignant features from protein marker data, despite the presence of benign-appearing cells under standard light microscopy, physicians could be alerted to early malignant alterations. In current clinical practice, FNA diagnoses are used simply to report on the presence or absence of malignant cells within a given specimen. Now, with the ability to quantify and profile cancer cells from FNA samples, this procedure could serve as a powerful adjunct to the semiquantitative data obtained from immunohistochemistry, and thus enhance diagnostic accuracy.

The ability to perform multiplexed, platform-independent, and rapid analyses of biological specimens remains a major research bottleneck. Many current nanosensing techniques are time-consuming, require operator input, and have prolonged processing times. Furthermore, they can generally only commit to specific markers (e.g., EpCAM). DMR, however, offers a biosensor technology with unique advantages in a clinical setting. This novel technology requires minimal sample preparation, can perform assays in turbid/obscure media, can be adapted to profile different targets (DNA, protein, metabolites, cells), and has the capability for high-throughput operation.

At the cellular level, malignant tumors are characterized by a complex mixture of cell types: benign, malignant, stromal, vasculature, and inflammatory cells (e.g., macrophages). Thus, given the limited quantity of biological sample possible, specificity becomes an increasingly important metric for successful detection of cancer. A common drawback faced by all current diagnostic strategies is that as sensitivity is increased, background also increases. This then compromises the SNR necessary for optimal detection. Existing technologies therefore continue to rely on negatively selecting (purifying) competing cells that are not of interest. This is especially true with circulating tumor cells. To further enhance cancer cell detection, we have similarly explored fast leukocyte depletion protocols, particularly in whole blood where leukocytes are highly abundant. We previously noted that DMR can detect, in its current iteration, two cancer cells per microliter volume of sample. While impressive for most purposes, this would translate into 2000 cells/mL. In the CTC detection realm, the clinical reality is more consistent with roughly 10–100 cells/5–7 mL of blood. Purification strategies to improve capturing or "enrichment" of cancer cells therefore become paramount for DMR and any other CTC detection method. Enrichment approaches exploit various parameters including size, morphology, or surface markers. Our group is actively working on optimizing and advancing such strategies to achieve improved CTC detection. The enriched milieu of biopsies or surgically resected tissue does not require enrichment strategies when analyzed by our sensitive DMR assay.

DMR leverages numerous advantages over existing detection technologies. They include the following:

Versatility: DMR methods can be used to measure virtually any type of target including whole cells, proteins, enzymes, nucleic acid sequences, and drugs.

No sample purification required: DMR uses magnetic fields for signal generation and detection (i.e., magnetic resonance). Because magnetic fields pass through biological samples regardless of their optical properties, assays can be performed in diverse media. Light-based assay methods, for example, fluorescence, bioluminescence, absorption, or colorimetry, are sensitive to materials in the sample that scatter light, absorb light, or fluoresce.

Sensitivity: DMR assays are of very high sensitivity (10^{-14} M) and enzymatic amplification is usually not required in contrast to PCR- or ELISA-based methods.

Multiplexed measurements: DMR's exquisite sensitivity technology is ideally suited for evaluating several CTC markers simultaneously thus increasing the detection sensitivity.

Rapid clinical answers: The technology can potentially be used at the point-of-care. While not all assay types may require fast clinical answers, there are clearly scenarios where such information is desirable. For example, having rapid answers during image-guided FNA could reduce procedure time and cost.

True quantitation: NMR measurements are inherently quantitative. Parallel processing of control samples allows accurate measurements of cell number as well as biomarker densities.

Homogeneous assay format: DMR employs neither solid phase attachment nor separation of bound and free analytes. For example, there is no immobilization of the biomolecules (e.g., protein, DNA, small molecules) onto glass slides, resulting in faster hybridization kinetics. These features make the technology particularly suitable for miniaturization and/or for microfluidic applications.

Potential for use as implantable devices: DMR is based on MNPs that are nontoxic to mammalian cells and have been widely used in clinical trials. Many other nanoparticle-based sensors (e.g., Q-dots, gold particles) use materials that are potentially toxic, not used in natural metabolic processes, or lack a pathway of degradation and elimination, all of which make their clinical use problematic.

The clinical application of our innovative customized tool would fulfill many (if not all) of the key requirements desired of biodiagnostic technologies, namely: assay sensitivity, selectivity, versatility, low cost, and portability [4]. Manufacturing costs from reusable NMR components (<$200) and the disposable microfluidic chip (<$1) also render DMR technology a practical and scalable option. These features thus further support our conviction that this platform will soon become widely adopted into both the clinical and research arenas.

42.5 Future Trends

We anticipate that the expanded possibilities of scientific inquiry, created by the DMR technology, will synergize interactions between the bench and the clinic. DMR chip customization, shaped by emerging science, should ensure that the platform remains current and responsive to researchers' inquiries. Given the recent success with human FNA specimens, our next goal will be to shift from using primary tumor specimens to alternative clinical samples such as proximal fluids (e.g., ascites) or peripheral blood. This will no doubt add further layers of complexity, in terms of both heightened sensitivity and specificity needs. For example, in peripheral blood, the ratio of leukocytes to circulating tumor cells is far greater than that observed in primary tumors. Our group therefore seeks to overcome such challenges by developing and optimizing cancer-specific assays (based on our novel chemistries and tools) to amplify signal and to lower detection limits. Amplifying molecular cancer signatures by associating more sensors (in our case, nanoparticles) to each biomarker will be critical for the development of rare cell detection strategies.

DMR is not a niche platform and has multiple implications for all facets of research and care. For example, it could be adapted to advancing proteomic or genomic data, including the Cancer Genome Atlas. For emerging molecules of interest, for which antibodies are not yet available, DMR would allow versatile use of different recognition moieties, such as thioaptamers or peptides. We are also currently exploring additional technologies for combining sensitive DMR measurements with higher throughput purification chips. In parallel, we are seeking to further miniaturize our devices and to increase the number of channels for simultaneous measurements. Finally, we are exploring various approaches for achieving single-cell analysis of targeted nanoparticles, using technologies such as Hall sensors. It is our belief that the current plateau in cancer management could benefit greatly from leveraging such promising innovative rare cell detection and profiling technologies.

ACKNOWLEDGMENTS

We would like to acknowledge Drs. Jered Haun and David Issadore, and many others, for their invaluable contributions to the optimization of the DMR technology and in the assay development over the years.

REFERENCES

1. Allegra, C. J., J. M. Jessup, M. R. Somerfield, S. R. Hamilton, E. H. Hammond, D. F. Hayes, P. K. McAllister, R. F. Morton, and R. L. Schilsky. 2009. American Society of Clinical Oncology provisional clinical opinion: Testing for KRAS gene mutations in patients with metastatic colorectal carcinoma to predict response to anti-epidermal growth factor receptor monoclonal antibody therapy. *J Clin Oncol* 27:2091–2096.
2. Devaraj, N. K., R. Upadhyay, J. B. Haun, S. A. Hilderbrand, and R. Weissleder. 2009. Fast and sensitive pretargeted labeling of cancer cells through a tetrazine/*trans*-cyclooctene cycloaddition. *Angew Chem Int Ed Engl* 48:7013–7016.
3. Devaraj, N. K., R. Weissleder, and S. A. Hilderbrand. 2008. Tetrazine-based cycloadditions: Application to pretargeted live cell imaging. *Bioconjug Chem* 19:2297–2299.
4. Giljohann, D. A. and C. A. Mirkin. 2009. Drivers of biodiagnostic development. *Nature* 462:461–464.

5. Harisinghani, M. G., J. Barentsz, P. F. Hahn, W. M. Deserno, S. Tabatabaei, C. H. van de Kaa, J. de la Rosette, and R. Weissleder. 2003. Noninvasive detection of clinically occult lymph-node metastases in prostate cancer. *N Engl J Med* 348:2491–2499.

6. Haun, J. B., N. K. Devaraj, S. A. Hilderbrand, H. Lee, and R. Weissleder. 2010. Bioorthogonal chemistry amplifies nanoparticle binding and enhances the sensitivity of cell detection. *Nat Nanotechnol* 5:660–665.

7. Heath, J. R. and M. E. Davis. 2008. Nanotechnology and cancer. *Annu Rev Med* 59:251–265.

8. Koh, I., R. Hong, R. Weissleder, and L. Josephson. 2008. Sensitive NMR sensors detect antibodies to influenza. *Angew Chem Int Ed Engl* 47:4119–4121.

9. Lee, H., Y. Liu, D. Ham, and R. M. Westervelt. 2007. Integrated cell manipulation system—CMOS/ microfluidic hybrid. *Lab Chip* 7:331–337.

10. Lee, H., E. Sun, D. Ham, and R. Weissleder. 2008. Chip–NMR biosensor for detection and molecular analysis of cells. *Nat Med* 14:869–874.

11. Lee, H. and R. Weissleder. 2009. Rapid detection and profiling of cancer cells in fine-needle aspirates. *Proc Natl Acad Sci USA* 106:12459–12464.

12. Lee, H., T. J. Yoon, and R. Weissleder. 2009. Ultrasensitive detection of bacteria using core-shell nanoparticles and an NMR-filter system. *Angew Chem Int Ed Engl* 48:5657–5660.

13. Shaw, A. T., B. Y. Yeap, M. Mino-Kenudson, S. R. Digumarthy, D. B. Costa, R. S. Heist, B. Solomon, H. Stubbs, S. Admane, U. McDermott, J. Settleman, S. Kobayashi, E. J. Mark, S. J. Rodig, L. R. Chirieac, E. L. Kwak, T. J. Lynch, and A. J. Iafrate. 2009. Clinical features and outcome of patients with non-small-cell lung cancer who harbor EML4-ALK. *J Clin Oncol* 27:4247–4253.

14. Tsao, M. S., A. Sakurada, J. C. Cutz, C. Q. Zhu, S. Kamel-Reid, J. Squire, I. Lorimer, T. Zhang, N. Liu, M. Daneshmand, P. Marrano, G. da Cunha Santos, A. Lagarde, F. Richardson, L. Seymour, M. Whitehead, K. Ding, J. Pater, and F. A. Shepherd. 2005. Erlotinib in lung cancer—Molecular and clinical predictors of outcome. *N Engl J Med* 353:133–144.

15. Haun, J. B., C. M. Castro, R. Wang, V. M. Peterson, B. S. Mari-nelli, H. Lee, and R. Weissleder. 2011. Micro-NMR for rapid molecular analysis of human tumor samples. Sci Transl Med 3:71ra16.

16. Issadore, D., C. Min, M. Liong, J. Chung, R. Weissleder and H. Lee. 2011. Miniature magnetic resonance system for point-of-care diagnostics. LabChip 11:2282–2287.

43

Field Effect Transistor Nanosensor for Breast Cancer Diagnostics

Pritiraj Mohanty, Yu Chen, Xihua Wang, Mi K. Hong, Carol L. Rosenberg, David T. Weaver, and Shyamsunder Erramilli

CONTENTS

43.1 Introduction

A biosensor, described succinctly by Ziegler and Gopel (1998), is "a device integrated with a biological sensing element, which is a product derived from a living system and a signal transducer which can provide a recognition signal of the presence of a specific substance". The biological sensing element, which determines the specificity of the biosensor, can be composed of an enzyme, an antibody, a nucleic acid, or another analyte detecting molecule. The specific binding or reaction between the target and the receptor (or the biological sensing element) introduces a signal that is then transduced and measured. Because of these fundamental properties, biosensors can be configured for macromolecular recognition, such as with human cells of different types, viruses, and pathogenic organisms. Therefore, there is a far-reaching diagnostic utility in these devices in applications ranging from human health to food safety, drug response, and personalized medicine.

Biosensors may be categorized by the operational mechanism of the sensors (Byfield and Abuknesha, 1994). Although optical biosensors using colorimetric, fluorescence, luminescence, and absorbance are industry and diagnostics standards, these strategies necessitate target labeling and amplification. Also, the instrumentation footprint necessary to sensitively read optics-based signals is large compared with that achievable in devices incorporating nanotechnologies and microelectronics. Thus, emerging technologies that improve the sensitivity, cost, instrumentation, or field applicability of biosensors are beginning to be implemented. Mechanical biosensors utilize mass loading during the recognition process, introducing a dynamic resonance frequency change or static deflection of the device. Electronic biosensors measure the change of the capacitance or conductance of the device due to biological recognition.

43.1.1 Nanoelectrical Sensors

A field effect transistor (FET) uses an electric field to control the electrical channel of conduction and hence the conductivity of the charge carriers in the channel. The flow of charge carriers between the source and the drain can be tuned by modifying the size and the shape of the conducting channel by applying an electric field to the gate. In the biosensor configuration, the FET consists of a nanowire channel between the source and the drain terminals. The nanowire surface can be biofunctionalized so that a biomolecular binding event can create an electric field, similar to the control electric field applied to a conventional FET (Figure 43.1). In devices that use the FET principle, a designated, physically separated sensor surface is formed by precision manufacturing. The FET sensor is connected to an electronic circuit to monitor the specific conductance of this sensor surface.

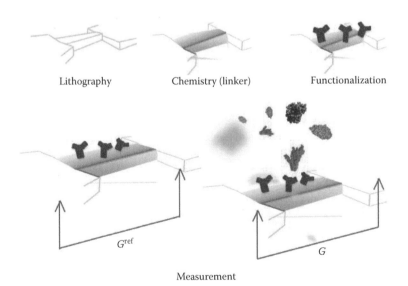

FIGURE 43.1 FET biosensor manufacturing and measurement. The diagram indicates the stepwise manufacturing of FET devices for diagnostic tests. Highly sensitive nanosensors are formed with precise dimensionality and surface area. Following a process of lithography and chemistry, antibodies are conjugated to the surface of the sensor. Analyte measurements are conducted with samples containing heterogeneous mixtures of disease-relevant proteins, such as those that occur in blood, saliva, and other fluids. Specific analyte binding contributes to a surface charge differential, detected electrically as a change in conductance (ΔG) across the nanosensor surface.

FET biosensors are adapted for the measurement of biomolecules interacting with such a sensor surface (Figure 43.1). As with other forms of sensing, the surface of a FET biosensor is modified to selectively recognize specific analytes. In the illustration, antibodies are conjugated to the surface as part of the manufacturing process. These antibodies are selected for specific detection of a protein of interest. Molecular binding events between the analyte and the antibodies on the biosensor surface cause changes to the biosensor surface charge density and/or surface potential. In this manner, precision manufacturing of FET biosensors allows sensitive analyte recognition. The differential conductance amplitude is correlated to the analyte concentration in the sample solution.

The FET was first introduced as a sensor for analyte ion concentration measurements in 1970 (Bergveld, 1970), and was later adapted for many different sensor applications (Chung et al., 2006; Hammond and Cumming, 2005). In 1997, detection of DNA hybridization was reported using a FET sensor (Souteyrand et al., 1997). When these technologies were introduced, the operational dimensions of the device were in the micron–millimeter range, with recording buffer concentrations on the order of a millimole. Others described DNA hybridization detection with a FET sensor that monitored nanomolar DNA concentrations, but a similar operational area was still required (Fritz et al., 2002).

Important innovations have now been introduced which allow these measurements to be taken to the nanoscale. Although the molecular binding events are identical to those in larger devices, the nanoscale has the advantage of increased sensitivity of detection. In their pioneering work, Lieber and coworkers used chemically grown silicon nanowire FETs for sensing (Cui et al., 2001b). These nanochannel FET sensors demonstrated significant advantages of real-time, label-free, and highly sensitive detection of a wide range of analytes, including proteins (Wang et al., 2005), nucleic acids (Hahm and Lieber, 2004), small molecules (Chen et al., 2008; Wang et al., 2005, 2008), and viruses (Patolsky et al., 2004) in single-element or multiplex format (Zheng et al., 2005).

A significant breakthrough has been achieved in the utilization of CMOS-compatible materials, such as silicon, to develop FET biosensors. Common nanoscale architectures include nanowires, nanotubes, nanopores, micro- or nanocantilevers, and nanoparticles (including quantum dots). Together with carbon nanotubes, silicon nanowires are considered to be the fundamental building blocks for scaling microelectronics down to the nanolevel. However, current carbon nanotube synthetic methods produce mixtures of metallic and semiconductor nanotubes, indicating the difficulty of fabricating scalable, homogeneous devices reproducibly.

Silicon technology is one of the most developed and well-studied technologies for the microelectronics industry. Therefore, many manufacturing, miniaturization, and multiplexing capabilities are advantageously adaptable to silicon-based nanosensors. Further, the doping of silicon can be well-controlled, ensuring that silicon nanowires can be consistently made semiconducting, whether synthesized by chemical methods (chemical vapor deposition or laser ablation) or fabricated by a lithography process. Since silicon nanowires are configurable to ultrasmall dimensions (20 nm in diameter and 2 μm in length), these devices promise ultrahigh sensitivities sufficient to measure a single bonding event (such as a single virus particle detection).

There are several technical improvements which illustrate the importance of size effects and justify the development of nanoscale FET biosensors. First, biological recognition at the sensor surface is reflective of the entirety of the physical signal across the whole sensor device. Thus, the sensitivity of the device, relating to its signal-to-noise ratio, is improved by increasing the surface-to-volume ratio. Whereas attachment of a biomolecule to a macroscale planar device affects the "surface" characteristics, attachment to a nanoscale wire affects the "bulk" characteristics. Second, many of the biological molecules of interest in biomedical, environmental, and food applications are known to have sizes on the micro- or nanoscale. Thus, a device with submicron or nanoscale dimensions will have the best efficiency as a biological sensor. Third, as the device size is reduced to the nanoscale, properties such as the mobility of the charge carrier change (Elfstrom et al., 2007) finite size effects dominate the device character, and nanodevice performance is enhanced.

Nanowires have many applications, including sensitivity enhancement for surface plasmon resonance biosensors (Byun et al., 2005, 2007; Kim, 2006). Surface-modified silicon nanochannels bind to charged groups and ions and serve as nanoscale pH sensors (Chen et al., 2006). Nanowires composed of conducting polymers (polyaniline) can be used to coat the electrode for voltammetric detection of DNA hybridization (Chang et al., 2007; Zhu et al., 2006) as well as for the detection of the food-borne pathogen, *Bacillus cereus* (Kim et al., 2006). Nanotechnologies are driving the development of low-cost, portable FET devices for clinical applications that will greatly improve diagnostic efficiency. The promise of the nanoscale semiconductor technology lies in its potential for highly parallel detection of thousands of target molecules in tandem, and at very low concentrations relevant to their normal and diseased state concentrations in blood and other human bodily fluids. Therein lies a fundamental advantage over traditional screening methods that ordinarily detect one type of protein or biomarker per assay. Screening and diagnosis by conventional techniques does not allow for the check of a large number of biomarkers due to technological and cost limitations. Array chip techniques for both proteomics and genomics often require several stages of amplification for reliable analysis. The time and the number of cells needed for these techniques prove to be bottlenecks for the future molecular applications in medical science. The nanosensor technologies described here are designed to overcome these challenges, as they require short analysis time with small sample volume.

Perhaps the most promising application is the use of these sensors for monitoring and detecting specific molecules associated with a particular disease, such as in cancer patient diagnostics. The FET nanosensor may logically be used for early-stage detection of cancer, a strategy crucial to the prevention of cancer-related death.

43.1.2 Breast Cancer Diagnostics

Human breast cancer is a clonal disease, presumed to develop when a cell acquires sufficient germ line or somatic abnormalities to be transformed and to express full malignant potential (Balmain et al., 2003). Although originally considered to be a single disease of the breast, extensive research and clinical studies have led to the realization that breast tumors are biologically heterogeneous. Hence, an individual patient is known to require therapies that are tailored to her specific cancer. The development of microarray RNA techniques, combined with a search for appropriate biomarkers, has led to the realization that associated with each breast cancer there may be changes in the expression level in hundreds of genes (Branca, 2003; Van den Eynden et al., 2004). Traditional medical treatment protocols are unable to properly consider tumor heterogeneity and thus may be prescribing treatments that may not be effective for a particular patient (Garber, 2004; Ring and Kroetz, 2002).

Currently, the major biological distinctions between individual tumors are made by evaluating expression of estrogen receptor (ER), progesterone receptor (PR), and Her2/*neu*, the product of the c-erbB-2 oncogene. Classification of tumors using a set of biomarkers (Weigelt et al., 2003) suggests that each patient's tumor may be associated with a unique "fingerprint." Individually or in clusters, these biomarkers may have prognostic value (i.e., estimating outcome) and/or predictive value (i.e., estimating response to specific treatments, or in predicting survival in metastasized breast cancers; Bild et al., 2006; Minn et al., 2005; Weigelt et al., 2003). This new understanding has the potential to lead to truly patient-specific treatment protocols (Garber, 2004), but it must be supported by accurate, immediate, and comprehensive evaluation of biomarkers by new diagnostics strategies.

43.2 Materials

43.2.1 Wafer-Level Fabrication in Silicon-Based Materials

43.2.1.1 Silicon Wafers

Silicon nanowires are fabricated from standard silicon-on-insulator (SOI) wafers using nanolithography and surface micromachining. The starting wafer has a typical silicon thickness of 30–100 nm on top of a sacrificial oxide layer varying in thickness from 100 to 500 nm. SOI wafers are used to replace the traditional silicon wafers in order to reduce parasitic device capacitance and thus improve performance of the devices in microelectronics (Celler and Cristoloveanu, 2003). The wafer used in our experiment is an SOI wafer (100) (University Wafer), which includes a device layer (100 nm thick), an insulation layer

(380 nm thick), and a boron-doped substrate (around 600 μm thick). The volume resistivity of the device layer is 10–20 Ωcm for these experiments.

43.2.1.2 Design Features

We refer to the manufactured surface consisting of one or more silicon nanowires as a "chip." Typically, the nanosensors are designed with 10–100 nanowires per chip. Silicon nanowires are fabricated by exposure on three sides to provide three-dimensional relief. The associated benefit to this geometry is an increase in surface-to-volume ratio of the silicon nanowire, which enhances the dielectric effect of molecular binding. Second, we vary the system size to determine optimal structure size. Both the silicon nanowire cross-sectional dimensions and the proximity of the gates are changed in a series of devices. The main manufacturing variance parameter to tune the sensitivity range is the dimensional width and length of the nanowire between the source and drain. The nanosensor thickness, which will be nearly identical for all wires on a single chip, is chosen to be 30–100 nm in order to maintain high sensitivity, nanosensor reproducibility, and stability. The bias voltage applied on each nanosensor is also a contributor to modulating sensitivity.

We also fabricate nanowire sensors with multiple gates. Additional gates enable better sensitivity by providing a means to fine-tune the nanowire conductance channel. In addition, we fabricate arrays of nanowires for differential measurement in anticipation of diagnostic need. In the differential configuration, the detected signal is the difference between the signal from a functionalized nanowire and an unfunctionalized nanowire. This technique allows a more accurate determination of the biomolecular signal.

43.2.2 Nanochannel Surface Chemistry and Functionalization

The silicon-based nanochannels contain an oxide layer (e.g., Al_2O_3 grown by atomic layer deposition [ALD]) added during the fabrication process. This layer can be used for functionalization of the analyte detection molecules, which may be proteins, receptors, antibodies, nucleic acids, ligands, or small molecules, using chemical modification protocols extensively published in the literature.

43.2.2.1 Silane Treatment

The silicon oxide surface is modified using 3-aminopropyltriethoxysilane (APTES; Sigma-Aldrich, St. Louis, MO). The amine group of APTES reacts with glutaraldehyde to provide an aldehyde group that can then form an imine linkage to the primary amine group on the detector receptor proteins. Other ligands, polymers, and chemicals may be conjugated to the surface by the same or related chemistries. Alternatively, the surface of the silicon nanowire is exposed to an ethanol solution of 3-(trimethoxysilyl) propyl aldehyde to yield an aldehyde-terminated nanosensor surface.

Silanization is conducted after the Al_2O_3 coating steps. The nanowire device is placed into a hot water bath for a short time to clean the surfaces. The device is then blown dry with N_2 gas. The device is silanized by 5% APTES (Gelest, Inc. Morrisville, PA) solution in 95% ethanol, 5% DI water for 2 h. APTES is removed by rinsing with ethanol (>3 times) and DI water (>3 times) and then dried with N_2 gas following a bake at 120°C for 30 min. The conjugation-ready receptors (Section 43.2.2.2) are then directly coupled to these aldehyde surfaces.

43.2.2.2 Antibody Conjugation

Immobilizing a selected antibody or receptor to the modified surface contributes the desired specificity for molecular interactions to be studied. These molecules are prepared for linking to the aldehyde surfaces by cross-linking chemistry. First, EDC and sulfo-NHS (Sigma-Aldrich) are used to cross-link to the detector protein by standard conjugation protocols. For the studies reported here, the biotinylated monoclonal CA15.3 from Fitzgerald (host species—mouse) recognizing the CA15.3 tumor antigen is used. Second, after the aldehyde-linked surface has been bound to the antibodies to CA15.3 or other analyte detections of interest, it is important to make sure that there are no free (i.e., unreacted) aldehyde surface groups that remain. A potential artifact can result from nonspecific binding of other proteins and peptides to the unbound aldehyde links. Passivation of the surface with ethanolamine is sufficient, and it is routinely checked using fluorescently labeled peptide markers. Functionalization is optimized and checked by using an immunoglobulin assay (human IgG). The same procedure is operational for other molecules with primary amine groups.

43.3 Methods

The invention of the transistor was a defining point for the modern electronics revolution. An accelerator of this technology occurred with the realization of techniques to shrink the physical dimensions of the transistors to the nanoscale. This capability to manufacture at the nanoscale is also expected to facilitate an important breakthrough in diagnostics. Fabrication of the nanowire sensor involves a series of processes, including pattern definition by electron beam lithography, material deposition by thermal evaporation or electron beam evaporation, plasma cleaning, liftoff, and a number of dry (reactive ion plasma) and wet (acid) etch processes. We have succeeded in developing manufacturing processes for a number of materials, including insulating GaAs, semi-insulating or doped silicon, silicon oxide, and metallic suspended structures.

43.3.1 Nanochannel Device Fabrication

43.3.1.1 Photolithography

Photolithography is currently the most widely used fabrication method in industrial settings for submicron or nanoscale structures. During the stepwise photolithography process, a mask is used to define the structure on the wafer. The wafer is first spin-coated with a polymer (photoresist) that is sensitive to the UV light. After soft baking, the wafer is placed beneath the photomask and intensely exposed to the UV light. The more common type of resist, positive photoresist, becomes more soluble in developer after exposure to UV light (negative photoresist, in contrast, becomes less soluble).The developer is then applied to remove the photoresist pattern that has been exposed to the UV light.

The resolution of photolithography is limited by the wavelength of the UV light. In current technology, UV light with a wavelength of 183 nm is used to pattern features as small as 37 nm. Smaller wavelengths are able to create smaller structures, but as demand dictates further miniaturization, new optics which utilize reflection strategies are required. Other technologies, such as phase shifting masks or immersion of high-refractive efficient media between the wafer and apertures, can be used to pattern structures smaller than 50 nm.

43.3.1.2 Electron Beam Lithography

Electron beam lithography, laser beam lithography, and focused ion beam lithography are serial processes used to pattern micro/nanostructures. Because the throughput is low, electron beam lithography is used primarily to produce photomasks or in research laboratory settings. It is sufficient, however, to develop and test design fabrication prototypes for the nanochannel FET devices. The typical process is diagrammed in a stepwise manner (Figure 43.2). First, electron beam lithography is used to define source–drain electrodes. Two layers of polymethylmethacrylate (PMMA) are spun on the wafer surface, and exposure to a high-energy electron beam breaks the PMMA bonds, making it soluble in developer. We use a developer composed of an MIBK: IPA solution for PMMA. The resolution of electron beam lithography depends on the acceleration voltage of the electron beam, and also the secondary electron travel in the resist (PMMA). Electron beam evaporation is used for metallization. Titanium and gold are typical electrode materials, with titanium acting as an adhesive layer. After liftoff, another lithography process is used to define a nanochannel, and chromium is used as the mask for the subsequent etching process.

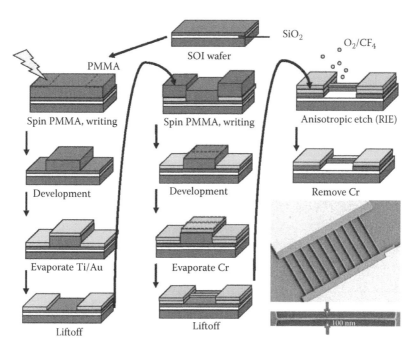

FIGURE 43.2 The stepwise fabrication process. Ti, Au, and Cr denote the metals titanium, gold, and chromium, respectively. Reactive ion etching (RIE) is used to give vertical relief to the wires. Lower right photograph: scanning electron micrograph image of a suite of 10 nanosensors with critical dimensions of 100 nm, connected to a common source/drain.

43.3.1.3 Reactive Ion Etching

Reactive ion etching (RIE) is a typical method for both isotropic and anisotropic etching. Chemically reactive plasma is used to remove the material on the wafer surface. The material to be removed dictates choice of plasma chemistry. For example, to etch silicon, plasmas based upon either CF_4 or SF_6 are used. While RIE is mainly a chemical reaction, the process also depends on the power and pressure, as the ions bombard the surface and transfer kinetic energy to the wafer in a process akin to sputtering. By appropriate power and pressure adjustments, we achieve a uniform anisotropic etch and straight side walls.

Wet etching can also be used in nanochannel device fabrication. Stern et al. (2007a) applied a wet anisotropic etching technique combined with a lithography process. They have demonstrated that high-quality nanochannel FET sensors can be created using "top-down" fabrication techniques. These devices can be used as highly sensitive, specific label-free sensors down to 100 femtomolar concentrations of antibody-based detection, as well as for utilities in real-time monitoring of the cellular immune response. In this multistep process, tetramethylammonium hydroxide is used to etch Si(111) planes around 100 times more slowly than the other planes, such that all imperfections not aligned to the (111) plane are smoothed. In their work, the oxide of the nanobar is removed and a direct silicon–carbon bond is formed by a photochemical hydrosilation reaction. Yields of successful selective functionalization by this approach have been reported to be low (around 2%), although improvement may be expected.

Following etching, a thin layer of Al_2O_3 (5–10 nm) is deposited on the nanowire(s) by ALD to form a high-quality insulation layer. Thin films grown by ALD are chemically bonded to the substrate and are pinhole-free. The Al_2O_3 insulation layer has almost perfect conformity compared with other methods and can be used to deposit many types of films. A possible advantage of the Al_2O_3 insulation layer is that the Al_2O_3 surface has an increased capacity for silanization compared to silicon. Al_2O_3 is also used to protect the electrodes in the solution.

43.3.2 Measurement with a Standard Nanochannel FET Device

Nanowires are particularly attractive as biosensors due to the critical dimensions of the nanostructures. The detection sensitivity is greatly enhanced as a result of the large surface-to-volume ratio achieved with nanoscale conductors, where the measured conductance is dominated by surface contributions. Therefore, the presence of charged proteins bound to the surface induces a large fractional change in the nanowire conductance, and enables relatively easy detection.

A conventional electrical circuit is used to measure the functionalized nanosensors. To establish the circuitry, of particular interest is the location of a control gate in the devices (Figure 43.3), which is the subject of future improvements in design, fabrication, and sensitivity. This gate has allowed us to enhance the sensitivity and to control the labeling of the nanowire, leading to a new method of using nanosensors for biomarkers. Without any gate biasing, we are able to measure the presence of the model marker at a concentration of <1 ng/mL. By adjusting the gate voltage, the sensitivity is increased by at least one order of magnitude.

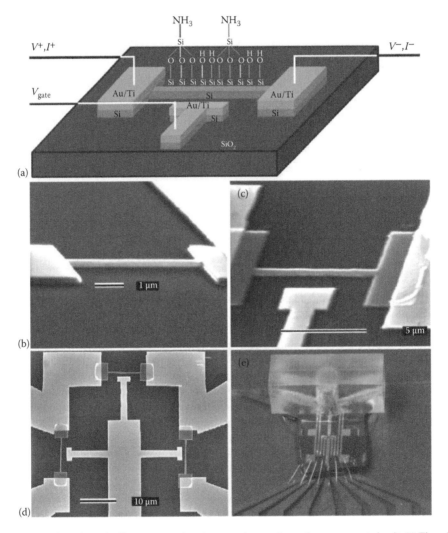

FIGURE 43.3 Device schematic diagram, scanning electron micrographs, and measurement circuit. (a) The schematic diagram of the silicon nanowire with side gates and electrodes. The nanowire is exposed on three sides along the longitudinal directions. (b) The nanowire shown here is 300 nm wide, 230 nm thick, and 8 μm long. (c) A silicon nanowire with a side gate. (d) The scanning electron micrograph displays three silicon nanowire devices on the same chip. (e) An optical micrograph shows the flow chamber sealed on top of the devices on the interface board.

The measurement circuit includes a small AC (alternating current) modulation (provided by an EG&G 5210 lock-in amplifier), superimposed on the DC bias across the nanowire (provided by a Keithley 2400 source meter) (Figure 43.3d). The AC modulation voltage and the DC bias voltage are added by a noninverting summing circuit integrated with the preamplifier circuit. Differential conductance measurements are taken by sweeping the DC bias at constant AC modulation amplitude and measuring the response with the lock-in amplifier, which is referenced to the AC signal frequency.

The measurement of interest is G, the change in the differential conductance due either to a change in the reference gate voltage ΔV_g, or to a change in molecular concentration Δm.

PDMS gel is used to seal the device and surround the nanowire, which is bathed in a fluid volume of 20–30 μL, connected to a syringe pump. Unlike in microfluidic cells, where laminar flow prevents complete mixing, a larger cell shortens the response time of diffusion-controlled processes.

43.4 Results

43.4.1 Breast Cancer Biomarkers and Clinical Need

Breast cancer is the most common life-threatening malignancy in women of most developed countries today, with approximately 180,000 new cases diagnosed every year. Roughly half of these newly diagnosed patients are node-negative. However, 30% of these cases progress to metastatic disease. Breast cancer biomarkers have emerged as an important determinant of diagnosis, staging, progression, and therapy decision making.

The CA15.3 tumor marker is produced from the mucin 1 (*MUC1*) gene product and it is an epithelial marker. CA15.3 is highly overexpressed in breast, ovarian, and other cancers compared with the normal epithelium from these tissues. The protein also has the property of being cleaved from the cell surface as a soluble protein fragment, and it is therefore in the bloodstream. Many studies have demonstrated that breast cancer patients have elevated CA15.3 levels.

43.4.2 Protein Sensing and Signal Amplification—General Features

A change in conductance is primarily due to the contribution of surface states to the conductance, which for larger sensors is dominated by volume effects. The fractional change is greatest for the smallest sensors, due to the increased surface-to-volume ratio.

Small changes in the conductance of the device (related to the inverse of the source–drain resistance) are best measured by considering the differential conductance with the derivative taken at constant V_t.

This method yields measurements at higher signal-to-noise ratio compared to digital method of taking derivatives. The differential conductance G depends on top gate voltage V_t or analyte concentration m in solution as well as bias voltage V_{ds}. The quantity of interest is ΔG, the change in conductance either due to a change in the top gate voltage V_t or due to change in concentration Δm. A higher signal ΔG can be obtained in the region of negative V_{ds} or positive V_t for our nanosensor.

To demonstrate that the FET nanosensors are a robust platform for sensitive protein concentration determinations, we showed that there was a corresponding change in the differential conductance of the device that was dependent on the protein concentration of the analyte. As a demonstration of this principle, we used a monoclonal antibody association with its hapten, as a model for a molecular binding interaction. Typically, antibodies will bind selected epitopes with high specificities and affinity. The antibiotin antibody (antibiotin) is known to specifically recognize biotin in either free or conjugated forms by many independent studies (Sigma-Aldrich, B3640).

A nanosensor device composed of 20 silicon wires in parallel (6 μm long, 250 nm wide, 100 nm thick) was formed. The silicon nanowires were covered by 5 nm of an Al_2O_3 insulating layer as specified earlier, and then chemically modified by biotinamidocaproyl-labeled bovine serum albumin (BSA), providing a biotin-conjugate surface. Protein binding experiments were conducted in 1 mM phosphate containing 1 mM NaCl, pH = 7.4.

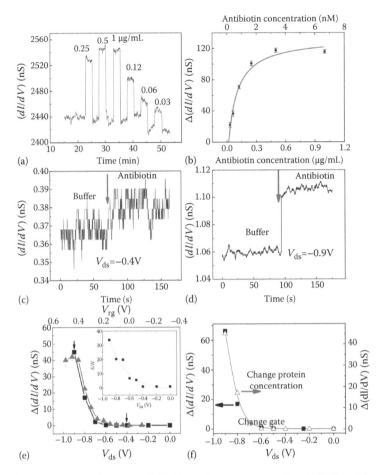

FIGURE 43.4 The FET nanosensor detects protein binding to the nanosensor surface. (a) Differential conductance versus concentration of antibiotin in phosphate buffer for a device composed of 20 wires of 250 nm width, $V_{ds} = -0.5$ V, $V_{rg} = 0$ V. (b) Change of conductance (with random error) versus antibiotin concentration. The solid line is fit by a two-state model. (c) and (d) Device response to antibiotin at different bias $V_{ds} = -0.4$ V and -0.9 V. (e) Device sensitivity change with bias and gate voltage; inset, signal-to-noise ratio. (f) Comparison of conductance change introduced by 5 mV of reference gate voltage change (black dots, left axis) and 100 ng/mL of antibiotin solution (red triangles, right axis).

To determine the protein binding contribution to conductance changes, solutions bearing different concentrations of the antibiotin are introduced via a syringe pump into the FET nanosensor chamber. Less than 2 min is required to register a stable reading of the conductance (Figure 43.4c). Changes in the differential conductance G are observed consistent with different antibiotin concentrations as these solutions are pumped to the nanosensor surface (Figure 43.4a). The nanosensor accurately monitors antibiotin concentrations despite changes to order of additions of the antibiotin concentrations (Figure 43.4a; data not shown). Further, these data are delineated in plots of differential conductance as a function of protein concentration, showing that a typically shaped binding curve can be formed (Figure 43.4b). At higher concentrations of 0.5 μg/mL, the device response saturates, presumably corresponding to the coating of the entire silicon nanowire surface. At low concentrations, the device response is rapid and precise.

Protein binding and concentration-dependent measurements with the FET nanowire device are demonstrated by the antibiotin experiments diagrammed in Figure 43.4a and b. Silicon nanochannels, functionalized with biotinylated BSA, are used to detect antibiotin in 1 mM NaCl and 1 mM phosphate buffer solution. At the concentration of salt used in solution, the Debye screening length λ_D at room temperature is about 9.6 nm. λ_D is sufficiently large that the surface potential is sensitive to protein binding, but short enough to screen out biomolecules in solution (Stern et al., 2007b). All solutions used in the measurements are dialyzed in order to keep constant ionic strength. The mechanism of ligand receptor binding can be simply explained by a two-state model. Assuming the dissociation constant of biotin–antibiotin binding is K_{eq}, then,

$$\text{Antibiotin} + \text{Biotin} \xrightarrow{\;K_{eq}\;} \text{Antibiotin} - \text{biotin complex} \tag{43.1}$$

The fractional occupancy of the binding sites is related to the biotin solution concentration m by a Langmuir adsorption isotherm. Assume the total bonding sites on the silicon wires are N and there are a_p protein molecules attached on the surface:

$$a_p = \frac{N}{K_{eq}/m + 1} \tag{43.2}$$

By assuming the conductance change of the device is proportional to the charge concentration of the silicon nanowire surface, the data can be fit to the aforementioned two-state model, and a dissociation constant of 5.2×10^{-10} M is calculated. This value compares favorably with an experimental K_d value of 10^{-9} M reported in previous studies (Cui et al., 2001b). Furthermore, the protein detection limit is 3 ng/mL (around 20 pM), when the signal-to-noise ratio is set to 1.

We also demonstrated that by adjusting the source–drain bias voltage to the subthreshold region, the device sensitivity can be improved. The differential conductance change Δg contributed by 100 ng/mL antibiotin is 0.02 ± 0.01 nS at $V_{ds} = -0.4$ V (Figure 43.4c), while Δg is 45 ± 0.1 nS at $V_{ds} = -0.9$ V (Figure 43.4d). The signal-to-noise ratio increases from 2 to 34 (see Figure 43.4e, inset). The signal due to 100 ng/mL antibiotin injection is plotted as a function of V_{ds} and V_{rg} (Figure 43.4e).

This plot clearly shows the effect of reverse bias amplification. The change Δg, due to concentration change at fixed reference gate voltage, can be compared to the change caused by varying the reference gate voltage while keeping the concentration fixed. Equivalence between the surface potential change and concentration change is then established. Δg introduced by 100 ng/mL (660 pM) of the antibiotin monoclonal antibody is equivalent to a gate voltage change of 7.2 ± 0.3 mV (see Figure 43.4f).

These experiments illustrate the FET nanosensor capability to sensitively measure analytes in solution. Many applications are possible for this technology, because it combines high sensitivity with speed, reproducibility, and miniaturization.

43.4.3 Breast Cancer Tumor Antigen Determinations with FET Nanosensors

Although image methods (mammograms) are routinely used in breast cancer treatment decisions, clinicians are seeking reliable early indicators from simpler and more cost-efficient procedures, such as blood tests. There are two major utilities for measuring the CA15.3 tumor antigens in the circulation of breast cancer patients. First, tests that yield the greatest level of sensitivity are of most importance for CA15.3 and early detection of the cancer. Current methods operate in at a baseline detection cutoff of >38 U/mL. Therefore, tests that increase the sensitivity to lower CA15.3 levels are of great value. Second, additional measurements of CA15.3 are valuable in determining progression status for breast cancer patients on therapy. Here the CA15.3 biomarker is monitored as a prognostic marker as an indicator of late-stage disease and/or metastasis. If CA15.3 levels rise, it is an indicator of progression, and the clinical team may redirect the treatment plan. CA15.3 in the 500–2500 U/mL range is monitored.

The specification of the analyte being detected on the nanosensor is determined by the tumor antigens that can be recognized. In the examples shown here, we selected tumor antigens where there are detection methods and reagents currently available as a proof-of-concept test.

Silicon nanowires were functionalized (Chen et al., 2010) with the antibody CA15.3 specific to the soluble portion of the protein that is shed from the tumor and enters the bloodstream. As discussed earlier, the conjugated selected antibody on the silicon nanowire FET device surface contributes to the desired specificity of the measurement. The silicon nanowire FET devices undergo a change in its electrical conductance (or differential conductance G) depending on the binding of CA15.3 to the respective antibody

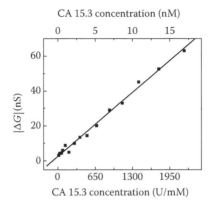

FIGURE 43.5   Measurements of CA15.3 binding events on FET biosensors functionalized with CA15.3 antibody.

(Chen et al., 2010). The ability of a nanoscale CA15.3 FET device to detect CA15.3 at the clinical relevant levels is shown in the following experiments.

The CA15.3 cancer antigen used for detection is dialyzed in solution with $150\,\mu M$ NaCl and $10\,\mu M$ phosphate buffer. Upon binding to antigen, the conductance of the nanosensor probe ("nanowire") changes (differential conductance G) in a CA15.3 concentration–dependent manner (Figure 43.5). A measurement with an independently calibrated CA15.3 is also demonstrated. A real-time signal fluctuation is indicated when the syringe pump changes from buffer to $36\,U/mL$ CA15.3 in buffer. Furthermore, a plot of differential conductance versus concentration is a straight line at over three logarithms of concentrations range as shown.

43.4.3.1 Detection Sensitivity

The detection limit of the conductance change is $1.68\,nS$, setting a limit in the detection of determined of $7.7\,U/mL$ ($\sim 60\,pM$). The relationship of the differential conductance G and CA15.3 concentration m is shown in Figure 43.5. Over the range of clinical interest, the experimental data can be fit into a straight line from which we can derive a linear relationship between G and m:

$$G\left(V_{ds},V_{rg},m\right) = G_1 + G_2 m \tag{43.3}$$

Here, G_1 and G_2 are device-dependent parameters and $G_1 = 307 \pm 2\,nS$, $G_2 = -(28 \pm 1) \times 10^{-3}\,nS/(U/mL)$ from Figure 43.5. (At higher concentrations, the response is nonlinear and can be fit into an adsorption isotherm described earlier.) With fixed CA15.3 concentration, differential conductance measurement can be done with varying reference gate voltage as shown in Figure 43.5:

$$G\left(V_{ds},V_{rg},m\right) = G_3 + G_4 V_{rg} \tag{43.4}$$

where G_3 and G_4 are device-dependent parameters and $G_3 = 304 \pm 2\,nS$, $G_4 = 2.3 \pm 0.1\,nS/mV$. For a given device, the parameter values serve to characterize the charge and equilibrium binding of a protein. An equivalence between the reference gate voltage V_{rg} and the free CA 15.3 biomarker concentration m is established from Equations 43.3 and 43.4.

Additionally, the device response time is $<60\,s$, indicating that the measurement is essentially instantaneous. This important feature of the FET nanosensor configuration may be primarily limited by the passive diffusion of the analyte to the nanosensor surface.

43.4.3.2 Specificity of the Nanosensor to CA15.3

In addition, measurement of the unrelated monoclonal antibody, antibiotin, is used as a specificity control. Antibiotin, at differing protein concentrations below 20 nM, shows no differential conductance

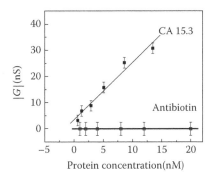

FIGURE 43.6 Selectivity of nanosensors based on antibody recognition. The differential conductance *G* is determined from a CA15.3 nanosensor exposed to CA 15.3 or a protein control, antibiotin, at differing protein concentrations below 20 nM shows no differential conductance *G* different from baseline.

G different from baseline. The data indicate that the conductance change observed is primarily due to specific binding of the biomarker to its cognate antibody (Figure 43.6).

In summary, the nanosensor response was determined to operate as a function of CA15.3 concentration, demonstrated here over a span of 0–2300 U/mL, underscoring the wide dynamic range of this assay. Importantly, fluctuations in this dynamic range may be useful to a variety of clinical situations, for both early-stage diagnosis and late-stage prognosis, without reconfiguration.

When the control solution includes similar concentration of antibiotin, there is no detectable conductance change, which proves that the conductance change is due to selective binding between CA15.3 antibody and CA15.3 antigen. With a biotin concentration that is 30 times higher than the selective target (CA15.3 antigen), a specific conductance change can still be observed. The CA15.3 nanosensor responds as a function of CA15.3 concentration, demonstrated here over a span of 17–2200 U/mL, and underscores the broad dynamic range that is achievable, which has utility for both early-stage diagnosis (low concentration) and late-stage prognosis and monitoring to treatment (high concentration) without reconfiguration of the device and/or spectrum of nanosensors on a single chip.

The fundamental advantage of the label-free nanosensor device architectures is the combination of high detection sensitivity and a standard semiconductor-based CMOS-compatible fabrication technique, suitable for scalable manufacturing. The unique FET configuration allows device-level signal amplification and the required degree of control to enable large-scale parallel architecture for detection of multiple target molecules. This combination is important for the screening, diagnosis, and prognosis of diseases in a clinical setup, requiring fast analysis time, small sample volume, and low cost. Further development of our array-based technique with high-sensitivity protein assay is likely to be generally important for genomics and proteomics.

43.5 Discussion

The approach described here combines recent advances in the areas of nanoelectronics and biosensing. Eventual application of this work will be to enable highly parallel assay of protein biomarkers and signaling peptide molecules. In principle, the techniques can be used to detect any type of protein biomarkers, such as may occur in an air or liquid environment. Compared to the current assay methodologies, the proposed techniques facilitate a highly parallel approach for biomolecular recognition in small sample volumes. The key technology advantages—ultrahigh sensitivity, sensor flexibility, low sample volume, and highly scalable manufacturability—will facilitate point-of-care testing, and significantly reduce costs per test.

Recent advances in semiconductor technology, materials chemistry of surface conjugation, cancer diagnosis of tumor markers, and semiconductor device engineering are converging to produce greater diagnostic opportunities. In the semiconductor industry, the CMOS-compatible scalable architecture has enabled the creation and use of integrated chips with multimillion processing elements. Our nanoscale FET elements leverage the use of CMOS-compatible chip architecture (Stern et al., 2007a) for ultrasensitive high-speed assay of multiple analytes. Although, for demonstration purposes, we have focused on the CA15.3 breast cancer biomarker, the device can be used for any tumor biomarker by the suitable adaptation of other specific antibodies on the device surface.

Others have also demonstrated cancer antigen detection at femtomolar sensitivities with devices fabricated by a "bottom-up" method (Cui et al., 2001a), using vapor–liquid–solid nanowire growth strategy. In this manufacturing method, silicon wires with feature widths between 5 and 30 nm are formed by modulating the nanocluster catalyst size on prepatterned wafers containing electrodes. Just as with our nanosensors, these other approaches offer real-time, label-free, and highly sensitive detection of a wide range of analytes, including proteins (Wang et al., 2005), nucleic acids (Hahm and Lieber, 2004), small molecules (Chen et al., 2008; Wang et al., 2005, 2008), and viruses (Patolsky et al., 2004) in single-element or multiplex formats (Zheng et al., 2005). However, the scalability of these methods is significantly impeded by the tedious chemically grown processes and assembly steps required with ultrasmall diameters of the silicon wires. For industry applications, top-down manufacturing strategies, such as ours, are much preferred.

There are several directions for improvements of CA15.3 testing that incorporate nanosensor strategies. First, it will be important to continue to adjust the assay sensitivity. The nanosensor platform depends on our ability to detect ultrasmall currents in the nanowire in the sub-nanoampere range. The corresponding antigen detection sensitivity is estimated to be on the order of 1 U/µL, which is a 1000-fold increase on the detection sensitivity of standard techniques such as chemiluminescent immunoassay. It should be noted that the current sensitivity is adequate for baseline determinations of the patients by comparison with the available clinical diagnostic tests. Not only does the small nanoscale size of the device contribute to the high sensitivity (as reflected in the corresponding change in the nanowire conductance), but it also provides a small sensing area or binding site for the antigens. Since these antigens are large protein molecules (300–450 kDa), the change in conductance due to the binding of a single protein to the functionalized nanowire surface could be measured, providing the ultimate detection sensitivity. Refinements that investigate the transistor amplification of the detector sensitivity as a function of gate voltage promise to be an important development area. In addition, continuing a refinement of the surface dimensions toward ultrahigh sensitivity may also be of value.

Second, we expect that these nanosensors will be active over much broader dynamic ranges than was illustrated in this report. Since the FET response is a function of CA15.3 concentration, demonstrated here over a span of 0–2300 U/mL, it underscores the wide dynamic range in the necessary clinically significant levels. Modifications that extend the dynamic range are clearly of interest for other settings are actively being investigated. In cancer diagnostic applications, it may be favorable to utilize common nanosensor devices for CA15.3 and other tumor antigens, without device reconfiguration, but providing differential sensitivity ranges for different settings. These attributes may be preferred in the clinical context of providing early-stage diagnosis or late-stage prognosis and monitoring from a common platform.

Third, our nanosensor device response time is <60 s with a CA15.3 detection limit of ~1–2 U/mL (~10 pM). Experiments are underway to determine the factors driving the response time for antibody–antigen interactions, but a likely parameter is the diffusion process. Adjustments to the response times and sample processing functions are expected to be of assistance toward a point-of-care platform.

43.5.1 Multiplexing with Other Breast Cancer Markers

Multiplexed antibody arrays may be formed utilizing the same strategy, and also by using commercially available reagents. Additional breast cancer biomarkers may be valuable to examine in tandem with CA15.3. CA27.29 is also a mucin-associated antigen, which is detected by the monoclonal antibody B27.29, specific for the protein core of the MUC1 gene product. CA27.29 has been evaluated in comparison to other tumor markers including CA15.3 (Gion et al., 1999). Specifically, in a detailed comparison of the diagnostic accuracy of CA27.29 and CA15.3, Gion et al. (1999) show that CA27.29 discriminates primary breast cancer better than CA15.3, especially in patients with limited disease. In healthy subjects, CA15.3 was found to be significantly higher than CA27.29. In breast cancer patients, CA27.29 was found to be considerably higher than CA15.3, suggesting that CA27.29 is a better marker. The CA27.29 antigen is detected by monoclonal antibody B27.29, specific for the protein core of the MUC1 product. This antibody is already in use in fully automated versions of the CA27.29 assay such as the ACS:180 BR assay (Chiron Diagnostics) based on a conventional chemiluminescent immunoassay. In this technique, a mouse monoclonal antibody is incubated with both the patient sample and purified CA27.29 coupled covalently to paramagnetic particles in a solid phase. Both the solid-phase CA27.29 and the antigen in the sample compete for binding to the antibody. The resulting chemiluminescent signal is therefore found to be inversely proportional to the amount of antigen in the sample.

The American Society of Clinical Oncology (ASCO) recommended the use of a set of tumor markers for breast cancer in the 2007 clinical practice guideline (Harris et al., 2007). In particular, CA15.3 and CA27.29 were evaluated for their ability to determine diagnosis and prognosis, monitor therapy, and predict recurrence of breast cancer after curative surgery and radiotherapy. CA27.29 is elevated in 29%, 36%, and 59% stages I, II and III cancers, respectively. In comparison, the incidence of elevation for CA15.3 in the same samples was 15%, 23%, and 54.5% for these patients, respectively (Gion et al., 1999). Although a greater sensitivity toward earlier detection of disease progression may be evident from CA27.29 assays, it is clear that examining both markers in tandem may be beneficial. Thus, a technique capable of simultaneous detection of multiple biomarkers with high sensitivity in small volumes of samples is desirable.

Additional serum breast cancer markers, including CA125, CA19.9, and CEA, may also be productively examined. Further, circulating cell-bound markers may also be considered for eventual evaluation with nanosensors, provided that these assays achieve acceptable sensitivities and specificities. Investigations are underway with other cancer markers, such as HER2, EGFR, and other general epithelial cancer markers, in this regard. Advances in multiplexing with FET nanosensors are expected to allow device-level signal amplification and the required controllability to enable large-scale parallel processing architecture for detection of multiple target molecules.

The fundamental advantage of our label-free device architecture is the combination of high detection sensitivity and a standard semiconductor-based CMOS-compatible fabrication technique, suitable for scalable manufacturing. The combination of these nanosensor technology advances is important for the development of high-quality screening, diagnosis, and prognosis of diseases in the clinic, requiring fast analysis time, small sample volumes, and low cost. Further development of our array-based technique with high-sensitivity protein assay may be broadly important to genomics and proteomics.

43.6 Future Trends

It is attractive to consider these nanosensor platforms in light of future trends toward personalized medicine, because these technologies offer advantages of high sensitivity and speed at low cost.

In this chapter, we have discussed how nanochannel FET devices can be fabricated using CMOS-compatible lithography processes, familiar to the semiconductor industry. It is possible with nanochannel FET devices to build multiple sensor arrays integrated with current semiconductor technology. By addressing refinements to the design circuitry and the etching process, and multiplexing the functionalization steps, these devices open promising avenues to detecting many analytes relevant to cancer diagnostics with high sensitivity.

43.6.1 Multiplexing

It is well recognized in cancer diagnostics and therapy selection that many tumor markers are present that may contribute to a better definition of the disease state. Yet, clinical access to these markers will often be limited by costs, as physicians do not want to order large numbers of individual tests. Thus, an avenue of great utility will be the ability to create a platform that provides highly multiplexed information, such as with the FET biosensor arrays depicted here. Implementing these FET biosensors with an appropriate circuit design provides the combined benefit of coordinating individual signal measurements as part of the management of highly multiplexed reads and high information content in tandem. Many innovations in the recording and interpretation of this multiplexed information for cancer diagnostics will follow.

The tumor antigen CA15.3 is typically reported as a single analyte test in clinical diagnostics practice. According to product labeling for different assays, these CA15.3 tests are indicated for the serial measurement of CA15.3-reactive antigenic determinants as an aid in the monitoring of patients previously diagnosed with breast cancer. CA15.3 measurements are approved for disease progression or response to therapy in conjunction with other clinical methods, such as imaging (Harris et al., 2007). The CA15.3 assay can also be used as an aid in the detection of recurrence in previously treated stages II and III breast cancer patients. Improvements to the CA15.3 measurements are likely to include the incorporation of additional serum markers in multimarker algorithms. Further, changes to CA15.3 in response to chemotherapy treatment may provide patient benefit (Kim et al., 2009). Tests with high sensitivity and specificity are required to make clinical decisions derived from biomarker levels during kinetic responses to chemotherapy.

In addition to utilities in breast cancer diagnostics, serum marker testing may also be beneficial in other cancer monitoring, such as for ovarian cancer. It has been demonstrated that multimarker algorithms of four or more serum markers (panel of CA125, HE4, CEA, and VCAM-1 in this example) outperform single prognostic markers (Yurkovetsky et al., 2010). Also, it is evident that patient baseline level variations of many serum markers are an important constraint, and utilizing multiple independent markers in tandem is thought to aid a more robustly applicable test outcome.

43.6.2 Miniaturization

A striking benefit of these nanosensor technologies is the ability to implement them with dramatically reduced footprints. It is estimated that thousands of separate tests may be run in single chip devices with external dimensions of the chip of $1\,cm^2$.

43.6.3 Point of Care

Device portability and reliability are the cornerstones of developing point-of-care diagnostics. Today, many clinical tests in the cancer diagnostics arena operate from highly sophisticated instrumentation and optics, and necessarily require highly skilled operators. Ostensibly, the use of nanosensors allows for the transition to a user interface of great simplicity. In operation, the devices of the future will have complex algorithms that read and process the individual biomarker outputs. However, these algorithms will function in the background of the device, making it possible to transform the diagnostic to a user-friendly interface that is clinically actionable.

REFERENCES

Balmain, A., Gray, J., and Ponder, B. (2003). The genetics and genomics of cancer. *Nat Genet* 33 (Suppl), 238–244.

Bergveld, P. (1970). Development of an ion-sensitive solid-state device for neurophysiological measurements. *IEEE Trans Biomed Eng* 17, 70–71.

Bild, A. H., Yao, G., Chang, J. T., Wang, Q., Potti, A., Chasse, D., Joshi, M.-B., Harpole, D., Lancaster, J. M., Berchuck, A., Olson, J. A., Jr., Marks, J. R., Dressman, H. K., West, M., and Nevins, J. R. (2006). Oncogenic pathway signatures in human cancers as a guide to targeted therapies. *Nature* 439, 353–357.

Branca, M. (2003). Genetics and medicine. Putting gene arrays to the test. *Science* 300, 238.

Byfield, M. P. and Abuknesha, R. A. (1994). Biochemical aspects of biosensors. *Biosens Bioelectron* 9, 373–400.

Byun, K. M., Kim, S. J., and Kim, D. (2005). Design study of highly sensitive nanowire-enhanced surface plasmon resonance biosensors using rigorous coupled wave analysis. *Opt Express* 13, 3737–3742.

Byun, K. M., Yoon, S. J., Kim, D., and Kim, S. J. (2007). Sensitivity analysis of a nanowire-based surface plasmon resonance biosensor in the presence of surface roughness. *J Opt Soc Am A Opt Image Sci Vis* 24, 522–529.

Celler, G. K. and Cristoloveanu, S. (2003). Frontiers of silicon-on-insulator. *J Appl Phys* 93, 4955–4978.

Chang, Z., Zhu, N. N., Zhao, K., Fan, H., He, P.-G., and Fang, Y. Z. (2007). Study of polyaniline nanowire modified electrode for electrochemical DNA sensor. *Acta Chim Sin* 65, 135–139.

Chen, Y., Wang, X. H., Erramilli, S., Mohanty, P., and Kalinowski, A. (2006). Silicon-based nanoelectronic field-effect pH sensor with local gate control. *Appl Phys Lett* 89, 223512.

Chen, Y., Wang, X. H., Hong, M., Erramilli, S., and Mohanty, P. (2008). Surface-modified silicon nano-channel for urea sensing. *Sens Actuators B Chem* 133, 593–598.

Chen, Y., Wang, X., Hong, M. K., Rosenberg, C. L., Reinhard, B. M., Erramilli, S., and Mohanty, P. (2010). Nanoelectronic detection of breast cancer biomarker. *Appl Phys Lett* 97, 233702–233704.

Chung, W. Y., Yang, C. H., Pijanowska, D. G., Grabiec, P. B., and Torbicz, W. (2006). ISFET performance enhancement by using the improved circuit techniques. *Sens Actuators B Chem* 113, 555–562.

Cui, Y., Lauhon, L. J., Gudiksen, M. S., Wang, J. F., and Lieber, C. M. (2001a). Diameter-controlled synthesis of single-crystal silicon nanowires. *Appl Phys Lett* 78, 2214–2216.

Cui, Y., Wei, Q., Park, H., and Lieber, C. M. (2001b). Nanowire nanosensors for highly sensitive and selective detection of biological and chemical species. *Science* 293, 1289–1292.

Elfstrom, N., Juhasz, R., Sychugov, I., Engfeldt, T., Karlstrom, A. E., and Linnros, J. (2007). Surface charge sensitivity of silicon nanowires: Size dependence. *Nano Lett* 7, 2608–2612.

Fritz, J., Cooper, E. B., Gaudet, S., Sorger, P. K., and Manalis, S. R. (2002). Electronic detection of DNA by its intrinsic molecular charge. *Proc Natl Acad Sci U S A* 99, 14142–14146.

Garber, K. (2004). Genomic medicine. Gene expression tests foretell breast cancer's future. *Science* 303, 1754–1755.

Gion, M., Mione, R., Leon, A. E., and Dittadi, R. (1999). Comparison of the diagnostic accuracy of CA27.29 and CA15.3 in primary breast cancer. *Clin Chem* 45, 630–637.

Hahm, J. and Lieber, C. M. (2004). Direct ultrasensitive electrical detection of DNA and DNA sequence variations using nanowire nanosensors. *Nano Lett* 4, 51–54.

Hammond, P. A. and Cumming, D. R. S. (2005). Performance and system-on-chip integration of an unmodified CMOS ISFET. *Sens Actuators B Chem* 111, 254–258.

Harris, L., Fritsche, H., Mennel, R., Norton, L., Ravdin, P., Taube, S., Somerfield, M. R., Hayes, D. F., and Bast, R. C., Jr. (2007). American Society of Clinical Oncology 2007 update of recommendations for the use of tumor markers in breast cancer. *J Clin Oncol* 25, 5287–5312.

Kim, D. (2006). Effect of resonant localized plasmon coupling on the sensitivity enhancement of nanowire-based surface plasmon resonance biosensors. *J Opt Soc Am A Opt Image Sci Vis* 23, 2307–2314.

Kim, J. S., Park, W. I., Lee, C.-H., and Yi, G.-C. (2006). ZnO nanorod biosensor for highly sensitive detection of specific protein binding. *J Kor Phys Soc* 49, 1635–1639.

Kim, H. S., Park, Y. H., Park, M. J., Chang, M. H., Jun, H. J., Kim, K. H., Ahn, J. S., Kang, W. K., Park, K., and Im, Y. H. (2009). Clinical significance of a serum CA15-3 surge and the usefulness of CA15-3 kinetics in monitoring chemotherapy response in patients with metastatic breast cancer. *Breast Cancer Res Treat* 118, 89–97.

Minn, A. J., Gupta, G. P., Siegel, P. M., Bos, P. D., Shu, W., Giri, D. D., Viale, A., Olshen, A. B., Gerald, W. L., and Massagué, J. (2005). Genes that mediate breast cancer metastasis to lung. *Nature* 436, 518–524.

Patolsky, F., Zheng, G., Hayden, O., Lakadamyali, M., Zhuang, X., and Lieber, C. M. (2004). Electrical detection of single viruses. *Proc Natl Acad Sci U S A* 101, 14017–14022.

Ring, H. Z. and Kroetz, D. L. (2002). Candidate gene approach for pharmacogenetic studies. *Pharmacogenomics* 3, 47–56.

Souteyrand, E., Cloarec, J. P., Martin, J. R., Wilson, C., Lawrence, I., Mikkelsen, S., and Lawrence, M. F. (1997). Direct detection of the hybridization of synthetic homo-oligomer DNA sequences by field effect. *J Phys Chem B* 101, 2980–2985.

Stern, E., Klemic, J. F., Routenberg, D. A., Wyrembak, P. N., Turner-Evans, D. B., Hamilton, A. D., LaVan, D. A., Fahmy, T. M., and Reed, M. A. (2007a). Label-free immunodetection with CMOS-compatible semiconducting nanowires. *Nature* 445, 519–522.

Stern, E., Wagner, R., Sigworth, F. J., Breaker, R., Fahmy, T. M., and Reed, M. A. (2007b). Importance of the Debye screening length on nanowire field effect transistor sensors. *Nano Lett* 7, 3405–3409.

Van den Eynden, G. G., Van der Auwera, I., Van Laere, S., Colpaert, C. G., van Dam, P., Merajver, S., Kleer, C. G., Harris, A. L., Van Marck, E. A., Dirix, L. Y., and Vermeulen, P. B. (2004). Validation of a tissue microarray to study differential protein expression in inflammatory and non-inflammatory breast cancer. *Breast Cancer Res Treat* 85, 13–22.

Wang, X. H., Chen, Y., Gibney, K. A., Erramilli, S., and Mohanty, P. (2008). Silicon-based nanochannel glucose sensor. *Appl Phys Lett* 92, 013903-1–013903-3.

Wang, W. U., Chen, C., Lin, K. H., Fang, Y., and Lieber, C. M. (2005). Label-free detection of small-molecule-protein interactions by using nanowire nanosensors. *Proc Natl Acad Sci U S A* 102, 3208–3212.

Weigelt, B., Glas, A. M., Wessels, L. F., Witteveen, A. T., Peterse, J. L., and van't Veer, L. J. (2003). Gene expression profiles of primary breast tumors maintained in distant metastases. *Proc Natl Acad Sci U S A* 100, 15901–15905.

Yurkovetsky, Z., Skates, S., Lomakin, A., Nolen, B., Pulsipher, T., Modugno, F., Marks, J., Godwin, A., Gorelik, E., Jacobs, I., Menon, U., Lu, K., Badgwell, D., Bast, R. C., Jr., and Lokshin, A. E. (2010). Development of a multimarker assay for early detection of ovarian cancer. *J Clin Oncol* 28, 2159–2166.

Zheng, G., Patolsky, F., Cui, Y., Wang, W. U., and Lieber, C. M. (2005). Multiplexed electrical detection of cancer markers with nanowire sensor arrays. *Nat Biotechnol* 23, 1294–1301.

Zhu, N. N., Chang, Z., He, P.-G., and Fang, Y. Z. (2006). Electrochemically fabricated polyaniline nanowire-modified electrode for voltammetric detection of DNA hybridization. *Electrochim Acta* 51, 3758–3762.

Ziegler, C. and Gopel, W. (1998). Biosensor development. *Curr Opin Chem Biol* 2, 585–591.

44

Measuring the Electric Field in Skin to Detect Malignant Lesions

Richard Nuccitelli, KaYing Lui, Kevin Tran, Brian Athos, Mark Kreis, and Pamela Nuccitelli

CONTENTS

44.1 Introduction

Skin is the largest body organ, and it has an amazing regenerative capacity. It is our first line of defense against environmental hazards that we are exposed to everyday such as ultraviolet radiation and pathogens. The long-term exposure to these hazards has been shown to often give rise to malignant lesions such as melanoma and basal cell carcinoma most commonly at 60–70 years of age in humans. In fact, basal cell carcinoma is the most common form of cancer in the American population with 1.8 million cases in the United States alone every year. Despite the prevalence of this disease, and the easy access to the skin, the clinician has no medical device capable of detecting this form of cancer on a routine basis.

We have developed a new instrument called the Dermacorder® that can measure the electric field generated by skin tumors noninvasively with a 1 min scan of the suspicious lesion. The Dermacorder's sensor vibrates at around 850 Hz and hovers above the skin. It does not have to penetrate the stratum corneum to determine the surface potential of the epidermis below it. Instead, it forms a capacitor with the epidermal layer and the vibration results in an oscillating current on the sensor. This current is converted to voltage to generate an oscillating signal. Since we can control the voltage on the sensor, V_b, we can determine the value of V_b for which the oscillating output voltage goes to 0. That will only happen when V_b is equal to the unknown surface potential, V_s. More details of this method are provided in Section 44.3. First, it will be helpful to describe the electrical properties of mammalian skin.

While the outermost layer of skin, the stratum corneum, is an inert, nonconductive dielectric, the epidermis below it is composed of many polarized epithelial layers that generate a voltage difference or transepithelial potential of 20–50 mV, inside positive, across the epidermis. This voltage difference will drive ionic current out of any low-resistance pathways in the epidermis such as those formed by wounds or skin lesions that cause disruptions in the tight junctions that are responsible for the high transepithelial resistance. Dubois–Reymond first measured these ionic currents that exit skin wounds 167 years ago using a galvanometer [1]. This was confirmed by Illingworth and Barker [2] using the more modern vibrating probe technique [3]. They measured up to $10 \mu A/cm^2$ exiting accidentally amputated fingertips in children during a 2 week period following the injury. Such a wound current must generate an electric field in the skin bordering the wound. This field has been measured in the skin of guinea pigs [4,5], newts [6], and most recently in humans [7].

44.1.1 How Is This Lesion Field Generated?

The ultimate driving force for all lesion currents is the voltage across the epidermis. The epidermis of the skin normally generates a voltage across itself, termed the transepithelial potential, by pumping positive ions from its apical to its basal side. This is accomplished by segregating Na^+ channels to the apical end and K^+ channels to the basal end of the epithelial cells, while utilizing an Na^+/K^+-ATPase to lower intracellular $[Na^+]$ and raise intracellular $[K^+]$ (Figure 44.1A). This low intracellular $[Na^+]$ (combined with the negative membrane potential) results in Na^+ movement into the cell on the apical end where the channels are localized, and the high intracellular $[K^+]$ results in K^+ efflux on the basal side where the K^+ channels are localized. This transepithelial ion flux creates a transepidermal potential of between 20 and 50 mV, inside positive, in mammalian skin [4,8], and has been termed the "skin battery" [6] (Figure 44.1B). At the site of a lesion that disrupts the tight junctions between epidermal cells, a low-resistance pathway will form and this transepidermal voltage will drive current out along this pathway (Figure 44.1C). Since this positive wound current flows toward the lesion on the basal side of the epidermis, and then away from the lesion on the apical side, a lateral electric field will be generated by this flow of current on both sides of the epidermis but will exhibit opposite directions of current flow on the two sides (Figure 44.1C). This model predicts that the field just beneath the stratum corneum will be oriented with the positive pole at the lesion and the negative pole away from the lesion. In contrast to this, the field deeper in the epidermal multilayer will have the opposite polarity, with the negative pole at the lesion site.

44.2 Materials

44.2.1 Animals

Hairless, albino, SKH-1 mice, and immunodeficient nude (Nu/Nu) mice between 1 and 6 months old were used in this study. All measurements were done under inhalation anesthesia using 1.4% isoflurane in O_2. Melanomas were generated by injecting 10^6 B16-F10 murine melanoma cells just beneath the skin on the back of the mouse. A similar number of SCC VII/SF squamous cells were used to generate subdermal squamous cell carcinomas. All animal procedures were approved by the institutional animal care and use committees either at Eastern Virginia Medical School, Norfolk, VA, or at In-Vivo Technologies, Burlingame, CA.

44.2.2 Cell Culture

Murine melanoma B16-F10 cells were obtained from ATCC (Manassas, VA). Murine squamous cell carcinoma cells (SCC VII/SF) were obtained from Dr. J. R. Grandis at the University of Pennsylvania. All cells were stored frozen in liquid nitrogen until needed. They were then thawed in a 37°C water bath and were transferred to a culture flask containing Dulbecco's modified Eagle's medium (DMEM) supplemented with 10% fetal bovine serum (FBS; Atlanta Biologicals), 4 mM L-glutamine (Cellgro), and 2% penicillin–streptomycin solution (Cellgro). The cells were grown in a 5% CO_2 incubator at 37°C

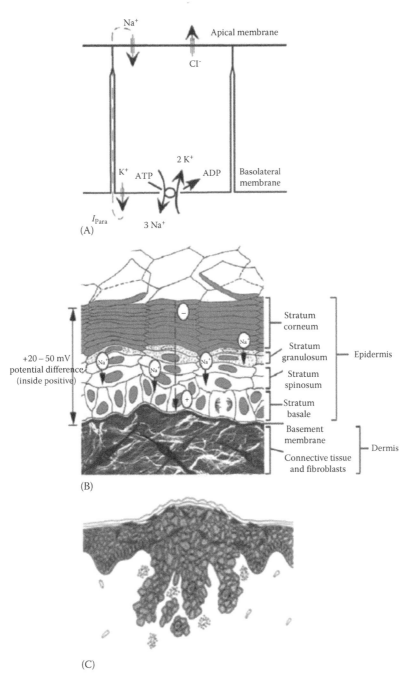

FIGURE 44.1 Generation of skin lesion electric fields. (A) Diagram of a typical epithelial cell in a monolayer with Na^+ channels localized on the apical plasma membrane and K^+ channels localized on the basolateral membranes along with the Na^+/K^+-ATPase. This asymmetric distribution of ion channels generates a transcellular flow of positive current that must flow back between the cells through the paracellular pathway (I_{para}). This current flow generates a transepithelial potential that is positive on the basolateral side of the monolayer. (B) Unbroken skin maintains this "skin battery" or transepithelial potential of 20–50 mV. (C) Diagram of a squamous cell carcinoma with the expected current flow out of the low resistance pathway. The developing squamous cell carcinoma disrupts tight junctions between the epidermal cells and this reduces the resistance to current flow in that region. The TEP could drive current through this newly formed low-resistance pathway in a current path shown. This could generate a lateral electric field whose negative vector points toward the lesion center at the lower portion of the epidermis and away from the wound on the upper portion just beneath the stratum corneum.

and passaged up to 17 times. Prior to injection, cells were pelleted by centrifugation and resuspended in HBSS at a concentration of 10^5 cells/μL.

44.2.3 Tumor Induction

Tumors were generated in female SKH-1 mice and Nu/Nu mice (Charles River, Wilmington, MA) by injecting $10\,\mu$L containing 10^6 B16-F10 murine melanoma cells or 10^6 SCC VII/SF squamous carcinoma cells under the skin using a $50\,\mu$L syringe (Hamilton, NV) with a 30-gauge hypodermic needle. This results in the appearance of visible tumor nodules within 2–3 days. The melanomas grow fairly rapidly from 2 to $3\,mm^3$ in volume within 4–7 days to about $6\,mm^3$ by 10 days after injection. We chose to use the hairless SKH-1 mouse and nude mouse strains because their skin is pigment-free, with very few hairs. The pigment-free skin facilitates transillumination photography to image the tumors and associated blood vessels. The reduced hair density facilitates measurement of the surface potential with the BFI and Dermacorder.

44.3 Methods

44.3.1 Dermacorder®

The fundamental principle of this technique, originally proposed by Lord Kelvin [9], is that the unknown surface potential of an object can be determined by forming a parallel-plate capacitor using a piece of metal with known potential, connecting the two surfaces and measuring the current flow between them. Zisman [10] introduced an oscillating approach in which one plate is vibrated up and down to vary the capacitance and Bluh and Scott [11] adapted it for measuring bioelectric potentials. Baikie et al. [12] refined this approach and used it to measure the electric field in a corn coleoptile.

In our application, the Dermacorder (Figure 44.2A through D) vibrates a small flat metal electrode $500\,\mu$m in diameter above the skin in air, forming a parallel-plate capacitor between this electrode and the opposing conductive surface of the epidermis (Figure 44.2E). The signal on the current-to-voltage converter attached to the sensing tip will normally oscillate due to the vibration (Figure 44.2F) but this oscillation falls to zero when the voltage value equal and opposite to the local epidermal voltage is applied to the stratum corneum. This provides a simple method for detecting the surface potential of the epidermis without touching it with the probe itself.

A built-in video camera (Figure 44.2D) aids the operator in positioning the sensor over the region of interest and records a snapshot of the probe position at each measurement. These images can then be played back like a slow motion movie with the surface potential data displayed simultaneously. The distance between the sensor and the surface is extracted from the current signal and fed back to the Z-axis stepper motor to maintain a constant distance between the probe and the skin surface as the probe is scanned over the uneven surface of the skin. This information is also plotted along with the surface potential so that the surface topology can be correlated with the electric field (Figure 44.3, middle right graph).

44.3.2 Measurement of the Lateral Electric Field of the Tumors

The mice were immobilized by inhalation anesthesia (4% isoflurane in O_2 to render unconscious, followed by 1.4% isoflurane in O_2 to maintain anesthesia). Tumors were generated on the mouse's posterior back region where movement due to respiration is minimized. As a control, we injected either dead melanoma cells or PBS alone. Each of the regions was scanned daily to determine if there was a consistent significant difference between them.

The general method for scanning the tumors was as follows. Before scanning, the skin above each tumor was wiped with propanol. The mouse was positioned so that the tumor and 2–3 mm of adjacent skin were in the horizontal plane. The skin over and around the tumors was covered with a thin ($30\,\mu$m thick) piece of clear plastic wrap and the surface of the wrap was cleaned with alcohol. The probe head was positioned over the center of the melanoma to be scanned. The operation of the Dermacorder scan was controlled from a computer interface.

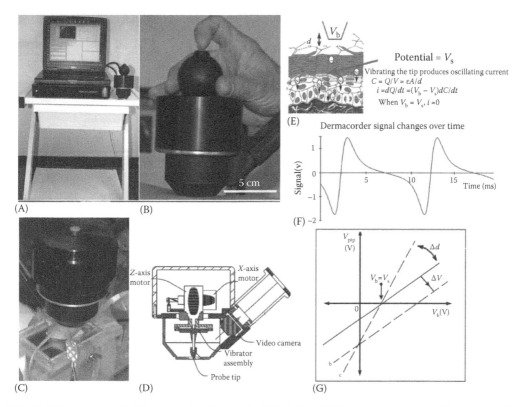

$C = Q/V = \varepsilon A/d$

$i = dQ/dt = (V_b - V_s)dC/dt$

When $V_b = V_s$, $i = 0$

FIGURE 44.2 Dermacorder® for measuring the electric field of skin. (A) Dermacorder and control box on portable cart. (B) Close-up of Dermacorder being used to measure the electric field on an arm. (C) Bench-mounted Dermacorder measuring the electric field on a mouse lesion. (D) Schematic illustration of the main components of the Dermacorder. (E) Description of how the signal is generated when oscillating the sensor over a distance "d" causing the charge on the capacitor to oscillate. (F) Typical sensor output voltage when the biasing voltage is not equal to the surface potential. (G) When the peak-to-peak potential (V_{ptp}) from (F) is plotted versus the biasing voltage, a straight line is generated. The value of V_b at $V_{ptp} = 0$ gives the surface potential and the slope of that line is inversely proportional to the distance between the probe and the skin.

44.4 Results

In order to confirm that the Dermacorder is functioning properly we routinely scan an 8 mm long carbon film resistor with 3 V across it driving current through the resistor. This should generate a uniform voltage gradient of ~375 mV/mm depending on the thickness of the carbon film. A screen shot of the computer after two scans of this resistor is shown in Figure 44.3. The slanted line on the output graph indicates a linear voltage gradient of 397 mV/mm and the horizontal line indicates that the surface scanned is quite flat. This result is exactly what we expect and indicates that the Dermacorder is functioning properly.

44.4.1 Preparation of the Device and Active Height Adjustment

In order to obtain a reliable two-dimensional surface potential distribution, it is critical that all measurements be made at the same distance from the surface being scanned. We use a Z-axis stepper motor (Figure 44.2D) to adjust the height of the probe before each measurement to compensate for the uneven topology of the skin. We obtain the distance measurement from the probe signal because when we plot the peak-to-peak voltage (V_{ptp}) (Figure 44.2F) versus the applied voltage, we obtain a straight line with

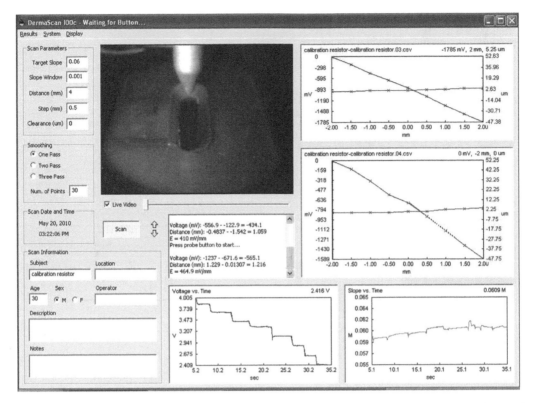

FIGURE 44.3 Dermascan computer screen just after the Dermacorder® completed two scans over the central 4 mm of an 8 mm long carbon film resistor with current passing through it. The current flow generates a linear voltage gradient of about 400 mV/mm along the resistor. The steeply sloped line in the plot indicates this gradient at 397 mV/mm. All of these data are automatically stored in an Excel file along with photographs of the probe position in which each measurement was taken. Two scans are always shown: the current scan (center right) and the one taken just before it (top right). The image shown is of the sensor vibrating over a slot in the transparent window that would normally rest on the skin. Here, the slot is positioned over a carbon film resistor.

a slope that is inversely proportional to the distance between the probe and the skin (Figure 44.2G). As this distance changes, the slope of the line shown in Figure 44.2G changes; so we adjust the probe height to maintain a constant slope before accepting each measurement of the surface potential.

In order to determine the unknown surface potential of the epidermis, the Dermascan software plots V_{ptp} versus the applied voltage (V_b). We plot two values of the peak-to-peak voltage output; one when the applied voltage is −10 V relative to the probe and the other when the applied voltage is 10 V relative to the probe. The value of V_b for which V_{ptp} equals zero is the intercept on the graph and this is the unknown surface potential we seek (Figure 44.2G).

44.4.2 Measurement of Malignant Skin Lesions

When we scan a skin region without any lesions present, we obtain the expected very flat response (upper trace in Figure 44.4E). However, when we scan over a murine melanoma (Figure 44.4D) or squamous cell carcinoma (Figure 44.4F) with the Dermacorder, the surface potential of the epidermis above these lesions is usually 100–300 mV more negative than that of the epidermal regions surrounding the lesion. Eighty-eight percent of the malignant lesions scanned exhibited this negative surface potential. When heat-killed melanoma cells were injected, no significant change in surface potential was detected. This current pattern detected above malignant lesions is consistent with positive current leaving the skin above the lesion. While it is clear that the Dermacorder can detect malignant lesions, the mystery is how these lesions located far below the epidermis influence the current flow through the epidermis.

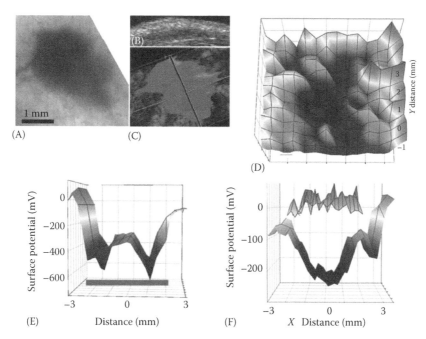

FIGURE 44.4 Imaging murine melanomas and squamous cell carcinomas. (A) Transillumination photograph of a 3 day old melanoma. (B) High-resolution ultrasound image of a slice through the same melanoma imaged in (A). (C) Three-dimensional reconstruction of hundreds of ultrasound slices through the same tumor imaged in (A). (D) Two-dimensional Dermacorder® scan of the melanoma imaged in (A)–(C). (E) Cross section through the x–z plane (where z is surface potential) of (D) at $y=0$ showing the surface potential profile over the melanoma. The top scan is from control lesion-free skin. (F) Cross section through the x–z plane at $y=0$ from a scan of a squamous cell carcinoma.

44.4.3 Measurement of Bacterial Infections

We suspected that inflammation might be involved in this signaling between the lesion and the epidermis and tested this hypothesis by injecting bacteria beneath the skin in the same manner that was used to generate malignant lesions. We found that bacterial injections also generated a negative surface potential on the epidermis above the bacterial infection (Figure 44.5) (Nuccitelli et al., in preparation). The surface potential of the epidermis over the bacterial infection looks very similar to that over malignant lesions. This potential profile was usually most negative at the edges of the infection and often was less negative in the middle, resulting in a "w"-shaped potential profile. We hypothesize that the white blood cells release a signaling factor that can trigger ion efflux from the epidermis.

44.5 Discussion

The Dermacorder is the first instrument designed to noninvasively detect the electric field in skin. Several unique design features make this possible: (1) an internal video camera provides a real-time image to facilitate positioning of the Dermacorder on the lesion of interest and to capture exact positional information for each stored potential measurement; (2) by constantly monitoring the distance between the probe and the skin, instructions can be relayed to a stepper motor to adjust to changes in surface topology and thus maintain a constant separation distance between the probe and the skin; (3) a second motor moves the probe parallel to the skin to provide a linear scan of the surface potential along that axis; and (4) each scan can traverse up to 10 mm with a 0.5 mm resolution.

The Dermacorder opens up a new research domain in the physiology of mammalian skin by allowing us to correlate the endogenous electric field in skin with its many functions. These electric fields near inflamed lesions were unexpected because the location of these lesions is usually beneath the dermis

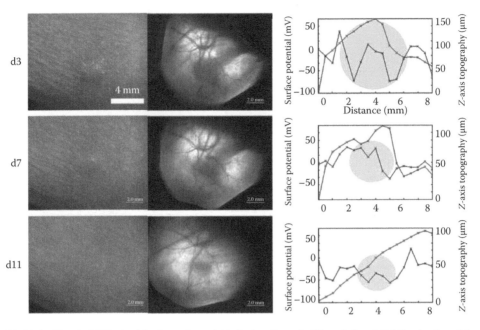

FIGURE 44.5   Electric field in the skin above a bacterial infection. On each of Days 3, 7, and 11 following the subdermal injection of *Staphylococcus epidermidis*, a reflected light photograph, transilluminated photo, and a Dermacorder® scan were recorded. The position of the infected region on the scan is indicated by the shadow and was taken from the transillumination image. The scale bar shown in upper left applies to all of the photos. The upper curve in each graph is the z-axis topography and the lower curve is the surface potential.

and far from the outer epidermal layer that is being scanned by the Dermacorder. In order for these deep lesions to generate a change in the electric field between the stratum corneum and the outer epidermal layer, they must somehow influence ion flow through the epidermis.

One way that this could occur is via messenger molecules secreted by the immune system. White blood cells involved in inflammation could release a molecule that could diffuse to the epidermis and influence the permeability of that layer. For example, proinflammatory cytokines such as interferon-gamma (IFN-γ) induce endocytosis of tight junction proteins in epithelial cells and removal of IFN-γ from the culture medium results in re-accumulation of tight junction proteins in the tight junctions [13]. Another cytokine, interleukin-1β (IL-1β), has been found to change the resistance across a monolayer of primary human keratinocytes by first increasing it at 24 h and then decreasing it at 72 and 96 h [14]. Local reductions in transepithelial resistance will lead to current flow across the epidermis in that region driven by the transepithelial potential and this current flow between the epidermis and stratum corneum will generate the local electric fields that we are detecting with the Dermacorder.

44.6 Future Trends

The Dermacorder technology is new, so much more work needs to be done to better understand these first observations using this approach to study mammalian skin. If the immune system really is involved, we should be able to identify the components of this system that are critical for generating the electric field. Experiments designed to test the effect of IFN-γ and IL-1β on the local electric field would be a good place to start.

Since all malignant lesions will be inflamed, the Dermacorder should be able to detect malignant lesions as a diagnostic technique. However, it remains to be seen if it will be possible to distinguish malignant lesions from inflamed benign lesions by the magnitude or shape of the signal. That ability to distinguish benign from malignant lesions would be required for the Dermacorder to be a useful diagnostic tool for the detection of skin cancer.

REFERENCES

1. DuBois-Reymond E. Vorlaufiger abrifs einer untersuchung uber den sogenannten froschstrom und die electomotorischen fische. *Ann Phy U Chem* 58:1, 1843.
2. Illingworth CM, Barker AT. Measurement of electrical currents emerging during the regeneration of amputated fingertips in children. *Clin Phys Physiol Meas* 1:87–89, 1980.
3. Jaffe LF, Nuccitelli R. An ultrasensitive vibrating probe for measuring extracellular currents. *J Cell Biol* 63:614–628, 1974.
4. Barker AT, Jaffe LF, Vanable JW, Jr. The glabrous epidermis of cavies contains a powerful battery. *Am J Physiol* 242:R358–R366, 1982.
5. Jaffe LF, Vanable JW. Electric fields and wound healing. *Clin Dermatol* 2:34–44, 1984.
6. McGinnis ME, Vanable JW, Jr. Electrical fields in *Notophthalmus viridescens* limb stumps. *Dev Biol* 116:184–193, 1986.
7. Nuccitelli R, Nuccitelli P, Ramlatchan S, Sanger R, Smith PJS. Imaging the electric field associated with mouse and human skin wounds. *Wound Repair Regen* 16:432–441, 2008.
8. Foulds IS, Barker AT. Human skin battery potentials and their possible role in wound healing. *Br J Dermatol* 109:515–522, 1983.
9. Kelvin L. Contact electricity of metals. *Philos Mag* 46:82–120, 1898.
10. Zisman WA. A new method of measuring contact potential differences in metals. *Rev Sci Instrum* 3:367–370, 1932.
11. Bluh O, Scott BIH. Vibrating probe electrometer for the measurement of bioelectric potentials. *Rev Sci Instrum* 21:867–868, 1950.
12. Baikie ID, Smith PJS, Porterfield DM, Estrup PJ. Multitip scanning bio-Kelvin probe. *Rev Sci Instrum* 70:1842–1850, 1999.
13. Utech M, Mennigen R, Bruewer M. Endocytosis and recycling of tight junction proteins in inflammation. *J Biomed Biotechnol* 2010:484987, 2010. Epub.:484987.
14. Kirschner N, Poetzl C, von den Driesch P, Wladykowski E, Moll I, Behne MJ, Brandner JM. Alteration of tight junction proteins is an early event in psoriasis: Putative involvement of proinflammatory cytokines. *Am J Pathol* 175:1095–1106, 2009.

Part X

Thermometric Sensing

45

Next Generation Calorimetry Based on Nanohole Array Sensing

Gregory J. Kowalski, Mehmet Sen, and Dale Larson

CONTENTS

45.1 Introduction

Thermodynamic properties such as the enthalpy of reaction and entropy change of reaction are important characteristics of binding interactions between drugs and their targets. Calorimetry is a valuable method for studying the thermodynamics of binding reactions and can be used by pharmaceutical researchers to accelerate the pace and reduce the cost of developing new drugs and result in therapeutics with improved efficacy. Unlike the rest of the material in this book, this chapter presents technology used to support the development of targeted therapies, rather than directly advance diagnosis or prognosis of cancer. This is done by addressing key limitations associated with isothermal titration calorimetry (ITC), namely, that too much compound and protein are required to perform a calorimetry experiment and the experimental throughput is too low and incompatible with the scale of the drug development process.

In this chapter we discuss the uses of calorimetry, present detailed thermodynamic analysis of the calorimetric measurement and its sensitivity to errors, and describe a novel calorimeter in development which can reduce the required sample size by orders of magnitude without loss of accuracy or precision.

The detection and characterization of binding interactions (protein–protein, protein–DNA, small molecule–protein, and peptide–protein) are central to basic biological research and pharmaceutical R&D. As a result, many analytical techniques have been developed that are used to study various types and aspects of binding interactions. These techniques include enzyme-linked immunosorbent assay (ELISA), surface plasmon resonance (SPR), mass spectrometry, fluorescence resonance energy transfer, fluorescence correlation spectroscopy, fluorescence anisotropy, protein arrays, nucleic acid microarrays, and calorimetry. All of these methods have strengths and weaknesses that depend on the experimental objectives, the specific reactants, their concentrations, and the throughput that is

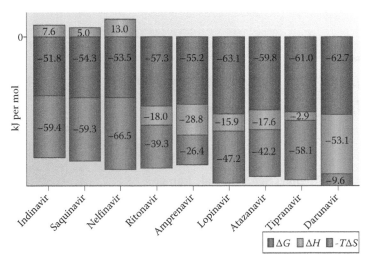

FIGURE 45.1 The thermodynamic characteristics of best-in-class compounds for pharmaceutics developed over an 11-year span for HIV treatment are illustrated. Over this development, the enthalpy of reaction of the compounds initially increases and then becomes more negative indicating a greater reactivity. The improvement in the effectiveness of the therapy also increased with the introduction of the new pharmaceuticals. (Taken from Ladbury, J.E., et al., *Nat. Rev. Drug Discov.*, 9, 23, 2010.)

required. Calorimetry is a particularly advantageous method if one is interested in studying the thermodynamics of binding interactions.

Ladbury et al. [14] discuss the significance of the thermodynamics characteristics (enthalpy of reaction, ΔH, change in entropy, ΔS, and the change in Gibbs free energy, ΔG) of compound–protein interactions and their relationship to the best-in-class drugs used in targeted therapies. In Figure 45.1, taken from Ladbury et al. [14], the history of an HIV target therapy drugs development over an 11-year span and the thermodynamic relationships of the binding reaction is summarized for nine different compounds. For effective drug therapy it is desired to have a large, negative enthalpy of reaction and a low change in entropy value. The enthalpy of reaction is related to the number of binding sites and the nature of the bonds, while the change in entropy is related to the geometry and irreversibility of the reaction. In pharmaceutical development, it is easier to manipulate the entropy change of the reaction than to modify the enthalpy of reaction or the number and type of binding sites. Current screening technologies are unable to distinguish between compounds with different contributions from enthalpy and entropy. Calorimetry is uniquely able to make this distinction and thereby start the pharmaceutical development process with a compound that is more likely to succeed. It is evident in Figure 45.1 that development of the HIV therapy with compounds having large entropy changes and small, positive enthalpy of reaction changes occurred and that as the therapy developed, the best-in-class compound had characteristics of large negative ΔH and small values of ΔS. As Ladbury states relating to Figure 45.1, "It suggests that, had initial compounds been selected based on the criterion of ΔH, best-in-class compounds could have been identified more quickly" [14]. This information strongly suggests that the ability to identify classes of pharmaceutical compounds that have an inherently large, negative enthalpy of reaction early in the therapy development will shorten the time to an effective treatment.

In a calorimetry experiment the heat of reaction, enthalpy of reaction, is measured and from this measurement and knowledge of the experimental conditions the Gibbs free energy, entropy, affinity constant, and stoichiometry are determined. Calorimetry is not currently used for screening because the sample throughput is too low and the amount of sample required is too large; rather it is used to gain a deep understanding into the nature of a particular binding reaction. Figure 45.2 summarizes the number of compounds under consideration as a function of development time in pharmaceutical R&D. Initially, there are a large number of compounds that are investigated using high-throughput screening and calorimetry is typically used later in the developmental process, at a point when the number of compounds

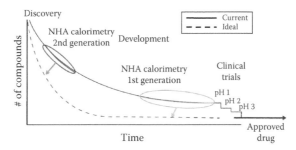

FIGURE 45.2 Illustration demonstrating the trends between number of compounds and time in pharmaceutical research and development.

being pursued has been pruned to a select few. Developing a sensitive, fast, low sample volume calorimeter could contribute to shift the trend illustrated in Figure 45.2 from the solid to dotted line, improving the productivity of pharmaceutical R&D and shortening the development times.

The ability to measure enthalpy and determine entropy allows the biochemist or medicinal chemist to assess the relative contributions of enthalpy, which is driven by the number and type of bonds in the binding reaction, and entropy, which is driven by the geometry of the ligand and the binding site. Understanding these contributions is critical in drug development because it allows for the selection of compounds that are more readily optimized. Campoy and Freire [2] describe the importance of the knowledge about the relative contributions by writing:

> enthalpically driven ligands possess better potentiality towards optimization than entropically driven ligands, because it is easier to improve entropy than enthalpy. Besides, a favorable enthalpy contribution is an indication of specific interactions between binding partners and a good way to ensure ligand specificity, selectivity and adaptability. On the other hand, an unfavorable enthalpy contribution is an indication of non-specific interactions between the binding partners, making it very difficult to provide specificity, selectivity and adaptability.

John Ladbury [12] wrote:

> The real strength of ITC is to add a further level of information to the decision-making process. In a case where two or more compounds have similar affinities, it has been suggested that the compounds with the most favorable ΔH terms are likely to make the best lead compounds for further chemical modification. This is because the ΔH term corresponds to the energy associated with the net change in non-covalent bonds. A more favorable enthalpy of reaction term, ΔH, suggests a better complementarily of bonds in the interface. Since improving the bonding in the biomolecule-drug interface is a major challenge, the initial enthalpy data is a useful tool to discriminate between compounds.

Calorimetry has other advantages as an analytical method, but also has some important limitations at the present time. One advantage is the ability to measure the effects of pH and electrostatics on complex formation [12]. Another is that there is no need for chemical modification (e.g., labeling) or immobilization of any of the components. The limitations of the current generation of calorimeters include:

- Inadequate sensitivity for low ΔH reactions.
- Large amount of protein required (0.5–5 mg).
- Low experimental throughput because of the combination of experiment run times of 60–90 min and the need to sequentially run controls to assess the significance of the confounding effects. The potential confounding effects primarily include heating due to the mixing of dissimilar sample media [17] (buffers with different pH, ionic strength, solvents), the presence of DMSO from compound storage, and solvation [22].

- Frequently compounds with poor solubility generate hits in high-throughput screens. Unfortunately, the concentrations of these compounds required to meet the mass requirement for reactants (protein and its ligand) are often above the solubility limit. As a result, calorimetry studies on the interactions of these compounds with their targets cannot be done. Paradoxically, additional synthetic/medicinal chemistry is required before calorimetry can be used, but this chemistry work cannot be justified without the calorimetry data. The outcome of this is that potentially promising compounds are not pursued. The ability to analyze smaller amounts of reactants reduces this need for concentration.

To be successful, any next generation technology needs to match or exceed the measurement performance of the current generation of calorimetry instrumentation and address these limitations.

There are several groups pursuing miniaturized calorimeters described in the literature. From among these approaches, nanohole array technology is particularly promising. The nanohole array uses optical transmission through a nanoscale aperture that is smaller than the wavelength of the light to measure local temperature in the reaction cell (Section 45.3.6). The key advantage that this technology has over other methods is the extremely small size of the sensor, which, at $0.15\,\mu m \times 1.0\,\mu m$, is much smaller than the thermopiles and thermistors that the other technologies use. From discussions with calorimetrists, compound usage is the dominant limitation to today's calorimeters and this technology will require the least compound among the chip scale calorimeter technologies found in the literature. A short summary of these developmental stage calorimetry technologies and their limitations is provided below.

- Enthalpy arrays described by Torres et al. [20] and commented on by Salemme [17] are based on the use of amorphous silicon thermistors made using photolithography with $50\,\mu m$ design rules, where each sensor consists of two thermistors connected in a Wheatstone bridge configuration. The sample consists of two $250\,nL$ drops of liquid, one for each reactant, that are merged electrostatically using a $100\,V$ potential. One of the sensors in the Wheatstone bridge is for the sample and the other is for a reference sample. The difference between the temperatures of these two sensors indicates the heat of the reaction. In Salemme's review [17] of this chapter, he states that due to the need for controls and signal-to-noise concerns this technology will require ~20% of the amount of protein required by conventional ITC.

- Calorimetric biosensors with integrated microfluidics are described by Zhang and Tadigadapa [25]. Their technology uses calorimetry in a mode that somewhat resembles how conventional SPR experiments are done. They have replaced the SPR phenomenon with a measurement of temperature change using a microthermopile. In their assay, one of the reactants is immobilized on the chip surface and the other is delivered by flow. This represents a very interesting approach that seems to be a hybrid combining conventional label-free detection with calorimetry. On the plus side, it provides thermodynamic data that is not provided by other label-free methods and it should be able to provide the off-rate (if its sensitivity is adequate), but it does not have the advantage provided by calorimetry of not needing to immobilize reagents and studying the binding interaction in solution. The potential for steric hindrance, due to the immobilization, altering the binding interaction is a concern.

- Lerchner et al. [15] describe a microfluidic chip calorimeter that is very similar to the calorimeter described by Zhang and Tadigadapa [25] with a significant improvement in temperature control. Lerchner et al. report a temperature controller that stabilizes the temperature to $100\,\mu K$, which is very impressive, and they introduce both reactants via flow with no immobilization of either reactant required. The work by Lerchner et al. focuses on the use of their technology to monitor microbial growth, not as an analytical calorimeter, although it is reasonable to expect that the basic technology could be developed as an analytical instrument.

- Johannessen et al. [7–9] describe a calorimeter for the analysis of very low cell volumes and for cell-based assays. The sample chamber used has a $200\,\mu m$ diameter. Similarly, Verhaegen et al. [21] describe a microphysiometer that is used to perform functional cell-based assays using calorimetry. They also use lithography and micromachining to fabricate their chip.

All of these technologies are interesting and the range of uses that are described (cell-based assays, low volume binding interactions, monitoring of microbial growth, high-throughput screening) speak to the value of calorimetry. The fundamental advantage in small temperature sensor size possible with nanohole arrays is the possibility of creating instrumentation systems that will transform calorimetry into a routinely used analytical method.

45.2 Materials

Experimental setup consisted of a constant wavelength light source, a flow cell assembly, a nanohole array chip, and a sensor for the transmitted light intensity measurements. A photograph of a typical experimental set up is shown in Figure 45.3. The key components are labeled in the figure. The monochromatic, polarized light source (1, numbers in curly brackets refer to components in the figure) is directed through the beam expander (6) to the sample flow cell detector holder (4) that is mounted on the sample detector stage. The thermal control unit (3) and computer connections boards (7) for the data acquisition system are also shown. Not shown are the syringe pumps and microfluidics connections. Figure 45.4 is a close-up of the sample holder with highlights on the NAS chip input aperture and microfluidics tubing connections, NAS chip, and the heater. An exploded view of the sample holder with the NAS chip, heater, and detector is given in Figure 45.5. Details of the components used in the experiment are given later.

The metallic film was excited by a 632 nm helium–neon (He–Ne) laser light source (Melles Griot, 05-LHP-991 Albuquerque, NM) and the total transmitted light through the sub-wavelength nanoholes was measured by a photomultiplier tube (PMT) detector (Hamamatsu H6780, Bridgewater, NJ).

The nanohole array chip was fabricated by first depositing 25 nm chromium and 105 nm Au film on a 4 in. diameter, 0.650 mm thick glass wafer (Mark Optics Santa Ana, CA) using E-Beam Evaporator. Coated wafer was then sliced into rectangular slides. Nanohole perforation on the metallic film layer was

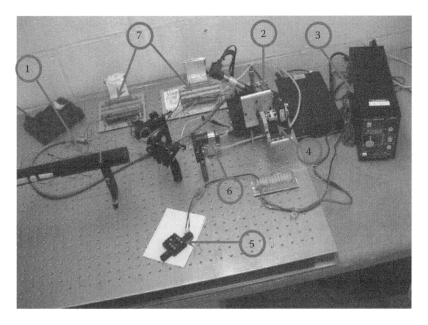

FIGURE 45.3 A photograph of the flow cell nanohole sensor experiment with photomultiplier tube (PMT) detector option. The number devices are (1) He–Ne polarized laser, (2) flow cell detector stage, (3) thermal control unit, (4) PMT detector flow cell assembly, (5) assembled sample holder, (6) beam expander, and (7) computer data acquisition cable connectors. The PMT voltage output is connected to the computer data acquisition cables. The thermal control unit is connected to heater (white square visible above the flow cell assembly) to control the temperature of the flow cell fluid.

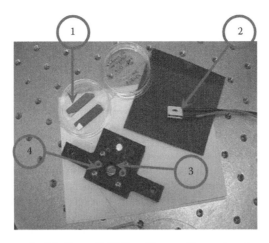

FIGURE 45.4 Close-up photograph of the flow cell, NAS, and heaters. The numbered components are (1) two-nanohole array sensor chips, (2) heater used to control temperature, (3) inlet aperture for the incident light passed through sample and NAS, and (4) microfluidic port.

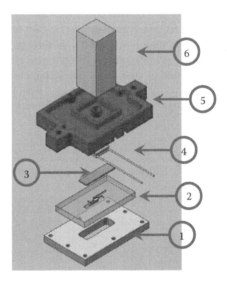

FIGURE 45.5 Exploded view flow cell detector assembly. The numbered components are (1) flow cell outer support surface, (2) PDMS flow cell, (3) NAS chip, (4) heater, (5) flow cell NAS chip holder, and (6) transmitted light detector, either a photomultiplier tube or CCD camera.

performed using Zeiss CrossBeam focused ion beam (FIB). These procedures were done at the Harvard Center for Nanoscale Systems.

The flow cell was created using micromolding process. The mold design, with the feature details of the flow cell, were designed using CAD (computer-aided design) software. These designs were fabricated by a 3D prototype printer (3D Systems Viper Si 2, SLA Acura 40 Resin material, Rock Hill, SC). Channels and the reaction chamber were fabricated by casting polydimethylsiloxane (PDMS, Dow Corning Sylgard 184 Silicone elastomer base and curing agent, 10:1 ratio, Midland, MI) using the printed molds. Remaining parts of the clamp assembly were fabricated using 3D prototype ABS plastic printer (Dimension bst, ABSplus plastic model material, Eden Prairie, MN).

To control the temperature of the system, a thermoelectric heater (Melcor, SH 1.4-15-045, Trenton, NJ) was controlled by a temperature-controlling unit (Wavelength Electronics LFI-3751, Bozeman, MT) and a thermistor (U.S. Sensor Corp., 30 kΩ microthermistor, USP 5506, Orange, CA).

The injections were performed with 10 mm diameter syringes connected to the syringe pumps (Harvard Apparatus 11P and Harvard Apparatus 4400-001, Holliston, MA). Syringes were connected to the flow cell by polyethylene tubing (Intramedic, Clay Adams, PE240, #427450, Kongens Lyngby, Denmark).

The responses from the sensors and the system were measured in real time and recorded using LabVIEW software (National Instruments, Austin, TX).

Saturated salt solution used in the experiments was prepared using sodium chloride, NaCl, (Crystalline/ Certified ACS, Fisher Scientific, Fisher Chemical, S271-1, Pittsburgh, PA), and deionized (DI) water.

45.3 Methods

45.3.1 Calorimetry

In this section, the thermodynamics of a binding reaction is discussed as it is applied to the calorimetric process. The simplest view of a calorimeter is that it provides a means to quantitatively measure the energy released as a heat transfer effect. This section will culminate in a sensitivity analysis to illustrate the sources and significance of experiment errors and of design strategies to reduce them.

45.3.2 General Thermodynamic Relationships

Consider a control volume that is initially filled with a compound M and will have an injection of a fixed mass of compound L. This control volume represents the sample cell in a typical ITC experiment. In an ITC experiment, between 8 and 14 injections are used to determine the thermodynamic properties of the binding reaction. Both compounds M and L are initially at a known concentration in a buffer, B. The general situation is shown in Figure 45.6.

The reaction equation for this example assumes a single binding site per molecule and is

$$[L] + [M] \leftrightarrow [ML] \tag{45.1}$$

Not all of the injected compound L may be consumed in forming the compound ML and the ratio of the species in the control volume at equilibrium is described by the equilibrium binding constant which is the reciprocal of the disassociation constant, K_D:

$$K_D = \left[\frac{[ML]}{[L][M]} \right] \tag{45.2}$$

The equilibrium binding constant can be used to determine the change in the Gibbs free energy and the enthalpy of formation of the reaction. The Gibbs free energy is a measure of the thermodynamic potential between stable equilibrium states of a reaction under constant temperature and pressure:

$$\Delta G = RT \ln(K_D) \tag{45.3}$$

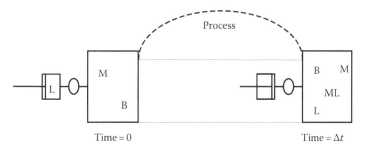

FIGURE 45.6   Schematic illustration of the ITC process that will be used in the thermodynamic analysis.

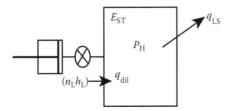

FIGURE 45.7   The calorimeter control volume illustrating the energy flows.

where R and T represent the universal gas constant and experimental temperature, respectively. These parameters can be used to determine the entropy change of the reaction:

$$\Delta S = \frac{(\Delta H - \Delta G)}{T} \tag{45.4}$$

As shown in Figure 45.6, the end state may contain unreacted molecules of L because of its inability to locate a binding site or because of the reverse chemical reaction. The energy flows that occur during this titration process are shown in Figure 45.7.

The energy balance on the control volume for the constant pressure process from the start of the reaction to the return of the control volume to an equilibrium state is

$$\int_{t_o}^{t_f} n_L h_L\, dt = \int_{t_o}^{t_f} \dot{E}_{ST}\, dt + \int_{t_o}^{t_f} q_{dil}\, dt + \int_{t_o}^{t_f} q_{LS}\, dt - \int_{t_o}^{t_f} P_H\, dt \tag{45.5}$$

In Equation 45.5, n_L represents the rate of the number of moles of L added, h_L represents enthalpy of L, q_{LS} is the heat loss from the control volume, and E_{ST} is the energy storage rate in the control volume. The energy flows include terms representing a heater power, P, which is commonly used in ITC systems to maintain a constant temperature (isothermal). ITC systems typically use a reference and sample cell. The term q_{dil} represents either the folding or solvation energy released and is included to represent cases where the energy change observed can occur from more than one chemical source. These terms will be discussed later in relation to the analysis of sensitivity and error sources of calorimetric measurements. The inlet term on the left hand side (LHS) of Equation 45.5 includes the flow work term associated with the injection process. For an incompressible material undergoing a constant pressure process, the energy storage term, $\dot{E}_{ST}$, is

$$\dot{E}_{ST} = \frac{\partial}{\partial t}\left[n_M h_M + n_{ML} h_{ML} + n_L h_L \right] \tag{45.6}$$

The injectant L, as shown in Figure 45.6, is held at a constant temperature and pressure, and its enthalpy is constant and can be factored out of the integral on the LHS of Equation 45.5. With Equation 45.6, the first integral on the right hand side (RHS) of Equation 45.5 can be performed and the energy balance reduces to the following form:

$$h_L \Delta n_L = \left[n_M h_M + n_{ML} h_{ML} + n_L h_L \right]|_{t_f} - \left[n_M h_M + n_{ML} h_{ML} + n_L h_L \right]|_{t_o}$$

$$+ \int_{t_o}^{t_f} q_{dil}\, dt + \int_{t_o}^{t_f} q_{LS}\, dt - \int_{t_o}^{t_f} P_H\, dt \tag{45.7}$$

where $\Delta n_L = \int_{t_o}^{t_f} n_L\, dt$ is the number of moles of L injected into the control volume.

When applying Equation 45.7 to a control volume undergoing a reaction, the absolute enthalpy should be used. The absolute enthalpy is the sum of the enthalpy of formation of the compound at the reference temperature and the change in enthalpy produced by the temperature difference between the control volume and the reference temperature. The reference temperature is 298.15 K:

$$(\overset{\circ}{h}_L + \Delta h_L\mid_i)\Delta n_L = \left[n_M\mid_{t_f}(\overset{\circ}{h}_M + \Delta h_M\mid_{t_f}) + n_{ML}\mid_{t_f}(\overset{\circ}{h}_{ML} + \Delta h_{ML}\mid_{t_f}) + n_L\mid_{t_f}(\overset{\circ}{h}_L + \Delta h_L\mid_{t_f}) \right]$$

$$- \left[n_M\mid_{t_o}(\overset{\circ}{h}_M + \Delta h_M\mid_{t_o}) + n_{ML}\mid_{t_o}(\overset{\circ}{h}_{ML} + \Delta h_{ML}\mid_{t_o}) + n_L\mid_{t_o}(\overset{\circ}{h}_L + \Delta h_L\mid_{t_o}) \right] + \int_{t_o}^{t_f} q_{dil}\, dt$$

$$+ \int_{t_o}^{t_f} q_{LS}\, dt - \int_{t_o}^{t_f} P_H\, dt \tag{45.8}$$

where

$\overset{\circ}{h}_x$ is the enthalpy of formation of compound x at state t_f, x = compound M, L, or ML

$\Delta h_x\mid_j = c_{px}(T_f - T_R)$ is the enthalpy change for compound x between the reference state and specific control volume state t_f

To use Equation 45.8 to identify the enthalpy of reaction, it is rearranged to group the enthalpy of formations and is typically divided by the number of moles of reactant, Δn_L. The enthalpy of reaction, ΔH_R, term is introduced to shorten Equation 45.8. The enthalpy of reaction is one of the thermodynamic parameters of interest in describing the protein–drug interaction that is used to identify inherently more efficient therapies. In this chapter, the enthalpy of reaction refers to the energy released only by the reaction at the temperature of the control volume as shown in Equation 45.8:

$$\Delta H_R = \left[\overset{\circ}{h}_L\left(1 - \frac{n_L\mid_{t_f}}{\Delta n_L} + \frac{n_L\mid_{t_o}}{\Delta n_L}\right) + \left(\frac{n_M\mid_{t_o}}{\Delta n_L} - \frac{n_M\mid_{t_f}}{\Delta n_L}\right)\overset{\circ}{h}_M + \left(\frac{n_{ML}\mid_{t_o}}{\Delta n_L} - \frac{n_{ML}\mid_{t_f}}{\Delta n_L}\right)\overset{\circ}{h}_{ML} \right]$$

$$+ \left[\frac{n_M\mid_{t_o}}{\Delta n_L}\Delta h_M\mid_{t_o} + \frac{n_{ML}\mid_{t_o}}{\Delta n_L}\Delta h_{ML}\mid_{t_o} + \frac{n_L\mid_{t_o}}{\Delta n_L}\Delta h_L\mid_{t_o} \right]$$

$$- \left[\frac{n_M\mid_{t_f}}{\Delta n_L}\Delta h_M\mid_{t_f} + \frac{n_{ML}\mid_{t_f}}{\Delta n_L}\Delta h_{ML}\mid_{t_f} + \frac{n_L\mid_{t_f}}{\Delta n_L}\Delta h_L\mid_{t_f} \right] - \Delta h_L\mid_i$$

$$= \left(\frac{1}{\Delta n_L}\right)\left[\int_{t_o}^{t_f} q_{dil}\, dt + \int_{t_o}^{t_f} q_{LS}\, dt - \int_{t_o}^{t_f} P_H\, dt \right] \tag{45.9}$$

The enthalpy changes in Equation 45.9 can be expressed in terms of the control volume temperature and the reference temperature. The rate of heat transfer from the control volume is expressed in terms of the temperature difference between the control volume and the surroundings and the overall heat transfer conductance, C_{HT}. The overall heat transfer conductance is a design parameter of the control volume, or cell, and is necessary to remove the energy released between injections:

$$\Delta H_R = \left[\overset{\circ}{h}_L\left(1 - \frac{n_L\mid_{t_f}}{\Delta n_L} + \frac{n_L\mid_{t_o}}{\Delta n_L}\right) + \left(\frac{n_M\mid_{t_o}}{\Delta n_L} - \frac{n_M\mid_{t_f}}{\Delta n_L}\right)\overset{\circ}{h}_M + \left(\frac{n_{ML}\mid_{t_o}}{\Delta n_L} - \frac{n_{ML}\mid_{t_f}}{\Delta n_L}\right)\overset{\circ}{h}_{ML} \right]$$

$$+ \left[\frac{n_M\mid_{t_o}}{\Delta n_L}c_{pM}(T_o - T_R) + \frac{n_{ML}\mid_{t_o}}{\Delta n_L}c_{pML}(T_o - T_R) + \frac{n_L\mid_{t_o}}{\Delta n_L}c_{pL}(T_o - T_R) \right]$$

$$- \left[\frac{n_M\mid_{t_f}}{\Delta n_L}c_{pM}(T_f - T_R) + \frac{n_{ML}\mid_{t_f}}{\Delta n_L}c_{pML}(T_f - T_R) + \frac{n_L\mid_{t_f}}{\Delta n_L}c_{pL}(T_f - T_R) \right] - c_{pL}(T_i - T_R)$$

$$= \left(\frac{1}{\Delta n_L}\right)\left[\int_{t_o}^{t_f} q_{dil}\, dt + C_{HT}\int_{t_o}^{t_f}(T_{cv} - T_\infty)\, dt - \int_{t_o}^{t_f} P_H\, dt \right] \tag{45.10}$$

In an ITC calorimeter, there is typically a 5°C–10°C temperature difference between the control volume and the surroundings.

Equation 45.10 represents the energy flows during the reaction, and the common, practical assumptions used in calorimetry which simplifies this relationship have not yet been introduced. For example, it is commonly assumed that the process begins at an equilibrium condition and then the reaction and subsequent heat transfer are allowed to proceed to a new equilibrium condition. Usually, the new state is at the same temperature as the starting state. In Equation 45.10, a potential temperature difference is included to assess its effects on the errors and sensitivity of the calorimetric device. Another common assumption is to equate the temperature of the injectant and that of the control volume, $T_i = T_o$. In practice, this condition is not automatically satisfied and requires careful experimental design and control to be true. If one assumes that $T_i = T_o$ when it is not, there is an error in both the temperature and observed energy results:

$$\Delta T_{in} \mid_{error} = T_o - T_i \tag{45.11}$$

And Equation 45.10 can be rewritten as

$$\Delta H_R = \left(\frac{1}{\Delta n_L} \right) \left[\int_{t_o}^{t_f} q_{dil}\, dt + C_{HT} \int_{t_o}^{t_f} (T_{cv} - T_\infty)\, dt - \int_{t_o}^{t_f} P_H\, dt - \{ c_{pL} (\Delta T_{in} \mid_{error} - T_R) \} \right] \tag{45.12}$$

The curly bracket term is the energy released error for the analysis if the actual temperature difference is not taken into account. Not accounting for this error term will increase or decrease the observed ΔH_R term.

The aforementioned development also illustrates the need to accurately account for the mass flow of each species in order to accurately determine the enthalpy of reaction and the equilibrium binding constant. In standard calorimetric practice, a fixed amount of one species [M] is introduced into the control volume at the beginning of the experiment:

$$[M] = [M]_{TOT} = n_{M_{tot}} \tag{45.13}$$

During the measurement a portion of [M] is converted to [ML] because of the injected [L]. The amount of the injectant per injection is Δn_L as defined in Equation 45.7. After a number of injections, N_i, the total amount of [L] in the control volume is

$$n_L \mid_i = N_i \Delta n_L = n_L \mid_o + \Delta n_L + n_{ML} \mid_o \tag{45.14}$$

A mass balance per species in the control volume after the N_ith injection yields the following relationships:

$$n_M \mid_i = n_M \mid_{TOT} - n_{ML} \mid_i \tag{45.15}$$

$$n_L \mid_i = (N_i \Delta n_L) - n_{ML} \mid_i \tag{45.16}$$

However, as described by Equation 45.2, not all the injected [L] will be consumed during the reaction. After any given injection

$$n_L \mid_i = \left(\frac{1}{K_D} \right) \left(\frac{n_M \mid_i}{n_{ML} \mid_i} \right) \tag{45.17}$$

Equation 45.17 provides a measure of the completeness of the reaction for an injection. Under equilibrium conditions, the reaction proceeds in both directions and the control volume will contain a mixture of the compounds [L], [M], and [ML]. From Equation 45.17, it can be observed that if $n_M \mid_i$ is much larger than the amount of reacted compound, $n_{ML} \mid_i$; the number of moles of [L] will approach zero; all of the injected compound will be consumed. This case closely corresponds to the complete reaction limit. At this limit, the measured energy released will equal the enthalpy of formation of the reaction, ΔH_R^o.

Under complete reaction limit:

$$n_L \mid_f = 0 = n_L \mid_o \tag{45.18}$$

And the reaction equation (Equation 45.1) can be written as

$$\Delta n_L + a(n_M) + b(n_{ML}) \leftrightarrow c(n_M) + d(n_{ML}) \tag{45.19}$$

The reaction equation (Equation 45.1) can also be interpreted as a relationship of compound molar ratios and the LHS of Equation 45.19 is at an equilibrium condition until the addition of Δn_L:

$$\frac{n_M \mid_o - n_M \mid_f}{\Delta n_L} = 1 = \frac{n_{ML} \mid_o - n_{ML} \mid_f}{\Delta n_L} \tag{45.20}$$

Doing a species mass balance yields the following four relationships that can be solved for the unknown constants, "c" and "d" in Equation 45.19:

$$a = \text{known} = \frac{n_M \mid_o}{\Delta n_L} \tag{45.21a}$$

$$b = \text{known} = \frac{n_{ML} \mid_o}{\Delta n_L} \tag{45.21b}$$

$$c = \frac{n_M \mid_o}{\Delta n_L} - 1 \tag{45.21c}$$

$$d = 1 + \frac{n_{ML} \mid_o}{\Delta n_L} \tag{45.21d}$$

Once the coefficients in Equation 45.19 are known, they can be introduced into the LHS of Equation 45.9 and the enthalpy of reaction per injection can be determined in terms of the enthalpy of formation of each compound and the enthalpy changes between the control volume and the reference temperature:

$$\Delta H_R = \left[h_L^\circ + (a-c)h_M^\circ + (b-d)h_{ML}^\circ \right] + \left[a\Delta h_M \mid_{t_o} + b\Delta h_{ML} \mid_{t_o} + \Delta h_L \mid_{t_o} \right]$$
$$- \left[c\Delta h_M \mid_{t_f} + d\Delta h_{ML} \mid_{t_f} + \Delta h_L \mid_{t_f} \right] - \Delta h_L \mid_i \tag{45.22}$$

In practice and under normal circumstances, the conditions for complete reaction may not exist and one needs to include the equilibrium binding constant, K_D, in the calculations. The equilibrium binding constant is determined using calorimetry by performing the previous experiment using a sequence of injections. There will be a significant amount of the injectant, [L], in the control volume for the case when either the equilibrium binding constant, K_D, is small or the ratio ($n_{ML}\mid_i/n_M\mid_i$) is approaching or is greater than 1. In calorimetry, this condition is used to determine the equilibrium binding constant through a nonlinear iterative solution. The energy released in this case, ΔH_{ER}, is less than the enthalpy of the reaction:

$$\Delta H_{ER} = \Delta H_R F_{CR} = \text{Energy released during the reaction} \tag{45.23}$$

where F_{CR} is the fraction of compound [L] that has not been converted to compound [ML]

$$= \left(\frac{n_{ML} \mid_f - n_{ML} \mid_o}{\Delta n_L} \right) \tag{45.24}$$

Using Equation 45.19 and the definition of the equilibrium binding constant (Equation 45.2) F_{CR}, can be expressed in terms of the molar fraction and K_D:

$$F_{CR} = 1 + \left(\frac{1}{K_D \Delta n_L} \right) \left[n_{MTOT} \left(\frac{1}{n_M \mid_o} - \frac{1}{n_M \mid_f} \right) \right] \tag{45.25}$$

Equation 45.25 is used to extract the enthalpy of reaction parameter from the measured energy released using the relationship in Equation 45.10. To correctly interpret Equation 45.10, it is necessary to know the amount of injected compound [L] converted to [ML]. In standard practice, Equations 45.15 through 45.17 are combined to form the following nonlinear relationship:

$$\left(n_{ML} \right)^2 - \left[N_i \Delta n_L + n_{MTOT} + K_D \right] n_{ML} + \left(N_i \Delta n_L \right) n_{MTOT} = 0 \tag{45.26}$$

The equilibrium binding constant is determined by performing a number of sequential injections under conditions where the ratio $(n_{ML} \mid_i / n_M \mid_i)$ is approaching 1 and iteratively solving Equations 45.26 and 45.10. In this calculation, the relationship between the energy released per mole of injectant and the molar ratio is used with Equation 45.26. The behavior of the completeness of the reaction described by Equation 45.17 plays a critical role in this calculation. If the equilibrium binding constant is large or the ratio $(n_{ML} \mid_i / n_M \mid_i)$ is small, there are a large number of [M] molecules available which can react with the [L] molecules and the energy released per mole of injectant will be constant and equal to the enthalpy of reaction. The iterative technique will fail under these conditions. This behavior implies that for the evaluation of the enthalpy of reaction and the equilibrium binding constant the calorimeter experiment must be carefully planned using the appropriate concentrations and sample cell size.

45.3.3 ITC Operation

The most common approach to using Equation 45.10, the control volume energy balance, for calorimetric measurements is the ITC method in which the temperature of the sample and reference cells is held constant by varying the power during the reaction. The preceding analysis of the control volume applies to the sample and reference cell. Both the sample and reference cell are 5°C–10°C above the temperature of the surrounding bath. The bath acts as a controlled temperature heat reservoir that isolates the reference and sample cell from random changes in the ambient environment. There will be a heat flow from both cells and it is assumed that it is the same because of similar construction and that both cells are controlled to the same temperature:

$$\int_{t_o}^{t_f} q_{ls,s} \, dt = C_{HT,S} \int_{t_o}^{t_f} \left(T_{cl} - T_\infty \right) dt \tag{45.27}$$

$$\int_{t_o}^{t_f} q_{ls,r} \, dt = C_{HT,R} \int_{t_o}^{t_f} \left(T_{cl,R} - T_\infty \right) dt \tag{45.28}$$

In the reference cell, no reaction takes place and its energy balance (Equation 45.10) reduces to

$$C_{HT,R} \int_{t_o}^{t_f} (T_{cl,R} - T_\infty) \, dt - \int_{t_o}^{t_f} P_{H,R} \, dt = 0 \tag{45.29}$$

All the enthalpy terms go to zero since there is no reaction caused by an injected compound and the temperature is held constant. In practice, both the temperature of the reference cell and the power are monitored and the reference cell's temperature is controlled to a set value. In this presentation, the temperature and the power of the reference are treated as separate parameters for the purpose of doing an error and sensitivity analysis on the device.

In ITC mode, the temperature of the sample cell is maintained constant by adjusting its heater power, $P_{H,S}$, while the reaction is occurring. To describe this control process, the change in the sample heater power is introduced:

$$P_{H,S} = P_{H,S_o} - \Delta P_H = P_{H,R} - \Delta P_H \tag{45.30}$$

The last term on the RHS applies under the assumption that the sample and reference cells are exactly alike. For constant temperature operation $T_o = T_f = T_{cl} = T_{ITC}$, Equation 45.10 for the sample cell becomes

$$\Delta H_{ER} = \left(\frac{1}{\Delta n_L}\right) \left[C_{HT,S} \int_{t_o}^{t_f} (T_{ITC} - T_\infty) dt - \int_{t_o}^{t_f} (P_{H,S_o} - \Delta P_H) dt \right] \tag{45.31}$$

The enthalpy change for each compound caused by the temperature difference between the operating and reference temperature has been moved to the left side of the equation since they will correspond to the energy released as observed in the experiment. In practice these terms are included in the enthalpy of reaction, which differentiates this measured value from the enthalpy of formation of the reaction; it is the negative of the commonly used heating value parameter used in some fields. These statements are based on the assumption that the injected compound is at the same temperature as the sample cell, $T_i = T_{ITC}$, as stated earlier. If this condition is not satisfied, then the temperature difference clearly produces an error in the enthalpy of reaction measurement. A similar interpretation is made concerning the folding or solvation term, $q_{dil,I}$, if it is not zero. However, in the case of the folding or solvation terms the error is one of the interpreting accurately measured data, while the temperature difference error is one of the poor operating procedures or experimental designs.

The ITC device can be used to evaluate the energy released from a reaction by monitoring the change in the heater power of the sample cell. As stated by Ladbury et al. [14], the thermodynamic information of the protein–compound interaction can provide information on selecting the class of compounds with the largest enthalpy of reaction magnitude which could shorten the time to effective therapies. If all assumptions are satisfied and, in practice, the well-designed device will satisfy the following assumptions, $C_{HT,R} = C_{HT,S}$ and $T_{ITC|sample} = T_{ITC|reference} = T_{ITC|set}$. Equation 45.29 can be solved for the heat transfer in terms of the reference cell's heater power which is equal to the set value of that for the sample cell. Substituting this result into Equation 45.32 for the sample cell's set value, heat transfer term yields

$$\Delta H_{ER} = \left(\frac{1}{\Delta n_L}\right) \left[\int_{t_o}^{t_f} P_{H,R} \, dt - \int_{t_o}^{t_f} P_{H,S_o} \, dt + \int_{t_o}^{t_f} \Delta P_H \, dt \right] = \left(\frac{1}{\Delta n_L}\right) \int_{t_o}^{t_f} \Delta P_H \, dt \tag{45.32}$$

Equation 45.32 demonstrates that in a well-designed and controlled ITC calorimeter the basic measurement is the power change. Using a differential measurement of the constant temperature relative to the set point and between the reference and sample cell and of using the reference cell as a compensation device for the heat loss from the sample cell requires careful design and quality manufacturing. In order to understand and to include all potential sources of error in the analysis, the following parameters are introduced:

$$\Delta C_{HT} = C_{HT,S} - C_{HT,R} \tag{45.33}$$

$$\Delta T_{set} = T_{ITC}\mid_{sample} - T_{ITC}\mid_{set} \tag{45.34}$$

$$\Delta P_{set} = P_{H,sample} - P_{H,reference} \tag{45.35}$$

Solving Equation 45.32 for the observed power difference term, ΔP_H, combined with Equations 45.33 through 45.35 and 45.10 and setting the folding or salvation term to zero yields the following form of the energy balance. The folding and solvation energy release terms are neglected in this analysis because the error and sensitivity of the calorimeter design are being investigated:

$$\Delta P_T = \int_{t_o}^{t_f} \Delta P_H \, dt = \left[h_L^\circ \left(1 - \frac{n_L \mid_{t_f}}{\Delta n_L} + \frac{n_L \mid_{t_o}}{\Delta n_L} \right) + \left(\frac{n_M \mid_{t_o}}{\Delta n_L} - \frac{n_M \mid_{t_f}}{\Delta n_L} \right) h_M^\circ + \left(\frac{n_{ML} \mid_{t_o}}{\Delta n_L} - \frac{n_{ML} \mid_{t_f}}{\Delta n_L} \right) h_{ML}^\circ \right]$$

$$+ \left[\left(\frac{n_M \mid_{t_o}}{\Delta n_L} c_{pM} + \frac{n_{ML} \mid_{t_o}}{\Delta n_L} c_{pML} + \frac{n_L \mid_{t_o}}{\Delta n_L} c_{pL} \right) - \left(\frac{n_M \mid_{t_f}}{\Delta n_L} c_{pM} + \frac{n_{ML} \mid_{t_f}}{\Delta n_L} c_{pML} + \frac{n_L \mid_{t_f}}{\Delta n_L} c_{pL} \right) - c_{pL} \right]$$

$$\times \left(T_{ITC,set} - \Delta T_{set} - T_R \right) - \left(\frac{1}{\Delta n_L} \right)$$

$$\times \left[(C_{HT,R} - \Delta C_{HT}) \int_{t_o}^{t_f} (T_{ITC,set} + \Delta T_{set} - T_\infty) dt + \int_{t_o}^{t_f} (P_{H,ref} + \Delta P_{set}) dt \right] \qquad (45.36)$$

Equation 45.36 has been expanded so that all parameters that are expected to contribute to the experimental uncertainty in the measure power change are identified for use with the RMS formulation (Equation 45.37) for estimating the uncertainty, ΔE, ΔP_T, and the sensitivity of the procedure. The terms containing such ΔC_{HT}, $T_{ITC,set}$, ΔT_{set}, ΔP_{set}, and $T_{in} \mid_{error}$ are direct uncertainties that are summed in the RMS expected uncertainty formulation:

$$\Delta E_{\Delta PT} = \left[\sum_i \left(\frac{\partial (\Delta P_T)}{\partial X_i} \Delta X_i \right)^2 + \Delta \left\{ \Delta P_T \mid_{\Delta C_{HT}, \Delta T_{set}, \Delta P_{set}, T_{in} \mid_{error}} \right\} \right] \qquad (45.37)$$

The variables that are controlled or observed are listed in Table 45.1 as are the expected uncertainties for them in terms of the enthalpy of formation of the reaction. For the ITC calorimeter process, Equation 45.37 expands to

$$\Delta E_{\Delta PT} = \begin{bmatrix} \left(\frac{\partial (\Delta P_T)}{\partial (\Delta H_R)} \Delta (\Delta H_R) \right)^2 + \left(\frac{\partial (\Delta P_T)}{\partial (\Delta C_{HT,R})} \Delta C_{HT,R} \right)^2 \\ + \left(\frac{\partial (\Delta P_T)}{\partial (\Delta T_{ITC,set})} \Delta T_{ITC,set} \right)^2 + \left(\frac{\partial (\Delta P_T)}{\partial (\Delta T_\infty)} \Delta T_\infty \right)^2 \\ + \left(\frac{\partial (\Delta P_T)}{\partial (\Delta P_{H,ref})} \Delta P_{H,ref} \right)^2 \end{bmatrix}^{(1/2)}$$

$$+ \Delta \left(\Delta P_T \mid_{\Delta C_{HT}} \right) + \Delta \left(\Delta P_T \mid_{\Delta T_{set}} \right) + \Delta \left(\Delta P_T \mid_{\Delta P_{set}} \right) + \Delta \left(\Delta P_T \mid_{T_{in} \mid_{error}} \right) \qquad (45.38a)$$

where the uncertainty in the enthalpy of formation of the reaction is given by

$$\Delta (\Delta H_R) = \left[\left(\frac{\partial (\Delta H_R)}{\partial (\Delta n_L)} \Delta (\Delta n_L) \right)^2 + \left(\frac{\partial (\Delta H_R)}{\partial (n_{MTOT})} \Delta (n_{MTOT}) \right)^2 + \left(\frac{\partial (\Delta P_T)}{\partial (\Delta T_{ITC,set})} \Delta T_{ITC,set} \right)^2 \right]^{(1/2)} \qquad (45.39)$$

The expected uncertainty in the enthalpy of reaction measurement for the ITC approach developed in Equation 45.39 and the values listed in Table 45.1 was determined to be ±1.55 kJ/mol or ±2.5% of the reported enthalpy of reaction (60.88 KJ/mol). For the ITC approach, the observed uncertainty is ±1.73 kJ/mol or 2.8% of the enthalpy of reaction. The calculated expected uncertainty is consistent with the experimentally

TABLE 45.1

Summary of Parameters and Values Used in Estimating the
Experimental Uncertainty for an ITC Type of Calorimeter

Parameter Category	Parameter	Value
Controlled variables	Concentration of [L]	$200\,\mu M$
	$V_{injectant}$	$10\,\mu L$
	Δn_L	$2 \times 10^{-9}\,mol$
	Concentration of [M]	$10\,\mu M$
	V_{cell}	$1\,mL$
	n_{MTOT}	$1 \times 10^{-8}\,mol$
	T_{ITC}	$30°C$
	$C_{HT,R}$	$5.41 \times 10^{-10}\,kW/K$
	ΔH_R	$-1455\,kcal/mol = 60.88\,kJ/mol$
	ΔP_H	$2.71 \times 10^{-9}\,kW$
	T_{ITC}	$30°C$
	T_∞	$25°C$
Expected uncertainty	$\Delta(\Delta P_H)$	$0.005\,P_H = 1.35 \times 10^{-12}\,kW$
	ΔT_{ITC}	$0.005\,K$
	ΔT_∞	$0.005\,K$
	$\Delta C_{HT,R}$	$0.01\,C_{HT,R} = 5.41 \times 10^{-12}\,kW/K$
	$\Delta P_{(\Delta P_{set})}$	$0.304\,kJ/mol$
	$\Delta P_{(\Delta T_{set})}$	$0.005\,K$
	$\Delta P_{(\Delta C_H)}$	$3.04\,kJ/mol$
	$\Delta P_{(\Delta T_{in})}$	$0.001\,K$

Source: Ladbury, J.E. and Doyle, M. (eds.), *Biocalorimetry*, Vol. 2, Wiley &
  Sons, Chichester, U.K., 2004.

Note: The example is based on a standard reference reaction of 2′-CMP and
  ribonuclease A.

reported value. The largest contributions to the uncertainty are related to the uncertainties of the heat transfer conductance and differences between the heat transfer from the reference and sample cells.

45.3.4 Alternative Calorimeter Method

An alternative calorimetry method is also suggested by Equation 45.31. In this method, the temperature change as a function of time during the reaction and the return to the new equilibrium state is monitored instead of the change in power of the sample cell, and the reference cell is eliminated. The energy change is determined using the observed heat loss which is measured using a calibrated value of the overall conductance and the temperature of the surrounding bath. Equation 45.31 reduces to the following form:

$$\Delta H_{ER} = \left(\frac{1}{\Delta n_L}\right)\left[C_{HT,S}\int_{t_o}^{t_f}(T_{ITC} - T_\infty)\,dt - \int_{t_o}^{t_f}q_{dil}\,dt\right] \tag{45.40}$$

where ΔH_{ER} is defined by Equations 45.22 and 45.23. The heater power term is zero in this method and has been eliminated from the relationship. The surrounding bath temperature, T_∞, is controlled to a set value as in the ITC method and has a much larger thermal capacitance than the sample cell in order to absorb the energy release without significantly increasing the bath temperature. As stated previously, the overall heat transfer conductance, $C_{HT,S}$, is calibrated for a specific device. The operating assumptions are that the sample cell reaches equilibrium with the surroundings at the beginning and the end of each injection. Focusing on the design and control of the alternative method, the folding and solvation energy release, q_{dil}, is set to zero and Equation 45.40 is solved for the observed heat loss:

$$q_{loss,obs} = \left(\frac{1}{\Delta n_L}\right)\left[C_{HT,S}\int_{t_o}^{t_f}(T_C - T_\infty)dt\right] = \Delta H_{ER}$$

$$= \left[h_L^\circ\left(1 - \frac{n_L\mid_{t_f}}{\Delta n_L} + \frac{n_L\mid_{t_o}}{\Delta n_L}\right) + \left(\frac{n_M\mid_{t_o}}{\Delta n_L} - \frac{n_M\mid_{t_f}}{\Delta n_L}\right)h_M^\circ + \left(\frac{n_{ML}\mid_{t_o}}{\Delta n_L} - \frac{n_{ML}\mid_{t_f}}{\Delta n_L}\right)h_{ML}^\circ\right]$$

$$+ \left[\begin{array}{c}\left(\dfrac{n_M\mid_{t_o}}{\Delta n_L}c_{pM} + \dfrac{n_{ML}\mid_{t_o}}{\Delta n_L}c_{pML} + \dfrac{n_L\mid_{t_o}}{\Delta n_L}c_{pL}\right)\\[2mm] -\left(\dfrac{n_M\mid_{t_f}}{\Delta n_L}c_{pM} + \dfrac{n_{ML}\mid_{t_f}}{\Delta n_L}c_{pML} + \dfrac{n_L\mid_{t_f}}{\Delta n_L}c_{pL}\right) - c_{pL}\end{array}\right](T_C - T_R) \qquad (45.41)$$

Equation 45.41 is simpler than Equation 45.36 because it has fewer set point and controlled variables and no controlled power terms as used in the ITC approach. It does require a fast and accurate temperature sensor and an accurate determination of the heat transfer conductance to record the heat flow during the reaction. The expected uncertainty in the observed heat loss is due to expected uncertainties in the measured temperatures, the calibrated overall heat transfer conductance, and the energy released term. The uncertainty of the enthalpy of formation of the reaction is calculated using Equation 45.39:

$$\Delta E_{q_{loss,obs}} = \left[\begin{array}{c}\left(\dfrac{\partial(q_{loss,obs})}{\partial(\Delta H_R)}\Delta H_R\right)^2 + \left(\dfrac{\partial(q_{loss,obs})}{\partial(\Delta C_{HT,S})}\Delta C_{HT,S}\right)^2\\[3mm] +\left(\dfrac{\partial(q_{loss,obs})}{\partial(\Delta T_{ITC,set})}\Delta T_C\right)^2\\[3mm] +\left(\dfrac{\partial(q_{loss,obs})}{\partial(\Delta T_\infty)}\Delta T_\infty\right)^2\end{array}\right]^{(1/2)} + \Delta P_{T1inerror} \qquad (45.42)$$

For the alternative method, the expected uncertainty in the enthalpy of reaction as calculated with Equations 45.42 and 45.39 and the values listed in Table 45.1 is found to be $\pm 0.686\,kJ/mol$ or a $\pm 1.13\%$ of the enthalpy of reaction. This reduction in the expected uncertainty is 39.7% of that reported for the ITC method. This estimate is based on the same expected uncertainties in the heat transfer conductance and measured temperatures used in the ITC approach. The heat transfer conductance terms are the largest contributors to the expected uncertainty. These results illustrate that the alternative method is inherently more accurate than the ITC because of the reduced number of parameters.

45.3.5 Nanohole Array Sensor Calorimeter

The uncertainty of the temperature measurements is characteristic of a thermopile type of measurement. Using instead the nanohole array sensor as a small, fast sensitive temperature would reduce this uncertainty by a factor of 5 to 0.001°C. The index of refraction resolution of the nanohole array sensor as reported by Stark et al. [18] is 1×10^{-7} index of refraction units.

For constant concentration and pressure, the change in the index of refraction is related to the temperature:

$$\Delta n_R = \left(\frac{dn}{dT} \right) \Delta T_R \tag{45.43}$$

The expected change in the temperature for the minimum index of refraction change is determined using Lorenz–Lorentz law and equation of state to determine (dn/dT) and using properties of water at 20°C. ΔT_R is found to be 6.6×10^{-4} K. This change in the expected temperature corresponds to a change in energy resolution per 20 pL sample (water) of 52 pJ. The expected uncertainty in the measured enthalpy of reaction using the nanohole array temperature sensor and the alternative calorimeter method is found to be 1.12%; almost no change from the previous estimate. The reason for this small change is that most of the expected uncertainty is related to the heat transfer conductance. Using a small, fast temperature sensor will allow more accurate determination of both the heat transfer conductance and measured temperature response of the calorimeter. It is estimated that the volume of the calorimeter using nanohole arrays for temperature sensing will be 1000 times smaller than the current ITC sample cell allowing calorimetry to be used earlier in the pharmaceutical development process and having ΔH and ΔS will shorten the therapy development time.

45.3.6 EOT Sensing

In 1998, Ebbesen et al. [3] reported that his team was seeing extraordinary optical transmission (EOT) through periodically perforated optically thick* monometallic films. According to diffraction theory, the transmission through these holes should have been proportional to fourth power of the hole diameter. They found that the throughput was substantially higher than that predicted by diffraction theory. Not only were photons transmitted efficiently but photons which were incident on the surface between the holes were transmitted, and that transmission fractions when normalized to hole area were larger than unity. It was determined that when surface plasmon (SP) were excited at their resonance, the photons were collected and the energy was funneled down through the holes. Surface plasmons are fluctuations in electronic density at the surface of the chip. The mechanism by which the power is transmitted through the holes is still the subject of research. What is clear, however, is that this transmission is affected by the dielectric properties of the metallic film and those of the material in direct contact with the metallic film.

The effect of the dielectric property of the material in contact with the metallic film was shown by Krishnan et al. [11] when he measured the transmission through a nanohole array constructed in a gold film over a glass substrate for different wavelengths. The results clearly show a shift in the spectral pattern of the transmitted light for the different dielectric values. To illustrate the dependence of the transmitted light on the dielectric property, this data was graphed for three different wavelengths as shown in Figure 45.8. In this figure, the transmission is plotted as a function of the dielectric properties which is consistent with the monochromatic nanohole array sensors developed in [16,17,22]. Three different behaviors were observed. At a wavelength of 750 nm, the transmission increases, reaches a peak, and then decreases. The behavior for wavelengths of 1010 and 1100 nm show two different, monotonic increases in transmission as a function of dielectric constant. These two wavelengths provide evidence that the nanohole array technology could be developed as a sensor that measures changes in the dielectric or other properties associated with it.

This dependency of the transmission on the dielectric properties has led several groups [1,4–6,16,18,23,24] to study the utility of this phenomenon as an affinity biosensor. In nanohole array sensing (NAS), analyte-specific capture reagents are immobilized on the surface of the metallic film, and when the sensor is exposed to sample the desired analytes bind to their respective capture agents. This binding changes the effective dielectric function of the metallic film and hence, the transmission spectrum. This type of sensing enables the measurement of analyte concentrations and binding kinetics (both on- and off-rates) in a multiplexed detection scheme. The experiments that can be done with this technology are very similar to those done with conventional SPR. It is well known in the SPR literature that there are bulk index of refraction effects and temperature effects that must be addressed to achieve accurate results and this is also the case with NAS. The effect of temperature can either be controlled for or measured and a correction applied.

---

* At least two times the surface plasmon (SP) depth. Optically thick implies that there can be no cross talk from one surface to another due to SP–SP interaction.

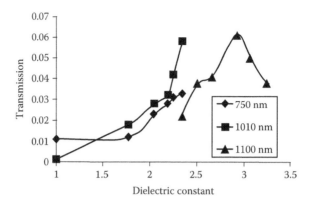

FIGURE 45.8 Using the data reported by Krishnan [11], the behavior of a nanohole array in a gold film over a glass substrate and the behavior of the transmitted light as a function of dielectric constant are illustrated for three different wavelengths. The graph illustrates how the transmission varies when excited with a monochromatic light source.

It is this temperature effect that led to our work in developing a calorimeter using nanohole array technology. At a fundamental level, a nanohole array calorimeter is very similar to the technology used for affinity detection using NAS technology. The key difference is that, instead of either controlling or correcting for temperature changes, the temperature change is the desired signal. Another important difference is that with nanohole array calorimetry (NAC) no capture agents are immobilized and all binding is done in solution.

The ability of nanohole arrays to serve as a temperature sensor was shown by Kowalski et al. [10]. A schematic of the experiment is shown in Figure 45.9 and the illustrated components were described in Section 45.2. A NAS was constructed using a 100 nm gold film over a 0.762 mm thick silicone dioxide slide and the array consisted of 81 holes, a 9×9 pattern of nanoholes, with a diameter of 150 nm spaced 350 nm apart. The NAS was mounted into a PDMS flow cell (described in Section 45.4) that contains distilled water. A 0.450 mm thermistor was installed in the flow cell to provide the temperature measurements as the flow cell was heated from a temperature 31°C to 43°C. The transmission of the incident He–Ne laser light was measured using a PMT. The dielectric constant of the distilled water is a function of temperature and the transmitted light will change as the temperature of the flow cell changes.

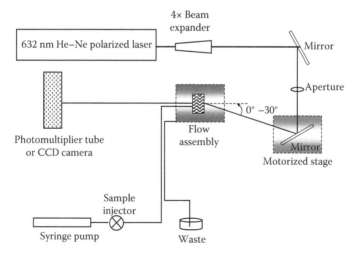

FIGURE 45.9 Schematic drawing of the experimental apparatus showing the light source (laser) being directed to the flow cell assembly where the fluid and reactants are combined to create EOT that is measured using the photomultiplier tube or a CCD camera.

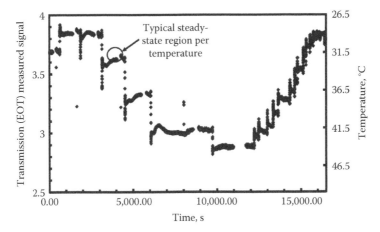

FIGURE 45.10 Measured EOT as a function of time using thermal controller for temperature–light transmission (EOT) calibration.

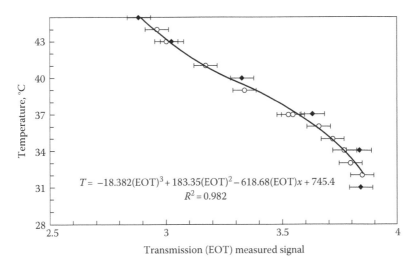

FIGURE 45.11 Summary of the steady state measured EOT signal as a function of the temperature of distilled water in the fluid cell for the EOT data shown in Figure 45.6 and the corresponding temperatures. The steady-state regime are the last 50 data points or observed constant values per set to avoid scatter related to the thermal response of the flow cell and thermal control devices and to the differences in the response time of the nanohole array sensor and the thermistor.

This ability to sense temperature when combined with the inherently small size of nanohole arrays (as small as $0.100\,\mu m \times 1.0\,\mu m$) enables the NAC technology to detect temperature changes from reactions with very low reactant masses, thus addressing the key limitation of the current generation of calorimeters. Figure 45.10 shows the measured EOT signal as a function of time as the temperature controller steps up through a temperature range and then back down through the same range. Figure 45.11 shows the same EOT signal plotted versus the control temperature and includes the equation for the relationship between EOT and T.

45.4 Results

A preliminary experimental device was fabricated to test the use of NAS and EOT to measure thermal response as fluids were injected into a microscale reaction chamber. The four-port flow cell used in the

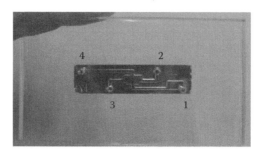

FIGURE 45.12 Nanohole array chip (gold colored on the back), inlet (1 and 2) and outlet ports (3 and 4), and the PDMS layer.

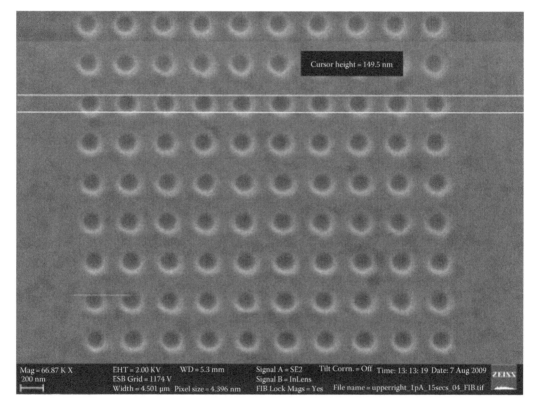

FIGURE 45.13 FIB image of a 10 × 10 nanohole array. Hole diameter is 150 nm and pitch distance is 350 nm. (Photo by Mehmet Sen, Center for Nanoscale Systems, Harvard University, Cambridge, MA, Summer 2009.)

experimental setup is shown in Figure 45.12. The flow cell was created using micromolding process. The flow cell assembly consisted of a glass chip with chromium bonding layer and gold film with the nanohole array coated on top and a PDMS layer on top with the channel and reaction chamber details inside. An electron microscope photograph of a typical nanohole array chip is shown in Figure 45.13. Each chip contains individual sensors, each with a 10 × 10 array of 150 nm diameter holes with a pitch distance of 350 nm. The thermoelectric heater surface was kept at a constant temperature (30°C in this study) using the temperature controller and a thermistor placed on the heater surface for sending feedback to the heater. The bottom surface of the sensing chip was in contact with the constant temperature surface and the upper surface of the chip was covered with the flow cell material (PDMS), which has a low thermal conductivity ($k = 0.15$ W/m·K). This configuration minimizes the effect of the environmental temperature variation on the sensing region. A second thermistor was inserted through the first inlet port and measures the incoming liquid temperature. The second thermistor was used for response comparison purposes.

The light transmission through the nanohole arrays was measured by the PMT which converts the number of incident photon to a voltage output. Due to the properties of the EOT explained in the previous sections, the light transmission through the nanohole arrays is dependent on the temperature and concentration of the medium above the surface. The output from the PMTs (transient response of EOT signal), temperature information from the two thermistors, current and the voltage (hence the power) response of the thermoelectric heater were measured in real time and recorded using LabView software and National Instruments data acquisition hardware. Post-processing and analysis of the experimental data were performed in MATLAB®.

The following experimental procedure was used to check the repeatability and the continuous monitoring capability of the EOT signal due to temperature change. The system was filled with DI water and brought to steady state at the set temperature. Room temperature water was injected with a volume of 16 μL and at a speed of 0.1 mL/min. The injections were performed with 10 mm diameter syringes connected to the syringe pumps. The room temperature injection was repeated several times after the system has reached steady-state conditions. The results show that the EOT signal is repeatable and maximum and minimum deviations from 30°C mean temperature are +0.15°C and −0.20°C, respectively.

Figure 45.14 illustrates the transient response of EOT signal compared to the 450 μm diameter thermistor. The temperature measured from the thermistor and EOT signal were normalized to make this comparison. The shorter width in the EOT signal's pulse suggests its faster transient behavior compared to the thermistor used in the experiment. The duration of each injection was 9.6 s. Once the injection ends, the EOT signal goes back to the steady state after 13 s and the thermistor signal goes back to steady state after 140 s.

For the next experiment, the flow cell was filled with DI water initially and the inlet channel was filled with saturated salt solution and the system was brought to 30°C. Saturated salt solution was introduced from the inlet channel, similar to a salt dilution experiment. Measurement of heat of dilution of NaCl solution requires high sensitivity and has been used for calibrating industrial ITC devices [19]. Each injection consisted of 4 μL of saturated salt solution introduced at a rate of 0.1 mL/min. After each injection, transient EOT signal was recorded. In this experiment, the dielectric constant of the fluid in the flow cell is changing as a function of temperature and of the salt concentration. The temperature changes as a result of the energy released by the salt dilution process and as covered in Section 45.2 is a transient process. The transmitted light through the NAS will also change as the dielectric constant changes. Once the system reaches a new steady state, the next injection was done. The results of this experiment are summarized in Figure 45.15.

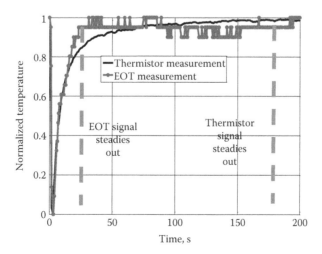

FIGURE 45.14 Transient response of the thermistor and the EOT signal for the third injection.

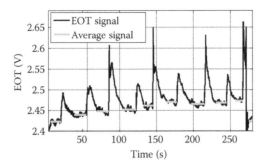

FIGURE 45.15 EOT response of saturated salt solution injections. Each pulse in the plot represents an injection of 4 µL saturated salt solution at a rate of 0.1 mL/min.

Each injection in Figure 45.15 represents an addition of saturated salt solution. Light gray lines represent the averaged steady signal values for concentration calculations. The last injection is the pure water injection. It can be observed that EOT level goes back to the initial level. The spiked peaks observed in Figure 45.15 between the light gray lines correspond to the temperature change of the solution caused by the energy released due to the dilution of the salt injection and the concentration change due to diffusion. The injections result in a temperature change smaller than 0.015°C on the heater surface. The reaction due to the salt injection ended after 20 s for the first and 15 s for the second injection. This duration is governed by the mass diffusion. The signal goes to a steady level after each injection. The steady level increase after each injection is due to the increasing concentration in the flow cell.

A titration type of result can be obtained by monitoring the steady-state EOT level as the concentration changes. For this purpose, the salt concentration in the flow cell at each injection is calculated using the following equation:

$$m_{NaCl}^{t+1} = m_{NaCl}^{t} + m_{NaCl}^{injected} - \frac{m_{NaCl}^{t}}{V_{flow\ cell}} V_{injection} \tag{45.44}$$

where
  t represents the time
  m represents the mass of the NaCl present in the flow cell
  V represents the volume of the reaction chamber

This approach is similar to the concentration calculations in the reaction chamber [19] for the ITC calorimetry.

The theoretical curve obtained from Equation 45.44 and the averaged steady EOT signals after each injection signals were normalized and plotted and the results are presented in Figure 45.16. These results

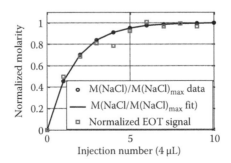

FIGURE 45.16 Comparison of steady-state concentration values with theoretically calculated concentration values.

suggest that a titration type of experiment can be performed using this device and that small values of energy released during the dilution process can be measured.

45.5 Discussion

Calorimetry and its potential as a screening tool in the earlier stages of the drug research and current limitations of the available methods were discussed. Unlike the rest of the material in this book, this chapter does not directly advance diagnosis or prognosis of cancer, rather it presents technology used to support the development of targeted therapies. Possible contributions of nanohole array technology to the calorimetric research field, such as faster response time and multiplexing capability, were presented. A detailed uncertainty analysis has been done on a widely used calorimetric technique and proposed nanohole array calorimeter approach. This uncertainty analysis shows that a calorimeter using nanohole array temperature sensors can be at least as precise as a conventional ITC system.

The preliminary experimental results show that continuous temperature changes can be monitored using the changes in the EOT signal. A transient response comparison is done experimentally with an industrial thermistor device and it has been shown that EOT has faster response capability (results shown above in Figure 45.14). Tests show that EOT signal is repeatable if similar conditions are present. It should be noted that the flow cell volume used in this study can be further reduced by photolithographic methods since the footprint of a sensor is only $3.3\,\mu m \times 3.3\,\mu m$ and can be made even smaller. For these experiments, the volume of the flow cell was $8.8\,\mu L$, which was limited by the prototype printer resolution. However, this value is still below the current industrial values, $200\,\mu L$ to $1.5\,mL$. Salt dilution tests showed that EOT signal can continuously monitor the temperature and concentration changes and hence can be utilized as a calorimeter device for chemical reactions. However, since during a reaction the concentration is not constant and the light source travels through the mixture, the EOT signal should be interpreted with caution since it is affected by temperature change and concentration change.

45.6 Future Trends

Methods are under development to separate the concentration and temperature effects on the EOT signal. The analysis and experiments described demonstrate that a calorimeter based on NAS can provide a small, low compound use device that can be used early in the pharmaceutical development process. Future tests using biological reactants and other standardized reactant pairs are being performed. Future developments to further reduce the calorimeter test cell volume from $8.8\,\mu L$ to the nanoliter regime are in process and are expected to be tested in the near future. As the volume is reduced, corresponding improvements in controlling the microfluidic system for delivering the reactants as well as thermally controlling them during delivery are also under development.

REFERENCES

1. Brolo, A. G., R. Gordon, B. Leathem, and K. L. Kavanaugh. 2004. Surface plasmon sensor based on the enhanced light transmission through arrays of nanoholes in gold films. *Langmuir* 20:4813–4815.
2. Campoy, A. V. and E. Freire. 2005. ITC in the post-genomic era …? Priceless. *Biophysical Chemistry* 115:115–124.
3. Ebbesen, T. W., H. J. Lezec, H. F. Ghaemi, T. Thio, and P. A. Wolff. 1998. Extraordinary optical transmission through sub-wavelength hole arrays. *Nature* 391:667–669.
4. Im, H., A. Lesuffleur, N. C. Lindquist, and S.-H. Oh. 2009. Plasmonic nanoholes in a multichannel microarray format for parallel kinetic assays and differential sensing. *Analytical Chemistry* 81:2854–2859.
5. Ji, J., J. O'Connell, D. Carter, and D. Larson. 2008. High-throughput nanohole array based system to monitor multiple binding events in real time. *Analytical Chemistry* 80:2491–2498.
6. Ji, J., J. Yang, and D. Larson. 2009. Nanohole arrays of mixed designs and microwriting for simultaneous and multiple protein binding studies. *Biosensors and Bioelectronics* 24:2847–2852.

7. Johannessen, E. A., J. M. R. Weaver, L. Bourova, P. Svoboda, P. H. Cobbold, and J. M. Cooper. 2002. Micromachined nanocalorimetric sensor for ultra-low-volume cell-based assays. *Analytical Chemistry* 74:2190–2197.

8. Johannessen, E. A., J. M. R. Weaver, P. H. Cobbold, and J. M. Cooper. 2002. Heat conduction nanocalorimeter for pl-scale single cell measurements. *Applied Physics Letters* 80:2029–2031.

9. Johannessen, E. A., J. M. R. Weaver, P. H. Cobbold, and J. M. Cooper. 2002. A suspended membrane nanocalorimeter for ultralow volume bioanalysis. *IEEE Transactions on Nanobioscience* 1:29–36.

10. Kowalski, G. J., A. Talakoub, J. Ji, J. G. O'Connell, M. Sen, and D. Larson. 2009. Fast temperature sensing using changes in extraordinary optical transmission through an array of subwavelength apertures. *Optical Engineering* 48:104402–104409.

11. Krishnan, A., T. Thio, T. J. Kim, H. J. Lezec, T. W. Ebbesen, P. A. Wolff, J. Pendry, L. Martin-Moreno, and F. J. Garcia-Vidal. 2001. Evanescently coupled resonance in surface plasmon enhanced transmission. *Optics Communications* 200:1–7.

12. Ladbury, J. E. 2004. Application of isothermal titration calorimetry in the biological sciences: Things are heating up! *BioTechniques* 37:885–887.

13. Ladbury, J. E. and M. Doyle (eds.). 2004. *Biocalorimetry*, Vol. 2. Wiley & Sons, Chichester, U.K.

14. Ladbury, J. E., G. Klebe, and E. Freire. 2010. Adding calorimetric data to decision making in lead discovery: A hot tip. *Nature Reviews Drug Discovery* 9:23–27.

15. Lerchner, J., A. Wolf, G. Wolf, V. Baier, E. Kessler, M. Nietzsch, and M. Krugel. 2006. A new micro-fluid chip calorimeter for biochemical applications. *Thermochimica Acta* 445:144–150.

16. Lindquist, N. C., A. Lesuffleur, H. Im, and S.-H. Oh. 2009. Sub-micron resolution surface plasmon resonance imaging enabled by nanohole arrays with surrounding Bragg mirrors for enhanced sensitivity and isolation. *Lab on a Chip* 9:382–387.

17. Salemme, F. R. 2004. High-throughput biochemistry heats up. *Nature Biotechnology* 22:1100–1101.

18. Stark, P., A. E. Halleck, and D. N. Larson. 2005. Short order nanohole arrays in metals for highly sensitive probing of local indices of refraction as the basis for a highly multiplexed biosensor technology. *Methods: Companion to Methods in Enzymology* 37:37–47.

19. Tellinghuisen, J. 2007. Calibration in isothermal titration calorimetry: Heat and cell volume from heat of dilution of NaCl (aq). *Analytical Biochemistry* 360:47.

20. Torres, F. E., P. Kuhn, D. D. Bruyker, A. G. Bell, M. V. Wolkin, E. Peeters, J. R. Williamson, G. B. Anderson, G. P. Schmitz, M. I. Recht, S. Schweizer, L. G. Scott, J. Ho, S. A. Elrod, P. G. Schultz, R. A. Lerner, and R. H. Bruce. 2004. Enthalpy arrays. *Proceedings of the National Academy of Sciences* 101:9517–9522.

21. Verhaegen, K., K. Baert, J. Simaels, and W. V. Driessche. 2000. A high-throughput silicon microphysiometer. *Sensors and Actuators A* 82:186–190.

22. Whitesides, G. M. and V. M. Krishnamurthy. 2006. Designing ligands to bind proteins. *Quarterly Reviews of Biophysics* 38:385–395.

23. Yang, J., J. Ji, J. Hogle, and D. Larson. 2008. Metallic nanohole arrays on fluoropolymer substrates as small label-free real-time bioprobes. *Nano Letters* 8:2718–2724.

24. Yang, J., J. Ji, J. Hogle, and D. Larson. 2009. Multiplexed plasmonic sensing based on small-dimension nanohole arrays and intensity interrogation. *Biosensors and Bioelectronics* 24:2334–2338.

25. Zhang, Y. and S. Tadigadapa. 2004. Calorimetric biosensors with integrated microfluidic channels. *Biosensors and Bioelectronics* 19:1733–1743.

Part XI

Cantilever-Based Technology

46

Microcantilever Biosensor Array for Cancer Research: From Tumor Marker Detection to Protein Conformational State Analysis

Riccardo Castagna and Carlo Ricciardi

CONTENTS

46.1 Introduction

In the last few years, there has been a growing interest in bio- and immunosensors based on functionalized micromachined cantilevers to detect the presence of specific compounds [1]. Microcantilevers (MCs) are single-side-clamped suspended beams that can operate either as surface stress sensor or as microbalance (Figure 46.1). Changes in the surface properties of the MC through binding or hybridization of analytes to the receptors on the MC will directly influence its surface stress, causing the MC to deflect proportionally to the analytes concentration. Otherwise, the mass of the binding entities anchored on the MC surfaces can be detected due to the decrease in the resonant frequency. In this way, using the following well-known equation, it is possible to estimate the mass increment Δm looking at the frequency shift value Δf:

$$\frac{\Delta f}{f_0} = -\frac{1}{2}\frac{\Delta m}{m_0} \tag{46.1}$$

where
    f_0 is the resonant frequency
    m_0 is the mass loading on the MC before binding

FIGURE 46.1 Picture of the core of the experimental setup: 1. vacuum chamber; 2. Peltier cell; 3. piezo-disk; 4. micro-cantilever arrays; and 5. optical micrograph of one array.

This straightforward relationship holds if the added mass is uniformly deposited on cantilever surface and if the beam spring constant k remains essentially constant after molecule binding. Furthermore, it has been previously proved that the adsorbate properties such as thickness, stiffness, and surface stress have a neglecting influence on the vibrational characteristics of the Si resonator [2–4].

In recent years, it has been demonstrated that with the dynamic mode, vibrating cantilever beams can be used to detect ultrasmall masses, ranging from femtogram to the attogram range [1]. Determination of a single virus particle was performed in vacuum [5], and quantitative detection in clinical ranges of prostate-specific antigen (PSA) and specific C-reactive protein (CRP) has been reported [6]. Similarly, in this work, we report about the development of an MC-based biosensor suitable to perform in vacuum quantitative and precise detection of specific proteins.

The identification and quantification of proteins involved in disease pathogenesis and physiological pathway have brought many advantages in the recent boom of proteomics field and the analysis of the proteome dynamics. In particular, in the field of medicine, proteomics activities might accelerate the discovery of new drug targets and protein disease markers useful for in vitro diagnostics [7]. Moreover, the understanding of the molecular pathways at the basis of diseases is mandatory for the elucidation of target-specific remedy. In fact, while the causes of many documented clinical problems vary greatly in their nature and origin, in some cases, the cause is found at the protein level, involving protein function, protein regulation, or protein–protein interactions. This will lead to new pharmaceutical treatment opportunities and to reliable diagnostic information, which is essential for choosing the appropriate therapy. Thus, providing reliable information about a person's condition and supporting therapeutic approach, proteomics plays a fundamental role in basic research, especially for a better comprehension of disease mechanisms and evolution [8]. For instance, in order to accurately diagnose and closely monitor complex diseases, such as cancer, the quantitative detection of multiple proteins is necessary. The development of label-free biosensors for sensitive and specific detection of protein analytes represents then an imperative opportunity, but still an open challenge. We focused on the tyrosine kinase receptor (TKR) Tie-2 and its primary activating ligand angiopoietin-1 (Ang-1), an oligomeric-secreted glycoprotein and a member, with Ang-2 and Ang-3/4, of the Ang family of growth factors (GFs), crucial molecules of angiogenic processes. Angiogenesis is the process that leads to the formation of new blood vessel growth from existing ones and plays an essential role in the development of tissues in the vertebrate embryo. It is also involved in a wide variety of physiological and pathological conditions in adults, including wound repair, metabolic diseases, inflammation, cardiovascular disorders, and tumor progression. It is well established that the Ang-Tie-2 pathway is involved in tumor angiogenesis, even if the exact roles of Angs on tumor angiogenesis are still under debate [9]. Nevertheless, experimental and clinical studies have demonstrated that increased expression of Ang-1 and Ang-2, respectively, promotes or inhibits tumor angiogenesis, and correlates with a reduced or extended survival time of patients and with a declined or improved clinical outcome. In general, these studies suggest that Ang-1 is a proangiogenic factor that promotes endothelial cell survival and tumor angiogenesis, especially in the presence of vascular endothelial growth factor (VEGF) [6].

In the present work, we performed Ang-1 binding experiments by hybridization with the anti-Ang-1 monoclonal antibody (mAb). Thanks to the combination of results coming from both the first and second modes of vibration, we achieved measurements with relative uncertainty lower than 0.5%. Such a precision allowed us to evaluate the protein surface density (number of molecules per cm^2) that revealed interesting features concerning the multimerization state of Ang-1.

46.2 Materials

46.2.1 Reagents

3-Aminopropyltriethoxysilane (APTES, anhydrous, 99%), glutaraldehyde (GA, 25%, v/v, water solution), and toluene (anhydrous, 99.8%) were obtained from Sigma-Aldrich (St. Louis, MO). Sulfuric acid (95%–97%, w/w), hydrogen peroxide (30%, w/w), orthoboric acid, and sodium chloride used to prepare borate buffer were ACS reagents (essay ≥99.5%) also purchased from Sigma-Aldrich.

BioWhittaker/Cambrex (East Rutherford, NJ) provided Dulbecco's phosphate-buffered saline (PBS; BE17-512F). Recombinant Protein G (PtG), purified from *Streptococcus*, was from PIERCE (Rockford, IL). Human recombinant proteins Ang-1 and VEGF-A$_{165}$ were obtained from R&D Systems (Minneapolis, MN); Ang-1 was provided as 6XHIStagged. Sigma provided the anti-Ang-1 (A0604) mAb; the anti-6XHIS (552565) mAb was from BD Pharmingen (San Diego, CA); goat anti-mouse IgG (H+L) HRP-conjugated Ab and goat anti-rabbit IgG Fc HRP-conjugated Ab were from CHEMICON International (Temecula, CA). PtG Sepharose 4 Fast Flow, and 3,3′,5,5′-tetramethylbenzidine (TMB) were purchased from GE Healthcare Bio-Sciences (Uppsala, Sweden).

46.3 Methods

46.3.1 In Vitro Biochemical Assays Preparatory for MC Protein Binding Experiments

46.3.1.1 Enzyme-Linked Immunosorbent Assay

To verify the protein-binding specificity on the silicon surface and the evaluation of immobilized protein activity, we performed enzyme-linked immunosorbent assay (ELISA) experiments. Functionalized and unfunctionalized macrosilicon slides were placed in microtiter plates and incubated with PBS solution in the presence or absence of PtG (0.05 mg/mL) for 2 h at room temperature (RT) using an orbital shaker. In parallel, PtG standards were diluted in PBS to obtain a range of 8–1000 ng for the microtiter plate calibration curves, and then incubated for 2 h at RT. The plates were washed four times with PBS and then treated using the following protocols designed for PtG detection. Plates were incubated first with anti-Ang-1 Ab and then with the secondary anti-mouse HRP-conjugated Ab using the same aforementioned conditions of washing and incubation. Plates were rinsed four times with PBS. Subsequently, TMB was used to generate color reaction that was measured at 405 nm in a spectrophotometer for microtiter plate (HTS 7000 Bio Assay Reader, PerkinElmer [Bucking-hamshire, UK]). The amount of bound PtG in the samples was calculated from the standard curves.

46.3.1.2 Immunoprecipitation Assays

Immunoprecipitation (IP) experiments were used to establish the optimal experimental conditions for the Ab–antigen (Ag) biodesign. For IP experiments, Ang-1 (500 ng) was incubated with anti-Ang-1 Ab (0.1–1.5 mg) in the presence of PtG Sepharose in PBS for 2 h at RT. After three washes in PBS, solubilized proteins were resolved on 8% sodium dodecyl sulfate–polyacrylamide gel electrophoresis (SDS-PAGE), immunoblotted with anti-6XHIS or anti-Ang-1 Abs, and detected by enhanced chemiluminescence technique (PerkinElmer). The optimal amount for anti-Ang-1 Ab was found to be 1 μg.

46.3.2 MC Fabrication and Measurement Setup

Arrays of MCs with dimensions of 400–800 μm in length, 50–120 μm in width, and 2–3 μm in thickness were fabricated using a combination of surface and bulk micromachining of silicon-on-insulator (SOI) wafers. The process flow, described in detail elsewhere [10], is composed of the following main steps:

1. Photolithography and wet etching in buffered oxide etch (BOE) solution for the patterning of the mask layer on the backside
2. Front side protection with a polymeric coating (Protek B2 from Brewer Science [Rolla, MO])
3. Membrane definition thanks to backside wet etching in KOH solution
4. Cantilever geometry definition thanks to photolithography and reactive ion etching (RIE) on the front side of the membrane
5. Removal of the buried oxide layer in BOE and cleaning by piranha solution

A small piezoelectric disk (PI Ceramic [Lederhose, Germany]) was used for the excitation of MCs. The sinusoidal signal was produced by a function generator (HP 33120A [Palo Alto, CA]), amplified, and sent to the piezoelectric actuator, to give us the possibility to control the frequency of the oscillation. The actuator was linked to a holding cell, the cantilevers were attached to the actuator with double-sided tape, and the cell was evacuated to a minimum pressure of 5×10^{-4} mbar by a series of a membrane and turbomolecular pump (MINI-Task System, Varian Inc. Vacuum Technologies [Palo Alto, CA]). Cantilever vibrational characteristics were measured by the optical lever technique. The position of a focused laser beam reflected off the top side of the cantilever onto a position-sensitive detector (PSD) was monitored. The current output of the PSD was amplified and converted into a voltage output, sent to a lock-in amplifier (EG&G 7260 [Trenton, NJ]) for signal extraction and filtering, and stored to a personal computer, together with the stimulus signal for the function generator (Figure 46.2). The procedure and the data analysis, fitting with a Lorentzian curve, were controlled in a LABVIEW® (National Instruments, Austin, TX) environment.

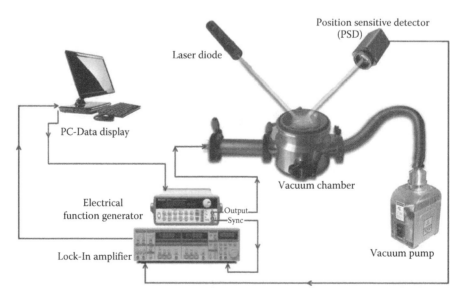

FIGURE 46.2 Schematic representation of the entire experimental setup for monitoring of microcantilever vibrational characteristics.

46.3.3 Statistical Data Analysis

Since MCs with slightly different geometrical dimensions (due to unavoidable tolerances in the realization process) were used, and first and second flexural modes were monitored for each MC, bioexperiments are better compared in terms of relative frequency deviation $\Delta f/f$ rather than absolute frequency shift. For each MC, we calculated the arithmetic mean of relative frequency deviation $(\Delta f/f)_{12}$ over the modes and the uncertainty as half-deviation of the modes (using standard deviation for few data can lead to an underestimation of the uncertainty). When more than one MC were considered, we used the "weighted average" method to have the best estimation of the true value. If N is the number of MCs, x_i a single measurement, σ_i its related uncertainty, and $w_i = 1/\sigma_i^2$ the *weight* of each uncertainty, the weighted average $\bar{x}$ and its related accuracy $\sigma_{\bar{x}}$ are defined as

$$\bar{x} = \frac{\sum_{i=1}^{N} w_i x_i}{\sum_{i=1}^{N} w_i} \tag{46.2}$$

$$\sigma_{\bar{x}} = \frac{1}{\sqrt{\sum_{i=1}^{N} w_i}} \tag{46.3}$$

For further details concerning data statistics, please refer to [11].

46.3.4 Experimental Procedure

The experimental design, optimized for MC experiments, consisted of following functionalization steps that can be summarized as:

1. Thermal oxidation of the cantilever, to obtain a silicon oxide flat surface, followed by a piranha solution cleaning step (70% H_2SO_4:30% H_2O_2) allowing to obtain an active surface which exhibit hydroxyl (–OH) groups.
2. Surface silanization of the cantilever. A self-assembled monolayer (SAM) of the APTES is produced through a wet process. Active amino groups (–NH₂) are exposed on the surface after the treatment.
3. Surface activation with GA, which allowed to obtain the exposition of aldehyde (–CHO) groups.

After that, the following sequence of protein bindings on the activated surface has been exploited:

4. PtG binding
5. Antibody (Ab) binding
6. Exposure of the functionalized cantilever to the target analyte solution (Ag, Ang-1)

After each functionalization step, the MC arrays were dried with a nitrogen flow (dip-and-dry approach) and placed in the vacuum chamber to measure the resonance frequencies.

46.3.4.1 MC Functionalization

A silicon oxide film on silicon MCs was grown by thermal oxidation at 1100°C in O_2 atmosphere for 3 h. Before exposure to organosilanes, the SiO_2/Si MC were soaked into piranha solution (70% H_2SO_4:30% H_2O_2) for 15 min to remove organic contaminants, rinsed with doubly distilled water, and dried in a stream of nitrogen. Freshly cleaned MCs were then incubated with 1% (v/v) solution of APTES in toluene at 60°C for 10 min. Silane-coated MCs were rinsed several times with toluene and then dried in a stream

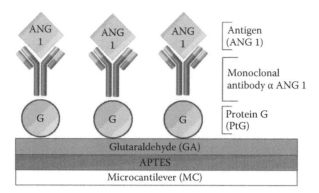

FIGURE 46.3 Schematic representation of microcantilever-based antibody–antigen design: two chemical functionalization steps (3-aminopropyltriethoxysilane [APTES] and glutaraldehyde [GA]), and three biomolecule-binding steps (Protein G, mAb, and Ang-1).

of nitrogen. Freshly silanized MCs were then incubated in a 1% GA solution (in borate buffer, 0.1 M; pH 8.8) for 1 h using an orbital shaker at 100 rpm. The relatively high pH ensures that mostly amino groups at MC surface are deprotonated ($-NH_2$) and thus able to react with aldehyde groups. After 15 min of incubation, 300 µL of a solution of sodium cyanoborohydride (5 M) in NaOH has been added to reduce the imine ($-N CH-$) formed from the reaction between $-NH_2$ and $-CHO$ moieties and thus to form a more stable bond. After incubation, MCs have been rinsed several times with doubly distilled water and dried in a stream of nitrogen.

46.3.4.2 "Ab–Ag" Hybridization Experiment

The pre-activated MCs were incubated with a PtG solution in PBS (50 µg/mL) for 2 h at RT using an orbital shaker. Afterward, MCs were washed (2 × 5 min and 2 × 10 min) in PBS-Tween 20 (0.05%) to remove nonspecifically bound proteins, and then in doubly distilled water (2 × 15 min) to remove possible salt residuals. After PtG binding, 5 µL of anti-Ang-1 Ab (Sigma Aldrich, A0604) solution in PBS (200 µg/mL) was dropped on MC's surface and the incubation was performed for 2 h at RT under controlled humidity conditions. After this, MCs were washed with PBS and doubly distilled water as reported earlier. Hybridization reaction was carried out by dropping 20 µL of Ang-1 solution in PBS (25 µg/mL) and performing the incubation for 2 h at RT under controlled humidity conditions. After this, MCs were washed with PBS and doubly distilled water as reported earlier. Hybridization experiment on anti-Ang-1 Ab-bound MCs was also carried out with a VEGF-A_{165} Ag, nonspecific for A0604 Ab, by using the same incubation conditions employed for Ang-1 hybridization reaction (Figure 46.3).

46.4 Results

As described earlier, the Ab–Ag design is composed of two functionalization steps (APTES and GA), and three biomolecule-binding steps (PtG, mAb, and Ang-1). Figure 46.4 reports a representative plot of the resonance curves of first and second flexural modes of an MC after all the aforementioned steps, starting from the oxidized cantilever, denominated as "bare." The curves are normalized to one to facilitate the graphical comparison: In fact, we frequently experienced variations in the intensity of the PSD signal, mainly due to the operator focusing of the laser beam. This fact also influences the Q-factor evaluation that we found to vary in the range 5,000–20,000, totally covering any possible trend linked to the variation of stiffness related to adsorption of soft materials like biomolecules [11].

Combining first and second flexural modes, as described elsewhere [11], we were able to quantify the mAb-immobilized molecules on the MC surfaces, obtaining an average mass of $\Delta m_{12} = (0.23 \pm 0.03)$ ng. Moreover, we calculated the protein surface density (number of molecules per cm^2) as follows, to compare cantilever with different dimensions:

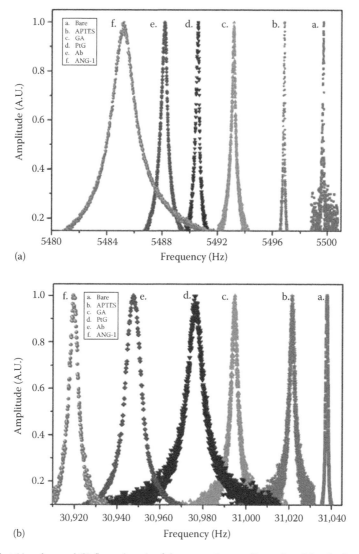

FIGURE 46.4 First (a) and second (b) flexural mode of the same microcantilever, describing the frequency response of the biosensor after each binding step.

$$\delta = -\frac{2m_0 \Delta f}{f_0} \frac{N_A}{MS} \tag{46.4}$$

where

N_A is the Avogadro constant
S is the exposed surface of the beam
M is the molecular mass of the protein

The evaluated weighted average Ab density resulted in $\overline{\delta}_{12} = (2.50 \pm 0.05) \times 10^{12}$, a value that is comparable with the theoretical maximum density estimated from the known size of an IgG molecule ($23.5\,\text{nm} \times 2.5\,\text{nm}$), which is 1.7×10^{12} molecules/cm^2 [12]. The slightly higher value we obtained is probably related to the fact that cantilever surface with immobilized Ab is not nanometrically smooth. In fact, the functionalization and protein immobilization steps result in a three-dimensional corrugated surface and the related density of active sites is larger with respect to that of a flat surface. For the following Ang-1 binding, we detected an average mass of $\Delta m_{12} = (0.411 \pm 0.001)$ ng. We then calculated the weighted average of Ang-1 surface densities, obtaining $\overline{\delta}_{12} = (5.14 \pm 0.02) \times 10^{12}$. The relative uncertainty of the entire set is smaller than 0.5%, indicating that the Ab–Ag biodesign can perform precise and accurate measurements.

It is remarkable to note that this technique can also be used to investigate how many target molecules bind to each probe: from our data, the ratio of Ag/Ab surface density is 2.06 ± 0.05, perfectly consistent with the two binding sites of the Ab. Interestingly, as shown elsewhere [11], we found a pretty similar surface density ratio, 2.0 ± 0.2, using Tie-2, the natural receptor of Ang-1, instead of the mAb. It is notable to remark that Tie-2 receptor has a single binding site for Ang-1 [13], implying that Ang-1 is not present in its monomeric form. This suggestion is confirmed by other research groups that recently characterized its modular structure, evidencing the presence of variable-size multimers [13]. From our measurements, the dimer seems to be the preferred immobilized conformation, but further studies are needed to deeply investigate the competitive kinetics of hybridization and the influence of washings. Anyhow, these results support that MC biosensors can be successfully applied as a quantitative helpful tool in solving the multimeric state of a target molecule as well as the stoichiometry of protein–protein interaction.

We also performed negative control experiments, in which the protein-immobilized MCs are dipped into PBS solution to evaluate the influence of nonspecific adsorption/desorption on resonance curves. We found that typical relative frequency shifts after PBS dipping are on the order of $(\Delta f/f)_{PBS} = 1 \times 10^{-4}$, at least one order of magnitude lower than typical variations due to specific protein binding ($\approx 10^{-3}$). Since nonspecific binding is often addressed as the major limit on ultimate cantilever-based sensitivity [1], we used $(\Delta f/f)_{PBS}$ as our experimental limit of detection (LOD): frequency variation below this limit was labeled as statistical fluctuation. We finally performed a specificity test, dipping five MCs in a solution containing a "false" Ag, VEGF-A$_{165}$: in this way, we wanted to check the system response due to the nonspecific adsorption of proteins similar to the target molecule, Ang-1. We found that the average relative frequency shift $(\Delta f/f)_{VEGF} = (-1.1 \pm 0.8) \times 10^{-5}$ was lower than our experimental LOD and nearly two orders of magnitude lower than typical shifts due to specific recognition. Figure 46.5 reports the graphical comparison between resonance curves of one cantilever dipped in Ang-1 solution (denominated "true" Ag) versus one dipped in VEGF-A$_{165}$ solution ("false" Ag). The same relative scale on x-axis is used to clearly evidence the large difference in the frequency shift induced by protein adsorption.

46.5 Discussion

In the last few years, MC biosensors have been used to quantify cells, nucleic acids, and proteins obtaining promising results. Nevertheless, a small amount of works has investigated the robustness and reproducibility of their results. We developed and tested an MC biosensor array focusing on the possibility to achieve a statistically representative amount of data to validate our results. We particularly focused on the precision and accuracy of our cantilever-based analysis, proposing the combination of results coming from both the first and second modes of vibration. We compared the cantilever vibrational modes in terms of precision (i.e., the variability of a measurement around its average value) rather than sensitivity. As a matter of fact, more than one research group has recently demonstrated significant advantages in mass sensitivity when monitoring higher modes of vibration respect to the fundamental one [1]. However, such works usually focus just on the improved detector sensitivity (essentially given by the increasing of resonance curve Q-factor), while a careful analysis of repeatability and reproducibility of frequency measurement is needed, in our opinion, to corroborate such advantages. In such a way, our MC-based system was successfully used to detect Ang-1 masses on the order of few hundreds of picograms with <0.5% of relative uncertainty.

Moreover, to clarify the specificity of the measurements, we tested our MC array biosensor with a "false" Ag, VEGF-A$_{165}$, obtaining a result comparable with our negative controls. These results underlined that such biosensor could represent a precious tool for complex systems analysis, able to determinate the modulation of proteins concentration that can lead to the activation of different cellular pathways.

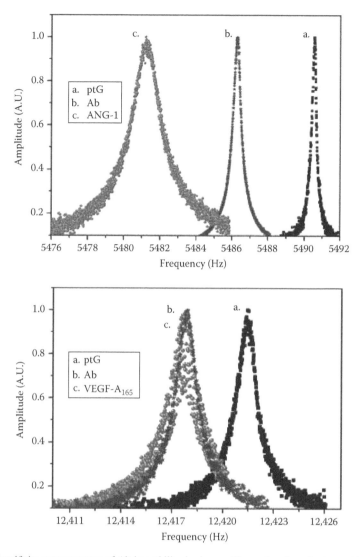

FIGURE 46.5 Specificity measurements of Ab-immobilized microcantilever: Ang-1 on the top and VEGF-A$_{165}$ on the bottom.

46.6 Future Trends

The main goal of a biosensor is to overcome the evident limits of nowadays laboratory technologies, such as time-consuming, expensiveness, difficult automation, low sensitivity, accuracy, and precision for quantitative methods. Therefore, the development of innovative biosensors such as Ab-immobilized MCs can represent new "easy-to-use" and inexpensive biological tools for the quantification of target biomolecules. We reported here about the development and testing of an MC biosensor array for in vacuum quantitative and precise protein detection. Anyway, despite their low LOD, the dip-and-dry technique described here still demands the employment of trained operators, since the handling of micromachined biosensors requires considerable skill to avoid cantilever damage or breakage. Current issues involved in MC biosensors may be addressed through the use of integrated lab-on-chip (LOC) microfluidic technologies. In such a way, it will be possible to overcome the employment of trained operators and to further reduce reagent volumes used for the detection. Nevertheless, the LOC integration of MC biosensor can lead to the ability of liquid measurements that will overcome the unnatural drying step that can unfold and damage the biomolecules used for the functionalization of the biosensor. Furthermore, liquid environment is essential in those applications where the observation of binding and unbinding kinetics is needed as well as for in situ and real-time measurement. An online measurement in liquid could also in principle reduce false-positive and false-negative responses.

In summary, we suggest that integration of microfluidic and cantilever technology would be a suitable solution to supply in situ portable devices for automation of routine diagnostic tests, such as point-of-care (POC) devices.

Abbreviations

Ab	Antibody
Ag	Antigen
Ang-1	Angiopoietin 1
APTES	3-Aminopropyltriethoxysilane
BOE	Buffered oxide etch
CRP	C-reactive protein
ELISA	Enzyme-linked immunosorbent assay
GA	Glutaraldehyde
GF	Growth factor
LOC	Lab-on-chip
LOD	Limit of detection
MC	Microcantilever
PBS	Phosphate-buffered saline
POC	Point of care
PSA	Prostate-specific antigen
PSD	Position-sensitive detector
PtG	Protein G
RIE	Reactive ion etching
SOI	Silicon on insulator
TKRs	Tyrosine kinase receptors
TMB	3,3′,5,5′-Tetramethylbenzidine
VEGF	Vascular endothelial growth factor

REFERENCES

1. Waggoner, P. S. and Craighead, H. G., Micro- and nanomechanical sensors for environmental, chemical, and biological detection. *Lab on a Chip*, 2007; 7(10): 1238–1255.
2. Lu, P., Lee, H. P., Lu, C., and O' Shea, S. J., Surface stress effects on the resonance properties of cantilever sensors. *Physical Reviews B*, 2005; 72(8): 085405.
3. Oliviero, G., Bergese, P., Canavese, G., Chiarib, M., Colombia, P., Cretich, M., Damin, F., Fiorilli, S., Marasso, S. L., Ricciardi, C., Rivolo, P., and Depero, L. E., A biofunctional polymeric coating for microcantilever molecular recognition. *Analytica Chimica Acta*, 2008; 630(2): 161–167.
4. Tamayo, J., Ramos, D., Mertens, J., and Calleja, M. J., Effect of the adsorbate stiffness on the resonance response of microcantilever sensors. *Applied Physics Letters*, 2006; 89(22): 224104.
5. Gupta, A., Akin, D., and Bashir, R., Single virus particle mass detection using microresonators with nanoscale thickness. *Applied Physics Letters*, 2004; 84(11): 1976–1978.
6. Lee, J. H., Yoon, K. H., Hwang, K. S., Park, J., Ahn, S., and Kim, T. S., Label free novel electrical detection using micromachined PZT monolithic thin film cantilever for the detection of C-reactive protein. *Biosensors and Bioelectronics*, 2004; 20(2): 269–275.
7. Vitzthum, F., Behrens, F., Anderson, N. L., and Shaw, J. H., Proteomics: From basic research to diagnostic application. A review of requirements & needs. *Journal of Proteome Research*, 2005; 4(4): 1086–1097.

8. Emmert-Buck, M. R., Strausberg, R. L., Krizman, D. B., Bonaldo, M. F., Bonner, R. F., Bostwick, D. G., Brown, M. R., Buetow, K. H., Chuaqui, R. F., Cole, K. A., Duray, P. H., Englert, C. R., Gillespie, J. W., Greenhut, S., Grouse, L., Hillier, L. W., Katz, K. S., Klausner, R. D., Kuznetzov, V., Lash, A. E., Lennon, G., Linehan, W. M., Liotta, L. A., Marra, M. A., Munson, P. J., Omstein, D. K., Prabhu, V. V., Prange, C., Schuler, G. D., Soares, M. B., Tolstoshev, C. M., Vocke, C. D., and Waterston, R. H., Molecular profiling of clinical tissue specimens: Feasibility and applications. *American Journal of Pathology*, 2000; 156(4): 1109–1115.

9. Metheny-Barlow, L. J. and Li, L. Y., The enigmatic role of angiopoietin-1 in tumor angiogenesis. *Cell Research*, 2003; 13(5): 309–317.

10. Canavese, G., Marasso, S. L., Quaglio, M., Cocuzza, M., Ricciardi, C., and Pirri, C. F., Polymeric mask protection for alternative KOH silicon wet etching. *Journal of Micromechanical Microengineering*, 2007; 17: 1387–1393.

11. Ricciardi, C., Fiorilli, S., Bianco, S., Canavese, G., Castagna, R., Ferrante, I., Digregorio, G., Marasso, S. L., Napione, L., and Bussolino, F., Development of microcantilever-based biosensor array to detect angiopoietin-1, a marker of tumor angiogenesis. *Biosensors and Bioelectronics*, 2010; 25(5): 1193–1198.

12. Gupta, A. K., Nair, P. R., Akin, D., Ladisch, M. R., Broyles, S., Alam, M. A., and Bashir, R., Anomalous resonance in a nanomechanical biosensor. *Proceedings of the National Academy of Sciences of the United States of America*, 2006; 103(36): 13362–13367.

13. Davis, S., Papadopoulos, N., Aldrich, T. H., Maisonpierre, P. C., Huang, T., Kovac, L., Xu, A., Leidich, R., Radziejewska, E., Rafique, A., Goldberg, J., Jain, V., Bailey, K., Karow, M., Fandl, J., Samuelsson, S. J., Ioffe, E., Rudge, J. S., Daly, T. J., Radziejewski, C., and Yancopoulos, G. D., Angiopoietins have distinct modular domains essential for receptor binding, dimerization and superclustering. *Nature Structural Biology*, 2003; 10(1): 38–44.

Index

A

Acute promyelocytic leukemia (APL), *see* Cancer, types, acute promyelocytic leukemia (APL)
AFM, *see* Atomic force microscope (AFM)
AgCl reference electrode, 521, 523, 544–551, 599, 647
ALCAM, *see* Biomarkers, ALCAM (CD166)
Alpha fetoprotein (AFP), *see* Biomarkers, AFP
3-Aminopropyltriethoxysilane (APTES), 716–718, 751, 805–812
Angiogenesis, 5–13, 56, 171, 187, 201, 804–805
Antibodies
    anti-fluorescein, 598, 601
    AP-anti IgG, 545, 551–552
    auto (*see* autoantibodies)
    biotinylated, 74, 159, 648–652
    concentrations, 133, 213, 459–460, 552–553, 722
    detecting, 461–462
    immobilized, 77–78, 120, 543, 546–548, 552
    methylcytosine, 720–725
    monoclonal, 17, 66, 102, 277, 376, 474, 488, 498, 512, 551, 580, 728, 739
    multiple, 450–451, 555
    primary, 259, 262, 499, 657–666, 740
    rapid detection of, 461–462
    unbound, 312, 445, 460
Antigen
    microarrays, 488–494
    tumor-associated (TAA), 484
Arrays
    acoustic sensor, 506
    antibody, 23
    antigen, 492
    cancer marker, 646
    CD, 423
    detector, 113–118, 191, 296–300
    electrochemical, 544
    electrode, 522
    multielectrode, 563–565, 567
    optical, 116
    protein, 44, 491
    96-sensor, 591, 594, 599
    tissue, 202
Asparaginase (ASNase), 90–104
    antigen, 94–104
Assays
    biochemical, 158, 174, 293, 421–423
    biomarker, 15, 604–605
    biosensor, 93, 173, 593
    cell based, 158, 174
    cellular, 141–151, 740
    chemical, 27, 425, 586
    colorimetric (*see* biosensor, colorimetric)

ELISA, 82–83, 92–105, 113, 126–133, 156–158, 430–460, 542, 582–593, 633–638, 654, 657–665, 732, 777, 805
    molecular pincer, 455–462
    type
        competitive, 43–52, 417–439, 546–553, 719
        direct binding, 25–26, 151, 369, 569, 682–687, 714, 729
        sandwich, 43, 136, 163–164, 390, 430, 435, 522
Atomic force microscope (AFM), 114, 360, 378–380
AuNPs, *see* Nanoparticles, gold
Autoantibodies, 461, 483–487, 497–498

B

BCR, *see* Breakpoint cluster region (BCR)
Biomarkers, 15–29, 156–167, 216–217, 257–277, 456, 463, 466, 760
    AFP, 434–451, 545–555, 672
    ALCAM (CD166), 71–84
    CA15-3, 125–137, 752–762
    CA 19-9, 16, 126, 156, 442, 542–555
    CA 27-29, 15, 442, 760–761
    CA 125, 16, 442, 542–555, 647–653, 761–762
    carcinoembryonic antigen (CEA), 16, 155–167, 270, 442–462, 542–555, 573–585, 658–672, 761–763
    EpCAM, 258–270, 294, 356–365, 376–382, 738–744
    epidermal growth factor receptor (EGFR), 8–18, 112, 150, 171–182, 217, 277, 309–316, 740–743, 761
    ERCC, 258–269
    estrogen receptor (ER), 15–17, 156, 257–269, 442, 750, 787–792
    FBP, 61–65
    HER2, 8–17, 126, 156, 200–201, 257–270, 277, 376, 442, 467–478, 592, 738–743, 750–761
    prostatic acid phosphatase (PAP), 15, 346, 517–531
    PSA, 14–15, 126, 156, 337, 347, 442–451, 473, 475, 518, 532, 575–585, 643–654, 657–666, 804–812
    salivary, 631–639
Biopsies, 231–251, 324–333, 346–347, 441
Biosensor
    CD-based, 419–430
    cell assay plate, 144–145
    colorimetric, 419, 441–451, 748
    disposable, 591–603
    electrochemical
        CV, 521–522, 648–662, 671, 673–687, 696–697
        DNA, 591–605, 671–688, 694, 701–706
        enzyme, 517–531
        immunosensor, 541–556, 573–587, 643–654, 657–667